Improve Your Grade.

Access included with any new book.

REGISTER NOW!

Registration will let you:

- Prepare for exams using chapter quizzes
- Grasp difficult concepts with animations and activities
- Brush up on basic skills, such as interpreting data and graphs with GraphIt! exercises
- Master key terms and vocabulary
- Explore additional case studies with InvestigateIt!

www.myecologyplace.com

TO REGISTER

1. Go to www.myecologyplace.com.
2. Click "Register."
3. Follow the on-screen instructions to create your login name and password.

Your Access Code is:

Note: If there is no silver foil covering the access code, it may already have been redeemed, and therefore may no longer be valid. In that case, you can purchase online access using a major credit card. To do so, go to www.myecologyplace.com and click on "Buy Access" and follow the on-screen instructions.

TO LOG IN

1. Go to www.myecologyplace.com.
2. Click "Log In."
3. Pick your book cover.
4. Enter your login name and password.

Hint:
Remember to bookmark the site after you log in.

Technical Support:
http://247pearsoned.custhelp.com

Elements of Ecology

Seventh Edition

THOMAS M. SMITH

UNIVERSITY OF VIRGINIA

ROBERT LEO SMITH

WEST VIRGINIA UNIVERSITY

Benjamin Cummings

San Francisco Boston New York
Cape Town Hong Kong London Madrid Mexico City
Montreal Munich Paris Singapore Sydney Tokyo Toronto

Editor-in-Chief: Beth Wilbur
Executive Director of Development: Deborah Gale
Acquisitions Editor: Star MacKenzie
Project Editor: Leata Holloway
Managing Editor: Michael Early
Senior Production Supervisor: Shannon Tozier
Production Management: Marsha Hall, Progressive Publishing Alternatives
Compositor: Progressive Publishing Alternatives
Design Manager: Marilyn Perry
Interior and Cover Designer: Hespenheide Design
Illustrators: Scientific Illustrators
Photo Researcher: Kristin Piljay
Director, Image Resource Center: Melinda Patelli
Image Rights and Permissions Manager: Zina Arabia
Image Permissions Coordinator: Elaine Soares
Manufacturing Buyer: Michael Penne
Executive Marketing Manager: Lauren Harp
Text and Cover Printer: Courier/Kendallville
Cover Photo Credits: Nature Picture Library/Tony Heald: Hippopotamus surrounded by water lettuce, Kruger NP, South Africa

ISBN 10-Digit 0-321-56147-3
       13-Digit 978-0-321-56147-3

Benjamin Cummings
is an imprint of

Contents

Preface

The first edition of *Elements of Ecology* appeared in 1976 as a short version of *Ecology and Field Biology*. Since that time, *Elements of Ecology* has evolved into a textbook intended for use in a one-semester introduction to ecology course. Although the primary readership will be students majoring in the life sciences, in writing this text we were guided by our belief that ecology should be part of a liberal education. We believe that students who major in such diverse fields as economics, sociology, engineering, political science, law, history, English, languages, and the like should have some basic understanding of ecology for the simple reason that it impinges on their lives.

Structure and Content

The structure and content of the text is guided by our basic belief that: 1) the fundamental unit in the study of ecology is the individual organism, and 2) the concept of adaptation through natural selection provides the framework for unifying the study of ecology at higher levels of organization: populations, communities, and ecosystems. A central theme of the text is the concept of tradeoffs— that the set of adaptations (characteristics) that enable an organism to survive, grow, and reproduce under one set of environmental conditions inevitably impose constraints on its ability to function (survive, grow, and reproduce) equally well under different environmental conditions. These environmental conditions include both the physical environment as well as the variety of organisms (both the same and different species) that occupy the same habitat. This basic framework provides a basis for understanding the dynamics of populations at both an evolutionary and demographic scale.

The text begins with an introduction to the science of ecology in Chapter 1 (The Nature of Ecology). The remainder of the text is divided into eight parts. Part One examines the constraints imposed on living organisms by the physical environment, both aquatic and terrestrial. Part Two begins by examining how these constraints imposed by the environment function as agents of change through the process of natural selection, the process through which adaptations evolve. The remainder of Part Two explores specific adaptations of organisms to the physical environment, considering both organisms that derive their energy from the sun (autotrophs) and those that derive their energy from the consumption and break-down of plant and animal tissues (heterotrophs).

Part Three examines the properties of populations, with an emphasis on how characteristics expressed at the level of the individual organisms (life history characteristics examined in Part Two) ultimately determine the collective dynamics of the population. Part Four extends our discussion from interactions among individuals of the same species to interactions among populations of different species (interspecific interactions). In these chapters we expand our view of adaptations to the environment from one dominated by the physical environment, to the role of species interactions in the process of natural selection and on the dynamics of populations.

Part Five explores the topic of ecological communities. This discussion draws upon topics covered in Parts Two to Four to examine the factors that influence the distribution and abundance of species across environmental gradients, both spatial and temporal.

Part Six combines the discussions of ecological communities (Part Five) and the physical environment (Part One) to develop the concept of the ecosystem. Here the focus is on the flow of energy and matter through natural systems. Part Seven continues the discussion of communities and ecosystems in the context of biogeography, examining the broad-scale distribution of terrestrial and aquatic ecosystems, as well as regional and global patterns of biological diversity.

Part Eight focuses on the interactions between humans and ecological systems. It is here that we examine the important current environmental issues relating to population growth, sustainable resource use, declining biological diversity, and global climate change. The objective of these chapters is to explore the role of the science of ecology in both understanding and addressing these critical environmental issues.

Throughout the text we explore this range of topics by drawing upon current research in the various fields of ecology, providing examples that enable the reader to develop an understanding of species natural history, the ecology of place (specific ecosystems), and the basic process of science.

New for the Seventh Edition

For those familiar with the sixth edition of this text, you will notice a number of changes in this new edition of *Elements of Ecology*. In addition to updating many of the examples and topics to reflect the most recent research and results in the field of ecology, we have made a number of changes in the organization and content of the text.

The objective of Chapter 1 (The Nature of Ecology) is to introduce students to the science of ecology and to provide a framework for understanding the organization of the remainder of the book. In keeping with this objective, we have expanded our discussion of the hierarchical structure of ecological systems. This framework helps the student to both understand the way ecologists view and study nature, as well as providing a framework for understanding why the text is structured the way it is—from individuals to ecosystems. In addition, we have expanded the current discussion of the scientific method and use it as a framework for presenting the discussion of the different approaches used by ecologists to understand pattern and process in natural systems (direct observation, experimentation, modeling, etc.).

In the sixth edition, the text's second chapter, Adaptation and Evolution, provided an introduction to the concepts of natural selection, inheritance, and speciation. In initially developing this chapter, we assumed that students using the text would fall into two general categories. The first category included students who had not taken a college biology course. For those students, the chapter was meant to provide a basic review of Mendelian genetics as well as the concepts of natural selection and speciation. For the second category, students who had a background in biology, it was hoped that the representation of the materials would function as a review and place the material into a new "ecological context." From my use of the textbook in the classroom, as well as comments from reviewers, we now believe that the chapter fell short of its objective, particularly in light of the advances and growing importance of genetics in ecological research. In response, we have reduced the emphasis on reviewing concepts of basic genetics, and placed the emphasis on natural selection and adaptation, providing a framework for addressing the role of the physical (aquatic and terrestrial) and biological (species interactions) environment as agents of natural selection presented in later chapters, emphasizing the role of tradeoffs and constraints in the evolution of adaptations.

In addition to the changes in content, the discussion of natural selection and adaptation has been moved from the opening of the text, where it was presented before the discussion of the physical environment, to its current place as Chapter 5 (Ecological Genetics: Adaptation and Natural Selection) in Part Two, where it now provides a framework for the discussion of specific adaptations of plants and animals in Chapters 6 (Plant Adaptations to the Environment), 7 (Animal Adaptations to the Environment), and 8 (Life History Patterns).

Although the majority of this new edition retains the general structure of the sixth edition, we have added additional and expanded coverage of a wide variety of topics throughout the text including: water balance in plants, density-dependent population regulation, allometry and animal metabolism, long-term dynamics of decomposition, role of the rhizosphere in decomposition and nutrient cycling, and the relationship between species diversity and ecosystem productivity.

A historical feature of the *Elements of Ecology* text is our focus on applying the science of ecology to current environmental issues, providing students with a first-hand understanding of the importance of ecology in the relationship between the human population and the natural environment. Since publication of the sixth edition, advances have been made in our understanding of the issues that are presented in Part Eight: Human Ecology (Chapter 26: Large-Scale Patterns of Biological Diversity; Chapter 27: Population Growth, Resource Use, and Sustainability; Chapter 28: Habitat Loss, Biodiversity and Conservation; and Chapter 29: Global Climate Change). In response we have updated many of the research results presented in these chapters to reflect the most current understanding of these issues, such as the Climate Change 2007 (Fourth Assessment) report of the IPCC (Intergovernmental Panel on Climate Change).

A feature new to the sixth edition was the inclusion of Researcher Profiles associated with specific chapters. This feature serves two functions. First, it introduces students to the new generation of ecologists pursuing questions relating directly to concepts presented within the chapter. Secondly, each functions as a case studies, allowing a more detailed presentation of methods, analyses, and results. Each profile integrates the work from a number of research papers, which together address a broader research question. This feature remains in the seventh edition under a new name: Field Studies. We believe this title more accurately reflects the emphasis of these short essays in providing students with a view of the new frontiers in ecological research and the new generation of ecologists that are defining the future directions for the science of ecology.

Ecology is a science rich in concepts, yet as with all science, it is quantitative. As such, a major objective of any science course should be the development of basic skills relating to the analysis and interpretation of empirical data. As with the sixth edition, the Quantifying Ecology feature in this new edition functions to provide students with an understanding of how concepts introduced in the chapters are quantified. In many chapters, the Quantifying Ecology boxes focus on assisting the reader with the interpretation of graphs, mathematical models, or quantitative methods that we have introduced within the main body of the text. In the seventh edition, however, we have added an additional feature to assist students in the development of quantitative skills. Interpreting Ecological Data is associated directly with various figures and tables in the text. The feature consists of two or more questions relating directly to the interpretation of data and analyses presented in the associated figure or table. It has become a common practice in many new textbooks to embellish figures and tables with annotations that function to provide the reader with an interpretation of the graph or data. Although we also use this technique for a number of complex graphics, our annotations are meant only as an extension of the figure or table captions. Rather, we believe that it is better to ask specific questions that will both encourage and assist the reader in the interpretation and understanding of the data and analyses that are presented. In doing so we hope to assist the reader in building the basic skills that are necessary to move beyond the examples presented in the text and begins to explore the wealth of ecological studies published in the books and journals that are referenced throughout the text. It is our belief that the development of these basic quantitative and interpretative skills are as important as understanding the body of concepts presented in the text that form the framework of the science of ecology. The answers to the questions presented in the Interpreting Ecological Data features are provided at the associated website.

Associated Materials

- Instructor's Resource DVD (0-3215-6786-2)
- Instructor's Guide/Test Bank (0-3215-6788-9)
- Computerized Test Bank: (0-3215-6787-0)
- Ecology Place Companion Website: (www.ecologyplace.com)
- Ecology on Campus Lab Manual: (0-8053-8214-3)
- Course Management Options (All CourseCompass and Blackboard courses offer pre-loaded content including tests, quizzes, and more.)

Acknowledgments

No textbook is a product of the authors alone. The material this book covers represents the work of hundreds of ecological researchers who have spent lifetimes in the field and the laboratory. Their published experimental results, observations, and conceptual thinking provide the raw material out of which the textbook is fashioned. We particularly acknowledge and thank the fourteen ecologists that are featured in the Field Studies boxes. Their cooperation in providing artwork and photographs is greatly appreciated.

Revision of a textbook depends heavily on the input of users who point out mistakes and opportunities. We took these suggestions seriously and incorporated most of them. We are deeply grateful to the following reviewers for their helpful comments and suggestions on how to improve this edition:

Peter Alpert, *University of Massachusetts*
Paul Bartell, *Texas A&M University*
Judith Bramble, *DePaul University*
William Brown, *SUNY Fredonia*
Chris Brown, *Tennessee Tech University*
Steve Blumenshine, *Fresno State University*
David Bybee, *Brigham Young University, Hawaii*
Dan Capuano, *Hudson Valley Community College*
Brian Chabot, *Cornell University*
Darren Divine, *Community College of Southern Nevada*
Curt Elderkin, *The College of New Jersey*
James Gould, *Princeton University*
Mark C. Grover, *Southern Utah University*

William Hallahan, *Nazareth College*
Floyd Hayes, *Pacific Union College*
John Jaenike, *University of Rochester*
Doug Keran, *Central Lakes Community College*
Jamie Kneitel, *California State University, Sacramento*
Ned J. Knight, *Linfield College*
James Lewis, *Fordham University*
Ken R. Marion, *University of Alabama, Birmingham*
Chris Migliaccio, *Miami Dade Community College*
Don Miles, *Ohio University*
L. Maynard Moe, *California State University, Bakersfield*
David Pindel, *Corning Community College*
James Refenes, *Concordia University*
Seith Reice, *University of North Carolina*
Ryan Rehmeir, *Simpson College*
Thomas Rosburg, *Drake University*
Tatiana Roth, *Coppin State College*
Irene Rossell, *University of North Carolina, Asheville*
Nathan Sanders, *University of Tennessee*
Alan Stam, *Capital University*
Mitch Wagener, *Western Connecticut State University*

Reviewers of Previous Editions:

John Anderson, *College of the Atlantic*
Morgan Barrows, *Saddleback College*
Christopher Beck, *Emory University*
Nancy Broshot, *Linfield College*
Evert Brown, *Casper College*
Mitchell Cruzan, *Portland State University*
Robert Curry, *Villanova University*
Richard Deslippe, *Texas Tech University*
Lauchlan Fraser, *University of Akron*
Sandi Gardner, *Triton College*
E. O. Garton, *University of Idaho*
Frank Gilliam, *Marshall University*
Brett Goodwin, *University of North Dakota*
Mark Gustafson, *Texas Lutheran University*
Greg Haenel, *Elon University*
Douglas Hallett, *Northern Arizona University*
Gregg Hartvigsen, *State University of New York at Geneseo*
Michael Heithaus, *Florida International University*
Jessica Hellman, *Notre Dame University*
Jason Hoeksema, *University of California at Santa Cruz*
John Jahoda, *Bridgewater State University*
Stephen Johnson, *William Penn University*
Jeff Klahn, *University of Iowa*
Frank Kuserk, *Moravian College*
Kate Lajtha, *Oregon State University*
Vic Landrum, *Washburn University*
Richard Lutz, *Rutgers University*
Richard MacMillen, *University of California at Irvine*
Ken Marion, *University of Alabama at Birmingham*
Deborah Marr, *Indiana University at South Bend*
Sherri Morris, *Bradley University*
Steve O'Kane, *University of Northern Iowa*

Matthew Parris, *University of Memphis*
Rick Relyea, *University of Pittsburgh*
Carol Rhodes, *College of San Mateo*
Eric Ribbens, *Western Illinois University*
Robin Richardson, *Winona State University*
Rowan Sage, *University of Toronto*
Thomas Sarro, *Mount Saint Mary College*
Maynard Schaus, *Virginia Wesleyan College*
Erik Scully, *Towson University*
Wendy Sera, *University of Maryland*
Mark Smith, *Chaffey College*
Paul Snelgrove, *Memorial University of Newfoundland*
Amy Sprinkle, *Jefferson Community College Southwest*
Barbara Shoplock, *Florida State University*
Christopher Swan, *University of Maryland*
Alessandro Tagliabue, *Stanford University*
Charles Trick, *University of Western Ontario*
Peter Turchin, *University of Connecticut*
Neal Voelz, *St. Cloud State University*
Joe von Fischer, *Colorado State University*
David Webster, *University of North Carolina at Wilmington*
Jake Weltzin, *University of Tennessee*

The publication of a modern textbook requires the work of many editors to handle the specialized tasks of development, photography, graphic design, illustration, copy editing, and production, to name only a few. We'd like to thank acquisitions editor Star MacKenzie for her editorial guidance. Her ideas and efforts have help to shape this edition. We'd also like to thank the rest of the editorial team—Leata Holloway, Project Editor, Laura Tomassi, Media Producer, and Erin Mann, Editorial Assistant. We also appreciate the efforts of production supervisor Shannon Tozier and the team at Progressive Information Technologies, especially Marsha Hall. Finally, we are indebted to Brian Morris at Scientific Illustrators, for all his efforts on the art program.

Through it all our families, especially our spouses Nancy and Alice, had to endure the throes of book production. Their love, understanding, and support provide the balanced environment that makes our work possible.

Thomas M. Smith
Robert Leo Smith

CHAPTER 1

The Nature of Ecology

Researcher Anker Nillsen taking morphological measurements from an Atlantic puffin chick (*Fratercula arctica*) as part of a study of this bird species being conducted in Rost, Norway.

The color photograph of Earthrise, taken by *Apollo 8* astronaut William A. Anders on December 24, 1968, is a powerful and eloquent image (Figure 1.1). One leading environmentalist has rightfully described it as "the most influential environmental photograph ever taken." Inspired by the photograph, economist Kenneth E. Boulding summed up the finite nature of our planet as viewed in the context of the vast expanse of space in his metaphor "spaceship Earth." What had been perceived throughout human history as a limitless frontier had suddenly become a tiny sphere: limited in its resources, crowded by an ever-expanding human population, and threatened by our use of the atmosphere and the oceans as repositories for our consumptive wastes.

A little more than a year later, on April 22, 1970, as many as 20 million Americans participated in environmental rallies, demonstrations, and other activities as part of the first Earth Day. The *New York Times* commented on the astonishing rise in environmental awareness, stating that "Rising concern about the environmental crisis is sweeping the nation's campuses with an intensity that may be on its way to eclipsing student discontent over the war in Vietnam." At the core of this social movement was a belief in the need to redefine our relationship with nature, and the particular field of study called upon to provide the road map for this new course of action was *ecology*.

1.1 | Ecology Is the Study of the Relationship between Organisms and Their Environment

With the growing environmental movement of the late 1960s and early 1970s, ecology—until then familiar only to a relatively small number of academic and applied biologists—was suddenly thrust into the limelight. Hailed as a framework for understanding the relationship of humans to their environment, *ecology* became a household word that appeared in newspapers, magazines, and books—although the term was often misused. Even now, people confuse it with terms such as *environment* and *environmentalism*. Ecology is neither. Environmentalism is activism with a stated aim of protecting the natural environment, particularly from the negative impacts of human activities. This activism often takes the form of public education programs, advocacy, legislation, and treaties.

So what is ecology? Ecology is a science. According to one accepted definition, **ecology** *is the scientific study of the relationships between organisms and their environment.* That definition is satisfactory so long as one considers *relationships* and *environment* in their fullest meanings. Environment includes the physical and chemical conditions as well as the biological or living components of an organism's surroundings. Relationships include interactions with the physical world as well as with members of the same and other species.

The term *ecology* comes from the Greek words *oikos,* meaning "the family household," and *logy,* meaning "the study of." It has the same root word as *economics,* meaning "management of the household." In fact, the German zoologist Ernst Haeckel, who originally coined the term *ecology* in 1866, made explicit reference to this link when he wrote:

> By ecology we mean the body of knowledge concerning the economy of nature—the investigation of the total relations of the animal both to its inorganic and to its organic; including above all, its friendly and inimical relations with those animals and plants with which it comes directly or indirectly into contact—in a word, ecology is the study of all those complex interrelationships referred to by Darwin as the conditions of the struggle for existence.

Haeckel's emphasis on the relation of ecology to the new and revolutionary ideas put forth in Charles Darwin's *The Origin of Species* (1859) is important. Darwin's theory of natural selection (Haeckel called it "the struggle for existence") is a cornerstone of the

Figure 1.1 | Photograph of Earthrise taken by *Apollo 8* astronaut William A. Anders on December 24, 1968.

science of ecology. It is a mechanism allowing the study of ecology to go beyond descriptions of natural history and examine the processes that control the distribution and abundance of organisms.

1.2 | Organisms Interact with the Environment in the Context of the Ecosystem

Organisms interact with their environment at many levels. The physical and chemical conditions surrounding an organism—such as ambient temperature, moisture, concentrations of oxygen and carbon dioxide, and light intensity—all influence basic physiological processes crucial to survival and growth. An organism must acquire essential resources from the surrounding environment and in doing so must protect itself from becoming food for other organisms. It must recognize friend from foe, differentiating between potential mates and possible predators. All of this effort is an attempt to succeed at the ultimate goal of all living organisms: to pass their genes on to successive generations.

The environment in which each organism carries out this "struggle for existence" is a place—a physical location in time and space. It can be as large and stable as an ocean or as small and transient as a puddle on the soil surface after a spring rain. This environment includes both the physical conditions and the array of organisms that coexist within its confines. This entity is what ecologists refer to as the ecosystem.

Organisms interact with the environment in the context of the **ecosystem.** The *eco–* part of the word relates to the environment. The *–system* part implies that the ecosystem functions as a collection of related parts that function as a unit. The automobile engine is an example of a system: components, such as the ignition and fuel pump, function together within the broader context of the engine. The ecosystem likewise consists of interacting components that function as a unit. Broadly, the ecosystem consists of two basic interacting components: the living, or **biotic,** and the nonliving (physical and chemical), or **abiotic.**

Consider a natural ecosystem, such as a forest (Figure 1.2). The physical (abiotic) component of the forest consists of the atmosphere, climate, soil, and water. The biotic component includes the many different organisms—plants, animals, and microbes—that inhabit the forest. Relationships are complex in that each organism not only responds to the abiotic environment but also modifies it and, in doing so, becomes part of the broader environment itself. The trees in the canopy of a forest intercept the sunlight and use this energy to fuel the process of photosynthesis. In doing so, the trees modify the environment of the plants below them, reducing the sunlight and lowering air temperature. Birds foraging on insects in the litter layer of fallen leaves reduce insect numbers and

Figure 1.2 | The interior of a forest ecosystem in coastal southeastern Alaska. Note the vertical structure within this forest. The Sitka spruce (*Picea sitchensis*) trees form the canopy, intercepting direct sunlight, and various species of mosses cover the surfaces of dead branches extending from the canopy to the ground. A diversity of shrub and herbaceous plant species make up the understory, and another layer of mosses covers the forest floor, accessing the nutrients made available by the community of bacteria and fungi that function as decomposers at the soil surface. This forest is also home to a wide variety of vertebrate and invertebrate animals, including larger species such as the bald eagle, black-tailed deer, and Alaskan brown bear.

modify the environment for other organisms that depend on this shared food resource. By reducing the populations of insects they feed on, the birds are also indirectly influencing the interactions among different insect species that inhabit the forest floor. We will explore these complex interactions between the living and the nonliving environment in greater detail in succeeding chapters.

1.3 | Ecological Systems Form a Hierarchy

The various kinds of organisms that inhabit our forest make up populations. The term *population* has many uses and meanings in other fields of study. In ecology, a **population** is a group of individuals of the same species that occupy a given area. Populations of plants and animals in an ecosystem do not function independently of each other. Some populations compete with other populations for limited resources, such as food, water, or space. In other cases, one population is the food resource for another. Two populations may mutually benefit each other, each doing better in the presence of the other. All populations of different species living and interacting within an ecosystem are referred to collectively as a **community.**

We can now see that the ecosystem, consisting of the biotic community and the physical environment, has many levels (Figure 1.3). On one level, individual organisms both respond to and influence the abiotic environment. At the next level, individuals of the same species form populations, such as a population of white oak trees or gray squirrels within a forest. Further, individuals of these populations interact among themselves and with individuals of other species to form a community. Herbivores consume plants, predators eat prey, and individuals compete for limited resources. When individuals die, other organisms consume and break down their remains, recycling the nutrients contained in their dead tissues back into the soil.

Organisms interact with the environment in the context of the ecosystem, yet all communities and ecosystems exist in the broader spatial context of the **landscape**—an area of land (or water) composed of a patchwork of communities and ecosystems. At the spatial scale of the landscape, communities and ecosystems are linked through such processes as the dispersal of organisms and the exchange of materials and energy.

Although each ecosystem on the landscape is distinct in that it is composed of a unique combination of physical conditions (such as topography and soils) and associated sets of plant and animal populations (community), the broad-scale patterns of climate and geology characterizing our planet (see Chapter 2) give rise to regional patterns in the geographic distribution of ecosystems. Geographic regions having similar geological and climate conditions (patterns of temperature, precipitation, and seasonality) support similar types of communities and ecosystems. For example, warm temperatures, high rates of precipitation, and a lack of seasonality characterize the world's equatorial regions. These warm, wet conditions year-round support vigorous plant growth and highly productive, evergreen forests known as tropical rain forests (see Chapter 23). These broad-scale regions dominated by similar types of ecosystems, such as tropical rain forest, grasslands, and deserts, are referred to as **biomes.**

The highest level of organization of ecological systems is the **biosphere**—the thin layer about the Earth that supports all of life. In the context of the biosphere, all ecosystems, both on land and in the water, are linked through their interactions—exchanges of materials and energy—with the other components of the Earth system: atmosphere, hydrosphere, and geosphere. Ecology is the study of the complex web of interactions between organisms and their environment at all levels of organization—from the individual organism to the biosphere.

Individual

What characteristics allow the Echinacea to survive, grow, and reproduce in the environment of the prairie grasslands of central North America?

Population

Is the population of this species increasing, decreasing, or remaining relatively constant from year to year?

Community

How does this species interact with other species of plants and animals in the prairie community?

Ecosystem

How do yearly variations in rainfall influence the productivity of plants in this prairie grassland ecosystem?

Landscape

How do variations in topography and soils across the landscape influence patterns of species composition and diversity in the different prairie communities?

Biome

What features of geology and regional climate determine the transition from forest to prairie grassland ecosystems in North America?

Biosphere

What is the role of the grassland biome in the global carbon cycle?

Figure 1.3 | The hierarchy of ecological systems.

1.4 | Ecologists Study Pattern and Process at Many Levels

As we shift our focus across the different levels in the hierarchy of ecological systems—from the individual organism to the biosphere—a different and unique set of patterns and processes emerges; and subsequently, a different set of questions and approaches for studying these patterns and processes is required (see Figure 1.3). The result is that the broader science of ecology is composed of a range of subdisciplines—from physiological ecology, which focuses on the functioning of individual organisms, to the perspective of Earth's environment as an integrated system forming the basis of global ecology.

Ecologists who focus on the level of the individual examine how features of morphology (structure), physiology, and behavior influence that organism's ability to survive, grow, and reproduce in its environment. Conversely, how do these same characteristics (morphology, physiology, and behavior) function to constrain the organism's ability to function successfully in other environments? By contrasting the characteristics of different species that occupy different environments, these ecologists gain insights into the factors influencing the distribution of species.

At the individual level, birth and death are discrete events; yet when we examine the collective of individuals that make up a population, these same processes are continuous as individuals are born and die. At the population level, birth and death are expressed as rates, and the focus of study shifts to examining the numbers of individuals in the population and how these numbers change through time. Populations also have a distribution in space, leading to such questions as how are individuals spatially distributed within an area, and how do the population's characteristics (numbers, rates of birth and death) change from location to location?

As we expand our view of nature to include the variety of plant and animal species that occupy an area, the ecological community, a new set of patterns and processes emerges. At this level of the hierarchy, the primary focus is on factors influencing the relative abundances of various species coexisting within the community. What is the nature of the interactions among the species, and how do these interactions influence the dynamics of the different species' populations?

The diversity of organisms comprising the community modify as well as respond to their surrounding physical environment, and so together the biotic and abiotic components of the environment interact to form an integrated system—the ecosystem. At the ecosystem level, the emphasis shifts from species to the collective properties characterizing the flow of energy and nutrients through the combined physical and biological system. At what rate are energy and nutrients converted into living tissues (termed *biomass*)? In turn, what processes govern the rate at which energy and nutrients in the form of organic matter (living and dead tissues) are broken down and converted into inorganic forms? What environmental factors limit these processes governing the flow of energy and nutrients through the ecosystem?

As we expand our perspective even further, the landscape can be viewed as a patchwork of ecosystems whose boundaries are defined by distinctive changes in the underlying physical environment and/or species composition. At the landscape level, questions focus on identifying factors that give rise to the spatial extent and arrangement of the various ecosystems that make up the landscape, and ecologists explore the consequences of these spatial patterns on such processes as the dispersal of organisms, the exchange of energy and nutrients between adjacent ecosystems, and the propagation of disturbances such as fire or disease.

At a continental to global scale, the questions focus on the broad-scale distribution of different ecosystem types or biomes. How do patterns of biological diversity (the number of different types of species inhabiting the ecosystem) vary geographically across the different biomes? Why do tropical rain forests support a greater diversity of species than do forest ecosystems in the temperate regions? What environmental factors determine the geographic distribution of the different biome types (e.g., forest, grassland, and desert)?

Finally, at the biosphere level, the emphasis is on the linkages between ecosystems and other components of the Earth system, such as the atmosphere. For example, how does the exchange of energy and materials between terrestrial ecosystems and the atmosphere influence regional and global climate patterns? Certain processes, such as movement of the element carbon between ecosystems and the atmosphere, operate at a global scale and require ecologists to collaborate with oceanographers, geologists, and atmospheric scientists.

In this textbook, we have used this hierarchical view of nature and the unique set of patterns and process associated with each level—the individual population, community, ecosystem, landscape, biome, and biosphere—as an organizing framework for studying the science of ecology. In fact, the science of ecology is functionally organized into subdisciplines based on these different levels of organization, each utilizing an array of specialized approaches and methodologies to address the unique set of questions that emerge at these different levels of ecological organization. The patterns and processes at these different levels of organization are linked, however, and identifying these linkages is a key objective of this text. For example, at the individual organism level, characteristics such as size, longevity, age at reproduction, and degree of parental care will directly influence rates of birth and survival for the collective of individuals comprising the species' population. At the community level, the same population will be influenced both positively and negatively through its interactions with populations of other species. In turn, the relative mix of species that make up the community will influence the collective properties of energy and nutrient exchange at the ecosystem level. As we shall see in the following chapters, patterns and processes at each level—from individuals to ecosystems—are intrinsically linked in a web of cause and effect with the patterns and processes operating at the other levels of this organizational hierarchy.

1.5 | Ecologists Investigate Nature Using the Scientific Method

Although each level in the hierarchy of ecological systems has a unique set of questions on which ecologists focus their research, all ecological studies have one thing in common: they involve the process known as the scientific method (Figure 1.4). This method demonstrates the power and limitations of science, and taken individually each step of the scientific method involves commonplace procedures. Yet taken together, these procedures form a powerful tool for understanding nature.

All science begins with observation. In fact, this first step in the process defines the domain of science: if something cannot be observed, it cannot be investigated by science. The observation need not be direct, however. For example, scientists cannot directly observe the nucleus of an atom, yet its structure can be explored indirectly through a variety of methods. Secondly, the observation must be repeatable—able to be made by multiple observers. This constraint helps to minimize unsuspected bias, when an individual might observe what they "want" or think they "ought" to observe.

The second step in the scientific method is to define a problem—a question regarding the observation that has been made. For example, an ecologist working in the prairie grasslands of North America might observe that the growth and productivity of grasses varies across the landscape. From this observation the ecologist may formulate the question, what environmental factors result in the observed variations in grassland productivity across the landscape? The question typically focuses on seeking an explanation for the observed patterns.

Once a question (problem) has been established, the next step is to develop a hypothesis. A **hypothesis** is an "educated guess" about what the answer to the question may be. The process of developing a hypothesis is guided by experience and knowledge, and it should be a statement of cause and effect that can be tested. For example, based on her knowledge that nitrogen availability varies

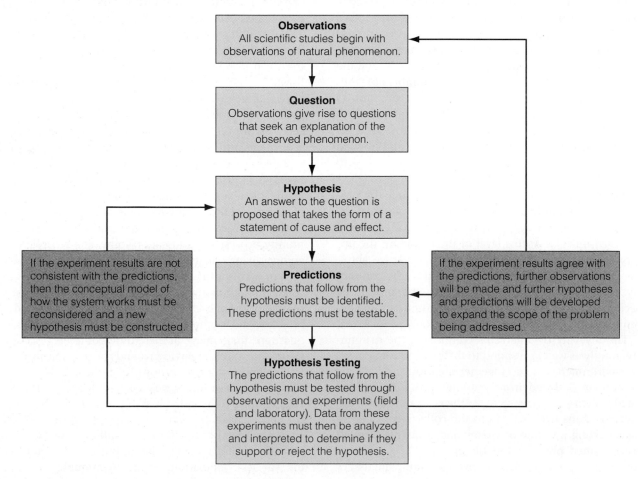

Figure 1.4 | A simple representation of the scientific method.

Classifying Ecological Data

All ecological studies involve collecting data—observations and measurements for testing hypotheses and drawing conclusions about a population. The term *population* in this context refers to a **statistical population.** An investigator is highly unlikely to gather observations on all members of a total population, so the part of the population actually observed is referred to as a **sample.** From this sample data, the investigator will draw his or her conclusions about the population as a whole. However, not all data are of the same type; and the type of data collected in a study directly influences the mode of presentation, types of analyses that can be performed, and interpretations that can be made.

At the broadest level, data can be classified as either categorical or numerical. **Categorical data** are *qualitative*—observations that fall into separate and distinct categories. The resulting data are labels or categories, such as the color of hair or feathers, sex, or reproductive status (pre-reproductive, reproductive, post-reproductive). Categorical data can be further subdivided into two categories: nominal and ordinal. **Nominal data** are categorical data in which objects fall into unordered categories, such as the previous examples of hair color or sex. In contrast, **ordinal data** are cate-

gorical data in which order is important, such as the example of reproductive status. In the special case where only two categories exist, such as in the case of presence or absence of a trait, categorical data are referred to as **binary.** Both nominal and ordinal data can be binary.

With **numerical data,** objects are "measured" based on some *quantitative* trait. The resulting data are a set of numbers, such as height, length, or weight. Numerical data can be subdivided into two categories: discrete and continuous. For **discrete data,** only certain values are possible, such as with integer values or counts. Examples include the number of offspring, number of seeds produced by a plant, or number of visits to a flower by a hummingbird during the course of a day. With **continuous data,** theoretically, any value within an interval is possible, limited only by the ability of the measurement device. Examples of this type of data include height, weight, or concentration.

1. What type of data does the variable "available nitrogen" (the *x*-axis) represent in Figure 1.5?

2. How might you transform this variable (available nitrogen) into categorical data? Would it be considered ordinal or nominal?

across the different soil types found in the region, and that nitrogen is an important nutrient limiting plant growth, the ecologist might hypothesize that *the observed variations in the growth and productivity of grasses across the prairie landscape are a result of differences in the availability of soil nitrogen.* As a statement of cause and effect, certain predictions follow from the hypothesis. If soil nitrogen is the factor limiting the growth and productivity of plants in the prairie grasslands, than grass productivity should be greater in areas with higher levels of soil nitrogen than in areas with lower levels of soil nitrogen. The next step is testing the hypothesis to see if the predictions that follow from the hypothesis do indeed hold true. This step requires gathering data (see Quantifying Ecology 1.1: Classifying Ecological Data).

To test this hypothesis, the ecologist can gather data in several ways. The first approach might be a field study to examine how patterns of soil nitrogen and grass productivity covary (vary together) across the landscape. If nitrogen is controlling grassland productivity, productivity should increase with increasing soil nitrogen. The ecologist would measure nitrogen availability and grassland productivity at various sites across the landscape.

Then, the relationship between these two variables, nitrogen and productivity, could be expressed graphically (see Quantifying Ecology 1.2: Displaying Ecological Data: Histograms and Scatter Plots on pages 10 and 11 to learn more about working with graphical data.). Go to QUANTIFYit! at www.ecologyplace.com to work with histograms and scatter plots.

After you've become familiar with scatter plots, you'll see the graph of Figure 1.5 shows nitrogen availability on the horizontal or *x*-axis and plant productivity on the vertical or *y*-axis. This arrangement is important. The scientist is assuming that nitrogen is the cause and that plant productivity is the effect. Because nitrogen (*x*) is the cause, we refer to it as the independent variable. Because it is hypothesized that plant productivity (*y*) is influenced by the availability of nitrogen, we refer to it as the dependent variable. (Go to GRAPHit! at www.ecologyplace.com for a tutorial on reading and interpreting graphs.)

From the observations plotted in Figure 1.5, it is apparent that grass productivity does in fact increase with increasing availability of nitrogen in the soil. Therefore, the data support the hypothesis. Had the data shown no relationship between grass productivity and nitrogen, the

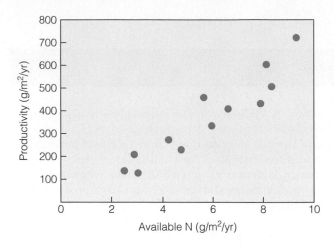

Figure 1.5 | The response of grassland production to nitrogen availability. Nitrogen, the independent variable, goes on the *x*-axis; grassland productivity, the dependent variable, goes on the *y*-axis.

ecologist would have rejected the hypothesis and sought a new explanation for the observed differences in grass productivity across the landscape. However, although the data suggest that grassland production does increase with increasing soil nitrogen, they do not prove that nitrogen is the factor controlling grass growth and production. Some other factor that varies with nitrogen availability, such as soil moisture or acidity, may actually be responsible for the observed relationship. To test the hypothesis another way, the ecologist may choose to do an experiment. An experiment is a test under controlled conditions performed to examine the validity of a hypothesis. In designing the experiment, the scientist will try to isolate the presumed causal agent—in this case, nitrogen availability.

The scientist may decide to do a field experiment (Figure 1.6), adding nitrogen to some field sites and not to others. The investigator controls the independent variable (levels of nitrogen) in a predetermined way, to reflect observed variations in soil nitrogen availability across the landscape, and monitors the response of the dependent variable (plant growth). By observing the differences in productivity between the grasslands fertilized with nitrogen and those that were not, the investigator tries to test whether nitrogen is the causal agent. However, in choosing the experimental sites, the ecologist must try to locate areas where other factors that may influence productivity, such as moisture and acidity, are similar. Otherwise, she cannot be sure which factor is responsible for the observed differences in productivity among the sites.

Finally, the ecologist might try a third approach—a series of laboratory experiments. Laboratory experiments give the investigator much more control over the environmental conditions. For example, she can grow the native grasses in the greenhouse under conditions of controlled temperature, soil acidity, and water availability

(Figure 1.7). If the plants exhibit increased growth under higher nitrogen fertilization, the investigator has further evidence in support of the hypothesis. Nevertheless, she faces a limitation common to all laboratory experiments;

Figure 1.6 | Field experiment at the Cedar Creek Long Term Ecological Research (LTER) site in central Minnesota; operated by the University of Minnesota. Experimental plots like those in the photograph are used to examine the effects on ecosystem functioning of elevated nitrogen deposition, increased concentrations of atmospheric carbon dioxide, and loss of biodiversity.

Figure 1.7 | These *Eucalyptus* seedlings are being grown in the greenhouse as part of an experiment examining the response of plant growth to varying levels of nutrient availability. The researcher shown here is using a portable instrument to measure the photosynthesis rates of plants that have received different levels of nitrogen during their growth period.

the results are not directly applicable in the field. The response of grass plants under controlled laboratory conditions may not be the same as their response under natural conditions in the field. There, the plants are part of the ecosystem and interact with other plants, animals, and the physical environment. Despite this limitation, the ecologist has accumulated additional data describing the basic growth response of the plants to nitrogen availability.

Having conducted several experiments that confirm the link between patterns of grass productivity to nitrogen availability, the ecologist may now wish to explore this relationship further, to see how the relationship between productivity and nitrogen is influenced by other environmental factors that vary across the prairie landscape. For example, how do differences in rainfall and soil moisture across the region influence the relationship between grass production and soil nitrogen? Once again hypotheses are developed, predictions made, and experiments conducted. As the ecologist develops a more detailed understanding of how various environmental factors interact with soil nitrogen to control grass production, a more general theory of the influence of environmental factors controlling grass production in the grassland prairies may emerge. A **theory** is an integrated set of hypotheses that together explain a broader set of observations than any single hypothesis—such as a general theory of environmental controls on productivity of the prairie grassland ecosystems of North America.

1.6 | Models Provide a Basis for Predictions

Scientists use the understanding derived from observation and experiments to develop models. Data are limited to the special case of what happened when the measurements were made. Like photographs, data represent a given place and time. Models use the understanding gained from the data to predict what will happen in some other place and time.

Models are abstract, simplified representations of real systems. They allow us to predict some behavior or response using a set of explicit assumptions, and as with hypotheses, these predictions should be testable through further observation or experiments. Models can be mathematical, like computer simulations, or they can be verbally descriptive, like Darwin's theory of evolution by natural selection (see Chapter 5). Hypotheses are models, although the term *model* is typically reserved for circumstances in which the hypothesis has at least some limited support through observations and experimental results. For example, the hypothesis relating grass production to nitrogen availability is a model. It predicts that plant productivity will increase with increasing nitrogen availability. However, this prediction is qualitative—it does not predict *how much* plant productivity will

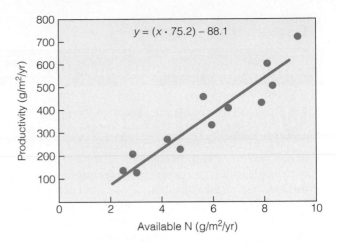

Figure 1.8 | A simple linear regression model to predict plant productivity (y-axis) from nitrogen availability (x-axis). The general form of the equation is $y = (x \times b) + a$, where b is the slope of the line (75.2) and a is the y-intercept (–88.1), or the value of y where the line intersects the y-axis (when $x = 0$).

increase. In contrast, mathematical models usually offer quantitative predictions. For example, from the data in Figure 1.5, we can develop a regression equation, a form of statistical model, to predict the amount of plant productivity per unit of nitrogen in the soil (Figure 1.8). (See QUANTIFYit! at www.ecologyplace.com to review regression analysis.)

All of the approaches just discussed—observation, experimentation, hypothesis testing, and development of models—appear in the following chapters to illustrate basic concepts and relationships. They are the basic tools of science. In each chapter, an array of figures and tables present the observations, experimental data, and model predictions used to test specific hypotheses regarding pattern and process at the different levels of ecological organization. Being able to analyze and interpret the data presented in these figures and tables is essential to your understanding of the science of ecology. To help you develop these skills, we have annotated certain figures and tables to guide you in their interpretation. In other cases, we pose questions that ask you to interpret, analyze, and draw conclusions from the data presented. These figures and tables are labeled as "Interpreting Ecological Data." (See Figure 2.16 on page 28 for the first example.)

1.7 | Uncertainty Is an Inherent Feature of Science

Collecting observations, developing and testing hypotheses, and constructing predictive models all form the backbone of the scientific method (see Figure 1.4). It is a continuous process of testing and correcting concepts to arrive at explanations for the variation we observe in the world around us, thus unifying observations that on first inspection seem unconnected. The

Displaying Ecological Data: Histograms and Scatter Plots

Whichever type of data an observer collects (see Quantifying Ecology 1.1), the process of interpretation typically begins with graphically displaying the set of observations. The most common method of displaying a single data set is constructing a **frequency distribution.** A frequency distribution is a count of the number of observations (frequency) having a given score or value. For example, consider this set of observations regarding flower color in a sample of 100 pea plants:

Flower color	Purple	Pink	White
Frequency	50	35	15

These data are categorical and nominal since the categories have no inherent order.

Frequency distributions are likewise used to display continuous data. This set of continuous data represents body lengths (in centimeters) of 20 sunfish sampled from a pond:

8.83, 9.25, 8.77, 10.38, 9.31, 8.92, 10.22, 7.95, 9.74, 9.51, 9.66, 10.42, 10.35, 8.82, 9.45, 7.84, 11.24, 11.06, 9.84, 10.75

With continuous data, the frequency of each value is often a single instance since multiple measurements are unlikely to be exactly the same. Therefore, continuous data are normally grouped into discrete categories, with each category representing a defined range of values. Each category must be nonoverlapping so that each observation belongs to only one category. For example, the body length data could be grouped into discrete categories:

Body length (intervals, cm)	Number of individuals
7.00–7.99	2
8.00–8.99	4
9.00–9.99	7
10.00–10.99	5
11.00–11.99	2

Once the observations have been grouped into categories, the resulting frequency distribution can then be displayed as a **histogram** (type of bar graph; Figure 1a). The x-axis represents the discrete intervals of body length, and the y-axis represents the number of individuals whose body length falls within each given interval.

In effect, the continuous data are transformed into categorical data for the purposes of graphical display. Unless there are previous reasons for defining categories,

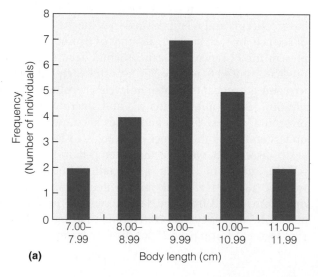

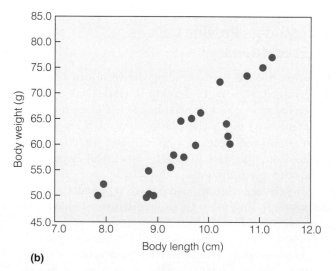

Figure 1 | (a) An example of a histogram relating the number of individuals belonging to different categories of body length from a sample of the sunfish population. **(b)** Scatter plot relating body length (x-axis) and body weight (y-axis) for the sample of sunfish presented in (a).

defining intervals is part of the data interpretation process—the search for pattern. For example, how would the pattern represented by the histogram in Figure 1a differ if the intervals were in units of 1 but started with 7.50 (7.50–8.49, 8.50–9.49, etc.)?

Often, however, the researcher is examining the relationship between two variables or sets of observations. When both variables are numerical, the most common method of graphically displaying the data is by using a scatter plot. A **scatter plot** is constructed by defining two axes (*x* and *y*), each representing one of the two variables being examined. For example, suppose the researcher who collected the observations of body length for sunfish netted from the pond also measured their weight in grams. The investigator might be interested in whether there is a relationship between body length and weight in sunfish.

In this example, body length would be the *x*-axis, or independent variable (Section 1.5), and body weight would be the *y*-axis, or dependent variable. Once the two axes are defined, each individual (sunfish) can be plotted as a point on the graph, with the position of the point being defined by its respective values of body length and weight (Figure 1b).

Scatter plots can be described as belonging to one of three general patterns, as shown in Figure 2. In plot (a) there is a general trend for *y* to increase with increasing values of *x*. In this case the relationship between *x* and *y* is said to be positive (as with the example of body length and weight for sunfish). In plot (b) the pattern is reversed, and *y* decreases with increasing values of *x*. In this case the relationship between *x* and *y* is said to be negative, or inverse. In plot (c) there is no apparent relationship between *x* and *y*.

You will find many types of graphs throughout the text, but most will be histograms and scatter plots. No matter which type of graph is presented, ask yourself the same set of questions—listed below—to help interpret the results. Review this set of questions by applying them to the graphs in Figure 1. What do you find out?

1. What type of data do the observations represent?

2. What variables do each of the axes represent, and what are their units (cm, g, color, etc.)?

3. How do values of *y* (the dependent variable) vary with values of *x* (the independent variable)?

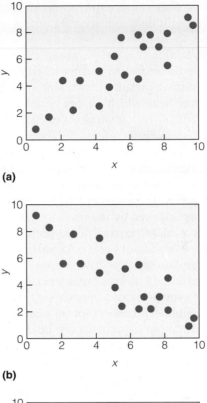

(a)

(b)

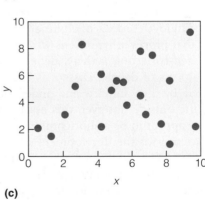

(c)

Figure 2 | Three general patterns for scatter plots.

Go to QUANTIFYit! and GRAPHit! at www.ecologyplace.com to further explore how to display data graphically.

difference between science and art is that, although both pursuits involve creation of concepts, in science the exploration of concepts is limited to the facts. In science, there is no test of concepts other than their empirical truth.

However, scientific concepts have no permanence, because they are only our interpretations of natural phenomena. We are limited to inspecting only a part of nature because to understand, we have to simplify. As discussed in Section 1.5, in designing experiments we control the pertinent factors and try to eliminate others that may confuse the results. Our intent is to focus on a subset of nature from which we can establish cause and effect. The trade-off is that whatever cause and effect we succeed in identifying represents only a partial connection to the nature we hope to understand. For that reason, when experiments and observations support our hypotheses, and when the predictions of the models are verified, our job is still not complete. We work to loosen the constraints imposed by the need to simplify so that we can understand. We expand our hypothesis to cover a broader range of conditions and once again begin testing its ability to explain our new observations.

It may sound odd at first, but science is a search for evidence that proves our concepts wrong. Rarely is there only one possible explanation for an observation. As a result, any number of hypotheses can be developed that may be consistent with an observation, and determining that experimental data are consistent with a hypothesis does not prove that the hypothesis is true. The real goal of hypothesis testing is to eliminate incorrect ideas. Thus, we must follow a process of elimination, searching for evidence that proves a hypothesis wrong. Science is essentially a self-correcting activity, dependent on the continuous process of debate. Dissent is the activity of science, fueled by free inquiry and independence of thought. To the outside observer, this essential process of debate may appear to be a shortcoming. After all, we depend on science for the development of technology and the ability to solve problems. For the world's current environmental issues, the solutions may well involve difficult ethical, social, and economic decisions. In this case, the uncertainty inherent to science is discomforting. However, we must not mistake uncertainty for confusion, nor should we allow disagreement among scientists to become an excuse for inaction. Instead, we need to understand the uncertainty so that we may balance it against the costs of inaction.

1.8 | Ecology Has Strong Ties to Other Disciplines

The complex interactions taking place within ecological systems involve all kinds of physical, chemical, and biological processes. To study these interactions, ecologists must draw on other sciences. This dependency makes ecology an interdisciplinary science.

Although in upcoming chapters we explore topics that are typically the subject of disciplines such as biochemistry, physiology, and genetics, we do so only in the context of understanding the interplay of organisms with their environment. The study of how plants take up carbon dioxide and lose water (see Chapter 6), for example, belongs to plant physiology. Ecology looks at how these processes respond to variations in rainfall and temperature. This information is crucial to understanding the distribution and abundance of plant populations and the structure and function of ecosystems on land. Likewise, we must draw upon many of the physical sciences, such as geology, hydrology, and meteorology. They will help us chart other ways organisms and environments interact. For instance, as plants take up water, they influence soil moisture and the patterns of surface water flow. As they lose water to the atmosphere, they increase atmospheric water content and influence regional patterns of precipitation. The geology of an area influences the availability of nutrients and water for plant growth. In each example, other scientific disciplines are crucial to understanding how individual organisms both respond to and shape their environment.

In the 21st century, ecology is entering a new frontier, one that requires expanding our view of ecology to include the dominant role of humans in nature. Among the many environmental problems facing humanity, four broad and interrelated areas are crucial: human population growth, biological diversity, sustainability, and global climate change. As the human population increased from approximately 500 million to more than 6.5 billion in the past two centuries, dramatic changes in land use have altered Earth's surface. The clearing of forests for agriculture has destroyed many natural habitats, resulting in a rate of species extinction that is unprecedented in Earth's history. In addition, the expanding human population is exploiting natural resources at unsustainable levels. Due to growing demand for energy from fossil fuels that is needed to sustain economic growth, the chemistry of the atmosphere is changing in ways that are altering Earth's climate. These environmental problems are ecological in nature, and the science of ecology is essential to understanding their causes and identifying ways to mitigate their impacts (see Ecological Issues: The Human Factor, and Part Eight: Human Ecology). Addressing these issues, however, requires a broader interdisciplinary framework to better understand their historical, social, legal, political, and ethical dimensions. That broader framework is known as environmental science. Environmental science examines the impact of humans on the natural environment and as such covers a wide range of topics including agronomy, soils, demography, agriculture, energy, and hydrology, to name but a few.

Ecologists tend to distinguish between the basic science of ecology—the study of the interaction of organisms with their environment—and the application of ecology to understand human interactions with the environment. The former concept is typically associated with study of the "natural world," the environment apart from humans, whereas the latter focuses on the effects of human activities on the natural environment. This distinction extends even to the professional journals that report research results. Journals such as *Ecology* (Ecological Society of America) and the *Journal of Ecology* (British Ecological Society) report on studies of the natural world, whereas *Ecological Applications* (Ecological Society of America) and the *Journal of Applied Ecology* (British Ecological Society) report on the influence of human activities on the environment. This traditional distinction, however, is becoming increasingly difficult to maintain, both in theory and practice. Ecologists find themselves having to expand the very definition of what constitutes the "natural world."

Our species has an ever-growing influence on Earth's environment. The human population now exceeds 6.5 billion—and, like our population, our collective impact on the planet's environment continues to grow. We use more than 50 percent of all freshwater resources, and our activities have transformed 30 to 40 percent of the terrestrial surface to produce food, fuel, and fiber (see Chapter 27). Although air pollution has long been an issue, changes in the atmosphere resulting from burning fossil fuels now have the potential to change Earth's climate (see Chapter 29).

In the title of his 1989 book, environmental writer Bill McKibben declared *The End of Nature*. His point was that humans have so altered Earth's environment that nature, "the separate and wild province, the world apart from man," no longer exists. Although many of us may not agree with McKibben, it has become increasingly difficult to study the natural world without considering the influence of human activities, past or present, on the ecological systems that are the focus of our research. For example, by the end of the 19th century the forests of eastern North America had been cleared for settlement and agricultural production (crops and/or pasture). In the 1930s and 1940s many of these lands were abandoned; agricultural production moved westward, leading to the regrowth of forests in eastern North America. Ecologists cannot study these ecosystems without explicitly considering their history. We cannot understand the distribution and abundance of tree species across the region without understanding past patterns of land use. We cannot study the cycling of nutrients within these forested watersheds without understanding how rapidly nitrogen and other nutrients are being deposited from atmospheric pollutants (see Chapter 22). Nor can we understand the causes of population decline in bird species inhabiting the forests of eastern North America without understanding how the fragmentation of forested lands from rural and urban development has restricted patterns of movement, susceptibility to predation or disease, and availability of habitat. Some leading questions facing ecologists today are directly related to the potential effects of human activities on terrestrial and aquatic ecosystems and on the diversity of life they support. Throughout the text, we will highlight these questions and topics in the feature Ecological Issues to illustrate the importance of the science of ecology in better understanding the human relationship with the environment—an environment that we are a part of.

1. How would you define nature? Does your definition include the human species? Why?

2. What would you consider as the most important environmental issue of our time? What role might the science of ecology (as you know it) play in helping us understand this issue?

1.9 | The Individual Is the Basic Unit of Ecology

As we noted earlier, ecology encompasses a broad area of investigation—from the individual organism to the biosphere. Our study of the science of ecology uses this hierarchical framework in the chapters that follow. We begin with the individual organism, examining the processes it uses and constraints it faces in maintaining life under varying environmental conditions. The individual organism forms the basic unit in ecology. The individual senses and responds to the prevailing physical environment. The collective properties of birth and death of individuals drive the dynamics of populations, and individuals of different species interact with each other in the context of the community. But perhaps most importantly, the individual through the process of reproduction passes genetic information to successive individuals, defining the nature of individuals that will compose future populations, communities, and ecosystems. At the individual level we can begin to understand the mechanisms that give rise to the diversity of life and ecosystems on Earth—mechanisms that are governed by the process of natural selection. But before embarking on our study of ecological systems, in Part One (The Abiotic Environment) we examine characteristics of the abiotic (physical and chemical) environment that function to sustain and constrain the patterns of life on our planet.

Summary

Ecology | 1.1

Ecology is the scientific study of the relationships between organisms and their environment. Environment includes physical and chemical conditions as well as biological or living components of an organism's surroundings. Relationship includes interactions with the physical world as well as with members of the same and other species.

Ecosystems | 1.2

Organisms interact with their environment in the context of the ecosystem. Broadly, the ecosystem consists of two components, the living (biotic) and the physical (abiotic), interacting as a system.

Hierarchical Structure | 1.3

Ecological systems can be viewed in a hierarchical framework, from individual organisms to the biosphere. Organisms of the same species that inhabit a given physical environment make up a population. Populations of different kinds of organisms interact with members of their own species as well as with individuals of other species. These interactions range from competition for shared resources to predation to mutual benefit. Interacting populations make up a biotic community. Community plus the physical environment make up an ecosystem.

All communities and ecosystems exist in the broader spatial context of the landscape—an area of land (or water) composed of a patchwork of communities and ecosystems. Geographic regions having similar geological and climatic conditions support similar types of communities and ecosystems, referred to as biomes. The highest level of organization of ecological systems is the biosphere—the thin layer around Earth that supports all of life.

Ecological Studies | 1.4

At each level in the hierarchy of ecological systems—from the individual organism to the biosphere—a different and unique set of patterns and processes emerges; and subsequently, a different set of questions and approaches for studying these patterns and processes is required.

Scientific Method | 1.5

All ecological studies are conducted by using the scientific method. All science begins with observation, from which questions emerge. The next step is the development of a hypothesis—a proposed answer to the question. The hypothesis must be testable through observation and experiments.

Models | 1.6

From research data, ecologists develop models. Models allow us to predict some behavior or response using a set of explicit assumptions. They are abstractions and simplifications of natural phenomena. Such simplification is necessary to understand natural processes.

Uncertainty in Science | 1.7

An inherent feature of scientific study is uncertainty; it arises from the limitation that we can focus on only a small subset of nature, and it results in an incomplete perspective. Because we can develop any number of hypotheses that may be consistent with an observation, determining that experimental data are consistent with a hypothesis is not sufficient to prove that the hypothesis is true. The real goal of hypothesis testing is to eliminate incorrect ideas.

An Interdisciplinary Science | 1.8

Ecology is an interdisciplinary science because the interactions of organisms with their environment and with each other involve physiological, behavioral, and physical responses. The study of these responses draws upon such fields as physiology, biochemistry, genetics, geology, hydrology, and meteorology.

Individuals | 1.9

The individual organism forms the basic unit in ecology. The individual responds to the environment. The collective birth and death of individuals defines the dynamics of populations, and the interactions among individuals of the same and different species define communities. The individual passes genes to successive generations.

Study Questions

1. How do ecology and environmentalism differ? In what way does environmentalism depend on the science of ecology?
2. Define the terms *population, community, ecosystem, landscape, biome,* and *biosphere.*
3. How might including the abiotic environment within the framework of the ecosystem help ecologists achieve the basic goal of understanding the interaction of organisms with their environment?
4. What is a hypothesis? What is the role of hypotheses in science?
5. An ecologist observes that the diet of a bird species consists primarily of large grass seeds (as opposed to smaller grass seeds or the seeds of other herbaceous plants found in the area). He hypothesizes that the birds are choosing the larger seeds because they have a higher concentration of nitrogen than do other types of seeds at the site. To test the hypothesis, the ecologist compares the large grass seeds with the other types of seeds, and the results clearly show that the large grass seeds do indeed have a much higher concentration of nitrogen. Did the ecologist prove the hypothesis to be true? Can he conclude that the birds select the larger grass seeds due to their higher concentration of nitrogen? Why or why not?
6. What is a model? What is the relationship between hypotheses and models?
7. Given the importance of ecological research in making political and economic decisions regarding current environmental issues such as global warming, how do you think scientists should communicate uncertainties in their results to policy makers and the public?

Further Readings

Bates, M. 1956. *The nature of natural history.* New York: Random House.

A lone voice in 1956, Bates shows us that environmental concerns have had a long history before the modern environmental movement emerged. A classic that should be read by anyone interested in current environmental issues.

Bronowski, J. 1956. *Science and human values.* New York: Harper & Row.

Written by a physicist and poet, this short book is a wonderful discussion of the scientific process as a human undertaking. The book explores the implication of science as a human endeavor. Highly recommended.

Cronon, W. 1996. "The trouble with wilderness; or, getting back to the wrong nature." In *Uncommon ground: Rethinking the human place in nature* (W. Cronon, ed.), 69–90. New York: Norton.

This is a great paper for students to read; it stirred up an incredible amount of debate among ecologists, environmentalists, and conservationists. Cronon describes how the idea of pristine nature is a human construct and says that, consequently, ecology is not natural without considering humans an integral part of it.

McKibben, W. 1989. *The end of nature.* New York: Random House.

In this provocative book, McKibben explores the philosophies and technologies that have brought humans to their current relationship with the natural world.

McIntosh, R. P. 1985. *The background of ecology: Concept and theory.* Cambridge: Cambridge University Press.

MacIntosh provides an excellent history of the science of ecology from a scientific perspective.

Worster, D. 1994. *Nature's economy.* Cambridge: Cambridge University Press.

This history of ecology was written from the perspective of a leading figure in environmental history.

PART ONE

The Physical Environment

In January 2004, two small, robotic vehicles named the Mars Exploration Rovers landed on the Martian surface. Their mission was to explore the surface for evidence of whether life ever arose on Mars. The rovers were not looking for living organisms or even for fossils within the rocks littering the terrain. These robotic vehicles were exploring the geology of the Martian surface. Their task was to determine the history of water on Mars. Although there is no liquid water on the surface of Mars today, the record of past water activity on Mars can be found in the planet's rocks, minerals, and geologic landforms, particularly in those that can form only in the presence of water.

Why search for water as evidence of life? Because life as we understand it is impossible without conditions that allow for the existence of water in a liquid form. The history of water on Mars is crucial to determining if the Martian environment was ever conducive to life. The mission of the Mars Exploration Rovers was not to search for direct evidence of life, but rather to provide us with information on the "habitability" of the environment.

Although studying the physical environment is the central mission of disciplines such as geology, meteorology, and hydrology, the concept of **habitability**—the ability of the physical environment to support life—links the physical sciences with the discipline of ecology. To illustrate this connection between ecology and the study of the physical environment, we move from the surface of Mars to a small chain of islands off the western coast of South America— the Galápagos Islands, which so influenced the thinking of young Charles Darwin.

When we think of penguins, the frozen landscape of the Antarctic generally comes to mind. Yet the Galápagos Islands, which lie on the equator, are home to the smallest of the penguin species: *Spheniscus mendiculus*, or the Galápagos penguin. These penguins stand only 16 to 18 inches (40 to 45 cm) tall and weigh only 5 pounds (2 to 2.5 kg). Found only on the Galápagos Islands, they live the farthest north of all the penguin species.

Galápagos penguins eat mostly small fish such as mullet and sardines; they depend on ocean currents flowing from the cooler waters of the south to bring these fish to their feeding grounds. Darwin himself noted the importance of currents flowing north from the South Pacific Ocean to the Galápagos environment: "Considering that these islands are placed directly under the equator, the climate is far from being excessively hot; this seems chiefly caused by the singularly low temperature of the surrounding water, brought here by the great southern Polar current" (*Voyage of the* Beagle). However, the prevailing winds giving rise to the cool waters that bathe these tropical islands are not always predictable. Periodically, the trade winds flowing westward in this region stall, and the waters around the Galápagos Islands warm. The warm waters dramatically reduce the fish populations upon which the penguins depend on. Such an occurrence caused a severe shortage of food about 20 years ago—more than 70 percent of the Galápagos penguins died. Since then their numbers have increased, and the population currently is estimated at about 800 breeding pairs. As far back as the 16th century, the region's fishermen have recorded periods of warming such as the one that occurred 20 years ago, and these periods of severe food shortage no doubt have affected the penguin population since their ancestors first arrived on the islands.

Even from the short example just given, you can see that understanding the ecology of Galápagos penguins very much depends on understanding the physical environment of these islands—namely, features of the environment that influence the islands' habitability for the penguin population. Ecologists use two very different timescales in viewing the interaction between organisms and their surrounding physical environment. Over a period of many generations, the physical environment represents a guiding force in the process of natural selection, favoring individuals with certain characteristics over others (see Chapter 5). Over a shorter period, the physical environment influences the physiological performance of individuals as well as the availability of essential resources, both of which directly influence the survival, growth, and reproduction of individuals within the population. The example of the Galápagos penguin illustrates both of these timescales.

Over evolutionary time, the physical environment of the Galápagos Islands has influenced the characteristics and behavior of Galápagos penguins through the process of natural selection. First, this penguin's small body size (smallest of all penguin species) enables it to dissipate heat easily to the surrounding air—an important characteristic in this tropical environment (see Chapter 7). Second, unlike most other penguins, Galápagos penguins have no particular breeding season and may have as many as three clutches in a single year. This adaptation allows them to cope with their variable and unreliable food resource, which is a direct result of unpredictable changes in the patterns of prevailing winds and ocean currents. An understanding of these characteristics requires an understanding of the physical environment that has shaped this species over evolutionary time through the process of natural selection.

Over the short term, yearly variations in currents and water temperatures around the islands are crucial to understanding the population dynamics of this species. Periodic increases and decreases in the population of Galápagos penguins are a direct response to variations in the availability of food resources.

In the chapters that make up Part One, we will look at features of Earth's physical environment that directly influence its habitability. We will explore features of the two dominant environments characterizing our planet—land and water. We begin by examining Earth's climate (Chapter 2); the broad-scale patterns of temperature, winds, precipitation, seasonality, and ocean currents. We then turn our attention to the dominant physical characteristics of aquatic (Chapter 3) and terrestrial (Chapter 4) environments. These chapters will set the stage for our discussion in Part Two of the adaptations of plants and animals to these physical environments.

The Galápagos penguin (*Spheniscus mendiculus*).

CHAPTER 2

Climate

As the sun rises, warming the morning air in this tropical rain forest on the island of Borneo, fog that has formed in the cooler night air begins to evaporate.

What determines whether a particular geographic region will be a tropical forest, a grassy plain, or a barren landscape of sand dunes? The aspect of the physical environment that most influences a particular ecosystem by placing the greatest constraint on organisms is climate. *Climate* is a term we tend to use loosely. In fact, people sometimes confuse climate with weather. **Weather** is the combination of temperature, humidity, precipitation, wind, cloudiness, and other atmospheric conditions occurring at a specific place and time. **Climate** is the long-term average pattern of weather and may be local, regional, or global.

The structure of terrestrial ecosystems is largely defined by the dominant plants, which in turn reflect the prevailing physical environmental conditions, namely climate (see Chapter 23). Geographic variations in climate, primarily temperature and precipitation, govern the large-scale distribution of plants and therefore the nature of terrestrial ecosystems (Figure 2.1). In this chapter, we learn how climate determines the availability of heat and water on Earth's surface and influences the amount of solar energy that plants can harness.

2.1 | Earth Intercepts Solar Radiation

The outer edge of Earth's atmosphere intercepts solar radiation. The resulting molecular interactions create heat and cause thermal patterns that, coupled with Earth's rotation and movement, generate the prevailing winds and ocean currents. These movements of air and water in turn influence Earth's weather patterns, including the distribution of rainfall.

Solar radiation, the electromagnetic energy emanating from the Sun, travels more or less unimpeded through the vacuum of space until it reaches Earth's atmosphere. Scientists conceptualize solar radiation as a stream of photons, or packets of energy, that—in one of the great paradoxes of science—behave either as waves or as particles, depending on how they are observed. Scientists characterize waves of energy in terms of their wavelength (λ), or the physical distances between successive crests, and their frequency (ν), or the number of crests that pass a given point per second. All objects emit radiant energy, typically across a wide range of wavelengths. The exact nature of the energy emitted, however, depends on the object's temperature (Figure 2.2). The hotter the object is, the more energetic the emitted photons and the shorter the wavelength. A very hot surface such as that of the Sun ($\sim 5800°C$) gives off primarily **shortwave radiation.** In contrast, cooler objects such as Earth's surface (average temperature of 15°C) emit radiation of longer wavelengths, or **longwave radiation.**

Of the solar radiation that reaches the top of Earth's atmosphere, only 51 percent makes it to the surface. What happens to all the incoming energy? If we say the amount of solar radiation reaching the top of the atmosphere equals 100 units, then on average, clouds and the atmosphere reflect and scatter 26 units, and Earth's surface reflects an additional 4 units, for a total of 30 units being reflected back to space (Figure 2.3). Together the atmosphere and clouds absorb another 19 units (for a total of 49 units), leaving 51 units of direct and indirect solar radiation to be absorbed by Earth's surface.

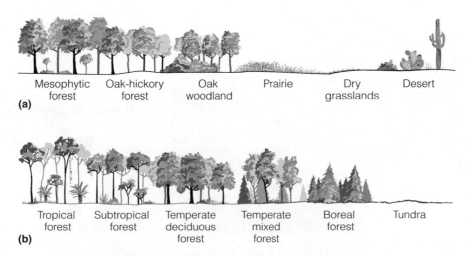

Figure 2.1 | Gradients of vegetation in North America from east to west and south to north.
(a) The east-west gradient reflects a decrease in annual precipitation (does not cut across the Rocky Mountains). **(b)** The south-north gradient reflects a decrease in mean annual temperature. Note that with decreasing precipitation and temperatures, the stature of the vegetation decreases. See Chapter 23 for a detailed discussion of the characteristics and distribution of terrestrial ecosystems.

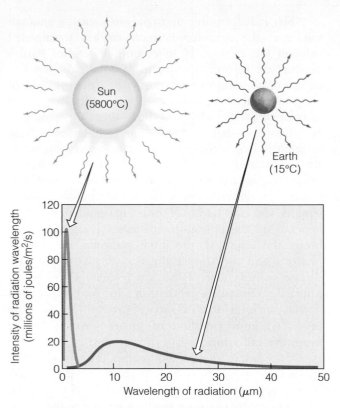

Figure 2.2 | The wavelength of radiation emitted by an object is a function of its temperature. The Sun, with an average surface temperature of 5800°C, emits relatively shortwave radiation as compared to Earth, with an average surface temperature of 15°C, which emits relatively longwave radiation. When comparing these two graphs, note the difference in scale (1 J/s = 1 W).

Of the 51 units that reach Earth's surface, 23 units are used to evaporate water and another 7 units heat the air next to the surface, leaving 21 units to heat the landmasses and oceans. The landmasses and oceans in turn emit radiation back to the atmosphere in the form of longwave (thermal) radiation. The amount of longwave radiation emitted by Earth's surface exceeds the 21 units of solar radiation that are absorbed. Actually, some 117 units in total are emitted. How is this possible? Because, although the surface receives solar (shortwave) radiation only during the day, it constantly emits longwave radiation during both day and night. Additionally, the atmosphere above allows only a small fraction of this energy (6 units) to pass through into space. Most of the energy (111 units) is absorbed by the water vapor and CO_2 in the atmosphere and by clouds. Much of this energy (96 units) is radiated back to Earth, producing the **greenhouse effect** (see Chapter 29), which is crucial to maintaining the planet's surface warmth. As a result, Earth's surface receives nearly twice as much longwave radiation from the atmosphere as it does shortwave radiation from the Sun. In all of these exchanges (see Figure 2.3), the energy lost at Earth's surface (30 units + 117 units = 147 units) is exactly balanced by the energy gained (51 units + 96 units = 147 units). The radiation budget of Earth is in balance.

Electromagnetic radiation emitted by the Sun covers a wide range of wavelengths. Of the total range of solar radiation reaching Earth's atmosphere, the wavelengths of approximately 400 to 700 nm (a nanometer is 1 billionth of a meter) make up **visible light** (Figure 2.4). Collectively, these wavelengths are also known as **photosynthetically**

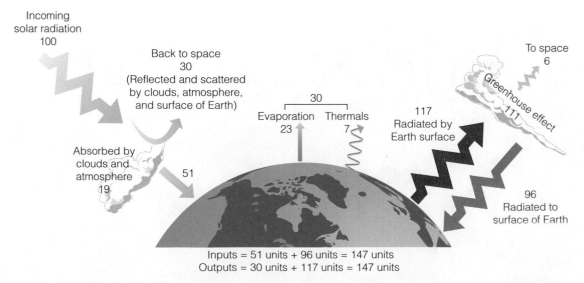

Inputs = 51 units + 96 units = 147 units
Outputs = 30 units + 117 units = 147 units

Figure 2.3 | Disposition of solar energy reaching Earth's atmosphere. Inputs include incoming solar (shortwave) radiation and longwave radiation returning to Earth as a function of the greenhouse effect. Outputs include heat from surface evaporation and thermals and longwave energy radiated by Earth's surface.

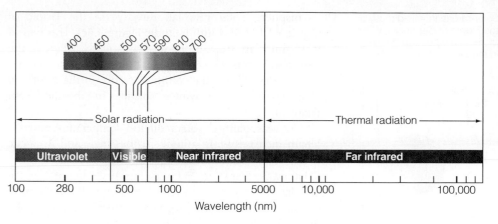

Figure 2.4 | A portion of the electromagnetic spectrum, separated into solar and thermal radiation. Ultraviolet, visible, and infrared light waves represent only a small part of the spectrum. To the left of ultraviolet radiation are X-rays and gamma rays (not shown).

(Adapted from Halverson and Smith 1979.)

active radiation (PAR) because they include the wavelengths that plants use as a source of energy in the process of photosynthesis (see Chapter 6). Wavelengths shorter than the visible range are ultraviolet (UV) light. There are two types of ultraviolet light: UV-A, with wavelengths from 315 to 380 nm; and UV-B, with wavelengths from 280 to 315 nm. Radiation with wavelengths longer than the visible range is known as infrared radiation. Near-infrared radiation includes wavelengths of approximately 740 to 4000 nm, and far-infrared or thermal radiation includes wavelengths from 4000 to 100,000 nm.

2.2 | Intercepted Solar Radiation Varies Seasonally

The amount of solar energy intercepted at any point on Earth's surface varies markedly with latitude (Figure 2.5). Two factors influence this variation. First, at higher latitudes, radiation hits the surface at a steeper angle, spreading sunlight over a larger area. Second, radiation that penetrates the atmosphere at a steep angle must travel through a deeper layer of air. In the process, it encounters more particles in the atmosphere, which reflect more of it back into space.

Although the variation in solar radiation reaching Earth's surface with latitude can explain the gradient of decreasing temperature from the equator to the poles, it does not explain the systematic variation occurring over the course of a year. What gives rise to the seasons on Earth? Why do the hot days of summer give way to the changing colors of fall, or the freezing temperatures and snow-covered landscape of winter to the blanket of green signaling the onset of spring? The explanation is quite simple—because Earth does not stand up straight but rather tilts on its side.

Earth, like all planets, is subjected to two distinct motions. While it orbits around the Sun, Earth rotates about an axis that passes through the North and South Poles, giving rise to the brightness of day followed by the darkness of night (the diurnal cycle). Earth travels about the Sun in a plane called the ecliptic (a plane traveled by all other planets except Pluto). By chance, Earth's axis of spin is not perpendicular to the ecliptic but tilted at an angle of 23.5° (Figure 2.6). This tilt (inclination) is responsible for the seasonal variations in temperature and day length. Only at the equator are there exactly 12 hours of daylight and darkness every day of the year. At the vernal equinox (approximately March 21) and autumnal equinox (approximately September 22), solar radiation falls directly on the equator (Figure 2.6b). At this time, the equatorial region is heated most intensely, and every place on Earth receives the same 12 hours each of daylight and night.

At the summer solstice (approximately June 22) in the Northern Hemisphere, solar rays fall directly on the Tropic of Cancer (23.5° north latitude; Figure 2.6a). This is when days are longest in the Northern Hemisphere, and the Sun heats the surface most intensely. In contrast, the Southern Hemisphere experiences winter at this time. At winter solstice (about December 22) in the Northern

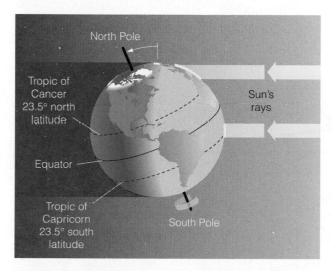

Figure 2.5 | Solar radiation striking Earth at high latitudes arrives at an oblique angle and spreads over a wide area. Therefore, it is less intense than energy arriving vertically at the equator.

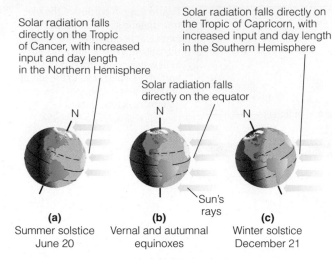

Solar radiation falls directly on the Tropic of Cancer, with increased input and day length in the Northern Hemisphere

Solar radiation falls directly on the Tropic of Capricorn, with increased input and day length in the Southern Hemisphere

Solar radiation falls directly on the equator

Sun's rays

(a)
Summer solstice
June 20

(b)
Vernal and autumnal equinoxes

(c)
Winter solstice
December 21

Figure 2.6 | Angle of the Sun and circle of illumination at the equinoxes and the winter and summer solstices.

Hemisphere, solar rays fall directly on the Tropic of Capricorn (23.5° south latitude; Figure 2.6c). This period is summer in the Southern Hemisphere, whereas the Northern Hemisphere is enduring shorter days and colder temperatures. Thus, the summer solstice in the Northern Hemisphere is the winter solstice in the Southern Hemisphere.

The seasonality of solar radiation, temperature, and day length increases with latitude. At the Arctic and Antarctic circles (66.5° north and south latitudes, respectively), day length varies from 0 to 24 hours over the course of the year. The days shorten until the winter solstice, a day of continuous darkness. The days lengthen with spring, and on the day of the summer solstice, the Sun never sets.

Figure 2.7 shows how annual, seasonal, and daily solar radiation vary over Earth. Although in theory every location on Earth receives the same amount of daylight during a year, in high latitudes where the Sun is never

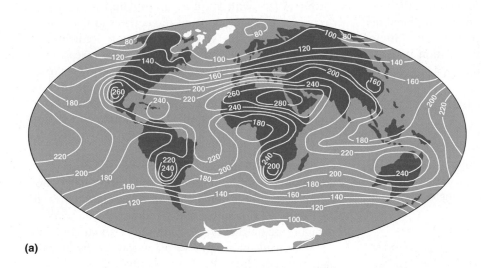

(a)

Figure 2.7 | Annual variation in solar radiation on Earth. **(a)** Global mean solar radiation at Earth's surface. **(b)** Seasonal variation of the daily extraterrestrial solar radiation incident on a horizontal surface at the top of Earth's atmosphere in the northern hemisphere.

(Adapted from Barry and Chorley 1992.)

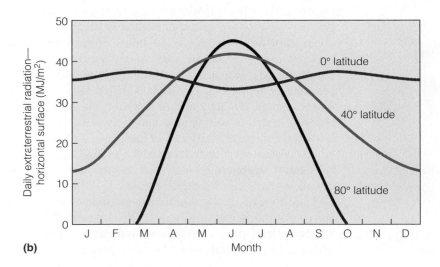

(b)

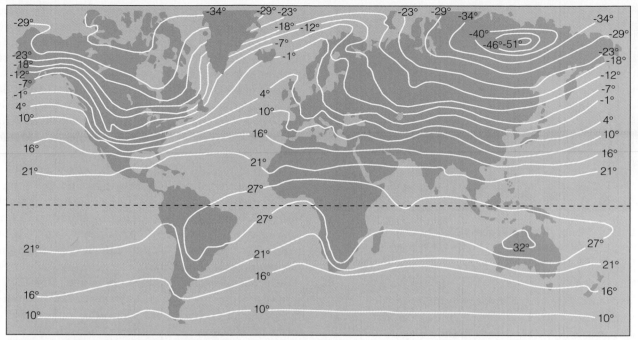

(a) January isotherms (lines of equal temperature) around the Earth

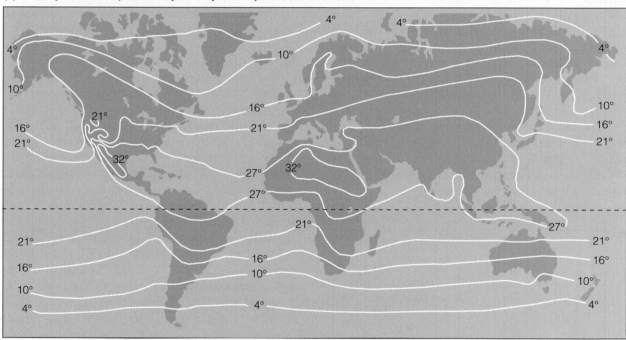

(b) July isotherms (lines of equal temperature) around the Earth

Figure 2.8 | Mean annual global temperatures change with latitude and season. **(a)** Mean sea-level temperatures (°C) in January. **(b)** Mean sea-level temperatures (°C) in July. Note the colder temperatures during the Northern Hemisphere winter (January), warmer temperatures during the Southern Hemisphere summer, and the reversal in temperature patterns with the shift to summer in the Northern Hemisphere and winter in the Southern Hemisphere (July).

positioned directly overhead, the annual input of solar radiation is the lowest. This pattern of varying exposures to solar radiation controls mean annual temperature around the globe (Figure 2.8). Like annual solar radiation, mean annual temperatures are highest in tropical regions and decline as one moves toward the poles.

2.3 | Air Temperature Decreases with Altitude

Whereas varying degree and length of exposure to solar radiation may explain changes in latitudinal, seasonal, and daily temperatures, they do not explain why air gets

Figure 2.9 | Although near the equator, Mount Kilimanjaro in Africa is snowcapped and supports tundra-like vegetation near its summit. Global warming is causing a rapid melting of this snowcap.

cooler with increasing altitude. Mount Kilimanjaro, for example, rises from the hot plain of tropical East Africa; but its peak is capped with ice and snow (Figure 2.9). The explanation of this apparent oddity of snow in the tropics lies in the physical properties of air.

The weight of all the air molecules surrounding Earth is a staggering 5600 trillion tons. The air's weight acts as a force on Earth's surface, and the amount of force exerted over a given area of surface is called **atmospheric pressure,** or air pressure. Envision a vertical column of air. The pressure at any point in the column can be measured in terms of the total mass of air above that point. As we climb in elevation, the mass of air above us decreases, and therefore pressure declines. Although atmospheric pressure decreases continuously, the rate of decline slows with increasing altitude (Figure 2.10).

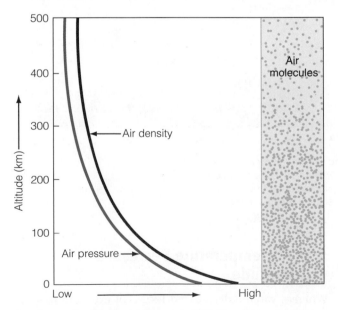

Figure 2.10 | Both air pressure and air density decrease with increasing altitude above sea level.

Because of the greater air pressure at Earth's surface, the density of air (the number of molecules per unit volume) is high, decreasing in parallel with air pressure as we climb in altitude. As altitude above sea level increases, both air pressure and density decrease. Although by an altitude of 50 km, air pressure is only 0.1 percent of that measured at sea level, the atmosphere continues to extend upward for many hundreds of kilometers, gradually becoming thinner and thinner until it merges into outer space.

Air pressure and density decrease systematically with height above sea level, but air temperature has a more complicated vertical profile. Air temperature normally decreases from Earth's surface up to an altitude of approximately 11 km (nearly 36,000 feet). The rate at which temperature decreases with altitude is called the **environmental lapse rate.**

The decrease in air temperature as one moves farther from Earth's surface is caused by two factors. Because of the greater air pressure at the surface, air molecules move more quickly. Since temperature is a measure of the average speed of the air molecules (average kinetic energy of the particles), the average speed of the molecules and the temperature are both higher near the surface. The decrease in air pressure with altitude results in a reduced rate of motion of the molecules and therefore a decline in temperature. The primary reason for the decrease in air temperature with increasing altitude, however, is the corresponding decline in the "warming effect" of Earth's surface. The absorption of solar radiation warms Earth's surface. Energy (longwave radiation) is emitted upward from the surface, heating the air above it. This process of transfer continues upward as heat flows spontaneously from warmer to cooler areas but at a continuously declining rate as the energy emitted from the surface is dissipated.

Unlike air pressure and density, air temperature does not decline continuously with increasing height above Earth's surface. In fact, at certain heights in the atmosphere, a change in altitude can result in an abrupt change in temperature. Atmospheric scientists use these specific altitudes to distinguish different regions in the atmosphere (Figure 2.11). Beginning at Earth's surface, the regions are called the troposphere, the stratosphere, the mesosphere, and the thermosphere. The boundary zones between these four regions of the atmosphere are the tropopause, stratopause, and mesopause, respectively. The two most important regions for climate, and therefore life on Earth, are the troposphere and stratosphere.

So far, this discussion of the change in air temperature with increasing altitude has assumed no vertical movement of air from the surface to the top of the atmosphere. When a volume of air at the surface warms, however, it becomes buoyant and rises (like a hot-air balloon does). As the volume of air (known as a parcel of air) rises, decreasing pressure makes it expand and cool.

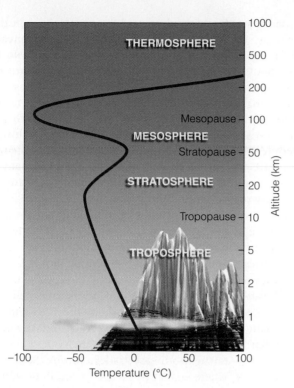

Figure 2.11 | Changes in atmospheric temperature (global average) with altitude above sea level. Regions of the atmosphere are labeled, and Mount Everest (the highest mountain peak on Earth) is drawn for perspective.

(Adapted from Graedel and Crutzen 1995.)

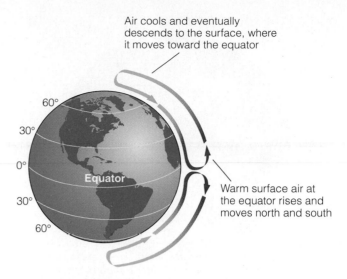

Figure 2.12 | Circulation of air cells and prevailing winds on an imaginary, nonrotating Earth. Air heated at the equator rises and moves north and south. After cooling at the poles, it descends and moves back toward the equator.

The decrease in air temperature through expansion, rather than through heat loss to the surrounding atmosphere, is called **adiabatic cooling.** The same process works in an air conditioner, where a coolant is compressed. As the coolant moves from the compressor to the coils, the drop in pressure makes it expand and cool.

The rate of adiabatic cooling depends on how much moisture is in the air. The adiabatic cooling of dry air is approximately 10°C per 1000 m elevation. Moist air cools more slowly ($\sim$6°C per 1000 m; see Section 3.2 for discussion of differences in the specific heat of water and air). The rate of temperature change with elevation is called the **adiabatic lapse rate.**

2.4 | Air Masses Circulate Globally

The blanket of air surrounding Earth—the atmosphere—is not static. It is in a constant state of movement, driven by rising and sinking air masses and by Earth's rotation on its axis. The equatorial region receives the largest annual input of solar radiation. Warm air rises because it is less dense than the cooler air above it. Air heated at the equatorial region rises to the top of the troposphere, establishing a zone of low pressure down at the surface (Figure 2.12). More air rising beneath it forces the air mass to spread north and south toward the poles. As air

masses move poleward, they cool, become heavier, and sink. The sinking air raises surface air pressure (high-pressure zone). The cooled, heavier air then flows toward the low pressure zone at the equator, replacing the warm air rising over the tropics and closing the pattern of air circulation.

If Earth were stationary and without irregular landmasses, the atmosphere would circulate as shown in Figure 2.12. Earth, however, spins on its axis from west to east. Although each point on Earth's surface makes a complete rotation every 24 hours, the speed of rotation varies with latitude (and circumference). At a point on the equator (its widest circumference at 40,176 km), the speed of rotation is 1674 km per hour. In contrast, at 60° north or south, Earth's circumference is approximately half that at the equator (20,130 km), and the speed of rotation is 839 km per hour. According to the law of angular motion, the momentum of an object moving from a greater circumference to a lesser circumference will deflect in the direction of the spin, and an object moving from a lesser circumference to a greater circumference will deflect in the direction opposite that of the spin. As a result, air masses and all moving objects in the Northern Hemisphere are deflected to the right (clockwise motion) and in the Southern Hemisphere to the left (counterclockwise motion). This deflection in the pattern of air flow is the **Coriolis effect,** named after the 19th-century French mathematician G. C. Coriolis, who first analyzed the phenomenon (Figure 2.13).

Due to the deflection of air masses, the Coriolis effect prevents a direct, simple flow of air from the equator to the poles. Instead, it creates a series of belts of prevailing winds, named for the direction they come from. These belts break the simple flow of surface air toward the

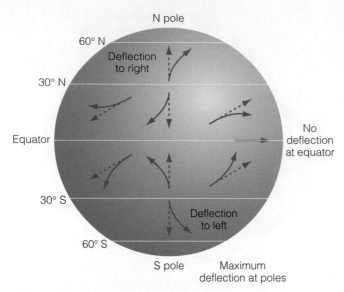

Figure 2.13 | Effect of the Coriolis force on wind direction. The effect is absent at the equator, where the linear velocity is the greatest, 465 m/s (1040 mi/h). Any object on the equator is moving at the same rate. The Coriolis effect increases regularly toward the poles. If an object, including an air mass, moves northward from the equator at a constant speed, it speeds up because Earth moves more slowly (403 m/s at 30° latitude, 233 m/s at 60° latitude, and 0 m/s at the poles) than the object does. As a result, the object's path appears to deflect to the right or east in the Northern Hemisphere and to the left or west in the Southern Hemisphere.

equator and the flow aloft to the poles into a series of six cells, three in each hemisphere. They produce areas of low and high pressure as air masses ascend from and descend toward the surface, respectively (Figure 2.14). To trace the flow of air as it circulates between the equator and poles, we begin at Earth's equatorial region, which receives the largest annual input of solar radiation.

Air heated in the equatorial zone rises upward, creating a low-pressure zone near the surface—the **equatorial low.** This upward flow of air is balanced by a flow of air

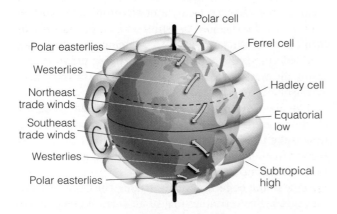

Figure 2.14 | Belts and cells of air circulation about a rotating Earth. This circulation gives rise to the trade, westerly, and easterly winds.

from the north and south toward the equator (equatorial low). As the warm air mass rises it begins to spread, diverging northward and southward toward the North and South Poles, cooling as it goes. In the Northern Hemisphere, the Coriolis effect forces air in an easterly direction, slowing its progress north. At about 30° north latitude, the now cool air sinks, closing the first of the three cells—the Hadley cells, named for the Englishman George Hadley, who first described this pattern of circulation in 1735. The descending air forms a semipermanent high-pressure belt at the surface and encircling Earth—the **subtropical high.** Having descended, the cool air warms and splits into two currents flowing over the surface. One moves northward toward the pole, diverted to the right by the Coriolis effect to become the prevailing **westerlies.** Meanwhile, the other current moves southward toward the equator. Also deflected to the right, this southward-flowing stream becomes the strong, reliable winds that were called **trade winds** by the 17th-century merchant sailors who used them to reach the Americas from Europe. In the Northern Hemisphere, these winds are known as the northeast trades. In the Southern Hemisphere, where similar flows take place, these winds are known as the southeast trades.

As the mild air of the westerlies moves poleward, it encounters cold air moving down from the pole (approximately 60° N). These two air masses of contrasting temperature do not readily mix. They are separated by a boundary called the polar front—a zone of low pressure (the **subpolar low**) where surface air converges and rises. Some of the rising air moves southward until it reaches approximately 30° latitude (region of the subtropical high), where it sinks back to the surface and closes the second of the three cells—the Ferrel cell, named after the American meteorologist William Ferrel.

As the northward-moving air reaches the pole, it slowly sinks to the surface and flows back (southward) toward the polar front, completing the last of the three cells—the polar cell. This southward-moving air is deflected to the right by the Coriolis effect, giving rise to the **polar easterlies.** Similar flows occur in the Southern Hemisphere (see Figure 2.14).

2.5 | Solar Energy, Wind, and Earth's Rotation Create Ocean Currents

The global pattern of prevailing winds plays a crucial role in determining major patterns of surface water flow in Earth's oceans. These systematic patterns of water movement are called **currents.** In fact, until they encounter one of the continents, the major ocean currents generally mimic the movement of the wind currents above.

Each ocean is dominated by two great circular water motions, or **gyres.** Within each gyre, the ocean current moves clockwise in the Northern Hemisphere and counterclockwise in the Southern Hemisphere (Figure 2.15).

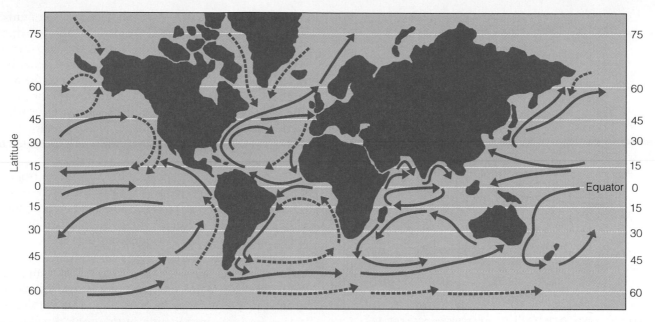

Figure 2.15 | Ocean currents of the world. Notice how the circulation is influenced by the Coriolis force (clockwise movement in the Northern Hemisphere and counterclockwise movement in the Southern Hemisphere) and continental landmasses and how oceans are connected by currents. Dashed arrows represent cool water, and solid arrows represent warm water.

Along the equator, trade winds push warm surface waters westward. When these waters encounter the eastern margins of continents, they split into north- and south-flowing currents along the coasts, forming north and south gyres. As the currents move farther from the equator, the water cools. Eventually, they encounter the westerly winds at higher latitudes (30–60° N and 30–60° S), which produce eastward-moving currents. When these eastward-moving currents encounter the western margins of the continents, they form cool currents that flow along the coastline toward the equator. Just north of the Antarctic continent, ocean waters circulate unimpeded around the globe.

2.6 | Temperature Influences the Moisture Content of Air

Air temperature plays a crucial role in the exchange of water between the atmosphere and Earth's surface. Whenever matter, including water, changes from one state to another, energy is either absorbed or released. The amount of energy released or absorbed (per gram) during a change of state is known as **latent heat** (from the Latin *latens*, "hidden"). In going from a more ordered state (liquid) to a less ordered state (gas), energy is absorbed (energy required to break bonds between molecules). While going from a less ordered to a more ordered state, energy is released. **Evaporation,** the transformation of water from a liquid to a gaseous state, requires 2260 joules (J) of energy per gram of liquid water converted to water vapor (1 joule is the equivalent energy of 1 watt of power radiated or dissipated for 1 second). **Condensation,** the transformation of water vapor to a liquid state, releases an equivalent amount of energy. When air comes into contact with liquid water, water molecules are freely exchanged between the air and the water's surface. When the evaporation rate equals the condensation rate, the air is said to be saturated.

In the air, water vapor acts as an independent gas that has weight and exerts pressure. The amount of pressure water vapor exerts independent of the pressure of dry air is called **vapor pressure.** Vapor pressure is typically defined in units of pascals (Pa). The water vapor content of air at saturation is called the **saturation vapor pressure.** The saturation vapor pressure, also known as the water vapor capacity of air, cannot be exceeded. If the vapor pressure exceeds the capacity, condensation occurs and reduces the vapor pressure. Saturation vapor pressure varies with temperature, increasing as air temperature increases (due to increasing molecular kinetic energy; Figure 2.16). Warm air has a greater capacity for water vapor than does cold air.

The amount of water in a given volume of air is its absolute humidity. The most familiar measure is **relative humidity,** or the amount of water vapor in the air expressed as a percentage of the saturation vapor pressure. At saturation vapor pressure, the relative humidity is 100 percent. If air cools while the actual moisture content (water vapor pressure) remains constant, then relative humidity increases as the value of saturation vapor pressure declines. If the air cools to a point where the actual vapor pressure exceeds the saturation vapor pressure,

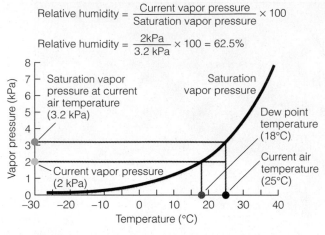

$$\text{Relative humidity} = \frac{\text{Current vapor pressure}}{\text{Saturation vapor pressure}} \times 100$$

$$\text{Relative humidity} = \frac{2\,\text{kPa}}{3.2\,\text{kPa}} \times 100 = 62.5\%$$

Figure 2.16 | Saturation vapor pressure (VP) as a function of air temperature (saturation VP increases with air temperature). For a given air temperature, the relative humidity is the ratio of current VP to saturation VP: (current VP/saturation VP) × 100. For a given VP, the temperature at which saturation VP occurs is called the dew point. (Go to QUANTIFYit! at www.ecology.place.com to review functions.)

Interpreting Ecological Data

Q1. Assume that the actual (current) water vapor pressure remains the same over the course of the day, and that the current air temperature of 25°C in the above graph represents the air temperature at noon (12:00 p.m.). How would you expect the relative humidity to change from noon to 5:00 p.m.? Why?

Q2. What is the approximate relative humidity at 35°C? (Assume that actual water vapor pressure remains the same as in the above figure: 2 kPa.)

moisture in the air will condense and form clouds. As soon as particles of water or ice in the air become too heavy to remain suspended, precipitation falls.

For a given water content of a parcel of air (vapor pressure), the temperature at which saturation vapor pressure is achieved is called the **dew point temperature.** Think about finding dew or frost on a cool fall morning. As nightfall approaches, temperatures drop and relative humidity rises. If cool night air temperatures reach the dew point, water condenses and dew forms, lowering the amount of water in the air. As the sun rises, air temperature warms and the water vapor capacity (saturation vapor pressure) increases. As a result, the dew evaporates, increasing vapor pressure in the air.

2.7 | Precipitation Has a Distinctive Global Pattern

By bringing together patterns of temperature, winds, and ocean currents, we are ready to understand the global pattern of precipitation. Precipitation is not evenly distributed across Earth (Figure 2.17). At first the global map of annual precipitation in Figure 2.17 may seem to have no discernible pattern or regularity. But if we exam-

ine the simpler pattern of variation in average rainfall with latitude (Figure 2.18), a general pattern emerges. Precipitation is highest in the region of the equator, declining as one moves north and south. The decline, however, is not continuous. Two peaks occur in the midlatitudes followed by a further decline toward the poles. The sequence of peaks and troughs seen in Figure 2.18 corresponds to the pattern of rising and falling air masses associated with the belts of prevailing winds presented in Figure 2.14.

As the warm trade winds move across the tropical oceans, they gather moisture. Near the equator, the northeasterly trade winds meet the southeasterly trade winds. This narrow region where the trade winds meet is the **intertropical convergence zone (ITCZ),** characterized by high amounts of precipitation. Where the two air masses meet, air piles up, and the warm humid air rises and cools. When the dew point is reached, clouds form, and precipitation falls as rain. This pattern accounts for high precipitation in the tropical regions of eastern Asia, Africa, and South and Central America (see Figure 2.17).

Having lost much of its moisture, the ascending air mass continues to cool as it splits and moves northward and southward. In the region of the subtropical high (approximately 30° north and south), where the cool air descends, two belts of dry climate encircle the globe (the two troughs at the midlatitudes seen in Figure 2.18). The descending air warms. Because the saturation vapor pressure rises, it draws water from the surface through evaporation, causing arid conditions. In these belts, the world's major deserts have formed (see Chapter 23).

As the air masses continue to move north and south, they once again draw moisture from the surface, but to a lesser degree because of the cooler surface conditions. Moving poleward, they encounter cold air masses originating at the poles (approximately 60° north and south). Where the surface air masses converge and rise, the ascending air mass cools and precipitation occurs (seen as the two peaks in precipitation between 50° and 60° north and south in Figure 2.18). From this point on to the poles, the cold temperature and associated low-saturation vapor pressure function to restrict precipitation.

One other pattern is worth noting in Figure 2.18. In general, rainfall is greater in the Southern Hemisphere than in the Northern Hemisphere (note the southern shift in the rainfall peak associated with the ITCZ). This is because the oceans cover a greater proportion of the Southern Hemisphere, and water evaporates more readily from the water's surface than from the soil and vegetation. This is also why the interior of continents generally experiences less precipitation than the coastal regions do. As air masses move inland from the coast, water vapor lost from the atmosphere through precipitation is not recharged (from surface evaporation) as readily over land

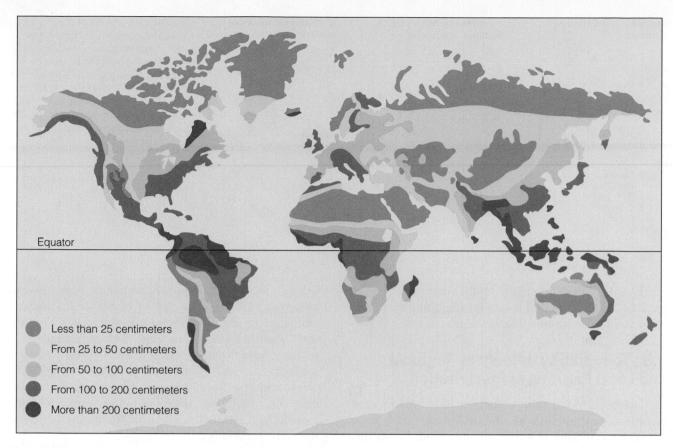

Figure 2.17 | Annual world precipitation. Relate the wettest and driest areas to mountain ranges, ocean currents, and winds.

Legend:
- Less than 25 centimeters
- From 25 to 50 centimeters
- From 50 to 100 centimeters
- From 100 to 200 centimeters
- More than 200 centimeters

Equator

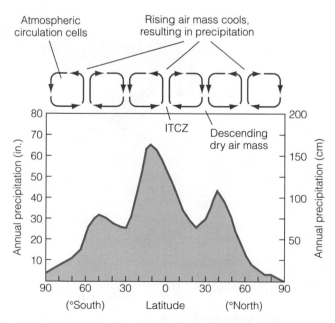

Figure 2.18 | Variation in mean annual precipitation with latitude. The peaks in rainfall correspond to rising air masses, such as that of the intertropical convergence zone, whereas the troughs are associated with descending dry air masses.

as it is over the open waters of the ocean (note the gradients of precipitation from the coast to the interiors of North America and Europe/Asia in Figure 2.17).

Missing from our discussion thus far is the temporal variation of precipitation over Earth. The temporal variation is directly linked to the seasonal changes in the heating of Earth and its effect on the movement of global pressure systems and air masses. This is illustrated in seasonal movement north and south of the ITCZ, which follows the apparent migration of the direct rays of the Sun (see Figure 2.19).

The ITCZ is not stationary but tends to migrate toward regions of the globe with the warmest surface temperature. Although tropical regions about the equator are always exposed to warm temperatures, the Sun is directly over the geographical equator only twice a year, at the spring and fall equinoxes. At the northern summer solstice, the Sun is directly over the Tropic of Cancer; at the winter solstice (which is summer in the Southern Hemisphere), the Sun is directly over the Tropic of Capricorn. As a result, the ITCZ moves poleward and invades the subtropical highs in northern summer; in the winter it moves southward, leaving clear, dry weather behind. As the ITCZ migrates

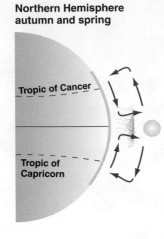

Northern Hemisphere summer

Equator

Northern Hemisphere autumn and spring

Tropic of Cancer

Tropic of Capricorn

Northern Hemisphere winter

Equator

Figure 2.19 | Shifts of the intertropical convergence zone, producing rainy seasons and dry seasons. As the distance from the equator increases, the dry season is longer and the rainfall is less. These oscillations result from changes in the Sun's altitude between the equinoxes and the solstices, as diagrammed in Figure 2.6. Patterns of air circulation are shown in Figure 2.14.

southward, it brings rain to the southern summer. Thus, as the ITCZ shifts north and south, it brings on the wet and dry seasons in the tropics (Figure 2.20).

2.8 | Topography Influences Regional and Local Patterns of Precipitation

Mountainous topography influences local and regional patterns of precipitation. Mountains intercept air flow. As an air mass reaches a mountain, it ascends, cools, becomes saturated with water vapor (because of lower saturation vapor pressure), and releases much of its moisture at upper altitudes of the windward side. As the now cool, dry air descends the leeward side, it warms again and gradually picks up moisture. As a result, the windward

side of a mountain supports denser, more vigorous vegetation and different species of plants and associated animals than does the leeward side, where in some areas dry, desert-like conditions exist. This phenomenon is called a **rain shadow** (Figure 2.21). Thus in North America, the westerly winds that blow over the Sierra Nevada and the Rocky Mountains, dropping their moisture on west-facing slopes, support vigorous forest growth. By contrast, the eastern slopes exhibit semidesert or desert conditions.

Some of the most pronounced effects of this same phenomenon occur in the Hawaiian Islands. There, plant cover ranges from scrubby vegetation on the leeward side of an island to moist, forested slopes on the windward side (Figure 2.22).

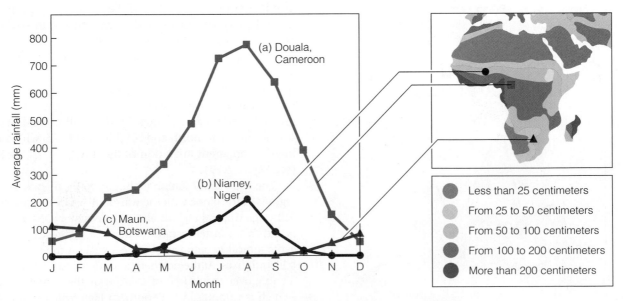

Figure 2.20 | Seasonal variations in precipitation at three sites within the intertropical convergence zone. Although site **(a)** shows a seasonal variation, precipitation exceeds 50 mm each month. Sites **(b)** and **(c)** are in the ITCZ regions that experience a distinct wet (summer) and dry (winter) season. The rainy season is 6 months out of phase for these two sites, reflecting the difference in the timing because the summer months occur at different times in the two hemispheres.

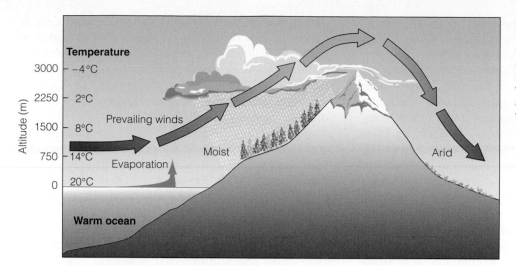

Figure 2.21 | Formation of a rain shadow. Air is forced to go over a mountain. As it rises, the air mass cools and loses its moisture as precipitation on the windward side. The descending air, already dry, picks up moisture from the leeward side.

2.9 | Irregular Variations in Climate Occur at the Regional Scale

The patterns of temporal variation in climate that we have discussed thus far occur at regular and predictable intervals: seasonal changes in temperature with the rotation of Earth about the Sun, and migration of the intertropical convergence zone with the resulting seasonality of rainfall in the tropics and monsoons in Southeast Asia. Not all features of the climate system, however, occur so regularly. Earth's climate system is characterized by variability at both the regional and global scales.

The Little Ice Age, a period of cooling that lasted from approximately the mid-14th to the mid-19th century, brought bitterly cold winters to many parts of the Northern Hemisphere, affecting agriculture, health, politics, economics, emigration, and even art and literature. In the mid-17th century, glaciers in the Swiss Alps advanced, gradually engulfing farms and crushing entire villages. In

1780, New York Harbor froze, allowing people to walk from Manhattan to Staten Island. In fact, the image of a white Christmas evoked by Charles Dickens and the New England poets of the 18th and 19th centuries is largely a product of the cold and snowy winters of the Little Ice Age. But the climate has since warmed to the point where a white Christmas in these regions is an anomaly.

The Great Plains region of central North America has undergone periods of drought dating back to the mid-Holocene period some 5000 to 8000 years ago, but the homesteaders of the early 20th century settled the Great Plains at a time of relatively wet summers. They assumed these moisture conditions were the norm, and they employed the agricultural methods they had used in the East. So they broke the prairie sod for crops; but the cycle of drought returned, and the prairie grasslands became a dust bowl.

These examples reflect the variability in Earth's climate systems, which operate on timescales ranging from

(a)

(b)

Figure 2.22 | Rain shadow on the mountains of Maui, Hawaiian Islands. **(a)** The windward, east-facing slopes intercept the trade winds and are cloaked with wet forest. **(b)** Low-growing, shrubby vegetation is found on the dry side.

decades to tens of thousands of years, driven by changes in the input of energy to Earth's surface (see Chapter 29: Global Climate Change). Earth's orbit is not permanent. Changes occur in the tilt of the axis and the shape of the yearly path about the Sun. These variations affect climate by altering the seasonal inputs of solar radiation. Occurring on a timescale of tens of thousands of years, these variations are associated with the glacial advances and retreats throughout Earth's history (see Chapter 18).

Variations in the level of solar radiation to Earth's surface are also associated with sunspot activity—huge magnetic storms on the Sun. These storms are associated with strong solar emissions and occur in cycles, with the number and size reaching a maximum approximately every 11 years. Researchers have related sunspot activity, among other occurrences, to periods of drought and winter warming in the Northern Hemisphere.

Interaction between two components of the climate system, the ocean and the atmosphere, are connected to some major climatic variations that occur at a regional scale. As far back as 1525, historic documents reveal that fishermen off the coast of Peru recorded periods of unusually warm water. The Peruvians referred to these as El Niño, because they commonly appear at Christmastime, the season of the Christ Child (Spanish: *El Niño*). El Niño is the phenomenon in the waters of the Galápagos Islands described in the introduction of Part One (pp. 16–17). Now referred to by scientists as the **El Niño– Southern Oscillation (ENSO),** this phenomenon is a global event arising from large-scale interaction between the ocean and the atmosphere. The Southern Oscillation, a more recent discovery, refers to an oscillation in the surface pressure (atmospheric mass) between the southeastern tropical Pacific and the Australian-Indonesian regions. When the waters of the eastern Pacific are abnormally warm (an El Niño event), sea level pressure drops in the eastern Pacific and rises in the west. The reduction in the pressure gradient is accompanied by a weakening of the low-latitude easterly trades.

Although scientists still do not completely understand the cause of the ENSO phenomenon, its mechanism has been well documented. Recall from Section 2.4 that the trade winds blow westward across the tropical Pacific (see Figure 2.14). As a consequence, the surface currents within the tropical oceans flow westward (see Figure 2.15), bringing cold, deeper waters to the surface off the coast of Peru in a process known as upwelling (see Section 3.8). This pattern of upwelling, together with the cold-water current flowing from south to north along the western coast of South America, results in this region of the ocean being normally colder than one would expect given its equatorial location (Figure 2.23a).

As the surface currents move westward the water warms, giving the water's destination, the western Pacific, the warmest ocean surface on Earth. The warmer water of the western Pacific causes the moist maritime air to rise and cool, bringing abundant rainfall to the region (Figure 2.23a; also see Figure 2.17). In contrast, the cooler waters of the eastern Pacific result in relatively dry conditions along the Peruvian coast.

During an El Niño event, the trade winds slacken, reducing the westward flow of the surface currents (Figure 2.23b). The result is a reduced upwelling and a warming of the surface waters in the eastern Pacific. Rainfall

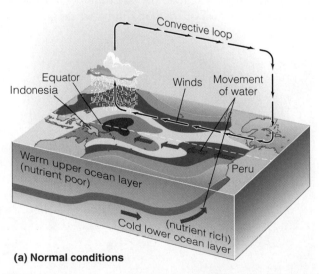

(a) Normal conditions

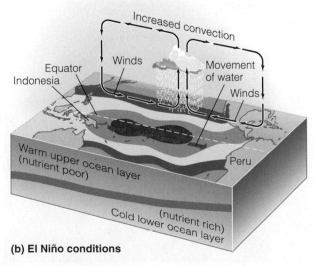

(b) El Niño conditions

Figure 2.23 | Schematic of the El Niño–Southern Oscillation (ENSO) that occurs off the western coast of South America. **(a)** Under normal conditions, strong trade winds move surface waters westward. As the surface currents move westward, the water warms. The warmer water of the western Pacific causes the moist maritime air to rise and cool, bringing abundant rainfall to the region. **(b)** Under ENSO conditions, the trade winds slacken, reducing the westward flow of the surface currents. Rainfall follows the warm water eastward, with associated flooding in Peru and drought in Indonesia and Australia.

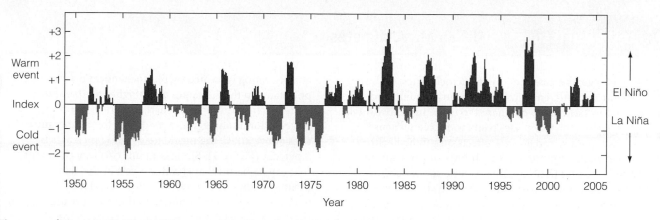

Figure 2.24 | Record of El Niño–La Niña events during the second half of the 20th century. Numbers at the left of the diagram represent the ENSO index, which includes a combination of six factors related to environmental conditions over the tropical Pacific Ocean: air temperature, surface water temperature, sea-level pressure, cloudiness, and wind speed and direction. Warm episodes are in red; cold episodes are in blue. An index value greater than +1 represents an El Niño. A value less than −1 represents a La Niña.

follows the warm water eastward, with associated flooding in Peru and drought in Indonesia and Australia.

This eastward displacement of the atmospheric heat source (latent heat associated with the evaporation of water; see Section 3.2) overlaying the warm surface waters results in large changes in the global atmospheric circulation, in turn influencing weather in regions far removed from the tropical Pacific.

At other times, the injection of cold water becomes more intense than usual, causing the surface of the eastern Pacific to cool. This variation is referred to as **La Niña** (Figure 2.24). It results in droughts in South America and heavy rainfall, even floods, in eastern Australia.

2.10 | Most Organisms Live in Microclimates

Most organisms live in local conditions that do not match the general climate profile of the larger region surrounding them. For example, today's weather report may state that the temperature is 28°C and the sky is clear. However, your weather forecaster is painting only a general picture. Actual conditions of specific environments will be quite different depending on whether they are underground versus on the surface, beneath vegetation or on exposed soil, on mountain slopes or at the seashore. Light, heat, moisture, and air movement all vary greatly from one part of the landscape to another, influencing the transfer of heat energy and creating a wide range of localized climates. These **microclimates** define the conditions organisms live in (see Ecological Issues: Urban Microclimates).

On a sunny but chilly day in early spring, flies may be attracted to sap oozing from the stump of a maple tree. The flies are active on the stump despite the near-freezing air temperature because, during the day, the surface of the

stump absorbs solar radiation, heating a thin layer of air above the surface. On a still day, the air heated by the tree stump remains close to the surface, and temperatures decrease sharply above and below this layer. A similar phenomenon occurs when the frozen surface of the ground absorbs solar radiation and thaws. On a sunny, late winter day, the ground is muddy even though the air is cold.

By altering soil temperatures, moisture, wind movement, and evaporation, vegetation moderates microclimates, especially areas near the ground. For example, areas shaded by plants have lower temperatures at ground level than do places exposed to the Sun. On fair summer days in locations 25 mm (1 inch) aboveground, dense forest cover can reduce the daily range of temperatures by 7°C to 12°C below soil temperature in bare fields. Under the shelter of heavy grass and low plant cover, the air at ground level is completely calm. This calm is an outstanding feature of microclimates within dense vegetation at Earth's surface. It influences both temperature and humidity, creating a favorable environment for insects and other ground-dwelling animals.

Topography, particularly aspect (the direction that a slope faces), influences the local climatic conditions. In the Northern Hemisphere, south-facing slopes receive the most solar energy, whereas north-facing slopes receive the least. At other slope positions, energy varies between these extremes, depending on their compass direction.

Different exposure to solar radiation at south- and north-facing sites has a marked effect on the amount of moisture and heat present. Microclimate conditions range from warm, dry, variable conditions on the south-facing slope to cool, moist, more uniform conditions on the north-facing slope. Because high temperatures and associated high rates of evaporation draw moisture from soil and plants, the evaporation rate at south-facing slopes is often 50 percent higher, the average temperature is higher, and soil moisture is lower. Conditions are driest on the

Urban areas create their own microclimates, with significant differences in temperature, rainfall, and wind flow patterns as compared to those in nearby rural areas. Urban microclimates result in high energy use, poor air quality, and adverse effects on public health.

On warm summer days with little or no wind, the air temperature in urban areas can be several degrees hotter than in the surrounding countryside. Scientists thus refer to these urban areas as "urban heat islands." Urban areas are warmer than surrounding rural environments because of their energy balance, or the difference between the amounts of energy gained and lost. In rural environments, the solar energy absorbed by the vegetation and ground is partially dissipated by water evaporation from the vegetation and soil. In urban areas there is less vegetation, so the buildings, streets, and sidewalks absorb most of the solar energy. In areas with narrow streets and tall buildings, the building walls radiate heat toward one another instead of skyward. Further, because the man-made surfaces of asphalt, cement, and brick are not porous, most rainfall is lost as runoff to storm drains before evaporation can cool the air. Waste heat from cars, buses, and city buildings also contributes to the input of energy. Although this waste heat eventually makes its way into the atmosphere, it can contribute as much as one-third of the heat received from solar energy. Adding to the problem are construction materials (asphalt, concrete, bricks, and tar), which are better conductors of heat than is the vegetation that dominates the landscapes of surrounding rural areas. At night, these materials slowly give off heat that was stored during the day.

The heat island effect can raise temperatures from 6°C to 8°C above those in the surrounding countryside. In Baltimore, Phoenix, Tucson, and Washington, D.C., for example, scientific data show that July's maximum temperatures during the past 30 to 80 years have been steadily increasing at a rate of 0.5 to 1 degree Fahrenheit every decade. The highest temperatures are in areas of highest population density and activity, whereas temperatures decline markedly toward the periphery of the city. Although temperature differences are detectable throughout the year, the heat island effect is most pronounced during the summer and early winter—particularly at night, when heat stored by pavement and buildings reradiates into the air.

The heat island effect also has an impact on air quality within urban areas. Throughout the year, urban areas are blanketed with particulate matter and pollutants from the combustion of fossil fuels and industrial activity. Smog is created by photochemical reactions of pollutants in the atmosphere. At higher temperatures, these reactions occur at an increasing rate. In Los Angeles, for example, for every degree the temperature rises above 20°C, the incidence of smog increases by 3 percent (Figure 1). Smog contains ozone, a pollutant that can be harmful at elevated levels in the air we breathe. Ozone also adversely affects vegetation both within the urban environment and in the surrounding rural areas. The heat island effect exacerbates these effects on air quality, as higher ambient temperatures during the summer months increase air-conditioning energy use. As power plants burn more fossil fuels, they increase both pollution levels and energy costs.

Particulate matter has other microclimatic effects. Because of a city's low evaporation rate and the lack of vegetation, relative humidity is lower in urban areas than in surrounding rural areas. However, the particulates act as condensation nuclei for water vapor in the air, producing fog and haze. Fog is much more frequent in urban areas than in the country, especially in winter.

1. What simple steps could be taken to decrease the heat island effect in urban areas?

2. How might the urban heat island effect influence surrounding rural environments?

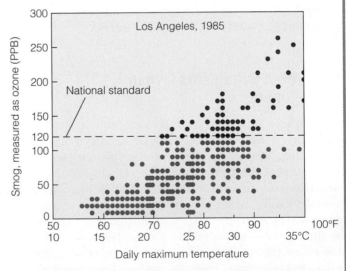

Figure 1 | Relationship between daily maximum temperature and smog concentration (measured as ozone concentration in parts per billion—PPB) for the city of Los Angeles (1985). The dashed horizontal line at 120 PPB represents the national standard as defined by the U.S. Environmental Protection Agency.

top of south-facing slopes, where air movement is greatest, and dampest at the bottom of north-facing slopes.

The same microclimatic conditions occur on a smaller scale on north- and south-facing slopes of large ant hills, mounds of soil, dunes, and small ground ridges in otherwise flat terrain, as well as on the north- and south-facing sides of buildings, trees, and logs. The south-facing sides of buildings are always warmer and drier than the north-facing sides—a consideration for landscape planners, horticulturists, and gardeners. North sides of tree trunks are cooler and moister than south sides, as reflected by more vigorous growth of moss on the north sides. In winter, the temperature of the north-facing side of a tree may be below freezing while the south side, heated by the Sun, is warm. This temperature difference may cause frost cracks in the bark as sap, thawed by day, freezes at night. Bark beetles and other wood-dwelling insects that seek cool, moist areas for laying their eggs prefer north-facing locations. Flowers on the south side of tree crowns often bloom sooner than those on the north side.

Microclimatic extremes also occur in depressions in the ground and on the concave surfaces of valleys, where the air is protected from the wind. Heated by sunlight during the day and cooled by terrestrial vegetation at night, this air often becomes stagnant. As a result, these sheltered sites experience lower nighttime temperatures (especially in winter), higher daytime temperatures (especially in summer), and higher relative humidity. If the temperature drops low enough, frost pockets form in these depressions. The microclimates of the frost pockets often display the same phenomenon, supporting different kinds of plant life than found on surrounding higher ground.

Although the global and regional patterns of climate discussed in this chapter constrain the large-scale distribution and abundance of plants and animals, the localized patterns of microclimate define the actual environmental conditions sensed by the individual organism. This localized microclimate thus determines the distribution and activities of organisms in a particular region.

Summary

Interception of Solar Radiation | 2.1

Earth intercepts solar energy in the form of shortwave radiation, which easily passes through the atmosphere, and emits much of it back as longwave radiation. However, energy of longer wavelengths cannot readily pass through the atmosphere and so is returned to Earth, producing the greenhouse effect.

Seasonal Variation | 2.2

The amount of solar radiation intercepted by Earth varies markedly with latitude. Tropical regions near the equator receive the greatest amount of solar radiation, and high latitudes receive the least. Because Earth tilts on its axis, parts of Earth receive seasonal differences in solar radiation. These differences give rise to seasonal variations in temperature and rainfall. There is a global gradient in mean annual temperature; it is warmest in the tropics and declines toward the poles.

Altitude and Temperature | 2.3

Heating and cooling, influenced by energy emitted from Earth's surface and by atmospheric pressure, cause air masses to rise and sink. This movement of air masses involves an adiabatic process in which heat is neither gained from nor lost to the outside.

Atmospheric Circulation | 2.4

Vertical movements of air masses give rise to global patterns of atmospheric circulation. The spin of Earth on its axis deflects air and water currents to the right in the Northern Hemisphere and to the left in the Southern Hemisphere. Three cells of global air flow occur in each hemisphere.

Ocean Currents | 2.5

The global pattern of winds and the Coriolis effect cause major patterns of ocean currents. Each ocean is dominated by great circular water motions, or gyres. These gyres move clockwise in the Northern Hemisphere and counterclockwise in the Southern Hemisphere.

Atmospheric Moisture | 2.6

Atmospheric moisture is measured in terms of relative humidity. The maximum amount of moisture the air can hold at any given temperature is called the saturation vapor pressure, which increases with temperature. Relative humidity is the amount of water in the air, expressed as a percentage of the maximum amount the air could hold at a given temperature.

Precipitation | 2.7

Wind, temperature, and ocean currents produce global patterns of precipitation. They account for regions of high precipitation in the tropics and belts of dry climate at approximately 30° N and S latitude.

Topography | 2.8

Mountainous topography influences local and regional patterns of precipitation. As an air mass reaches a mountain, it ascends, cools, becomes saturated with water vapor, and releases much of its moisture at upper altitudes of the windward side.

Irregular Variation | 2.9

Not all temporal variation in regional climate occurs at a regular interval. Irregular variations in the trade winds give rise to periods of unusually warm waters off the coast of western South America. Referred to by scientists as the El Niño, this phenomenon is a global event arising from large-scale interaction between the ocean and the atmosphere.

Microclimates | 2.10

The actual climatic conditions that organisms live in vary considerably within one climate. These local variations, or microclimates, reflect topography, vegetative cover, exposure, and other factors on every scale. Angles of solar radiation cause marked differences between north- and south-facing slopes, whether on mountains, sand dunes, or ant mounds.

Study Questions

1. Why does the equator receive more solar radiation than the polar regions? What is the consequence of latitudinal patterns of temperature?
2. What is the greenhouse effect, and how does it influence the energy balance (and temperature) of Earth?
3. The 23.5° tilt of Earth on its north–south axis gives rise to the seasons (review Figure 2.5). How would the pattern of seasons differ if the Earth's tilt were 90°? How would this influence the diurnal (night–day) cycle?
4. Why are the coastal waters of the southeastern United States warmer than the coastal waters off the southwestern coast? (Assume the same latitude.)
5. The air temperature at noon on January 20 was 45°F, and the air temperature at noon on July 20 at the same location was 85°F. The relative humidity on both days was 75 percent. On which of these two days was there more water vapor in the air?
6. How might the relative humidity of a parcel of air change as it moves up the side of a mountain? Why?
7. What is the intertropical convergence zone (ITCZ), and why does it give rise to a distinct pattern of seasonality in precipitation in the tropical zone?
8. What feature of global atmospheric circulation gives rise to the desert zones of the midlatitudes?
9. Which aspect, south-facing or north-facing slopes, would receive the most solar radiation in the mountain ranges of the Southern Hemisphere?
10. Spruce Knob (latitude 38.625° N) in eastern West Virginia is named for the spruce trees dominating the forests at this site. Spruce trees are typically found in the colder forests of the more northern latitudes (northeastern United States and Canada). What does the presence of spruce trees at Spruce Knob tell you about this site?

Further Readings

Ahrens, C. D. 2003. *Meteorology today: An introduction to weather, climate, and the environment.* 6th ed. Belmont, CA: Brooks/Cole.
    An excellent introductory text on climate, clearly written and well illustrated.

Fagen, B. 2001. *The Little Ice Age: How climate made history, 1300–1850.* New York: Basic Books.
    An enjoyable book that gives an overview of the effects of the Little Ice Age on human history.

Graedel, T. E., and P. J. Crutzen. 1997. *Atmosphere, climate and change.* New York: Scientific American Library.

A short introduction to climate written for the general public. Provides an excellent background for those interested in topics relating to air pollution and climate change.

Philander, G. 1989. El Niño and La Niña. *American Scientist* 77:451–459.

Suplee, C. 1999. El Niño, La Niña. *National Geographic* 195:73–95.
    These two articles provide a general introduction to the El Niño–La Niña climate cycle.

CHAPTER 3

The Aquatic Environment

A rainstorm over the ocean—a part of the water cycle.

Water is the essential substance of life, the dominant component of all living organisms. About 75–95 percent of the weight of all living cells is water, and there is hardly a physiological process in which water is not fundamentally important.

Covering some 75 percent of the planet's surface, water is also the dominant environment on Earth. A major feature influencing the adaptations of organisms that inhabit aquatic environments is water salinity (see Section 3.5). For this reason, aquatic ecosystems are divided into two major categories: saltwater (or marine) and freshwater. These two major categories are further divided into a variety of aquatic ecosystems based on the depth and flow of water, substrate, and the type of organisms (typically plants) that dominate. We will explore the diversity of aquatic environments and the organisms that inhabit them later in Chapters 24 and 25. In this chapter, we will examine the unique physical and chemical characteristics of water and how those characteristics interact to define the different aquatic environments and constrain the evolution of organisms that inhabit them.

3.1 | Water Cycles between Earth and the Atmosphere

All marine and freshwater aquatic environments are linked, either directly or indirectly, as components of the **water cycle** (also referred to as the **hydrologic cycle;** Figure 3.1)—the process by which water travels in a sequence from the air to Earth and returns to the atmosphere.

Solar radiation, which heats Earth's atmosphere (see Chapter 2) and provides energy for the evaporation of water, is the driving force behind the water cycle. **Precipitation** sets the water cycle in motion. Water vapor, circulating in the atmosphere, eventually falls in some form of precipitation. Some of the water falls directly on the soil and bodies of water. Some is intercepted by vegetation, dead organic matter on the ground, and urban structures and streets (**interception**).

Because of interception, which can be considerable, various amounts of water never infiltrate the ground but evaporate directly back to the atmosphere. Precipitation that reaches the soil moves into the ground by **infiltration.** The rate of infiltration depends on the type of soil (see Section 4.8), slope, vegetation, and intensity of the precipitation. During heavy rains when the soil is saturated, excess water flows across the surface of the ground as **surface runoff** or overland flow. At places, it concentrates into depressions and gullies, and the flow changes from sheet to channelized flow—a process that can be observed on city streets as water moves across the pavement into gutters. Because of low infiltration, runoff from urban areas might be as much as 85 percent of the precipitation.

Some water entering the soil seeps down to an impervious layer of clay or rock to collect as **groundwater** (see Figure 3.1 also see Ecological Issues: Groundwater

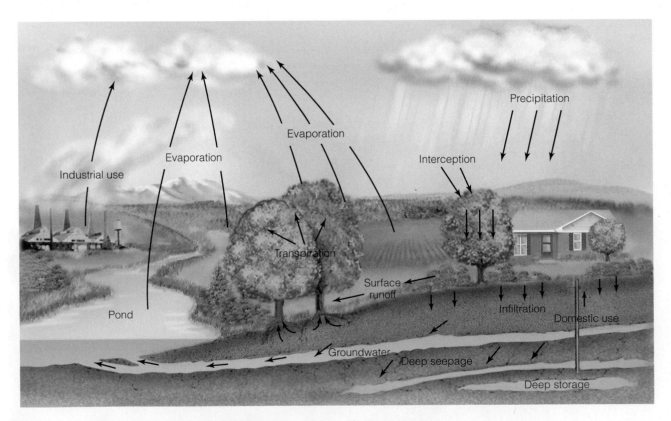

Figure 3.1 | The water cycle on a local scale, showing major pathways of water movement.

The addition of water to agricultural fields, known as the process of irrigation, is an ancient and important practice. In dry environments, irrigation is essential. However, even in regions where average precipitation is adequate for agricultural production, irrigation often provides a steady supply of water to fields, making plant growth less dependent on the effects of weather. Although only 15 percent of the world's cultivated land is irrigated, it accounts for 35–40 percent of the global food harvest.

In most cases, groundwater supplies the necessary water for irrigation. Approximately 30 percent of the groundwater used for irrigation in the United States comes from a single source: the High Plains–Ogallala aquifer (Figure 1). An aquifer is a layer of water-bearing permeable rock, sand, or gravel capable of providing significant amounts of water. The buried sand and gravel

that hold the aquifer water originated from rivers that have flowed eastward from the Rockies over the past several million years. The thickness of the aquifer, averaging 65 m, varies from more than 1000 m in Nebraska to less than 20 m in New Mexico. The water it contains dates to the last ice age, more than 10,000 years ago.

The high permeability of the High Plains–Ogallala aquifer enables a large volume of water to be pumped from the ground, but it also means the cumulative effect of pumping at any location is to draw down the water level over an extensive area of the aquifer. Water is now being pumped from more than 200,000 wells in the region as much as 50 times faster than water in the aquifer is being recharged. For decades, it was assumed that the aquifer's water supply was endless. Intensive pumping since the 1940s, however, has steadily lowered the water level in many areas. Withdrawals in 1990 for irrigation alone exceeded 14 billion gallons per day. During the period of drought in the western states that extended from mid-1992 to late 1996, the decline in the water level (depth) of the aquifer averaged 40 cm per year.

One consequence of decreased water levels in the aquifer is a large increase in the costs of pumping, which is becoming prohibitively expensive in some places. Pumping has also decreased the volume of water discharged from the aquifer into streams and springs. For example, pumping from the aquifer in Colorado decreased flow in the Arkansas River, which flows through Kansas.

Besides the impacts on agriculture, the decline of the High Plains–Ogallala aquifer affects the supply of drinking water in the region. Although 95 percent of the water pumped from the aquifer is used for irrigation, upward of 80 percent of the people in the region depend on the aquifer for drinking water. Water in some areas of the aquifer does not meet the U.S. Environmental Protection Agency's (EPA) standards for human consumption, and declining water levels reduce the quality of remaining waters as the concentrations of salts and other solutes increase.

1. In many areas of the High Plains–Ogallala aquifer, there is no possibility of developing additional water sources and supplies. In these areas, what steps could possibly be taken to help conserve existing water supplies?

2. Because the use of water in one state can influence the long-term availability of water throughout the region (other states), how would you propose that these water resources be managed?

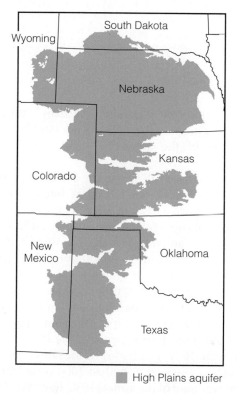

High Plains aquifer

Figure 1 | The High Plains–Ogallala aquifer underlies an area of approximately 174,000 square miles that extends through eight states. This aquifer is the principal source of water in one of the major agricultural regions of the United States.

Resources). From there, water finds its way into springs and streams. Streams coalesce into rivers as they follow the topography of the landscape. In basins and floodplains, lakes and wetlands form. Rivers eventually flow to

the coast, forming the transition from freshwater to marine environments.

Water remaining on the surface of the ground, in the upper layers of the soil, and collected on the surface of

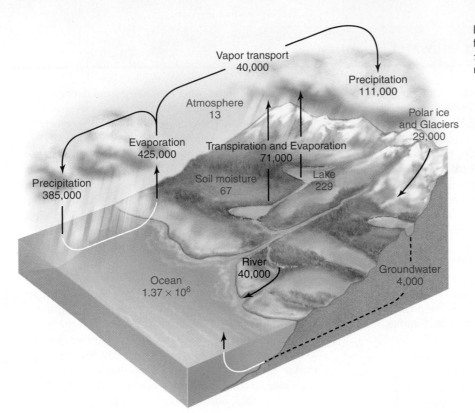

Vapor transport
40,000

Atmosphere
13

Precipitation
111,000

Polar ice
and Glaciers
29,000

Evaporation
425,000

Transpiration and Evaporation
71,000

Soil moisture
67

Lake
229

Precipitation
385,000

River
40,000

Groundwater
4,000

Ocean
1.37×10^6

Figure 3.2 | Global water cycle. Values for reservoirs (shown in blue) are in 10^8 km³. Values for fluxes (shown in red) are in km³ per year.

vegetation—as well as water in the surface layers of streams, lakes, and oceans—returns to the atmosphere by evaporation. The rate of evaporation is governed by how much water vapor is in the air relative to the saturation vapor pressure (see Section 2.6). Plants cause additional water loss from the soil. Through their roots, they take in water from the soil and lose it through the leaves and other organs in a process called transpiration. **Transpiration** is the evaporation of water from internal surfaces of leaves, stems, and other living parts (see Chapter 6). The total amount of evaporating water from the surfaces of the ground and vegetation (surface evaporation plus transpiration) is called **evapotranspiration.**

Figure 3.2 is a diagram of the global water cycle showing the various reservoirs (bodies of water) and fluxes (exchanges between reservoirs). The total volume of water on Earth is approximately 1.4 billion cubic kilometers (km³), of which more than 97 percent resides in the oceans. Another 2 percent of the total is found in the polar ice caps and glaciers, and the third largest active reservoir is groundwater (0.3 percent). Over the oceans, evaporation exceeds precipitation by some 40,000 km³. A significant proportion of the water evaporated from the oceans is transported by winds over the land surface in the form of water vapor, where it is deposited as precipitation. Of the 111,000 km³ of water that falls as precipitation on the land surface, only some 71,000 km³ is returned to the atmosphere as evapotranspiration. The remaining 40,000 km³ is

carried as runoff by rivers and eventually returns to the oceans. This amount balances the net loss of water from the oceans to the atmosphere through evaporation (Figure 3.2).

The relatively small size of the atmospheric reservoir (only 13 km³) does not reflect its importance in the global water cycle. In Figure 3.2, note the large fluxes between the atmosphere and the oceans and land surface relative to the amount of water residing in the atmosphere at any given time (reservoir size). The importance of the atmosphere in the global water cycle is better reflected by the turnover time of this reservoir. The turnover time is calculated by dividing the size of the reservoir by the rate of output (flux out). For example, the turnover time for the ocean is the size of the reservoir (1.37×10^6 km³) divided by the rate of evaporation (425 km³ per year), or more than 3000 years. In contrast, the turnover time of the atmospheric reservoir is approximately 0.024 year. That is to say, the entire water content of the atmosphere is replaced on average every nine days.

3.2 | Water Has Important Physical Properties

The physical arrangement of its component molecules makes water a unique substance. A molecule of water consists of two atoms of hydrogen (H) joined to one atom of oxygen (O), represented by the chemical symbol

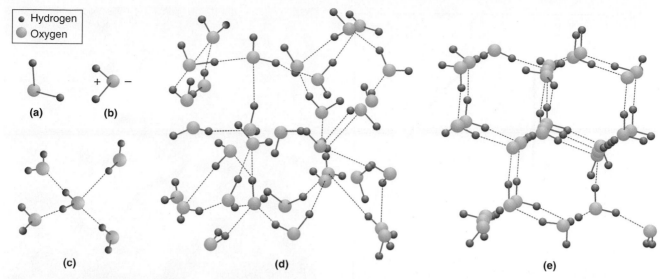

Figure 3.3 | The structure of water. **(a)** An isolated water molecule, showing the angular arrangement of the hydrogen atoms. **(b)** Polarity of water. **(c)** Hydrogen bonds to one neighboring molecule of water. **(d)** The structure of liquid water. **(e)** The open lattice structure of ice.

H_2O. The H atoms are bonded to the O atoms asymmetrically, such that the two H atoms are at one end of the molecule and the O atom at the other (Figure 3.3a). The bonding between the two hydrogen atoms and the oxygen atom is via shared electrons (called a covalent bond), so that each H atom shares a single electron with the oxygen. The shared hydrogen atoms are closer to the oxygen atom than they are to each other. As a result, the side of the water molecule where the H atoms are located has a positive charge, and the opposite side where the oxygen atom is located has a negative charge, thus polarizing the water molecule (termed a *polar covalent bond*; Figure 3.3b).

Because of its polarity, each water molecule becomes weakly bonded with its neighboring molecules (Figure 3.3c). The positive (hydrogen) end of one molecule attracts the negative (oxygen) end of the other. The angle between the hydrogen atoms encourages an open, tetrahedral-like arrangement of water molecules. This situation, when hydrogen atoms act as connecting links between water molecules, is **hydrogen bonding.** The simultaneous bonding of a hydrogen atom to the oxygen atoms of two different water molecules gives rise to a lattice arrangement of molecules (Figure 3.3d). These bonds, however, are weak in comparison to the bond between the hydrogen and oxygen atoms. As a result, they are easily broken and reformed.

Water has some unique properties related to its hydrogen bonds. One property is high **specific heat**—the number of calories necessary to raise 1 gram of water 1 degree Celsius. The specific heat of water is defined as a value of 1, and other substances are given a value relative to water. Water can store tremendous quantities of heat energy with a small rise in temperature. As a result, great quantities of heat must be absorbed before the temperature of natural waters, such as ponds, lakes, and seas, rises just 1°C. These waters warm up slowly in spring and cool off just as slowly in the fall. This process prevents the wide seasonal fluctuations in the temperature of aquatic habitats so characteristic of air temperatures and moderates the temperatures of local and worldwide environments (see Chapter 2). The high specific heat of water also is important in the thermal regulation of organisms. Because 75–95 percent of the weight of all living cells is water, temperature variation is also moderated relative to changes in ambient temperature.

Due to the high specific heat of water, large quantities of heat energy are required to change its state between solid (ice), liquid, and gaseous (water vapor) phases. Collectively, the energy released or absorbed in transforming water from one state to another is called latent heat (see Section 2.6). Removing only 1 calorie (4.184 joules) of heat energy will lower the temperature of a gram of water from 2°C to 1°C, but approximately 80 times as much heat energy (80 calories per gram) must be removed to convert that same quantity of water at 1°C to ice (freezing point of 0°C). Likewise, it takes 536 calories to overcome the attraction between molecules and convert 1 g of water at 100°C into vapor, the same amount of heat needed to raise 536 g of water 1°C.

The lattice arrangement of molecules gives water a peculiar density–temperature relationship. Most liquids become denser as they are cooled. If cooled to their freezing temperature they become solid, and the solid phase is denser than the liquid. This description is not true for water. Pure water becomes denser as it is cooled until it reaches 4°C (Figure 3.4). Cooling below this temperature results in a decrease in density. When 0°C is reached,

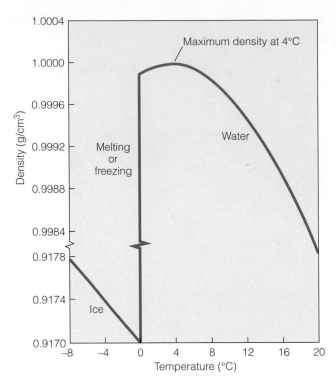

Figure 3.4 | Density of pure water (and ice) as a function of temperature. The maximum density of water is at 4°C, and it declines dramatically as water changes from a liquid to a solid (ice).

Figure 3.5 | The property of surface tension allows the water strider (*Gerris remigis*) to glide across the water surface.

freezing occurs and the lattice structure is complete—each oxygen atom is connected to four other oxygen atoms by means of hydrogen atoms. The result is a lattice with large, open spaces and therefore decreased density (see Figure 3.3e). Water molecules when frozen occupy more space than they would in liquid form. Because of its reduced density, ice is lighter than water and floats on it. This property is crucial to life in aquatic environments. The ice on the surface of water bodies insulates the waters below, helping to keep larger bodies of water from freezing solid during the winter months.

Due to hydrogen bonding, water molecules tend to stick firmly to each other, resisting external forces that would break these bonds. This property is called **cohesion.** In a body of water, these forces of attraction are the same on all sides. At the water's surface, however, conditions are different. Below the surface, molecules of water are strongly attracted to one another. Above the surface is the much weaker attraction between water molecules and air. Therefore, molecules on the surface are drawn downward, resulting in a surface that is taut like an inflated balloon. This condition, called **surface tension,** is important in the lives of aquatic organisms.

For example, the surface of water is able to support small objects and animals, such as the water striders (*Gerridae* spp.) and water spiders (*Dolomedes* spp.) that run across a pond's surface (Figure 3.5). To other small organisms, surface tension is a barrier, whether they wish to penetrate the water below or escape into the air above. For some, the surface tension is too great to break; for others, it is a trap to avoid while skimming the surface to feed or to lay eggs. If caught in the surface tension, a small insect may flounder on the surface. The nymphs of mayflies (*Ephemeroptera spp.*) and caddis flies (*Trichoptera spp.*) that live in the water and transform into winged adults are hampered by surface tension when trying to emerge from the water. While slowed down at the surface, these insects become easy prey for fish.

Cohesion is also responsible for the viscosity of water. **Viscosity** is the property of a material that measures the force necessary to separate the molecules and allow an object to pass through the liquid. Viscosity is the source of frictional resistance to objects moving through water. This frictional resistance of water is 100 times greater than that of air. The streamlined body shape of many aquatic organisms, for example most fish and marine mammals, helps to reduce frictional resistance. Replacement of water in the space left behind by the moving animal increases drag on the body. An animal streamlined in reverse, with a short, rounded front and a rapidly tapering body, meets the least water resistance. The perfect example of such streamlining is the sperm whale (*Physeter catodon*; Figure 3.6).

Water's high viscosity relative to air is due largely to its greater density. The density of water is about 860 times greater than that of air (pure water has a density of 1000 kg/m³). Although the resulting viscosity of water limits the mobility of aquatic organisms, it also benefits them. If a body submerged in water weighs less than the water it displaces, it is subjected to an upward force called **buoyancy.** Because most aquatic organisms

Figure 3.6 | The body of the sperm whale (*Physeter catodon*) is streamlined in reverse, with a short, rounded front and a rapidly tapering body. This shape meets the least water resistance.

(plants and animals) are close to neutral buoyancy (their density is similar to that of water), they do not require structural material such as skeletons or cellulose to hold themselves erect against the force of gravity. Similarly, when moving on land, terrestrial animals must raise their mass against the force of gravity for each step they take. Such movement requires significantly more energy than swimming movements do for aquatic organisms.

But water's greater density can profoundly affect the metabolism of marine organisms inhabiting the deeper waters of the ocean. Because of its greater density, water also experiences greater changes in pressure with depth than does air.

At sea level, the weight of the vertical column of air from the top of the atmosphere to the sea surface is 1 kg/cm^2, or 1 atmosphere (atm). In contrast, pressure increases 1 atm for each 10 m in depth. Because the deep ocean varies in depth from a few hundred meters down to the deep trenches at more than 10,000 m, the range of pressure at the ocean bottom is from 20 atm to more than 1000 atm. Recent research has shown that both proteins and biological membranes are strongly affected by pressure and must be modified to work in animals living in the deep ocean.

3.3 | Light Varies with Depth in Aquatic Environments

When light strikes the surface of water, a certain amount is reflected back to the atmosphere. The amount of light reflected from the surface depends on the angle at which the light strikes the surface. The lower the angle, the larger the amount of light reflected. As a result, the amount of light reflected from the water surface will vary both diurnally and seasonally as a person moves from the equator to the poles (see Section 2.2 for a complete discussion).

The amount of light entering the water surface is reduced further by two additional processes. First, suspended particles, both alive and dead, intercept the light and either absorb or scatter it. The scattering of light increases its path of travel through the water and results in further attenuation. Second, water itself absorbs light (Figure 3.7). Even in perfectly clean, clear water, only about 40 percent of shortwave radiation reaches a depth

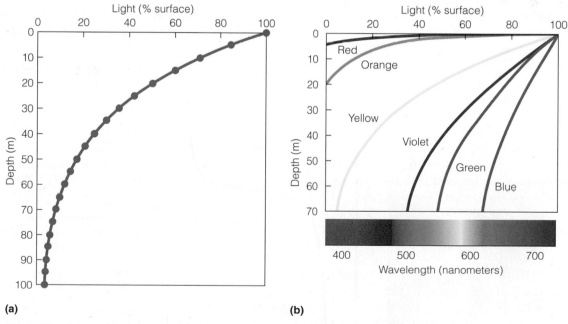

(a)   (b)

Figure 3.7 | **(a)** Attenuation of incident light with water depth (pure water), expressed as a percentage of light at the water surface. Estimates assume a light extinction coefficient of $k_w = 0.035$ (see Quantifying Ecology 4.1, pp. 63–64). **(b)** The passage of light through water reduces the quantity of light and modifies its spectral distribution (see Figure 2.4). Red wavelengths are attenuated more rapidly than green and blue wavelengths.

of 1 m. Moreover, water absorbs some wavelengths more than others. First to be absorbed are visible red light and infrared radiation in wavelengths greater than 750 nm. This absorption reduces solar energy by one-half. Next, in clear water yellow disappears, followed by green and violet, leaving only blue wavelengths to penetrate deeper water. A fraction of blue light is lost with increasing depth. In the clearest seawater, only about 10 percent of blue light reaches to more than 100 m in depth.

These changes in the quantity and quality of light have important implications to life in aquatic environments, both directly by influencing the quantity and distribution of productivity (see Section 20.4 and Chapter 24) and indirectly by influencing the vertical profile of temperature with water depth. The lack of light in deeper waters of the oceans has resulted in various adaptations. Organisms of the deeper ocean (200–1000 m deep) are typically silvery gray or deep black, and organisms living in even deeper waters (below 1000 m) often lack pigment. Another adaptation is large eyes, which give these organisms maximum light-gathering ability. Many organisms have adapted organs that produce light through chemical reactions referred to as bioluminescence (see Section 24.10).

3.4 | Temperature Varies with Water Depth

As we discussed in Chapter 2 (Section 2.1), surface temperatures reflect the balance of incoming and outgoing radiation. As solar radiation is absorbed in the vertical water column, the temperature profile with depth might be expected to resemble the vertical profile of light shown in Figure 3.7—that is, decreasing exponentially with depth. However, the physical characteristic of water

density discussed earlier (Section 3.2, Figure 3.4) plays an important role in modifying this pattern.

As sunlight is absorbed in the surface waters, it heats up (Figure 3.8). Winds and surface waves mix the surface waters, distributing the heat vertically. As a result, the decline in water temperature with depth will lag the decline in solar radiation. Below this mixed layer, however, temperatures drop rapidly. This region of the vertical depth profile where the temperature declines most rapidly is called the **thermocline.** The depth of the thermocline will depend on the input of solar radiation to the surface waters and on the degree of vertical mixing (wind speed and wave action). Below the thermocline, water temperatures continue to fall with depth but at a much slower rate. The result is a distinct pattern of temperature zonation with depth.

The difference in temperature between the warm, well-mixed surface layer and the cooler waters below the thermocline causes a distinctive difference in water density in these two vertical zones. The thermocline is located between an upper layer of warm, lighter (less dense) water called the **epilimnion** and a deeper layer of cold, denser water called the **hypolimnion** (Figure 3.8; also see Section 21.8 and Figure 21.18). The density change at the thermocline acts as a physical barrier that prevents mixing of the upper (epilimnion) and lower (hypolimnion) layers.

Just as seasonal variation in the input of solar radiation to Earth's surface results in seasonal changes in surface temperatures (see Section 2.2), seasonal changes in the input of solar radiation to the water surface give rise to seasonal changes in the vertical profile of temperature in aquatic environments (Figure 3.9). Due to the relatively constant input of solar radiation to the water surface throughout the year, the thermocline is a permanent

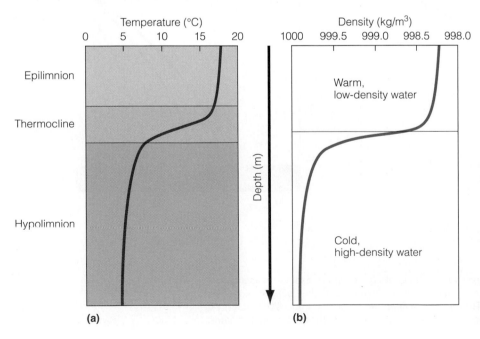

(a)　　　　(b)

Figure 3.8 | Temperature and density profiles with water depth for an open body of water such as a lake or pond. **(a)** The vertical profile of temperature might be expected to resemble the profile of light presented in Figure 3.7, but vertical mixing of the surface waters transports heat to the waters below. Below this mixed layer, temperatures decline rapidly in a region termed the thermocline. Below the thermocline, temperatures continue declining at a slower rate. The vertical profile can therefore be divided into three distinct zones: epilimnion, thermocline, and hypolimnion. **(b)** The rapid decline in temperature in the thermocline results in a distinct difference in water density (see Figure 3.4) in the warmer epilimnion as compared to the cooler waters of the hypolimnion, leading to a two-layer density profile—warm, low-density surface water and cold, high-density deep water.

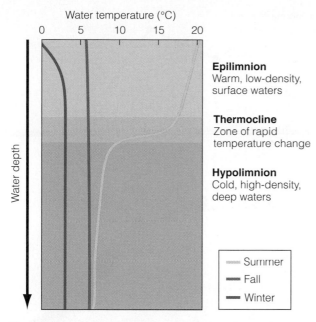

Water temperature (°C)

Epilimnion
Warm, low-density, surface waters

Thermocline
Zone of rapid temperature change

Hypolimnion
Cold, high-density, deep waters

Summer
Fall
Winter

Figure 3.9 | Seasonal changes in the vertical temperature profile (with water depth) for an open body of water such as a lake or pond. As air temperatures decline during the fall months, the surface water cools and sinks so that the temperature is uniform with depth. With the onset of winter, surface water further cools and ice may form on the surface. When spring arrives, the process reverses and the thermocline once again forms.

feature of tropical waters. In the waters of the temperate zone, a distinct thermocline exists during the summer months. By fall, conditions begin to change, and a turnabout takes place. Air temperatures and sunlight decrease, and the surface water of the epilimnion starts to cool. As it does, the water becomes denser and sinks, displacing the warmer water below to the surface, where it cools in turn. As the difference in water density between the epilimnion and hypolimnion continues to decrease, winds are able to mix the vertical profile to greater depths. This process continues until the temperature is uniform throughout the basin (see Figure 3.9). Now, pond and lake water circulate throughout the basin. This process of vertical circulation, called the fall turnover, is an important component of nutrient dynamics in open-water ecosystems (see Chapter 21). Stirred by wind, the process of vertical mixing may last until ice forms at the surface.

Then comes winter, and as the surface water cools to below 4°C, it becomes lighter again and remains on the surface. (Remember, water becomes lighter above and below 4°C; see Figure 3.4.) If the winter is cold enough, surface water freezes; otherwise, it remains close to 0°C. Now the warmest place in the pond or lake is on the bottom. In spring, the breakup of ice and heating of surface water with increasing inputs of solar radiation to the surface again causes the water to stratify.

Because not all bodies of water experience such seasonal changes in stratification, do not consider this phenomenon as characteristic of all deep bodies of water. In some very deep lakes and the oceans, the thermocline simply descends during periods of turnover and does not disappear at all. In such bodies of water, the bottom water never becomes mixed with the top layer. In shallow lakes and ponds, temporary stratification of short duration may occur; in other bodies of water, stratification may exist, but the depth is not sufficient to develop a distinct thermocline. However, some form of thermal stratification occurs in all open bodies of water.

The temperature of a flowing body of water (stream or river), on the other hand, is variable (Figure 3.10). Small, shallow streams tend to follow, but lag behind, air temperatures. They warm and cool with the seasons but rarely fall below freezing in winter. Streams with large areas exposed to sunlight are warmer than those shaded by trees, shrubs, and high banks. That fact is ecologically important because temperature affects the stream community, influencing the presence or absence of cool-water and warm-water organisms. For example, the dominant predatory fish shift from species such as trout and smallmouth bass, which require cooler water and more oxygen, to species such as suckers and catfish, which require warmer water and less oxygen (see Figure 24.13).

3.5 | Water Functions as a Solvent

As you stir a spoonful of sugar into a glass of water, it dissolves, forming a homogeneous, or uniform, mixture. A liquid that is a homogeneous mixture of two or more substances is called a **solution.** The dissolving agent of a solution is the **solvent,** and the substance that is dissolved is referred to as the **solute.** A solution in which water is the solvent is called an **aqueous solution.**

Water is an excellent solvent that can dissolve more substances than can any other liquid. This extraordinary ability makes water a biologically crucial substance. Water provides a fluid that dissolves and transports molecules of nutrients and waste products, helps to regulate temperature, and preserves chemical equilibrium within living cells.

The solvent ability of water is due largely to the bonding discussed in Section 3.2. Because the H atom is bonded to the O atoms asymmetrically (see Figure 3.3), one side of every water molecule has a permanent positive charge and the other side has a permanent negative charge; such a situation is called a permanent dipole (*dipole* refers to oppositely charged poles). Because opposite charges attract, water molecules are strongly attracted

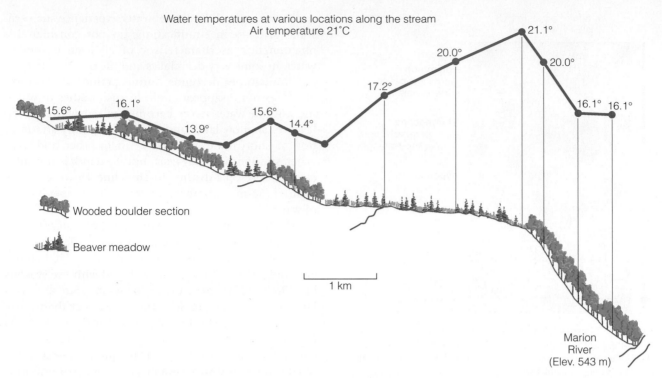

21.1°

20.0°

20.0°

17.2°

16.1°   16.1°

15.6°

16.1°

15.6°

14.4°

13.9°

Wooded boulder section

Beaver meadow

1 km

Marion
River
(Elev. 543 m)

Figure 3.10 | Profile of Bear Brook (Adirondack Mountains, New York) and a graph of its water temperatures. Water temperatures rise as the stream moves through the open beaver meadows, once again declining as it flows into the shaded cover of the wooded forest.

to each other; and they also attract other molecules carrying a charge.

Compounds that consist of electrically charged atoms or groups of atoms are called **ions.** Sodium chloride (table salt), for example, is composed of positively charged sodium ions (Na^+) and negatively charged chloride ions (Cl^-) arranged in a crystal lattice. When placed in water, the attractions between negative (oxygen atom) and positive (hydrogen atoms) charges on the water molecule (see Figure 3.3) and those of the sodium and chloride atoms are greater than the forces (ionic bonds) holding the salt crystals together. Consequently, the salt crystals readily dissociate into their component ions when placed into contact with water; that is, they dissolve.

The solvent properties of water are responsible for most of the minerals (elements and inorganic compounds) found in aquatic environments. When water condenses to form clouds, it is nearly pure except for some dissolved atmospheric gases. In falling to the surface as precipitation, water acquires additional substances from particulates and dust particles suspended in the atmosphere. Water that falls on land flows over the surface and percolates into the soil, obtaining more solutes. Surface waters, such as streams and rivers, pick up more solvents from the substances through and over which they flow. The waters of most rivers and lakes contain 0.01–0.02 percent dissolved

minerals. The relative concentrations of minerals in these waters reflect the substrates over which the waters flow. For example, waters that flow through areas where the underlying rocks consist largely of limestone (composed primarily of calcium carbonate; $CaCO_3$) will have high concentrations of calcium (Ca^{2+}) and bicarbonate (HCO_3^-).

In contrast to freshwaters, the oceans have a much higher concentration of solutes. In effect, the oceans function as a large still. The flow of freshwaters into the oceans continuously adds to the solute content of the waters, as pure water evaporates from the surface to the atmosphere. The concentration of solutes, however, cannot continue to increase indefinitely. When the concentration of specific elements reaches the limit set by the maximum solubility of the compounds they form (grams per liter), the excess amounts will precipitate and be deposited as sediments. Calcium, for example, readily forms calcium carbonate ($CaCO_3$) in the waters of the oceans. The maximum solubility of calcium carbonate, however, is only 0.014 gram per liter of water, a concentration that was reached early in the history of the oceans. As a result, calcium ions continuously precipitate out of solution and are deposited as limestone sediments on the ocean bottom.

In contrast, the solubility of sodium chloride is very high (360 grams per liter). In fact, these two elements, sodium and chlorine, make up some 86 percent of sea salt. Sodium and chlorine—along with other major elements

Table 3.1 | Composition of Seawater of 35 Practical Salinity Units (psu)

Elements	g/kg	Milli-moles/kg	Milli-equivalents/kg
Cations			
Sodium	10.752	467.56	467.56
Potassium	0.395	10.10	10.10
Magnesium	1.295	53.25	106.50
Calcium	0.416	10.38	20.76
Strontium	0.008	0.09	0.18
			605.10
Anions			
Chlorine	19.345	545.59	545.59
Bromine	0.066	0.83	0.83
Fluorine	0.0013	0.07	0.07
Sulfate	2.701	28.12	56.23
Bicarbonate	0.145	2.38	—
Boric acid	0.027	0.44	—
			602.72

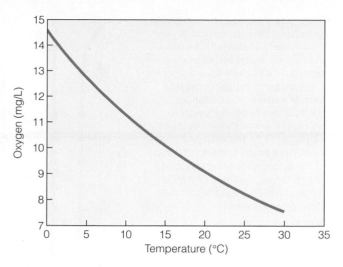

Figure 3.11 | The solubility of oxygen in pure water as a function of temperature. Solubility decreases with increasing water temperature.

such as sulfur, magnesium, potassium, and calcium, whose relative proportions vary little—compose 99 percent of sea salts (Table 3.1). Determination of the most abundant element, chlorine, is used as an index of salinity. Salinity is expressed in **practical salinity units (psu),** represented as ‰ and measured as grams of chlorine per kilogram of water. The salinity of the open sea is fairly constant, averaging about 35‰. In contrast, the salinity of freshwater ranges from 0.065 to 0.30‰. However, over geologic timescales (hundreds of millions of years), the salinity of the oceans has increased and continues to do so.

3.6 | Oxygen Diffuses from the Atmosphere to the Surface Waters

Water's role as a solvent is not limited to dissolving solids. The surface of a body of water defines a boundary with the atmosphere. Across this boundary, gases are exchanged through the process of diffusion. **Diffusion** is the general tendency of molecules to move from a region of high concentration to one of lower concentration. The process of diffusion results in a net transfer of two metabolically important gases, oxygen and carbon dioxide, from the atmosphere (higher concentration) into the surface waters (lower concentration) of aquatic environments.

Oxygen diffuses from the atmosphere into the surface water. The rate of diffusion is controlled by the solubility of oxygen in water and the steepness of the diffusion gradient (the difference in concentration between the air

and the surface waters where diffusion is occurring). The solubility of gases in water is a function of temperature, pressure, and salinity. The saturation value of oxygen is greater for cold water than warm water because the solubility (ability to stay in solution) of a gas in water decreases as the temperature rises (Figure 3.11). However, solubility increases as atmospheric pressure increases and decreases as salinity increases, which is not significant in freshwater.

Once oxygen enters the surface water, the process of diffusion continues, and oxygen diffuses from the surface to the waters below (lower concentration). Water, with its greater density and viscosity relative to air, limits how quickly gases diffuse through water. Gases diffuse some 10,000 times slower in water than in air. In addition to the process of diffusion, oxygen absorbed by surface water is mixed with deeper water by turbulence and internal currents. In shallow, rapidly flowing water and in wind-driven sprays, oxygen may reach and maintain saturation and even supersaturated levels due to the increase of absorptive surfaces at the air–water interface. Oxygen is lost from the water as temperatures rise, decreasing solubility, and through the uptake of oxygen by aquatic life.

During the summer, oxygen, like temperature (see Section 3.4), may become stratified in lakes and ponds. The amount of oxygen usually is greatest near the surface, where an interchange between water and atmosphere, further stimulated by the stirring action of the wind, takes place (Figure 3.12). Besides entering the water by diffusion from the atmosphere, oxygen is also a product of photosynthesis, which is largely restricted to the surface waters due to the limitations of available light (see Figure 3.7 and Chapter 6). The quantity of oxygen decreases with depth because of the oxygen demand by

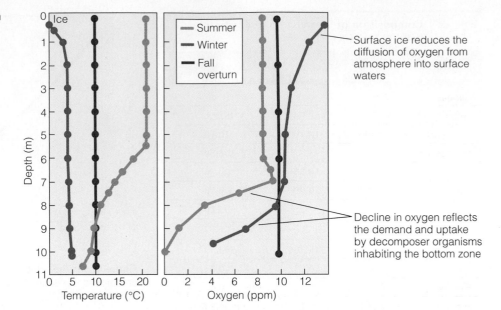

Figure 3.12 | Oxygen stratification in Mirror Lake, New Hampshire, in winter, summer, and late fall. The late fall turnover results in uniform temperature as well as uniform distribution of oxygen throughout the lake basin. In summer, a pronounced stratification of both temperature and oxygen exists. Oxygen declines sharply in the thermocline and is nonexistent on the bottom due to its uptake by decomposer organisms in the sediments. In winter, oxygen is also stratified; but its concentration is low in deep water.

(Adapted from Likens 1985.)

Surface ice reduces the diffusion of oxygen from atmosphere into surface waters

Decline in oxygen reflects the demand and uptake by decomposer organisms inhabiting the bottom zone

decomposer organisms living in the bottom sediments (Chapter 21). During spring and fall turnover, when water recirculates through the lake, oxygen becomes replenished in deep water. In winter, the reduction of oxygen in unfrozen water is slight because the demand for oxygen by organisms is reduced by the cold, and oxygen is more soluble at low temperatures. Under ice, however, oxygen depletion may be serious due to the lack of diffusion from the atmosphere to the surface waters.

As with ponds and lakes, oxygen is not distributed uniformly within the depths of the oceans (Figure 3.13). A typical oceanic oxygen profile shows a maximum amount in the upper 10–20 m, where photosynthetic activity and diffusion from the atmosphere often lead to saturation. With increasing depth, the oxygen content

declines. In the open waters of the ocean, concentrations reach a minimum value of 500–1000 m, a region referred to as the *oxygen minimum zone*. Unlike lakes and ponds, where the seasonal breakdown of the thermocline and resulting mixing of surface and deep waters result in a dynamic gradient of temperature and oxygen content, the limited depth of surface mixing in the deep oceans maintains the vertical gradient of oxygen availability year-round.

The availability of oxygen in aquatic environments characterized by flowing water is quite different. The constant churning and swirling of stream water over riffles and falls gives greater contact with the atmosphere; therefore, the oxygen content of the water is high, often near saturation for the prevailing temperature. Only in deep holes or in polluted waters does dissolved oxygen show any significant decline.

Even under ideal conditions, however, gases are not very soluble in water. For example, rarely is oxygen limited in terrestrial environments. In aquatic environments, however, the supply of oxygen, even at saturation levels, is meager and problematic. Compared with its concentration of 0.21 liter per liter in the atmosphere (21 percent by volume), oxygen in water reaches a maximum solubility of 0.01 liters per liter (1 percent) in freshwater at a temperature of 0°C. As a result, the concentration of oxygen in aquatic environments often limits respiration and metabolic activity (see Section 6.1).

3.7 | Acidity Has a Widespread Influence on Aquatic Environments

The solubility of carbon dioxide is somewhat different from that of oxygen in its chemical reaction with water. Water has a considerable capacity to absorb carbon

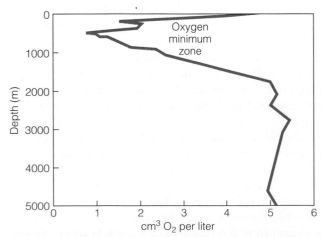

Figure 3.13 | Vertical profile of oxygen with depth in the tropical Atlantic Ocean. The oxygen content of the waters declines to a depth known as the oxygen minimum zone. Oxygen increase below this zone is thought to be due to the influx of cold, oxygen-rich waters that originally sank in the polar waters (see Section 3.8).

dioxide, which is abundant in both freshwater and salt-water. Upon diffusing into the surface, carbon dioxide reacts with water to produce carbonic acid (H_2CO_3):

$$CO_2 + H_2O \longleftrightarrow H_2CO_3$$

Carbonic acid further dissociates into a hydrogen ion and a bicarbonate ion:

$$H_2CO_3 \longleftrightarrow HCO_3^- + H^+$$

Bicarbonate may further dissociate into another hydrogen ion and a carbonate ion:

$$HCO_3^- \longleftrightarrow H^+ + CO_3^{2-}$$

The carbon dioxide–carbonic acid–bicarbonate system is a complex chemical system that tends to stay in equilibrium. (Note that the arrows in the preceding equations go in both directions.) Therefore, if CO_2 is removed from the water, the equilibrium is disturbed and the equations will shift to the left, with carbonic acid and bicarbonate producing more CO_2 until a new equilibrium is produced.

The chemical reactions just described result in the production and absorption of free hydrogen ions (H^+). The abundance of hydrogen ions in solution is a measure of **acidity.** The greater the number of H^+ ions, the more acidic the solution will be. **Alkaline** solutions are those that have a large number of OH^- (hydroxyl ions) and few H^+ ions. The measurement of acidity and alkalinity is pH, calculated as the negative logarithm (base 10) of the concentration of hydrogen ions in solution. In pure water, a small fraction of molecules dissociates into ions: $H_2O \rightarrow H^+ + OH^-$, and the ratio of H^+ ions to OH^- ions is 1:1. Because both occur in a concentration of 10^{-7} moles per liter, a neutral solution has a pH of 7 [$-\log(10^{-7}) = 7$]. A solution departs from neutral when one ion increases and the other decreases. Customarily, we use the negative logarithm of the hydrogen ion to describe a solution as an acid or a base. Thus, a gain of hydrogen ions to 10^{-6} moles per liter means a decrease of OH^- ions to 10^{-8} moles per liter, and the pH of the solution is 6. The negative logarithmic pH scale goes from 1 to 14. A pH greater than 7 denotes an alkaline solution (greater OH^- concentration) and a pH of less than 7 an acidic solution (greater H^+ concentration).

Although pure water is neutral in pH, because the dissociation of the water molecule produces equal numbers of H^+ and OH^- ions, the presence of CO_2 in the water will alter this relationship. The preceding chemical reactions result in the production and absorption of free hydrogen ions (H^+). Because the abundance of hydrogen ions in solution is the measure of acidity, the dynamics of the carbon dioxide–carbonic acid–bicarbonate system directly affects the pH of aquatic ecosystems. In general,

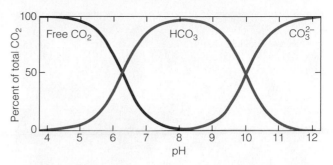

Figure 3.14 | Theoretical percentages of CO_2 in each of the three forms of water in relation to pH. At low values of pH (acidic conditions), most of the CO_2 is in its free form. At intermediate values (neutral conditions) bicarbonate dominates, whereas under alkaline conditions most of the CO_2 is in the form of carbonate ions.

the carbon dioxide–carbonic acid–bicarbonate system functions as a buffer to keep the pH of water within a narrow range. It does this by absorbing hydrogen ions in the water when they are in excess (producing carbonic acid and bicarbonates) and producing them when they are in short supply (producing carbonate and bicarbonate ions). At neutrality (pH 7), most of the CO_2 is present as HCO_3^- (Figure 3.14). At a high pH, more CO_2 is present as CO_3^{2-} than at a low pH, where more CO_2 occurs in the free condition. Addition or removal of CO_2 affects pH, and a change in pH affects CO_2.

The pH of natural waters ranges between 2 and 12. Waters draining from watersheds dominated geologically by limestone will have a much higher pH and be well buffered as compared to waters from watersheds dominated by acid sandstone and granite. The presence of the strongly alkaline ions sodium, potassium, and calcium in ocean waters results in seawater being slightly alkaline, usually ranging from 7.5 to 8.4.

The pH of aquatic environments can exert a powerful influence on the distribution and abundance of organisms. Increased acidity can affect organisms directly, by influencing physiological processes, and indirectly, by influencing the concentrations of toxic heavy metals. Tolerance limits for pH vary among plant and animal species, but most organisms cannot survive and reproduce at a pH below about 4.5. Aquatic organisms are unable to tolerate low pH conditions largely because acidic waters contain high concentrations of aluminum. Aluminum is highly toxic to many species of aquatic life and thus leads to a general decline in aquatic populations.

Aluminum is insoluble when the pH is neutral or basic. Insoluble aluminum is present in very high concentrations in rocks, soils, and river and lake sediments. Under normal pH conditions, the aluminum concentrations of lake water are very low; however, as the pH drops and becomes more acidic, aluminum begins to dissolve, raising the concentration in solution.

(a)

(b)

Figure 3.15 | (a) A fast mountain stream. The elevation gradient is steep, and fast-flowing water scours the stream bottom, leaving largely bedrock material. (b) In contrast, a slow-flowing stream meanders through a growth of willows. The relatively flat topography reduces the flow rate and allows finer sediments to build up on the stream bottom. These two streams represent very different environmental conditions and subsequently support very different forms of aquatic life (see Chapter 24).

3.8 | Water Movements Shape Freshwater and Marine Environments

The movement of water—currents in streams, and waves in an open body of water or breaking on a shore—determines the nature of many aquatic environments. The velocity of a current molds the character and structure of a stream. The shape and steepness of the stream channel; its width, depth, and roughness of the bottom; and the intensity of rainfall and rapidity of snowmelt all affect velocity. In fast streams, velocity of flow is 50 cm per second or higher (see Quantifying Ecology 24.1: Streamflow). At this velocity, the current removes all particles less than 5 mm in diameter and leaves behind a stony bottom. High water volume increases the velocity; it moves bottom stones and rubble, scours the streambed, and cuts new banks and channels. As the gradient decreases and the width, depth, and volume of water increase, silt and decaying organic matter accumulate on the bottom. Thus, the stream's character changes from fast water to slow (Figure 3.15).

Wind generates waves on large lakes and on the open sea (Figure 3.16). The frictional drag of the wind on the surface of smooth water causes ripples. As the wind continues to blow, it applies more pressure to the steep side of the ripple, and wave size begins to grow. As the wind becomes stronger, short, choppy waves of all sizes appear; and as they absorb more energy, they continue to grow. When the waves reach a point where the energy supplied by the wind equals the energy lost by the breaking waves, they become whitecaps. Up to a certain point, the stronger wind, the higher the waves.

The waves breaking on a beach do not contain water driven in from distant seas. Each particle of water remains largely in the same place and follows an elliptical orbit with the passage of the wave. As a wave moves forward, it loses energy to the waves behind and disappears, its place taken by another. The swells breaking on a beach are distant descendants of waves generated far out at sea.

As the waves approach land, they advance into increasingly shallow water. The height of each wave rises until the wave front grows too steep and topples over. As the waves break onshore, they dissipate their energy, pounding rocky shores or tearing away sandy beaches in one location and building up new beaches elsewhere.

Figure 3.16 | Waves breaking on a rocky shore.

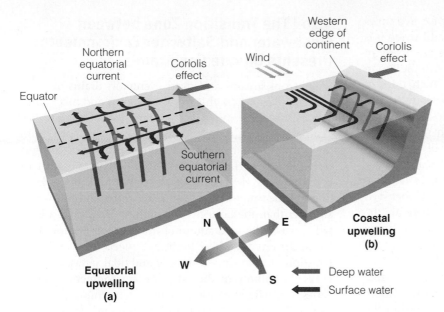

Equator

Northern equatorial current

Coriolis effect

Wind

Western edge of continent

Coriolis effect

Southern equatorial current

N

E

W

S

**Equatorial upwelling
(a)**

**Coastal upwelling
(b)**

← Deep water

← Surface water

Figure 3.17 | **(a)** Along the equator, the Coriolis effect acts to pull the westward-flowing currents to the north and south (purple solid arrows), resulting in an upwelling of deeper cold waters to the surface. **(b)** Along the western margins of the continents, the Coriolis effect causes the surface waters to move offshore (purple solid arrows). Movement of the surface waters offshore results in an upwelling of deeper, colder waters to the surface. Example shown is for the Northern Hemisphere.

In Chapter 2, we discussed the patterns of ocean currents, influenced by the direction of the prevailing winds and the Coriolis effect (see Section 2.5). However, deep waters of the oceans move quite differently from waters of the surface currents. Because deep waters are isolated from the wind, their motion does not depend on it. Movement of deep waters does, however, result from changes occurring at the surface. As discussed earlier, seawater increases in density as a result of decreasing temperature and increasing salinity. When seawater increases in density, it sinks. As the warm, highly saline surface currents of the tropical waters move north and southward, they cool (see map of surface currents in Figure 2.15). As these waters cool, they increase in density and sink. Because these cold, dense waters originated at the surface, they contain high concentrations of oxygen. As these waters sink, they begin the return trip to the tropics—in the form of deep-water currents. When these deep-water currents meet in the equatorial waters of the ocean, they form a region of **upwelling** where the deep waters move up to the surface, closing the pattern of ocean circulation (Figure 3.17a).

In coastal regions, winds blowing parallel to the coast move the surface waters offshore. Water moving upward from the deep replaces this surface water, creating a pattern of coastal upwelling (Figure 3.17b).

3.9 | Tides Dominate the Marine Coastal Environment

Tides profoundly influence the rhythm of life on ocean shores. Tides result from the gravitational pulls of the Sun and the Moon, each of which causes two bulges (tides) in the waters of the oceans. The two bulges caused by the Moon occur at the same time on opposite sides of

Earth on an imaginary line extending from the Moon through the center of Earth (Figure 3.18). The tidal bulge on the Moon side is due to gravitational attraction; the bulge on the opposite side occurs because the gravitational force there is less than at the Earth's center. As Earth rotates eastward on its axis, the tides advance westward. Thus, Earth will in the course of one daily rotation pass through two of the lunar tidal bulges, or high tides, and two of the lows, or low tides, at right angles (90° longitude difference) to the high tides.

The Sun also causes two tides on opposite sides of Earth, and these tides have a relation to the Sun like that of the lunar tides to the Moon. Because the Sun has a weaker gravitational pull than the Moon does, solar tides

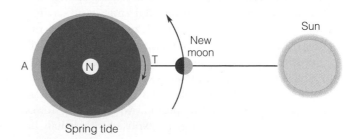

A

N

T

New moon

Sun

Spring tide

Figure 3.18 | Tides result from the gravitational pull of the Moon on Earth. Centrifugal force applied to a kilogram of mass is 3.38 mg. This centrifugal force on a rotating Earth is balanced by gravitational force, except at those (moving) points on Earth's surface that directly aligned with the moon. Thus the centrifugal force at point N, the center of the rotating Earth, is 3.38 mg. Point T on Earth is directly aligned with the Moon. At this point the Moon's gravitational force is 3.49 mg, a difference of 0.11 mg. Because the Moon's gravitational force is greater than the centrifugal force at T, the force is directed away from the Earth and causes a tidal bulge. At point A on the opposite side of the Earth from T, the Moon's gravitational force is 3.27 mg, 0.11 mg less than the centrifugal force at N. This difference causes a tidal bulge on the opposite side of Earth.

are partially masked by lunar tides—except for two times during the month: when the Moon is full and when it is new. At these times, Earth, Moon, and Sun are nearly in line, and the gravitational pulls of the Sun and the Moon are additive. This combination makes the high tides of those periods exceptionally large, with maximum rise and fall. These are the fortnightly spring tides, a name derived from the Saxon word *sprungen*, which refers to the brimming fullness and active movement of the water. When the Moon is at either quarter, its pull is at right angles to the pull of the Sun, and the two forces interfere with each other. At those times, the differences between high and low tides are exceptionally small. These are the neap tides, from an old Scandinavian word meaning "barely enough."

Tides are not entirely regular, nor are they the same all over Earth. They vary from day to day in the same place, following the waxing and waning of the Moon. They may act differently in several localities within the same general area. In the Atlantic, semidaily tides are the rule. In the Gulf of Mexico and the Aleutian Islands of Alaska, the alternate highs and lows more or less cancel each other out, and flood and ebb follow one another at about 24-hour intervals to produce one daily tide. Mixed tides in which successive or low tides are of significantly different heights through the cycle are common in the Pacific and Indian oceans. These tides are combinations of daily and semidaily tides in which one partially cancels out the other.

Local tides around the world are inconsistent for many reasons. These reasons include variations in the gravitational pull of the Moon and the Sun due to the elliptical orbit of Earth, the angle of the Moon in relation to the axis of Earth, onshore and offshore winds, the depth of water, the contour of the shore, and wave action.

The area lying between the water lines of high and low tide, referred to as the **intertidal zone,** is an environment of extremes. The intertidal zone undergoes dramatic shifts in environmental conditions with the daily patterns of inundation and exposure. As the tide recedes, the uppermost layers of life are exposed to air, wide temperature fluctuations, intense solar radiation, and desiccation for a considerable period, whereas the lowest fringes of the tidal zone may be exposed only briefly before the high tide submerges them again. Temperatures on tidal flats may rise to 38°C when exposed to direct sunlight and drop to 10°C within a few hours when the flats are covered by water.

Organisms living in the sand and mud do not experience the same violent temperature fluctuations as those living on rocky shores do. Although the surface temperature of the sand at midday may be 10°C (or more) higher than that of the returning seawater, the temperature a few centimeters below the sand's surface remains almost constant throughout the year.

3.10 | The Transition Zone between Freshwater and Saltwater Environments Presents Unique Constraints

Water from streams and rivers eventually drains into the sea. The place where this freshwater joins and mixes with the saltwater is called an estuary (see Chapter 24). Temperatures in estuaries fluctuate considerably, both daily and seasonally. Sun and inflowing and tidal currents heat the water. High tide on the mudflats may heat or cool the water, depending on the season. The upper layer of estuarine water may be cooler in winter and warmer in summer than the bottom—a condition that, as in a lake, will cause spring and autumn turnovers (see Section 3.4).

In the estuary, where freshwater meets the sea, the interaction of inflowing freshwater and tidal saltwater influences the salinity of the estuarine environment. Salinity varies vertically and horizontally, often within one tidal cycle (Figure 3.19). Salinity may be the same from top to bottom; or it may be completely stratified, with a layer of freshwater on top and a layer of dense, salty water on the bottom. Salinity is homogeneous when currents are strong enough to mix the water from top to bottom. The salinity in some estuaries is homogeneous at low tide, but at high tide a surface wedge of seawater moves upstream more rapidly than the bottom water. Salinity is then unstable, and density is inverted. The seawater on the surface tends to sink as lighter freshwater rises, and mixing takes place from the surface to the bottom. This phenomenon is known as **tidal overmixing.** Strong winds, too, tend to mix saltwater with freshwater in some estuaries; but when the winds are still, the river water flows seaward on a shallow surface over an upstream movement of seawater, more gradually mixing with the salt.

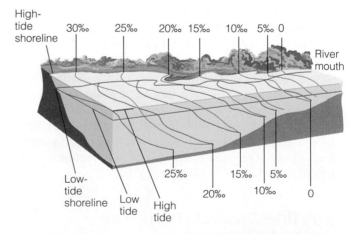

Figure 3.19 | Vertical and horizontal stratification of salinity (‰; practical salinity units; see Section 3.5) from the river mouth to the estuary at high tide (brown lines) and low tide (blue lines). At high tide, the incoming seawater increases the salinity toward the river mouth. At low tide, salinity is reduced. Salinity increases with depth because the lighter freshwater flows over the denser saltwater.

Horizontally, the least saline waters are at the river mouth and the most saline at the sea (see Figure 3.19). Incoming and outgoing currents deflect this configuration. In all estuaries of the Northern Hemisphere, outward-flowing freshwater and inward-flowing seawater are deflected to the right (relative to the axis of water flow from the river to ocean) because of Earth's rotation. As a result, salinity is higher on the left side; the concentration of metallic ions carried by rivers varies from drainage to drainage; and salinity and chemistry differ among estuaries. The portion of dissolved salts in the estuarine waters remains about the same as that of seawater, but the concentration varies in a gradient from freshwater to sea.

To survive in estuaries, aquatic organisms must have evolved physiological or behavioral adaptations to changes in salinity. Many oceanic species of fish are able to move inward during periods when the flow of freshwater from rivers is low and the salinity of estuaries increases. Conversely, freshwater fish move into the estuarine environment during periods of flood when salinity levels drop. Due to the stressful conditions that organisms face in the mixed zones of estuaries, there is often a relatively low diversity of organisms despite the high productivity found in these environments (see Chapter 24).

Summary

The Water Cycle | 3.1

Water follows a cycle, traveling from the air to Earth and returning to the atmosphere. It moves through cloud formation in the atmosphere, precipitation, interception, and infiltration into the ground. It eventually reaches groundwater, springs, streams, and lakes from which evaporation takes place, bringing water back to the atmosphere in the form of clouds. The various aquatic environments are linked, either directly or indirectly, by the water cycle.

The largest reservoir in the global water cycle is the oceans, which contain more than 97 percent of the total volume of water on Earth. In contrast, the atmosphere is one of the smallest reservoirs but has a very fast turnover time.

Properties of Water | 3.2

Water has a unique molecular structure. The hydrogen atoms are located on the side of the water molecule that has a positive charge. The opposite side, where the oxygen atom is located, has a negative charge, thus polarizing the water molecule. Because of their polarity, water molecules become coupled with neighboring water molecules to produce a lattice like structure with unique properties.

Depending on its temperature, water may be in the form of a liquid, solid, or gas. It absorbs or releases considerable quantities of heat with a small rise or fall in temperature. Water has a high viscosity that affects its flow. It exhibits high surface tension, caused by a stronger attraction of water molecules for each other than for the air above the surface. If a body is submerged in water and weighs less than the water it displaces, it is subjected to the upward force of buoyancy. These properties are important ecologically and biologically.

Light | 3.3

Both the quantity and quality of light change with water depth. In pure water, red and infrared light are absorbed first, followed by yellow, green, and violet; blue penetrates the deepest.

Temperature in Aquatic Environments | 3.4

Lakes and ponds experience seasonal shifts in temperature. In summer there is a distinct vertical gradient of temperature, resulting in a physical separation of warm surface waters and the colder waters below the thermocline. When the surface waters cool in the fall, the temperature becomes uniform throughout the basin and water circulates throughout the lake. A similar mixing takes place in the spring when the water warms. In some very deep lakes and the oceans, the thermocline simply descends during turnover periods and does not disappear at all.

Temperature of flowing water is variable, warming and cooling with the season. Within the stream or river, temperatures vary with depth, amount of shading, and exposure to sun.

Water as a Solvent | 3.5

Water is an excellent solvent with the ability to dissolve more substances than any other liquid can. The solvent properties of water are responsible for most of the minerals found in aquatic environments. The waters of most rivers and lakes contain a relatively low concentration of dissolved minerals, determined largely by the underlying bedrock over which the water flows. In contrast, the oceans have a much higher concentration of solutes. As pure water evaporates from the surface to the atmosphere, the flow of freshwaters into the oceans continuously adds to the solute content of the waters.

The solubility of sodium chloride is very high; together with chlorine, it makes up some 86 percent of sea salt. The concentration of chlorine is used as an index of salinity. Salinity is expressed in practical salinity units, or psu (represented as ‰), measured as grams of chlorine per kilogram of water.

Oxygen | 3.6

Oxygen enters the surface waters from the atmosphere through the process of diffusion. The amount of oxygen water can hold depends on temperature, pressure, and salinity. In lakes, oxygen absorbed by surface water mixes with deeper water by turbulence. During the summer, oxygen may become stratified, decreasing with depth because of decomposition in bottom sediments. During spring and fall turnover, oxygen becomes replenished in deep water. Constant swirling of stream water gives it greater contact with the atmosphere and thus maintains a high oxygen content.

Acidity | 3.7

The measurement of acidity is pH, the negative logarithm of the concentration of hydrogen ions in solution. In aquatic environments, a close relationship exists between the diffusion of carbon dioxide into the surface waters and the degree of acidity and alkalinity. Acidity influences the availability of nutrients and restricts the environment of organisms sensitive to acid situations.

Water Movement | 3.8

Currents in streams and rivers as well as waves in open sea and breaking on ocean shores determine the nature of many aquatic and marine environments. Velocity of currents shapes the environment of flowing water. Waves pound rocky shores and tear away and build up sandy beaches. Movement of water in surface currents of the ocean affects the patterns of deep-water circulation. As the equatorial currents move northward and southward, deep waters move up to the surface, forming regions of upwelling. In coastal regions, winds blowing parallel to the coast create a pattern of coastal upwelling.

Tides | 3.9

Rising and falling tides shape the environment and influence the rhythm of life in coastal intertidal zones.

Estuaries | 3.10

Water from all streams and rivers eventually drains into the sea. The place where this freshwater joins and mixes with the salt is called an estuary. Temperatures in estuaries fluctuate considerably, both daily and seasonally. The interaction of inflowing freshwater and tidal saltwater influences the salinity of the estuarine environment. Salinity varies vertically and horizontally, often within one tidal cycle.

Study Questions

1. Draw a simple diagram of the water cycle and then describe the processes involved.
2. How does the physical structure of water influence its ability to absorb and release heat energy?
3. What property of water allows aquatic organisms to function with far fewer supportive structures (tissues) than terrestrial organisms have?
4. What is the fate of visible light in water?
5. What is the thermocline? What causes the development of a thermocline?
6. Explain why seasonal stratification of temperature and oxygen takes place in deep ponds and lakes.
7. Increasing the carbon dioxide concentration of water will have what effect on its pH?
8. The concentration of which element is used to define the salinity of water?
9. What causes the upwelling of deeper, cold waters in the equatorial zone of the oceans?
10. What causes the tides?

Further Readings

Garrison, T. 2001. *Oceanography: An invitation to marine science.* Belmont, CA: Brooks/Cole.
    A clearly written and well-illustrated introductory text for those interested in more detail on the subject matter.

Hutchinson, G. E. 1957–1967. *A treatise on limnology. Vol. 1: Geography, physics, and chemistry.* New York: Wiley.
    A classic reference.

Hynes, H. B. N. 2001. *The ecology of running waters.* Caldwell, NJ: Blackburn Press.
    A reprint of a classic and valuable work, this major reference continues to be influential.

McLusky, D. S. 1989. *The estuarine ecosystem.* 2nd ed. New York: Chapman & Hall.
    Clearly describes the structure and function of estuarine ecosystems.

Nybakken, J. W. 2005. *Marine biology: An ecological approach.* 6th ed. San Francisco: Benjamin Cummings.
    Chapters 1 and 6 provide an excellent introduction to the physical environment of the oceans.

CHAPTER 4

The Terrestrial Environment

The input and decomposition of dead organic matter is a key factor in the development of forest soils.

Our introduction of aquatic environments in Chapter 3 was dominated by discussion of the physical and chemical properties of water—characteristics such as depth, flow rate, and salinity. When considering the term *terrestrial environment*, however, people typically do not think of the physical and chemical characteristics of a place. What we most likely visualize is the vegetation: the tall, dense forests of the wet tropics; the changing colors of autumn in a temperate forest; or the broad expanses of grass that characterize the prairies. Animal life depends on the vegetation within a region to provide the essential resources of food and cover—and as such, the structure and composition of plant life constrain the distribution and abundance of animal life. But ultimately, as with aquatic environments, the physical and chemical features of terrestrial environments set the constraints for life. Plant life is a reflection of the climate and soils (as discussed in Chapter 6); and regardless of the suitability of plant life for providing essential resources, the physical conditions within a region impose the primary constraints on animal life as well (Chapter 7).

In this chapter, we will explore key features of the terrestrial environment that directly influence life on land. Life emerged from the water to colonize the land more than a billion years ago. The transition to terrestrial environments posed a unique set of problems for organisms already adapted to an aquatic environment. To understand the constraints imposed by the terrestrial environment, we must start by looking at the physical differences between the terrestrial and aquatic environments and at the problems these differences create for organisms making the transition from water to land.

4.1 | Life on Land Imposes Unique Constraints

The transition from life in aquatic environments to life on land brought with it a variety of constraints. Perhaps the greatest constraint imposed by terrestrial environments is desiccation. Living cells, both plant and animal, contain about 75–95 percent water. Unless the air is saturated with moisture (see Section 2.6), water readily evaporates from the surfaces of cells via the process of diffusion. The water that is lost to the air must be replaced if the cell is to remain hydrated and continue to function. Maintaining this balance of water between organisms and their surrounding environment (referred to as an organism's **water balance**) has been a major factor in the evolution of life on land. For example, in adapting to the terrestrial environment, plants have evolved extensively specialized cells for different functions. Aerial parts of most plants, such as stems and leaves, are coated with a waxy cuticle that prevents water loss. While it reduces water loss, the waxy surface also prevents gas exchange

(carbon dioxide and oxygen) from occurring. As a result, terrestrial plants have evolved pores on the leaf surface (stomata; see Chapter 6) that allow gases to diffuse from the air into the interior of the leaf.

To stay hydrated, an organism must replace water that it has lost to the air. Terrestrial animals can acquire water by drinking and eating. For plants, however, the process is passive. Early in their evolution, land plants evolved vascular tissues consisting of cells joined into tubes that transport water and nutrients throughout the plant body. The topic of water balance and the array of characteristics that plants and animals have evolved to overcome the problems of water loss are discussed in more detail in Chapters 6 and 7.

Desiccation is not the only constraint imposed by the transition from water to land. Because air is less dense than water, it results in a much lower drag (frictional resistance) on the movement of organisms; but it greatly increases the constraint imposed by gravitational forces. The upward force of buoyancy due to the displacement of water helps organisms in aquatic environments overcome the constraints imposed by gravity (see Section 3.2). In contrast, the need to remain erect against the force of gravity in terrestrial environments results in a significant investment in structural materials such as skeletons (for animals) or cellulose (for plants). The giant kelp (*Macrocystis pyrifera*) inhabiting the waters off the coast of California is an excellent example (Figure 4.1a). It grows in dense stands called kelp forests. Anchored to the bottom sediments, these kelp (macroalgae) can grow 100 feet or more toward the surface. The kelp are kept afloat by gas-filled bladders attached to each blade, yet when the kelp plants are removed from the water, they collapse into a mass. Lacking supportive tissues strengthened by cellulose and lignin, the kelp cannot support its own weight under the forces of gravity. In contrast, a tree of equivalent height inhabiting the coastal forest of California (Figure 4.1b) must allocate more than 80 percent of its total mass to supportive and conductive tissues in the trunk (bole), branches, leaves, and roots.

Another characteristic of terrestrial environments is their high degree of variability, both in time and space. Temperature variations on land (air) are much greater than in water. The high specific heat of water (see Section 3.2) prevents wide daily and seasonal fluctuations in the temperature of aquatic habitats; in contrast, such fluctuations are a characteristic of air temperatures (see Chapter 2). Likewise, the timing and quantity of precipitation received at a location constrain the availability of water for terrestrial plants and animals as well as their ability to maintain water balance. These fluctuations in temperature and moisture have both a short-term effect on metabolic processes and a long-term influence on the evolution and distribution of terrestrial plants and animals (see

(a)

(b)

Figure 4.1 | **(a)** The giant kelp (*Macrocystis pyrifera*) inhabits the waters off the coast of California. Anchored to the bottom sediments, these kelp plants can grow 100 feet or more toward the surface despite their lack of supportive tissues. These kelp plants are kept afloat through the buoyancy of gas-filled bladders attached to each blade, yet when the kelp plants are removed from the water, they collapse into a mass. **(b)** In contrast, a redwood tree (*Sequoia sempervirens*) of comparable height allocates more than 80 percent of its biomass to supportive and conductive tissues that help the tree resist gravitational forces.

Chapters 6 and 7). Ultimately, the geographic variation in climate (see Chapter 2) governs the large-scale distribution of plants and therefore the nature of terrestrial ecosystems (see Figure 2.1 and Chapter 23).

4.2 | Plant Cover Influences the Vertical Distribution of Light

In contrast to aquatic environments, where the absorption of solar radiation by the water itself results in a distinct vertical gradient of light, the dominant factor influencing the vertical gradient of light in terrestrial environments is the absorption and reflection of solar radiation by plants. When walking into a forest in summer, you will observe a decrease in light (Figure 4.2a). You can observe much the same effect if you examine the lowest layer in grassland or an old field (Figure 4.2b). The quantity and quality (spectral composition) of light that does penetrate the canopy of vegetation to reach the ground varies with both the quantity and orientation of the leaves.

The amount of light at any depth in the canopy is affected by the number of leaves above. As we move down through the canopy, the number of leaves above increases, so the amount of light decreases. However, because leaves vary in size and shape, the number of leaves is not the best measure of quantity.

The quantity of leaves, or foliage density, is generally expressed as the leaf area. Because most leaves are flat, the leaf area is the surface area of one or both sides of the leaf. When the leaves are not flat, the entire surface area is sometimes measured. To quantify the changes in light environment with increasing area of leaves, we need to define the area of leaves per unit ground area (m^2 leaf area/m^2 ground area). This measure is the **leaf area index (LAI)** (Figure 4.3). A leaf area index of 3 (LAI = 3) indicates a quantity of 3 m^2 of leaf area over each 1 m^2 of ground area.

The greater the leaf area index above any surface, the lower the quantity of light reaching that surface. As you move from the top of the canopy to the ground in a forest, the cumulative leaf area and LAI increase. Correspondingly, light decreases. The general relationship between available light and leaf area index is described by Beer's law (see Quantifying Ecology 4.1: Beer's Law and the Attenuation of Light).

In addition to the quantity of light, the spectral composition (quality) of light varies through the plant canopy. Recall from Chapter 2 (Section 2.1) that the wavelengths of approximately 400 to 700 nm make up visible light (see Figure 2.4). These wavelengths are also known as photosynthetically active radiation (PAR) because they include the wavelengths used by plants as a source of energy in photosynthesis (see Chapter 6). The transmittance of PAR is typically less than 10 percent, whereas the transmittance of far-red radiation (730 nm) is much greater. As a result, the ratio of red (660 nm) to far-red radiation (the R/FR ratio) decreases through the canopy. This shift in the spectral quality of light affects the production of phytochrome (a pigment that allows a plant to perceive shading by other plants), thus influencing patterns of growth and allocation (see Chapter 6, Section 6.7).

Besides the quantity of leaves, the orientation of leaves on the plant influences the attenuation of light through the canopy. The angle at which a leaf is oriented relative to

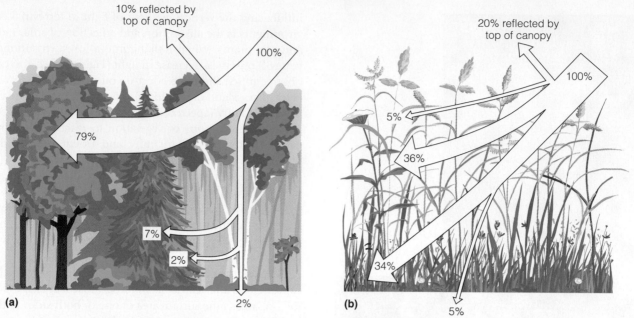

Figure 4.2 | Absorption and reflection of light by the plant canopy. **(a)** A mixed conifer–deciduous forest reflects about 10 percent of the incident photosynthetically active radiation (PAR) from the upper canopy, and it absorbs most of the remaining PAR within the canopy. **(b)** A meadow reflects 20 percent of the PAR from the upper surface. The middle and lower regions, where the leaves are densest, absorb most of the rest. Only 2–5 percent of PAR reaches the ground.

(Adapted from Larcher 1980.)

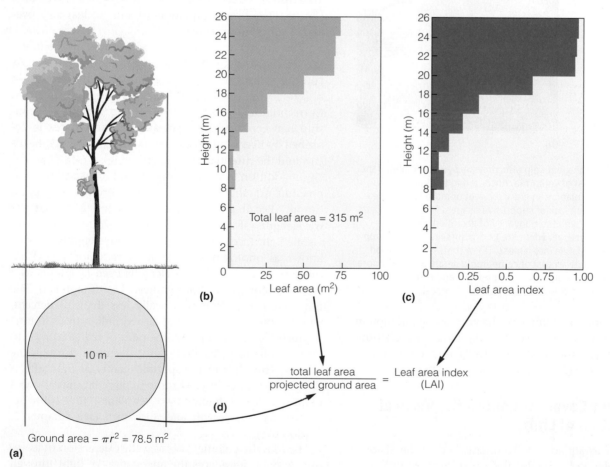

Figure 4.3 | The concept of leaf area index (LAI). **(a)** A tree with a crown 10 m wide projects a circle of the same size on the ground. **(b)** Foliage density (area of leaves) at various heights above the ground. **(c)** Contributions of layers in the crown to the leaf area index. **(d)** Calculation of LAI. The total leaf area is 315 m². The projected ground area is 78.5 m². The LAI is 4.

Beer's Law and the Attenuation of Light

Due to the absorption and reflection of light by leaves, there is a distinct vertical gradient of light availability from the top of a plant canopy to the ground. The greater the surface area of leaves, the less light will penetrate the canopy and reach the ground. The vertical reduction, or attenuation, of light through a stand of plants can be estimated using Beer's law, which describes the attenuation of light through a homogeneous medium. The medium in this case is the canopy of leaves. Beer's law can be applied to the problem of light attenuation through a plant canopy using the following relationship:

Light reaching any vertical position i, expressed as a proportion of light reaching the top of the canopy

Leaf area index above height i

$$AL_i = e^{-LAI_i \times k}$$

Light extinction coefficient

The subscript i refers to the vertical height of the canopy. For example, if i were in units of meters, a value of $i = 5$ refers to a height of 5 m above the ground. The value e is the natural logarithm (2.718). The light extinction coefficient, k, represents the quantity of light attenuated per unit of leaf area index (LAI) and is a measure of the degree to which leaves absorb and reflect light. The extinction coefficient will vary as a function of leaf angle (see Figure 4.4) and the optical properties of the leaves. Although the value of AL_i is expressed as a proportion of the light reaching the top of the canopy, the quantity of light at any level can be calculated by multiplying this value by the actual quantity of light (or photosynthetically active radiation) reaching the top of the canopy (units of $\mu mol/m^2/s$).

For the example presented in Figure 4.3, we can construct a curve describing the available light at any height in the canopy. In Figure 1, the light extinction coefficient has a value of $k = 0.6$, an average value for a temperate deciduous forest. We label vertical positions from the top of the canopy to ground level on the curve. Knowing the amount of leaves (LAI) above any position in the canopy (i), we can use the equation to calculate the amount of light there.

The availability of light at any point in the canopy will directly influence the levels of photosynthesis (see Figure 6.2). The light levels and rates of light-limited

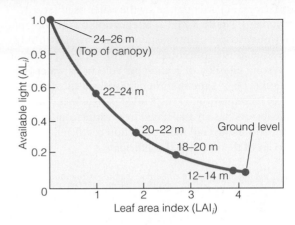

Figure 1 | Relationship between leaf area index above various heights in the canopy (LAI_i) and the associated values of available light (AL_i), expressed as a proportion of PAR at the top of the canopy.

photosynthesis for each of the vertical canopy positions are shown in the curve in Figure 2. Light levels are expressed as a proportion of values for fully exposed leaves at the top of the canopy ($1500\ \mu mol/m^2/s$). As one moves from the top of the canopy downward, the amount of light reaching the leaves and the corresponding rate of photosynthesis decline.

Beer's law can also be used to describe the vertical attenuation of light in aquatic environments, but applying the light extinction coefficient (k) is more complex. The

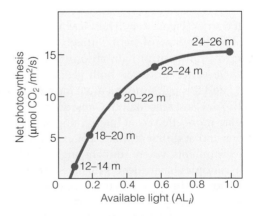

Figure 2 | Relationship between available light (PAR) and rate of net photosynthesis at various heights in the canopy. Available light is expressed as the proportion of PAR at the top of the canopy (assumed to be $1500\ \mu mol/m^2/s$).

continued on page 60

reduction of light with water depth is a function of various factors: (1) attenuation by the water itself (see Section 3.3, Figure 3.7); (2) attenuation by phytoplankton (microscopic plants suspended in water), typically expressed as the concentration of chlorophyll (the light-harvesting pigment of plants) per volume of water (see Section 6.1); (3) attenuation by dissolved substances; and (4) attenuation by suspended particulates. Each of these factors has an associated light extinction coefficient, and the overall light extinction coefficient (k_T) is the sum of the individual coefficients:

Whereas the light extinction coefficient for leaf area expresses the attenuation of light per unit of LAI, these values of k are expressed as the attenuation of light per unit of water depth (such as centimeter, meter, inches, or feet). Beer's law can then be used to estimate the quan-

tity of light reaching any depth (z) by using the following equation:

$$AL_z = e^{-k_T z}$$

If the ecosystem supports submerged vegetation, such as kelp (see Figure 4.1), seagrass, or other plants that are rooted in the bottom sediments, the preceding equation can be used to calculate the available light at the top of the canopy. The equation describing the attenuation of light as a function of LAI can then be applied (combined) to calculate the further attenuation from the top of the plant canopy to the sediment surface.

1. If we assume that the value of k used to calculate the vertical profile of light in Figure 1 ($k = 0.6$) is for a plant canopy where the leaves are positioned horizontally (parallel to the forest floor), how would the value of k differ (higher or lower) for a forest where the leaves were oriented at a 60-degree angle (see example in Figure 4.4)?

2. In shallow-water ecosystems, storms and high wind can result in bottom sediments (particulates) being suspended in the water for some time before once again settling to the bottom. How would this situation affect the value of k_T and the attenuation of light in the water profile?

the Sun changes the amount of light it absorbs. If a leaf that is perpendicular to the Sun absorbs 1.0 unit of light energy (per unit leaf area/time), the same leaf displayed at a 60-degree angle to the Sun will absorb only 0.5 unit. The reason is that the same leaf area represents only half the projected surface area and therefore intercepts only half as much light energy (Figure 4.4). Thus, leaf angle influences the vertical distribution of light through the canopy as well as the total amount of light absorbed and reflected. The sun angle varies, however, both geographically (see Section 2.1) and through time at a given location (over the course of the day and seasonally). Consequently, different leaf angles are more effective at intercepting light in different locations and at different times. For example, in high-latitude environments, where sunlight angles are low (see Figure 2.5), canopies having leaves that are displayed at an angle will absorb light more effectively. Leaves that are displayed at an angle rather than perpendicular to the Sun are also typical of arid tropical environments. In these hot and dry environments, angled leaves reduce light interception during midday, when temperatures and demand for water are at their highest.

Although light decreases downward through the plant canopy, some direct sunlight does penetrate

openings in the crown and reaches the ground as sunflecks. Sunflecks can account for 70–80 percent of solar energy reaching the ground in forest environments (Figure 4.5).

In many environments, seasonal changes strongly influence leaf area. For example, in the temperate regions of the world, many forest tree species are deciduous, shedding their leaves during the winter months. In these cases, the amount of light that penetrates a forest canopy varies with the season (Figure 4.6 on page 62). In early spring in temperate regions, when leaves are just expanding, 20–50 percent of the incoming light may reach the forest floor. In other regions characterized by distinct wet and dry seasons (see Chapter 2), a similar pattern of increased light availability at the ground level occurs during the dry season.

4.3 | Soil Is the Foundation upon Which All Terrestrial Life Depends

Soil is the medium for plant growth; the principal factor controlling the fate of water in terrestrial environments; nature's recycling system, which breaks down the waste products of plants and animals and transforms them into

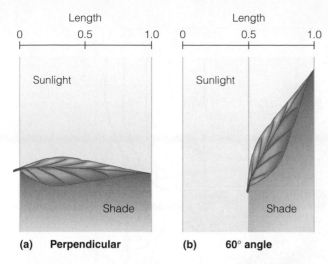

(a) Perpendicular **(b) 60° angle**

Figure 4.4 | Influence of leaf orientation (angle) on the interception of light energy. If a leaf that is perpendicular to the source of light **(a)** intercepts 1.0 unit of light energy, the same leaf at an angle of 60° relative to the light source will intercept only 0.5 unit **(b)**. The reduction in intercepted light energy is a result of the angled leaf projecting a smaller surface area relative to the light source.

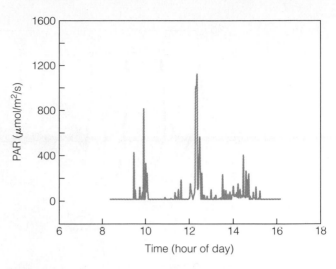

Figure 4.5 | Changes in the availability of light (photosynthetically active radiation; PAR) at ground level in a redwood forest in northern California over the course of a day. The spikes result from sunflecks in an otherwise low-light environment (average PAR of 30 μmol/m²/s). The median sunfleck length on this day was 2 seconds.

(Adapted from Pfitsch and Pearcy 1989.)

their basic elements (see Chapter 21); and habitat to a diversity of animal life, from small mammals to countless forms of microbial life.

As familiar as it is, soil is hard to define. One definition says that soil is a natural product formed and synthesized by the weathering of rocks and the action of living organisms. Another states that soil is a collection of natural bodies of earth, composed of mineral and organic matter and capable of supporting plant growth. Indeed, one eminent soil scientist, Hans Jenny—a pioneer of modern soil studies—will not give an exact definition of soil. In his book *The Soil Resource*, he writes:

> Popularly, soil is the stratum below the vegetation and above hard rock, but questions come quickly to mind. Many soils are bare of plants, temporarily or permanently, or they may be at the bottom of a pond growing cattails. Soil may be shallow or deep, but how deep? Soil may be stony, but surveyors (soil) exclude the larger stones. Most analyses pertain to fine earth only. Some pretend that soil in a flowerpot is not a soil, but soil material. It is embarrassing not to be able to agree on what soil is. In this pedologists are not alone. Biologists cannot agree on a definition of life and philosophers on philosophy.

Of one fact we are sure. Soil is not just an abiotic environment for plants. It is teeming with life—billions of minute and not so minute animals, bacteria, and fungi. The interaction between the biotic and the abiotic makes the soil a living system.

Soil scientists recognize soil as a three-dimensional unit, or body, having length, width, and depth. In most places on Earth's surface, exposed rock has crumbled and broken down to produce a layer of unconsolidated debris overlaying the hard, unweathered rock. This unconsolidated layer, called the **regolith,** varies in depth from virtually nonexistent to tens of meters. This interface between rock and the air, water, and living organisms that characterizes the surface environment is where soil is formed.

4.4 | The Formation of Soil Begins with Weathering

Soil formation begins with the weathering of rocks and their minerals. Weathering includes the mechanical destruction of rock materials into smaller particles as well as their chemical modification. **Mechanical weathering** results from the interaction of several forces. When exposed to the combined action of water, wind, and temperature, rock surfaces flake and peel away. Water seeps into crevices, freezes, expands, and cracks the rock into smaller pieces. Wind-borne particles, such as dust and sand, wear away at the rock surface. Growing roots of trees split rock apart.

Without appreciably influencing their composition, physical weathering breaks down rock and minerals into smaller particles. Simultaneously, these particles are chemically altered and broken down through the process of **chemical weathering.** The presence of water, oxygen, and acids resulting from the activities of soil organisms and the continual addition of organic matter (dead plant and animal tissues) enhance the chemical weathering process. Rainwater falling on and filtering through this

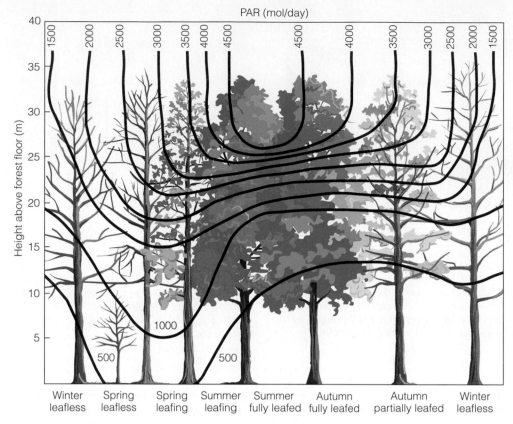

Figure 4.6 | Levels of photosynthetically active radiation (PAR) within and above a yellow poplar (*Liriodendron tulipifera*) stand over a year. The lines (isopleths) define the gradient of PAR. Solar radiation is most intense in summer, but since the canopy intercepts most of the PAR, little reaches the forest floor. The greatest amounts of PAR reach the forest floor in spring, when trees are still leafless. The forest receives the least PAR in winter, with its lower solar elevations and shorter day lengths. The amount of PAR reaching the forest floor in winter is little more than that in midsummer.

(Adapted from Hutchinson and Matt 1977.)

organic matter and mineral soil sets up a chain of chemical reactions that transform the composition of the original rocks and minerals.

4.5 | Soil Formation Involves Five Interrelated Factors

Five interdependent factors are important in soil formation: parent material, climate, biotic factors, topography, and time.

Parent material is the material from which soil develops. The original parent material could originate from the underlying bedrock; from glacial deposits (till); from sand and silt carried by the wind (eolian); from gravity moving material down a slope (colluvium); and from sediments carried by flowing water (fluvial), including water in floodplains. The physical character and chemical composition of the parent material are important in determining soil properties, especially during the early stages of development.

Biotic factors—plants, animals, bacteria, and fungi—all contribute to soil formation. Plant roots can function to break up parent material, enhancing the process of weathering, as well as to stabilize the soil surface and thus reduce erosion. Plant roots pump nutrients up from soil depths and add them to the surface. In doing so, plants recapture minerals carried deep into the soil by weathering processes. Through photosynthesis, plants capture the Sun's energy and add some of this energy to the soil in the form of organic carbon. On the soil surface, microorganisms break down the remains of dead plants and animals that eventually become organic matter incorporated into the soil (see Chapter 21).

Climate influences soil development both directly and indirectly. Temperature, precipitation, and winds directly influence the physical and chemical reactions responsible for breaking down parent material and the subsequent **leaching** (movement of solutes through the soil) and movement of weathered materials. Water is essential for the process of chemical weathering, and the greater the depth of water percolation, the greater the depth of weathering and soil development. Temperature controls the rates of biochemical reactions, affecting the balance between the accumulation and breakdown of organic materials. Consequently, under conditions of warm temperatures and abundant water, the processes of weathering, leaching, and plant growth (input of organic matter) are maximized. In contrast, under cold, dry conditions, the influence of these processes is much more modest. Indirectly, climate influences a region's plant and animal life, both of which are important in soil development.

Topography, the contour of the land, can affect how climate influences the weathering process. More water runs off and less enters the soil on steep slopes than on level land, whereas water draining from slopes enters the soil on low and flat land. Steep slopes are also subject to soil erosion and soil creep—the downslope movement of soil material that accumulates on lower slopes and lowlands.

Time is a crucial element in soil formation: all of the factors just listed assert themselves through time. The weathering of rock material; the accumulation, decomposition, and mineralization of organic material; the loss of minerals from the upper surface; and the downward movement of materials through the soil all require considerable time. Forming well-developed soils may require 2000 to 20,000 years.

4.6 | Soils Have Certain Distinguishing Physical Characteristics

Soils are distinguished by differences in their physical and chemical properties. Physical properties include color, texture, structure, moisture, and depth. All may be highly variable from one soil to another.

Color is one of the most easily defined and useful characteristics of soil. It has little direct influence on the function of a soil but can be used to relate chemical and physical properties. Organic matter (particularly humus) makes soil dark or black. Other colors can indicate the chemical composition of the rocks and minerals from which the soil was formed. Oxides of iron give a color to the soil ranging from yellowish-brown to red, whereas manganese oxides give the soil a purplish to black color. Quartz, kaolin, gypsum, and carbonates of calcium and magnesium give whitish and grayish colors to the soil. Blotches of various shades of yellowish-brown and gray indicate poorly drained soils or soils saturated by water. Soils are classified by color using standardized color charts.

Soil texture is the proportion of different-sized soil particles. Texture is partly inherited from parent material and partly a result of the soil-forming process. Particles are classified on the basis of size into gravel, sand, silt, and clay. Gravel consists of particles larger than 2.0 mm. They are not part of the fine fraction of soil. Soils are classified based on texture by defining the proportion of sand, silt, and clay.

Sand ranges from 0.05 to 2.0 mm, is easy to see, and feels gritty. Silt consists of particles from 0.002 to 0.05 mm in diameter that can scarcely be seen by the naked eye; it feels and looks like flour. Clay particles are less than 0.002 mm, too small to be seen under an ordinary

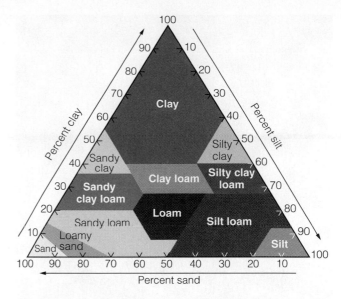

Figure 4.7 | A soil texture chart showing the percentages of clay (below 0.002 mm), silt (0.002–0.05 mm), and sand (0.05–2.0 mm) in the basic soil texture classes. For example, a soil with 60 percent sand, 30 percent silt, and 10 percent clay would be classified as a sandy loam.

Interpreting Ecological Data

Q1. What is the texture classification for a soil with 60 percent silt, 35 percent clay, and 5 percent sand?

Q2. What is the texture classification for a soil with 60 percent clay and 40 percent silt?

microscope. Clay controls the most important properties of soils, including its water-holding capacity (see Section 4.8) and the exchange of ions between soil particles and soil solution (see Section 4.9). A soil's texture is the percentage (by weight) of sand, silt, and clay. Based on proportions of these components, soils are divided into texture classes (Figure 4.7).

Soil texture affects pore space in the soil, which plays a major role in the movement of air and water in the soil and the penetration by roots. In an ideal soil, particles make up 50 percent of the soil's total volume; the other 50 percent is pore space. Pore space includes spaces within and between soil particles, as well as old root channels and animal burrows. Coarse-textured soils have large pore spaces that favor rapid water infiltration, percolation, and drainage. To a point, the finer the texture, the smaller the pores, and the greater the availability of active surface for water adhesion and chemical activity. Very fine-textured or heavy soils, such as clays, easily become compacted if plowed, stirred, or walked on. They are poorly aerated and difficult for roots to penetrate.

Soil depth varies across the landscape, depending on slope, weathering, parent materials, and vegetation. In grasslands, much of the organic matter added to the soil is from the deep, fibrous root systems of the grass plants. By contrast, tree leaves falling on the forest floor are the principal source of organic matter in forests. As a result, soils

Figure 4.8 | The pattern of horizontal layering or soil horizons is easily visible where a recent cut has been made along a road bank. This soil is relatively shallow, with the parent material close to the surface.

developed under native grassland tend to be several meters deep, and soils developed under forests are shallow. On level ground at the bottom of slopes and on alluvial plains, soils tend to be deep. Soils on ridgetops and steep slopes tend to be shallow, with bedrock close to the surface.

4.7 | The Soil Body Has Horizontal Layers, or Horizons

Initially, soil develops from undifferentiated parent material. Over time, changes occur from the surface down, through the accumulation of organic matter near the surface and the downward movement of material. These changes result in the formation of horizontal layers that are differentiated by physical, chemical, and biological characteristics. Collectively, a sequence of horizontal layers constitutes a **soil profile.** This pattern of horizontal layering, or **horizons,** is easily visible where a recent cut has been made along a road bank or during excavation for a building site (Figure 4.8).

The simplest general representation of a soil profile consists of four horizons: O, A, B, and C (Figure 4.9). The surface layer is the **O horizon,** or organic layer. This horizon is dominated by organic material, consisting of partially decomposed plant materials such as leaves, needles, twigs, mosses, and lichens. This horizon is often subdivided into a surface layer composed of undecomposed leaves and twigs (Oi), a middle layer composed of partially decomposed plant tissues (Oe), and a bottom layer consisting of dark brown to black, homogeneous organic

Organic layer: dominated by organic material, consisting of undecomposed or partially decomposed plant materials, such as dead leaves

Topsoil: largely mineral soil developed from parent material; organic matter leached from above gives this horizon a distinctive dark color

Subsoil: accumulation of mineral particles, such as clay and salts leached from topsoil; distinguished based on color, structure, and kind of material accumulated from leaching

Unconsolidated material derived from the original parent material from which the soil developed

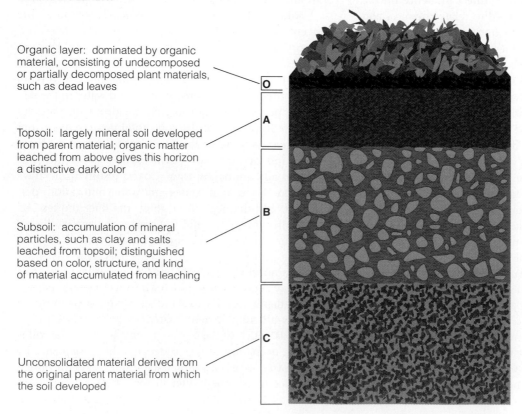

Figure 4.9 | A generalized soil profile. Over time, changes occur from the surface down, through the accumulation of organic matter near the surface and the downward movement of material. These changes result in the formation of horizontal layers, or horizons.

material—the humus layer (Oa). This pattern of layering is easily seen by carefully scraping away the surface organic material on the forest floor. In temperate regions, the organic layer is thickest in the fall, when new leaf litter accumulates on the surface. It is thinnest in the summer after decomposition has taken place.

Below the organic layer is the **A horizon,** often referred to as the topsoil. This is the first of the layers that are largely composed of mineral soil derived from the parent materials. In this horizon, organic matter (humus) leached from above accumulates in the mineral soil. The accumulation of organic matter typically gives this horizon a darker color, distinguishing it from lower soil layers. Downward movement of water through this layer also results in the loss of minerals and finer soil particles, such as clay, to lower portions of the profile—sometimes giving rise to an **E horizon,** a zone or layer of maximum leaching, or eluviation (from Latin *ex* or *e*, "out," and *lavere*, "to wash") of minerals and finer soil particles to lower portions of the profile. Such E horizons are quite common in soils developed under forests, but because of lower precipitation they rarely occur in soils developed under grasslands.

Below the A (or E) horizon is the **B horizon,** also called the subsoil. Containing less organic matter than the A horizon, the B horizon shows accumulations of mineral particles such as clay and salts due to leaching from the topsoil. This process is called illuviation (from the Latin *il*, "in," and *lavere*, "to wash"). The B horizon usually has a denser structure than the A horizon, making it more difficult for plants to extend their roots downward. B horizons are distinguished on the basis of color, structure, and the kind of material that has accumulated as a result of leaching from the horizons above.

The **C horizon** is the unconsolidated material that lies under the subsoil and is generally made of original material from which the soil developed. Because it is below the zones of greatest biological activity and weathering and has not been sufficiently altered by the soil-forming processes, it typically retains much of the characteristics of the parent materials from which it was formed. Below the C horizon lies the bedrock.

4.8 | Moisture-Holding Capacity Is an Essential Feature of Soils

If you dig into the surface layer of a soil after a soaking rain, you should discover a sharp transition between wet surface soil and the dry soil below. As rain falls on the surface, it moves into the soil by infiltration. Water moves by gravity into the open pore spaces in the soil, and the size of the soil particles and their spacing determine how much water can flow in. Wide pore spacing at the soil surface increases the rate of water infiltration, so coarse soils have a higher infiltration rate than fine soils do.

If there is more water than the pore space can hold, we say that the soil is **saturated,** and excess water drains freely from the soil. If water fills all the pore spaces and is held there by internal capillary forces, the soil is at **field capacity.** Field capacity is generally expressed as the percentage of the weight or volume of soil occupied by water when saturated compared to the oven-dried weight of the soil at a standard temperature. The amount of water a soil holds at field capacity varies with the soil's texture—the proportion of sand, silt, and clay. Coarse, sandy soil has large pores; water drains through it quickly. Clay soils have small pores and hold considerably more water. Water held between soil particles by capillary forces is **capillary water.**

As plants and evaporation from the soil surface extract capillary water, the amount of water in the soil declines. When the moisture level decreases to a point where plants can no longer extract water, the soil has reached the **wilting point.** The amount of water retained by the soil between field capacity and wilting point (or the difference between FC and WP) is the **available water capacity (AWC),** as shown in Figure 4.10. The AWC provides an estimate of the water available for uptake by plants.

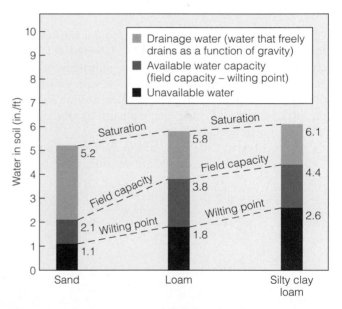

Figure 4.10 | Water content of three different soils at wilting point (WP), field capacity (FC), and saturation. The three soils differ in texture from coarse-textured sand to fine-textured silty clay loam (see soil texture chart of Figure 4.7). Available water capacity (AWC) is defined as the difference between FC and WP. Both FC and WP increase from coarse- to fine-textured soils, and the highest AWC is in the intermediate-textured soils.

Interpreting Ecological Data

Q1. Although fine-textured soils (silty clay loam) have a greater available water capacity, for this value to be achieved, the soil must be at or above field capacity. In arid regions, low and infrequent precipitation may keep soil water content below field capacity for most of the growing season. If the measured value of soil water content at a site is 1.9 inches per foot of soil, which of the three soil types represented in Figure 4.10 would have the greatest soil water available for uptake by plants?

Q2. What if the value of soil water was 2.9 inches per foot of soil?

Although water still remains in the soil—filling up to 25 percent of the pore spaces—soil particles hold it tightly, making it difficult to extract.

Both the field capacity and wilting point of a soil are heavily influenced by soil texture. Particle size of the soil directly influences the pore space and surface area onto which water adheres. Sand has 30 percent to 40 percent of its volume in pore space, whereas clays and loams (see soil texture chart in Figure 4.7) range from 40 percent to 60 percent. As a result, fine-textured soils have a higher field capacity than sandy soils, but the increased surface area results in a higher value of the wilting point as well (see Figure 4.10). Conversely, coarse-textured soils (sands) have a low field capacity and a low wilting point. Thus, AWC is highest in intermediate clay loam soils.

The topographic position of a soil affects the movement of water both on and in the soil. Water tends to drain downslope, leaving soils on higher slopes and ridgetops relatively dry and creating a moisture gradient from ridgetops to streams.

4.9 | Ion Exchange Capacity Is Important to Soil Fertility

Chemicals within the soil dissolve into the soil water to form a solution (see Section 3.5). Referred to as exchangeable nutrients, these chemical nutrients in solution are the most readily available for uptake and use by plants (see Chapter 6). They are held in soil by the simple attraction of oppositely charged particles and are constantly interchanging with the soil solution.

As described in Chapter 3, an **ion** is a charged particle. Ions carrying a positive charge are **cations,** whereas ions carrying a negative charge are **anions.** Chemical elements and compounds exist in the soil solution both as cations, such as calcium (Ca^{2+}), magnesium (Mg^{2+}), and ammonium (NH_4^+), and as anions, such as nitrate (NO_3^-) and sulfate (SO_4^{2-}). The ability of these ions in soil solution to bind to the surface of soil particles depends on the number of negatively or positively charged sites within the soil. The total number of charged sites on soil particles within a volume of soil is called the **ion exchange capacity.** In most soils of the temperate zone, cation exchange predominates over anion exchange due to the prevalence of negatively charged particles in the soil, referred to as **colloids.** The total number of negatively charged sites, located on the leading edges of clay particles and soil organic matter (humus particles), is called the **cation exchange capacity (CEC).** These negative charges enable a soil to prevent the leaching of its positively charged nutrient cations. Because in most soils there are far fewer positively charged than negatively charged sites, anions such as nitrate (NO_3^-) and phosphate (PO_3^{4-}) are not retained on exchange sites in soils but tend to leach away quickly if not taken up by plants. The CEC is a basic measure of soil quality and increases with higher clay and organic matter content.

Cations occupying the negatively charged particles in the soil are in a state of dynamic equilibrium with similar cations in the soil solution (Figure 4.11). Cations in soil solution are continuously being replaced by or exchanged with cations on the clay and humus particles. The relative abundance of different ions on exchange sites is a function of their concentration in the soil solution and the relative affinity of each ion for the sites. In general, the physically smaller the ion and

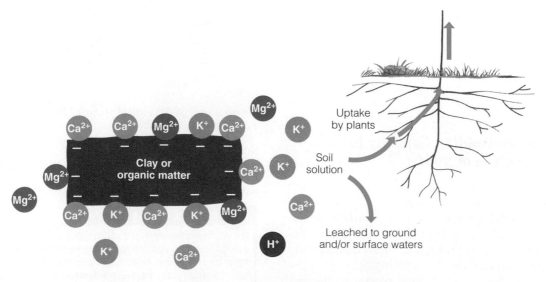

Figure 4.11 | The process of cation exchange in soils. Cations occupying the negatively charged particles in the soil are in a state of dynamic equilibrium with similar cations in the soil solution. Cations in soil solution are continuously being replaced by or exchanged with cations on clay and humus particles. Cations in the soil solution are also taken up by plants and leached to ground and surface waters.

the greater its positive charge, the more tightly it is held. The lyotropic series places the major cations in order of their strength of bonding to the cation exchange sites in the soil:

$$Al^{3+} > H^+ > Ca^{2+} > Mg^{2+} > K^+ = NH^{4+} > Na^+$$

However, higher concentrations in the soil solution can overcome these differences in affinity.

Hydrogen ions added by rainwater, by acids from organic matter, and by metabolic acids from roots and microorganisms increase the concentration of hydrogen ions in the soil solution and displace other cations, such as Ca^{2+}, on the soil exchange sites. As more and more hydrogen ions replace other cations, the soil becomes increasingly acidic (see Section 3.5). Acidity is one of the most familiar of all chemical conditions in the soil. Typically, soils range from pH 3 (extremely acid) to pH 9 (strongly alkaline). Soils of just over pH 7 (neutral) are considered basic, and those of pH 5.6 or below are acid. As soil acidity increases, the proportion of exchangeable Al^{3+} increases and Ca^{2+}, Na^+, and other cations decrease. High aluminum (Al^{3+}) concentrations in soil solution can be toxic to the plants. Aluminum toxicity damages the root system first, making the roots short, thick, and stubby. The result is reduced nutrient uptake.

4.10 | Basic Soil Formation Processes Produce Different Soils

Broad regional differences in geology, climate, and vegetation give rise to characteristically different soils. The broadest level of soil classification is the order. Each order has distinctive features, summarized in Figure 4.12, and

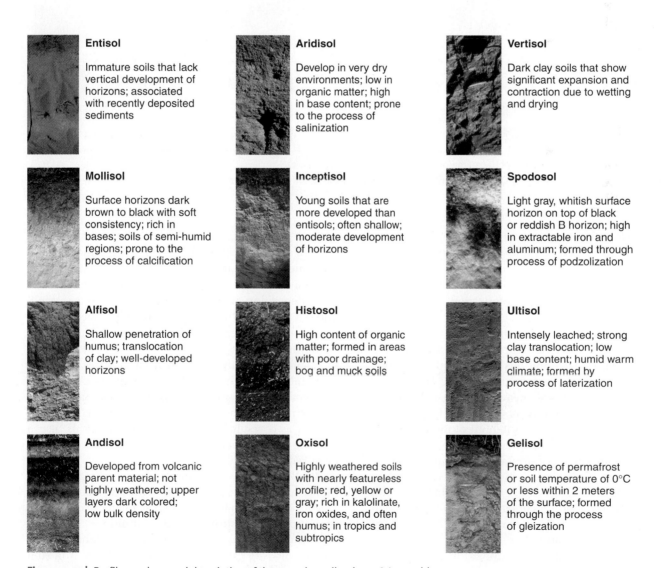

Entisol
Immature soils that lack vertical development of horizons; associated with recently deposited sediments

Aridisol
Develop in very dry environments; low in organic matter; high in base content; prone to the process of salinization

Vertisol
Dark clay soils that show significant expansion and contraction due to wetting and drying

Mollisol
Surface horizons dark brown to black with soft consistency; rich in bases; soils of semi-humid regions; prone to the process of calcification

Inceptisol
Young soils that are more developed than entisols; often shallow; moderate development of horizons

Spodosol
Light gray, whitish surface horizon on top of black or reddish B horizon; high in extractable iron and aluminum; formed through process of podzolization

Alfisol
Shallow penetration of humus; translocation of clay; well-developed horizons

Histosol
High content of organic matter; formed in areas with poor drainage; bog and muck soils

Ultisol
Intensely leached; strong clay translocation; low base content; humid warm climate; formed by process of laterization

Andisol
Developed from volcanic parent material; not highly weathered; upper layers dark colored; low bulk density

Oxisol
Highly weathered soils with nearly featureless profile; red, yellow or gray; rich in kalolinate, iron oxides, and often humus; in tropics and subtropics

Gelisol
Presence of permafrost or soil temperature of 0°C or less within 2 meters of the surface; formed through the process of gleization

Figure 4.12 | Profiles and general description of the 12 major soil orders of the world.

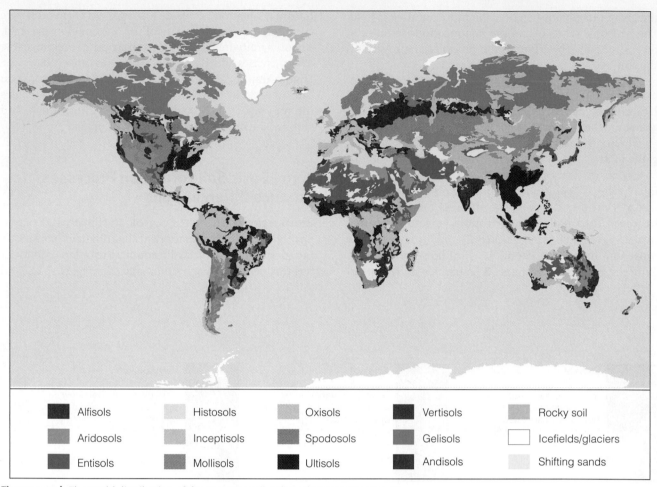

Figure 4.13 | The world distribution of the 12 major soil orders shown in Figure 4.12.

(Adapted from USGS, Soil Conservation Service.) Go to **GRAPHit!** at **www.ecologyplace.com** to graph global soil degradation.

Legend:
- Alfisols
- Aridosols
- Entisols
- Histosols
- Inceptisols
- Mollisols
- Oxisols
- Spodosols
- Ultisols
- Vertisols
- Gelisols
- Andisols
- Rocky soil
- Icefields/glaciers
- Shifting sands

its own distribution, mapped in Figure 4.13. Although a wide variety of processes are involved in soil formation (pedogenesis), soil scientists recognize five main soil-forming processes that give rise to these different classes of soils. These processes are laterization, calcification, salinization, podzolization, and gleization.

Laterization is a process common to soils found in humid environments in the tropical and subtropical regions. The hot, rainy conditions cause rapid weathering of rocks and minerals. Movements of large amounts of water through the soil cause heavy leaching, and most of the compounds and nutrients made available by the weathering process are transported out of the soil profile if not taken up by plants. The two exceptions to this process are compounds of iron and aluminum. Iron oxides give tropical soils their unique reddish coloring (see Ultisol profile in Figure 4.12 as an example of soil formed through the process of laterization). Heavy leaching also causes these soils to be acidic due to the loss of other cations (other than H^+).

Calcification occurs when evaporation and water uptake by plants exceed precipitation. The net result is an upward movement of dissolved alkaline salts, typically calcium carbonate ($CaCO_3$), from the groundwater. At the same time, the infiltration of water from the surface causes a downward movement of the salts. The net result is the deposition and buildup of these deposits in the B horizon (subsoil). In some cases, these deposits can form a hard layer called caliche (Figure 4.14a).

Salinization is a process that functions similarly to calcification, only in much drier climates. It differs from calcification in that the salt deposits occur at or very near the soil surface (Figure 4.14b). Saline soils are common in deserts but may also occur in coastal regions as a result of sea spray. Salinization is also a growing problem in agricultural areas where irrigation is practiced.

Podzolization occurs in cool, moist climates of the midlatitude regions where coniferous vegetation (e.g., pine forests) dominates. The organic matter of coniferous vegetation creates strongly acidic conditions. The acidic soil solution enhances the process of leaching, causing the removal of cations and compounds of iron and aluminum from the A horizon (topsoil). This process

(a)

(b)

Figure 4.14 | **(a)** In arid regions, salinization occurs when salts (the white crust at the center of the photo) accumulate near the soil surface because of surface evaporation. **(b)** Calcification occurs when calcium carbonates precipitate out from water moving downward through the soil or from capillary water moving upward from below. The result is an accumulation of calcium in the B horizon (seen as the white soil layer in the photo).

creates a sublayer in the A horizon that is composed of white- to gray-colored sand (see Spodosol profile in Figure 4.12 as an example of soil formed through the process of podzolization).

Gleization occurs in regions with high rainfall or low-lying areas associated with poor drainage (waterlogged). The constantly wet conditions slow the breakdown of organic matter by decomposers (bacteria and fungi), allowing the matter to accumulate in upper layers of the soil. The accumulated organic matter releases or-

ganic acids that react with iron in the soil, giving the soil a black to bluish-gray color (see Gelisol profile in Figure 4.12 as an example of soil formed through the process of gleization).

These five processes represent the integration of climate and edaphic (relating to the soil) factors on the formation of soils, giving rise to the geographic diversity of soils that influence the distribution, abundance, and productivity of terrestrial ecosystems. We will explore these topics further in Chapters 20, 21, and 23.

Summary

Life on Land | 4.1

Maintaining the balance of water between organisms and their surrounding environment has been a major influence on the evolution of life on land. The need to remain erect against the force of gravity in terrestrial environments results in a significant investment in structural materials. Variations in temperature and precipitation have both a short-term effect on metabolic processes and a long-term influence on the evolution and distribution of terrestrial plants and animals. The result is a distinct pattern of terrestrial ecosystems across geographic gradients of temperature and precipitation.

Light | 4.2

Light passing through a canopy of vegetation becomes attenuated. The density and orientation of leaves in a plant canopy influence the amount of light reaching the ground. Foliage density is expressed as leaf area index (LAI), the area of leaves per unit of ground area. The amount of light reaching the ground in terrestrial vegetation varies with the

season. In forests, only about 1–5 percent of light striking the canopy reaches the ground. Sunflecks on the forest floor enable plants to endure shaded conditions.

Soil Defined | 4.3

Soil is a natural product of unconsolidated mineral and organic matter on Earth's surface. It is the medium for plant growth; the principal factor controlling the fate of water in terrestrial environments; nature's recycling system, which breaks down the waste products of plants and animals and transforms them into their basic elements; and habitat to a diversity of animal life.

Weathering | 4.4

Soil formation begins with the weathering of rock and minerals. In mechanical weathering, water, wind, temperature, and plants break down rock. In chemical weathering, the activity of soil organisms, the acids they produce, and rainwater break down primary minerals.

Soil Formation | 4.5

Soil results from the interaction of five factors: parent material, climate, biotic factors, topography, and time. Parent material provides the substrate from which soil develops. Climate shapes soil development through temperature, precipitation, and its influence on vegetation and animal life. Biotic factors—vegetation, animals, bacteria, and fungi—add organic matter and mix it with mineral matter. Topography influences the amount of water entering the soil and the rates of erosion. Time is required to fully develop distinctive soils.

Distinguishing Characteristics | 4.6

Soils differ in the physical properties of color, texture, and depth. Although color has little direct influence on soil function, it can be used to relate chemical and physical properties. Soil texture is the proportion of different-sized soil particles—sand, silt, and clay. A soil's texture is largely determined by the parent material but is also influenced by the soil-forming process. Soil depth varies across the landscape, depending on slope, weathering, parent materials, and vegetation.

Soil Horizons | 4.7

Soils develop in layers called horizons. Four horizons are commonly recognized, although not all of them are necessarily present in any one soil: the O or organic layer; the A horizon, or topsoil, characterized by accumulation of organic matter; the B horizon, or subsoil, in which mineral materials accumulate; and the C horizon, the unconsolidated material underlying the subsoil and extending downward to the bedrock.

Moisture-Holding Capacity | 4.8

The amount of water a soil can hold is one of its important characteristics. When water fills all pore spaces, the soil is saturated. When a soil holds the maximum amount of water it can retain, it is at field capacity. Water held between soil particles by capillary forces is capillary water. When the moisture level is at a point where plants cannot extract water, the soil has reached wilting point. The amount of water retained between field capacity and wilting point is the available water capacity. The available water capacity of a soil is a function of its texture.

Ion Exchange | 4.9

Soil particles, particularly clay particles and organic matter, are important to nutrient availability and the cation exchange capacity of the soil—the number of negatively charged sites on soil particles that can attract positively charged ions. Cations occupying the negatively charged particles in the soil are in a state of dynamic equilibrium with similar cations in the soil solution. Percent base saturation is the percentage of sites occupied by ions other than hydrogen.

Soil Formation Processes Form Different Soils | 4.10

Broad regional differences in geology, climate, and vegetation give rise to characteristically different soils. The broadest level of soil classification is the order. Each order has distinctive features. Soil scientists recognize five main soil-forming processes that give rise to these different classes of soils. These processes are laterization, podzolization, calcification, salinization, and gleization.

Study Questions

1. Name two constraints imposed on organisms in the transition of life from aquatic to terrestrial environments.
2. Assume that two forests have the same quantity of leaves (leaf area index). In one forest, however, the leaves are oriented horizontally (parallel to the forest floor). In the other forest, the leaves are positioned at an angle of 60 degrees. How would the availability of light at the forest floor differ for these two forests at noon? In which forest would the leaves at the bottom of the canopy (lower in the tree) receive more light at mid-morning?
3. What is the general shape of the curve that describes the vertical attenuation of light through the plant canopy based on Beer's law? Why is it not a straight line (linear)?
4. What five major factors affect soil formation?
5. What role does weathering play in soil formation? What factors are involved in the process of weathering?
6. Use Figure 4.10 to answer this question: Which soil holds more moisture at field capacity: clay or sand? Which soil holds more moisture at wilting point: clay or sand? Which soil type has a greater availability of water for plant uptake when the water content of the soil is 3.0 in./ft (value on y-axis)?
7. What is the major factor distinguishing the O and A soil horizons?
8. Why do clay soils typically have a higher cation exchange capacity than do sandy soils?
9. How does pH influence the base saturation of a soil?
10. Why is the process of salinization more prevalent in arid areas? How does irrigation increase the process of salinization in agricultural areas?
11. What soil-forming process is dominant in the wet tropical regions? How does this process influence the availability of nutrients to plant roots in the A horizon?

Further Readings

Brady, N. C., and R. W. Weil. 1999. *The nature and properties of soils*. 12th ed. Upper Saddle River, NJ: Prentice Hall.

The classic introductory textbook on soils. Used for courses in soil science.

Jenny, H. 1994. *Factors of soil formation*. Mineola, NY: Dover Publications.

A well-written and accessible book by one of the pioneers in soil science.

Kohnke, H., and D. P. Franzmeier. 1994. *Soil science simplified*. Prospect Heights, IL: Wavelength Press.

A well-written and illustrated overview of concepts and principles of soil science for the general reader. Provides many examples and applications of basic concepts.

Patton, T. R. 1996. *Soils: A new global view*. New Haven, CT: Yale University Press.

Presents a new view and approach to studying soil formation at a global scale.

PART TWO

The Organism and Its Environment

The Namib Desert, stretching for 1200 miles along the southwest coast of Africa, is home to the highest sand dunes in the world. Rainfall is a rare event in the Namib. But each morning as the Sun rises, the cool, moist air of this coastal desert begins to warm, and the Namib becomes shrouded in fog. And each morning, black thumbnail-sized beetles perform one of nature's more bizarre behaviors (Figure 1). These tenebrionid beetles (*Stenocara* spp.) upend their bodies into a handstand. The beetle stays in this position as fog droplets collect on its back and then gradually roll down the wing case (called the elytra) into its mouth. By viewing the bumps on its back (Figure 2) through an electron microscope, we can see a wax-coated carpet of tiny nodules covering the sides of the bumps as well as the valleys between them that aid in channeling water from the beetle's back to its mouth.

The tenebrionid beetles of the Namib Desert illustrate two important concepts: the relationship between structure and function and how that relationship reflects adaptations of the organism to its environment. The structure of the beetle's back and the beetle's behavior of standing on its head in the morning fog serve the function of acquiring water, a scarce and essential resource in this arid environment. This same set of characteristics, however, are unlikely to be efficient for acquiring water in the desert regions of the continental interior, where morning fog does not form; or in wet environments, such as a tropical rain forest, where standing pools of water are readily available.

Fundamental questions for the ecologist are, What controls the distribution and abundance of species? What enables a species to succeed in one environment but not another? To the ecologist, the link between structure and function provides the first clue. The characteristics that an organism exhibits—its physiology, morphology (physical structure), behavior, and lifetime pattern of development and reproduction (life history)—reflect adaptations to its particular environment. Each environment presents a different set of constraints on processes relating to survival, growth, and reproduction. The set of characteristics that enable an organism to succeed in one environment typically preclude it from doing equally well under a different set of environmental conditions. The different evolutionary solutions to life in various environments represent the products of trade-offs. In nature, one size does not fit all. Earth's diverse environments are inhabited by 1.5 million known species, which is 1.5 million different ways that life exists on this planet.

Despite the diversity of species, all organisms (from single-celled bacteria to the largest of all animals, the blue whale) represent solutions to the same three basic functions shared by all living organisms: assimilation, reproduction, and the ability to respond to external stimuli. Organisms must acquire energy and matter from the external environment for the synthesis of new tissue through the process of assimilation. To maintain the continuity of life, some of the assimilated resources (energy and matter) must be allocated to reproduction—the production of new individuals. Finally, organisms must be able to respond to external stimuli relating to both the physical (such as heat and humidity) and the biotic (such as recognition of potential mates or predators) environments.

Perhaps the most fundamental constraint on life is the acquisition of energy. It takes energy to acquire and assimilate essential nutrients and perform the processes associated with life—synthesis, growth, reproduction, and maintenance. Chemical energy is generated in the breakdown of carbon compounds in all living cells, by the process called respiration. But the ultimate source of energy for life on Earth is the Sun (solar energy; see Chapter 2). Solar energy fuels photosynthesis, the process of assimilation in green plants. By consuming plant and animal tissues, all other organisms use energy that comes directly or indirectly from photosynthesis. The source that an organism derives its energy from is one of the most basic distinctions in ecology. Organisms that derive their energy from sunlight are referred to as **autotrophs,** or **primary producers.** Organisms that derive energy from consuming plant and animal tissues and then breaking down assimilated carbon compounds are called **heterotrophs,** or **secondary producers.** These two modes of acquiring energy impose fundamentally different evolutionary constraints; thus, our discussion of adaptations to the environment in Part Two is divided into the subjects of autotrophs (Chapter 6: Plant Adaptations to the Environment) and heterotrophs (Chapter 7: Animal Adaptations to the Environment). These chapters will focus on adaptations relating to the exchange of energy and matter between organisms and their environment: the processes of energy, carbon, nutrient, and water balances that govern the survival and growth of individual organisms. The final chapter (Chapter 8: Life History Patterns) treats adaptations relating to reproduction—how resources are used to assure the continuity of life. The concept of trade-offs is a central theme in each of these discussions, linking the constraints imposed by different environments (and resources) and the "solutions" reflected in the traits that characterize each species. But first we will turn our attention to the topic of natural selection (Chapter 5: Ecological Genetics: Adaptation and Natural Selection), the unifying concept that provides a foundation for understanding the evolution of adaptations as well as linking the patterns and processes we will explore at the hierarchical levels of ecological study: organisms, populations, communities, and ecosystems.

Figure 1 | A tenebrionid beetle (*Stenocara* spp.) perched upon a sand dune in the Namib Desert of southwestern Africa.

Figure 2 | Fog droplets can be seen on the wing case (elytra) of this tenebrionid beetle.

CHAPTER 5

Ecological Genetics: Adaptation and Natural Selection

Examples of variations in the structure of the hind legs of different beetles. Although most beetles use their legs for walking, legs have been variously modified and adapted for other uses. **(a)** In diving beetles of the family Dytiscidae, the last pair of legs bear rows of long hairs that aid in swimming. **(b)** Dung beetles of the family Scarabaeidae have legs that are widened and spined, which they use to fashion freshly laid dung into huge circular structures in which the female lays a single egg. **(c)** The hind legs of flea beetles (Chrysomelidae) are enlarged and designed for jumping.

(a)

(b)

(c)

An unusual fact of natural history is that one of every four animals on the planet is a beetle. When the eminent 19th-century British biologist J.B.S. Haldane was asked what conclusions he could draw regarding this fact, he is reported to have commented that God must have "an inordinate fondness for beetles."

There are over 330,000 species of beetles in the order Coleoptera (the name means "sheathed wing" and refers to the hardened and thickened forewings that function as a sheath to cover the more delicate, membranous hind wings). Beetles have been found almost everywhere on Earth, from the Arctic tundra to the sand dunes of the world's deserts (see example of tenebrionid beetles in Part Two introduction). They exploit almost every habitat and type of food.

The beetle's diversity of habitats and food resources is paralleled by an equal diversity of morphology and behaviors. For example, while most beetles use their legs for walking, legs have been variously modified and adapted for other uses. Among aquatic beetles, such as the diving beetles of the family Dytiscidae, the last pair of legs bear rows of long hairs that aid in swimming. Other beetles have legs that are widened and spined for digging. One group of beetles in the family Scarabaeidae, the dung beetles, use their legs to fashion freshly laid dung into huge circular structures and then roll them into an underground nest. The female then lays a single egg in each ball of dung and then covers the nest with more dung and soil. The hind legs of some beetles, such as flea beetles (Chrysomelidae) and flea weevils (Curculionidae), are enlarged and designed for jumping. These variations in the leg structure represent adaptations—characteristics that enable an organism to exploit a particular resource or thrive in a given environment.

As the evolutionary ecologist Ernst Mayr so poetically wrote, in the early 19th century, examples such as these served to illustrate the "wise laws that brought about the perfect adaptation of all organisms one to another and to their environment." Adaptation, after all, implied design—and design, a designer. Natural history was the task of cataloging the creations of the divine architect. By the mid-1800s, however, a revolutionary idea emerged that would forever change our view of nature.

> In considering the origin of species, it is quite conceivable that a naturalist . . . might come to the conclusion that species had not been independently created, but had descended, like varieties, from other species. Nevertheless, such a conclusion, even if well founded, would be unsatisfactory, until it could be shown how the innumerable species, inhabiting this world, have been modified, so as to acquire that perfection of structure and coadaptation which justly excites our admiration.

The pages that followed in Charles Darwin's *The Origin of Species*, first published on November 24, 1859, al-

Figure 5.1 | Charles Darwin (1809–1882).

tered the history of science and brought into question a view of the world that had been held for millennia (Figure 5.1). Charles Darwin put forward in those pages a mechanism to explain how the diversity of organisms inhabiting our world have acquired the features seemingly designed to enable them to survive and reproduce. He called it the theory of natural selection. Its beauty lay in its simplicity: the mechanism of natural selection is the simple elimination of "inferior" individuals.

5.1 | Adaptations Are a Product of Natural Selection

Stated more precisely, **natural selection** is the differential success (survival and reproduction) of individuals within the population that results from their interaction with their environment. As outlined by Darwin, natural selection is a product of two conditions: (1) that variation occurs among individuals within a population in some "heritable" characteristic, and (2) that this variation results in differences among individuals in their survival and reproduction. Natural selection is a numbers game. Darwin wrote:

> Among those individuals that do reproduce, some will leave more offspring than others. These individuals are considered more fit than the others because they contribute the most to the next generation. Organisms that leave few or no offspring contribute little or nothing to the succeeding generations and so are considered less fit.

The **fitness** of an individual is measured by the proportionate contribution it makes to future generations. Under a given set of environmental conditions, individuals having certain characteristics that enable them to survive and reproduce, eventually passing those characteristics on to the next generation, are selected for. Individuals without those traits are selected against. In this way, the process of natural selection results in changes in the properties of populations of organisms over the course of generations, by a process known as **evolution.**

An **adaptation** is any heritable behavioral, morphological, or physiological trait of an organism that has evolved over a period of time by the process of natural selection such that it maintains or increases the fitness (long-term reproductive success) of an organism under a given set of environmental conditions. The concept of adaptation by natural selection is central to the science of ecology. The study of the relationship between organisms and their environment is the study of adaptations. Adaptations represent the characteristics (traits) that enable an organism to survive, grow, and reproduce under the prevailing environmental conditions. Adaptations likewise govern the interaction of the organism with other organisms, both of the same and different species. How adaptations enable an organism to function in the prevailing environment—and conversely, how those same adaptations limit its ability to successfully function in other environments—is the key to understanding the distribution and abundance of species, the ultimate objective of the science of ecology.

5.2 | Genes Are the Units of Inheritance

By definition, adaptations are traits that are inherited—passed from parent to offspring. So to understand the evolution of adaptations, we must first understand the basis of inheritance: how characteristics are passed from parent to offspring, and what forces bring about changes in those same characteristics through time (from generation to generation).

At the root of all similarities and differences among organisms is the information contained within the molecules of DNA (deoxyribonucleic acid). You will recall

from basic biology that DNA is organized into discrete subunits—genes, which form the informational units of the DNA molecule. A **gene** is a stretch of DNA coding for a polypeptide chain (sequence of amino acids), where one or more polypeptides make up a protein. The alternate forms of a gene are called **alleles** (derived from the term *allelomorphs*, which in Greek means "different form"). The process of creating proteins from the genetic code in DNA is called **gene expression.** All of the DNA in a cell is collectively called the **genome.**

Genes are arranged in linear order along microscopic, threadlike bodies called **chromosomes.** The position occupied by a gene on the chromosome is called the **locus** (Latin for *place*). In most multicellular organisms, each individual cell contains two copies of each type of chromosome, one inherited from its mother through the ovum and one inherited through its father through the sperm. At any locus, therefore, every diploid individual contains two copies of the gene—one at each corresponding (homologous) position in the maternal and paternal chromosomes (termed homologous chromosomes). These two copies are the alleles of the gene in that individual. If the two copies of the gene are the same, then the individual is **homozygous** at that given locus. If the two alleles at the locus are different, then the individual is **heterozygous** at the locus. The pair of alleles present at a given locus defines the **genotype** of an individual; therefore, homozygous and heterozygous are the two main categories of genotypes.

5.3 | The Phenotype Is the Physical Expression of the Genotype

The outward appearance of an organism for a given characteristic is its **phenotype.** The phenotype is the external, observable expression of the genotype. When an individual is heterozygous, the two different alleles may produce an individual with intermediate characteristics; or one allele may mask the expression of the other (Figure 5.2). In the case where one allele masks the expression of the other, the allele that is expressed is referred to as the **dominant allele,** while the allele that is masked is called the **recessive allele.** If the

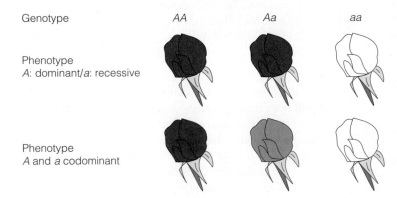

Genotype   *AA*   *Aa*   *aa*

Phenotype
A: dominant/*a*: recessive

Phenotype
A and *a* codominant

Figure 5.2 | Example of different modes of gene expression. In this example, flower color is controlled by a single locus having two alternate alleles, *A* and *a*. The *A* allele codes for the production of red pigment, while the *a* allele does not (absence of pigment). In the first case, heterozygous individuals (*Aa*) exhibit the same phenotype as homozygous *AA* individuals, indicating that *A* is the dominant allele. The recessive allele (*a*) is expressed only in homozygous recessive (*aa*) individuals. In the second case, the heterozygous individuals are intermediate in form to the homozygotes. In this case the alleles are said to be codominant, and each allele has a proportional effect on the phenotype.

physical expression of the heterozygous individual is intermediate between those of the homozygotes, the alleles are said to be **codominant,** and each allele has a specific value (proportional effect) that it contributes to the phenotype.

Phenotypic characteristics that fall into a limited number of discrete categories, such as the example of flower color presented in Figure 5.2, are referred to as **qualitative traits.** Even though all genetic variation is discrete (in the form of alleles), most phenotypic traits have a continuous distribution. These traits, such as height or weight, are referred to as **quantitative traits.** The continuous distribution of most phenotypic traits occurs for two reasons: (1) most traits have more than one gene locus affecting them, and (2) most traits are affected by the environment. For example, if the phenotypic characteristic of flower color illustrated in Figure 5.2 is controlled by two loci rather than a single locus (each with two alleles—A:a and B:b), there are nine possible genotypes (Figure 5.3). In contrast to the three distinct flower colors (phenotypes) produced in the case of a single locus, there is now a range of flower colors varying in hue between dark red and white depending on the number of alleles coding for the production of red pigment (see Figure 5.2). The greater the number of loci, the greater the range of possible phenotypes.

Genotype	# of alleles for red pigment	Phenotype (flower color)
AABB	4	
AABb	3	
AaBB	3	
AAbb	2	
AaBb	2	
aaBB	2	
Aabb	1	
aaBb	1	
aabb	0	

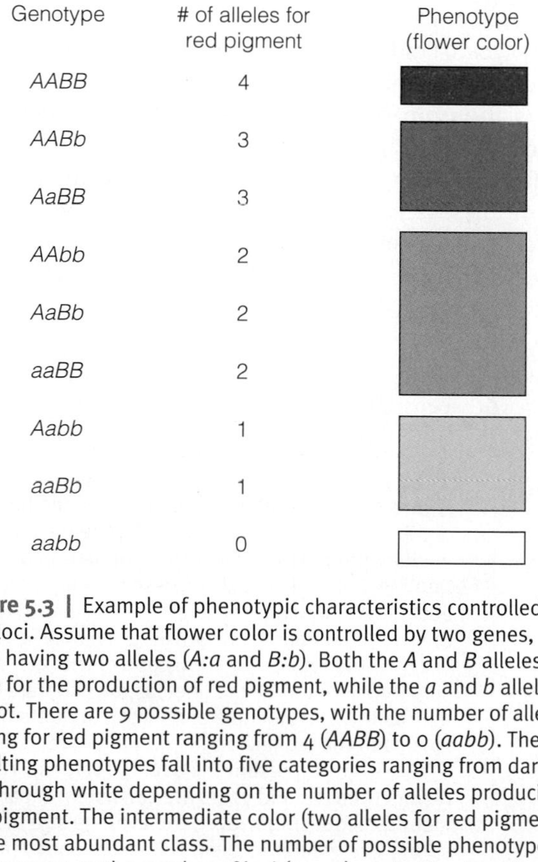

Figure 5.3 | Example of phenotypic characteristics controlled by two loci. Assume that flower color is controlled by two genes, each having two alleles (A:a and B:b). Both the A and B alleles code for the production of red pigment, while the a and b alleles do not. There are 9 possible genotypes, with the number of alleles coding for red pigment ranging from 4 (AABB) to 0 (aabb). The resulting phenotypes fall into five categories ranging from dark red through white depending on the number of alleles producing red pigment. The intermediate color (two alleles for red pigment) is the most abundant class. The number of possible phenotypes will increase as the number of loci (genes) controlling the phenotype increases.

The second factor influencing phenotypic variation is the environment. The expression of most phenotype traits is affected to varying degrees by the environment. Since environmental factors themselves usually vary continuously—temperature, rainfall, sunlight, level of predation, and so on—the environment can cause the phenotype produced by a given genotype to vary continuously. To illustrate this point, we can use the example of flower color controlled by two loci presented earlier (and in Figure 5.3). Pigment production during flower development can be affected by temperature. If temperatures below some optimal value or range function to reduce the expression of the A and B alleles in the production of red pigment, fluctuations in temperatures over the period of flower development in the population of plants will function to further increase the range of flower colors (shades between red and white) produced by the nine genotypes. We will examine the role of gene–environment interactions in greater detail later in Section 5.9.

5.4 | Genetic Variation Occurs at the Level of the Population

Adaptations are the characteristics of individual organisms—a reflection of the interaction of the genes and the environment. They are the product of natural selection. Although the process of natural selection is driven by the success or failure of individuals, the population—the collective of individuals and their alleles—changes through time, as individuals either succeed or fail to pass their genes to successive populations. For this reason, to understand the process of adaptation through natural selection, we must first understand how genetic variation is organized within the population.

A species is rarely represented by a single continuous interbreeding population. Instead, the population of a species is typically composed of a group of subpopulations—local populations of interbreeding individuals, linked to each other in varying degrees by the movement of individuals (see Chapter 12: Metapopulations). Thus genetic variation can occur at two hierarchical levels: within subpopulations and among subpopulations. When genetic variation occurs among subpopulations of the same species, it is called **genetic differentiation.**

The sum of genetic information (alleles) across all individuals in the population is referred to as the **gene pool.** The gene pool represents the total genetic variation within a population. Genetic variation within a population can be quantified in several ways. The most fundamental measures are **allele frequency** and **genotype frequency.** The word *frequency* in this context refers to the proportion of a given allele or genotype among all the alleles or genotypes present at the locus in the population.

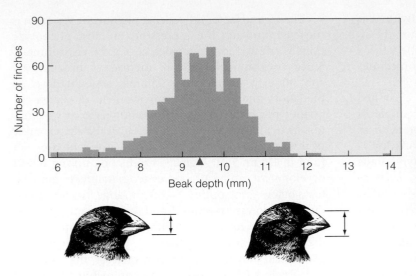

Figure 5.4 | Variation in beak size (as measured by depth) in the population of Galápagos medium ground finch (*Geospiza fortis*) on the island of Daphne Major as estimated by individual birds sampled during 1976. The histogram represents the number of individuals that were sampled (*y*-axis) in each category (0.2 mm) of bill depth (*x*-axis). The estimate of the population mean is marked by the blue triangle. (Adapted from Grant 1999 after Boag and Grant 1984.)

5.5 | Adaptation Is a Product of Evolution by Natural Selection

Earlier we defined evolution as changes in the properties of populations of organisms over the course of generations (Section 5.1). More specifically, phenotypic evolution can be defined as a change in the mean or variance (see QUANTIFYit! for discussion of statistics for estimating measures of central tendency and variation within populations) of a phenotypic trait across generations due to changes in allele frequencies. In favoring one phenotype over another, the process of natural selection acts directly on the phenotype. But in doing so, natural selection changes allele frequencies within the population. Changes in allele frequencies from parental to offspring generations are a product of differences in relative fitness (survival and reproduction) of individuals in the parental generation.

The work of Peter and Rosemary Grant provides an excellent documented example of natural selection. The Grants have spent more than two decades studying the birds of the Galápagos Islands, the same islands whose diverse array of animals so influenced the young Charles Darwin when he was the naturalist aboard the expeditionary ship HMS *Beagle*. Among other events, the Grants'

research documented a dramatic shift in a physical characteristic of finches inhabiting some of these islands during a period of extreme climate change.

Recall from our initial discussion in Section 5.1 that natural selection is a product of two conditions: (1) that variation occurs among individuals within a population in some "heritable" characteristic, and (2) that this variation results in differences among individuals in their survival and reproduction. Figure 5.4 shows variation in beak size in Darwin's medium ground finch (*Geospiza fortis*) on the 40-ha islet of Daphne Major, one of the Galápagos Islands off the coast of Ecuador. Heritability of beak size in this species was established by examining the relationship between the beak size of parents and their offspring (Figure 5.5).

Beak size is a trait that influences the feeding behavior of these seed-eating birds. Individuals with large beaks can feed on a wide range of seeds, from small to large, whereas individuals with smaller beaks are limited to feeding on smaller seeds (Figure 5.6).

During the early 1970s, the island received regular rainfall (127–137 mm per year), supporting an abundance of seeds and a large finch population (1500 birds). In 1977, however, a periodic shift in the climate

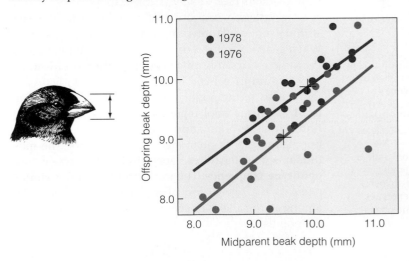

Figure 5.5 | The relationship between the beak depth (size) of offspring and their parents in the medium ground finch (*G. fortis*) population on Daphne Major. The *x*-axis represents midparent beak depth, which is the average beak depth for the two parent birds. The *y*-axis is the average beak depth of their offspring. The slope of the relationship (represented by the lines) is the estimate of heritability. The blue line and circles are data from 1976, and the red line and circles are data from 1978 (+ signs represent the average values). The results from the two years are consistent (nearly identical slopes); however, the average size of offspring was greater in 1978. Data from both years show a strong relationship between the beak depth of parents and their offspring. (Adapted from Grant 1999 after Boag 1983.)

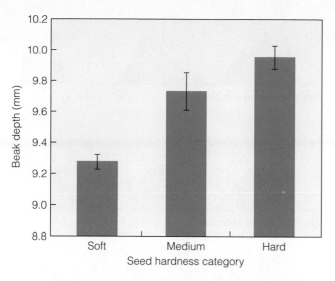

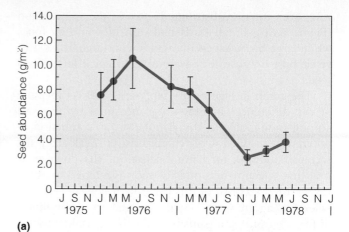

(a)

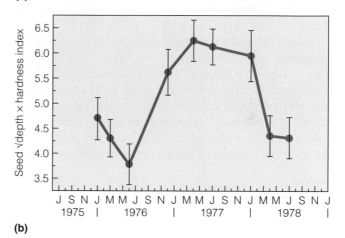

(b)

Figure 5.6 | Bill depth of medium ground finches (*G. fortis*) feeding on soft, medium, and hard seeds on Daphne Major in 1977. The bars represent the mean bill depth for birds feeding on the corresponding class of seeds, and the lines represent ±1 standard error. As can be seen, beak size has a direct influence on the hardness and size of seeds selected by individual birds.

(After Boag and Grant 1984.)

of the eastern Pacific Ocean—called La Niña (see Chapter 2, Section 2.9)—altered weather patterns over the Galápagos, causing a severe drought. That season only 24 mm of rain fell. During the drought, seed production declined drastically. Small seeds declined in abundance faster than large seeds did, increasing the average size and hardness of seeds available (Figure 5.7). The decline in food (seed) resources resulted in an 85 percent decline in the finch population due to mortality and possible emigration (Figure 5.8a). Mortality, however,

Figure 5.7 | Changes in **(a)** seed abundance and **(b)** seed size and hardness on Daphne Major for the period of July 1975 to July 1978. Points represent mean values, and associated lines represent the 95 percent confidence intervals. Seed size and hardness index is the square root of the product of seed depth and hardness.

(Adapted from Grant 1999 after Boag and Grant 1981.)

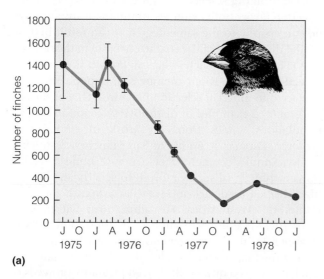

(a)

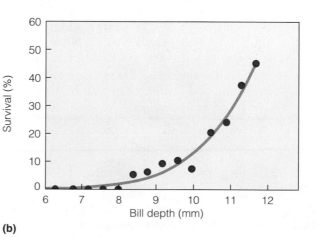

(b)

Figure 5.8 | (a) Decline of the population of medium ground finch on Daphne Major during the 1977 drought. Points represent mean estimates, and associated lines represent the 95 percent confidence interval. The population declined in the face of seed scarcity during a prolonged drought (Figure 5.7a). **(b)** Birds with larger beak size had a much greater rate of survival due to their ability to feed on the larger, harder seeds that comprised the majority of food resources during the drought period (Figure 5.7b).

(Adapted from Grant 1999 after Boag and Grant 1981.)

was not equally distributed across the population (Figure 5.8b). Small birds had difficulty finding food, while large birds, especially males with large beaks, survived best because they were able to crack large, hard seeds.

The graph in Figure 5.8b represents a direct measure of the differences in fitness (as measured by survival) among individuals in the population as a function of differences in phenotypic characteristics (beak size), the second condition for natural selection. The phenotypic trait that selection acts directly upon is referred to as the **target of selection;** in this example it is beak size. The **selective agent** is the environmental cause of fitness differences among organisms with different phenotypes, or in this case, the change in food resources (abundance and size distribution of seeds).

The increased survival rate of individuals with larger beaks resulted in a shift in the distribution of beak size (phenotypes) in the population (Figure 5.9). This type of natural selection, where the mean value of the trait is shifted toward one extreme over another (Figure 5.10a), is called **directional selection.** In other cases, natural selection may favor individuals near the population mean at the expense of the two extremes; this is referred

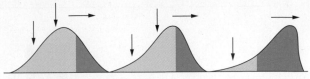

(a) Directional selection

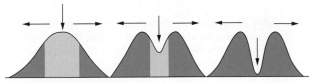

(b) Stabilizing selection

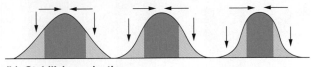

(c) Disruptive selection

Figure 5.10 | Three types of selection: The curves represent the distribution of characteristics (phenotypes) within the population. **(a)** Directional selection moves the mean of the population toward one extreme. **(b)** Stabilizing selection favors organisms with values close to the population mean. **(c)** Disruptive selection increases the frequencies of both extremes. Downward arrows represent selection pressures; horizontal arrows represent the direction of change.

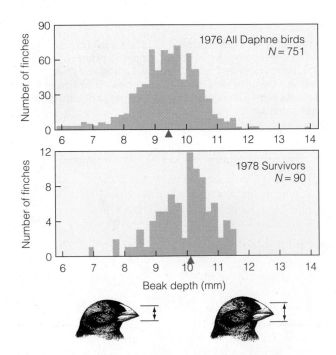

Figure 5.9 | Distribution of beak depth for the population of medium ground finches inhabiting Daphne Major **(a)** before and **(b)** after natural selection. The estimate of mean beak depth for both census periods is shown by the blue triangles. Note the increase in the mean beak depth for the population due to the differential survival of individuals related to beak size as shown in Figure 5.8.

(Adapted from Grant 1999 after Boag and Grant 1984.)

to as **stabilizing selection** (Figure 5.10b). When natural selection favors both extremes simultaneously, although not necessarily to the same degree, it can result in a bimodal distribution of the characteristic(s) in the population (Figure 5.10c). Such selection, known as **disruptive selection,** occurs when members of a population are subjected to different selection pressures. One of the few documented examples of disruptive selection in natural populations comes from the work of evolutionary ecologist Thomas B. Smith of San Francisco State University. Smith documented evidence of non-sex-related polymorphism (distinct morphological types) in the black-bellied seedcracker (*Pyrenestes ostrinus*), found in Cameroon, West Africa. For adult birds within both sexes, there is a distinct bimodal distribution in bill size (Figure 5.11) that results from disruptive selection.

For the black-bellied seedcracker, disruptive selection is related to seed quality. Birds feed primarily on seeds of two species of sedge; these seeds are similar in size but differ dramatically in hardness. Individuals with large bills feed more effectively on the hard-seeded sedge species, whereas individuals with small bills feed more effectively on the soft-seeded species.

(a)

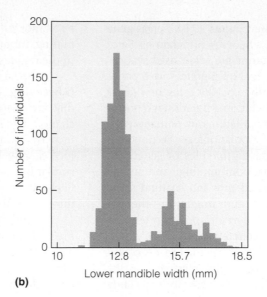

Figure 5.11 | Bimodal distribution of bill size (as measured by the width of the lower section of bill, i.e., mandible: x-axis) for adult black-bellied seedcrackers (*Pyrenestes ostrinus*) **(a).** The bimodal distribution of bill size is found for both males and females and is a result of disruptive selection related to seed quality. Smaller-billed individuals feed on soft-shelled seeds, whereas large-billed individuals are more successful at feeding on hard-shelled seeds **(b).** These two types of seeds are the primary food source for the species.

(Adapted from Smith 1993.) (Nature Publishing Group.)

(b)

5.6 | Several Processes Can Function to Alter Patterns of Genetic Variation

Natural selection is the only process that leads to adaptation because it is the only one in which the changes in allele frequency from one generation to the next are a product of differences in the relative fitness (survival and reproduction) of individuals in the population. Yet not all phenotypic characteristics represent adaptations, and processes other than natural selection can be important factors influencing changes in genetic variation (allele and genotype frequencies) within populations. For example, mutation is the ultimate source of the genetic variation that natural selection acts upon. **Mutations** are heritable changes in a gene or a chromosome. The word *mutation* refers to the process of altering a gene or chromosome as well as to the product, the altered state of the gene or chromosome. Mutation is a random force in evolution that produces genetic variation. Any altered phenotypic characteristic resulting from mutation may be beneficial, neutral, or harmful. Whether a mutation is beneficial or not depends on environment. A mutation that enhances an organism's fitness in one environment could harm it in another. Most of the mutations that have significant effect, however, are harmful; but the harmful mutations do not survive long. Natural selection eliminates most deleterious genes from the gene pool, leaving behind only genes that enhance (or at least do not harm) an organism's ability to survive, grow, and reproduce in its environment.

Another factor that can directly influence patterns of genetic variation within a population is a change in allele frequencies due to random chance—a process known as **genetic drift.** Recall from basic biology that the recombination of alleles in sexual reproduction is a random process. The offspring produced in sexual reproduction, however, represent only a subset of the parents' alleles. If the parents have only a small number of offspring, then not all of the parents' alleles will be passed on to their progeny due to chance assortment of chromosomes at meiosis (the process of recombination). In effect, genetic drift is the evolutionary equivalent of sampling error, with each successive generation representing only a subset or sample of the gene pool from the previous generation.

In a large population genetic drift will not affect each generation much, because the random nature of the process will tend to average out. But in a small population the effect could be rapid and significant. To illustrate this point, we can use the analogy of tossing a coin. With a single toss of the coin, the probability of each of the two possible outcomes, heads or tails, is equal, or 50 percent. Likewise, with a series of four coin tosses, the probability of the outcome being two heads and two tails is 50 percent. But each individual outcome in the coin tosses is independent; and therefore, in a series of four coin tosses, there is also a probability of 0.0625, or 6.25 percent, that the outcome will be four heads. The probability of the outcome being all heads drops to 9.765×10^{-4} if the number of tosses is increased to 10, and this probability drops to 8.88×10^{-16} for 50 tosses. Likewise, the probability of heterozygous (*Aa*) individuals in the population producing only homozygous (either *aa* or *AA*) offspring under a system of random mating decreases with increasing population size.

Patterns of genetic variation within a population can also be influenced by the movement of individuals into, or out of, the population. Recall from the discussion of genetic variation in Section 5.4 that the population of a species is typically composed of a group of subpopulations—local populations of interbreeding individuals that are linked to each other in varying degrees by the movement of individuals (see Chapter 12: Metapopulations). **Migration** is defined as the movement of individuals between local populations (see Chapter 12), whereas **gene flow** is the movement of genes between populations. Since individuals carry genes, the terms are often used synonymously; however, if an individual immigrates into a population but does not successfully reproduce, the new genes are not established in the population. Migration is a potent force in reducing the level of population differentiation (genetic differences among local populations; see Section 5.4).

One of the most important principles of genetics is that under conditions of random mating, and in the absence of the factors discussed thus far—natural selection, mutation, genetic drift, and migration—the frequency of alleles and genotypes in a population remains constant from generation to generation. In other words, no evolutionary change occurs through the process of sexual reproduction itself. This principle, referred to as the **Hardy–Weinberg principle,** is named for Godfrey Hardy and Wilhelm Weinberg, who each independently published the model in 1908 (see Quantifying Ecology 5.1: Hardy–Weinberg Principle). Mating is random when the chance that an individual mates with another individual of a given genotype is equal to the frequency of that genotype in the population. When individuals choose mates nonrandomly with respect to their genotype—or more specifically, select mates based on some phenotypic trait—the behavior is referred to as **assortative mating.** Perhaps the most recognized and studied form of assortative mating is female mate choice. Discussed in detail in Chapter 8 (Life History Patterns), female mate choice is the behavior in which females exhibit a bias toward certain males as mates based on specific phenotypic traits (often secondary sex characteristics), such as body size or coloration.

Positive assortative mating occurs when mates are phenotypically more similar to each other than expected by chance. Positive assortative mating is common, and one of the most widely reported examples relates to the timing of reproduction. Plants mate assortatively based on flowering time. In populations of plants with an extended flowering time, early flowering plants are often no longer flowering when late flowering plants are in bloom.

The genetic effect of positive assortative mating is to increase the frequency of homozygotes while decreasing the frequency of heterozygotes in the population. Think of a locus where *AA* individuals tend to be larger than *Aa*, which in turn are larger than *aa* individuals. With positive assortative mating, *AA* will mate with *AA*, and *aa* with other *aa*. All of these matings will produce only homozygous offspring. Even mating between *Aa* individuals will result in half of the offspring being homozygous. The genetic effects of positive assortative mating are only at the loci that affect the phenotypic characteristic by which the organisms are selecting mates.

Negative assortative mating occurs when mates are phenotypically less similar to each other than expected

by chance. Though not as common as positive assortative mating, negative assortative mating results in an increase in the frequency of heterozygotes.

A special case of nonrandom mating is inbreeding. **Inbreeding** is the mating of individuals in the population that are more closely related than expected by random chance. Unlike positive assortative mating, inbreeding increases homozygosity at all loci. Inbreeding affects all loci equally because related individuals are genetically similar by common ancestry, and they are therefore more likely than unrelated individuals are to share alleles throughout the genome.

Inbreeding can be detrimental. Offspring are more likely to inherit rare, recessive, deleterious genes. These genes can cause decreased fertility, loss of vigor, reduced fitness, reduced pollen and seed fertility in plants, and even death. These consequences are referred to as **inbreeding depression.**

As we have seen from the preceding discussion, nonrandom mating changes genotypic frequencies from one generation to the next, but assortative mating does not directly result in a change of allele frequencies within a population. The other three processes discussed—mutation, migration, and genetic drift, together with natural selection—alter the allele frequencies, therefore resulting in a shift in the distribution of genotypes (and potentially phenotype) within the population. As such, all four processes function as agents of evolution. However, natural selection is special among the four evolutionary processes because it is the only one that leads to adaptation. The other three can only speed up or slow the development of adaptations.

5.7 | Natural Selection Can Result in Genetic Differentiation

The example of natural selection in the population of medium ground finches as described earlier represents a shift in the distribution of phenotypes in the population inhabiting the island of Daphne Major. This shift in the mean phenotype (beak size) reflects a change in genetic variation (allele and genotype frequencies) within the population. Natural selection can also function to alter genetic variation among populations through the process of genetic differentiation (see Section 5.4).

Species having a wide geographic distribution often encounter a broader range of environmental conditions than do those species whose distribution is more restricted. The variation in environmental conditions often gives rise to a corresponding variation in morphological, physiological, and behavioral characteristics. Significant differences often exist among local populations of a single species inhabiting different regions. The greater the distance between populations, the more pronounced the differences often become as each population adapts to the locality it inhabits. Geographic variation within a

Hardy–Weinberg Principle

The Hardy–Weinberg principle states that both allele and genotype frequencies will remain the same in successive generations of a sexually reproducing population if certain criteria are met: (1) mating is random; (2) mutations do not occur; (3) the population is large, so that genetic drift is not a significant factor; (4) there is no migration; and (5) natural selection does not occur.

If we have only two alleles at a locus, designated as A and a, then the usual symbols for designating their frequencies are p and q, respectively. Since frequencies (proportions) must sum to 1, then:

$$p + q = 1 \text{ or } q = 1 - p$$

Genotypic frequencies are typically designated by uppercase letters. In the case of a locus with two alleles, P is the frequency of AA, H is the frequency of Aa, and Q is the frequency of aa. As with gene frequencies, genotype frequencies must sum to 1:

$$P + H + Q = 1$$

Given a population having the genotypic frequencies of

$$P = 0.64, H = 0.32, \text{ and } Q = 0.04$$

we can calculate the allele frequencies as follows:

$$p = P + H/2 = 0.64 + (0.32/2) = 0.8$$
$$q = Q + H/2 = 0.04 + (0.32/2) = 0.2$$

The frequency of heterozygous individuals (H: Aa) is divided by 2 because only one of the allele pair is A or a.

With a population consisting of the three genotypes just described (AA, Aa, and aa), there are six possible types of mating (Table 1). For example, the mating $AA \times AA$ occurs only when an AA female mates with an AA male, with the frequency of occurrence being $P \times P$ (or P^2) under the conditions of random mating. Similarly, an $AA \times Aa$ mating occurs when an AA female mates with an Aa male (proportion $P \times H$) or when an Aa female mates with an AA male (proportion $H \times P$). Therefore the overall proportion of $AA \times Aa$ matings is $PH + HP = 2\,PH$. The frequencies of these and the other four types of matings are given in the second column of Table 1.

To calculate the offspring genotypes produced by these matings, we must first examine the offspring produced by each of the six possible pairings of parental genotypes (Table 1). Since homozygous AA genotypes produce only A-bearing gametes (egg or sperm), and homozygous aa genotypes produce only a-bearing gametes, the mating of $AA \times AA$ individuals will produce only AA offspring, and likewise, the mating of $aa \times aa$ individuals will produce only offspring with genotype aa. In addition, the mating of $AA \times aa$ individuals will produce only heterozygous offspring (Aa).

In contrast to homozygous individuals, heterozygous individuals produce both A- and a-bearing gametes.

Table 1 | **Calculation of Offspring Genotype Frequencies for a Randomly Mating Population**

(a) Parental Genotype Frequencies			(b) Parental Allele Frequencies	
P	H	Q	p	q
(AA)	(Aa)	(aa)	0.8	0.2
0.64	0.32	0.04		

(c) Mating Genotype		Frequency of Mating		Offspring Frequencies		
				AA	Aa	aa
$AA \times AA$	P^2	$.64 \times .64 =$	.4096	1 .4096	0	0
$AA \times Aa$	$2PH$	$2 \times .64 \times .32 =$	.4096	$\frac{1}{2}$.2048	$\frac{1}{2}$.2048	0
$AA \times aa$	$2PQ$	$2 \times .64 \times .04 =$	.0512	0	1 .0512	0
$Aa \times Aa$	H^2	$.32 \times .32 =$	.1024	$\frac{1}{4}$.0256	$\frac{1}{2}$.0512	$\frac{1}{4}$.0256
$Aa \times aa$	$2HQ$	$2 \times .32 \times .04 =$	.0256	0	$\frac{1}{2}$.0128	$\frac{1}{2}$.0128
$aa \times aa$	Q^2	$.04 \times .04 =$	.0016	0	0	1 .0016
Total (frequency of mating):			1.0000			
				.64	.32	.04
Totals (next generation):				P'	H'	Q'

continued on page 84

Therefore, the mating of a heterozygous individual with a homozygous (either *AA* or *aa*) or another heterozygous individual will produce offspring of all three possible genotypes (*AA*, *Aa*, and *aa*). The relative frequencies of offspring genotypes depend on the specific combination of parents (Figure 1). The offspring frequencies presented in Figure 1 are based on the assumption that an *Aa* heterozygote individual produces an equal number of *A*- and *a*-bearing gametes (referred to as Mendelian segregation).

Using the data presented in Figure 1 and the frequencies of the different types of matings in column 2 of Table 1, the genotype frequencies of the offspring, denoted as P' (*AA*), H' (*Aa*), and Q' (*aa*), are presented in column 3 of Table 1. The new genotype frequencies are calculated as the sum of the products shown at the bottom of Table 1. For each genotype, the frequency of each mating producing the genotype is multiplied by the fraction of the genotypes produced by that mating.

We can now calculate the allele frequencies (p and q) for the generation of offspring (designated as p' and q') using the formula presented earlier:

$$p' = P' + H'/2 = 0.64 + (0.32/2) = 0.8$$

$$q' + Q' + H'/2 = 0.04 + (0.32/2) = 0.2$$

Note that both the genotype and allele frequencies of the offspring generation are the same as those of the parental generation.

In natural populations the assumptions of the Hardy–Weinberg principle are never fully met. Mating is not random, mutations do occur, individuals move between local populations, and natural selection does occur. All of these circumstances change the frequencies of genotypes and alleles from generation to generation, acting as evolutionary forces in a population. The beauty of the Hardy–Weinberg principle is that it functions as a null model, where deviations from the expected frequencies can provide insight into the evolutionary forces at work within a population.

1. How would the frequency of heterozygotes in the population change if the frequency for the *A* allele in the example described was $p = 0.5$?

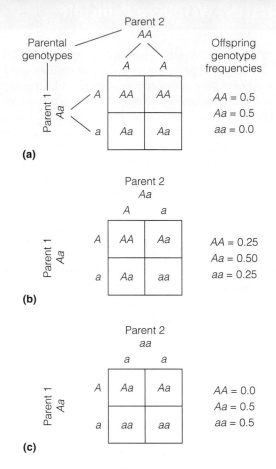

(a)

(b)

(c)

Figure 1 | Proportion of offspring genotypes produced by heterozygous individuals mating with (a) homozygous *AA* individual, (b) another heterozygous individual, and (c) homozygous *aa* individual. The frequencies of genotypes produced from each of these matings are shown.

2. When the frequency of an allele is greater than 0.8, most of these alleles are contained in homozygous individuals (as illustrated in the example). When the frequency of an allele is less than 0.1, in which genotype are most of these alleles?

species can result in clines, ecotypes, and geographic isolates.

A **cline** is a measurable, gradual change over a geographic region in the average of some phenotypic character, such as size and coloration; or it can be a gradient in genotypic frequency. Clines are usually associated with an environmental gradient such as temperature, moisture, light, or altitude. Continuous variation results from gene flow from one population to another along the gra-

dient. Because environmental constraints influencing natural selection vary along the gradient, any one population along the gradient will differ genetically to some degree from another, the difference increasing with the distance between the populations.

Clinal differences exist in size, body proportions, coloration, and physiological adaptations among animals. For example, the white-tailed deer (*Odocoileus virginianus*) in North America exhibits clinal variation in body weight

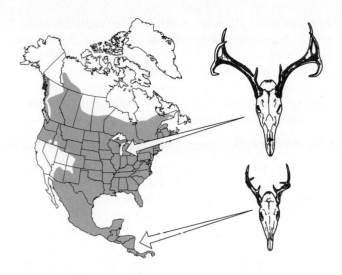

Figure 5.12 | Comparative skull sizes for white-tailed deer in the northern and southern areas of their geographic distribution. (Adapted from Baker 1984.)

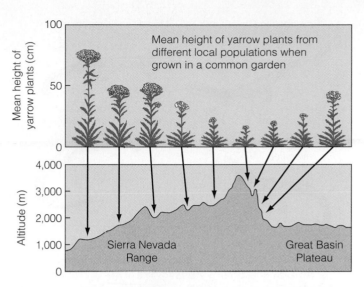

Figure 5.13 | Local population of yarrow plants (*Achillea millefolium*) inhabiting the Sierra Nevada mountains of western North America exhibit an inverse relationship between altitude and plant height, with local populations inhabiting lower altitudes being much taller in stature than montane (high-altitude) populations. When seeds from local populations were planted in a common garden at Stanford University, the differences in stature (mean plant height) observed in their native habitats (altitudes) were maintained, demonstrating the role of genetic differentiation in the observed phenotypic differences among local populations. (From Campbell and Reece 2005 after Clausen et al. 1948.)

(Figure 5.12). White-tailed deer in Canada and the northern United States are heaviest, weighing on average more than 136 kg. Individuals of the species weigh on average 93 kg in Kansas, 60 kg in Louisiana, and 46 kg in Panama. The smallest individuals of this species, the Key deer in Florida, weigh less than 23 kg.

Clinal variation can show marked discontinuities. Such abrupt changes, or step clines, often reflect abrupt changes in local environments. Such variants are called ecotypes. An **ecotype** is a population adapted to its unique local environmental conditions. For example, a population inhabiting a mountaintop may differ from a population of the same species in the valley below. When several habitats (unique environments) that the species is adapted to recur throughout the species' range, ecotypes often are scattered like a mosaic across the landscape.

Yarrow (*Achillea millefolium*) is a plant species of the temperate and subarctic regions of the Northern Hemisphere with an exceptional number of ecotypes. It exhibits considerable variation, an adaptive response to different climates at various altitudes. Populations exhibit an inverse relationship between altitude and plant height; local populations inhabiting lower altitudes are much taller than montane ("high-altitude") populations.

In a classic study, plant ecologists J. Clausen, D. D. Keck, and W. M. Hiesey of Stanford University collected seeds from local populations of yarrows at different elevations in a transect across the Sierra Nevada mountains. To remove the effects of environmental differences at the different elevations, seeds from the various local populations were planted in a common garden at Stanford University (Figure 5.13). Results of the experiment revealed that the local populations maintained the differences in stature and other associated traits that are observed in their native habitats (altitudes), demonstrating

the role of genetic as well as environmental factors in the phenotypic variation that occurs in populations at different altitudes. Later experiments demonstrated that the observed differences among local populations in stature, and leaf shape and morphology, represent physiological adaptations to the constraints imposed by differences in altitude.

The southern Appalachian Mountains are noted for their diversity of salamanders. This diversity is fostered in part by a rugged terrain, an array of environmental conditions, and the limited ability of salamanders to disperse (Figure 5.14). Populations become isolated from one another, preventing a free flow of genes. One species of salamander, *Plethodon jordani*, breaks into a number of semi-isolated populations, each characteristic of a particular part of the mountains. These subpopulations make up **geographic isolates,** in which the free flow of genes among subpopulations is prevented by some extrinsic barrier—in the case of the salamanders, rivers and mountain ridges. The degree of isolation depends on the efficiency of the extrinsic barrier, but rarely is the isolation complete. These geographic isolates are often classified as **subspecies,** a taxonomic term for populations of a species that are distinguishable by one or more characteristics. Unlike clines, for subspecies we can draw a geographic line separating the subpopulations into subspecies. Nevertheless, it is often difficult to draw the line between species and subspecies.

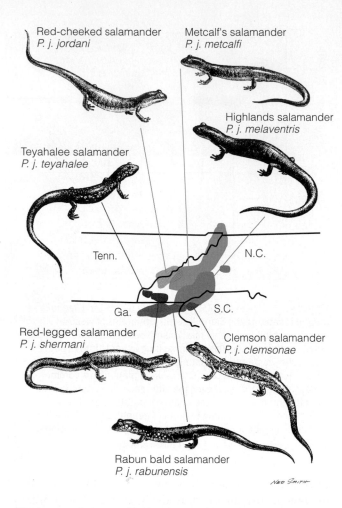

Red-cheeked salamander
P. j. jordani

Metcalf's salamander
P. j. metcalfi

Highlands salamander
P. j. melaventris

Teyahalee salamander
P. j. teyahalee

Tenn.

N.C.

Ga.

S.C.

Red-legged salamander
P. j. shermani

Clemson salamander
P. j. clemsonae

Rabun bald salamander
P. j. rabunensis

NED SMITH

Figure 5.14 | Geographical isolates in *Plethodon jordani* of the Appalachian highlands. These salamanders originated when a population of the salamander *P. yonahlossee* became separated by the French Broad River valley. The eastern population developed into Metcalf's salamander, which spread northward, the only direction in which any group member could find suitable ecological conditions. To the south, southwest, and northwest, the mountains end abruptly, limiting the remaining *jordani*. Metcalf's salamander is the most specialized and ecologically divergent and the least competitive. Next, the red-cheeked salamander became isolated from the red-legged and the rest of the group by the deepening of the Little Tennessee River. Remaining members are still somewhat connected.

5.8 | Adaptations Reflect Trade-offs and Constraints

If Earth were one large homogeneous environment, perhaps a single phenotype, a single set of characteristics might bestow upon all living organisms the ability to survive, grow, and reproduce. But this is not the case. As we have seen in Part One (The Physical Environment), environmental conditions that directly influence life vary in both space and time. Patterns of temperature, precipitation, and seasonality vary across Earth's surface (Chapter 2), producing a diversity of unique terrestrial

environments. Likewise, variations in depth, salinity, pH, and dissolved oxygen define an array of freshwater and marine habitats (Chapter 3). Each combination of environmental conditions presents a unique set of constraints on the organisms that inhabit them—constraints on their ability to maintain basic metabolic processes that are essential to survival and reproduction. Therefore, as features of the environment change, so will the set of traits (phenotypic characteristics) that increase the ability of individuals to survive and reproduce. Natural selection will favor different phenotypes under different environmental conditions. This principle was clearly illustrated in the example of Darwin's medium ground finch in Section 5.5, where a change in the resource base (abundance, size, and hardness of seeds) resulted in a shift in the distribution of phenotypes within the population. Simply stated, the fitness of any phenotype is a function of the prevailing environmental conditions; and the characteristics that enable an individual to maximize fitness under one set of environmental conditions generally limit its ability to do so under a different set of conditions. The limitations on the fitness of a phenotype under different environmental conditions are a function of trade-offs imposed by constraints that can ultimately be traced to the laws of physics and chemistry.

This general but important concept of adaptive trade-offs is illustrated in the example of natural selection for beak size in the population of Darwin's medium ground finch (*Geospiza fortis*) presented in Section 5.5. Recall from Figure 5.6 that the ability to utilize different seed resources (size and hardness) is related to beak size. Individuals with small beaks feed on the smallest and softest seeds, and individuals with larger bills feed on the largest and hardest seeds. These differences in diet as a function of beak size reflect a trade-off in morphological characteristics (the depth and width of the bill) that allow for the effective exploitation of different seed resources. This pattern of trade-offs is even more apparent if we compare differences in beak morphology and the utilization of seed resources for the three most common species of Darwin's ground finch that inhabit Santa Cruz island in the Galápagos.

The distributions of beak size (phenotypes) for individuals of the three most common species of Darwin's ground finches are shown in Figure 5.15a. As their common names suggest, the mean value of beak size increases from the small (*G. fuliginosa*) to the medium (*G. fortis*) and large (*G. magnirostris*) ground finch. In turn, the proportions of various seed sizes in their diets (Figure 5.15b, on page 89) reflect these differences in beak size, with the average size and hardness of seeds in the diets of these three populations increasing as a function of beak size. Small beak size restricts the ability of the smaller finch species (*G. fuliginosa*) to feed on larger, harder seed resources. In contrast, large beak size allows individuals

Department of Zoology University of Guelph, Guelph, Ontario, Canada

Most of the lakes scattered over the northern regions of North America were formed after the immense ice sheets that covered the region retreated about 15,000 years ago. One group of fish inhabiting these new environments, the three-spine sticklebacks (*Gasterosteus* spp.), have undergone a rapid period of speciation, with the species inhabiting the lakes of coastal British Columbia being among the youngest species on Earth. No more than two species occur in any one lake, and pairs of species in different lakes seem to have evolved completely independently of other pairs.

In every lake where a pair of threespine stickleback species is found, the two species have different patterns of habitat use and diet. One of the species feeds on plankton in the open-water zone (limnetic form), and the other species exploits larger prey from the sediments and submerged vegetation of the shallower, nearshore waters (benthic form). Morphological differences between the species correlate with these differences in habitat use and diet (Figure 1). Whether constraints imposed by these distinct environments are partly responsible for the evolution of the pairs of species is a central question in the research of Beren Robinson of the Department of Zoology at the University of Guelph.

In lakes where only a single species is found, the species of threespine stickleback tend to be intermediate in morphology and habit to the limnetic and benthic species. In sampling the population of one such

species—*Gasterosteus aculeatus,* which occurs in Cranby Lake in the coastal region of British Columbia—Robinson found that individuals sampled from the open-water habitat differed morphologically from those individuals sampled from the shallower nearshore waters. In addition, these differences in morphology parallel the differences observed between species that occupy these two habitats in lakes where species pairs are found (Figure 1). Robinson hypothesized that these individuals represented distinct phenotypes that are products of natural selection promoting divergence within the population. Divergent (or disruptive) selection can potentially occur when individuals within a population face trade-offs involving the performance of different tasks. Trade-offs can occur when performance on one task (such as feeding on plankton in the open water) results in a cost to performance and fitness on a second task (feeding on the sediments in the shallow water of the shoreline). Severe trade-offs can result in divergent selection that favors resource specialization.

To test his hypothesis, Robinson needed to establish that two conditions were met. First, that morphological differences between the two forms were heritable, rather than an expression of phenotypic plasticity in response to the two different habitats or diets. Second, that the observed morphological differences between the two forms influence their foraging efficiency—that they do in fact represent a trade-off.

To test the first condition, Robinson reared offspring of the two forms under identical laboratory conditions (environmental conditions and diet). The results showed that although there was some degree of phenotypic plasticity, differences in most characteristics remained between the two forms when raised under identical conditions. On average, the benthic form (BF) had (1) shorter overall body length, (2) deeper body, (3) wider mouth, (4) more dorsal spines, and (5) fewer gill rakers than did the limnetic form (LF). Therefore, the phenotypic difference in morphology between these two forms is indeed heritable.

To test the second condition, Robinson conducted feeding trials in the laboratory to test for trade-offs in the

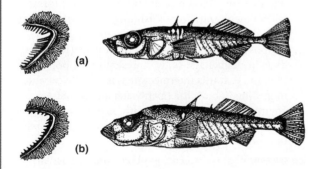

Figure 1 | Morphological differences in the gill rakers of the stickleback species based on **(a)** limnetic form and **(b)** benthic form feeding habits.

Illustration by Laura Nagel, Queen's University. (From Schulter 1993.)

continued on page 88

foraging efficiency of the two forms. The foraging success of individual fish was assessed in two artificial habitats, mimicking conditions in the limnetic and benthic environments.

Two food types were used in the trials. Brine shrimp larvae (*Artemia*), a common prey found in open water, were placed in the artificial limnetic habitats. Larger amphipods, fast-moving arthropods with hard exoskeletons that forage on dead organic matter on the sediment surface, were placed in the artificial benthic habitats.

An experimental trial consisted of releasing a single fish into the test aquarium and observing it for a period of time. At the end of the observation period, the total number of prey items eaten was determined and the data were converted into two measures of foraging success: intake rate (number of prey items consumed per minute) and capture effort (mean number of bites per prey item).

Results of the foraging trials revealed distinct differences in the foraging success of the two morphological forms (phenotypes; Figure 2). The LF individuals were most successful at foraging on the brine shrimp larvae. They had a higher consumption rate and required only half the number of bites to consume as compared to the BF individuals. In contrast, BF individuals had a higher intake rate for amphipods and on average consumed larger amphipods than did LF individuals.

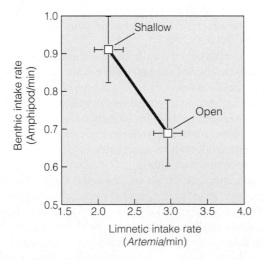

Figure 2 | Mean intake rate (and standard errors) of limnetic (open) and benthic (shallow) forms in open-water (food source: *Artemia* [brine shrimp larvae]) and shallow-water (food source: amphipod) feeding trials.

(Adapted from Robinson 2000.)

Robinson was able to determine that the higher intake rate of brine shrimp larvae by LF individuals was related to this form's greater number of gill rakers, and greater mouth width was related to the higher intake rate of amphipods by BF individuals. Therefore, foraging efficiency was found to be related to morphological differences between the two forms, suggesting trade-offs in characteristics related to the successful exploitation of these two distinct habitats and associated food resources.

These and other experiments by Robinson strongly suggest that divergent natural selection is occurring in these populations and may represent the early stages of speciation. Previous studies of divergent species of sticklebacks (species pairs) inhabiting the lakes of British Columbia have suggested that opposing selective pressures in open-water and nearshore (shallow water) environments have been a major factor in the evolution of these species. Yet, Beren Robinson's work is unique in its illustration that natural selection is at work within a single population, resulting in different morphological phenotypes that inhabit these two distinct environments. It is one of few studies that have actually quantified the trade-offs faced by individuals within a population, relating to the exploitation of different resources. Robinson's work offers us important insights into the mechanisms at work in the evolution of diversity in this closely related group of fishes.

Bibliography

Robinson, B. W. 2000. Trade offs in habitat-specific foraging efficiency and the nascent adaptive divergence of sticklebacks in lakes. *Behaviour* 137:865–888.

Robinson, B. W., and S. Wardrop. 2002. Experimentally manipulated growth rate in threespine sticklebacks: Assessing trade-offs with developmental stability. *Environmental Biology of Fishes* 63:67–78.

1. How do you think that the differences in morphology, diet, and habitat of the two phenotypic forms within the population of fish might possibly lead to reproductive isolation and eventual speciation?

2. Suppose that an intermediate phenotype (intermediate in characteristics) was present in the population and that this intermediate was equally capable of feeding in both the open-water and nearshore environments (and as efficiently as both of the other phenotypes). How would this change the interpretation of the results?

See QUANTIFYit! at **www.ecologyplace.com** to calculate summary statistics using stickleback data.

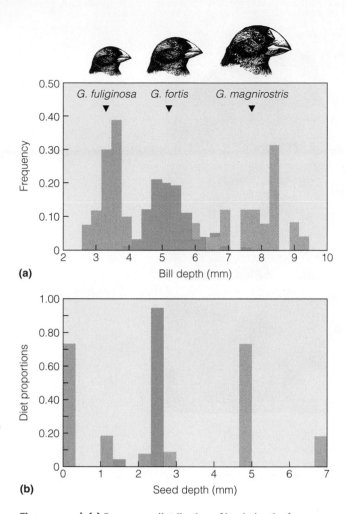

Figure 5.15 | (a) Frequency distribution of beak depths (upper mandible) of adult males in populations of the small (*Geospiza fuliginosa*), medium (*G. fortis*), and large (*G. magnirostris*) ground finch. Mean values are indicated by the solid triangles. The differences in beak size for the three species result in differences in the size of seeds in their diet, expressed as the proportion in the diet of various seed sizes **(b)**.

(Adapted from Grant 1999 after Schluter 1982.)

of the largest species, *G. magnirostris*, to feed on a range of seed resources from small, softer seeds to larger, harder seeds. However, because they are less efficient at exploiting the smaller seed resources, these larger-beaked individuals restrict their diet to the larger, harder seeds. The profitability—defined as the quantity of food energy gained per unit of time spent handling these small seeds (see Section 14.6)—is extremely low for the larger birds and makes feeding on smaller seeds extremely inefficient for these individuals. This inefficiency is directly related to the greater metabolic (food energy) demands of the larger birds, which illustrates a second important concept regarding the role of constraints and trade-offs in the process of natural selection: individual phenotypic characteristics (such as bill size) often are components of a larger adaptive complex involving multiple traits and loci.

The phenotypic trait of beak size in Darwin's ground finches is but one in a complex of interrelated morphological characteristics that determine the foraging behavior and diet of these birds. Larger beak size is accompanied by increased body size (length and weight) as well as specific changes in components of skull architecture and head musculature (Figure 5.16), all of which are directly related to feeding functions.

In summary, the beak size of a bird sets the potential range of seed types in the diet. This relationship between morphology and diet represents a basic trade-off that constrains the evolution of adaptations in Darwin's finches relating to their acquisition of essential food resources. In addition, the example of Darwin's ground finches illustrates how natural selection operates on genetic variation at the three levels we initially defined in Section 5.4: within a population, among subpopulations of the same species, and among different species. Natural selection operated in the local population of medium ground finches on Daphne Major island during the period of drought in the mid-1970s, increasing the mean beak size of birds in this population in response to the shift in the abundance and quality of seed resources. In addition, natural selection has resulted in differences in mean beak size between populations of the medium ground finch inhabiting the islands of Daphne Major and Santa Cruz (see Chapter 13, Figure 13.20). The larger mean beak size for the population on Santa Cruz is believed to be a result of competition from the population of small ground finch present on the island (*G. fuliginosa* does not occupy Daphne Major). The presence of the smaller species on the island has the effect of reducing the availability of smaller, softer seeds (see Figure 5.15) and increasing the relative fitness of *G. fortis* individuals with larger beaks that can feed on the larger, harder seeds (see Figure 5.6).

Population genetic studies have also shown that natural selection is the evolutionary force that has resulted in the genetic differentiation of various species of Darwin's finches that inhabit the Galápagos Islands (Figure 5.17). The process in which one species gives rise to multiple species that exploit different features of the environment, such as food resources or habitats, is called **adaptive radiation.** The different features of the environment exert the selection pressures that push the populations in various directions; and reproductive isolation, the necessary condition for speciation to occur, is often a by-product of the changes in morphology, behavior, or habitat preferences that are the actual targets of selection. In this way, the differences in beak size and diet among the three species of ground finch are magnified versions of the differences observed within a population, or among populations inhabiting different islands.

In the remaining chapters of Part Two, we will examine this basic principle of trade-offs as it applies to the adaptation of species and explore how the nature of

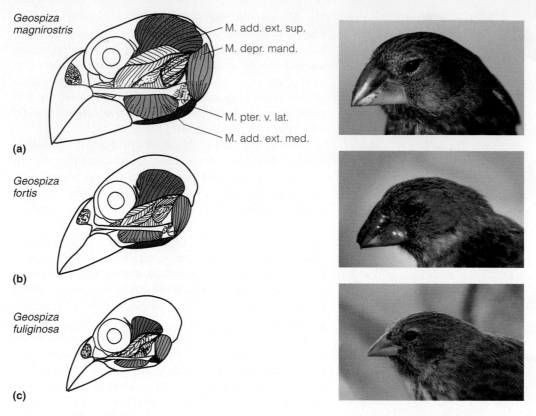

Geospiza
magnirostris

M. add. ext. sup.

M. depr. mand.

M. pter. v. lat.

M. add. ext. med.

(a)

Geospiza
fortis

(b)

Geospiza
fuliginosa

(c)

Figure 5.16 | Beak structure and superficial jaw muscles of **(a)** large, **(b)** medium, and **(c)** small ground finches. Muscles identified by abbreviation are *M. adductor mandibulae externus* (superficial and medialis), *M. depressor mandibulae*, and *M. pterygoideus ventralis lateralis*. An individual's ability to exploit different seed resources (size, hardness, etc.) is influenced by a complex of morphological and behavioral characteristics including those illustrated.

(Adapted from Grant 1999 after Bowman 1961.)

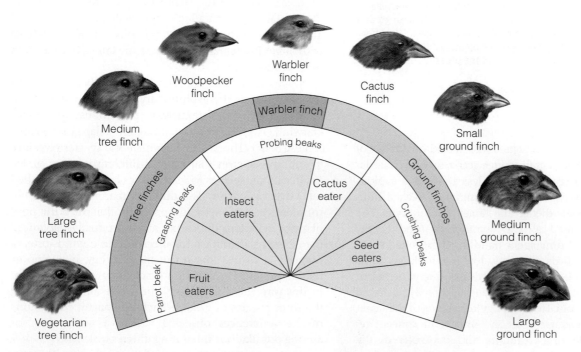

Figure 5.17 | The concept of adaptive radiation is illustrated by the diversity of Darwin's ground finches inhabiting the Galápagos Islands. Genetic studies show that the various species of finches arise from a single ancestral species, but they have evolved a rich diversity of beak sizes and shapes.

(From Patel 2006.)

adaptations changes with changing environmental conditions. In Chapters 6 and 7 we will explore various adaptations of plant and animal species, respectively, to key features of the physical environment that directly influence the basic processes of survival and assimilation. In Chapter 8 we will explore trade-offs involved in the evolution of life history characteristics (adaptations) relating to reproduction. The role of species interactions as selective agents in the process of natural selection will be examined later, in Part Four (Species Interactions).

Throughout the text, adaptation by natural selection is a unifying concept, a mechanism for understanding the distribution and abundance of species. We will explore the selective forces giving rise to the adaptations that define the diversity of species as well as the advantages and constraints arising from those adaptations under different environmental conditions. Finally, we will examine how the trade-offs in adaptations to different environmental conditions give rise to the patterns and processes observed in communities and ecosystems as environmental conditions change in space and time.

5.9 | Organisms Respond to Environmental Variation at the Individual and Population Levels

Organisms respond to changes in the environment (both spatial and temporal) in several different ways that operate at the levels of the population and the individual. At both these levels, the responses reflect that interaction between genes and the environment.

Over the timescale of generations, natural selection can alter the distribution of phenotypes in the population (mean and variance) in response to a change in environmental conditions. As environmental conditions change, selection will favor certain phenotypes over others, resulting in a shift in the distribution of phenotypes (as well as allele and genotype frequencies) within the population (see Figure 5.10). The result will be an increase in the average fitness of individuals in the population. This is an example of adaptation to environmental change through natural selection at the population level.

This same process can occur spatially. As we have seen in the examples of Section 5.7, local populations of the same species inhabiting different environments and experiencing different selective agents will undergo genetic differentiation. This differentiation results in a divergence in the distribution of genotypes and associated phenotypes. The result can be distinctive ecotypes—populations adapted to their unique local environmental conditions.

Individuals can also respond to both temporal and spatial changes in the environment. If an organism is mobile, and alternative environments (habitats) are available on the landscape, the simplest response to a change in environmental conditions is to move to a more suitable location. As we will see in Chapter 7, this is a

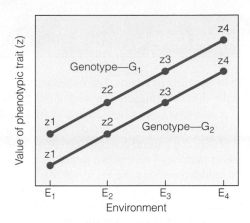

Figure 5.18 | An example of the norm of reaction: the set of phenotypes expressed by a genotype in different environments. The norms of reaction for two genotypes, G1 and G2, in four distinct environments: E1—E4 are shown. The two lines represent the phenotypic characteristics (z_1—z_4) exhibited by two genotypes (G1 and G2) in the four different environments. The ability of a genotype to express different phenotypic characteristics under different environmental conditions is called phenotypic plasticity.

Interpreting Ecological Data

Q1. Is there any environment in which the two genotypes will express the same phenotype?

Q2. Is it possible for the two genotypes to exhibit the same phenotype?

Q3. Suppose rather than the two lines representing the norms of reaction for the two genotypes being parallel, they crossed each other to form an *X*. How would this change your answers to questions 1 and 2?

common behavior in many animals. Yet this option is not available to sessile (nonmobile) organisms; and even for mobile organisms, distance or other obstacles may limit the individual's ability to move to more suitable habitat.

A second mechanism allowing individual organisms to respond to changes in environmental conditions is the direct influence of the environment on gene expression. Recall from Section 5.3 that most phenotypic traits are influenced by the environment; that is to say, the phenotypic expression of the genotype is influenced by the environment. The ability of a genotype to give rise to different phenotypic expressions under different environmental conditions is termed **phenotypic plasticity.** The set of phenotypes expressed by a single genotype across a range of environmental conditions is referred to as the **norm of reaction** (Figure 5.18). Note that we are not talking about different genotypes adapted to different environmental conditions, but about a single genotype (set of alleles) capable of altering the development or expression of a phenotypic trait to suit the conditions encountered by the individual organism. The result is to improve the individual's ability to survive, grow, and reproduce under the prevailing environmental conditions (i.e., increase fitness).

When we think of adaptation and evolution, we typically think of timescales involving millions of years. Yet, as we have seen from examples such as the shift in bill size in Darwin's ground finch (see Figure 5.9), evolution by natural selection can occur over relatively short periods of time when populations are exposed to sudden shifts in the environment. In one group of organisms, evolution by natural selection has been accelerated by our attempts at eradication, with potentially grave consequences to our own species.

Antibiotics are a class of compounds, natural or synthetic, that destroy or inhibit the growth of microorganisms. Their discovery and successful use against disease-causing bacteria is one of modern medicine's greatest success stories. Since these compounds first became widely used in the mid-20th century, they have saved countless lives through their ability to fight infections and disease-causing bacteria. But just four years after drug companies began mass-producing penicillin—the first of these compounds—microbes began appearing that could resist it.

Antibiotic resistance spreads fast. Between 1979 and 1987, for example, a survey by the U.S. Centers for Disease Control and Prevention (CDC) found that 0.02 percent of bacterial strains that cause pneumonia were penicillin resistant. Today, 6.6 percent of these bacterial strains are resistant. The CDC also reports that in 1992, more than 13,000 hospital patients died of bacterial infections that were resistant to antibiotic treatment.

The increased prevalence of antibiotic resistance is an outcome of evolution, specifically the accelerated process of natural selection in bacterial populations brought about by the widespread use of antibiotics. All populations of organisms, including bacteria, exhibit genetic variations that influence traits among individuals—in this case, the ability of the bacterium to withstand the effects of the antibiotic. Whenever antibiotics are used, there is selective pressure for resistance to occur. When a person takes an antibiotic, the drug kills the vast majority of bacteria that are defenseless. If any of the bacteria are resistant, however, they will survive. These resistant bacteria then multiply, increasing their numbers—up to a million-fold in a day. This process builds on itself, with more and more bacteria developing resistance to more and more antibiotic compounds. Natural selection favors the antibiotic-resistant individuals, whose increased fitness results in a greater proportion of the population being antibiotic resistant in each succeeding generation.

An individual can develop an antibiotic-resistant infection either by becoming initially infected with a resistant strain of bacteria or by having a resistant bacterial strain emerge in the body once antibiotic treatment begins. A bacterium develops antibiotic resistance as a result of acquiring genes that confer resistance to the action caused by the compound (i.e., the actual effects of the drug). Bacteria acquire genes that confer resistance in one of three ways: (1) Bacterial DNA may mutate spontaneously, which is likely the most common cause of resistance. (2) In a process called *transformation*, one bacterium may take up DNA from another bacterium. (3) Resistance may be acquired from a small circle of DNA called a *plasmid*, which can be transmitted from one type of bacterium to another. A single plasmid can provide a variety of resistances. In 1968, more than 12,000 people in Guatemala died in an epidemic of bacteria-caused diarrhea. The strain of bacteria harbored a plasmid that carried resistances to four different antibiotics.

Although bacterial antibiotic resistance is a natural phenomenon, societal factors such as inappropriate antibiotic use also contribute to the problem. For example, doctors sometimes prescribe antibiotics for a cold, cough, or the flu, all of which are viral and do not respond to antibiotics. Also, patients who are prescribed antibiotics but do not take the complete course (correct dose for the entire period of the prescription) can contribute to resistance.

Another much-publicized concern is the use of antibiotics in livestock. The drugs are used in healthy animals to prevent disease or administered at low levels in feed for long durations to increase the rate of weight gain. Although the U.S. Food and Drug Administration (FDA) Center for Veterinary Medicine limits the amount of antibiotic residue allowed in poultry and other meats, these drugs can cause bacteria to become resistant to drugs used to treat humans, ultimately making some human illnesses harder to treat.

1. In being treated for potentially deadly bacterial infections, patients are often given a series of different antibiotics. Can you think why this treatment might be utilized?

2. Why does the development of antibiotic-resistant populations of bacteria represent an example of evolution by natural selection?

Some of the best examples of phenotypic plasticity occur among plants. The size of the plant, the ratio of reproductive tissue to vegetative tissue, and even the shape of the leaves may vary widely at different levels of nutrition, light, moisture, and temperature. An excellent illustration of phenotypic plasticity in plants is the work of colleagues Sonia Sultan of Wesleyan University and Fakhri Bazzaz of Harvard University. Sultan and Bazzaz

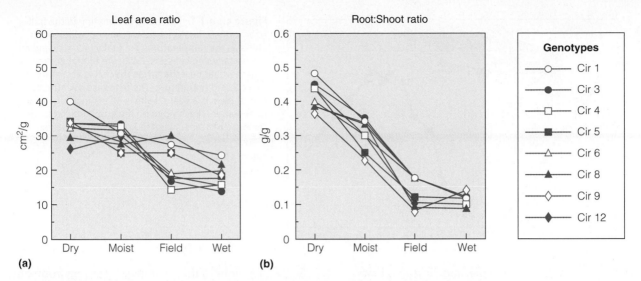

Figure 5.19 | Norms of reaction for eight genotypes (designated as CIR *x*) of spotted Lady's thumb (*Polygonum persicaria*) at four soil moisture levels (water content measured as mean percentage of soil dry weight: dry—11%, moist—14%, field capacity—26%, and wet—74%). Each point represents the mean of four plants (replicates). **(a)** Leaf area ratio is the ratio of leaf area (cm²) to plant weight (g). **(b)** Root:shoot is the ratio of root mass (g) to shoot mass (g), where shoot mass includes both the leaf and stem (aboveground) tissues. Both indices are measures of the allocation of biomass to different plant tissues.

(Adapted from Sultan and Bazzaz 1993.)

examined the response of eight different genotypes of spotted lady's thumb (*Polygonum persicaria*), an annual herbaceous plant species, to different soil water environments. Individuals of each of the eight genotypes were grown in the greenhouse under four different water treatments. Other environmental factors (e.g., light, nutrients, temperature, etc.) were identical for the plants. The resulting reaction norms for two phenotypic characteristics, leaf area ratio and root-to-shoot ratio, for each of the eight different genotypes are presented in Figure 5.19. All eight genotypes exhibit a wide variation in morphological responses to the different soil moisture environments (treatments). These differences in phenotypic traits for a given genotype under different environmental conditions reflect differences in the allocation of biomass to different tissues (leaves, stem, and roots) during the growth and development of the individual plant. This response is referred to as **developmental plasticity.** As such, these changes are irreversible. After the adult plant develops, these patterns of biomass allocation (proportions of leaf, stem, and root) will remain largely unchanged, regardless of any changes in the moisture environment (e.g., drought or flooding).

In contrast to developmental plasticity, other forms of phenotypic plasticity in response to prevailing environmental conditions are reversible. For example, fish have an upper and lower limit of tolerance to temperature (Figure 5.20). They cannot survive at water temperatures above and below these limits. However, these upper

and lower limits change seasonally as water temperatures warm and cool. During the warmer months of summer, the bullhead catfish *Ictalurus nebulosus* has an upper lethal temperature of 36°C. In winter this upper lethal temperature drops to 28°C. As water temperatures change seasonally, shifts in enzyme and membrane structure allow the individual's physiology to adjust slowly over a period of time, influencing heart rate, metabolic rate, neural activity, and enzyme reaction rates. These reversible phenotypic changes in an individual organism in response to changing environmental conditions are referred to as **acclimation.**

Acclimation is a common response in both plant and animal species, involving adjustments relating to biochemical, physiological, morphological, and behavioral traits.

As we reviewed in Part One (The Physical Environment), the physical environment varies on several different timescales that directly affect the functioning of organisms. Light, temperature, and moisture vary both diurnally and seasonally in a regular and predictable way, yet interannual variations in these same features can be less predictable, giving rise to periods of drought or flooding or to unusual periods of cold or heat. Over a much longer timescale, geological processes and periodic variations in climate can change environmental conditions at a regional to global scale. And at all of these timescales, organisms have evolved mechanisms for responding to these changes in environmental conditions.

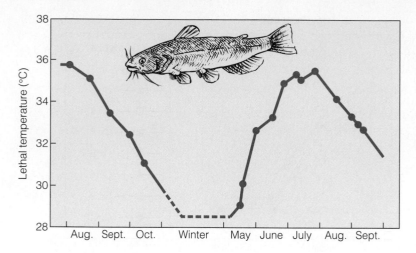

Figure 5.20 | Temperature acclimation in the bull-head catfish. Tolerance for warmer or colder water shifts as the temperatures gradually increase and decrease between seasons. Exposure to higher temperatures would be lethal when the fish is acclimatized to a colder temperature. The response over the period shown as a dashed line is an estimate (no direct observation). Each point on the graph indicates the temperature that is fatal to 50 percent of the fish after an exposure of 15 hours.

(Adapted from Fry 1947.)

Biochemical changes in enzyme and hormonal systems allow individual organisms to adjust physiological processes and behavior on a timescale of seconds or less in response to changes in environmental conditions that occur during a daily cycle. Seasonal adjustments of these same biological systems (biochemical, physiological, and behavioral) allow the individual to alter its basic metabolism to function under these periodic and predictable environmental changes through the process of acclimation. Finally, the combination of acclimation and developmental plasticity can allow an individual of a given genotype to alter the expression of its phenotypic traits during the development to suit the conditions that the individual organism encounters. All of these processes represent responses that are stimulated by signals from the environment that are received and processed at the level of individual cells and ultimately under genetic control. As such, they are all examples of phenotypic plasticity—the ability of a single genotype to express different phenotypic responses as a function of environmental conditions.

Some alterations of the phenotype by the environment (phenotypic plasticity) are not adaptive, such as stunted growth in response to inadequate nutrition. Other environmental alterations of the phenotype, such as those discussed earlier, represent adaptive responses to environmental stimuli and have evolved by natural selection. In effect, phenotypic plasticity—the ability of a genotype to respond to external environmental stimuli—represents an adaptation controlled by the process of natural selection. Ultimately all adaptations, phenotypic traits that enable an organism to survive and reproduce under the prevailing environmental conditions, are the product of natural selection operating on genetic variation within the population over a timescale of generations.

In the three chapters that follow, we will explore a range of adaptations by both plants and animals that allow them to survive, grow, and reproduce in the diversity of environments that occur on our planet. These adaptations include phenotypic responses at all of the levels discussed in this chapter.

Summary

Adaptation | 5.1

Characteristics that enable an organism to thrive in a given environment are called adaptations. Adaptations are a product of natural selection. Natural selection is the differential fitness of individuals within the population that results from their interaction with their environment, where the fitness of an individual is measured by the proportionate contribution it makes to future generations. The process of natural selection results in changes in the properties of populations of organisms over the course of generations, by a process known as evolution.

Genes | 5.2

The units of heredity are genes, which are linearly arranged on threadlike bodies called chromosomes. The alternative forms of a gene are alleles. The pair of alleles

present at a given locus defines the genotype. If both alleles at the locus are the same, the individual is homozygous. If the alleles are different, the individual is heterozygous. The sum of heritable information carried by the individual is the genome.

Phenotype | 5.3

The phenotype is the physical expression of the genotype. The manner in which the genotype affects the phenotype is termed the mode of gene action. When heterozygous individuals exhibit the same phenotype as one of the homozygotes, the allele that is expressed is termed dominant and the masked allele is termed recessive. If the physical expression of the heterozygote is intermediate between the homozygotes, the alleles are said to be codominant.

Even though all genetic variation is discrete, most phenotypic traits have a continuous distribution because (1) most traits are affected by more than one locus, and (2) most traits are affected by the environment.

Genetic Variation | 5.4

Genetic variation occurs at three levels: within subpopulations, among subpopulations of the same species, and among different species. The sum of genetic information across all individuals in the population is the gene pool. The fundamental measures of genetic variation within a population are allele frequency and genotype frequency.

Natural Selection | 5.5

Natural selection acts on the phenotype, but in doing so it alters both genotype and allele frequencies within the population. There are three general types of natural selection: directional selection, stabilizing selection, and disruptive selection. The target of selection is the phenotypic trait that natural selection acts upon, while the selective agent is the environmental cause of fitness differences among individuals in the population.

Processes Influencing Genetic Variation | 5.6

Natural selection is the only evolutionary process that can result in adaptations; however, some processes can function to alter patterns of genetic variation from generation to generation. These include mutation, migration, genetic drift, and nonrandom mating. Mutations are heritable changes in a gene or chromosome. Migration is the movement of individuals between local populations, resulting in the transfer of genes between local populations. Genetic drift is a change in allele frequency due to random chance.

Nonrandom mating on the basis of phenotypic traits is referred to as assortative mating. Assortative mating can be either positive (mates are more similar than expected by chance) or negative (dissimilar). A special case of nonrandom mating is inbreeding—the mating of individuals that are more closely related than expected by chance.

Genetic Differentiation | 5.7

Natural selection can function to alter genetic variation between populations; this result is referred to as genetic differentiation. Species having a wide geographic distribution often encounter a broader range of environmental conditions than do species whose distribution is more restricted. The variation in environmental conditions often gives rise to a corresponding variation in many morphological, physiological, and behavioral characteristics due to different selective agents in the process of natural selection.

Trade-offs and Constraints | 5.8

The environmental conditions that directly influence life vary in both space and time. Likewise, the objective of selection changes with environmental circumstances in both space and time. The characteristics enabling a species to survive, grow, and reproduce under one set of conditions limit its ability to do equally well under different environmental conditions.

Phenotypic Plasticity | 5.9

The ability of a genotype to give rise to a range of phenotypic expression under different environmental conditions is termed phenotypic plasticity. The range of phenotypes expressed under different environmental conditions is termed the norm of reaction. If the phenotypic plasticity occurs during the growth and development of the individual and represents an irreversible characteristic, it is referred to as developmental plasticity. Reversible phenotypic changes in an individual organism in response to changing environmental conditions are referred to as acclimation.

Study Questions

1. What is natural selection? What conditions are necessary for natural selection to occur?
2. Distinguish between the terms *gene* and *allele*.
3. What is the relationship between an individual's genotype and a phenotype?
4. If the phenotype trait of an *Aa* heterozygous individual is the same as that of an *AA* homozygous individual, which allele is recessive?
5. David Reznick, an ecologist at the University of California at Riverside, studied the process of natural selection in populations of guppies (small freshwater fish) on the island of Trinidad. Reznick found that populations at lower elevations face the assault of predatory fish, whereas the populations at higher elevations live in peace because few predators can move upstream past the waterfalls. The average size of individuals in the higher-elevation waters is larger than the average size of guppies in the lower-elevation populations. Reznick hypothesized that the smaller size of individuals in the lower-elevation populations was a result of increased rates of predation on larger individuals; in effect, predation was selecting for smaller individuals in the population. To test this hypothesis, Reznick moved individuals from the lower elevations to (unoccupied) pools upstream, where predation was not a factor. Eleven years in these conditions produced a population of individuals that were on average larger than the individuals of the downstream populations. Is the study by Reznick an example of natural selection (does it

meet the necessary conditions)? If so, what type of selection does it represent (directional, stabilizing, or disruptive)? Can you think of any alternative hypotheses to explain why the average size of individuals may have shifted through time as a result of moving the population to the upstream (higher-elevation) environment?

6. Why are small populations more prone to variations in allele frequency from generation to generation as a result of genetic drift than are large populations?

7. How might genetic drift and inbreeding be important processes in the conservation of endangered species?

8. Why is natural selection the only process that can result in adaptation?

9. Does the example of variation in body size of guppies from upstream and downstream populations presented in question 5 represent a trade-off similar to that of variation in beak size of medium ground finches presented in Section 5.8? What is the selective agent in the example in question 5?

10. What is phenotypic plasticity?

11. Contrast developmental plasticity and acclimation.

Further Readings

Conner, J. K., and D. L. Hartl. 2004. *A primer of ecological genetics.* Sunderland, MA: Sinauer.

An excellent introduction to population and quantitative genetics for the ecologist.

Desmond, A., and J. Moore. 1991. *Darwin: The life of a tormented evolutionist.* New York: Norton.

Written by two historians, this book is an anthology providing an introduction to the man and his works.

Gould, S. J. 1992. *Ever since Darwin: Reflections in natural history.* New York: Norton.

A collection of Gould's essays written for scientific journals. The first in a series of humorous and fun reading. See other collections, including *The Panda's Thumb, The Flamingo's Smile,* and *Dinosaur in a Haystack.*

Gould, S. J. 2002. *The structure of evolutionary theory.* Cambridge, MA: Belknap Press of Harvard University Press.

The last book written by Gould, the best, most respected popular science writer in the field of evolution. Although the book is more technical than others, Chapter 1 gives an excellent overview of current thinking in evolutionary theory.

Grant, P. R., and B. R. Grant. 2000. "Non-random fitness variation in two populations of Darwin's finches." *Proceedings of the Royal Society of London Series B* 267:131–138.

An excellent source of additional information for those intrigued by the example of natural selection in Darwin's ground finch presented in the chapter.

Mayr, E. 2001. *What evolution is.* New York: Basic Books.

An excellent primer on the topics of natural selection and evolution by a leading figure in evolution. Wonderfully written and accessible to the general reader.

Reznick, D. N., F. H. Shaw, F. H. Rodd, and R. G. Shaw. 1997. "Evaluation of the rate of evolution in natural populations of guppies *(Poecilia reticulata)." Science* 275:1934–1937.

A beautifully designed experiment for evaluating the role of natural selection in the evolution of life history characteristics.

Weiner, J. 1994. *The beak and the finch: A story of evolution in our time.* New York: Knopf.

Winner of the Pulitzer Prize. Gives the reader a firsthand view of scientific research in action.

CHAPTER 6

Plant Adaptations to the Environment

These Kokerboom trees in the desert region of Bloedkoppie, Namibia (in southwestern Africa), use the CAM photosynthetic pathway to conserve water in this harsh environment.

All life on Earth is carbon based. This means all living creatures are made up of complex molecules built on a framework of carbon atoms. The carbon atom is able to bond readily with other carbon atoms, forming long, complex molecules. The carbon atoms needed to construct these molecules—the building blocks of life—are derived from various sources. The means by which organisms acquire and use carbon represent some of the most basic adaptations required for life. Humans, like all other heterotrophs, gain their carbon by consuming other organisms. However, the ultimate source of carbon from which life is constructed is carbon dioxide (CO_2) in the atmosphere.

Not all living organisms can use this abundant form of carbon directly. Only autotrophs can transform carbon in the form of CO_2 into organic molecules and living tissue. Autotrophs can be subdivided into chemoautotrophs and photoautotrophs, depending on how they derive the energy for their metabolism. Chemoautotrophs convert carbon dioxide into organic matter using the oxidation of inorganic molecules (such as hydrogen gas or hydrogen sulfide) or methane as a source of energy. Chemoautotrophs are the dominant primary producers in oxygen-deficient environments, such as the hydrothermal vents of the deep ocean floor (see Chapter 24). The dominant form of autotrophs, photoautotrophs, uses the Sun's energy to drive the process of converting CO_2 into simple organic compounds. That process, carried out by green plants, algae, and some types of bacteria, is photosynthesis. Photosynthesis is essential for the maintenance of life on Earth.

Although all green plants derive their carbon from photosynthesis, how organisms, from the most minute of flowering plants (members of the duckweed family—Lemnaceae) to the largest of trees, allocate the products of photosynthesis to the basic processes of growth and maintenance varies immensely. These differences represent a diversity of evolutionary solutions to the problem of being a plant—acquiring the essential resources of carbon, light, water, and mineral nutrients necessary to support the process of photosynthesis. In this chapter, we will examine the variety of adaptations that plants have evolved to successfully survive, grow, and reproduce across virtually the entire range of environmental conditions found on Earth.

First, let us review the process so essential to life on Earth—or as the author John Updike so poetically phrased it, "the lone reaction that counterbalances the vast expenditures of respiration, that reverses decomposition and death."

6.1 | Photosynthesis Is the Conversion of Carbon Dioxide into Simple Sugars

Photosynthesis is the process by which energy from the Sun, in the form of shortwave radiation (photosynthetically active radiation, or PAR; see Section 2.1), is harnessed to drive a series of chemical reactions that result in the fixation of CO_2 into carbohydrates (simple sugars) and the release of oxygen (O_2) as a by-product.

The process can be expressed in the simplified form shown here:

$$6CO_2 + 12H_2O \longrightarrow C_6H_{12}O_6 + 6O_2 + 6H_2O$$

The net effect of this chemical reaction is the utilization of 6 molecules of water (H_2O) and the production of 6 molecules of oxygen (O_2) for every 6 molecules of CO_2 that are transformed into 1 molecule of sugar ($C_6H_{12}O_6$). The synthesis of various other carbon-based compounds—such as complex carbohydrates, proteins, fatty acids, and enzymes—from these initial products occurs in the leaves as well as other parts of the plant.

Photosynthesis, a complex sequence of metabolic reactions, can be separated into two processes, often referred to as the light and dark reactions. The light reactions begin with the initial photochemical reaction where chlorophyll (light-absorbing pigment) molecules within the chloroplasts absorb light energy. The absorption of a photon of light raises the energy level of the chlorophyll molecule. The excited molecule is not stable, and the electrons rapidly return to their ground state, thus releasing the absorbed photon energy. This energy is transferred to another acceptor molecule, resulting in a process called photosynthetic electron transport. This process results in the synthesis of ATP (adenosine triphosphate) from ADP (adenosine diphosphate) and of NADPH (the reduced form of NADP) from $NADP^+$ (nicotinamide adenine dinucleotide phosphate). The high-energy substance ATP and the strong reductant NADPH produced in the light reactions are essential for the second step in photosynthesis—the dark reactions.

In the dark reactions, CO_2 is biochemically incorporated into simple sugars. The dark reactions do not directly require the presence of sunlight. They are, however, dependent on the products of the light reactions and therefore ultimately depend on the essential resource of sunlight.

The process of incorporating CO_2 into simple sugars begins in most plants when the five-carbon molecule RuBP (ribulose biphosphate) combines with CO_2 to form two molecules of a three-carbon compound called 3-PGA (phosphoglycerate).

CO_2	+	RuBP	$\longrightarrow$	2 3-PGA
1-carbon molecule		5-carbon molecule		3-carbon molecule

This reaction, called carboxylation, is catalyzed by the enzyme **rubisco** (ribulose biphosphate carboxylase-oxygenase). The plant quickly converts the 3-PGA formed in this process into the energy-rich sugar molecule G3P (glyceraldehyde 3-phosphate). The synthesis of G3P from 3-PGA requires both ATP and NADPH, the high-energy molecule and reductant that are formed in the light reactions. Some of this G3P is used to produce simple sugars ($C_6H_{12}O_6$), starches, and other carbohydrates required for plant growth and maintenance; the remainder is used to synthesize new RuBP to continue the process. The synthesis of new RuBP from G3P requires additional ATP. In this way, the availability of light energy (solar radiation) can limit the dark reactions of photosynthesis through its control on the production of ATP and NADPH required for the synthesis of G3P and the regeneration of RuBP. This photosynthetic pathway involving the initial fixation of CO_2 into the three-carbon PGAs is called the Calvin–Benson cycle, or C_3 cycle, and plants employing it are known as **C_3 plants** (Figure 6.1).

The C_3 pathway has one major drawback. The enzyme rubisco driving the process of carboxylation also acts as an oxygenase—rubisco can catalyze the reaction between O_2 and RuBP. The oxygenation of RuBP results in the eventual release of CO_2. This competitive reaction to the carboxylation process reduces the efficiency of C_3 photosynthesis.

Some of the carbohydrates produced in photosynthesis are used in the process of cellular respiration—the harvesting of energy from the chemical breakdown of simple sugars and other carbohydrates. Cellular respiration (also referred to as aerobic respiration) involves the oxidation of carbohydrates to generate energy in the form of ATP and takes place exclusively in the mitochondria.

$$C_6H_{12}O_6 + 6O_2 \longrightarrow 6CO_2 + 6H_2O + ATP$$

In the absence of oxygen, the process of fermentation (also referred to as anaerobic respiration) enables some cells to convert glucose into lactic acid and ATP ($C_6H_{12}O_6 \rightarrow 2C_3H_6O_3 + 2\,ATP$). Respiration occurs in all living cells, both plant and animal.

Because leaves use CO_2 during photosynthesis and produce CO_2 during respiration, the difference in the rates of these two processes is the net gain of carbon, referred to as **net photosynthesis.**

$$\text{Net photosynthesis} = \text{Photosynthesis} - \text{Respiration}$$

The rates of photosynthesis and respiration, and therefore net photosynthesis, are typically measured in moles CO_2 per unit leaf area (or mass) per unit time (e.g., $\mu mol/m^2/s$).

6.2 | The Light a Plant Receives Affects Its Photosynthetic Activity

Solar radiation provides the energy required to convert CO_2 into simple sugars. Thus, the availability of light (PAR) to the leaf directly influences the rate of photosynthesis (Figure 6.2). As the amount of light declines, the rate of carbon uptake in photosynthesis will eventually decline to a level where it equals the rate of carbon loss in respiration. At that point, the rate of net photosynthesis is zero. The light level (value of PAR) at which this occurs is called the **light compensation point (LCP).** At light levels below the compensation point, the rate of carbon loss due to respiration exceeds the rate of uptake in the process of photosynthesis, resulting in a net loss of carbon dioxide from the leaf to the atmosphere. The rate of carbon loss when the value of PAR is zero provides an estimate of the rate of respiration (point at which photosynthesis = 0).

As light levels exceed the light compensation point, the rate of photosynthesis increases with PAR; and the light reactions limit the rate of photosynthesis. Eventually, photosynthesis becomes light saturated, and now the dark reactions limit the rate of photosynthesis. The value of PAR, above which no further increase in photosynthesis occurs, is referred to as the **light saturation point.** In some plants adapted to extremely shaded environments, photosynthetic rates decline as light levels

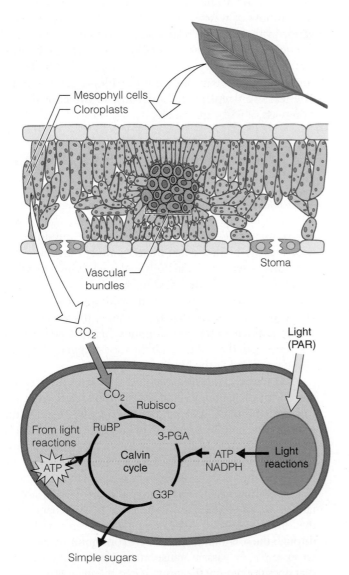

Figure 6.1 | The process of photosynthesis occurs in the chloroplast within the mesophyll cells. A simple representation of the C_3 photosynthetic pathway, or Calvin cycle, is shown. Note the link between the light and dark reactions, since the products of the light reactions (ATP and NADPH) are required to synthesize the energy-rich sugar molecule G3P and to regenerate RuBP (dark reactions). (PAR = photosynthetically active radiation.)

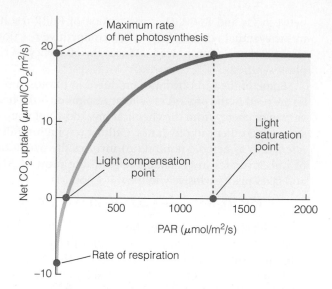

Figure 6.2 | Response of photosynthetic activity (*y*-axis) to available light (*x*-axis; PAR = photosynthetically active radiation). The plant increases its rate of photosynthesis as the light level increases up to a maximum rate known as the light saturation point. The light compensation point is the value of PAR at which the uptake of CO_2 for photosynthesis equals the loss of CO_2 in respiration. The value of net CO_2 exchange at PAR = 0 (dark) provides an estimate of the rate of respiration.

exceed saturation. This negative effect of high light levels, called **photoinhibition,** can be the result of "overloading" the processes involved in the light reactions.

6.3 | Photosynthesis Involves Exchanges between the Plant and Atmosphere

The process of photosynthesis occurs in specialized cells within the leaf that are called **mesophyll** cells (see Figure 6.1). For photosynthesis to take place within the mesophyll cells, CO_2 must move from the outside atmosphere into the leaf. In terrestrial (land) plants, CO_2 enters the leaf by diffusing through openings on its surface called **stomata** (Figure 6.3).

Diffusion is the movement of a substance from areas of higher to lower concentration. Concentrations of CO_2 are often described in units of parts per million (ppm) of air. A CO_2 concentration of 355 ppm would be 355 molecules of CO_2 for every 1 million molecules of air. Substances diffuse from areas of high concentration to areas of low concentration until the concentrations in the two areas are equal. As long as the concentration of CO_2 in the air outside the leaf is greater than that inside the leaf, CO_2 will continue to diffuse through the stomata.

As CO_2 diffuses into the leaf through the stomata, why don't the concentrations of CO_2 inside and outside the leaf come into equilibrium? As CO_2 is transformed into sugar during the photosynthesis, the concentration inside the leaf declines. As long as photosynthesis occurs, the gradient remains. If photosynthesis stopped and the

stomata remained open, CO_2 would diffuse into the leaf until the internal CO_2 equaled the outside concentration.

When photosynthesis and the demand for CO_2 are reduced for any reason, the stomata tend to close, thus reducing flow into the leaf. The stomata close because they play a dual role. As CO_2 diffuses into the leaf through the stomata, water vapor inside the leaf diffuses out through the same openings. This water loss through the stomata is called **transpiration.**

How rapidly the water can move from inside the leaf, through the stomata, and into the surrounding outside air depends on the diffusion gradient of water vapor from inside to outside the leaf. Like CO_2, water vapor diffuses from areas of high concentration to areas of low concentration—from wet to dry. For all practical purposes, the air inside the leaf is saturated with water, so the outflow of water is affected by the amount of water vapor in the air—the relative humidity (see Section 2.6; Figure 2.16). The drier the air (lower the relative humidity), the more rapidly the water inside the leaf will diffuse through the stomata into the outside surrounding air. The leaf must replace the water lost to the atmosphere; otherwise, it will wilt and die.

6.4 | Water Moves from the Soil, Through the Plant, to the Atmosphere

The force exerted outward on a cell wall by the water contained in the cell is called **turgor pressure.** The growth rate of plant cells and the efficiency of their physiological processes are highest when the cells are at maximum turgor—that is, when they are fully hydrated. When the water content of the cell declines, turgor pressure drops and water stress occurs, ranging from wilting to dehydration. For leaves to maintain maximum turgor, the water lost to the atmosphere in transpiration must be replaced by water taken up from the soil through the root system of the plant and transported to the leaves.

You may recall from basic physics that work—the displacement of matter, such as transporting water from the soil into the plant roots and to the leaves—requires the transfer of energy. The measure of energy available ("free") to do work is called Gibbs free energy (*G*), named for the American physicist Willard Gibbs, who first developed the concept in the 1870s. In the process of active transport, such as transporting water from the ground to an elevated storage tank using an electric pump, the input of free energy is in the form of electricity to the pump. The movement of water through the soil–plant–atmosphere continuum, however, is an example of passive transport, a spontaneous reaction that does not require the input of free energy.

All reactions, either chemical or mechanical, will proceed spontaneously from a state of high free energy to one of lower free energy. For example, if we define the difference in free energy (ΔG) between two states, A and B, as

$$\Delta G = G_A - G_B$$

where G_A and G_B are the quantities of free energy in states A and B respectively, we can predict the direction that the reaction between these two states will take. If the value of ΔG is positive (e.g., $G_A > G_B$), the reaction will proceed spontaneously from state A to state B. If we wish to reverse the direction of this reaction (from B to A), we must add free energy to B such that ΔG is negative (e.g., $G_A < G_B$). If the free energy of A and B are the same ($\Delta G = 0$), the reaction is at equilibrium.

In the example, the transition from states A and B can represent either a chemical or physical reaction. Photosynthesis is an example of a chemical reaction requiring the input of free energy in the form of solar radiation (PAR), because the quantity of free energy in products of the reaction (state B) is greater than that of the reactants (state A; see Section 6.1). As we shall see, the movement of water from the soil, through the plant, and to the atmosphere is an example of a spontaneous physical reaction progressing from a state of higher to lower free energy.

So if the movement of water from the soil into the roots, from the roots to the leaf, and from the leaf to the atmosphere is a spontaneous reaction, what is the source of free energy that allows this work to be accomplished? To answer this question, let's begin by considering the movement of water from the leaf to the atmosphere—transpiration. Recall that transpiration is driven by the process of diffusion—the movement of a substance (in this case, water molecules) down a concentration gradient from areas of high concentration (within the leaf) to areas of low concentration (outside air). The free energy that allows this work to be accomplished is the kinetic energy associated with the random movement (and collision) of the water molecules.

The movement of a substance through diffusion depends on the existence of a concentration gradient, which also represents a gradient of free energy. The region of higher concentration, having a greater density of water molecules, will have a greater value of free energy resulting from the random motion and collision of molecules than will the region of lower concentration. The result will be the net movement of water molecules from the region of high free energy to the region of low free energy until the level of free energy is the same in both regions ($\Delta G = 0$). When the concentration is the same in both regions (homogeneous), the net exchange of energy (and molecules) between them is zero, and the reaction is in a state of equilibrium.

The measure used to describe the free energy of water at any point along the soil–plant–atmosphere continuum is called water potential (Ψ). For any system, the value of Gibbs free energy is expressed relative to some arbitrary zero—pure water (no solute content), which has a high amount of free energy, is arbitrarily assigned a water potential of zero ($\Psi = 0$) units pressure: MPa.

When relative humidity of the atmosphere is 100 percent, the atmospheric water potential (Ψ_{atm}) is zero. As values drop below 100 percent, the value of free energy declines and (Ψ_{atm}) becomes negative (Figure 6.3a).

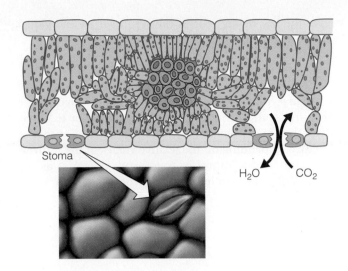

Stoma

H_2O CO_2

Figure 6.3 | Cross section of a C_3 leaf, showing epidermal cells and stomata. In terrestrial (land) plants the stomata—openings on the surface of the leaf—allow for the exchange of CO_2 and water between the atmosphere and the leaf interior through the process of diffusion.

Under most physiological conditions, the air within the leaf is at or near saturation (relative humidity ~100 percent). As long as the relative humidity of the air is below 100 percent, a steep gradient of water potential between the leaf (Ψ_{leaf}) and the atmosphere (Ψ_{atm}) will be driving the process of diffusion. Water vapor will move from the region of higher water potential (interior of the leaf) to the region of lower water potential (atmosphere), from a state of high to low free energy.

As water is lost to the atmosphere through the stomata, the water content of the cells decreases (turgor pressure drops) and in turn increases the concentration of solutes in the cell. This decrease in the cell's water content (and corresponding increase in solute concentration) decreases the water potential of the cells (lowers free energy). Unlike the water potential of the atmosphere, which is determined only by relative humidity, several factors determine water potential within the plant. Turgor pressure (positive pressure) in the cell increases the plant's water potential (free energy). Therefore, a decrease in turgor pressure associated with water loss functions to decrease water potential. The component of plant water potential due to turgor pressure is represented as ψ_p.

Increasing concentrations of solutes in the cells are associated with water loss and will lower the water potential (free energy). This component of plant water potential is termed **osmotic potential** (ψ_π), because the difference in solute content inside and outside the cell results in the movement of water through the process of osmosis.

The surfaces of larger molecules, such as those in the cell walls, exert an attractive force on water. This tendency for water to adhere to surfaces reduces the free energy of the water molecules, reducing water potential. This component of water potential is called **matric potential** (ψ_m). The total water potential Ψ at any point in the plant, from the leaf to the root, is the sum of these individual components:

$$\Psi = \psi_p + \psi_\pi + \psi_m$$

The osmotic and matric potentials will always have a negative value, while the turgor pressure (hydrostatic pressure) component can be either positive or negative. Thus the total potential (Ψ) can be either positive or negative, depending on the relative values of the individual components. Values of total water potential at any point along the continuum (soil, root, leaf, and atmo-

sphere), however, are typically negative; and the movement of water will be from areas of higher (zero or less negative) to lower (more negative) potential (from the region of higher to the region of lower free energy). Therefore, the movement of water from the soil to the root, from the root to the leaf, and from the leaf to the atmosphere depends on maintaining a gradient of increasingly negative water potential at each point along the continuum (Figure 6.4).

$$\Psi_{atm} < \Psi_{leaf} < \Psi_{root} < \Psi_{soil}$$

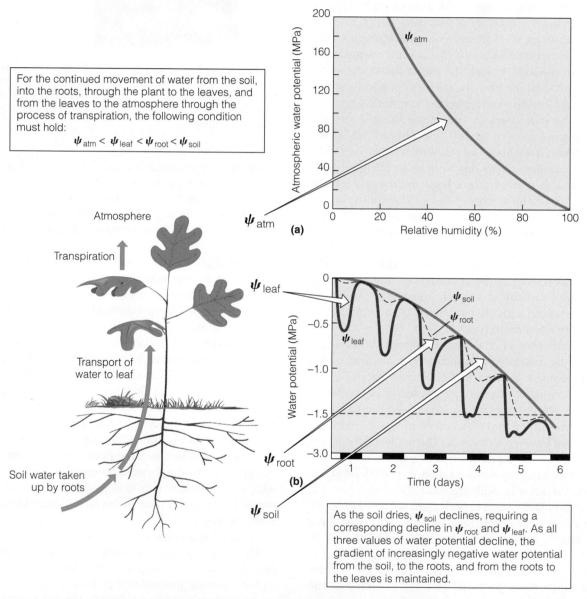

For the continued movement of water from the soil, into the roots, through the plant to the leaves, and from the leaves to the atmosphere through the process of transpiration, the following condition must hold:

$$\psi_{atm} < \psi_{leaf} < \psi_{root} < \psi_{soil}$$

As the soil dries, ψ_{soil} declines, requiring a corresponding decline in ψ_{root} and ψ_{leaf}. As all three values of water potential decline, the gradient of increasingly negative water potential from the soil, to the roots, and from the roots to the leaves is maintained.

Figure 6.4 | Transport of water along a gradient of water potential (Ψ) from soil to leaves to air. Water will move from a region of high to low (more negative) water potential. **(a)** As the relative humidity of the air drops below 100 percent, a steep decline in atmospheric water potential maintains the gradient from leaf to the air. **(b)** As the soil dries, the soil water potential becomes increasingly negative. This decline in soil water potential requires a corresponding decline in the water potential of the roots and leaves to maintain the gradient and flow of water. The graph depicts changes in the plant's leaf and root water potential in response to declining soil moisture (declining soil water potential) over a period of six days. Note the diurnal changes in leaf water potential. Leaf water potential is highest (less negative) at night when the stomata are closed.

Drawn by the low water potential of the atmosphere, water from the surface of and between the mesophyll cells within the leaf evaporates and escapes through the stomata. This gradient of water potential is transmitted into the mesophyll cells and on to the water-filled xylem (hollow conducting tubes throughout the plant) in the leaf veins. The gradient of increasingly negative water potential extends down to the fine rootlets in contact with soil particles and pores. As water moves from the root and up through the stem to the leaf, the root water potential declines so that more water moves from the soil into the root.

Water loss through transpiration continues as long as (1) the amount of energy striking the leaf is enough to supply the necessary latent heat of evaporation, (2) moisture is available for roots in the soil, and (3) the roots are capable of maintaining a more negative water potential than that of the soil. At field capacity (see Section 4.8), water is freely available, and soil water potential (Ψ_{soil}) is at or near zero. As water is drawn from the soil, the water content of the soil declines, and the soil water potential becomes more negative. As the water content of the soil declines, the remaining water adheres more tightly to the surfaces of the soil particles, and the matric potential becomes more negative. For a given water content, the matric potential of soil is influenced strongly by its texture (see Figure 4.10). Soils composed of fine particles, such as clays, have a higher surface area (per soil volume) for water to adhere to than sandy soils do. Clay soils therefore maintain more negative matric potentials for the same water content.

As soil water potential becomes more negative, the root and leaf water potentials must decline (become more negative) to maintain the potential gradient. If precipitation does not recharge soil water and soil water potentials continue to decline, eventually the plant will not be able to maintain the potential gradient. At that point the stomata will close to stop further water loss through transpiration. However, this closure also results in stopping further uptake of CO_2. The soil water potential at which its stomata close is determined by the plant's ability to reduce leaf water potentials further without disrupting basic physiological processes. The value of leaf water potential at which stomata close and net photosynthesis ceases varies among plant species (Figure 6.5) and reflects basic differences in their biochemistry, physiology, and morphology.

The rate of water loss varies with daily environmental conditions, such as humidity and temperature, and with the characteristics of plants. Opening and closing the stomata is probably the plant's most important means of regulating water loss. The trade-off between CO_2 uptake and water loss through the stomata results in a direct link between water availability in the soil and the plant's ability to carry out photosynthesis. To carry out photosynthesis, the plant must open its stomata; but when it does, it will lose water, which it must replace. If water is scarce,

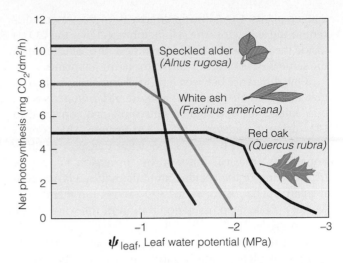

Figure 6.5 | Changes in net photosynthesis (y-axis) as a function of leaf water potential for three tree species from the northeastern United States. The decline in net photosynthesis with declining leaf water potentials (more negative values) results primarily from stomatal closure. As the water content of the soil declines, the plant must reduce the leaf water potential to maintain the gradient so that water can move from the soil to the roots and from the roots to the leaves. The plant eventually reaches a point where it cannot further reduce leaf water potentials, and the stomata will close to reduce the loss of water.

the plant must balance the opening and closing of the stomata, taking up enough CO_2 while minimizing the loss of water. The ratio of carbon fixed (photosynthesis) per unit of water lost (transpiration) is called the **water-use efficiency.**

We can now appreciate the trade-off faced by terrestrial plants. To carry out photosynthesis, the plant must open the stomata to take up CO_2. But at the same time, the plant will lose water through the stomata to the outside air—water that must be replaced through the plant's roots. If its access to water is limited, the plant must balance the opening and closing of stomata to allow for the uptake of CO_2 while minimizing water loss through transpiration. This balance between photosynthesis and transpiration is an extremely important constraint that has governed the evolution of terrestrial plants and directly influences the productivity of ecosystems under differing environmental conditions (see Chapter 20).

6.5 | The Process of Carbon Uptake Differs for Aquatic and Terrestrial Plants

A major difference in CO_2 uptake and assimilation by aquatic versus terrestrial plants is the lack of stomata in submerged aquatic plants. CO_2 diffuses from the atmosphere into the surface waters and is then mixed into the water column. Once dissolved, CO_2 reacts with the water to form bicarbonate (HCO_3^-). This reaction is reversible, and the concentrations of CO_2 and bicarbonate tend toward a dynamic equilibrium (see Section 3.7). In aquatic plants, CO_2 diffuses directly from the waters adjacent to the leaf across the cell membrane. Once the CO_2 is inside the aquatic plant, photosynthesis proceeds in much the same way as outlined earlier for terrestrial plants.

One difference between terrestrial and aquatic plants is that some aquatic plants can also use bicarbonate as a carbon source. However, the plants must first convert it to CO_2 using the enzyme carbonic anhydrase. This conversion can occur in two ways: (1) active transport of bicarbonate into the leaf followed by conversion to CO_2 or (2) excretion of the enzyme into adjacent waters and subsequent uptake of converted CO_2 across the membrane. As CO_2 is taken up, its concentration in the waters adjacent to the leaf will decline. Because the diffusion of CO_2 in water is 10^4 times slower than in the air, it can easily become depleted (low concentrations) in the waters adjacent to the leaf, reducing rates of plant uptake and photosynthesis. This constraint can be particularly important in still waters such as dense sea-grass beds or rocky intertidal pools.

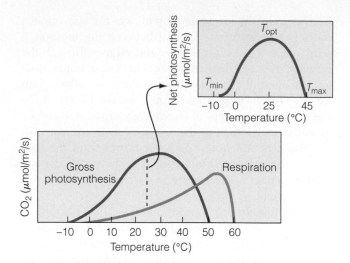

Figure 6.6 | General response of the rates of photosynthesis and respiration to temperature. At any temperature, the difference between these two rates is the rate of net photosynthesis (net uptake rate of CO_2). Here, the optimal temperature for net photosynthesis is between 20°C and 30°C.

6.6 | Plant Temperatures Reflect Their Energy Balance with the Surrounding Environment

Both photosynthesis and respiration respond directly to variations in temperature (Figure 6.6). As temperatures rise above freezing, both photosynthesis and respiration rates increase. Initially, photosynthesis increases faster than respiration. As temperatures continue to rise, the photosynthetic rate reaches a maximum related to the temperature response of the enzyme rubisco. As temperatures continue to rise, photosynthetic rate declines and respiration rate continues to increase. As temperatures rise further, even respiration declines as temperatures reach critical levels. The temperature response of net photosynthesis is the difference between the rate of carbon uptake in photosynthesis and the rate of carbon loss in respiration (see Figure 6.6). Three values describe the temperature response curve: T_{min}, T_{opt}, and T_{max}. The values T_{min} and T_{max} are, respectively, the minimum and maximum temperatures at which net photosynthesis approaches zero, meaning no net carbon uptake. T_{opt} is the temperature, or range of temperatures, over which net carbon uptake is at its maximum.

The temperature of the leaf, not the air, controls the rate of photosynthesis and respiration; and leaf temperature depends on the exchange of energy (radiation) between the leaf and its surrounding environment. Plants absorb both shortwave (solar) and longwave (thermal) radiation (see Section 2.1). Plants reflect some of this solar radiation and emit longwave radiation back to the atmosphere. The difference between the radiation a plant receives and the radiation it reflects and emits back to the surrounding environment is the net energy balance of the plant (R_n). The net energy balance of a plant is analogous to the concept of the energy balance of Earth presented in Chapter 2 (see Fig-

ure 2.3). Of the net radiation absorbed by the plant, some is used in metabolic processes and stored in chemical bonds—namely, in the processes of photosynthesis and respiration. This quantity is quite small, typically less than 5 percent of R_n. The remaining energy heats the leaves and the surrounding air. On a clear, sunny day, the amount of energy plants absorb can raise internal leaf temperatures well above ambient (air or water temperature). Internal leaf temperatures may go beyond the optimum for photosynthesis and possibly reach critical levels (Figure 6.6).

To maintain internal temperatures within the range of tolerance (positive net photosynthesis), plants must exchange heat with the surrounding environment. Terrestrial plants exchange heat by convection and evaporation; aquatic plants do so primarily by convection (see Quantifying Ecology 7.1: Heat Exchange and Temperature Regulation). Recall from Chapter 2 that convection is the transfer of heat energy between a solid and a moving fluid. Convective loss depends on the difference between the temperature of the leaf and the surrounding fluid (air or water). If the leaf temperature is higher than that of the surrounding air, there is a net transfer of heat from the leaf to the surrounding air. Evaporation occurs in the process of transpiration. As plants transpire water from their leaves to the surrounding atmosphere through the stomata, the leaves lose energy and their temperature declines through evaporative cooling (see Section 3.2).

The transfer of heat from the plant to the surrounding environment is influenced by the existence of the **boundary layer,** a layer of still air (or water) adjacent to the surface of each leaf. The environment of the boundary layer differs from that of the surrounding environment (air or water) because it is modified by the diffusion of heat, water, and CO_2 from the plant surface. As water is transpired from the stomata, the humidity of the air within the boundary layer increases, reducing further transpiration. Likewise, as heat is transferred from the leaf surface to the boundary layer, the air temperature of the boundary layer increases, further reducing heat transfer from the leaf surface. Under still conditions (no air or water flow), the boundary layer will increase in thickness, further reducing the transfer of heat and materials (water and CO_2) between the leaf and the atmosphere (or water).

Wind or water flow functions to reduce the size of the boundary layer, allowing for mixing between the boundary layer and the surrounding air (or water) and reestablishing the diffusion gradient between the leaf surface and the bulk air.

Leaf size and shape influences the thickness and dynamics of the boundary layer, and therefore, the ability of plants to exchange heat through convection (see Quantifying Ecology 7.1: Heat Exchange and Temperature Regulation). Air tends to move more smoothly (laminar flow) over a larger surface than a smaller one, and as a result the boundary layer tends to be thicker and more intact in larger leaves. Deeply lobed leaves, like those of some oaks, and small, compound leaves (Figure 6.7) tend to disrupt the flow of air, causing turbulence that functions to reduce the boundary layer and increase the exchange of heat and water.

The ability of terrestrial plants to dissipate heat by evaporation is dependent on the rate of transpiration. The transpiration rate is in turn influenced by the relative air humidity and by the availability of water to the plant (see Section 6.3).

6.7 | Carbon Gained in Photosynthesis Is Allocated to the Production of Plant Tissues

Because leaves take in carbon dioxide during photosynthesis as well as produce carbon dioxide during respiration, a simple, economic approach can be used to explore the balance of these two processes. This approach, referred to as the **carbon balance,** focuses on the balance between uptake of CO_2 in photosynthesis and its loss through the process of respiration. Thus far, our discussion of plant carbon balance has focused on net photosynthesis—the balance between the processes of photosynthesis and respiration in the leaves (photosynthetic tissues) of green plants. However, plants are not composed only of leaves; they also have roots and supportive tissues such as stems. In keeping with our simple, economic model, the net uptake of carbon by the whole plant will be the difference between the uptake of carbon in photosynthesis minus the loss of carbon in respiration (Figure 6.8a). The total carbon uptake, or gain per unit time, will be the product of the average rate of carbon uptake in photosynthesis per unit of leaf area (photosynthetic surface) multiplied by the total surface area of leaves. Because all living cells respire, the total loss of carbon in respiration per unit time will be a function of the

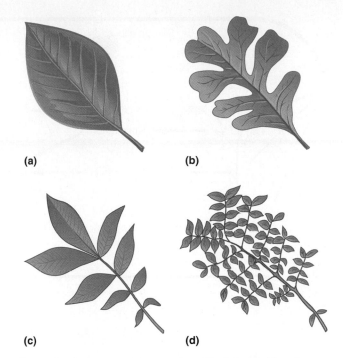

Figure 6.7 | Four general categories of leaf shape: **(a)** entire, **(b)** lobed, **(c)** simple compound, and **(d)** double compound. Smaller and more lobed leaves function to reduce the boundary layer and increase the exchange of heat energy through convection (see Quantifying Ecology 7.1: Heat Exchange and Temperature Regulation).

total mass of living tissue; that is, the sum of leaf, stem, and root tissues. The net carbon gain for the whole plant per unit time is the difference between these two values (carbon gain and carbon loss). This net carbon gain is then allocated to a variety of processes. Some of the carbon is used in maintenance and the rest in synthesis of new tissues for plant growth and reproduction.

How the net carbon gain is allocated will have a major influence on plant survival, growth, and reproduction. Acquiring the resources essential to supporting photosynthesis and growth involves different plant tissues. Leaf tissue is the photosynthetic surface, providing access to the essential resources of light and CO_2. Root tissue provides access to belowground resources such as water and nutrients in the soil; this tissue also anchors the plant to the soil. Stem tissue provides vertical support, elevating leaves above the ground and increasing access to light by reducing the chance of being shaded by taller plants (see Section 4.2). Stem tissue also provides the conductive tissue necessary to move water and nutrients from the roots to other parts of the plant. As we discuss in the following sections, the availability of these essential resources for plant growth influences the allocation of carbon to the production of various tissues.

Under ideal conditions (no resource limitations), the allocation of carbon to the further production of leaf tissue will promote the fastest growth. Increased allocation to leaf tissue increases the photosynthetic surface,

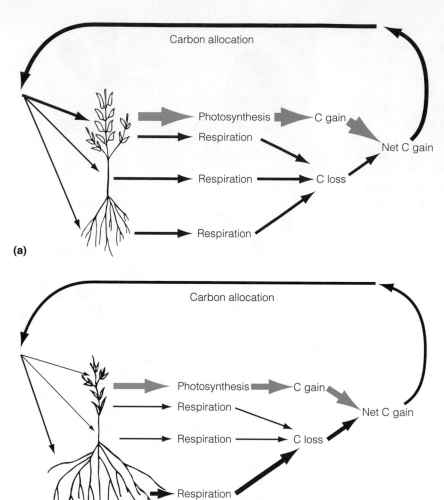

Figure 6.8 | The net carbon gain of a plant is the difference between carbon gain in photosynthesis and carbon loss in respiration. Because leaves (or more generally, photosynthetic tissues) are responsible for carbon gain, while all plant tissues respire (leaves, stem, and roots), the net carbon gain is directly influenced by the pattern of carbon allocation to the production of different plant tissues. **(a)** Increased allocation to leaves increases carbon gain (photosynthesis) relative to carbon loss (respiration) and therefore increases net carbon gain. **(b)** Increased allocation to roots has the opposite effect, decreasing carbon gain relative to carbon loss. The result is a lower net carbon gain by the plant.

which increases the rate of carbon uptake as well as carbon loss due to respiration. Allocation to all other tissues, such as stem and root, increases the respiration rate but does not directly increase the capacity for carbon uptake through photosynthesis. The consequence is the reduction of net carbon gain by the plant (Figure 6.8b). However, the allocation of carbon to the production of stem and root tissue is essential for acquiring the key resources necessary to maintain photosynthesis and growth. As these resources become scarce, it becomes increasingly necessary to allocate carbon to the production of these tissues—at the expense of leaf production. The implications of these shifts in patterns of carbon allocation are addressed in the following sections.

6.8 | Constraints Imposed by the Physical Environment Have Resulted in a Wide Array of Plant Adaptations

In Part One (Chapters 2–4), we explored variation in the physical environment over Earth's surface: the salinity, depth, and flow of water; spatial and temporal patterns in climate (precipitation and temperature); variations in geology and soils. In all but the most extreme of these environments, autotrophs harness the energy of the Sun to fuel the conversion of carbon dioxide into glucose in the process of photosynthesis. To survive, grow, and reproduce, plants must maintain a positive carbon balance, converting enough carbon dioxide into glucose to offset the expenses of respiration (photosynthesis > respiration). To accomplish this, a plant must acquire the essential resources of light, carbon dioxide, water, and mineral nutrients as well as tolerate other features of the environment that directly affect basic plant processes such as temperature, salinity, and pH. Although often discussed—and even studied as though they are independent of each other—the adaptations exhibited by plants to these features of the environment are not independent, for reasons relating to the physical environment and to the plants themselves.

Many features of the physical environment that directly influence plant processes are interdependent. For example, the light, temperature, and moisture environments are all linked through a variety of physical processes as discussed in Chapters 2–4. The amount of solar radiation not only influences the availability of light (PAR) required for photosynthesis but also directly influences the temperature of the leaf and its surroundings. In addition, air temperature directly affects the relative humidity, a key feature influencing the rate of transpiration and evaporation of water from the soil (see Section 2.6, Figure 2.16). For this reason, we see a correlation in the adaptations of plants to variations in these environmental factors. Plants adapted to dry, sunny environments must be able to deal with the higher demand for water associated with higher temperatures and lower relative humidity, and they tend to have characteristics such as smaller leaves and increased production of roots.

In other cases, there are trade-offs in the ability of plants to adapt to limitations imposed by multiple environmental factors, particularly resources. One of the most important of these trade-offs involves the acquisition of above- and belowground resources. Allocating carbon to the production of leaves and stems increases the plant's access to the resources of light and carbon dioxide, but it is at the expense of allocating carbon to the production of roots. Likewise, allocating carbon to the production of roots increases access to water and soil nutrients but limits carbon allocation to the production of leaves. The set of characteristics (adaptations) that allow a plant to successfully survive, grow, and reproduce under one set of environmental conditions inevitably limits its ability to do equally well under different environmental conditions. We explore the consequences of this simple premise, first introduced in Chapter 5, in the following sections.

6.9 | Species of Plants Are Adapted to Different Light Environments

As we have seen in Chapter 2, the amount of solar radiation reaching Earth's surface varies diurnally, seasonally, and geographically (Section 2.2). However, a major factor influencing the amount of light (PAR) a plant receives is the presence of other plants through shading (see Section 4.2, and Quantifying Ecology 4.1: Beer's Law and the Attenuation of Light). Although the amount of light that reaches an individual plant varies continuously as a function of the area of leaves above it, plants live in one of two qualitatively different light environments—sun or shade—depending on whether they are overtopped by other plants. Plants have evolved a range of physiological and morphological adaptations that allow individuals to survive, grow, and reproduce in these two different light environments.

The relationship between the availability of light and the rate of photosynthesis varies among plants (Figure 6.9). Plants growing in shaded environments tend to have a lower light compensation point, a lower light saturation point, and a lower maximum rate of photosynthesis than plants growing in high-light environments.

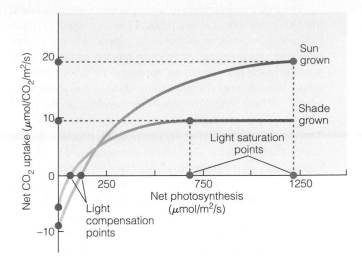

(a)

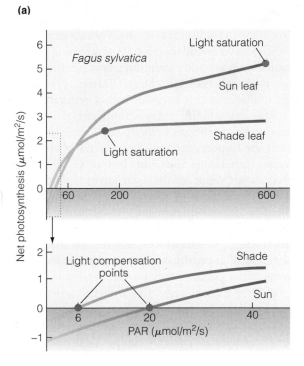

(b)

Figure 6.9 | (a) General patterns of photosynthetic response to light availability (photosynthetically active radiation; PAR) for plants grown in sun and shade environments. Shade-grown plants typically have a lower light compensation point and a lower light saturation point than do sun-grown plants. These same differences in photosynthetic response to light are also seen when comparing shade-tolerant (shade-adapted) and shade-intolerant (sun-adapted) plant species. (b) Example of shift in photosynthetic characteristics of sun and shade leaves for copper beech (*Fagus sylvatica*). Sun leaf is from the periphery of the canopy, where leaves are exposed to full sunlight. Shade leaf is from the shaded interior of the canopy. Insert shows the regions of the response curves below 50 μmol/m²/s, illustrating the differences in the light compensation points.

(Adapted from Larcher 1995.)

These differences relate in part to lower concentrations of the photosynthetic enzyme rubisco (see Section 6.1) found in shade-grown plants. Plants must expend a large amount of energy and nutrients to produce rubisco and other components of the photosynthetic apparatus. In shaded environments, low light, not the availability of rubisco to catalyze the fixation of CO_2, limits the rate of photosynthesis. The shade plant produces less rubisco as a result. By contrast, production of chlorophyll, the light-harvesting pigment in the leaves, often increases. The reduced energy cost of producing rubisco and other compounds involved in photosynthesis lowers the rate of leaf respiration. Because the light compensation point is the value of PAR necessary to maintain photosynthesis at a rate that exactly offsets the loss of CO_2 in respiration (net photosynthesis = 0), the lower rate of respiration can be offset by a lower rate of photosynthesis, requiring less light. The result is a lower light compensation point. However, the same reduction in enzyme concentrations limits the maximum rate at which photosynthesis can occur when light is abundant (high PAR), lowering both the light saturation point and maximum rate of photosynthesis.

The work of ecologist Stuart Davies of Harvard University illustrates this trade-off in leaf respiration rate, light compensation point, and maximum rate of net photosynthesis under conditions of high and low light. Davies examined the response of nine tree species of the genus *Macaranga* (inhabiting the rain forests of Borneo in Malaysia) to variations in the light environment. Davies grew seedlings of the nine species in a greenhouse under two light regimes: high light (total daily PAR of 21.4 mol/m²/day), and low light (7.6 mol/m²/day). The shaded (low light) treatment was created using shade cloth. After a period of approximately six months, he took measurements of net photosynthesis under varying light levels (values of PAR) for seedlings of each species grown under high and low light. The resulting light response curves, such as the one presented in Figure 6.2, were used to estimate values of leaf respiration, light compensation point, and the maximum rate of net photosynthesis at light saturation. A comparison of the nine species is presented in Figure 6.10.

The rate of leaf respiration for seedlings grown under low-light conditions was significantly lower than the corresponding value for individuals of the same species grown under high-light conditions (Figure 6.10a). This reduction in leaf respiration rate was accompanied by both a decrease in the light compensation point (Figure 6.10b) and the maximum rate of net photosynthesis at light saturation (Figure 6.10c). These shifts in the photosynthetic characteristics of leaves from plants grown in high- and low-light environments were also accompanied by changes in leaf morphology (structure).

The ratio of surface area (cm²) to weight (g) for a leaf is called the specific leaf area (SLA; cm²/g). The value of SLA represents the surface area of leaf produced per gram of biomass (or carbon) allocated to the production of leaves. Plants grown under low-light conditions typically produce leaves with a greater specific leaf area. In the

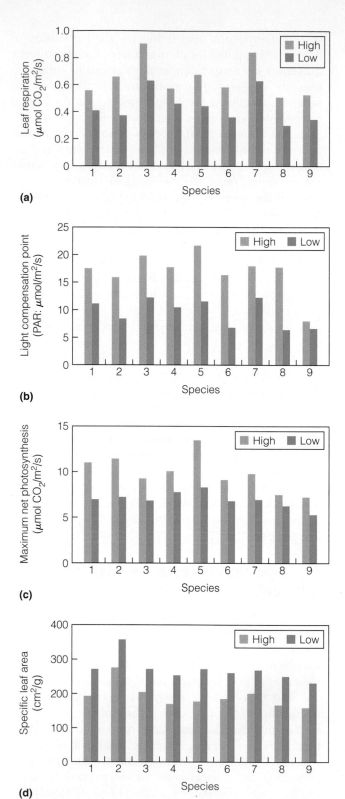

Figure 6.10 | Variation in **(a)** leaf respiration rate, **(b)** light compensation point, **(c)** light saturated rate of net photosynthesis, and **(d)** specific leaf area for seedlings of nine tree species of the genus *Macaranga* grown in contrasting light environments (high light: 21.4 mol/m²/day, low light: 7.6 mol/m²/day). The nine species of *Macaranga* inhabit the forests of Borneo, Malaysia. Species codes: (1) *M. hosei*, (2) *M. winkleri*, (3) *M. gigantea*, (4) *M. hypoleuca*, (5) *M. beccariana*, (6) *M. triloba*, (7) *M. trachyphylla*, (8) *M. hullettii*, and (9) *M. lamellata*.

(Adapted from Davies 1998.)

Relative Growth Rate

When we think of growth rate, what typically comes to mind is a measure of change in size during some period of time, such as change in weight during the period of a week (grams weight gain/week). However, this conventional measure of growth is often misleading when comparing individuals of different sizes or tracking the growth of an individual through time. Although larger individuals may have a greater absolute weight gain when compared with smaller individuals, this may not be the case when weight gain is expressed as a proportion of body weight (proportional growth). A more appropriate measure of growth is the mass-specific or relative growth rate. **Relative growth rate (RGR)** expresses growth during an observed period of time as a function of the size of the individual. This calculation is found by dividing the increment of growth during some observed time period (grams weight gain) by the size of the individual at the beginning of that time period (grams weight gain/total grams weight at the beginning of observation period) and then dividing the period of time to express the change in weight as a rate (g/g/time).

> **Relative growth rate** = weight gain during the period of observation per plant weight at the beginning of the observation period per time interval of observation period $(g\ g^{-1}\ time^{-1})$

$$RGR\ (g/g/time) = NAR\ (g/cm^2/time) \times LAR\ (cm^2/g)$$

> **Net assimilation rate** = weight gain during the period of observation per total area of leaves (leaf area) per time interval of observation period $(g/cm^2/time)$

> **Leaf area ratio** = total area of leaves (leaf area) per total plant weight (cm^2/g)

> **Leaf area ratio** = total area of leaves (leaf area) per total plant weight (cm^2/g)

> **Leaf weight ratio** = total weight of leaves per total plant weight (g/g)

$$LAR\ (cm^2/g) = LWR \times SLA$$

> **Specific leaf area** = total area of leaves (leaf area) per total weight of leaves (cm^2/g)

Using RGR to evaluate the growth of plants has an additional value; it can be partitioned into components reflecting the influences of assimilation (photosynthesis) and allocation on growth—the assimilation of new tissues per unit leaf area $(g/cm^2/time)$ called the **net assimilation rate (NAR),** and the leaf area per unit of plant weight (cm^2/g), called the **leaf area ratio (LAR).**

The NAR is a function of the total gross photosynthesis of the plant minus the total plant respiration. It is the net assimilation gain expressed on a per unit leaf area basis. The LAR is a function of the amount of that assimilation that is allocated to the production of leaves—more specifically, leaf area—expressed on a per unit plant weight basis.

The LAR can be further partitioned into two components that describe the allocation of net assimilation to leaves, the **leaf weight ratio (LWR),** and a measure of leaf density or thickness, the **specific leaf area (SLA).** The LWR is the total weight of leaves expressed as a proportion of total plant weight (g leaves/g total plant weight), whereas the SLA is the area of leaf per gram of leaf weight. For the same tissue density, a thinner leaf would have a greater value of SLA.

The real value of partitioning the estimate of RGR is to allow for comparison, either among individuals of different species or among individuals of the same species grown under different environmental conditions. For example, the data presented in the table on the next page are the results of a greenhouse experiment where seedlings of *Acacia tortilis* (a tree that grows on the savannas of southern Africa) were grown under two different light environments: full sun and shaded (50 percent full sun). Individuals were harvested at two times (4 and 6 weeks), and the total plant weight, total leaf weight, and total leaf area were measured. The mean values of these measures are shown in the table. From these values, estimates of RGR, LAR, LWR, and SLA were calculated. The values of RGR are calculated using the total plant weights at 4 and 6 weeks. NAR was then calculated by dividing the RGR by LAR. Because LAR varies through time (between weeks 4 and 6), the average of LAR at 4 and 6 weeks was used to characterize LAR in estimating RGR. Note that the average size (weight and leaf area) of seedlings grown in the high-light environment is approximately twice that of seedlings grown in the shade. Despite this difference in size, and the lower light levels to support photosynthesis, the difference in RGR between sun- and shade-grown seedlings is only about 20 percent. By examining the components of RGR, we can see how the shaded individuals are able to accomplish this task. Low-light conditions reduced rates of photosynthesis, subsequently reducing NAR for the individuals grown in the shade. The

continued on page 110

plants compensated, however, by increasing the allocation of carbon (assimilates) to the production of leaves (higher LWR) and producing thinner leaves (higher SLA) than did the individuals grown in full sun. The result is that individuals grown in the shade have a greater LAR (photosynthetic surface area relative to plant weight), offsetting the lower NAR and maintaining comparable RGR.

	Week 4		Week 6	
	Sun	Shade	Sun	Shade
Leaf area (cm^2)	18.65	12.45	42	24
Leaf weight (g)	0.056	0.032	0.126	0.061
Stem weight (g)	0.090	0.058	0.283	0.138
Root weight (g)	0.099	0.043	0.239	0.089
Total weight (g)	0.245	0.133	0.648	0.288
LAR (cm^2/g)	75.998	93.750	64.854	83.304
SLA (cm^2/g)	334	392		
LWR (g/g)	0.228	0.239	0.194	0.213
RGR (g/g/week)			0.471	0.382
NAR (g/cm^2/week)			0.007	0.004

These results illustrate the value of using the RGR approach for examining plant response to varying environmental conditions, either among individuals of the same species or among individuals of different species adapted to different environmental conditions. By partitioning the components of plant growth into measures directly related to morphology, carbon allocation, and photosynthesis, we can begin to understand how plants both acclimate and adapt to differing environmental conditions.

1. When plants are grown under dry conditions (low water availability), there is an increase in the allocation of carbon to the production of roots at the expense of leaves. How would this shift in allocation influence the plant's leaf area ratio (LAR)?

2. Nitrogen availability can directly influence the rate of net photosynthesis. Assuming no change in the allocation of carbon or leaf morphology, how would an increase in the rate of net photosynthesis resulting from an increase in nitrogen availability influence relative growth rate? Which component of RGR would be influenced by the increase in net photosynthesis?

experiment conducted by Davies, SLA increased for all nine species when grown under low-light conditions (Figure 6.10d). In general, leaves grown under reduced light conditions are larger (in surface area) and thinner than those grown under high light levels (Figure 6.11). The shift in leaf structure effectively increases the surface

area for the capture of light (the limiting resource) per unit of biomass allocated to the production of leaves.

In addition to producing broader, thinner leaves, plants grown under shaded conditions allocate a greater proportion of their net carbon gain to leaf production and less to the roots (Figure 6.12). Just as with the changes in leaf morphology (thinner, broader leaves), this shift in allocation from roots to leaves increases the surface area for the interception of light. The increase in leaf area (photosynthetic surface area) can partially offset the decrease in photosynthetic rates per unit leaf area under low-light conditions, allowing the plant to maintain a positive carbon balance and continued growth (see Quantifying Ecology 6.1: Relative Growth Rate).

The patterns in photosynthetic and morphological response to variations in the light environment described earlier can occur among individuals of the same species grown under different light conditions (as in Figures 6.11 and 6.12) and even among leaves on the same plant with different exposures to light (as in Figures 6.9b and Figure 6.11). Such change in the physiology or form of an individual organism (genotype) in response to changes in environmental conditions is an example of phenotypic plasticity discussed in Chapter 5 (see Section 5.9). These differences, however, are most pronounced between species of plants adapted to high- and low-light environments (see Field Studies: Kaoru Kitajima). These adaptations represent genetic differences in the potential

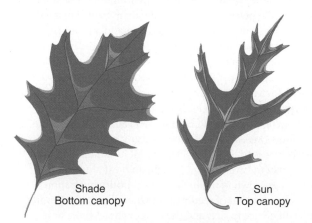

Shade
Bottom canopy

Sun
Top canopy

Figure 6.11 | Example of the response of leaf shape (morphology) to the light environment under which it developed. Red oak (*Quercus rubra*) leaves vary in size and shape from the top to the bottom of the tree. Leaves at the top of the canopy receive higher levels of solar radiation and experience higher temperatures than those growing at the bottom. Upper leaves are smaller, thicker, and more lobed than those growing at the bottom. This morphology aids in heat loss through convection. Conversely, the larger, thinner leaves at the bottom increase the surface area for capture of light.

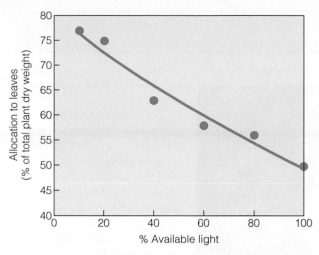

Figure 6.12 | Changes in allocation to leaves for broadleaf peppermint (*Eucalyptus dives*) seedlings grown under different light environments in the greenhouse. Allocation to leaves is expressed as a percentage of the total dry weight of the plant at harvest. Each point represents the average response of five seedlings. Light availability is expressed as a percentage of full sunlight. Levels of shading were controlled by shade cloth of varying density. The increased allocation to leaves, together with the production of thinner leaves, functions to increase the photosynthetic surface for the capture of light.

Interpreting Ecological Data

Q1. What is the average allocation to leaves for seedlings grown in full-sunlight conditions in the greenhouse?

Q2. How does the average allocation to leaves vary with decreasing availability of light?

Q3. Let us assume that allocation to stems does not vary with changes in the availability of light during seedling growth. Using the simple model of carbon allocation presented in Figure 6.8, how would the graph of allocation to roots under the different light environments differ from the graph of allocation to leaves presented in Figure 6.12?

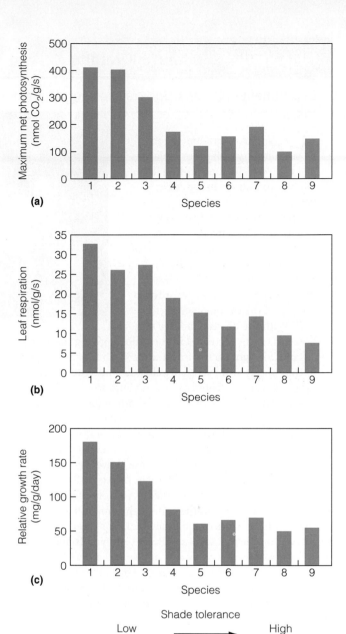

Figure 6.13 | Differences in the rates of **(a)** light-saturated photosynthesis, **(b)** leaf respiration, and **(c)** relative growth rate for nine tree species that inhabit the forests of northeastern North America (boreal forest). Species are ranked from highest (shade intolerant) to lowest (shade tolerant) in tolerance to shade. Species codes: (1) *Populus tremuloides*, (2) *Betula papyrifera*, (3) *Betula allegheniensis*, (4) *Larix laricina*, (5) *Pinus banksiana*, (6) *Picea glauca*, (7) *Picea mariana*, (8) *Pinus strobus*, and (9) *Thuja occidentalis*.

Interpreting Ecological Data

Q1. In general, how does net photosynthesis and leaf respiration vary with increasing shade tolerance for the nine boreal tree species? What does this imply about the corresponding pattern of gross photosynthesis with increasing shade tolerance for these species?

Q2. Based on the data presented in graphs (a) and (b), how would you expect the light compensation point to differ between *Populus tremuloides* and *Picea glauca*?

response of the plant species (between-species genetic variation; Chapter 5) to the light environment. Plant species adapted to high-light environments are called **shade-intolerant** species, or sun-adapted species. Plant species adapted to low-light environments are called **shade-tolerant** species, or shade-adapted species.

The variations in photosynthesis, respiration, and growth rate that characterize plant species adapted to different light environments are illustrated in the work of plant ecologist Peter Reich and colleagues at the University of Minnesota. They examined the characteristics of nine tree species that inhabit the cool temperate forests of northeastern North America (boreal forest). The species differ widely in shade tolerance from very tolerant of shaded conditions to very intolerant. Seedlings of the nine species were grown in the greenhouse, and measurements of maximum net photosynthetic rate at light saturation, leaf respiration rate, and relative growth rate (growth rate per unit plant biomass; see Quantifying Ecology 6.1: Relative Growth Rate) were made over the course of the experiment (Figure 6.13).

Department of Botany, University of Florida, Gainesville, Florida

A major factor influencing the availability of light to a plant is its neighbors. By intercepting light, taller plants shade individuals below, influencing rates of photosynthesis, growth, and survival. Nowhere is this effect more pronounced than on the forest floor of a tropical rain forest (Figure 1), where light levels are often less than 1 percent of those recorded at the top of the canopy (see Section 5.2). With the death of a large tree, however, a gap is created in the canopy, giving rise to an "island" of light on the forest floor. With time, these gaps in the canopy will eventually close, as individuals grow up to the canopy from below or neighboring trees expand their canopies, once again shading the forest floor.

How these extreme variations in availability of light at the forest floor have influenced the evolution of rain forest plant species has been the central research focus of plant ecologist Kaoru Kitajima of the University of Florida. Kitajima's work in the rain forests of Barro Colorado Island in Panama presents a story of plant adaptations to variations in the light environment that includes all life stages of the individual, from seed to adult.

Within the rain forests of Barro Colorado Island, the seedlings of some tree species survive and grow only in the high-light environments created by the formation of canopy gaps, whereas the seedlings of other species can survive for years in the shaded conditions of the forest floor.

Figure 1 | The dense canopy and understory in a tropical rain forest results in little light reaching the forest floor.

In a series of experiments designed to determine shade tolerance based on patterns of seedling survival in sun and shade environments (see Figure 6.10), Kitajima noted that seed mass (weight) is negatively correlated with seedling mortality rates. Interestingly, large-seeded species not only had higher rates of survival in the shade but also exhibited slower rates of growth after germination. Intuitively, one might think that larger reserves of energy and nutrients within the seed (larger mass) would allow for a faster rate of initial development, but this was not the case. What role does seed size play in the survival and growth of species in different light environments? An understanding of these relationships requires close examination of how seed reserves are used.

The storage structure(s) within a seed are called the cotyledon. Upon germination, cotyledons transfer reserve materials (lipids, carbohydrates, mineral nutrients) into developing shoots and roots. The cotyledons of some species serve strictly as organs to store and transfer seed reserves throughout their life span and are typically positioned at or below the ground level (Figure 2a). The cotyledons of other species develop a second function— photosynthetic carbon assimilation. In these species, the cotyledons function as "seed leaves" and are raised above the ground (Figure 2b). As Kitajima's research has revealed, the physiological function of cotyledons is crucial in determining the growth response of seedlings to the light environment.

Kitajima conducted an experiment involving tree species that differed in cotyledon function (photosynthetic and storage) and shade tolerance. Seedlings were raised from germination under two light levels: sun (23 percent full sun) and shade (1 percent full sun).

Changes in biomass and leaf area were recorded during a period of 40 days postgermination to determine when light began to affect seedling growth (determined for each species by the difference between individuals grown in shade and sun environments).

The smaller seeds of the shade-intolerant species had photosynthetic cotyledons and developed leaves earlier than did shade-tolerant species with their larger storage cotyledons. These differences reflect two different "strategies" in the use of initial seed reserves. Shade-

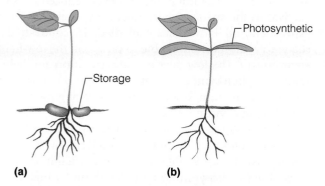

Figure 2 | Two contrasting functional forms of cotyledons found in tropical rain forest trees: **(a)** storage and **(b)** photosynthetic.

intolerant species invested reserves into the production of leaves to bring about a rapid return (carbon uptake in photosynthesis), whereas shade-tolerant species kept seed reserves as storage for longer periods at the expense of growth rate.

Having used their limited seed reserves for the production of leaves, the shade-intolerant species responded to light availability earlier than did the shade-tolerant species. And without sufficient light, mortality was generally the outcome.

So the experiments revealed that larger seed storage in shade-tolerant species does not result in a faster initial growth under shaded conditions. Rather, these species (shade tolerant) exhibit a conservative strategy of slow use of reserves over a prolonged period.

Whether shade tolerant or intolerant, once seedlings use up seed reserves, they must maintain a positive net carbon gain as a prerequisite for survival (see Section 6.7). What suites of seedling traits allow some species to survive better than others in the shade? To answer this question, Kitajima grew seedlings in the experimental sun and shade environments for an extended period beyond the reserve phase.

Individuals of both shade tolerant and shade intolerant species shifted allocation of carbon from producing roots to producing leaves. They also produced broader, thinner leaves under shaded conditions (higher specific leaf area: SLA; see Quantifying Ecology 6.1: Relative Growth Rate).

Despite the similarity between shade-tolerant and shade-intolerant species in their phenotypic responses to reduced available light, these two types exhibit different overall patterns of morphology and carbon allocation. Under both sun and shade conditions, shade-tolerant species had a greater proportional allocation to roots

(relative to leaves), thicker leaves (lower SLA), and, as a result, a lower photosynthetic surface area than did shade-intolerant species. As a result, the relative growth rates of shade-intolerant species were consistently greater than those of the shade-tolerant species, both in sun and shaded conditions (Figure 3).

Whereas the characteristics exhibited by the shade-intolerant species reflect strong natural selection for fast growth within light gaps, shade-tolerant species appear adapted to survive for many years in the understory, where their ability to survive attacks by pathogens and herbivores is enhanced by their well-developed reserves within the root system.

Bibliography

Kitajima, K. 1996. Ecophysiology of tropical tree seedlings. In *Tropical forest plant ecophysiology* (S. Mulkey, R. L. Chazdon, and A. P. Smith, eds.), 559–596. New York: Chapman & Hall.

Kitajima, K. 2002. Do shade-tolerant tropical tree seedlings depend longer on seed reserves? *Functional Ecology* 16:433–444.

1. What processes might create gaps in the forest canopy?
2. How might seed size influence the method of seed dispersal from the parent plant?

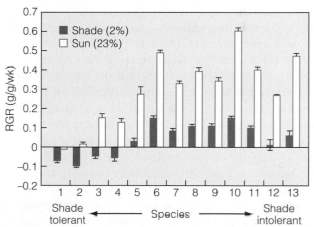

Figure 3 | Relative growth rates (RGRs) of 13 tropical tree species grown as seedlings from germination to 10 weeks under controlled shade and sun conditions. Species are ranked based on their shade tolerance (survival in shade). Species codes: (1) *Aspidosperma cruenta*, (2) *Tachigalia versicolor*, (3) *Bombacopsis sessilis*, (4) *Platypodium elegans*, (5) *Lonchocarpus latifolius*, (6) *Lafoensia punicifolia*, (7) *Terminalia amazonica*, (8) *Cordia alliodora*, (9) *Pseudobombax septenatum*, (10) *Luehea seemannii*, (11) *Ceiba pentandra*, (12) *Cavanillesia platanifolia*, (13) *Ochroma pyramidale* (Adapted from Kitajima 1994.)

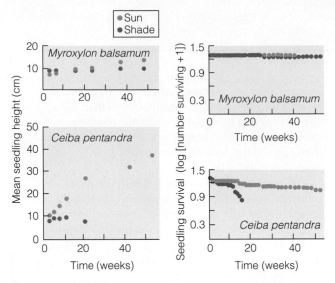

Figure 6.14 | Seedling survival and growth over a period of one year for seedlings of two tree species on Barro Colorado Island, Panama, grown under sun and shade conditions. *Ceiba pentandra* is a shade-intolerant species; *Myroxylon balsamum* is shade tolerant.

(Adapted from Augspurger 1982.)

Species adapted to lower light environments (shade tolerant) are characterized by lower maximum rates of net photosynthesis, leaf respiration, and relative growth rate than are species adapted to higher light environments (shade intolerant).

The difference in the photosynthetic characteristics between shade-tolerant and shade-intolerant species influences not only rates of net carbon gain and growth but also ultimately the ability of individuals to survive in low-light environments. This relationship is illustrated in the work of Caroline Augspurger of the University of Illinois. She conducted a series of experiments designed to examine the influence of light availability on seedling survival and growth for a variety of tree species, both shade tolerant and intolerant, that inhabit the tropical rain forests of Panama. Augspurger grew tree seedlings of each species under two levels of light availability. These two treatments mimic the conditions found either under the shaded environment of a continuous forest canopy (shade treatment) or in the higher light environment in openings or gaps in the canopy caused by the death of large trees (sun treatment). She continued the experiment for a year, monitoring the survival and growth of seedlings on a weekly basis. Figure 6.14 presents the results for two contrasting species, shade tolerant and shade intolerant.

The shade-tolerant species (*Myroxylon balsamum*) showed little difference in survival and growth rates under sunlight and shade conditions. In contrast, the survival and growth rates of shade-intolerant species (*Ceiba pentandra*) were dramatically reduced under shade conditions. These observed differences are a direct result of the difference in the adaptations relating to photosynthesis and

carbon allocation discussed earlier. The higher rate of light-saturated photosynthesis results in a high growth rate for the shade-intolerant species in the high-light environment. The associated high rate of leaf respiration and light compensation point, however, reduced rates of survival in the shaded environment. By week 20 of the experiment, all individuals had died. In contrast, the shade-tolerant species was able to survive in the low-light environment. The low rates of leaf respiration and light-saturated photosynthesis that allow for the low-light compensation point, however, limit rates of growth even in high-light environments.

The dichotomy in adaptations between shade-tolerant and shade-intolerant species reflects a trade-off between characteristics that enable a species to maintain high rates of net photosynthesis and growth under high-light conditions and the ability to continue survival and growth under low-light conditions. The changes in biochemistry, physiology, leaf morphology, and carbon allocation exhibited by shade-tolerant species enable them to reduce the amount of light required to survive and grow. However, these same characteristics limit their ability to maintain high rates of net photosynthesis and growth when light levels are high. In contrast, plants adapted to high-light environments (shade-intolerant species) can maintain high rates of net photosynthesis and growth under high-light conditions but at the expense of continuing photosynthesis, growth, and survival under shaded conditions.

6.10 | The Link between Water Demand and Temperature Influences Plant Adaptations

As with the light environment, terrestrial plants have evolved a range of adaptations in response to variations in precipitation and soil moisture. As we saw in the earlier discussion of transpiration (see Section 6.3), however, the demand for water is linked to temperature. As air temperature rises, the saturation vapor pressure will likewise rise (see Section 2.6), increasing the gradient of water vapor between the inside of the leaf and the outside air and subsequently the rate of transpiration. As a result, the amount of water required by the plant to offset losses from transpiration will likewise increase with temperature.

When the atmosphere or soil is dry, plants respond by partially closing the stomata and opening them for shorter periods of time. In the early period of water stress, a plant closes its stomata during the hottest part of the day when relative humidity is the lowest. It resumes normal activity in the afternoon. As water becomes scarcer, the plant opens its stomata only in the cooler, more humid conditions of morning. Closing the stomata reduces the loss of water through transpiration, but it also reduces CO_2 diffusion into the leaf and the dissipation of heat through evaporative cooling. As a result, the photosynthesis rate declines and leaf temperatures may rise. Some plant species, such as evergreen rhododendrons, respond to moisture stress by an inward curling of the leaves. Others show it in a wilted appearance caused by a lack of turgor in the leaves. Leaf curling and wilting allow leaves to reduce water loss and heat gain by reducing the surface area exposed to solar radiation.

Prolonged moisture stress inhibits the production of chlorophyll, causing the leaves to turn yellow or, later in the summer, to exhibit premature autumn coloration. As conditions worsen, deciduous trees may prematurely shed their leaves—the oldest ones dying first. Such premature shedding can result in dieback of twigs and branches.

In tropical regions with distinct wet and dry seasons (see Section 2.7), some species of trees and shrubs have evolved the characteristic of dropping their leaves at the onset of the dry season. These plants are termed *drought deciduous*. In these species, leaf senescence occurs as the dry season begins, and new leaves are grown just before the rainy season begins.

Some species of plants, referred to as C_4 plants or CAM plants, have evolved a modified form of photosynthesis that increases water-use efficiency in warmer and drier environments. The modification involves an additional step in the conversion (fixation) of CO_2 into sugars.

In C_3 plants, the capture of light energy and the transformation of CO_2 into sugars occur in the mesophyll cells (see Section 6.1). The products of photosynthesis move into the vascular bundles, part of the plant's transport system, where they can be transported to other parts of the plant. In contrast, plants possessing the **C_4 photosynthetic pathway** have a leaf anatomy different from that of C_3 plants (see Figure 6.3). C_4 plants have two distinct types of photosynthetic cells: the mesophyll cells and the **bundle sheath cells.** The bundle sheath cells surround the veins or vascular bundles (Figure 6.15). C_4 plants divide photosynthesis between the two types of cells: the mesophyll and the bundle sheath cells.

In C_4 plants, CO_2 reacts with PEP (phosphoenolpyruvate), a three-carbon compound, within the mesophyll cells. This is in contrast to the initial reaction with RuBP in C_3 plants. This reaction is catalyzed by the enzyme **PEP carboxylase,** producing OAA (oxaloacetate) as the initial product. The OAA is then rapidly transformed into the four-carbon molecules of malic and aspartic acids, from which the name C_4 photosynthesis is derived. These organic acids are then transported to the bundle sheath cells (see Figure 6.15). There, enzymes break down the organic acids to form CO_2, reversing the process that is carried out in the mesophyll cells. In the bundle sheath cells, the CO_2 is transformed into sugars using the C_3 pathway involving RuBP and rubisco.

The extra step in the fixation of CO_2 gives C_4 plants certain advantages. First, PEP does not interact with oxygen, as does RuBP. This eliminates the inefficiency that

occurs in the mesophyll cells of C_3 plants when rubisco catalyzes the reaction between O_2 and RuBP, leading to the production of CO_2 and a decreased rate of net photosynthesis (see Section 6.1). Second, the conversion of malic and aspartic acids into CO_2 within the bundle sheath cells acts to concentrate CO_2. The concentration of CO_2 within the bundle sheath cells can reach much higher concentrations than in either the mesophyll cells or the surrounding atmosphere. The higher concentrations

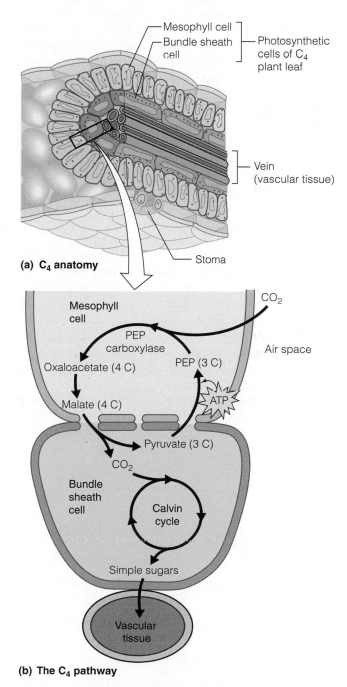

(a) **C_4 anatomy**

(b) **The C_4 pathway**

Figure 6.15 | The pathway of photosynthesis. Different reactions take place in the mesophyll and bundle sheath cells. Compare to the C_3 pathway (see Figure 6.1). (PEP, phosphoenolpyruvate; OAA, oxaloacetate.)

of CO_2 in the bundle sheath cells increase the efficiency of the reaction between CO_2 and RuBP catalyzed by rubisco. The net result is generally a higher maximum rate of photosynthesis in C_4 plants than in C_3 plants.

To understand the adaptive advantage of the C_4 pathway, we must go back to the trade-off in terrestrial plants between the uptake of CO_2 and the loss of water through the stomata. Due to the higher photosynthetic rate, C_4 plants exhibit greater water-use efficiency. That is, for a given degree of stomatal opening and water loss, C_4 plants typically fix more carbon. This increased water-use efficiency can be a great advantage in hot, dry climates where water is a major factor limiting plant growth. However, it comes at a price. The C_4 pathway has a higher energy expenditure because of the need to produce PEP and the associated enzyme, PEP carboxylase.

The C_4 photosynthetic pathway is not found in algae, bryophytes, ferns, gymnosperms (includes conifers, cycads, and ginkgos), or the more primitive flowering plants (angiosperms). C_4 plants are mostly grasses native to tropical and subtropical regions and some shrubs characteristic of arid and saline environments, such as *Larrea* (creosote bush) and *Atriplex* (saltbush) that dominate regions of the desert southwest in North America. The distribution of C_4 grass species in North America reflects the advantage of the C_4 photosynthetic pathway under warmer and drier conditions (Figure 6.16). The proportion of grass species that are C_4 increases from north to south, reaching a maximum in the southwest.

In the hot deserts of the world, environmental conditions are even more severe. Solar radiation is high, and water is scarce. To counteract these conditions a small group of desert plants, mostly succulents in the families Cactaceae (cacti), Euphorbiaceae, and Crassulaceae, use a third type of photosynthetic pathway—crassulacean acid

metabolism (CAM). The **CAM pathway** is similar to the C_4 pathway in that CO_2 is first transformed into the four-carbon compounds by using the enzyme PEP carboxylase. The four-carbon compounds are later converted back into CO_2, which is transformed into glucose using the C_3 cycle. Unlike C_4 plants, however, in which these two steps are physically separate (in mesophyll and bundle sheath cells), both steps occur in the mesophyll cells but at separate times (Figure 6.17).

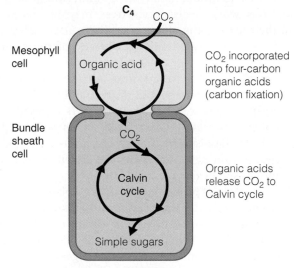

(a) Spatial separation of steps.
In C_4 plants, carbon fixation and the Calvin cycle occur in different types of cells.

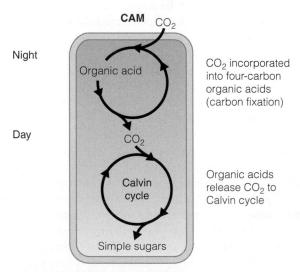

(b) Temporal separation of steps.
In CAM plants, carbon fixation and the Calvin cycle occur in the same cells at different times.

Figure 6.17 | Photosynthesis in CAM plants. **(a)** At night, the stomata open, the plant loses water through transpiration, and CO_2 diffuses into the leaf. CO_2 is stored as malate in the mesophyll, to be used in photosynthesis by day. **(b)** During the day, when stomata are closed, the stored CO_2 is refixed in the mesophyll cells using the C_3 cycle.

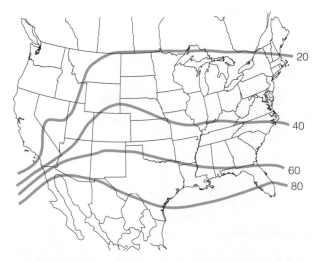

Figure 6.16 | Percentage of total grass species that are C_4. Values on isopleths (lines) represent percentages.

(Adapted from Teeri and Stone 1976.)

CAM plants open their stomata at night, taking up CO_2 and converting it to malic acid using PEP, which accumulates in large quantities in the mesophyll cells. During the day, the plant closes its stomata and reconverts the malic acid into CO_2, which it then fixes using the C_3 cycle. Relative to both C_3 and C_4 plants, the CAM pathway is slow and inefficient in the fixation of CO_2. But by opening their stomata at night when temperatures are lowest and relative humidity is highest, CAM plants dramatically reduce water loss through transpiration and increase water-use efficiency.

Plants may also respond to a decrease in available soil water by increasing the allocation of carbon to the production of roots while decreasing the production of leaves (Figure 6.18). By increasing its production of roots, the plant can explore a larger volume and depth of soil for extracting water. The reduction in leaf area decreases the amount of solar radiation the plant intercepts as well as the surface area that is losing water through transpiration. The combined effect is to increase the uptake of water per unit leaf area while reducing the total amount of water that is lost to the atmosphere through transpiration.

The decline in leaf area with decreasing water availability is actually a combined effect of reduced allocation of carbon to the production of leaves and changes in leaf morphology (size and shape). The leaves of plants grown under reduced water conditions tend to be smaller and thicker (lower specific leaf area: see Section 6.9) than those of individuals growing in more mesic

(wet) environments. In some plants, the leaves are small, the cell walls are thickened, the stomata are tiny, and the vascular system for transporting water is dense. Some species have leaves covered with hairs that scatter incoming solar radiation, whereas others have leaves coated with waxes and resins that reflect light and reduce its absorption. All these structural features reduce the amount of energy striking the leaf and thus the loss of water through transpiration.

Although the decrease in leaf area and corresponding increase in biomass allocated to roots (Figure 6.18) observed for plants growing under reduced water availability functions to reduce transpiration and increase the plant's ability to acquire water from the soil, this shift in patterns of allocation has consequences for plant growth. The reduced leaf area decreases carbon gain from photosynthesis relative to the loss of carbon from respiration (see Figure 6.8). The result is a reduction in net carbon gain as well as plant growth rate.

As with the shifts in plant morphology under varying light environments (see Section 6.9), the observed changes in leaf shape and carbon allocation of individuals in response to the availability of water represents phenotypic plasticity (as in Figure 6.18). Differences, however, are most pronounced between species of plants adapted to wet (mesic) and dry (xeric) environments (Figure 6.19).

6.11 | Plants Vary in Their Response to Environmental Temperatures

The photosynthetic temperature response curves (see Figure 6.6) for several terrestrial plant species are shown in Figure 6.20. These species vary in the range of temperatures over which net photosynthesis is at its maximum, T_{opt}. In fact, the differences in T_{opt} for the species correspond to differences in the thermal environments that the species inhabit. Species found in cooler environments typically have a lower T_{min}, T_{opt}, and T_{max} than species that inhabit warmer climates. These differences in the temperature response of net photosynthesis are directly related to a variety of biochemical and physiological adaptations that act to shift the temperature responses of photosynthesis and respiration toward the prevailing temperatures in the environment. These differences are most pronounced between plants using the C_3 and C_4 photosynthetic pathways (see Section 6.10). C_4 plants inhabit warmer, drier environments and exhibit higher optimal temperatures for photosynthesis (generally between 30°C and 40°C) than do C_3 plants (Figure 6.21).

Although species from different thermal habitats exhibit different temperature responses for photosynthesis and respiration, these responses are not fixed. When individuals of the same species are grown under different thermal conditions in the laboratory or greenhouse, divergence in the temperature response of net

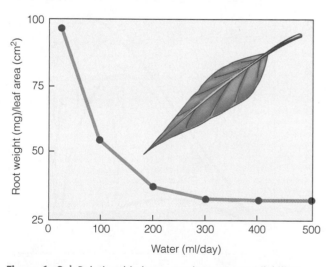

Figure 6.18 | Relationship between plant water availability and the ratio of root mass (mg) to leaf area (cm²) for broadleaf peppermint (*Eucalyptus dives*) seedlings grown in the greenhouse. Each point on the graph represents the average value for plants grown under the corresponding water treatment. As water availability decreases, plants allocate more carbon to producing roots than to producing leaves. This increased allocation to roots increases the surface area of roots for the uptake of water, while the decline in leaf area decreases water loss through transpiration.

(Adapted from Smith et al. 2002.)

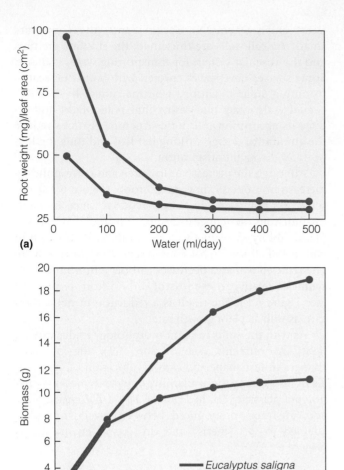

(a)

(b)

Figure 6.19 | Comparison of patterns of carbon allocation and growth rate for two species of *Eucalyptus* along an experimental gradient of water availability. **(a)** Although both species exhibit the same patterns of response to declining water availability, the xeric species (*E. dives*) exhibits a consistently higher ratio of root mass (mg) to leaf area than does the mesic species (*E. saligna*) across the water gradient. **(b)** *E. saligna*'s growth rate (biomass gain over period of experiment) continues to increase with increasing water availability. *E. dives* reaches maximum growth rate at intermediate water treatments.

(Adapted from Smith et al. 2002.)

photosynthesis is often observed (Figure 6.22). In general, the range of temperatures over which net photosynthesis is at its maximum shifts in the direction of the thermal conditions under which the plant is grown. That is to say, individuals grown under cooler temperatures exhibit a lowering of T_{opt}, whereas those individuals grown under warmer conditions exhibit an increase in T_{opt}. This same shift in the temperature response can be observed in individual plants in response to seasonal shifts in temperature (Figure 6.23). These modifications in the temperature response of net photosynthesis are a

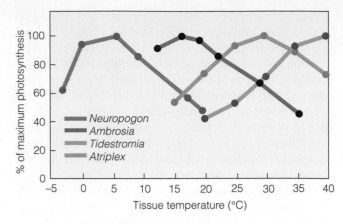

Figure 6.20 | Relationship between net photosynthesis and temperature for various terrestrial plant species from dissimilar thermal habitats: *Neuropogon acromelanus* (Arctic lichen), *Ambrosia chamissonis* (cool, coastal dune plant), *Atriplex hymenelytra* (evergreen desert shrub), and *Tidestromia oblongifolia* (summer-active desert perennial).

(Adapted from Mooney et al. 1976.)

result of the process of acclimation—reversible phenotypic changes in response to changing environmental conditions (see Section 5.9).

In addition to the influence of temperature on plant carbon balance, periods of extreme heat or cold can directly damage plant cells and tissues. Plants that inhabit seasonally cold environments, where temperatures drop below freezing for periods of time, have evolved several

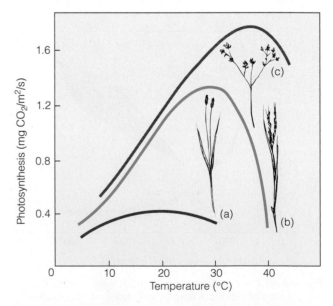

Figure 6.21 | Effect of change in leaf temperature on the photosynthetic rates of C_3 and C_4 plants. **(a)** A C_3 plant, the north temperate grass *Sesleria caerulea*, exhibits a decline in the rate of photosynthesis as the temperature of the leaf increases. **(b)** A C_4 north temperate grass, *Spartina anglica*. **(c)** A C_4 shrub of the North American hot desert, *Tidestromia oblongifolia* (Arizona honeysweet). The maximum rate of photosynthesis for the C_4 species occurs at higher temperatures than for the C_3 species.

(Adapted from Bjorkman 1973.)

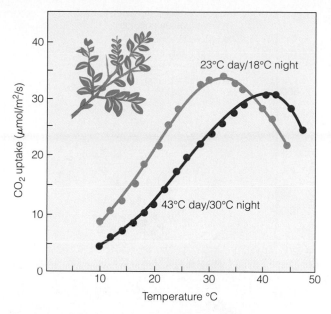

Figure 6.22 | Relationship between temperature and net photosynthesis for cloned plants of big saltbush (*Atriplex lentiformis*) grown under two different day/night temperature regimes. The shift in T_{opt} corresponds to the temperature conditions under which the plants were grown.

(Adapted from Pearcy 1977.)

adaptations for survival. The ability to tolerate extreme cold, referred to as frost hardening, is a genetically controlled characteristic that varies among species as well as among local populations of the same species. In seasonally changing environments, plants develop frost hardening through the fall and achieve maximum hardening in winter. Plants acquire frost hardiness—the turning of cold-sensitive cells into hardy ones—through the formation or addition of protective compounds in the cells. Plants synthesize and distribute substances such as

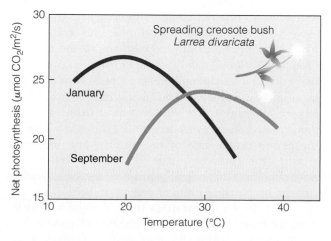

Figure 6.23 | Seasonal shift in the relationship between net photosynthesis and temperature for creosote bush (*Larrea divaricata*) shrubs growing in the field. Note that T_{opt} shifts to match the prevailing temperatures.

(Adapted from Mooney et al. 1978.)

sugars, amino acids, and other compounds that function as antifreeze, lowering the temperature at which freezing occurs. Once growth starts in spring, plants lose this tolerance quickly and are susceptible to frost damage in late spring.

Producing the protective compounds that allow leaves to survive freezing temperatures requires a significant expenditure of energy and nutrients. Some species avoid these costs by shedding their leaves before the cold season starts. These plants are termed *winter deciduous,* and their leaves senesce during the fall. The leaves are replaced during the spring, when conditions are once again favorable for photosynthesis. In contrast, needle-leaf evergreen species—such as pine (*Pinus* spp.) and spruce (*Picea* spp.) trees—contain a high concentration of these protective compounds, allowing the needles to survive the freezing temperatures of winter.

6.12 | Plants Exhibit Adaptations to Variations in Nutrient Availability

Plants require a variety of chemical elements to carry out their metabolic processes and to synthesize new tissues (Table 6.1). Thus, the availability of nutrients has many direct effects on plant survival, growth, and reproduction. Some of these elements, known as **macronutrients,** are needed in large amounts. Other elements are needed in lesser, often minute quantities. These elements are called **micronutrients,** or trace elements. The prefixes *micro–* and *macro–* refer only to the quantity of nutrients needed, not to their importance to the organism. If micronutrients are lacking, plants fail as completely as if they lacked nitrogen, calcium, or any other macronutrient.

Of the macronutrients, carbon (C), hydrogen (H), and oxygen (O) form the majority of plant tissues. These elements are derived from CO_2 and H_2O and are made available to the plant as glucose through photosynthesis. The remaining six macronutrients—nitrogen (N), phosphorus (P), potassium (K), calcium (Ca), magnesium (Mg), and sulfur (S)—exist in varying states in the soil and water, and their availability to plants is affected by different processes depending on their location in the physical environment (see Chapters 3 and 4). In terrestrial environments, plants take up nutrients from the soil. In aquatic environments, plants take up nutrients from the substrate or directly from the water.

The best example of the direct link between nutrient availability and plant performance involves nitrogen. Nitrogen plays a major role in photosynthesis. In Section 6.1, we examined two important compounds in photosynthesis: the enzyme rubisco and the pigment chlorophyll. Rubisco catalyzes the transformation of carbon dioxide into simple sugars, and chlorophyll absorbs light energy. Nitrogen is a component of both compounds. In fact, more than 50 percent of the nitrogen content of a leaf is in some way involved directly with the process of

Table 6.1 | Essential Elements in Plants

Element	Major Functions
Macronutrients	
Carbon (C) Hydrogen (H) Oxygen (O)	Basic constituents of all organic matter.
Nitrogen (N)	Used only in a fixed form: nitrates, nitrites, ammonium. Component of chlorophyll and enzymes (such as rubisco); building block of protein.
Calcium (Ca)	In plants, combines with pectin to give rigidity to cell walls; activates some enzymes; regulates many responses of cells to stimuli; essential to root growth.
Phosphorus (P)	Component of nucleic acids, phospholipids, ATP, and several enzymes.
Magnesium (Mg)	Essential for maximum rates of enzymatic reactions in cells. Integral part of chlorophyll; involved in protein synthesis.
Sulfur (S)	Basic constituent of protein.
Potassium (K)	Involved in osmosis and ionic balance; activates many enzymes.
Micronutrients	
Chlorine (Cl)	Enhances electron transfer from water to chlorophyll in plants.
Iron (Fe)	Involved in the production of chlorophyll; is part of the complex protein compounds that activate and carry oxygen and transport electrons in mitochondria and chloroplasts.
Manganese (Mn)	Enhances electron transfer from water to chlorophyll and activates enzymes in fatty-acid synthesis.
Boron (B)	Fifteen functions are ascribed to boron in plants, including cell division, pollen germination, carbohydrate metabolism, water metabolism, maintenance of conductive tissue, translation of sugar.
Copper (Cu)	Concentrates in chloroplasts, influences photosynthetic rates, activates enzymes.
Molybdenum (Mo)	Essential for symbiotic relationship with nitrogen-fixing bacteria.
Zinc (Zn)	Helps form growth substances (auxins); associated with water relationships; active in formation of chlorophyll; component of several enzyme systems.
Nickel (Ni)	Necessary for enzyme functioning in nitrogen metabolism.

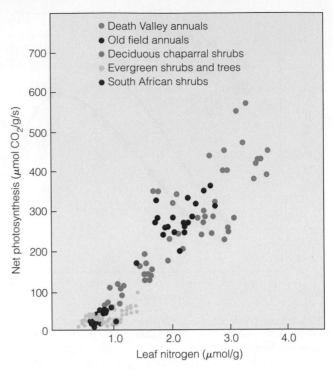

Figure 6.24 | Influence of leaf nitrogen concentrations on maximum observed rates of net photosynthesis for a variety of species from differing habitats.

(Adapted from Field and Mooney 1983.)

photosynthesis, with much of it tied up in these two compounds. As a result, the maximum (light saturated) rate of photosynthesis for a species is correlated with the nitrogen content of its leaves (Figure 6.24).

The uptake of a nutrient depends on supply and demand. Figure 6.25a illustrates the typical relationship between the uptake of a nutrient and its concentration in soil. Notice that the uptake rate increases with the concentration until some maximum rate is achieved. No further increase occurs above this concentration, because the plant meets its demands. In the case of nitrogen, low concentration in the soil or water restricts the plant to low uptake rates. In turn, a lower uptake rate decreases the concentrations of nitrogen in the leaf (Figure 6.25b) and, consequently, the concentrations of rubisco and chlorophyll. Therefore, lack of nitrogen limits the maximum rate of photosynthesis and growth. A similar pattern holds for other essential nutrients.

We have seen that geology, climate, and biological activity alter the availability of nutrients in the soil (see Chapter 4). Consequently, some environments are relatively rich in nutrients, and others are poor. How do plants from low-nutrient environments succeed? Because plants require nutrients for synthesizing new tissue, a plant's growth rate influences its demand for a nutrient. In turn, the plant's uptake rate of the nutrient also influences growth. This relationship may seem circular, but the important point is that not all plants have the same inherent (maximum potential) rate of growth. In Section 6.9 (see Figure 6.13), we saw how shade-tolerant plants have an inherently lower rate of photosynthesis and growth than shade-intolerant plants do, even under high-light conditions. This lower rate of photosynthesis and growth means a lower demand for resources, including nutrients. The same pattern of reduced photosynthesis occurs among plants that are characteristic of low-nutrient environments.

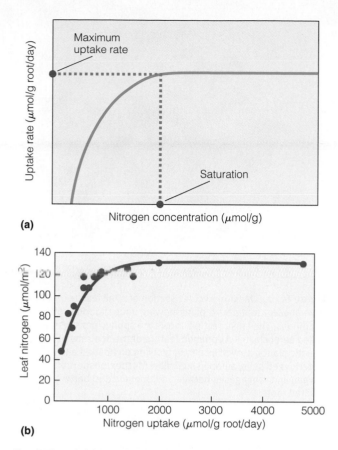

(a)

(b)

Figure 6.25 | **(a)** Uptake of nitrogen by plant roots increases with concentration in soil until the plant arrives at maximum uptake. **(b)** Influence of root nitrogen uptake on leaf nitrogen concentrations.

(Adapted from Woodward and Smith 1994.)

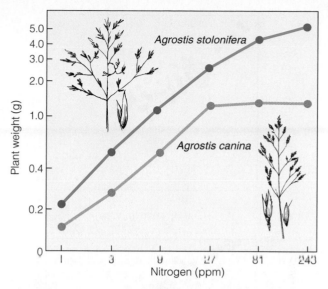

Figure 6.26 | Growth responses of two species of grass—creeping bentgrass (*Agrostis stolonifera*), found in high-nutrient environments, and velvet bentgrass (*Agrostis canina*), found in low-nutrient environments—to the addition of different levels of nitrogen fertilizer. *A. canina* responds to nitrogen fertilizer up to a certain level only.

(Adapted from Bradshaw et al. 1974.)

Figure 6.26 shows the growth responses of two grass species when soil is enriched with nitrogen. The species that naturally grows in a high-nitrogen environment keeps increasing its rate of growth with increasing availability of soil nitrogen. The species native to a low-nitrogen environment reaches its maximum rate of growth at low to medium nitrogen availability. It does not respond to further additions of nitrogen.

Some plant ecologists suggest that a low maximum growth rate is an adaptation to a low-nutrient environment. One advantage of slower growth is that the plant can avoid stress under low-nutrient conditions. A slow-growing plant can still maintain optimal rates of photosynthesis and other metabolic processes crucial for growth under low-nutrient availability. In contrast, a plant with an inherently high rate of growth will show signs of stress.

A second hypothesized adaptation to low-nutrient environments is leaf longevity (Figure 6.27). Leaf production has a cost to the plant. This cost can be defined in terms of the carbon and other nutrients required to grow the leaf. At a low rate of photosynthesis, a leaf needs a longer time to "pay back" the cost of its production. As a result, plants inhabiting low-nutrient environments tend to have

longer-lived leaves. A good example is the dominance of pine species on nutrient poor, sandy soils in the coastal region of the southeastern United States. In contrast to deciduous tree species, which shed their leaves every year, these pines have needles that live for up to three years.

Like water, nutrients are a belowground resource of terrestrial plants. Their ability to exploit this resource is related to the amount of root mass. One way that plants growing in low-nutrient environments compensate is by increasing root production. This increase is one cause of their low growth rates. As in the case of water limitation, carbon is allocated to root production at the cost of leaf production. The reduced leaf area reduces the rate of carbon uptake in photosynthesis relative to the rate of carbon loss in respiration. The result is a lower net carbon gain and growth rate by the plant.

6.13 | Wetland Environments Present Unique Constraints on Plant Adaptations

In wetland environments (see Chapter 25), where soils are saturated with water for most or all of the year, or in areas along waterways that are prone to flooding, too much water can place as much stress on plants as too little water does. Plant species differ in their ability to deal with the stresses of flooding and waterlogged soils. Symptoms exhibited by plants intolerant of such conditions are similar to those of drought and include closing stomata, yellowing and premature loss of leaves, wilting, and rapid reduction in photosynthesis. The causes, however, are different.

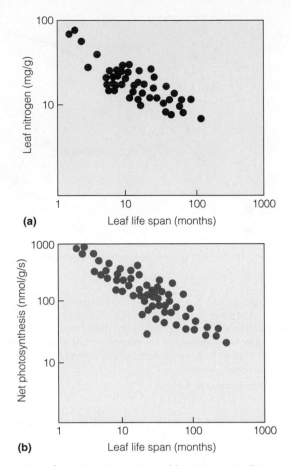

(a)

(b)

Figure 6.27 | Relationship between **(a)** leaf longevity (life span) and leaf nitrogen concentration and **(b)** leaf longevity and net photosynthetic rate (maximum) for a wide variety of plants from different environments. Each data point represents a single species. Species having longer-lived leaves tend to have lower leaf nitrogen concentrations and subsequently lower rates of photosynthesis.

(Adapted from Reich et al. 1996.)

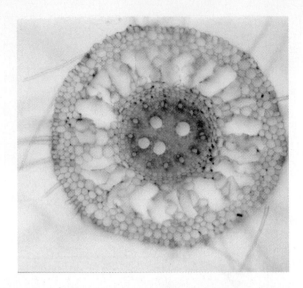

Figure 6.28 | Magnified cross section of a tall fescue (*Festuca arundinacea*) root from a plant growing under flooded soil conditions. The cross section shows the spongy passages called aerenchyma. A common feature of the anatomy of aquatic plants, as well as plants growing on flooded and waterlogged soils, aerenchyma allow for the transport of oxygen and other gases between submerged and better-aerated roots.

Growing plants need both sufficient water and a rapid gas exchange (carbon dioxide and oxygen) with their environment. Much of this exchange occurs between the roots and the air spaces within the soil. When water fills soil pores, such gas exchange cannot take place. The plants, experiencing depressed O_2 levels, in effect asphyxiate; they drown. Roots depend on gaseous O_2 to carry on aerobic respiration. Lacking oxygen, flooded roots are forced to shift to an alternative anaerobic metabolism (see Section 6.1 for discussion of aerobic and anaerobic respiration). Further, anaerobic conditions in the soil can lead to the production of substances toxic to plants.

In response to these anaerobic conditions, some species of plants accumulate ethylene in their roots. Ethylene gas, a growth hormone, is highly insoluble in water and is normally produced in small amounts in the roots. Under flooded conditions ethylene diffusion from the roots slows, and oxygen diffusion into the roots virtually stops. Ethylene then accumulates to high levels in the roots. Ethylene stimulates cells in the roots to self-destruct and separate to form interconnected, gas-filled chambers called **aerenchyma** (Figure 6.28). These chambers, typical of aquatic plants, allow some exchange of gases between submerged and better-aerated roots. Aerenchyma also allow oxygen to diffuse between the plant parts above the water and the submerged tissues. In plants such as water lilies, the aerenchyma ventilate the entire plant, with interconnected gas spaces throughout the leaves and roots occupying nearly 50 percent of the plant tissue.

In other plants—especially woody ones—the original roots, deprived of oxygen, die. In their place, adventitious roots (roots arising in positions where roots normally would not grow) emerge from the submerged part of the stem. Replacing the functions of the original roots, adventitious roots spread horizontally along the soil surface where oxygen is available. Some plants found in poorly drained soils—for example, red maple (*Acer rubrum*) and white pine (*Pinus strobus*)—develop shallow, horizontal root systems to cope with flooding. These shallow root systems, however, make them highly susceptible to drought and windthrow.

Some woody species can grow on permanently flooded sites. Among them are bald cypress (*Taxodium distichum*), mangroves, willows, and water tupelo (*Nyssa aquatica*). Bald cypresses growing on sites with fluctuating water tables develop knees or **pneumatophores,** specialized growths of the root system (Figure 6.29). These growths may be beneficial, although they are not necessary for survival. Pneumatophores on mangrove trees

Figure 6.29 | Pneumatophores, or "knees," are typical features of a bald cypress swamp.

help with gas exchange and provide oxygen to roots during tidal cycles (see Chapter 25).

Plants of salt marshes and other saline habitats grow in a physiologically dry environment. Salinity limits the amount of water they can absorb (see Chapter 3). Known as **halophytes,** the plants take in water containing high levels of solutes. Characteristically, halophytes accumulate high levels of ions within their cells, especially in the leaves. The high solute concentration, which may equal or exceed that of seawater, allows halophytes to maintain a high cell water content in the face of a low external osmotic potential (see Chapter 3, Section 3.6).

After taking up water heavy in sodium and chloride, some halophytes dilute it with water they have stored in their tissues. Some plants have salt-secreting glands that deposit excess salt on the leaves so it will be washed away by rain. Others remove salts mechanically at the root membranes.

Halophytes, however, vary in their degree of salt tolerance. Some plants, such as salt marsh hay grass (*Spartina patens*), grow best at low salinities. Others, such as salt marsh cordgrass (*S. alternifolia*), do best at moderate levels of salinity (see Figures 25.6, 25.7, and 25.9 for examples of salt marsh halophytes). A few, such as glasswort (*Salicornia* spp.), tolerate high salinities (see photograph in Figure 25.8).

Summary

Photosynthesis and Respiration | 6.1

Photosynthesis harnesses light energy from the Sun to convert CO_2 and H_2O into glucose. A nitrogen-based enzyme, rubisco, catalyzes the transformation of CO_2 into sugar. Because the first product of the reaction is a three-carbon compound, this photosynthetic pathway is called C_3 photosynthesis. Cellular respiration releases energy from carbohydrates to yield energy, H_2O, and CO_2. The energy released in this process is stored as the high-energy compound ATP. Respiration occurs in the living cells of all organisms.

Photosynthesis and Light | 6.2

The amount of light reaching a plant influences its photosynthetic rate. The light level at which the rate of carbon dioxide uptake in photosynthesis equals the rate of carbon dioxide loss due to respiration is called the light compensation point. The light level at which a further increase in light no longer produces an increase in the rate of photosynthesis is the light saturation point.

CO₂ Uptake and Water Loss | 6.3

Photosynthesis involves two key physical processes: diffusion and transpiration. CO_2 diffuses from the atmosphere to the leaf through leaf pores, or stomata. As photosynthesis slows down during the day and demand for CO_2 lessens, stomata close to reduce loss of water to the atmosphere. Water loss through the leaf is called transpiration. The amount of water lost depends on the humidity. Water lost through transpiration must be replaced by water taken up from the soil.

Water Movement | 6.4

Water moves from the soil into the roots, up through the stem and leaves, and out to the atmosphere. Differences in water potential along a water gradient move water along this route. Plants draw water from the soil, where the water potential is the highest, and release it to the atmosphere, where it is the lowest. Water moves out of the leaves through the stomata in transpiration, and this reduces water potential in the roots so that more water moves from the soil through the plant. This process continues as long as water is available in the soil. This loss of water by transpiration creates moisture conservation problems for plants. Plants need to open their stomata to take in CO_2, but they can conserve water only by closing the stomata.

Aquatic Plants | 6.5

A major difference between aquatic and terrestrial plants in CO_2 uptake and assimilation is the lack of stomata in submerged aquatic plants. In aquatic plants, there is a direct diffusion of CO_2 from the waters adjacent to the leaf across the cell membrane.

Plant Energy Balance | 6.6

Leaf temperatures affect both photosynthesis and respiration. Plants have optimal temperatures for photosynthesis beyond which photosynthesis declines. Respiration increases with temperature. The internal temperature of all plant parts is influenced by heat gained from and lost to the environment. Plants absorb longwave and shortwave radiation. They reflect some of it back to the environment. The difference is the plant's net radiation balance. The plant uses some of the absorbed radiation in photosynthesis. The remainder must be either stored as heat in the plant and surrounding air or dissipated through the processes of evaporation (transpiration) and convection.

Net Carbon Gain and Carbon Allocation | 6.7

The net carbon gain (per unit time) of a plant is the difference between carbon uptake in photosynthesis and carbon loss through respiration. The net carbon gain is then allocated to a variety of plant processes, including the production of new tissues. Because, in general, only leaves (photosynthetic tissues) are able to photosynthesize, yet all living tissues respire, the net carbon gain (and subsequently the growth) of a plant is influenced by the patterns of carbon allocation.

Interdependence of Plant Adaptations | 6.8

Plants have evolved a wide range of adaptations to variations in environmental conditions. The adaptations exhibited by plants to these features of the environment are not independent, for reasons relating to the physical environment and to the plants themselves.

Plant Adaptations to High and Low Light | 6.9

Plants exhibit a variety of adaptations and phenotypic responses (phenotypic plasticity) in response to different light environments. Shade-adapted (shade-tolerant) plants have low photosynthetic, respiratory, metabolic, and growth rates. Sun plants (shade-intolerant) generally have higher photosynthetic, respiratory, and growth rates but lower survival rates under shaded conditions. Leaves in sun plants tend to be small, lobed, and thick. Shade-plant leaves tend to be large and thin.

Alternative Pathways of Photosynthesis | 6.10

The C_4 pathway of photosynthesis involves two steps and is made possible by leaf anatomy that differs from C_3 plants. C_4 plants have vascular bundles surrounded by chlorophyll-rich bundle sheath cells. C_4 plants fix CO_2 into malate and aspartate in the mesophyll cells. They transfer these acids to the bundle sheath cells, where they are converted into CO_2.

Photosynthesis then follows the C_3 pathway. C_4 plants have a high water-use efficiency (the amount of carbon fixed per unit of water transpired). Succulent desert plants, such as cacti, have a third type of photosynthetic pathway, called CAM. CAM plants open their stomata to take in CO_2 at night, when the humidity is high. They convert CO_2 to malate, a four-carbon compound. During the day, CAM plants close their stomata, convert malate back to CO_2, and follow the C_3 photosynthetic pathway.

Adaptations to Heat and Cold | 6.11

Plants exhibit a variety of adaptations to extremely cold as well as hot environments. Cold tolerance is mostly genetic and varies among species. Plants acquire frost hardiness through the formation or addition of protective compounds in the cell, where these compounds function as antifreeze. The ability to tolerate high air temperatures is related to plant moisture balance.

Plant Adaptations to Nutrient Availability | 6.12

Terrestrial plants take up nutrients from soil through the roots. As roots deplete nearby nutrients, diffusion of water and nutrients through the soil replaces them. Availability of nutrients directly affects a plant's survival, growth, and reproduction. Nitrogen is important because rubisco and chlorophyll are nitrogen-based compounds essential to photosynthesis. Uptake of nitrogen and other nutrients depends on availability and demand. Plants with high nutrient demands grow poorly in low-nutrient environments. Plants with lower demands survive and grow, slowly, in low-nutrient environments. Plants adapted to low-nutrient environments exhibit lower rates of growth and increased longevity of leaves. The lower nutrient concentration in their tissues means lower-quality food for decomposers.

Plant Adaptations to Wetland Environments | 6.13

Plants intolerant of flooding exhibit symptoms similar to those of drought. Plants also experience reduced oxygen availability to the roots, disrupting metabolic processes and changing patterns of root growth. In waterlogged environments ethylene—a growth hormone in roots—may increase, stimulating root cells to form interconnected, gas-filled chambers. These chambers permit the exchange of oxygen between submerged and aerated roots. Plants adapted to waterlogged environments have gas-filled chambers that carry oxygen from the leaves to the roots.

Study Questions

1. What does it mean to say that life on Earth is carbon based?
2. Distinguish between photosynthesis and assimilation. How are they related?
3. What is the function of respiration?

4. What is the role of light (PAR) in the process of photosynthesis?
5. In the relationship between net photosynthesis and available light (PAR) shown in Figure 6.2, there is a net loss of carbon dioxide by the leaf at levels of light

below the light compensation point. Why does this occur? Based on this relationship, how do you think net photosynthesis varies over the course of the day?

6. How does diffusion control the uptake of carbon dioxide and the loss of water from the leaf?

7. How does the availability of water to a plant constrain the rate of photosynthesis?

8. What is the advantage of the C_4 photosynthetic pathway as compared to the conventional C_3 pathway? How might these advantages influence where these plant species are found?

9. What is the advantage of a lower light compensation point (LCP) for plant species adapted to low-light environments? What is the cost of maintaining a low LCP?

10. How do plants growing in shaded environments respond to increase their photosynthetic surface area?

11. How does a decrease in water availability influence the allocation of carbon (photosynthates) in the process of growth?

12. What is the basis for the relationship between leaf nitrogen concentration and rate of net photosynthesis shown in Figure 6.24?

13. How could increased leaf longevity (longer-lived leaves) function as an adaptation to low-nutrient environments?

Further Readings

Dale, J. E. 1992. How do leaves grow? *Bioscience* 42:423–432.
How do environmental conditions influence leaf development? This paper gives the reader a basic understanding of the processes involved in leaf growth and the background necessary to answer this question.

Grime, J. 1971. *Plant strategies and vegetative processes.* New York: Wiley.
An excellent, integrated overview of plant adaptations to the environment. This book describes how the various features of a plant's life history, from seed to adult, reflect adaptations to different habitats and the constraints imposed on plant survival, growth, and reproduction.

Lambers, H., F. S. Chapin III, and T. L. Pons, 1998. *Plant physiological ecology.* New York: Springer.
For more information, read this technical (but well written, illustrated, and organized) book that delves further into the processes presented in this chapter.

Larcher, W. 1996. *Physiological plant ecology.* 3rd ed. New York: Springer-Verlag.
An excellent reference on plant ecophysiology. Like the previous text, this is a fine reference book for more information on the materials presented in the chapter. Less technical, but also less comprehensive.

Schulze, E. D., R. H. Robichaux, J. Grace, P. W. Randel, and J. R. Ehleringer. 1987. Plant water balance. *Bioscience* 37:30–37.
A good introduction to plant water balance that is well written, well illustrated, and not too technical. For students who wish to expand their understanding of the topic, it complements the materials presented in this chapter.

Walker, D. 1992. *Energy, plants and man.* East Sussex, UK: Packard.
This book is a humorous, well-written presentation of plant biology. The sections on photosynthesis are easy to read, well illustrated, and an excellent introduction to the topic. Highly recommended.

Woodward, F. I. 1987. *Climate and plant distribution.* Cambridge, UK: Cambridge University Press.
An excellent overview of plant energy and water balance, as well as plant adaptations to climate. It is easy to read, well referenced, and concise.

C H A P T E R 7

Animal Adaptations to the Environment

A male white-fronted brown lemur (*Eulemur fulvus albifrons*) feeds on fruits in the canopy of a rain forest in Madagascar (Africa). This omnivore feeds on fruits, leaves, and insects in the forest canopy, where it lives in small, cohesive groups of 3–12 individuals.

All autotrophs, whether microscopic phytoplankton or the giant sequoia trees of the western United States, derive their energy from the same process—photosynthesis. The story is quite different for animals. Because heterotrophic organisms derive their energy, and most of their nutrients, from consuming organic compounds contained in plants and animals, they encounter literally hundreds of thousands of different types of potential food items—packaged as the diversity of plant and animal species inhabiting Earth. For this reason alone, animal adaptation is a much more complex and diverse topic than that of plants (presented in Chapter 6). However, several key processes are common to all animals: acquiring and digesting food, absorbing oxygen, maintaining body temperature and water balance, and adapting to systematic variation in light and temperature (the diurnal and seasonal cycles). The aquatic and terrestrial environments also impose several fundamentally different constraints on animal adaptation. In this chapter, we examine the variety of adaptations animals have evolved to maintain the basic metabolic processes that allow them to successfully survive, grow, and reproduce in the diversity of environments existing on Earth. In doing so, we will focus on the benefits and constraints imposed by specific adaptations and how the trade-offs involved influence the organisms' success under different environmental conditions.

7.1 | Size Imposes a Fundamental Constraint on the Evolution of Organisms

Living organisms occur in a wide range of sizes (Figure 7.1). The smallest animals are around 2–10 μg, while the largest living animals are mammals: the blue whale weighing more than 100,000 kg in marine environments, and the African elephant at 5000 kg on land. Each taxonomic group of animals has its own particular size range, largely due to design constraints. Some groups such as Bryozoa (aquatic colonial animals: see Figure 7.1) contain species all within one or two orders of magnitude, while mammals are hugely variable in size. The smallest mammal is a species of shrew (see Figure 7.1) weighing only about 2 gm fully grown—or about 100 million (10^8) times lighter than the blue whale.

Size has consequences for structural and functional relationships in animals, and as such, presents a fundamental constraint on adaptation. Most morphological and physiological features change as a function of body size in a predictable way—by a process known as **scaling.**

Geometrically similar objects, such as cubes or spheres, are referred to as isometric (Greek for "having equal measurement"). The surface area (*SA*) and volumes (*V*) of isometric objects are related to their linear dimensions (length = *l*) to the second and third power,

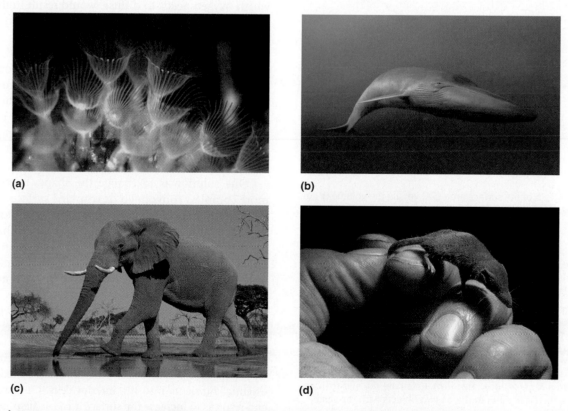

(a)

(b)

(c)

(d)

Figure 7.1 | Living organisms occur in a wide range of body size. **(a)** Bryozoa are aquatic colonial animals, with individuals as small as 0.5 mm and 5 μg. Mammals represent a huge range of body size. The largest mammal in marine environments is the **(b)** blue whale, weighing over 100,000 kg. On land, mammals range in size from the **(c)** African elephant at 5000 kg to the **(d)** pygmy shrew, weighing only 2 g fully grown.

respectively. Since surface areas (SA) are l^2 and volumes (V) are l^3, it follows that surface area is $V^{2/3}$ (or $V^{0.67}$).

For example, the surface area of a square is l^2, where l is the length of each side. Therefore the surface area of a cube of length l is $6l^2$ (six sides). In contrast, the volume of the cube is l^3. Therefore, if the surface area (SA) of a cube is plotted against its volume (V) using logarithmic coordinates (Figure 7.2), a straight line is obtained that corresponds to the equation:

$$\log SA = \log 6 + 0.67 \log V$$

The value $\log 6$ ($= 0.78$) represents the intercept, while the value 0.67 is the slope of the relationship, describing the change in SA per unit change in V. For other isometric shapes the value of the intercept will change, but the isometric scaling exponent of 0.67 will remain the same.

An important consequence of the relationship of isometric scaling is the relationship between the surface area and volume. If surface area per unit volume (SA/V) is plotted against volume (V), a straight line is produced with a slope of -0.33 (see Figure 7.2). Thus smaller bodies have a larger surface area relative to their volume than do larger objects of the same shape.

This relationship between surface area and volume imposes a critical constraint on the evolution of animals. The range of biochemical and physiological processes associated with basic metabolism (assimilation and respiration) require the transfer of materials and energy between the organism's interior and its exterior environment. For example, most organisms depend on oxygen (O_2) to maintain the process of cellular respiration (see Section 6.1). Every living cell in the body therefore requires that oxygen diffuse into it to function and survive. Oxygen is a relatively small molecule that readily diffuses across the cell surface; in a matter of seconds, it can penetrate into a millimeter of living tissue. So if we imagine a spherically shaped organism that is 1 mm in radius, the center of this tiny spherical organism is close enough to the surface that as oxygen is used up in the process of respiration, it will be replenished by a steady diffusion from the surface in contact with the external environment (air or water).

Now imagine a spherical organism with the radius of a golf ball: approximately 21 cm. It would now take over an hour for oxygen to diffuse into the center. Although the layers of cells just below the surface would receive adequate oxygen, the continuous depletion of oxygen as it diffuses toward the center, and the greater distance over which oxygen needs to diffuse, would result in the death of the interior cells (and eventually the organism) due to oxygen depletion.

The problem is that, as the size (radius) of the organism increases, the surface area of the body across which oxygen diffuses into the organism decreases relative to the interior volume of the body that requires the oxygen (the SA/V ratio decreases as shown in Figure 7.2). So how can this constraint be overcome so that larger organisms can maintain an adequate flow of oxygen to the entire interior of the body?

One solution is to change the shape. By creating a more complex, convoluted or wrinkled surface as shown in Figure 7.3, we can increase the surface area for an object having the same volume as the golf-ball-shaped organism. The difference is that now (1) no point on the interior of the organism is greater than a few millimeters from the surface, and (2) the total surface area over which oxygen can diffuse is much greater.

Another solution to the problem is to actively transport oxygen into the interior of the body. Many of the smallest animals have a tubelike shape with a central chamber (Figure 7.4). These animals draw water into their interior chamber (tube), allowing for the diffusion of oxygen and essential nutrients into the interior cells. Once again, the end result is to increase the surface area for absorption (diffusion) relative to the volume (SA/V), assuring that every point (cell) in the interior is close enough to the surface to allow for the diffusion of oxygen. This is a workable

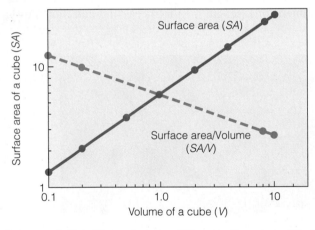

Figure 7.2 | Plot of the surface area (SA) versus the volume (V) of a cube (solid line). Both surface area and volume are plotted on a logarithmic scale ($\log_{10}$). The slope of the line is 0.67. The dashed line shows the ratio of surface area to volume (SA/V), with a slope of -0.33. Note that SA/V decreases with increasing size (volume) of the cube. This relationship holds true for any isometric object and constrains the ability of organisms to exchange energy and matter with the external environment with increasing body size.

Interpreting Ecological Data

Q1. The volume of a cube of length 4 is 64 (or 4^3), and the surface area is 96 (or 6×4^2). Now consider a 3-D rectangle having the following dimensions of length (l), height (h), and width (w): $l = 16$, $h = 2$, and $w = 2$. The volume is $l \times h \times w = 64$. The surface area is $4(l \times h) + 2(w \times h)$. Calculate the surface area. How does the SA/V differ for these two objects with the same volume?

Q2. Which of the two objects (cube or 3-D rectangle) would be a more efficient design for exchanging substances between the surface and the body interior?

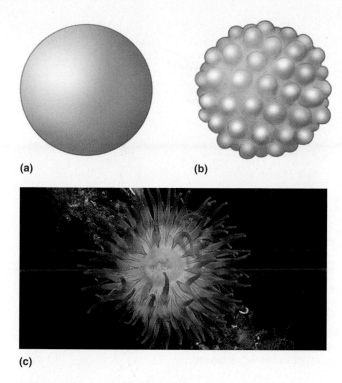

(a)   (b)

(c)

Figure 7.3 | One way to increase the ratio of surface area to volume (SA/V) for an object of a given size (volume) is to alter the shape. Objects **(a)** and **(b)** have the same volume; but by creating a more complex, convoluted surface, object (b) has a much greater surface area (and SA/V) than does object (a). Note the similarity between the body form of the sea anemone **(c)** and object (b).

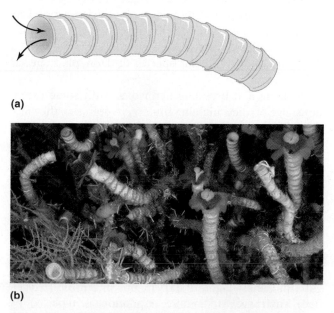

(a)

(b)

Figure 7.4 | **(a)** A tube-shaped body with a central chamber greatly increases the surface area of an organism, providing a greater exchange surface with the external environment and reducing the distance from the surface to any point within the body interior. Note the similarity in design of (a) with that of tube-shaped organisms such as the tube worms shown in **(b)**.

solution for smaller organisms; but as size increases, a more complex network of transport vessels (tubes) is needed for oxygen to reach every point in the body.

Much of the shape of larger organisms is governed by the transport of oxygen and other essential substances to cells in the interior of the body. To achieve this end, animals have evolved a complex set of anatomical structures. Lungs function as interior chambers that bring oxygen close to blood vessels, where it can be transferred to molecules of hemoglobin for transport throughout the body. A circulatory system with a heart functioning as a pump assures that oxygen-containing blood is actively transported into the minute vessels or capillaries that permeate all parts of the body. These complex systems increase the surface area for exchange, assuring that all cells in the body are well within the maximum distance over which oxygen can diffuse at the rate necessary to support cellular respiration.

The same constraints on body size apply to the wide range of metabolic processes that require the exchange of materials and energy between the external environment and the interior of the organism. Carbon and other essential nutrients must be taken in through a surface. The food canal (digestive system) in most animals is a tube in which the process of digestion occurs and through which dissolved substances must be absorbed into the circulatory system for transport throughout the body. In the smallest of animals, such as the Bryozoa (see Figure 7.1) or tube worms (Figure 7.4b), the central chamber into which water is drawn also functions as the food canal, where digestion occurs and substances are absorbed. Waste products then exit through the opening as water is expelled. In larger animals the food canal is a tube extending from the mouth to the anus (Section 7.2). As food travels through the tube it is broken down, and essential nutrients and amino acids are absorbed and transported into the circulatory system. The greater the surface area of the food canal, the greater its ability to absorb food. Because surface area increases as the square of length, the larger the animal (which increases as a cube), the greater the surface area of its food canal must be to maintain a constant ratio of surface area to volume.

From these simple examples, it should be clear that increasing body size requires complex changes in the organism's structure. These changes represent adaptations that maintain the relationship between the volume (or mass) of living cells that must be constantly supplied with essential resources from the outside environment and the surface area through which these exchanges occur.

In the following sections, we examine various adaptations relating to the ability of animals to maintain the exchange of essential nutrients (food), oxygen, water, and thermal energy (heat) with the external environment. We also consider how those adaptations are constrained by both body size and the physical environments in which the animals live.

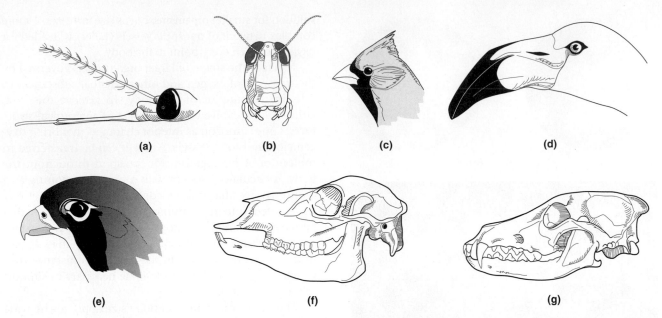

Figure 7.5 | Mouthparts reflect how organisms obtain their food. **(a)** Piercing mouthparts of a mosquito. **(b)** Chewing mouthparts of a grasshopper. **(c)** Strong, conical bill of a seed-eating bird. **(d)** Straining bill of a flamingo. **(e)** Tearing beak of a hawk. **(f)** Grinding molars of an herbivore, a deer. **(g)** Canine and shearing teeth of a carnivorous mammal, the coyote.

7.2 | Animals Have Various Ways of Acquiring Energy and Nutrients

The diversity of potential energy sources in the form of plant and animal tissues requires an equally diverse array of physiological, morphological, and behavioral characteristics that enable animals to acquire (Figure 7.5) and assimilate these resources. There are many ways to classify animals, based on the resources they use and how they exploit them. The most general of these classifications is the division based on how animals use plant and animal tissues as sources of food. Animals that feed exclusively on plant tissues are classified as **herbivores.** Those that feed exclusively on the tissues of other animals are classified as **carnivores,** whereas those that feed on both plant and animal tissues are called **omnivores.** In addition, animals that feed on dead plant and animal matter, called detritus, are detrital feeders, or **detritivores** (see Chapter 21). Each of these four feeding groups has characteristic adaptations allowing it to exploit its particular diet.

Herbivory

Because plants and animals have different chemical compositions, the problem facing herbivores is how to convert plant tissue to animal tissue. Animals are high in fat and proteins, which they use as structural building blocks. Plants are low in proteins and high in carbohydrates—much of it in the form of cellulose and lignin in cell walls—which have a complex structure and are difficult to break down (see Chapter 21). Nitrogen is a major constituent of protein. In plants, the ratio of carbon to nitrogen is about 50 to 1. In animals, the ratio is about 10 to 1.

Herbivores are categorized by the type of plant material they eat. Grazers feed on leafy material, especially grasses. Browsers feed mostly on woody material. Granivores feed on seeds, and frugivores eat fruit. Other types of herbivorous animals, such as avian sapsuckers (*Sphyrapicus* spp.) and sucking insects such as aphids, feed on plant sap; and hummingbirds, butterflies, and a variety of moth and ant species feed on plant nectar (nectivores).

Grazing and browsing herbivores, with some exceptions, live on diets high in cellulose. In doing so, they face several dietary problems. Their diets are rich in carbon, but low in protein. Most of the carbohydrates are locked in indigestible cellulose, and the proteins exist in chemical compounds. Lacking the enzymes needed to digest cellulose, herbivores depend on specialized bacteria and protozoa living in their digestive tracts. These bacteria and protozoans digest cellulose and proteins, and they synthesize fatty acids, amino acids, proteins, and vitamins. For most vertebrates, bacteria and protozoans concentrate in the foregut (from mouth to intestines) or hindgut (intestines; Figure 7.6a). In this anaerobic (oxygen free) environment, aerobic respiration is replaced by fermentation, a form of anaerobic respiration (see Section 6.1 for a discussion of aerobic and anaerobic respiration). **Fermentation,** which converts sugars to inorganic acids and alcohols in the absence of oxygen, is a less efficient process than aerobic respiration.

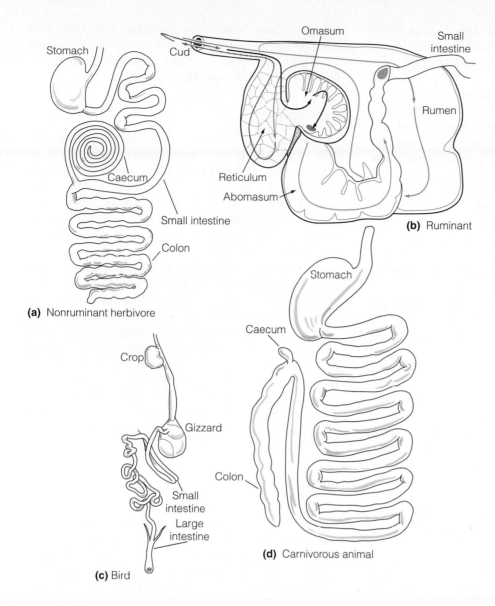

Figure 7.6 | **(a)** Digestive tract of a nonruminant herbivore, characterized by a long intestine and well-developed caecum. **(b)** The ruminant's four-compartment stomach consists of the rumen, reticulum, omasum, and abomasum. Food enters the rumen and the reticulum. The ruminant regurgitates fermented material (cud) and rechews it. Finer material enters the reticulum and then the omasum and abomasum. Coarser material reenters the rumen for further fermentation. **(c)** The digestive tract of a bird with a crop and gizzard (the stomach). **(d)** The relatively simple digestive tract of a carnivorous mammal consists of the esophagus, stomach (collectively the foregut), small intestine, and a small caecum and large intestine (collectively the hindgut).

In herbivorous insects, bacteria and protozoans inhabit the hindgut. Some species of cellulose-consuming wood beetles and wasps depend on fungi. These insects carry fungal spores with them externally when they invade new wood.

Ruminants, such as cattle and deer, are exemplary cases of herbivores anatomically specialized for digesting cellulose. They have a highly complex digestive system consisting of a four-compartment stomach—the rumen (from which the group gets its name), reticulum, omasum, and abomasum (or true stomach; Figure 7.6b)—and a long intestine. The rumen and reticulum, inhabited by anaerobic bacteria and protozoans, function as fermentation vats. These microbes break down the cellulose into usable nutrients. Ruminants have highly developed salivary glands that excrete substances to regulate acidity (pH) and chemistry in the rumen.

As ruminants graze, they chew their food hurriedly. The food material descends to the rumen and reticulum, where it is softened to a pulp by the addition of water, kneaded by muscular action, and fermented by bacteria. At leisure, the animals regurgitate the food, chew it more thoroughly to reduce plant particle size (thus making it more accessible to microbes), and swallow it again. The mass again enters the rumen. Finer material moves into the reticulum. Contractions force the material into the third compartment, or omasum, where it is further digested and finally forced into the abomasum, or true glandular stomach.

The digestive process carried on by the microorganisms in the rumen produces fatty acids. These acids rapidly absorb through the wall of the rumen into the bloodstream, providing the ruminant with a major source of food energy. Part of the material in the rumen converts to methane, which is expelled from the body, and part is converted into compounds that can be used directly as food energy. To recapture still more of the energy and nutrients, the ruminant digests many of the microbial cells in the abomasum. Further bacterial action breaks down complex carbohydrates into sugars. In addition to carrying on fermentation, the bacteria synthesize B-complex vitamins and amino acids.

Most of the digestion in ruminants occurs in the foregut. Among nonruminants, such as rabbits and horses, digestion takes place—less efficiently—in the hindgut. Nonruminant vertebrate herbivores, such as horses, have simple stomachs, long intestinal tracts that increase surface area and slow the passage of food through the gut, and a well-developed caecum, which is a pouch attached to the colon of the intestine and is where fermentation takes place (Figure 7.6a).

Lagomorphs—rabbits, hares, and pikas—resort to a form of coprophagy, the ingestion of fecal material for further extraction of nutrients. Part of ingested plant material enters the caecum (see Figure 7.6a), and part enters the intestine to form dry pellets. In the caecum, microorganisms process the ingested material and expel it into the large intestine and then out of the body as soft, green, moist pellets. Lagomorphs reingest the soft pellets, which are much higher in protein and lower in crude fiber than the hard fecal pellets. The coprophagy recycles 50 to 80 percent of feces. The reingestion is important because the pellets, functioning as "external rumen," provide bacterially synthesized B vitamins and ensure a more thorough digestion of dry material and better use of protein. Coprophagy is widespread among the detritus-feeding animals, such as wood-eating beetles and millipedes that ingest pellets after they have been enriched by microbial activity.

Seed-eating birds—gallinaceous (chicken-like) birds, pigeons, doves, and many species of songbirds—have three separate chambers. The first chamber is a pouch in the esophagus called the crop, which is a reservoir for food that passes on to the stomach (Figure 7.6c). The stomach secretes enzymes to begin digestion. The food then passes to the gizzard, which functions as a powerful grinding organ. Birds assist the grinding action of the gizzard by swallowing small pebbles and gravel, or grit.

Among marine fish, herbivorous species are small and typically inhabit coral reefs. Characterized by high diversity (many different species), they make up about 25 to 40 percent of the fish biomass about the reefs. These herbivorous fish feed on algal growth that, unlike the food of terrestrial herbivores, lacks lignin and other structural carbon compounds that are more difficult to digest. The fish gain access to the nutrients inside the algal cells through one or more of four basic types of digestive mechanisms. In some fish with thin-walled stomachs, low stomach pH (acidic) weakens algal cell walls and allows digestive enzymes access to the cell contents. Fish having gizzardlike stomachs can ingest inorganic material that mechanically breaks down algal cells to release nutrients. Some reef fish have specialized jaws that shred or grind algal material before it reaches the intestine. Other fish depend on microbial fermentation in the hindgut to assist in breaking down algal cells. These four types are not mutually exclusive. Some marine herbivores may combine low stomach pH or grinding and shredding with microbial fermentation in the hindgut.

Carnivory

Herbivores are the energy source for carnivores—the flesh eaters. Unlike herbivores, carnivores are not faced with problems relating to digesting cellulose or to the quality of food. Because the chemical composition of the flesh of prey and the flesh of predators is quite similar, carnivores encounter no problem in digesting and assimilating nutrients from their prey. Their major problem is obtaining enough food.

Lacking the need to digest complex cellulose compounds, carnivores have short intestines and simple stomachs (Figure 7.6d). In mammalian carnivores, the stomach is little more than an expanded hollow tube with muscular walls. It stores and mixes foods, adding mucus, enzymes, and hydrochloric acid to speed digestion. In carnivorous birds such as hawks and owls, the gizzard is little more than an extendable pocket with reduced muscles in which digestion, started in the anterior stomach, continues. In hawks and owls, the gizzard acts as a barrier against the hair, bones, and feathers that these birds regurgitate and expel from the mouth as pellets.

Omnivory

Omnivory includes animals that feed on both plants and animals. The food habits of many omnivores vary with the seasons, stages in the life cycle, and their size and growth rate. The red fox (*Vulpes vulpes*), for example, feeds on berries, apples, cherries, acorns, grasses, grasshoppers, crickets, beetles, and small rodents. The black bear (*Ursus americanus*) feeds heavily on vegetation—buds, leaves, nuts, berries, tree bark—supplemented with bees, beetles, crickets, ants, fish, and small- to medium-sized mammals.

7.3 | Animals Have Various Nutritional Needs

Animals require a variety of mineral elements (Table 7.1) and amino acids. Of the 20 amino acids required to make proteins, most animal species can synthesize about half of them, as long as their diet includes organic nitrogen. The remaining ones, termed essential amino acids, must be supplied by the diet. These nutritional needs differ little among vertebrates and invertebrates. Insects, for example, have the same dietary requirements as vertebrates, although they need more potassium, phosphorus, and magnesium and less calcium, sodium, and chlorine than vertebrates do. The ultimate source of most of these nutrients is plants. For this reason, the quantity and quality of plants affect the nutrition of herbivorous consumers.

Table 7.1 | **Essential Minerals in Animals**

Element	Role
Carbon (C) Hydrogen (H) Oxygen (O)	Basic constituents of all organic matter.
Nitrogen (N)	Building block of protein.
Calcium (Ca)	Needed for acid–base relationships, clotting of blood, contraction and relaxation of heart muscles. Controls movement of fluid through cells; gives rigidity to skeletons of vertebrates; forms shells of mollusks, arthropods, and one-celled Foraminifera.
Phosphorus (P)	Necessary for energy transfer; major component of nuclear material of cells; acid–base balance; bone and tooth formation.
Magnesium (Mg)	Essential for maximum rates of enzymatic reactions in cells; enzyme activation.
Sulfur (S)	Basic constituent of protein.
Sodium (Na)	Maintenance of acid–base balance, osmotic homeostasis, formation and flow of gastric and intestinal secretions, nerve transmission, lactation, growth, and maintenance of body weight.
Potassium (K)	Involved in synthesis of protein, growth, and carbohydrate metabolism.
Chlorine (Cl)	Role is similar to that of sodium, which it is associated with in salt (NaCl).
Fluorine (F)	Maintenance of tooth (and probably bone) structure.
Iron (Fe)	Component of respiratory pigment hemoglobin in blood of vertebrates and hemolymph of insects; electron carriers in energy metabolism.
Manganese (Mn)	Enzyme systems.
Selenium (Se)	Closely related to vitamin E in function.
Cobalt (Co)	Required by ruminants for the synthesis of vitamin B_{12} by bacteria in the rumen.
Copper (Cu)	Involved in iron metabolism; melanin synthesis; electron transport.
Molybdenum (Mo)	Enzyme systems.
Zinc (Zn)	Functions in several enzyme systems, especially the respiratory enzyme carbonic anhydrase in red blood cells, and in certain digestive enzymes.
Iodine (I)	Involved in thyroid metabolism.
Chromium (Cr)	Involved in glucose and energy metabolism.

When food is scarce, consumers may suffer from acute malnutrition, leave the area, or starve. In other situations there may be enough food to relieve hunger, but its low quality affects reproduction, health, and longevity.

The highest-quality plant food for herbivores, vertebrate and invertebrate, is high in nitrogen in the form of protein. As the nitrogen content of their food increases, the animals' assimilation of plant material improves, increasing growth, reproductive success, and survival. Nitrogen is concentrated in the growing tips, new leaves, and buds of plants. Its content declines as leaves and twigs mature and become senescent. Herbivores have adapted to this period of new growth. Herbivorous insect larvae are most abundant early in the growing season, and they complete their growth before the leaves mature. Many vertebrate herbivores, such as deer, give birth to their young at the start of the growing season, when the most protein-rich plant foods will be available for their growing young.

Although availability and season strongly influence food selection, both vertebrate and invertebrate herbivores do show some preference for the most nitrogen-rich plants, which they probably detect by taste and odor. For example, beavers show a strong preference for willows (*Salix* spp.) and aspen (*Populus* spp.), two species that are high in nitrogen content. Chemical receptors in the nose and mouth of deer encourage or discourage consumption of certain foods. During drought, nitrogen-based compounds are concentrated in certain plants, making them more attractive and vulnerable to herbivorous insects. However, preference for certain plants means little if they are unavailable. Food selection by herbivores is an interaction among quality, preference, and availability (see Field Studies: Martin Wikelski).

The need for quality foods differs among herbivores. Ruminant animals, as already pointed out, can subsist on rougher or lower-quality plant materials because bacteria in the rumen can synthesize such requirements as vitamin B_1 and certain amino acids from simple nitrogen-based compounds. Therefore, the caloric content and the nutrients in a certain food might not reflect its real nutritive value for the ruminant. Nonruminant herbivores require a larger amount of complex proteins in their foods. Seed-eating herbivores exploit the concentration of nutrients in

Department of Ecology and Evolutionary Biology, Princeton University, Princeton, New Jersey

The isolated archipelago of the Galápagos Islands off the western coast of South America is known for its amazing diversity of animal and plant life. It was the diversity of life on these islands that so impressed the young Charles Darwin and laid the foundations for his theory of natural selection (see Chapter 5). However, one inhabitant of the Galápagos fauna has consistently been met with revulsion by historic visitors: the marine iguana, *Amblyrhynchus cristatus* (Figure 1). Indeed, even Darwin himself commented on this "hideous-looking creature."

Marine iguanas are widely distributed throughout the Galápagos Islands, and different populations vary dramatically in size (both length and weight). Due to these variations, many of the iguana populations were long considered separate species, yet modern genetic studies have confirmed that all of the populations are part of a single species. What could possibly account for this marked variation in body size among populations? This question has been central to the research of Princeton University ecologist Martin Wikelski. Studies by Wikelski and his colleagues during the past decade have revealed an intriguing story of the constraints imposed by variations in the environment of the Galápagos on the evolution of these amazing creatures.

In a series of studies, Wikelski and his colleagues have examined differences in body size between two populations of marine iguanas that inhabit the islands of Santa Fe and Genovesa. The populations of these two islands differ markedly in body size, with an average body length of 25 cm (maximum body weight of 900 g) for adult males on Genovesa as compared to 40 cm (maximum body weight of 3500 g) for adult males on the island of Santa Fe. Wikelski hypothesized that these differences reflected energetic constraints on the two populations in the form of food supply.

Marine iguanas are herbivorous reptiles that feed on submerged intertidal and subtidal algae (seaweed) along the rocky island shores, referred to as algae pastures. To determine the availability of food for iguana populations, Wikelski and colleagues measured the standing biomass and productivity of pastures in the tidal zones of these two islands. Their results show that the growth of algae pastures correlates with sea surface temperatures. Waters in the tidal zone off Santa Fe (the more southern island) are cooler than those off Genovesa, and as a result both the length of algae plants and the productivity of pastures are much higher off Santa Fe.

By examining patterns of food intake and growth of marked individuals on the two islands, Wikelski was able to demonstrate that food intake limits growth rate and subsequent body size in marine iguanas, which in turn depends on the availability of algae (Figure 2). Body size differences between the two island populations can be explained by differences in food availability.

Temporal variations in climate and sea surface temperatures also influence food availability for the marine iguanas across the Galápagos Islands. Marine iguanas can live for up to 30 years, and environmental conditions can change dramatically within an individual's lifetime. El Niño events (see Section 3.8) usually recur at intervals of 3–7 years, but more were prevalent in the decade of the 1990s. During El Niño years in the Galápagos, sea surface temperatures increase from an average of 18°C to a maximum of 32°C as cold ocean currents and cold-rich upwellings are disrupted. As a result, green and red algal species—the preferred food of marine iguanas—disappear and are replaced by the brown algae, which the iguanas find hard to digest. Up to 90 percent of marine iguana populations on islands can die of starvation due to these environmental changes.

Figure 1 | The Galápagos marine iguana (*Amblyrhynchus cristatus*).

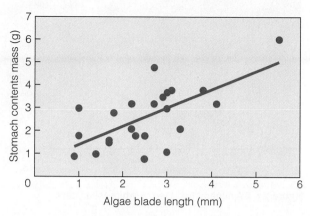

Figure 2 | The food intake (dry mass in stomach) for iguanas of a given length (200–250 mm) from both study islands increased with increasing length of the algae pasture in the intertidal zone.
(Adapted from Wikelski et al. 1997.)

In studying patterns of mortality during the El Niño events of the 1990s, Wikelski observed the highest mortality rate for larger individuals. This higher mortality rate directly related to observed differences in foraging efficiency with body size. Wikelski and colleagues determined that although larger individuals have a higher daily intake of food, smaller individuals have a higher food intake per unit body mass, a result of higher foraging efficiency (food intake per bite per gram body mass). Large iguanas on both islands showed a marked decline in body mass during the El Niño events. The result is a strong selective pressure against large body size during these periods of food shortage (Figure 3).

Perhaps the most astonishing result of Wikelski's research is that the marine iguanas exhibit an unusual

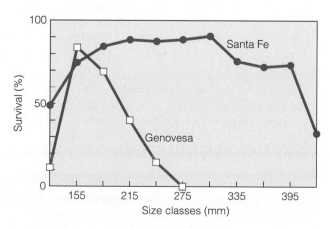

Figure 3 | Survival of individually marked animals on Genovesa (squares) and Santa Fe (dots).
(Adapted from Wikelski and Trillmich 1997.)

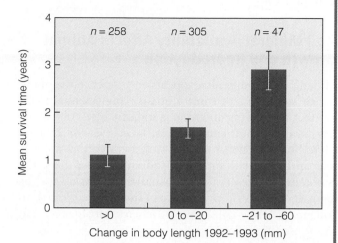

Figure 4 | Relation between change in body length and survival time for adult iguanas on the island of Santa Fe during the 1997–1998 El Niño (ENSO) cycle. Values of *n* refer to sample size.
(Adapted from Wikelski 2000.) (Nature Publishing Group.)

adaptation to the environmental variations caused by El Niño. Change in body length is considered to be unidirectional in vertebrates, but Wikelski repeatedly observed shrinkage of up to 20 percent in the length of individual adult iguanas. This shrinking coincided with low food availability resulting from El Niño events.

Shrinking did not occur equally across all size classes. Wikelski found an inverse relationship between the initial body size of individuals and the observed change in body length during the period of food shortage—larger individuals shrank less than smaller individuals.

Shrinkage was found to influence survival. Large adult individuals that shrank more survived longer because their foraging efficiency increased and their energy expenditure decreased (Figure 2).

Bibliography

Wikelski, M., and C. Thom. 2000. Marine iguanas shrink to survive El Niño. *Nature* 403:37–38.

Wikelski, M., V. Carrillo, and F. Trillmich. 1997. Energy limits on body size in a grazing reptile, the Galápagos marine iguana. *Ecology* 78:2204–2217.

1. Does the mortality of iguanas during El Niño events represent a case of natural selection? Which of the three models of selection best describes the pattern of natural selection?
2. If the iguanas could not "shrink" during the period of resource shortage, how do you think the El Niño events would influence natural selection?

Go to **QUANTIFYIT!** at www.ecologyplace.com to explore how to use a scatter plot to display data.

the seeds. Among the carnivores, quantity is more important than quality. Carnivores rarely have a dietary problem because they consume animals that have resynthesized and stored protein and other nutrients from plants in their tissues.

7.4 | Mineral Availability Affects Animal Growth and Reproduction

Mineral availability also appears to influence the abundance and fitness of some animals. One essential nutrient that is receiving attention is sodium, often one of the least available nutrients in terrestrial ecosystems. In areas of sodium deficiency in the soil, herbivorous animals face an inadequate supply of sodium in their diets. The problem has been noted in Australian herbivores such as kangaroos, in African elephants (*Loxodonta africana*), in rodents, in white-tailed deer (*Odocoileus virginianus*), and in moose (*Alces alces*).

Sodium deficiency can influence the distribution, behavior, and physiology of mammals, especially herbivores. The spatial distribution of elephants across the Wankie National Park in central Africa appears to correlate closely with the sodium content of drinking water. The most elephants are found at water holes with the highest sodium content. Three herbivorous mammals— the European rabbit (*Oryctolagus cuniculus*), the moose, and the white-tailed deer—experience sodium deficiencies in parts of their range. In sodium-deficient areas in southwestern Australia, the European rabbit builds up reserves of sodium in its tissues during the nonbreeding season. These reserves become exhausted near the end of the breeding season, forcing the rabbits to graze selectively on sodium-rich plants, often to the point of depleting these plant populations.

Ruminants face severe mineral deficiencies in spring. Attracted by the flush of new growth, deer, bighorn sheep (*Ovis canadensis*), mountain goats (*Oreamnos americanus*), elk (*Cervus elaphus*), and domestic cattle and sheep feed on new succulent growth of grass, but with high physiological costs. Vegetation is much higher in potassium relative to calcium and magnesium in the spring than during the rest of the year. This high intake of potassium stimulates the secretion of aldosterone, the principal hormone that promotes sodium retention by the kidney. Although aldosterone stimulates sodium retention, it also facilitates the excretion of potassium and magnesium. Because concentrations of magnesium in soft tissues and skeletal stores are low in herbivores, these animals experience magnesium deficiency. This deficiency results in a rapid onset of diarrhea and often muscle spasms (tetany). The deficiency comes when mineral demands are high—late in gestation for females, and at the beginning of antler growth for male deer and elk.

To counteract this mineral imbalance in the spring, large herbivores seek mineral licks—places in the land-

Figure 7.7 | A mineral lick used by white-tailed deer.

scape where animals concentrate to satisfy their mineral needs by eating mineral-rich soil (Figure 7.7). Although sodium chloride is associated with mineral licks, animal physiologists hypothesize that it is not sodium the animals seek but magnesium—and in the case of bighorn sheep, mountain goats, and elk, calcium as well.

The size of deer, their antler development, and their reproductive success all relate to nutrition. Other factors being equal, only deer obtaining high-quality foods grow large antlers. Deer on diets low in calcium, phosphorus, and protein show stunted growth, and males develop only thin spike antlers.

7.5 | Animals Require Oxygen to Release Energy Contained in Food

Animals obtain their energy from organic compounds in the food they eat; and they do so primarily through aerobic respiration, which requires oxygen (see Section 6.1). Oxygen is easily available in the atmosphere for terrestrial animals. However, for aquatic animals, oxygen may be limiting and its acquisition problematic (see Section 3.6).

Differences between terrestrial and aquatic animals in the means of acquiring oxygen reflect the availability of oxygen in the two environments. Minute terrestrial organisms take in oxygen by diffusion across the body surface. Insects have tracheal tubes that open to the outside through openings (or spiracles) on the body wall (Figure 7.8a). The tracheal tubes carry oxygen directly to the body cells.

Unable to meet demands through the direct diffusion of oxygen across the body surface, larger terrestrial animals (mammals, birds, and reptiles) have some form of lungs. Unlike tracheal systems that branch throughout the insect body, lungs are restricted to one location. Structurally, lungs have innumerable small sacs that increase surface area across which oxygen readily diffuses

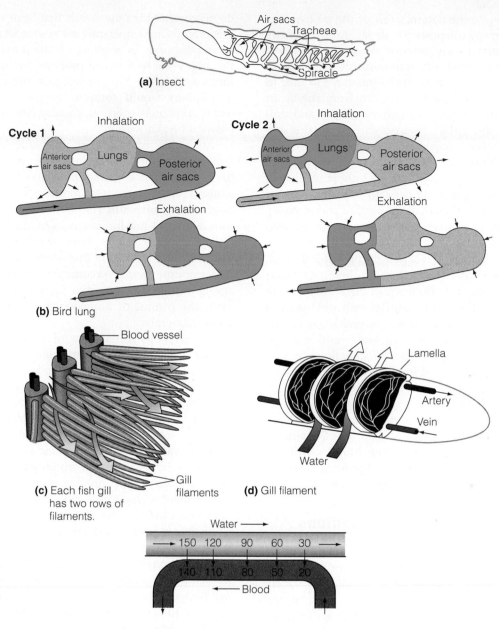

(a) Insect

(b) Bird lung

(c) Each fish gill has two rows of filaments.

(d) Gill filament

(e) Water and blood flow through lamellae

Figure 7.8 | Respiratory systems. **(a)** The tracheal system and spiracles of an insect (grasshopper). Air enters the tracheal tubes through spiracles—openings on the body wall. **(b)** Gas exchange in the bird lung requires two cycles, each involving inhalation and exhalation. During the first inhalation, most of the air flows past the lungs into a posterior air sac. That air passes through the lungs upon exhalation, and the next inhalation ends up in the anterior air sac. At the same time, the posterior sacs draw in more air. This flow pattern allows oxygenated blood to leave the lungs with the highest possible amount of oxygen. **(c)** Fish obtain oxygen from water by their gills. Gill filaments have flattened plates called lamellae. **(d)** Blood flowing through capillaries within the lamella pick up oxygen from the water through a countercurrent exchange. Water flows across the lamellae in a direction opposite to the blood flow. **(e)** Blood entering the gills is low in oxygen. As it flows through the lamellae, it picks up more and more oxygen from the water. The water flowing in the opposite direction gradually loses more and more of its oxygen (numbers refer to O_2 concentration).

into the bloodstream. Amphibians take in oxygen through a combination of lungs and vascularized skin. Lungless salamanders are an exception; they live in a moist environment and take in oxygen directly through the skin.

In addition to lungs, birds have accessory air sacs that act as bellows to keep air flowing through the lungs as they inhale and exhale (Figure 7.8b). Air flows one way only, forming a continuous circuit through the interconnected system regardless of whether the bird is inhaling or exhaling. During inhalation, most of the air bypasses the lungs and enters the posterior air sac. Air then passes through the lungs to the anterior air sac; at the same time, the posterior air sac draws in more air.

In aquatic environments, organisms have to take in oxygen from the water or gain oxygen from the air in some way. Marine mammals such as whales and dolphins come to the surface to expel carbon dioxide and take in air containing oxygen to the lungs. Some aquatic insects rise to the surface to fill the tracheal system with air. Others, like diving beetles, carry a bubble of air with them when submerged. Held beneath the wings, the air bubble contacts the spiracles of the beetle's abdomen.

Minute aquatic animals, zooplankton, take up oxygen from the water by diffusion across the body surface. Fish, the major aquatic vertebrates, pump water through their mouth. The water passes through slits in the pharynx, flows over gills, and exits through the back of the gill covers (Figure 7.7c). The close contact with and the rapid flow of water over the gills allows for exchanges of oxygen and carbon dioxide between water and the gills (Figure 7.7d). Water passing over the gills flows in a direction opposite to that of blood, setting up a countercurrent exchange. As the blood flows through capillaries, it acquires more and more oxygen. It also encounters water that is more and more concentrated with oxygen because water is just beginning to pass over the gills. As water continues its flow, it encounters blood with lower oxygen concentration, aiding the uptake of oxygen through the process of diffusion (Figure 7.7e).

7.6 | Regulation of Internal Conditions Involves Homeostasis and Feedback

In an ever-changing physical environment, organisms must maintain a fairly constant internal environment within the narrow limits required by their cells, organs, and enzyme systems. They need some means of regulating their internal environment relative to external conditions including body temperature, water balance, pH, and the amounts of salts in fluids and tissues. For example, the human body must maintain internal temperatures within a narrow range around 37°C. An increase or decrease of only a few degrees from this range could prove fatal. The maintenance of a relatively constant internal environment in a varying external environment is called **homeostasis.**

Whatever the processes involved in regulating an organism's internal environment, homeostasis depends on negative feedback—meaning that when a system deviates from the normal or desired state, mechanisms function to restore the system to that state. The thermostat that controls the temperature in your home is an example of a negative feedback system. If we wish the temperature of the room to be 20°C (68°F), we set that point on the thermostat. When the temperature of the room air falls below that point, a temperature-sensitive device within the thermostat trips the switch that turns on the furnace. When the room temperature reaches the set point, the thermostat responds by shutting off the furnace. Should the thermostat fail to function properly and not shut off the furnace, then the furnace would continue to heat, the temperature would continue to rise, and the furnace would ultimately overheat, causing either a fire or a mechanical breakdown.

A key difference between mechanical and living systems is that in living systems the set point is not firmly fixed, as it often is in mechanical systems. Instead, organisms have a limited range of tolerances, called **homeostatic plateaus.** Homeostatic systems work within minimum and maximum values by using negative feedback to regulate activity above or below the set point. If the system deviates from that set point, a negative feedback response ensues—a control mechanism inhibits any strong movement away from the set point. Among animals, the control of homeostasis is both physiological and behavioral.

An example is temperature regulation in humans (Figure 7.9). The normal temperature, or set point, for humans is 37°C. When the temperature of the environment rises, sensory mechanisms in the skin detect the change. They send a message to the brain, which automatically relays the message to receptors that increase blood flow to the skin, induce sweating, and stimulate behavioral responses. Water excreted through the skin evaporates, cooling the body. When the environmental temperature falls below a certain point, another reaction takes place. This time, it reduces blood flow and causes shivering, an involuntary muscular exercise that produces more heat.

If the environmental temperature becomes extreme, the homeostatic system breaks down. When it gets too warm, the body cannot lose heat fast enough to maintain normal temperature. Metabolism speeds up, further raising body temperature, until death results from heatstroke. If the environmental temperature drops too low, metabolic processes slow down, further decreasing body temperature, until death by freezing ensues.

7.7 | Animals Exchange Energy with Their Surrounding Environment

In principle, an animal's energy balance is the same as that described for a plant in Chapter 6 (Section 6.6). Animals, however, differ significantly from plants in their thermal relations with the environment. Animals can produce significant quantities of heat by metabolism, and their mobility allows them to seek out or escape heat and cold.

Body structure influences the exchange of heat between animals and the external environment. Consider a simple thermal model of an animal body (Figure 7.10). The interior or core of the body must be regulated within a defined range of temperature. In contrast, the temperature of the environment surrounding the animal's body

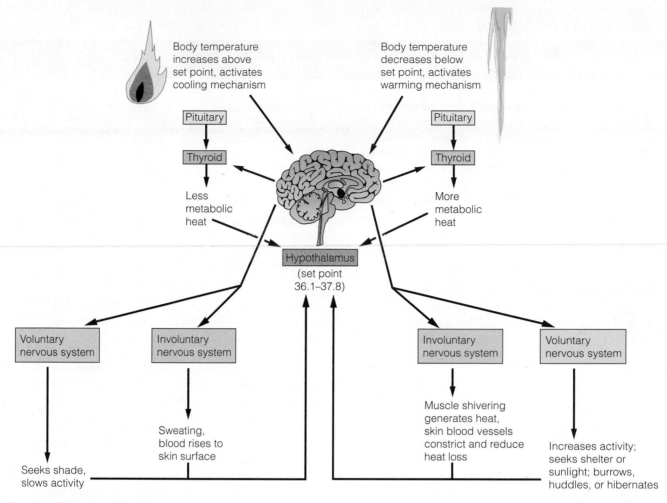

Body temperature increases above set point, activates cooling mechanism

Body temperature decreases below set point, activates warming mechanism

Pituitary

Thyroid

Less metabolic heat

Pituitary

Thyroid

More metabolic heat

Hypothalamus
(set point 36.1–37.8)

Voluntary nervous system

Involuntary nervous system

Involuntary nervous system

Voluntary nervous system

Sweating, blood rises to skin surface

Muscle shivering generates heat, skin blood vessels constrict and reduce heat loss

Seeks shade, slows activity

Increases activity; seeks shelter or sunlight; burrows, huddles, or hibernates

Figure 7.9 | Thermoregulation is an example of homeostasis in action. The hypothalamus in one's brain receives feedback or senses the temperature of blood arriving from the body core. If body core temperature rises, it responds accordingly in two ways, activating the autonomic (or involuntary) and voluntary nervous systems and the endocrine system.

varies. The temperature at the body's surface, however, is not the same as the air or water temperature in which the animal lives. Rather, it is the temperature at a thin layer of air (or water) called the boundary layer, which lies at the surface (Section 6.6) just above and within hair, feathers, and scales.

Therefore, body surface temperature differs from both the air (or water) and the core body temperatures. Separating the body core from the body surface are layers of muscle tissue and fat, across which the temperature gradually changes from the core temperature to the body surface temperature. This layer of insulation influences the organism's thermal **conductivity**; that is, the ability to conduct or transmit heat (see Quantifying Ecology 7.1: Heat Exchange and Temperature Regulation).

To maintain its core body temperature, the animal has to balance gains and losses of heat to the external environment. It does so by changes in metabolic rate and by heat exchange. The core area exchanges heat (produced by metabolism and stored in the body) with the surface area by conduction, the transfer of heat through a

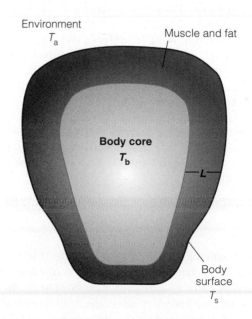

Environment T_a

Muscle and fat

Body core T_b

L

Body surface T_s

Figure 7.10 | Temperatures in an animal body. Body core temperature is T_b, environmental temperature is T_a, surface temperature is T_s, and L is the thickness of the outer layer of the body.

Heat Exchange and Temperature Regulation

Internal temperature regulation is a crucial process in all animals. Whether endothermic or ectothermic, this regulation is a function of balancing inputs of heat, from both the external environment and metabolic processes (particularly in endothermic animals), with losses to the external environment. The total heat stored by the body of an organism (H_{stored}) can be represented by the sum of these inputs and losses:

$$H_{stored} = H_{metabolism} + H_{conduction} + H_{convection} + H_{radiation} + H_{evaporation}$$

The value of heat energy from metabolic processes ($H_{metabolism}$) will always be positive, representing a gain of heat energy from respiration. In contrast, the transfer of heat between the organism and the external environment through the processes of conduction, convection, radiation, and evaporation can be either positive (gain of heat energy) or negative (loss of heat energy). Physical laws governing the transfer of energy determine these inputs and outputs of heat energy, in which energy always travels from hot to cold—from regions of higher energy to regions of lower energy content.

Heat transfer through the process of evaporation (latent heat exchange) is discussed in Chapter 3 (see Section 3.2), and the process of radiative heat transfer is covered in Chapter 2. The other two forms of energy transfer, conduction and convection, are particularly important in the energy balance and thermal regulation of animals in both terrestrial and aquatic environments.

Conductive heat transfer is the movement of heat through solids or between two solids that are in direct contact. Conduction occurs when heat is transferred between the core and the surface of an organism's body (see Section 7.7 and Figure 7.9). As in all forms of heat transfer, energy flows from the region of high temperature to the region of low temperature. The rate of conductive heat transfer ($H_{conduction}$) through a solid is described by the following equation:

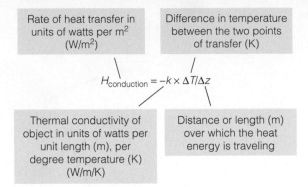

The symbol Δ refers to "a difference." Therefore, ΔT is the difference in temperature between the two regions (such as the body core and surface), and Δz is the difference in position (z)—the length or distance between the two points of transfer. The **thermal conductivity** of an object (k) describes its ability to transfer heat. Various factors, including its density and specific heat, will influence an object's thermal conductivity. Forms of insulation—such as fat, fur, or feathers—will decrease an organism's thermal conductivity with the surroundings.

As an example of conduction, consider the transfer of heat energy through an organism. In an endothermic animal, the core body temperature is maintained by metabolic processes, and heat will be transferred from the core to the body surface, where temperatures are lower. If we assume that the thermal conductivity (k) of the body is 1.25 (W/m/K) and the distance Δz from the core to the surface is 10 cm, we can calculate the heat transfer between the core and the surface in Figure 1.

If the value of $H_{conduction}$ is positive, then the direction of flow is outward. If the value is negative, the flow of heat would be inward, from the surface to the body core. This would be the case for a reptile basking on a warm surface such as a rock heated by the Sun. We can use the same approach to calculate the transfer of heat between two objects in contact, such as the reptile and the rock.

solid. Influencing this exchange are the thickness and conductivity of fat and the movement of blood to the surface. The surface layer exchanges heat with the environment by convection, conduction, radiation, and evaporation—all influenced by the characteristics of skin and body covering (see Quantifying Ecology 7.1: Heat Exchange and Temperature Regulation for a quantitative discussion of heat exchange and thermal regulation).

External environmental conditions heavily influence how animals confront thermal stress. Because air has a lower specific heat and absorbs less solar radiation than water does, terrestrial animals face more radical and dan-

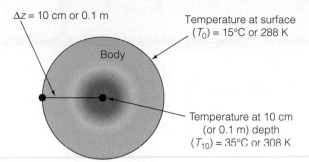

Δz = 10 cm or 0.1 m

Temperature at surface
(T_0) = 15°C or 288 K

Body

Temperature at 10 cm
(or 0.1 m) depth
(T_{10}) = 35°C or 308 K

$$H_{conduction} = -k \times (T_{10} - T_0)/length$$
$$H_{conduction} = -1.25 \times (308 - 288)/0.1$$
$$H_{conduction} = 250 \ W/m^2$$

Figure 1 | Daily variation in human body temperature.

The transfer of heat energy between a solid and a moving fluid (air or water) is called **convection.** As with conduction, the rate of convective heat transfer ($H_{convection}$) is a function of the temperature gradient between the object and the surrounding environment, in this case the fluid:

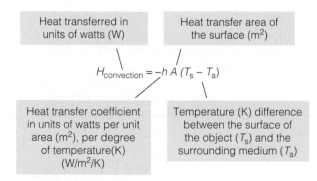

Heat transferred in units of watts (W)

Heat transfer area of the surface (m^2)

$$H_{convection} = -h \ A \ (T_s - T_a)$$

Heat transfer coefficient in units of watts per unit area (m^2), per degree of temperature(K) (W/m^2/K)

Temperature (K) difference between the surface of the object (T_s) and the surrounding medium (T_a)

The **heat transfer coefficient** represents how easily heat can move through the fluid. The value will depend on the type of fluid (gas or liquid) and its flow (velocity and viscosity) and thermal (specific heat) properties. For example, for a given gas or liquid, the heat transfer coefficient increases with the flow rate (such as wind speed).

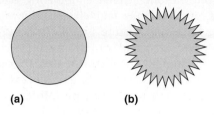

(a)　　　　(b)

Figure 2 |

An object's surface area relative to its volume plays an important role in convective heat transfer. Because heat energy being emitted by an object is transferred to the surrounding fluid across its surface area, the greater the surface area per volume (or mass), the faster heat is transferred. As discussed in Section 7.11, the result is that smaller bodies with their greater ratio of surface area to volume will exchange heat more readily than will larger bodies having a lower surface area per volume. Shape as well as size, however, influences the exchange of heat between a body and the surrounding environment. For example, the cross sections of the two objects in Figure 2 are the same size when measured as volume. But object (b) has a much greater surface area than (a) does, so object (b) has a greater capacity for exchanging heat energy with the surrounding fluid. As discussed in both Chapters 6 and 7, this relationship between shape and heat transfer is important in the energy and thermal balance of organisms, especially poikilotherms. It represents an important adaptive constraint in the evolution of both plants and animals.

1. Figure 6.11 illustrates the differences observed in the size and shape of leaves on the same plant that are exposed to full sunlight and shade. How might the differences in leaf morphology shown in Figure 6.11 influence the ability of these two leaves to dissipate heat through convection?

2. Given the importance of conduction and convection in regulating the body temperature of poikilotherms, how do you think general patterns of body shape might differ between homeotherms and poikilotherms of the same body mass (size)?

gerous changes in their thermal environment than do aquatic animals. Incoming solar radiation can produce lethal heat. The loss of radiant heat to the air, especially at night, can result in deadly cold. Aquatic animals live in a more stable energy environment (see Chapter 3), but they have a lower tolerance for temperature changes.

7.8 | Animals Fall into Three Groups Relative to Temperature Regulation

To regulate temperature, some groups of animals generate heat metabolically. This internal heat production is **endothermy,** meaning "heat from within." The result is homeothermy (from the Greek *homeo*, "the same"), or maintenance of a fairly constant temperature independent of external temperatures. Another group of animals acquires heat primarily from the external environment. Gaining heat from the environment is **ectothermy**, meaning "heat from without." Unlike endothermy, ectothermy results in a variable body temperature. This means of maintaining body temperature is **poikilothermy** (from the Greek *poikilos*, "manifold" or "variegated").

Birds and mammals are notable **homeotherms**, popularly called warm blooded. Fish, amphibians, reptiles, insects, and other invertebrates are **poikilotherms**, often called cold blooded because they can be cool to the touch. A third group regulates body temperature by endothermy at some times and ectothermy at other times. These animals are **heterotherms** (from *hetero*, "different"). Heterotherms employ both endothermy and ectothermy, depending on environmental situations and metabolic needs. Bats, bees, and hummingbirds belong to this group.

The terms *homeotherm* and *endotherm* are often used synonymously, as are *poikilotherm* and *ectotherm*; but there is a difference. *Ectotherm* and *endotherm* emphasize the mechanisms that determine body temperature. The other two terms, *homeotherm* and *poikilotherm*, represent the nature of body temperature (either constant or variable).

7.9 | Poikilotherms Depend on Environmental Temperatures

Poikilotherms, such as amphibians, reptiles, and insects, gain heat easily from the environment and lose it just as fast. Environmental sources of heat control the rates of metabolism and activity among most poikilotherms. Rising temperatures increase the rate of enzymatic activity, which controls metabolism and respiration (Figure 7.11). For every 10°C rise in temperature, the rate of metabolism in poikilotherms approximately doubles. They become ac-

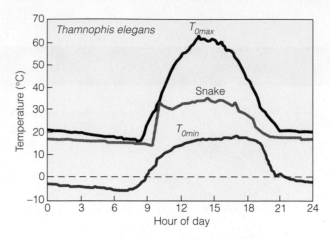

Figure 7.12 | Daily temperature variation in the western terrestrial garter snake (*Thamnophis elegans*) within its operative temperature range (T_{omin} and T_{omax}). The snake maintains a fairly constant temperature during the daylight hours.

(Adapted from Peterson et al. 1993.)

tive only when the temperature is sufficiently warm. Conversely, when ambient temperatures fall, metabolic activity declines and poikilotherms become sluggish.

Poikilotherms have an upper and lower thermal limit that they can tolerate. Most terrestrial poikilotherms can maintain a relatively constant daytime body temperature by behavioral means, such as seeking sunlight or shade. Lizards and snakes, for example, may vary their body temperature by no more than 4°C to 5°C (Figure 7.12); and body temperature of amphibians may vary by 10°C when active. The range of body temperatures at which poikilotherms carry out their daily activities is the **operative temperature range.**

Poikilotherms have a low metabolic rate and a high ability to exchange heat between body and environment (high conductivity). During normal activities, poikilotherms carry out aerobic respiration. Under stress and while pursuing prey, the poikilotherms' inability to supply sufficient oxygen to the body requires that much of their energy production come from anaerobic respiration, in which oxygen is not used. This process depletes stored energy and accumulates lactic acid in the muscles. (Anaerobic respiration can occur in the muscles of marathon runners and other athletes, causing leg cramps.) Anaerobic respiration metabolism limits poikilotherms to short bursts of activity and results in rapid physical exhaustion.

Aquatic poikilotherms, when completely immersed, maintain no appreciable difference between their body temperature and the surrounding water. Aquatic poikilotherms are poorly insulated. Any heat produced in the muscles moves to the blood and on to the gills and skin, where heat transfers to the surrounding water by convection. Exceptions are sharks and tunas, which possess a **rete,** a blood circulation system that allows them to keep internal temperatures higher than external ones (see

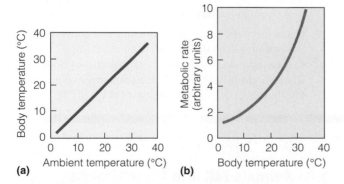

Figure 7.11 | Relationship among body temperature, resting metabolic rate, and ambient temperature in poikilotherms. **(a)** Body temperature is a function of ambient temperature. **(b)** Resting metabolism is a function of body temperature.

(Adapted from Hill and Wyse 1989.)

Section 7.14). Because seasonal water temperatures are relatively stable, fish and aquatic invertebrates maintain a constant temperature within any given season. They adjust seasonally to changing temperatures by acclimation, or physiological adjustment to a change in environmental conditions (see Chapter 5, Section 5.14 and Figure 5.30). They undergo these physiological changes over a period of time. Upper and lower limits of tolerance to temperature vary among the poikilotherm species. If they live at the upper end of their tolerable thermal range, poikilotherms will adjust their physiology at the expense of being able to tolerate the lower range. Similarly, during periods of cold, the animals shift physiological functions to a lower temperature range that would have been debilitating before. Because water temperature changes slowly through the year, aquatic poikilotherms can make adjustments slowly. Fish are highly sensitive to rapid change in environmental temperatures. If they are subjected to a sudden temperature change (faster than biochemical and physiological adjustments can occur), they may die of thermal shock.

To maintain a tolerable and fairly constant body temperature during active periods, terrestrial and amphibious poikilotherms rely on behavioral thermoregulation. They seek out appropriate microclimates (Figure 7.13). Insects such as butterflies, moths, bees, dragonflies, and damselflies bask in the sun to raise their body temperature to the level necessary to become highly active. When they become too warm, these animals seek the shade. Semiterrestrial frogs, such as bullfrogs (*Rana catesbeiana*) and green frogs (*Rana clamitans*), exert considerable control over their body temperature. By basking in the sun, frogs can raise their body temperature as much as 10°C above ambient temperature. Because of associated evapo-

rative water losses, such amphibians must be either near or partially submerged in water. By changing position or location or by seeking a warmer or cooler substrate, amphibians can maintain body temperatures within a narrow range.

Most reptiles are terrestrial and exposed to widely fluctuating surface temperatures. The simplest way for a reptile to raise body temperature is to bask in the sun. Snakes, for example, heat up rapidly in the morning sun (see Figure 7.12). When they reach the preferred temperature, the animals move on to their daily activities, retreating to the shade to cool when necessary. In this way, they maintain a stable body temperature during the day. In the evening, the reptile experiences a slow cooling. Its body temperature at night depends on its location.

Lizards raise and lower their bodies and change body shape to increase or decrease heat conduction between themselves and the rocks or soil they rest on. They also seek sunlight or shade or burrow into the soil to adjust their temperatures. Desert beetles, locusts, and scorpions exhibit similar behavior. They raise their legs to reduce contact between their body and the ground, minimizing conduction and increasing convection by exposing body surfaces to the wind. Thus, body temperatures of poikilotherms do not necessarily follow the general ambient temperature.

7.10 | Homeotherms Escape the Thermal Restraints of the Environment

Homeothermic birds and mammals meet the thermal constraints of the environment by being endothermic. They maintain body temperature by oxidizing glucose and other energy-rich molecules in the process of respiration. The process of oxidation is not 100 percent efficient,

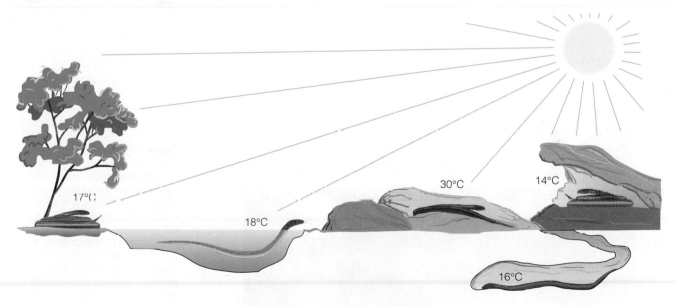

Figure 7.13 | Microclimates a snake typically uses to regulate body temperature during the summer.
(Adapted from Pearson et al. 1993.)

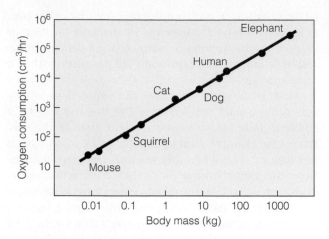

Figure 7.14 | Relationship between metabolic rate, as measured by oxygen consumption per hour, and body mass (kg) for a range of mammal species. Both variables (metabolic rate and body mass) are plotted on a logarithmic scale ($\log_{10}$).

(After Schmidt-Nielsen 1997.)

and in addition to the production of chemical energy in the form of ATP (see Section 6.1), some energy is converted to heat energy. Because oxygen is used in the process of respiration, an organism's basal metabolic rate is typically measured by the rate of oxygen consumption. Recall from Section 6.1 that all living cells respire; therefore, the rate of respiration for homeothermic animals is proportional to their body mass (grams body mass$^{0.75}$; however, the exponent varies across different taxonomic groups, ranging from 0.6 to 0.9; Figure 7.14).

For homeotherms, the **thermoneutral zone** is a range of environmental temperatures within which the metabolic rates are minimal (Figure 7.15). Outside this zone, marked by upper and lower **critical temperatures,** metabolic rate increases.

Maintenance of a high body temperature is associated with specific enzyme systems that operate optimally within a high temperature range, with a set point of about 40°C. Because efficient cardiovascular and respiratory systems bring oxygen to their tissues, homeotherms can maintain a high level of energy production through aerobic respiration (high metabolic rates). Thus, they can sustain high levels of physical activity for long periods. Independent of external temperatures, homeotherms can exploit a wider range of thermal environments. They can generate energy rapidly when the situation demands, escaping from predators or pursuing prey.

To regulate the exchange of heat between the body and the environment, homeotherms use some form of insulation—a covering of fur, feathers, or body fat (see Figure 7.10). For mammals, fur is a major barrier to heat flow; but its insulation value varies with thickness, which is greater on large mammals than on small ones. Small mammals are limited in the amount of fur they can carry, because a thick coat could reduce their ability to move. Mammals change the thickness of their fur with the season, by a form of acclimation (Section 5.14). Aquatic mammals—especially of Arctic regions—and Arctic and Antarctic birds such as auklets (Alcidae) and penguins have a heavy layer of fat beneath the skin. Birds reduce heat loss by fluffing the feathers and drawing the feet into them, making the body a round, feathered ball. Some Arctic birds, such as ptarmigan (*Lagopus* spp.), have feathered feet—unlike most birds, which have scaled feet that function to lose heat.

Although the major function of insulation is to keep body heat in, it also keeps heat out. In a hot environment, an animal has to either rid itself of excess body heat or prevent heat from being absorbed in the first place. One way is to reflect solar radiation from light-colored fur or feathers. Another way is to grow a heavy

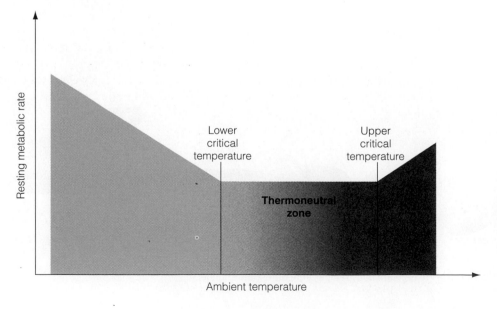

Figure 7.15 | General resting metabolic response of homeotherm to changes in ambient temperature. For temperatures within the thermal neutral zone, resting metabolic rate changes little with a change in ambient temperature. Beyond these limits, however, metabolic rate increases markedly with either an increase or decrease in ambient temperature as a result of feedback mechanisms (see Section 7.5 and Figure 7.9).

(After Schmidt-Nielsen 1997.)

coat of fur that heat does not penetrate. Large mammals of the desert, notably the camel, use this method. The outer layers of hair absorb heat and return it to the environment.

Some insects—notably moths, bees, and bumblebees—have a dense, furlike coat over the thoracic region that serves to retain the high temperature of flight muscles during flight. The long, soft hairs of caterpillars, together with changes in body posture, act as insulation to reduce convective heat exchange.

When insulation fails, many animals resort to shivering, which is a form of involuntary muscular activity that increases heat production. Many species of small mammals increase heat production without shivering by burning (oxidizing) highly vascular brown fat. Found about the head, neck, thorax, and major blood vessels, brown adipose tissue (fat) is particularly prominent in hibernators, such as bats and groundhogs (*Marmota monax*).

Many species employ evaporative cooling to reduce the body heat load. Birds and mammals lose some heat by evaporation of moisture from the skin. When their body heat is above the upper critical temperature, they accelerate evaporative cooling by sweating and panting. Only certain mammals have sweat glands—in particular, horses and humans. Panting in mammals and gular fluttering in birds increase the movement of air over moist surfaces in the mouth and pharynx. Many mammals, such as pigs, wallow in water and wet mud to cool down.

7.11 | Endothermy and Ectothermy Involve Trade-offs

A prime example of the trade-offs involved in the adaptations of organisms to their environment are endothermy and ectothermy, the two alternative approaches to regulation of body temperature in animals. Each strategy has advantages and disadvantages that enable the organisms to excel under different environmental conditions. For example, endothermy allows animals to remain active regardless of environmental temperatures, whereas environmental temperatures largely dictate the activity of poikilotherms. However, the freedom of activity enjoyed by homeotherms comes at a great energy cost. To generate heat through respiration, homeotherms must take in calories (food). Of the food energy that is assimilated, a minimum goes to growth (most goes to respiration).

The metabolic heat produced by homeothermy can be lost to the surrounding environment (see Quantifying Ecology 7.1: Heat Exchange and Temperature Regulation), and this heat must be replaced by additional heat generated through respiration. As a result, metabolic costs weigh heavily against homeotherms. In contrast, ectotherms can allocate more of their energy intake to biomass production than to metabolic needs. Not needing to burn calories to provide metabolic heat, ectotherms require fewer calories (food) per gram of body weight.

Because they do not depend on internally generated body heat, ectotherms can curtail metabolic activity in times of food and water shortage and temperature extremes. Their low energy demands enable some terrestrial poikilotherms to colonize areas of limited food and water.

One of the most important features influencing its ability to regulate body temperature is an animal's size. A body exchanges heat with the external environment (either air or water) in proportion to the surface area exposed. In contrast, it is the entire body mass (or volume) that is being heated (see Figure 7.10 and discussion in Section 7.1).

Cold-blooded organisms (ectotherms) absorb heat across their surface but must absorb enough energy to heat the entire body mass (volume). Therefore, the ratio of surface area to volume (SA/V) is a key factor in controlling the uptake of heat and the maintenance of body temperature. As an organism's size increases, the SA/V ratio decreases (see Figure 7.2). Because the organism has to absorb sufficient energy across its surface to warm the entire body mass, the amount of energy and/or the period of time required to raise body temperature likewise increases. For this reason, ectothermy imposes a constraint on maximum body size for cold-blooded animals and restricts the distribution of the larger poikilotherms to the warmer, aseasonal regions of the subtropics and tropics. For example, large reptiles such as alligators, crocodiles, iguanas, komodo dragons, anacondas, and pythons are all restricted to warm tropical environments.

The constraint that size imposes on warm-blooded animals (homeotherms) is opposite that for cold-blooded animals. For homeotherms, it is the body mass (or volume) that produces heat through respiration, while heat is lost to the surrounding environment across the body surface. The smaller the organism, the larger the SA/V ratio; therefore the greater the relative heat loss to the surrounding environment. To maintain a constant body temperature, the heat loss must be offset by increased metabolic activity (respiration). Thus, small homeotherms have a higher mass-specific metabolic rate (metabolic rate per unit body mass; Figure 7.16) and consume more food energy per unit of body weight than do large ones. Small shrews (*Sorex* spp.), for example, ranging in weight from 2 to 29 g (see Figure 7.1), require daily an amount of food (wet weight) equivalent to their own body weight. Therefore, small animals are forced to spend most of their time seeking and eating food. The mass-specific metabolic rate (respiration rate per gram of body weight) of small endotherms rises so rapidly that below a certain size, they could not meet their energy demands. On average, 2 g is about as small as an endotherm can be and still maintain a metabolic heat balance, although this minimum constraint depends on the thermal environment. Some shrews and hummingbirds undergo daily torpor (see Section 7.13) to reduce their

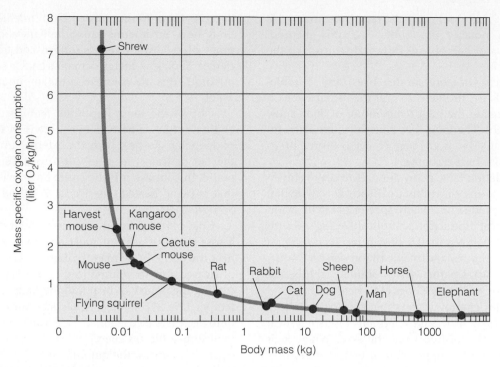

Figure 7.16 | Observed relationship between metabolic rate (oxygen consumption) per unit body mass (mass-specific metabolic rate) and body mass for various mammal species. Mass-specific metabolic rate increases with decreasing body mass. Body mass is plotted on a logarithmic scale ($\log_{10}$). (Adapted from Schmidt-Nielson 1979.)

Interpreting Ecological Data

Q1. How does the variable plotted on the *y*-axis of this graph (mass-specific metabolic weight) differ from the variable plotted on the *y*-axis of Figure 7.16?

Q2. What does the graph imply about the rates of cellular respiration for a mouse compared to a horse?

Q3. How would the graph differ if the *y*-axis was plotted on a logarithmic scale ($\log_{10}$)?

metabolic needs. Due to the conflicting metabolic demands of body temperature and growth, most young birds and mammals are born in an altricial state (see Section 8.8), meaning they are blind, naked, and helpless, beginning life as ectotherms. They depend on the body heat of their parents to maintain their body temperature, which allows these young animals to allocate most of their energy to growth.

7.12 | Heterotherms Take on Characteristics of Ectotherms and Endotherms

Species that sometimes regulate their body temperature and sometimes do not are called *temporal heterotherms*. At different stages of their daily and seasonal cycle or in certain situations, these animals take on characteristics of endotherms or ectotherms. They can undergo rapid, drastic, repeated changes in body temperature.

Insects are ectothermic and poikilothermic; yet in the adult stage, most species of flying insects are heterothermic. When flying, they have high rates of metabolism, with heat production as great as or greater than that of

homeotherms. They reach this high metabolic state in a simpler way than do homeotherms, because they are not constrained by the uptake and transport of oxygen through the lungs and vascular system. Insects take in oxygen by demand through openings in the body wall and transport it throughout the body in a tracheal system (see Section 7.5).

Temperature is crucial to the flight of insects. Most cannot fly if the temperature of the body muscles is below 30°C, nor can they fly if muscle temperature is above 44°C. This constraint means that an insect has to warm up before it can take off, and it has to get rid of excess heat in flight. With wings beating up to 200 times per second, flying insects can produce a prodigious amount of heat.

Some insects, such as butterflies and dragonflies, warm up by orienting their bodies and spreading their wings to the sun. Most warm up by shivering their flight muscles in the thorax. Moths and butterflies vibrate their wings to raise thoracic temperatures above ambient. Bumblebees pump their abdomens without any external wing movements. They do not maintain any physiological set point, and they cool down to ambient temperatures when not in flight.

7.13 | Torpor Helps Some Animals Conserve Energy

To reduce metabolic costs during periods of inactivity, some small homeothermic animals become heterothermic and enter into torpor daily. Daily **torpor** is the dropping of body temperature to approximately ambient temperature for a part of each day, regardless of season.

Some birds, such as hummingbirds (Trochilidae) and poorwills (*Phalaenoptilus nuttallii*), and small mammals, such as bats, pocket mice, kangaroo mice, and white-footed mice, experience daily torpor. Such daily torpor seems to have evolved as a way to reduce energy demands over that part of the day when the animals are inactive. Nocturnal mammals, such as bats, go into torpor by day; and diurnal animals, such as hummingbirds, go into torpor by night. As the animal goes into torpor, its body temperature falls steeply. With the relaxation of homeothermic responses, the body temperature declines to within a few degrees of ambient. Arousal returns the body temperature to normal rapidly as the animal renews its metabolic heat production.

To escape the rigors of long, cold winters, many terrestrial poikilotherms and a few heterothermic mammals go into a long, seasonal torpor called **hibernation.** Hibernation is characterized by the cessation of activity. Hibernating poikilotherms experience such physiological changes as decreased blood sugar, increased liver glycogen, altered concentration of blood hemoglobin, altered carbon dioxide and oxygen content in the blood, altered muscle tone, and darkened skin.

Hibernating homeotherms become heterotherms and invoke controlled hypothermia (reduction of body temperature). They relax homeothermic regulation and allow the body temperature to approach ambient temperature. Heart rate, respiration, and total metabolism fall, and body temperature sinks below 10°C. Associated with hibernation are high levels of CO_2 and an associated decrease in blood pH (increased acidity). This state, called acidosis, lowers the threshold for shivering and reduces the metabolic rate. Hibernating homeotherms, however, are able to rewarm spontaneously using only metabolically generated heat.

Among homeotherms, entrance into hibernation is a controlled process difficult to generalize from one species to another. Some hibernators, such as the groundhog (*Marmota monax*), feed heavily in late summer to store large fat reserves, from which they will draw energy during hibernation. Others, like the chipmunk (*Tamias striatus*), lay up a store of food instead. All hibernators, however, convert to a means of metabolic regulation different from that of the active state. Most hibernators rouse periodically and then drop back into torpor. The chipmunk, with its large store of seeds, spends much less time in torpor than do species that store large amounts of fat.

Although popularly said to hibernate, black bears, grizzly bears, and female polar bears do not. Instead, they enter a unique winter sleep from which they easily rouse. They do not enter extreme hypothermia but allow body temperatures to decline only a few degrees below normal. The bears do not eat, drink, urinate, or defecate, and females give birth to and nurse young during their sleep; yet they maintain a metabolism that is near normal. To do so, the bears recycle urea, normally excreted in urine, through the bloodstream. The urea is degraded into amino acids that are reincorporated in plasma proteins.

Hibernation provides selective advantages to small homeotherms. For them, maintaining a high body temperature during periods of cold and limited food supply is too costly. It is far less expensive to reduce metabolism and allow the body temperature to drop. Doing so eliminates the need to seek scarce food resources to maintain higher body temperatures.

7.14 | Some Animals Use Unique Physiological Means for Thermal Balance

Due to an animal's limited tolerance for heat, storing body heat does not seem like a sound option to maintain thermal balance in the body. But certain mammals, especially the camel, oryx, and some gazelles, do just that. The camel, for example, stores body heat by day and dissipates it by night, especially when water is limited. Its temperature can fluctuate from 34°C in the morning to 41°C by late afternoon. By storing body heat, these animals of dry habitats reduce the need for evaporative cooling and thus reduce water loss and food requirements.

Many ectothermic animals of temperate and Arctic regions withstand long periods of below-freezing temperatures in winter through supercooling and developing a resistance to freezing. **Supercooling** of body fluids takes place when the body temperature falls below the freezing point without actually freezing. The presence of certain solutes in the body that function to lower the freezing point of water (see Chapter 3) influences the amount of supercooling that can take place. Some Arctic marine fish, certain insects of temperate and cold climates, and reptiles exposed to occasional cold nights employ supercooling by increasing solutes, notably glycerol, in body fluids. Glycerol protects against freezing damage, increasing the degree of supercooling. Wood frogs (*Rana sylvatica*), spring peepers (*Hyla crucifer*), and gray tree frogs (*Hyla versicolor*) can successfully overwinter just beneath the leaf litter because they accumulate glycerol in their body fluids.

Some intertidal invertebrates of high latitudes and certain aquatic insects survive the cold by freezing and then thawing out when the temperature moderates. In some species, more than 90 percent of the body fluids may freeze, and the remaining fluids contain highly concentrated solutes. Ice forms outside the shrunken cells, and muscles and organs are distorted. After thawing, they quickly regain normal shape.

To conserve heat in a cold environment and to cool vital parts of the body under heat stress, some animals

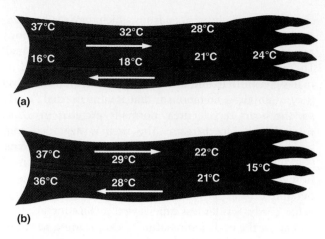

(a)

(b)

Figure 7.17 | A model of countercurrent flow in the limb of a mammal, showing hypothetical temperature changes in the blood **(a)** in the absence and **(b)** in the presence of countercurrent heat exchange.

have evolved **countercurrent heat exchange** (Figure 7.17). For example, the porpoise (*Phocaena* spp.), swimming in cold Arctic waters, is well insulated with blubber. It could experience an excessive loss of body heat, however, through its uninsulated flukes and flippers. The porpoise maintains its body core temperature by exchanging heat between arterial (coming from the lungs) and venous (returning to the lungs) blood in these structures (Figure 7.18). Veins completely surround arteries, which carry warm blood from the heart to the extremities.

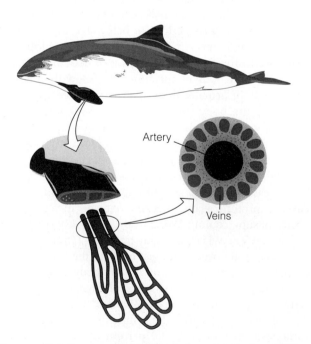

Artery

Veins

Figure 7.18 | The porpoise and its relatives, the whales, use flippers and flukes as temperature-regulating devices. Several veins in the appendages surround the arteries. Venous blood returning to the body core is warmed through heat transfer, retaining body heat.

(Adapted from Schmidt-Nielson 1977.)

Warm arterial blood loses its heat to the cool venous blood returning to the body core. As a result, little body heat passes to the environment. Blood entering the flippers cools, whereas blood returning to the deep body warms. In warm waters, where the animals need to get rid of excessive body heat, blood bypasses the heat exchangers. Venous blood returns unwarmed through veins close to the skin's surface to cool the body core. Such vascular arrangements are common in the legs of mammals and birds as well as in the tails of rodents, especially the beaver (*Castor canadensis*).

Many animals have arteries and veins divided into small, parallel, intermingling vessels that form a discrete vascular bundle or net known as a rete. In a rete, the principle is the same as in the blood vessels of the porpoise's flippers. Blood flows in opposite directions, and heat exchange takes place.

Countercurrent heat exchange can also keep heat out. The oryx (*Oryx beisa*), an African desert antelope exposed to high daytime temperatures, can experience elevated body temperatures yet keep the highly heat-sensitive brain cool by a rete in the head. The external carotid artery passes through a cavernous sinus filled with venous blood that is cooled by evaporation from the moist mucous membranes of the nasal passages (Figure 7.19). Arterial blood passing through the cavernous sinus cools on the way to the brain, reducing the temperature of the brain 2°C to 3°C lower than that of the body core.

Countercurrent heat exchangers are not restricted to homeotherms. Certain poikilotherms that assume some degree of endothermism employ the same mechanism. The swift, highly predaceous tuna (*Thunnus* spp.) and the mackerel shark (*Isurus tigris*) possess a rete in a band of dark muscle tissue used for sustained swimming effort. Metabolic heat produced in the muscle warms the venous blood, which gives up heat to the adjoining newly oxygenated blood returning from the gills. Such a countercurrent heat exchange increases the power of the muscles because warm muscles contract and relax more rapidly. Sharks and tuna maintain fairly constant body temperatures, regardless of water temperatures.

7.15 | Maintenance of Water Balance for Terrestrial Animals Is Constrained by Uptake and Conservation

Living cells, both plant and animal, contain about 75 to 95 percent water. Water is essential for virtually all biochemical reactions within the body, and it functions as a medium for excreting metabolic wastes and for dissipating excess heat through evaporative cooling. To stay properly hydrated, an organism must offset these water losses by the uptake of water from the external environment. Maintaining this balance between the uptake and loss of water with the surrounding environment is referred to as an organism's water balance (Section 4.1).

(a)

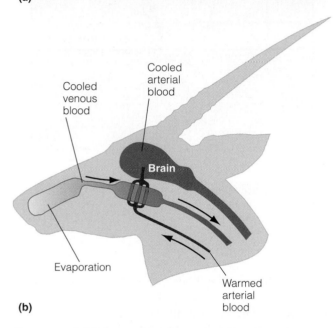

Cooled arterial blood

Cooled venous blood

Brain

Evaporation

(b)

Warmed arterial blood

Figure 7.19 | (a) The oryx (*Oryx beisa*), an African desert antelope, can keep a cool head despite a high body core temperature by means of a rete. (b) Arterial blood passes in small arteries through a pool of venous blood that is cooled by evaporation as it drains from the nasal region and into the pool.

Terrestrial animals have three major ways of gaining water and solutes: directly by drinking and eating and indirectly by producing metabolic water in the process of respiration (see Section 6.1). They lose water and solutes through urine, feces, evaporation from the skin, and from the moist air they exhale. Some birds and reptiles have a salt gland, and all birds and reptiles have a cloaca—a common receptacle for the digestive, urinary, and reproductive tracts. They reabsorb water from the cloaca back into the body proper. Mammals have kidneys capable of producing urine with high ion concentrations.

In arid environments, animals, like plants, face a severe problem of water balance. They can solve the problem in either of two ways: by evading the drought or by avoiding its effects. Animals of semiarid and desert regions may evade drought by leaving the area during the dry season and moving to areas where permanent water

is available. Many of the large African ungulates (Figure 7.20) and many birds use this strategy. During hot, dry periods the spadefoot toad (*Scaphiopus couchi*) of the southern deserts of the United States remains belowground in a state of dormancy and emerges when the rains return. Some invertebrates inhabiting ponds that dry up in summer, such as the flatworm *Phagocytes vernalis,* develop hardened casings and remain in them for the dry period. Other aquatic or semiaquatic animals retreat deep into the soil until they reach the level of groundwater. Many insects undergo diapause, a stage of arrested development in their life cycle from which they emerge when conditions improve (see Section 7.19).

Other animals remain active during the dry season but reduce respiratory water loss. Some small desert rodents lower the temperature of the air they breathe out. Moist air from the lungs passes over cooled nasal membranes, leaving condensed water on the walls. As the rodent inhales, this water humidifies and cools the warm, dry air.

There are other approaches to the problem. Some small desert mammals reduce water loss by remaining in burrows by day and emerging by night. Many desert mammals, from kangaroos to camels, extract water from the food they eat—either directly from the moisture content of the plants or from metabolic water produced during respiration—and produce highly concentrated urine and dry feces. Some desert mammals can tolerate a certain degree of dehydration. Desert rabbits may withstand water losses of up to 50 percent and camels of up to 27 percent of their body weight.

7.16 | Animals of Aquatic Environments Face Unique Problems in Maintaining Water Balance

Aquatic animals face the constant exchange of water with the external environment through the process of osmosis. As in the discussion of passive transport of water in plants (Section 6.4), osmotic pressure moves water through cell membranes from the side of greater water concentration to the side of lesser water concentration. Aquatic organisms living in freshwater are **hyperosmotic**—they have a higher salt concentration in their bodies than does the surrounding water. Consequently, water moves inward into the body, while salts move outward across their gills. Their problem is to prevent uptake, or rid themselves of excess water, and replace salts lost to the external environment. Freshwater fish maintain osmotic balance by absorbing and retaining salts in special cells in the gills and by producing copious amounts of watery urine. Amphibians balance the loss of salts through the skin by absorbing ions directly from the water and transporting them across the skin and gill membranes. In the terrestrial phase, amphibians store water from the kidneys in the bladder. If circumstances demand it, they can reabsorb the water through the bladder wall.

(a)

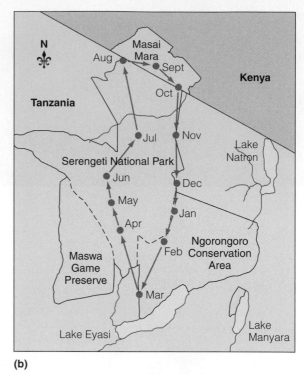

(b)

Figure 7.20 | (a) Many of the large ungulate species in the semiarid regions of Africa, such as the wildebeest shown here, migrate over the course of the year, following the seasonal shift in rainfall. (b) The changing distribution of wildebeest populations in the contiguous Serengeti, Masai Mara, and Ngorongoro Conservation areas in East Africa. This seasonal pattern of migration gives these species consistent access to food (grass production) and water.

Marine fish face problems opposite to those in freshwater. These organisms are **hypoosmotic,** having a lower salt concentration in their bodies than does the surrounding water. When the concentration of salts is greater outside the body than within, organisms tend to dehydrate. Osmosis draws water out of the body into the surrounding environment. In marine and brackish [term defined in Chapter 3] environments, organisms have to inhibit water loss by osmosis through the body wall and prevent an accumulation of salts in the body.

There are many solutions to this problem. Invertebrates solve it by being **isosmotic**—their body fluids have the same osmotic pressure as seawater. Marine bony (teleost) fish absorb saltwater into the gut. They secrete magnesium and calcium through the kidneys and pass these ions off as a partially crystalline paste. In general, fish excrete sodium and chloride, major ions in seawater, by pumping the ions across special membranes in the gills. This pumping process is one type of active transport, moving salts against the concentration gradient, but it has a high energy cost. Sharks and rays retain enough urea to maintain a slightly higher concentration of solute in the body than exists in surrounding seawater. Birds of the open sea and sea turtles can consume seawater because they possess special salt-secreting nasal glands. Seabirds of the order Procellariiformes (e.g., albatrosses, shearwaters, and petrels) excrete fluids in excess of 5 percent salt from these glands. Petrels forcibly eject the fluids through the nostrils;

other species drip the fluids out of the internal or external nares. In marine mammals, the kidney is the main route for elimination of salt. Porpoises have highly developed kidneys to eliminate salt loads rapidly.

7.17 | Buoyancy Helps Aquatic Organisms Stay Afloat

Aquatic animals have adapted various mechanisms to help them stay afloat in water. Most aquatic animals inhabiting the oceans have densities very close to that of seawater. Because living tissues are generally denser (heavier) than water, larger animals cannot maintain buoyancy unless their bodies have lower-density areas that counter the higher density of most tissues.

Most fish have a gas or swim bladder (Figure 7.21a) that typically accounts for 5 to 10 percent of total body volume. These fish can control the degree of buoyancy by regulating the amount of gas in the bladder. Lungs in air-breathing animals sustain neutral buoyancy.

Some marine animals, such as the squid, maintain neutral buoyancy by replacing heavy chemical ions in the body fluids with lighter ions. Squids have body cavities in which lighter ammonium ions replace the heavier sodium ions. As a result, an equal amount of body fluid is less dense than the same volume of seawater. Another mechanism is increased storage of lipids (fats and oils). Lipids are less dense than seawater. Large amounts of

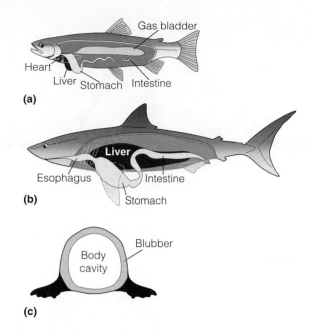

Figure 7.21 | Buoyancy adaptation in fish and seal. **(a)** Gas bladder in a fish. **(b)** Large, fat-filled liver of a shark. **(c)** Blubber surrounding the body of a seal.

lipids are present in fishes that lack swim bladders (such as sharks, mackerels, bluefish, and bonito). Lipid deposits are located in muscles, internal organs, and the body cavity (Figure 7.21b). In marine mammals, lipids are typically deposited as a layer of fat just below the skin (blubber; Figure 7.21c). Besides aiding in buoyancy, blubber functions as insulation and energy storage.

Maintaining neutral buoyancy in open-water environments releases organisms from many of the structural constraints imposed by gravity in terrestrial environments. Organisms such as the jellyfish, squid, and octopus quickly lose their graceful forms when removed from the water. Beached whales die from suffocation when they are no longer able to support their body weight through neutral buoyancy. It is no coincidence that the largest vertebrate and invertebrate organisms on Earth inhabit the oceans.

7.18 | Daily and Seasonal Light and Dark Cycles Influence Animal Activity

The major influence of light on animals is its role in timing daily and seasonal activities including feeding, food storage, reproduction, and migratory movements. The internal mechanisms in organisms used to control the periodicity of various functions or activities are **biological clocks.** An internal biological clock is fundamental to all living organisms, influencing hormones that play a role in the sleep cycle, metabolic rate, and body temperature.

Biological processes fluctuate in cycles, or rhythms, that range from minutes to months or even years. Biological processes that cycle in 24-hour intervals are called daily rhythms. When a daily rhythm results from a physiological response to the diurnal environmental

cycle, it is called a **circadian rhythm** (Latin *circa,* "about," and *dia,* "day"). Many involuntary (autonomic) processes of individual organisms exhibit a circadian rhythm, including the control of body temperature, cardiovascular function, melatonin (hormone related to the sleep–wake cycle) secretion, cortisol (primary stress hormone) secretion, and metabolism.

The circadian rhythm, with its sensitivity to light and dark, is the major mechanism for the biological clock—that timekeeper of physical and physiological activity in living things. Where is such a clock located in organisms? Its position must expose it to its time-setter: light.

Circadian rhythms are believed to include at least three elements: (1) *input pathway* that transmits environmental signals (light) to the clock, (2) *pacemaker* (clock), and (3) *output pathway* through which the pacemaker regulates the rhythms. In mammals, the pacemaker is the suprachiasmic nucleus (SCN), a distinct group of cells located in the hypothalamus region of the brain. The input pathway to the pacemaker in mammals begins with photoreceptors in the eye. The retina of the eyes contains "classical" photoreceptors as well as photo-responsive cells. These cells, which contain a photosensitive pigment, follow a pathway leading to the SCN. As the SCN receives information about illumination from the retina, it is interpreted and transmitted to the pineal gland, a small organ shaped like a pine cone (hence its name) and located in the epithalamus region of the brain. The pineal gland provides the output pathway through the secretion of melatonin. Melatonin is a structurally simple hormone that communicates information about the light environment to various parts of the body. Secretion of melatonin peaks at night and declines during the day. Ultimately, melatonin has the ability to entrain biological rhythms and has important effects on reproductive function in many animals.

The eye is the only photosensitive organ in mammals. In other vertebrates, such as birds and some lizards, the pineal gland is on the surface of the brain, directly under the skull. The pineal gland contains photoreceptors that function as the input pathway and allow the gland to function directly as the pacemaker. Some animal species have a parietal eye, a photosensory organ connected to the pineal gland (Figure 7.22). Parietal eyes are found in lizards, frogs, and lampreys as well as in some species of fish, such as tuna and pelagic sharks, where it is visible as a light-sensitive spot on top of the head. A poorly developed version of the parietal eye, often called the parapineal gland, occurs in salamanders.

Ecologists are particularly interested in the adaptive value of biological clocks. One adaptive value is that the biological clock gives the organism a time-dependent mechanism. It enables the organism to prepare ahead of time for periodic changes in the environment. Circadian rhythms help organisms with physical aspects of the environment other than light or dark. For example, a rise in humidity and a drop in temperature accompany the

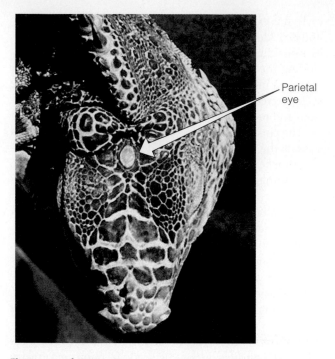

Figure 7.22 | Some animal species, such as the iguana pictured here, have a parietal eye—a photosensory organ connected to the pineal gland. Parietal eyes are found in lizards, frogs, and lampreys as well as some species of fish. In tuna and pelagic sharks, the eye is visible as a light-sensitive spot on top of the head.

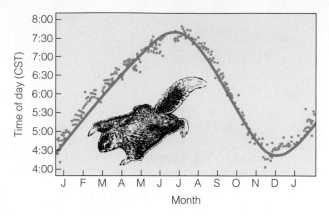

Figure 7.23 | Seasonal variation in the time of day when flying squirrels become active.

(Adapted from Decoursey 1960.)

transition from light to dark. Wood lice, centipedes, and millipedes, which lose water rapidly in dry air, spend the day in the darkness and damp under stones, logs, and leaves. At dusk they emerge, when the humidity of the air is more favorable. These animals show an increased tendency to escape from light as the length of time they spend in darkness increases. On the other hand, the strength of their response to low humidity decreases with darkness. Thus, these invertebrates come out at night into places too dry for them during the day, and they quickly retreat to their dark hiding places as light comes.

The circadian rhythms of many organisms relate to biotic aspects of their environment. Predators such as insectivorous bats must match their feeding activity to the activity rhythm of their prey. Moths and bees must seek nectar when flowers are open. Flowers must open when insects that pollinate them are flying. The circadian clock lets insects, reptiles, and birds orient themselves by the Sun's position. Organisms make the most economical use of energy when they adapt to the periodicity of their environment.

7.19 | Critical Day Lengths Trigger Seasonal Responses

In the middle and upper latitudes of the Northern and Southern Hemispheres, the daily periods of light and dark lengthen and shorten with the seasons (see Section 2.2). The activities of animals are geared to the changing seasonal rhythms of night and day. The flying squirrel (*Glaucomys volans*), for example, starts its daily activity with nightfall, regardless of the season. As the short days of winter turn to the longer days of spring, the squirrel begins its activity a little later each day (Figure 7.23).

Most animals of temperate regions have reproductive periods that closely follow the changing day lengths of the seasons. For most birds, the height of the breeding season is the lengthening days of spring; for deer, the mating season is the shortening days of fall.

The signal for these responses is **critical day length.** When the duration of light (or dark) reaches a certain proportion of the 24-hour day, it inhibits or promotes a photoperiodic response. Critical day length varies among organisms, but it usually falls somewhere between 10 and 14 hours. Through the year, plants and animals compare critical day length with the actual length of day or night and respond appropriately. Some organisms can be classified as **day neutral**; they are not controlled by day length, but by some other influence such as rainfall or temperature. Others are short-day or long-day organisms. **Short-day** organisms are those whose reproductive or other seasonal activity is stimulated by day lengths shorter than their critical day length. **Long-day** organisms are those whose seasonal responses, such as reproduction, are stimulated by day lengths longer than the critical day length.

Many organisms exhibit both long- and short-day responses. Because the same duration of dark and light occurs two times a year, in spring and fall, the organisms could get their signals mixed. For them, the distinguishing cue is the direction in which the critical day length is changing—that is, whether the critical day length is reached as long days become shorter, or as short days become longer.

Diapause, a stage of arrested growth over winter in insects of the temperate regions, is controlled by photoperiod. The time measurement in such insects is precise, usually between 12 and 13 hours of light. A quarter-hour difference in the light period can determine whether an insect goes into diapause or not. The shortening days of late summer and fall forecast the coming of winter and call for diapause. The lengthening days of late winter and early spring are signals for the insect to resume development, pupate, emerge as an adult, and reproduce.

Increasing day length induces spring migratory behavior, stimulates gonadal development, and brings on the reproductive cycle in birds. After the breeding season, the gonads of birds regress spontaneously. During this time, light cannot induce gonadal activity. The short days of early fall hasten the termination of this period. The progressively shorter days of winter then become stimulatory. The lengthening days of early spring bring the birds back into the reproductive stage.

In mammals, photoperiod influences activity such as food storage and reproduction, too. Consider, for example, such seasonal breeders as sheep and deer. Melatonin initiates their reproductive cycle. More melatonin is produced when it is dark, so these animals receive a higher concentration of melatonin as the days become shorter in the fall. This increase in melatonin reduces the sensitivity of the pituitary gland to negative feedback effects of hormones from the ovaries and testes. Lacking this feedback, the anterior pituitary releases pulses of another hormone (called *luteinizing hormone*) that stimulates growth of ova in the ovaries and sperm production in the testes.

Activities of animals through the year reflect this seasonal response to changing day length, referred to as circannual rhythms. The reproductive cycle of the white-tailed deer for example, begins in fall, and the young are born in spring when the highest-quality food for lactating mothers and young is available. In tropical Central America, home of many species of fruit-eating (frugivorous) bats, the reproductive periods track the seasonal production of food. The birth periods of frugivorous bats coincide with the peak period of fruiting. Young are born when both females and young will have adequate food. Insects and other arthropods reach their greatest biomass early in the rainy season in the Costa Rican forests. At that time, the insectivorous bats give birth to their young.

7.20 | Activity Rhythms of Intertidal Organisms Follow Tidal Cycles

Along the intertidal marshes, fiddler crabs (*Uca* spp.; the name refers to the enormously enlarged claw of the male, which he waves incessantly) swarm across the exposed mud of salt marshes and mangrove swamps at low tide. As high tide inundates the marsh, fiddler crabs retreat to their burrows, where they await the next low tide. Other intertidal organisms—from diatoms, green algae, sand beach crustaceans, and salt-marsh periwinkles to intertidal fish such as blennies (*Malacoctenus*) and cottids (*Cottidae*)—also respond to both daily and tidal cycles.

J. D. Palmer and his associates of the University of Massachusetts, Amherst, brought fiddler crabs into the laboratory and held them under constant temperature and light, devoid of tidal cues. Palmer found that the crabs exhibit the same tidal rhythm in their activity as they do in the marsh (Figure 7.24). This tidal rhythm

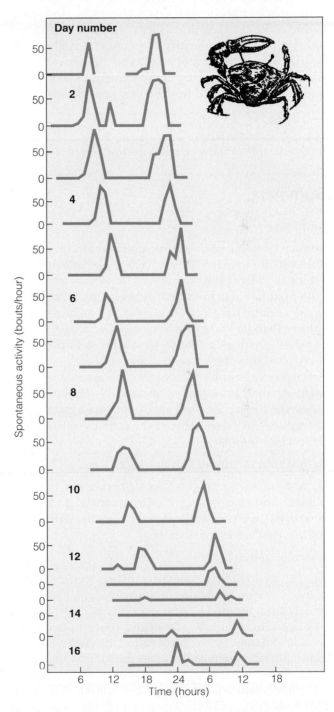

Figure 7.24 | Tidal rhythm of a fiddler crab in the laboratory in constant light at a constant temperature of 22°C for 16 days. Because the lunar day is 51 minutes longer than the solar day, the tides occur 51 minutes later each solar day; thus, peaks of activity appear to move to the right.

(Adapted from Palmer 1990.)

mimics the ebb and flow of tides every 12.4 hours, one-half of the lunar day of 24.8 hours, the interval between successive moonrises. Under the same constant conditions, fiddler crabs exhibit a circadian rhythm of changes in body color, turning dark by day and light by night.

Is the clock in this case unimodal, with a 12.4-hour cycle, or is it bimodal, with a 24.8-hour cycle, close to the period of the circadian clock? Does one clock keep a solar-day rhythm of approximately 24 hours and another clock keep a lunar-day rhythm of 24.8 hours? Palmer and his associates conducted a series of experiments to address this question. The evidence suggests that one solar-day clock synchronizes daily activities, while two strongly coupled lunar-day clocks synchronize tidal activity. Each lunar-day clock drives its own tidal peak. If one clock quits running in the absence of environmental cues, the other one still runs. This feature enables tidal organisms to synchronize their activities in a variable tidal environment. Day–night cycles reset solar-day rhythms, and tidal changes reset tidal rhythms. Even at the cellular level, organisms do not depend on one clock any more than most of us keep a single clock at home. Organisms have built-in redundancies. Such redundancies enable the various clocks to run at differing speeds, governing different processes with slightly differing periods.

Summary

Body Size | 7.1

Size has consequences for structural and functional relationships in animals; as such, it is a fundamental constraint on adaptation. For objects of similar shape, the ratio of surface area to volume decreases with size; that is, smaller bodies have a larger surface area relative to their volume than do larger objects of the same shape. The decreasing surface area relative to volume with increasing body size limits the transfer of materials and energy between the organism's interior and its exterior environment. Animals have evolved an array of solutions to increase the surface area and enable adequate exchange of energy and materials between the interior cells and the external environment.

Acquisition of Energy and Nutrients | 7.2

To acquire energy and nutrients, herbivores consume plants; carnivores consume other animals; and omnivores feed on both plant and animal tissues. Detritivores feed on dead organic matter.

Nutritional Needs | 7.3

Directly or indirectly, animals get their nutrients from plants. Low concentrations of nutrients in plants can adversely affect the growth, development, and reproduction of plant-eating animals. Herbivores convert plant tissue to animal tissue. Among plant eaters, the quality of food, especially its protein content and digestibility, is crucial. Carnivores must secure a sufficient quantity of nutrients already synthesized from plants and converted into animal flesh.

Mineral Requirements | 7.4

Three essential nutrients influencing the distribution, behavior, growth, and reproduction of grazing animals are sodium, calcium, and magnesium. Grazers seek these nutrients from mineral licks and from the vegetation they eat.

Oxygen Uptake | 7.5

Animals generate energy by breaking down organic compounds primarily through aerobic respiration, which requires oxygen. Differences between terrestrial and aquatic animals in their means of acquiring oxygen reflect the availability of oxygen in the two environments. Most terrestrial animals have some form of lungs, whereas most aquatic animals use gills to transfer gases between the body and the surrounding water.

Regulation of Internal Conditions | 7.6

To confront daily and seasonal environmental changes, organisms must maintain some equilibrium between their internal and external environment. Homeostasis is the maintenance of a relatively constant internal environment in a variable external environment through negative feedback responses. Through various sensory mechanisms, an organism responds physiologically and/or behaviorally to maintain an optimal internal environment relative to its external environment. Doing so requires an exchange between the internal and external environments.

Energy Exchange | 7.7

Animals maintain a fairly constant internal body temperature, known as the body core temperature. They use behavioral and physiological means to maintain their heat balance in a variable environment. Layers of muscle fat and surface insulation of scales, feathers, and fur insulate the animal body core against environmental temperature changes. Terrestrial animals face a more dynamic and often more threatening thermal environment than do aquatic animals.

Thermal Regulation | 7.8

Animals fall into three major groups relative to temperature regulation: poikilotherms, homeotherms, and heterotherms. Poikilotherms, so named because they have variable body temperatures influenced by ambient temperatures, are ectothermic. Animals that depend on internally produced heat to maintain body temperatures are endothermic. They are called homeotherms because they maintain a rather constant body temperature independent of the environment. Many animals are heterotherms that function either as endotherms or ectotherms, depending on external circumstances.

Poikilotherms | 7.9

Poikilotherms gain heat from and lose heat to the environment. Poikilotherms have low metabolic rates and high thermal conductance. Environmental temperatures control their rates of metabolism. Poikilotherms are active only when environmental temperatures are moderate; they are sluggish when temperatures are cool. They have, however, upper and lower limits of tolerable temperatures. Most aquatic poikilotherms maintain no appreciable difference between body temperature and water temperature.

Poikilotherms resort to behavioral means of regulating body temperature. They exploit variable microclimates by moving into warm, sunny places to heat up and by seeking shaded places to cool. Many amphibians move in and out of water. Insects and desert reptiles raise and lower their bodies to reduce or increase conductance from the ground or convective cooling. Desert animals resort to shade or spend the heat of day in underground burrows.

Homeotherms | 7.10

Homeotherms maintain high internal body temperature by oxidizing glucose and other energy-rich molecules. They have high metabolic rates and low thermal conductance. Body insulation of fat, fur, feathers, scales, and fur-like covering on many insects reduces heat loss from the body. A few desert mammals employ heavy fur to keep out desert heat and cold. When insulation fails during cold, many homeotherms resort to shivering and burning fat reserves. For homeotherms, evaporative cooling by sweating, panting, and wallowing in mud and water is an important way of dissipating body heat.

Trade-offs in Thermal Regulation | 7.11

The two approaches to maintaining body temperature, ectothermy and endothermy, involve trade-offs. Unlike poikilotherms, homeotherms are able to remain active regardless of environmental temperatures. For homeotherms, a high rate of aerobic metabolism comes at a high energy cost. This cost places a lower limit on body size. Due to the low metabolic cost of ectothermy, poikilotherms can curtail metabolic activity in times of food and water shortage and temperature extremes. Their low energy demands enable some terrestrial poikilotherms to colonize areas of limited food and water.

Heterotherms | 7.12

Based on environmental and physiological conditions, heterotherms take on the characteristics of endotherms or ectotherms. Some normally homeothermic animals become ectothermic and drop their body temperature under certain environmental conditions. Many poikilotherms, notably insects, need to increase their metabolic rate to generate heat before they can take flight. Most accomplish this feat by vibrating their wings or wing muscles or by basking in the sun. After flight, their body temperatures drop to ambient temperatures.

Torpor | 7.13

During environmental extremes, some animals enter a state of torpor to reduce the high energy costs of staying warm or cool. They slow their metabolism, heartbeat, and respiration and lower their body temperature. Birds such as hummingbirds and mammals such as bats undergo daily torpor, the equivalent of deep sleep, without the extensive metabolic changes of seasonal torpor. Hibernation (seasonal torpor over winter) involves a complete rearrangement of metabolic activity to run at a very low level. Heartbeat, breathing, and body temperature are all greatly reduced.

Unique Physiological Means to Maintain Thermal Balance | 7.14

Many homeotherms and heterotherms employ countercurrent circulation, the exchange of body heat between arterial and venous blood reaching the extremities. This exchange reduces heat loss through body parts or cools blood flowing to such vital organs as the brain.

Some desert mammals use hyperthermia to reduce the difference between body and environmental temperatures. They store up body heat by day, then release it to the cool desert air by night. Hyperthermia reduces the need for evaporative cooling and thus conserves water. Some cold-tolerant poikilotherms use supercooling, the synthesis of glycerol in body fluids, to resist freezing in winter. Supercooling takes place when the body temperature falls below freezing without freezing body fluids. Some intertidal invertebrates survive the cold by freezing, then thawing with warmer temperatures.

Water Balance for Terrestrial Animals | 7.15

Terrestrial animals must offset water loss from evaporation, respiration, and waste excretion by consuming and/or conserving water. Terrestrial animals gain water by drinking, eating, and producing metabolic water. Animals of arid regions may reduce water loss by becoming nocturnal, producing highly concentrated urine and feces, becoming hyperthermic during the day, using only metabolic water, and tolerating dehydration.

Water Balance in Aquatic Animals | 7.16

Aquatic animals need to prevent the uptake of or rid themselves of excess water. Freshwater fish maintain osmotic balance by absorbing and retaining salts in special cells in the body and by producing copious amounts of watery urine. Many marine invertebrates maintain in their body cells the same osmotic pressure as that in seawater. Marine fish secrete excess salt and other ions through kidneys or across gill membranes.

Buoyancy | 7.17

Aquatic animals use a wide variety of mechanisms to maintain buoyancy, including gas-filled bladders, lungs, and lipid deposits.

Daily and Seasonal Changes in Day Length | 7.18

Living organisms, except bacteria, have an innate circadian rhythm of activity and inactivity. Under natural conditions, the circadian rhythm is set to the 24-hour day by external time cues, notably light and dark (day and night). This setting synchronizes the activity of plants and animals with the environment. The onset and cessation of activity depend on whether the organisms are light-active or dark-active.

Circadian rhythms operate the biological clocks in the cells of plants and in the brains of multicelled animals. Animals produce more of a special hormone,

melatonin, in the dark than in the light. Thus, melatonin becomes a device for measuring day length.

Critical Day Length | 7.19

Seasonal changes in activity are based on day length. Lengthening days of spring and the shortening days of fall stimulate migration in animals and reproduction and food storage as well. These seasonal or circannual rhythms bring living organisms into a reproductive state at the time of year when offspring probability of survival is highest. The rhythms synchronize within a population such activities as mating and migration.

Tidal Cycles | 7.20

Intertidal organisms are influenced by two environmental rhythms: day length and tidal cycles of 12.4 hours. Intertidal organisms appear to have two lunar-day clocks that set tidal rhythms and one solar-day clock that sets circadian rhythms.

Study Questions

1. What constraints are imposed by a diet of plants as compared to one of animal tissues?
2. What is homeostasis?
3. In Figure 7.16, why does mass-specific metabolic rate (metabolic rate per unit weight) of mammals increase with decreasing body mass?
4. Why are the largest species of poikilotherms found in the tropical and subtropical regions?
5. How might you expect the average size of mammal species to vary from the tropics to the polar regions? Why?
6. Why might it be easier to capture a snake in the early morning rather than the afternoon?
7. Why do homeotherms typically have a greater amount of body insulation than do poikilotherms?
8. What behaviors help poikilotherms maintain a fairly constant body temperature during their season of activity?
9. How does the size and shape of an animal's body influence its ability to exchange heat with the surrounding environment?
10. Contrast the problem of maintaining water balance for freshwater and marine fishes.
11. How does supercooling enable some insects, amphibians, and fish to survive freezing conditions?
12. How does countercurrent circulation work, and what function does it serve?
13. Distinguish between hibernation and torpor.
14. Consider a population of fish living below a power plant that is discharging heated water. The plant shuts down for 3 days in the winter. How would that affect the fish?
15. Define circadian rhythm.
16. How does day length influence the seasonal activity of plants and animals?
17. What is the adaptive value of seasonal synchronization of animal activity?

Further Readings

Bonner, J. T. 2006. *Why size matters*. Princeton: Princeton Univ. Press.

  This short book provides an excellent overview of the constraints imposed by body size in the evolution and ecology of animals from one of the leading figures in the field.

French, A. R. 1988. The patterns of mammalian hibernation. *American Scientist* 76:569–575.

  This article provides an excellent, easy-to-read, and well-illustrated overview of the concept of hibernation.

Heinrich, B. 1996. *The thermal warriors: Strategies of insect survival*. Cambridge: Harvard University Press.

  This enjoyable book describes the variety of strategies insects use to heat their bodies. It is full of strange and wonderful examples of evolution in the world of insects.

Johnson, C. H., and J. W. Hastings. 1986. The elusive mechanisms of the circadian clock. *American Scientist* 74:29–36.

  This article discusses the history of the search for the mechanisms controlling circadian rhythm and gives an easily understood overview of our current understanding.

Lee, R. E., Jr. 1989. Insect cold-hardiness: To freeze or not to freeze. *Bioscience* 39:308–313.

This article offers an excellent overview of the diversity of adaptations that represent evolutionary solutions to the problems faced by insects in dealing with seasonal variations in temperature. It includes a good discussion of supercooling and cold hardening in insects.

Palmer, J. D. 1996. Time, tide, and living clocks of marine organisms. *American Scientist* 84:570–578.

An easy-to-read, well-illustrated article on the adaptation of marine organisms to the tidal cycle.

Schmidt-Neilsen, K. 1997. *Animal physiology: Adaptation and environment*. 5th ed. New York: Cambridge University Press.

This comprehensive text gives you more information about animal physiology and the range of adaptations that allow animals to cope with various terrestrial and aquatic environments.

Storey, K. B., and J. M. Storey. 1996. Natural freezing survival in animals. *Annual Review of Ecology and Systematics* 27:365–386.

An excellent if somewhat technical review of the adaptations that allow animals to cope with freezing temperatures.

Takahashi, J. S., and M. Hoffman. 1995. Molecular biological clocks. *American Scientist* 83:158–165.

A great, readable article that provides an overview of the mechanisms involved in biological clocks.

CHAPTER 8

Life History Patterns

A female Alaskan brown bear (*Ursus arctos*) with her three cubs. Litter sizes range from 1 to 4 cubs, born between the months of January and March. The cubs remain with their mothers for at least 2 1/2 years, so the most frequently females can breed is every 3 years.

An organism's life history is its lifetime pattern of growth, development, and reproduction. Life histories combine a rich array of adaptations relating to physiology, morphology, and behavior. They involve adaptations to the prevailing physical environment, such as those discussed in Chapters 6 and 7. Life histories also include adaptations relating to the organism's interactions with other organisms—the biological environment. Perhaps the most important of these life history characteristics are the adaptations relating to reproduction. The true measure of an organism's reproductive success is fitness (see Chapter 5). Evolution is the product of differential reproduction of individuals and the process of natural selection.

If reproductive success (the number of offspring that survive to reproduce) is the measure of fitness, imagine designing an organism with the objective of maximizing its fitness. It would reproduce as soon as possible after birth, and it would reproduce continuously, producing large numbers of large offspring that it would nurture and protect. Yet such an organism is not possible. Each individual has a limited amount of resources that can be allocated to specific tasks, and allocation to one task reduces the resources available for the others. Allocation to reproduction reduces the amount of resources available for growth. Should an individual reproduce early in life or delay reproduction? For a given allocation of resources to reproduction, should an individual produce many small or fewer, but larger offspring? Thus, organisms face trade-offs in life history characteristics related to reproduction, just as they do in the adaptations related to carbon, water, and energy balance discussed in Chapters 6 and 7. These trade-offs involve modes of reproduction; age at reproduction; allocation to reproduction; timing of reproduction; number and size of eggs, young, or seeds produced; and parental care. These trade-offs are imposed by constraints of physiology, energetics, and the prevailing physical and biotic environment—the organism's habitat. In this chapter, we explore these trade-offs and the diversity of solutions that have evolved to assure success at the one essential task for continuation of life on our planet: reproduction.

8.1 | Reproduction May Be Sexual or Asexual

In Chapter 5, we explored how genetic variation among individuals within a population arises from the shuffling of genes and chromosomes in sexual reproduction. In sexual reproduction between two diploid individuals, the individuals produce haploid (one-half the normal number of chromosomes) gametes—egg and sperm—that combine to form a diploid cell, or zygote, that has a full complement of chromosomes. Because the possible number of gene recombinations is enormous, recombination is an immediate and major source of genetic vari-

Figure 8.1 | The freshwater hydra reproduces asexually by budding.

ability among offspring. However, not all reproduction is sexual. Many organisms reproduce asexually. Asexual reproduction produces offspring without the involvement of egg and sperm. It takes many forms; but in all cases, the new individuals are genetically the same as the parent. Strawberry plants spread by rhizomes, the underground stems from which new roots and vertical stems sprout. The one-celled paramecium reproduces by dividing in two. Hydras, cnidarians that live in freshwater (Figure 8.1), reproduce by budding, a process by which a bud pinches off as a new individual. In spring, wingless female aphids emerge from eggs that have survived the winter and give birth to wingless females through **parthenogenesis** (Greek *parthenos*, "virgin"; Latin *genesis*, "to be born"), a form of asexual reproduction in which the ovum develops without fertilization by a male.

Organisms that rely heavily on asexual reproduction occasionally revert to sexual reproduction. Many of these reversions to sexual reproduction are induced by an environmental change at some time in the organism's life cycle. During warmer parts of the year, hydras turn to sexual reproduction to produce eggs that lie dormant over the winter and from which young hydras emerge in the spring to mature and reproduce asexually. After giving birth to several generations of wingless females, aphids produce a generation of winged females. These winged females migrate to different food plants, become established, and reproduce parthenogenetically. Later in the summer, these same females move back to the original food plants and give birth to true sexual forms, winged males and females that lay eggs rather than give birth to young. After mating, each female lays one or more overwintering eggs. Because the males produce sperm with the X chromosome only, the eggs that hatch in spring will produce wingless females that give birth to female young.

Each form of reproduction, asexual and sexual, has its trade-offs. The ability to survive, grow, and reproduce indicates that an organism is well adapted to the prevailing environmental conditions. Asexual reproduction produces offspring that are genetically identical to the parent and are therefore well adapted to the local environment. Because all individuals are capable of reproducing, asexual reproduction results in a potential for high population growth. However, the cost of asexual reproduction is the loss of genetic recombination that increases variation among offspring. Low genetic variability among individuals in the population means that the population responds more uniformly to a change in environmental conditions than does a sexually reproducing population. If a change in environmental conditions is detrimental, the effect on the population can be catastrophic.

In contrast, the mixing of genes and chromosomes that occurs in sexual reproduction produces genetic variability to the degree that each individual in the population is genetically unique. This genetic variability produces a broader range of potential responses to the environment, increasing the probability that some individuals will survive environmental changes. But this variability comes at a cost. Only half the genetic material of each individual offspring comes from each parent. Compared with asexually produced offspring, which have only the genes of their single parent, each progeny

from a sexually reproducing pair of individuals contributes only one half as much to the evolutionary fitness of either parent (see Chapter 5). In addition, sexual reproduction requires specialized reproductive organs that, aside from reproduction, have no direct relationship to an individual's survival. Production of gametes (egg and sperm), courtship activities, and mating are energetically expensive. The expense of reproduction is not shared equally by both sexes. The eggs (ovum) produced by females are much larger and energetically much more expensive than sperm produced by males. As we will examine in the following sections, this difference in energy investment in reproduction between males and females has important implications in the evolution of life history characteristics.

8.2 | Sexual Reproduction Takes a Variety of Forms

Sexual reproduction takes a variety of forms. The most familiar involves separate male and female individuals. It is common to most animals. Plants with that characteristic are called **dioecious** (Greek *di*, "two," and *oikos*, "home"); examples are holly trees (*Ilex* spp.) and stinging nettle (*Urtica* spp.; Figure 8.2a).

In some species, individual organisms possess both male and female organs. They are **hermaphroditic**

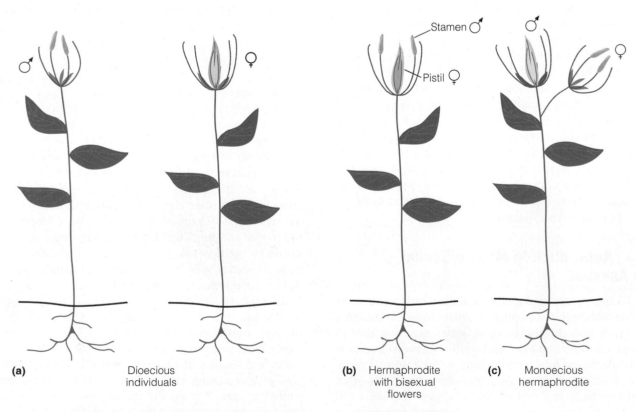

(a) Dioecious individuals

(b) Hermaphrodite with bisexual flowers

(c) Monoecious hermaphrodite

Figure 8.2 | Floral structure in **(a)** dioecious plant (separate male and female individuals), **(b)** hermaphroditic plant possessing bisexual flowers, and **(c)** monoecious plant possessing separate male and female flowers.

Figure 8.3 | Hermaphroditic earthworms mating.

Figure 8.4 | Parrotfishes (Scaridae) that inhabit coral reefs exhibit sex change. When a large dominant male mating with a harem of females is removed (by a predator or experimenter), within days, the largest female in the harem becomes a dominant male and takes over the missing male's function.

(Greek *hermaphroditos*). In plants, individuals can be hermaphroditic by possessing bisexual flowers with male organs (stamens) and female organs (ovaries), such as lilies (*Lilium* spp.) and buttercups (*Ranunculus* spp.; Figure 8.2b). Such flowers are termed *perfect*. Asynchronous timing of the maturation of pollen and ovules often occurs, reducing the chances of self-fertilization. Other plants are **monoecious** (Greek *mon*, "one," and *oikos*, "house"). They possess separate male and female flowers on the same plant, as do birch (*Betula* spp.) and hemlock (*Tsuga* spp.) trees (Figure 8.2c). Such flowers are called *imperfect*. This strategy of sexual reproduction can be an advantage in the process of colonization. A single self-fertilized hermaphroditic plant can colonize a new habitat and reproduce, establishing a new population, as do self-fertilizing annual weeds that colonize disturbed sites.

Among animals, hermaphroditic individuals possess the sexual organs of males and females (both testes and ovaries), a condition common in invertebrates such as earthworms (Figure 8.3). In these species, referred to as **simultaneous hermaphrodites,** the male organ of one individual is mated with the female organ of the other, and vice versa. The result is that a population of hermaphroditic individuals has the potential to produce twice as many offspring as a population of unisexual individuals does.

Some species are **sequential hermaphrodites.** Animals—such as some mollusks and echinoderms—and some plants may be males first during one part of their life cycle and females in another part. Some fish may be females first, then males. Sex change usually takes place as individuals mature or grow larger. A change in the sex ratio of the population stimulates sex change among some animals. Removing individuals of the other sex initiates sex reversal among some species of marine fish (Figure 8.4). Among some coral reef fish, removal of females from a social group stimulates males to change sex and become females. In other species, removal of males stimulates a one-to-one replacement of males by sex-reversing females. Among the mollusks, the Gastropoda (snails and slugs) and Bivalvia (clams and mussels) have sex-changing species. Almost all of these species change from male to female.

Plants also can undergo sex change. One such plant is jack-in-the-pulpit (*Arisaema triphyllum*), a clonal herbaceous plant found in the woodlands of eastern North America (Figure 8.5). Jack-in-the-pulpit may produce male flowers one year, an asexual vegetative shoot the next, and female flowers the next. Over its life span, a jack-in-the-pulpit may produce both sexes as well as an asexual vegetative shoot, but in no particular sequence. Usually an asexual stage follows a sex change. Sex change in jack-in-the-pulpit appears to be triggered by the large

Figure 8.5 | The jack-in-the-pulpit becomes asexual, male, or female depending on energy reserves. The plant gets its name from the flower stem, or spadix, enclosed in a hoodlike sheath. This fruiting plant is the female stage.

energy cost of producing female flowers. Jack-in-the-pulpit plants generally lack sufficient resources to produce female flowers in successive years; male flowers and pollen are much cheaper to produce than female flowers and subsequent fruits.

8.3 | Mating Systems Describe the Pairing of Males and Females

On a brushy rise of ground at the edge of a forest, a pair of red foxes (*Vulpes vulpes*) has dug a deep burrow. Inside are the female and her litter of pups. Outside at the burrow entrance are scattered bits of fur and bones, the leftovers of meals carried to the den by the male for his mate and pups. Back in the woods, a female white-tailed deer (*Odocoileus virginianus*) has hidden a young dappled fawn in a patch of ferns on the forest floor. The fawn's father is gone and has no knowledge of this offspring's existence. His only interaction with the mother was during the previous fall, when she was one of several females he mated with during a short period of several days.

The fox and the deer represent two extremes in what is referred to as the **mating system**—the pattern of mating between males and females in a population (also see Chapter 5). The structure of mating systems ranges from **monogamy,** which involves the formation of a lasting pair bond between one male and one female, to **promiscuity,** in which males and females mate with one or many of the opposite sex and form no pair bond.

Monogamy is most prevalent among birds and rare among mammals, except for several carnivores such as foxes (*Vulpes* spp.) and weasels (*Mustela* spp.), and a few herbivores such as the beaver (*Castor* spp.), muskrat (*Ondatra zibethica*), and prairie vole (*Microtus ochrogaster*). Monogamy exists mostly among species in which cooperation by both parents is needed to raise the young successfully. Most species of birds are seasonally monogamous (during the breeding season), because most young birds are helpless at hatching and need food, warmth, and protection. The avian mother is no better suited than the father to provide these needs. Instead of seeking other mates, the male can increase his fitness more by continuing his investment in the young. Without him, the young carrying his genes may not survive. Among mammals, the situation is different. The females lactate (produce milk), providing food for the young. Males often can contribute little or nothing to the survival of the young, so it is to their advantage to mate with as many females as possible. Some exceptions are the foxes, wolves, and other canids, among which the male provides for the female and young and defends the territory (area defended for exclusive use and access to resources; see Section 11.10 for discussion). Both males and females regurgitate food for the weaning young.

Monogamy, however, has another side. Among many species of monogamous birds, the female or male may "cheat" by engaging in extra-pair copulations while maintaining the reproductive relationship with the primary mate and caring for the young. By engaging in extra-pair relationships, the female may increase her fitness by rearing young sired by two or more males. The male increases his fitness by producing offspring with several females.

Polygamy is the acquisition by an individual of two or more mates. It can involve one male and several females or one female and several males. A pair bond exists between the individual and each mate. The individual having multiple mates (be it male or female) is generally not involved in caring for the young. Freed from parental duty, the individual can devote more time and energy to competition for more mates and resources. The more unevenly such crucial resources as food or quality habitat are distributed, the greater the opportunity for a successful individual to control the resource and several mates.

The number of the opposite sex an individual can monopolize depends on the degree of synchrony in sexual receptivity. For example, if females in the population are sexually active for only a brief period, as with the white-tailed deer, the number a male can monopolize is limited. However, if females are receptive over a long period of time, as with elk (*Cervus elaphus*), the size of a harem a male can control depends on the availability of females and the number of mates the male has the ability to defend.

Environmental and behavioral conditions result in various types of polygamy. In **polygyny,** an individual male pairs with two or more females. In **polyandry,** an individual female pairs with two or more males. Polyandry is interesting because it is the exception rather than the rule. This system is best developed in three groups of birds, the jacanas (Jacanidae; Figure 8.6), phalaropes (*Phalaropus* spp.), and some sandpipers (Scolopacidae). The female competes for and defends

Figure 8.6 | An example of polyandry. The male African jacana (*Actophilornis africanus*) is shown defending the young. After the female lays a clutch, the male incubates the eggs and cares for the young while the female seeks additional mates.

resources essential for the male. In addition, females compete for available males. As in polygyny, this mating system depends on the distribution and defensibility of resources, especially quality habitat. The female produces multiple clutches of eggs, each with a different male. After the female lays a clutch, the male begins incubation and becomes sexually inactive.

The nature and evolution of male–female relationships are influenced by environmental conditions, especially the availability and distribution of resources and the ability of individuals to control access to resources. If the male has no role in feeding and protecting young and defends no resources available to them, the female gains no advantage by remaining with him. Likewise, the male gains no increase in fitness by remaining with the female. If the habitat were sufficiently uniform, so that little difference existed in the quality of territories held by individuals, selection would favor monogamy because female fitness in all habitats would be nearly the same. However, if the habitat is diverse, with some parts more productive than others, competition may be intense, and some males will settle on poorer territories. Under such conditions, a female may find it more advantageous to join another female in the territory of the male defending a rich resource than to settle alone with a male in a poorer territory. Selection under those conditions will favor a polygamous relationship, even though the male may provide little aid in feeding the young.

8.4 | Acquisition of a Mate Involves Sexual Selection

The flamboyant plumage of the peacock (Figure 8.7) presented a troubling problem for Charles Darwin. Its tail feathers are big and clumsy and require a considerable allocation of energy to grow. They are also very conspicuous and present a hindrance when a peacock is

Figure 8.7 | Male peacock in courtship display.

Figure 8.8 | This bull elk is bugling a challenge to other males in a contest to control a harem.

trying to escape predators. In the theory of natural selection (see Chapter 5), what could account for the peacock's tail? Of what possible benefit could it be?

In his book *The Descent of Man and Selection in Relation to Sex*, published in 1871, Darwin observed that the elaborate and often outlandish plumage of birds as well as the horns, antlers, and large size of polygamous males seemed incompatible with natural selection. To explain why males and females of the same species often differ greatly in body size, ornamentation, and color (referred to as *sexual dimorphism*), Darwin developed a theory of sexual selection. He proposed two processes to account for these differences between the sexes: intrasexual selection and intersexual selection.

Intrasexual selection involves male-to-male (or in some cases, female-to-female) competition for the opportunity to mate. It leads to exaggerated secondary sexual characteristics, such as large size, aggressiveness, and organs of threat such as antlers and horns (Figure 8.8) that aid in competition for access to mates.

Intersexual selection involves the differential attractiveness of individuals of one sex to another (see Field Studies: Alexandra L. Basolo). In the process of intersexual selection, the targets of selection are characteristics in males such as bright or elaborate plumage used in sexual displays, as well as the elaboration of some of the same characteristics related to intrasexual selection (such as horns and antlers). It is a form of assortative mating (see Section 5.6), where the female selects a mate based on specific phenotypic characteristics. There is intense rivalry among males for female attention. In the end, the female determines the winner, selecting an individual as a mate. The result is an increase in relative fitness for those males that are chosen, shifting the distribution of male phenotypes in favor of the characteristics on which female choice is based (see Chapter 5). But do such characteristics as bright coloration, elaborate plumage, and size

School of Biological Sciences, University of Nebraska–Lincoln, Lincoln, Nebraska

The elaborate and often flamboyant physical traits exhibited by males of many animal species—bright coloration, exceedingly long feathers or fins—have always presented a dilemma to the traditional theory of natural selection. Because females in the process of mate selection often favor these male traits, sexual selection (see Section 8.4) will reinforce these characteristics. However, male investment in these traits may also reduce the amount of energy available for other activities that are directly related to individual fitness, such as reproduction, foraging, defense, predator avoidance, and growth. The effect of such trade-offs in energy allocation on the evolution of animal traits is the central focus of ecologist Alexandra Basolo's research, which is changing how behavioral ecologists think about the evolution of mate selection.

Basolo's research focuses on the small freshwater fishes of the genus *Xiphophorus* that inhabit Central America. One group of species within this genus, the swordtail fish, exhibits a striking sexual dimorphism in the structure of the caudal fin. Males have a colorful, elongated caudal appendage, termed the *sword* (Figure 1),

that is absent in females. This appendage appears to play no role other than as a visual signal to females in the process of mate selection. To test the hypothesis that this trait results partly from female choice (intersexual selection), Basolo undertook a series of laboratory experiments to determine if females exhibited a preference for male sword length. Her test subject was the green swordtail, *Xiphophorus helleri*, shown in Figure 1. These experiments allowed females to choose between a pair of males differing in sword length. Five tests with different pairs of males were conducted in which the sword differences between paired males varied. Female preference was measured by scoring the amount of time a female spent in association with each male.

Results of the experiments revealed that females preferred males with longer swords. The greater the difference in sword length between two males, the greater was the difference in time that the female spent with them (Figure 2). The results suggest that sexual selection through female choice will influence the relative fitness of males. The benefit of having a long sword is the increased probability of mating. But what is the cost? Locomotion accounts for a large part of the energy budget of fish, and the elongated caudal fin (sword) of the male swordtails may well influence the energetic cost of

Figure 1 | The green swordtail, *Xiphophorus helleri*. Males (shown in the photo) have a colorful, elongated appendage on the lower caudal fin (tail), termed the *sword*, which functions as a visual cue to attract females during courtship.

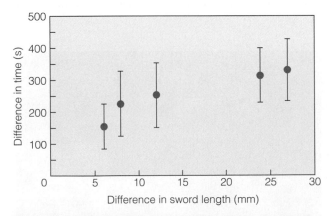

Figure 2 | Relationship between the difference in sword length between the two test males and the difference in the time spent with the male having the longer sword.

(Adapted from Basolo 1990b.)

swimming. The presence of the sword increases mating success (via female choice) but may well negatively affect swimming activities.

To evaluate the costs associated with sword length, Basolo undertook a series of experiments using another species of swordtail, the Montezuma swordtail (*X. montezumae*) found in Mexico. Like the green swordtail, the Montezuma swordtail males have an asymmetric caudal fin as a result of an extended sword, and the presence of this sword increases mating success. The experiments were designed to quantify the metabolic costs of the sword fins during two types of swimming—routine and courtship—for males with and without sword fins. Males were chosen from the population having average-length sword fins. For some of these males, the sword was surgically removed (excised). Comparisons were then made between males with and without swords for both routine (no female present) and courtship (female present) swimming. Male courtship behavior involves a number of active maneuvers. Routine swimming by males occurs in the absence of females, whereas the presence of females elicits courtship–swimming behavior.

Basolo placed test males into a respirometric chamber—a glass chamber instrumented to measure the oxygen content of the water continuously. For a trial where a female was present, the female was suspended in the chamber in a cylindrical glass tube having a separate water system. During each trial, water was sampled from the chamber for oxygen content to determine the rate of respiration. Higher oxygen consumption indicates a higher metabolic cost (respiration rate).

Results of the experiments show a significant energy cost associated with courtship behavior (Figure 3). A 30 percent increase in net cost was observed when females were present for both groups (males with and without swords) as a result of increased courtship behavior. However, the energy cost for males with swords was significantly higher than that for males without swords for both routine and courtship swimming behavior (Figure 3). Thus, although sexual selection via female choice favors long swords, males with longer swords experience higher metabolic costs during swimming, suggesting that sexual and natural selection have opposing effects on sword evolution.

The cost of a long sword to male swordtails extends beyond the energetics of swimming. Other studies have shown that more conspicuous males are more likely to be attacked by predators than are less conspicuous individuals. In fact, in green swordtail populations that occur sympatrically (together) with predatory fish, the average sword

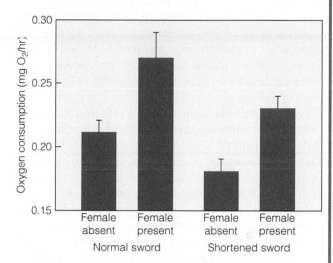

Figure 3 | Mean oxygen consumption for males with intact (normal) swords and shortened (excised) swords in the absence and presence of females. Female-absent results represent routine swimming. Female-present results represent male courtship swimming.
(Adapted from Basolo 2003.)

length of males in the population is significantly shorter than in populations where predators are not present.

Despite the cost, both in energy and probability of survival, the sword fin of the male swordtails confers an advantage in the acquisition of mates that must offset the energy and survival costs in terms of natural selection.

Bibliography

Basolo, A. L. 1990a. Female preference predates the evolution of the sword in swordtail fish. *Science* 250:808–810.

Basolo, A. L. 1990b. Female preference for male sword length in the green swordtail, *Xiphophorus helleri*. *Animal Behaviour* 40:332–338.

Basolo, A. L., and G. Alcaraz. 2003. The turn of the sword: Length increases male swimming costs in swordtails. *Proceedings of the Royal Society of London* 270:1631–1636.

1. In her experiments examining the costs and benefits of swords on the caudal fins of male fish, did Prof. Basolo actually quantify differences in fitness associated with this characteristic?
2. Often, sexual selection favors characteristics that appear to reduce the probability of survival for the individual. Doesn't this run counter to the idea of natural selection presented in Chapter 2?

Go to **QUANTIFYit!** at **www.ecologyplace.com** to perform regression analysis.

really influence the selection of males by females of the species?

Marion Petrie of the University of Newcastle, England, conducted some experiments to examine intersexual selection in peacocks (*Pavo cristatus*). She measured the tail length of male peacocks chosen by females as mates over the breeding season. Her results show that females selected males with larger tail feathers. She then removed tail feathers from a group of large-tailed males and found that reduced tail size led to a reduction in mating success. In the case of the peacock's tail, size does matter.

However, tail size itself is not what is important; it is what tail size implies about the individual. The large, colorful, and conspicuous tail makes the male more vulnerable to predation, or in many other ways reduces the male's probability of survival. A male that can carry these handicaps and survive shows his health, strength, and genetic superiority. Females showing preference for males with large tail feathers will produce offspring that will carry genes for high viability. Thus, the driving force behind the evolution of exaggerated secondary sexual characteristics in males is selection by females. In fact, the offspring of female peacocks that mated with males having large tail feathers had higher rates of survival and growth than did the offspring of those paired with males having short tails. A similar mechanism may be at work in the selection of male birds with bright plumage. One hypothesis proposes that only healthy males can develop bright plumage. There is evidence from some species that males with low parasitic infection have the brightest plumage. Females selecting males based on differences in the brightness of plumage are in fact selecting males that are the most disease resistant, therefore increasing their fitness (see Section 15.7 for detailed discussion).

8.5 | Females May Acquire Mates Based on Resources

A female exhibits two major approaches in choosing a mate. In the sexual selection just discussed, the female selects for phenotypic characteristics such as exaggerated plumage or displays that are indirectly related to the male's health and quality as a mate. The second approach is that the female can base her choice on the potential mate's ability to provide access to essential resources, such as habitat or food, that will function to improve her fitness.

For monogamous females, the criterion for mate selection often appears to be acquisition of a resource, usually a high-quality habitat or territory defended by the male (see Section 11.10). Does the female select the male and accept the territory that goes with him, or does she select the territory and accept the male that goes with it? There is some evidence from laboratory and field studies that female songbirds base their choice, in part, on the variety of the male's song. In aviary studies, female great

tits (*Parus major*) were more receptive of males with more varied or elaborate songs (Figure 8.9). In the field, male sedge warblers (*Acrocephalus schoenobaenus*) with more complex song repertoires appear to hold the higher-quality territories, so the more complex song may convey that fact to the female. None of these studies, however, determined the fitness of females attracted to these males.

Among polygamous species, the question becomes more complex. In those cases in which females acquire a resource along with the male, the situation is similar to that of monogamous relationships. Among polygamous birds, for example, the females show strong preference for males with high-quality territories. On territories with superior nesting cover and an abundance of food, females can attain reproductive success even if they share the territory and the male with other females.

When polygamous males offer no resources, it would seem that the female has limited information upon which to act. She might select a winner from among males that defeat others in combat—as in bighorn sheep, elk, and seals. She might select mates by intensity of courtship display or some morphological feature that may reflect a male's genetic superiority and vitality.

Among some polygamous species, it is hard to see female choice at work. In elk and seals, a dominant male takes charge of a harem of females (see Figure 8.8 on page 163). Nevertheless, the females express some choice. Protestations by female elephant seals over the attention of a dominant male may attract other large males nearby, who attempt to replace the male. Such behavior ensures that females will mate only with the highest-ranking male.

In other situations, females seem clearly in control. Males put on an intense display for them. Such advertising

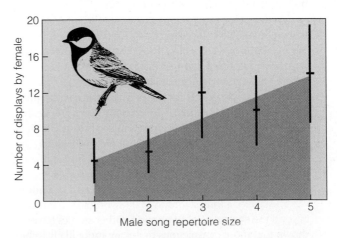

Figure 8.9 | Mean number of copulation–solicitation displays given by 11 female great tits (*Parus major*) as a function of the complexity of the male song. Complexity of song was measured in terms of repertoire size, ranging from one to five song types. Females responded more frequently to male courtship songs that included a greater number of song types.

(Adapted from M. C. Baker et al. 1986).

Figure 8.10 | Male greater sage grouse (*Centrocercus urophasianus*) exhibiting courtship display on a communal courtship ground called a lek. Females visit these leks of displaying males, select a male, mate, and move on.

can be costly for courting polygamous males. Because of conspicuous behavior and inattention, they may be subject to intense predation. Outstanding examples of female choice appear in lek species. These animals aggregate into groups on communal courtship grounds called **leks** or arenas (Figure 8.10). Males on the lek defend small territories that hold no resources and advertise their presence by colorful vocal and visual displays. Females visit the leks of displaying males, select a male, mate, and move on. Although few species engage in this type of mating system, it is widespread in the animal world, from insects to frogs to birds and mammals. Males defend small, clustered mating territories, whereas females have large overlapping ranges that the males cannot economically defend. Leks provide an unusual opportunity for females to choose a mate among the displaying males. Congregating about dominant males with the most effective displays, subdominant and satellite males may steal mating opportunities. Most of the matings on the lek, however, are by a small percentage of the males in a dominance hierarchy formed in the absence of females.

8.6 | Organisms Budget Time and Energy to Reproduction

Like plants, which must allocate carbon to the production of various plant tissues (Section 6.7), all organisms, both plant and animal, must allocate their limited energy to meet many demands. Some energy must go to growth, to maintenance, to acquiring food, to defending a territory, and to escaping predators. Energy must also be allocated to reproduction. The time and energy allocated to reproduction make up an organism's **reproductive effort.**

As with the allocation of carbon in plants, in animals the allocation of energy to various processes involves trade-offs. The more energy an organism expends on reproduction, the less it can allocate for growth and maintenance. For example, reproducing females of the terrestrial isopod *Armadillidium vulgare* have a lower rate of growth than do nonreproducing females. Nonreproducing females devote as much energy to growth as reproducing females devote to both growth and reproduction. Likewise, there is a negative relationship between annual growth and the size of the cone crop for Douglas-fir trees (*Pseudotsuga menziesii;* Figure 8.11). The greater the number of cones produced per tree, the smaller the increment of radial growth. Life histories represent trade-offs—compromises between competing objectives. By increasing the current allocation to reproduction, the organism—isopod or fir tree—is decreasing its growth, which may have consequences for its ability to compete for resources, survive, or reproduce in the future.

The amount of energy organisms invest in reproduction varies. Herbaceous perennials invest between 15 and 20 percent of annual net production to reproduction, including vegetative propagation. Wild annuals that reproduce only once expend 15 to 30 percent; most grain crops, 25 to 30 percent; and corn and barley, 35 to 40 percent. The common lizard (*Lacerta vivipara*) invests 7 to 9 percent of its annual energy assimilation to reproduction. The female Allegheny Mountain salamander (*Desmognathus ochrophaeus*) spends 48 percent of its annual energy budget on reproduction.

An organism's investment in reproduction includes the costs of care and nourishment as well as the production of offspring. This investment in reproduction

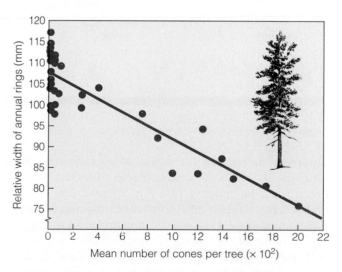

Figure 8.11 | Douglas-fir trees (*Pseudotsuga menziesii*) exhibit an inverse relationship between the allocation to reproduction (number of cones produced) and annual growth (as measured by radial growth).

(Adapted from Eis 1965.)

Interpreting Trade-offs

Many of the life history characteristics discussed in this chapter involve trade-offs, and understanding the nature of trade-offs involves the analysis of costs and benefits for a particular trait.

One trade-off in reproductive effort discussed in Section 8.6 involves the number and size of offspring produced. The graph in Figure 1 is similar to the one presented in Figure 8.11, showing the trade-off relationship between seed size and the number of seeds produced per plant. The example assumes a fixed allocation (100 units); and therefore the number of seeds produced per plant declines with increasing seed size.

Based on this information alone, it would appear that the best strategy for maximizing reproductive success would be to produce small seeds, thereby increasing the number of offspring produced. However, we must also consider any benefits to reproductive success that might vary as a function of seed size. The reserves

of energy and nutrients associated with large seed size have been shown to increase the probability of successful establishment, particularly for seedlings in low-resource environments (see example of the relationship between seed size and seedling survival in shade in the Chapter 6 Field Studies: Kaoru Kitajima). A generalized relationship between seed size and seedling survival for two different environments (wet and dry) is plotted in Figure 2. In both environments, survival increases with seed size; however, in dry environments, the probability of survival declines dramatically with decreasing seed size.

By multiplying the number of seeds produced by the probability of survival, we can now calculate the expected reproductive success (the number of surviving offspring produced per plant) for plants producing seeds of a given size in both the wet and dry environments (Figure 3).

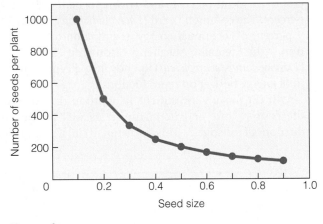

Figure 1 |

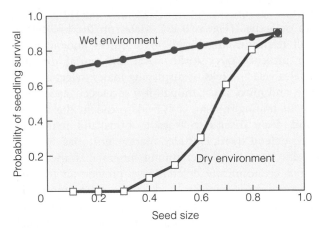

Figure 2 |

ultimately determines the fitness of an individual (see Chapter 5). An individual's fitness, however, is not defined by the number of offspring produced over a lifetime, but rather by the number of offspring that survive to reproduce. How a given investment in reproduction is allocated, the timing of reproduction, the number and size of offspring produced, and the care and defense provided will all interact in the context of the environment to determine the return to the individual in terms of increased fitness (see Quantifying Ecology 8.1: Interpreting Trade-offs).

8.7 | Species Differ in the Timing of Reproduction

How should an organism invest its allocation in reproduction through time? One approach is initially to invest all energy into growth, development, and energy storage, followed by one massive reproductive effort and then death. In this strategy, an organism sacrifices future prospects by expending all its energy in one suicidal act of reproduction. This mode of reproduction is called **semelparity.**

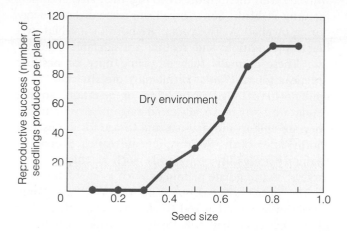

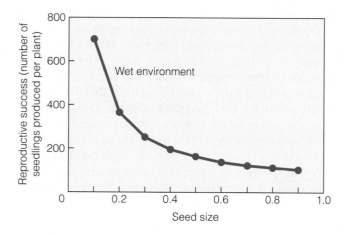

Figure 3 |

In wet environments, where all seed sizes have comparable probabilities of survival, the strategy of producing many small seeds results in the highest reproductive success and fitness. In contrast, the greater probability of survival makes the strategy of producing large seeds the most successful in dry environments, even though far fewer seeds are produced.

Interpreting the trade-offs observed in life history characteristics, such as the one illustrated between seed size and the number of seeds produced, requires understanding how those trade-offs function in the context of the environment (both biotic and abiotic) in which the species lives. Costs and benefits of a trait can change as the environmental conditions change. The diversity of life history traits exhibited by species is testimony that there is no single "best" solution for all environmental conditions.

1. In the example just presented, natural selection should favor plants producing small seeds in wet environments and plants that produce larger seeds in dry environments, resulting in a difference in average seed size in these two environments. What might you expect in an environment where annual rainfall is relatively high during most years (wet) but in which periods of drought (dry) commonly persist for several years?

2. The seeds of shade-tolerant plant species are typically larger than those of shade-intolerant species. How might this difference reflect a trade-off in life history characteristics relating to successful reproduction in sun and shade environments? See the discussion of shade tolerance in Chapter 6 and the Field Studies feature in that chapter.

Semelparity is employed by most insects and other invertebrates, by some species of fish (notably salmon), and by many plants. It is common among annuals, biennials, and some species of bamboos. Many semelparous plants, such as ragweed (*Ambrosia* spp.), are small, short-lived, and found in ephemeral or disturbed habitats. For them, it would not pay to hold out for future reproduction, for the chances are slim. They gain their maximum fitness by expending all their energies in one bout of reproduction.

Other semelparous organisms, however, are long-lived and delay reproduction. Mayflies (Ephemeroptera) may spend several years as larvae before emerging from the surface of the water for an adult life of several days devoted to reproduction. Periodical cicadas spend 13 to 17 years belowground before they emerge as adults to stage a single outstanding exhibition of reproduction. Some species of bamboo delay flowering for 100 to 120 years, produce one massive crop of seeds, and die. Hawaiian silverswords (*Argyroxiphium* spp.) live 7 to 30 years before flowering and dying. In general, species that evolve semelparity have to increase fitness enough to compensate for the loss of repeated reproduction.

Organisms that produce fewer young at one time and repeat reproduction throughout their lifetime are called **iteroparous.** Iteroparous organisms include most vertebrates, perennial herbaceous plants, shrubs, and trees. For an iteroparous organism, the problem is timing reproduction—early in life or later. Whatever the choice, it involves trade-offs. Early reproduction means less growth, earlier maturity, reduced survivorship, and reduced potential for future reproduction. Later reproduction means increased growth, later maturity, and increased survivorship, but less time for reproduction. In effect, if an organism is to make a maximum contribution to future generations, it has to balance the profits of immediate reproduction against future prospects, including the cost to fecundity (total offspring produced) and its own survival.

8.8 | Parental Investment Depends on the Number and Size of Young

In theory, a given allocation to reproduction can potentially produce many small young or fewer large ones. The number of offspring affects the parental investment each receives. If the parent produces a large number of young, it can afford only minimal investment in each one. In such cases, animals provide no parental care, and plants store little food energy in seeds (Figure 8.12). Such organisms usually inhabit disturbed sites, unpredictable environments, or places such as the open ocean where opportunities for parental care are difficult at best. By dividing energy for reproduction among as many young

as possible, these parents increase the chances that some young will successfully settle somewhere and reproduce in the future.

Parents that produce few young are able to expend more energy on each individual. The amount of energy will vary with the number of young, their size, and their maturity at birth. Some organisms expend less energy during incubation. The young are born or hatched in a helpless condition and require considerable parental care. These animals, such as young mice or nestling American robins (*Turdus migratorius*), are **altricial.** Other animals have longer incubation or gestation, so the young are born in an advanced stage of development. They are able to move about and forage for themselves shortly after birth. Such young are called **precocial.** Examples are gallinaceous birds, such as chickens and turkeys, and ungulate mammals, such as cows and deer.

The degree of parental care varies widely. Some species of fish, such as cod (*Gadus morhua*), lay millions of floating eggs that drift freely in the ocean with no parental care. Other species, such as bass, lay eggs in the hundreds and provide some degree of parental care. Among amphibians, parental care is most prevalent among tropical toads and frogs and some species of salamanders. Among reptiles, which rarely exhibit parental care, crocodiles are an exception. They actively defend the nest and young for a considerable time. Invertebrates exhibit parental care to varying degrees. Octopus; crustaceans such as lobsters, crayfish, and shrimp; and certain amphipods such as millipedes brood and defend eggs. Parental care is best developed among the social insects: bees, wasps, ants, and termites. Social insects perform all functions of parental care, including feeding, defense, heating and cooling, and sanitation.

8.9 | Fecundity Depends on Age and Size

For many species, the number of offspring produced varies with the age and size of the parent. Many plants and ectothermic (cold-blooded) animals (fish, reptiles, amphibians, and invertebrates) do not have a characteristic adult size. These species can continue to grow throughout their adult lives (although typically at a continuously declining relative rate). This condition is referred to as **indeterminate growth** (for example, see Chapter 7 Field Studies: Martin Wikelski).

Among plants, perennials delay flowering until they reach a sufficiently large size (and leaf area) to support seed production. Many biennials in poor environments also delay flowering beyond the usual 2-year life span until environmental conditions become more favorable. Annuals show no relationship between size and the percentage of energy devoted to reproductive output; as a result, size differences among annuals result in differences in the number of seeds produced. Small plants produce fewer seeds, even though the plants may be

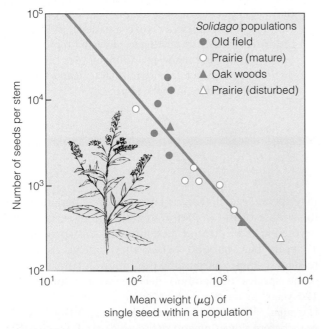

Figure 8.12 | Inverse relationship between mean weight of individual seeds and the number of seeds produced per stem for populations of goldenrod (*Solidago* spp.) in a variety of habitats. (Adapted from Werner and Platt 1976.)

contributing the same proportion of energy to reproduction as larger plants do.

Similar patterns exist among poikilothermic (cold-blooded) animals. Production of offspring (fecundity) in fish increases with size, which increases with age. Because early fecundity reduces both growth and later reproductive success, fish obtain a selective advantage by delaying sexual maturation until they grow larger. Gizzard shad (*Dorosoma cepedianum*) reproducing at 2 years of age produce about 59,000 eggs. Those delaying reproduction until the third year produce about 379,000 eggs. Among the gizzard shad, only about 15 percent spawn at 2 years of age and about 80 percent at 3 years. The number of eggs produced by loggerhead sea turtles (*Caretta caretta*) is constrained by the female's egg-carrying capacity, which is related to body size. Similarly, there is a positive relationship between body size and the number of young produced for female big-handed crabs (*Heterozius rohendifrons*) inhabiting the coastal environments of New Zealand (Figure 8.13a).

An apparent relationship also exists between body size and fecundity for some endothermic (warm-blooded) animals. Heavier females are more successful in reproduction, and more of their young survive. For example, the body weight of female European red squirrels (*Sciurus vulgaris*) in Belgian forests is strongly correlated with lifetime reproductive success (Figure 8.13b). Few squirrels weighing less than 300 g reproduce.

8.10 | Food Supply Affects the Production of Young

Within a given region, production of young may reflect the abundance of food. In species with indeterminate growth, the availability of food can directly influence body size and therefore reproductive effort (see Chapter 7 Field Studies: Martin Wikelski).

In environments where the availability of food resources is highly variable through time, the number of offspring that can physiologically be produced may be greater than the number that can be provided for during certain years. Under these circumstances, it may be necessary to reduce the number of young. In many species of birds, asynchronous hatching and **siblicide,** the killing of one sibling by another, reduce the number of offspring.

In asynchronous hatching, the young are of several ages. The older siblings beg more vigorously for food, forcing the harried parents to ignore the calls of the younger, smaller sibling, which perishes. For example, the common grackle (*Quiscalus quiscula*) begins incubation before the entire clutch of five eggs has been laid. The eggs laid last are heavier, and the young from them grow fast. However, if food is scarce, the parents fail to feed these late offspring because of more vigorous begging by the larger, older siblings. The last-hatched young then die of starvation. Thus, asynchronous hatching favors the early-hatched young, ensuring the survival of some siblings under adverse conditions.

In other situations, the older or more vigorous young simply kill their weaker siblings. Several birds, including raptors, herons, egrets, gannets, boobies, and skuas, practice siblicide. The parents normally lay two eggs, possibly to insure against infertility of a single egg. The larger of the two hatchlings kills the smaller sibling, or runt, and the parents redirect all resources to the surviving chick (Figure 8.14). These birds are not alone in siblicidal tendencies. The females of some parasitic wasps lay

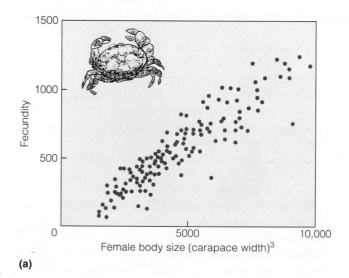

(a)

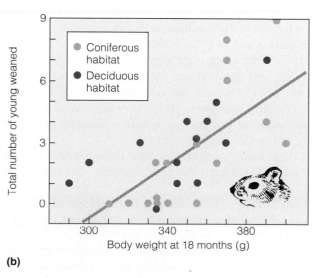

(b)

Figure 8.13 | **(a)** Annual production of young (fecundity) by female big-handed crabs (native to New Zealand) increases with female body size (as measured by the width of the carapace, or shell). **(b)** Lifetime reproductive success of the European red squirrel is correlated with body weight in the first winter as an adult.

(a) (Adapted from Jones 1978.) (b) (Adapted from Wauters and Dhondf 1989.)

Figure 8.14 | Example of siblicide in the masked booby (*Sula dactylatra*), in which the larger of two offspring kills the smaller sibling. Masked booby parents collaborate by ignoring the battle between the offspring.

two or more eggs in a host, and the larvae fight each other until only one survives.

8.11 | Reproductive Effort May Vary with Latitude

Birds in temperate regions have larger clutch sizes (the number of eggs produced) than do those in the tropics (Figure 8.15), and mammals at higher latitudes have larger litters than do those at lower latitudes. Lizards exhibit a similar pattern. Those living at lower latitudes have smaller clutches, reproduce at an earlier age, and experience higher adult mortality than those living at higher latitudes.

Insects, too, such as the milkweed beetle (*Oncopeltus* spp.), exhibit a latitudinal pattern in reproductive effort. Temperate and tropical milkweed beetles have a similar duration of the egg stage, egg survivorship, developmental rate, and age at sexual maturity. Although the clutch sizes are the same, the temperate species lay more clutches and therefore a greater number of eggs. Total egg production of the tropical species is only 60 percent of that of the temperate species.

Plants also follow the general principle of allocation on a latitudinal basis. For example, three species of cattail grow in North and Central America. The common cattail (*Typha latifolia*) grows under a broad range of climates, with a distribution extending from the Arctic Circle to the equator. The narrow-leaved cattail (*T. angustifolia*) is restricted to the northern latitudes. The southern cattail (*T. domingensis*) grows only in the southern latitudes. These three cattails show a climatic gradient in the allocation of energy to vegetative (asexual) reproduction. *T. angustifolia* and the northern populations of *T. latifolia* grow

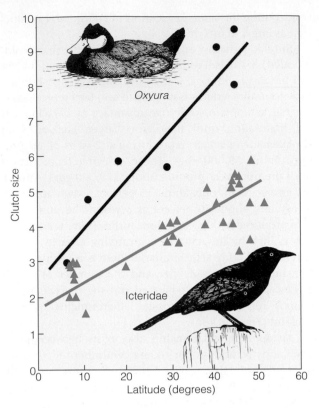

Figure 8.15 | Relationship between clutch size and latitude in birds. Represented are the family Icteridae (blackbirds, orioles, and meadowlarks) in North and South America and the worldwide genus *Oxyura* (ruddy and masked ducks) of the family Anatidae. (Adapted from Cody 1966.)

earlier and faster than *T. domingensis* and produce a greater number of rhizomes (ramets). The southern populations produce fewer but larger rhizomes.

Why might this pattern of geographic variation in reproductive allocation occur? David Lack, an English ornithologist, proposed that clutch size in birds has evolved to equal the largest number of young the parents can feed. Thus, clutch size is an adaptation to food supply. Temperate species, he argued, have larger clutches because increasing day length during the spring provides a longer time to forage for food to support large broods. In the tropics, where day length is roughly 12 hours, foraging time is more limited.

Martin Cody at University of California, Los Angeles, modified Lack's ideas by proposing that clutch size results from different allocations of energy to egg production, avoidance of predators, and competition. In temperate regions, periodic local climatic catastrophes (such as a harsh winter or a summer drought) can hold a population below the level that the resources could support. Organisms respond with larger clutches and a higher rate of population increase. In tropical regions, with predictable climates and increased probability of survival, there is no need for extra young.

A third hypothesis, proposed by N. Philip Ashmole of the University of Edinburgh (Scotland), states that clutch size varies in direct proportion to the seasonal variation in resources, especially food. Population density is regulated primarily by mortality in winter, when resources are scarce. Greater winter mortality means more food for the survivors during the breeding season. This abundance is reflected in larger clutches. Thus, geographical variation in mean clutch size and the size of the breeding population is inversely related to winter food supply. Although reproductive output among organisms does appear to be greater at higher latitudes, the number of comparable species for which there are data is too small to confirm any of these hypotheses. Many more studies along a latitudinal gradient are needed.

8.12 | Habitat Selection Influences Reproductive Success

Reproductive success depends heavily on choice of habitat—the environment (place) in which an organism lives. Habitat provides access to essential resources, nesting sites, and cover from potential predators. It can also influence the ability to attract a mate. Settling on a less-than-optimal habitat can result in reproductive failure. The process by which organisms actively choose a specific location to inhabit is called **habitat selection.** Given the importance of habitat selection on an organism's fitness, how are they able to assess the quality of an area where they settle? What do they seek in a living place? Such questions have been intriguing ecologists for many years. Understanding the relationship between habitat selection, reproductive success, and population dynamics has become particularly important due to the loss of habitats for many species as a result of human impacts on the landscape (see Chapter 28).

Habitat selection has been most widely studied in birds, particularly in species that defend breeding territories (see Section 11.10). Territories can be delineated, and features of the habitat can be described and contrasted with the surrounding environment. Of particular importance in the study of habitat selection is the ability to contrast areas that have been chosen as habitats with adjacent areas that have not. Using this approach, a wide variety of studies examining the process of habitat selection in birds have demonstrated a strong correlation between the selection of an area as habitat and the structural features of the vegetation. These studies suggest that habitat selection most likely involves a hierarchical approach. Birds appear initially to assess the general features of the landscape: the type of terrain; presence of lakes, ponds, streams, and wetlands; gross features of the vegetation such as open grassland and shrubby areas; and types and extent of forests. Once in a broad general area, the birds respond to more specific features, such as the structural configuration of vegetation, particularly the presence or absence of various vertical layers such as shrubs, small trees, tall canopy, and degree of patchiness (Figure 8.16). Frances James, an avian ecologist at Florida State University, coined the term *niche gestalt* (German *gestalt*, meaning "shape or form") to describe the vegetation profile associated with the breeding territory of a particular species.

Besides the physical structure of the vegetation, the actual plant species present can be important. Certain species of plants might produce preferred food items, such as seeds and fruits, or influence the type and quantity of insects available as prey for insectivorous birds.

The structural features of the vegetation that define its suitability for a given species may be related to a variety of specific needs, such as food, cover, and nesting sites.

  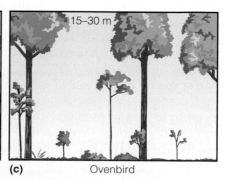

(a) Yellowthroat   **(b)** Hooded warbler   **(c)** Ovenbird

Figure 8.16 | Vegetation structure characterizing the habitat of three warbler species. **(a)** Common yellowthroat (*Geothlypis trichas*), a bird of shrubby margins of woodland and wetlands and brushy fields. **(b)** Hooded warbler (*Wilsonia citrina*), a bird of small forest openings. **(c)** Ovenbird (*Seiurus aurocapillus*), an inhabitant of deciduous or mixed conifer–deciduous forests with open forest floor. The labels 9–10 m, 18–30 m, and 15–30 m refer to height of vegetation.

(Adapted from James 1971.)

The lack of song perches may prevent some birds from colonizing an otherwise suitable habitat. An adequate nesting site is another requirement. Animals require sufficient shelter to protect themselves and young against enemies and adverse weather. Cavity-nesting animals require suitable dead trees or other structures with cavities. In areas without such sites, populations of birds and squirrels can be increased dramatically by providing nest boxes and den boxes.

Habitat selection is a common behavioral characteristic of a wide variety of vertebrates other than birds; fish, amphibians, reptiles, and mammals furnish numerous examples. Garter snakes (*Thamnophis elegans*) living along the shores of Eagle Lake in the sagebrush–ponderosa pine country of northeastern California select rocks of intermediate thickness (20 to 30 cm) over thinner and thicker rocks as their retreat sites. Shelter under thin rocks becomes lethally hot; shelter under thicker rocks does not allow the snakes to warm to their preferred range of body temperature (T_b; see Chapter 7). Under the rocks of intermediate thickness, snakes are able to achieve and maintain their preferred body temperature for a long period. Insects, too, cue in on habitat features. Thomas Whitham, a plant ecologist at Northern Arizona University, studied habitat selection by the gall-forming aphid (*Pemphigus betae*), which parasitizes the narrow-leaf cottonwood (*Populus angustifolia*). He found that aphids select the largest leaves to colonize and discriminate against small leaves. Beyond that, they select the best positions on the leaf. Occupying this particular habitat, which provides the best food source, produces individuals with the highest fitness.

Even though a given habitat may provide suitable cues, it still may not be selected. The presence or absence of others of the same species may influence individuals to choose or avoid a particular site. In social or colonial species like herring gulls (*Larus argentatus*), an animal will choose a site only if others of the same species are already nearby. On the other hand, the presence of predators and human activity may discourage a species from occupying an otherwise suitable habitat (see Section 14.8).

Most species exhibit some flexibility in habitat selection. Otherwise, these animals would not settle in what appears to us as a less-suitable habitat, nor would they colonize new habitats. Often, individuals are forced to make this choice. Available habitats range from optimal to marginal; the optimal habitats, like good seats at a concert, fill up fast. The marginal habitats go next; and latecomers or subdominant individuals are left with the poor habitats, where they may have little chance of reproducing successfully.

Do plants select habitats, and if so how? Plants can hardly get up and move about to find a suitable site. Plants, like animals, fare better in certain habitat types, characterized by such environmental factors as light, moisture, nutrients, and presence of herbivores. The only recourse plants have in habitat selection is to evolve dispersal strategies that influence the probability that a seed will arrive at a place suitable for germination and seedling survival. Habitat selection for plants involves their ability (and often, an element of chance) to disperse with the aid of wind, water, or animal agents to preferred patches of habitat.

8.13 | Environmental Conditions Influence the Evolution of Life History Characteristics

The life history characteristics exhibited by a species are the product of evolution and should reflect adaptations to the prevailing environmental conditions under which natural selection occurred (see Ecological Issues: The Life History of Maize: A Story of Unnatural Selection). If this is the case, do species inhabiting similar environments exhibit similar patterns of life history characteristics? Do life history characteristics exhibit patterns related to the habitats that species occupy?

One way of classifying environments (or species habitats) relates to their variability in time. We can envision two contrasting types of habitats: (1) those that are variable in time or short-lived and (2) those that are relatively stable (long-lived and constant), with little random environmental fluctuations. The ecologists Robert MacArthur of Princeton University, E. O. Wilson of Harvard University, and later E. Pianka of the University of Texas used this dichotomy to develop the concept of *r*- and *K*-selection.

The theory of *r*- and *K*-selection predicts that species adapted to these two different environments will differ in life history traits such as size, fecundity, age at first reproduction, number of reproductive events during a lifetime, and total life span. Species popularly known as **r-strategists** are typically short-lived. They have high reproductive rates at low population densities, rapid development, small body size, large number of offspring (but with low survival), and minimal parental care. They make use of temporary habitats. Many inhabit unstable or unpredictable environments that can cause catastrophic mortality independent of population density. For these species, environmental resources are rarely limiting. They exploit noncompetitive situations. Some *r*-strategists, such as weedy species, have means of wide dispersal, are good colonizers, and respond rapidly to disturbance.

K-strategists are competitive species with stable populations of long-lived individuals. They have a slower growth rate at low populations, but they maintain that

Corn is one of the major global food crops and perhaps one of the most generally recognized plant species on Earth. Current annual global production approaches 600 million metric tons, or approximately 100 kg for every person on the planet. The plant that we know as modern corn, however, has undergone an amazing transformation from its ancestral form. Its transformation has involved the alteration of life history characteristics through selective breeding—a process of human guided evolution.

Corn, or maize as it is called in most parts of the world, is a domesticated plant of the Americas. Besides adopting many other indigenous plants like beans, squash, melons, and tobacco, European colonists in America quickly adopted maize agriculture from Native Americans. Maize as well as other crop species developed by Native Americans then quickly spread to other parts of the world, changing the nature of global agriculture. The original maize plant encountered by early colonists, however, already had a long history of transformation. Maize was developed about 7000 years ago from teosinte, a wild grass growing in Central America (southern Mexico). Teosinte is a group of large, Central and South American grasses of the genus *Zea*. There are five recognized species of teosinte, including *Zea mays*, from which maize has been developed. The ancestral kernels of teosinte looked very different from today's corn (see Figure 1). These kernels were small and were not fused together like the kernels on the husked ear of early maize and modern corn.

By systematically collecting and cultivating those plants best suited for human consumption, Native Americans encouraged the formation of ears or cobs on early maize plants. The first ears of maize were only a few inches long and had only eight rows of kernels. Cob length and size of early maize grew over the next several thousand years, leading to gradual increases in the yields of each crop. This transformation involved a major shift in the pattern of resource (carbon and other nutrients) allocation. The major criterion used in selecting plants for breeding was increased allocation of resources to those parts of the plant that are used as food. For corn, this translated into an increased allocation of resources to reproduction—the production of seeds. By selectively breeding those plants that allocated the greatest amount of resources to the production of seeds, humans have altered the species' life history to meet our needs for food. The increased allocation to reproduction is at the expense of allocation to the production of other structures, primarily roots. The well-developed root system characteristic of grasses allows for water uptake in the regions of limited rainfall where these species grow naturally. Modern varieties of maize often need to be irrigated to compensate for their reduced ability to tolerate periods of low soil moisture. To avoid the need for irrigation, varieties have been specifically developed for regions where soil moisture is depleted by midsummer. These varieties grow and mature more quickly, avoiding the drought stress associated with late summer.

The original ancestors of modern corn can still be found in Central America today, but virtually all populations of teosinte are either threatened or endangered. The Mexican government has taken action in recent years to protect wild teosinte populations, and there is currently a large amount of scientific interest in using wild populations in the future development of domesticated corn, including conferring beneficial teosinte traits, such as insect resistance and perennialism, to cultivated maize lines.

1. Can you think of another example where the life history of a plant or animal species has been actively manipulated for human purposes? Which characteristics have been manipulated?

2. Can you think of any life history characteristics of humans that have been altered by the development of technology or modern culture? Think of life history in the broadest sense to include behavioral and cultural factors that influence lifetime patterns of growth, development, and reproduction.

Figure 1 | Domestication of the corn plant involved selecting for increased allocation to reproduction—the production of seed. Ancestral kernels of teosinte were small and not fused together like the kernels on the husked ear of modern corn.

growth rate at high densities. *K*-strategists can cope with physical and biotic pressures. They possess both delayed and repeated reproduction and have a larger body size and slower development. They produce few seeds, eggs, or young. Among animals, parents care for the young; among plants, seeds possess stored food that gives the seedlings a strong start. The mortality of *K*-strategists relates more to density than to unpredictable environmental conditions. They are specialists, efficient users of a particular environment, but their populations are at or near carrying capacity (maximum sustainable population size) and are resource limited. These qualities, combined with their lack of means for wide dispersal, make *K*-strategists poor colonizers of new and empty habitats.

The terms *r* and *K* used to characterize these two contrasting strategies relate to the parameters of the logistic model of population growth presented later (in Chapter 11): *r* is the per capita rate of growth, and *K* is the carrying capacity (maximum size a population can sustain). Using the classification of *r*- and *K*- strategies for comparing species across a wide range of sizes is of limited value. For example, the correlation among body size, metabolic rate, and longevity in warm-blooded organisms (see Chapter 7) results in species with small body size generally being classified as *r* species and those with large body size as *K* species. The concept of *r* species and *K* species is most useful in comparing organisms that are either taxonomic or functionally similar. The spotted (*Ambystoma maculatum*) and redback (*Plethodon cinereus*) salamanders found in eastern North America provide such an example of contrasting life history strategies (Figure 8.17).

The spotted salamander, an *r*-strategist, is found under logs and piles of damp leaves in deciduous forest habitats. During the month of February, individuals migrate to ponds and other small bodies of water to reproduce. After mating, females lay up to 250 eggs in large, compact, gelatinous masses that are attached to twigs just below the water surface. After mating, adults leave the water and provide no parental care of eggs or young. In contrast, the redback salamander, a *K*-strategist, occupies similar habitats in mixed coniferous–deciduous forests. After mating, females lay 4 to 10 eggs, which are deposited in a cluster within the crevice of a rotting log or stump. The female then curls about the egg cluster, guarding them until the larvae hatch.

The plant ecologist J. Phillip Grime of Sheffield, England, used a framework similar to that used by MacArthur and Wilson to develop a life history classification for plants. He based it on three primary strategies—*R*, *C*, and *S*—that relate plant adaptations to different habi-

(a)

(b)

Figure 8.17 | The spotted (*Ambystoma maculatum*) and redback (*Plethodon cinereus*) salamanders found in eastern North America provide an example of contrasting life history strategies. **(a)** The spotted salamander lays a large number of eggs that form an egg mass, which it then abandons. **(b)** In contrast, the redback lays only few eggs, which it guards until they hatch.

tats (Figure 8.18a). Species exhibiting the *R*, or ruderal, strategy rapidly colonize disturbed sites. These species are typically small in stature and short-lived. Allocation of resources is primarily to reproduction, with characteristics allowing for a wide dispersal of seeds to newly disturbed sites. Predictable habitats with abundant resources favor species that allocate resources to growth, favoring resource acquisition and competitive ability (*C* species). Habitats where resources are limited favor stress-tolerant species (*S* species) that allocate resources to maintenance. These three strategies form the end points of a triangular classification system that allows for intermediate strategies, depending on such environmental factors as resource availability and frequency of disturbance (Figure 8.18b).

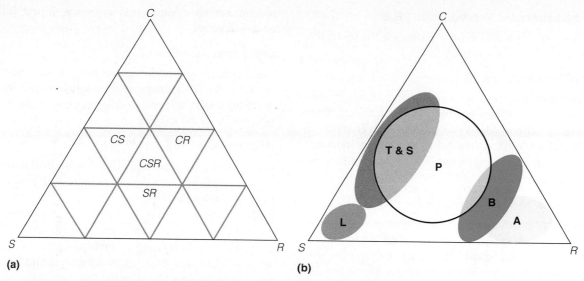

Figure 8.18 | Grime's model of life history variation in plants based on three primary strategies: ruderals (*R*), competitive (*C*), and stress-tolerant (*S*). **(a)** These primary strategies define the three points of the triangle. Intermediate strategies are defined by combinations of these three (e.g., *CS, CR, CSR,* and *SR*). **(b)** Grime's assessment of life history strategies of most trees and shrubs (T & S), lichens (L), biennial herbs (B), perennial herbs (P), and annual herbs (A).

(Adapted from Grime 1977.)

Summary

Asexual and Sexual Reproduction | 8.1

Fitness is an organism's ability to leave behind reproducing offspring. Organisms that contribute the most offspring to the next generation are the fittest. Reproduction can be asexual or sexual. Asexual reproduction, or cloning, results in new individuals that are genetically the same as the parent. Sexual reproduction combines egg and sperm in a diploid cell, or zygote. Sexual reproduction produces genetic variability among offspring.

Forms of Sexual Reproduction | 8.2

Sexual reproduction takes a variety of forms. Plants with separate males and females are called dioecious. An organism with both male and female sex organs is hermaphroditic. Plant hermaphrodites have bisexual flowers or, if they are monoecious, separate male and female flowers on the same individual. Some plants and animals change gender.

Mating Systems | 8.3

Mating systems include two basic types: monogamy and polygamy. In polygyny, the male acquires more than

one female; in polyandry, the female acquires more than one male. The potential for competitive mating and sexual selection is higher in polygamy than in monogamy.

Sexual Selection | 8.4

Selection of a proper mate is essential if an organism is to contribute to the next generation. An important component of mating strategy is sexual selection. In general, males compete with males for the opportunity to mate with females, but females finally choose mates. Sexual selection favors traits that enhance mating success, even if they handicap the male by making him more vulnerable to predation. Male competition is intrasexual selection, whereas intersexual selection involves the differential attractiveness of individuals of one sex to the other. By choosing the best males, females ensure their own fitness.

Resources and Mate Selection | 8.5

Females may also choose mates based on the acquisition of resources, usually a defended territory or habitat. By choosing a male with a high-quality territory, the female may increase her fitness.

Energy Investment in Reproduction | 8.6

The amount of time and energy parents allot to reproduction is reproductive effort.

Timing of Reproduction | 8.7

To maximize fitness, an organism has to balance immediate reproductive efforts against future prospects. One alternative, semelparity, is to invest maximum energy in a single reproductive effort. The other alternative, iteroparity, is to allocate less energy each time to repeated reproductive efforts.

Number and Size of Offspring | 8.8

Organisms that produce many offspring have a minimal investment in each offspring. They can afford to send a large number into the world with a chance that a few will survive. By so doing, they increase parental fitness but decrease the fitness of the young. Organisms that produce few young invest considerably more in each one. Such organisms increase the fitness of the young at the expense of the fitness of the parents.

Age and Size | 8.9

A direct relationship between size and fecundity exists among plants and ectotherms. The larger the size, the more young are produced. Among both endotherms and ectotherms, heavier females are more successful in producing a greater number of offspring.

Food Supply | 8.10

Production of young often reflects food availability. When food is scarce, parents may fail to feed some off-spring. In other situations, vigorous young kill their weaker siblings.

Latitudinal Variation | 8.11

In general, clutch and litter sizes increase from the tropics to the poles. This gradient may reflect either length of daylight, which influences foraging time, or the more stable climate in the tropics.

Habitat Selection | 8.12

Reproductive success depends heavily on the choice of habitats. Habitat selection is genetic and partly behavioral. Most studies of habitat selection have focused on birds that defend breeding territories. Results suggest that habitat selection involves a hierarchical approach, with the initial selection based on general features of the landscape; within this area, individuals respond to specific features of the vegetation or habitat.

Environmental Influences | 8.13

Organisms living in variable or ephemeral environments or facing heavy predation produce many offspring, ensuring that some will survive. Annual plants, short-lived mammals, insects, and semelparous species produce large numbers of young. Long-lived species produce few young. Iteroparous species may adjust the number of young in response to environmental conditions and the availability of resources.

Study Questions

1. Why might you expect sexual reproduction to be an advantage in a variable environment?
2. Contrast dioecious and monoecious flowering plants.
3. What are some advantages of hermaphroditism?
4. Distinguish among monogamy, polygamy, polygyny, and polyandry.
5. Why is monogamy more common in birds than it is in mammals?
6. How might female preference for a male trait (sexual selection), such as coloration or body size, drive selection in a direction counter to that of natural selection?
7. Contrast intrasexual selection and intersexual selection.
8. Discuss the trade-off in the number of offspring produced and the degree of parental care.
9. What conditions favor semelparity over iteroparity?
10. For a given allocation to reproduction, there is a trade-off between the number and size of offspring produced (see Figure 8.12). What type(s) of environment would favor plant species with a strategy of producing many small seeds rather than few large ones?
11. What is the difference between r-selected and K-selected organisms? Which strategy would you expect to be more prevalent in unpredictable environments (high stochastic variation in conditions)?

Further Readings

Alcock, J. 2001. *Animal behavior: An evolutionary approach.* 7th ed. Sunderland, MA: Sinauer.

   This text is an excellent treatment of topics covered in this chapter. It is a good reference for students who want to pursue specific topics relating to behavioral ecology.

Andersson, M., and Y. Iwasa. 1996. Sexual selection. *Trends in Ecology and Evolution* 11:53–58.

   An excellent but technical review of sexual selection.

Buss, D. M. 1994. The strategies of human mating. *American Scientist* 82:238–249.

   This article, an application of sexual selection theory to humans, is a fun read for students. It explores the question of whether mate selection by females has influenced male characteristics in humans.

Krebs, J. R., and N. D. Davies. 1993. *An introduction to behavioral ecology.* 3rd ed. Oxford: Blackwell Scientific.

   This text provides a comprehensive discussion of behavioral topics that are covered in this chapter.

Policansky, D. 1982. Sex change in plants and animals. *Annual Review of Ecology and Systematics* 13:471–495.

   A review article that explores the variety of examples, in both plants and animals, of individuals that change sex over the course of their lifetime. It explores our understanding of cues that result in the shift as well as mechanisms by which the shift occurs.

Stearns, S. C. 1992. *The evolution of life histories.* Oxford: Oxford University Press.

   This book explores the link between natural selection and life history. It does an excellent job of illustrating how both biotic and abiotic factors interact to influence the evolution of specific life history traits.

PART THREE | Populations

The Copper River in southeastern Alaska originates at the Copper Glacier on Mount Wrangell, in what is now the Wrangell–St. Elias National Park. From its headwaters at the Copper Glacier, the river flows 300 miles westward, dropping some 3600 feet through the Alaskan wilderness before reaching the Gulf of Alaska. On its 300-mile journey to the ocean, it flows past 12 major glaciers, each further contributing to the volume of water and sediments that give the Copper River its distinctive opaque gray color. Its color can invoke a sense of lifelessness, but each spring the Copper River comes alive as millions of Chinook salmon (*Oncorhynchus tshawytscha*) enter the mouth of the river from the Gulf of Alaska and make their way upriver to spawn (breed; see photo).

The salmon stop eating once they enter freshwater to spawn. Their bodies darken, losing their shiny, silvery color. In battling against the current, the condition of the Chinooks deteriorates. Their sheer abundance in the waters and weakened state make the Chinook easy prey for predators such as the Alaskan brown bear and bald eagle, and most of the salmon die before their task is completed. The survivors continue their trek to the headwaters and then to the streams that feed the river along its journey. Once the Chinooks arrive, the female chooses a site and digs a nest with her tail (nests are referred to as *redds*). She deposits from 3000 to 5000 eggs that are then fertilized by one or more males. Within a week of breeding, the adults will die.

The life cycle of Chinook salmon begins when the young emerge from the redds. Hatchlings live in freshwater for up to a year before heading downriver to the coastal waters. Once in the sea, the salmon move across an area of thousands of miles while they mature. Within 2 to 5 years, adults return to the Copper River, and the cycle begins again.

This cycle has repeated itself for thousands of years, and the population of Chinook salmon in the Copper River has persisted. If on average, each individual that spawns is successful in producing one offspring that also survives to reproduce successfully, the population will persist. The Copper River population depends on a variety of factors that influence survival patterns over the Chinook life cycle, such as the abundance of food and the environmental conditions in the river and streams that are home to the hatchlings and in the open oceans that are home to the adults. Likewise, environmental conditions influence reproductive success. In years that are unusually warm and dry, the streams and tributaries of the Copper River can become so shallow that access to spawning areas becomes severely restricted, reducing reproduction.

This story is repeated along the coastline of the Pacific Northwest. In Alaska alone, Chinook return to the Yukon, Kuskokwim, Nushagak, Susitna, Kenai, Alsek, Taku, and Stikine rivers, as well as countless small streams that flow to the coast. Because salmon return at maturity to their parent streams to spawn, the larger population that inhabits the waters of the northern Pacific Ocean is actually a collection of local populations, each of which is genetically unique. Individuals return to the specific locations in the streams where they were born and thus breed within specific subpopulations. The dynamics of the larger coastal population, both demographic and genetic, are therefore a product of the collective dynamics of these local breeding populations.

The story of the Chinook salmon is unique to this species, a product of its life history characteristics (see Chapter 8). Yet this salmon's story illustrates a much broader concept of the processes governing population dynamics. For each individual, birth and death are discrete events; but for the population, birth and death are collective properties, and these collective properties govern the dynamics of the population through time. The population is an ever-changing entity as individuals are born and die. If the number of individuals being born exceeds the number dying, the size of the population will increase. Should the reverse situation occur (births < deaths), the population size will decline.

Various factors can influence the survival and reproductive success of individuals. The net effect of these factors is to regulate the population's growth. Finally, the collective of all Chinook salmon in the Pacific Northwest does not form one homogeneous, interbreeding population. Rather, the larger regional population consists of a collection of local populations that are connected by the occasional exchange of individuals among them. Therefore, the larger regional population is governed by processes operating at two spatial scales: the dynamics of local populations and the larger dynamic of the collective of local populations. In Part Three, we explore these basic concepts as they relate to the dynamics of natural populations. In Chapter 9 we will examine the properties characterizing the collective of individuals that define the population. In Chapter 10 we will examine how the collective properties of birth and death govern the dynamics of local populations. In Chapter 11 we will examine factors, both environmental and behavioral, that function to regulate the growth of populations. Finally, in Chapter 12, we will explore the processes that influence the interactions among local subpopulations and, subsequently, the dynamics of the larger regional population.

Chinook salmon move against the current to reach the upper reaches of the river, where they will spawn (reproduce).

CHAPTER 9

Properties of Populations

This field of daisies represents a population—a group of individuals of the same species inhabiting a given area.

As an individual, how do you perceive the world? Most of us regard a friend, a neighborhood maple tree, a daisy in a field, a squirrel in the park, or a bluebird nesting in the backyard as individuals. Rarely do we consider such individuals as part of a larger unit—a population. Although the term *population* has many different meanings and uses, for biologists and ecologists it has a very specific definition. A population is a group of individuals of the same species that inhabit a given area. This definition has two important features. First, by requiring that individuals be of the same species, the definition suggests the potential (in sexually reproducing organisms) for interbreeding among members of the population. As such, the population is a genetic unit. It defines the gene pool, the focus of evolution (see Chapter 5). Second, the population is a spatial concept, requiring a defined spatial boundary—for example, the breeding population of Chinook salmon inhabiting the Copper River described in the introduction to Part Three or the population of Darwin's ground finch inhabiting the Island of Daphne Major (Galápagos Islands) described in Chapter 5 (see Section 5.4).

Populations have unique features because they are an aggregate of individuals. Populations have structure, which relates to characteristics such as density (the number of individuals per unit area), proportion of individuals in various age classes, and spacing of individuals relative to each other. Populations also exhibit dynamics—a pattern of continuous change through time that results from the birth, death, and movement of individuals. In this chapter, we explore the basic features used to describe the structure of populations, and in doing so we set the foundation for examining the dynamics of population structure in subsequent chapters (Chapters 10–12).

9.1 | Organisms May Be Unitary or Modular

A population is considered to be a group of individuals, but what constitutes an individual? For most of us, defining an individual would seem to be no problem. We are individuals, and so are dogs, cats, spiders, insects, fish, and so on throughout much of the animal kingdom. What defines us as individuals is our unitary nature. Form, development, growth, and longevity of unitary organisms are predictable and determinate from conception on. The zygote, formed through sexual reproduction (see Section 8.1), grows into a genetically unique organism. There is no question about recognizing an individual. This simplistic view of an individual breaks down, however, when the organism is modular rather than unitary.

In modular organisms, the zygote develops into a unit of construction (a module) that then produces further, similar modules. Most plants are modular. A tree, shrub, or herbaceous plant grown from a seed is an individual

with its own genetic characteristics (see Chapter 5). Once established, some species of trees—such as black locust (*Robinia pseudoacacia*) and quaking aspen (*Populus tremuloides*)—shrubs, and many perennial herbaceous plants grow root extensions to send up new shoots or suckers that may remain attached to root extensions or break off to live independently (Figure 9.1). These new modules (or clones) may cover a considerable area and appear to be individuals. The individual tree or plant produced by sexual reproduction and thus arising from a zygote is a genetic individual, or **genet.** Modules produced asexually by the genet are **ramets.** The ramets may remain physically linked to the parent genet or they may separate, functioning independently. These modules can produce seeds and their own lateral extensions or ramets.

Whether living independently or physically linked to the original individual, all ramets possess the same genetic constitution as the original product of sexual reproduction. Thus, by producing ramets, the genet can cover a relatively large area and considerably extend its life. Some modules die, others live, and new ones appear.

Plants are the most obvious group of modular organisms. Many other modular animals, such as corals, sponges, and bryozoans (Figure 9.2), also grow by repeated production of modules.

Technically, to study populations of modular organisms, we must recognize two levels of population structure: the module (ramet) and the individual (genet). For practical purposes, however, ramets are often counted as—and function as—individual members of the population.

9.2 | The Distribution of a Population Defines Its Spatial Location

The **distribution** of a population describes its spatial location, the area over which it occurs. Distribution is based on the presence and absence of individuals. If we assume that each red dot in Figure 9.3 represents an individual's position within a population on the landscape, we can draw a line (shown in blue) defining the population distribution—a spatial boundary within which all individuals in the population reside. When the defined area encompasses all the individuals of a species (total area occupied by a population), the distribution describes the population's **geographic range.**

Population distribution is influenced by the occurrence of suitable environmental conditions. The red maple (*Acer rubrum*), for example, is the most widespread of all deciduous trees of eastern North America (Figure 9.4). The northern limit of its geographic range coincides with the area in southeastern Canada where minimum winter temperatures drop to −40°C. Its southern limit is the Gulf Coast and southern Florida. Dry conditions halt its westward range. Within this geographic range, the tree grows under a wide variety of soil types, soil moisture, acidity, and elevations—from wooded swamps to dry

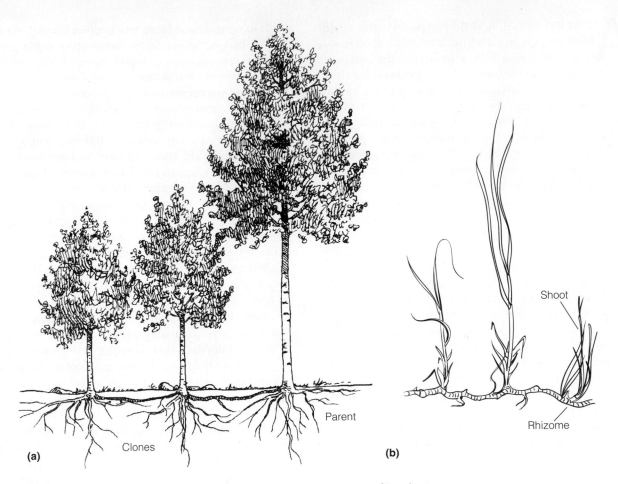

Figure 9.1 | Two examples of modular growth in plants. **(a)** Growth in an aspen tree (*Populus tremuloides*) involves modules of roots and root buds, which give rise to clones. These clones are various ages, with the youngest individuals forming the leading edge of growth away from the parent. **(b)** Growth pattern of sea grass *Halodule beaudettei* (shoal grass) along rhizomes. The rhizomes grow below bottom sediments, and new short shoots are produced at regular intervals as the rhizome grows laterally.

Figure 9.2 | Examples of clonal animal species: **(a)** corals and **(b)** sponges.

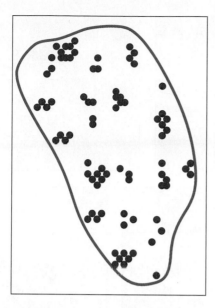

Figure 9.3 | A hypothetical population. Each red dot represents an individual organism. The blue line defines the population distribution, or the area in which the population occurs.

ridges. Thus, the red maple exhibits high tolerance to temperature and other environmental conditions. In turn, this high tolerance allows a widespread geographic range.

Figure 9.4 | Red maple (*Acer rubrum*), one of the most abundant and widespread trees in eastern North America, thrives on a wider range of soil types, texture, moisture, acidity, and elevation than does any other forest species in North America. The northern extent of its range coincides with the minimum winter temperature in southeastern Canada.

The example of red maple illustrates another important factor limiting the distribution of a population: geographic barriers. Although this tree species occupies several islands south of mainland Florida (see map in Figure 9.4), the southern and eastern limits of its geographic range correspond to the Gulf and Atlantic coastline. Although environmental conditions may be suitable for establishment and growth in other geographic regions of the world (such as Europe and Asia), the red maple is restricted in its ability to colonize those areas. Other barriers to dispersal (movement of individuals), such as mountain ranges or extensive areas of unsuitable habitat, may likewise restrict the spread and therefore the geographic range of a species.

Later, in Part Four, we will explore another factor that can restrict the distribution of a population: interactions, such as competition and predation, with other species.

Within the geographic range of a population, individuals are not distributed equally throughout the area. Individuals occupy only those areas that can meet their requirements. An organism responds to a variety of environmental factors, and it can inhabit a location only when those factors are all within the range of tolerance (see Section 5.8). As a result, we can describe the distribution of a population at various spatial scales. For example, in Figure 9.5, the distribution of the moss *Tetraphis pellucida* is described at several different spatial scales, ranging from its geographic distribution at a global scale to the location of individuals within a single clump occupying the stump of a dead conifer tree. This species of moss can grow only in areas where the temperature, humidity, and pH are suitable, and different factors may be limiting at different spatial scales. At the continental scale, the suitability of climate (temperature and humidity) is the dominant factor. Within a particular area, distribution of the moss is limited to microclimates along stream banks, where conifer trees are abundant. Within a particular locality, it occupies the stumps of conifer trees where the pH is sufficiently acidic.

Due to environmental heterogeneity, most populations are divided into subpopulations, each occupying a patch of suitable habitat that is separated from other subpopulations by areas that are unsuitable. In the example of *Tetraphis* presented in Figure 9.5, distribution of individuals within a region is limited to stream banks, where temperature and humidity are within its range of tolerance, and stands of conifers are present to provide a substrate for growth. As a result, the population is divided into **local subpopulations,** each associated with a watershed. Ecologists refer to the collective of local subpopulations as a **metapopulation,** which we will address in detail in Chapter 12. Ecologists typically study these local, or subpopulations, rather than the entire population of a species over its geographic range. For this

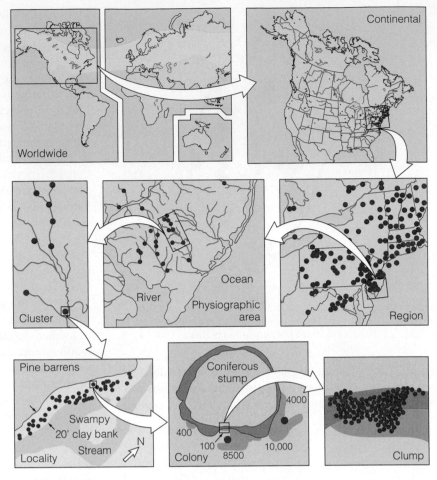

Figure 9.5 | Distribution of the moss *Tetraphis pellucida* at various spatial scales, from its global geographic range to the location of individual colonies on a tree stump.
(Adapted from Firman 1964; as illustrated in Krebs 2001.)

reason, it is important when referring to a population to define explicitly its boundaries (spatial extent). For example, an ecological study might refer to the population of red maple trees in the Three Ridges Wilderness Area of the George Washington–Jefferson National Forest in Virginia, or to the population of Chinook salmon spawning in the Copper River of Alaska (see introduction to Part Three).

9.3 | Abundance Reflects Population Density and Distribution

Whereas distribution defines the spatial extent of a population, **abundance** defines its size—the number of individuals in the population. In Figure 9.3, the population abundance is the total number of red dots (individuals) within the blue line that defines the population distribution.

Abundance is a function of two factors: (1) the population density and (2) the area over which the population is distributed. **Population density** is the number

of individuals per unit area (per square kilometer, hectare, or square meter), or per unit volume (per liter or m³). By placing a grid over the population distribution shown in Figure 9.3, as is done in Figure 9.6, we can calculate the density for any given grid cell by counting the number of red dots that fall within its boundary. Density measured simply as the number of individuals per unit area is referred to as **crude density.** The trouble with this measure is that individuals are typically not equally numerous over the geographic range of the population (see Section 9.2). Individuals do not occupy all the available space within the population's distribution, because not all areas are suitable. As a result, density can vary widely from location to location (see Figure 9.6).

How individuals are distributed within a population—in other words, their spatial position relative to each other—has an important bearing on density. Individuals of a population may be distributed randomly, uniformly, or in clumps (Figure 9.7). Individuals may be distributed randomly if each individual's position is independent of those of the others.

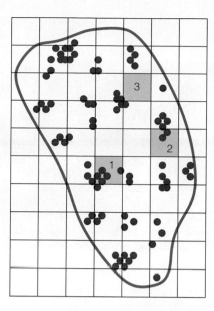

Figure 9.8 | Shrubs in the Kara Kum Desert of central Asia conform to a uniform distribution. The root systems of these shrubs extend laterally up to eight times the diameter of their canopy, and competition for water in this arid environment is intense.

Figure 9.6 | We can use the hypothetical population shown in Figure 9.3 to define the distinction between abundance and density. Abundance is defined as the total number of individuals in the population (red dots). Population density is defined as the number of individuals per unit area. The grid divides the distribution into quadrats of equal size. If we assume that each grid cell is 1 m², the density of grid cell 1 is 5 individuals/m², the density of grid cell 2 is 2 individuals/m², and the density of the third grid cell is zero (unoccupied).

By contrast, individuals distributed uniformly are more or less evenly spaced. A uniform distribution usually results from some form of negative interaction among individuals, such as competition (see Chapter 11), which functions to maintain some minimum distance among members of the population. Uniform distributions are common in animal populations where individuals defend an area for their own exclusive use (territoriality; see Section 11.10) or in plant populations where severe competition exists for belowground resources such as water or nutrients (Figure 9.8).

The most common spatial distribution is clumped, in which individuals occur in groups. Clumping results from a variety of factors. For example, suitable habitat or other resources may be distributed as patches on the larger landscape. Some species form social groups, such as fish that move in schools or birds in flocks. Plants that reproduce asexually form clumps, as ramets extend

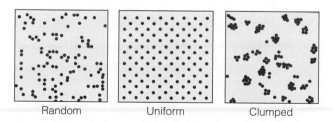

| Random | Uniform | Clumped |

Figure 9.7 | Patterns of the spatial distribution for individuals within a population: random, uniform, and clumped.

outward from the parent plant (see Figure 9.1). The distribution of humans is clumped because of social behavior, economics, and geography, reinforced by the growing development of urban areas during the past century (see the introduction to Part Eight). In the example presented in Figure 9.6, the individuals within the population are clumped; as a result, the density varies widely between grid cells.

As with geographic distribution (see Section 9.2), the spatial distribution of individuals within the population can also be described at multiple spatial scales. For example, the distribution of the shrub *Euclea divinorum*, found in the savanna ecosystems of Southern Africa, is clumped (Figure 9.9a). The clumps of *Euclea* are associated with the canopy cover of another plant that occupies the savanna: trees of the genus *Acacia* (Figure 9.9b). The clumps, however, are uniformly spaced, reflecting the uniform distribution of *Acacia* trees on the landscape. The regular distribution of trees is a function of severe competition among neighboring individuals for water (see Section 11.10). In the example of *Tetraphis* presented in Figure 9.5, the spatial distribution of individuals is clumped at two different spatial scales. Populations are concentrated in long bands or strips along the stream banks, leaving the rest of the area unoccupied. Within these patches, individuals are further clumped in patches corresponding to the distribution of conifer stumps.

To account for patchiness, ecologists often refer to **ecological density,** the number of individuals per unit of available living space. For example, in a study of bobwhite quail (*Colinus virginianus*) in Wisconsin, biologists expressed density as the number of birds per mile of hedgerow (the birds' preferred habitat), rather than as birds per hectare. Ecological densities are rarely estimated, because determining what portion of a habitat represents living space is typically a difficult undertaking.

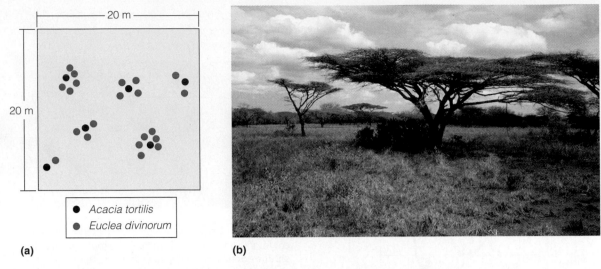

(a)

● *Acacia tortilis*
● *Euclea divinorum*

(b)

Figure 9.9 | Spatial distribution of the shrub *Euclea divinorum*, inhabiting the savannas of southern Africa. Individuals are clumped under the canopies of *Acacia tortilis* trees, as seen in both **(a)** the mapping of individuals within a sample plot and in **(b)** the accompanying photograph. The clumps, however, are uniformly spaced due to the uniform spacing of *A. tortilis* trees on the landscape.

9.4 | Determining Density Requires Sampling

Population size (abundance) is a function of population density and the area that is occupied (geographic distribution). In other words, population size = density × area. But how is density determined? When both the distribution (spatial extent) and abundance are small, as in the case of many rare or endangered species (see Chapter 28), a complete count may be possible. Likewise, in some habitats that are unusually open, such as antelope living on an open plain or waterfowl concentrated in a marsh, density may be determined by a direct count of all individuals. In most cases, however, density must be estimated by sampling the population.

A method of sampling used widely in the study of populations of plants and sessile (attached) animals involves quadrats, or sampling units. Researchers divide the area of study into subunits, in which they count animals or plants of concern in a prescribed manner, usually counting individuals in only a subset or sample of the subunits (as in Figure 9.6). From these data, they determine the mean density of the units sampled. Multiplying the mean value by the total area provides an estimate of population size (abundance). The accuracy of sampling can be influenced by the manner in which individuals are spatially distributed over the landscape (Section 9.3). The estimate of density can also be influenced by the choice of boundaries or sample units. If a population is clumped—concentrated into small areas—and the population density is described in terms of individuals per square kilometer, you receive a false impression (Figure 9.10). In this case, it is important to report an estimate of variation or provide a confidence interval for the estimate of density. In cases where clump-

ing is a result of habitat heterogeneity (habitat is clumped), ecologists may choose to use the index of ecological density for the specific areas (habitats) in which the species is found (for example, stream banks in Figure 9.5).

For mobile populations, animal ecologists must use other sampling methods. Capturing, marking, and recapturing individuals within a population—known generally as mark-recapture—is the most widely used technique to estimate animal populations. There are many variations of this technique, and entire books are devoted to various methods of application and statistical analysis. Nevertheless, the basic concept is simple.

Capture-recapture or mark-recapture methods are based on trapping, marking, and releasing a known number of marked animals (M) into the population (N). After giving the marked individuals an appropriate period of time to once again mix with the rest of the population, some individuals are again captured from the

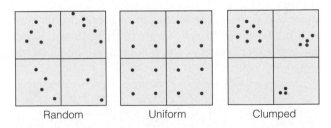

Random   Uniform   Clumped

Figure 9.10 | The difficulty of sampling. Each area contains a population of 16 individuals. We divide each area into four sampling units and choose one at random. Our estimates will be quite different, depending on which unit we select. For the random population, the estimates are 20, 20, 16, and 8. For the uniform population, any sampling unit gives a correct estimate (16). For the clumped population, the estimates are 32, 20, 0, and 12.

population (n). Some of the individuals caught in this second period will be carrying marks (recaptured, R), and others will not. If we assume that the ratio of marked to sampled individuals in the second sample (n/R) represents the ratio for the entire population (N/M), we can compute an estimate of the population using the following relationship:

$$\frac{N}{M} = \frac{n}{R}$$

The only variable that we do not know in this relationship is N. We can solve for N by rearranging the equation as follows:

$$N = \frac{nM}{R}$$

For example, suppose that in sampling a population of rabbits, a biologist captures and tags 39 rabbits from the population. After their release, the ratio of the number of rabbits in the entire population (N) to the number of tagged or marked rabbits (M) is N/M. During the second sample period, the biologist captures 15 tagged rabbits (R) and 19 unmarked ones—a total of 34 (n). The estimate of population size, N, is calculated as

$$N = \frac{nM}{R} = \frac{(34 \times 39)}{15} = 88$$

This simplest method, the single mark–single recapture, is known as the Lincoln index or Petersen index of relative population size.

For work with most animals, ecologists find that a measure of relative density or abundance is sufficient. Methods involve observations relating to the presence of organisms rather than to direct counts of individuals. Techniques include counts of vocalizations, such as recording the number of drumming ruffed grouse heard along a trail; counts of animal scat seen along a length of road traveled; or counts of animal tracks, such as the number of opossums crossing a certain dusty road. If these observations have some relatively constant relationship to total population size, you can then convert the data to the number of individuals seen per kilometer or heard per hour. Such counts, called indices of abundance, cannot function alone as estimates of actual density. However, a series of such index figures collected from the same area over a period of years depicts trends in abundance. Counts obtained from different areas during the same year provide a comparison of abundance between different habitats. Most population data on birds and mammals are based on indices of relative abundance rather than on direct counts.

9.5 | Populations Have Age Structures

Abundance describes the number of individuals in the population but provides no information on their characteristics—that is, how individuals within the population may differ from each other. Unless each generation reproduces and dies in a single season, not overlapping the next generation (such as annual plants and many insects), the population will have an age structure: the number or proportion of individuals in different age classes. Because reproduction is restricted to certain age classes and mortality is most prominent in others, the relative proportions of each age group bear on how quickly or slowly populations grow (see Chapter 10).

Populations can be divided into three ecologically important age classes or stages: prereproductive, reproductive, and postreproductive. We might divide humans into young people, working adults, and senior citizens. How long individuals remain in each stage depends largely on the organism's life history (see Chapter 8). Among annual species, the length of the prereproductive stage has little influence on the rate of population growth. In organisms with variable generation times, the length of the prereproductive period has a pronounced effect on the population's rate of growth. Short-lived organisms often increase rapidly, with a short span between generations. Populations of long-lived organisms, such as elephants and whales, increase slowly and have a long span between generations.

Determining a population's age structure requires some means of obtaining the ages of its members. For humans, this task is not a problem; but it is for wild populations. Age data for wild animals can be obtained in several ways, and the method varies with the species (Figure 9.11). The most accurate, but most difficult, method is to mark young individuals in a population and follow their survival through time. This method requires a large number of marked individuals and a lot of time. For this reason, biologists may use other, less-accurate methods. These methods include examining a representative sample of individual carcasses to determine their ages at death. A biologist might look for the wear and replacement of teeth in deer and other ungulates, growth rings in the cementum of the teeth of carnivores and ungulates, or annual growth rings in the horns of mountain sheep. Among birds, observations of plumage changes and wear in both living and dead individuals can separate juveniles from subadults (in some species) and adults. Aging of fish is most commonly accomplished by counting rings deposited annually (annuli) on hard parts including scales, otoliths (ear bones), and spines.

Studies of the age structure of plant populations are few. The major reason is the difficulty of determining the age of plants and whether the plants are genetic individuals (genets) or ramets.

Foresters have tried to use age structure as a guide to timber management (see Chapter 27). They employ size (diameter of the trunk at breast height, or dbh) as an indicator of age on the assumption that diameter increases with age—the greater the diameter, the older the tree.

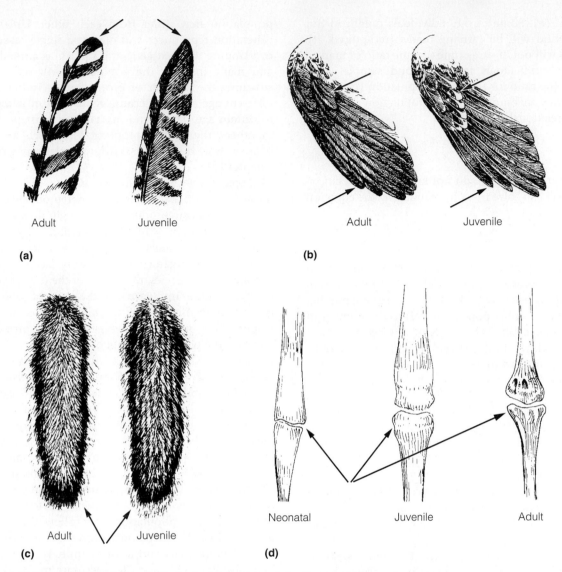

Adult   **Juvenile**

(a)

Adult   **Juvenile**

(b)

Adult   **Juvenile**

(c)

Neonatal   **Juvenile**   **Adult**

(d)

Figure 9.11 | Examples of methods used by ecologists for determining the age of birds and mammals. **(a)** The leading primary feather on an adult wild turkey is rounded, whereas that of a juvenile is pointed. **(b)** Besides having a pointed leading primary feather, the juvenile bobwhite quail has buff-colored primary coverts (see arrows). **(c)** Differences in the color bars on the tail of gray squirrels distinguish adults from juveniles. Juveniles have much more distinctive bands of white and black along the tail edge. **(d)** Researchers can detect differences in the bone structure of bat wings, and determine age, by feeling the wing bones of living individuals.

Such assumptions, foresters discovered, were valid for dominant canopy trees; but with their growth suppressed by lack of light, moisture, or nutrients, smaller understory trees, seedlings, and saplings add little to their diameters. Although their diameters suggest youth, small trees often are the same age as large individuals in the canopy.

You can determine the approximate ages of trees in which growth is seasonal by counting annual growth rings (Figure 9.12). Attempts to age nonwoody plants have not been successful. The most accurate method of determining the age structure of short-lived herbaceous plants is to mark individual seedlings and follow them through their lifetimes.

Age pyramids (Figure 9.13) are snapshots of the age structure of a population at some period in time, providing a picture of the relative sizes of different age groups in the population. As we shall see in Chapter 10, the age structure of a population is a product of the age-specific patterns of mortality and reproduction. In plant populations, the distribution of age classes is often highly skewed (Figure 9.14, p. 192), because dominant overstory trees can inhibit the establishment of seedlings and growth and survival of juvenile trees. One or two age classes dominate the site until they die or are removed, allowing young age classes access to resources such as light, water, and nutrients so they can grow and develop.

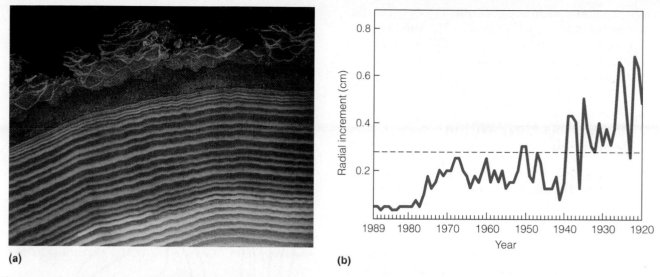

(a)   **(b)**

Figure 9.12 | **(a)** Cross section of tree trunk showing annual growth rings. By measuring the width of each ring, a pattern of radial growth through time can be established. **(b)** Example of time series of radial increments for an American beech tree (*Fagus grandifolia*) in central Virginia. Dashed line is overall average for the tree over time.

9.6 | Sex Ratios in Populations May Shift with Age

Populations of sexually reproducing organisms in theory tend toward a 1:1 sex ratio (the proportion of males to females). The primary sex ratio (the ratio at conception) also tends to be 1:1. This statement may not be universally true, and it is, of course, difficult to confirm.

In most mammalian populations, including humans, the secondary sex ratio (the ratio at birth) is often weighted toward males, but the population shifts toward females in the older age groups. Generally, males have a shorter life span than females do. The shorter life expectancy of males can be a result of both physiological and behavioral factors. For example, in many animal species, rivalries among males occur for

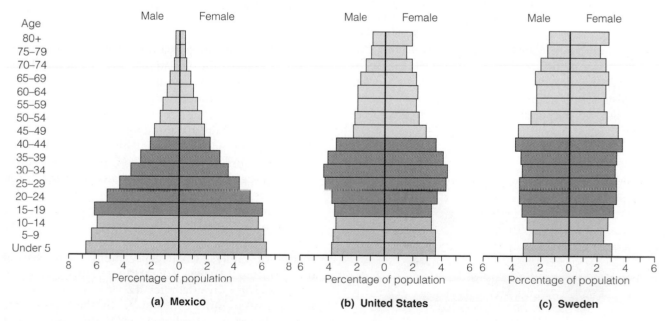

Figure 9.13 | Age pyramids for three human populations in 1989. **(a)** Mexico shows an expanding population. A broad base of young will enter the reproductive age classes. **(b)** The United States has a less tapered age pyramid. The youngest age classes are no longer the largest. **(c)** The age pyramid for Sweden is characteristic of a population that is approaching zero growth.

Go to **GRAPHIT!** at www.ecologyplace.com to explore island biogeography theory.

dominant positions in social hierarchies or for the acquisition of mates (see Section 8.4). Among birds, males tend to outnumber females because of increased mortality of nesting females, which are more susceptible to predation and attack.

9.7 | Individuals Move Within the Population

At some stage in their lives, most organisms are to some degree mobile. The movement of individuals directly influences their local density. The movement of individuals in space is referred to as **dispersal**, although the term *dispersal* most often refers to the more specific movement of individuals away from each other. When individuals move out of a subpopulation, it is referred to as **emigration.** When an individual moves from another location into a subpopulation, it is called **immigration.** The movement of individuals among subpopulations within the larger geographic distribution is a key process in the dynamics of metapopulations (see Chapter 12) and in maintaining the flow of genes between these subpopulations (see Chapter 5).

Many organisms, especially plants, depend on passive means of dispersal involving gravity, wind, water, and animals. The distance these organisms travel depends on the agents of dispersal. Seeds of most plants fall near the parent, and their density falls off quickly with distance (Figure 9.15). Heavier seeds, such as the acorns of oaks (*Quercus* spp.), have a much shorter dispersal range than do the lighter wind-carried seeds of maples (*Acer* spp.), birch (*Betula* spp.), milkweed (Asclepiadaceae), and dandelions (*Taxaxacum officinale*). Some plants, such as cherries and viburnums (*Viburnum* spp.), depend on active carriers such as particular birds and mammals to disperse their seeds by eating the fruits and carrying the seeds to some distant point. These seeds pass through the animals' digestive tract and are deposited in their feces.

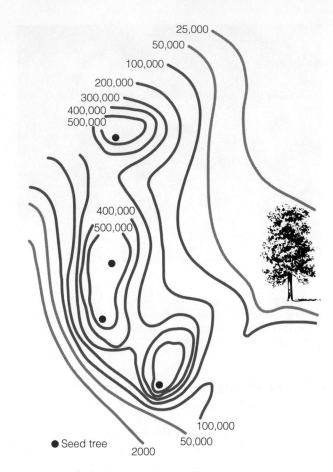

Figure 9.15 | Pattern of annual seedfall of yellow poplar (*Liriodendron tulipifera*). Lines define areas of equal density of seeds. With this wind-dispersed species, seedfall drops off rapidly away from the parent trees.
(Adapted from Engle 1960.)

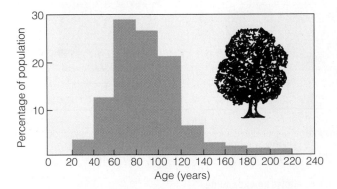

Figure 9.14 | An oak (*Quercus*) forest in Sussex, England, is dominated by trees in the 60- to 120-year age classes. There has been no recruitment of new trees (younger age classes) for 20 years.

Other plants possess seeds armed with spines and hooks that catch on the fur of mammals, the feathers of birds, and the clothing of humans. In the example of the clumped distribution of *E. divinorum* shrubs (see Figure 9.9), birds disperse seeds of this species. The birds feed on the fruits and deposit the seeds in their feces, as they perch atop the *Acacia* trees. In this way, the seeds are dispersed across the landscape, and the clumped distribution of the *E. divinorum* is associated with the use of *Acacia* trees as bird perches.

For mobile animals, dispersal is active; but many depend on a passive means of transport, such as wind and moving water. Wind carries the young of some species of spiders, larval gypsy moths, and cysts of brine shrimp (*Artemia salina*). In streams, the larval forms of some invertebrates disperse downstream in the current to suitable habitats. In the oceans, the dispersal of many organisms is tied to the movement of currents and tides.

Dispersal among mobile animals may involve young and adults, males and females; there is no hard-and-fast rule about who disperses. The major dispersers among

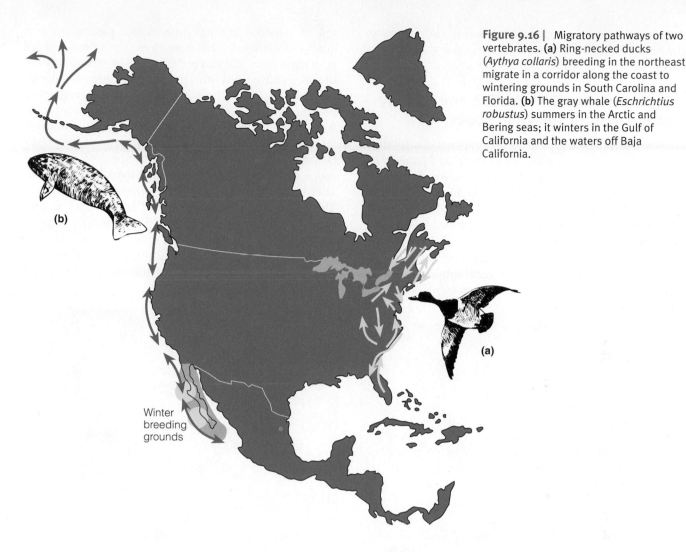

Figure 9.16 | Migratory pathways of two vertebrates. **(a)** Ring-necked ducks (*Aythya collaris*) breeding in the northeast migrate in a corridor along the coast to wintering grounds in South Carolina and Florida. **(b)** The gray whale (*Eschrichtius robustus*) summers in the Arctic and Bering seas; it winters in the Gulf of California and the waters off Baja California.

Winter breeding grounds

(b)

(a)

birds are usually the young. Among rodents, such as deer mice (*Peromyscus maniculatus*) and meadow voles (*Microtus pennsylvanicus*), subadult males and females make up most of the dispersing individuals. Crowding, temperature change, quality and abundance of food, and photoperiod all have been implicated in stimulating dispersal in various animal species.

Often, the dispersing individuals are seeking vacant habitat to occupy. As a result, the distance they travel will depend partly on the density of surrounding populations and the availability of suitable unoccupied areas.

Unlike the one-way movement of animals in the processes of emigration and immigration, **migration** is a round-trip. The repeated return trips may be daily or seasonal. Zooplankton in the oceans, for example, move down to lower depths by day and move up to the surface by night. Their movement appears to be a response to light intensity. Bats leave their daytime roosting places in caves and trees, travel to their feeding grounds, and return by daybreak. Other migrations are seasonal, either short range or long range. Earthworms annually make a vertical migration deeper into the soil to spend the winter below the freezing depths and move back to the

upper soil when it warms in spring. Elk (*Cervus canadensis*) move down from their high mountain summer ranges to lowland winter ranges. On a larger scale, caribou (*Rangifer tarandus*) move from the summer calving range in the arctic tundra to the boreal forests for the winter, where lichens are their major food source. Gray whales (*Eschrichtius robustus*) move down from the food-rich arctic waters in summer to their warm wintering waters of the California coast, where they give birth to young (Figure 9.16). Similarly, humpback whales (*Megaptera novaeangliae*) migrate from northern oceans to the central Pacific off the Hawaiian Islands. Perhaps the most familiar of all are long-range and short-range migrations of waterfowl, shorebirds, and neotropical migrants in spring to their nesting grounds and in fall to their wintering grounds (see Field Studies: T. Scott Sillett, pp. 230–231).

Another type of migration involves only one return trip. Such migrations occur among Pacific salmon (*Oncorhynchus* spp.) that spawn in freshwater streams (see the introduction to Part Three). The young hatch and grow in the headwaters of freshwater coastal streams and rivers and travel downstream and out to sea, where they reach sexual maturity. At this stage, they return to the home stream to spawn (reproduce) and then die.

9.8 | Population Distribution and Density Change in Both Time and Space

Dispersal has the effect of shifting the spatial distribution of individuals and consequently the localized patterns of population density. Emigration may cause density in some areas to decline, while immigration into other areas will increase the density of subpopulations or even establish new subpopulations in habitats that were previously unoccupied.

In some instances, dispersal can result in the shift or expansion of a population's geographic range. The role of dispersal in range expansion is particularly evident in populations that have been introduced to a region where they did not previously exist. A wide variety of species have been introduced, either intentionally or unintentionally, into regions outside their geographic distribution. As the initial population becomes established, individuals disperse into areas of suitable habitat, expanding the geographic distribution as the population grows. A map showing the spread of the gypsy moth (*Lymantria dispar*) in the eastern United States after its introduction in 1869 is shown in Figure 9.17. The story of the introduction of this species is presented in detail in Ecological Issues: Human-Assisted Dispersal.

In other cases, the range expansion of a population has been associated with temporal changes in environmental

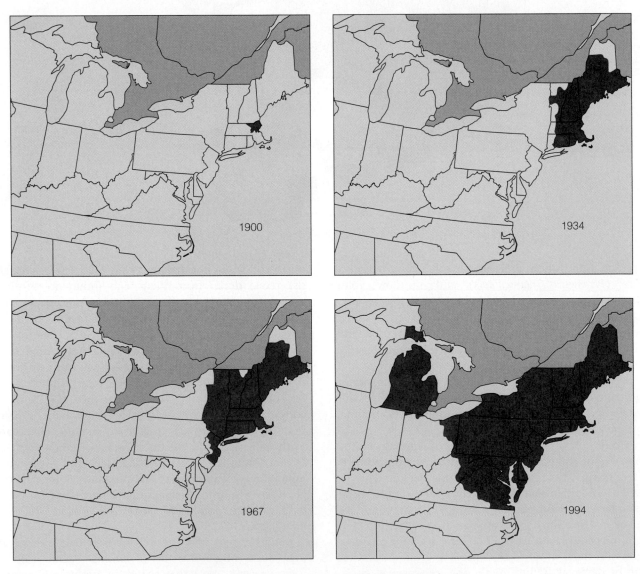

Figure 9.17 | Spread of the gypsy moth (*Lymantria dispar*) in the United States after its introduction in Massachusetts in 1869. Note how dispersal since its original introduction has expanded the geographic distribution of this population. See Ecological Issues: Human-Assisted Dispersal for details.

(Adapted from Liebhold 1992; as illustrated in Krebs 2001.)

Dispersal is a key feature of the life histories of all species, and a diversity of mechanisms have evolved to allow plant and animal species to move across the landscape and seascape. In plants, seeds and spores can be dispersed by wind, water, or through active dispersal by animals (see Section 15.15). In animals, the dispersal of fertilized eggs, particularly in aquatic environments, can result in dispersal of offspring across significant distances. But dispersal typically involves the movement—either active or passive—of individuals, both juvenile and adult. In recent centuries, however, a new source of long-distance dispersal has led to the redistribution of species at a global scale: dispersal by humans.

Humans are increasingly moving about the world. As they do so, they may either accidentally or intentionally introduce organisms, including plants, into places where they have never occurred. Sometimes these introductions are harmless, but often the introduced organisms negatively affect native species and ecosystems. In the past few centuries, many plants have been introduced accidentally by being imported with agricultural products. The seeds of weed species are unintentionally included in shipments of imported crop seeds or on the bodies of domestic animals. Or, seed-carrying soil from other countries is often loaded onto ships as ballast and then dumped in another country in exchange for cargo. Humans have also introduced non-native plants intentionally for ornamental and agricultural purposes. Most introduced plants do not become established and reproduce, but many do form extensive colonies. These invasive plants compete with native species for resources such as light, water, nutrients, pollinators, and seed dispersers. They may also alter the functioning of ecosystems by changing patterns of water use or through the natural frequency of disturbances such as fire—all of which can severely affect native species.

The gypsy moth (*Lymantria dispar*) is native to Europe and Asia and is the major introduced pest of eastern United States hardwood forests. The gypsy moth is found mainly in the temperate regions of the world, including central and southern Europe, northern Africa, central and southern Asia, and Japan. Leopold Trouvelot, a French astronomer with an interest in insects, originally introduced the species into Medford, Massachusetts, in 1869. As part of an effort to begin a commercial silk industry, Trouvelot wanted to develop a strain of silk moth that was resistant to disease. However, several gypsy moth caterpillars escaped from Trouvelot's home and established themselves in the surrounding areas. Some 20 years later, the first outbreak of gypsy moths occurred; and despite all control efforts since that time, the gypsy moth has persisted and extended its range (see Figure 9.17). In the United States, the gypsy moth has rapidly moved north to Canada, west to Wisconsin, and south to North Carolina. Gypsy moth caterpillars defoliate millions of acres of trees annually in the United States (Figure 1). In the forests of eastern North America, losses to European gypsy moths in 1981 were $764 million, and the Asian strain that has invaded the Pacific Northwest has already necessitated a $20 million eradication campaign.

Kudzu (*Pueraria montana*), a species of vine native to Asia, was originally introduced into the United States as an ornamental vine at the Philadelphia Centennial Exposition of 1876. By the early part of the 20th century, kudzu was being enthusiastically promoted as a fodder crop, and rooted cuttings were sold to farmers through the mail. In the 1930s and 1940s, kudzu was propagated and promoted by the Soil Conservation

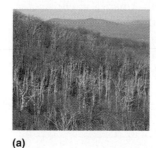
(a)

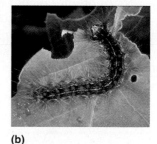
(b)

Figure 1 | An oak forest **(a)** in summer has been completely defoliated by gypsy moth caterpillars **(b)**.

continued on page 196

Figure 2 | Kudzu vines have blanketed the ground, shrubs, and trees.

Service as a means of holding soil on the swiftly eroding gullies of the deforested southern landscape, especially in the Piedmont regions of Alabama, Georgia, and Mississippi. By the 1950s, however, kudzu was recognized as a pest and removed from the list of species acceptable for use under the Agricultural Conservation Program, and in 1998 it was listed by the U.S. Congress as a Federal Noxious Weed. Although it spreads slowly, kudzu completely covers all other vegetation, blanketing trees with a dense canopy through which very little light can penetrate (Figure 2). Estimates of kudzu infestation in the southeastern United States vary greatly, from as low as 2 million to as high as 7 million acres.

1. In looking around your campus or local community, can you identify any examples of plant or animal species that owe their presence to active dispersal by humans?

2. How is agriculture an example of human-assisted dispersal?

---

conditions, shifting the spatial distribution of suitable habitats. Such is the case of the shift in the distribution of tree populations in eastern North America as climate has changed during the past 20,000 years (see Section 18.9, Figure 18.19). Examples of predicted changes in the distribution of plant and animal populations resulting from future human-induced changes in Earth's climate are discussed later, in Chapter 29.

Although the movement of individuals within the population results in a changing pattern of distribution and density through time, the primary factors driving the dynamics of population abundance are the demographic processes of birth and death. The processes of birth and death, and the resulting changes in population structure, are the focus of our attention in the following chapter.

Summary

Unitary and Modular Organisms | 9.1

A population is a group of individuals of the same species living in a defined area. We characterize populations by distribution, abundance, density, and age structure. Most animal populations are made up of unitary individuals with a definitive growth form and longevity. In most plant populations, however, organisms are modular. These plant populations may consist of sexually produced parent plants and asexually produced stems arising from roots. A similar population structure occurs in animal species that exhibit modular growth.

Distribution | 9.2

The distribution of a population describes its spatial location, or the area over which it occurs. The distribution of a population is influenced by the occurrence of suitable environmental conditions. Within the geographic range of a population, individuals are not distributed equally throughout the area. Therefore, the distribution of individuals within the population can be described at a range of different spatial scales.

Individuals within a population are distributed in space. If the spacing of each individual is independent of the others, then the individuals are distributed randomly;

if they are evenly distributed, with a similar distance among individuals, it is a uniform distribution. In most cases, individuals are grouped together in a clumped or aggregated distribution. These groups of individuals can function as subpopulations.

Abundance | 9.3

The number of individuals in a population defines the abundance. Abundance is a function of two factors: (1) the population density and (2) the area over which the population is distributed. Population density is the number of individuals per unit area or volume. Because landscapes are not homogeneous, not all of the area is suitable habitat. The number of organisms in available living space is the true or ecological density.

Sampling Populations | 9.4

Determination of density and dispersion requires careful sampling and appropriate statistical analysis of the data. For sessile organisms, researchers often use sample plots. For mobile organisms, researchers use capture-recapture techniques or determine relative abundance using indicators of animal presence, such as tracks or feces.

Age Structure | 9.5

The age number or proportion of individuals within each age class defines the age structure of a population.

Individuals making up the population are often divided into three ecological periods: prereproductive, reproductive, and postreproductive.

Sex Ratios | 9.6

Sexually reproducing populations have a sex ratio that tends to be 1:1 at conception and birth but often shifts as a function of sex-related differences in mortality.

Dispersal | 9.7

At some stage of their life cycles, most individuals are mobile. For some organisms, such as plants, dispersal is passive and dependent on various dispersal mechanisms. For mobile organisms, dispersal can occur for a variety of reasons, including the search for mates and unoccupied habitat. For some species, dispersal is a systematic process of movement between areas by a process called migration.

Population Dynamics | 9.8

Dispersal has the effect of shifting the spatial distribution of individuals and as a result the localized patterns of population density. Although the movement of individuals within the population results in a changing pattern of distribution and density through time, the primary factors driving the dynamics of population abundance are the demographic processes of birth and death.

Study Questions

1. How does asexual reproduction make it difficult to define what constitutes an individual within a population?

2. Suppose you were given the task of estimating the density of two plant species in a field. Based on the life histories of the two species, you expect that the spatial distribution of one of the species is approximately uniform, whereas the other is likely to be clumped. How might your approach to estimating the density of these two species differ?

3. The age structure of a population can provide insight into whether the population is growing or declining. A large number of individuals in the young age classes relative to the older age classes often indi-

cates a growing population. In contrast, a large proportion of individuals in the older age classes relative to the young age classes suggests a population in decline (see Figure 9.13). What factors might invalidate this interpretation? When might a large number of individuals in the young age classes relative to the older age classes not indicate a growing population?

4. Modern humans are a highly mobile species. Think of three locations in your local community that might be used as areas for estimating the population density. How might the daily movement pattern of people in your community change the estimate of density at these locations during the course of the day?

Further Readings

Brown, J. H., D. W. Mehlman, and G. C. Stevens. 1995. Spatial variation in abundance. *Ecology* 76:2028–2043.

An excellent overview of the factors influencing geographic patterns of abundance in populations.

Cook, R. E. 1983. Clonal plant populations. *American Scientist* 71:244–253.

An introduction to the nature of modular growth in plants and its implications for the study of plant populations.

Gaston, K. J. 1991. How large is a species' geographic range? *Oikos* 61:434–438.

> This paper explores the methods used to define a species' geographic range and how range size is influenced by various aspects of a species' life history.

Gompper, M. E. 2002. Top carnivores in the suburbs? Ecological and conservation issues raised by colonization of Northeastern North America by coyotes. *Bioscience* 52:185–190.

> Fascinating story of the coyote's dispersal and enormous range expansion.

Krebs, C. J. 1999. *Ecological methodology.* 2nd ed. San Francisco: Benjamin Cummings.

> Essential reading for those interested in the sampling of natural populations. The text provides an excellent introduction with many illustrative examples.

Laliberte, A. S., and W. J. Ripple, 2004. Range contractions of North American carnivores and ungulates. *Bioscience* 54:123–138.

> In contrast to the Gompper paper (above) that presents a case of range expansion, this paper explores examples of range contraction of many of the large mammals that were once widely distributed across North America.

Mack, R., and W. M. Lonsdale. 2001. Humans as global plant dispersers: Getting more than we bargained for. *Bioscience* 51:95–102.

> Tells the story of how humans are acting as agents of species dispersal and illustrates some of the unexpected consequences. A good follow-up to the Ecological Issues essay in this chapter.

CHAPTER 10

Population Growth

Elephant (*Loxodonta africana*) herd in Amboseli National Park, Kenya, Africa. This small group is composed of adult females as well as juveniles of various ages. Because African elephants are largely restricted to national parks and conservation areas, these local populations function as closed populations, with no immigration or emigration.

The term *population growth* refers to how the number of individuals in a population increases or decreases with time. This growth is controlled by the rate at which new individuals are added to the population through the processes of birth and immigration and the rate at which individuals leave the population through the processes of death and emigration. We refer to populations in which immigration and/or emigration occur as open populations. Those in which movement into and out of the population does not occur (or is not a significant influence on population growth) are referred to as closed populations.

In this chapter, we will explore the process of population growth under the conditions where population dynamics are a function only of demographic processes relating to birth and death: the conditions where either the population is "closed" (no immigration or emigration) or the rates of immigration and emigration are equal. Later, in Chapters 11 and 12, we will relax this assumption to examine how the interactions among subpopulations through the process of dispersal influence the dynamics of local populations as well as the overall dynamics of the larger metapopulation.

10.1 | Population Growth Reflects the Difference between Rates of Birth and Death

Suppose we were to monitor a population of an organism that has a very simple life history, such as a population of freshwater hydra (see Section 8.1, Figure 8.1) growing in an aquarium in the laboratory. We define the population size as $N(t)$, where t refers to time. Let us assume that the initial population is small, so that the food supply within the aquarium is much more than is needed to support the current population. How will the population change through time?

Because no emigration or immigration is allowed by the lab setting, the population is closed. The number of hydra will increase as a result of new "births" (recall from the photograph in Figure 8.1 that hydra reproduce asexually by budding). Additionally, the population will decrease as a result of some hydra dying. Because the processes of birth and death in this population are continuous (no defined period of synchronized birth or death), we can define the proportion of hydra producing a new individual per unit of time as b, and the proportion of hydra dying per unit of time can be d. If we start with $N(t)$ hydra at time t, then to calculate the total number of hydra reproducing over a given time period, Δt (the symbol Δ refers to a "change" in the associated variable; in this case, a change in time or time interval), we simply need to multiply the proportion reproducing per unit time by the total number of hydra and the length of the time period: $bN(t)\Delta t$. Because each reproducing hydra will add only one individual to the total

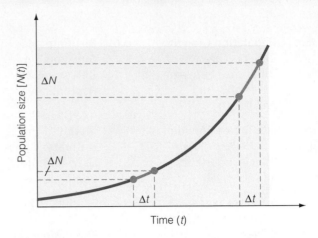

Figure 10.1 | Change in the size of a hypothetical population of hydra through time (green line). The change in population size, ΔN, for a given time interval, Δt, differs as a function of time (t), as indicated by the slope of the line segments shown in orange. (See QUANTIFY it! at www.ecologyplace.com to review functions.)

population (see Figure 8.1), the number of births is $B(t) = bN(t)\Delta t$. Note that B and N are functions of time (because they change as time goes along), but the per capita birthrate b is a constant. For this reason, we write $B(t)$ and $N(t)$, but simply b. The number of deaths, $D(t)$, is calculated in a similar manner, so that $D(t) = dN(t)\Delta t$.

The population size at the next time period ($t + \Delta t$) would then be

$$N(t + \Delta t) = N(t) + B(t) - D(t)$$

or

$$N(t + \Delta t) = N(t) + bN(t)\Delta t - dN(t)\Delta t$$

The resulting pattern of population size as a function of time is shown in Figure 10.1.

We can define the change in population over the time interval (Δt) by rearranging the equation just presented. First, we move $N(t)$ to the left-hand side of the equation, and then divide both sides by Δt:

$$\frac{N(t + \Delta t) - N(t)}{\Delta t} = bN(t) - dN(t)$$

$$= (b - d)N(t)$$

If we substitute ΔN for $[N(t + \Delta t) - N(t)]$, we can rewrite the equation as

$$\frac{\Delta N}{\Delta t} = (b - d)N(t)$$

The term on the left-hand side of the equation defines the unit change in population size per unit change in time, or the slope of the relationship between $N(t)$ and t (the "rise" over the "run") presented in Figure 10.1. Because the relationship between $N(t)$ and t is nonlinear

Derivatives and Differential Equations

Suppose we wish to measure the rate of change in a population through time. Let us assume that population size, $N(t)$, is a linear function of time, t. The resulting graph will be a straight line, which may look something like Figure 1.

Let us select two points on the graph at $t = t_1$ and $t = t_2$. Then we call $\Delta t = t_2 - t_1$ the "run" and $\Delta N = N(t_2) - N(t_1)$ the "rise." The rate of population change ΔN over the time interval Δt is given by the slope of the line (s) defined as the rise per unit of run:

$$s = \frac{\Delta N}{\Delta t}$$

If we choose a different pair of points, then we will still have the same value of the slope (rate of population change), because the slope of a linear function (straight line) does not depend on t. The slope of a nonlinear function (curve), however, does depend on the value of t (Figure 2).

Suppose we hold t_1 constant, and move t_2 closer to t_1 (as in graphs (a) and (b) in Figure 2). As we do so, the slope (indicated by the orange dashed line) will vary; but as the two points get progressively closer together, the slopes will vary by smaller and smaller amounts and will in fact approach a constant "limiting value." When this happens—as in graph (c)—we call the limiting value the slope of the tangent to the curve at point t_1, or simply the slope of the curve at point t_1 (note the tangent intersects the function at the point $[N(t_1), t_1]$ only). Mathematically, we can express this value as

$$\text{slope at } t_1 = \lim_{\Delta t \to 0} \frac{\Delta N}{\Delta t} = \lim_{\Delta t \to 0} \frac{N(t_1 + \Delta t) - N(t_1)}{\Delta t}$$

The slope of the function $N(t)$ at t_1 is known as the derivative of $N(t)$, written as $dN(t)/dt$. Thus, the definition of the derivative is

$$\frac{dN(t)}{dt} = \lim_{\Delta t \to 0} \frac{N(t_1 + \Delta t) - N(t_1)}{\Delta t}$$

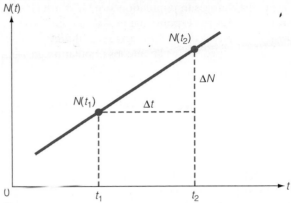

Figure 1

It is important to remember that when the derivative of the function $N(t)$ is evaluated at point t, t is held constant, whereas Δt is varied (approaches zero).

For those who have not taken a course in calculus, this process of taking limits may sound confusing. But in practice, we can think of making the time interval Δt so small that decreasing it further will not noticeably affect the slope of the tangent (as shown in Figure 2)—our estimate of the rate of population change.

An equation where the derivative appears on the left-hand side is referred to as a differential equation, as with the exponential model of population growth: $dN/dt = rN$. We will encounter many other examples of differential equations in our discussions of population dynamics in Chapters 11–15.

1. Why does the size of the time interval (Δt) matter when estimating the slope of a nonlinear function (curve)?

2. What errors would arise in estimating the rate of population growth if the time interval (Δt) were too large, as in graph (a) in Figure 2?

Go to **QUANTIFYit!** at **www.ecologyplace.com** to practice differential equations.

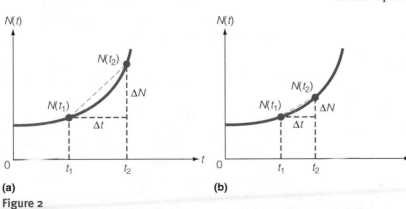

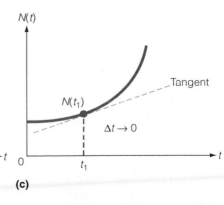

(a) (b) (c)

Figure 2

Exponential Model of Population Growth

To predict changes in population size (N) through time (t) using the exponential model of population growth, it is first necessary to integrate the differential equation presented for instantaneous population growth:

$$\frac{dN}{dt} = rN$$

The first step is to collect terms with N in them on the same side of the equation:

$$\frac{dN}{N} = rdt$$

We then take the integral of each side. The integral of $dN/N = \ln(N) + c_1$, where c_1 is a constant of integration. The integral of $rdt = rt + c_2$, where c_2 is a constant of integration.

$$\ln N + c_1 = rt + c_2$$

We then combine the constants of integration into a single constant, c. Let $c = c_2 - c_1$:

$$\ln N = rt + (c_2 - c_1) = rt + c$$

Next take each side of the equation and raise e to that power, to get rid of the logarithms:

$$e^{\ln N} = e^{rt+c}$$

We can then simplify the equation by applying the rules of exponents:

$$N = e^c e^{rt}$$

We now need to determine what e^c means. Because e and c are both constants, e^c must also be some constant—its value will not change with changing time or population size. We can solve for the value of e^c at a specific time. The simplest solution is for $t = 0$, the initial time when we observe our population. The value of N will be $N(0)$. Now we solve the equation for $t = 0$ and get

$$N(0) = e^c e^{r(0)} \text{ or } N(0) = e^c e^{(0)}$$

Because anything raised to the 0 power is 1:

$$N(0) = e^c$$

So, e^c is the initial population size.

If we now substitute $N(0)$ for e^c in the equation $N = e^c e^{rt}$ above and $N(t)$ for N to indicate that it is the population size at time t, the resulting equation is

$$N(t) = N(0)e^{rt}$$

This equation can now be used to predict the population size at any time in the future.

1. Why is setting $t = 0$ the simplest solution for determining the value of e^c?

2. For readers who have taken a course in calculus, what are you actually estimating (in Figures 10.1 or 10.3) when you integrate the equation representing the exponential growth model: $dN/dt = rN$? Assume that $N(0)$ is the value of $N(t)$ where the curve intersects the y-axis.

(curve), the slope changes as a function of time, so that the rate of change depends on the time interval being evaluated (Figure 10.1). For this reason, the rate of change is best described by the derivative (see Quantifying Ecology 10.1: Derivatives and Differential Equations), written as:

$$\frac{dN}{dt} = (b - d)N$$

The term $\Delta N/\Delta t$ is replaced by dN/dt to express that Δt (the time interval) approaches a value of zero, and the rate of change is instantaneous. The values of b and d represent the instantaneous (per capita) rates of birth and death; and because they are constants, we can define $r = (b - d)$, and rewrite the equation for continuous population growth as

$$\frac{dN}{dt} = rN$$

The value r is the instantaneous per capita rate of growth (sometimes called the intrinsic rate of population growth), and the resulting equation is referred to as the model of **exponential population growth.**

The model of exponential growth ($dN/dt = rN$) predicts the rate of population change through time (Figure 10.2). If we wish to define the equation to predict population size, $N(t)$, under conditions of exponential growth, it is necessary to integrate the differential equation presented earlier. The result is

$$N(t) = N(0)e^{rt}$$

where $N(0)$ is the initial population size at $t = 0$, and e is the base of the natural logarithms; its value is approximately 2.72. For readers with a background in calculus, the proof is presented in Quantifying Ecology 10.2: Exponential Model of Population Growth.

Examples of exponential growth for differing values of r are shown in Figure 10.2. Note that when $r = 0$, there is no change in population size. For values of $r > 0$ the population increases exponentially, whereas values of $r < 0$ result in an exponential decline in the population.

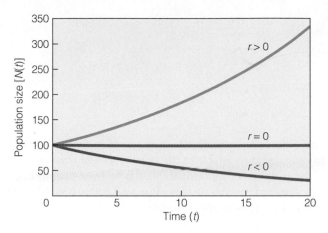

Figure 10.2 | Examples of exponential growth under differing values of r, the instantaneous per capita growth rate. When $r > 0$ ($b > d$), the population size increases exponentially; for values of $r < 0$ ($b > d$), there is an exponential decline. When $r = 0$ ($b = d$), there is no change in population size through time.

Exponential growth results in a continuously accelerating rate of population increase (or decelerating rate of decrease) as a function of population size (Figure 10.3).

Exponential growth is characteristic of populations inhabiting favorable environments at low population densities, such as during the process of colonization and establishment in new environments. An example of a population undergoing exponential growth is the rise of the reindeer herd introduced on St. Paul, one of the Pribilof Islands, Alaska (Figure 10.4). After being introduced on St. Paul in 1910, reindeer expanded rapidly from 4 males and 22 females to a herd of 2000 in only 30 years. The whooping crane (*Grus americana*) is another example of a population exhibiting exponential growth (Figure 10.5). This species was endangered and on the brink of extinction until recently. It was first protected in 1916, but only 15 birds existed in 1941. The whooping crane breeds in

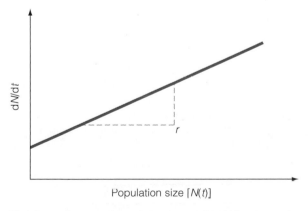

Figure 10.3 | Rate of population growth (dN/dt) expressed as a function of population size [$N(t)$] for the exponential growth model: $dN/dt = rN$. The growth rate increases continuously with $N(t)$. The slope of the line, defined as (dN/dt)/$N(t)$, is the instantaneous population growth rate, r.

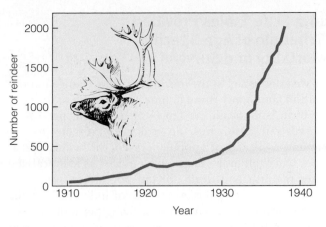

Figure 10.4 | Exponential growth of the St. Paul reindeer (*Rangifer tarandus*) herd following introduction in 1910.
(Adapted from Scheffer 1951.)

the Northwestern Territories of Canada and migrates to overwinter on the Texas coast at the Aransas National Wildlife Refuge. Counts of the entire population since 1938 have provided the data presented in Figure 10.5.

(a)

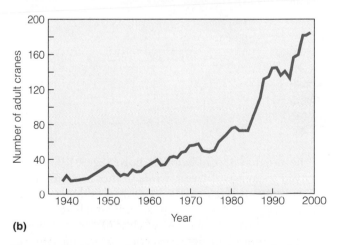

(b)

Figure 10.5 | Pattern of exponential population growth of the whooping crane, (a) an endangered species that has recovered from near extinction in 1941. Population estimates (b) are based on an annual count of adult birds on the wintering grounds at Aransas National Wildlife Refuge.
(Data from Binkley and Miller 1983, and Cannon 1996.)

10.2 | Life Tables Provide a Schedule of Age-Specific Mortality and Survival

As we established in the previous section, change in population abundance through time is a function of the rates of birth and death, as represented by the per capita growth rate r. But how do ecologists estimate the per capita growth rate of a population? For the hydra population, where all individuals can be treated as identical, the rates of birth and death for the population were estimated by counting the proportion of individuals in the population either giving birth or dying per unit of time. But when birth and death rates vary with age, a different approach must be used.

To obtain a clear and systematic picture of mortality and survival within a population, ecologists use an approach involving the construction of life tables. The **life table** is simply an age-specific account of mortality. This technique, first developed by students of human populations, is used by life insurance companies as the basis for evaluating age-specific mortality rates. Now, however, population ecologists are using life tables to examine systematic patterns of mortality and survivorship within animal and plant populations.

The construction of a life table begins with a **cohort,** a group of individuals born in the same period of time. For example, data presented in the following table represent a cohort of 530 gray squirrels (*Sciurus carolinensis*) from a population in northern West Virginia that was the focus of a decade-long study. The fate of these 530 individuals was tracked until all had died some 6 years later. The first column of numbers, labeled x, represents the age classes; in this example, the age classes are in units of years. The second column, n_x, represents the number of individuals from the original cohorts that are alive at the specified age (x).

x	n_x
0	530
1	159
2	80
3	48
4	21
5	5

Of the original 530 individuals born (age 0), only 159 survived to an age of 1 year, while of those 159 individuals, only 80 survived to age 2. Only 5 individuals survived to age 5, and none of those individuals survived to age 6 (that is why there is no age class 6).

It is common practice when constructing life tables to express the number of individuals surviving to any given age as a proportion of the original cohort size. This value, represented as l_x, represents the probability at birth of surviving to any given age.

x	n_x	l_x	
0	530	1.00	$n_0 / n_0 = 530 / 530$
1	159	0.30	$n_1 / n_0 = 159 / 530$
2	80	0.15	$n_2 / n_0 = 80 / 530$
3	48	0.09	
4	21	0.04	
5	5	0.01	

The difference between the number of individuals alive for any age class (n_x) and the next older age class (n_{x+1}) is the number of individuals that have died during that time interval. We define this value as d_x, which gives us a measure of age-specific mortality.

x	n_x	d_x	
0	530	371	$n_0 - n_1 = 530 - 159$
1	159	79	$n_1 - n_2 = 159 - 80$
2	80	32	
3	48	27	
4	21	16	
5	5	5	

The number of individuals that died during any given time interval (d_x) divided by the number alive at the beginning of that interval (n_x) provides an **age-specific mortality rate, q_x.**

x	n_x	d_x	q_x	
0	530	371	0.70	$d_0 / n_0 = 371 / 530$
1	159	79	0.50	$d_1 / n_1 = 79 / 159$
2	80	32	0.40	
3	48	27	0.55	
4	21	16	0.75	
5	5	5	1.00	

A complete life table for the cohort of gray squirrels, including all of the preceding calculations, is presented in Table 10.1. In addition, the calculation of age-specific life expectancy, e_x, the average number of years into the future that an individual of a given age is expected to live, is presented in Quantifying Ecology 10.3: Life Expectancy.

Table 10.1 | Gray Squirrel Life Table

x	n_x	l_x	d_x	q_x
0	530	1.0	371	0.7
1	159	0.3	79	0.5
2	80	0.15	32	0.4
3	48	0.09	27	0.55
4	21	0.04	16	0.75
5	5	0.01	5	1.0

Life Expectancy

Most of us are not familiar with the concept of life tables; however, almost everyone has heard or read statements like, "The average life expectancy for a male in the United States is 72 years." What does this mean? What is life expectancy? Life expectancy (e) typically refers to the average number of years an individual is expected to live from the time of its birth. Life tables, however, are used to calculate age-specific life expectancies (e_x), or the average number of years that an individual of a given age is expected to live into the future. We can use the life table for the cohort of female gray squirrels presented in Table 10.1 to examine the process of calculating age-specific life expectancies for a population.

The first step in estimating e_x is to calculate L_x using the n_x column of the life table. L_x is the average number of individuals alive during the age interval x to $x + 1$. It is calculated as the average of n_x and n_{x+1}. This estimate assumes that mortality within any age class is distributed evenly over the year.

x	n_x	L_x	
0	530	344.5	$= (n_0 + n_1) / 2 = (530 + 159) / 2 = 344.5$
1	159	119.5	
2	80	64.0	$= (n_2 + n_3) / 2 = (80 + 48) / 2 = 64$
3	48	34.5	
4	21	13.0	
5	5	2.5	$= (n_5 + n_6) / 2 = (5 + 0) / 2 = 2.5$

Next, the L_x values are used to calculate T_x, the total years lived into the future by individuals of age class x in the population. This value is calculated by summing the values of L_x cumulatively from the bottom of the column to age x.

x	L_x	T_x	
			$= L_0 + L_1 + L_2 + L_3 + L_4 + L_5$
			$= 344.5 + 119.5 + 64 + 34.5 + 13 + 2.5 = 578$
0	344.5	578.0	
1	119.5	233.5	
2	64.0	114.0	
3	34.5	50.0	$= L_4 + L_5 = 13 + 2.5 = 15.5$
4	13.0	15.5	
5	2.5	2.5	$= L_5 = 2.5$

In the example of the gray squirrel, the value of T_0 is 578. This means that the 530 individuals in the cohort lived a total of 578 years (some only 1 year, while others lived to age 5).

The life expectancy for each age class (e_x) is then calculated by dividing the value of T_x by the corresponding value of n_x. In other words, divide the total number of years lived into the future by individuals of age x by the total number of individuals in that age group.

x	n_x	T_x	e_x	
0	530	578.0	1.09	$= T_0 / n_0 = 578 / 530 = 1.09$
1	159	233.5	1.47	
2	80	114.0	1.43	$= T_2 / n_2 = 114 / 80 = 1.43$
3	48	50.0	1.06	
4	21	15.5	0.75	
5	5	2.5	0.50	

Note that life expectancy changes with age. On average, gray squirrel individuals born can expect to live for only 1.09 years. However, for those individuals that survive past their first birthday, life expectancy increases to 1.47. Life expectancy remains high for age class 2 and then declines for the remainder of the age classes.

1. Why does life expectancy increase for those individuals that survive to age 1 (1.47 as compared to 1.09 for newborn individuals)?

2. Which would have a greater influence on the life expectancy of a newborn (age 0)—a 20 percent decrease in mortality rate for individuals of age class $0(x = 0)$, or a 20 percent decrease in the mortality rate of age class $4(x = 4)$ individuals? Why?

Go to **GRAPHit!** at www.ecologyplace.com to learn more about age structure diagrams.

10.3 | Different Types of Life Tables Reflect Different Approaches to Defining Cohorts and Age Structure

There are two basic kinds of life tables. The first type is the **cohort** or **dynamic life table.** This is the approach used in constructing the gray squirrel life table presented in Table 10.1. The fate of a group of individuals, born at a given time, is followed from birth to death—for example, a group of individuals born in the year 1955. A modification of the dynamic life table is the **dynamic composite life table.** This approach constructs a cohort from individuals born over several time periods instead of just one. For example, you might follow the fate of individuals born in 1955, 1956, and 1957.

Table 10.2 | Life Table of a Sparse Gypsy Moth Population in Northeastern Connecticut

x	n_x	l_x	d_x	q_x
Eggs	450	1.000	135	0.300
Instars I–III	315	0.700	258	0.819
Instars IV–VII	57	0.127	33	0.582
Prepupae	24	0.053	1	0.038
Pupae	23	0.051	7	0.029
Adults	16	0.036	0	1.000

Source: Data from R. W. Campbell 1969.

The second type is the **time-specific life table.** It is constructed by sampling the population in some manner to obtain a distribution of age classes during a single time period. Although it is much easier to construct, this type of life table requires some crucial assumptions. First, you assume that you sampled each age class in proportion to its numbers in the population. Second, you must assume that the age-specific mortality rates (and birthrates) have been constant over time.

Most life tables have been constructed for long-lived vertebrate species having overlapping generations (such as humans). Many animal species, especially insects, live through only one breeding season. Because their generations do not overlap, all individuals belong to the same age class. We obtain the values of n_x by observing a natural population several times over its annual season, estimating the size of the population at each time. For many insects, the n_x values can be obtained by estimating the number surviving from egg to adult. If records are kept of weather, abundance of predators and parasites, and the occurrence of disease, death from various causes can also be estimated.

Table 10.2 represents the fate of a cohort from a single gypsy moth egg mass. The age interval, or x column, indicates the life history stages, which are of unequal duration. The n_x column indicates the number of survivors at each stage. The d_x column gives an accounting of deaths in each stage.

Table 10.3 | Life Table for a Natural Population of *Sedum smallii*

x	l_x	d_x	q_x
Seed produced	1.000	0.16	0.160
Available	0.840	0.630	0.750
Germinated	0.210	0.177	0.843
Established	0.033	0.009	0.273
Rosettes	0.024	0.010	0.417
Mature plants	0.014	0.014	1.000

Source: Data from Sharitz and McCormick 1973.

In plant demography, the life table is most useful in studying three areas: (1) seedling mortality and survival, (2) population dynamics of perennial plants marked as seedlings, and (3) life cycles of annual plants. An example of the third type is Table 10.3, showing a life table for the annual elf orpine (*Sedum smallii*). The time of seed formation is the initial point in the life cycle. The l_x column indicates the proportion of plants alive at the beginning of each stage; the d_x column indicates the proportion dying, rather than the actual number of individuals (as in the other examples).

10.4 | Life Tables Provide Data for Mortality and Survivorship Curves

Although we can graphically display data from any of the columns in a life table, the two most common approaches are the construction of (1) a mortality curve based on the q_x column, and (2) a survivorship curve based on the l_x column. A mortality curve plots mortality rates in terms of q_x against age. Mortality curves for the life tables presented in Table 10.1 (gray squirrel) and Table 10.3 (*S. smallii*) are shown in Figure 10.6. For the gray squirrel cohort (Figure 10.6a), the curve consists of two parts: a juvenile phase, in

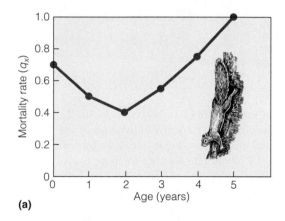
(a)

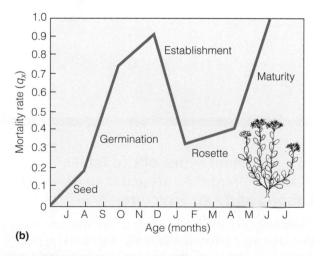
(b)

Figure 10.6 | Examples of mortality curves for **(a)** gray squirrel population, based on Table 10.1, and **(b)** *Sedum smallii*, based on Table 10.3.

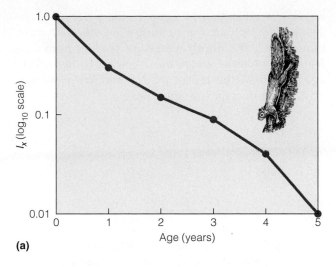

(a)

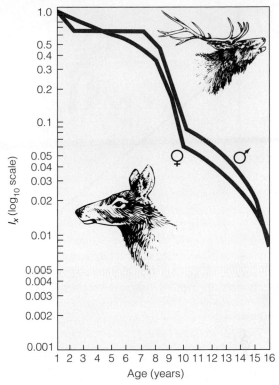

Figure 10.8 | Comparison of survivorship curves for male and female red deer (*Cervus elaphus*).

(Adapted from Lowe 1969.)

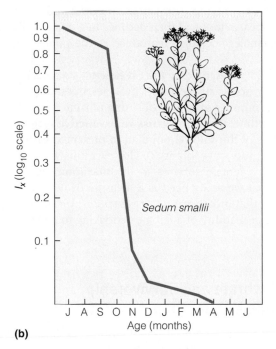

(b)

Figure 10.7 | Survivorship curve for **(a)** gray squirrel based on Table 10.1 and **(b)** *Sedum smallii*, based on Table 10.3.

which the rate of mortality is high, and a post-juvenile phase, in which the rate decreases with age until mortality reaches some low point, after which it increases again. For plants, the mortality curve may assume various patterns, depending on whether the plant is annual or perennial and how we express the age structure. Mortality rates for the *Sedum* population (Figure 10.6b) are initially high, declining once seedlings are established.

Survivorship curves plot the l_x from the life table against time or age class (x). The time interval is on the horizontal axis, and survivorship is on the vertical axis. Survivorship (l_x) is plotted on a $\log_{10}$ scale. Survivorship curves for the life tables presented in Table 10.1 (gray squirrel) and Table 10.3 (*S. smallii*) are shown in Figure 10.7.

Life tables and survivorship curves are based on data obtained from one population of the species at a particular

time and under certain environmental conditions. They are like snapshots. For this reason, survivorship curves are useful for comparing one time, area, or sex with another (Figure 10.8).

Survivorship curves fall into three general idealized types (Figure 10.9). When individuals tend to live out their physiological life span, survival rate is high

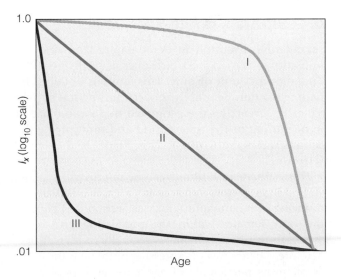

Figure 10.9 | The three basic types of survivorship curves.

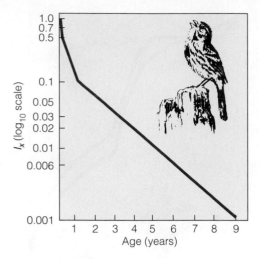

l_x (log$_{10}$ scale)

Age (years)

Figure 10.10 | Survivorship curve for the song sparrow (*Melospiza melodia*). The curve is typical of birds. After a period of high juvenile mortality (type III), the curve becomes linear (type II).

(Adapted from Johnson 1956.)

throughout the life span, followed by heavy mortality at the end. With this type of survivorship pattern, the curve is strongly convex, or type I. Such a curve is typical of humans and other mammals and has also been observed for some plant species. If survival rates do not vary with age, the survivorship curve will be straight, or type II. Such a curve is characteristic of adult birds, rodents, and reptiles, as well as many perennial plants. If mortality rates are extremely high in early life—as in oysters, fish, many invertebrates, and many plant species, including most trees—the curve is concave, or type III. These generalized survivorship curves are idealized models to which survivorship of a species can be compared. Many survivorship curves show components of these three generalized types at different times in the life history of a species (Figure 10.10).

10.5 | Birthrate Is Age-Specific

A standard convention in demography (the study of populations) is to express birthrates as births per 1000 population per unit of time. This figure is obtained by dividing the number of births over some period of time (typically a year) by the estimated population size at the beginning of the time period and multiplying the resulting number by 1000. This figure is the **crude birthrate.**

This estimate of birthrate can be improved by taking two important factors into account. First, in a sexually reproducing (nonhermaphroditic; see Section 8.2) population, only females within the population give birth. Second, the birthrate of females generally varies with age. Therefore, a better way of expressing birthrate is the number of births per female of age x. Because population

increase is a function of reproduction by females, the age-specific birthrate can be further modified by determining only the mean number of females born to a female in each age group, b_x. Following is the table of **age-specific birthrates** for the gray squirrel population used to construct the life table (Table 10.1):

x	b_x
0	0
1	2
2	3
3	3
4	2
5	0
Σ	10

At age 0, females produce no young; thus the value of b_x is 0. The average number of female offspring produced by a female of age 1 is 2. For females of ages 2 and 3, the b_x value increases to 3 and then declines to 2 at age 4. By age 5 the females no longer reproduce, so the value of b_x is 0.

The sum (represented by the Greek letter sigma, Σ) of the b_x values across all age classes provides an estimate of the average number of female offspring born to a female over her lifetime; this is the **gross reproductive rate.** In the example of the squirrel population presented earlier, the gross reproductive rate is 10. However, this value assumes that a female survives to the maximum age of 5 years. What we really need is a measure of net reproductive rate that incorporates the age-specific birthrate as well as the probability of a female's surviving to any specific age.

10.6 | Birthrate and Survivorship Determine Net Reproductive Rate

We can use the gray squirrel population as the basis for constructing a fecundity or fertility table (Table 10.4). The **fecundity table** uses the survivorship column, l_x, from the life table together with the age-specific birthrates (b_x) described earlier. Although b_x may initially increase with age, survivorship (l_x) in each age class declines. To adjust for mortality, we multiply the b_x

Table 10.4 | Gray Squirrel Fecundity Table

x	l_x	b_x	$l_x b_x$
0	1.0	0.0	0.00
1	0.3	2.0	0.60
2	0.15	3.0	0.45
3	0.09	3.0	0.27
4	0.04	2.0	0.08
5	0.01	0.0	0.00
Σ		10.0	1.40

values by the corresponding l_x, the survivorship values. The resulting value, $l_x b_x$, gives the mean number of females born in each age group, adjusted for survivorship.

Thus, for 1-year-old females, the b_x value is 2; but when adjusted for survival (l_x), the value drops to 0.6. For age 2 the b_x is 3; but $l_x b_x$ drops to 0.45, reflecting poor survival of adult females. The values of $l_x b_x$ are summed over all ages at which reproduction occurs. The result represents the **net reproductive rate, R_0,** defined as the average number of females that will be produced during a lifetime by a newborn female. If the R_0 value is 1, females will on average replace themselves in the population (produce one daughter). If the R_0 value is less than 1, the females are not replacing themselves. If the value is greater than 1, females are more than replacing themselves. For the gray squirrel, an R_0 value of 1.4 suggests a growing population of females. Note the significant difference between the gross and net reproductive rates (10 and 1.4, respectively). The difference reflects the fact that only a small proportion of the females born will survive to the maximum age and produce 10 female offspring.

Because the value of R_0 is a function of the age-specific patterns of birth and survivorship, it is a product of the life history characteristics discussed in Chapter 8: the allocation to reproduction, the timing of reproduction, the trade-off between the size and number of offspring produced, and the degree of parental care. The net reproductive rate (R_0) therefore provides a means of evaluating both the individual (fitness) and population consequences of specific life history characteristics.

10.7 | Age-Specific Mortality and Birthrates Can Be Used to Project Population Growth

Age-specific mortality rates (q_x) from the life table together with the age-specific birthrates (b_x) from the fecundity table can be combined to project changes in the population into the future. To simplify the process, the values for age-specific mortality are converted to age-specific survival. If q_x is the proportion of individuals alive at the beginning of an age class that die before reaching the next age class, then $1 - q_x$ is the proportion that survive to the next age class (Table 10.5)—designated as s_x. With age-specific values of s_x and b_x, we can project the growth of a population by constructing a population projection table.

We will illustrate the construction of a **population projection table** by using data from Table 10.5 and a hypothetical population of squirrels introduced into an unoccupied oak forest. Because females form the reproductive units of the population, we follow only the females in constructing the table. The year of establishment is designated as year 0. The introduced population of female

Table 10.5 | Age-Specific Survival and Birthrates for Squirrel Population

x	l_x	q_x	s_x	b_x
0	1.0	0.7	0.3	0.0
1	0.3	0.5	0.5	2.0
2	0.15	0.4	0.6	3.0
3	0.09	0.55	0.45	3.0
4	0.04	0.75	0.25	2.0
5	0.01	1.0	0.0	0.0

squirrels consists of 20 juveniles (age 0) and 10 adults (age 1), giving a total population of $N(0) = 30$. The following table gives age-specific birthrates (b_x), survival rates (s_x), and number of females in each age class (x) at year 0. These values can now be used to project the population at year 1.

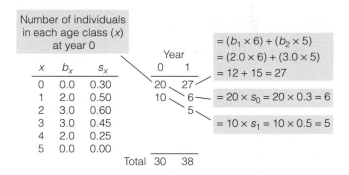

Not all of the squirrels in the initial population (year 0) will survive into the next year (year 1). The survival of these two age groups is obtained by multiplying the number of each by the s_x value. Because the s_x of the 1-year-old females is 0.5, we know that 5 individuals ($10 \times 0.5 = 5$) survive to year 1 and reach age 2. The s_x value of age 0 is 0.3, so that only 6 of the 20 females less than 1 year of age in year 0 survive ($20 \times 0.3 = 6$) to year 1 and become 1-year-olds. In year 1, we now have six 1-year-olds and the five 2-year-olds, with both these age classes reproducing. The b_x value of the six 1-year-olds is 2, so they produce 12 offspring. The five 2-year-olds have a b_x value of 3, so they produce 15 offspring. Together, the two age classes produce 27 young for year 1, and they now make up age class 0. The total population for year 1, $N(1)$, is 38. Survivorship and fecundity are determined in a similar manner for each succeeding year (Table 10.6). Survivorship is tabulated year by year diagonally down the table to the right through the years, while new individuals are being added each year to age class 0.

From such a population projection table, we can calculate **age distribution** for each successive year (see Section 9.5)—the proportion of individuals in the

Table 10.6 | Population Projection Table, Squirrel Population

Age	Year (t) 0	1	2	3	4	5	6	7	8	9	10
0	20	27	34.1	40.71	48.21	58.37	70.31	84.8	101.86	122.88	148.06
1	10	6	8.1	10.23	12.05	14.46	17.51	21.0	25.44	30.56	36.86
2	0	5	3.0	4.05	5.1	6.03	7.23	8.7	10.50	12.72	15.28
3	0	0	3.0	1.8	2.43	3.06	3.62	4.4	5.22	6.30	7.63
4	0	0	0	1.35	0.81	1.09	1.38	1.6	1.94	2.35	2.83
5	0	0	0	0	0.33	0.20	0.27	0.35	0.40	0.49	0.59
Total $N(t)$	30	38	48.2	58.14	68.93	83.21	100.32	120.85	145.36	175.30	211.25
Lambda	λ	1.27	1.27	1.21	1.19	1.21	1.20	1.20	1.20	1.20	1.20

various age classes for any one year—by dividing the number in each age class (x) by the total population size for that year [$N(t)$]. In comparing the age distribution of the squirrel population through time (Table 10.7), we observe that the population attains an unchanging or **stable age distribution** by year 7. From that year on, the proportions of each age group in the population remain the same year after year, even though the population [$N(t)$] is increasing.

Another piece of information that can be derived from the population projection shown in Table 10.6 is an estimate of population growth. By dividing the total number of individuals in year $t + 1$, $N(t + 1)$, by the total number of individuals in the previous year, $N(t)$, we can arrive at the **finite multiplication rate,** λ (Greek letter lambda), for each time period.

$$\frac{N(t + 1)}{N(t)} = \lambda$$

The rate λ has been calculated for each time interval and is shown at the bottom of each column (year) in Table 10.6. Note that initially λ varies between years, but once the population has achieved a stable age distribution, the value of λ remains constant. Values of λ greater than 1.0 indicate a population that is growing, whereas values less than 1.0 indicate a population in decline. A value of

$\lambda = 1.0$ indicates a stable population size, neither increasing nor decreasing through time.

The population projection table demonstrates two important concepts of population growth: (1) the rate of population growth, as estimated by λ, is a function of the age-specific rates of survival (s_x) and birth (b_x), and (2) the constant rate of increase of the population from year to year and the stable age distribution are results of survival and birth rates for each age class that are constant through time.

Given a stable age distribution in which λ does not vary, λ can be used as a multiplier to project population size into the future ($t + 1$). This can be shown very simply by multiplying both sides of the equation for λ shown earlier by the current population size, $N(t)$:

$$N(t + 1) = N(t)\lambda$$

We can predict the population size at year 1 by multiplying the initial population size $N(0)$ by λ, and for year 2 by multiplying $N(1)$ by λ:

$$N(1) = N(0)\lambda$$
$$N(2) = N(1)\lambda$$

Note that by substituting $N(0)\lambda$ for $N(1)$ (see above), we can rewrite the equation predicting $N(2)$ as

$$N(2) = [N(0)\lambda]\lambda = N(0)\lambda^2$$

Table 10.7 | Approximation of Stable Age Distribution, Squirrel Population

Age	Proportion in Each Age Class for Year 0	1	2	3	4	5	6	7	8	9	10
0	.67	.71	.71	.71	.69	.70	.70	.70	.70	.70	.70
1	.33	.16	.17	.17	.20	.17	.17	.18	.18	.18	.18
2		.13	.06	.07	.06	.07	.07	.07	.07	.07	.07
3			.06	.03	.03	.04	.04	.03	.03	.03	.03
4				.02	.01	.01	.01	.01	.01	.01	.01
5					.01	.01	.01	.01	.01	.01	.01

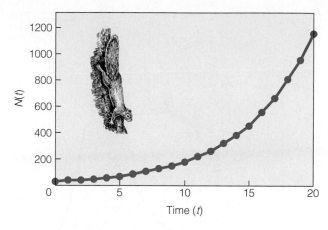

Figure 10.11 | Change in population size [$N(t)$] with time for the gray squirrel population represented in Table 10.1. Population estimates are based on the discrete model of geometric population growth [$N(t) = N(0)\lambda t$] with an initial population density [$N(0)$] of 30 and a finite growth multiplier of $\lambda = 1.2$ from the population projection table (Table 10.6).

In fact, we can use λ to project the population at any year into the future using the following general form of the relationship developed previously:

$$N(t) = N(0)\lambda^t$$

For our squirrel population, we can multiply the population size at year 0, $N(0) = 30$, by $\lambda = 1.20$, the value derived from the population projection table, to obtain a population size of 36 for year 1. If we again multiply 36 by 1.20, or the initial population size 30 by λ^2 (1.20^2), we get a population size of 43 for year 2; and if we multiply the initial population size of 30 by λ^{10}, we arrive at a projected population size of 186 for year 10 (Figure 10.11). These population sizes do not correspond exactly to the population sizes calculated in the population projection table, because λ fluctuates above and below the eventual value attained at stable age distribution. Only after the population achieves a stable age distribution does the λ value of 1.20 project future population size.

The equation $N(t) = N(0)\lambda_t$ describes a pattern of population growth (Figure 10.11) similar to that presented for the exponential growth model developed in Section 10.1. When described over discrete time intervals, however, the pattern of growth is termed **geometric population growth.** In this example, the time interval (Δt) is 1 year, the interval (x) used in constructing the life and fecundity tables from which λ is derived.

Note that the equation predicting population size through time using the finite growth multiplier λ is similar to the corresponding equation describing conditions of exponential growth developed in Section 10.1:

$$N(t) = N(0)\lambda^t \quad \text{(geometric population growth)}$$

$$N(t) = N(0)e^{rt} \quad \text{(exponential population growth)}$$

In fact, the two equations (finite and continuous) illustrate the relationship between λ and r:

$$\lambda = e^r \quad \text{or} \quad r = \ln\lambda$$

For the gray squirrel population, we can calculate the value of $r = \ln(1.20)$, or 0.18.

Unlike the original calculation of r for the hydra population in Section 10.1, this estimate of the per capita population growth rate does not assume that all individuals in the population are identical. It is derived from λ, which as we have seen is an estimate of population growth based on the age-specific patterns of birth and death for the population. This estimate does, however, assume that the age-specific rates of birth and death for the population are constant; that is, they do not change through time. It is this assumption that results in the population converging on a stable age distribution and constant value of λ.

The geometric and exponential models developed thus far provide an important theoretical framework for understanding the demographic processes governing the dynamics of populations. But nature is not constant; systematic and stochastic (random) processes, both internal (demographic) and external (environmental), can influence population dynamics.

10.8 | Stochastic Processes Can Influence Population Dynamics

Thus far, we have considered population growth as a deterministic process. Because the rates of birth and death are assumed to be constant, for a given set of initial conditions—values of r or λ, and $N(0)$—both the exponential and geometric models of population growth will predict only one exact outcome. Recall, however, that the age-specific values of survival and birth in the life and fecundity tables (Tables 10.1 and 10.4) represent probabilities and averages derived from the cohort or population under study. For example, the values of b_x are the average number of females produced by a female of that age group. For the 1-year-old females, the average value is 2.0; however, some female squirrels in this age class may have given birth to four female offspring, whereas others may not have given birth at all. The same holds true for the age-specific survival rates (s_x), which represent the probability of a female of that age surviving to the next age class. For example, in Table 10.5 the probability of survival for a 1-year-old female gray squirrel is 0.5—the same probability of getting a heads or tails in a coin toss. Although survival (and mortality) is expressed as a probability, for any individual it is a discrete event—you either survive to the next year or not, just as the outcome of a single coin toss will be either heads or tails. If we toss a coin 10 times, however, we expect to get on average an outcome of 5 heads and 5 tails. This is in fact what we assume when we multiply the probability of survival (0.5) by the number of females in an age class (10) to project the number surviving to the next year (5) in Table 10.6. But each individual outcome in the 10 coin tosses is independent, and there is a possibility of getting 4 heads and 6 tails (probability $p = 0.2051$), or even 0 heads and 10 tails ($p = 9.765 \times 10^{-4}$). The same is true for the probability of survival when applied to individuals in a specific age class. The realization that population dynamics represent the combined outcome of many individual probabilities has led to the development of probabilistic, or stochastic models of population growth. These models allow the rates of birth and death to vary about the mean estimate represented by the values of b_x and s_x.

The stochastic (or random) variations in birth and death rates occurring in populations from year to year are called **demographic stochasticity,** and they cause populations to deviate from the predictions of population growth based on the deterministic models discussed in this chapter.

Besides demographic stochasticity, random variations in the environment, such as annual variations in climate (temperature and precipitation) or the occurrence of natural disasters such as fire, flood, and drought, can directly influence birth and death rates within the population. Such variation is referred to as **environmental stochasticity.**

10.9 | A Variety of Factors Can Lead to Population Extinction

When deaths exceed births, populations decline: R_0 becomes less than 1.0; r becomes negative. Unless the population can reverse the trend, it may become so low that it declines toward extinction (Figure 10.2).

A variety of factors can lead to population extinction. Extreme environmental events, such as droughts, floods, or extreme temperature events (heat or cold), can increase mortality rates and reduce population size. Should the environmental conditions exceed the bounds of tolerance for the species (for examples, see Figures 5.20 and 6.20), the event could well lead to extinction.

A severe shortage of resources, caused by either environmental extremes (as discussed earlier) or overexploitation, could result in a severe population decline and possible extinction, should the resource base not recover in time to allow for adequate reproduction by survivors. In the example of exponential growth in the population of reindeer introduced on St. Paul in 1910 (Figure 10.4), the reindeer overgrazed their range so severely that the herd plummeted from its high of more than 2000 in 1938 to 8 animals in 1950 (Figure 10.12). The decline produced a curve typical of a population that exceeds the

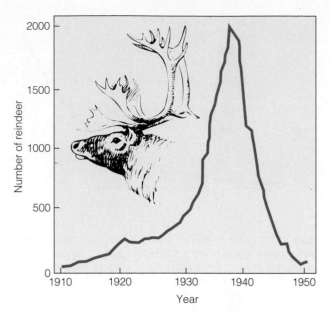

Figure 10.12 | After the initial period of exponential growth of the St. Paul reindeer herd presented in Figure 10.4, overexploitation of resources degraded the habitat, resulting in a sharp decline in population numbers.
(Adapted from Scheffer 1951.)

resources of its environment. Growth stops abruptly and declines sharply in the face of environmental deterioration. From a low point, the population may recover to undergo another phase of exponential growth; or it may decline to extinction.

When a new species is introduced to an ecosystem, either through the natural process of immigration (see Section 9.5) or through human activity (see Ecological Issues: Human-Assisted Dispersal in Chapter 9), the resulting interactions with species in the community can often be detrimental. The introduction of a novel predator, competitor, or parasite (disease) can have a devastating effect on the target population (see Part Four: Species Interactions).

A leading cause of current population extinctions is the loss of habitat due to human activities. The cutting of forests and clearing of lands for agriculture and development have resulted in a significant decline in the available habitat for many species and is currently the leading cause of species extinctions at a global scale (see Chapter 28).

Not all species are equally susceptible to extinction, and various factors influencing the vulnerability of species to extinction are discussed in detail in Chapter 28 (see Section 28.3). However, regardless of the species, small populations—due to their greater vulnerability to demographic and environmental stochasticity (see Section 10.8) and loss of genetic variability—are more susceptible to extinction than larger populations are.

10.10 | Small Populations Are Susceptible to Extinction

Small populations can be susceptible to a variety of factors that directly influence the rates of survival and birth. Small populations are more susceptible to both demographic and environmental stochasticity (see Section 10.8). If only a few individuals make up the population, the fate of each individual can be crucial to the survival of the population. Declining population size also may directly influence birthrates as a result of life history characteristics related to mating and reproduction (Chapter 8).

Among species that are widely dispersed, such as large cats, finding a mate may be impossible once the population density falls below a certain point. Many insect species use chemical odors or pheromones to communicate with and attract mates. As population density falls, there is less probability that an individual's chemical message will reach a potential mate, and reproductive rates may decrease. Similarly, as a plant population declines and individuals are more widely scattered, the distance between plants increases and pollination may become less likely.

Ecologists Erine Hackney and James McGraw of West Virginia University examined the reproductive limitations imposed by small population size on American ginseng (*Panax quinquefolius*), a perennial herbaceous plant species inhabiting the deciduous forest ecosystems of eastern North America. Wild populations of American ginseng have historically been harvested for their medicinal value, and at some locations it has been reduced to populations of only a few dozen individuals. Hackney and McGraw established experimental populations of varying density using cultivated plants. Fruit production per plant declined with decreasing population size (Figure 10.13). The reduced pattern of per-plant fertility was due to reduced visitation by pollinators. Similar studies have confirmed that smaller populations of flowering plants are less apparent to potential pollinator species and therefore have a lower frequency of visitation (see Chapter 15).

In other cases, small population size may result in the breakdown of social structures in species that practice facilitation or cooperative behaviors relating to mating, foraging, or defense. Species that aggregate into groups on communal courtship grounds (called leks; see Section 8.5) are particularly susceptible to disruption of mating behavior and reproduction with declining population size. Many gregarious species live in herds or packs that enable the individuals to defend themselves from predators or find food. Once the population is too small to sustain an effective herd or pack, the population may decline from increased mortality due to predation or starvation.

(a)

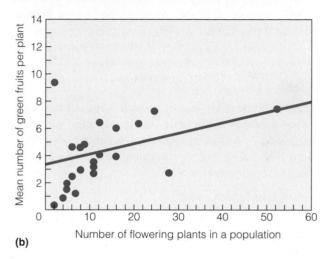

(b)

Figure 10.13 | Example of the Allee effect of reduced fecundity rate at low population density in the perennial herbaceous plant (a) American ginseng. (b) There is a reduction in fruit production per plant at small population sizes in the 26 planted experimental populations. Line represents the general trend as defined by the linear regression $y = 3.36 + 0.077\,x$, where x is the number of flowering plants in population and y is the mean number of green fruits per plant.

(Adapted from Hackney and McGraw 2001.)

The decline in either reproduction or survival under conditions of low population density is known as the **Allee effect**—named for the ecologist W. C. Allee, who first proposed this mechanism of population regulation.

Another factor that has been suggested as a potential cause of extinction in a small population is reduced genetic diversity. Just as a matter of chance, small populations will support less genetic variation than will larger populations. This reduced genetic variation may influence the population's ability to adapt to exposure to a new disease, predator, or changes in the physical environment, such as climate. Two mechanisms that operate in small populations can function to reduce genetic variation—genetic drift and inbreeding, both of which are discussed in detail in Chapter 5 (Section 5.6).

Summary

Population Growth | 10.1

In a population with no immigration or emigration, the rate of change in population size through time over a defined time interval is a function of the difference between the rates of birth and death. When the birthrate exceeds the death rate, the rate of population change increases with population size. As the time interval over which population change is evaluated decreases, approaching zero, the change in population size is expressed as a continuous function, and the resulting pattern is termed exponential population growth. The difference in the instantaneous per capita rates of birth and death is defined as r, the instantaneous per capita growth rate.

Life Table | 10.2

Mortality and its complement, survivorship, are best analyzed by means of a life table—an age-specific summary of mortality. By following the fate of a cohort of individuals until all have died, we can calculate age-specific estimates of mortality and survival.

Types of Life Tables | 10.3

We can construct a cohort or dynamic life table by following one or more cohorts of individuals through time. A time-specific life table is constructed by sampling the population in some manner to obtain a distribution of age classes during a single time period.

Mortality and Survivorship Curves | 10.4

From the life table, we derive both mortality curves and survivorship curves. They are useful for comparing demographic trends within a population and among populations under different environmental conditions and for comparing survivorship among various species. Survivorship curves fall into three major types: type I, in which individuals tend to live out their physiological life span; type II, in which mortality and thus survivorship are constant through all ages; and type III, in which survival of young is low.

Birthrate | 10.5

Birth is the greatest influence on population increase. Like mortality rate, birthrate is age-specific. Certain age classes contribute more to the population than others do.

Net Reproductive Rate | 10.6

The fecundity table provides data on the gross reproduction, b_x, and survivorship, l_x, of each age class. The sum of these products gives the net reproductive rate, R_0, defined as the average number of females that will be produced during a lifetime by a newborn female.

Population Projection Table | 10.7

We can use age-specific estimates of survival and birth rates from the fecundity table to project changes in population density. The procedure involves using the age-specific survival rates to move individuals into the next age class and age-specific birthrates to project recruitment into the population. The resulting population projection table provides future estimates of both population density and age structure. Estimates of changes in population density can be used to calculate λ (lambda), a discrete estimate of population growth rate. This estimate can be used to predict changes in population size through time (geometric growth model). In addition, l can be used to estimate r, the instantaneous per capita growth rate. The estimate of r based on λ accounts for differences in the age-specific rates of birth and death.

Stochastic Processes | 10.8

Because the age-specific values of survival and birth derived from the life and fecundity tables represent average values (probabilities), actual values for individuals within the population can vary. The random variations in birth and death rates that occur in populations from year to year are called demographic stochasticity. Random variations in the environment that directly influence rates of birth and death are termed environmental stochasticity.

Extinction | 10.9

A variety of factors can result in a population declining to extinction, including environmental stochasticity, the introduction of new species, and habitat destruction.

Small Populations | 10.10

Small populations (low density) are more susceptible to extinction than are larger populations. This susceptibility is due to various factors, including demographic stochasticity; environmental stochasticity; disruption of social structures that influence mating, feeding, and defense; and loss of genetic diversity.

Study Questions

1. What is the difference between a discrete ($\Delta N/\Delta t$) and continuous (dN/dt) model of population growth? What is the difference between geometric and exponential growth?

2. What is a life table, and what information (data) is required to construct one?

3. How do the gross reproductive rate and the net reproductive rate differ?

4. How can the value of net reproductive rate for a population (R_0) be used to assess whether a population is increasing or declining?

5. To use a life table to project population growth, what assumption must be made regarding the age-specific rates of survival (s_x)?

6. How is the finite growth multiplier (λ) calculated from the population projection table?

7. How can the finite growth multiplier be used to predict future values of population density, $N(t)$?

8. What environmental factors might result in random yearly variations in the rates of survival and birth within a population?

9. Identify two factors that could possibly cause a population to decline to extinction. How might these factors be influenced by population size?

10. Why are small populations more susceptible to extinction from demographic stochasticity than are larger populations?

Further Readings

Begon, M., and M. Mortimer. 1996. *Population ecology: A unified study of animals and plants.* 3rd ed. Oxford: Blackwell Scientific Publications.

>This well-written and organized introductory text on the ecology of populations is an excellent resource for those wishing to read further about the structure and dynamics of natural populations.

Carey, J. R. 2001. Insect biodemography. *Annual Review of Entomology* 46:79–110.

>A review of life tables in insect populations.

Deevey, E. S., Jr. 1947. Life tables for natural populations of animals. *Quarterly Review of Biology* 22:283–314.

>A classic, pioneering paper on life tables.

Gibbons, J. W., D. E. Scott, and T. J. Ryan, et al. 2000. The global decline of reptiles, déjà vu amphibians. *Bioscience* 50:653–666.

>The decline in amphibian populations has received much attention. Now it appears that reptile populations are also declining. This paper reviews significant threats to reptile populations that may account for this observed decline.

Gotelli, N. J. 2001. *A primer of ecology.* 3rd ed. Sunderland, MA: Sinauer Associates.

>This text explains in detail the most common mathematical models in population and community ecology.

Kaufman, L., and K. Mallory, eds. 1986. *The last extinction.* Cambridge, MA: MIT Press.

>An excellent overview of causes of global extinction.

Levin, D. A. 2002. Hybridization and extinction. *American Scientist* 90:254–261.

>Discusses how interbreeding between species can lead to the extinction of a rare species.

CHAPTER 11

Intraspecific Population Regulation

The pygmy sweep (*Parapriacanthus ransonetti*) form densely packed schools in small openings in the coral knolls of the South Pacific. This species feeds at night on small plankton by using bioluminescent organs located at the back of their pectoral fins.

N

o population continues to grow indefinitely. In particular, populations that exhibit exponential growth eventually confront the limits of the environment. As a population's density changes, interactions mediated by the environment occur among members of the population and tend to regulate the population's size. These interactions include a wide variety of mechanisms relating to physiological, morphological, and behavioral adaptations.

11.1 | The Environment Functions to Limit Population Growth

The exponential model of population growth that we developed in Chapter 10 is based on several assumptions about the environment in which the population is growing. The model assumes that essential resources (space, food, etc.) are unlimited and that the environment is constant (no seasonal or annual variations that might influence the probabilities of birth or death). In the real world, the environment is not constant, and resources are limited. As the density of a population increases, demand for resources increases. If the rate of consumption exceeds the rate at which resources can be resupplied, then the resource base will shrink. Shrinking resources and the potential for an unequal distribution of those resources will result in increased mortality, decreased fecundity, or both. The simplest form of representing changes in birth and death rates with increasing population is a straight line (linear function). The graph in Figure 11.1 presents an example where

the per capita birthrate (b) decreases with increasing population size (N). Conversely, the per capita death rate (d) increases with population size. We can describe the line representing the change in birthrate as a function of population size as

$$b = b_0 - aN$$

In this equation, b_0 is the intercept (value of b when N is equal to zero), and a is the slope of the line ($\Delta b/\Delta N$; see Figure 11.1). The intercept, b_0, represents the birthrate achieved under ideal (uncrowded with no resource limitation) conditions, whereas b is the actual birthrate, which is reduced as a function of crowding. The ideal birthrate, b_0, is the value used in the exponential model of population growth (see Section 10.1). Likewise, we can represent the change in death rate as a function of population size as

$$d = d_0 + cN$$

Again, the constant d_0 is the death rate when the population size is close to zero (no crowding or resource limitation), and the constant c represents the increase in death rate with increasing population size (slope of the line shown in Figure 11.1).

We can now rewrite the exponential model of population growth developed in Section 10.1 ($dN/dt = (b - d)N$) to include the variations in the rates of birth and death as a function of population size presented above:

$$\frac{dN}{dt} = [(b_0 - aN) - (d_0 + cN)]N$$

The pattern of population growth now differs from that of the original exponential model. As N increases, the birthrate ($b_0 - aN$) declines, the death rate ($d_0 + cN$) increases, and the result is a slowing of the rate of population growth. If the value of d exceeds that of b, population growth is negative, and population size declines (Figure 11.1). When the birthrate (b) is equal to the death rate (d), the rate of population change is zero ($dN/dt = 0$). The value of population size where the birthrate is equal to the death rate ($b = d$) represents the maximum sustainable population size under the prevailing environmental conditions. We can solve for this value by setting the equation for population growth developed above equal to zero and solving for N:

$$\frac{dN}{dt} = [(b_0 - aN) - (d_0 + cN)]N = 0$$

$$(b_0 - aN)N - (d_0 + cN)N = 0$$
(move the term for death rate to the right side of the equation)

$$(b_0 - aN)N = (d_0 + cN)N$$
(then divide both sides by N)

$$b_0 - aN = d_0 + cN$$
(move d_0 to the left side of the equation and aN to the right side)

$$b_0 - d_0 = aN + cN$$
(rearrange the right-hand side of the equation)

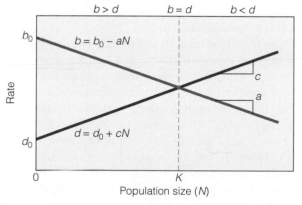

Figure 11.1 | Rates of birth (b) and death (d), represented as a linear function of population size N. The values b_0 and d_0 represent the ideal birth and death rates (respectively) under conditions where the population size is near zero and resources are not limiting. The values a and c represent the slopes of the lines describing changes in birth and death rates as a function of N (respectively). The population density where $b = d$ and population growth is zero is defined as K, the carrying capacity. For values of N above K, b is less than d and the population growth rate is negative. For values of N below K, b is greater than d, and the population growth rate is positive. (Go to **www.ecologyplace.com** to review linear functions.)

The Logistic Model of Population Growth

We can derive the logistic population growth by beginning with the equation that allows the rates of birth (b) and death (d) to vary as a function of population size, as outlined in Section 11.1:

$$\frac{dN}{dt} = (b - d)N$$

Because $b = b_0 - aN$ and $d = d_0 + cN$, we rewrite the equation as follows:

$$\frac{dN}{dt} = [(b_0 - aN) - (d_0 + cN)]N$$

After rearranging the terms (see p. 217), we have

$$\frac{dN}{dt} = [(b_0 - d_0) - (a + c)N]N$$

Next, we multiply by $(b_0 - d_0)/(b_0 - d_0)$. This term is equal to 1.0, so it only simplifies the equation further:

$$\frac{dN}{dt} = \frac{(b_0 - d_0)}{(b_0 - d_0)}[(b_0 - d_0) - (a + c)N]N$$

$$\frac{dN}{dt} = [(b_0 - d_0)]\left[\frac{(b_0 - d_0)}{(b_0 - d_0)} - \frac{(a + c)}{(b_0 - d_0)}N\right]N$$

Because we have defined $r = (b_0 - d_0)$ in Section 11.1, we have

$$\frac{dN}{dt} = rN\left[1 - \frac{(a + c)}{(b_0 - d_0)}N\right]$$

Note that $(a + c)/(b_0 - d_0) = 1/K$, as shown in left column below.

Making the appropriate substitution, we have

$$\frac{dN}{dt} = rN\left[1 - N\left(\frac{1}{K}\right)\right]$$

$$\frac{dN}{dt} = rN\left(1 - \frac{N}{K}\right)$$

This is the equation for the logistic model of population growth.

1. The equation is sometimes presented in an alternative but equivalent form: $dN/dt = rN[(K - N)/K]$. Show algebraically how this equation is equivalent to the one presented above.

Go to **QUANTIFYit!** at www.ecologyplace.com to review derivatives and working with differential equations.

$$(b_0 - d_0) = (a + c)N$$

[divide both sides by $(a + c)$]

$$\frac{(b_0 - d_0)}{(a + c)} = N$$

Because b_0, d_0, a, and c are constants, we can define a new constant:

$$K = \frac{(b_0 - d_0)}{(a + c)}$$

The constant K is the **carrying capacity**—the maximum sustainable population size for the prevailing environment (see Ecological Issues: The Human Carrying Capacity). It is a function of the supply of resources (e.g., food, water, space, etc.).

We can now rewrite the equation for population growth that includes the rates of birth (b) and death (d) that vary with population size using the value of carrying capacity, K, defined above:

$$\frac{dN}{dt} = rN\left(1 - \frac{N}{K}\right)$$

In this form, referred to as the **logistic model of population growth,** the per capita growth rate, r, is defined

as $b_0 - d_0$. The derivation of the logistic equation is presented in Quantifying Ecology 11.1: The Logistic Model of Population Growth.

The logistic model effectively has two components: the original exponential term (rN) and a second term ($1 - N/K$) that functions to reduce population growth as the population size approaches the carrying capacity. When the population density (N) is low relative to the carrying capacity (K), the term ($1 - N/K$) is close to 1.0, and population growth follows the exponential model (rN). However, as the population grows and N approaches K, the term ($1 - N/K$) approaches zero, slowing population growth. Should the population density exceed K, population growth becomes negative and population density declines back toward carrying capacity.

As with the exponential growth model, we can use the rules of calculus (see Quantifying Ecology 10.2: Exponential Model of Population Growth) to integrate the logistic growth equation and express population size as a function of time:

$$N(t) = \frac{K}{1 + \left\{\dfrac{[K - N(0)]}{N(0)}\right\}e^{-rt}}$$

Demographers estimate that the current human population of over 6.5 billion will most likely double by the year 2200. Can Earth support such an increase in the number of human inhabitants? Or, to ask more directly: What is the human carrying capacity of Earth? To address this question, we must qualify it by asking two important questions: (1) at what standard of living (nutritional, cultural, technological, etc.) and (2) at what cost to the environment? In his book *How Many People Can the Earth Support?* the population biologist Joel E. Cohen of Rockefeller University reviews eight estimates of human carrying capacity that use different assumptions relating to these two questions. The earliest estimate was made by E. G. Ravenstein and presented in a paper to the British Association for the Advancement of Science in 1891. The remaining seven are products of the latter part of the 20th century. Remarkably, the estimates vary by a thousandfold.

In October 1966, C. T. DeWit of the Institute for Biological and Chemical Research on Field Crops and Herbage in Wageningen, The Netherlands, presented his estimate of Earth's carrying capacity at the symposium "Harvesting the Sun: Photosynthesis in Plant Life." DeWit framed the question as follows: "How many people can live on Earth if photosynthesis is the limiting process?" In his answer, he estimated the potential global productivity of green plants (crops). After examining the geographic variation in the constraints on photosynthesis imposed by light (solar radiation), length of growing season, and temperature, DeWit calculated Earth's potential productivity in the absence of constraints by water and nutrients (assuming sufficient water and nutrients are made available through irrigation and fertilization). Based solely on estimates of primary production and the average annual per capita energy (caloric) requirement, DeWit estimated that 1000 billion people could live from Earth if photosynthesis is the limiting factor. This estimate is for the population that could live *from* Earth, not *on* Earth, since the estimate assumes that the entire land area is dedicated to the production of plants (food). To account for the land area needed to support activities other than food production, DeWit goes on to assume that each person would require a minimum of 750 m^2 of land for urban use, plus the area required for growing food (approximately 830 m^2). This requirement of approximately 1500 m^2 per person brings the carrying capacity down to 146 billion people. Countless problems are associated with the assumptions embedded in DeWit's calculations, and it is unlikely that the effects on Earth's environment would allow for human survival at such a density.

At the other extreme is a study published in 1970 by H. R. Hulett, who at that time was in the Department of Genetics of the Stanford University Medical School. Hulett set out to calculate the "optimum world population." He used the standard of living and resource consumption (food and other raw materials) for the average U.S. citizen as his definition of optimal lifestyle. Using the ratio of current (1970) world production of these materials to the average per capita American consumption, Hulett calculated an upper limit for the optimal world population. The U.S. population was approximately 200 million in 1970 when Hulett reached the conclusion that approximately 1 billion people represented the maximum world population supportable at the then-current agricultural and industrial levels of production.

So there we have it. Given the range of estimates (1 to 146 billion), the answer no doubt is somewhere in between. Technology will undoubtedly allow our species to further increase agricultural productivity, and necessity will require us to move from fossil fuels to alternative, renewable sources of energy in the future. Perhaps the real issue of Earth's carrying capacity lies not in our ability to feed, clothe, and shelter additional people but rather in the "kind of Earth" that we hope to inhabit.

1. What are some factors (environmental, economic, etc.) that influence the carrying capacity of your region or community?

2. Extreme environmental events in the United States—hurricanes in Florida, freezing temperatures in southern California, drought in the Midwest—dramatically affect agricultural production. What does this imply about the carrying capacity of the United States (or for that matter any region or country)? How might a population respond to unpredictable variations in essential resources through time?

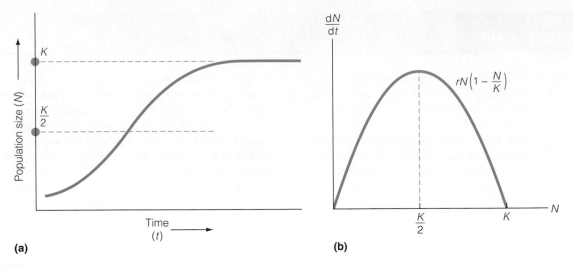

Figure 11.2 | **(a)** Change in population size (*N*) through time as predicted by the logistic model of population growth [$dN/dt = rN(1 - N/K)$]. Initially (low values of *N*) the population grows exponentially; but as *N* increases, the rate of population growth decreases, eventually reaching zero as the population size approaches the carrying capacity (*K*). **(b)** The relationship between the rate of population growth, dN/dt, and population size, *N*, takes the form of a parabola, reaching a maximum value at a population size of $N = K/2$.

The graph of population size (*N*) through time for the logistic model is shown in Figure 11.2a. When the population is small, it increases rapidly, at a rate slightly lower than that predicted by the exponential model. The rate of population growth (dN/dt) is at its highest when $N = K/2$ (called the inflection point) and then decreases as it approaches the carrying capacity (*K*; Figure 11.2b). This is in contrast to the exponential model, in which the population growth rate increases linearly with population size (see Figure 10.2).

As an illustration of logistic growth, we can return to the example of the gray squirrel population presented in Chapter 10. Let us assume a carrying capacity of $K = 200$ individuals. Given the value of $r = 0.18$ calculated from the life table and an initial population size of 30 individuals, the predicted patterns of population growth under both the exponential and logistic models of population growth are shown in Figure 11.3.

11.2 | Population Regulation Involves Density Dependence

The concept of carrying capacity suggests a negative feedback between population increase and resources available in the environment. As population density increases, the per capita availability of resources declines. The decline in per capita resources eventually reaches some crucial level at which it functions to regulate population growth. Implicit in this model of population regulation is **density dependence.**

Density-dependent effects influence a population in proportion to its size. They function by slowing the rate of population growth with increasing population den-

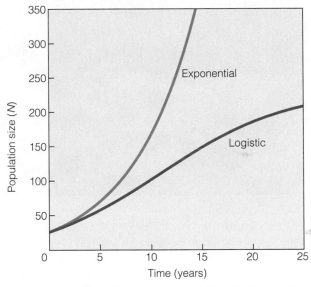

Figure 11.3 | Predictions of the exponential and logistic population growth models for the gray squirrel population from Tables 10.1 and 10.6; $r = 0.18$, $K = 200$, and $N (0) = 30$.

sity, by increasing the rate of mortality (termed **density-dependent mortality**), decreasing the rate of fecundity (**density-dependent fecundity**), or both. In the case of the logistic growth model, this is done by varying the rates of birth (*b*) and death (*d*), expressed through the value of the carrying capacity, *K* (Figure 11.4).

Mechanisms of density-dependent population regulation may include factors other than the direct effects of resource availability. For example, population density can influence patterns of predation (see Chapter 14), or the spread of disease and parasites (see Chapter 15).

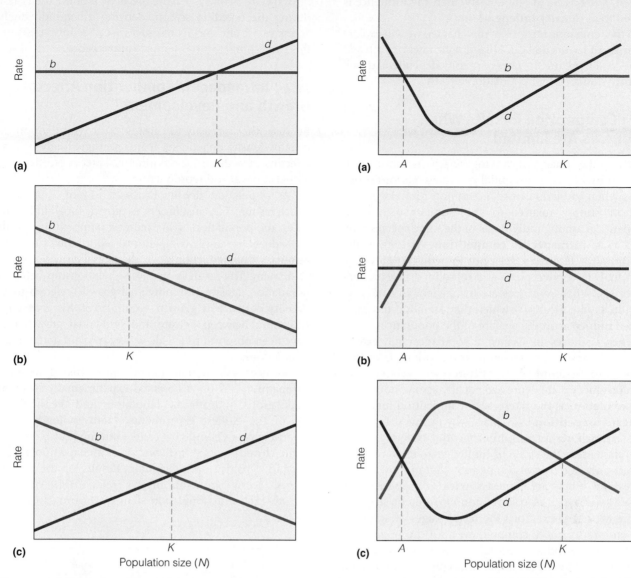

Figure 11.4 | Regulation of population size in three situations. **(a)** Birthrate (*b*) is independent of population density, as indicated by the horizontal line. Only the death rate (*d*) increases with population size. At *K*, equilibrium is maintained by increasing mortality. **(b)** The situation is reversed. Mortality is independent, but birthrate declines with population size. At *K*, a decreasing birthrate maintains equilibrium. **(c)** Full density-dependent regulation. Both birthrate and mortality are density dependent. Fluctuations in either one hold the population at or near *K*.

Figure 11.5 | Under the model of density dependence presented in Figure 11.4, birthrate (*b*) increases and/or death rate (*d*) declines as population size decreases, maintaining the population at carrying capacity (*K*). Under conditions of an Allee effect, below some minimum population density, the mortality rate increases **(a)**, birthrate decreases **(b)**, or both **(c)**. The result is a negative rate of population growth for population densities below *A*. Unlike the negative feedback that occurs when *N* exceeds *K* (returning population size to values at or below the carrying capacity), when values of *N* fall below *A*, a positive feedback occurs, resulting in a further decline in *N* and leading to extinction.

Density-dependent mechanisms have also been identified that reduce rates of birth and survival at low population densities, referred to as the Allee effect. Examples of the Allee effect are discussed in Section 10.10, where low population densities restrict the ability of individuals to find potential mates or result in the breakdown of social structures in species that practice facilitative or cooperative behaviors relating to mating, foraging, or defense (see Figure 10.13). The result is a form of inverse density dependence, where at low population size, birthrate declines; mortality increases; and below some minimum population density, the rate of population growth is negative (Figure 11.5).

Other factors that can directly influence rates of birth and death function independently of population density. If some environmental factor such as adverse weather conditions affects the population regardless of the number of individuals, or if the proportion of individuals affected is the same at any density, then the influence is referred to as **density independent.**

In the remaining sections of this chapter, we will explore the variety of factors that can influence the rates of birth and death—and therefore the rates of density-dependent as well as density-independent population growth.

11.3 | Competition Results When Resources Are Limited

Implicit in the concept of carrying capacity is competition among individuals for essential resources. **Competition** occurs when individuals use a common resource that is in short supply relative to the number seeking it. Competition among individuals of the same species is referred to as **intraspecific competition.** As long as the availability of resources does not impede the ability of individuals to survive, grow, and reproduce, no competition exists. When resources are insufficient to satisfy all individuals, the means by which they are allocated has a marked influence on the welfare of the population.

When resources are limited, a population may exhibit one of two responses: scramble competition or contest competition. **Scramble competition** occurs when growth and reproduction are depressed equally across individuals in a population as the intensity of competition increases. **Contest competition** takes place when some individuals claim enough resources while denying others a share. Generally, under the stress of limited resources, a species exhibits only one type of competition. Some are scramble species and others are contest species. One species may practice both types of competition at different stages in the life cycle. For example, the larval stages of some insects endure scramble competition until the population declines, and the adult stages face contest competition.

The outcomes of scramble and contest competition vary. In its extreme, scramble competition can lead to all individuals receiving insufficient resources for survival and reproduction, resulting in local extinction. In contest competition, only a fraction of the population suffers—the unsuccessful individuals. The survival, growth, and reproduction of individuals that successfully compete for the limited resources all function to sustain the population.

In many cases, competing individuals do not directly interact with one another. Instead, individuals respond to the level of resource availability that is depressed by the presence and consumption of other individuals in the population. For example, large herbivores such as zebras grazing on the savannas of Africa may influence each other not through direct interactions but by reducing the amount of grass available as food. Similarly, as a tree in the forest takes up water through its roots, it decreases the remaining amount of water in the soil for other trees. In these cases, competition is termed **exploitation.**

In other situations, however, individuals interact directly with each other, preventing others from occupying a habitat or accessing resources within it. For example, most bird species actively defend the area around their nest during the breeding season, denying other individuals access to the site and its resources (see Section 11.10). In this case, competition is termed **interference.**

11.4 | Intraspecific Competition Affects Growth and Development

Because the intensity of intraspecific competition is usually density dependent, it increases gradually, at first affecting growth and development. Later it affects individual survival and reproduction.

As population density increases toward a point at which resources are insufficient to provide for all individuals in the population, some (contest competition) or all individuals (scramble competition) reduce their intake of resources. That reduction slows the rate of growth and development. The result is an inverse relationship between population density and individual growth, referred to as **density-dependent growth.** Examples of this inverse relationship between density and individual growth rate have been observed in a wide variety of plant and animal populations.

Reduced growth rate under conditions of resource competition has been observed experimentally in plant populations in both the laboratory and the field. In one of the earliest experiments, plant ecologist J. N. Chatworthy of Oxford University examined the growth of white clover plants (*Trifolium repens*) grown in pots with varying densities of individuals. Results of the experiments clearly show an inverse relationship between growth rate and population density (Figure 11.6). The

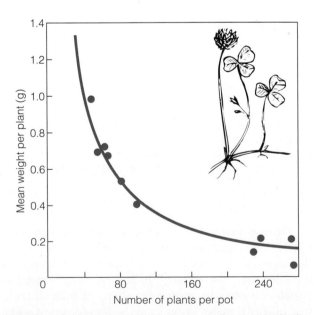

Figure 11.6 | Effect of population density on the growth of individuals. The growth rate and subsequent weight of white clover (*Trifolium repens*) plants declines markedly with increasing density of individuals planted in the pot.

(Adapted from Chatworthy 1960.)

mean weight of individual plants declines with increasing density of individuals in the pot. This decline is a direct consequence of resource limitation. At low densities, all individuals are able to acquire sufficient resources to meet demands for growth. As the density is increased (more individuals planted per pot), demand exceeds the supply of resources in the pot, and both growth rate and plant size decline.

In an experiment to examine the effects of intraspecific competition on growth and photosynthesis of the salt marsh species *Atriplex prostrata* (spear-leaved orache), Li-Wen Wang and colleagues at Ohio University grew plants in pots at varying densities that correspond to the range of densities observed in natural populations. Plants were grown under controlled environmental conditions in growth chambers. After 4 weeks, measurements of

photosynthesis were made, and plants were harvested to provide measures of tissue dry weight, leaf area, and plant height.

As with the results of the earlier experiment by Chatworthy, Wang and colleagues observed an inverse relationship between density and individual plant growth (Figure 11.7). High plant density caused an 80 percent reduction in mean plant weight (Figure 11.7a) and a 72 percent reduction in leaf area (Figure 11.7b) over values observed at the lowest plant densities. A 50 percent reduction in net photosynthetic rate paralleled the observed growth inhibition with increasing plant density (Figure 11.7c), indicating that the growth inhibition caused by intraspecific competition is mainly due to a decline in the net rate of carbon uptake. In addition to the decline in plant weight and leaf area with increasing density, the data

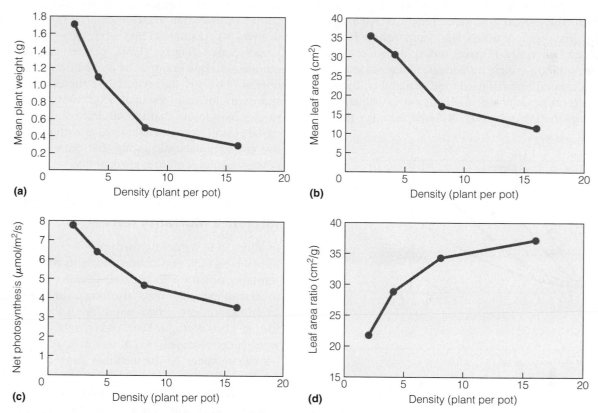

Figure 11.7 | Effects of intraspecific competition on growth and photosynthesis of *Atriplex prostrata*. Mean values of **(a)** plant dry weight (accumulated biomass), **(b)** leaf area, and **(c)** net photosynthetic rate plotted as a function of varying densities of individuals per pot in the experiments. Besides the general pattern of reduced photosynthesis and growth experienced under increasing competition, individual plants exhibit a shift in patterns of carbon allocation, as illustrated by the increase in leaf area ratio (leaf area cm²/plant weight g) with increasing density of plants per pot **(d)**. The increase in leaf area ratio is characteristic of reduced light availability during plant development.

(Data from Wang et al. 2005).

Interpreting Ecological Data

Q1. What is the approximate difference in mean plant weight between individuals grown at the density of 2 plants per pot and those grown at 8 plants per pot? How does this difference in mean plant weight compare to that between plants grown at 8 and 16 plants per pot?

Q2. Why is the decline in mean plant weight with increasing density not linear (straight line)? (Hint: Calculate the total biomass in each pot by multiplying the density [2, 4, 8, and 16 plants per pot] by mean plant weight [1.7, 1.1, 0.5 and 0.3 g] for each treatment.)

illustrate a corresponding shift in carbon allocation and plant morphology. The average leaf area ratio of individual plants (leaf area/total plant weight) increased with increasing plant density (Figure 11.7d), reflecting a relative increase in the allocation of photosynthates (total plant mass) to the production of leaf area (photosynthetic surface). You may recall from the discussion of plant response to variations in the light environment presented in Chapter 6 (Section 6.2 and Quantifying Ecology 6.1: Relative Growth Rate), an increase in leaf area ratio is indicative of reduced light availability, strongly suggesting that the observed reductions in photosynthesis and growth with increasing plant density are a result of shading and competition for light resources.

Similar patterns of density-dependent growth to those reported for plants have been observed among populations of ectothermic (cold-blooded) vertebrates. In a series of experiments to examine the effects of population density on the growth and development of *Rana tigrina* tadpoles, ecologists Madhab Dash and Ashok Hota of Sambalpur University (Orissa, India) reared *R. tigrina* larvae in isolation (single individual) and at various densities under conditions of fixed food availability. Tadpoles reared experimentally at high densities experienced slower growth (Figure 11.8a), required a longer time to change from tadpoles to frogs (metamorphosis), and had a lower probability of completing this transformation. Those that did reach threshold size were smaller than those living in less dense populations (Figure 11.8b).

In a similar study to that of Dash and Hota, Rick Relyea of the University of Pittsburgh examined the effects of intraspecific competition on experimental populations of wood frog tadpoles (*Rana sylvatica*). In addition to the influence of density on individual growth rate, Relyea examined the impacts of intraspecific competition on phenotypic variations in behavior and morphology (phenotypic plasticity: see Section 5.9 for discussion). Like Dash and Hota, Relyea found a decrease in the average individual growth rate with increasing population density (Figure 11.9a). The reductions in growth rate, however, were accompanied by distinct changes in behavior and morphology in response to the competitive environment. Individuals reared in higher-density populations exhibited an increase in activity (time spent in movement) (Figure 11.9b) and generally developed longer bodies (Figure 11.9c), shorter tails, and wider mouths—a clear example of competition-induced phenotypic plasticity (see Section 5.9). Increased activity was shown to increase resource acquisition, and previous studies have found that the observed shifts in morphology are associated with enhanced growth under competitive environments, suggesting that both behavioral and developmental responses are adaptive.

11.5 | Intraspecific Competition Can Influence Mortality Rates

In addition to reduced growth rate, a common response of plants to high population density is reduced survival. Mortality functions to increase resource availability for the remaining individuals, allowing for increased growth. In an experiment aimed at exploring this relationship, the late plant ecologist Kyoji Yoda planted seeds of horseweed (*Erigeron canadensis*) at a density of 100,000 seeds per square meter. As the seedlings grew, competition for the limited resources ensued (Figure 11.10a). The number of seedlings surviving declined within months to a density of approximately 1000 individuals. The death of individuals increased the per capita resource availability, and the average size of the surviving individuals increased as population density declined. The inverse relationship between population density and average plant size during the experiment can be seen more easily in Figure 11.10b. This progressive decline in density and increase in biomass (growth) of remaining individuals caused by the combined effects of density-dependent mortality and growth within a population is known as **self-thinning.**

Originally described in forest trees, self-thinning has been widely documented in plant populations and identified in sessile animals such as barnacles and mussels

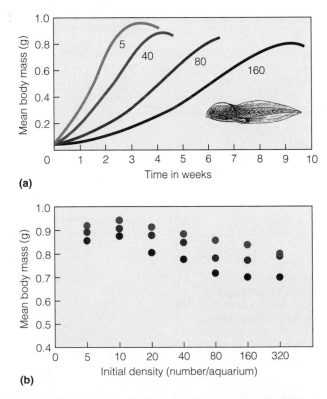

(a)

(b)

Figure 11.8 | Effect of population density on the growth and development of individuals. **(a)** The growth rate of the tadpole *Rana tigrina* declines swiftly as density increases from 5 to 160 individuals confined in the same space. **(b)** As a result, the mean body mass at metamorphosis declines as a function of the population density.

(Adapted from Dash and Hota 1980.)

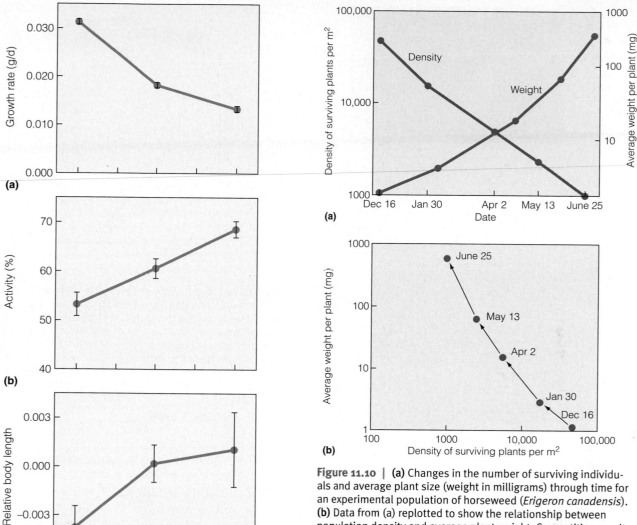

Figure 11.9 | Mean (**a**) growth rate, (**b**) activity, and (**c**) relative body length (±SE) of larval wood frogs as a function of density of intraspecific competitors. Activity was measured by determining the proportion (%) of observed individuals in motion during the observation period. The measure of relative body length is the residual of the regression of body length as a function of body mass. This approach removes differences in body length among individuals due purely to differences in overall size (body mass). Mean values (and estimates of variance) for the residuals were then calculated for each treatment, as presented in graph (c).
(Adapted from Relyea 2002.)

Figure 11.10 | (**a**) Changes in the number of surviving individuals and average plant size (weight in milligrams) through time for an experimental population of horseweed (*Erigeron canadensis*). (**b**) Data from (a) replotted to show the relationship between population density and average plant weight. Competition results in mortality, which in turn increases the per capita availability of resources, resulting in increased growth for the survivors.
(Adapted from Yoda 1963.)

(Figure 11.11). More recent studies have presented evidence of self-thinning in mobile animals.

Most evidence of self-thinning in mobile organisms comes from studies of stream-dwelling fish populations. Thomas Jenkins and colleagues at the University of California (Santa Barbara) undertook a multiyear experimental study to examine the effects of population density on individual growth of brown trout (*Salmo trutta*) in two stream ecosystems in the Sierra Nevada mountains (Sierra Nevada Aquatic Research Laboratory). Their data reveal an inverse relationship between average mass of individuals and density of surviving brook trout over the study period (Figure 11.12) that is similar to the general patterns shown in Figures 11.10 and 11.11 for populations of plants and sessile animals. The reduction in density of brook trout through time was due to intraspecific competition for limited food resources as mean body size (and associated demand for food resources) increases.

Ernest Keeley of the University of British Columbia conducted experiments in artificial stream channels, manipulating density of competitors and food abundance to examine density-dependent growth and mortality in populations of steelhead trout (*Oncorhynchus mykiss*). His results illustrate a pattern of decreasing growth and increasing mortality with increasing levels of per capita food competition, resulting either from an increase in

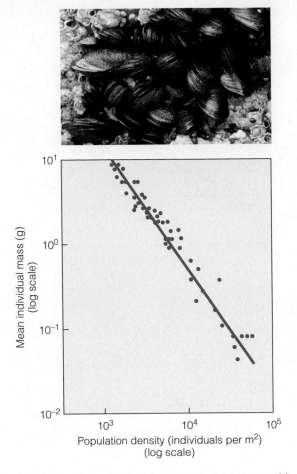

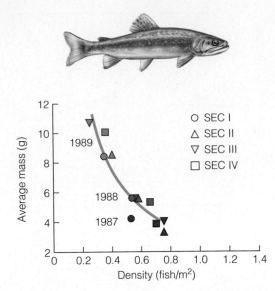

Figure 11.12 | Self-thinning in a population of brook trout inhabiting Convict Creek in the Sierra Nevada Mountain range. Average body mass of surviving brown trout is expressed as a function of the total density of individuals in different stream sections as measured during the fall censuses 1987–1989. Data points represent means of from 118 to 699 individuals. Each symbol represents the subpopulation occupying a specific stream section. The color red denotes data from 1987; the color blue, from 1988; the color green, from 1989.

(Adapted from Jenkins et al. 1999.)

Figure 11.11 | Relationship between mean individual mass (g) and population density (individuals per m²) for the black mussel, *Choromytilus meridionalis*.

(Adapted from Hughes and Griffiths 1988.)

population density for a given level of food abundance or from reduced food abundance for a given population density (Figure 11.13).

11.6 | Intraspecific Competition Can Reduce Reproduction

Besides directly influencing the survival and growth of individuals, competition within a population can reduce fecundity. The timing of the response depends on the nature of the population, and the mechanisms by which competition influences reproductive rate can vary with species. Harp seals (*Phoca groenlandica*) become sexually mature when they reach approximately 87 percent of their mature body weight of about 120 kg. Reduced growth rates (weight gain) under high population densities increase the mean age at which females become reproductive (Figure 11.14a). The result is that fertility in harp seals, as measured by the number of females giving birth to young, is density dependent, inversely related to the population density of the previous year (Figure 11.14b). Similar patterns of density-dependent fecundity

have been observed in bird populations (see Figure 3 of Field Studies: T. Scott Sillett).

For animal species that exhibit indeterminate rates of growth and development (see Field Studies: Martin Wikelski, pg. 134), density-dependent growth is a potentially powerful mechanism of population regulation, because fecundity is typically related to body size (see Chapter 8, Section 8.9). Population density has been shown to affect patterns of growth, age at maturity, and fecundity in fish populations, suggesting density dependence as an important mechanism of population control. Amy Schueller and colleagues at the University of Wisconsin–Stevens Point examined the effects of population density on maturity and fecundity of walleyes (*Sander vitreus*) in Big Crooked Lake, Wisconsin, over a period of 6 years. The researchers found a significant positive relationship between age of maturity (reproduction) and adult walleye population density (Figure 11.15c) similar to that observed for the harp seal population shown in Figure 11.14. The onset of sexual maturity is size dependent (Figure 11.15b), and the delayed maturity observed with increasing population density is a direct result of the inverse relationship between individual growth rates and population density (Figure 11.15a). In addition to delayed maturity, Schueller observed a decrease in mean fecundity rate (egg production) with increasing population density.

Density-dependent controls on fecundity are a common observation in plant populations. The amount of

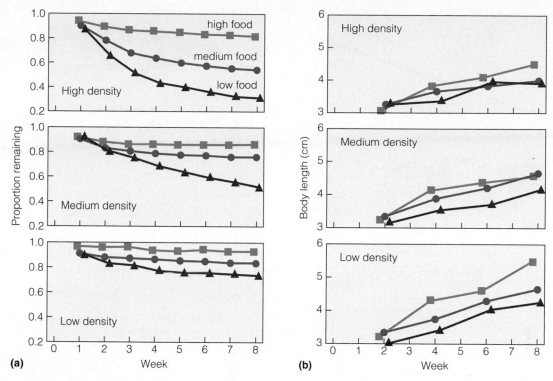

Figure 11.13 | Patterns of **(a)** mortality (as measured by proportion of initial individuals remaining) and **(b)** growth (as measured by body length) of juvenile steelhead trout in stream channels over an 8-week experimental period. Experiments examined the influence of intraspecific competition by manipulating density of competitors and abundance of food. Initial densities of individuals in the experimental stream channels were set at 32 (low density), 64 (medium density), or 128 individuals per m² (high density). Food abundance was varied by providing either 0.3 (low food), 0.6 (medium food), or 1.2 grams/m²/day (high food).

Interpreting Ecological Data

Q1. How does mortality differ for the high-density experimental populations under low, medium, and high food abundance? How do these differences among the three food abundance treatments change as population density is reduced (medium and low population densities)?

Q2. How does growth (body length) differ for the high-density experimental populations under low, medium, and high food abundance? How do these differences among the three food abundance treatments change as population density is reduced (medium and low population densities)?

Q3. Why do the greatest differences in mortality among the three food abundance treatments occur at high population density, while the greatest differences in growth rate among the three food abundance treatments occur at low population density?

grain produced by individual corn plants is reduced dramatically when planted at higher densities (Figure 11.16a). Similarly, the number of seeds produced per plant declines with increasing density of individuals in populations of the annual herb glasswort (*Salicornia europaea*; Figure 11.16b) inhabiting coastal salt marshes.

11.7 | High Density Is Stressful to Individuals

As a population reaches a high density, individual living space can become restricted. Often, aggressive contacts among individuals increase. One hypothesis of population regulation in animals is that increased crowding and

social contact cause stress. Such stress triggers hormonal changes that can suppress growth, curtail reproductive functions, and delay sexual activity. They may also suppress the immune system and break down white blood cells, increasing vulnerability to disease. In mammals, social stress among pregnant females may increase mortality of the young in the fetal stage (unborn) and cause inadequate lactation, stunting the growth and development of nursing young. Thus, stress results in decreased births and increased infant mortality. Such population-regulating effects have been confirmed in confined laboratory populations of several species of mice and to a lesser degree in enclosed wild populations of woodchucks (*Marmota monax*) and rabbits (*Oryctolagus cuniculus*). Evidence of

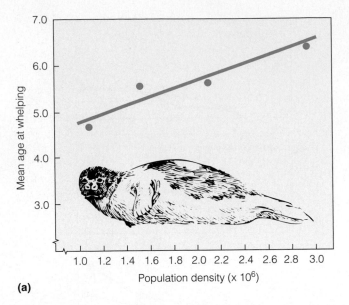

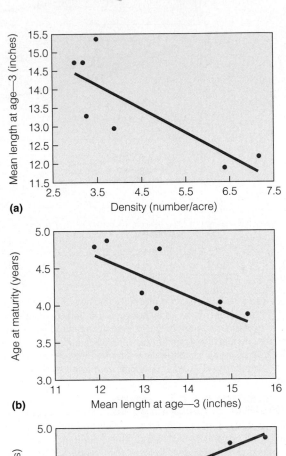

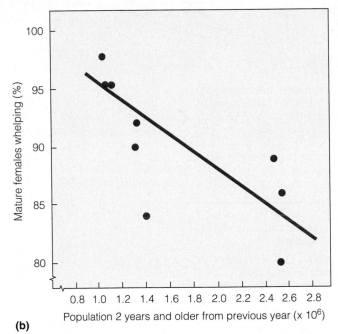

Figure 11.14 | **(a)** The mean age of sexual maturity for female harp seals (*Phoca groenlandica*) is related to weight more than to age. Seals arrive at sexual maturity when they reach 87 percent of average adult body weight. Seals attain this weight at an earlier age when population density is low. **(b)** As a result, fertility is density dependent. As the seal population (measured by the number of individuals 2 years and older during the previous year) increases, the percentage of females giving birth to young decreases markedly.

(Adapted from Lett et al. 1981.)

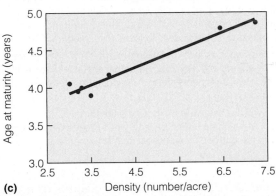

Figure 11.15 | Influence of population density on growth rate and age at sexual maturity for female walleye in Big Crooked Lake, Wisconsin. **(a)** Individual walleye growth, as measured by the mean length at 3 years of age, is inversely related to population density. **(b)** The age at which walleye achieve sexual maturity is size dependent, therefore increasing with decreasing growth rate. **(c)** The result is a positive relationship between age at maturity and population density for female walleye in the lake ecosystem.

(Adapted from Schueller et al. 2005).

the effects of stress in free-ranging wild animals, however, is difficult to obtain.

Pheromones are perfume-like chemical signals that function to communicate between individuals, influencing behavior and body function much like a hormone does. In social insects, pheromones released by the queen have been identified as an important mechanism in controlling the development and reproduction of colony members. Pheromones present in the urine of adult rodents can encourage or inhibit reproduction. One study by Adrianne Massey and John Vendenbergh of North Carolina State University involved wild female

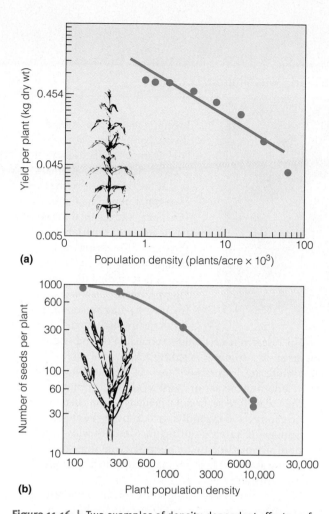

(a)

Population density (plants/acre × 10³)

Yield per plant (kg dry wt)

(b)

Plant population density

Number of seeds per plant

Figure 11.16 | Two examples of density-dependent effects on fecundity in plants. **(a)** Grain (seed) production of individual corn plants declines with increasing density of plants per acre. **(b)** Number of seeds produced per plant declines with increasing population density in the marsh shrub *Salicornia europaea*.

(a) (Adapted from Fery and Janick 1971.) (b) (Adapted from Watkinson and Davy 1985.)

house mice (*Mus musculus*) confined to grassy areas surrounded by roadways that prohibited dispersal. One group lived in a high-density population and the other in a low-density population. Urine from females of each population was absorbed onto filter paper and placed in laboratory cages with juvenile test females. Exposing juvenile females to urine from high-density populations delayed puberty, whereas exposing females to urine from low-density populations did not. The results suggest that pheromones contained in the urine of adult females in high-density populations function to delay puberty and help slow further population growth.

11.8 | Dispersal Can Be Density Dependent

Instead of coping with stress, some animals disperse, leaving the population to seek vacant habitats. Although dispersal is most apparent when population density is high, it goes on all the time. Some individuals leave the parent population whether it is crowded or not. There is no hard-and-fast rule about who disperses.

When a lack of resources resulting from high population density forces some individuals to disperse, the ones to go are usually subadults driven out by adult aggression. The odds are that such individuals will perish, although a few may arrive at some suitable area and successfully become established. Because dispersal under conditions of high population density is a response to overpopulation, this type of dispersal does not function as an effective means of population regulation. More important to population regulation is dispersal when density is low or increasing, well before the local population reaches the point of overexploiting resources.

For example, ecologists Dominique Berteaux and Stan Boutin of McGill University (Quebec, Canada) studied dispersal in a red squirrel (*Tamiasciurus hudsonicus*) population in the Yukon in Canada. They found that every year, a fraction of older reproductive females left their home areas during the summer months when food availability was high. The dispersal of female adults increased the survival of their juvenile offspring that remained on the home area during the winter months, when food resources are scarce.

Some dispersing individuals, especially juveniles, can maximize their probability of survival and reproduction only if they leave their birthplace. When intraspecific competition at home is intense, dispersers can relocate in habitats where resources are more accessible, breeding sites more available, and competition less intense. Further, the disperser reduces the risk of inbreeding (see Section 5.6). At the same time, dispersers incur risks that come with living in unfamiliar terrain.

Does dispersal actually regulate a population? Although dispersal is often positively correlated with population density, no relationship that can be generalized exists between the proportion of the population leaving and its increase or decrease. Dispersal may not function as a regulatory mechanism, but it contributes strongly to population expansion (see Section 9.8) and aids in the persistence of local populations (see Chapter 12).

11.9 | Social Behavior May Function to Limit Populations

Intraspecific competition can express itself in social behavior, or the degree to which individuals of the same species tolerate one another. Social behavior appears to be a mechanism that limits the number of animals living in a particular habitat, having access to a common food supply, and engaging in reproductive activities.

Many species of animals live in groups with some kind of social organization. In some populations, the group structure is crucial to acquiring resources (as with predators that hunt in packs) and maintaining defense. The organization, however, is often based on aggressiveness, intolerance, and the dominance of one individual

Smithsonian Migratory Bird Center, National Zoological Park, Washington, D.C.

One of the most studied and publicized conservation issues today is the decline of migratory songbirds, particularly neotropical migrants—species that migrate between breeding grounds in the temperate zone of North America and that winter in the tropics of Central and South America. Determining the factors responsible for the decline of these species, however, is difficult because birds spend parts of their annual cycle in different geographic locations. Furthermore, events during one period of the annual cycle are likely to influence populations in subsequent stages.

Understanding the factors influencing the population dynamics of migratory bird species is the research focus of avian ecologist T. Scott Sillett of the Smithsonian Migratory Bird Center. Sillett's research focuses on the population dynamics of the black-throated blue warbler (Figure 1), a migratory songbird that breeds in the forested regions of eastern North America and overwinters in the Greater Antilles (Jamaica and Cuba). The species is territorial and feeds largely on insects, primarily the larvae of Lepidoptera (butterflies and moths). Individuals exhibit strong site fidelity in both breeding and wintering grounds—meaning that individuals return to the same locations (often the same territorial areas) each year.

In a series of studies, Sillett and his colleagues quantified the demography of black-throated blue warbler populations from 1986 to 2000 at two locations during their annual cycle: the overwinter period at Copse Mountain, near Bethel Town in northwestern Jamaica, and the breeding season at Hubbard Brook Experimental Forest, New Hampshire. The Jamaican site was visited twice annually, first at the beginning of the overwinter period in autumn (October) and again at the end of the winter (March), before the start of the spring migration, in the following calendar year. Sillett studied the New Hampshire site each year during the breeding season from mid-May through August.

The research team made estimates of fecundity during the spring–summer breeding season at the New Hampshire site. Estimates of survival are much more difficult, requiring the use of mark-recapture sampling over multiple years (see Section 9.4). Sillett's team used mist nets to capture and tag individuals at both sites to allow for estimates of survival during the 6-month winter and 3-month summer stationary periods. From these data, the team was also able to estimate survivorship for the 3-month migratory period.

Results of the decade-long study reveal an interesting pattern of variations in fecundity and survival that are influenced by both density-dependent and density-independent factors. Average survival rates for birds at both locations (Jamaica and New Hampshire) did not differ significantly. Average monthly values during the summer (0.99, New Hampshire) and winter (0.93, Jamaica) stationary periods were high. In contrast, monthly survival probability during the periods of migration ranged from 0.77 to 0.81. Thus, apparently mortality rates are at least 15 times higher during the migration compared to the stationary periods, and more than 85 percent of annual mortality occurred during migration.

Although mean monthly values of survival did not vary between the sites, interannual patterns of survival were markedly different (Figure 2). The annual survival rate of warblers breeding in New Hampshire was relatively constant over the period of the study; however, the probability of survival on the wintering grounds in Jamaica varied more than threefold. Sillett and his colleagues found that the difference is related to different impacts of climate variation on the birds at the two locations. The El Niño–Southern Oscillation (ENSO) discussed in Chapter 2 (see Section 2.9) influences patterns of climate variation at both the Jamaica and New Hampshire sites, but it influences population dynamics in very different ways. Sillett used annual mean monthly values of the standardized

Figure 1 | Black-throated blue warbler feeding young.

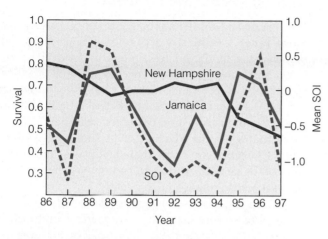

Figure 2 | Comparison of annual survival estimates for Jamaica and New Hampshire to mean monthly values of the Southern Oscillation Index (SOI).

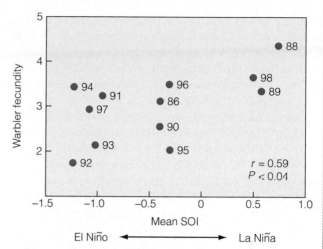

Figure 3 | Relationship between the mean monthly Southern Oscillation Index (SOI) and warbler fecundity, measured as the mean number of offspring fledged per warbler pair per year at the New Hampshire site. Numbers at each point represent the calendar year.

Southern Oscillation Index (SOI) to represent ENSO conditions for each calendar year. High, positive values of SOI indicate La Niña conditions, and low, negative values indicate El Niño conditions (Figure 2). Survival at the Jamaica site was low in El Niño years and high in La Niña years (Figure 2). During El Niño years in Jamaica, reduced rainfall leads to decreased food available for warblers in the winter dry season and, hence, to lower survival. La Niña years, in contrast, tend to be wetter and with increased food availability and higher survival.

In contrast to Jamaica, the ENSO cycle had little effect on survival rates at the New Hampshire site during the breeding season. The ENSO cycle did, however, significantly influence fecundity rates.

Both the mean number of young fledged per warbler pair (Figure 3) and the mass at fledging were lower in El Niño years relative to La Niña years. The influence of ENSO on fecundity was due to the influence of climate variation on food abundance. Prey biomass (lepidopteran larvae) was low during El Niño (dry) years and high during La Niña (wet) years.

Besides the influence of the ENSO cycle, Sillett and his colleagues observed density-dependent regulation of fecundity rates at the New Hampshire site over the period of study. They observed an inverse relationship between fecundity and population density. The result is a combination of density-independent (ENSO) and density-dependent factors that functions to control the black-throated blue warbler population dynamics. In El Niño years, high mortality (low survival) in the wintering grounds (Jamaica) and low fecundity in the breeding grounds (New Hampshire) function to reduce the density of the breeding population. In subsequent La Niña years, increasing rates of survival during the winter, combined with high rates of fecundity, cause the population to rise. As the density of the breeding population increases, density-dependent factors (crowding) control fecundity rates and population growth.

The work of Scott Sillett and his colleagues has greatly expanded our understanding of neotropical bird species—and of all migratory species whose dynamics depend on environmental conditions over a broad geographic region.

Bibliography

Rodenhouse, N. L., T. S. Sillett, P. J. Doran, and R. T. Holmes. 2003. Multiple density-dependence mechanisms regulate a migratory bird population during the breeding season. *Proceedings of the Royal Society of London* 270:2105–2110.

Sillett, T. S., R. T. Holmes, and T. W. Sherry. 2000. Impacts of a global climate cycle on population dynamics of a migratory songbird. *Science* 288:2040–2042.

Sillett, T. S., and R. T. Holmes. 2002. Variation in survivorship of a migratory songbird throughout its annual cycle. *Journal of Animal Ecology* 71:296–308.

1. Suppose that environmental conditions associated with El Niño resulted in a decreased survival in Jamaica during the winter but increased fecundity rates in New Hampshire. How would the population dynamics of the species differ from that discussed above?

2. How does population density influence the birth rate at the New Hampshire site?

over another. Two opposing forces are at work. One is mutual attraction of individuals; the other is a negative reaction against crowding—the need for individual space.

Each individual occupies a position in the group based on dominance and submissiveness. In its simplest form, the group includes an alpha individual dominant over all others, a beta individual dominant over others except the alpha, and an omega individual subordinate to all others. Individuals settle social rank by fighting, bluffing, and threatening at initial encounters between any given pair of individuals or at a series of such encounters. Once individuals establish social rank, they maintain it by habitual subordination of those in lower positions. Threats and occasional punishment handed out by those of higher rank reinforce this relationship. Such organization stabilizes and formalizes intraspecific (contest) competitive relationships and resolves disputes with a minimum of fighting and wasted energy.

Social dominance plays a role in population regulation when it affects reproduction and survival in a density-dependent manner. An example is the wolf. Wolves live in small groups, called packs, of 6 to 12 or more individuals. The pack is an extended kin group consisting of a mated pair, one or more juveniles from the previous year who do not become sexually mature until the second year, and several related nonbreeding adults.

The pack has two social hierarchies, one headed by an alpha female and the other headed by an alpha male—the leader of the pack, to whom all other members defer. Below the alpha male is the beta male—closely related, often a full brother, who has to defend his position against pressure from males below him in the social hierarchy.

Mating within the pack is rigidly controlled. The alpha male (occasionally the beta male) mates with the alpha female. She prevents lower-ranking females from mating with the alpha and other males, while the alpha male inhibits other males from mating with her. Therefore, each pack has one reproducing pair and one litter of pups each year. They are reared cooperatively by all members of the pack.

The size of packs, which is heavily influenced by the availability of food, governs the level of the wolf population in a region. Priority for food goes to the reproducing pair. At high pack density and decreased availability of food, individuals may be expelled or leave the pack. Unless they have the opportunity to settle successfully in a new area and form a pack, they may not survive. Thus, at high wolf densities, mortality increases and birthrates decline. When the population of wolves is low, sexually mature males and females leave the pack, settle in unoccupied habitat, and establish their own packs that have an alpha (reproducing) female. In this case, nearly every sexually mature female reproduces, and the wolf population increases. At very low densities, however, females may have difficulty locating males to establish a pack with, and so they fail to reproduce or even survive.

11.10 | Territoriality Can Function to Regulate Population Growth

The area that an animal normally uses during a year is its **home range.** Overall size of the home range varies with the available food resources, mode of food gathering, body size, and metabolic needs. Among mammal species, the home range size is related to body size (Figure 11.17), reflecting the link between body size and energy requirements (food resources; see Chapter 7 and Field Studies: Martin Wikelski). In general, carnivores require a larger home range than herbivores and omnivores of the same size. Males and adults have larger home ranges than do females and juveniles.

Although the home range is not defended, aggressive interactions may influence the movements of individuals within another's home range. Some species, however, defend a core area of the home range against others. If the animal defends any part of its home range, we define that part as a **territory**—a defended area (Figure 11.18). If the animal defends its entire home range, its home range and territory are the same.

By defending a territory, the individual secures sole access to an area of habitat and the resources it contains. The defense of a territory involves well-defined behavioral patterns: song and call, intimidation displays such as spreading wings and tail in birds and baring fangs in mammals, attack and chase, and marking with scents that evoke escape and avoidance in rivals. As a result, territorial individuals tend to occur in more or less regular

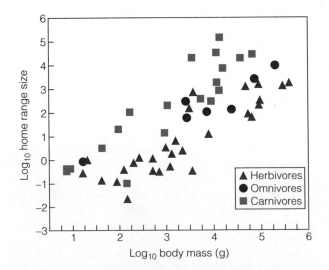

Figure 11.17 | Relationship between the size of home range and body weight of North American mammals. For a given body mass, the home range of carnivores is larger than that of herbivores because the home range of a carnivore must be large enough to support a population of the prey (other animals) that it feeds upon.

(Adapted from Harastad and Bunnell 1979.)

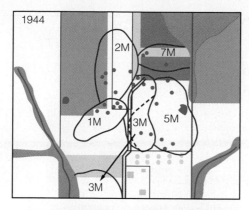

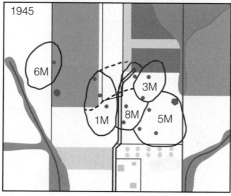

Figure 11.18 | Territories of the grasshopper sparrow (*Ammodramus savannarum*) as determined by observations of male banded birds, indicated by 1M, 2M, and so forth. Dots indicate song perches. Note how they are distributed near the territorial boundaries. Shaded areas represent crop field. Dashed lines indicate territory shifts before second nesting within the same breeding season. The return of the same males to nearly the same territory the second year is termed *philopatry*.

(Adapted from Smith 1963.)

patterns of distribution (see Section 9.3 and Figures 11.18 and 11.19).

The total area available divided by the average size of the territory determines the number of territorial owners a habitat can support. When the available area is filled, owners evict excess individuals, denying them access to resources and potential mates. These individuals make up a floating reserve of potential breeders. The existence of such a reserve of potentially breeding adults has been described for several bird species, including the red grouse (*Lagopus lagopus*) in Scotland, the Australian magpie (*Gymnorhina tibicen*), Cassin's auklet (*Ptychoramphus aleuticus*) of Alaska, and the white-crowned sparrow (*Zonotrichia leucophrys*) of California. Studies of a banded (marked for identification) population of white-crowned sparrows by Lewis Petrinvich and Thomas Patterson of the University of California, Riverside, indicated a surplus of potential breeding individuals. In fact, 24 percent of the individuals holding territories had been floaters (no territory) for a period ranging from 2 to 5 years before acquiring a territory. Floaters quickly replaced territory holders that disappeared during the breeding season.

Ecologist John Krebs of the University of Oxford conducted a field experiment in which breeding pairs of great tits (*Parus major*) were removed from their territories in an English oak woodland (Figure 11.19). The pairs were quickly replaced by new birds, largely first-year individuals that moved into the vacated territories from adjacent areas of suboptimal habitat, such as hedgerows.

Territoriality functions to limit access to the defended area by other individuals in the population, but under what conditions can territoriality function as a mechanism of population regulation? If all pairs that settle on an area

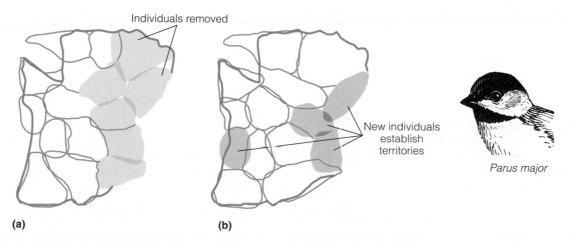

Figure 11.19 | Replacement of removed individuals and settlement of vacant territories in an oak woodland. Lines represent the territory boundaries of breeding pairs of great tits (*Parus major*). Six pairs were removed between March 19 and March 24 (1969), shown as the shaded areas in map **(a)**. Within 3 days, some of the resident pairs had shifted (and in some cases expanded) the boundaries of their territories, and four new pairs had taken up the new territories shown in map **(b)**. After this short period of adjustment, the territories again formed a complete mosaic over the woodland.

(Adapted from Krebs 1971.)

get a territory, territoriality only influences the spatial distribution of individuals within the population and does not regulate it. By contrast, if territorial size has a lower limit, the number of pairs that can settle on an area is limited, and individuals that fail to do so have to leave. In that situation, territoriality might regulate the local population; but only if an excess of nonterritorial males and females of reproductive age occurs, as is the case for examples presented earlier. Then, reproduction is limited by territoriality, and we have density-dependent population regulation.

11.11 | Plants Preempt Space and Resources

Plants are not territorial in the same sense that animals are, but plants can capture and hold on to space. This phenomenon in the plant world is analogous to territoriality in animals, especially if one accepts an alternative definition of territoriality: individual organisms spaced out more than we would expect from a random occupancy of suitable habitat. In fact, the presence of a uniform distribution is often used as an indication that competition is occurring within plant populations (see Figure 9.8).

Plants from dandelions to trees do capture a certain amount of space and exclude individuals of their own and other species. When a dandelion plant spreads its rosettes of leaves on the ground, it eliminates all other plants from the area it covers. Plants also establish zones of resource depletion associated with their canopy (leaves) and root systems. Taller individuals intercept light (see Section 4.2), shading the ground below and limiting successful establishment to species that can tolerate the reduced availability of light (see Section 6.9). Likewise, the uptake of water and nutrients from the soil limits availability to individuals with overlapping rooting zones.

Plant ecologist James Cahill of the University of Alberta has conducted a variety of experiments that have quantified the relationship between neighboring root biomass and belowground competition in plants using root exclusion tubes made of PVC pipe. The exclusion tubes are placed vertically into the soil to separate roots of the target plant, which is planted inside the tube, from the roots of other individuals in the population that naturally occur surrounding it. Differing quantities of neighboring roots are then allowed to access the soil within the tubes by placing differing numbers of holes through the sides of the tube before installing it. Results of these experiments have demonstrated a clear relationship between the growth of target plants and their overlap with the biomass of neighboring individuals (Figure 11.20).

Because of their longevity, some plants, especially trees, occupy space for a long time, preventing invasion by individuals of the same or other species. Plants successful in capturing space increase their fitness at the expense of others.

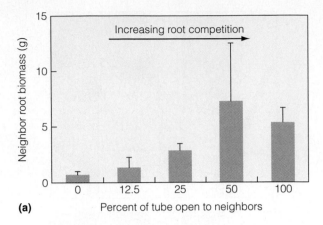

(a)

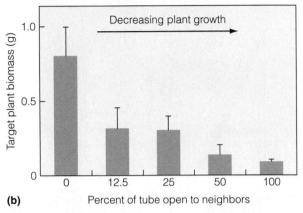

(b)

Figure 11.20 | Results of an experiment controlling root competition among neighboring plants by using root exclusion tubes made out of PVC pipe. By drilling holes, the investigator controlled access to soil in the tube by neighboring plants. **(a)** The amount of root biomass from neighboring plants that grew into the tubes increased as a function of the number of openings in the tube (percent of tube open to neighbors: *x*-axis). **(b)** The increase in belowground (root) competition resulted in a decline in the mean biomass (dry weight in grams) of target plants grown in the exclusion tubes.

(Adapted from Cahill 2000.)

11.12 | Density-Independent Factors Can Influence Population Growth

We have seen that population growth and fecundity are heavily influenced by density-dependent responses. But there are other, often overriding influences on population growth that do not relate to density. These influences are termed **density independent.** Factors such as temperature, precipitation, and natural disasters (fire, flood, and drought) may influence the rates of birth and death within a population (see Field Studies: T. Scott Sillett) but do not regulate population growth, because regulation implies feedback.

If environmental conditions exceed an organism's limits of tolerance, the result can be disastrous, affecting growth, maturation, reproduction, survival, and movement. The resulting increase in mortality rates can even lead to the extinction of local populations.

Pronounced changes in population growth often correlate directly with variations in moisture and temperature. For example, outbreaks of spruce budworm (*Choristoneura fumiferana*) are usually preceded by 5 or 6 years of low rainfall and drought. Outbreaks end when wet weather returns. Such density-independent effects can occur on a local scale where topography and microclimate influence the fortunes of local populations.

In desert regions, a direct relationship exists between precipitation and rate of increase in certain rodents and birds. Merriam's kangaroo rat (*Dipodomys merriami*) occupies lower elevations in the Mojave Desert. The kangaroo rat has the physiological capacity to conserve water and survive long periods of aridity. However, it does require the prevailing patterns of seasonal moisture availability to be sufficient to stimulate the growth of herbaceous desert plants in fall and winter. The kangaroo rat becomes reproductively active in January and February when plant growth, stimulated by fall rains, is green and succulent. Herbaceous plants provide a source of water, vitamins, and food for pregnant and lactating females. If rainfall is scanty, annual plants fail to develop, and reproduction by kangaroo rats is low. This close relationship between population dynamics and seasonal rainfall and success of winter annuals is also apparent in other rodents and birds occupying similar desert habitats.

In more northern regions of the temperate zone, winters can be harsh, and the accumulation of snow can directly (physiological) and indirectly (food availability) affect many animal species. L. D. Mech and colleagues at the Patuxent Wildlife Research Center (U.S. Fish and Wildlife Service) examined the relationship between winter snow accumulation and dynamics of the white-tailed deer population that inhabits northeastern Minnesota. Their studies reveal that the average number of offspring (fawns) produced per female (doe) in the spring (Figure 11.21a), and subsequently the annual change in the population (Figure 11.21b), is inversely related to the previous winter's snow accumulation.

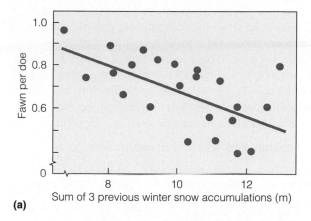

(a)

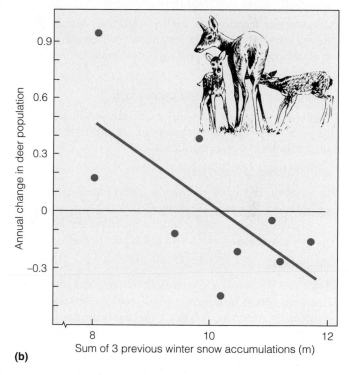

(b)

Figure 11.21 | Relationship between the sum of the previous three winter monthly snow accumulations in northeastern Minnesota and the population of white-tailed deer. **(a)** Fecundity (fawn:doe ratio). **(b)** Percent annual change in next winter's population. (Adapted from Mech et al. 1987.)

Summary

Logistic Population Growth | 11.1

Because resources are limited, exponential growth cannot be sustained indefinitely. The maximum population size that can be sustained for a particular environment is termed the carrying capacity (*K*). The logistic model of population growth incorporates the concept of carrying capacity into the previously developed model of exponential growth. The result is a decrease in the rate of population growth as the population size approaches the carrying capacity.

Density-Dependent Regulation | 11.2

Populations do not increase indefinitely. As resources become less available to an increasing number of individuals, birthrates decrease, mortality increases, and population growth slows. If the population declines, mortality decreases, births increase, and population growth speeds up. Positive and negative feedback function to regulate the population.

Competition | 11.3

Intraspecific competition occurs when resources are in short supply. Competition can take two forms: scramble and contest. In scramble competition, growth and reproduction are depressed equally across individuals as competition increases. In contest competition, dominant individuals claim sufficient resources for growth and reproduction. Others produce no offspring or perish. Competition can involve interference among individuals or indirect interactions via exploitation of resources.

Competition, Growth, and Development | 11.4

Competition for scarce resources can decrease or retard growth and development. Up to a point, plants respond to competition by modifying form and size.

Competition and Mortality | 11.5

A common response to high population density is reduced survival. Mortality functions to increase resource availability for the remaining individuals, allowing for increased growth.

Competition and Reproduction | 11.6

High population density and competition can also function to delay reproduction in animals and reduce fecundity in both plants and animals.

Density and Stress | 11.7

In animals, the stress of crowding may cause delayed reproduction, abnormal behavior, and reduced ability to resist disease and parasitic infections; in plants, crowding results in reduced growth and production of seeds.

Dispersal | 11.8

Dispersal is a constant phenomenon in populations. When dispersal occurs in response to the overexploitation of resources, or crowding, it is not a mechanism of population regulation. It can, however, regulate populations if the rate of dispersal increases in response to population growth.

Social Behavior | 11.9

Intraspecific competition may be expressed through social behavior. The degree of tolerance can limit the number of animals in an area and access of some animals to essential resources. A social hierarchy is based on dominance. Dominant individuals secure most of the resources. Shortages are borne by subdominant individuals. Such social dominance may function as a mechanism of population regulation.

Territoriality | 11.10

Social interactions influence the distribution and movement of animals. The area an animal normally covers in its life cycle is its home range. The size of a home range is influenced by body size.

If the animal or a group of animals defends a part or all of its home range as its exclusive area, it exhibits territoriality. The defended area is its territory. Territoriality is a form of contest competition in which part of the population is excluded from reproduction. These nonreproducing individuals act as a floating reserve of potential breeders, available to replace losses of territory holders. In such a manner, territoriality can act as a population-regulating mechanism.

Plants Preempt Space | 11.11

Plants are not territorial in the same sense as animals are, but they do hold on to space, excluding other individuals of the same or smaller size. Plants capture and hold space by intercepting light, moisture, and nutrients.

Density-Independent Factors | 11.12

Density-independent influences, such as weather, affect but do not regulate populations. They can reduce populations to the point of local extinction. Their effects, however, do not vary with population density. Regulation implies feedback.

Study Questions

1. What is the difference between the exponential and logistic models of population growth?
2. What is the definition of carrying capacity (K)? How is it used in the framework of the logistic model to introduce a mechanism of density-dependent population regulation?
3. We have seen from many examples in Chapter 11 that competition among individuals within a population can result in an inverse relationship between population density and the growth (see Figures 11.8 and 11.9) and reproduction (see Figures 11.14 and 11.15) of individuals within the population. Besides the inverse correlation between population density and growth/reproduction, what condition must hold for the researcher to establish that competition is responsible for these relationships?
4. Competition can function as a mechanism of density-dependent population regulation. How might scramble and contest competition differ in their effect on population growth (regulation)?
5. Distinguish between home range and territory.

6. How does the relationship between body size and home range size influence the estimates of population density and abundance discussed in Chapter 9?
7. What condition must hold true for territoriality to function as a density-dependent mechanism regulating population growth (and density)?
8. How might social dominance within a population function to regulate population growth?
9. Years of below-average rainfall in Kruger National Park (South Africa) cause a decline in the growth and productivity of grasses. Mortality rates in populations of herbivores such as the African buffalo then increase, resulting in a decline in population density. Based on the information that you have been provided, would you conclude the annual variations in rainfall within Kruger function as a density-dependent mechanism regulating growth of the buffalo population? Is there any additional information that would cause you to change your answer?

Further Readings

Barrett, G. W., and E. P. Odum. 2000. The twenty-first century: The world at carrying capacity. *Bioscience* 50:363–368.

A discussion on the relationship between global human population growth, economic growth, and world carrying capacity.

Lack, D. 1954. *The natural regulation of animal numbers.* London: Oxford University Press.

A classic text on population regulation that should be read by those interested in the historical development of ideas in population ecology.

Murdoch, W. W. 1994. Population regulation in theory and practice. *Ecology* 75:271–287.

This excellent paper contrasts the theory of population regulation as understood by ecologists against the difficulties of experimentally detecting regulation in natural populations.

Newton, I. 1998. *Population limitation in birds.* London: Academic Press Limited.

Chapters 2 through 5 present an excellent, readable overview of population regulation.

Sinclair, A. R. E. 1977. *The African buffalo: A study of resource limitations of populations.* Chicago: University of Chicago Press.

A classic study of intraspecific population regulation. Provides an example of the interaction between various factors that together function to regulate the population of this large herbivore.

Stephens, P. A., and W. J. Sutherland. 1999. Consequences of the Allee effect for behavior, ecology and conservation. *Trends in Ecology and Evolution* 14:401–405.

An extensive, but technical, review of the concept of the Allee effect in population regulation.

Turchin, P. 1999. Population regulation: A synthetic view. *Oikos* 84:160–163.

An excellent discussion of the broader issues relating to the regulation of natural populations.

Wolff, J. O. 1997. Population regulation in mammals: An evolutionary perspective. *Journal of Animal Ecology* 66:1–13.

This is an excellent discussion of the broader issues relating to the regulation of natural populations.

CHAPTER 12

Metapopulations

Caribou herd moving on winter grounds in Alaska. Caribou (*Rangifer tarandus*) are distributed in 32 herds based on the location of their calving areas, but the different herds may mix together on winter range.

n Chapters 10 and 11, we explored the ecology of populations that form a single spatial unit in which individuals share the same environment. Within the geographic distribution of a species, however, the environmental conditions are typically not uniformly favorable for the successful survival, growth, and reproduction of individuals. Rather, suitable habitat forms a network of patches of various shape and size within the larger landscape of unsuitable habitat (Figure 12.1). Where these habitat patches are large enough to support local breeding populations, the population of a species consists of a group of spatially discrete subpopulations. In 1970, population ecologist Richard Levins of Harvard University coined the term **metapopulation** to describe a population consisting of many local populations—a population of populations.

Just as we defined a population as a group of interacting individuals of the same species occupying a given habitat, the metapopulation is a collection of local populations interacting within the larger area or region. How might the dynamics of this collective of local populations differ from that of a single, continuous population? The answer depends on the degree to which the dynamics of the various local populations are connected. The models of population dynamics presented in Chapters 10 and 11 assume that populations are closed—no immigration or emigration (see the introduction to Chapter 10)—and therefore that population growth is solely a function of the processes of birth and death. Using this framework, the dynamics of each local population would be independent, and the dynamics of the metapopulation (the collective of local populations within the region) would simply be the sum of the dynamics exhibited by the local subpopulations.

If movement of individuals to and from local populations has a significant influence on their dynamics, then a new and broader framework is necessary. For example, Paul Ehrlich and colleagues at Stanford University in California have been studying (for three decades) the population dynamics of checkerspot butterflies (genus *Euphydryas*) occupying areas of serpentine soils in the vicinity of San Francisco Bay. The most extensively studied population is that of the bay checkerspot butterfly (*Euphydryas editha*) on Stanford University's Jasper Ridge Biological Preserve. The colony of butterflies was found to consist of three localized populations, sufficiently isolated to have independent dynamics. The dynamics of the three local populations on Jasper Ridge have been documented since 1960 and are presented in Figure 12.2. Note that one of the populations (area G) went extinct in 1964, was reestablished in 1966, and then went extinct again in 1974. The reestablishment of the population in area G after local extinction (extirpation) was a result of emigration from the other two populations occupying Jasper Ridge. Had the subpopulations been closed (no immigration), area G would have remained unoccupied. In this example, however, the local populations did not function in isolation; they interacted via dispersal (immigration and emigration) with the other subpopulations within the larger area of Jasper Ridge. In other

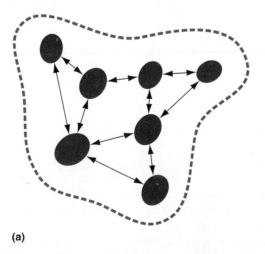

(a)

(b)

Figure 12.1 | (a) In the metapopulation concept, the distribution of a species (defined by the dashed line) is composed of a group of sub- or local populations (red circles) that are linked by dispersal (movement of individuals represented by arrows). (b) The ponds dotting the tundra landscape of Siberia (Russia) are an example of the metapopulation concept. The ponds represent a network of habitat patches of various size and shape for aquatic organisms. These habitat patches exist within a larger landscape of unsuitable habitat (land). The populations of aquatic organisms inhabiting these ponds represent distinct, spatially discrete local populations that form part of the larger metapopulation over the region.

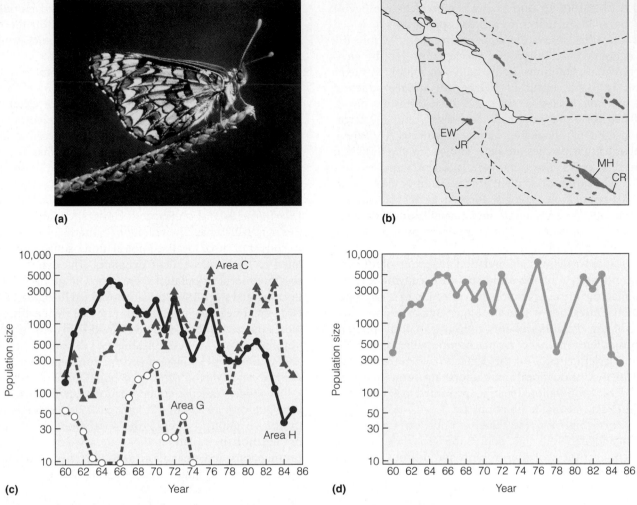

Figure 12.2 | Dynamics of three local populations of **(a)** the bay checkerspot butterfly (*Euphydryas editha*) on Jasper Ridge. **(b)** Distribution of serpentine soils in the San Francisco Bay area, including grassland habitats of the checkerspot butterfly. EW = Edgewood Park, JR = Jasper Ridge, CR = Coyote Reservoir, and MH = Morgan Hill. **(c)** Population size changes of the local populations in three areas (C, G, and H) of Jasper Ridge. Note the extinction and recolonization events in area G. **(d)** The dynamics of population size for the Jasper Ridge population, had the separate subpopulations not been identified (collective dynamics of the metapopulation at Jasper Ridge).

(Adapted from Ehrlich 1987.)

words, they functioned as a metapopulation, and an understanding of the regional dynamics of the species requires an understanding of processes operating at the level of the local population(s) as well as at the scale of the metapopulation.

12.1 | Four Conditions Define a Metapopulation

Many populations exhibit a patchy spatial distribution (see Section 9.3). But Ilkka Hanski of the University of Helsinki, a leading ecologist in the area of metapopulation biology, has suggested four necessary conditions for the term *metapopulation* to be applicable to a system of local

populations: (1) the suitable habitat occurs in discrete patches that may be occupied by local breeding populations; (2) even the largest populations have a substantial risk of extinction; (3) habitat patches must not be too isolated to prevent recolonization after local extinction; and (4) the dynamics of the local populations are not synchronized. Not all uses of the term *metapopulation* by ecologists, however, are limited to the strict set of conditions originally outlined by Hanski. In many applications of the concept, the metapopulation consists of a larger *core population* that functions as the main source of emigrants to smaller *satellite populations*. Under this set of conditions, the probability of local extinction for the core population may be extremely small (see condition 2). We will explore application of this broader use of the term *metapopulation* later, in Section 12.5.

Metapopulation dynamics differs from our discussion of population dynamics thus far in that it is governed by two sets of processes operating at two distinctive spatial scales. At the **local** or **within-patch scale,** individuals move and interact with each other in the course of their routine feeding and breeding activities. Population growth and regulation at the local scale are governed by the demographic processes (birth and death) presented in Chapter 10.

The second scale is the **metapopulation** or **regional scale,** which encompasses the set of local populations (patches) that compose the metapopulation. At this scale, dynamics are governed by the interaction of local populations, namely the process of dispersal and colonization. Because all local populations have a probability of extinction, the long-term persistence of the metapopulation depends on the process of (re)colonization.

Colonization involves the movement of individuals from occupied patches (existing local populations) to unoccupied patches to form new local populations. Individuals moving from one patch (population) to another typically move across habitat types that are not suitable for their feeding and breeding activities and often face substantial risk of failing to locate another suitable habitat patch to settle in. This dispersal of individuals between local populations is a key feature of metapopulation dynamics. If no individuals move between habitat patches, the local populations act independently. If the movement of individuals between local populations is sufficiently high, then the local populations will function as a single large population. Under this scenario, the dynamics of the various local populations may be synchronized and equally susceptible to factors that can lead to possible extinction (see condition 4).

At intermediate levels of dispersal, a dynamic emerges where the processes of local extinction and recolonization achieve some balance, where the metapopulation exists as a shifting mosaic of occupied and unoccupied habitat patches. The metapopulation concept is therefore closely linked with the processes of population turnover—extinction and establishment of new populations—and the study of metapopulation dynamics is essentially the study of the conditions under which these two processes are in balance.

12.2 | Metapopulation Dynamics Is a Balance between Colonization and Extinction

The fundamental idea of metapopulation persistence is a dynamic balance between the extinction of local populations and recolonization of empty habitat patches. In 1970, Levins proposed a simple model of metapopulation dynamics, where metapopulation size is defined by the fraction of (discrete) habitat patches (P) occupied at any given time (t). Within a given time interval, each subpopulation occupying a habitat patch has a probability of going extinct (e). Therefore, if P is the fraction of patches that is occupied during the time interval, the rate at which subpopulations will go extinct (E) is defined as

$$E = eP$$

The rate of colonization of empty patches (C) depends on the fraction of empty patches ($1 - P$) available for colonization and the fraction of occupied patches providing colonists (P), multiplied by the probability of colonization (m), a constant that reflects the rate of movement (dispersal) of individuals between habitat patches. Therefore, the colonization rate will be

$$C = [mP(1 - P)]$$

We can think of metapopulation growth in a manner analogous to our discussion of population growth in Chapter 10, where the change in the population (ΔN) over a given time interval (Δt) can be expressed as the difference between the rates of birth and death ($\Delta N/\Delta t = b - d$). The change in metapopulation, defined as the fraction of habitat patches occupied by local populations through time ($\Delta P/\Delta t$), can therefore be defined as the difference between the rates of colonization C and extinction E:

$$\Delta P/\Delta t = C - E$$

or

$$\Delta P/\Delta t = [mP(1 - P)] - eP$$

The model of metapopulation growth functions in a manner similar to the logistic model presented in Chapter 11, in that growth is regulated in a density-dependent fashion. This characteristic is probably not immediately apparent (at least to most of us) just by looking at the above equation, but it is much easier to see if we examine the equation graphically.

For any given values of e and m, we can plot the rates of extinction (E) and colonization (C) as a function of the proportion of habitat patches occupied (P; Figure 12.3). The extinction rate increases linearly with P, and the colonization rate forms a convex curve, initially rising with the proportion of patches occupied and then declining as the proportion approaches 1.0 (all patches are occupied). The value of P where the lines cross represents the equilibrium value, **P.** At this value of P, the extinction and colonization rates are equal ($E = C$) and the metapopulation growth rate is zero ($\Delta P/\Delta t = 0$). It is an equilibrium value because when the fraction of patches occupied (P) is below this value (**P**), the rate of colonization exceeds the rate of extinction and the number of occupied habitat patches increases. Conversely, if the value of P exceeds **P,** the rate of extinction exceeds the rate of colonization and the size of the

Equilibrium Proportion of Occupied Patches

When the rate of local population extinction (E) is equal to the rate at which unoccupied habitat patches are being colonized (C), the metapopulation size, as measured by the proportion of habitat patches occupied (the number of local population), will be at equilibrium (see Figure 12.3). By setting $\Delta P/\Delta t$ equal to zero, we can solve for the equilibrium value P:

$$\frac{\Delta P}{\Delta t} = 0$$

$$[mP(1 - P)] - eP = 0$$

Using simple algebraic substitution, we can solve for P (the equilibrium value P)

$$mP(1 - P) = eP$$

by dividing both sides of the equation by P

$$m(1 - P) = e$$

then dividing both sides of the equation by m

$$1 - P = \frac{e}{m}$$

and subtracting 1 from both sides

$$-P = -1 + \frac{e}{m}$$

then multiplying both sides of the equation by -1

$$P = 1 - \frac{e}{m}$$

For the metapopulation to persist, the equilibrium value P must be greater than zero, so the probability of extinction (e) cannot exceed the probability of colonization m.

1. In what way is the concept of P (the equilibrium proportion of occupied patches) similar to the concept of carrying capacity (K) presented in Chapter 11?

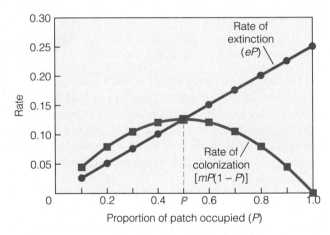

Figure 12.3 | Rates of extinction and colonization as a function of patch occupancy (P, the proportion of available habitat patches occupied) under the Levins model of metapopulation dynamics: $\Delta P/\Delta t = [mP(1 - P)] - eP$. Values of m (probability of colonization) and e (probability of extinction) were set at 0.5 and 0.25, respectively. Note that the equilibrium value of patch occupancy (P) is 0.5. At this value, the rate of extinction equals the rate of colonization. At values of P above 0.5, the rate of change is negative, and the value of P declines. When values of P are below the equilibrium value ($<$0.5), the rate of change is positive, and P increases with time.

Interpreting Ecological Data

Q1. Why is the rate of change ($\Delta P/\Delta t$) negative for values of P above the equilibrium (P)? (Answer in terms of the influence on the rates of extinction and colonization.)

Q2. Conversely, why is the rate of change ($\Delta P/\Delta t$) positive for values of P below the equilibrium (P)?

metapopulation (number of occupied patches) declines. So just as in the logistic model—in which the population density (N) tends to the equilibrium population size represented by the carrying capacity (K)—in the metapopulation model, the metapopulation density, P (proportion of patches occupied), will tend to the equilibrium metapopulation size represented by P.

The equilibrium value of P is a function of the probabilities of extinction (e) and colonization (m):

$$P = 1 - \frac{e}{m}$$

We calculate the equilibrium value mathematically from the metapopulation growth equation by setting the growth rate equal to zero: $\Delta P/\Delta t = 0$ (see Quantifying Ecology 12.1: Equilibrium Proportion of Occupied Patches).

The model of Levins just presented makes several assumptions for the sake of mathematical simplicity. It assumes that all patches are equal in size and quality as habitat. Regarding the process of colonization, the model assumes that each patch contributes equally to the pool of emigrants and that these emigrants have an equal probability of colonizing any of the unoccupied habitat patches. Finally, it assumes that the probability of extinction of any local population is independent of all other local populations; that is, the dynamics of local populations are asynchronous.

In reality, each of these assumptions may be (and most likely are) unrealistic for naturally occurring metapopulations. Local populations often differ in their

Human activity is having an ever-increasing impact on patterns of land cover. The conversion of naturally occurring ecosystems, such as forest and grassland, into highly managed agricultural fields and forest plantations is resulting in the large-scale loss of habitats for countless species of plants and animals (see Chapters 19 and 28). A result of this widespread conversion of land cover is the fragmentation of habitats (see Figure 28.4). With habitat fragmentation, formerly continuous populations are likewise fragmented into a network of local populations with varying degrees of isolation and risk of extinction. The Iberian lynx (*Lynx pardinus*) provides an example.

The Iberian lynx (Figure 1a), considered the most endangered felid species in the world, is restricted to the Iberian Peninsula (Spain and Portugal) in southwestern Europe. This medium-size cat (9–15 kg in weight), whose world population is below 1000 individuals, is distributed in nine spatially and genetically isolated populations (Figure 1b). Each of these populations is fragmented into separate local populations connected by dispersal. As a result, these populations fit the broader definition of metapopulation. The most studied of the Iberian lynx metapopulations is in the region of the Donana National Park in southwestern Spain (Figure 1c).

(a)

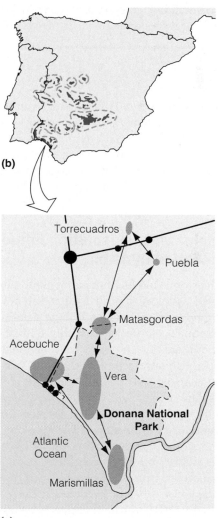

(b)

(c)

Figure 1 | Current distribution of the Iberian lynx (*Lynx pardinus*) **(a)**. **(b)** The distribution in Spain consists of nine regional populations (within dashed lines). Each of these populations consists of a group of local populations linked by dispersal. **(c)** Spatial distribution of the local populations (green shading) making up the lynx metapopulation of the Donana National Park area. Arrows indicate migration between local populations. A dashed line marks the national park boundaries. Black circles represent human settlements, and thick lines represent high-traffic roads.

(Adapted from Rodriguez 1992 and Gaona 1998.)

continued on page 244

This lynx population is distributed patchily on an area of approximately 1500 km² and has been isolated from the rest of the lynx distribution for at least the past 50 years. Before the 1950s, the lynx had a continuous distribution throughout the Donana areas, but during the past half-century human activities have caused fragmentation of the original distribution. Agriculture and tourist development now dominate the areas around the Donana National Park, including highly traveled roads and urban areas, greatly restricting patterns of movement.

Understanding that this population of cats functions as a metapopulation is crucial to management efforts aimed at preserving this species. The persistence of the species hinges on maintaining exchange among the local subpopulations. Currently, efforts are

under way to create natural habitat corridors that will allow for the safe movement of individuals between habitat patches, thus decreasing lynx traffic fatalities and promoting genetic exchange among the local populations.

We will examine the role of the metapopulation concept as it applies to habitat fragmentation in much more detail in Chapter 28 when we explore current topics in conserving species diversity.

1. Think of a plant or animal species that is indigenous (native) to your local area or region and whose population has been fragmented due to land-use changes. What processes have resulted in the fragmentation of habitat? What is the capacity of the species to disperse between habitat patches?

---

susceptibility to extinction. Habitat patches differ in size and spatial position relative to each other, and large-scale features of the regional environment often function to synchronize local population dynamics. In the following sections we will explore the consequences to metapopulation dynamics of relaxing each of these assumptions.

12.3 | Patch Area and Isolation Influence Metapopulation Dynamics

Although the simple model of metapopulation dynamics just presented assumes that all habitat patches are the same (size and quality) and contribute equally to the pool of potential colonizers, in reality habitat patches vary in size and quality. The ability of individuals to disperse between habitat patches is directly related to their spatial arrangement on the landscape—namely, their degree of isolation (see Figure 12.1). The work of Oskar Kindvall and Ingemar Ahlen of the Swedish University of Agricultural Sciences illustrates the importance of the location and size of habitat patches on metapopulation dynamics. Kindvall and Ahlen conducted a study of the metapopulation dynamics of the bush cricket (*Metrioptera bicolor*) in the Vomb valley of Sweden. The bush cricket, a medium-sized (12–19 mm), flightless katydid (Figure 12.4a), inhabits grass and heathland patches of varying size and isolation within a landscape dominated by pine forest. The metapopulation was surveyed in 1986, 1989, and 1990. Figure 12.4b shows the distribution of potential habitat patches within the valley. During the study period, the proportion of available patches occupied varied between 72 percent and 79 percent. Patterns of

occupancy were directly related to characteristics of the habitat patches and their influence on patterns of extinction and colonization.

Altogether, 18 habitat patches were colonized (or recolonized after local extinction) during the survey period. These patches were less isolated than patches that were not colonized (Figure 12.4c). The chance of colonization decreases dramatically when the interpatch distance exceeds about 100 m. The longest interpatch distance recorded for colonization by the crickets was 250 m over agricultural fields. Colonized and uncolonized patches did not differ significantly with respect to patch size; however, patch size did influence the probability of extinction.

From 1986 to 1990, a total of 18 local populations became extinct. Habitat destruction or alteration, such as grazing, house building, or the application of pesticides, caused six of these extinctions (see Ecological Issues: The Metapopulation Concept in Conservation Ecology). The other 12 extinctions occurred on patches displaying no noticeable change in habitat. These 12 patches were significantly smaller than those with persisting populations (Figure 12.4d). The risk of local extinction apparently increases with decreasing patch size. The probability of local population extinction, however, appeared not to be influenced by patch isolation.

The influence of patch size on the persistence of local populations of bush cricket was found to be indirect, through the influence of patch area on the size of local populations. The researchers found a significant positive relationship between patch size and local population size (Figure 12.5). Data revealed that the risk of local extinction increases for a patch size of less than one-half

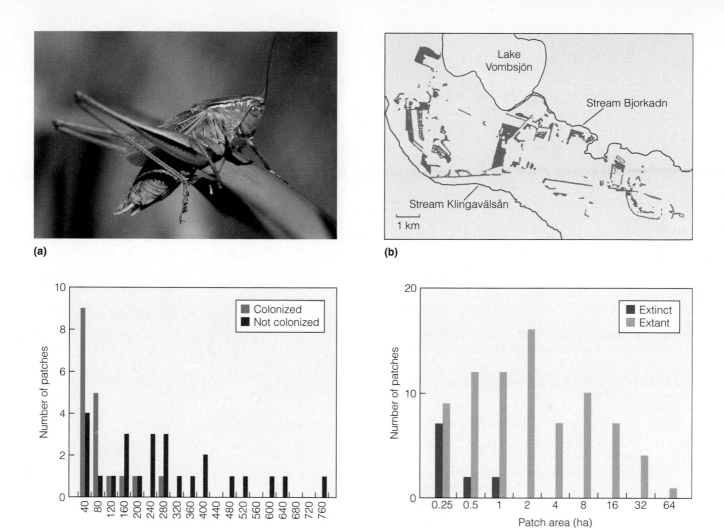

(a)

(b)

(c)

(d)

Figure 12.4 | The Swedish metapopulation of bush cricket (*Metrioptera bicolor*) **(a)** is located south of Lake Vombsjön, which is surrounded by two streams **(b)**. Patches of grassland with local populations that did not become extinct during the period of study are shown in brown. Other areas in tan were occupied either occasionally or not at all during the survey period. **(c)** The frequency distribution of interpatch distance between colonized and uncolonized patches during the survey period. **(d)** The frequency distribution of areas of patches with existing (extant) populations and patches where the local population became extinct during the survey period.

(Adapted from Kindvall 1992.)

Interpreting Ecological Data

Q1. In Figure 12.4c, interpatch distance is a measure of the spatial isolation of the habitat patches from other patches of suitable habitat on the landscape. What is the maximum interpatch distance beyond which dispersal and colonization of bush crickets does not appear to be possible?

Q2. In Figure 12.4d, the total number of habitat patches in each size category (*x*-axis) that were occupied by bush cricket populations during the study period can be calculated by adding the values of occupied (extant) and extinct. Which size category of habitat patch has the highest value for occupancy during the study period? Which size category has the largest existing value for occupancy? What patch size appears to represent a threshold for the long-term persistence of local cricket populations?

hectare (Figure 12.4d), which corresponds to a critical population size of about 12 males (see Sections 10.10 and 28.6 for discussion of minimum viable population concept). In 1990, only 67 percent of the suitable patches were larger than one-half hectare; however, 79 percent of the habitat patches were occupied. The fraction of habitat patches occupied (*P*) was maintained at a higher value than would be expected under isolation of patches as a result of recolonization after local extinction. The fraction of occupied patches (*P*) was fairly constant

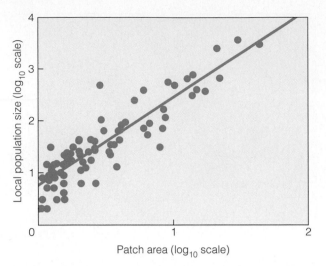

Figure 12.5 | Relationship between local population size and patch area in the metapopulation of bush crickets, *Metrioptera bicolor* (see Figure 12.4 for map of site). Values for both population size and patch have been log-transformed. The line represents the general trend as defined by the simple linear regression: $y = 1.70x + 0.74$ ($n = 83, r^2 = 0.81$).

(Adapted from Kindvall 1992.) Go to **QUANTIFYit!** at www.ecologyplace.com to review linear regression.

during the survey period (5 years); however, different patches were occupied each year.

As seen in the preceding example, both patch size and isolation influence local population dynamics. That is to say, metapopulation persistence depends simultaneously on patch size and isolation. The interaction between patch area and isolation is illustrated in the work of C. D. Thomas and T. M. Jones of Imperial College, England. Thomas and Jones examined patterns of extinction and colonization of patches of grassland habitat in the North and South Downs of southern England by the skipper butterfly (*Hesperia comma*; Figure 12.6a). The probability of local extinction declined with increasing patch area and increased with isolation. Conversely, the probability of colonization increased with patch area and declined with isolation from other local populations. The result, shown in Figure 12.6b, is a clear pattern of increasing probability of patch occupancy with declining isolation and increasing patch area, with increasing patch size compensating for increasing degree of isolation from neighboring populations.

To explore how the factors of patch size and isolation interplay to influence the dynamics of patch occupancy (*P*), we can use the graphical model of metapopulation growth presented in Figure 12.3. Isolation (distance from neighboring patches) decreases the probability of colonization (*m*) and therefore decreases the rate of colonization (*C*). Increasing patch size decreases the probability of extinction (*e*) and therefore the rate of extinction (*E*). Shifts in the rates of colonization and extinction function

(a)

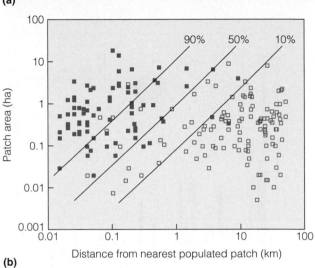

(b)

Figure 12.6 | Distribution of occupied (solid) and vacant (open) habitat patches of the skipper butterfly (*Hesperia comma*) **(a)** in southern Britain (1991) in relation to patch area and isolation from the nearest populated patch. **(b)** Lines give the 90%, 50%, and 10% probabilities of occupancy. Note the compensatory effect of isolation and patch area on occupancy.

(Adapted from Thomas 1993.)

Interpreting Ecological Data

The three lines in Figure 12.6b define the combined values of patch isolation (distance from nearest populated patch) and patch size (area) that include 90%, 50%, and 10% of the occupied habitat patches (solid squares) on the landscape during the study period. Use these lines to answer the following questions.

Q1. What is the approximate probability that a 10-ha habitat patch 1 km from the nearest populated patch will be occupied by skipper butterflies?

Q2. What is the approximate area necessary to provide a 50% probability that a habitat patch 1 km from any neighboring populated patches will be colonized and occupied by a local population of skipper butterflies?

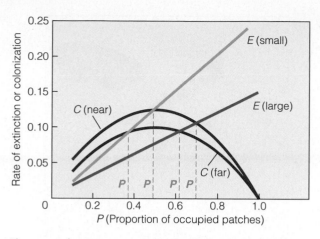

Figure 12.7 | Changes in the equilibrium proportion of habitat patches occupied (**P**) for habitat patches of differing size (small and large) and isolation (near and far). Increasing patch size functions to decrease the rate of extinction (*E* large), and increasing distance from neighboring patches functions to decrease the rate of colonization (*C* far).

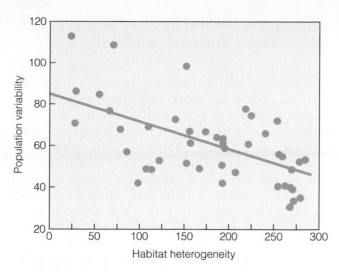

Figure 12.8 | Relationship between temporal variability of population size and habitat heterogeneity for local populations of the bush cricket (*Metrioptera bicolor*) in the Lake Vombsjön metapopulation (see Figure 12.4). Population variability was measured by the coefficient of variation in population size through time. Habitat heterogeneity is an index measuring the spatial heterogeneity of vegetation cover types (e.g., dense, low grassland; tall grassland; sparse grassland on sandy soils).

(Adapted from Kindvall 1996.)

to change the equilibrium value of P (proportion of patches occupied; Figure 12.7). An increase in isolation reduces the equilibrium value of P, and an increase in patch size increases the equilibrium value of P.

The compensatory effect between patch size and isolation can also be seen in Figure 12.7. Increasing isolation of habitat patches can be offset by an increase in the average patch size; and conversely, a decline in the average patch size can be compensated by a reduced degree of isolation.

12.4 | Habitat Heterogeneity Influences Local Population Persistence

Besides supporting larger local populations (greater carrying capacity), increasing patch size may influence the persistence of local populations by increasing the potential for environmental heterogeneity. Numerous studies have demonstrated that larger patches have the potential to be spatially more heterogeneous than smaller patches. As a result, larger patches may include habitat types lacking from smaller areas. An important consequence of such environmental heterogeneity is that when environmental conditions change, the essential resources (habitat or food) for a particular species might disappear entirely from small patches (though perhaps only temporarily). In larger areas, however, favorable conditions are more likely to remain somewhere within the patch. This hypothesis therefore predicts that increasing area reduces the risk of extinction by supporting greater habitat heterogeneity and thus reducing the impact of environmental stochasticity (see Section 10.8).

In his work on the dynamics of bush cricket metapopulation dynamics in Sweden (see previous sec-

tion), Kindvall explored the role of habitat heterogeneity as a possible mechanism in reducing fluctuations in population densities within habitat patches. The habitat requirements of the bush cricket are known to change in relation to weather conditions. Wet conditions have a depressing effect on both survival rate and fecundity, so that in extremely rainy years it is better for crickets to live in sites with sparse vegetation on sandy soils that dry quickly. But during periods of drought, the dense, tall grassland vegetation is beneficial for the survival of juveniles. As Figure 12.8 demonstrates, Kindvall found that habitat heterogeneity significantly affects temporal variability of local populations (as measured by the variation in population size over the study period). Populations inhabiting habitat patches that supported various types of vegetation cover (e.g., dense, low grassland; tall grassland; sparse grassland on sandy soils) exhibited less fluctuation in population densities and subsequently a lower probability of local extinction than did populations inhabiting more homogeneous patches.

The work of Oskar Kindvall shows that habitat heterogeneity can influence both the temporal variability of local populations and the probability of local extinction. For species susceptible to weather fluctuation, habitat heterogeneity becomes an important factor that determines the relative strength of environmental stochasticity acting on different local populations, and therefore the risk of extinction (see Section 10.11).

12.5 | Some Habitat Patches May Function as the Major Source of Emigrants

The ecologists James Brown and Astrid Kodric-Brown of the University of New Mexico introduced the concept of **rescue effect** into the metapopulation literature. By *rescue effect* they meant the increasing population size, and hence the decreasing risk of extinction (see Section 10.11), that occurs with an increasing rate of immigration. The focus of their work was the colonization of islands by individuals from a mainland population. Although migration between islands occurs, the major source of immigrants to island populations is emigration from the mainland population. As with our discussion of colonization in the context of metapopulations, the rate of immigration to an island is influenced by its degree of isolation and declines with distance from the mainland. Assuming the mainland population is extremely large relative to the island populations, species will not go extinct from the network of islands as long as there is some dispersal (emigration) from the mainland.

The **mainland–island metapopulation structure** is unique in that a single habitat patch (the mainland) is the dominant source of individuals emigrating to the other habitat patches within the metapopulation network. Many populations have an essentially similar mainland and island structure, owing to high variation in the sizes of habitat patches or populations. For a system to have mainland–island dynamics, there need not be a single mainland of extreme size. Substantial variation in patch or population size can produce the same effect. An example is the Morgan Hill population of the bay checkerspot butterfly (*Euphydryas editha bayensis*) in central California, studied by Susan Harrison and colleagues at Stanford University (Figure 12.9). One very large and apparently persistent "mainland" population dominates this metapopulation, with transient "island" populations found nearby in small habitat patches. Local extinctions will tend to strike the smallest populations, which have the least impact on the overall persistence of the metapopulation. Differences in habitat quality among patches, rather than patch or population size, may produce a similar rescue effect. Ronald Pulliam of the University of Georgia proposed the idea that dispersal from **source populations** in high-quality habitat may permit **sink populations** to exist in inferior habitat. Source populations are defined by their ability to maintain a positive growth rate ($r > 0$), whereas sink populations cannot support positive population growth ($r < 0$), because of their poor quality.

Ecologists also refer to **source and sink habitats**. The source–sink scenario reinforces the point that the presence and size of a population in a habitat patch might give an entirely misleading idea of environmental

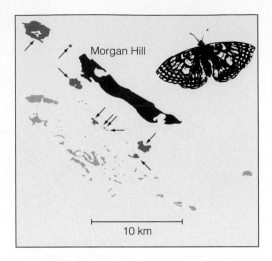

Figure 12.9 | Habitat for the bay checkerspot butterfly (*Euphydryas editha bayensis*) metapopulation is fragmented due to both natural and human-induced factors. The shaded areas are patches of serpentine grasslands: the butterfly's habitat. The largest population occupies the Morgan Hill area. Only those patches closest to the Morgan Hill area (marked by arrows) are usually occupied, suggesting that the butterfly is a poor disperser. Extinctions and recolonizations of smaller patches are common. (Adapted from Harrison et al. 1988.)

conditions in the patch. Sink populations may be persistent or even large due to high rates of immigration from the source populations, even though they cannot sustain a positive growth rate through reproduction.

An extreme example of a source–sink metapopulation is the metapopulation of the checkered white butterfly (*Pieris protodice*) in the Central Valley of California, studied by Arthur Shapiro of the University of California, Davis. The source population occupies a unique riparian habitat along the east side of the Sacramento Valley (patches of vegetation that occur on dredge tailings created by mining activity in the valley during the early 1900s), which is the only part of the regional geographic range where the butterfly can overwinter successfully. In most years, breeding colonies may be found far from the source population; but most breeding colonies do not persist into a second season. Reestablishment of colonies the following spring is dependent on emigration from the source population inhabiting the dredge tailings.

12.6 | Certain Factors Can Function to Synchronize the Dynamics of Local Populations

Asynchronous dynamics of local populations is a key factor in the persistence of a metapopulation (see Section 12.1). When the chance of extinction is completely independent in each local population, the probability of metapopulation extinction rapidly declines with increasing number of local populations. If extinction probabilities are correlated, however, metapopulations with even large numbers of local populations may be susceptible to extinction.

A number of demographic and environmental factors can potentially synchronize the dynamics of local populations. Environmental stochasticity often operates at a regional scale (see Section 10.11). Weather, for example, causes insect populations to fluctuate in synchrony over broad geographic areas, and a single drought or freeze may eliminate multiple local populations. Paul Ehrlich and colleagues at Stanford University documented the effects of a regional drought during the years 1975–1977 on butterfly populations in central California. The researchers found the synchronized extinction of local populations in response to a dramatic decline in a particular species of annual flowering plant that is an essential food resource for the butterflies. Odette Sutcliffe and colleagues of the University of Leeds observed a similar synchronization of local population extinction for the ringlet butterfly (*Aphantopus hyperantus*) in eastern England (Monk Wood National Nature Reserve, Cambridgeshire) during a period of severe regional drought. Such regional stochasticity reduces the likelihood of persistence for metapopulations. In turn, it increases the potential importance of other mechanisms that enable persistence, such as refuge habitats (as in the case of habitat heterogeneity and the bush cricket; see Section 12.4) or resistant or dormant life stages (see Section 7.12).

During longer time periods, landscape and habitat changes that cause deterministic extinctions have often been relatively synchronized over larger areas. The work of C. D. Thomas of Imperial College provides an example of regional scale synchronization of local populations resulting from a large-scale change in habitat availability. At the beginning of the 20th century, the skipper butterfly (*Hesperia comma*) inhabited most of the chalk grasslands of southern England (Figure 12.10). Despite its extensive geographic distribution, the skipper butterfly has specific ecological requirements. It occurs only on heavily grazed, calcareous grasslands where the turf is short and sparse. The conversion of lands to agriculture and a reduction in the stocking of lands with domestic animals led to a gradual decline of the skipper butterfly in the first half of the century, with a retraction of the distribution southward (Figure 12.10). But grazing by previously introduced rabbits assured that many of the remaining grasslands did not become overgrown, and the species survived throughout the southern portion of its distribution. When the myxomatosis virus was introduced in England in 1954 to control rabbit populations, the calcareous grasslands became overgrown and the shortgrass habitats disappeared throughout England. The skipper butterfly showed a period of rapid decline and is now found only in those areas (habitat refuges) where domestic grazing continues (Figure 12.10).

Changes in land-use practices, such as the conversion of natural ecosystems to agricultural or forest production,

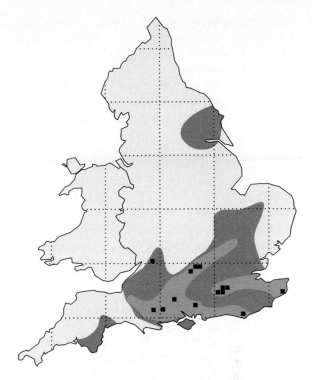

Figure 12.10 | Decline of the skipper butterfly (*Hesperia comma*) in England during the 20th century. Brown shading represents the pre-1920 range. The skipper butterfly was recorded in all calcareous outcrops within this range. Green shading represents the known range between 1920 and 1961. The current distribution, shown as black squares, consists of a complex of refuges (as of 1982 census). The decline in distribution is due to the conversion of native grasslands to pasture and agricultural fields and the reduction of grazing in remaining native grasslands after the introduction of the myxomatosis virus (used to control rabbit populations) in the mid-1950s.

(Adapted from Thomas 1993.)

usually occur over very large areas in a relatively short time, causing widespread changes in the fortunes of associated species (Chapters 19 and 28 discuss habitat fragmentation and species extinction in more detail).

12.7 | Species Differ in Their Potential Rates of Colonization and Extinction

Thus far, we have explored the influence of the physical environment on the potential rates of extinction and colonization of habitat patches within a metapopulation. Species, however, differ widely in their susceptibility to local extinction and their ability to colonize available habitats. An important component determining a species' ability to colonize new areas is dispersal rate, which varies greatly among species and is clearly the result of natural selection.

General life history classifications, such as *r*- and *K*-species or the ruderal–competitive–stress tolerant classification of plants put forward by Philip Grime (see Section 8.13), consider dispersal ability as a key characteristic of species adapted to ephemeral or disturbance-prone habitats. Frequent dispersal is common in insects that occupy temporary habitats or where variation in local carrying capacity is large. In contrast, less emigration occurs in stable or isolated habitats. In analyzing the evolution of flightlessness in insects, Derek Roff of McGill University in Montreal showed that flightlessness is associated with decreased environmental heterogeneity and habitat persistence.

High fecundity has also been associated with higher rates of dispersal. In a survey of ecological characteristics of British angiosperms, Helen Peat and Alastair Fitter of the University of York (England) found that species with wind-dispersed seeds produced more seeds than did species with no specialized dispersal mechanisms. Wind dispersal is an adaptation that permits long-distance dispersal, whereas mean dispersal distances tend to be small—often measured in centimeters—for species with unspecialized dispersal.

Mode of reproduction can also influence rates of dispersal and colonization. Plant species that reproduce asexually tend to have relatively high rates of growth and subsequent lower rates of local extinction. However, the asexual mode of reproduction through the production of ramets (see Section 9.1, Figure 9.1) can reduce dispersal ability and thus the ability to colonize new habitats.

The relationship between body size and home range size, presented in Chapter 11 (Section 11.9, Figure 11.10), directly influences population density (individuals/unit area). Smaller organisms can potentially support higher population densities per unit area than can species having a larger body size. Because the most likely cause for frequent local extinctions is simply that many local populations are small and hence vulnerable to stochastic extinction (see Section 10.10), small body size might well reduce the probability of extinction by allowing for a larger potential population size for a given area of habitat (patch size). This relationship also directly influences the minimum patch size necessary to support a viable local population (patch occupancy). Larger species require a larger minimum patch size to support a viable population.

In other circumstances, body size can increase the susceptibility of a species to extinction. In endothermic species, mass-specific metabolic rate increases exponentially with decreasing body mass (higher mass-specific metabolic rate in smaller species; see Section 7.10, Figure 7.15). The high mass-specific metabolic rate results in a shorter period to starvation for smaller species during periods of resource shortage, making smaller species potentially more susceptible to environmental stochasticity. Anu Peltonen and Ilkka Hanski of the University of Helsinki surveyed the presence of three species of shrews (*Sorex araneus*, *S. caecutiens*, and *S. minutus*) on 108 islands in 3 lakes in Finland. Their results demonstrated that on islands less than 8 ha in size, the probability of extinction varied significantly between species and was negatively correlated with body size. The smallest species, *S. minutus*, had the highest rate of extinction as a result of increased sensitivity to environmental stochasticity and variations in food supply.

Correlations have also been identified between mode of temperature regulation and susceptibility to extinction from environmental stochasticity. Generally, smaller ectothermic animals, such as insects, are more affected by environmental anomalies (e.g., drought, extreme temperature events, ice storms, etc.) than larger endothermic animals, giving rise to greater temporal variations in population size. In a review of the studies examining the population dynamics of 91 species of terrestrial vertebrates (mammals, birds, and lizards) and 99 species of terrestrial arthropods (moths, aphids, hoverflies, grasshoppers, etc.), Hanski found that vertebrates have generally less-variable populations and a greater degree of density dependence than do the invertebrate populations examined.

In general, virtually all aspects of a species' life history can influence demographic processes, either directly or indirectly, and can subsequently influence metapopulation dynamics. We revisit this topic later, in Chapter 19, when we explore patterns of species diversity at the landscape scale.

12.8 | The Concept of Population Is Best Approached by Using a Hierarchical Framework

The classic definition of a *population* as "a group of organisms of the same species occupying a particular area at a particular time" (see the introduction to Part Three) is so general that it could describe everything from a group of individuals occupying a single patch of habitat (local population) to all members of the species within its geographic range. Ecologist Edward O. Garton has proposed a more useful framework for the concept of population; it is based on a series of hierarchical spatial units containing groupings of individuals significant for our understanding, estimation, and prediction of future population conditions. Four levels of spatial aggregation that we have discussed thus far in the text provide an understanding of the role of population in the processes of demography, movement, geographic distribution, and evolution.

The smallest group of individuals that we have discussed is the local population. The local population is the spatial unit for which it is reasonable to estimate the rates of birth, death, emigration, and immigration—the focus of our discussion in Chapter 10. Individuals in this grouping are ideally distributed continuously within a single, contiguous patch of habitat.

The second spatial unit is the metapopulation—the focus of this chapter. This spatial unit represents a collection of local populations within sufficient proximity that dispersing individuals can readily colonize empty habitat patches resulting from local population extinction. Local populations within a single metapopulation may or may not show correlations in demographic rates; but the low rates of dispersal are sufficient to maintain substantial gene flow among populations and, therefore, to preserve genetic similarity.

The third spatial unit is the subspecies, presented in Chapter 5 (see Section 5.7). This spatial unit represents a collection of metapopulations in a geographic region, where local and metapopulations occupy habitat patches that may be separated by large distances or by large areas of unsuitable habitat, resulting in substantial demographic independence among metapopulations. The physical isolation of the metapopulations can result in a divergence in natural selection and the evolution of distinct subspecies (a taxonomic term for populations of a species that are distinguishable by one or more characteristics). Rare dispersal, however, maintains some gene flow and genetic similarity among these potentially interbreeding populations. The salamander *Plethodon jordani* in the southern Appalachian Mountains is an example of a population unit at this spatial scale (see Figure 5.14). This species is broken into several subspecies that form semi-isolated populations in this landscape of rugged terrain.

The fourth spatial unit is the species: the collection of subspecies encompassing the entire distribution and geographic range of the species (see Section 9.2).

Differentiating among these different spatial scales is essential to our understanding of the processes driving population dynamics, both demographically and genetically.

Summary

Conditions for Metapopulation | 12.1

When the population of a species consists of spatially discrete subpopulations, the collective of subpopulations is termed a *metapopulation*. Four conditions are necessary for the term *metapopulation* to be applicable to a system of local populations: (1) discrete habitat patches, (2) substantial risk of extinction, (3) migration and recolonization after local extinction, and (4) asynchronous dynamics. Under this set of conditions, processes operating at two distinct spatial scales govern the dynamics of the metapopulation. At the metapopulation or regional scale, which encompasses the set of local populations (patches) that compose the metapopulation, population dynamics are governed by the interaction of local populations—namely, the process of migration and colonization.

Extinction and Colonization | 12.2

Metapopulation persistence is a dynamic balance between the extinction and recolonization of empty habitat patches. For local extinction to occur, populations on separate patches must be reasonably isolated from one another, with most recruitment coming from within the patch (birth) rather than from immigration. Colonization involves the movement of individuals from occupied patches to unoccupied patches to form new local populations. The persistence of the metapopulation, therefore, depends on the balance between the rates of local extinction and the (re)colonization of unoccupied patches. If rates of immigration are high in comparison to local rates of extinction, the local populations will function as a single continuous population. In contrast, if the colonization rate is not sufficient to offset the local extinction rate, the metapopulation will not persist.

Patch Size and Isolation | 12.3

The ability of individuals to disperse between habitat patches is directly related to their spatial arrangement on the landscape. The rate of colonization declines with increasing isolation (distance from adjacent patches). The risk of local extinction increases with decreasing patch size. A major cause of the inverse relationship between patch size and probability of local extinction is that small patches support smaller populations, and small population size increases the probability of extinction due to demographic and genetic stochasticity.

Patch area and isolation can have a compensatory effect on the probability of patch occupancy (size of metapopulation). Reduced isolation can compensate for patch size through increased rates of recolonization. Conversely, larger patch size and associated decrease in probability of extinction can function to compensate for patch isolation (reduced probability of colonization).

Habitat Heterogeneity | 12.4

Increasing patch size may also influence the persistence of local populations by increasing the potential for environmental heterogeneity. An important consequence of such environmental heterogeneity is that when environmental conditions change, the essential resources for a particular species might disappear entirely from small patches; in larger areas, however, favorable conditions are more likely to remain somewhere within the patch.

Rescue Effect | 12.5

The role of immigration in maintaining the persistence of local populations that would otherwise go extinct has been termed the rescue effect, most noted in island populations. Assuming the mainland population is extremely large relative to the island populations, species will not go extinct from the network of islands as long as there is some migration from the mainland. Many populations have an essentially similar mainland and island structure owing to high variation in the sizes of habitat patches or populations.

Differences in habitat quality among patches, rather than patch or population size, may produce a similar rescue effect, giving rise to source and sink populations (or habitat patches). Source populations are defined by their ability to maintain a positive growth rate $(r > 0)$, whereas sink populations cannot support positive population growth $(r > 0)$, because of their poor quality.

Synchronized Dynamics | 12.6

Asynchronous dynamics of local populations is a key factor in the persistence of a metapopulation. When the chance of extinction is completely independent in each local population, the probability of metapopulation extinction rapidly declines with increasing number of local populations. If extinction probabilities are correlated, however, metapopulations with even large numbers of local populations may be susceptible to extinction. Some demographic and environmental factors, including weather and changes in patterns of land use, can potentially synchronize the dynamics of local populations.

Species Characteristics | 12.7

Species differ widely in their susceptibility to local extinction and their ability to colonize available habitats. Species life history characteristics such as mode of reproduction, fecundity rate, dispersal mechanism, and longevity can influence rates of dispersal and colonization. Likewise, characteristics such as body size and mode of thermal regulation have been correlated with probability of local extinction, particularly under conditions of environmental stochasticity.

Hierarchical Concept of Population | 12.8

Four levels of spatial aggregation provide an understanding of the role of population in the processes of demography, movement, geographic distribution, and evolution: (1) local population (demography), (2) metapopulation (movement), (3) subspecies (evolution), and (4) the species' geographic distribution (geography).

Study Questions

1. Define *metapopulation*. How does this concept relate to the spatial distribution (dispersion) of the population over a geographic region?
2. What four conditions are necessary for the term *metapopulation* to be applied to a system of local populations?
3. Discuss the role of local extinction and colonization on metapopulation dynamics. How does the rate of immigration influence the degree to which the dynamics of the local populations are synchronized? What effect does the synchronicity of local dynamics have on metapopulation persistence?
4. How do the size and spatial arrangement of habitat patches (local populations) influence metapopulation dynamics? Discuss in terms of probabilities of extinction and colonization.
5. How does habitat heterogeneity influence the dynamics of local populations? What does this imply about the importance of habitat heterogeneity among habitat patches?
6. What is the rescue effect? Under what set of conditions might this effect be important?
7. Contrast the terms *source population* and *sink population*. What feature(s) of a population define it as being a potential source or sink?
8. How does the concept of source and sink population relate to habitat quality?
9. What environmental factors might function to synchronize local population dynamics over a region?

Further Readings

den Boer, P. 1981. On the survival of populations in a heterogeneous and variable environment. *Oecologia* 50:39–53.

An early and important paper that examines the importance of the metapopulation concept in understanding the persistence of populations in variable environments.

Hanski, I., and M. Gilpin. 1991. Metapopulation dynamics: Brief history and conceptual domain. *Biological Journal of the Linnean Society* 42:3–16.

This paper is an excellent overview of the research areas relating to the study of metapopulations.

Hanski, I. 1999. *Metapopulation ecology*. New York: Oxford University Press.

The definitive reference text written by a leading scientist in the study of metapopulations. Provides a wealth of illustrated examples relating to topics covered in this chapter.

Holyoak, M., and C. Ray. 1999. A roadmap for metapopulation research. *Ecology Letters* 2:273–275.

Excellent discussion of the role of metapopulation research in conservation ecology.

Pulliam, R. 1988. Sources, sinks, and population regulation. *American Naturalist* 132:652–661.

The initial work that developed the idea of source and sink populations (and habitats).

The peacock bass (*Cichla ocellaris*), native to South America's Amazon–Oronoco watershed, is one of the most sought-after freshwater game fish in the world. Bright yellow with black vertical bars, this fish derives its common name from its basslike shape and from the conspicuous ocellus (eyespot)— black encircled by a gold ring—located at the base of the caudal fin. Its reputation among sport fishermen has resulted in this species being introduced to fresh-water lakes and rivers from South and Central America to southern Florida in the United States. In 1967, this species was introduced to Gatun Lake (42,315 hectares) in Panama. Since then it has become a popular sport fish in the region, attracting fishermen from around the world. Besides providing a source of tourist revenue, it is the only species of freshwater fish that is sold for consumption in this region.

Although at first the introduction of the peacock bass to Gatun Lake may seem like an unqualified success, there is a darker side to this story. As the species spread throughout the lake, its voracious predatory habits had a devastating effect on the native fish populations. In the lake, the peacock bass feeds mostly on adult fish of the genus *Melaniris,* thus decreasing its populations. Other predatory species that feed on *Melaniris* are Atlantic tarpon, black terns, and herons—all of which have sharply declined. A complex community structure has been highly simplified (see Section 28.2, Figure 28.7). Six of eight common fish species within the community have been eliminated or seriously reduced, all by the introduction of a single species of fish into the lake ecosystem.

The impact of peacock bass on the Gatun Lake ecosystem is only one of countless stories about the unintended consequences of introducing a nonindigenous species into an existing natural community. These problems arise from failing to understand the processes that influence population dynamics and ultimately structure ecological communities—species interactions.

Thus far in the text, we have examined interactions among organisms only as they pertain to individuals within the same population or species (intraspecific interactions). In Chapter 8, we examined interactions among individuals within the population that relate to reproduction (selecting mates; care and defense of offspring). In Chapter 11, we examined how competition for limited resources can function to limit population growth. But species cannot be viewed in isolation. Species occupying the same physical area—be it a lake, stream, forest, or field—interact in many ways, and central to these interactions is the need to acquire the basic resources necessary for growth and reproduction. Although different plant species that co-occur within a habitat may differ in their specific needs for certain essential nutrients or in their ability to continue the process of assimilation as resources become limiting (see Chapter 6), all plants require the same resources of water, light, carbon dioxide, and other essential nutrients (see Table 6.1). As a result, competition for these resources within a habitat can become intense, with their acquisition by individuals of one species reducing availability to individuals of other species.

Among heterotrophic organisms, the range of potential interactions expands. Because heterotrophic organisms derive their energy and nutrients from consuming organic matter, the very act of feeding involves interaction between species: interaction between predator (the consumer) and prey (the consumed). When different predator species feed on the same species as prey, there is also the potential for competition.

For some species, other organisms provide habitat as well as nourishment. Many microorganisms, such as bacteria and fungi, take up residence in or on other organisms. They draw their energy and nutrient needs from the host, hopefully without killing the source of resources that they depend on. This is the interaction between parasite and host.

Not all interactions among species are negative (i.e., involving winners and losers). Interactions that are mutually beneficial to the parties involved are ubiquitous and relate to nutrition, shelter, defense, and reproduction.

By influencing the demographic processes of birth and death, interspecific interactions play a central role in population dynamics. By differentially influencing the survival and reproduction of individuals within the population, these same interactions can function as agents of natural selection. Certain phenotypes within the population may be superior to others in either competing—or avoiding competition—with individuals of other species, resulting in an increase in their relative fitness. Some phenotypes within the population may be superior to others at capturing prey or in avoiding capture by predators. In other circumstances, a specific characteristic may change the nature of an interaction from negative or neutral to positive. In functioning as agents of natural selection,

species interactions play a central role in the process of evolution.

In Part Four, we will examine the diversity of species interactions, their influence on demographic processes, and their role as agents of natural selection. In Chapter 13, we will explore the process of interspecific competition, building on the conceptual and mathematical models of competition within populations that we developed in Chapter 11. In Chapter 14, we will examine the process of predation, including carnivory (animals eating animals) as well as herbivory (animals eating plants and algae). We will develop a model of the interaction between predator and prey that suggests a mutual regulation of populations, giving rise to a pattern of predictable population cycles through time. Finally, in Chapter 15 we will explore the relationship between parasites and their hosts and examine how, through the process of natural selection, antagonistic relationships sometimes evolve into interactions that are mutually beneficial to the species involved. The discussion of species interactions in these chapters will form the foundation for our examination of ecological communities in Part Five.

Gatun Lake, Panama

CHAPTER 13

Interspecific Competition

A yellow-necked mouse (*Apodemus flavicollis*) feeds on an acorn on a forest floor in South Lower Saxony, Germany. This mouse is only one of an array of species within the forest that depend on acorns as a food resource.

n Chapter 3 of *The Origin of Species,* Charles Darwin wrote: "as more individuals are produced than can possibly survive, there must in every case be a struggle for existence, either one individual with another of the same species, or with the individuals of distinct species, or with the physical conditions of life." The concept of interspecific competition is a cornerstone of evolutionary ecology. Darwin based his idea of natural selection on competition, the "struggle for existence." Because it is advantageous for individuals to avoid this struggle, competition has been regarded as the major force behind species divergence and specialization (Chapter 5).

13.1 | Interspecific Competition Involves Two or More Species

A relationship that affects the populations of two or more species adversely (– –) is interspecific competition. In interspecific competition, as in intraspecific competition (see Chapter 11), individuals seek a common resource in short supply. But in interspecific competition, the individuals are of two or more species. Both kinds of competition may take place simultaneously. Gray squirrels, for example, compete among themselves for acorns during a year when oak trees produce fewer acorns. At the same time, white-footed mice, white-tailed deer, wild turkey, and blue jays vie for the same resource. Because of competition, one or more of these species may broaden the base of their foraging efforts. Populations of these species may be forced to turn away from acorns to food that is less in demand.

Like intraspecific competition, interspecific competition takes two forms: exploitation and interference (see Section 11.3). As an alternative to this simple dichotomous classification of competitive interactions, Thomas Schoener of the University of California at Davis proposed that six types of interactions are sufficient to account for most instances of interspecific competition: (1) consumption, (2) preemption, (3) overgrowth, (4) chemical interaction, (5) territorial, and (6) encounter.

Consumption competition occurs when individuals of one species inhibit individuals of another by consuming a shared resource, such as the competition among various animal species for acorns described earlier. Preemptive competition occurs primarily among sessile organisms, such as barnacles, where the occupation by one individual precludes establishment (occupation) by others. Overgrowth competition occurs when one organism literally grows over another (with or without physical contact), inhibiting access to some essential resource. An example of this interaction is when a taller plant shades those individuals below, reducing available light (as discussed in Section 4.2 of Chapter 4). In chemical interactions, chemical growth inhibitors or toxins released by an individual inhibit or kill other species. Allelopathy in plants, in which chemicals produced by some plants inhibit germination and establishment of other species, is an example of this type of species interaction. Territorial competition results from the behavioral exclusion of others from a specific space that is defended as a territory (see Section 11.10). Encounter competition results when nonterritorial meetings between individuals negatively affect one or both of the participant species. Various species of scavengers fighting over the carcass of a dead animal provide an example of this type of interaction.

13.2 | There Are Four Possible Outcomes of Interspecific Competition

In the early 20th century, two mathematicians—the American Alfred Lotka and the Italian Vittora Volterra—independently arrived at mathematical expressions to describe the relationship between two species using the same resource. Both men began with the logistic equation for population growth that we developed in Chapter 11:

$$\frac{dN}{dt} = rN\frac{(K - N)}{K}$$

Next, they both modified the logistic equation for each species by adding to it a term to account for the competitive effect of one species on the population growth of the other. For species 1, this term is αN_2, where N_2 is the population size of species 2, and α is the competition coefficient that quantifies the per capita effect of species 2 on species 1. Similarly, for species 2, the term is βN_1, where β is the per capita competition coefficient that quantifies the per capita effect of species 1 on species 2. The competition coefficients can be thought of as factors for converting an individual of one species into the equivalent number of individuals of the competing species, based on their shared use of the resources that define the carrying capacities. In resource use, an individual of species 1 is equal to β individuals of species 2. Likewise, an individual of species 2 is equivalent to α individuals of species 1. These terms, in effect, convert the population density of the one species into the equivalent density of the other. Now we have a pair of equations that consider both intraspecific and interspecific competition.

$$\text{Species 1: } \frac{dN_1}{dt} = r_1 N_1\left(\frac{K_1 - N_1 - \alpha N_2}{K_1}\right) \quad (1)$$

$$\text{Species 2: } \frac{dN_2}{dt} = r_2 N_2\left(\frac{K_2 - N_2 - \beta N_1}{K_2}\right) \quad (2)$$

As you can see, in the absence of interspecific competition—either α or $N_2 = 0$ in Equation (1) and β or $N_1 = 0$ in Equation (2)—the population of each species grows logistically to equilibrium at K, the carrying capacity. In the presence of competition, the picture changes.

For example, the carrying capacity for species 1 is K_1, and as N_1 approaches K_1, the population growth (dN_1/dt) approaches zero. However, species 2 is also vying for the limited resource that determines K_1, so we must consider the impact of species 2. Because α is the per capita effect of species 2 on species 1, the total effect of species 2 on species 1 is αN_2. We must consider the effects of both species in calculating population growth. As the combined population effect ($N_1 + \alpha N_2$) approaches K_1, the growth rate of species 1 will approach zero. The greater the population size of the competing species (N_2), the greater the reduction in the growth rate of species 1 (dN_1/dt).

Depending on the combination of values for the Ks and for α and β, the Lotka–Volterra equations predict four different outcomes. Figure 13.1 graphically depicts all the possible outcomes. In two situations (Figures 13.1c and 13.1d), one species is the superior competitor and wins out over the other. In one case, species 1 inhibits the population growth of species 2 while continuing to increase. In this case, species 2 is driven to extinction. In the other case, species 2 inhibits further increase in species 1 while continuing to increase, and species 1 is driven to extinction. In the third situation (Figure 13.1e), each species, when abundant, inhibits the growth of the other species more than it inhibits its own growth. Eventually, one or the other species wins, driving the other to extinction. The outcome (which species wins) depends on the initial population densities of the two species (which species is most abundant). In the fourth situation (Figure 13.1f), neither population can achieve a density capable of eliminating the other. Each species inhibits its own population growth more than that of the other species.

In each graph of Figure 13.1, the x-axis represents the population size of species 1 (N_1) and the y-axis represents the population size of species 2 (N_2). Two lines are plotted on each graph, and each line represents one of the two species. The diagonal line for species 1 represents the combined population densities of species 1 and 2 that equal K_1 and therefore $dN/dt = 0$. The line for species 2 represents the combined population densities of species 1 and 2 that equal K_2. For any point on the species 1 line, $N_1 + \alpha N_2 = K_1$. When $N_1 = K_1$, then N_2 must be zero. Because α is the per capita effect of species 2 on species 1, the population density of species 2 that is exactly equivalent to the carrying capacity of species 1 ($\alpha N_2 = K_1$) will be $N_2 = K_1/\alpha$. Therefore, when $N_2 = K_1/\alpha$, then N_1 must equal zero. For the line describing species 2, the values where the line crosses the axes will be K_2 and $N_1 = K_2/\beta$, the population density of species 1 that is equal to the carrying capacity of species 2.

The diagonal lines are called **zero-growth isoclines.** The zero-growth isoclines for species 1 and 2 are shown in Figures 13.1a and 13.1b, respectively. In the region below the line, or isocline, for species 1 (Figure 13.1a), combinations of N_1 and N_2 are below carrying capacity ($N_1 + \alpha N_2 < K_1$); therefore, the population is increasing (dN_1/dt greater than zero). This increase is represented by the arrows that are parallel to the x-axis and point in the direction of increasing values of N_1. In the region above the isocline, combinations of N_1 and N_2 are greater than carrying capacity; therefore, the population is declining (dN_1/dt is negative). In this region, the arrows point in the direction of decreasing population size. Figure 13.1b depicts the analogous situation for species 2. By combining the zero isoclines for the two species, we can interpret the combined dynamics of the two populations (see Quantifying Ecology 13.1: Interpreting Population Isoclines).

Figures 13.1c–13.1f depict the possible outcomes when the two isoclines are combined. In Figure 13.1c, the isocline of species 1 is parallel to, and lies outside, the isocline of species 2. In this case, even when the population of species 2 is at its carrying capacity (K_2), its density cannot stop the population of species 1 from increasing ($K_2 < K_1/\alpha$). As species 1 continues to increase, species 2 eventually becomes extinct. In Figure 13.1d, the situation is reversed. Species 2 wins, leading to the exclusion of species 1.

In Figures 13.1e and 13.1f, the isoclines cross; but the outcomes of competition are quite different. In Figure 13.1e, note that along the x-axis, the value of K_2/β is less than the value of K_1. Recall that the value $N_1 = K_2/\beta$ is the density of species 1 exactly equal to the carrying capacity of species 2 (K_2). Because K_2/β is less than K_1, species 1 can achieve population densities that would exceed the density required to drive the population of species 2 to extinction ($>K_2$). Likewise, because K_1/α is less than K_2 along the y-axis, species 2 can achieve population densities high enough to drive species 1 to extinction ($>K_1$). Which species "wins" the competition depends on the initial populations of the two species.

In Figure 13.1f, the result of competition differs in that neither species can exclude the other. The result is coexistence. Note that in this case, K_1 is less than K_2/β. The population of species 1 can never reach a density sufficient to eliminate species 2. For this to happen, the population of species 1 would have to reach $N_1 = K_2/\beta$. Likewise, K_2 is less than K_1/α, so species 2 can never achieve a population density high enough to eliminate species 1. Intraspecific competition (K) inhibits the growth of each population more than interspecific competition inhibits the growth of the other species.

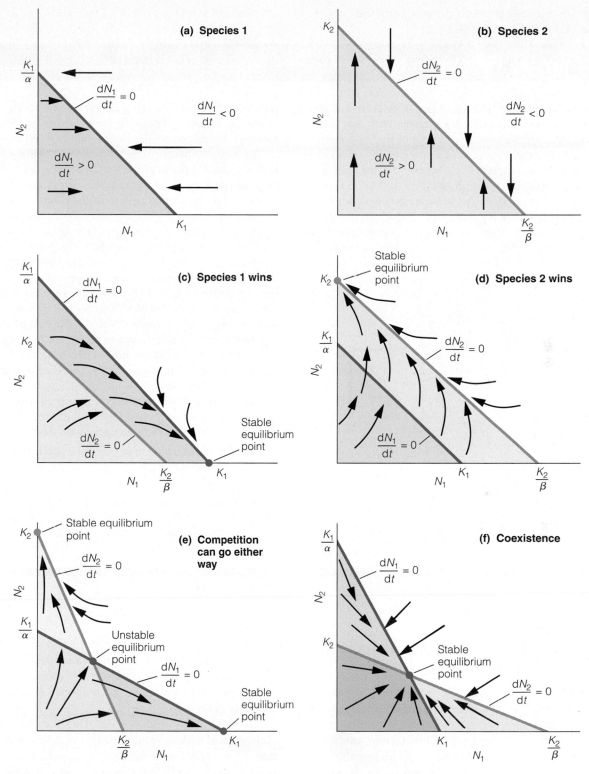

Figure 13.1 | The Lotka–Volterra model of competition between two species. **(a, b)** The zero isocline for each species is defined as the combinations of (N_1, N_2) for which $dN/dt = 0$ (zero population growth). In the shaded area (combined values of N_1, N_2) below the line, population growth is positive and the population increases (as indicated by the arrows); for combined values of (N_1, N_2) above the line, the population decreases. **(c)** The isocline of species 1 falls outside the isocline of species 2. Species 1 always wins, leading to the extinction of species 2. **(d)** The situation is the reverse of (c). **(e)** The isoclines cross. Each species inhibits the growth of the other more than its own growth. The more abundant species often wins. **(f)** Each species inhibits the growth of its own population more than that of the other by intraspecific competition. The species coexist.

The zero isoclines for species 1 and 2 are presented in Figures 13.1a and 13.1b, respectively. These isoclines (lines) represent the combined values of species 1 and species 2 under which the population growth rate is zero ($dN/dt = 0$; see graph below). For combined values of N_1 and N_2 that fall below the isocline (the region from the line to the x–y origin), population growth is positive. For values of N_1 and N_2 above the isocline, population growth is negative. Arrows are used to represent the direction of population change for any point. For the zero isocline of species 1, the arrows are parallel to the x-axis (N_1 axis), showing the direction of population change,

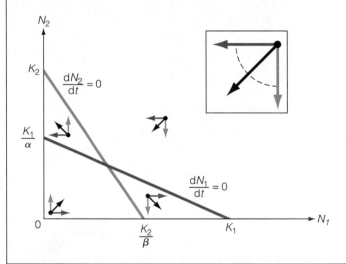

increasing if pointing in the direction away from the origin and declining if pointing toward the origin. For species 2, the arrows are parallel to the y-axis (N_2 axis).

To interpret the dynamics for any combination of α, β, K_1, and K_2, the isoclines for the two species must be drawn on the same x–y graph. Now when any point is plotted on the graph that represents the combined values of species 1 and 2 (N_1, N_2), two arrows must be plotted to represent the direction of change for both populations. In the graph shown here, the green arrow represents the change in species 1, and the yellow arrow shows the corresponding change in species 2 at each of the four points (values of (N_1, N_2), that are plotted. The future (predicted) values of N_1 and N_2 will therefore lie in the direction of the arrow. Therefore, the next point representing the combined values of N_1 and N_2 must lie somewhere between the two arrows (in the region defined by the dashed line—see insert) and is represented by the black arrow. In Figures 13.1c–13.1f, only the black arrows are shown, and the arrows are sometimes bent (curved) to show in which general direction the combined populations will move through time.

1. What is the outcome of competition for the case presented in the graph?

2. What parameter(s) in the Lotka–Volterra equations ($\alpha, \beta, K_1, K_2, r_1$, and r_2) will influence the actual projected values of N_1 and N_2 (N_1, N_2) within the region defined by the two arrows (green and yellow)?

13.3 | Laboratory Experiments Support the Lotka–Volterra Equations

The theoretical Lotka–Volterra equations stimulated studies of competition in the laboratory, where under controlled conditions an outcome is more easily determined than in the field. One of the first to study the Lotka–Volterra competition model experimentally was the Russian biologist G. F. Gause. In a series of experiments published in the mid-1930s, he examined competition between two species of *Paramecium: P. aurelia* and *P. caudatum*. *P. aurelia* has a higher rate of population growth than *P. caudatum* and can tolerate a higher population density. When Gause introduced both species into one tube containing a fixed amount of bacterial food, *P. caudatum* died out (Figure 13.2). In another experiment, Gause reared the losing species, *P. caudatum*, with

another species, *P. bursaria*. These two species coexisted because *P. caudatum* fed on bacteria suspended in solution, whereas *P. bursaria* confined its feeding to bacteria at the bottom of the tubes. Each species used food unavailable to the other.

In the 1940s and 1950s, Thomas Park at the University of Chicago conducted several classic competition experiments with laboratory populations of flour beetles. He found that the outcome of competition between *Tribolium castaneum* and *T. confusum* depended on environmental temperature, humidity, and fluctuations in the total number of eggs, larvae, pupae, and adults. Often, the outcome of competition was not determined until many generations had passed.

In a much later study, ecologist David Tilman of the University of Minnesota grew laboratory populations of two species of diatoms: *Asterionella formosa* and *Synedra*

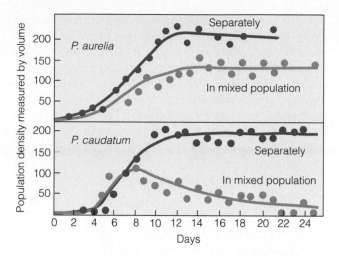

Figure 13.2 | Competition experiments with two ciliated protozoans, *Paramecium aurelia* and *P. caudatum*, grown separately and in a mixed culture. In a mixed culture, *P. aurelia* outcompetes *P. caudatum*, and the result is competitive exclusion.

(Adapted from Gause 1934.)

ulna. Both species require silica for the formation of cell walls. The researchers monitored population growth and decline as well as the level of silica in the water. When grown alone in a liquid medium to which silica was continually added, both species kept silica at a low level because they used it to form cell walls. However, when grown together, the use of silica by *S. ulna* reduced the concentration to a level below that necessary for *A. formosa* to survive and reproduce (Figure 13.3). By reducing resource availability, *Synedra* drove *Asterionella* to extinction.

13.4 | Studies Support the Competitive Exclusion Principle

In three of the four situations predicted by the Lotka–Volterra equations, one species drives the other to extinction. The results of the laboratory studies just presented tend to support the mathematical models. These and other observations have led to the concept called the **competitive exclusion principle,** which states that "complete competitors" cannot coexist. Complete competitors are two species (non-interbreeding populations) that live in the same place and have exactly the same ecological requirements. Under this set of conditions, if population A increases the least bit faster than population B, then A will eventually outcompete B, eventually leading to its local extinction.

Competitive exclusion, then, invokes more than competition for a limited resource. The competitive exclusion principle involves assumptions about the species involved as well as the environment in which they exist. First, this principle assumes that the competitors have exactly the same resource requirement. Second, it assumes that environmental conditions remain constant. Such

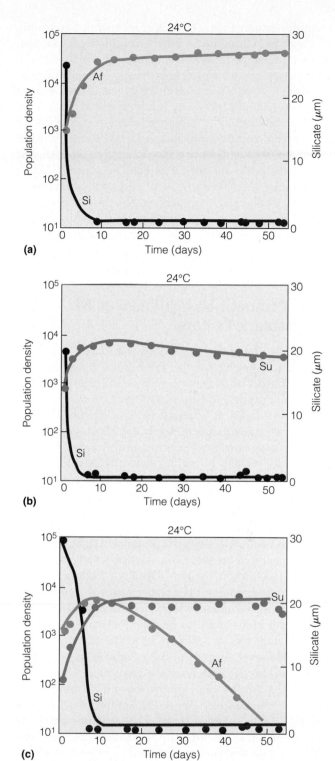

Figure 13.3 | Competition between two species of diatom, *Asterionella formosa* (Af) and *Synedra ulna* (Su), for silicate (Si). **(a, b)** Grown alone in a culture flask, both species reach a stable population that keeps silicate at a constant low level. *Synedra* draws silicate lower. **(c)** When the two are grown together, *Synedra* reduces silicate to a point at which *Asterionella* dies out.

(Adapted from Tilman et al. 1981.)

conditions rarely exist. The idea of competitive exclusion, however, has stimulated a more critical look at competitive relationships in natural situations. How similar can two species be and still coexist? What ecological conditions are necessary for coexistence of species that share a common resource base? The resulting research has identified a wide variety of factors affecting the outcome of interspecific competition, including environmental factors that directly influence a species' survival, growth, and reproduction but are not consumable resources (such as temperature or pH); spatial and temporal variations in resource availability; competition for multiple limiting resources; and resource partitioning. In the following sections, we examine each topic and consider how it functions to influence the nature of competition.

13.5 | Competition Is Influenced by Nonresource Factors

Interspecific competition involves individuals of two or more species vying for the same limited resource. An array of factors influences the relative abilities of different species to compete for the shared resource. For example, when a managed agricultural field or pasture is abandoned (see photograph in Figure 1 of Ecological Issues: The Changing Forests of Eastern North America, Chapter 18), a wide variety of native plant species colonize the site and compete for essential resources, including space. Species that have high rates of photosynthesis and allocate carbon to height growth and leaf production will overgrow and shade other species, preempting space and gaining access to the essential resource of light. At first, it would seem that the set of characteristics (adaptation) allowing fast growth under high light conditions (see Section 6.9) would be the decisive factor in determining the superior competitor(s). Field experiments have shown this assumption to be true; the shade-intolerant, fast-growing species quickly dominate these environments. However, features of the environment other than light also directly influence the germination, establishment, and growth rates of species—and thus the outcome of competition (see Field Studies: Katherine N. Suding). Environmental factors such as temperature, soil or water pH, relative humidity, and salinity directly influence physiological processes related to plant growth and reproduction, but they are not consumable resources that species compete over.

Fakhri Bazzaz of Harvard University examined the responses of a variety of annual plants that dominate during the early stages of old-field succession (colonization following disturbance; see Section 18.3) to variations in air temperature. Species differed significantly in the range of temperatures over which maximum rates of seed germination occurred (Figure 13.4). These differences among species in germination rates directly affect

patterns of seedling establishment, subsequent competition for resources, and the structure of old-field communities during the early stages of colonization and recovery. Differences in the temperature responses of these species can result in year-to-year variation in species composition within an old field as a result of differences in temperatures during the early growing season. Similar results have been observed in salt-marsh communities, where variations in salinity (both spatial and temporal) can shift the relative growth rates and competitive ability of plant species based on their salt tolerance.

13.6 | Temporal Variation in the Environment Influences Competitive Interactions

One species excludes another when it exploits a shared, limiting resource more effectively; the result is an increase in that species' population at the expense of its competitor. However, as in the example of the annual plant species studied by Bazzaz, each species does best under particular environmental conditions (see Figure 13.4). Low temperature may favor one competitor; if temperatures rise over time, the advantage may shift to another species. When environmental conditions vary through time, the competitive advantages also change. As a result, no one species will reach sufficient density to displace its competitors. In this manner, environmental variation allows competitors to coexist where under constant conditions one would exclude the other.

A similar example is the work of Peter Dye in the grasslands of southern Africa. He examined annual variations in the relative abundance of grass species occupying a savanna community in southwest Zimbabwe. From 1971 to 1981, the dominant grass species shifted from *Urochloa mosambicensis* to *Heteropogon contortus* (Figure 13.5a). This observed shift in dominance is a result of yearly variations in rainfall (Figure 13.5b). Rainfall during the 1971–72 and 1972–73 rainy seasons was much lower than average. *Urochloa mosambicensis* can maintain higher rates of survival and growth under dry conditions than can *H. contortus*, making it a better competitor under conditions of low rainfall. With the return to higher rainfall during the remainder of the decade, *H. contortus* became the dominant grass species. Annual rainfall in this semiarid region of southern Africa is highly variable, and fluctuations in species composition such as those shown in Figure 13.5 are a common feature of the community.

Besides shifting the relative competitive abilities of species, variation in climate can function as a density-independent limitation on population density (see Section 11.12). Periods of drought or extreme temperatures may depress populations below carrying capacity. If these events are frequent enough relative to the time required

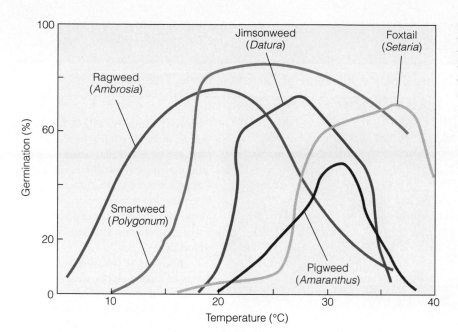

Figure 13.4 | Patterns of seed germination of five annual plant species along a gradient of temperature. These species dominate the early stages of secondary succession in field communities of the midwestern United States.

(Adapted from Bazzaz 1996.)

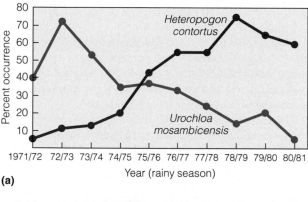

(a)

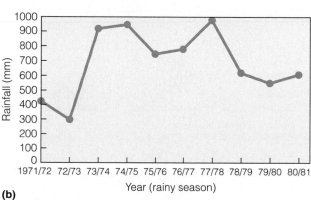

(b)

Figure 13.5 | (a) Shift in the dominant grass species in a savanna community in southwest Zimbabwe during the period 1971–81. The shift is in response to changing patterns of precipitation during the same period (b). *Urochloa mosambicensis* was able to compete successfully under the drier conditions during the 1971–72 and 1972–73 rainy seasons. With the increase in rainfall beginning in the 1973–74 season, *Heteropogon contortus* came to dominate the site.

(Adapted from Dye and Spear 1982.)

for the population to recover (approach carrying capacity), resources may be sufficiently abundant during the intervening periods to reduce or even eliminate competition.

13.7 | Competition Occurs for Multiple Resources

In many cases, competition between species involves multiple resources; and competition for one resource often influences an organism's ability to access other resources. One such example is the practice of interspecific territoriality, where competition for space influences access to food and nesting sites (see Section 11.10).

A wide variety of bird species in the temperate and tropical regions exhibits interspecific territoriality. Most often, this practice involves the defense of territories against closely related species, such as the gray (*Empidonax wrightii*) and dusky (*E. oberholseri*) flycatchers of the western United States. Some bird species, however, defend their territories against a much broader range of potential competitors. For example, the acorn woodpecker (*Melanerpes formicivorus*) defends territories against jays and squirrels as well as other species of woodpeckers. Strong interspecific territorial disputes likewise occur among brightly colored coral reef fish.

Competition among plants provides many examples of how competition for one resource can influence an individual's ability to exploit other essential resources, leading to a combined effect on growth and survival. R. H. Groves and J. D. Williams examined competition between populations of subterranean clover (*Trifolium subterraneum*) and skeletonweed (*Chondrilla juncea*) in a series of greenhouse experiments. Plants were grown

Department of Ecology and Evolutionary Biology, University of California, Irvine, California

Natural disturbances such as fire, wind, floods, or extreme weather conditions typically result in partial or complete destruction of plant life within an affected area. By altering the local environment, such disturbances can influence the relative rates of establishment, growth, and survival of different plant species within a community; and in doing so, they influence the outcome of interspecific competition. Plant ecologists have long appreciated that natural disturbances can alter the composition of plant communities, favoring some species while reducing the abundance of others. But the mechanisms responsible for the observed changes have rarely been explored. For ecologist Katherine Suding of the University of California, Irvine, however, the question of how disturbance influences the nature of competition within plant communities has been a central focus of her research in the grassland ecosystems of North America.

Natural disturbances can potentially influence the nature of competition by changing environmental conditions in various ways. By removing the competitively dominant species or simply reducing overall plant density, disturbance can reduce the intensity of competition within an affected area. Disturbances can also alter the abiotic environment. For example, within prairie grassland ecosystems, small (10- to 70-cm diameter) soil mounds created by the burrowing of small mammals significantly affect the surrounding biotic and abiotic environments. The formation of burrows reduces vegetation cover in the area surrounding the opening and influences soil compaction and soil moisture in the area of the mound. Plant species that inhabit the resulting mounds tend to differ from those that characterize the surrounding prairie.

To understand how this disturbance influences interspecific competition, Suding conducted a series of experiments in the tallgrass prairie ecosystem at the Resthaven Wildlife Area near Castalia, Ohio. Suding developed an experimental design to separate the effects of reduced competition (resulting from the removal of plants) and changes in the abiotic environment (mound formation) on the outcome of competition after disturbance. Two soil environment treatments were applied. On some plots, the soil environment was left undisturbed (no

mound). On a second set of plots, mounds similar to those formed by small mammals were simulated by extracting existing plants with a bulber, turning over the soil to a depth of 15 cm, removing the remaining plant roots and crowns, and then loosening the soil.

Besides the two soil treatments, three neighborhood treatments were applied. On some plots, the vegetation was left unaltered (adult neighborhood). On a second set of plots, the vegetation was removed, and seedlings of colonizing species were planted to simulate the conditions found in the early states after disturbance (juvenile neighborhood). On a third set of plots, all vegetation was removed (no neighbors).

The combination of two soil and three neighborhood treatments gives a total of six treatments (Figure 1). The combination of mound and juvenile neighborhood is the "gap treatment," and the combination of no mound and adult neighborhood is the "control treatment."

Three plant species, the C_4 grass *Andropogon gerardii*, and two forbs, *Ratibida pinnata* and *Coreopsis tripteris*, were included in the experiment. In each plot, three individuals

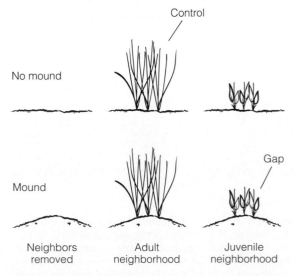

Figure 1 | Combinations of soil (mound and no mound) and neighborhood treatments used in the disturbance experiment.

of the same target species were planted, giving a total of 18 treatments (3 species × 2 soil × 3 neighborhood). The experiment lasted 2 years, and Suding harvested the six treatments annually and calculated estimates of relative growth rate (RGR: see Quantifying Ecology 6.1) for the intervening period. She used the natural logarithm (ln) of the ratio of the relative growth rates of target individuals measured with and without neighbors (ln RR) as an index of competition:

$$\ln RR = \ln (RGR_{\text{with neighbors}}/RGR_{\text{without neighbors}})$$

The competition index for the three species under the control (adult and no mound) treatment (RGR control/RGR neighbors removed) showed a clear competitive hierarchy, with neighboring plants having the least competitive effect on *Coreopsis*, an intermediate effect on *Ratibida*, and the greatest effect (reduction in RGR) on *Andropogon* (Figure 2a). The gap treatment (juvenile and mound) resulted in a reduction in the effects of competition for all three species, eliminating the competitive hierarchy observed under the control treatment (Figure 2a). The relative influence of decreasing intensity of competition (removal of neighboring plants) and changes in abiotic factors (mound formation) on the breakdown of the competitive hierarchy can be evaluated by comparing the results of the control treatment with both the mound treatment and the reduced neighborhood treatment (the two components that make up the gap treatment; see Figure 1). The comparison between the control and the reduced neighborhood treatments (Figure 2b) reveals a reduction in the intensity of competition, but the competitive hierarchy observed under the control treatment is not altered. In contrast, the mound treatment shifted the relative competitive abilities of the three species (Figure 2c), decreasing the competitive effect of neighboring plants on *Andropogon*

but increasing the competitive effect of neighboring plants on *Coreopsis*. The treatment did not influence the competitive index of *Ratibida*, and it retained its intermediate value in the hierarchy.

Suding's results clearly show that elimination of the competitive hierarchy existing in the established prairie (the control treatment) after the creation of gap environments by the burrowing behavior of small mammals is a result of abiotic changes (specifically, decreased compaction and lower soil moisture associated with mound formation) rather than changes in neighborhood characteristics such as biomass.

Katherine Suding's work provides valuable insights into the interactions between biotic and abiotic factors that occur after disturbances. It also demonstrates how those interactions can influence the relative competitive abilities of plant species and, ultimately, the composition and dynamics of plant communities.

Bibliography

Suding, K. 2001. The effect of spring burning on competitive rankings of prairie species. *Journal of Vegetation Science* 12:849–856.

Suding, K., and D. Goldberg. 2001. Do disturbances alter competitive hierarchies? Mechanisms of change following gap formation. *Ecology* 82:2133–2149.

1. What exactly is the index of competition, RR, measuring?
2. Based on the discussion of plant adaptations in Chapter 6, what characteristics might make the forb species *Coreopsis* and *Ratibida* better competitors than the grass *Andropogon* in the absence of disturbance?

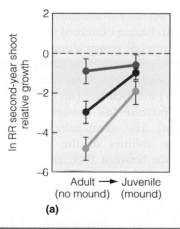
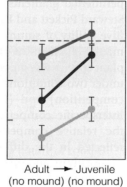
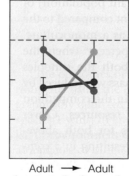

(a) Adult ⟶ Juvenile
(no mound) (mound)

(b) Adult ⟶ Juvenile
(no mound) (no mound)

(c) Adult ⟶ Adult
(no mound) (mound)

Figure 2 | Effects of selected treatments (see Figure 1) on the competitive response of *Andropogon* (green), *Coreopsis* (blue), and *Ratibida* (magenta). Each graph (a, b, and c) presents a comparison between the control (Adult, no mound) and one of the other 6 treatments outlined in Figure 1. Go to QUANTIFY it! at www.ecologyplace.com to use data to calculate standard deviation.

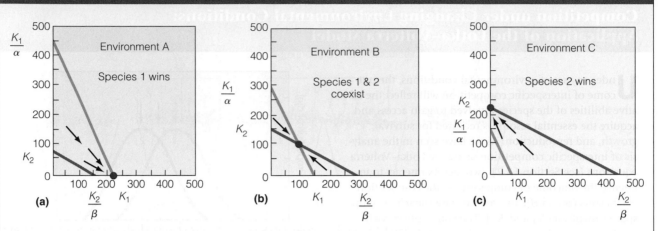

Figure 2 | Outcomes for the competitive interactions of the two species in the three different environments (parameters of Lokta–Volterra equations given in text).

1. Suppose that the competition coefficients were not the same for both species, but rather $\alpha = 0.5$ and $\beta = 0.25$. How would this influence the outcome of competition in environment A?

2. What factor or factors might cause the relative values of the competition coefficients (α and β) to change in the three different environments?

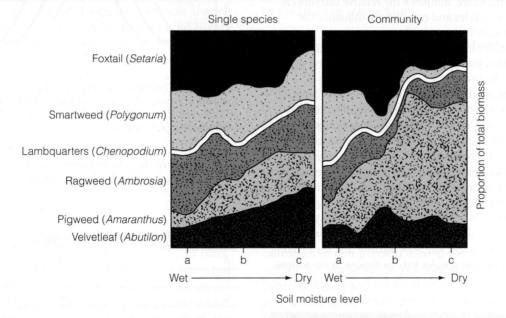

Figure 13.7 | Differential response of six summer annuals along an experimental moisture gradient (wet to dry soil conditions). The experiment involved two competitive environments. In the first, individuals were grown in monoculture (single species). In the second, individuals were grown in a mixture of all six species (community). The various shaded regions on each of the two graphs represent the relative contribution of each species to the total biomass at that point along the moisture gradient (soil moisture level). (Data from Pickett and Bazzaz 1978.)

Interpreting Ecological Data

Q1. Which of the six plant species accounts for the greatest proportion of total biomass at each of the three moisture levels (a, b, and c) when grown as a monoculture?

Q2. Which of the six plant species shows the largest increase in relative biomass (proportion of total) from the monoculture to the mixture experience under the dry soil conditions (level c)?

Q3. Which plant species shows the largest decrease in relative biomass from the monoculture to the mixture experiments under the wet soil conditions (level a)?

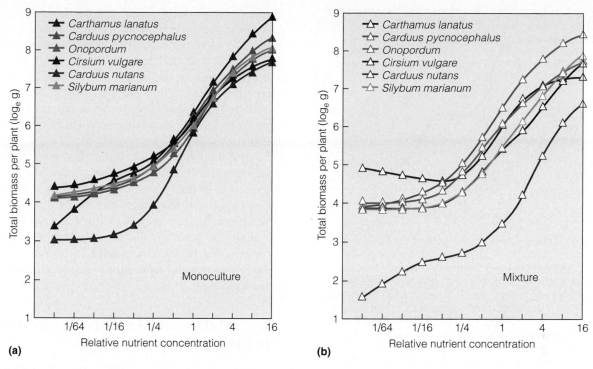

Figure 13.8 | Response of six thistle species to an experimental gradient of nutrient availability: **(a)** single-species populations (monocultures) and **(b)** mixed populations.

(Adapted from Austin et al. 1986.)

several greenhouse studies to explore the changing nature of interspecific competition among plant species across experimental gradients of nutrient availability. In one such experiment, the researchers examined the response of six species of thistle along a gradient of nutrient availability (application of nutrient solution). Plants were grown both in monoculture (single species) and mixture (all six species) under 11 different nutrient treatments, ranging from 1/64 to 16 times the recommended concentration of standard greenhouse nutrient solution. After 14 weeks, the plants were harvested, and their dry weights were determined. Responses of the six species along the nutrient gradient for monoculture and mixture experiments are shown in Figure 13.8.

Two important results emerge from the thistle experiment. First, as with Pickett and Bazzaz's experiment (see Figure 13.7), when grown in mixture, the response (accumulated biomass) of each species along the resource gradient differs from the pattern observed when grown in isolation (monoculture). Second, the relative competitive abilities of the species change along the nutrient gradient. This result is more easily seen when the response of each species in the mixed-species experiments is expressed on a relative basis. This expression is achieved by dividing the biomass (dry weight) value for each species at each nutrient level by the value of the species that achieved the highest biomass at that level. The relative performance of each species at each nutrient level then

ranges from 0 to 1.0. Relative responses of the three dominant (greatest biomass accumulation) thistle species along the nutrient gradient are shown in Figure 13.9. Note that *Carthamus lanatus* is the superior competitor under low nutrient concentrations, *Carduus pycnocephalus* at intermediate values, and *Silybum marianum* at the highest nutrient concentrations.

Field studies designed to examine the influence of interspecific competition across an environmental gradient often reveal that multiple environmental factors interact to influence the response of organisms across the landscape. In the coastal regions of New England, interspecific competition for nutrients has been shown to be an important factor determining the patterns of salt-marsh plant zonation (see Section 17.5). However, the relative competitive abilities of species for nutrients are influenced by the ability of plant species to tolerate a gradient of increasing physical stress relating to waterlogging, salinity, and oxygen availability in the soil and sediments (Figure 13.10). Upper boundaries (toward shoreline) of species distribution are set by interspecific competition for nutrients, whereas lower boundaries are set by the ability to tolerate the physical stress associated with increasing water depth (see Section 17.5).

Chipmunks furnish a striking example of the interaction of competition and tolerance to physical stress in determining species distribution along an environmental gradient. In this case, physiological tolerance,

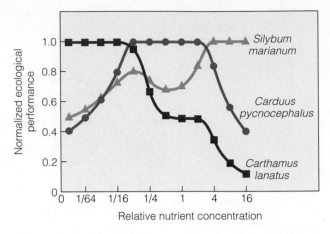

Figure 13.9 | Differences in the response of three thistle species grown at different relative nutrient levels. Nutrient levels (*x*-axis) represent 11 different nutrient treatments, ranging from 1/64 to 16 times the recommended concentration of standard greenhouse nutrient solution. Response (*y*-axis) is measured as the normalized ecological response of the three dominant thistle species when grown as a mixture (interspecific competition). This expression is achieved by dividing the biomass (dry weight) value for each species at each nutrient level by the value of the species that achieved the highest biomass at that level (see results of experiments presented in Figure 13.8). The relative performance of each species at each nutrient level then ranges from 0 to 1.0.

(Adapted from Austin et al. 1986.)

Interpreting Ecological Data

Q1. Which of the three species of thistle included in the graph had the highest biomass production under the 1/64 nutrient treatment? What does this imply about this species' competitive ability under low nutrient availability relative to other thistle species?

Q2. Using relative biomass production at each treatment level as an indicator of competitive ability, which thistle species is the superior competitor under the standard concentration of nutrient solution (1.0)?

Q3. At which nutrient level is the relative biomass of the three species most similar (smallest difference in the biomass of the three species)?

aggressive behavior, and restriction to habitats in which one organism has competitive advantage all play a part. On the eastern slope of the Sierra Nevada live four species of chipmunks: the alpine chipmunk (*Tamias alpinus*), the lodgepole chipmunk (*T. speciosus*), the yellow-pine chipmunk (*T. amoenus*), and the least chipmunk (*T. minimus*). Three or more of these species have strongly overlapping food requirements, and each species occupies a different altitudinal zone (Figure 13.11).

The line of contact between chipmunks is determined partly by interspecific aggression. Aggressive behavior by the dominant yellow-pine chipmunk determines the upper range of the least chipmunk. Although the least chipmunk can occupy a full range of habitats from sage-

brush desert to alpine fields, it is restricted in the Sierra Nevada to sagebrush habitat. Physiologically it is more capable of handling heat stress than the others, enabling it to inhabit extremely hot, dry sagebrush. If the yellow-pine chipmunk is removed from its habitat, the least chipmunk moves into vacated open pine woods. However, if the least chipmunk is removed from the sagebrush habitat, the yellow-pine chipmunk does not invade the habitat. The aggressive behavior of the lodgepole chipmunk in turn determines the upper limit of the yellow-pine chipmunk. The lodgepole chipmunk is restricted to shaded forest habitat because it is vulnerable to heat stress. Most aggressive of the four, the lodgepole chipmunk also may limit the downslope range of the alpine chipmunk. Thus, the range of each chipmunk is determined both by aggressive exclusion and by its ability to survive and reproduce in a habitat hostile to the other species.

13.9 | Interspecific Competition Influences the Niche of a Species

An organism free from interference by other species could use the full range of conditions and resources under which it can survive and reproduce. We call this range the **fundamental niche** of a species. Competition from other species often restricts a species to a portion of its fundamental niche. The portion of the fundamental niche that a species actually exploits as a result of interactions with other species (such as competition) is its **realized niche.** An example is provided by the work of J. B. Grace and R. G. Wetzel of the University of Michigan. Two species of cattail (*Typha*) occur along the shoreline of ponds in Michigan. One species, *Typha latifolia*, dominates in the shallower water, whereas *Typha angustifolia* occupies the deeper water farther from shore. When these two species were grown along the water depth gradient in the absence of the other species, a comparison of the results with their natural distributions revealed how competition influences their realized niche (Figure 13.12). Both species can survive in shallow waters; but only the narrow-leaved cattail, *T. angustifolia*, can grow in water deeper than 80 cm. When the two species grow together along the same gradient of water depth, their distributions, or realized niches, change. Even though *T. angustifolia* can grow in shallow waters (0 to 20 cm depth) and above the shoreline (−20 to 0 cm depth), in the presence of *T. latifolia* it is limited to depths of 20 cm or deeper. Individuals of *T. latifolia* outcompete individuals of *T. angustifolia* for the resources of nutrients, light, and space, limiting the distribution of *T. angustifolia* to the deeper waters. Note that the maximum abundance of *T. angustifolia* occurs in the deeper waters, where *T. latifolia* is not able to survive.

Although each of the two species of cattail has exclusive use of a subset of the range of habitats along the

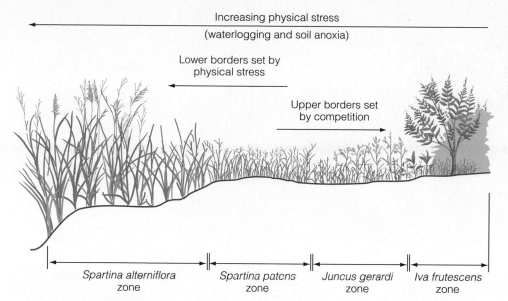

Increasing physical stress
(waterlogging and soil anoxia)

Lower borders set by
physical stress

Upper borders set
by competition

Spartina alterniflora zone | *Spartina patens* zone | *Juncus gerardi* zone | *Iva frutescens* zone

Figure 13.10 | Zonation of the dominant perennial plant species in a New England salt-marsh community. Upper borders of species distribution are a function of competition, whereas lower borders are a function of the species' ability to tolerate the physical stress associated with salinity, waterlogging, and low oxygen concentrations in the sediments.

(Adapted from Emery et al. 2001.)

shoreline, the species coexist at intermediate depths. When two or more organisms use a portion of the same resource simultaneously, be it food or habitat, the interaction is referred to as **niche overlap.** The amount of niche overlap is assumed to be proportional to the degree of competition for that resource. Niche overlap, however, does not always mean high competitive interaction. In

fact, resources may not be in short supply. Extensive niche overlap may indicate that little competition exists and that resources are abundant.

As preceding examples have illustrated, competition may force species to restrict their use of space, range of foods, or other resource-oriented activities. As a result, species do not always occupy that part of their

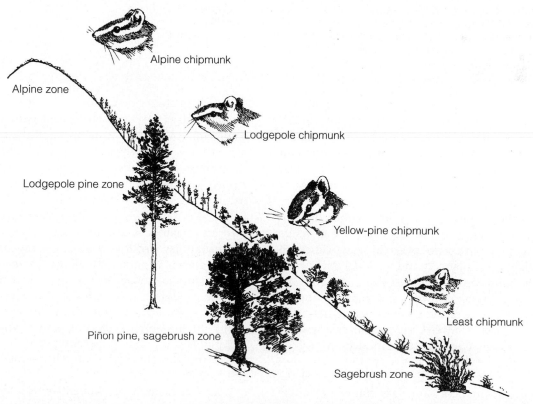

Alpine chipmunk

Alpine zone

Lodgepole chipmunk

Lodgepole pine zone

Yellow-pine chipmunk

Piñon pine, sagebrush zone

Least chipmunk

Sagebrush zone

Figure 13.11 | A transect of the Sierra Nevada in California, 38° north latitude, showing vegetation zonation and the altitudinal ranges of four species of chipmunks (*Eutamias*) on the east slope.

(Adapted from Heller-Gates 1971.)

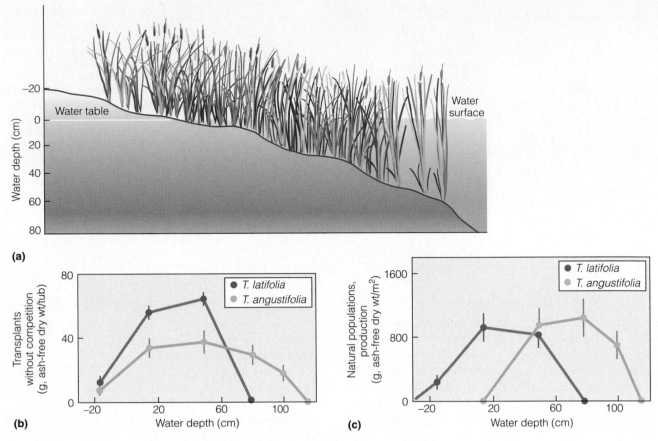

Figure 13.12 | Distribution of two species of cattail (*Typha latifolia* and *Typha angustifolia*) **(a)** along a gradient of water depth; **(b)** grown separately in an experiment; **(c)** growing together in natural populations. The response of the two species in the absence of competition (b) reflects their fundamental niche (physiological tolerances). The response of each species is altered by the presence of the other (c). They are forced to occupy only their realized niches.

(Adapted from Grace and Wetzel 1981.)

fundamental niche where conditions yield the highest growth rate, reproduction, or fitness (Figure 13.13). Jessica Gurevitch, of the University of New York at Stony Brook, examined the role of interspecific competition on the local distribution of *Stipa neomexicana*, a C₃ perennial grass found in the semiarid grassland communities of southeastern Arizona. *Stipa* is found only on the dry ridge crests where grass cover is low, rather than in moister, low-lying areas below the ridge crests where grass cover is greater. In a series of experiments, Gurevitch removed neighboring plants from individual *Stipa* plants in ridge-crest, midslope, and lower-slope habitats. She compared the survival, growth, and reproduction of these plants with control individuals (whose neighboring plants were not removed). Her results clearly show that *Stipa* has a higher growth rate, produces more flowers per plant, and has higher rates of seedling survival in midslope and lower-slope habitats (Figure 13.14). Its population density in these habitats is limited by competition with more successful grass species. *Stipa* distribution (or realized niche) is limited to suboptimal habitats due to interspecific competition.

Much of the evidence for competition comes from studies, such as the two just presented, demonstrating the contraction of a fundamental niche in the presence of a competitor. Conversely, when a species expands its niche in response to the removal of a competitor, the result is termed **competitive release.** Competitive release may occur when a species invades an island that is free of potential competitors, moves into habitats it never occupied on a mainland, and increases its abundance. Such expansion may also follow when a competing species is removed from a community, allowing remaining species to move into microhabitats they previously could not occupy. Such was the case with the distribution of cattails along the gradient of water depth presented in Figure 13.12. In the absence of competition from *T. latifoli*, *T. angustifolia* was able to expand its distribution to areas above the shoreline (expressed as negative values of water depth).

Another example of competitive release comes from the waters of the Antarctic Ocean. It is estimated that a century ago, the number of baleen whales (members of the Balaenidae, Balaenopteridae, and Eschrichtiidae

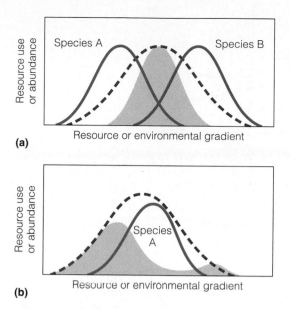

(a)

(b)

Figure 13.13 | Two possible relationships between fundamental and realized niche for a hypothetical species. The fundamental niche of the focal species is shown as a dashed red line, and the realized niche is shaded orange. **(a)** Species A and B compete for the resource and cause the realized niche of the focal species to compress to the central optimum. Note that the curves are symmetrical and bell-shaped. **(b)** A dominant species A (blue line) excludes the focal species from its optimum into the peripheral part of its fundamental niche, causing a bimodal, asymmetrical, realized niche shown in red.

(Adapted from Austin 1999.)

families of whale) feeding in the Antarctic during summer totaled about 1 million, with a biomass of 43 million tons. These whales are filter feeders, feeding primarily on Antarctic krill (small euphausiid shrimps; see Figure 24.19) and consuming approximately 4 percent of their body weight per day. By the 1930s, commercial whaling had reduced the whale population to about 340,000 individuals, and today the current biomass probably does not exceed 7 million tons, or about one-sixth of the initial population. Documented increases in the abundance of krill-dependent predators, such as seals and penguins, following the dramatic decline in baleen whale populations have been attributed to the effects of competitive release, resulting from the increased availability of krill.

13.10 | Coexistence of Species Often Involves Partitioning Available Resources

All terrestrial plants require light, water, and essential nutrients such as nitrogen and phosphorus. Consequently, competition between various co-occurring species ought to be common. The same should be true for the variety of insect-feeding bird species inhabiting the canopy of a forest, large mammalian herbivores feeding on the grasslands of east Africa, and predatory fish species that make the coral reef their home. How is it that these diverse arrays of potential competitors can coexist in the same community? The competitive exclusion principle introduced in Section 13.4 suggests that if two species have identical resource requirements, then one species will eventually displace the other. But how different do two species have to be in their use of resources before competitive exclusion does not occur (or conversely, how similar can two species be in their resource requirements and still coexist)?

We have already seen from the variety of examples presented thus far in the chapter that the coexistence of competitors is associated with some degree of "niche differentiation"—differences in the range of resources used or environmental tolerances. Observations of similar species sharing the same habitat suggest that they coexist by partitioning available resources. Animals use different kinds and sizes of food, feed at different times, or forage in different areas. Plants require different proportions of nutrients or have different tolerances for light and shade. Each species exploits a portion of the resources unavailable to others, resulting in differences among co-occurring species that would not be expected purely due to chance.

Field studies provide many reports of apparent resource partitioning. One example involves three species of annual plants growing together on prairie soil abandoned 1 year after plowing. Each plant exploits a different part of the soil resource (Figure 13.15). Bristly foxtail (*Setaria faberii*) has a fibrous, shallow root system that draws on a variable supply of moisture. It recovers rapidly from drought, takes up water rapidly after a rain, and carries on a high rate of photosynthesis even when partially wilted. Indian mallow (*Abutilon theophrasti*) has a sparse, branched taproot extending to intermediate depths, where moisture is adequate during the early part of the growing season but is less available later on. The plant is able to carry on photosynthesis at low water availability (Section 6.10). The third species, smartweed (*Polygonum pensylvanicum*), has a taproot that is moderately branched in the upper soil layer and develops mostly below the rooting zone of other species, where it has a continuous supply of moisture. The different responses of these three plant species to variation in water availability can be seen in the results of the water gradient experiment shown in Figure 13.7.

Apparent resource partitioning is also common among related animal species that share the same habitat and draw upon a similar resource base. Tamar Dayan, at Tel Aviv University, examined possible resource partitioning in a group of coexisting species of wild cats inhabiting the Middle East. Dayan and colleagues examined differences among species in the size of canine teeth, which are crucial to wild cats in capturing and killing their prey. For these cats, there is a general relationship between the size of canine and the prey species selected. Dayan found clear evidence of systematic differences in the size of the

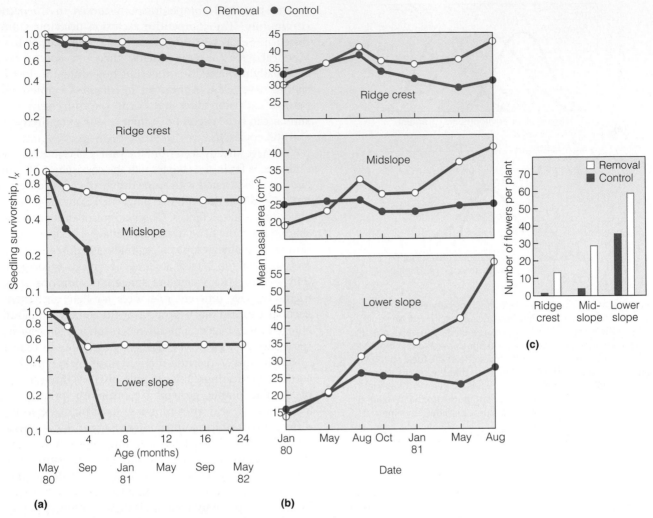

Figure 13.14 | Response of *Stipa neomexicana* plants in three different habitats (ridge crest, midslope, and lower slope). Results of both treatment (neighboring plants removed) and control (neighbors *not* removed) plants are shown for **(a)** seedling survival, **(b)** mean growth rate, and **(c)** flowers produced per plant. Under natural conditions, distribution of *Stipa* is restricted to the ridge-crest habitats due to competition from other grass species.

(Adapted from Gurevitch 1986.)

Interpreting Ecological Data

Q1. How does the influence of interspecific competition on seedling survival of *Stipa* differ between the ridge-crest and lower-slope habitats?

Q2. Experiment results show that *Stipa* individuals can effectively grow at the lower slope even under conditions of interspecific competition (as indicated by values of mean basal area in part b). Based on the results in Figure 13.14, what part(s) of the *Stipa* life cycle are most heavily influenced by interspecies competition, and how would these limitations affect distribution of the species on the landscape?

canine teeth, not only between male and female individuals within each of the species (sexual dimorphism; see Chapter 8) but also among the three coexisting cat species (Figure 13.16). The pattern observed suggests an exceptional regularity in the spacing of species along the axis defined by the average size of canine teeth (*x*-axis in Figure 13.16). Dayan and colleagues hypothesize that intra- and interspecific competition for food has resulted in natural selection favoring the observed differences,

thereby reducing the overlap in the types and sizes of prey that are taken.

Understanding the relationship between patterns of resource use and the coexistence of species is further complicated because the species' niche involves a wide variety of resource and other environmental factors. For simplicity, we often represent the niche as one-dimensional (as in Figure 13.13). In reality, a species' niche includes using many types of resources (such

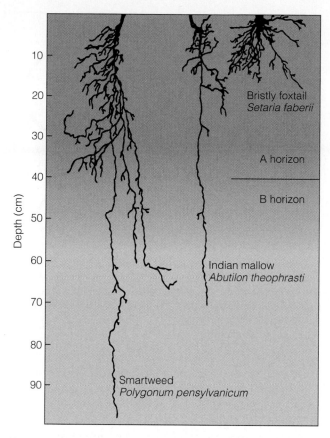

Figure 13.15 | Vertical partitioning of the prairie soil resource at different levels by three species of annual plants, one year after disturbance.

as food, a place to feed, cover, space, etc.), as well as responding to a wide variety of nonresource environmental factors that directly influence physiological performance and fitness (such as temperature, relative humidity, salinity, etc.). The limnologist G. E. Hutchinson proposed the idea of the niche as a multidimensional response called a *hypervolume*, where each axis is defined by a variable relating to the specific resource need or environmental factor that is essential for species survival, growth, and reproduction. We can

begin to visualize a multidimensional niche by creating a three-dimensional one. Consider three niche-related variables for a hypothetical organism: temperature, humidity, and food size (Figure 13.17).

Rarely do two or more species have exactly the same combination of requirements. Species may overlap on one dimension of the niche (such as size of insects selected as prey) but not on another (such as foraging height in the canopy). The total competitive interaction may be less than that suggested by the niche overlap on one gradient alone (Figure 13.18).

13.11 | Competition Can Influence Natural Selection

The patterns of resource partitioning discussed in the previous section have the effect of reducing competition among co-occurring species. By dividing the resource, each species reduces direct competition with the others. Resource partitioning results from specific physiological, morphological, or behavioral adaptations (see Chapter 5) that allow individuals access to essential resources and therefore influence fitness. Because these differences also reduce competition among co-occurring species, they are often regarded as an outcome of interspecific competition in the past. Competition is at the heart of Darwin's theory of natural selection (see Section 5.1). Characteristics that enable an organism to reduce competition will increase fitness, therefore influencing the evolution of characteristics related to acquiring resources.

Consider two bird species that feed on seeds. The pattern of seed selection for both species (A and B) can be expressed as a bell-shaped curve on a graph, with seed size as the *x*-axis and proportion of total diet as the *y*-axis (Figure 13.19). When two species show considerable overlap in the size of seeds they select, there is the potential for competition. If we assume that interspecific competition reduces the fitness of individuals of both species, then in theory, those individuals selecting seeds from the tails of the distributions where overlap is minimal will encounter less competition and increased fitness. The

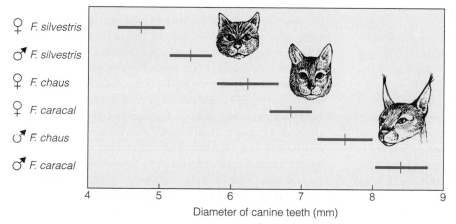

Figure 13.16 | Size (diameter) of canine teeth for small cat species that co-occur in Israel. Note the regular pattern of differences in size between species. Size is correlated with the size of prey selected by the different species.

(Adapted from Dayan et al. 1990.)

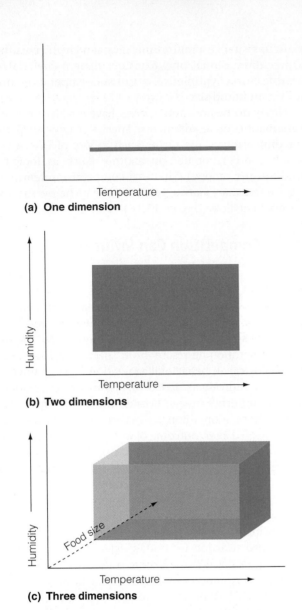

(a) One dimension

Temperature ⟶

(b) Two dimensions

Humidity ⟶

Temperature ⟶

(c) Three dimensions

Humidity ⟶

Food size

Temperature ⟶

Figure 13.17 | An illustration of niche dimension. Consider three variables composing the species' niche: temperature, humidity, and food size. **(a)** A one-dimensional niche involving only temperature. **(b)** A second dimension, humidity, has been added. Enclosing that space we have a two-dimensional niche. **(c)** Adding a third dimension, food size, and enclosing all three gives a three-dimensional niche space, or volume. A fourth variable would create a hypervolume.

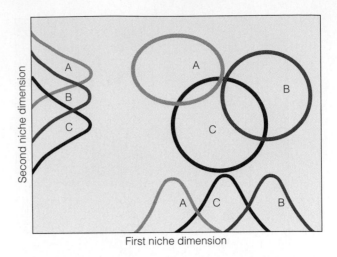

Second niche dimension

First niche dimension

Figure 13.18 | Niche relationships based on two environmental gradients. The three species (A, B, and C) exhibit considerable overlap on one gradient and little or none on the other. When we sum niche dimensions (in the circles), we find niche overlap reduced considerably.

(Adapted from Pianka 1978.)

range of seeds selected by an individual is constrained by beak size and shape (see example in Sections 5.5 and 5.8). If competition between the two species favors the selection of smaller (species A) and larger (species B) seeds, natural selection will favor those individuals in population A with small beaks while favoring those individuals with larger beaks in population B. Ultimately, the two species will diverge in both the range of beak size and the size of seeds selected as food. Direct interspecific competition will be reduced, promoting the coexistence of the two species.

Although the scenario just described is consistent with patterns of resource partitioning observed in nature (such as the example of differences in canine teeth presented in Figure 13.16), it is difficult to prove. Differences among species may relate to adaptation for the ability to exploit a certain environment or range of resources independent of competition. Differences among species have evolved over a long period of time, and we have limited or no information about resources and potential competitors that may have influenced natural selection. This issue led Joseph Connell, an ecologist at the University of California at Santa Barbara, to refer to the theory as the "ghosts of competition past." Some of the strongest evidence supporting this theory comes from studies examining differences in the characteristics of (sub)populations of a species that face different competitive environments. A good example is the work of Peter and Rosemary Grant, of Princeton University, involving two Darwin's finches of the Galápagos Islands. The Grants studied the medium ground finch (*Geospiza fortis*) and the small ground finch (*G. fuliginosa*), both of which feed on an overlapping array of seed sizes (for further discussion and illustrations, see Section 5.8). On the large island of Santa Cruz, where the two species of finch coexist, the distribution of beak sizes (phenotypes) for the two species does not overlap. Average beak size is significantly larger for *G. fortis* than for the smaller *G. fuliginosa* (Figure 13.20a). On the adjacent—and much smaller—islands of Los Hermanos and Daphne Major, the two species do not coexist; and the distributions of beak sizes for the two species are distinctively different from the patterns observed on Santa Cruz. The medium ground finch is

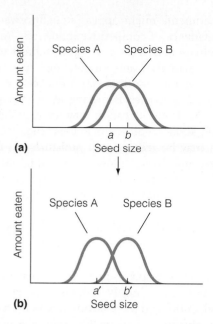

Figure 13.19 | An example of the possible influence of interspecific competition on patterns of resource use in two species (A and B) of seed-eating birds. **(a)** In the absence of interspecific competition, the two species feed on an overlapping range of seed sizes. The average (and modal) seed sizes selected by the two species are labeled as *a* and *b* respectively. **(b)** Due to interspecific competition for intermediate-size seeds, the diets of species A and B shift, reducing overlap in the range of seed sizes chosen as food. The new average (and modal) seed sizes selected by the two species are now labeled as *a'* and *b'* respectively.

allopatric (lives separately) on the island of Daphne Major, and the small ground finch is allopatric on Los Hermanos. Populations of each species on these two islands possess intermediate and overlapping distributions of beak sizes (Figures 13.20b and 13.20c). These patterns suggest that on islands where the two species coexist, competition for food results in natural selection favoring medium ground finch individuals with a large beak size that can effectively exploit larger seeds while also favoring small ground finch individuals that feed on smaller seeds. The outcome of this competition was a shift in feeding niches. When the shift involves features of the species' morphology, behavior, or physiology, it is referred to as **character displacement.**

The preceding example, as well as many others in Section 13.10, suggests that competing species can have evolutionary effects on each other that result in character displacement—that is, divergence in phenotypic traits relating to the exploitation of a shared and limited resource. However, until recently, the process of character displacement has never been documented by direct observational data. The first direct evidence of character displacement is provided by the recently published work of Peter and Rosemary Grant on the population of *G. fortis* inhabiting the small island of Daphne Major.

Before 1982, *G. fortis* (medium ground finch) was the only species of ground finch inhabiting the island of Daphne Major. The situation changed in 1982 when a new competitor species emigrated from the larger adjacent islands—the large ground finch, *G. magnirostris* (see Section 5.8 and Figure 5.16). *G. magnirostris* is a potential competitor on the island as a result of diet overlap with *G. fortis*. *G. magnirostris* feeds primarily on seeds of the herbaceous forb, Jamaican feverplant (*Tribulus cistoides*). The seeds are contained within a hard seed coat and exposed when a finch cracks or tears away the woody outer coating. Large-beaked members of the *G. fortis* population also feed on these seeds; in fact, during the 1976–77 drought, the survival of the population depended on this seed resource (see Section 5.5 for a discussion of natural selection in this population).

Initially the population of *G. magnirostris* on Daphne Major was too small in relation to the food supply to have anything but a small competitive effect on *G. fortis*. From 1982 to 2003, however, the population increased. Then little rain fell on the island during 2003 and 2004, and populations of both finch species declined dramatically as a result of declining food resources. During this period, *G. magnirostris* depleted the supply of large seeds

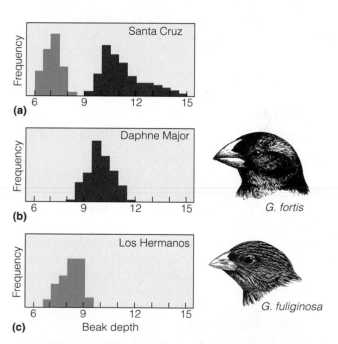

Figure 13.20 | Apparent character displacement in beak size in two populations of Galápagos finches—the medium ground finch (*Geospiza fortis*) and the small ground finch (*G. fuliginosa*). **(a)** On the large island of Santa Cruz, where the two finch species coexist, distribution of beak sizes (phenotypes) for these species does not overlap. Average beak size is significantly larger for *G. fortis* than for the smaller *G. fuliginosa*. In contrast, on the smaller islands of **(b)** Daphne Major and **(c)** Los Hermanos, where the two species do not co-occur, distribution of beak sizes for these species is intermediate and overlapping.

(Adapted from Grant 1986.)

from the Jamaican feverplant, causing the G. fortis population to depend on the smaller seed resources on the island. The result of this shift in resource availability due to competition from G. magnirostris was that during 2004 and 2005, G. fortis experienced strong directional selection against individuals with large beaks. The resulting decrease in the average beak size of the G. fortis population provides a clear example of character displacement.

13.12 | Competition Is a Complex Interaction Involving Biotic and Abiotic Factors

Demonstrating interspecific competition in laboratory "bottles" or the greenhouse is one thing, and demonstrating competition under natural conditions in the field is another. In the field, researchers (1) have little control over the environment, (2) have difficulty knowing whether the populations are at or below carrying capacity, and (3) lack full knowledge of the life history requirements or the subtle differences between the species.

In the previous sections, we reviewed an array of studies examining the role of competition in the field. Perhaps the most common are removal experiments, in which one of the potential competitors is removed and the response of the remaining species is monitored.

These experiments might appear straightforward, yielding clear evidence of competitive influences. But removing individuals may have direct and indirect effects on the environment that are not intended or understood by the investigators and that can influence the response of the remaining species. For example, removing (neighboring) plants from a location may increase light reaching the soil surface, soil temperatures, and evaporation. The result may be reduced soil moisture and increased rates of decomposition, influencing the abundance of belowground resources. These sometimes "hidden treatment effects" can hinder the interpretation of experimental results.

As we have seen in previous sections, competition is a complex interaction that seldom involves the interaction between two species for a single limiting resource. Interaction between species involves a variety of environmental factors that directly influence survival, growth, and reproduction; and these factors vary in both time and space. The outcome of competition between two species for a specific resource under one set of environmental conditions (temperature, salinity, pH, etc.) may differ markedly from the outcome under a different set of environmental conditions. As we shall see in the following chapters, competition is only one of many interactions occurring between species—interactions that ultimately influence population dynamics and community structure.

Summary

Interspecific Competition | 13.1

In interspecific competition, individuals of two or more species share a resource in short supply, thus reducing the fitness of both. As with intraspecific competition, competition between species can involve either exploitation or interference. Six types of interactions account for most instances of interspecific competition: (1) consumption, (2) preemption, (3) overgrowth, (4) chemical interaction, (5) territorial, and (6) encounter.

Competition Model | 13.2

The Lotka–Volterra equations describe four possible outcomes of interspecific competition. Species 1 may succeed over species 2; species 2 may succeed over species 1. Both of these outcomes represent competitive exclusion. The other two outcomes involve coexistence. One is unstable equilibrium, in which the species that was most abundant at the outset usually succeeds. A final possible outcome is stable equilibrium, in which two species coexist but at a lower population level than if each existed without the other.

Experimental Tests | 13.3

Laboratory experiments with species interactions support the Lotka–Volterra model.

Competitive Exclusion | 13.4

Experiment results led to formulation of the competitive exclusion principle—two species with exactly the same ecological requirements cannot coexist. This principle has stimulated critical examinations of competitive relationships outside the laboratory, especially how species coexist and how resources are partitioned.

Nonresource Factors | 13.5

Environmental factors such as temperature, soil or water pH, relative humidity, and salinity directly influence physiological processes related to growth and reproduction but are not consumable resources that species compete over. By differentially influencing species within a community, these nonresource factors can influence the outcome of competition.

Environmental Variability | 13.6

Environmental variability may give each species a temporary advantage. It allows competitors to coexist, where under constant conditions one would exclude the other.

Multiple Factors | 13.7

In many cases, competition between species involves multiple resources. Competition for one resource often influences an organism's ability to access other resources.

Environmental Gradients | 13.8

As environmental conditions change, so may the relative competitive ability of species. Shifts in competitive ability can result either from changes in the carrying capacities related to a changing resource base or from changes in the physical environment that interact with resource availability. Natural environmental gradients often involve the covariation of multiple factors, both resource and nonresource factors, such as salinity, temperature, and water depth.

Interspecific Competition and Natural Selection | 13.9

Interspecific competition can reduce the fitness of individuals. If certain phenotypes within the population function to reduce competition with individuals of other species, those individuals will encounter less competition and increased fitness. The result is a shift in the distribution of phenotypes (characteristics) within the competing population(s).

Resource Partitioning | 13.10

Many species that share the same habitat coexist by partitioning available resources. When each species exploits a portion of the resources unavailable to others, competition is reduced.

Niche | 13.11

The niche is the functional role of an organism in the community. That role might be constrained by interspecific competition. In the absence of interactions with other species, an organism occupies its fundamental niche. In the presence of interspecific competition, the fundamental niche is reduced to a realized niche—the conditions under which an organism actually exists. When two different organisms use a part of the same resource, such as food, their niches overlap. Overlap may or may not indicate competitive interaction. A species compresses or shifts its niche when competition forces it to restrict its type of food or habitat. In some cases, the realized niche may not provide optimal conditions for the species. In the absence of competition, the species experiences competitive release, thus expanding its niche.

Complexity of Competition | 13.12

Competition is a complex interaction that seldom involves the interaction between two species for a single limiting resource. Competition involves a variety of environmental factors that directly influence survival, growth, and reproduction—factors that vary in both time and space.

Study Questions

1. What condition(s) must be established before a researcher can definitively state that two species are competing for a resource? Is establishing that two species overlap in their use of a resource enough to determine that interspecific competition is occurring?

2. In analyzing the Lotka–Volterra model of interspecific competition in Section 13.2, three outcomes of competition between two species were identified. In three of the cases, one species outcompetes the other, driving its population to zero. In the fourth case, the two species coexist. What condition is necessary for this outcome?

3. Environmental factors such as temperature and salinity are not consumable resources, but species that occupy the same habitat often differ in their response to these factors. How might environmental factors such as temperature and salinity influence the outcome of competition between two species that occupy the same habitat?

4. Figure 13.19 presents a hypothetical case where two species overlap in their use of food resources (seeds). Let us assume that due to interspecific competition, the two species cannot co-occur in the same habitat (say species A succeeds). Now assume that the distribution of seed sizes in their habitat changes from year to year depending on rainfall, and that the average seed size varies from a to b (Figure 13.19). How might this temporal variation in resources influence the outcome of competition between the two species?

5. In the experiment conducted by Austin and colleagues (Figures 13.8 and 13.9), relative competitive abilities of the three dominant thistle species change under different nutrient availability. Based on the discussion of plant adaptations to nutrient availability in Chapter 6 (see Section 6.12), how do you think these three species might differ (morphology, maximum rates of photosynthesis, etc.)?

6. Resource partitioning due to natural selection is often interpreted as the cause for observed differences in the characteristics of closely related species that occupy the same area, such as the example of the cats in Figure 13.16, or the Galápagos finches in Figure 13.20. What other factors unrelated to competition might account for the observed differences?

Think about possible differences in the use of habitats within the area.

7. Distinguish between niche shift and character displacement. Which of these two concepts best describes the example of the Galápagos finches in Figure 13.20? Why?

Further Readings

Bazzaz, F. A. 1996. *Plants in changing environments: Linking physiological, population, and community ecology.* New York: Cambridge University Press.

An excellent overview of competition in plant populations, linking species characteristics and competitive interactions. Also illustrates the shifting nature of competitive interactions as environmental conditions change in time and space.

Connell, J. H. 1983. On the prevalence and relative importance of interspecific competition: Evidence from field experiments. *American Naturalist* 122:661–696.

This paper provides a review of field experiments that have examined the role of competition.

Grace, J. B., and D. Tilman. 1990. *Perspectives on plant competition.* San Diego: Academic Press.

An excellent overview of the concept of competition in plant communities. Provides a wealth of illustrated examples from laboratory, greenhouse, and field experiments.

Gurevitch, J. L., L. Murrow, A. Wallace, and J. J. Walch. 1992. Meta-analysis of competition in field experiments. *American Naturalist* 140:539–572.

An update of the paper by Connell (listed earlier).

Keddy, P. 1989. *Competition.* London: Chapman and Hall.

An outstanding review of theoretical developments in the area of competition. Also provides a wide array and review of experimental research aimed at examining the importance of competition in structuring ecological communities.

Schoener, T. W. 1983. Field experiments on interspecific competition. *American Naturalist* 122:240–285.

A companion paper to the one by Connell (listed earlier).

Werner, E. E., and J. D. Hall. 1976. Niche shifts in sunfishes: Experimental evidence and significance. *Science* 191:404–406.

A classic work in the study of how competition influences the species' niche.

Wiens, J. A. 1977. On competition and variable environments. *American Scientist* 65:590–597.

This paper discusses the influence of temporal variation in environmental conditions on the outcome of competition. This paper provides a contrast to the papers by Connell and Schoener (listed earlier).

CHAPTER 14

Predation

Lioness attacking a female Kudu in Etosha National Park, Namibia, Africa.

When the poet Alfred, Lord Tennyson wrote "Tho' Nature, red in tooth and claw," he was no doubt seeking to evoke savage images of predation. The very term *predator* brings to mind images of lions on the African savannas or the great white shark cruising coastal waters; however, *predation* is defined more generally as "the consumption of one living organism by another." Although all heterotrophic organisms acquire their energy by consuming organic matter, predators are distinguished from scavengers and decomposers (see Chapter 21) in that they feed on living organisms. As such, they function as agents of mortality with the potential to regulate prey populations. Likewise, being the food resource, the prey population has the potential to influence the growth rate of the predator population. These interactions between predator and prey species can have consequences on community structure and serve as agents of natural selection, influencing the evolution of both predator and prey.

14.1 | Predation Takes a Variety of Forms

The broad definition of predation as the consumption of one living organism (the prey) by another (the predator) excludes scavengers and decomposers. Nevertheless, this definition results in the potential classification of a wide variety of organisms as predators. The simplest classification of predators would be the categories of heterotrophic organisms presented in Chapter 7, which are based on their use of plant and animal tissues as sources of food: carnivores (carnivory—consumption of animal tissue), herbivores (herbivory—consumption of plant or algal tissue), and omnivores (omnivory—consumption of both plant and animal tissues). Predation, however, is more than a transfer of energy. It is a direct and often complex interaction of two or more species: the eater and the eaten. As a source of mortality, the predator population has the potential to reduce, or even regulate, the growth of prey populations. In turn, as an essential resource, the availability of prey may function to regulate the predator population. For these reasons ecologists recognize a functional classification, based on the specific interactions between predator and prey, that provides a more appropriate framework for understanding the interconnected dynamics of predator and prey populations.

In this functional classification of predators we reserve the term *predator*, or *true predator*, for species that kill their prey more or less immediately upon capture. These predators typically consume multiple prey organisms and continue over their lifetime to function as agents of mortality on prey populations. In contrast, most herbivores (grazers and browsers) consume only part of the individual plant. Although this activity may harm the plant, it usually does not result in mortality. Seed predators and planktivores (aquatic herbivores that feed on phytoplankton) are exceptions; these herbivores function as true predators. Like herbivores, parasites feed on the prey organism (the host) while it is still alive; and although harmful, their feeding activity is generally not lethal in the short term. However, the association between parasites and their host organisms has an intimacy that is not seen in true predators and herbivores, because many parasites live on or in their host organisms for some portion of their life cycle. The last category in this functional classification, the parasitoids, consists of a group of insects classified based on the egg-laying behavior of adult females and the development pattern of their larvae. The parasitoid attacks the prey (host) indirectly by laying its eggs on the host's body. When the eggs hatch, the larvae feed on the host, slowly killing it. As with parasites, parasitoids are intimately associated with a single host organism, and they do not cause the immediate death of the host.

In this chapter we will use the preceding functional classification, focusing our attention on the two categories of true predators and herbivores. (From this point forward, the term *predator* is used in referring to the category of true predator). We will discuss the interactions of parasites and parasitoids and their hosts later, in Chapter 15, focusing on the intimate relationship between parasite and host that extends beyond the feeding relationship between predator and prey.

We will begin by exploring the connection between the hunter and the hunted, using a mathematical model to define the link between the populations of predator and prey. We will then examine the wide variety of subjects and questions that emerge from this simple mathematic abstraction of predator–prey interactions.

14.2 | Mathematical Model Describes the Basics of Predation

In the 1920s, Lotka and Volterra turned their attention from competition (see Section 13.2) to the effects of predation on population growth. Independently, they proposed mathematical statements to express the relationship between predator and prey populations. They provided one equation for the prey population and another for the predator population.

The population growth equation for the prey population consists of two components: the exponential model of population growth presented in Chapter 10 ($dN/dt = rN$) and a mortality term that represents removal of the prey from the population by the predator. The per capita rate at which predators consume prey is assumed to increase linearly with the number of prey (Figure 14.1a) and can therefore be represented as cN_{prey}, where c represents the efficiency of predation, defined by the slope of the relationship shown in Figure 14.1a. The total rate of predation (number of prey

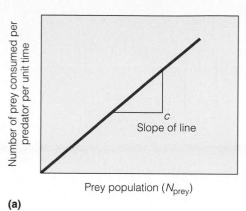

(a)

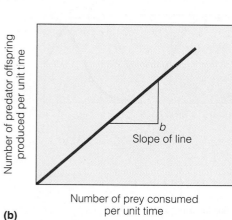

(b)

Figure 14.1 | **(a)** Relationship between prey population (*x*-axis) and the per capita consumption rate of predation (*y*-axis). The slope of the relationship ($\Delta y/\Delta x$), defined as *c*, represents the "efficiency of predation." **(b)** Relationship between the number of prey consumed (*x*-axis) and the number of predator offspring produced (*y*-axis). The slope of the relationship, *b*, represents the efficiency with which food is converted into predator population growth (reproduction).

captured per unit time) will be the product of the per capita rate of consumption (cN_{prey}) and the number of predators (N_{pred}), or $cN_{prey}N_{pred}$. This value represents a source of mortality for the prey population and must be subtracted from the rate of population increase represented by the exponential model of growth. The resulting equation representing the rate of change in the prey population (dN_{prey}/dt) is:

$$\frac{dN_{prey}}{dt} = rN_{prey} - cN_{prey}N_{pred}$$

The equation for the predator population consists of a term representing the birthrate and a second term representing the predator mortality rate. The birthrate is assumed to be a function of the amount of food consumed, increasing linearly with the rate at which prey are being captured and consumed: $cN_{prey}N_{pred}$ (Figure 14.1b). The birthrate is therefore the product of *b*, the efficiency with which food is converted into population growth (reproduction), which is defined by the slope of

the relationship shown in Figure 14.1b and the rate of predation ($cN_{prey}N_{pred}$), or $b(cN_{prey}N_{pred})$.

The predator mortality rate is assumed to be a constant proportion of the predator population and is therefore represented as dN_{pred}, where *d* is the probability of mortality. The resulting equation representing the rate of change in the predator population is

$$\frac{dN_{pred}}{dt} = b(cN_{prey}N_{pred}) - dN_{pred}$$

The Lotka–Volterra equations for predator and prey population growth therefore explicitly link the two populations, each functioning as a density-dependent regulator on the other. Predators regulate the growth of the prey population by functioning as a source of density-dependent mortality. The prey population functions as a source of density-dependent regulation on the birthrate of the predator population. To understand how these two populations interact, we can use the same graphical approach used to examine the outcomes of interspecific competition in Chapter 13 (see Section 13.2). In the absence of predators (or at very low predator density), the prey population will grow exponentially ($dN_{prey}/dt = rN_{prey}$). As the predator population increases, prey mortality will increase until eventually the mortality rate due to predation ($cN_{prey}N_{pred}$) is equal to the inherent growth rate of the prey population (rN_{prey}), and the net population growth for the prey species is zero ($dN_{prey}/dt = 0$). We can solve for the predator population density at which this occurs:

$$cN_{prey}N_{pred} = rN_{prey}$$
$$cN_{pred} = r$$
$$N_{pred} = \frac{r}{c}$$

Simply put, the growth rate of the prey population is zero when the number of predators is equal to the per capita growth rate of the prey population divided by the efficiency of predation. If the predator population exceeds this value, the growth rate of the prey becomes negative, and the population size declines. The mortality due to predation ($cN_{prey}N_{pred}$) is greater than the inherent growth rate of the prey population (rN_{prey}).

Likewise, we can examine the influence of prey population size on the growth rate of the predator population. The growth rate of the predator population will be zero ($dN_{pred}/dt = 0$) when the rate of predator increase (resulting from the consumption of prey) is equal to the rate of mortality:

$$b(cN_{prey}N_{pred}) = dN_{pred}$$
$$bcN_{prey} = d$$
$$N_{prey} = \frac{d}{bc}$$

The growth rate of the predator population is zero when the size of the prey population equals the mortality rate

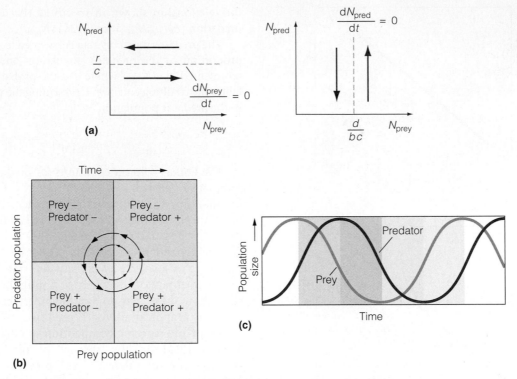

Figure 14.2 | Patterns predicted by the Lotka–Volterra model of predator–prey interaction. **(a)** Zero isoclines ($dN/dt = 0$) for the prey and predator populations. Note that the zero isocline for the prey population is defined by a fixed number of predators, and the zero isocline for the predator population is defined by a fixed number of prey. Arrows show the direction of change in population size relative to the isoclines. **(b)** The combined zero isoclines provide a means of examining the combined population trajectories of the predator and prey populations. Arrows represent the combined population trajectory. A minus sign indicates population decline, and a plus sign indicates population increase. This trajectory shows the cyclic nature of the predator–prey interaction. **(c)** By plotting the implied changes in size for both the predator and prey populations through time, we can see that the two populations continuously cycle out of phase with each other, and the density of predators lags behind that of prey. Shaded regions in the background of the graph relate to the corresponding regions in (b).

Interpreting Ecological Data

Q1. In Figure 14.2b, which of the four quadrants (regions of the graph) correspond to the following conditions?

$N_{prey} < d/bc$ and $N_{pred} > r/c$

Q2. In which of the four shaded regions of Figure 14.2c does the mortality rate of predators exceed their birthrate as a result of low prey density?

of the predator population divided by the product of the efficiency of predation and the conversion efficiency of captured prey into new predators (reproduction).

We can examine these results graphically (Figure 14.2a) using two axes that represent the population sizes of the prey (x-axis) and predator (y-axis). Having defined the size of the prey population at which the growth of the predator population is equal to zero ($N_{prey} = d/bc$) and the number of predators at which the growth rate of the prey population is zero ($N_{pred} = r/c$), we can draw the corresponding zero isoclines (see Section 13.2). If values of the prey population are to the right of the

predator isocline, the predator population increases; if they are to the left of the predator isocline, predator population declines. The same is true for the prey isocline. For values of predator population below the prey isocline, the prey growth rate is positive. For values of predator population above the prey isocline, the prey population declines. The two isoclines (Figure 14.2b) can be combined to examine changes in the growth rates of predator and prey populations for any combination of population sizes.

These paired equations, when solved, show that the two populations rise and fall in oscillations (Figure 14.2c). The oscillation occurs because as the predator population increases, it consumes more and more prey, until the prey population begins to decline. The declining prey population no longer supports the large predator population. The predators now face a food shortage, and many of them starve or fail to reproduce. The predator population declines sharply to a point where the reproduction of prey more than balances its losses through predation. The prey increases, eventually followed by an increase in the population of predators. The cycle may continue indefinitely. The prey is never quite destroyed; the predator never completely dies out.

14.3 | Model Suggests Mutual Population Regulation

The Lotka–Volterra model of predator–prey interactions assumes a mutual regulation of predator and prey populations. In the equations presented earlier, the growth of predator and prey populations are linked by a single term relating to the consumption of prey: $cN_{prey}N_{pred}$. For the prey population, this term serves to regulate population growth through mortality. In the predator population, it serves to regulate population growth through reproduction. Regulation of the predator population growth is a direct result of two distinct responses by the predator to changes in prey population. First, predator population growth depends on the rate at which prey are captured ($cN_{prey}N_{pred}$). The equation implies that the greater the number of prey, the more the predator eats. The relationship between the per capita rate of consumption and the number of prey is referred to as the predator's **functional response.** Second, this increased consumption of prey results in an increase in predator reproduction [$b(cN_{prey}N_{pred})$], referred to as the predator's **numerical response.**

This model of predator–prey interaction has been widely criticized for overemphasizing the mutual regulation of predator and prey populations. The continuing appeal of these equations to population ecologists, however, lies in the straightforward mathematical descriptions and in the oscillatory behavior that seems to occur in predator–prey systems. Perhaps the greatest value of this model is in stimulating a more critical look at predator–prey interactions in natural communities, including the conditions influencing the control of prey populations by predators. A variety of factors have emerged, including the availability of cover (refuges) for the prey, the increasing difficulty of locating prey as it becomes scarcer, choice among multiple prey species, and evolutionary changes in predator and prey characteristics (coevolution). In the following sections, we examine each of these topics and consider how they influence predator–prey interactions.

14.4 | Functional Responses Relate Prey Consumed to Prey Density

The English entomologist M. E. Solomon introduced the idea of functional response in 1949. A decade later, the ecologist C. S. Holling explored the concept in more detail, developing a simple classification based on three-general types of functional response. The functional response is the relationship between the per capita predation rate (number of prey consumed per unit time) and prey population size (see Figure 14.1a). How a predator's rate of consumption responds to changes in the prey population is a key factor influencing the predator's ability to regulate the prey population.

In developing the predatory prey equations in Section 14.2, we defined the per capita rate of predation as cN_{prey}, where c is the "efficiency" of predation, and N_{prey} is the number of prey (also used to represent the density of prey: number/unit area). Because the dynamics of the predator and prey populations are described as differential equations, this rate of predation is instantaneous (Δt approaches zero: see Quantifying Ecology 10.1: Derivatives and Differential Equations, p. 201). Alternatively, if we define the rate of predation as the number of prey eaten by a single predator during a period of search time (T_s) as N_e, we can express the per capita rate of predation as:

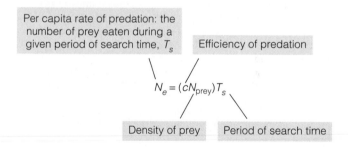

For a given period of search time (T_s), the relationship between the per capita rate of predation (N_e) and prey density (N_{prey}) for the equation just presented is linear; it was referred to by Holling as a **type I functional response** (Figure 14.3a). The rate of prey mortality due to predation (proportion of prey population eaten per unit time) is constant, a function of the efficiency of predation (c), as in Figure 14.3a.

The type I functional response is characteristic of passive predators, such as spiders that depend on prey happening upon their web or filter feeders that extract prey from a constant volume of water that washes over their filtering apparatus. The latter is a feeding behavior exhibited by a range of aquatic organisms from zooplankton to blue whales. A type I functional response can also occur when prey do not become sufficiently abundant to satiate the predators. Such is the case for the functional response of European kestrel (*Falco tinnunculus*) feeding

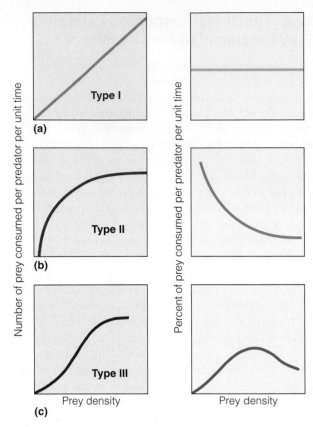

Figure 14.3 | Three types of functional response curves, which relate the per capita rate of predation to prey density. **(a)** Type I: The number of prey taken per predator increases linearly as prey density increases. Expressed as a proportion of the prey density, the rate of predation is constant—independent of prey density. **(b)** Type II: The predation rate rises at a decreasing rate to a maximum level. Expressed as a proportion of prey density, the rate of predation declines as the prey population grows. **(c)** Type III: The rate of predation is low at first and then increases in a sigmoid fashion, approaching an asymptote. Plotted as a proportion of prey density, the rate of predation is low at low prey density, rising to a maximum before declining as the rate of predation reaches its maximum.

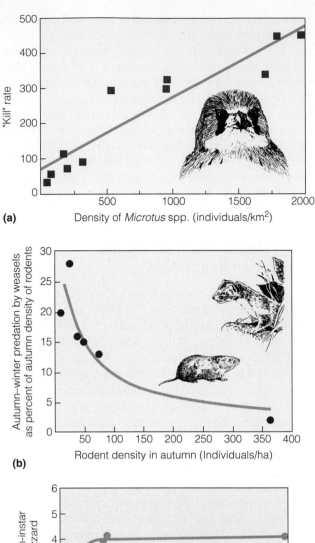

Figure 14.4 | Three examples of functional response curves relating the per capita rate of predation (*y*-axis) to prey density (*x*-axis). **(a)** Functional response of a pair of European kestrels to *Microtus* vole density during the breeding season. "Kill" rate (*y*-axis) is the number of prey taken during the breeding season. **(b)** Functional response of weasels preying on rodents in the deciduous forests of Bialowieza National Park, Poland. Note that the *y*-axis represents mortality rate rather than per capita consumption (see Figure 14.3b). **(c)** Functional response of bay-breasted warblers feeding on spruce budworm larvae.

((**a**) Adapted from Korpimaki and Norrdahl 1991; (**b**) Adapted from Jedrzejewski et al. 1995; (**c**) Adapted from Mook et al. 1963.)

on voles (*Microtus* spp.) presented in Figure 14.4a. These raptors take voles in proportion to their availability over the range of prey populations that are encountered in the field.

The type I functional response suggests a form of predation where all of the time allocated to feeding is spent searching (T_s). In general, however, the time available for searching will be shorter than the total time associated with consuming the N_e prey, because time is required to "handle" the prey item. Handling includes chasing, killing, eating, and digesting (type I functional response assumes no handling time below the maximum rate of ingestion). If we define T_h as the time required by a predator to handle an individual prey item, then the time spent handling N_e prey will be the product $N_e T_h$: The total time (T) spent searching and handling the prey is now:

$$T = T_s + (N_eT_h)$$

By rearranging the preceding equation, we can define the search time as:

$$T_s = T - N_eT_h$$

For a given total foraging time (T), search time now varies, decreasing with increasing allocation of time to handling.

We can now expand the original equation describing the type I functional response [$N_e = (cN_{prey})T_s$] by substituting the equation for T_s just presented. This includes the additional time constraint of handling the N_e prey items:

$$N_e = c(T - N_eT_h)N_{prey}$$

Note that N_e, the number of prey consumed during the time period T, appears on both sides of the equation, so to solve for N_e, we need to rearrange the equation.

$$N_e = c(N_{prey}T - N_{prey}N_eT_h)$$

Move c inside the brackets, giving:

$$N_e = cN_{prey}T - N_ecN_{prey}T_h$$

Add $N_ecN_{prey}T_h$ to both sides of the equation, giving:

$$N_e + N_ecN_{prey}T_h = cN_{prey}T$$

Rearrange the left-hand side of the equation, giving:

$$N_e(1 + cN_{prey}T_h) = cN_{prey}T$$

Divide both sides of the equation by $(1 + cN_{prey}T_h)$, giving:

$$N_e = \frac{cN_{prey}T}{(1 + cN_{prey}T_h)}$$

We can now plot the relationship between N_e and N_{prey} for a given set of values for c, T, and T_h (Figure 14.5). (Recall that the values of c, T, and T_h are constants.) In this relationship, referred to by Holling as a **type II functional response,** the per capita rate of predation (N_e) increases in a decelerating fashion only up to a maximum rate that is attained at some high prey density (see Figure 14.3b). The reason that the value of N_e approaches an asymptote is related to the predator's time budget. Recall that the total time (T) is divided into two activities: searching and handling ($T = T_s + N_eT_h$). As the number of prey captured during the total time period (T) increases, the handling time ($N_{prey}T_h$) also increases, decreasing the time available for further searching (T_s). The result is a declining mortality rate of prey with increasing prey density (see Figure 14.3b). The type II functional response is the most commonly reported for predators (see Figure 14.4b).

Holling also described a **type III functional response,** illustrated in Figures 14.3c and 14.4c. At high

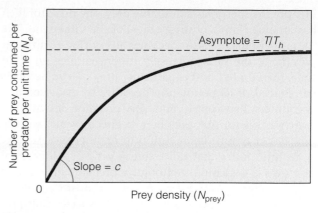

Figure 14.5 | Relationship between the density of prey population (x-axis) and the per capita rate of prey consumed (y-axis) for the model of predator functional response presented in Section 14.4 that includes both search (T_s) and handling N_eT_h time ($T = T_s + T_h$). At low prey density, the number of prey consumed is low, as is handling time. As prey density increases, the number of prey consumed increases; a greater proportion of the total foraging time (T) is spent handling prey, reducing time available for searching. As the handling time approaches the total time spent foraging, the per capita rate of prey consumed approaches an asymptote. The resulting curve is referred to as a type II functional response.

prey density, this functional response is similar to type II, and the explanation for the asymptote is the same. However, the rate at which prey are consumed is low at first, increasing in an S-shaped (sigmoid) fashion as the rate of predation approaches the maximum value. A type III functional response can potentially regulate a prey population because the initial rate of prey mortality increases with prey density (see Figure 14.4c). At low densities, mortality rate is negligible, but as prey density increases (as indicated by the upward sweep of the curve), prey mortality increases in a density-dependent fashion. However, the regulating effect of predators is limited to the interval of prey density where mortality increases. If prey density exceeds the upper limit of this interval, then mortality due to predation starts to decline.

Several factors may result in a type III response. Availability of cover (refuge) from which to escape predators may be an important factor. If the habitat provides only a limited number of hiding places, it will protect most of the prey population at low density; but the susceptibility of individuals will increase as the population grows.

Another reason for the sigmoidal shape of the type III functional response curve may be the predator's **search image,** an idea first proposed by the animal behaviorist L. Tinbergen. When a new prey species appears in the area, its risk of becoming selected as food by a predator is low. The predator has not yet acquired a search image—a way to recognize that species as a potential food item. Once the predator has captured an individual, it may identify the species as a desirable prey. The predator then has an easier time locating others of the same kind. The more adept the predator becomes at securing a particular prey item, the more intensely it concentrates on it. In time, the number of this particular prey species becomes so reduced or its population becomes so dispersed that encounters between it and the predator lessen. The search image for that prey item begins to wane, and the predator may turn its attention to another prey species.

A third factor that can result in a type III functional response is switching. Although a predator may have a strong preference for a certain prey, in most cases it can turn to another, more abundant prey species that provides more profitable hunting. If rodents, for example, are more abundant than rabbits and quail, foxes and hawks will concentrate on rodents.

Ecologists call the act of turning to more abundant, alternate prey *switching* (Figure 14.6). In **switching,** the predator feeds heavily on the more abundant species and pays little attention to the less abundant species. As the relative abundance of the second prey species increases, the predator turns its attention to that species.

At what point in prey abundance a predator switches depends considerably on the predator's food preference. A predator may hunt longer and harder for a palatable species before turning to a more abundant, less palatable alternate prey. Conversely, the predator may turn from the less desirable species at a much higher level of abundance than it would from a more palatable species.

Although simplistic, the model of functional response developed by Holling has been a valuable tool. It allows ecologists to explore how various behaviors, exhibited by both the predator and prey species, influence predation rate and subsequently predator and prey population dynamics. Because the model explicitly addresses the principle of time budget in the process of predation, this framework has been expanded to examine questions relating to the efficiency of foraging, a topic we will return to in Section 14.6.

14.5 | Predators Respond Numerically to Changing Prey Density

As the density of prey increases, the predator population growth rate should respond positively. A numerical response of predators can occur through reproduction by predators (as suggested by the conversion factor b in the Lotka–Volterra equation for predators) or through the movement of predators into areas of high prey density. The latter is referred to as an **aggregative response** (Figure 14.7). The response of predators to aggregate in areas of high prey density can be a crucial feature in determining a predator population's ability to regulate prey density. Aggregative response is important because most predator populations grow slowly in comparison to those of their prey. The numerical response of the bay-breasted warbler provides an example. L. J. Mook quantified the functional and numerical responses of bay-breasted warblers to an outbreak of prey: the spruce budworm. This bird species exhibits a type III functional response; the per capita rate of predation increases with

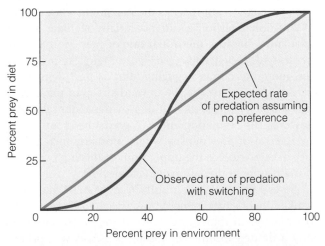

Figure 14.6 | A model of prey switching. The straight line represents the expected rate of predation assuming no preference by the predator. The prey are eaten in a fixed proportion to their relative availability (percent of total prey available to predator in environment). The curved line represents the change in predation rate observed in the case of prey switching. At low densities, the proportion of the prey species in the predator's diet is less than expected based on chance (based on its proportional availability to other prey species). Over this range of abundance, the predator is selecting alternative prey species. When prey density is high, the predator takes more of the prey than expected. Switching occurs at the point where the lines cross. The habit of prey switching results in a type III functional response between a predator and its prey species.

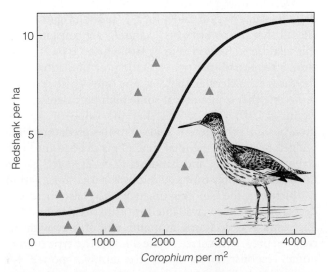

Figure 14.7 | Aggregative response in the redshank (*Tringa totanus*). The curve plots the density of the redshank in relation to the average density of its arthropod prey (*Corophium* spp.).
(Adapted from Hassel and May 1974.)

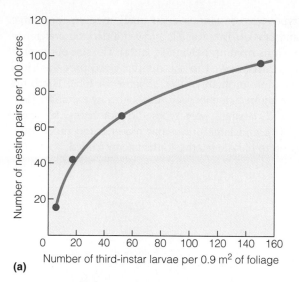

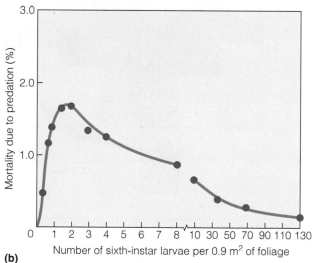

(a)

(b)

Figure 14.8 | **(a)** Numerical response of the bay-breasted warbler to changes in the population density of spruce budworms in New Brunswick, Canada. The x-axis represents the prey density, and the y-axis is the density of predators. **(b)** The predicted pattern of prey (y-axis) mortality rate as a function of prey density (x-axis). Mortality rate was estimated by combining the numerical response shown in (a) with the functional response of predation rate to prey density shown in Figure 14.4c.

(Adapted from Mook 1963.)

increasing prey density up to some maximum (see Figure 14.4c). In addition, the density of birds increases with increasing prey density (Figure 14.8a). This increase in density is a result of an increase in the number of nesting pairs occupying the area (breeding territories per hectare) and represents an aggregative response. The net result is that the percentage of prey organisms eaten per unit time by the entire predator population increases with the increase in prey density (Figure 14.8b). The combined functional and numerical responses by the bay-breasted warbler population interact to regulate the prey (spruce budworm) population at low population levels.

Other numerical responses by predator populations involve an increase in reproductive effort. The work of Wlodzimierz Jedrzejewski and colleagues at the Mammal Research Institute of the Polish Academy of Sciences is an example. Jedrzejewski examined the response of a weasel (*Mustela nivalis*) population to the density of two rodents, the bank vole (*Clethrionomys glareolus*) and the yellow-necked mouse (*Apodemus flavicollis*), in Biatowieza National Park in eastern Poland in the early 1990s. During that time, the rodents experienced a 2-year irruption in population size brought about by a heavy crop of oak, hornbeam, and maple seeds. The abundance of food stimulated the rodents to breed throughout the winter. The long-term average population density was 28–74 animals per hectare. During the irruption, the rodent population reached nearly 300 per hectare and then declined precipitously to 8 per hectare (Figure 14.9).

The weasel population followed the fortunes of the rodent population. At normal rodent densities, the winter weasel density ranged from 5 to 27 per 10 km^2, declining by early spring to 0 to 19. Following reproduction, the midsummer density rose to 42–47 weasels per 10 km^2. Because reproduction usually requires a certain minimal time (related to gestation period), a lag typically exists between an increase of a prey population and a numerical response by a predator population. No time lag, however, exists between increased rodent reproduction and weasel reproductive response. Weasels breed in the spring, and with an abundance of food they may have two litters or one larger litter. Young males and females breed during their first year of life. During the irruption,

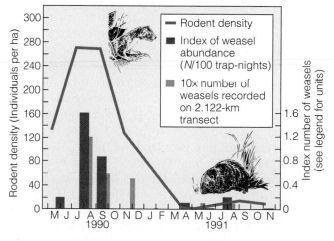

Figure 14.9 | Numerical response of weasels (predators) to an irruption and crash in the populations of forest rodents (prey). The x-axis represents time of year (month), and the y-axes represent the density of the rodent population (left axis) and an index of the predator population density (right axis). Brown bars represent weasel data from live trapping; orange bars represent data from captures, visual observations, and radio tracking. The reproductive rate of the weasel population is high enough to allow the population to track increases in the prey population during 1990.

(Adapted from Jedrzejewski et al. 1995.)

the number of weasels grew to 102 per 10 km^2, and during the crash the number declined to 8 per 10 km^2. The increase and decline in weasels was directly related to the spring rodent density.

14.6 | Foraging Involves Decisions about the Allocation of Time and Energy

Thus far, we have discussed the activities of predators almost exclusively in terms of foraging. But all organisms are required to undertake a wide variety of activities associated with survival, growth, and reproduction. Time spent foraging must be balanced against other time constraints such as defense, avoiding predators, searching for mates, or caring for young. This trade-off between conflicting demands has led ecologists to develop an area of research known as **optimal foraging theory.** At the center of optimal foraging theory is the hypothesis that natural selection should favor "efficient" foragers: individuals that maximize their energy or nutrient intake per unit of effort. The concept of efficient foraging involves an array of decisions involved in the process of foraging: what food to eat; where and how long to search; how to search. Optimal foraging theory approaches these decisions in terms of costs and benefits. Costs can be measured in terms of the time and energy expended in the act of foraging, and benefits should be measured in terms of fitness. However, it is extremely difficult to quantify the consequences of a specific behavioral choice on the probability of survival and reproduction. As a result, benefits are typically measured in terms of energy or nutrient gain, which is assumed to correlate with individual fitness.

One of the most active areas of research in optimal foraging theory has focused on the composition of animal diets—the process of choosing what to eat from among a variety of choices. We can approach this question using the framework of time allocation developed in the simple model of function response in Section 14.4, where the total time spent foraging (T) can be partitioned into two categories of activity: searching (T_s) and handling (T_h).

For simplicity, consider a predator hunting in a habitat that contains just two kinds of prey: P$_1$ and P$_2$. Assume that the two prey types yield E_1 and E_2 units of net energy gain (benefits), and they require T_{h_1} and T_{h_2} seconds to handle (costs). Profitability of the two prey types is defined as the net energy gained per unit handling time: E_1/T_{h_1} and E_2/T_{h_2}. Now suppose that P$_1$ is more profitable than P$_2$: $E_1/T_{h_1} > E_2/T_{h_2}$. Optimal foraging theory predicts that P$_1$ should be the preferred prey type because it has a greater profitability.

This same approach can be applied to a variety of prey items within a habitat. Behavioral ecologist Nicholas B. Davies of the University of Cambridge examined the feeding behavior of the pied wagtail (*Motacilla alba*) in a pasture near Oxford, England. The birds fed on various dung flies and beetles attracted to cattle droppings. Potential

prey types were of various sizes: small, medium, and large flies and beetles. The wagtails showed a decided preference for medium-sized prey (Figure 14.10a). The size of the prey selected corresponded to the optimal-sized prey the birds could handle profitably (E/T_h; Figure 14.10b). The birds virtually ignored smaller prey. Although easy to handle (low value of T_h), small prey types did not return sufficient energy (E), and large prey types required too much time and effort to handle relative to the energy gained.

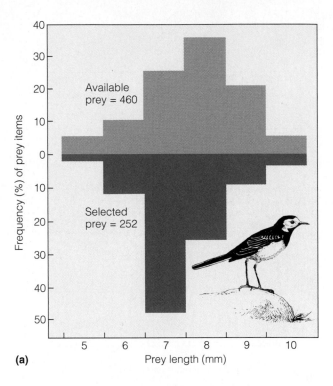

(a)

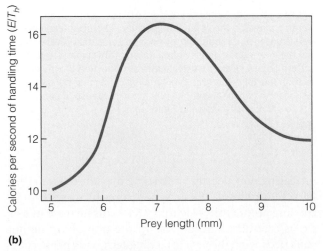

(b)

Figure 14.10 | (a) Relationship between prey length (*x*-axis) and frequency of prey in the diet of pied wagtails (*y*-axis) foraging in a pasture. Individuals show a definite preference for medium-sized prey, which are taken in disproportionate amounts compared to the frequencies of prey available. (b) The prey size (*x*-axis) chosen by pied wagtails represented the optimal size, providing maximum energy per handling time (E/T_h: *y*-axis).

(Adapted from Davies 1977.)

The simple model of optimal foraging presented here provides a means for evaluating which of two or more potential prey types is most profitable based on the net energy gain per unit of handling time. As presented, however, it also implies that the predator should always choose the most profitable prey item. Is there ever a situation where the predator should choose to eat the alternative, less profitable prey? To answer this question, we turn our attention to the second component of time involved in foraging: search time (T_s).

Suppose that while searching for P_1, the predator encounters an individual of P_2. Should it eat it or continue searching for another individual of P_1? The optimal choice will depend on the search time for P_1, defined as T_{s_1}. The profitability of consuming the individual of P_2 is E_2/T_{h_2}; the alternative choice of continuing to search, capture, and consume an individual of P_1 is $E_1/(T_{h_1} + T_{s_1})$, which now includes the additional time cost of searching for another individual of $P_1(T_{s_1})$. If $E_2/T_{h_2} > E_1/(T_{h_1} + T_{s_1})$, then according to optimal foraging theory, the predator should eat the individual of P_2. If this condition does not hold true, then the predator should continue searching for P_1. Testing this hypothesis requires the researcher to quantify the energy value and search and handling times of the various potential prey items. An example of this simple model of optimal prey choice is presented in Quantifying Ecology 14.1: A Simple Model of Optimal Foraging.

A wealth of studies examines the hypothesis of optimal prey choice in a wide variety of species and habitats, and patterns of prey selection generally follow the rules of efficient foraging. But the theory as presented here fails to consider the variety of other competing activities influencing a predator's time budget and the factors other than energy content that may influence prey selection. One reason that a predator consumes a varied diet is that it may not be able to meet its nutritional requirements (see Chapter 7) from a single prey species.

14.7 | Foragers Seek Productive Food Patches

Most animals live in a heterogeneous or patchy environment. Some patches might contain a high density of prey and others a low density or no prey at all. How should a predator allocate its time foraging in a patchy environment? Once again, optimal foraging theory would suggest that the decision is based on maximizing profitability, where profitability can now be defined as the energy gained per unit of time spent foraging in a given patch (rate of energy gain). The advantages to a predator of spending more time in highly profitable patches are obvious, but the question of when a predator should abandon one patch for a different patch is much subtler. This question has generated considerable interest and research and has given rise to another set of decision rules

in optimal foraging theory. The behavioral ecologist Eric Charnov of the University of New Mexico addressed this problem using an approach known as the **marginal value theorem.** The theorem predicts the length of time an individual should stay in a resource patch before leaving and seeking another. The length of stay is related to the richness of the food patch (prey density), the time required to get there (travel time), and the time required to extract the resource.

The marginal value theorem is illustrated in Figure 14.11. The x-axis represents time, and the y-axis represents the cumulative energy gain (G) from foraging activities in the patch. There is an initial time cost associated with traveling to the patch (t). Once foraging is initiated, the rate of energy gain, defined as the energy gain per unit time, is high. But as time progresses, the prey population becomes depleted (declining N_{prey}) and the rate declines, with the cumulative energy gain approaching an asymptote. The rate of return at any point in time is defined as the cumulative value of G (value on y-axis) divided by the combined travel time (t) and foraging time (T; value on x-axis). How long should the predator continue to forage in the patch before leaving? Optimal foraging theory predicts that the predator should abandon the patch when the rate of energy gain is at its maximum value, after which the rate of energy gain begins to decline (Figure 14.11c). The maximum rate of energy gain is represented by the line that is tangent to the curve of cumulative gain with time (Figure 14.11c). Before this time, the rate of return is still increasing; after this time, the rate of return declines.

The marginal value theorem also predicts that predators should remain in a rich food patch (higher prey density) for a longer time than they remain in a poorer one (Figure 14.12a), and that for patches of the same quality, time spent foraging (T) in a patch should increase with the amount of time spent traveling to the patch (increasing value of t; Figure 14.12b). Whether animals follow these predictions has been the object of experimental research both in the laboratory and in the field. Eric Wajnberg and colleagues at France's National Institute of Agronomy Research examined patterns of patch exploitation by females of the parasitoid wasp *Trichogramma brassicae* on habitat patches containing different host densities (eggs of the European corn borer, *Ostrinia nubilalis*). On average, *T. brassicae* females spent more time on patches of higher quality, and all patches were reduced to the same level of profitability before being left. The results appeared to agree with the optimal predictions of Charnov's marginal value theorem.

14.8 | Risk of Predation Can Influence Foraging Behavior

Most predators are also prey to other predatory species and therefore face the risk of predation while involved in their routine activities, such as foraging. Habitats and

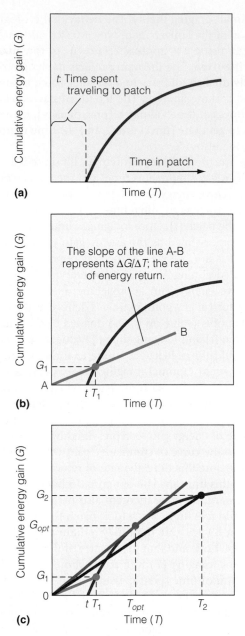

(a)

t: Time spent traveling to patch

Time in patch

Time (*T*)

(b)

The slope of the line A-B represents $\Delta G/\Delta T$; the rate of energy return.

B

G_1

A

$t\ T_1$

Time (*T*)

(c)

G_2

G_{opt}

G_1

0

$t\ T_1$ T_{opt} T_2

Time (*T*)

Figure 14.11 | How long should a predator remain in a habitat patch seeking food? The graphs shown above provide the answer to this question using Charnov's marginal value theorem, based on the assumption that the predator seeks to maximize the rate of energy return (energy gained per unit of time expended). **(a)** The total time spent foraging (*T*), which includes the time spent traveling to the patch (*t*), is shown on the *x*-axis. The cumulative energy gain (*G*) from the consumption of prey at any point in time is plotted on the *y*-axis. **(b)** For a given amount of foraging time (*T*₁) the predator acquires a corresponding amount of food energy (G_1). The slope of the line A-B that extends from the origin to the point on the energy gain curve that corresponds to (T_1G_1) represents the rate of energy gain for a predator that spends T_1 amount of time in the patch. **(c)** The theorem predicts that the optimal amount of time that a predator should spend in a patch is the value of *T* at which the rate of energy gain is maximum. The optimal solution (T_{opt}) is represented by the line that is tangent to the curve of cumulative gain with time [the blue line shown in **(c)**]. Note that for values of *T* below T_{opt}, the rate of energy gain continues to increase with increasing time spent in the patch. For values of *T* above T_{opt}, the rate of energy gain declines.

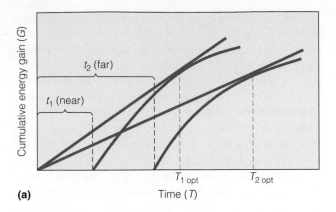

(a)

t_2 (far)

t_1 (near)

$T_{1\ opt}$ $T_{2\ opt}$

Time (*T*)

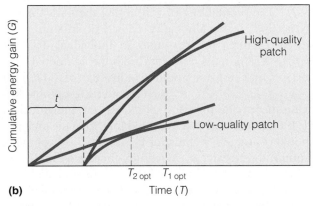

(b)

High-quality patch

t

Low-quality patch

$T_{2\ opt}$ $T_{1\ opt}$

Time (*T*)

Figure 14.12 | Under the marginal value theorem (see Figure 14.11), both the time spent traveling (*t*) to a habitat patch **(a)** and the quality of the patch **(b)** influence the time spent foraging (T_{opt}).

foraging areas vary in their foraging profitability and their risk of predation. In deciding whether it will feed, the forager must balance its energy gains against the risk of being eaten. If predators are about, then it may be to the forager's advantage not to visit a most profitable, but predator-prone, area and to remain in a less profitable but more secure part of the habitat. Many studies report how the presence of predators affects foraging behavior. In one such study, Jukka Suhonen of the University of Jyväskylä (Finland) examined the influence of predation risk on the use of foraging sites by willow tits (*Parus montanus*) and crested tits (*P. cristatus*) in the coniferous forests of central Finland. During the winter months, flocks of these two bird species forage in spruce, pine, and birch trees. Their major threat to survival is the Eurasian pygmy owl (*Glaucidium passerinum*). The owl is a diurnal ambush or sit-and-wait hunter that pounces downward upon its prey. Its major food is voles, and when vole populations are high, usually every 3 to 5 years, the predatory threat to these small passerine birds declines. When vole populations are low, however, the small birds become the owl's primary food. During these periods, the willow and crested tits forsake their preferred foraging sites on the outer branches and open parts of the trees, restricting their foraging activity to the

A Simple Model of Optimal Foraging

Faced with a variety of potential food choices, predators must make decisions. What types of food to eat? Where and how long to search for food? But how are these decisions made? Do predators function opportunistically, pursuing prey as they are encountered, or do they make choices and pass by potential prey of lesser food quality (energy content) while continuing the search for more preferred food types? If the objective is to maximize energy intake (energy gain per unit time), a predator should forage in a way that maximizes benefits (energy gained from consuming prey) relative to costs (energy expended). This concept of maximizing energy intake is the basis for models of optimal foraging.

Any food item has a benefit (energy content) and a cost (in time and energy involved in search and acquisition). The benefit–cost relationship determines how much profit a particular food item represents. The profitability of a prey item is the ratio of its energy content (E) to the time required for handling the item (T_h), or E/T_h.

Let us assume that a predator has two possible choices of prey, P_1 and P_2. The two prey types have energy contents of E_1 and E_2 (units of kJ) and take T_{h_1} and T_{h_2} seconds to handle. The searching time for the two prey types are T_{s_1} and T_{s_2} in seconds. We will define P_1 as the most profitable prey type (greater value of E/T_h).

As the predator searches for P_1, it encounters an individual of P_2. Should the predator capture and eat P_2 or continue to search for another individual of P_1? Which decision, capture P_2 or continue to search, would be the more profitable, maximizing the predator's energy intake? This is the basic question posed by optimal foraging theory, and the solution depends on the search time for P_1.

The profitability of capturing and eating P_2 is E_2/T_{h_2}; and the profitability of continuing the search, capturing, and eating another individual of P_1 is $E_1/(T_{h_1} + T_{s_1})$. Notice that the decision to ignore P_2 and continue the search carries the additional cost of the average search time for P_1, T_{s_1}. Therefore, the optimal solution, the

decision that will yield the greater profit, is based on the following conditions:

If: $$\frac{E_2}{T_{h_2}} > \frac{E_1}{(T_{h_1} + T_{s_1})}$$
then capture and eat P_2.

If: $$\frac{E_2}{T_{h_2}} < \frac{E_1}{(T_{h_1} + T_{s_1})}$$
then ignore P_2 and continue to search for P_1.

Therefore, if the search time for P_1 is short, the predator will be better off continuing the search. On the other hand, if the search time is long, the most profitable decision is to capture and consume P_2.

The benefit–cost trade-off for the optimal choice in prey selection is best understood through an actual example. David Irons and colleagues at Oregon State University examined the foraging behavior of glaucous-winged gulls (*Larus glaucescens*) that forage in the rock intertidal habitats of the Aleutian Islands, Alaska. Data on the abundance of three prey types (urchins, chitons, and mussels) in three intertidal zones (A, B, and C) are presented in the table below. Mean densities of the three prey types in numbers per m^2 are given for the three zones. Average energy content (E), handling time (T_h), and search time (T_s) for each of the three prey types are also listed in the table.

In feeding preference experiments, where search and handling time were not a consideration, chitons were the preferred prey type and are the obvious choice for maximizing energy intake. However, the average abundance of urchins across the three zones is greater than that of chitons. As a gull happens upon an urchin while hunting for chitons, should it capture and eat the urchin or continue to search for its preferred food, the chitons? Under conditions of optimal foraging, the decision depends on the conditions outlined earlier. The profit gained by capturing and consuming the urchin is $E/T_h = (7.45 \text{ kJ}/8.3 \text{ s})$, or 0.898. In contrast, the profit gained by ignoring the urchin and searching, capturing, and consuming another chiton is $E/(T_h + T_s) = [24.52 \text{ kJ}/(3.1 \text{ s} + 37.9 \text{ s})]$, or 0.598. Because the profit gained by consuming the urchin

Prey Type	Density Zone A	Density Zone B	Density Zone C	Energy (kJ/individual)	Handling Time (s)	Search Time (s)
Urchins	0.0	3.9	23.0	7.45	8.3	35.8
Chitons	0.1	10.3	5.6	24.52	3.1	37.9
Mussels	852.3	1.7	0.6	1.42	2.9	18.9

continued on page 294

is greater than the profit gained by ignoring it and continuing the search for chitons, it would make sense for the gull to capture and eat the urchin.

What about a gull foraging in intertidal zone A that happens upon a mussel? In this case, the profit gained by capturing and eating the mussel is (1.42/2.9), or 0.490, and the profit gained by continuing the search for a chiton remains [24.52 kJ/(3.1 s + 37.9 s)], or 0.598. In this case, the gull would be better off ignoring the mussel and continuing the search for chitons.

We now know what the gulls "should do," but do they in fact forage optimally as defined by this simple model of benefits and costs? If gulls are purely opportunistic, their selection of prey in each of the three zones would be in proportion to their relative abundances.

Irons and colleagues, however, found that the relative preferences for urchins and chitons were in fact related to their profitability (E/T_h); however, mussels were selected less frequently than predicted by their relative value of E.

1. How would reducing the energy content of chitons by half (to 12.26 kJ) influence the decision whether the gull should capture and eat the mussel or continue searching for a chiton in the example presented?

2. Because the gulls do not have the benefit of the optimal foraging model in deciding whether to select a prey item, how might natural selection result in the evolution of optimal foraging behavior?

---

denser inner parts of spruce trees that provide cover and to the tops of the more open pine and leafless birch trees.

14.9 | Coevolution Can Occur between Predator and Prey

By acting as agents of mortality, predators exert a selective pressure on prey species. That is to say, any characteristic that enables individual prey to avoid being detected and captured by a predator will increase its fitness. Natural selection should function to produce "smarter," more evasive prey (fans of the *Road Runner* cartoons should already understand this concept). However, failure to capture prey results in reduced reproduction and increased mortality of predators. Therefore, natural selection should also produce "smarter," more skilled predators. As prey species evolve characteristics to avoid being caught, predators evolve more effective means to capture them. To survive as a species, the prey must present a moving target that the predator can never catch. This view of the **coevolution** between predator and prey led the evolutionary biologist Van Valen to propose the Red Queen hypothesis. In Lewis Carroll's *Through the Looking Glass, and What Alice Found There*, there is a scene in the Garden of Living Flowers in which everything is continuously moving. Alice is surprised to see that no matter how fast she moves, the world around her remains motionless—to which the Red Queen responds, "Now, here, you see, it takes all the running you can do, to keep in the same place." So it is with prey species. To avoid extinction at the hands of predators, they must evolve means of avoiding capture; they must keep moving just to stay where they are.

14.10 | Animal Prey Have Evolved Defenses against Predators

Animal species have evolved a wide range of characteristics to avoid being detected, selected, and captured by predators. These characteristics are collectively referred to as **predator defenses.**

Chemical defense is widespread among many groups of animals. Some species of fish release alarm pheromones (chemical signals) that, when detected, induce flight reactions in members of the same and related species. Arthropods, amphibians, and snakes employ odorous secretions to repel predators. For example, when disturbed, the stinkbug (*Cosmopepla bimaculata*) discharges a volatile secretion from a pair of glands located on its back (Figure 14.13). The stinkbug can control the amount of fluid released and can reabsorb the fluid into the gland. In a series of controlled experiments, Bryan Krall and colleagues at Illinois State University have found that the secretion deters feeding by both avian and reptile predators.

Many arthropods possess toxic substances, which they acquired by consuming plants and then stored in their own bodies. Other arthropods and venomous snakes, frogs, and toads synthesize their own poisons.

Prey species have evolved numerous other defense mechanisms. **Cryptic coloration** includes colors and patterns that allow prey to blend into the background (Figure 14.14). Some animals are protectively colored, blending into the background of their normal environment. Such protective coloration is common among fish, reptiles, and many ground-nesting birds. **Object resemblance** is

Figure 14.13 | From a pair of glands located on its back, the stinkbug (*Cosmopepla bimaculata*) discharges a volatile secretion that has the effect of discouraging predators.

Figure 14.15 | The walking stick (Phasmatidae), which feeds on the leaves of deciduous trees, strongly resembles a twig.

common among insects. For example, walking sticks (Phasmatidae) resemble twigs (Figure 14.15), and katydids (Pseudophyllinae) resemble leaves. Some animals possess eyespot markings, which intimidate potential predators, attract the predators' attention away from the animal, or delude them into attacking a less vulnerable part of the body. Associated with cryptic coloration is **flashing coloration.** Certain butterflies, grasshoppers, birds, and ungulates, such as the white-tailed deer, display extremely visible color patches when disturbed and put to flight. The flashing coloration may distract and disorient predators; or, as in the case of the white-tailed deer, it may serve as a signal to promote group cohesion when confronted by a predator (Figure 14.16). When the

Figure 14.14 | Cryptic coloration in the flounder (*Parallchthys*), which inhabits the shallow coastal waters of eastern North America. Most flounders can change their color and pattern rapidly to match that of the bottom sediments, allowing them to avoid detection by both predators and potential prey.

animal comes to rest, the bright or white colors vanish, and the animal disappears into its surroundings.

Animals that are toxic to predators or use other chemical defenses often possess **warning coloration,** or **apoematism:** bold colors with patterns that may serve as warning to would-be predators. The black-and-white stripes of the skunk, the bright orange of the monarch butterfly, and the yellow-and-black coloration of many bees and wasps and some snakes may serve notice of danger to their predators (Figure 14.17). All their predators, however, must have an unpleasant experience with the prey before they learn to associate the color pattern with unpalatability or pain.

Some animals living in the same habitats with inedible species sometimes evolve a coloration that resembles or mimics the warning coloration of the toxic species. This type of mimicry is called **Batesian mimicry** after the English naturalist H. E. Bates, who described it when observing tropical butterflies. The mimic, an edible species, resembles the inedible species, called the model. Once the predator has learned to avoid the model, it avoids the mimic also. In this way, natural selection reinforces the characteristic of the mimic species that resembles that of the model species.

Most discussions of Batesian mimicry concern butterflies; but mimicry is not restricted to Lepidoptera and other invertebrates. Mimicry has also evolved in snakes with venomous models and nonvenomous mimics (Figure 14.18). For example, in eastern North America, the scarlet king snake (*Lampropeltis triangulum*) mimics the eastern coral snake (*Micrurus fulvius*); in southwestern North America, the mountain kingsnake (*Lampropeltis pyromelana*) mimics the western coral snake (*Micruroides euryxanthus*). Mimicry is not limited to color patterns. Some species of nonvenomous snakes are acoustic mimics of rattlesnakes. The fox snake (*Elaphe vulpina*) and the pine snake of eastern North America, the bull

Figure 14.16 | The white-tailed deer of North America received its name from the white tail patch that serves as an alarm and a distraction while the deer flees predators.

(a)

(b)

Figure 14.17 | The bright and distinctive coloration patterns of (a) the monarch butterfly (*Danaus plexippus*) and (b) the strawberry poison dart frog (*Dendrobates pumilio*) warn potential predators of these species' toxicity.

(a)

(b)

Figure 14.18 | The warning coloration of (a) the poisonous coral snake (*Micrurus fulvius*) is mimicked by (b) the nonvenomous scarlet king snake (*Lampropeltis triangulum*).

snake of the Great Plains, and the gopher snake of the Pacific States, all subspecies of *Pituophis melanoleucus*, rapidly vibrate their tails in leafy litter to produce a rattle-like sound.

Another type of mimicry is called Müllerian, after the 19th-century German zoologist Fritz Müller. With

Müllerian mimicry, many unpalatable or venomous species share a similar color pattern. Müllerian mimicry is successful because the predator has to be exposed to only one of the species before learning to stay away from all other species with the same warning color patterns. The black-and-yellow-striped bodies of social wasps,

(a)

(b)

(c)

Figure 14.19 | Example of Müllerian mimicry. The black and yellow striped bodies of **(a)** social wasps (Vespidae), **(b)** solitary digger wasps (Sphecidae), and **(c)** caterpillars of the cinnabar moths (*Callimorpha jacobaeae*) warn predators that these organisms are inedible. All are unrelated species with a shared color pattern that functions to keep predators away.

solitary digger wasps, and caterpillars of the cinnabar moths warn predators that the organism is inedible (Figure 14.19). All are unrelated species with a shared color pattern that functions to keep predators away.

Some animals employ **protective armor** for defense. Clams, armadillos, turtles, and many beetles all withdraw into their armor coats or shells when danger approaches. Porcupines, echidnas, and hedgehogs possess quills (modified hairs) that discourage predators.

Still other animals use **behavioral defenses,** which include a wide range of behaviors by prey species aimed at avoiding detection, fleeing, and warning others of the presence of predators. Animals may change their foraging behavior in response to the presence of predators, as in the example of the willow and crested tits (see Section 14.8). Some species give an alarm call when a predator is sighted. Because high-pitched alarm calls are not species specific, they are recognized by a wide range of nearby animals. Alarm calls often bring in numbers of potential prey that mob the predator. Other behavioral defenses include distraction displays, which are most common among birds. These defenses direct the predator's attention away from the nest or young.

For some prey, living in groups is the simplest form of defense. Predators are less likely to attack a concentrated group of individuals. By maintaining a tight, cohesive group, prey make it difficult for any predator to obtain a victim (Figure 14.20). Sudden, explosive group flight can confuse a predator, leaving it unable to decide which individual to follow.

A subtler form of defense is the timing of reproduction so that most of the offspring are produced in a short period. Prey are thus so abundant that the predator can take only a fraction of them, allowing a percentage of the young to escape and grow to a less-vulnerable size. This phenomenon is known as **predator satiation.** Periodic cicadas (*Magicicada* spp.) emerge as adults once every 13 years in the southern portion of their range in North America and once every 17 years in the northern portion of their range, living the remainder of the period as nymphs underground. Though these cicadas emerge only once every 13 or 17 years, a local population emerges somewhere within their range virtually every year. When emergence occurs, the local density of cicadas can number in the millions of individuals per hectare. Ecologist Kathy Williams, of San Diego State University, and her colleagues tested the effectiveness of predator satiation during the emergence of periodic cicadas in northwest Arkansas. Williams found that the first cicadas emerging in early May were eaten by birds, but avian predators quickly became satiated. Birds consumed 15–40 percent of the cicada population at low cicada densities but only a very small proportion as cicada densities increased (Figure 14.21). Williams's results demonstrated that, indeed, the synchronized, explosive emergences of periodic cicadas are an example of predator satiation.

Figure 14.20 | Musk ox (*Ovibos moschatus*) form a circle, each facing outward, to present a combined defense when threatened by predators.

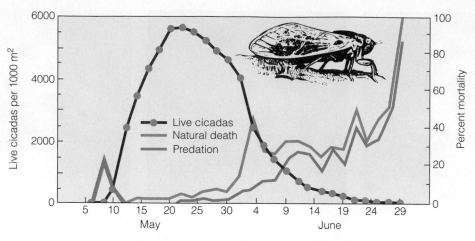

Figure 14.21 | Estimated daily population density of periodic cicadas (*Magicicada*) on a study site in Arkansas (left *y*-axis) and estimated daily mortality due to bird predation and natural causes (right *y*-axis). Maximum cicada density occurred around May 24, and maximum predation occurred around June 10. At the height of predation, most of the cicadas had already emerged and escaped bird predation.

(Adapted from Williams et al. 1993.)

The predator defenses just discussed fall into two broad classes: permanent and induced. Permanent, or **constitutive defenses,** are fixed features of the organism, such as object resemblance and warning coloration. In contrast, defenses that are brought about, or induced, by the presence or action of predators are referred to as **induced defenses.** Behavioral defenses are an example of induced defenses, as are chemical defenses such as alarm pheromones that, when detected, induce flight reactions. Induced defenses can also include shifts in physiology or morphology, representing a form of phenotypic plasticity (see Field Studies: Rick A. Relyea).

14.11 | Predators Have Evolved Efficient Hunting Tactics

As prey have evolved ways of avoiding predators, predators have evolved better ways of hunting. Predators have three general methods of hunting: ambush, stalking, and pursuit. Ambush hunting means lying in wait for prey to come along. This method is typical of some frogs, alligators, crocodiles, lizards, and certain insects. Although ambush hunting has a low frequency of success, it requires minimal energy. Stalking, typical of herons and some cats, is a deliberate form of hunting with a quick attack. The predator's search time may be great, but pursuit time is minimal. Pursuit hunting, typical of many hawks, lions, wolves, and insectivorous bats, involves minimal search time because the predator usually knows the location of the prey, but pursuit time is usually great. Stalkers spend more time and energy encountering prey. Pursuers spend more time capturing and handling prey.

Predators, like their prey, may use cryptic coloration to blend into the background or break up their outlines (Figure 14.22). Predators use deception by resembling the prey. Robber flies (*Laphria* spp.) mimic bumblebees, their prey (Figure 14.23). The female of certain species of fireflies imitates the mating flashes of other species to attract males of those species, which she promptly kills and eats. Predators may also employ chemical poisons, as do venomous snakes, scorpions, and spiders. They may form a group to attack large prey, as lions and wolves do.

14.12 | Herbivores Prey on Autotrophs

Although the term *predator* is typically associated with animals that feed on other animals, herbivory is a form of predation in which animals prey on autotrophs

Figure 14.22 | The alligator snapping turtle uses a combination of cryptic coloration and mimicry to avoid detection and attract prey. By lying motionless on the bottom with its mouth wide open, it wiggles its worm-shaped tongue (see bottom of mouth) to attract and ambush potential prey.

Robber fly   Bumblebee

Figure 14.23 | The robber fly (*Laphria* spp.) illustrates aggressive mimicry. It is the mimic of the bumblebee (*Megabombus pennsylvanicus*), on which it preys.

(plants and algae). Herbivory represents a special type of predation because herbivores typically do not kill the individuals they feed on. Because the ultimate source of food energy for all heterotrophs is carbon fixed by plants in the process of photosynthesis (see Chapter 6), autotroph–herbivore interactions represent a key component of all communities.

If you measure the amount of biomass actually eaten by herbivores, it may be small—perhaps 6–10 percent of total plant biomass present in a forest community or as high as 30–50 percent in grassland communities (see Chapter 20, Section 20.11). In years of major insect outbreaks, however, or in the presence of an overabundance of large herbivores, consumption is considerably higher (Figure 14.24). Consumption, though, is not necessarily the best measure of the importance of herbivory within a community. Grazing on plants can have a subtler impact on both plants and herbivores.

The removal of plant tissue—leaf, bark, stems, roots, and sap—affects a plant's ability to survive, even though the plant may not be killed outright. Loss of foliage and subsequent loss of roots will decrease plant biomass, reduce the vigor of the plant, place it at a competitive disadvantage with surrounding vegetation, and lower its reproductive effort. The effect is especially strong in the juvenile stage, when the plant is most vulnerable and least competitive with surrounding vegetation.

A plant may be able to compensate for the loss of leaves by increasing photosynthesis in the remaining leaves. However, it may be adversely affected by the loss of nutrients, depending on the age of the tissues removed. Young leaves are dependent structures—importers and consumers of nutrients drawn from reserves in roots and other plant tissues. Grazing herbivores, both vertebrate and invertebrate, often concentrate on younger leaves and shoots because they are lower in structural carbon compounds such as lignins, which are difficult to digest and provide little if any energy (see Section 21.4). By selectively feeding on younger tissues, grazers remove considerable quantities of nutrients from the plant.

Plants respond to defoliation with a flush of new growth that drains nutrients from reserves that otherwise would have gone into growth and reproduction. For ex-

(a)

(b)

Figure 14.24 | Examples of the impact of high rates of herbivory. (a) Intense predation on oaks by gypsy moths in the forests of eastern North America. (b) Contrast between heavily grazed grassland in southeast Africa and an adjacent area where large herbivores have been excluded.

ample, Anurag Agrawal of the University of Toronto found that herbivory by longhorn beetles (*Tetraopes* spp.) reduced fruit production and mass of milkweed plants (*Asclepias* spp.) by as much as 20–30 percent.

If defoliation of trees is complete (Figure 14.24a), as often happens during an outbreak of gypsy moths (*Lymantria dispar*) or fall cankerworms (*Alsophila pometaria*), leaves that regrow in their place are often quite different in form. The leaves are often smaller, and the total canopy (area of leaves) may be reduced by as much as 30–60 percent. In addition, the plant uses stored reserves to maintain living tissue until new leaves form, reducing reserves that it will require later. Regrown twigs and tissues are often immature at the onset of cold weather, reducing their ability to tolerate winter temperatures. Such weakened trees are more vulnerable to insects and disease. In contrast to deciduous tree species, defoliation kills coniferous species.

Department of Biological Sciences, University of Pittsburgh, Pittsburgh, Pennsylvania

Ecologists have long appreciated the influence of predation on natural selection. Predators select prey based on their sizes and shapes, thereby acting as a form of natural selection that alters the range of phenotypes within the population. In doing so, predators alter the genetic composition of the population (gene pool), which determines the range of phenotypes in future generations. Through this process, many of the mechanisms of predator avoidance discussed in Section 14.10 are selected for in prey populations. In recent years, however, ecologists have discovered that predators can have a much broader influence on the characteristics of prey species through nonlethal effects. For example, presence of a predator can change the behavior of prey, causing them to reduce activity (or hide) to avoid being detected. This change in behavior can reduce foraging activity. In turn, changes in the rate of food intake can influence prey growth and development, resulting in shifts in their morphology (size and shape of body). This shift in the phenotype of individual prey, induced by the presence and activity of predators, is termed *induction* and represents a form of phenotypic plasticity (see Section 5.9).

The discovery that predators can influence the characteristics (phenotype) of prey species through natural selection and induction presents a much more complex picture of the role of predation in evolution. Although ecologists are beginning to understand how natural selection and induction function separately, very little is known about how these two processes interact to determine the observed range of phenotypes within a prey

population. Thanks to the work of ecologist Rick Relyea, however, this picture is becoming much clearer.

Relyea's research is conducted in wading pools that are constructed to serve as experimental ponds. In one series of experiments, Relyea explored the nature of induced changes in behavior and morphology in prey (gray tree frog tadpoles, *Hyla versicolor*) by introducing caged predators (dragonfly larvae, *Anax longipes*) into the experimental ponds (Figure 1). The tadpoles can detect waterborne chemicals produced by the predators, allowing Relyea to simulate the threat of predation to induce changes in the tadpoles while preventing actual predation. By comparing the characteristics of tadpoles in control ponds (no predator present) and in ponds with caged predators, he was able to examine the responses induced by the presence of predators.

Results of the experiments reveal that induction by predatory chemical cues altered the tadpoles' behavior. They became less active in the presence of predators (Figure 2). Reduced activity makes prey less likely to encounter predators and improves their probability of survival. The predators' presence also induced a shift in the morphology of tadpoles—a form of phenotypic plasticity. Tadpoles raised in the experimental ponds in which predators were present have a greater tail depth and shorter overall body length than do individuals raised in

Figure 1 | Dragonfly larvae feeding on tadpole.

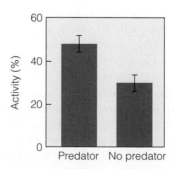

Figure 2 | Activity of gray tree frog tadpoles when reared either in the absence (green bar) or presence (blue bar) of caged predators.
Go to **QUANTIFYIT!** at **www.ecologyplace.com** to use data to calculate standard deviation.

(Adapted from Relyea 2002.)

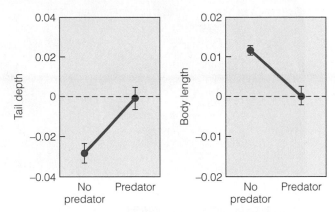

Figure 3 | Relative morphology of tree frog tadpoles when reared in either the presence or absence of caged predators. All values are relative to the mean, with negative values below and positive values above the mean value of the respective characteristic (see Figure 11.9 for complete description of statistical analyses).

(Adapted from Relyea 2002.)

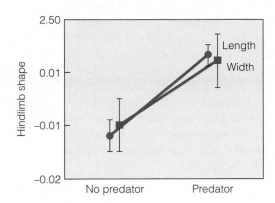

Figure 4 | Relative morphology of adult wood frog that developed from tadpoles reared in either the presence or absence of caged predators. All values are relative to the mean (see Figure 3).

(Adapted from Reylea 2001.)

the absence of predators (control ponds; Figure 3). Interestingly, previous studies showed that tadpoles with deeper tails and shorter bodies escape dragonfly predators better than tadpoles with the opposite morphology. Therefore, the induced morphological responses that were observed in Relyea's experiments are adaptive; they are a form of phenotypic plasticity that functions to increase the survival of individual tadpoles.

Relyea's experiments clearly show that predators can induce changes in prey phenotype and that the induced changes are of the same type as those that result from natural selection.

The experiments discussed here focus on only one life stage in the development of the tree frog: the larval (tadpole) stage. But how might these changes in morphology early in development affect traits later in life? As the tadpoles metamorphose into adult frogs, they have drastically different morphologies and occupy different habitats. To answer this question, Relyea conducted an experiment to examine how differences in the morphology of wood frog tadpoles (*Rana sylvatica*), induced by the presence of predators, subsequently affected the morphology of the adult frog later in development.

As in previous experiments, tadpoles reared with caged predators developed relatively deeper tail fins and had shorter bodies, lower mass, and longer developmental times than did tadpoles reared without predators. Adult frogs that emerged from the tadpoles exposed to predators (and exhibiting these induced changes during the larval stage) exhibited no differences in mass but developed relatively large hindlimbs

and forelimbs and narrower bodies as compared to individuals emerging from environments where predators were absent (Figure 4). These results clearly show that predator-induced shifts in traits early in development can subsequently alter traits later in development.

Bibliography

Relyea, R. 2001. The lasting effects of adaptive plasticity: Predator-induced tadpoles become long-legged frogs. *Ecology* 82:1947–1955.

Relyea, R. 2002a. The many faces of predation: How induction, selection, and thinning combine to alter prey phenotypes. *Ecology* 83:1953–1964.

Relyea, R. 2002b. Local population differences in phenotypic plasticity: Predator induced changes in wood frog tadpoles. *Ecological Monographs* 72:77–93.

1. Theory predicts that phenotypic plasticity should evolve when alternative phenotypes are favored in different environments. How does this prediction relate to the patterns of phenotypic plasticity observed by Relyea? What are the different environments that may function as a selective force to reinforce the benefit of phenotypic plasticity?
2. In the last experiment described, induced shifts in morphology observed during the larval stage (tadpole) influenced the morphology of these individuals at the adult stage. What type of experiment might determine if these induced changes in phenotype influence the fitness of adult frogs?

Browsing animals such as deer, rabbits, and mice selectively feed on the soft, nutrient-rich growing tips (apical meristems) of woody plants, often killing the plants or changing their growth form. Burrowing insects, like the bark beetles, bore through the bark and construct egg galleries in the phloem–cambium tissues. In addition to phloem damage caused by larval and adult feeding, some bark beetle species carry and introduce a blue stain fungus into a tree that colonizes sapwood and disrupts water flow to the tree crown, hastening tree death.

Some herbivores, such as aphids, do not consume tissue directly but tap plant juices instead, especially in new growth and young leaves. Sap-sucking insects can decrease growth rates and biomass of woody plants by as much as 25 percent.

Grasses have their meristems, the source of new growth, close to the ground. As a result, grazers first eat the older tissue and leave intact the younger tissue with its higher nutrient concentration. Therefore, grasses are generally tolerant of grazing; and, up to a point, most benefit from it. The photosynthetic rate of leaves declines with leaf age. Grazing stimulates production by removing older tissue functioning at a lower rate of photosynthesis, increasing the light availability to underlying young leaves. Some grasses can maintain their vigor only under the pressure of grazing, even though defoliation reduces sexual reproduction. Not all grasses, however, tolerate grazing. Species with vulnerable meristems or storage organs can be quickly eradicated under heavy grazing.

14.13 | Plants Have Evolved Characteristics That Deter Herbivores

Most plants are sessile; they cannot move. Thus, avoiding predation requires adaptations that discourage being selected by herbivores. The array of characteristics that plants employ to deter herbivores includes both structural and other defenses.

Structural defenses, such as hairy leaves, thorns, and spines, can discourage feeding (Figure 14.25), thereby reducing the amount of tissues removed by herbivores.

For herbivores, often the quality of food rather than the quantity is the constraint on food supply. Because of the complex digestive process needed to break down plant cellulose and convert plant tissue into animal flesh, high-quality forage rich in nitrogen is necessary (see Chapter 7, Section 7.3). If the nutrient content of the plants is not sufficient, herbivores can starve to death on a full stomach. Low-quality foods are tough, woody, fibrous, and indigestible. High-quality foods are young, soft, and green or they are storage organs such as roots, tubers, and seeds. Most plant tissues are relatively low in quality, and herbivores that have to live on such resources suffer high mortality or reproductive failure.

Plants contain various chemicals that are not involved in the basic metabolism of plant cells. Many of

Figure 14.25 | Thorns of this *Acacia* tree can deter herbivores or reduce levels of defoliation by browsers.

these chemicals, referred to as **secondary compounds,** either reduce the ability of herbivores to digest plant tissues or deter herbivores from feeding. Although these chemicals represent an amazing array of compounds, they can be divided into three major classes based on their chemical structure: nitrogen-based compounds, terpenoids, and phenolics. Nitrogen-based compounds include alkaloids such as morphine, atropine, nicotine, and cyanide. Terpenoids (also called isoprenoids) include a variety of essential oils, latex, and plant resins (many spices and fragrances contain terpenoids). Phenolics are a general class of aromatic compounds (i.e., contain the benzene ring) including the tannins and lignins.

Some secondary compounds are produced by the plant in large quantities and are referred to as **quantitative inhibitors.** For example, tannins and resins may constitute up to 60 percent of the dry weight of a leaf. In the vacuoles of their leaves, oaks and other species contain tannins that bind with proteins and inhibit their digestion by herbivores. From 5 to 35 percent of the carbon contained in the leaves of terrestrial plants is in the form of lignins—complex, carbon-based molecules that are impossible for herbivores to digest, making the nitrogen and other essential nutrients bound in these compounds unavailable to the herbivore. These types of compounds reduce digestibility and thus potential energy gain from food (see Sections 7.2 and 7.3).

Other secondary compounds that function as defenses against herbivory are present in small to minute quantities and are referred to as **qualitative inhibitors.** These compounds are toxic, often causing herbivores to avoid their consumption. This category of compounds includes cyanogenic compounds (cyanide) and alkaloids such as nicotine, caffeine, cocaine, morphine, and mescaline that interfere with specific metabolic pathways of physiological processes. Many of these compounds, such as pyrethrin, have become important sources of pesticides.

Although the qualitative inhibitors function to protect against most herbivores, some specialized herbivores have developed ways of breaching these chemical defenses. Some insects can absorb or metabolically detoxify the chemical substances. They even store the plant poisons to use them in their own defense, as the larvae of monarch butterflies do, or in the production of pheromones (chemical signals). Some beetles and certain caterpillars sever veins in leaves before feeding, stopping the flow of chemical defenses.

Some plant defenses are constitutive, such as structural defenses or quantitative inhibitors (tannins, resins, or lignins) that provide built-in physical or biological barriers against the attacker. Others are active, induced by the attacking herbivore. These induced responses can be local (at the site of the attack) or can extend systematically throughout the plant. Often, these two types of defenses are used in combination. For example, when attacked by bark beetles carrying an infectious fungus in their mouthparts, conifer trees release from the attack sites large amounts of resin (constitutive, quantitative defense) that flows out onto the attackers, entombing the beetles. Meanwhile, the tree mobilizes induced defenses against the pathogenic fungus that the intruder has deposited at the wound site.

In another kind of plant–insect interaction, it has been found that some plants appear to call for help, attracting the predators of their predators. Parasitic and predatory arthropods often prevent plants from being severely damaged by killing herbivores as they feed on the plants. Recent studies show that a variety of plant species, when injured by herbivores, emit chemical signals to guide natural enemies to the herbivores. It is unlikely that the herbivore-damaged plants initiate the production of chemicals solely to attract predators. The signaling role probably evolved secondarily from plant responses that produce toxins and deterrents against herbivores. For example, in a series of controlled laboratory studies, Ted Turlings and James Tumlinson, researchers at the Agricultural Research Service of the U.S. Department of Agriculture, found that corn seedlings under attack by caterpillars release several volatile terpenoid compounds. Consequently, the caterpillars become highly attractive to the parasitoid wasps that attack the herbivores. Experiment results showed that the induced emission of volatiles is not limited to the site of damage but occurs throughout the plant. The systematic release of volatiles by injured corn seedlings results in a significant increase in visitation by the parasitoid wasp *Cotesia marginiventris*.

Various hypotheses have been put forward to explain why plants differ in the types of defenses they have evolved to avoid herbivores. A feature common to all of these hypotheses is the trade-off between the costs and benefits of defense. The cost of defense in diverting energy and nutrients from other needs must be offset by the benefits of avoiding predation.

14.14 | Plants, Herbivores, and Carnivores Interact

In our discussion thus far, we have considered herbivory on plants and carnivory on animals as two separate topics, linked only by the common theme of predation. However, they are linked in another very important way. Plants are consumed by herbivores, which in turn are consumed by carnivores. Therefore we cannot really understand an herbivore–carnivore system without understanding plants and their herbivores, nor can we understand plant–herbivore relations without understanding predator–herbivore relationships. All three— plants, herbivores, and carnivores—are interrelated. Ecologists are beginning to understand these three-way relationships.

A classic case (Figure 14.26) is the three-level interaction of plants, the snowshoe hare (*Lepus americanus*), and its predators—lynx (*Felis lynx*), coyote (*Canis latrans*), and horned owl (*Bubo virginianus*). The snowshoe hare inhabits the high-latitude forests of North America. In winter, it feeds on the buds of conifers and the twigs of aspen, alder, and willow, which are termed *browse*. Browse consists mainly of smaller stems and young growth rich in nutrients. The hare–vegetation interaction becomes critical

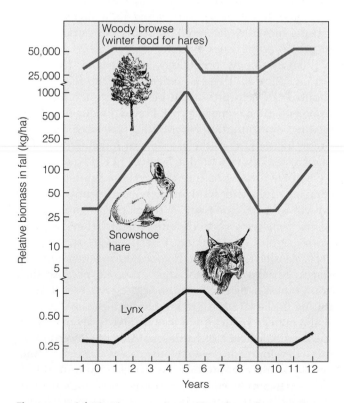

Figure 14.26 | The three-way interaction of woody vegetation, snowshoe hare, and lynx. Note the time lag between the cycles of the three populations.

(Adapted from Keith et al. 1974.)

when the amount of essential browse falls below that needed to support the population over winter (approximately 300 g per individual per day). Excessive browsing when the hare population is high reduces future woody growth, bringing on a food shortage.

The shortage and poor quality of food lead to malnutrition, parasite infections, and chronic stress. Those conditions and low winter temperatures weaken the hares, reducing reproduction and making them extremely vulnerable to predation. Intense predation causes a rapid decline in the number of hares. Now facing their own food shortage, the predators fail to reproduce, and populations decline. Meanwhile, upon being released from the pressures of browsing by hares, plant growth rebounds. As time passes, with the growing abundance of winter food as well as the decline in predatory pressure, the hare population starts to recover and begins another cycle. Thus, an interaction between predators and food supply (plants) produces the hare cycle, and in turn the hare cycle affects the population dynamics of its predators.

14.15 | Predators Influence Prey Dynamics through Lethal and Nonlethal Effects

As we have seen in the preceding sections, the ability of predators to suppress prey populations has been well documented. Predators can suppress prey populations through consumption; that is, they reduce prey population growth by killing and eating individuals. Besides causing mortality, however, predators can cause changes in prey characteristics by inducing defense responses in prey morphology, physiology, or behavior (see Field Studies: Rick A. Relyea). Predator-induced defensive responses can help prey avoid being consumed, but such responses often come at a cost. Prey individuals may lose feeding opportunities by avoiding preferred but risk-prone habitats, as in the example of foraging by willow and crested tits presented in Section 14.8. Reduced activity by prey in the presence of predators can reduce prey foraging time and food intake, subsequently delaying growth and development. A convincing demonstration of the long-term costs of antipredator behavior comes from studies of aquatic insects such as mayflies (*Ephemeroptera*), which do not feed during their adult life stages. Mayflies are ideal study subjects because their adult fitness depends on the energy reserves they develop during the larval stage. Thus, it has been possible to show that a marked reduction in feeding activity by mayfly larvae in the presence of predators leads to slower growth and development, which ultimately translates into smaller adults that produce fewer eggs (Figure 14.27).

Predator-induced defensive responses can potentially influence many aspects of prey population regulation and dynamics, given the negative reproductive consequences of antipredator behavior. Translating behavior decisions to population-level consequences, however,

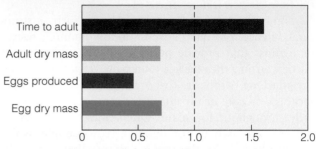

Figure 14.27 | Consequences of reduced activity in the mayfly *Baetis tricaudatus*. For each measure (time to adult, dry mass of adults, number of eggs produced, and dry mass of eggs), the ratio (*x*-axis) of its average value in the presence of predators (fish) relative to that in the absence of predators is given. A ratio smaller than 1.0 represents a reduced average value in the presence of predators, whereas values greater than 1.0 represent a greater average value in the presence of predators.

(Adapted from Scrimgeour and Culp 1994, as illustrated in Lima 1998.)

Interpreting Ecological Data

Q1. Based on the results of the experimental study presented in Figure 14.27, how does the reduced activity of larval mayflies in the presence of predators influence the time required for larvae to develop into adult mayflies?

Q2. How does the presence of predators and associated reduction in activity during the larval stage influence the fitness of adult mayflies? Explain the variables you used to draw your conclusions about adult fitness.

can be difficult. But research by Eric Nelson and colleagues at the University of California at Davis has clearly demonstrated an example of reduction in prey population growth resulting from predator-induced changes in prey behavior. Nelson and colleagues studied the interactions between herbivorous and predatory insects in fields of alfalfa (*Medicago sativa*). Pea aphids (*Acyrthosiphon pisum*) feed by inserting their mouthparts into alfalfa phloem tissue, and they reproduce parthenogenetically (see Section 8.1) at rates of 4 to 10 offspring per day. A suite of natural enemies attacks the aphids, including damsel bugs (*Nabis* spp.). The aphids respond to the presence of foraging predators by interrupting feeding and walking away from the predator or dropping off the plant. The costs suffered by the aphids due to their defensive behavior may include increased mortality or reduced reproduction.

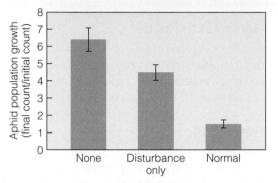

Figure 14.28 | Per capita population growth rate (*y*-axis) of pea aphids in field cages containing either no predatory damsel bugs (none), damsel bugs with proboscis removed (disturbance only), and normal damsel bugs (normal). Bars represent the mean population growth for each treatment, where population growth was measured as the ratio of final to initial population size. Vertical lines associated with each bar represent ±1 standard error of the mean.

Go to **QUANTIFYit!** at **www.ecologyplace.com** to perform confidence intervals and t-tests.

(Adapted from Nelson et al. 2004.)

Damsel bugs feed by piercing aphids with a long proboscis and ingesting the body contents. Damsel bugs, therefore, influence prey in two ways: first by con-suming aphids and second by disturbing their feeding behavior. In a series of controlled experiments, Nelson was able to distinguish between the effects of these two influences by surgically removing the mouthparts (proboscises) of some damsel bugs, therefore making them unable to kill and feed on aphids. By exposing aphids to these damsel bugs, the researchers were able to test the predators' ability to suppress aphid population growth through behavioral mechanisms only. The greatest reduction in aphid population growth was caused by normal predators that were able to consume and disturb the aphids, but aphid population growth was also strongly reduced by nonconsumptive predators (Figure 14.28). These field experiments clearly demonstrated that predators reduce population growth partly through predator-induced changes in prey behavior and partly through direct mortality (consuming prey individuals).

As we have seen in this chapter, the relationship between a predator and its prey is influenced by an array of specific behavioral, morphological, and physiological adaptations, making it difficult to generalize about the influence of predation on prey populations. Nonetheless, many laboratory and field studies offer convincing evidence that predators can significantly alter prey abundance. Whereas the influence of competition on community structure is somewhat obscure, the influence of predation is more demonstrable. Because all heterotrophs derive their energy and nutrients from consuming other organisms, the influence of predation can more readily be noticed throughout a community. As we shall see in Chapters 16 and 17, the direct influence of predation on the population density of prey species can have the additional impact of influencing the interactions among prey species, particularly competitive relationships.

Summary

Forms of Predation | 14.1

Predation is defined generally as the consumption of all or part of one living organism by another. Forms of predation include carnivory, parasitoidism, cannibalism, and herbivory.

Model of Predation | 14.2

A mathematical model that links the two populations through the processes of birth and death can describe interactions between predator and prey. Predation represents a source of mortality for the prey population, while the reproduction of the predator population is linked to the consumption of prey.

Mutual Population Regulation | 14.3

The models of predator–prey interactions predict oscillations of predator and prey populations, with the predator population lagging behind that of the prey population. The results of the models assume mutual regulation of predator and prey populations.

Functional Response | 14.4

The interaction between predator and prey involves a functional response and a numerical response. A functional response is one in which the number of prey taken increases with the density of prey. There are three types of functional responses. In type I, the number of prey affected increases linearly. In type II, the number of prey affected increases at a decreasing rate toward a maximum value. In type III, the number of prey taken increases sigmoidally as the density of prey increases. Only type III is important as a population-regulating mechanism.

Numerical Response | 14.5

A numerical response is the increase of predators with an increased food supply. Numerical response may involve an aggregative response: the influx of predators to a food-rich area. More important, a numerical response involves a change in the growth rate of a predator population through changes in fecundity.

Optimal Foraging | 14.6

Central to the study of predation is the concept of optimal foraging. This approach to understanding the foraging behavior of animals assumes that natural selection should favor "efficient" foragers: individuals that maximize their energy or nutrient intake per unit of effort. Decisions are based on the relative profitability of alternative prey types, defined as the energy gained per unit of handling time. An optimal diet includes the most efficient size of prey for handling and net energy return.

Foraging in a Heterogeneous Environment | 14.7

Optimal foraging concentrates activity in the most profitable areas or patches. After depleting the food to the average profitability of the area as a whole, the predator abandons that patch for one more profitable. The time a predator should spend foraging in a patch is influenced by the time required to travel to the patch and the quality of the patch in terms of prey density.

Foraging Behavior and Risk of Predation | 14.8

Most predators are also prey to other predatory species and thus face the risk of predation while involved in their routine activities, such as foraging. If predators are about, it may be to the forager's advantage not to visit a most profitable but predator-prone area and to remain in a less profitable but more secure part of the habitat.

Coevolution of Predator and Prey | 14.9

Prey species evolve characteristics to avoid being caught by predators. Predators have evolved their own strategies for overcoming these prey defenses. This process represents a coevolution of predator and prey in which each functions as an agent of natural selection on the other.

Predator Defenses | 14.10

Chemical defense in animals usually takes the form of distasteful or toxic secretions that repel, warn, or inhibit would-be attackers. Cryptic coloration and behavioral patterns enable prey to escape detection. Warning coloration declares that the prey is distasteful or disagreeable. Some palatable species mimic unpalatable species for protection. Armor and aggressive use of weapons defend some prey. Alarms and distraction displays help others. Another form of defense is predator satiation—prey species reproduce so many young at once that predators can take only a fraction of them. Predator defenses can be classified as permanent or induced.

Predator Evolution | 14.11

Predators have evolved different methods of hunting that include ambush, stalking, and pursuit. Predators also employ cryptic coloration for hiding and aggressive mimicry for imitating the appearance of prey.

Herbivory | 14.12

Herbivory is a form of predation. The amount of plant or algal biomass actually eaten by herbivores varies among communities. Plants respond to defoliation with a flush of new growth, which draws down nutrient reserves. Such drawdown can weaken plants, especially woody ones, making them more vulnerable to insects and disease. Moderate grazing may stimulate leaf growth in grasses up to a point. By removing older leaves less active in photosynthesis, grazing stimulates the growth of new leaves.

Herbivore Defenses | 14.13

Plants affect herbivores by denying them palatable or digestible food or by producing toxic substances that interfere with growth and reproduction. Certain herbivore specialists are able to breach the chemical defenses. They detoxify the secretions, block their flow, or sequester them in their own tissues as a defense against predators. Defenses can be either permanent (constitutive) or induced by damage inflicted by herbivores.

Vegetation–Herbivore–Carnivore Systems | 14.14

Plant–herbivore and herbivore–carnivore systems are closely related. An example of a three-level feeding interaction is the cycle of vegetation, hares, and their predators. Malnourished hares fall quickly to predators. Recovery of hares follows recovery of plants and decline in predators.

Lethal and Nonlethal Influences | 14.15

Besides influencing prey population directly through mortality, predators can cause changes in prey characteristics by inducing defense responses in prey morphology, physiology, or behavior. Reduced activity by prey in the presence of predators can reduce foraging time and food intake, subsequently delaying growth and development. The net result can be a reduction in the growth rate of the prey population.

Study Questions

1. The Lotka–Volterra model of predator–prey dynamics suggests mutual control between predator and prey populations that results in the two populations oscillating through time (see Figure 14.2). Why does the predator population lag behind the prey population?

2. What is a functional response in predation? What component of the Lotka–Volterra model of predator–prey dynamics presented in Section 14.2 represents the functional response?

3. Distinguish among type I, type II, and type III functional responses. Which type of functional response is included in the Lotka–Volterra model of predator–prey dynamics presented in Section 14.2?

4. In Hindu mythology, Brahma created a large and monstrous creature that grew rapidly and devoured everything in its path. In reality, predators are much more selective about what they eat. What factors appear to be important in determining what a predator selects to eat among the possible array of potential prey?

5. Optimal foraging theory suggests that a predator will select among possible prey based on their relative profitability (energy gained per unit of energy expended). Do you think that predators directly evaluate the profitability of potential prey items before selecting or rejecting them? If not, how might a foraging strategy evolve?

6. What is a numerical response? Which terms in the Lotka–Volterra model of predator–prey dynamics presented in Section 14.2 relate to the numerical response of predators to prey?

7. How can predators function as agents of natural selection in prey populations?

8. Distinguish between permanent and induced predator defenses. Provide an example of each in both plants and animals.

9. How can prey function as agents of natural selection in predator populations?

Further Readings

Hay, M. E. 1991. Marine-terrestrial contrasts in the ecology of plant chemical defenses against herbivores. *Trends in Ecology and Evolution* 6:362–365.

> This review article, an excellent overview of plant chemical defenses, contrasts the differences in plant chemical defenses that have evolved in terrestrial and aquatic plant species. Also considers how adaptations to the abiotic environment constrain the adaptation of plant defenses.

Krebs, C. J., R. Boonstra, S. Boutin, and A. R. E. Sinclair. 2001. What drives the 10-year cycle of snowshoe hares? *Bioscience* 51:25–35.

> An excellent, easy-to-read overview of factors involved in the predator–prey cycle discussed in Section 14.14.

O'Donoghue, M., S. Boutin, C. J. Krebs, and E. J. Hofer. 1997. Numerical responses of coyotes and lynx to the snowshoe hare cycle. *Oikos* 80:150–162.

> Another excellent paper; addresses the snowshoe hare–lynx predator–prey cycle as discussed in this chapter.

O'Donoghue, M., S. Boutin, C. J. Krebs, G. Zuleta, D. Murray, and E. J. Hofer. 1998. Functional responses of coyotes and lynx to the snowshoe hare cycle. *Ecology* 79:1193–1208.

> A sibling manuscript to the preceding reference.

Tollrian, R., and C. D. Harvell. 1999. *The ecology and evolution of inducible defenses.* Princeton, NJ: Princeton University Press.

> An extensive, excellent review of the defenses exhibited by prey species (both plant and animal) that are induced by the presence and activity of predators. The text provides a multitude of illustrated examples from recent research in the growing area of coevolution between predator and prey.

Williams, K., K. G. Smith, and F. M. Steven. 1993. Emergence of the 13-year periodic cicada: Phenology, mortality, and predator satiation. *Ecology* 74:1143–1152.

> A classic example of the consequences of predator satiation on the functional response of predators.

CHAPTER 1 5

Parasitism and Mutualism

Corn (*Zea mays*) infected with corn smut (*Ustilago maydis*).

n our discussion of predator–prey interactions in Chapter 14, we examined the concept of coevolution—the evolution of one species in response to interaction with another species. Prey have evolved means of defense against predators, and predators have evolved ways to breach those defenses: an evolutionary "game" of adaptation and counteradaptation. The process of coevolution is even more evident in the interactions between parasites and their hosts. The parasite lives on or in the host organism for some period of its life, in a relationship referred to as symbiosis (Greek *sym*, "together," and *bios*, "life"). **Symbiosis,** as defined by the eminent evolutionary biologist Lynn Margulis, is the "intimate and protracted association between two or more organisms of different species." This definition does not specify whether the result of the association between the species is positive, negative, or benign. It therefore includes a wide variety of interactions in which the fate of individuals of one species depends on their association with individuals of another. Some symbiotic relationships benefit both species involved, as in the case of mutualism. In other symbiotic relationships, however, one species benefits at the expense of the other. Such is the case of the parasitic relationship, in which the host organism is the parasite's habitat as well as its source of nourishment. For the parasite, this is an obligatory relationship; the parasite requires the host organism for its survival and reproduction. To minimize the negative impact of parasites, host species have evolved a variety of defense mechanisms.

15.1 | Parasites Draw Resources from Host Organisms

Parasitism is a type of symbiotic relationship between organisms of different species. One species, the parasite, benefits from a prolonged, close association with the other species—the host, which is harmed. Parasites increase their fitness by exploiting host organisms for food, habitat, and dispersal. Although they draw nourishment from the tissues of the host organism, parasites typically do not kill their hosts as predators do. However, the host may die from secondary infection or suffer reduced fitness due to stunted growth, emaciation, modification of behavior, or sterility. In general, parasites are much smaller than their hosts, are highly specialized for their mode of life, and reproduce more quickly and in greater numbers than their hosts.

The definition of parasitism just presented may appear unambiguous. But as with predation (see Section 14.1), the term *parasitism* is often used in a more general sense to describe a much broader range of interactions. Interactions between species frequently satisfy some but not all parts of this definition, because in many cases it is hard to demonstrate that the host is harmed. In other cases there may be no apparent specialization by the parasite, or the interaction between the organisms may be short-lived. For example, due to the episodic nature of their feeding habits, mosquitoes and hematophagic ("blood-feeding") bats are typically not considered parasitic. *Parasitism* can also be used to describe a form of feeding in which one animal appropriates food gathered by another (the host), a behavior termed *cleptoparasitism* (literally meaning "parasitism by theft"). An example is the brood parasitism practiced by many species of cuckoo (Cuculidae). Many cuckoos use other bird species as "babysitters"; they deposit their eggs in the nest of the host species, which raise the cuckoo young as one of their own. In the following discussion, we use the narrower definition of *parasite* as given in the previous paragraph. This definition includes a wide range of organisms—viruses, bacteria, protists, fungi, plants, and an array of invertebrates, among them arthropods. A heavy load of parasites is termed an **infection,** and the outcome of an infection is a **disease.**

Parasites are distinguished by size. Ecologically, parasites may be classified as microparasites and macroparasites. **Microparasites** include viruses, bacteria, and protozoans. They are characterized by small size and a short generation time. They develop and multiply rapidly within the host and are the class of parasites that we typically associate with the term *disease*. The infection generally lasts a short time relative to the host's expected life span. Transmission from host to host is most often direct, although other species may serve as carriers or vectors.

Macroparasites are relatively large. Examples include flatworms, acanthocephalans, roundworms, flukes, lice, fleas, ticks, fungi, rusts, and smuts. Macroparasites have a comparatively long generation time and typically do not complete an entire life cycle in a single host organism. They may spread by direct transmission from host to host or by indirect transmission, involving intermediate hosts and carriers.

Although the term *parasite* is most often associated with heterotrophic organisms such as animals, bacteria, and fungi, over 4000 species of parasitic plants derive some or all of their sustenance from another plant. Parasitic plants have a modified root—the haustorium—that penetrates the host plant and connects to the vascular tissues (xylem and/or phloem). Parasitic plants may be classified as holoparasites or hemiparasites based on whether they carry out the process of photosynthesis. Hemiparasites, such as most species of mistletoe (Figure 15.1), are photosynthetic plants that contain chlorophyll when mature and obtain water, with its dissolved nutrients, by connecting to the host xylem. Holoparasites, such as broomrape and dodder (Figure 15.2), lack chlorophyll and are thus nonphotosynthetic. These plants function as heterotrophs that rely totally on the host's xylem and phloem for carbon, water, and other essential nutrients.

Figure 15.1 | Mistletoe (*Phoradendron* spp.) with fruits on European birch (*Betula* spp.). Mistletoes are hemiparasites. Although capable of photosynthesis, they penetrate the host tree, extracting water and nutrients.

Parasites are extremely important in interspecific relations. In contrast with the species interactions of competition and predation, however, it was not until the late 1960s that ecologists began to appreciate the role of parasitism in population dynamics and community structure. Parasites have dramatic effects when they are introduced into host populations that have not evolved defenses against them. In such cases, diseases sweep through and decimate the population.

15.2 | Hosts Provide Diverse Habitats for Parasites

Hosts are the habitats for parasites, and the diverse arrays of parasites that have evolved exploit every conceivable habitat on and within their hosts. Parasites that live on the host's skin, within the protective cover of feathers and

Figure 15.2 | Squawroot (*Conopholis americana*), a member of the broomrape family, is a holoparasite on the roots of oak.

hair, are **ectoparasites.** Others, known as **endoparasites,** live within the host. Some burrow beneath the skin. They live in the bloodstream, heart, brain, digestive tract, liver, spleen, mucosal lining of the stomach, spinal cord, nasal tract, lungs, gonads, bladder, pancreas, eyes, gills of fish, muscle tissue, or other sites. Parasites of insects live on the legs, on the upper and lower body surfaces, and even on the mouthparts.

Parasites of plants also divide up the habitat. Some live on the roots and stems; others penetrate the roots and bark to live in the woody tissue beneath. Some live at the root collar, commonly called a crown, where the plants emerge from the soil. Others live within the leaves, on young leaves, on mature leaves, or on flowers, pollen, or fruits. A major problem for parasites, especially parasites of animals, is gaining access to and escaping from the host. Parasites can enter and exit host animals through various pathways including the mouth, nasal passages, skin, rectum, and urogenital system; they travel to their point of infection through the pulmonary, circulatory, or digestive systems.

For parasites, host organisms are like islands that eventually disappear (die). Because the host serves as a habitat enabling their survival and reproduction, parasites have to escape from one host and locate another—something they cannot do at will. Endo-macroparasites can escape only during a larval stage of their development, known as the infective stage, when they must make contact with the next host. The process of transmission from one host to another can occur either directly or indirectly and can involve adaptations by parasites to virtually all aspects of feeding, social, and mating behaviors in host species.

15.3 | Direct Transmission Can Occur between Host Organisms

Direct transmission occurs when a parasite is transferred from one host to another without the involvement of an intermediate organism. The transmission can occur by direct contact with a carrier, or the parasite can be dispersed from one host to another through the air, water, or other substrate. Microparasites are more often transmitted directly, as in the case of influenza (airborne) and smallpox (direct contact) viruses and the variety of bacterial and viral parasites associated with sexually transmitted diseases.

Many important macroparasites of animals and plants also move from infected to uninfected hosts by direct transmission. Among internal parasites, the roundworms (*Ascaris*) live in the digestive tracts of mammals. Female roundworms lay thousands of eggs in the host's gut that are expelled with the feces, where they are dispersed to the surrounding environment (water, soil, ground vegetation). If they are swallowed by a host of the correct species, the eggs hatch in the host's intestines, and the larvae bore their way into the blood vessels and come to rest in the lungs. From there they ascend to the mouth, usually by causing the host to cough, and are swallowed again to reach the stomach, where they mature and enter the intestines.

The most important debilitating external parasites of birds and mammals are spread by direct contact. They include lice, ticks, fleas, botfly larvae, and mites that cause mange. Many of these parasites lay their eggs directly on the host; but fleas lay their eggs and their larvae hatch in the host's nests and bedding, from which they leap onto nearby hosts.

Some parasitic plants also spread by direct transmission—notably those classified as holoparasites, such as members of the broomrape family (Orobanchaceae). Two examples are squawroot (*Conopholis americana*), which parasitizes the roots of oaks (see Figure 15.2), and beech-drops (*Epifagus virginiana*), which parasitizes mostly the roots of beech trees. Seeds of these plants are dispersed locally; upon germination, their roots extend through the soil and attach to the roots of the host plant.

Some fungal parasites of plants spread through root grafts. For example, *Fomes annosus*, an important fungal infection of white pine (*Pinus strobus*), spreads rapidly through pure stands of the tree when roots of one tree grow onto (and become attached to) the roots of a neighbor.

15.4 | Transmission between Hosts Can Involve an Intermediate Vector

Some parasites are transmitted between hosts by an intermediate organism, or vector. For example, the black-legged tick (*Ixodes scapularis*), functions as an arthropod vector in the transmission of Lyme disease, the major arthropod-borne disease in the United States. Named for its first noted occurrence at Lyme, Connecticut, in 1975, the disease is caused by a bacterial spirochete, *Borrelia burgdorferi*. It lives in the bloodstream of vertebrates, from birds and mice to deer and humans. The spirochete depends on the tick for transmission from one host to another.

Malaria parasites infect a wide variety of vertebrate species, including humans. The four species of protozoan parasites (*Plasmodium*) that cause malaria in humans are transmitted to the bloodstream by the bite of an infected female mosquito of the genus *Anopheles* (Figure 15.3). Mosquitoes are known to transmit more than 50 percent of the approximately 102 arboviruses (a contraction of "arthropod-borne viruses") that can produce disease in humans, including dengue and yellow fever.

Insect vectors are also involved in the transmission of parasites among plants. European and native elm bark beetles (*Scolytus multistriatus* and *Hylurgopinus rufipes*) carry spores of the fungi *Ophiostoma ulmi* that spreads the devastating Dutch elm disease from tree to tree. Mistletoes (*Phoradendron* spp.) belong to a group of plant parasites known as hemiparasites (see Figure 15.1) that, although photosynthetic, draw water and nutrients from

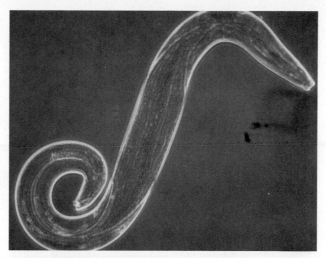

(a)

(b)

Figure 15.3 | Malaria is a recurring infection produced in humans by protozoan parasites **(a)** transmitted by the bite of an infected female mosquito of the genus *Anopheles* **(b)**. Today, more than 40 percent of the world's population is at risk, and more than 1 million are killed each year by malaria.

their host plant. Transmission of mistletoes between host plants is linked to seed dispersal. Birds feed on the mistletoe fruits. The seeds pass through the digestive system unharmed and are deposited on trees where the birds perch and defecate. The sticky seeds attach to limbs and send out rootlets that embrace the limb and enter the sapwood.

15.5 | Transmission Can Involve Multiple Hosts and Stages

In Chapter 10, we introduced the concept of life cycle—the phases associated with the development of an organism, typically divided into juvenile (or prereproductive), reproductive, and postreproductive. Some species of parasites cannot complete their entire life cycle in a single

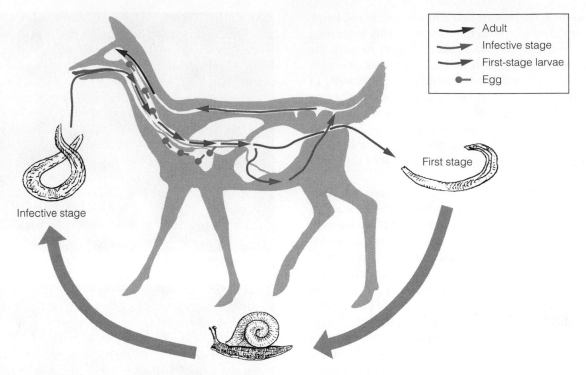

Figure 15.4 | The life cycle of a macroparasite, the meningeal worm *Parelaphostrongylus tenuis*, which infects white-tailed deer, moose, and elk. Transmission is indirect, involving snails as an intermediate host.

(Adapted from Anderson 1963.)

host species. The host species in which the parasite becomes an adult and reaches maturity is referred to as the **definitive host.** All others are **intermediate hosts,** which harbor some developmental phase. Parasites may require one, two, or even three intermediate hosts. Each stage can develop only if the parasite can be transmitted to the appropriate intermediate host. Thus, the dynamics of a parasite population are closely tied to the population dynamics, movement patterns, and interactions of the various host species.

Many parasites, both plant and animal, use this form of indirect transmission and spend different stages of the life cycle with different host species. Figure 15.4 shows the life cycle of the meningeal worm (*Parelaphostrongylus tenuis*), a parasite of the white-tailed deer in eastern North America. Snails or slugs that live in the grass serve as the intermediate host species for the larval stage of the worm. The deer picks up the infected snail while grazing. In the deer's stomach, the larvae leave the snail, puncture the deer's stomach wall, enter the abdominal membranes, and travel via the spinal cord to reach spaces surrounding the brain. Here, the worms mate and produce eggs. Eggs and larvae pass through the bloodstream to the lungs, where the larvae break into air sacs and are coughed up, swallowed, and passed out with the feces. The snails acquire the larvae as they come into contact with the deer feces on the ground. Once within the snail, the larvae continue to develop to the infective stage.

15.6 | Hosts Respond to Parasitic Invasions

Just as the coevolution of predators and prey has resulted in the adaptation of defense mechanisms by prey species, likewise host species exhibit a range of adaptations that minimize the impact of parasites. Some responses are mechanisms that reduce parasitic invasion. Other defense mechanisms aim to combat parasitic infection once it has occurred.

Some defensive mechanisms are behavioral, aimed at avoiding infection. Birds and mammals rid themselves of ectoparasites by grooming. Among birds, the major form of grooming is preening, which involves manipulating plumage with the bill and scratching with the foot. Both activities remove adults and nymphs of lice from the plumage. Deer seek dense, shaded places where they can avoid deerflies, which are common to open areas.

If infection should occur, the first line of defense involves the inflammatory response. The death or destruction (injury) of host cells stimulates the secretion of histamines (chemical alarm signals), which induce increased blood flow to the site and cause inflammation. This reaction brings in white blood cells and associated cells that directly attack the infection. Scabs can form on the skin, reducing points of further entry. Internal reactions can produce hardened cysts in muscle or skin that enclose and isolate the parasite. An example is the cysts that

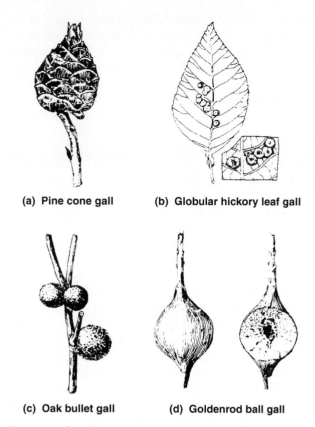

(a) Pine cone gall

(b) Globular hickory leaf gall

(c) Oak bullet gall

(d) Goldenrod ball gall

Figure 15.5 | Galls are a growth response to an alien substance in plant tissues. In this case, the presence of a parasitic egg stimulates a genetic transformation of the host's cells. **(a)** The pine cone gall, a bud gall on willows caused by the gall midge *Rhabdophaga strobiloides*. **(b)** The hickory leaf gall, induced by the gall aphid *Phylloxera caryaeglobuli*. **(c)** The oak bullet gall, caused by the gall wasp *Disholcaspis globulus*. **(d)** The goldenrod ball gall, a stem gall induced by a gallfly, *Eurosta solidaginis*.

encase the roundworm *Trichinella spiralis* (Nematoda) in the muscles of pigs and bears and that cause trichinosis when ingested by humans in undercooked pork.

Plants respond to bacterial and fungal invasion by forming cysts in the roots and scabs in the fruits and roots, cutting off contact by the fungus with healthy tissue. Plants react to attacks on leaf, stem, fruit, and seed by gall wasps, bees, and flies by forming abnormal growth structures unique to the particular gall insect (Figure 15.5). Gall formation exposes the larvae of some gall parasites to predation. For example, John Confer and Peter Paicos of Ithaca College (New York) reported that the conspicuous, swollen knobs of the goldenrod ball gall (Figure 15.5d) attract the downy woodpecker (*Picoides pubescens*), which excavates and eats the larva within the gall.

The second line of defense is the immune response (or immune system). When a foreign object such as a virus or bacteria, termed an *antigen* (a contraction of "antibody-generating"), enters the bloodstream, it elicits an immune response. White cells called lymphocytes

(produced by lymph glands) produce antibodies. The antibodies target the antigens present on the parasite's surface or released into the host, helping to counter their effects. These antibodies are energetically expensive to produce. They also are potentially damaging to the host's own tissues. Fortunately, the immune response does not have to kill the parasite to be effective. It only has to reduce the feeding, movements, and reproduction of the parasite to a tolerable level. The immune system is extremely specific, and it has a remarkable "memory." It can "remember" antigens it has encountered in the past and react to them more quickly and vigorously in subsequent exposures.

The immune response, however, can be breached. Some parasites vary their antigens more or less continuously. By doing so, they are able to keep one jump ahead of the host's response. The result is a chronic infection of the parasite in the host. Antibodies specific to an infection normally are composed of proteins. If the animal suffers from poor nutrition and its protein deficiency is severe, normal production of antibodies is inhibited. Depletion of energy reserves breaks down the immune system and allows viruses or other parasites to become pathogenic. The ultimate breakdown in the immune system occurs in humans infected with the human immunodeficiency virus (HIV), the causal agent of AIDS, which is transmitted sexually, through the use of shared needles, or by infected donor blood. The virus attacks the immune system itself, exposing the host to a range of infections that prove fatal.

15.7 | Parasites Can Affect Host Survival and Reproduction

Although host organisms exhibit a wide variety of defense mechanisms to prevent, reduce, or combat parasitic infection, all share the common feature of requiring resources that the host might otherwise have used for some other function. Given that organisms have a limited amount of energy, it is not surprising that parasitic infections function to reduce both growth and reproduction. J. J. Schall of the University of Vermont examined the impact of malaria on the western fence lizard (*Sceloporus occidentalis*) inhabiting California. Clutch size (number of eggs produced) is approximately 20 percent smaller in malaria-infected females compared with noninfected individuals. Reproduction is reduced because infected females are less able to store fat during the summer, so they have less energy for egg production the following spring.

Parasitic infection can reduce the reproductive success of males. Females of many species choose mates based on the secondary sex characteristics, such as bright and ornate plumage of male birds (see discussion of intrasexual selection in Chapter 8). Full expression of these characteristics can be limited by parasite infection, thus reducing the male's ability to successfully attract a mate.

(a)

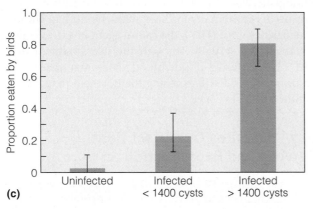

(b)

(c)

Figure 15.6 | Infection of the California killifish **(a)** by the trematode parasite causes abnormal behavior that increases the susceptibility of individuals to predation. **(b)** The frequency of conspicuous behaviors each fish displayed within a 30-minute period (*y*-axis) in relation to the intensity of parasitism (number of cysts per fish brain). In the parasitized population (squares), the number of conspicuous behaviors increased with parasite intensity. All unparasitized fish (circles) had smaller numbers of conspicuous behaviors that parasitized fish. **(c)** A comparison of the proportion of fish eaten by birds after 20 days (*y*-axis), showing that heavily parasitized fish were preyed on more frequently than unparasitized fish. Vertical lines in bars represent 95 percent confidence intervals.

(Adapted from Lafferty and Morris 1996.)

For example, the bright red color of the male zebra finch's beak depends on its level of carotenoid pigments, the naturally occurring chemicals that are responsible for the red, yellow, and orange coloration patterns in animals as well as in foods such as carrots. Birds cannot synthesize carotenoids and must obtain them through the diet. Besides being colorful pigments, carotenoids stimu-

late the production of antibodies and absorb some of the damaging free radicals that arise during the immune response. In a series of laboratory experiments, Jonathan Blount and colleagues from the University of Glasgow (Scotland) found that only those males with the fewest parasites and diseases can devote sufficient carotenoids to producing bright red beaks and therefore succeed in attracting mates and reproducing.

Although most parasites do not kill their host organisms, increased mortality can result from a variety of indirect consequences of infection. One interesting example is when the infection alters the behavior of the host, increasing its susceptibility to predation. Rabbits infected with the bacterial disease tularemia (*Francisella tularensis*), transmitted by the rabbit tick (*Haemaphysalis leporispalustris*), are sluggish and thus more vulnerable to predation. In another example, ecologists Kevin Lafferty and Kimo Morris of the University of California, Santa Barbara, observed that killifish (*Fundulus parvipinnis*; Figure 15.6a) parasitized by trematodes (flukes) display abnormal behavior such as surfacing and jerking. In a comparison of parasitized and unparasitized populations, the scientists found that the frequency of conspicuous behaviors displayed by individual fish is related to the intensity of parasitism (Figure 15.6b). The abnormal behavior of the infected killifish attracts fish-eating birds. Lafferty and Morris found that heavily parasitized fish were preyed upon more frequently than unparasitized individuals (Figure 15.6c). Interestingly, the fish-eating birds represent the trematodes' definitive host, so that by altering its intermediate host's (the killifish's) behavior, making it more susceptible to predation, the trematode ensures the completion of its life cycle.

15.8 | Parasites May Regulate Host Populations

For parasite and host to coexist under a relationship that is hardly benign, the host needs to resist invasion by eliminating the parasites or at least minimizing their effects. In most circumstances, natural selection has resulted in a level of immune response where the allocation of metabolic resources by the host species minimizes the cost of parasitism yet does not unduly impair its own growth and reproduction. Conversely, the parasite gains no advantage if it kills its host. A dead host means dead parasites. The conventional wisdom about host–parasite evolution is that virulence is selected against, so that parasites become less harmful to their hosts and thus persist. Does natural selection work this way in parasite–host systems?

Natural selection does not necessarily favor peaceful coexistence of hosts and parasites. To maximize fitness, a parasite should balance the trade-off between virulence and other components of fitness such as transmissibility. Natural selection may yield deadly (high virulence) or

Humans have always been bedeviled by parasites, but more so in recent stages of human history (see Part Eight Introduction). During our first 2 million years as hunter-gatherers, the most bothersome parasites were macroparasites such as roundworms, directly transmitted. Only microparasites with high transmission rates that produced no immunity could persist in such small groups of hosts.

Once humans became sedentary agriculturalists aggregated in villages, however, populations became large enough to sustain bacterial and viral parasites. Many of these parasites evolved from those causing diseases in domestic animals. Measles, for instance, evolved from canine distemper. Populations were too small at first to support disease continuously without reinfection from some neighboring settlement. Once a settlement grew into a sufficiently large city, the population was dense enough to maintain a reservoir of infection. As commerce developed between cities, people and goods began to move long distances. They introduced diseases from one part of the world to another where the populations lacked immunity. Periodic epidemics swept through the cities.

A classic example of the importation, direct transmission, and rapid spread of a disease is bubonic plague. Plague is caused by a rod-shaped bacillus, *Yersinia pestis*. It is transmitted from host to host mostly by the bite of its vector (fleas from infected rodents) and directly from person to person by mucus droplets spread by coughing. Infected individuals become ill in a few hours to a few days, showing symptoms of high fever and swollen lymph nodes. Death often follows within several days. Plague was called the Black Death because of the dark color on the faces of many dead victims.

Burrowing rodents are the reservoir of the bacillus. It is most closely associated with the black rat (*Rattus rattus*), a native of central India, also the original focal point of the disease. An agile climber, the black rat easily boarded cargo ships that carried it to port cities in Asia and the Mediterranean region. Hidden in the baggage of caravans, the black rat spread across the steppes of Asia, where undoubtedly it transferred the plague bacillus to burrowing rodents in the region. In 1331, an epidemic of the plague swept through China. Mongol armies carried the disease with them as they swept across Asia to the Mediterranean. At the siege of Caffa in 1346 on the Crimean peninsula, the Mongol army was devastated by

the plague and withdrew after catapulting the victims' corpses into town. Trade resumed, and ships carried infected black rats to ports of southern Europe.

Conditions were right for the spread of disease. Europe was experiencing huge population growth, climate was worsening, and crops were failing. By the end of December 1347, the disease that had decimated the Mongol army spread to Italy and southern France; by December 1348, it reached southern Germany and England; and by December 1350, it reached Scandinavia. Between 1348 and 1350, one-third of the European population, including entire villages, succumbed to the Black Death—upsetting the social, political, and economic stability of Europe. Later outbreaks occurred in 1630 in Milan, in 1665 in London, and in 1720–21 in Marseilles. Sporadic local outbreaks occurred throughout the world until 1944, when antibiotics quickly cured the disease if diagnosed early. Harbored by burrowing rodents, the plague bacillus still thrives worldwide—including North America.

Other diseases introduced into populations lacking immunity mimic the spread and devastation of the Black Death. Smallpox, measles, and typhus, carried to the New World by Spanish and English explorers and settlers, spread rampant through indigenous populations of South, Central, and North America. Disease devastated the Aztecs in Mexico and nearly exterminated the native American population of New England, allowing uncontested settlement of the region by the English. In more recent times, a massive flu epidemic (known as "Spanish Flu" or "La Grippe") spread worldwide in 1918 and killed 21 million people, including 500,000 people in the United States. Flu still remains a threat because of its high mutation rate. Strains evolve faster than resistance develops to the disease. As a result, flu returns in waves of different types.

Because many of the old plagues, like measles, have been contained by vaccinations and other health measures, many of us are complacent about disease. Even so, new diseases—Ebola in Zaire and AIDS everywhere—warn us that plagues are still with us. New mutant forms of old diseases such as tuberculosis, once considered conquered—combined with a massive increase in world population, changing global climate, and rapid transcontinental movement of people and goods—set the stage for future plagues.

benign (low virulence) parasites depending on the requirements for parasite reproduction and transmission. For example, the term *vertical transmission* is used to describe parasites transmitted directly from the mother to the offspring during the perinatal period (the period

immediately before or after birth). Typically, parasites that depend on this mode of transmission cannot be as virulent as those transmitted through other forms of direct contact between adult individuals, because the recipient (host) must survive until reproductive maturity to pass on the parasite. The host's condition is important to a parasite only as it relates to the parasite's reproduction and transmission. If the host species did not evolve, the parasite might well be able to achieve some optimal balance of host exploitation. But just as with the coevolution of predator and prey (see "Red Queen Hypothesis," Section 14.9), host species do evolve. The result is an "arms race" between parasite and host.

Parasites can have the effect of decreasing reproduction and increasing the probability of host mortality, but few studies have quantified the effect of a parasite on the dynamics of a particular plant or animal population under natural conditions. Parasitism can have a debilitating effect on host populations, a fact that is most evident when parasites invade a population with no evolved defenses (see Ecological Issues: Plagues Upon Us). In such cases, the spread of disease may be virtually density independent, reducing populations, exterminating them locally, or restricting distribution of the host species. The chestnut blight (*Endothia parasitica*), introduced into North America from Europe, nearly exterminated the American chestnut (*Castanea dentata*) and removed it as a major component of the forests of eastern North America. Dutch elm disease, caused by a fungus (*Ophiostoma ulmi*) spread by beetles, has nearly removed the American elm (*Ulmus americana*) from North America and the English elm (*U. glabra*) from Great Britain. Anthracnose (*Discula destructiva*), a fungal disease, is decimating flowering dogwood (*Cornus florida*), an important understory tree in the forests of eastern North America.

Rinderpest, a viral disease of domestic cattle, was introduced to East Africa in the late 19th century and subsequently decimated herds of African buffalo (*Syncerus caffer*) and wildebeest (*Connochaetes taurinus*). Avian malaria carried by introduced mosquitoes has eliminated most native Hawaiian birds below 1000 m, the altitude above which the mosquito cannot persist.

On the other hand, parasites may function as density-dependent regulators on host populations. Such incidents typically occur with directly transmitted endemic (native) parasites that are maintained in the population by a small reservoir of infected carrier individuals. Outbreaks of these diseases appear to occur when the host population density is high; and they tend to reduce host populations sharply, resulting in population cycles of host and parasite similar to those observed for predator and prey (see Section 14.2). Examples are distemper in raccoons and rabies in foxes, both significant diseases in controlling their host populations.

In other cases, the parasite may function as a selective agent of mortality, infecting only a subset of the population. Distribution of macroparasites, especially those with indirect transmission, is highly clumped. Some individuals in the host population carry a higher load of parasites than others do (Figure 15.7). These in-

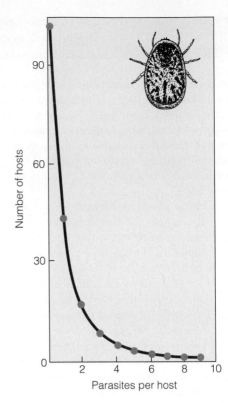

Figure 15.7 | Clumped distribution of the tick *Ixodes trianguliceps* (Birula) on a population of the European field mouse, *Apodemus sylvaticus*. Most individuals in the host population carry no ticks. A few individuals carry most of the parasite load. (Adapted from Randolph 1975.)

dividuals are most likely to succumb to parasite-induced mortality, suffer reduced reproductive rates, or both. Such deaths often are caused not directly by the macroparasites, but indirectly by secondary infection. In a study of reproduction, survival, and mortality of bighorn sheep (*Ovis canadensis*) in south-central Colorado, Thomas Woodard and colleagues at Colorado State University found that individuals may be infected with up to seven different species of lungworms (Nematoda). Highest infections occur in the spring when the lambs are born. Heavy lungworm infections in the young bring about a secondary infection—pneumonia—that kills the lambs. The researchers found that such infections can sharply reduce mountain sheep populations by reducing reproductive success.

15.9 | Parasitism Can Evolve into a Mutually Beneficial Relationship

Parasites and their hosts live together in a symbiotic relationship in which the parasite derives its benefit (habitat and food resources) at the expense of the host organism. Hosts have evolved a variety of defenses to minimize the negative impact of the parasite's presence. In the situation where adaptations have countered the negative impacts, the relationship may be termed **commensalism**— a relationship between two species in which one species benefits without significantly affecting the other. At some stage in host–parasite coevolution, the relationship may become beneficial to both species. For example, a host tolerant of parasitic infection may begin to exploit the relationship. At that point the relationship is termed **mutualism.** There are many examples of "parasitic relationships" in which there is an apparent benefit to the host organism. For example, rats infected with the intermediate stages of the tapeworm *Spirometra* grow larger than uninfected rats do. The tapeworm larva produces an analogue of vertebrate growth hormone. In this example, is the increased growth beneficial or harmful to the host? Similarly, many mollusks, when infected with the intermediate stages of digenetic flukes (Digenea), develop thicker, heavier shells that could be deemed an advantage. Some of the clearest examples of evolution from parasites to mutualists are parasites that are transmitted vertically from mother to offspring (see discussion in Section 15.8). Theory predicts that vertically transmitted parasites will be selected to increase host survival and reproduction because maximization of host reproductive success benefits both the parasite and host. This prediction has been supported by studies examining the effects of *Wolbachia*, a common group of bacteria that infect the reproductive tissues of arthropods. Investigations of the effects of *Wolbachia* on host fitness in the wasp *Nasonia vitripennis* have shown that infection increases host fitness and that infected females produce more offspring than do uninfected females. Similar increases in fitness have been reported for natural populations of *Drosophila*.

Mutualism is a relationship between members of two species that benefits both. Out of this relationship, individuals of both species enhance their survival, growth, or reproduction. Evidence, however, suggests that often this interaction is more a reciprocal exploitation than a cooperative effort between individuals. Many classic examples of mutualistic associations appear to have evolved from species interactions that previously reflected host–parasite or predator–prey interactions. In many cases of apparent mutualisms, the benefits of the interaction for one or both of the participating species may be environment dependent. For example, many tree species have associated with their roots fungal mycorrhizae (see Section 15.11). The fungi obtain organic nutrients from the plant via the phloem, and in nutrient-poor soil the trees seem to benefit by increased nutrient uptake, particularly phosphate by the fungus. In nutrient-rich soils, however, the fungi appear to be a net cost rather than benefit; and this seemingly mutualistic association appears much more like a parasitic invasion by the fungus. Depending on external conditions, the association switches between mutualism and parasitism.

15.10 | Mutualisms Involve Diverse Species Interactions

Mutualistic relationships involve many diverse interactions that extend beyond simply acquiring essential resources. Thus it is important to consider the different attributes of mutualistic relationships and how they affect the dynamics of the populations involved. Mutualisms can be characterized by a number of variables: the benefits received, the degree of dependency, the degree of specificity, and the duration of the intimacy.

Mutualism is defined as an interaction between members of two species that serves to benefit both parties involved, and the benefits received can include a wide variety of processes. Benefits may include provision of essential resources such as nutrients or shelter (habitat) and may involve protection from predators, parasites, and herbivores; or, they may reduce competition with a third species. Finally, the benefits may involve reproduction, such as dispersal of gametes or zygotes.

Mutualisms also vary in how much the species involved in the mutualistic interaction depend upon each other. Obligate mutualists cannot survive or reproduce without the mutualistic interaction, whereas facultative mutualists can. In addition, the degree of specificity of mutualism varies from one interaction to another, ranging from one-to-one, species-specific associations (termed specialists) to association with a wide diversity of mutualistic partners (generalists). The duration of intimacy in the association also varies among mutualistic interactions. Some mutualists are symbiotic, while others are free living (nonsymbiotic). In symbiotic mutualism, individuals coexist and their relationship is more often obligatory—at least one member of the pair becomes totally dependent on the other. Some forms of mutualism are so permanent and obligatory that the distinction between the two interacting organisms becomes blurred. Reef-forming corals of the tropical waters provide an example. These corals secrete an external skeleton composed of calcium carbonate. The individual coral animals, called *polyps*, occupy little cups, or corallites, in the larger skeleton that forms the reef (Figure 15.8 on p. 320). These corals have in their tissues single-celled, symbiotic algae called zooxanthellae. Although the coral polyps are carnivores, feeding on zooplankton suspended in the surrounding water, they acquire only about 10 percent of their daily energy requirement from zooplankton. They obtain the remaining 90 percent of their energy from carbon produced by the symbiotic algae through photosynthesis. Without the algae, these corals would not be able to survive and flourish in their nutrient-poor environment (see Field Studies: John J. Stachowicz). In turn, the coral provides the algae with shelter and mineral nutrients, particularly nitrogen in the form of nitrogenous wastes.

Section of Evolution and Ecology, Center for Population Biology, University of California, Davis, California

Facilitative, or positive, interactions are encounters between organisms that benefit at least one of the participants and cause harm to neither. Such interactions are considered "mutualisms" when both species derive benefit from the interaction. Ecologists have long recognized the existence of mutualistic interactions, but there is still far less research on positive interactions than on competition and predation. Now, however, ecologists are beginning to appreciate the ubiquitous nature of positive interactions and their importance in affecting populations and in the structuring of communities. The research of marine ecologist John Stachowicz has been at the center of this growing appreciation for the importance of facilitation.

Stachowicz works in the shallow-water coastal ecosystems of the southeastern United States. The large colonial corals and calcified algae that occupy the warm subtropical waters of this region provide a habitat for a diverse array of invertebrate and vertebrate species. In well-lit habitats, corals and calcified algae (referred to as coralline algae) grow slowly relative to the fleshy species of seaweed. The persistence of corals appears to be linked to the high abundance of herbivores that suppress the growth of the seaweeds, which grow on and over the coral and coralline algae and eventually cause their death. In contrast, the relative cover of corals is generally low in habitats such as reef flats and seagrass beds, where herbivory is less intense.

Stachowicz hypothesized that mutualism plays an influential role in the distribution of coral species. Although corals are typically associated with the colorful and diverse coral reef ecosystems of the tropical and subtropical coastal waters, many temperate and subarctic habitats support corals, and some tropical species occur where temperatures drop to 10°C or lower for certain months of the year. One such species is the coral *Oculina arbuscula*.

Oculina arbuscula occurs as far north as the coastal waters of North Carolina, forming dense aggregations in poorly lit habitats where seaweeds are rare or absent. In certain areas of the coastal waters, however, *O. arbuscula* does co-occur with seaweeds on natural and artificial reefs. It is the only coral in this region with a structurally complex branching morphology that provides shelter for a species-rich epifauna. More than 300 species of invertebrates are known to live among the branches of *Oculina* colonies.

How can *O. arbuscula* persist in these well-lit, shallow-water systems? In well-lit habitats, corals grow slowly relative to seaweeds, and the persistence of coral reefs appears to be tightly linked to high abundance of herbivores that

prevent seaweed from growing on and over the corals. When herbivorous fish or sea urchins are naturally or experimentally removed from tropical reefs, seaweed biomass increases dramatically and corals are smothered. In contrast, on the temperate reefs of North Carolina, herbivorous fish are less abundant than in the tropics, and the standing biomass of seaweed is typically much higher. On these reefs, herbivorous fish and urchins also alter the species composition of the seaweed community by selectively removing their preferred species; but they do not diminish the total seaweed biomass. The dependence of corals on positive interactions with herbivores may thus explain why corals are generally uncommon in temperate latitudes.

Stachowicz suspected the role of a key herbivore in these temperate reef ecosystems: the herbivorous crab *Mithrax forceps*. He hypothesized that the success of *O. arbuscula* on temperate reefs is derived from its ability to harbor symbiotic, herbivorous crabs that mediate competition with encroaching seaweeds. To evaluate the hypothesis, he conducted field experiments monitoring the fouling (overgrowth by seaweeds) and growth of corals in the presence and absence of crabs. Experiments were located at Radio Island Jetty near Beaufort, North Carolina.

In these experiments, metal stakes were driven into substrate, and one coral (which had previously been weighed) was fastened to each stake. A single crab was then placed on a subset of the corals, and the remainder was left vacant. At the end of the experiment, all seaweed (and other epiphytic growth) was removed from the

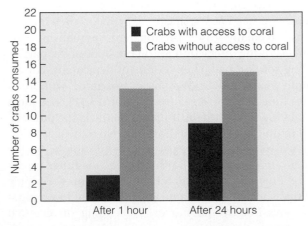

Figure 1 | Predation on *M. forceps* with and without access to corals.

(Adapted from Stachowicz 1999.)

corals, dried, and weighed. After removal of the seaweeds, the corals were reweighed to measure growth.

To determine if association with *O. arbuscula* reduced predation on *M. forceps*, Stachowicz tethered crabs both with and without access to coral. He checked each tether after 1 and 24 hours to see if crabs were still present.

The experiment results clearly demonstrated a mutually beneficial association between *O. arbuscula* and *M. forceps*. The coral shelters the crab from predators (Figure 1); in turn, the crab defends the coral from overgrowth by encroaching competitors, thus enhancing coral growth and survivorship (Figure 2).

Interactions between *O. arbuscula* and *M. forceps* have population and community implications extending beyond these two species. The crab directly alters the benthic community, enhancing the growth and survival of its host and ensuring the persistence of the diverse community associated with the coral structure.

Bibliography

Stachowicz, J. 2001. Mutualism, facilitation, and the structure of ecological communities. *BioScience* 51:235–246.

Stachowicz, J., and M. Hay. 1996. Facultative mutualism between an herbivorous crab and a coralline alga: Advantages of eating noxious seaweeds. *Oecologia* 105:337–387.

Stachowicz, J., and M. Hay. 1999. Mutualism and coral persistence: The role of herbivore resistance to algal chemical defense. *Ecology* 80:2085–2101.

1. Would you classify the relationship between corals and crabs described here as an example of facilitation or obligate mutualism? Can you determine this based on the information provided? If not, what additional information would you need?

2. In the results presented in Figures 1 and 2, has Stachowicz demonstrated that the mutualistic interactions directly influence the fitness of the two species involved?

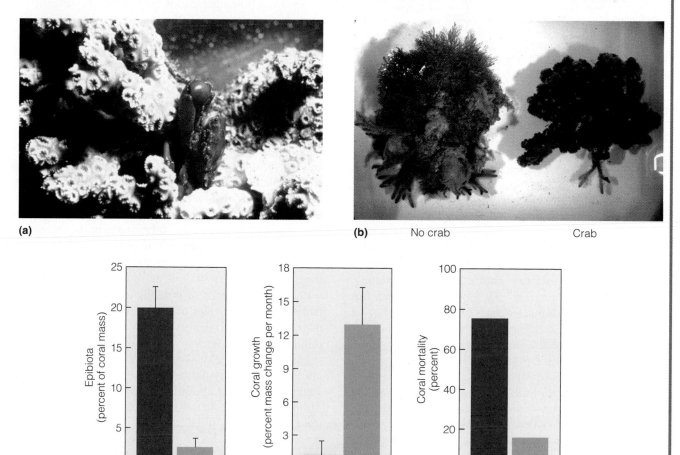

Figure 2 | (a) The crab *Mithrax forceps* hiding within the branches of the coral *Oculina arbuscula*.
(b) When corals are grown in the field without crabs, they are smothered by a dense cover of seaweed.
By reducing overgrowth (c), the crab increases coral growth (d) and survival (e).

(Adapted from Stachowicz 2001.)

(a)

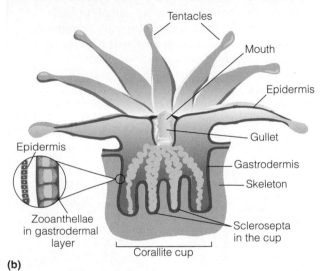

Tentacles

Mouth

Epidermis

Epidermis

Gullet

Gastrodermis

Skeleton

Zooanthellae
in gastrodermal
layer

Sclerosepta
in the cup

Corallite cup

(b)

Figure 15.8 | **(a)** Photograph showing individual polyps of the Great Star (*Montastrea cavernosa*) coral. **(b)** Anatomy of a coral polyp, showing the location of the symbiotic zooanthellae.

Figure 15.9 | Lichen consists of a fungus and an alga combined within a spongy body called a *thallus*.

two species, but rather involve a variety of plants, pollinators, and seed dispersers.

In the following sections we explore the diversity of mutualistic interactions. The discussion centers on the benefits derived by mutualists: acquisition of energy and nutrients, protection and defense, and reproduction and dispersal.

15.11 | Mutualisms Are Involved in the Transfer of Nutrients

The digestive system of herbivores is inhabited by a diverse community of mutualistic organisms that play a crucial role in the digestion of plant materials. The chambers of a ruminant's stomach (see Section 7.2) contain large populations of bacteria and protozoa that carry out the process of fermentation. Inhabitants of the rumen are primarily anaerobic, adapted to this peculiar environment. Ruminants are perhaps the best studied, but not the only, example of the role of mutualism in animal nutrition. The stomachs of virtually all herbivorous mammals and some species of birds and lizards rely on the presence of a complex microbial community to digest cellulose in plant tissues.

Mutualistic interactions are also involved in the uptake of nutrients by plants. Nitrogen is an essential constituent of protein, a building block of all living material. Although nitrogen is the most abundant constituent of the atmosphere, approximately 79 percent in its gaseous state, it is unavailable to most life. It must first be converted into a chemically usable form. One group of organisms that can use gaseous nitrogen (N_2) are the nitrogen-fixing bacteria of the genus *Rhizobium*. These bacteria (called rhizobia) are widely distributed in the soil, where they can grow and multiply. But in this free-living state, they do not fix nitrogen. Legumes—a group of

Lichens are involved in a symbiotic association in which the fusion of mutualists has made it even more difficult to distinguish the nature of the individual. Lichens (Figure 15.9) consist of a fungus and an alga (or in some cases cyanobacterium) combined within a spongy body called a *thallus*. The alga supplies food to both organisms, and the fungus protects the alga from harmful light intensities, produces a substance that accelerates photosynthesis in the alga, and absorbs and retains water and nutrients for both organisms. There are about 25,000 known species of lichens, each composed of a unique combination of fungus and alga.

In nonsymbiotic mutualism, the two organisms do not physically coexist; yet they depend on each other for some essential function. Although nonsymbiotic mutualisms may be obligatory, most are not. Rather, they are facultative, representing a form of mutual facilitation. Pollination in flowering plants and seed dispersal are examples. These interactions are generally not confined to

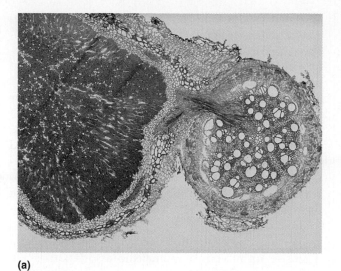

(a)

(b)

Figure 15.10 | Nitrogen-fixing *Rhizobium* bacteria (a) infect the plant roots, thus forming nodules (b). Bacteria derive carbon from the plant and in return provide the plant with nitrogen.

plant species that include clover, beans, and peas—attract the bacteria through the release of exudates and enzymes from the roots. Rhizobia enter the root hairs, where they multiply and increase in size. This invasion and growth results in swollen, infected root hair cells, which form root nodules (Figure 15.10). Once infected, rhizobia within the root cells reduce gaseous nitrogen to ammonia (a process referred to as nitrogen fixation). The bacteria receive carbon and other resources from the host plant; in return, the bacteria contribute fixed nitrogen to the plant, allowing it to function and grow independently of the availability of mineral (inorganic) nitrogen in the soil (see Chapter 6, Section 6.12).

Another example of a symbiotic relationship involving plant nutrition is the relationship between plant roots and mycorrhizal fungi. The fungi assist the plant with the uptake of nutrients and water from the soil. In return, the plant provides the fungi with carbon, a source of energy.

Endomycorrhizae have an extremely broad range of hosts; they have formed associations with over 70 percent of all plant species. Mycelia—masses of interwoven fungal filaments in the soil—infect the tree roots. They penetrate host cells to form a finely bunched network called an arbuscule (Figure 15.11a). The mycelia act as extended roots for the plant but do not change the shape or structure of the roots. They draw in nitrogen and

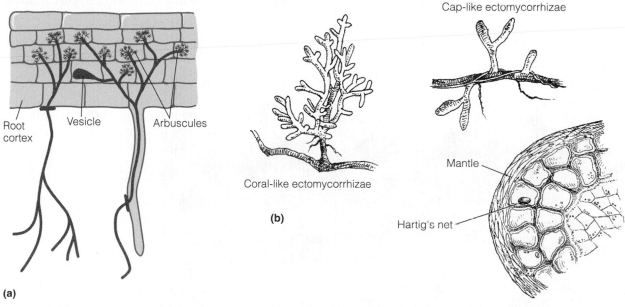

(a)

(b)

Figure 15.11 | (a) Endomycorrhizae grow within tree roots, and fungal hyphae enter the cells. (b) Ectomycorrhizae form a mantle of fungi about the tips of rootlets. Hyphae invade the tissues of rootlets between the cells. The network is called Hartig's net.

phosphorus at distances beyond those reached by the roots and root hairs. Another form, ectomycorrhizae, produces shortened, thickened roots that look like coral (Figure 15.11b). The threads of the fungi penetrate between the root cells. Outside the root, they develop into a network that functions as extended roots. Ectomycorrhizae have a more restricted range of hosts than do endomycorrhizae. They are associated with about 10 percent of plant families, and most of these species are woody.

Together, either ecto- or endomycorrhizae are found associated with the root systems of the vast majority of terrestrial plant species and are especially important in nutrient-poor soils. They aid in the decomposition of dead organic matter and the uptake of water and nutrients (see Sections 21.5 and 6.12 respectively), particularly nitrogen and phosphorus from the soil into the root tissue.

15.12 | Some Mutualisms Are Defensive

Other mutualistic associations involve defense of the host organism. A major problem for many livestock producers is the toxic effects of certain grasses, particularly perennial ryegrass and tall fescue. These grasses are infected by symbiotic endophytic fungi that live inside plant tissues (Figure 15.12). The fungi (Clavicipitaceae and Ascomycetes) produce alkaloid compounds in the tissue of the host grasses. The alkaloids, which impart a bitter taste to the grass, are toxic to grazing mammals, particularly domestic animals, and to a number of insect herbivores. In mammals, the alkaloids constrict small blood vessels in the brain, causing convulsions, tremors, stupor, gangrene of the extremities, and death. At the same time, these fungi seem to stimulate plant growth and seed production. This symbiotic relationship suggests a defensive mutualism between plant and fungi. The fungi defend the host plant against grazing. In return, the plant provides food to the fungi in the form of photosynthates (products of photosynthesis).

A group of Central American ant species (*Pseudomyrmex* spp.) that live in the swollen thorns of acacia (*Acacia* spp.) trees provides another example of defensive mutualism. Besides providing shelter, the plants supply a balanced and almost complete diet for all stages of ant development. In return, the ants protect the plants from herbivores. At the least disturbance, the ants swarm out of their shelters, emitting repulsive odors and attacking the intruder until it is driven away.

Perhaps one of the best-documented examples of a defensive or protective mutualistic association is the cleaning mutualism found in coral reef communities between cleaner shrimp or cleaner fishes and a large number of fish species. Cleaner fishes and shrimp obtain food by cleaning ectoparasites and diseased and dead tissue from the host fish (Figure 15.13a). In so doing, they benefit the host fish by removing harmful and unwanted materials.

Cleaning mutualism also occurs in terrestrial environments. The red-billed oxpecker (Figure 15.13b) of Africa is a bird that feeds almost exclusively by gleaning ticks and other parasites from the skin of large mammals such as antelope, buffalo, rhinoceros, or giraffe (also domestic cattle). It has always been assumed that these birds significantly reduce the number of ticks on the host animal, yet a recent study by ecologist Paul Weeks of Cambridge University brings into question whether this relationship is indeed mutualistic. In a series of field experiments, Weeks found that changes in adult tick load of cattle were unaffected by excluding the birds. In addition,

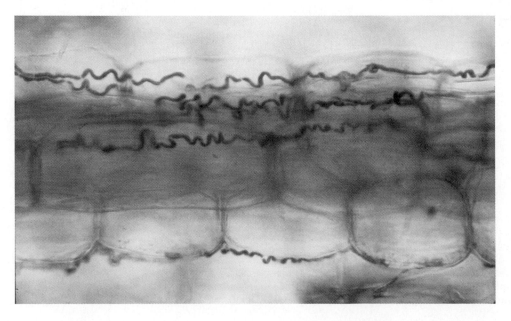

Figure 15.12 | Endophytic fungi in a blade of fescue (*Festuca*).

(a)

(b)

Figure 15.13 | Examples of cleaning mutualism. **(a)** Bluehead wrasse (small fish at eel's mouth) participating in cleaning symbiosis with a moray eel (*Muraenidae*). The cleaner fish obtains food by cleaning ectoparasites from the host fish. **(b)** The red-billed oxpecker of Africa feeds almost exclusively by gleaning ticks and other parasites from the skin of large mammals such as the impala shown here.

oxpeckers will peck a vulnerable area (often an ear) and drink blood when parasites are not available.

15.13 | Mutualisms Are Often Necessary for Pollination

The goal of cross-pollination is to transfer pollen from the anthers of one plant to the stigma of another plant of the same species (see Section 8.2). Some plants simply release their pollen in the wind. This method works well and costs little when plants grow in large homogeneous stands, such as grasses and pine trees often do. Wind dispersal can be unreliable, however, when individuals of the same species are scattered individually or in patches across a field or forest. In these circumstances, pollen transfer typically depends on insects, birds, and bats.

Plants entice certain animals by color, fragrances, and odors, dusting them with pollen and then rewarding them with a rich source of food: sugar-rich nectar, protein-rich pollen, and fat-rich oils. Providing such rewards is expensive for plants. Nectar and oils are of no value to the plant except as an attractant for potential pollinators.

They represent energy that the plant otherwise might expend in growth.

Nectivores (animals that feed on nectar) visit plants to exploit a source of food. While feeding, the nectivores inadvertently pick up pollen and carry it to the next plant they visit. With few exceptions, the nectivores are typically generalists that feed on many different plant species. Because each species flowers briefly, nectivores depend on a progression of flowering plants through the season.

Many species of plants, such as blackberries, elderberries, cherries, and goldenrods, are generalists themselves. They flower profusely and provide a glut of nectar that attracts a diversity of pollen-carrying insects, from bees and flies to beetles. Other plants are more selective, screening their visitors to ensure some efficiency in pollen transfer. These plants may have long corollas, allowing access only to insects and hummingbirds with long tongues and bills and keeping out small insects that eat nectar but do not carry pollen. Some plants have closed petals that only large bees can pry open. Orchids, whose individuals are scattered widely through their habitats, have evolved a variety of precise mechanisms for pollen transfer and

Figure 15.14 | Only males of a single wasp species (*Neozeleboria cryptoides*) pollinate the elbow orchid of southeastern Australia (*Chiloglottis trapeziformis*). A structure on the flower mimics the body of the smaller female wasp and emits an odor that imitates the female's pheromones (chemical signals). When a male wasp struggles to mate with the female decoy, pollen sticks to his body, which he then transfers to other elbow orchids.

reception so that pollen is not lost when the insect visits flowers of other species (Figure 15.14).

15.14 | Mutualisms Are Involved in Seed Dispersal

Plants with seeds too heavy to be dispersed by wind depend on animals to carry them some distance from the parent plant and deposit them in sites favorable for germination and seedling establishment. Some seed-dispersing animals upon which the plant depends may be seed predators as well, eating the seeds for their own nutrition. Plants depending on such animals produce a tremendous number of seeds during their reproductive lives. Most of the seeds are consumed, but the sheer number ensures that a few will be dispersed, come to rest on a suitable site, and germinate (see concept of predator satiation, Section 14.10).

For example, a very close mutualistic relationship exists between wingless-seeded pines of western North America [whitebark pine (*Pinus albicaulis*), limber pine (*Pinus flexilis*), southwestern white pine (*Pinus strobiformis*), and piñon pine (*Pinus edulis*)] and several species of jays [Clark's nutcracker (*Nucifraga columbiana*), piñon jay (*Gymnorhinus cyanocephalus*), western scrub jay (*Aphelocoma californica*), and Steller's jay (*Cyanocitta stelleri*)]. In fact, there is a close correspondence between the ranges of these pines and jays. The relationship is especially close between Clark's nutcracker and the whitebark pine. Research by ecologist Diana Tomback of the University of Colorado, Denver, has revealed that only Clark's nutcracker has the morphology and behavior appropriate to disperse the seeds successfully away from the tree. The bird carries as many as 50 seeds in cheek pouches for distances up to 20 km away from the source and caches them deep enough in the soil of forest and open fields to reduce their detection and predation by rodents.

Seed dispersal by ants is prevalent among a variety of herbaceous plants that inhabit the deserts of the southwestern United States, the shrublands of Australia, and the deciduous forests of eastern North America. Such plants, called **myrmecochores,** have an ant-attracting food body on the seed coat called an **elaiosome** (Figure 15.15). Appearing as shiny tissue on the seed coat, the elaiosome contains certain chemical compounds essential for the ants. The ants carry seeds to their nests, where they sever the elaiosome and eat it or feed it to their larvae. The ants discard the intact seed within abandoned galleries of the nest. The area around ant nests is richer in nitrogen and phosphorus than the surrounding soil, providing a good substrate for seedlings. Further, by removing seeds far from the parent plant, the ants significantly reduce losses to seed-eating rodents. Plants may enclose their seeds in a nutritious fruit attractive to fruit-eating animals—the frugivores (Figure 15.16). Frugivores are not seed predators. They eat only the tissue surrounding the seed and, with some exceptions, do not damage the seed. Most frugivores do not depend exclusively on fruits, which are only seasonally available and deficient in proteins.

To use frugivorous animals as agents of dispersal, plants must attract them at the right time. Cryptic coloration, such as green unripened fruit among green leaves, and unpalatable texture, repellent substances, and hard outer coats discourage consumption of unripe fruit. When seeds mature, fruit-eating animals are attracted by attractive odors, softened texture, increasing sugar and oil content, and "flagging" of fruits with colors.

Most plants have fruits that can be exploited by an array of animal dispersers. Such plants undergo quantity

Figure 15.15 | Bleeding hearts, trilliums, and several dozen other plants have appendages on their seeds that contain oils attractive to ants. The ants carry the seeds to their nest, where the elaiosomes are removed and consumed as food, leaving the seeds unharmed.

Figure 15.16 | The frugivorous cedar waxwing (*Bombycilla cedrorum*) feeds on the red berries of mountain ash (*Sorbus*).

dispersal; they scatter a large number of seeds to increase the chance that various consumers will drop some seeds in a favorable site. Such a strategy is typical of, but not exclusive to, plants of the temperate regions, where fruit-eating birds and mammals rarely specialize in one kind of fruit and do not depend exclusively on fruit for sustenance. The fruits are usually succulent and rich in sugars and organic acids. They contain small seeds with hard seed coats resistant to digestive enzymes, allowing the seeds to pass through the digestive tract unharmed. Such seeds may not germinate unless they have been conditioned or scarified by passage through the digestive tract. Large numbers of small seeds may be dispersed, but few are deposited on suitable sites.

In tropical forests, 50–75 percent of the tree species produce fleshy fruits whose seeds are dispersed by animals. Rarely are these frugivores obligates of the fruits they feed on, although exceptions include many tropical fruit-eating bats.

15.15 | Mutualism Can Influence Population Dynamics

Mutualism is easy to appreciate at the individual level. We grasp the interaction between an ectomycorrhizal fungus and its oak or pine host; we count the acorns dispersed by squirrels and jays; and we measure the cost of dispersal to oaks in terms of seeds consumed. Mutualism improves the growth and reproduction of the fungus, the oak, and the seed predators. But what are the consequences at the population and community levels?

Mutualism exists at the population level only if the growth rate of species A increases with the increasing density of species B, and vice versa (see Quantifying Ecology 15.1: A Model of Mutualistic Interactions). For symbiotic mutualists where the relationship is obligate, the influence is straightforward. Remove species A, and the population of species B no longer exists. If ectomycorrhizal spores fail to infect the rootlets of young pines,

the fungi will not develop. If the young pine invading a nutrient-poor field fails to acquire a mycorrhizal symbiont, it will not grow well, if at all.

Discerning the role of facultative (nonsymbiotic) mutualisms in population dynamics can be more difficult. As discussed in Sections 15.13 and 15.14, mutualistic relationships are common in plant reproduction, where plant species often depend on animal species for pollination, seed dispersal, or germination. Although some relationships between pollinators and certain flowers are so close that loss of one could result in the extinction of the other, in most cases the effects are subtler and require detailed demographic studies to determine the consequences on species fitness.

When the mutualistic interaction is diffuse, involving a number of species—as is often the case with pollination systems and seed dispersal by frugivores—the influence of specific species–species interactions is difficult to determine. In other situations, the mutualistic relationship between two species may be mediated or facilitated by a third species, much the same as for vector organisms and intermediate hosts in parasite–host interactions. Mutualistic relationships among conifers, mycorrhizae, and voles in the forests of the Pacific Northwest as described by ecologist Chris Maser of the University of Puget Sound (Washington) and his colleagues are one such example (Figure 15.17). To acquire nutrients from the soil, the conifers depend on mycorrhizal fungi associated with the root system. In return, the mycorrhizae depend on the conifers for energy in the form of carbon (see Section 15.10). The mycorrhizae also have a

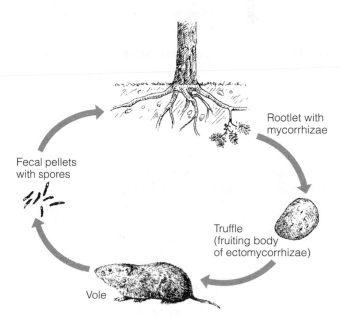

Figure 15.17 | A mutualism involving three species and both symbiotic and nonsymbiotic interactions. Voles eat truffles, the belowground fruiting bodies of some mycorrhizae. The spores become concentrated in fecal pellets. The voles disperse the spores to locations where they can infect new host plants.

A Model of Mutualistic Interactions

The simplest model for a mutualistic interaction between two species is similar to the basic Lotka–Volterra model as described in Chapter 13 for two competing species. The crucial difference is that rather than negatively influencing each other's growth rate, the two species have positive interactions. The competition coefficients α and β are replaced by positive interaction coefficients, reflecting the per capita effect of an individual of species 1 on species 2 (α_{12}) and the effect of an individual of species 2 on species 1 (α_{21}).

$$\text{Species 1: } \frac{dN_1}{dt} = r_1 N_1 \left(\frac{K_1 - N_1 + \alpha_{21} N_2}{K_1} \right)$$

$$\text{Species 2: } \frac{dN_2}{dt} = r_2 N_2 \left(\frac{K_2 - N_2 + \alpha_{12} N_1}{K_2} \right)$$

All of the terms are analogous to those used in the Lotka–Volterra equations for interspecific competition, except that $\alpha_{21} N_2$ and $\alpha_{12} N_1$ are added to the respective population densities (N_1 and N_2) rather than subtracted.

This model describes a facultative rather than obligate interaction because the carrying capacities of the two species are positive, and each species (population) can grow in the absence of the other. In this model, the presence of the mutualist offsets the negative effect of the species' population on the carrying capacity. In effect, the presence of the one species increases the carrying capacity of the other.

To illustrate this simple model, we can define values for the parameters $r_1, r_2, K_1, K_2, \alpha_{21},$ and α_{12}.

$$r_1 = 3.22, K_1 = 1000, \alpha_{12} = 0.5$$
$$r_2 = 3.22, K_2 = 1000, \alpha_{21} = 0.6$$

As with the Lotka–Volterra model for interspecific competition, we can calculate the zero isocline for the two mutualistic species that are represented by the equations presented two paragraphs above. The zero isocline for species 1 is solved by defining the values of N_1 and N_2, where $(K_1 - N_1 + \alpha_{21} N_2)$ is equal to zero. As with the competition model, because the equation is a linear function, we can define the line (zero isocline) by solving for only two points. Likewise, we can solve for the species 2 isocline. The resulting isoclines are shown in Figure 1.

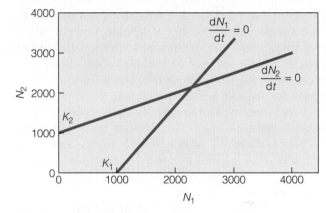

Figure 1 | Zero isoclines for species 1 (N_1) and species 2 (N_2) based on the modified Lotka–Volterra model for two mutualistic species presented in text.

mutualistic relationship with voles that feed on the fungi and disperse the spores, which then infect the root systems of other conifer trees.

Perhaps the greatest limitation in evaluating the role of mutualism in population dynamics is that many—if not most—mutualistic relationships arise from indirect interaction in which the affected species never come into contact. Mutualistic species influence each other's fitness or population growth rate indirectly through a third species or by altering the local environment (habitat modification)—topics we will revisit in Part Five. Mutualism may well be as significant as either competition or predation in its effect on population dynamics and community structure.

Note that, unlike the possible outcomes with the competition equations (see Figure 13.2), the zero isoclines extend beyond the carrying capacities of the two species (K_1 and K_2), reflecting that the carrying capacity of each species is effectively increased by the presence of the mutualist (other species). If we use the equations to project the density of the two populations through time (Figure 2), each species attains a higher density in the presence of the other species than when they occur alone (in the absence of the mutualist).

1. On the graph displaying the zero isoclines shown in Figure 1, plot the four points listed below and indicate the direction of change for the two populations.

$$(N_1, N_2) = 500, 500$$
$$(N_1, N_2) = 3500, 3000$$
$$(N_1, N_2) = 3000, 1000$$
$$(N_1, N_2) = 1000, 3000$$

2. What outcome do the isoclines indicate for the interaction between these two species?

Figure 2 | Population trajectories for a pair of facultative mutualists using the equations presented in text. Parameters for the Lotka–Volterra equations are presented in table form in the text. Population projections are shown for simulations of the equations both with and without the presence of the mutualist. (Adapted from Morin 1999.)

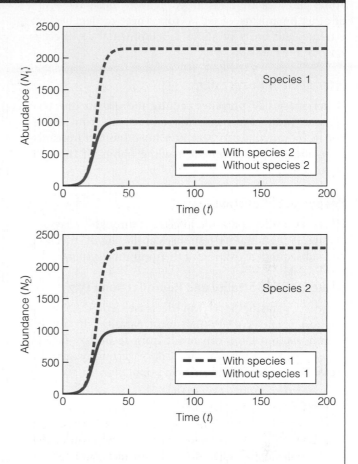

Summary

Characteristics of Parasites | 15.1

Parasitism is a symbiotic relationship between individuals of two species in which one benefits from the association while the other is harmed. Parasitic infection can result in disease. Microparasites include viruses, bacteria, and protozoa. They are small, have a short generation time, multiply rapidly in the host, tend to produce immunity, and spread by direct transmission. They are usually associated with dense populations of the host. Macroparasites are relatively large and include parasitic worms, lice, ticks, fleas, rusts, smuts, fungi, and other forms. They have a comparatively long generation time, rarely multiply directly in the host, persist with continual reinfection, and spread by both direct and indirect transmission.

Parasite–Host Relationships | 15.2

Parasites exploit every conceivable habitat in host organisms. Many are specialized to live at certain sites, such as in plant roots or an animal's liver. Parasites must (1) gain entrance into and (2) escape from the host. Their life cycle revolves about these two problems.

Direct Transmission | 15.3

Transmission for many species of parasites occurs directly from one host to another. It occurs either through direct physical contact or through the air, water, or another substrate.

Indirect Transmission | 15.4

Some parasites are transmitted between hosts by means of other organisms, called vectors. These carriers become intermediate hosts of some developmental or infective stage of the parasite.

Intermediate Hosts | 15.5

Other species of parasites require more than one type of host. Indirect transmission takes them from definitive to intermediate to definitive host. Indirect transmission often depends on the feeding habits of the host organisms.

Response to Infection | 15.6

Hosts respond to parasitic infections through behavioral changes, inflammatory responses at the site of infection, and subsequent activation of their immune systems.

Influence on Mortality and Reproduction | 15.7

A heavy parasitic load can decrease reproduction of the host organism. Although most parasites do not kill their hosts, mortality can result from secondary factors. Consequently, parasites can reduce fecundity and increase mortality rates of the host population.

Population Response | 15.8

Under certain conditions, parasitism can regulate a host population. When introduced to a population that has not developed defense mechanisms, parasites can spread quickly, leading to high rates of mortality and in some cases to virtual extinction of the host species.

Predation to Mutualism | 15.9

Mutualism is a positive reciprocal relationship between two species that may have evolved from predator–prey or host–parasite relationships. Where adaptations have countered the negative impacts, the relationship is termed commensalism. Where the interaction is beneficial to both species, the interaction is termed mutualism.

Mutualistic Relationships | 15.10

Mutualistic relationships involve diverse interactions. Mutualisms can be characterized by a wide number of variables relating to the benefits received, degree of dependency of the interaction, degree of specificity, and duration of the intimacy of the association.

Nutrient Uptake | 15.11

Symbiotic mutualisms are involved in the uptake of nutrients in both plants and animals. The chambers of a ruminant's stomach contain large populations of bacteria and protozoa that carry out the process of fermentation. Some plant species have a mutualistic association with nitrogen-fixing bacteria that infect and form nodules on their roots. The plants provide the bacteria with carbon, and the bacteria provide nitrogen to the plant. Fungi form mycorrhizal associations with the root systems of plants, assisting in the uptake of nutrients. In return, they derive energy in the form of carbon from the host plant.

Mutualisms Involving Defense | 15.12

Other mutualistic associations are associated with defense of the host organism.

Pollination | 15.13

Nonsymbiotic mutualisms are involved in the pollination of many species of flowering plants. While extracting nectar from the flowers, the pollinator collects and exchanges pollen with other plants of the same species. To conserve pollen, some plants have morphological structures that permit only certain animals to reach the nectar.

Seed Dispersal | 15.14

Mutualism is also involved in seed dispersal. Some seed-dispersing animals that the plant depends upon may be seed predators as well, eating the seeds for their own nutrition. Plants depending on such animals must produce a tremendous number of seeds to ensure that a few are dispersed, come to rest on a suitable site, and germinate. Alternatively, plants may enclose their seeds in a nutritious fruit attractive to frugivores (fruit-eating animals). Frugivores are not seed predators. They eat only the tissue surrounding the seed and, with some exceptions, do not damage the seed.

Population Dynamics | 15.15

Mutualistic relationships, both direct and indirect, may influence population dynamics in ways that we are just beginning to appreciate and understand.

Study Questions

1. In both predation and parasitism, one organism (species) derives its energy and nutrients from consuming another organism. How do these two processes differ?

2. For the parasite trematode discussed in Section 15.7, infection begins as snails grazing on algae incidentally ingest worm eggs. The eggs hatch into worms that prevent a snail's own reproduction. Instead, the infected snail nourishes the growing larval worms, which eventually develop into a free-swimming stage and leave the snails to seek their second, or intermediate, host—the California killifish. In traveling to the fish's brain, the worm causes the fish to behave differently from other killifish; it moves about jerkily near the water's surface. This behavior attracts predators like herons. The heron, in turn, becomes the host to the adult worm. The adult trematode takes up final residence in the bird's gut, releasing thousands of eggs that are deposited by way of bird droppings back into the salt marsh, completing the parasite's life cycle. How might such a complex life cycle have evolved?

3. How might a patchy or clumped distribution of hosts affect the spread of parasites? What spatial distribution of hosts (random, uniform, or clumped) would present the greatest difficulty in transmitting parasites from host to host?

4. What is mutualism? Look up some examples of mutualism and examine them critically. Are they in fact mutualistic?

5. Distinguish among symbiosis, obligate, and facultative mutualism.

6. Is fruit predation a chancy way of distributing seeds? Why?

7. Is mutualism reciprocal exploitation, or are the two species acting together for mutual benefit?

Further Readings

Barth, F. G. 1991. *Insects and flowers: The biology of a partnership.* Princeton, NJ: Princeton University Press.

  Although technical, this monograph is an excellent overview of the ecology of plant–pollinator interactions.

Boucher, D. H., ed. 1985. *The biology of mutualism.* London: Croom Helm.

  This volume provides a multitude of examples describing the range of mutualistic relationships covered in this chapter.

Dobson, A. P., and E. R. Carper. 1996. Infectious diseases and human population history. *BioScience* 46:115–125.

  An interesting and well-written review of the role of infectious disease in the history of the human population.

Futuyma, D. J., and M. Slatkin, eds. 1983. *Coevolution.* Sunderland, MA: Sinauer Associates.

  This edited volume provides an excellent overview of research in the area of coevolution.

Handel, S. N., and A. J. Beattie. 1990. Seed dispersal by ants. *Scientific American* 263:76–83.

  This review provides an excellent discussion of the mutualistic relationship between ants and plants.

Moore, J. 1984. Parasites that change the behavior of their host. *Scientific American* 250:108–115.

  An excellent introduction to this fascinating area of research.

Muscatine, L., and J. W. Porter. 1977. Reef corals: Mutualistic symbioses adapted to nutrient-poor environments. *BioScience* 27:454–460.

  An excellent introduction to the ecology of corals and the symbiotic relationship that is the foundation of the most diverse aquatic ecosystems on our planet.

Stachowicz, J. 2001. Mutualism, facilitation, and the structure of ecological communities. *BioScience* 51:235–246.

  An excellent review of the role of positive interactions on population dynamics and community structure. Contains many well-illustrated examples of current research in this growing area of research.

When European settlers to North America first explored the region west of the Mississippi River, they encountered a landscape on a scale unlike any they had known in Europe. The forested landscape of the east gave way to a vast expanse of grass and wildflowers. Grassy meadows are a common feature of the European landscape, but these grasslands seemed to extend endlessly westward. The prairies of North America once covered a large portion of the continent, ranging from Illinois and Indiana in the east into the Rocky Mountains of the west and extending from Canada in the north to Texas in the south. Today less than 1 percent of the prairie remains, mostly in small isolated patches, the result of a continental-scale transformation of this region to agriculture.

To reverse the loss of prairie ecosystems, efforts were begun in the 1960s in areas of the Midwest, such as Illinois, Minnesota, and Wisconsin, to reestablish native plant species on degraded areas of pasture land and abandoned croplands. These early efforts were in effect an attempt to reconstruct native prairie communities—the set of plant and animal species that once occupied these areas. But how does one start to rebuild an ecological community? Can a community be constructed by merely bringing together a collection of species in one place?

Many early reconstruction efforts met with failure. They involved planting whatever native plant species might be available in the form of seeds, often on very small plots surrounded by agricultural lands. The native plant species grew, but their populations often declined over time. Early efforts failed to appreciate the role of natural disturbances in maintaining these communities. Fire has historically been an important feature of the prairie, and many of the species were adapted to periodic burning. In the absence of fire, native species were quickly displaced by nonnative plant species from adjacent pastures.

Prairie communities are characterized by a diverse array of plant species that differ in the timing of germination, growth, and reproduction over the course of the growing season. The result is a shifting pattern of plant populations through time that together provide a consistent resource base for the array of animal species throughout the year. Attempts at restoration that do not include this full complement of plant species typically cannot attract and support the animal species that characterize native prairie communities.

The size of restoration projects was often a key factor in their failure. Small, isolated fragments tend to support species at low population levels and are thus prone to local extinction. These isolated patches were too distant from other patches of native grassland for the natural dispersal of other species, both plant and animal. Isolated patches of prairie often lacked the appropriate pollinator species required for successful plant reproduction.

Much has been learned from early attempts at restoring natural communities, and many restoration efforts have since succeeded. Restored prairie sites at Fermi National Accelerator Laboratory in northern Illinois are the product of 40 years of effort and now contain approximately 1000 acres—currently the largest restored prairie habitat in the world.

Attempts at reconstructing communities raise countless questions about the structure and dynamics of ecological communities, questions that in one form or another had been central to the study of ecological communities for more than a century. The ecological community is defined as the set of plant and animal species that occupy an area. Some species are abundant, supporting large populations within the area; others are rare, represented by only a small number of individuals. What controls the relative abundance of species within the community? Are all species equally important to the functioning and persistence of the community? How do the component species interact with each other? Do these interactions restrict or enhance the presence of other species? How do communities change through time? How does the community's size influence the number of species it can support? How do different communities on the larger landscape interact?

In Part Five, we will explore these questions, building on our understanding of species interactions developed in Part Four. We begin in Chapter 16 by examining the properties of ecological communities, their biological and physical structure, and how the structure of communities changes across the landscape. In Chapter 17, we will examine the factors influencing the structure of ecological communities. We start with the concept of the fundamental niche of a species and continue with how adaptations to the physical environment (see Part Two) influence the distribution and abundance of populations of species (see Part Three). We then examine how the interactions occurring

among species, both positive and negative, modify the potential distribution and abundance of species within an area, thus determining the structure of the community. In Chapter 18, we will explore the dynamic nature of community structure—how the relative abundance of species changes through time. Finally, in Chapter 19, we will examine how patterns in the physical environment (climate, geology, topography, and soils), agents of disturbance, and biotic processes interact to create a mosaic of communities on the broader landscape and how the spatial arrangement of these communities influences their dynamics.

As we shall see in the chapters that follow, ecological communities are more than an assemblage of species whose geographic distributions overlap. Ecological communities represent a complex web of interactions whose nature changes as environmental conditions vary in space and time.

The restoration project on the land surrounding the Fermi National Accelerator Laboratory west of Chicago, Illinois, is a leading success story among attempts to restore areas of the once vast midwestern prairie grasslands. The laboratory facility can be seen in the background.

CHAPTER 16

Community Structure

Coral reefs inhabiting the shallow coastal waters of subtropical and tropical oceans are among the most biologically diverse communities on our planet.

n walking through a forest or swimming along a coral reef, we see a collection of individuals of different species—plants and animals that form local populations. Sharing environments and habitats, these species of plants and animals interact in various ways. The group of species that occupy a given area, interacting either directly or indirectly, is called a **community.** This definition embraces the concept of community in its broadest sense. It is a spatial concept—the collective of species occupying a place within a defined boundary. Because ecologists generally do not study the entire community, the term *community* is often used in a more restrictive sense. It refers to a subset of the species, such as a plant, bird (avian), small mammal, or fish community. This use of *community* suggests relatedness or similarity among the members in their taxonomy, response to the environment, or use of resources.

The definition of community also recognizes that species living in close association may interact. They may compete for a shared resource, such as food, light, space, or moisture. One may depend on the other as a source of food. They may provide mutual aid, or they may not directly affect each other at all.

Like a population, a community has attributes that differ from those of its components and that have meaning only within the collective. These attributes include the number of species, their relative abundances, the nature of their interactions, and their physical structure (defined primarily by the growth forms of the plant components of the community). In this chapter, we examine the properties defining the structure of the community. In the chapters that follow, we will turn our attention to the processes influencing community structure and dynamics.

16.1 | The Number of Species and Their Relative Abundance Define Diversity

The mix of species, including their number and relative abundance, defines the biological structure of a community. The simplest measure of community structure is a count of the number of species occurring within the community, referred to as **species richness** and typically denoted by the symbol S.

Among the array of species that make up the community, however, not all are equally abundant. We can discover this characteristic by counting all the individuals of each species in a number of samples within a community and determining what percentage each species contributes to the total number of individuals of all species. This measure is known as **relative abundance.**

As an example of relative abundance, Tables 16.1 and 16.2 contain samples representing the tree species

Table 16.1 | Structure of One Mature Deciduous Forest Stand in Northern West Virginia

Species	Number of Individuals	Relative Abundance (Percentage of Total Individuals)
Yellow poplar (*Liriodendron tulipifera*)	76	29.7
White oak (*Quercus alba*)	36	14.1
Black oak (*Quercus velutina*)	17	6.6
Sugar maple (*Acer saccharum*)	14	5.4
Red maple (*Acer rubrum*)	14	5.4
American beech (*Fagus grandifolia*)	13	5.1
Sassafras (*Sassafras albidum*)	12	4.7
Red oak (*Quercus rubra*)	12	4.7
Mockernut hickory (*Carya tomentosa*)	11	4.3
Black cherry (*Prunus serotina*)	11	4.3
Slippery elm (*Ulmus rubra*)	10	3.9
Shagbark hickory (*Carya ovata*)	7	2.7
Bitternut hickory (*Carya cordiformis*)	5	2.0
Pignut hickory (*Carya glabra*)	3	1.2
Flowering dogwood (*Cornus florida*)	3	1.2
White ash (*Fraxinus americana*)	2	0.8
Hornbeam (*Carpinus carolinia*)	2	0.8
Cucumber magnolia (*Magnolia acuminata*)	2	0.8
American elm (*Ulmus americana*)	1	0.39
Black walnut (*Juglans nigra*)	1	0.39
Black maple (*Acer nigra*)	1	0.39
Black locust (*Robinia pseudoacacia*)	1	0.39
Sourwood (*Oxydendrum arboreum*)	1	0.39
Tree of heaven (*Ailanthus altissima*)	1	0.39
	256	100.00

Table 16.2 | Structure of a Second Deciduous Forest Stand in Northern West Virginia

Species	Number of Individuals	Relative Abundance (Percentage of Total Individuals)
Yellow poplar (*Liriodendron tulipifera*)	122	44.5
Sassafras (*Sassafras albidum*)	107	39.0
Black cherry (*Prunus serotina*)	12	4.4
Cucumber magnolia (*Magnolia acuminata*)	11	4.0
Red maple (*Acer rubrum*)	10	3.6
Red oak (*Quercus rubra*)	8	2.9
Butternut (*Juglans cinerea*)	1	0.4
Shagbark hickory (*Carya ovata*)	1	0.4
American beech (*Fagus grandifolia*)	1	0.4
Sugar maple (*Acer saccharum*)	1	0.4
	274	100.0

composition of two forest communities. The sample from the first forest consists of 24 species. Two species, yellow poplar and white oak, make up nearly 44 percent of the total number of individuals. The four next most abundant trees—black oak, sugar maple, red maple, and American beech—each make up a little over 5 percent of the total. Nine species range from 1.2 percent to 4.7 percent, and the nine remaining species as a group represent about 0.5 percent. The second forest presents a somewhat different picture. This community consists of 10 species, of which two—yellow poplar and sassafras—make up almost 84 percent of the total tree density.

Although both forest communities illustrate the pattern of a few common species associated with many rare ones, these two forests exhibit quite different patterns of species richness and relative abundance. A common method for comparing the patterns of species richness and abundance between communities involves plotting the relative abundance of each species against rank, where rank is defined by the order of species from the most to the least abundant. Thus, the most abundant species is plotted first along the *x*-axis, with the corresponding value on the *y*-axis being the value of relative abundance. This process is continued until all species are plotted. The resulting graph is called a

rank-abundance diagram. Figure 16.1 depicts the rank-abundance curves for the two forest stands presented in Tables 16.1 and 16.2.

As the rank-abundance diagram shows, these two forest communities differ in both species richness and how individuals are apportioned among the species (relative abundance). The first forest community has greater species richness and a more equitable distribution of individuals among the species; this characteristic is referred to as **species evenness.** The greater species richness is reflected by the greater length of the rank-abundance curve (number of species), and the more equitable distribution of individuals among the species (species evenness) is indicated by the more gradual slope of the rank-abundance curve.

Although the graphical procedure of rank-abundance diagrams can be used to visually assess (interpret) differences in the biological structure of communities, these diagrams offer no means of quantifying the observed differences. Ecologists have addressed this need by developing **diversity indexes,** which consider both the number and relative abundance of species within the community.

One of the simplest and most widely used indices of diversity is the Simpson's index. The term *Simpson's diversity index* can actually refer to any one of three closely related indexes.

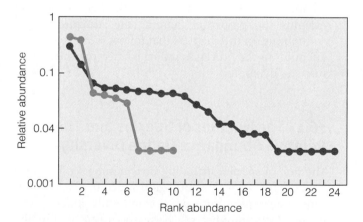

Figure 16.1 | Rank-abundance curves for the two forest communities described in Tables 16.1 and 16.2. Rank abundance is the species ranking based on relative abundance, ranked from the most to least abundant (*x*-axis). Relative abundance (*y*-axis) is expressed on a $\log_{10}$ axis. The forest community in Table 16.1 (brown line) has a higher species richness (length of curve) and evenness (slope of curve) than the forest community in Table 16.2 (orange line).

Interpreting Ecological Data

Q1. How does the slope of the rank-abundance curve vary with increasing species evenness? Why?

Q2. What would the rank-abundance curve look like for a forest community consisting of 10 species, where all of the 10 tree species are equally abundant?

Simpson's index (D) measures the probability that two individuals randomly selected from a sample will belong to the same species (category):

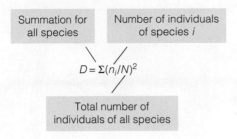

$$D = \Sigma(n_i/N)^2$$

The value of D ranges between 0 and 1. In the absence of diversity, where only one species is present, the value of D is 1. As both species richness and evenness increase, the value approaches 0.

Because the greater the value of D, the lower the diversity, D is often subtracted from 1 to give

$$\text{Simpson's index of diversity} = 1 - D$$

The value of this index also ranges between 0 and 1, but now the value increases with diversity. In this case, the index represents the probability that two individuals randomly selected from a sample will belong to different species.

Another method is to take the reciprocal of D:

$$\text{Simpson's reciprocal index} = \frac{1}{D}$$

This index, called the *Simpson's diversity index*, is most commonly used. The lowest possible value of this index is 1, representing a community containing only one species. The higher the value, the greater is the diversity. The maximum value is the number of species in the community (species richness: S). For example, there are 10 tree species in the forest community presented in Table 16.2, so the maximum possible value of the index is 10.

Because the Simpson's index actually refers to three related but different indexes, it is important to identify which is being used and reported.

Another widely used index of diversity that also considers both species richness and evenness is the **Shannon index** (also called the Shannon-Weiner index). To compute the index, the relative abundance of each species (p_i) is expressed as

$$p_i = n_i/N$$

where, as in the calculation of the Simpson's index (D), n_i is the number of individuals of species i and N is the total number of individuals of all species.

The Shannon index (H) is then computed as

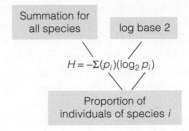

$$H = -\Sigma(p_i)(\log_2 p_i)$$

In the absence of diversity, where only one species is present, the value of H is 0. The maximum values of the index, which occurs when all species are present in equal numbers, is $H_{max} = \ln S$, where ln is the natural logarithm (approximately 2.718) and S is the total number of species (species richness).

16.2 | Numerical Supremacy Defines Dominance

Although the numbers of tree species occurring in the two forest communities (species richness) presented in Tables 16.1 and 16.2 differ more than twofold, the two communities share a common feature. Both communities are composed of a few common tree species with high population density, whereas the remaining tree species are rare and at low population density. This is a characteristic of most communities. When a single or few species predominate within a community, these species are referred to as **dominants.**

Dominance is the converse of diversity. In fact, the basic Simpson index, D, is often used as a measure of dominance. Recall that values of D range from 0 to 1, where 1 represents complete dominance—only one species present in the community.

Dominant species are usually defined separately for different taxonomic or functional groups of organisms within the community. For example, yellow poplar is a dominant tree species in both of the forest communities just discussed, but we could likewise identify the dominant herbaceous plant species within the forest or the dominant species of bird or small mammal.

Dominance typically is assumed to mean the greatest in number. But in populations or among species, where individuals can vary widely in size, abundance alone is not always a sufficient indicator of dominance. In a forest, for example, the small or understory trees can be numerically superior; yet a few large trees that overshadow the smaller ones will account for most of the biomass (living tissue). In such a situation, we may wish to define dominance based on some combination of characteristics that include both the number and size of individuals.

Because dominant species typically achieve their status at the expense of other species in the community, they are often the dominant competitors under the prevailing environmental conditions. For example, the American chestnut tree (*Castanea dentata*) was a dominant component of oak–chestnut forests in eastern North America until the early 20th century. At that time, the chestnut blight introduced from Asia decimated chestnut tree populations. Since then a variety of species—including oaks, hickories, and yellow poplar—have taken over the chestnut's position in the forest. As we shall see in Chapter 17, however, processes other than competition can also be important in determining dominance within communities.

16.3 | Keystone Species Influence Community Structure Disproportionately to Their Numbers

Relative abundance is just one measure, based only on numerical supremacy, of a species' contribution to the community. Other, less-abundant species, however, may play a crucial role in the function of the community. A species that has a disproportionate impact on the community relative to its abundance is referred to as a **keystone species.**

Keystone species function in a unique and significant manner through their activities, and their effect on the community is disproportionate to their numerical abundance. Their removal initiates changes in community structure and often results in a significant loss of diversity. Their role in the community may be to create or modify habitats or to influence the interactions among other species. An organism that functions as a keystone species by creating habitat is the coral *Oculina arbuscula*, which occurs along the eastern coast of the United States as far north as the coastal waters of North Carolina. It is the only coral in this region with a structurally complex, branching morphology that provides shelter for a species-rich epifauna (organisms that live on and among the coral). More than 300 species of invertebrates are known to live among the branches of *Oculina* colonies, and many more are reported to complete much of their life cycle within the coral (see Field Studies John J. Stachowicz, in Chapter 15).

In other cases, keystone herbivores may modify the local community by their feeding activities. An excellent example is the role of the African elephant in the savanna communities of southern Africa. This herbivore feeds primarily on a diet of woody plants (browse). Elephants are destructive feeders that often uproot, break, and destroy the shrubs and trees they feed on (Figure 16.2). Reduced density of trees and shrubs favors the growth and production of grasses. This change in the composition of the plant community is to the elephant's disadvantage, but other herbivores that feed on the grasses benefit from it.

Predators often function as keystone species within communities (see Section 17.3 for further discussion of

Figure 16.2 | Elephants, which consume large quantities of woody vegetation and uproot trees, are a keystone species in the savannas of southern Africa. By decreasing woody vegetation, elephants increase the productivity of grasses upon which a wide variety of grazing herbivores depend.

keystone predators). For example, sea otters (*Enhydra lutris*) are a keystone predator in the kelp bed communities found in the coastal waters of the Pacific Northwest. Sea otters eat urchins, which feed on kelp. The kelp beds provide habitat to a wide diversity of other species, and if sea otter populations decline in an area (as has happened in large areas of western Alaska since the early 1990s), the sea urchin population increases dramatically. The result is overgrazing of the kelp beds and a loss of habitat for the many species inhabiting these communities.

16.4 | Food Webs Describe Species Interactions

Perhaps the most fundamental process in nature is that of acquiring the energy and nutrients required for assimilation. The species interactions discussed in Part Four—predation, parasitism, competition, and mutualism—are all involved in acquiring these essential resources. For this reason, ecologists studying the structure of communities often focus on the feeding relationships among the component species, or how species interact in the process of acquiring the resources necessary for metabolism, growth and reproduction.

An abstract representation of feeding relationships within a community is the **food chain.** A food chain is a descriptive diagram—a series of arrows, each pointing from one species to another, representing the flow of food energy from prey (the consumed) to predator (the consumer). For example, grasshoppers eat grass; clay-colored sparrows eat grasshoppers; and marsh hawks prey on the sparrows. We write this relationship as follows:

grass → grasshopper → sparrow → hawk

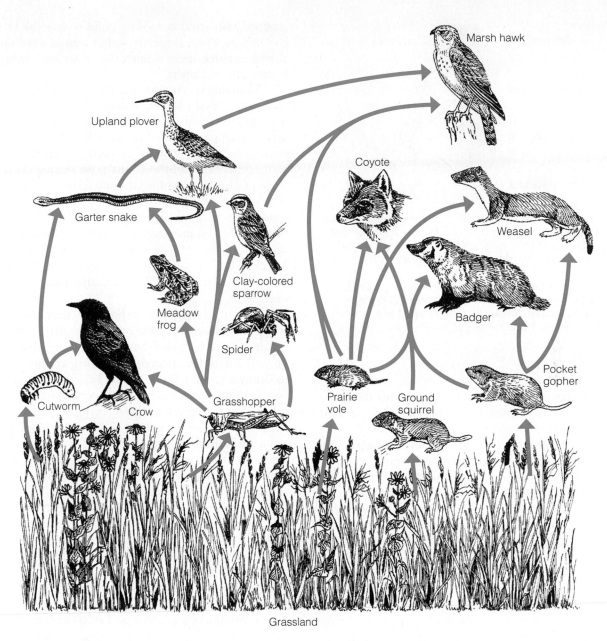

Figure 16.3 | A food web for a prairie grassland community in the midwestern United States. Arrows flow from prey (consumed) to predator (consumer).

Feeding relationships in nature, however, are not simple, straight-line food chains. Rather, they involve many food chains meshed into a complex food web with links leading from primary producers through an array of consumers (Figure 16.3). Such **food webs** are highly interwoven, with linkages representing a wide variety of species interactions.

A simple hypothetical food web is presented in Figure 16.4 to illustrate the basic terminology used to describe the structure of food webs. Each circle represents a species, and the arrows from the consumers to the species being consumed are termed **links.** The species in the webs are distinguished by whether they are basal species, intermediate species, or top predators. **Basal species**

feed on no other species but are fed upon by others. **Intermediate species** feed on other species, and they themselves are prey of other species. **Top predators** are not subjected to predators; they prey on intermediate and basal species. These terms refer to the structure of the web rather than to strict biological reality.

Although any two species are linked by only a single arrow representing the relationship between predator (the consumer) and prey (the consumed), the dynamics of communities cannot be understood solely in terms of direct interactions between species. For example, a predator may reduce competition between two prey species by controlling their population sizes below their respective carrying capacities. An analysis of the mechanisms

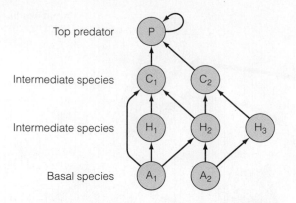

Figure 16.4 | Hypothetical food web illustrating the various categories of species. A_1 and A_2 feed on no other species in the food web and are referred to as basal species (typically autotrophs). H_1, H_2, and H_3 are herbivores. C_2 is a carnivore, and C_1 is defined as an omnivore, because it feeds on more than one trophic level. Species designated as H and C are all intermediate species because they function as both predators and prey within the food web. P is a top predator, because it is eaten by no other species within the food web. P also exhibits cannibalism, because this species feeds on itself.

controlling community structure must include these "indirect" effects represented by the structure of the food web; we will explore this topic in Chapter 17.

The simple designation of feeding relationships using the graphical approach of food webs can become incredibly complex in communities of even moderate diversity. For this reason, ecologists often simplify the representation of food webs by lumping species into broader categories that represent general feeding groups based on the source from which they derive energy. In Part Two, we defined organisms that derive their energy from sunlight as autotrophs, or primary producers. Organisms that derive energy from consuming plant and animal tissue are called heterotrophs, or secondary producers, and are further subdivided into herbivores, carnivores, and omnivores based on their consumption of plant tissues, animal tissues, or both. These feeding groups are referred to as **trophic levels,** after the Greek word *trophikos,* meaning "nourishment."

16.5 | Species within a Community Can Be Classified into Functional Groups

The grouping of species into trophic levels is a functional classification, defining groups of species that derive their energy in a similar manner. Another approach is to subdivide each trophic level into groups of species that exploit a common resource in a similar fashion; these groups are termed **guilds.** For example, hummingbirds and other nectar-feeding birds form a guild of species that exploit the common resource of flowering plants in a similar fashion. Likewise, seed-eating birds could be

grouped into another feeding guild within the broader community. Because species within a guild draw upon a shared resource, there is potential for strong interactions between the members.

Classifying species into guilds can simplify the study of communities, allowing researchers to focus on more manageable subsets of the community. Yet by classifying species into guilds based on their functional similarity, ecologists can also explore questions about the very organization of communities. Just as we can use the framework of guilds to explore the interactions of the component species within a guild, we can also use this framework to pose questions about the interactions between the various guilds that compose the larger community. At one level, a community can be a complex assembly of component guilds interacting with each other and producing the structure and dynamics that we observe.

In recent years, ecologists have expanded the concept of guilds to develop a more broadly defined approach of classifying species based on function rather than taxonomy. The term **functional type** is now commonly used to define a group of species based on their common response to the environment, life history characteristics, or role within the community. For example, plants may be classified into functional types based on their photosynthetic pathway (C_3, C_4, and CAM), which, as we have seen from Chapter 6, relates to their ability to photosynthesize and grow under different thermal and moisture environments. Similarly, plant ecologists use the functional classification of shade-tolerant and shade-intolerant to reflect basic differences in the physiology and morphology of plant species in response to the light environment (Section 6.9). Grouping plants or animals into the categories of iteroparous and semelparous, as presented in Chapter 8, also represents a functional classification based on the timing of reproductive effort.

As with the organization and classification of species into guilds, using functional groups allows ecologists to simplify the structure of communities into manageable units for study and to ask basic questions about the factors that structure communities, as we shall see later in the discussion of community dynamics in Chapter 18.

16.6 | Communities Have a Characteristic Physical Structure

Communities are characterized not only by the mix of species and by the interactions among them—the biological structure—but also by their physical features. The physical structure of the community reflects abiotic factors, such as the depth and flow of water in aquatic environments. It also reflects biotic factors, such as the spatial arrangement of the resident organisms. For example, the

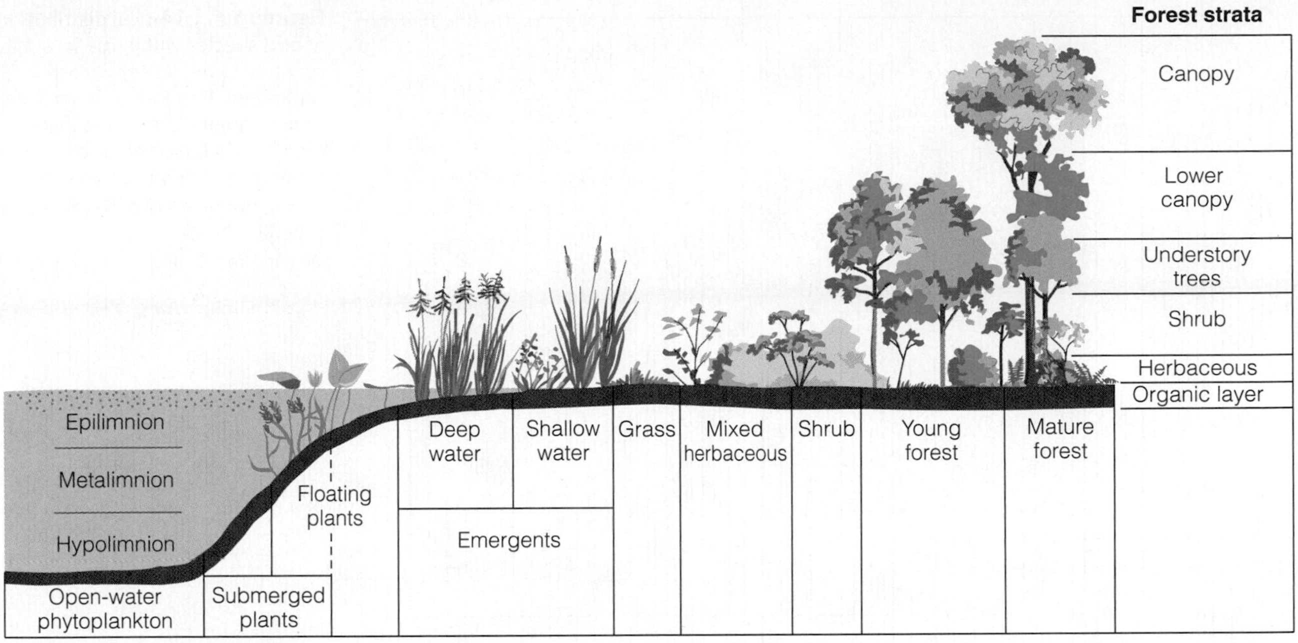

Forest strata

Canopy

Lower canopy

Understory trees

Shrub

Herbaceous

Organic layer

Epilimnion

Metalimnion

Hypolimnion

Open-water phytoplankton

Submerged plants

Floating plants

Deep water

Shallow water

Emergents

Grass

Mixed herbaceous

Shrub

Young forest

Mature forest

Figure 16.5 | A vertical view of communities from aquatic to terrestrial. In aquatic and terrestrial environments, the primary zone of decomposition and regeneration is the bottom stratum, and the zone of energy fixation is the upper stratum. From left to right in the figure, community stratification and complexity increase. Stratification in aquatic communities is largely physical, influenced by gradients of oxygen, temperature, and light. Stratification in terrestrial communities is largely biological. Dominant vegetation affects the physical structure of the community and the microclimatic conditions of temperature, moisture, and light. Because the forest has four or five strata, it supports a greater diversity of life than can grassland with two strata. Floating and emergent aquatic plant communities typically support greater diversity of life than can open water.

size and height of the trees and the density and spatial distribution of their populations help define the physical attributes of the forest community.

The form and structure of terrestrial communities are defined primarily by their vegetation. Plants may be tall or short, evergreen or deciduous, herbaceous or woody. Such characteristics can describe growth forms. Thus, we might speak of shrubs, trees, and herbs and further subdivide the categories into needle-leaf evergreens, broadleaf evergreens, broadleaf deciduous trees, thorn trees and shrubs, dwarf shrubs, ferns, grasses, forbs, mosses, and lichens. Ecologists often classify and name terrestrial communities based on the dominant plant growth forms and their associated physical structure: forests, woodlands, shrublands, or grassland communities (see Chapter 23).

In aquatic environments, the dominant organisms are also used to classify and name communities. Kelp forests, seagrass meadows, and coral reefs are examples. However, the physical structure of aquatic communities is more often defined by features of the abiotic environment, such as water depth, flow rate, or salinity (see Chapter 24).

Every community has an associated vertical structure (Figure 16.5). On land, the growth form of the plants largely determines vertical structure—their size, branching, and leaves—and this vertical structure in turn influences, and is influenced by, the vertical gradient of light (see Section 4.2). The vertical structure of the plant community also provides a physical framework in which many forms of animal life are adapted to live. A well-developed forest ecosystem, for example, has multiple layers of vegetation. From top to bottom, they are the canopy, the understory, the shrub layer, the herb or ground layer, and the forest floor.

The upper layer, the **canopy,** is the primary site of energy fixation through photosynthesis. The canopy structure has a major influence on the rest of the forest. If the canopy is fairly open, considerable sunlight will reach the lower layers. If ample water and nutrients are available, a well-developed **understory** and shrub strata will form. If the canopy is dense and closed, light levels are low, and the understory and shrub layers will be poorly developed.

In the forests of the eastern United States, the understory consists of tall shrubs such as witch hobble (*Viburnum alnifolium*), understory trees such as dogwood (*Cornus* spp.) and hornbeam (*Carpinus caroliniana*), and younger trees, some of which are the same species as those in the canopy. The nature of the **herb layer**

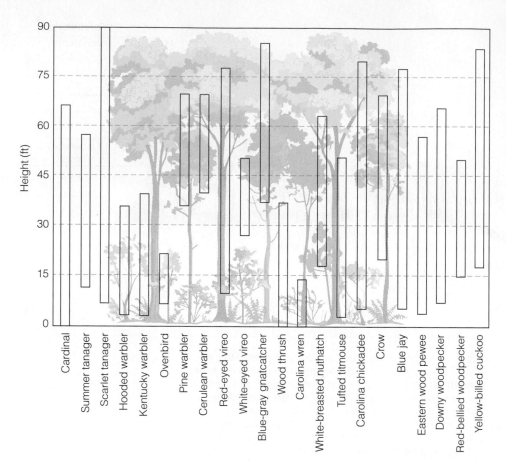

Figure 16.6 | Vertical distribution of bird species within the forest community on Walker Branch watershed, Oak Ridge, Tennessee. Height range represented by colored bars is based on total observations of birds during the breeding season regardless of activity.

(Adapted from Anderson and Shugart 1974.)

depends on the soil moisture and nutrient conditions, slope position, density of the canopy and understory, and exposure of the slope, all of which vary from place to place throughout the forest. The final layer, the **forest floor,** is where the important process of decomposition takes place and where microbial organisms feeding on decaying organic matter release mineral nutrients for reuse by the forest plants (see Chapter 21).

Aquatic ecosystems such as lakes and oceans have strata determined largely by light penetration through the water column (see Section 3.3). They also have distinctive profiles of temperature and oxygen (see Section 3.4). In the summer, well-stratified lakes have a layer of well-mixed water, the epilimnion; a second layer, the metalimnion, which is characterized by a thermocline (a steep and rapid decline in temperature relative to the waters above and below); the hypolimnion, a deep, cold layer of dense water at about 4°C (39°F), often low in oxygen; and a layer of bottom sediments. Two other structural layers are also recognized, based on light penetration: an upper zone, the **photic zone,** where the availability of light supports photosynthesis primarily by phytoplankton; and in deeper waters, the **aphotic zone,** an area without light. The bottom zone, where decomposition is most active, is referred to as the **benthic zone.**

Characteristic organisms inhabit each available vertical layer in a community. In addition to the vertical distribution of plant life already described, various types of consumers and decomposers occupy all levels of the community—although decomposers are typically found in greater abundance in the forest floor (soil surface) and sediment (benthic) layers. Considerable interchange takes place among the vertical strata, but many highly mobile animals confine themselves to only a few layers (Figure 16.6). The species occupying a given vertical layer may change during the day or season. Such changes reflect daily and seasonal variations in the physical environment such as humidity, temperature, light, and oxygen concentrations in the water; shifts in the abundance of essential resources such as food; or different requirements of organisms for the completion of their life cycles. For example, zooplankton migrate vertically in the water column during the course of the day in response to varying light (see Section 3.3).

16.7 | Zonation Is Spatial Change in Community Structure

As we move across the landscape, the physical and biological structure of the community changes. Often these changes are small, subtle ones in the species composition

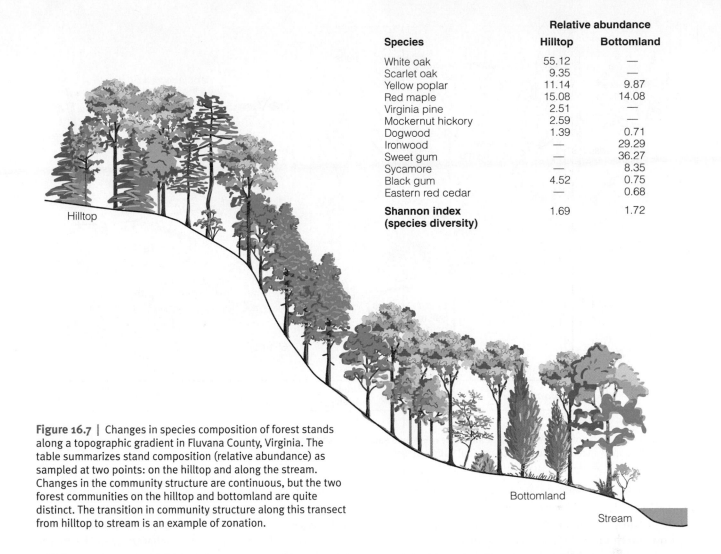

Species	Relative abundance	
	Hilltop	Bottomland
White oak	55.12	—
Scarlet oak	9.35	—
Yellow poplar	11.14	9.87
Red maple	15.08	14.08
Virginia pine	2.51	—
Mockernut hickory	2.59	—
Dogwood	1.39	0.71
Ironwood	—	29.29
Sweet gum	—	36.27
Sycamore	—	8.35
Black gum	4.52	0.75
Eastern red cedar	—	0.68
Shannon index (species diversity)	1.69	1.72

Hilltop

Bottomland

Stream

Figure 16.7 | Changes in species composition of forest stands along a topographic gradient in Fluvana County, Virginia. The table summarizes stand composition (relative abundance) as sampled at two points: on the hilltop and along the stream. Changes in the community structure are continuous, but the two forest communities on the hilltop and bottomland are quite distinct. The transition in community structure along this transect from hilltop to stream is an example of zonation.

or height of the vegetation. However, as we travel farther, these changes often become more pronounced. For example, central Virginia just east of the Blue Ridge Mountains is a landscape of rolling hills. The area is a mosaic of forest and field. As we walk through a forest, the physical structure of the community—the canopy, understory, shrub layer, and forest floor—appears much the same. But the biological structure, the mix of species that compose the community, may change quite dramatically. As we move from a hilltop to the bottomland along a stream (Figure 16.7), the mix of trees changes from oaks (*Quercus* spp.) and hickory (*Carya* spp.) to species associated with much wetter environments, such as sycamore (*Platanus occidentalis*), hornbeam (*Carpinus caroliniana*), and sweet gum (*Liquidambar styraciflua*). Besides changes in the vegetation, the animal species—insects, birds, and small mammals—that occupy the forest also change. These changes in the physical and biological structures of communities as one moves across the landscape are referred to as **zonation.**

Patterns of spatial variation in community structure or zonation are common to all environments, aquatic and terrestrial. Figure 16.8 provides an example of zonation in a salt marsh along the northeastern coastline of North America. In moving from the shore and through the marsh to the upland, notice the variations in the physical and biological structures of the communities. The dominant plant growth forms in the marsh are grasses and sedges. These growth forms give way to shrubs and trees as we move to dry land and the depth of the water table increases. In the zone dominated by grasses and sedges, the dominant species change as we move back from the tidal areas. These differences result from various environmental changes across a spatial gradient, including microtopography, water depth, sediment oxygenation, and salinity. The changes are marked by distinct plant communities that are defined by changes in dominant plants as well as in structural features such as height, density, and spatial distribution of individuals.

The intertidal zone of a sandy beach provides an example in which the zonation is dominated by

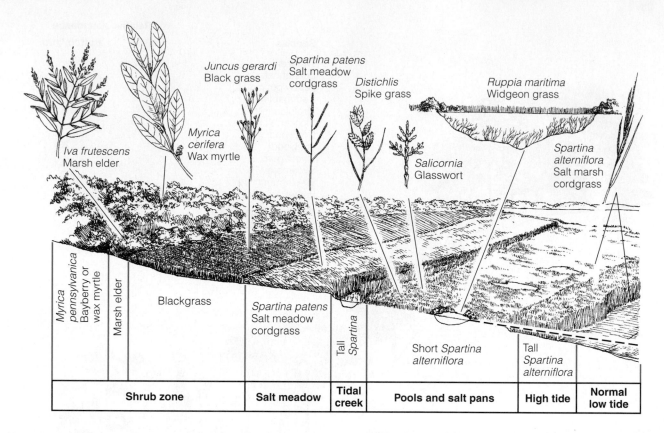

Figure 16.8 | Patterns of zonation in an idealized New England salt marsh, showing the relationship of plant distribution to microtopography and tidal submergence.

heterotrophic organisms rather than autotrophs (Figure 16.9). Patterns of species distribution relate to the tides. Sandy beaches can be divided into supratidal (above the high-tide line), intertidal (between the high- and low-tide lines), and subtidal (below the low-tide line; continuously inundated) zones, each home to a unique group of animal organisms. Pale, sand-colored ghost crabs (*Ocypode quadrata*) and beach fleas (*Talorchestia* and *Orchestia* spp.) occupy the upper beach, or supratidal zone. The intertidal beach is the zone where true marine life begins. An array of animal species adapted to the regular periods of inundation and exposure to the air are found within this zone. Many of these species, such as the mole crab (*Emerita talpoida*), lugworm (*Arenicola cristata*), and hard-shelled clam (*Mercenaria mercenaria*), are burrowing animals, protected from the extreme temperature fluctuations that can occur between periods of inundation and exposure. In contrast, the subtidal zone is home to a variety of vertebrate and invertebrate species that migrate into and out of the intertidal zone with the changing tides.

16.8 | Defining Boundaries between Communities Is Often Difficult

As previously noted, the community is a spatial concept involving the species that occupy a given area. Ecologists typically distinguish between adjacent communities or community types based on observable differences in their physical and biological structures: the different species assemblages characteristic of different physical environments. How different must two adjacent areas be before we call them separate communities? This is not a simple question. Consider the forest in Figure 16.7. Given the difference in species composition from hilltop to bottomland, most ecologists would define these two areas as different vegetation communities. As we walk between these areas, however, the distinction may not seem so straightforward. If the transition between the two communities is abrupt, it may not be hard to define community boundaries. But, if the species composition and patterns of dominance shift gradually, the boundary is not as clear.

Ecologists use various sampling and statistical techniques to delineate and classify communities. Generally, all employ some measure of community similarity or difference (see Quantifying Ecology 16.1: Community Similarity). Although it is easy to describe the similarities and differences between two areas in terms of species composition and structure, actually classifying areas into distinct groups of communities involves a degree of subjectivity that often depends on the study objectives and the spatial scale at which vegetation is being described.

The example of forest zonation presented in Figure 16.7 occurs over a relatively short distance of a few hundred meters, or an area of only a few hectares. As

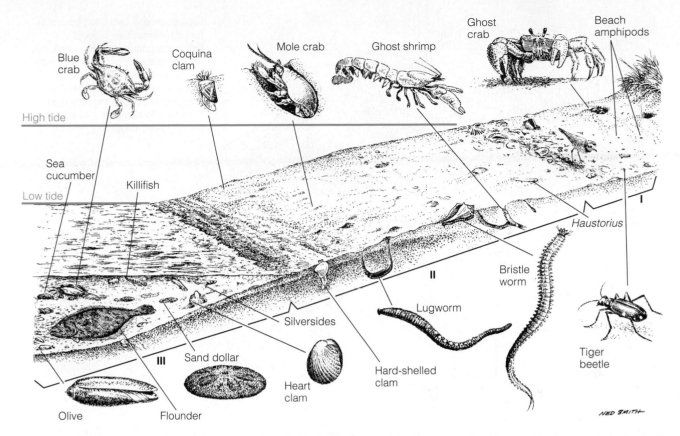

Figure 16.9 | Life on a sandy ocean beach along the Mid-Atlantic coast is an example of zonation dominated by changes in the fauna. The distribution of organisms changes along a gradient from land to sea as a function of the degree and duration of inundation during the tidal cycle. I—supratidal zone: ghost crabs and sand fleas. II—intertidal zone: ghost shrimp, bristle worms, clams, lugworms, mole crabs. III—subtidal zone: founder, blue crab, sea cucumber. The blue line indicates high tide.

we consider ever-larger areas, differences in community structure—both physical and biological—increase. An example is the pattern of forest zonation in Great Smoky Mountains National Park (Figure 16.10). The zonation is a complex pattern related to elevation, slope position, and exposure. Note that the description of the forest communities in the park contains few species names. Names like *hemlock forest* are not meant to suggest a lack of species diversity; they are just a shorthand method of naming communities for the dominant tree species. Each community could be described by a complete list of species, their population sizes, and their contributions to the total biomass (as with the communities in Tables 16.1 and 16.2). However, such lengthy descriptions are unnecessary to communicate the major changes in the structure of communities across the landscape. In fact, as we expand the area of interest to include the entire eastern United States, the nomenclature for classifying forest communities becomes even broader. In Figure 16.11 (on page 346), which is a broad-scale description of forest zonation in the eastern United States developed by Lucy Braun, all of Great Smoky Mountains National Park shown in Figure 16.10 (located in southeastern

Tennessee and northwestern North Carolina) is described as a single forest community type: oak–chestnut, a type that extends from New York to Georgia.

These large-scale examples of zonation make an important point that we return to when examining the processes responsible for spatial changes in community structure: our very definition of community is a spatial concept. Like the biological definition of population (Chapter 9), the definition of community refers to a spatial unit that occupies a given area. In a sense, the distinction among communities is arbitrary, based on the criteria for classification. As we shall see, the methods used in delineating communities as discrete spatial units have led to problems in understanding the processes responsible for patterns of zonation (see Chapter 17).

16.9 | Two Contrasting Views of the Community

At the beginning of this chapter, we defined the community as the group of species (populations) that occupy a given area, interacting either directly or indirectly. Interactions can have both positive and negative influences on species populations. How important are these interactions in

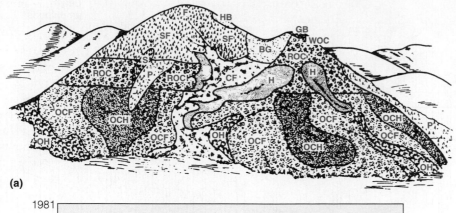

(a)

Figure 16.10 | Two descriptions of forest communities in Great Smoky Mountains National Park. **(a)** Topographic distribution of vegetation types on an idealized west-facing mountain and valley. **(b)** Idealized arrangement of community types according to elevation and aspect.

(Adapted from Whittaker 1954.)

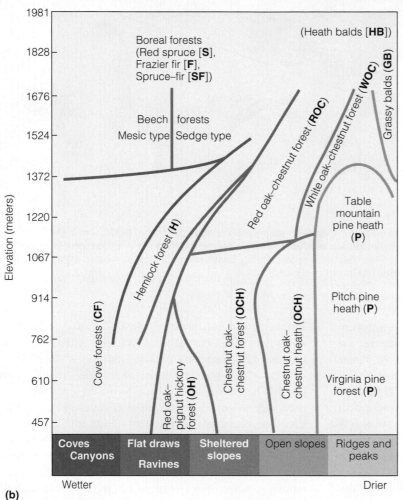

(b)

determining community structure? In the first half of the 20th century, this question led to a major debate in ecology that still influences our views of the community.

When we walk through most forests, we see a variety of plant and animal species—a community. If we walk far enough, the dominant plant and animal species will change (see Figure 16.7). As we move from hilltop to valley, the structure of the community will differ. But what if we continue our walk over the next hilltop and into the adjacent valley? We will most likely notice that although the communities on the hilltop

and valley are quite distinct, the communities on the two hilltops or valleys are quite similar. As a botanist might put it, they exhibit relatively consistent floristic composition. At the International Botanical Congress of 1910, botanists adopted the term *association* to describe this phenomenon. An association is a type of community with (1) relatively consistent species composition, (2) a uniform, general appearance (physiognomy), and (3) a distribution that is characteristic of a particular habitat, such as the hilltop or valley. Whenever the particular habitat or set of environmental

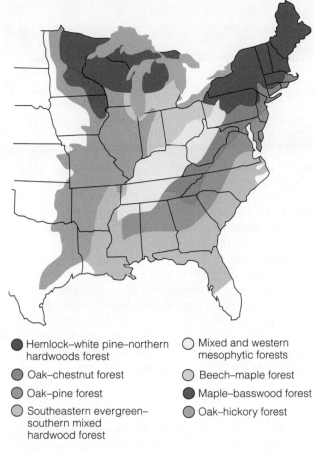

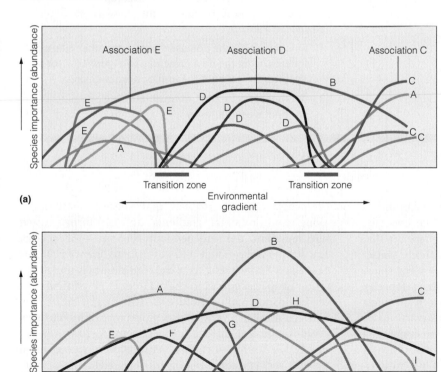

Legend for Figure 16.11:
- ● Hemlock–white pine–northern hardwoods forest
- ○ Mixed and western mesophytic forests
- ◐ Oak–chestnut forest
- ◔ Beech–maple forest
- ◔ Oak–pine forest
- ● Maple–basswood forest
- ○ Southeastern evergreen–southern mixed hardwood forest
- ◔ Oak–hickory forest

Figure 16.11 | Large-scale distribution of deciduous forest communities in the eastern United States is defined by eight regions. (Adapted from Braun 1950.)

(a) labels: Association E, Association D, Association C, Transition zone, Transition zone, Environmental gradient. Y-axis: Species importance (abundance)

(b) Y-axis: Species importance (abundance); X-axis: Environmental gradient

conditions repeats itself in a given region, the same group of species occurs.

Some scientists of the early 20th century thought that association implied processes that might be responsible for structuring communities. The logic was that if clusters or groups of species repeatedly associate together, that is indirect evidence for either positive or neutral interactions among them. Such evidence favors a view of communities as integrated units. A leading proponent of this thinking was the Nebraskan botanist Frederick Clements. Clements developed what has become known as the **organismic concept of communities.** Clements likened associations to organisms, with each species representing an interacting, integrated component of the whole. Development of the community through time (a process termed *succession* that we will explore in Chapter 18) was viewed as development of the organism.

As depicted in Figure 16.12a, in Clements' view the species in an association have similar distributional limits along the environmental gradient, and many of them rise to maximum abundance at the same point. Transitions between adjacent communities (or associations) are narrow, with few species in common. This view of the community suggests a common evolutionary history and similar fundamental responses and tolerances (see Section 2.9) for the component species. Mutualism and coevolution play an important role in the evolution of species that make up the association. The community has evolved as an integrated whole; species interactions are the "glue" holding it together.

Figure 16.12 | Two models of community. **(a)** The organismal, or discrete, view of communities proposed by Clements. Clusters of species (Cs, Ds, and Es) show similar distribution limits and peaks in abundance. Each cluster defines an association. A few species (e.g., A) have sufficiently broad ranges of tolerance that they occur in adjacent associations but in low numbers. A few other species (e.g., B) are ubiquitous. **(b)** The individualistic, or continuum, view of communities proposed by Gleason. Clusters of species do not exist. Peaks of abundance of dominant species, such as A, B, and C, are merely arbitrary segments along a continuum.

Community Similarity

When saying that a community's structure changes as we move across the landscape, we imply that the set of species that define the community differ from one place to another. But how do we quantify this change? How do ecologists determine where one community ends and another begins? Distinguishing between communities based on differences in species composition is important in understanding the processes that control community structure as well as in conservation efforts to preserve natural communities (see Chapter 28).

Various indexes have been developed that measure the similarity between two areas or sample plots based on species composition. Perhaps the most widely used is **Sorensen's coefficient of community (CC).** The index is based on species presence or absence. Using a list of species compiled for the two sites or sample plots that are to be compared, the index is calculated as

Number of species common
to both communities

$$CC = 2c/(s_1 + s_2)$$

Number of species in community 1

Number of species in community 2

As an example of this index, we can use the two forest communities presented in Tables 16.1 and 16.2:

$$s_1 = 24 \text{ species}$$
$$s_2 = 10 \text{ species}$$
$$c = 9 \text{ species}$$
$$CC = \frac{(2 \times 9)}{(24 + 10)} = \frac{18}{34} = 0.529$$

The value of the index ranges from 0, when the two communities share no species in common, to 1.0, when the species composition of the two communities is identical (all species in common).

The *CC* does not consider the relative abundance of species. It is most useful when the intended focus is the presence or absence of species. Another index of community similarity that is based on the relative abundance of species within the communities being compared is the **percent similarity (PS).**

To calculate *PS*, first tabulate species abundance in each community as a percentage (as was done for the two communities in Tables 16.1 and 16.2). Then add the lowest percentage for each species that the communities have in common. For the two forest communities, 16 species are exclusive to one community or the other. The lowest percentage of those 16 species is 0, so they need not be included in the summation. For the remaining nine species, the index is calculated as follows:

$$PS = 29.7 + 4.7 + 4.3 + 0.8 + 3.6 + 2.9$$
$$+ 0.4 + 0.4 + 0.4 = 47.2$$

This index ranges from 0, when the two communities have no species in common, to 100, when the relative abundance of the species in the two communities is identical. When comparing more than two communities, a matrix of values can be calculated that represents all pairwise comparisons of the communities—referred to as a *similarity matrix.*

1. Calculate both the Sorensen's and percent similarity indexes using the data presented in Figure 16.7 for the forests on the hilltop and bottomland sites.

2. Are these two forest communities more or less similar than the two sites in West Virginia?

In contrast to Clements' organismal view of communities was the botanist H. A. Gleason's view of community. Gleason stressed the individualistic nature of species distribution. His view became known as the **individualistic,** or **continuum concept.** The continuum concept states that the relationship among coexisting species (species within a community) is due to similarities in their requirements and tolerances, not to strong interactions or common evolutionary history. In fact, Gleason concluded that changes in species abundance along environmental gradients occur so gradually that it is not practical to divide the vegetation (species) into associations. Unlike Clements, Gleason asserted that species distributions along environmental gradients do not form clusters but represent the independent responses of species. Transitions are gradual and difficult to identify (Figure 16.12b). What we refer to as the community is merely the group of species found to coexist under any particular set of environmental conditions. The major difference between these two views is the importance of interactions, evolutionary and current, in the structuring of communities. It is tempting to choose between these views; but as we will see, current thinking involves elements of both perspectives.

Summary

Diversity | 16.1

A community is the group of species (populations) that occupy a given area and interact either directly or indirectly. The number of species in the community defines species richness. Species diversity involves two components: species richness and species evenness, which reflect how individuals are apportioned among the species (relative abundance).

Dominance | 16.2

When a single or a few species predominate within a community, they are referred to as dominants. The dominants are the most numerous.

Keystone Species | 16.3

Keystone species are species that function in a unique and significant manner through their activities, and their effect on the community is disproportionate to their numerical abundance. Their removal initiates changes in community structure and often results in a significant loss of diversity. Their role in the community may be to create or modify habitats or to influence the interactions among other species.

Food Webs | 16.4

Feeding relationships can be graphically represented as a food chain: a series of arrows, each pointing from one species to another that is a source of food. Within a community, many food chains mesh into a complex food web with links leading from primary producers through an array of consumers. Species that are fed upon but do not feed on others are termed basal species. Species that feed on others but are not prey for other species are termed top predators. Species that are both predators and prey are termed intermediate species.

Functional Groups | 16.5

Species that exploit a common resource in a similar fashion are termed guilds. Functional group or functional type is a more general term used to define a group of species based on their common response to the environment, life history characteristics, or role within the community.

Physical Structure | 16.6

Communities are characterized by physical structure. In terrestrial communities, the structure is largely defined by the vegetation. Vertical structure on land reflects the life-forms of plants. In aquatic communities, it is largely defined by physical features such as light, temperature, and oxygen profiles. All communities have an autotrophic and a heterotrophic layer. The autotrophic layer carries out photosynthesis. The heterotrophic layer uses carbon stored by the autotrophs as a food source. Vertical layering provides the physical structure in which many forms of animal life live.

Zonation | 16.7

Changes in the physical structure and biological communities across a landscape result in zonation. Zonation is common to all environments, aquatic and terrestrial. Zonation is most pronounced where sharp changes occur in the physical environment, as in aquatic communities.

Community Boundaries | 16.8

In most cases, transitions between communities are gradual, and defining the boundary between communities is difficult. The way we classify a community depends on the scale we use.

Concept of the Community | 16.9

Historically, there have been two contrasting concepts of the community. The organismal concept views the community as a unit, an association of species, in which each species is a component of the integrated whole. The individualistic concept views the co-occurrence of species as a result of similarities in requirements and tolerances.

Study Questions

1. Distinguish between species richness and species evenness.
2. Distinguish between a dominant and a keystone species.
3. Why are plants classified as basal species in the structure of a food web?
4. Are all carnivores top predators? What distinguishes a top predator in the structure of a food chain?
5. Contrast the vertical stratification of an aquatic community with that of a terrestrial community.
6. Define zonation.
7. In Figure 16.10, the vegetation of Great Smoky Mountains National Park is classified into distinct community types. Does this approach suggest the organismal or individualistic concept of communities? Why?

Further Readings

Brown, J. H. 1995. *Macroecology*. Chicago: University of Chicago Press.

> In this book, Brown presents a broad perspective for viewing ecological communities over large geographic and time scales.

Estes, J., M. Tinker, T. Williams, and D. Doak. 1998. Killer whale predation on sea otters linking oceanic and nearshore ecosystems. *Science* 282:473–476.

> An excellent example of the role of keystone species in the coastal marine communities of western Alaska.

Pimm, S. L. 1982. *Food webs*. New York: Chapman & Hall.

> Although first published over 20 years ago, this book remains the most complete and clearest introduction to the study of food webs.

Pimm, S. L. 1991. *The balance of nature*. Chicago: University of Chicago Press.

> An excellent example of the application of theoretical studies on food webs and the structure of ecological communities to current issues in conservation ecology.

Power, M. E., D. Tilman, J. Estes, B. Menge, W. Bond, L. Mills, G. Daily, J. Castilla, J. Lubchenco, and R. Paine. 1996. Challenges in the quest for keystones. *Bioscience* 46:609–620.

> This paper reviews the concept of keystone species as presented by many of the current leaders in the field of community ecology.

Ricklefs, R. E., and D. Schluter, eds. 1993. *Ecological communities: Historical and geographic perspectives*. Chicago: University of Chicago Press.

> This pioneering work examines biodiversity in its broadest geographical and historical contexts, exploring questions relating to global patterns of species richness and the historical events that shape both regional and local communities.

CHAPTER 1 7

Factors Influencing the Structure of Communities

Douglas-fir and western hemlock with an abundance of dead wood and decomposing logs—a setting characteristic of old-growth forests.

The community is a group of plant and animal species that inhabit a given area. As such, understanding the biological structure of the community depends on understanding the distribution and abundance of species. Thus far we have examined a wide variety of topics addressing this broad question, including the adaptation of organisms to the physical environment, the evolution of life history characteristics and their influence on population demography, and the interactions among different species. In Chapter 16, we examined characteristics that define both the biological and physical structure of communities and described the structure of community change as one moves across the landscape. However, the role of science is to go beyond description and to answer fundamental questions about the processes that give rise to these observed patterns. What processes shape these patterns of community structure? How will communities respond to the addition or removal of a species? Why are communities in some environments more or less diverse than others? In this chapter, we integrate our discussion of the adaptation of organisms to the physical environment presented in Part Two (Chapters 5–8) with the discussion of species interactions presented in Part Four (Chapters 13–15) to explain the processes that control community structure in a wide variety of communities.

17.1 | The Fundamental Niche Constrains Community Structure

In Chapter 5, we introduced the concept of environmental tolerances (see Section 5.9). All living organisms have a range of environmental conditions under which they can successfully survive, grow, and reproduce. This range of environmental conditions is not the same for all organisms. The conditions under which an organism can function successfully are the consequence of a wide variety of physiological, morphological, and behavioral adaptations. As well as allowing an organism to function under a specific range of environmental conditions, these same adaptations also limit its ability to do equally well under different conditions. In Part Two, we explored many examples of this premise. Plants adapted to high-light environments exhibit characteristics that preclude them from being equally successful under low-light conditions (see Chapter 6). Animals that regulate body temperature through ectothermy (cold-blooded animals—poikilotherms) are able to reduce energy requirements during periods of resource shortage. Dependence on external sources of energy, however, limits diurnal and seasonal periods of activity as well as the geographic distribution of poikilotherms (see Chapter 7). The number of offspring produced at any one time constrains the nature of parental care (see Chapter 8). Each set of adaptations

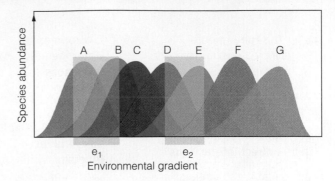

Figure 17.1 | Fundamental niches of seven hypothetical species along an environmental gradient (e.g., moisture, temperature, or elevation) in the absence of competition from other species. The species all have bell-shaped responses to the gradient, but each has different tolerance limits defined by a minimum and maximum value along the gradient. As conditions change, for example from e_1 to e_2, the set of species that can potentially occur in the community changes.

reflects a solution to a set of environmental conditions and, conversely, restricts or precludes adaptation to another. Such adaptations define the fundamental niche of a species (see Section 13.9).

Environmental conditions vary in both time and space (Part One). This observation, when combined with inherent differences in the fundamental niches of species, provides a starting point for exploring the processes that structure communities. We can represent the fundamental niches of various species with bell-shaped curves along an environmental gradient, such as availability of water or light for plants (Figure 17.1). The response of each species is defined in terms of its population density or abundance. Although the fundamental niches overlap, each species has limits beyond which it cannot survive. The distribution of fundamental niches along the environmental gradient represents a primary constraint on the structure of communities. For any given range of environmental conditions, only a subset of species can survive, grow, and reproduce. As environmental conditions change from location to location, the possible distribution and abundance of species will change—in turn changing the community's structure. For example, Figure 17.2 depicts geographic distributions of three of the tree species that are components of the forest communities in West Virginia as described in Tables 16.1 and 16.2. As we discussed in Chapter 9, these geographic distributions reflect the occurrence of suitable environmental conditions (within the range of environmental tolerances). Note that the geographic distributions of the three species are quite distinct, and the sites in West Virginia are part of a relatively small geographic region where the distributions of these three species overlap. As we move from this site in West Virginia to other regions of eastern North America, the set of tree species whose

Figure 17.2 | Geographic distribution of three tree species that are part of the two forest communities presented in Tables 16.1 and 16.2. The three tree species have distinct geographic distributions, and the site in West Virginia (shown as a red circle) falls within a very limited region where the distributions of the three species overlap. As you move across eastern North America, the set of tree species whose distributions overlap will change, and so will the species composition of the forest communities.

distributions overlap will change and, subsequently, so will the biological structure of the forest community.

This view of community represents what ecologists refer to as a **null model.** It assumes that the presence and abundance of the individual species found in a given community are solely a result of the independent responses of each individual species to the prevailing physical environment. Interactions among species have no significant influence on community structure. Considering the examples we have reviewed in the previous three chapters, this assumption must seem somewhat odd. However, it is helpful as a framework for comparing the actual patterns observed within the community. For example, this particular null model is the basis for comparisons in the experiments, examined in the previous chapters, in which the interactions between two species (competition, predation, parasitism, and mutualism) are explored by physically removing one species and examining the population response of the other. If the population of the remaining species does not differ from that observed previously in the presence of the removed species, we could assume that the apparent interspecific interaction has no influence on the remaining species' abundance within the community.

A great deal of evidence, however, indicates that species interactions do influence both the presence and abundance of species within a wide variety of communities. As we have seen in the examples presented in Part Four, species interactions modify the fundamental niche of both species involved, influencing their relative abundance and, in some cases, their distribution. The process of interspecific competition can reduce the abundance of or even exclude some species from a community, while positive interactions such as facilitation and mutualism can enhance the presence of a species or even extend a

species' distribution beyond that defined by its fundamental niche (see Field Studies: Sally Hacker, p. 352). Because studies that examine species interactions typically focus on only two (or at best a small subset) of the species found within a community, such studies most likely underestimate the importance of species interactions on the structure and dynamics of communities.

17.2 | Species Interactions Are Diffuse

One reason such experiments tend to underestimate the importance of species interactions in communities is that such interactions are often diffuse, involving a number of species. The work of Norma Fowler at the University of Texas provides an example. She examined competitive interactions within an old-field community by selectively removing species of plants from experimental plots and assessing the growth responses of remaining species. Her results showed that competitive interactions within the community tended to be rather weak and diffuse because removing a single species had relatively little effect. The response to removing groups of species, however, tended to be much stronger, suggesting that individual species compete with several other species for essential resources within the community. In diffuse competition, the direct interactions between any two species may be weak, making it difficult to determine the effect of any given species on another. Collectively, however, competition may be an important factor limiting the abundance of all species involved.

Diffuse interactions, where one species may be influenced by interactions with many different species, is not limited to competition. In the example of predator–prey cycles in Chapter 14 (see Section 14.16), a variety of predator species (including the lynx, coyote, and horned owl) are

Department of Zoology, Oregon State University, Corvallis, Oregon

Salt marsh plant communities are ideal for examining the forces that structure natural communities. They are typically dominated by a small number of plant species that form distinct zonation patterns (see Figure 16.8). Seaward distribution of marsh plant species is set by harsh physical conditions such as waterlogged soils and high soil salinities, whereas terrestrial borders are generally set by competitive interactions (see Figure 13.10). Yet marsh plants also have strong ameliorating effects on these harsh physical conditions. Shading by marsh plants limits surface evaporation and the accumulation of soil salts. In addition, the transport of oxygen to the rooting zone (rhizosphere) by marsh plants can alleviate anaerobic substrate conditions. How might these modifying effects of marsh plants on the physical environment influence the structure of salt marsh communities? This question has been central to the research of ecologist Sally Hacker of Oregon State University.

To examine the role of plant–physical environment interactions on salt marsh plant zonation, Hacker focused on the terrestrial border of New England marshes. In southern New England, terrestrial marsh borders (see Figure 13.10) are dominated by the perennial shrub *Iva frutescens* (marsh elder) mixed with the rhizomatous perennial rush *Juncus gerardi* (black grass), which also dominates the lower marsh elevations. The seaward border of the *Iva* zone is often characterized by low densities of stunted (35–50 cm) adult plants, whereas at higher elevations *Iva* are taller (up to 150 cm), more productive, and reach higher densities.

Previous studies had suggested that *Iva* is relatively intolerant of high soil salinities and waterlogged soil conditions. Given the potential role of marsh plants to modify the local environment, Hacker hypothesized that the modifying effects of *Juncus* on soil environment function to extend the seaward distribution of *Iva*. To test this hypothesis, Hacker and colleague Mark Bertness of Brown University applied one of three treatments to randomly selected adult *Iva* shrubs on the seaward border of the *Iva* zone. The three treatments were designed to examine the effects of *Juncus* neighbors on established adult *Iva*: (1) all *Juncus* within a 0.5-m radius of each *Iva* plant were regularly clipped to ground level (neighbor removal, or NR), (2) all *Juncus* were clipped (as in NR) and then the soil was covered with a water-permeable fabric (shaded neighbor removal, or SNR), and (3) control. The use of fabric in the SNR treatment mimics the effect of *Juncus* shading on soil salinities without the effects of *Juncus* transporting oxygen to the rhizosphere (increased soil oxygen).

Soil physical conditions (soil salinity and redox, a measure of oxygen content of soils) and *Iva* performance (photosynthetic rates and leaf production) were monitored in all treatments for a 2-year period.

Removing *Juncus* plants in the neighborhood of *Iva* shrubs strongly affected local physical conditions (Figure 1). Removing *Juncus* neighbors more than doubled soil salinities in contrast to other treatments and led to more than an order-of-magnitude drop in soil redox, suggesting that the presence of *Juncus* neighbors increases soil oxygen levels. Because shading plots without *Juncus* (SNR treatment) prevented salinity increases but did not influence soil redox, the NR and SNR treatments separated the effects caused by both salt buffering and soil oxidation from those caused only by soil oxidation.

Photosynthetic rate and leaf production of *Iva* individuals in the treatment where *Juncus* neighbors were removed (NR) declined significantly in comparison to either the shaded neighbor removal (SNR) or control (C)

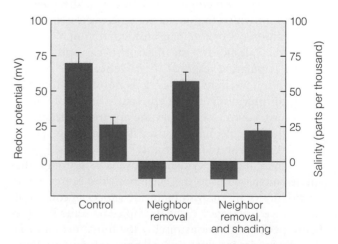

Figure 1 | Redox potential (blue) and surface salinity (green) of soil in the *Iva frutescens* neighbor manipulation treatments. The data are means (± standard error) of pooled monthly (June–September) measurements during 1991–92.

(Adapted from Bertness and Hacker 1994.)

treatments (Figure 2). Fourteen months after the experimental treatments were established, all *Iva* in the NR treatment were dead. These results show that soil salinity is the primary factor influencing the performance of *Iva* across the gradient. They also show that the presence of *Juncus,* with its superior ability to withstand waterlogging and salt stress, modifies physical conditions in such a manner as to create a hospitable environment for *Iva,* and allow this species to extend its distribution to lower intertidal habitats.

The *Iva–Juncus* interaction has interesting consequences for higher trophic levels in the marsh. The most common insects living on *Iva* are aphids (*Uroleucon ambrosiae*) and their predators, ladybird beetles (*Hippodamia convergens* and *Adalia bipunctata*). Interestingly, aphids are most abundant on short, stunted *Iva* in the lower intertidal zone, despite having far higher growth rates on the taller *Iva* shrubs in the upper intertidal zone. This reduced growth rate occurs because ladybird beetles prefer tall structures, and increased predation on tall plants restricts aphids to the poorer-quality *Iva* plants in the lower marsh. These findings prompted the investigators to hypothesize that the *Iva–Juncus* interaction is critical in maintaining aphid populations in the marsh.

To explore this hypothesis, Hacker examined the abundance of aphids and ladybird beetles on the *Iva* individuals with (treatment C) and without (treatment NR) *Juncus* neighbors. To determine how the aphid population growth rates were affected by the absence of

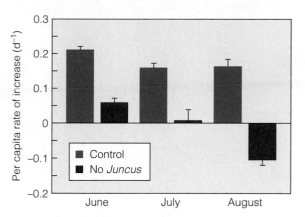

Figure 3 | The per capita rate of population increase for the period June–August (1993) on control and no *Juncus* plants (neighbors removed).

Juncus, Hacker calculated aphid population growth rates as the per capita rate of increase per day.

The overall percentage of plants with aphids and ladybird beetle predators was significantly higher for *Iva* plants without *Juncus* (NR) than for control (C) plants. This result suggests that *Juncus* neighbors influence stunted *Iva* by making it less noticeable for colonizing aphids as well as their predators. Although the removal of neighboring *Juncus* increased the proportion of *Iva* individuals colonized by aphids, growth rates for aphid populations (Figure 3) were lower on the stunted *Iva* without *Juncus* (NR) than for control individuals (C). Even though aphids are better at finding stunted *Iva* host plants when *Juncus* is removed, by late summer (August), population growth rates were negative on *Iva* individuals without neighbors—indicating that food quality of *Iva* host plants decreases such that aphids were unable to produce enough offspring to replace themselves.

Bibliography

Bertness, M. D., and S. D. Hacker. 1994. Physical stress and positive associations among marsh plants. *The American Naturalist* 144:363–372.

Hacker, S. D., and M. D. Bertness. 1996. Trophic consequences of a positive plant interaction. *The American Naturalist* 148:559–575.

1. How do the results presented in Figures 1 and 2 suggest that salinity is the factor limiting seaward distribution of *Iva* in the marsh?

2. How does the presence of *Juncus* function to maintain the aphid populations in the marsh?

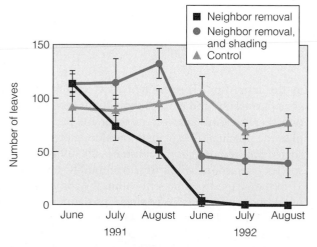

Figure 2 | Total number of leaves on adult *Iva frutescens* under experimentally manipulated conditions, with and without neighbors. All data are means (± standard error).

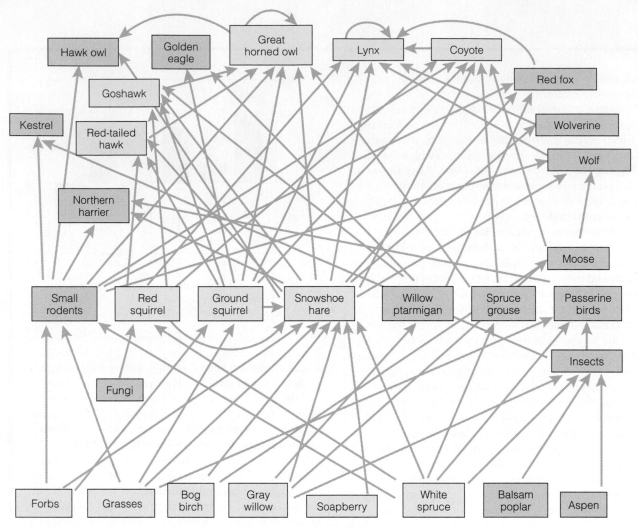

Figure 17.3 | A generalized food web for the boreal forests of northwestern Canada. Dominant species within the community are shown in green. Arrows link predator with prey species. Arrows that loop back to the same species (box) represent cannibalism.

(Adapted from Krebs 2001.)

responsible for periodic cycles observed in the snowshoe hare population. Examples of diffuse mutualisms relating to both pollination and seed dispersal are presented in Chapter 15 (see Sections 15.13 and 15.14), where a single plant species may depend on a variety of animal species for successful reproduction. Although food webs present only a limited view of species interactions within a community, they are an excellent means of illustrating the diffuse nature of species interactions (see Quantifying Ecology 17.1: Quantifying the Structure of Food Webs: Connectance). Charles J. Krebs of the University of British Columbia has developed a generalized food web for the boreal forest communities of northwestern Canada (Figure 17.3). This food web contains the plant–snowshoe hare–carnivore system discussed in Chapter 14 (see Figure 14.19). The arrows point from predator to prey, and an arrow that circles back

to the same box (species) represents cannibalism (e.g., great horned owl and lynx). Although this food web shows only the direct links between predator and prey, it also implies the potential for competition among predators for a shared prey resource; and it illustrates the diffuse nature of species interactions within this community. For example, 11 of the 12 predators present within the community prey upon snowshoe hares. Any single predator species may have a limited effect on the snowshoe hare population, but the combined impact of multiple predators regulates the snowshoe hare population. This same example illustrates the diffuse nature of competition within this community. Although the 12 predator species feed on a wide variety of prey species, snowshoe hares represent an important shared food resource for the three dominant predators: lynx, great horned owls, and coyotes.

Quantifying the Structure of Food Webs: Connectance

The use of food webs to describe community structure has raised many questions about the forces that structure ecological communities. Communities differ in the arrangement of species and feeding links, and these differences affect the population dynamics of component species. As we will see in Section 17.3, the manner in which species are connected within the food web has implications for community structure and dynamics beyond the direct interactions between predator and prey. One way that food webs of different communities differ is in their degree of connectance.

Connectance is a way of describing how many possible links in a food web are present. Links are simply the lines that link consumers and the consumed. One formula for connectance is

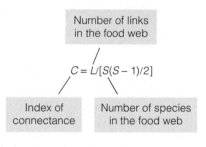

$$C = L/[S(S-1)/2]$$

Number of links in the food web

Index of connectance

Number of species in the food web

This formula is based on the notion that in a web consisting of S species, there are $S(S-1)/2$ possible unidirectional links (the link between any two species is in only one direction), excluding cannibalism.

Consider the simple food web just presented. There are four species (circles) and three links (arrows). The number of possible links is $4(4-1)/2$, and the connectance is 3/6 or 0.5. These possible links are shown in the following figure. No arrows are shown, because the possible links could represent either one of two possible interactions (direction of arrow).

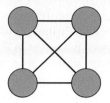

Linkage density, L/S, refers to the average number of feeding links per species. It is a function of the connectance and the number of species in the food web. In the example presented above, the value of linkage density is 3/4, or 0.75.

The density of interactions in a food web provides one measure of community complexity. Highly connected systems contain many links for a given number of species. One central question in the study of food webs is how connectance varies with diversity. Some recent studies suggest that the number of links in a food web increases with species richness, but connectance decreases. The reason for this pattern is not completely understood, but the connectance of a food web will be influenced by the abundance of generalists (predators that feed on a variety of prey species) as compared to the number of feeding specialists (predators that feed on only one or few prey species). Likewise, connectance will increase with the presence of omnivores as compared to strict herbivores or carnivores.

Use the food web shown here to answer the following questions.

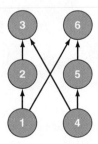

1. How many links are present in this food web?
2. How many possible links are there (using the formula presented earlier)?
3. What is the value of connectance for this food web?
4. Are any of the species in the food web omnivores?

17.3 | Food Webs Illustrate Indirect Interactions

Food webs also illustrate a second important feature of species interactions within the community: indirect effects. Indirect interactions occur when one species does not interact with a second species directly, but instead influences a third species that does directly interact with the second. For example, in the food web presented in Figure 17.3, lynx do not directly interact with white spruce; however, by reducing snowshoe hare and other herbivore populations that feed on white spruce, lynx predation can positively affect the white spruce population (survival of seedlings and saplings). The key feature of indirect interactions is that they can potentially arise throughout the entire community because of a single direct interaction between only two component species.

By affecting the outcome of competitive interactions among prey species, predation provides another example of indirect effects within food webs. Robert Paine of the University of Washington was one of the first ecologists to demonstrate this point. The intertidal zone along the rocky coastline of the Pacific Northwest is home to a variety of mussels, barnacles, limpets, and chitons (all invertebrate herbivores). All of these species are preyed upon by the starfish *Pisaster* (Figure 17.4). Paine conducted an experiment in which he removed the starfish from some areas (experimental plots) while leaving other areas undisturbed for purposes of comparison (controls). After the starfish were removed, the number of prey species in the experimental plots dropped from 15 at the beginning of the experiment to 8. In the absence of predation, several of the mussel and barnacle species that were superior competitors excluded the other species and reduced overall diversity in the community. This type of indirect interaction is called **keystone predation,** where the predator enhances one or more inferior competitors by reducing the abundance of the superior competitor (see discussion of keystone species in Section 16.3).

Ecologist Robert Holt of the University of Florida first described the conditions that might promote a type of indirect interaction he referred to as **apparent competition.** In its simplest form, apparent competition occurs when a single species of predator feeds on two prey species (Figure 17.5). When the predator species is absent, each population of the two prey species is regulated by purely

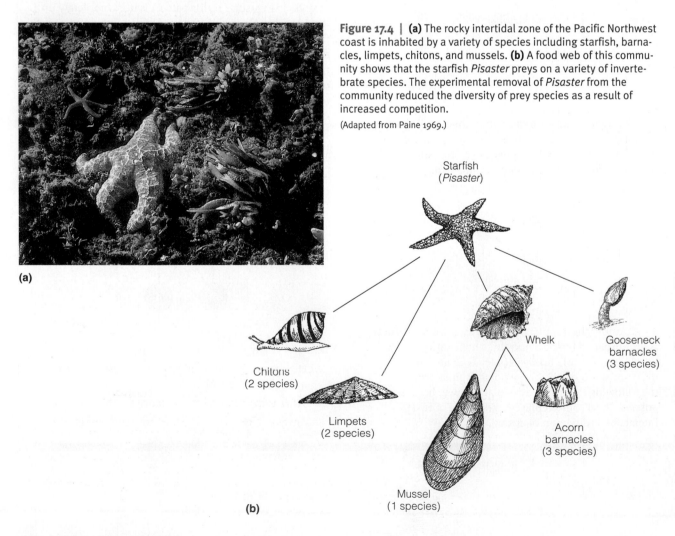

(a)

(b)

Figure 17.4 | **(a)** The rocky intertidal zone of the Pacific Northwest coast is inhabited by a variety of species including starfish, barnacles, limpets, chitons, and mussels. **(b)** A food web of this community shows that the starfish *Pisaster* preys on a variety of invertebrate species. The experimental removal of *Pisaster* from the community reduced the diversity of prey species as a result of increased competition.

(Adapted from Paine 1969.)

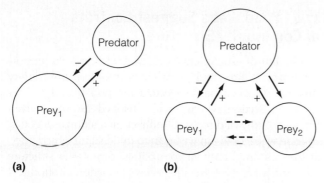

(a)   **(b)**

Figure 17.5 | Diagram illustrating the emergence of apparent competition between two prey species (Prey$_1$ and Prey$_2$) that have a common predator (Predator). Direct interactions are represented by a solid arrow, and indirect interactions between species are indicated by a dashed arrow. The size of the predator and prey populations is indicated by the size of the respective circles. The relationship between a single prey species (Prey$_1$) and its predator **(a)** is modified by the presence of a second prey species (Prey$_2$) upon which the predator also feeds **(b)**. The combined populations of the two prey species, however, function to increase the predator population to a level above what would be supported by either of the two prey species alone. The higher rates of predation thus reduce the populations of both prey species, even though the two prey species do not interact directly. The reduction in both prey populations gives the outward appearance of interspecific competition.

intraspecific, density-dependent mechanisms (see Chapter 11). Neither species competes, directly or indirectly, with the other. Now assume that the predator abundance depends on the combined abundance of all prey species (numerical response; discussed in Section 14.5). Under these conditions, the combined population abundance of the two prey species will support a higher predator density than in situations where only a single prey species occurs. Because increased predator abundance will increase the rate of prey consumption (see Section 14.4), a situation can occur in which both prey species occur at lower densities when they occur together than when they occur separately. The pattern of reduced population density in the presence of another species would have the outward appearance of interspecific competition (the prey species are less abundant when coexisting than when in the other species' absence), but in fact the two species do not compete either directly or indirectly (see Figure 17.5). The lower abundance of the two co-occurring prey species is caused entirely by the greater abundance of the predator population supported by both prey populations together than by either prey population alone.

Apparent competition is an interesting concept that can arise from the structure of food webs. But does it really occur in nature? Many studies have identified community patterns that are consistent with apparent competition, and there is convincing experimental evidence of apparent competition in intertidal, freshwater, and terrestrial communities. One such study was conducted by the ecologists Christine Müller and H. C. J. (Charles) Godfray of Imperial College in Berkshire, England. Müller and Godfray exam-

ined the role of apparent competition between two species of aphids that do not interact directly, yet share a common predator. The nettle aphid (*Microlophium carnosum*) feeds only on nettle plants (*Urtica* spp.), whereas the grass aphid (*Rhopalosiphum padi*) feeds on a variety of grass species. Although these two aphid species use different plant resources within the field community, they share a common predator, the ladybug beetle (Coccinellidae). In their study, the researchers placed potted nettle plants containing colonies of nettle aphids in plots of grass within the field community that contained natural populations of grass aphid. On a subset of the grass plots, they applied fertilizer that led to rapid grass growth and an increase in the local population of grass aphids. Nettle aphid colonies adjacent to the fertilized plots suffered a subsequent decline in population density when compared to colonies that were adjacent to unfertilized plots (control plots with low grass aphid populations). The reduced population of nettle aphids in the vicinity of high population densities of grass aphids (fertilized plots) thus was due to increased predation by ladybug beetles, attracted into the area by the large concentrations of grass aphids; it was not due to direct resource competition between the two aphid species.

Some indirect interactions have negative consequences for the affected species, as in the preceding case of apparent competition. In other cases, however, indirect interactions between species can be positive. An example comes from a study of subalpine ponds in Colorado by Stanley Dodson of the University of Wisconsin. It involves the relationships between two herbivorous species of *Daphnia* and their predators, a midge larva (*Chaoborus*) and a larval salamander (*Ambystoma*). The salamander larvae prey on the larger of the two *Daphnia* species, while the midge larvae prey on the small species (Figure 17.6). In a study of 24 pond communities

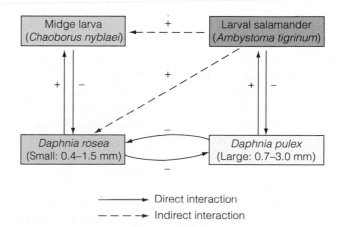

Direct interaction
Indirect interaction

Figure 17.6 | Diagram showing the relationship among the midge larva (*Chaoborus*), larval salamander (*Ambystoma*), and two species of *Daphnia* (*D. rosea* and the larger *D. pulex*) that inhabit pond communities in the mountains of Colorado. Removing salamander larvae from some ponds resulted in the competitive exclusion of *D. rosea* by *D. pulex* and the local extinction of the midge population that preyed on it.

(Adapted from Dodson 1974.)

in the mountains of Colorado, Dodson found that where salamander larvae were present, the number of large *Daphnia* was low and the number of small *Daphnia* high. However, in ponds where salamander larvae were absent, small *Daphnia* were absent and midges could not survive. The two species of *Daphnia* apparently compete for the same resources. When the salamander larvae are not present, the larger of the two *Daphnia* species can outcompete the smaller. With the salamander larvae present, however, predation reduces the population growth rate of the larger *Daphnia*, allowing the two species to coexist. In this example, two indirect positive interactions arise. The salamander larvae indirectly benefit the smaller species of *Daphnia* by reducing the population size of its competitor. Subsequently, the midge apparently depends on the presence of salamander larvae for its survival in the pond. The indirect interaction between the midge and the larval salamander is referred to as **indirect commensalism,** because the interaction is beneficial to the midge while neutral to the larval salamander. When the indirect interaction is beneficial to both species, the indirect interaction is termed **indirect mutualism.**

This role of indirect interactions can be demonstrated only in controlled experiments involving manipulations of the species populations involved. The importance of indirect interactions remains highly speculative; but experiments such as those just presented strongly suggest that indirect interactions among species, both positive and negative, can be an integrating force in structuring natural communities. There is a growing appreciation within ecology for the role of indirect effects in shaping community structure, and understanding these complex interactions is more than an academic exercise: it has direct implications for conservation and management of natural communities.

As with the example of starfish in the intertidal zone, removing a species from the community can have many unforeseen consequences. For example, Joel Berger of the University of Nevada and colleagues have examined how the local extinctions of grizzly bears (*Ursus arctos*) and wolves (*Canis lupus*) from the southern Greater Yellowstone ecosystem, resulting from decades of active predator control, have affected the larger ecological community. One unforeseen consequence of losing these large predators is the decline of bird populations that use the vegetation along rivers (riverine habitat) within the region. The elimination of large predators from the community resulted in an increase in the moose population (prey species). Moose selectively feed on willow (*Salix* spp.) and other woody species that flourish along the river shorelines. The increase in moose populations dramatically affected the vegetation in riverine areas that provide habitat for a wide variety of bird species and led to the local extinction of some populations.

17.4 | Food Webs Suggest Controls of Community Structure

The wealth of experimental evidence illustrates the importance of both direct and indirect interactions on community structure. On that basis, rejecting the null model as presented in Section 17.1 would be justified. However, given the complexity of direct and indirect interactions suggested by food webs, how can we begin to understand which interactions are important in controlling community structure and which are not? Are all species interactions important? Does some smaller subset of interactions exert a dominant effect, whereas most have little impact beyond those species directly involved? The hypothesis that all species interactions are important in maintaining community structure would suggest that the community is like a house of cards—that is, removing any one species may have a cascading effect on all others. The hypothesis that only a smaller subset of species interactions are controlling community structure suggests a more loosely connected assemblage of species.

These questions are at the forefront of conservation ecology, due to the dramatic decline in biological diversity that is a result of human activity. Certain species within the community can exert a dominant influence on its structure, such as the predatory starfish that inhabits the rocky intertidal communities. However, the relative importance of most species in the functioning of communities is largely a mystery. One approach being used to understand the influence of species diversity on the structure and dynamics of communities is to group species into functional categories based on criteria relating to their function within the community. For example, the concept of guilds presented in Section 16.5 is a functional grouping of species based on sharing similar functions within the community or exploiting the same resource (e.g., grazing herbivores, pollinators, cavity-nesting birds). By aggregating species into a smaller number of functional groups, researchers can explore the processes controlling community structure in more general terms. For example, what is the role of mammalian predators in boreal forest communities? This functional grouping of species can be seen in the food web presented in Figure 17.3, where the categories (boxes) of forbs, grasses, small rodents, insects, and passerine birds represent groups of functionally similar species.

One way to simplify food webs is to aggregate species into trophic levels, as discussed in Section 16.4. The food web presented in Figure 17.3 has been aggregated into three trophic levels: primary producers, herbivores, and carnivores (Figure 17.7). Although this is an obvious oversimplification, using this approach raises some fundamental questions concerning the processes that control community structure.

As with food webs, the arrows in a simple food chain based on trophic levels point in the direction of energy flow—from primary producers to herbivores, and from

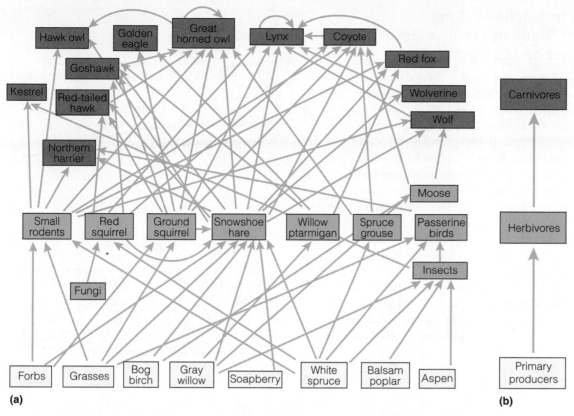

Figure 17.7 | **(a)** Aggregation of species forming the food web for Canadian boreal forests presented in Figure 17.3 into trophic levels (generalized feeding groups). **(b)** As with the food web, arrows point from prey to predator.

herbivores to carnivores. The structure of food chains suggests that the productivity and abundance of populations at any given trophic level are controlled (limited) by the productivity and abundance of populations in the trophic level below them. This phenomenon is called **bottom-up control.** Plant population densities control the abundance of herbivore populations, which in turn control the densities of carnivore populations in the next trophic level. However, as we have seen from the previous discussion of predation and food webs, **top-down control** also occurs when predator populations can control the abundance of prey species.

Work by Mary Power and her colleagues at the University of Oklahoma Biological Station suggests that the role of top predators (carnivores) on community structure can extend to lower trophic levels, influencing primary producers as well as herbivore populations. Power and colleagues showed that a top predator (largemouth bass, *Micropterus salmoides*) had strong indirect effects that cascaded through the food web to influence the abundance of benthic algae in stream communities of the midwestern United States. In these stream communities, herbaceous minnows (primarily *Campostoma anomalum*) graze on algae, and in turn largemouth bass (and two species of sunfish) feed on the minnows. During periods

of low flow, isolated pools form in the streams. As part of the experiment, bass were removed from some pools, and the populations of algae and minnows were monitored. Pools with bass had low minnow populations and a luxuriant growth of algae. In contrast, pools from which the bass were removed had high minnow populations and low populations (biomass) of algae. In this example, top predators (carnivores) were shown to control the abundance of plant populations (primary producers) indirectly through their direct control on herbivores (also see Field Studies, Brian Silliman, p. 432).

A now famous article written by Nelson Hairston, Fred Smith, and Larry Slobodkin first introduced the concept of top-down control with the frequently quoted "the world is green" proposition. These three ecologists proposed that the world is green (plant biomass accumulates) because predators keep herbivore populations in check. Although this proposition is supported by a growing body of experimental studies like those by Mary Power and her colleagues, experimental data required to test this hypothesis are still limited, particularly in terrestrial ecosystems. However, this proposition continues to cause great debate within the field of community ecology. We will return to the topic in Chapter 20 when discussing factors that control primary productivity.

17.5 | Species Interactions along Environmental Gradients Involve Both Stress Tolerance and Competition

We have now seen that the biological structure of a community is first constrained by the environmental tolerances of the species (fundamental niche). These tolerances are in turn modified through direct and indirect interactions with other species (realized niche). Competitors and predators, for example, can restrict a species from a community; conversely, mutualists can facilitate a species' presence and abundance within the community. As we move across the landscape, variations in the abiotic environment will alter these constraints on species distribution and abundance. Species differ in their range of environmental tolerances, and species interactions can change based on the environmental context (see Section 13.8).

The discussion of plant adaptations to environmental conditions in Chapter 6 provides some insight into how the relative competitive abilities of plant species may vary across environmental gradients of resource availability. Adaptations of plants to variations in the availability of light (see Section 6.9), water (see Section 6.10), and nutrients (see Section 6.12) result in a general pattern of trade-offs between the characteristics that enable a species to survive and grow under low resource availability and those that allow for high rates of photosynthesis and growth under high resource availability (Figure 17.8a). Competitive success in plants is often linked to their growth rate and the acquisition of resources (see Chapter 13). Species that have the highest growth rate and acquire most of the resources at any given point on the resource gradient have the competitive advantage there. The differences in adaptations to resource availability among the species in Figure 17.8a result in a competitive advantage for each species over the range of resource conditions under which they have the greatest growth rate relative to the other plant species present (see discussion of changing competitive ability along resource gradients in Section 13.8). The result is a pattern of zonation along the gradient (Figure 17.8b) that reflects the changing relative competitive abilities. The lower boundary of each species along the gradient is defined by its ability to tolerate (survive and maintain a positive carbon balance) resource limitation, while the upper boundary is defined by competition. Such a trade-off in tolerance and competitive ability can be seen in the examples of interspecific competition and the distribution of cattail species with water depth (see Figure 13.12) and in the distribution of grass species in the semiarid regions of southeastern Arizona (see Figure 13.14). The trade-off between tolerance and competitive ability is also evident in the patterns of zonation along gradients of soil moisture, such as the one reported by the plant ecologist Robert Whittaker (formerly of Cornell University) for lower elevations of the central Siskiyou Mountains of Oregon and California (Figure 17.9).

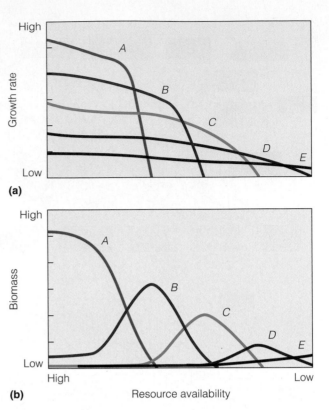

(a)

(b)

Resource availability

Figure 17.8 | (a) General pattern of trade-off between a species' ability to survive and grow under low resource availability and the maximum growth rate achieved under high resource availability as presented in Chapter 6 for the resources of light, water, and soil nutrients. The result is an inverse relationship between physiological maximum growth rate and the minimum resource requirement for hypothetical plant species A–E. Assuming that the superior competitor at any point along the resource gradient (x-axis) will be the species with the highest growth rate, the species' relative competitive ability changes with resource availability. **(b)** The outcome of competition will be a pattern of zonation in which lower boundaries (value of resource availability) for the species are due to differences in their tolerances for low resource availability and upper boundaries are a product of competition. (Adapted from Smith and Huston 1989.)

Interpreting Ecological Data

Q1. In the hypothetical example in graph (a), under what resource conditions (availability) is the growth rate of each plant species assumed to be optimal?

Q2. What is the rank order of the hypothetical plant species in graph (a), moving from most to least tolerant of resource limitation?

Q3. If species B were removed from the hypothetical community, how would the predicted distribution of species A along the resource gradient (x-axis) in graph (b) be altered? How would you expect the distribution of species C to change? Why?

Competition among plant species rarely involves a single resource, however (see Section 13.7). The experiments of R. H. Groves and J. D. Williams examining competition between populations of subterranean clover and skeletonweed (see Section 13.7 and Figure 13.6) clearly show that there is an interaction between competition for both aboveground (light) and belowground resources (water and nutrients). The differences in adaptations relating to the acquisition of above- and belowground resources when they are in short supply can result in changing patterns of competitive ability along gradients where these two classes of resources co-vary. Allocating carbon to the production of leaves and stems provides increased access to the resources of light, but at the expense of allocating carbon to the production of roots. Likewise, allocating carbon to the production of roots increases access to water and soil nutrients but limits the production of leaves and, therefore, the future rate of carbon gain through photosynthesis. As the availability of water (or nutrients) increases along a supply gradient (such as in Figure 17.9), the competitive advantage shifts from those species adapted to low availability of water (high root production) to those species that allocate carbon to leaf production and height growth but that require higher water availability to survive (Figure 17.10).

This framework of trade-offs between the set of characteristics that enable individuals of a plant species to survive and grow under low resource conditions as compared to the set of characteristics that would enable those same individuals to maximize growth and competitive

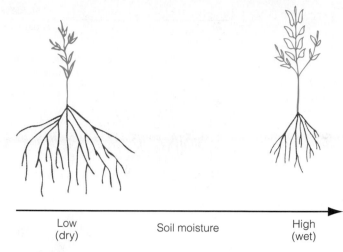

Low (dry)   Soil moisture   High (wet)

Figure 17.10 | General trends in plant adaptations (characteristics) that increase fitness along a soil moisture gradient. At low soil moisture, allocation to the production of roots at the expense of leaves aids in acquiring water and reducing transpiration, allowing the plant to survive (tolerance). As water availability increases, the overall increase in plant growth results in competition for light as some individuals overtop others. A shift in allocation to height growth (stems) and the production of leaves increases a plant's competitive ability.

ability under higher resource conditions is a powerful tool for understanding changes in the structure and dynamics of plant community structure along resource gradients. However, applying this simple framework of trade-offs in phenotypic characteristics can become more complicated when dealing with environmental gradients and communities where there are interactions of resource and nonresource factors (see Section 13.5).

The complex nature of competition along an environmental gradient involving both resource and nonresource factors is nicely illustrated in the pattern of plant zonation in salt marsh communities along the coast of New England (see Figure 13.10). Nancy Emery and her colleagues at Brown University conducted a number of field experiments to identify the factors responsible for patterns of species distribution in these coastal communities. Experiments included the addition of nutrients, removal of neighboring plants, and reciprocal transplants (i.e., planting species in areas where they are not naturally found to occur along the gradient). Experiment results indicate that the patterns of zonation are an interaction between the relative competitive abilities of species for nutrients and the ability of plant species to tolerate increasing physical stress. The low marsh is dominated by *Spartina alterniflora* (smooth cordgrass), a large perennial grass with extensive rhizomes. The upper edge of *S. alterniflora* is bordered by *Spartina patens* (salt-meadow cordgrass), a perennial turf grass, which is replaced at higher elevations in the marsh by *Juncus gerardi* (black needle rush), a dense turf grass (Figure 17.11). Although the low marsh experiences daily flooding by the tides, the *S. patens* and *J. gerardi* zones are inundated only

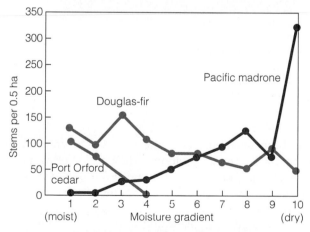

Figure 17.9 | Distribution of tree species along a soil-moisture gradient at low elevations in the central Siskiyou Mountains of Oregon and California. Data are from 50 stands sampled between 610 and 915 m in elevation. Species distributions reflect differences in competitive ability at the moist end of the gradient and the ability to tolerate dry conditions at the dry end of the gradient.

(Adapted from Whittaker 1960.)

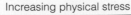

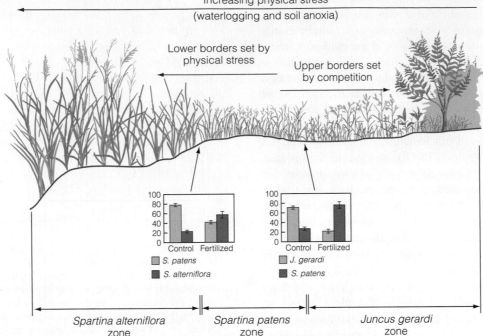

Figure 17.11 | Patterns of plant zonation and physical stress along an elevation gradient in a tidal salt marsh. The lower marsh experiences daily flooding by the tides, but the upper zones are inundated only during the high-tide cycles. The higher water levels of the lower marsh result in lower oxygen levels in the sediments and higher salinities. The lower (elevation) boundary of each species is determined by its tolerance to physical stress, and the upper boundaries are limited by competition. Bar graphs show shifts in percentage of cover by the two adjoining species within the border zones under normal (control) conditions and when fertilizer was added to increase nutrient availability. Note that the increase in nutrient availability resulted in the subordinate competitor under ambient conditions (control) becoming the dominant (superior competitor).

(Adapted from Emery et al. 2001.)

Interpreting Ecological Data

Q1. In the transition zone between the areas dominated by *Spartina alterniflora* and *Spartina patens*, which of the two species was the superior competitor (dominated) in the experimental plots under control conditions? Was the competitive outcome altered when nutrient availability was enhanced in the experimental plots (fertilized)? How so?

Q2. In the transition zone between the areas dominated by *Spartina patens* and *Juncus gerardi*, which of the two species was the superior competitor (dominated) in the experimental plots under control conditions? Was the competitive outcome altered when nutrient availability was enhanced in the experimental plots (fertilized)? How so?

Q3. What do the results of these experiments suggest about the role of nutrients in limiting the distribution of plant species along the gradient from low (sea side) to high (land side) marsh?

during high-tide cycles (see Chapter 3). These differences in the frequency and duration of tidal inundation establish a spatial gradient of increasing salinity, waterlogging, and reduced oxygen levels across the marsh (see Figure 17.11). Individuals of *S. patens* and *J. gerardi* that were transplanted into lower marsh positions exhibited stunted growth and increased mortality. Thus, the lower distribution of each species is determined by its physiological tolerance to the physical stress imposed by tidal inundation (its fundamental niche). In contrast, individuals of *S. alterniflora* and *S. patens* exhibited increased growth when transplanted onto higher marsh positions where the neighboring plants

had been removed. They were excluded by competition from higher marsh positions when neighboring plants were present (not removed). These results indicate that the upper distribution of each species in the marsh was limited by competition.

At first, this example would seem to be a clear case of the trade-off between adaptations for stress tolerance and competitive ability (high growth rate and resource use), as suggested in Figure 17.8. However, such was not the case. The experimental addition of nutrients to the marsh indeed changed the outcome of competition, but not in the manner that might be predicted. The addition of nutrients completely reversed the relative competitive abilities of the species, allowing *S. alterniflora* and *S. patens* to shift their distributions to higher marsh positions (see Figure 17.11).

J. gerardi, the dominant under ambient (low) nutrient conditions, allocates more carbon to root biomass than either species of *Spartina* does. That allows *Juncus* to be the superior competitor under conditions of nutrient limitation but limits its ability to tolerate the higher water levels of the lower marsh. In contrast, *S. alterniflora* allocates a greater proportion of carbon to aboveground tissues, producing taller tillers (stems and leaves)—an advantage in the high water levels of the lower marsh. The trade-off in allocation to belowground and aboveground tissues results in the competitive hierarchy and thus patterns of zonation observed under ambient conditions. When nutrients are not limiting (nutrient addition experiments), competition for light dictates the competitive outcome among marsh plants. The greater allocation of carbon to height growth by the *Spartina* species increased their competitive ability on the upper marsh.

In the salt marsh plant community, a trade-off between competitive ability belowground and the ability to tolerate the physical stress associated with the low oxygen and high salinity levels of the lower marsh appears to drive zonation patterns across the salt marsh landscape. In this environment, the stress gradient does not correspond to the resource gradient as in Figure 17.8, allowing the characteristics for stress tolerance to enhance competitive ability under high resource availability.

Spatial variation in resource and nonresource factors that directly influence physiological processes gives rise to patterns of zonation in terrestrial as well as aquatic communities. Patterns of temperature and moisture resulting from regional variations in climate (see Chapter 2) are the major determinant of regional and global patterns of vegetation distribution, and they form the basis of most vegetation classification systems (see Chapter 23). On a local scale, climate interacts with soils and topography to influence patterns of temperature and soil moisture (see Section 4.8). The underlying geology of an area interacts with climate to influence soil characteristics such as texture. In turn, texture directly affects soil moisture-holding capacity, cation exchange, and base saturation (see Section 4.9), influencing the moisture and nutrient environment of plants. In aquatic environments, water depth, flow rate, and salinity are the major environmental gradients that directly influence the distribution and dynamics of communities (see Chapter 24).

17.6 | Environmental Heterogeneity Influences Community Diversity

As we have seen thus far, the biological structure (species composition) of a community reflects both the direct response (survival, growth and reproduction) of the component species to the prevailing abiotic environmental conditions, as well as their interactions (directly and indirectly). In turn, as environmental conditions change from location to location, so will the set of species that can potentially occupy the area and the way they interact. This framework has helped us understand why the biological structure of a community changes as we move across a landscape from hilltop to valley or from the shoreline into the open waters of a lake or pond. However, environmental conditions are typically not homogeneous even within a given community. For example, ecologist Philip Robertson and colleagues quantified spatial variation in soil nitrogen and moisture across an abandoned agricultural field in southeastern Michigan. Once used for agriculture, the site was abandoned in the late 1920s and has since reverted to an old-field community composed of a variety of forb, grass, and shrub species. Detailed sampling of a 0.5-ha plot within the old field revealed considerable spatial variation (more than an order of magnitude) in soil moisture and nitrate at this spatial scale (Figure 17.12). Studies similar to that of Robertson and his colleagues have shown comparable patterns of fine-scale environmental variation within forest, intertidal, and benthic communities.

But how does environmental heterogeneity within a community influence patterns of diversity? Do variations in environmental conditions translate into an area's ability to support more species? Some examples that we have considered in previous chapters provide an answer to this question in plant communities. Heterogeneity in the light environment of the forest floor caused by the death of canopy trees (gap formation) has been shown to increase tree species diversity in forest ecosystems. The increase in available light below canopy gaps allows for the survival and growth of shade-intolerant species (see Section 6.9, Figure 6.14) that otherwise would be excluded from the community. Likewise, heterogeneity in the soil environment of prairie communities caused by the burrowing of small mammals results in small-scale variations in plant species composition (see Field Studies: Katherine N. Suding in Chapter 13).

A good example of the influence of environmental heterogeneity comes from the link between vegetation structure and bird species diversity. The structural features of the vegetation that influence habitat suitability (see Section 8.12) for a given bird species are related to a variety of species-specific needs relating to food, cover, and nesting sites. Because these needs vary among species, the structure of vegetation has a pronounced influence on the diversity of bird life within the community. Increased vertical structure means more resources and living space and a greater diversity of potential habitats (see Section 16.6, Figure 16.6). Grasslands, with their two strata, support 6 or 7 species of birds, all of which nest on the ground. A deciduous forest in eastern North America may support 30 or more species occupying

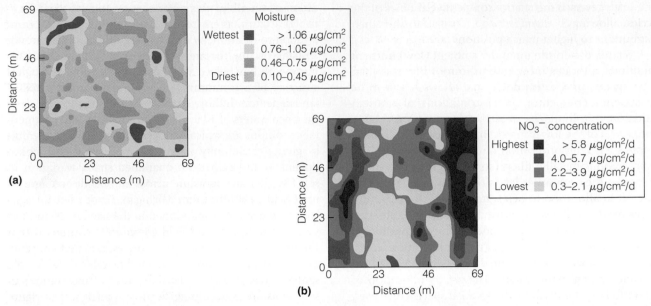

Figure 17.12 | Variations in **(a)** soil moisture and **(b)** nitrogen (nitrate; NO_3^-) production in an old-field community in Michigan (abandoned agricultural field).

(Adapted from Robertson et al. 1998.)

different strata. The scarlet tanager (*Piranga olivacea*) and wood pewee (*Contopus virens*) occupy the canopy; the hooded warbler (*Wilsonia citrina*) is a forest shrub species; and the ovenbird (*Seiurus aurocapillus*) forages and nests on the forest floor.

The late Robert MacArthur of Princeton University was the first ecologist to quantify the relationship between the structural heterogeneity of vegetation and the diversity of animal species that depend on the vegetation as habitat. He measured bird species diversity and the structural heterogeneity of vegetation in 13 communities in the northeastern United States. The communities represented a variety of structures, from grassland to deciduous forest. Bird species diversity in each community was measured using an index of species diversity (see Section 16.1). To quantify the structural heterogeneity of the vegetation, MacArthur developed an index of foliage height diversity. The value of the index increased with the number of vertical layers (and therefore the maximum height of the vegetation), as well as the relative abundance of vegetation (biomass) within the vertical layers. By comparing the two indexes, MacArthur found a strong relationship between bird species diversity and the index of foliage height diversity for the various communities (Figure 17.13). Since the publication of this pioneering work by MacArthur in the early 1960s, similar relationships between the structural diversity of habitats and the diversity of animal species within a community have been reported for a wide variety of taxonomic groups in both terrestrial and aquatic environments.

17.7 | Resource Availability Can Influence Plant Diversity within a Community

In Chapter 6 (Section 6.12), we examined the role of nutrient availability on plant processes. In general, increased availability of nutrients can support higher rates of photosynthesis, plant growth, and a higher density of plants per unit area. It might seem somewhat odd, therefore, that a variety of studies have shown an inverse relationship between nutrient availability and plant diversity in communities.

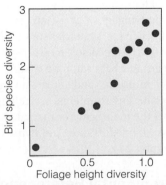

Figure 17.13 | Relationship between bird species diversity and foliage height diversity for deciduous forest communities in eastern North America. Foliage height diversity is a measure of the vertical structure of the forest. The greater the number of vertical layers of vegetation, the greater the diversity of bird species present in the forest.

(Adapted from MacArthur and MacArthur 1961.)

Michael Huston, an ecologist at Texas State University, examined the relationship between the availability of nutrients and species richness at 46 tropical rain forest sites in Costa Rica. Huston found an inverse relationship between species richness and a composite index of soil fertility (Figure 17.14). Tropical forest communities on soils with lower nutrient availability supported a greater number of tree species (species richness) than did communities on more fertile soils. Huston hypothesized that the inverse relationship results from reduced competitive displacement under low nutrient availability. Low nutrient availability reduces growth rates and supports a lower density and biomass of vegetation. Species that might dominate under higher nutrient availability cannot realize their potential growth rates and biomass and thus are unable to displace slower-growing, less competitive species.

A wide variety of field and laboratory experiments have supported the hypothesis put forward by Huston. In a series of competition experiments under controlled greenhouse conditions, Fakhri Bazzaz of Harvard University and the British ecologist John Harper found that two herbaceous plant species—white mustard (*Sinapis alba*) and cress (*Lepidium sativum*)—coexisted on less fertile soil, whereas *Lepidium* was driven to extinction by *Sinapis* under conditions of higher soil fertility.

The Park Grass experiment was begun at Rothamsted Experimental Station in Great Britain in 1859 to examine the effects of fertilizers on yield and quality of hay from permanently maintained grasslands. This experiment has continued for more than 140 years. Beginning with a uniform mixture of grass and other herbaceous species, various types, quantities, and schedules of fertilization have been applied to experimental plots within the field. Changes in species composition began as early as the second year and increased through time until a relatively stable community structure was achieved. The unfertilized

plots are the only ones retaining the original diversity of species that were planted. In all cases, the number of species was reduced by fertilization, and the most heavily fertilized plots became dominated by only a few grass species. Nearly identical results to those of the Park Grass experiment have been obtained in other fertilization experiments in both agricultural and natural grassland communities.

Greenhouse and field experiments both leave little doubt that changes in nutrient availability can greatly alter the composition and structure of plant species in communities. In all experimental studies to date, the effect of increasing nutrient availability has been to decrease diversity. But what processes cause this decrease in diversity, allowing some plant species to displace others under conditions of high nutrient availability? In a series of field experiments, ecologist James Cahill of the University of Alberta (Canada) examined how competition in grassland communities shifts along a gradient of nutrient availability. The experiments revealed a shift in the importance of belowground and aboveground competition and the nature of their interaction under varying levels of nutrient availability. Cahill's work indicates that competition for belowground and aboveground resources differs in an important way. Competition for belowground resources is *size symmetric* because nutrient uptake is proportional to the size of the plant's root system. Symmetric competition results when individuals compete in proportion to their size, so that larger plants cause a large decrease in the growth of smaller plants and small plants cause a small (but proportionate to their size) decrease in the growth of larger plants. In contrast, competition for light (aboveground) is generally one-sided or *size asymmetric*—larger plants have a disproportionate advantage in competition for light by shading smaller ones, resulting in initial size differences being compounded over time. Any factor that reduces the growth rate of a plant initiates a positive feedback loop that decreases the plant's likelihood of obtaining a dominant position in the developing size hierarchy.

Under low nutrient availability, plant growth rate, size, and density are low for all species. Competition primarily occurs belowground and therefore is symmetric. Competitive displacement is low and diversity is maintained. As nutrient availability increases, growth rate, size, and density increase. Species that are able to maintain higher rates of photosynthesis and growth exhibit a disproportionate increase in size. As faster-growing species overtop the others, creating a disparity in light availability, competition becomes strongly asymmetric. Species that achieve high rates of growth and stature under conditions of high soil fertility eventually outcompete and displace the slower-growing, smaller-stature species, thus reducing the species richness of the community.

In contrast to the inverse relationship between soil fertility (nutrient availability) and plant species richness

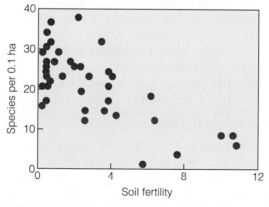

Figure 17.14 | Relationship of tree species richness (species per 0.1 ha) to a simple index of soil fertility for 46 forest communities in Costa Rica. The fertility index is the sum of the percentage values of phosphorus, potassium, and calcium obtained by dividing the value of each nutrient by the mean value of that nutrient for all 46 sites.

(Adapted from Huston 1980.)

observed in terrestrial plant communities, the pattern between nutrient availability and the species richness of autotrophic organisms in aquatic communities is quite different. Ecologist Helmut Hillebrand (University of Cologne, Germany) and colleagues reviewed the results of 44 published studies in which nutrient availability was experimentally manipulated in aquatic communities (19 freshwater and 23 marine). Their analyses reveal that, unlike the pattern observed in terrestrial plant communities, fertilization results in an increase in the species richness of autotrophs in both freshwater and marine communities. Although there is no consensus as to what causes the observed differences between terrestrial and aquatic communities, several factors have been suggested, including differences in the role of competition in terrestrial and marine environments. Unlike the competitive displacement of species under high nutrient availability discussed earlier for terrestrial plant communities, competitive exclusion among phytoplankton species in pelagic waters is less common. Reduced competition may result from rapid nutrient changes that occur in the water column (see discussion in Chapter 21 of nutrient cycling in aquatic ecosystems) and restrict competitive dominance for any given species. Or, different species in the assemblage of phytoplankton may be limited by more than one nutrient, so that no single species has a competitive advantage.

Summary

Fundamental Niche and Community Structure | 17.1

The range of environmental conditions tolerated by a species defines its fundamental niche. These constraints on the ability of species to survive and flourish will limit their distribution and abundance to a certain range of environmental conditions. Species differ in the range of conditions they tolerate. As environmental conditions change in both time and space, the possible distribution and abundance of species will change. This framework provides a null model against which to compare observed community patterns.

Diffuse Interactions | 17.2

Experiments that examine only two potentially interacting species tend to underestimate the importance of species interactions in communities because interactions are often diffuse, involving a number of species. In diffuse competition, direct interaction between any two species may be weak, making it difficult to determine the effect of any given species on another. Collectively, however, competition may be an important factor limiting the abundance of all species involved. Diffuse interactions involving predation and competition can be seen in the structure of food webs.

Indirect Interactions | 17.3

Food webs also illustrate indirect interactions among species within the community. Indirect interactions occur when one species does not interact with a second species directly, but influences a third species that does interact directly with the second. For example, a predator may increase the population density of one or more inferior competitors by reducing the abundance of the superior competitor it preys on. Indirect positive interactions result when one species benefits another indirectly through its interactions with others, reducing either competition or predation.

Controls on Community Structure | 17.4

To understand the role of species interactions in structuring communities, food webs are often simplified by placing species into functional groups based on their similarity in using resources or their role within the community. One such functional classification divides species into trophic levels based on general feeding groups (primary producers, herbivores, carnivores, etc.). The resulting food chains suggest the possibility of either bottom-up (primary producers) or top-down (top carnivores) control on community structure and function.

Species Interactions along Environmental Gradients | 17.5

The biological structure of a community is first constrained by the environmental tolerances of the species (fundamental niche), which are then modified through direct and indirect interactions with other species (realized niche). As we move across the landscape, variations in the physical environment will alter the nature of both these constraints on species distribution and abundance. There is a general trade-off between a species' stress tolerance and its competitive ability along gradients of resource availability. This trade-off can result in patterns of zonation across the landscape where variations in resource availability exist. The relationship between stress tolerance and competitive ability is more complex along gradients that include both resource and nonresource factors, such as temperature, salinity, or water depth.

Environmental Heterogeneity | 17.6

Environmental conditions are not homogeneous within a given community, and spatial variations in environmental conditions within the community can function to increase diversity by supporting a wider array of species. The structure of vegetation has a pronounced influence on the diversity of animal life within the community. Increased vertical structure means more resources and living space and a greater diversity of habitats.

Resource Availability | 17.7

A variety of studies have shown an inverse relationship between nutrient availability and plant diversity in communities. By reducing growth rates, low nutrient availability functions to reduce competitive displacement. As nutrient availability in terrestrial communities increases, competition shifts from belowground (symmetrical competition) to aboveground (asymmetrical competition). The net result is an increase in competitive displacement and a reduction in plant species diversity as the faster-growing, taller plant species dominate the light resource.

Unlike the pattern observed in terrestrial plant communities, fertilization results in an increase in the species richness of autotrophs in both freshwater and marine communities.

Study Questions

1. How does the species' fundamental niche function to constrain community structure?

2. The number of tree species that occur in a hectare of equatorial rain forest in eastern Africa can exceed 250. In contrast, the number of tree species occurring in a hectare of tropical woodland in southern Africa rarely exceeds three. In which forest community (rain forest or woodland) do you think diffuse competition would be the most prevalent? Why?

3. Define indirect interaction in the context of food webs.

4. Give an example of how predation can result in indirect positive interactions between species.

5. Contrast bottom-up and top-down control in the structure of a food web.

6. In the ecologist Mary Power's work, presented in Section 17.4, the top predators appear to control on plant productivity by controlling the abundance of herbivores (their prey). Now suppose we were to conduct a second experiment and reduce plant productivity by using some chemical that had no direct effect on the consumer organisms. If the results show that reduced plant productivity reduces herbivore populations, in turn leading to the decline of the top predator, what type of control would this imply? How might you reconcile the findings of these two experiments?

7. Using Chapter 6 (Section 6.12) as your resource, what characteristics might enable a plant species (species A) to tolerate low soil nutrient availability? How might these characteristics limit the maximum growth rates under high soil nutrient conditions? Conversely, what characteristics might enable a plant species (species B) to maintain high growth rates under high soil nutrient availability? How might these characteristics limit the plant species' ability to tolerate low soil nutrient conditions? Now predict the outcome of competition between species A and B in two plant communities, one with low soil nutrients and the other with abundant nutrients. Discuss in terms of tolerance and competition.

8. How does the structure of vegetation within a community influence the diversity of animal life?

9. Contrast symmetric and asymmetric competition. How does the availability of soil nutrients shift the nature of competition from symmetric to asymmetric?

Further Readings

Brown, J., T. Whitham, S. Ernest, and C. Gehring. 2001. Complex species interactions and the dynamics of ecological systems: Long-term experiments. *Science* 93:643–650.
    Review of long-term experiments that have revealed some of the complex interactions occurring within ecological communities.

Huston, M. 1994. *Biological diversity.* Cambridge, UK: Cambridge University Press.
    An essential resource for those interested in community ecology. Huston provides an extensive review of geographic patterns of species diversity and presents a framework for understanding the distribution of species in both space and time.

McPeek, M. 1998. The consequences of changing the top predator in a food web: A comparative experimental approach. *Ecological Monographs* 68:1–23.
    This paper presents a series of experiments directed at understanding the role of top predators in structuring care communities. An excellent example of the application of experimental manipulations to understanding species interactions in ecological communities.

Pimm, S. L. 1991. *The balance of nature.* Chicago: University of Chicago Press.
    An excellent example of the application of theoretical studies on food webs and the structure of ecological communities to current issues in conservation ecology.

Power, M. E. 1992. Top-down and bottom-up forces in food webs: Do plants have primacy? *Ecology* 73:733–746.
    This is an excellent discussion of the role of predation in structuring communities, including a contrast between bottom-up and top-down controls on the structure of food webs.

Rosenzweig, M. 1995. *Species diversity in space and time.* Cambridge, UK: Cambridge University Press.
    This book combines empirical studies and ecological theory to address various topics relating to patterns of species diversity. An excellent discussion of large-scale patterns of biological diversity over geological time.

CHAPTER 18

Community Dynamics

A diverse grassland community occupies a site once used for agriculture. The study of processes involved in the colonization of abandoned agricultural lands has given ecologists important insights into the dynamics of plant communities.

n Chapter 17 we examined the variety of processes that interact to influence the structure of communities. The changing nature of community structure across the landscape (zonation) reflects the shifting distribution of populations in response to changing environmental conditions, as modified by the interactions (direct and indirect) among the component species. Yet the structure (physical and biological) of the community also changes through time: it is dynamic, reflecting the population dynamics of the component species. The vertical structure of the community changes with time as plants establish themselves, mature, and die. The birth and death rates of species change in response to environmental conditions, resulting in a shifting pattern of species dominance and diversity through time. This changing pattern of community structure through time—community dynamics—is the topic of this chapter.

18.1 | Community Structure Changes Through Time

Community structure varies in time as well as in space. Suppose that rather than moving across the landscape, as with the examples of zonation in Chapter 16 (Section 16.7, Figures 16.7–16.9), we stand in one position and observe the area as time passes. For example, abandoned cropland and pasture land are common sights in agricultural regions in once forested areas in eastern North America (see Ecological Issues: American Forests). No longer tended, the land quickly grows up in grasses, goldenrod (*Solidago* spp.), and weedy herbaceous plants. In a few years, these same weedy fields are invaded by shrubby growth—blackberries (*Rubus* spp.), sumac (*Rhus* spp.), and hawthorn (*Crataegus* spp.). These shrubs are followed by fire cherry (*Prunus pennsylvanica*), pine (*Pinus* spp.), and aspen (*Populus* spp.). Many years later, this abandoned land supports a forest of maple (*Acer* spp.), oak (*Quercus* spp.), cherry, or pine (Figure 18.1). The process you would have observed, the gradual and seemingly directional change in community structure through time from field to forest, is called **succession.** Succession, in its most general definition, is the temporal change in community structure. Unlike zonation (changes in community structure across the landscape), succession refers to changes in community structure at a given location on the landscape through time.

The sequence of communities from grass to shrub to forest historically has been called a **sere** (from the word *series*), and each of the changes is a seral stage. Although each **seral stage** is a point in a continuum of vegetation through time, it is often recognizable as a distinct community. Each stage has its characteristic structure and species composition. A seral stage may last only 1 or 2 years, or it may last several decades. Some stages may be missed completely or may appear only in abbreviated or altered form. For example, when an abandoned field is

(a)

(b)

(c)

Figure 18.1 | Successional changes over 30 years in a western Pennsylvania field. **(a)** In 1942, it was moderately grazed. **(b)** The same area 21 years later in 1963. **(c)** In 1972, quaking aspen and maple have claimed some of the ground.

Old-field communities, such as the one shown in Figure 1, are a common sight in the eastern portion of the United States. These fields represent the early stages in the process of secondary succession, a process that began with the abandonment of agricultural lands (cropland or pasture) and that will eventually lead to forest. Although more than 50 percent of the U.S. land area east of the Mississippi River is currently covered by forest, the vast majority of these forest communities are less than 100 years old, the product of a continental-scale shift in land use that has occurred over the past 200 years.

When colonists first arrived on the eastern shores of North America in the 17th century, the landscape was dominated by forest. Native Americans had historically used fire to clear areas for planting crops or to create habitat for game species. However, their impact on the landscape was minor compared to what was to come as a result of the westward expansion of European colonization. The clearing of forest was driven by the need for agricultural lands and forest products, and as the population of colonists grew and expanded westward, so did the clearing of land (Figure 2). By the 19th century, most of the forests in eastern North America had been felled for agriculture. But by the early part of the 20th century, there would be a reversal of this trend.

The Dust Bowl period in the 1930s saw the beginning of the decline in small family farms in the agricultural regions west of the Mississippi. With the mechanization of agriculture, and the large-scale production of chemical fertilizers by the late 1940s, agriculture in the West would undergo a major transition, moving from small, family-owned farms to large commercial farms. The rise of large-scale commercial agriculture hastened the decline in agriculture east of the Mississippi River—a decline that began in the 1800s with the end of the large plantation farms in the Southern states. By the 1930s, the amount of agricultural land in the east had peaked, and it has been declining ever since (Figure 3). Since 1972 alone, more than 4500 square miles of farmland in the eastern United States has been abandoned and is currently reverting to forest. This is twice the amount of agricultural land that has been converted to commercial and residential development during the same period.

1. How might the shifts in land use and cover discussed here have changed the diversity of animal life in the eastern United States during the past 100 years?

2. How do the changes in biological diversity in the eastern part of the United States during the past century compare to the shifts that have occurred west of the Mississippi River because of the shift in agriculture and the decline of prairie ecosystems (see Part Five Introduction, p. 330)?

Figure 1 | A typical old-field community in the eastern portion of the United States. Once managed agricultural land, the site is now undergoing the process of secondary succession after abandonment. Eventually the site will revert to forest, like that seen in the background.

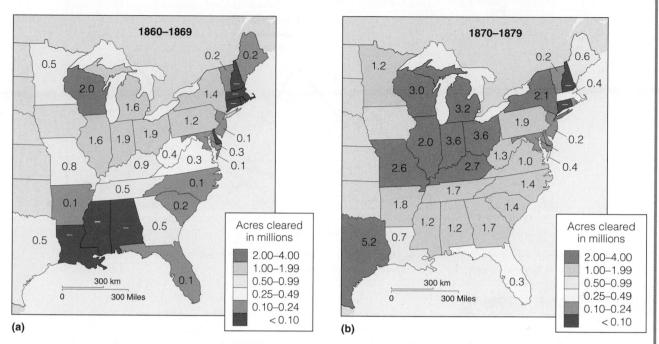

Figure 2 | Millions of acres of forest lands were cleared for agriculture during the decades of **(a)** 1860–69 and **(b)** 1870–79 in the eastern United States. Note the westward expansion and accelerating rate of clearing during this period.

(Adapted from Williams 1989.)

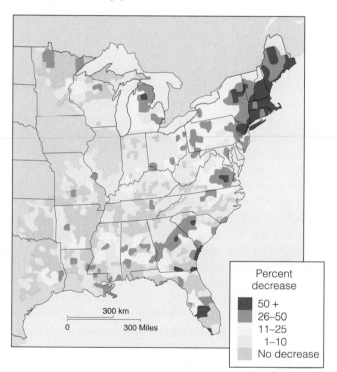

Figure 3 | Percent decrease in agricultural land in the eastern United States since its peak in 1930.

(Adapted from Williams 1989.)

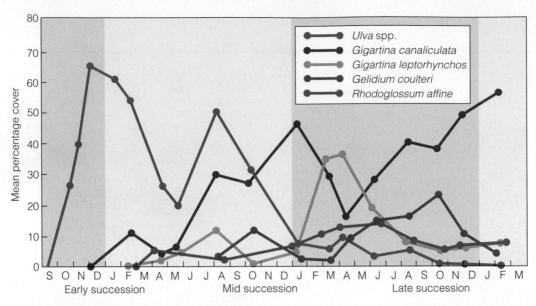

Figure 18.2 | Mean percentage of five algal species that colonized concrete blocks introduced into the rocky intertidal zone in September 1974. Note the change in species dominance over time.
(Adapted from Sousa 1979.)

Interpreting Ecological Data

Q1. At what time does the species *Gigartina canaliculata* first appear in the experiments (month and year)? At what period during the experiment does this species of algae dominate the community?

Q2. Which algal species dominates the community during the first year of succession? Which species dominates during the last year of the experiment?

Q3. Which algal species never dominate the community (greatest relative abundance)?

Q4. During which period of the observed succession (early, mid, or late) is overall species diversity the highest?

colonized immediately by forest trees (as in Figure 18.1), the shrub stage appears to have been bypassed; however, structurally, its place is taken by the incoming young trees.

Like zonation, the process of succession is generally common to all environments, both terrestrial and aquatic. The ecologist William Sousa carried out a series of experiments designed to examine the process of succession in a rocky intertidal algal community in southern California. A major form of natural disturbance in these communities is the overturning of rocks by the action of waves. Algal populations then recolonize these cleared surfaces. To examine this process, Sousa placed concrete blocks in the water to provide new surfaces for colonization. The study results show a pattern of colonization and extinction, with other species displacing populations that initially colonized the concrete blocks as time progresses (Figure 18.2). This is the process of succession. The initial, or **early successional species** (often referred to as **pioneer species**) are usually characterized by high growth rates, smaller size, high degree of dispersal, and high rates of per capita population growth. In contrast, the **late successional species** generally have lower rates of dispersal and colonization, slower per capita growth rates, and they are larger and longer-lived. As the terms *early* and *late succession* imply, the patterns of species

replacement with time are not random. In fact, if Sousa's experiment were to be repeated tomorrow, we would expect the resulting patterns of colonization and extinction (the successional sequence) to be very similar to those presented in Figure 18.2.

A similar pattern of succession occurs in terrestrial plant communities. Figure 18.3 depicts the patterns of woody plant species replacement after forest clearing (clear-cutting) at the Hubbard Brook Experimental Forest in New Hampshire. Before forest clearing, seedlings and saplings of beech (*Fagus grandifolia*) and sugar maple (*Acer saccharum*) dominated the understory. Large individuals of these two tree species dominated the canopy, and the seedlings represent successful reproduction of the parent trees. After the larger trees were removed by timber harvest in 1970, the numbers of beech and maple seedlings declined and were soon replaced by raspberry thickets and seedlings of sun-adapted (shade-intolerant), fast-growing, early successional tree species such as pin cherry (*Prunus pennsylvanica*) and yellow birch (*Betula alleghaniensis*). After many years, these species will be replaced by the late successional species of beech and sugar maple that previously dominated the site.

The two studies just presented point out the similar nature of successional dynamics in two very different

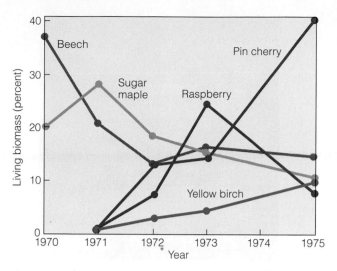

Figure 18.3 | Changes in the relative abundance (percentage of living biomass) of woody species in the Hubbard Brook Experimental Forest after a clear-cut.

(Adapted from Bormann and Likens 1979.)

environments. They also present examples of two different types of succession: primary and secondary. **Primary succession** occurs on a site previously unoccupied by a community—a newly exposed surface such as the cement blocks in a rocky intertidal environment. In contrast, the study at Hubbard Brook after forest clearing is an example of **secondary succession.** Unlike primary succession, secondary succession occurs on previously occupied (vegetated) sites after disturbance. In this case, disturbance is defined as any process that results in the removal (either partial or complete) of the existing vegetation (community). As seen in the Hubbard Brook example, the disturbance does not always result in the removal of all individuals. In these cases, the amount (density and biomass) and composition of the surviving community

will have a major influence on the proceeding successional dynamics. A more detailed discussion of disturbance and its role in structuring communities is presented in Chapter 19.

18.2 | Primary Succession Occurs on Newly Exposed Substrates

Primary succession begins on sites that never have supported a community, such as rock outcrops and cliffs, sand dunes, and newly exposed glacial till. For example, consider primary succession on a very inhospitable site: a sand dune. Sand, a product of pulverized rock, is deposited by wind and water. Where deposits are extensive, as along the shores of lakes and oceans and on inland sand barrens, sand particles may be piled in long, windward slopes to form dunes (Figure 18.4). Under the forces of wind and water, the dunes can shift, often covering existing vegetation or buildings. The establishment and growth of plant cover acts to stabilize the dunes. The late plant ecologist H. C. Cowles of the University of Chicago first described colonization of sand dunes and the progressive development of vegetation in his pioneering classic study (published in 1899) of plant succession on the dunes of Lake Michigan.

Grasses, especially beach grass (*Ammophila breviligulata*), are the most successful pioneering plants and function to stabilize the dunes with their extensive system of rhizomes (see Section 9.1). Once the dunes are stabilized, mat-forming shrubs invade the area. Subsequently, the vegetation shifts to dominance by trees—first pines and then oak. Due to low moisture reserves in the sand, oak is rarely replaced by more moisture-demanding (mesophytic) trees. Only on the more favorable leeward slopes and in depressions, where microclimate is more moderate and where moisture can accumulate, does succession proceed to more

Figure 18.4 | Primary succession on a coastal sand dune colonized by beach grass (*Ammophila breviligulata*).

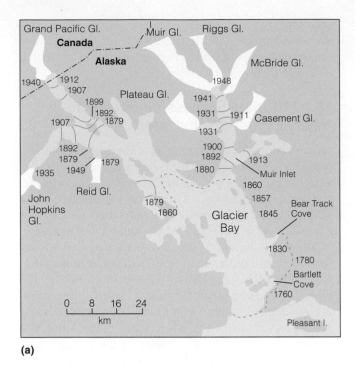

(a)

(b)

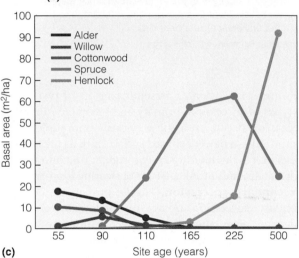

(c)

Figure 18.5 | **(a)** The Glacier Bay fjord complex in southeastern Alaska, showing the rate of ice recession since 1760. As the ice retreats, it leaves moraines along the edge of the bay in which primary succession occurs. **(b)** Primary succession along riverine environments of Glacier Bay National Park, Alaska. **(c)** Changes in community composition (basal area of woody plant species) with age for sites (time since sediments first exposed) at Glacier Bay.

(Adapted from Hobbie 1994.)

mesophytic trees such as sugar maple, basswood, and red oak. Because these trees shade the soil and accumulate litter on the soil surface, they act to improve nutrients and moisture conditions. On such sites, a mesophytic forest may become established without going through the oak and pine stages. This example emphasizes one aspect of primary succession: the colonizing species ameliorate the environment, paving the way for invasion of other species.

Newly deposited alluvial soil on a floodplain represents another example of primary succession. Over the past 200 years, the glacier that once covered the entire region of Glacier Bay National Park, Alaska, has been retreating (melting; Figures 18.5a and 18.5b). As the glacier retreats, the newly exposed landscape is initially colonized by a variety of species such as alder (*Alnus* spp.) and cottonwood (*Populus* spp.). Eventually, these early successional species are replaced by the later successional tree species of spruce (*Picea* spp.) and hemlock (*Tsuga* spp.), and the resulting forest (Figure 18.5c) resembles the forest communities in the surrounding landscape.

18.3 | Secondary Succession Occurs After Disturbances

A classic example of secondary succession in terrestrial environments is the study of old-field succession in the Piedmont region of North Carolina by the eminent plant ecologist Dwight Billings (Duke University) in the late 1930s. During the first year after a crop field has been abandoned, the ground is claimed by annual crabgrass (*Digitaria sanguinalis*); its seeds, lying dormant in the soil, respond to light and moisture and germinate. However, the crabgrass' claim to the ground is short-lived. In late summer, the seeds of horseweed (*Lactuca canadensis*), a winter annual, ripen. Carried by the wind, the seeds settle on the old field, germinate, and by early winter have produced rosettes. The following spring, horseweed, off to a head start over crabgrass, quickly claims the field. During the second summer, the field is invaded by other plants: white aster (*Aster ericoides*) and ragweed (*Ambrosia artemisiifolia*).

By the third summer, broomsedge (*Andropogon virginicus*), a perennial bunchgrass, invades the field.

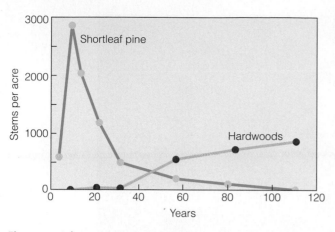

Figure 18.6 | Decline in the abundance of shortleaf pine (*Pinus echinata*) and increase in the density of hardwood species (oak, *Quercus*, and hickory, *Carya*, species) during secondary succession on abandoned farmland in the Piedmont region of North Carolina.

(Adapted from Billings 1938.)

Figure 18.7 | A kelp forest off the Aleutian Islands of Alaska: *Cymathera triplicata* (foreground); *Alaria fistulosa* (rear). Kelp forests in the eastern and northern Pacific commonly have complex three-dimensional structure, with many coexisting species. As in coral reefs, shading is a major mechanism of intraspecific and interspecific competition.

Abundant organic matter and ability to exploit soil moisture efficiently permits broomsedge to dominate the field. About this time, pine seedlings, finding room to grow in open places among the clumps of broomsedge, invade the field. Within 5 to 10 years, the pines are tall enough to shade the broomsedge. Eventually, hardwood species such as oaks and ash grow up through the pines and, as the pines die, take over the field (Figure 18.6). Development of the hardwood forest continues as shade-tolerant trees and shrubs—dogwood, redbud, sourwood, hydrangea, and others—fill the understory.

Similarly, studies of physical disturbance in marine environments have demonstrated secondary succession in seaweed, salt marsh, mangrove, seagrass, and coral reef communities. Ecologist David Duggins of the University of Washington examined the process of succession after disturbance in the subtidal kelp forests of Torch Bay, Alaska (Figure 18.7). One year after the removal of the kelp forest, a mixed canopy of kelp species (*Nereocystis luetkeana* and *Alaria fistulosa*) formed, and an understory of *Costaria costata* and *Laminaria dentigera* developed. During the second and third years, continuous stands of *Laminaria setchellii* and *Laminaria groenlandica* developed, and the community had returned to its original composition. A similar pattern has been observed in the subtidal kelp forests off the California coast.

Secondary succession in seagrass communities has been described for a variety of locations, including the shallow tropical waters of Australia and the Caribbean. Secondary succession in the seagrass communities of Florida Bay has been described in detail by ecologist Jay Zieman of the University of Virginia. Wave action associated with storms or heavy grazing by sea turtles and urchins creates openings in the grass cover, exposing the underlying sediments. Erosion on the down-current side

of these openings results in localized disturbances called blowouts (Figure 18.8). Initial colonizers on these disturbed sites are typically rhizophytic macroalgae, in which species of *Halimeda* and *Penicillus* are the most common (Figure 18.9). These algae have some sediment-binding capability, but their ability to stabilize the sediments is minimal, and their major function in the early successional stage seems to be the contribution of sedimentary particles as they die and decompose. *Halodule wrightii*, the local pioneer species of seagrass, colonizes readily from either seed or rapid vegetative branching. The carpet of cover laid by *Halodule* further stabilizes the sediment surface. Eventually, *Thalassia testudinum* will begin to colonize the region. Its larger leaves and extensive rhizome and root system effectively trap and retain particles, increasing the organic matter of the sediment. The once disturbed area again resembles the surrounding seagrass community.

18.4 | The Study of Succession Has a Rich History

The study of succession has been a focus of ecological research for more than a century. Early in the 20th century, botanists E. Warming in Denmark and Henry Cowles in the United States largely developed the concept of ecological succession. The intervening years have seen a variety of hypotheses attempting to address the processes that drive succession—the seemingly consistent directional change in species composition through time.

Frederick Clements (1916, 1936) developed a theory of plant succession and community dynamics known as the *monoclimax hypothesis*. The community is viewed as a

Figure 18.8 | Disturbed areas (light-colored areas) within seagrass communities in Florida Bay. These areas, called *blowouts,* undergo a process of recovery that involves a shift in species dominance from macroalgae to seagrass (see Figure 18.9). Insert shows a *Thalassia testudinum–* dominated community in Florida Bay.

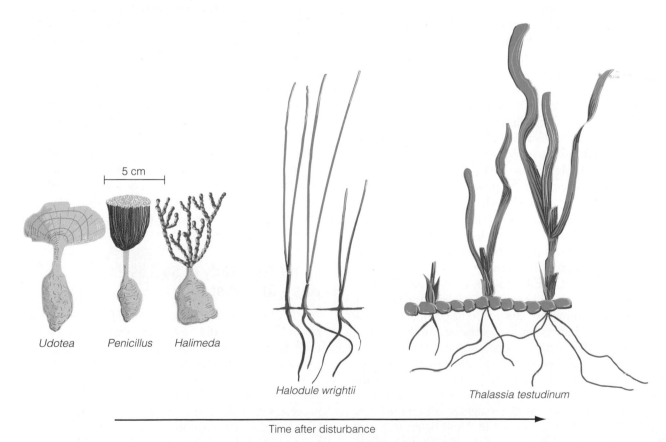

Udotea   Penicillus   Halimeda

Halodule wrightii

Thalassia testudinum

Time after disturbance

Figure 18.9 | Idealized pattern of secondary succession in seagrass communities of Florida Bay after disturbance (see Figure 18.8). The disturbed area is initially colonized by rhizophytic macroalgae. These algae species are soon displaced by the early successional seagrass species *Halodule wrightii*. This species is eventually displaced by the larger species *Thalassia testudinum*.

highly integrated superorganism (see Section 16.9), and the process of succession represents the gradual and progressive development of the community to the ultimate, or climax stage. The process was seen as analogous to the development of an individual organism.

In 1954, F. Egler proposed a hypothesis he termed *initial floristic composition*. In Egler's view, the process of succession at any site is dependent on which species get there first. Species replacement is not an orderly process, because some species suppress or exclude others from colonizing the site. No species is competitively superior to another. The colonizing species that arrive first inhibit any further establishment of newcomers. Once the original colonizers eventually die, the site then becomes accessible to other species. Succession is therefore very individualistic, dependent on the particular species that colonize the site and the order in which they arrive.

In 1977, ecologists Joseph Connell (University of California, Santa Barbara) and Ralph Slatyer (Australian National University) proposed a generalized framework for viewing succession that considers a range of species interactions and responses through succession. They offered three models.

The *facilitation model* states that early successional species modify the environment so that it becomes more suitable for later successional species to invade and grow to maturity. In effect, early-stage species prepare the way for late-stage species, facilitating their success (see discussion of facilitation, Chapter 15).

The *inhibition model* involves strong competitive interactions. No one species is completely superior to another. The first species to arrive hold the site against all invaders. They make the site less suitable for both early and late successional species. As long as they live and reproduce, the first species maintain their position. The species relinquish it only when they are damaged or die, releasing space to another species. Gradually, however, species composition shifts as short-lived species give way to long-lived ones.

A third model, the *tolerance model,* holds that later successional species are neither inhibited nor aided by species of earlier stages. Later-stage species can invade a newly exposed site, establish themselves, and grow to maturity independent of species that precede or follow them. They can do so because they tolerate a lower level of some resources. Such interactions lead to communities composed of those species most efficient in exploiting available resources. An example might be a highly shade-tolerant species that could invade, persist, and grow beneath the canopy because it is able to exist at a lower level of one resource: light. Ultimately, through time, one species would prevail.

Since the work of Connell and Slatyer, the search for a general model of plant succession has continued among ecologists. Although many hypotheses have been put forward in recent decades, a general trend in thinking has emerged. The current focus is on how the adaptations and life history traits of individual species influence species interactions and ultimately species distribution and abundance under changing environmental conditions.

18.5 | Succession Is Associated with Autogenic Changes in Environmental Conditions

The changes in environmental conditions that bring about shifts in the physical and biological structures of communities across the landscape are varied. They can, however, be grouped into two general classes: allogenic and autogenic. **Autogenic** environmental change is a direct result of the presence and activities of organisms within the community. For example, the vertical profile of light in a forest is a direct result of the interception and reflection of solar radiation by the trees (see Section 4.2 and Quantifying Ecology 4.1). In contrast, **allogenic** environmental change is a feature of the physical environment—it is governed by physical rather than biological processes. Examples are the decline in average temperature with elevation in mountainous regions (see Section 2.3), the decrease in temperature with depth in a lake or ocean (see Section 3.4), and the changes in salinity and water depth in coastal environments.

In Section 18.1, we defined succession as changes in community structure though time—specifically, changes in species dominance. One group of species initially colonizes an area, but as time progresses they decline and are replaced by other species. We observe this general pattern of changing species dominance as time progresses in most natural environments, thus suggesting a common underlying mechanism.

As reviewed in the previous section, plant succession has been a major focus of study since the birth of ecology as a science. Many hypotheses and models have been put forward to explain succession, but ecologists have achieved no general consensus. One obstacle is the diversity of environments and associated communities in which succession has been studied. No single cause fits all the examples. Despite this lack of consensus, a general model of plant succession has begun to emerge.

One feature common to all plant succession is autogenic environmental change. In both primary and secondary succession, colonization alters environmental conditions. One clear example is the alteration of the light environment. Leaves reflecting and intercepting solar radiation create a vertical profile of light within a plant community. In moving from the canopy to ground level, less light is available to drive the processes of photosynthesis. During the initial period of colonization, few if any plants are present. In the case of primary succession, the newly exposed site has never been occupied. In the case of secondary succession, plants have been killed or removed by some disturbance. Under these circumstances, the availability of light at the ground level is high, and seedlings are able to establish themselves. As plants grow, their leaves intercept sunlight, reducing the availability of light to shorter plants. This reduction in available light will decrease rates of photosynthesis, slowing the growth of these shaded individuals. Assuming that not all plant species can photosynthesize and grow at the same rate, plant species that can grow tall the fastest will have greater access to the light resource. They subsequently reduce the availability of light to the slower-growing species. This reduction in light enables the fast-growing species to outcompete the other species and dominate the site. However, in changing the availability of light below the canopy, the dominant species create an environment that is more suitable for the species that will later displace them as dominants.

Recall from Chapter 6 (Section 6.9) that not all plant species respond in the same way to variation in available light. Sun-adapted, shade-intolerant plants exhibit a different response to light than do shade-adapted, shade tolerant species. Shade-intolerant species exhibit high rates of photosynthesis and growth in high light environments. Under low light levels, they are not able to survive (see Figure 6.14). In contrast, shade-tolerant plant species exhibit much lower rates of photosynthesis and growth under high light conditions but are able to continue photosynthesis, growth, and survival under lower light. There is a fundamental physiological trade-off between the adaptations that enable high rates of growth under high light conditions and the ability to continue growth and survival under shaded conditions (see also Figure 6.9).

In the early stages of plant succession, shade-intolerant species can dominate because of their high growth rates. Shade-intolerant species grow above and shade the slower-growing, shade-tolerant species. As time progresses and light levels decline below the canopy, however, seedlings of the shade-intolerant species cannot grow and survive in the shaded conditions. At that time, although shade-intolerant species dominate the canopy, no new individuals are being recruited into their populations. In contrast, shade-tolerant species are able to germinate and grow under the canopy. As the shade-intolerant plants that form the canopy die, shade-tolerant species in the understory will replace them.

Figure 18.10 shows this pattern of changing population recruitment, mortality, and species composition through time in the forest community in the Piedmont region of North Carolina presented in Figure 18.6. Fast-growing, shade-intolerant pine species dominate in early succession. Over time, the number of new pine seedlings declines as the light decreases at the forest floor. Shade-tolerant oak and hickory species, however, are able to

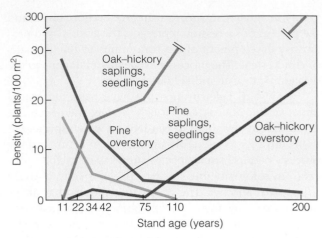

Figure 18.10 | Dominance shift of overstory and understory (seedlings and saplings) pines, oaks, and hickories during secondary succession in the Piedmont region of North Carolina (see Figure 18.6). Early successional pine species initially dominate the site. Pine seedling regeneration declines as the light decreases in the understory. Shade-tolerant oak and hickory seedlings establish themselves under the reduced light conditions. As pine trees in the overstory die, oak and hickory replace them as the dominant species in the canopy.

(Adapted from Billings 1938.)

establish seedlings in the shaded conditions of the understory. As the pine trees in the canopy die, the community shifts from a forest dominated by pine species to one dominated by oaks and hickories.

In this example, succession results from changes in the relative environmental tolerances and competitive abilities of the species under autogenically changing environmental conditions. Shade-intolerant species are able to dominate the early stages of succession because of their ability to grow quickly in the high light environment. However, as autogenic changes in the light environment occur, the ability to tolerate and grow under shaded conditions enables shade-tolerant species to rise to dominance.

Light is not the only environmental factor that changes during the course of succession, however. Other autogenic changes in environmental conditions can influence patterns of succession. The seeds of some plant species cannot germinate on the surface of mineral soil; these seeds require the buildup of organic matter on the soil surface before they can germinate and become established. In the example of secondary succession in seagrass communities (Figures 18.8 and 18.9), early colonizing species function to stabilize the sediments and add organic matter, allowing for later colonization by other plant species.

Consider the example of primary succession on newly deposited glacial sediments (see Figure 18.5). Due to the absence of a well-developed soil, little nitrogen is present in these newly exposed surfaces, thus restricting the establishment, growth, and survival of most plant species. However, those terrestrial plant species that have the

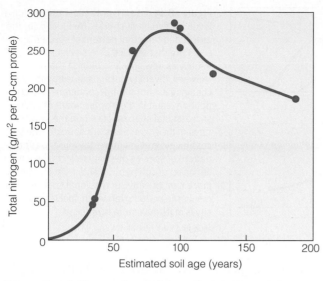

Figure 18.11 | Changes in soil nitrogen during primary succession in Glacier Bay National Park since the retreating glacier exposed the surface for colonization by plants. Initially, the virtual absence of nitrogen limits site colonization to alder, which has a mutualistic association with *Rhizobium* bacteria, allowing it access to atmospheric nitrogen. As plant litter decomposes, nitrogen is released to the soil through mineralization. With the buildup of soil organic matter and nitrogen levels, other plant species are able to colonize the site and displace alder (see Figure 18.5).

(Adapted from Oosting 1942.)

mutualistic association with nitrogen-fixing *Rhizobium* bacteria are able to grow and dominate the site (see Section 15.11, Figure 15.9). These plants provide a source of carbon (food) to the bacteria that inhabit their root systems. In return, the plants have access to the atmospheric nitrogen fixed by the bacteria. Alder, which colonizes the newly exposed glacial sediments in Glacier Bay, is one such plant species (see Figure 18.5c).

As individual alder shrubs shed their leaves or die, the nitrogen they contain is released to the soil through the processes of decomposition and mineralization (Figure 18.11; see also Chapter 21). Now other plant species can colonize the site. As nitrogen becomes increasingly available in the soil, species that do not have the added cost of mutualistic association and exhibit faster rates of growth and recruitment come to dominate the site. As in the Piedmont forest example, succession is a result of autogenic change in the environment and the relative competitive abilities of the species colonizing the site.

The exact nature of succession varies from one community to another, and it involves a variety of species interactions and responses that include facilitation, competition, inhibition, and differences in environmental tolerances. However, in all cases, the role of temporal, autogenic changes in environmental conditions and the differential response of species to those changes are key features of community dynamics.

18.6 | Species Diversity Changes During Succession

In addition to shifts in species dominance, patterns of plant species diversity change over the course of succession. Patterns of diversity through succession have been investigated by comparing sites within an area that are at different stages of succession (seral stages); such groups of sites are known as chronoseres or **chronosequences.** Studies of secondary succession in old-field communities have shown that plant species diversity typically increases with site age (time since abandonment). The plant ecologist Robert Whittaker of Cornell University, however, observed a different temporal pattern of species diversity for sites in New York (Figure 18.12). Species diversity increases into the late herbaceous stages and then decreases into shrub stages. Species diversity then increases again in young forest, only to decrease as the forest ages.

The processes of species colonization and replacement drive succession. To understand the changing patterns of species richness and diversity during succession, we need to understand how these two processes vary in time. Colonization by new species increases local species richness. Species replacement typically results from competition or an inability of a species to tolerate changing environmental conditions. Species replacement over time acts to decrease species richness.

During the early phases of succession, diversity increases as new species colonize the site. However, as time progresses, species become displaced—replaced as dominants by slower-growing, more shade-tolerant species. The peak in diversity during the middle stages of succession corresponds to the transition period, after the arrival of later successional species but before the decline (replacement) of early successional species. The two peaks

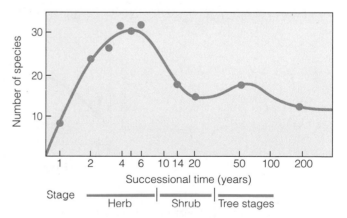

Figure 18.12 | Changes in plant diversity during secondary succession of an oak–pine forest in Brookhaven, New York. Diversity is reported as species richness in 0.3-ha samples. Species richness increases into the late herbaceous stages, declines into the shrub stage, increases once again into the early forest stages, and declines thereafter. The peaks in diversity correspond to periods of transition between these stages, where species from both stages are present at the site.

(Adapted from Whittaker 1975, Whittaker and Woodman 1968.)

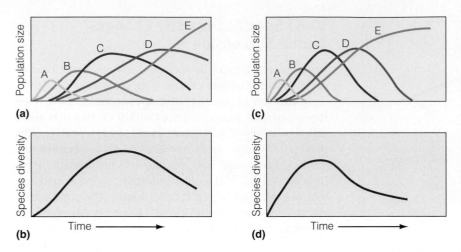

Figure 18.13 | **(a)** Hypothetical succession involving five plant species (A–E), and **(b)** the associated temporal pattern of species diversity. Note that species diversity increases initially as new species colonize the site. However, diversity declines as autogenically changing environmental conditions and competition result in the displacement of early successional species. **(c)** When the growth rates of the five species are doubled, the succession progresses more quickly, and **(d)** the pattern of species diversity reflects the earlier onset of competition and more rapid displacement of early successional species. As a result, the period over which species diversity is at its maximum is reduced.

(Adapted from Huston 1994.)

in diversity seen in Figure 18.12 correspond to the transition between the herbaceous- and shrub-dominated phases, when both groups of plants were present, and the transition between early and later stages of woody plant succession. Species diversity declines as shade-intolerant tree species displace the earlier successional trees and shrubs.

The rate of displacement is influenced by the growth rates of species involved in the succession. If growth rates are slow, the displacement process moves slowly; if growth rates are fast, displacement occurs more quickly. This observation led Michael Huston, an ecologist at Texas State University, to conclude that patterns of diversity through succession will vary with environmental conditions (particularly resource availability) that directly influence the rates of plant growth. By slowing the population growth rate of competitors that will eventually displace earlier successional species, the period of coexistence is extended, and species diversity can remain high (Figure 18.13). This hypothesis has the interesting consequence of predicting the highest diversity at low to intermediate levels of resource availability by extending the period over which species coexist (see Section 17.7).

Disturbance can have an effect similar to that of reduced growth rates by extending the period over which species coexist. In the simplest sense, disturbance acts to reset the clock in succession (Figure 18.14). By reducing

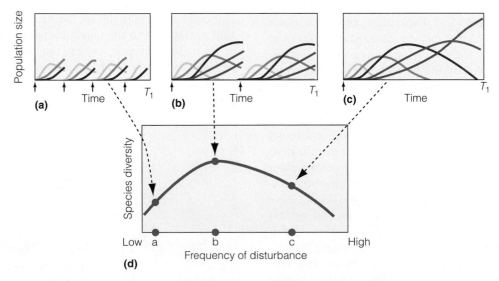

Figure 18.14 | Patterns of succession for the five hypothetical plant species shown in Figure 18.13 under three levels of disturbance frequency: frequent, intermediate, and no disturbance. Time of the disturbance is shown as an arrow on the *x*-axis. Note the differences in species composition and diversity at time T_1 on the *x*-axis. **(a)** Under a high frequency of disturbance, the absence of later successional species reduces overall diversity. **(b)** At an intermediate frequency of disturbance, all species coexist, and diversity is at a maximum. **(c)** When disturbance is absent, later successional species eventually displace the earlier successional species, and once again diversity is low. **(d)** The general form of the relationship between species diversity and the frequency of disturbance is a hump-shaped curve with maximum diversity at intermediate frequency and magnitude of disturbance. The three examples shown in (a)–(c) are labeled on the *x*-axis.

(Adapted from Huston 1994.)

or eliminating plant populations, the site is once again colonized by early successional species, and the process of colonization and species replacement begins again. If the frequency of disturbance (defined by the time interval between disturbances; see Chapter 19) is high, then later successional species will never have the opportunity to colonize the site. Under this scenario, diversity will remain low. In the absence of disturbance, later successional species will eventually displace earlier successional species, and species diversity will decline. At an intermediate frequency of disturbance, colonization can occur; but competitive displacement is held to a minimum. The pattern of high diversity at intermediate frequencies of disturbance was proposed independently by Michael Huston and by Joseph Connell of the University of California at Santa Barbara and is referred to as the **intermediate disturbance hypothesis.**

18.7 | Succession Involves Heterotrophic Species

Although our discussion and examples of succession have thus far focused on temporal changes in the autotrophic component of the community (plant succession), changes in the heterotrophic component also occur. A well-studied example is succession in the heterotrophic communities involved in decomposition. Dead trees, animal carcasses, and droppings form substrates on which communities of organisms involved in decomposition exist. Within these communities, groups of plants and animals succeed each other in a process of colonization and replacement. Succession is characterized by early dominance of fungi and invertebrates that feed on dead organic matter. Available energy and nutrients are most abundant in the early stages of succession and decline steadily as succession proceeds.

When a windstorm uproots or breaks a tree, the fallen tree becomes the stage for succession of plant and animal colonists until the log becomes part of the forest soil (Figure 18.15). The newly fallen tree, its bark and wood intact, is a ready source of shelter and nutrients. The first to exploit this resource are bark beetles and wood-boring beetles that drill through the bark and feed on the inner bark and the cambium, reducing it to frass (feces) and fragments, and tunnel galleries in which to lay eggs. Both adults and larvae drill more tunnels as they feed. *Ambrosia* beetles tunnel into the sapwood, creating galleries in which fungi grow. The tunnels provide a passageway, and the frass and softened wood form a substrate for bacteria. Loosened bark provides cover for predatory insects that are soon to follow: centipedes, mites, pseudoscorpions, and beetles. As decay proceeds, the softened wood holds more moisture; but the accessible nutrients have been depleted, leaving behind more complex, decay-resistant compounds. The pioneering arthropods leave for other logs. Fungi able to break down cellulose and lignin remain (see Chapter 21). Moss and lichens find the softened wood an

Figure 18.15 | A succession of plant and animal species are involved in the decomposition of a fallen log.

ideal habitat. Plant seedlings, too, take root on the softened logs, and their roots penetrate the heartwood, providing a pathway for fungal growth into the depths of the log.

Eventually, the log is broken into light brown, soft, blocky pieces, and the bark and sapwood are gone. At this advanced stage of decay, the log provides the greatest array of microhabitats and the highest species diversity. Invertebrates of many kinds find shelter in the openings and passages; salamanders and mice dig tunnels and move into the rotten wood. Fungi and other microorganisms abound, and many species of mites feed on decomposed wood and fungi. At last, the log crumbles into a red-brown, mulch-like mound of lignin materials resistant to decay, its nutrients and energy largely depleted. The tree is incorporated into the soil.

As plant succession advances, changes in the structure and composition of the vegetation result in changes in the animal life that depends on the vegetation as habitat (Figure 18.16). Each successional stage has its own distinctive fauna. Because animal life is often influenced more by structural characteristics than by species composition (see Section 8.12), successional stages of animal life may not correspond to the stages identified by plant ecologists.

During plant succession, animals can quickly lose their habitat by vegetation change. In eastern North America, grasslands and old fields support meadowlarks, meadow mice, and grasshoppers. When woody

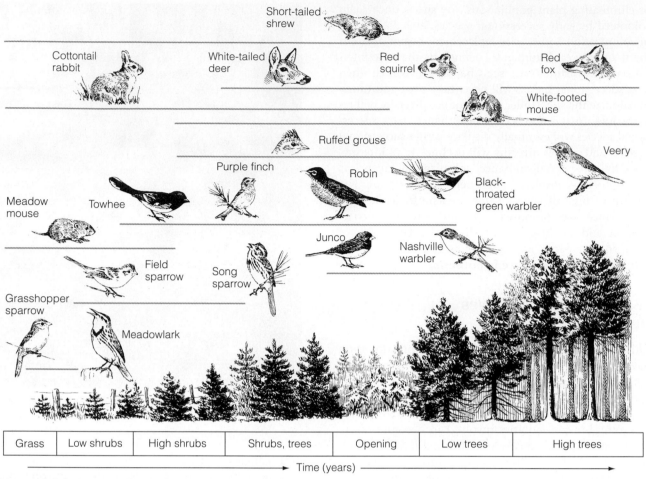

| Grass | Low shrubs | High shrubs | Shrubs, trees | Opening | Low trees | High trees |

Time (years)

Figure 18.16 | Changes in the composition of animal species inhabiting various stages of plant succession, from old-field to conifer forest, in central New York. Species appear or disappear as vegetation density and height change. Brown lines represent the range of vegetation (stages) inhabited by the associated species.

plants—both young trees and shrubs—invade, a new structural element appears. Grassland animals soon disappear, and shrubland animals take over. Towhees, catbirds, and goldfinches claim the thickets, and meadow mice give way to white-footed mice. As woody plant growth proceeds and the canopy closes, species of the shrubland decline and are replaced by birds and insects of the forest canopy. As succession proceeds, the vertical structure becomes more complex. New species appear, such as tree squirrels, woodpeckers, and birds of the forest understory, including hooded warblers and ovenbirds.

18.8 | Systematic Changes in Community Structure Are a Result of Allogenic Environmental Change at a Variety of Timescales

The focus on succession thus far has been on shifting patterns of community structure in response to autogenic changes in environmental conditions. Such changes occur at timescales relating to the establishment and growth of the organisms that make up the community. However, purely abiotic environmental (allogenic) change can produce patterns of succession over timescales ranging from days to millennia or longer. Environmental fluctuations that occur repeatedly during an organism's lifetime are unlikely to influence patterns of succession among species with that general life span. For example, annual fluctuations in temperature and precipitation will influence the relative growth responses of different species in a forest or grassland community (see Figure 13.4), but they will have little influence on the general patterns of secondary succession outlined in Figures 18.2 and 18.3. In contrast, shifts in environmental conditions that occur at periods as long as or longer than the organisms' life span are likely to result in successional shifts in species dominance. For example, seasonal changes in temperature, photoperiod, and light intensity produce a well-known succession of dominant phytoplankton in freshwater lakes that is repeated with very little

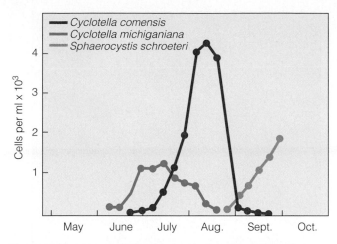

Figure 18.17 | Temporal changes in the abundance of dominant phytoplankton species during the period May through October in Lawrence Lake, Michigan (1979). Mean generation times of species range from 1 to 10 days.

(Adapted from Crumpton and Wetzel 1982.)

variation each year. Seasonal succession of phytoplankton in Lawrence Lake, a small temperate lake in Michigan, is presented in Figure 18.17. Periods of dominance are correlated with species' optimal temperature, nutrient, and light requirements, all of which systematically change over the growing season. Competition and seasonal patterns of predation by herbivorous zooplankton also interact to influence the temporal patterns of species composition.

Over a much longer timescale of decades to centuries, patterns of sediment deposition can have a major influence on the successional dynamics of coastal estuarine communities. The marshlands of the River Fal in Cornwall, England, have expanded seaward some 800 m during the past century. The seaward expansion is a result of silt deposition lowering water depths. On the landward side, woodland plant species invade the marshlands, leading to a successional sequence from marsh to woodland over time.

A similar pattern of long-term transition in community structure resulting from sediment deposition occurs in freshwater environments. Ponds and small lakes act as a settling basin for inputs of sediment from the surrounding watershed. These sediments form an oozy layer that provides a substrate for rooted aquatics such as the branching green algae, *Chara*, and pondweeds. These plants bind the loose matrix of bottom sediments and add materially to the accumulation of organic matter. Rapid addition of organic matter and sediments reduces water depth and increases the colonization of the basin by submerged and emergent vegetation. That, in turn, enriches the water with nutrients and organic matter. This enrichment further stimulates plant growth and sedimentation and expands the surface area available for colonization by larger species

of plants that root in the sediments. Eventually, the substrate, supporting emergent vegetation such as sedges and cattails, develops into a marsh. As drainage improves and the land builds higher, emergent plants disappear. Meadow grasses invade to form a marsh meadow in forested regions and wet prairie in grass country. Depending on the region, the area may pass into grassland, swamp woodland of hardwoods or conifers, or a peat bog.

Over an even longer timescale, changes in regional climate directly influence the temporal dynamics of communities. The shifting distribution of tree species and forest communities during the 18,000 years that followed the last glacial maximum in eastern North America is an example of how long-term allogenic changes in the environment can directly influence patterns of both succession and zonation at local, regional, and even global scales.

18.9 | Community Structure Changes Over Geologic Time

Since its inception some 4.6 billion years ago, Earth has changed profoundly. Landmasses emerged and broke into continents. Mountains formed and eroded, seas rose and fell, and ice sheets advanced to cover large expanses of the Northern and Southern hemispheres and then retreated. All these changes affected the climate and other environmental conditions from one region of Earth to another. Many species of plants and animals evolved, disappeared, and were replaced by others. As environmental conditions changed, so did the distribution and abundance of plant and animal species.

Records of plants and animals composing past communities lie buried as fossils: bones, insect exoskeletons, plant parts, and pollen grains. The study of the distribution and abundance of ancient organisms and their relationship to the environment is **paleoecology.** The key to present-day distributions of animals and plants can often be found in paleoecological studies. For example, paleoecologists have reconstructed the distribution of plants in eastern North America after the last glacial maximum of the Pleistocene.

The Pleistocene was an epoch of great climatic fluctuations throughout the world. At least four times in North America and three times in Europe, ice sheets advanced and retreated. With each movement, the biota retreated and advanced again with a somewhat different mix of species.

Each glacial period was followed by an interglacial period. The climatic oscillations in each interglacial period had two major stages: cold and temperate. During the cold stage, tundralike vegetation (see Section 23.9) dominated the landscape. As the glaciers retreated, light-demanding forest trees such as birch (*Betula* spp.) and

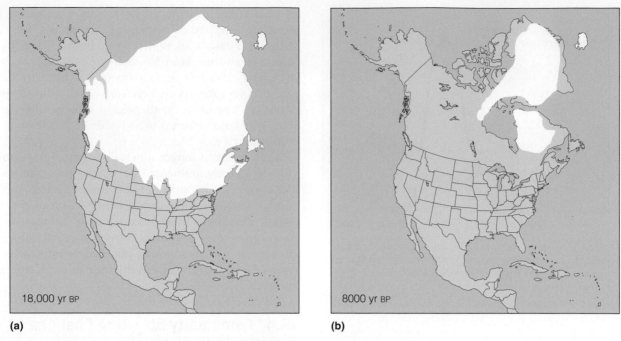

(a)   **(b)**

Figure 18.18 | Distribution of the glacial ice sheet on the North American continent at 18,000 **(a)** and 8000 **(b)** years before present (B.P.).

pine advanced. Then, as the soil improved and the climate warmed, these trees were replaced by more shade-tolerant species such as oak and ash (*Fraxinus* spp.). As the next glacial period began to develop, species such as firs (*Abies* spp.) and spruces (*Picea* spp.) took over the forest. At that time, both climate and soil began to deteriorate. Heaths began to dominate the vegetation, and forest species disappeared.

The last great ice sheet, the Laurentian, reached its maximum advance about 20,000 to 18,000 years before present (B.P.) during the Wisconsin glaciation stage in North America (Figure 18.18). Canada was under ice. A narrow belt of tundra about 60 to 100 km wide bordered the edge of the ice sheet and probably extended southward into the high Appalachians. Boreal forest, dominated by spruce and jack pine (*Pinus banksiana*), covered most of the eastern and central United States as far as western Kansas.

As the climate warmed and the ice sheet retreated northward, plant species invaded the glaciated areas. The maps in Figure 18.19 reflect the advances of four major tree genera in eastern North America after the retreat of the ice sheet. Margaret Davis of the University of Minnesota developed these maps from patterns of pollen deposition in sediment cores taken from lakes in eastern North America. By examining the presence and quantity of pollen deposited in sediment layers and dating the sediments with radiocarbon, she was able to obtain a picture of the spatial and temporal dynamics of tree communities over the past 18,000 years.

These analyses identify plants at the level of genus rather than species because, in many cases, we cannot identify species from pollen grains. Note that different genera, and probably species, expanded their distribution northward with the retreat of the glacier at markedly different rates. Differences in the rates of range expansion are most likely due to the differences in temperature responses of the species, distances and rates at which seeds can disperse, and interactions among species. The implication is that, during the past 18,000 years, the distribution and abundance of species and the subsequent structure of forest communities in eastern North America have changed dramatically (Figure 18.20).

18.10 | The Concept of Community Revisited

Our initial discussion of the processes influencing community structure and dynamics contrasted two views of the community (see Section 16.9). Through his organismal concept, Clements viewed the community as a quasi-organism made up of interdependent species. By contrast, in his individualistic or continuum concept, Gleason saw the community as an arbitrary concept and stated that each species responds independently to the underlying features of the environment. Research reveals that, as with most polarized debates, the reality lies somewhere in the middle, and our viewpoints are often colored by our perspective. The organismal community is

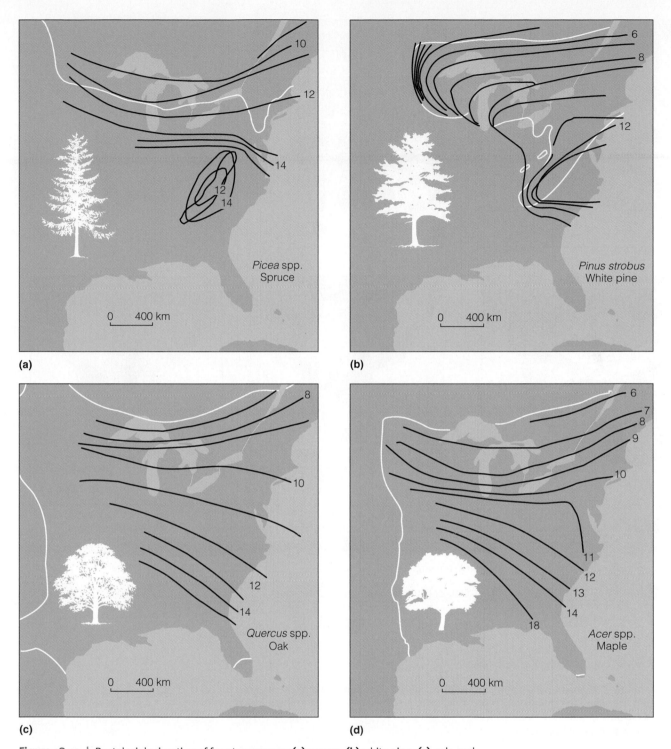

Figure 18.19 | Postglacial migration of four tree genera: **(a)** spruce, **(b)** white pine, **(c)** oak, and **(d)** maple. Dark lines represent the leading edges of the northward-expanding populations. White lines indicate the boundaries of the present-day ranges. The numbers are thousands of years before present (B.P.).

(Adapted from Davis 1981.)

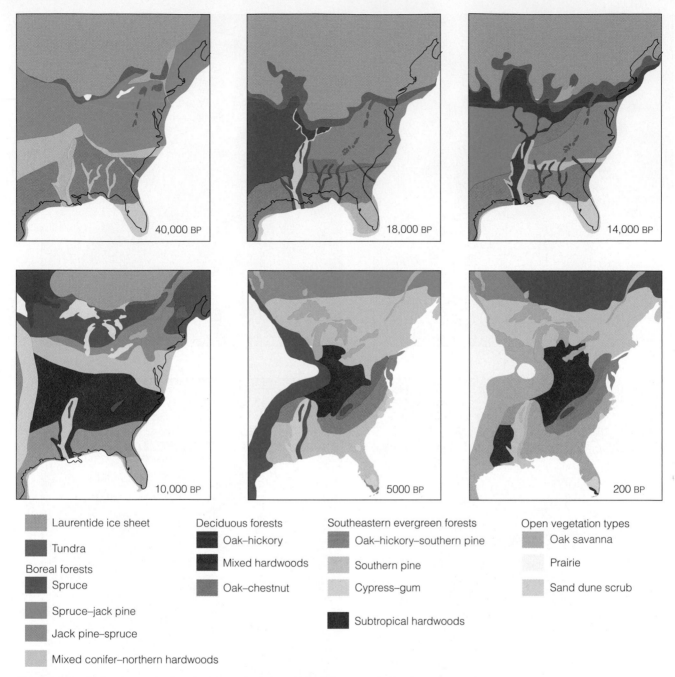

Figure 18.20 | Changes in the distribution of plant communities during and after the retreat of the Wisconsin ice sheet; the changes are reconstructed from pollen analysis at sites throughout eastern North America.

(Adapted from Delcourt and Delcourt 1981.)

Legend:

Laurentide ice sheet

Tundra

Boreal forests
- Spruce
- Spruce–jack pine
- Jack pine–spruce
- Mixed conifer–northern hardwoods

Deciduous forests
- Oak–hickory
- Mixed hardwoods
- Oak–chestnut

Southeastern evergreen forests
- Oak–hickory–southern pine
- Southern pine
- Cypress–gum
- Subtropical hardwoods

Open vegetation types
- Oak savanna
- Prairie
- Sand dune scrub

a spatial concept. As we stand in the forest, we see a variety of plant and animal species interacting and influencing the overall structure of the forest. The continuum view is a population concept, focusing on the response of the component species to the underlying features of the environment.

A simple example of the continuum concept is presented in Figure 18.21, which represents a transect up a mountain in an area with four plant species present. The distributions of the four plant species are presented in two ways. In one view, the species distributions are plotted as a function of altitude or elevation. Note that the four species exhibit a continuum of species regularly replacing each other in a sequence of A, B, C, and D with increasing altitude—very similar to the individualistic view of communities. The second

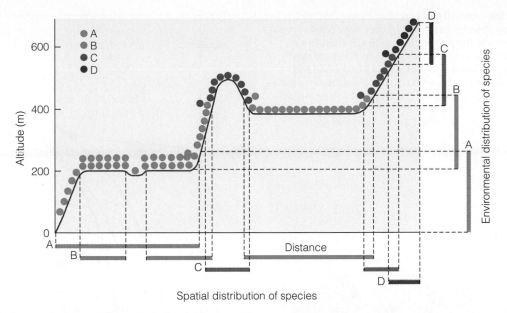

Figure 18.21 | Patterns of co-occurrence for four plant species on a landscape along a gradient of altitude. Environmental distributions of the four species are presented in two ways: (1) as the spatial distributions of the species along a transect on the mountainside, and (2) as a function of their response to altitude. Note that the species responses to the altitudinal gradient are continuous, but their spatial distributions along the transect are discontinuous. Patterns of species composition along the mountain gradient result from the spatial pattern of environmental conditions (altitude) and the individual responses of the species. Responses of the species to the environmental gradient are consistent with the individualistic or continuum view of communities proposed by Gleason (see Figure 16.11b). However, consistent patterns of species co-occurrence across the landscape are a function of the spatial distribution of environmental conditions. Repeatable patterns of species co-occurrence in similar habitats are consistent with the idea of plant associations supported by Clements (see Figure 16.11a).

(Adapted from Austin and Smith 1989.)

view of species distributions is a function of distance along the altitudinal gradient (mountainside). As we move up the mountainside, the distributions of the four species are not continuous. As a result, we might recognize several species associations as we walk along the transect. These associations are identified in Figure 18.21 by different symbols representing the combination of species. Communities composed of coexisting species are a consequence of the spatial pattern of the landscape.

The two views are quite different yet consistent. Each species has a continuous response along an environmental gradient: elevation. Yet the spatial distribution of that environmental variable across the landscape determines the overlapping patterns of species distributions—the composition of the community.

This same approach can be applied to the patterns of forest communities in Great Smoky Mountains National Park as first presented in Chapter 16 (see Figure 16.10). Different elevations and slope positions are characterized

by unique tree communities, identified by and named for the dominant tree species (Figure 18.22b). When presented in this fashion, the distributions of plant communities appear to support the organismal model of communities put forward by Clements. Yet if we plot the distributions of major tree species along a direct environmental gradient, such as soil moisture availability (Figure 18.22c), the species appear to be distributed independently of each other, thus supporting Gleason's view of the community.

The simple example in Figure 18.21 examines only one feature of the environment (elevation), yet the structure of communities is the product of a complex interaction of pattern and process. Species respond to a wide array of environmental factors that vary spatially and temporally across the landscape, and the interactions among organisms influence the nature of those responses. The product is a dynamic mosaic of communities that occupy the larger landscape—the topic we shall examine in Chapter 19.

Figure 18.22 | Two views of plant communities in **(a)** Great Smoky Mountains National Park. **(b)** Distribution of plant communities (associations) in the park in relation to elevation (*y*-axis) and slope position and aspect (*x*-axis). Communities are classified based on the dominant tree species. **(c)** Distribution and abundance of major tree species that make up vegetation communities in the park, plotted along the gradient of moisture availability (a function of slope position and aspect).

(b) (Adapted from Whittaker 1954.) (c) (Adapted from Whittaker 1956.)

(a)

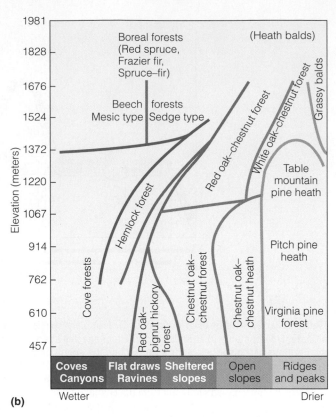

(b)

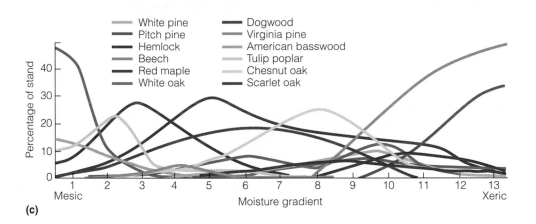

(c)

Summary

Succession | 18.1

With time, natural communities change. This gradual sequential change in the relative abundance of species in a community is succession. Opportunistic, early successional species yield to late successional species. Succession occurs in all environments. The similarity of successional patterns in different environments suggests a common set of processes.

Primary Succession | 18.2

Primary succession begins on sites devoid of or unchanged by organisms. Examples include newly formed sand dunes, lava flows, or newly exposed glacial sediments.

Secondary Succession | 18.3

Secondary succession begins after disturbance on sites where organisms are already present. Examples on land include abandoned agricultural lands or the reestablishment of vegetation after logging or fire. In aquatic ecosystems, disturbances caused by storms, wave action, or herbivory can initiate the process of secondary succession.

History | 18.4

The study of succession has been a focus of ecological research for more than a century. The intervening years have seen a variety of hypotheses attempting to address the processes that drive succession. These hypotheses include a variety of processes related to colonization, facilitation, competition, inhibition, and differences in environmental tolerances.

Autogenic Environmental Change | 18.5

Environmental changes can be autogenic or allogenic. Autogenic changes are a direct result of the activities of organisms in the community. Changes in environmental conditions independent of organisms are allogenic. Succession is the progressive change in community composition through time in response to autogenic changes in environmental conditions. One example is the changing light environment and the shift in dominance from fast-growing, shade-intolerant plants to slow-growing, shade-tolerant plants observed in terrestrial plant succession. Autogenic changes in nutrient availability, soil organic matter, and stabilization of sediments can likewise have a major influence on succession.

Species Diversity and Succession | 18.6

Patterns of species diversity change during the course of succession. Species colonization increases species richness, whereas species replacement acts to decrease the number of species present. Species diversity increases during the initial stages of succession as the site is colonized by new species. As early successional species are displaced by later arrivals, species diversity tends to decline. Peaks in diversity tend to occur during succession stages that correspond to the transition period, after the arrival of later successional species but before the decline of early successional species. Patterns of diversity during succession are influenced by resource availability and disturbance.

Heterotrophic Species | 18.7

Changes in the heterotrophic component of the community also occur during succession. The succession of organisms involved in the decomposition of fallen logs in a forest provides an example. Successional changes in vegetation affect the nature and diversity of animal life. Certain sets of species are associated with the structure of vegetation found during each successional stage.

Allogenic Environmental Change | 18.8

Fluctuations in the environment that occur repeatedly during an organism's lifetime are unlikely to influence patterns of succession among species with that general life span. Allogenic, abiotic environmental changes that occur over timescales greater than the longevity of the dominant organisms can produce patterns of succession over timescales ranging from days to millennia or longer.

Long-Term Changes | 18.9

The current pattern of vegetational distribution reflects the glacial events of the Pleistocene. Plants retreated and advanced with the movements of the ice sheets. The rates and distances of their advances are reflected in the present-day ranges of species and the distribution of plant communities.

Community Revisited | 18.10

The community is a spatial concept; the individual continuum is a population concept. Each species has a continuous response along an environmental gradient, such as elevation or moisture. Yet the spatial distribution of that environmental variable across the landscape determines the overlapping patterns of distribution—the composition of the community.

Study Questions

1. Distinguish between zonation and succession.
2. Distinguish between primary and secondary succession.
3. Defoliation of oak trees by gypsy moth larvae caused the death of extensive forest stands in the Blue Ridge Mountains of central Virginia. The recovery of these forest communities after defoliation includes the growth of existing trees and shrubs that have escaped defoliation, as well as colonization of the site by tree species outside the community. Is this an example of primary or secondary succession? Why?
4. Use the discussion of secondary succession in an abandoned agricultural field presented in Section 18.3 to differentiate among the models of facilitation, competition, and tolerance as they apply to the process of succession.
5. Classification of plant species into three primary life history strategies (R, C, and S) proposed by the plant ecologist Philip Grime was discussed in Chapter 8, Section 8.13. Which of these three plant strategies is most likely to be found colonizing a

newly disturbed site (early successional species)? Which of the three plant strategies is most likely to characterize later successional species?

6. Why is the ability to tolerate low resource availability often associated with plant species that dominate during the later stages of succession?

7. In Section 17.7, we discussed the difference between size-symmetrical and size-asymmetrical competition for resources. How might the nature of these two types of competition shift during the process of plant succession?

8. If the vertical structure of the vegetation increases during the process of terrestrial plant succession, how might the pattern of animal species diversity respond? (Hint: See Section 17.6.)

Further Readings

Bazzaz, F. A. 1979. The physiological ecology of plant succession. *Annual Review of Ecology and Systematics* 10:351–371.
    This now classic paper contrasts the physiology of plant species characteristic of different stages of succession. Bazzaz provides a framework for understanding the process of succession as the result of varying plants' adaptations to changing environmental conditions through time.

Bazzaz, F. A. 1996. *Plants in changing environments: Linking physiological, population, and community ecology.* New York: Cambridge University Press.
    This book integrates a variety of laboratory and field studies aimed at understanding the dynamics of plant communities.

Chapin, F. S., III, L. Walker, C. Fastie, and L. Sharman. 1994. Mechanisms of primary succession following deglaciation at Glacier Bay, Alaska. *Ecological Monographs* 64:149–175.
    A detailed study of the process of primary succession in Glacier Bay, Alaska.

Golley, F., ed. 1978. *Ecological succession.* Benchmark Papers. Stroudsburg, PA: Dowden, Hutchinson, and Ross.
    Succession as viewed over time. This edited volume presents a historical view of ecological theory relating to the process of succession. Golley presents a variety of original manuscripts with commentaries that trace the development of theoretical thinking on the subject.

Huston, M., and T. M. Smith. 1987. Plant succession: Life history and competition. *American Naturalist* 130:168–198.
    This manuscript presents a framework for understanding the process of plant succession as a result of the trade-offs in the evolution of plant life history characteristics and the shifting nature of competition as environmental conditions change through time.

Smith, T. M., and M. Huston. 1987. A theory of spatial and temporal dynamics of plant communities. *Vegetation* 3:49–69.
    Expands the framework first developed by Huston and Smith (see preceding article) to understand patterns of zonation and succession in plant communities from the perspective of evolutionary trade-offs in plant life history characteristics.

Tilman, D. 1988. *Plant strategies and the dynamics and structure of plant communities.* Princeton, NJ: Princeton University Press.
    In this book, Tilman draws upon a wide range of field and laboratory studies to develop a theoretical framework for understanding pattern and process in plant communities.

West, D. C., H. H. Shugart, and D. B. Botkin, eds. 1981. *Forest succession: Concepts and application.* New York: Springer-Verlag.
    An excellent reference on patterns and processes relating to succession in forest communities.

CHAPTER 19

Landscape Ecology

Landscape showing the contrast between highly managed agricultural fields in the foreground and native forest communities occupying the hills in the background.

Figure 19.1 | A view of a Virginia landscape showing a mosaic of patches consisting of different types of land cover: natural forest, plantations, fields, water, and rural development.

I n Chapter 16, we defined a community as a group of plant and animal species occupying a given area. Although ecological communities, by definition, have a spatial boundary (see Section 16.8), they likewise have a spatial context within the larger landscape. Consider the view of the Virginia countryside in Figure 19.1. It is a patchwork of forest, fields, golf course, hedgerows, pine plantation, pond, and human habitations. This patchwork of different types of land cover is called a **mosaic,** using the analogy of mosaic art in which an artist combines many small pieces of variously colored material to create a larger pattern or image (Figure 19.2). The artist creates the emerging pattern, defining boundaries by using different shapes and colors of materials to construct the mosaic. In a similar fashion, the landscape mosaic is a product of the boundaries defined by changes in the physical and biological structure of the distinct communities, called **patches,** which form its elements. In the artist's mosaic, the elements interact only visually to present the emerging image. In the landscape mosaic, patches and their boundaries make up the structural and functional components of the landscape that interact in a variety of ways depending on their size and spatial arrangement. The study of the causes behind the formation of patches and boundaries and the ecological consequences of these spatial patterns on the landscape is called **landscape ecology.**

Figure 19.2 | Mosaic in Khirbat al-Mafjar, Israel, dating from A.D. 710–750. Boundaries defining the different objects in the scene (lion, tree, antelope, etc.) are created using small pieces of stone that differ in shape and color.

19.1 | Environmental Processes Create a Variety of Patches in the Landscape

Patches, relatively homogeneous community types (such as the crop fields, forest, pond, or lawns shown in Figure 19.1) differ from their surroundings both in structure and in species composition. They vary in size, shape, and type and are embedded within a complex mosaic of surrounding patches from which they are often separated by distinct boundaries. The communities that surround a patch constitute its matrix. But how does the mosaic of patches develop on the landscape?

Patches making up the landscape mosaic result from the interactions of several environmental factors, including regional variations in geology, topography, soils, and climate. Upon this natural stage, human activity makes its mark on the broad-scale distribution of communities. For example, ongoing fragmentation of large tracts of land by human development eventually results in a mosaic of smaller, often isolated patches of forest, grassland, and shrubland (Figure 19.3). Many landscape patterns we observe today in the United States reflect early land-survey methods, developed in the United States, that divided land into sections, half sections, and quarter sections of 160 acres. Historically, these land surveys were set on straight lines (consider the Mason-Dixon Line and many state and county boundaries), with no attempt to follow topography or natural boundaries (Figure 19.4). The straight-line survey resulted in straight corners where woods and fields, croplands and developments, and other landscape elements met. This checkerboard pattern, often overlooked by the casual observer, has a lasting impact on the landscape.

Although human activity is often the most obvious force determining the size and shape of landscape patches, natural variations in geology and soil conditions (see Chapter 5) and natural events such as fire and grazing by herbivores all interact to govern the formation of patches. As a result, patches vary considerably in size and shape.

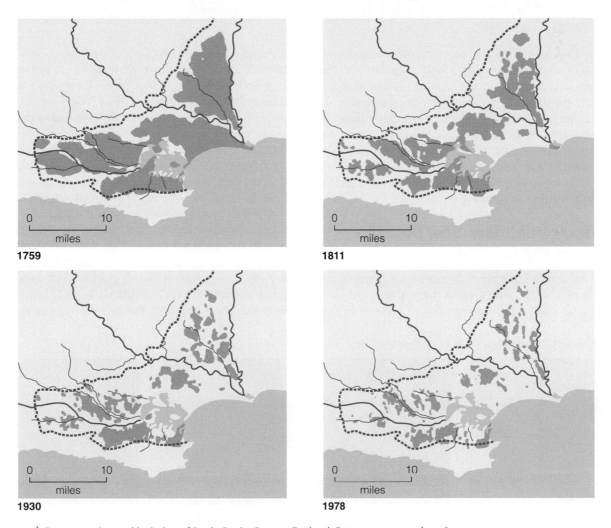

Figure 19.3 | Fragmentation and isolation of Poole Basin, Dorset, England. Between 1759 and 1978, the area lost 86 percent of its heathland (40,000 ha to 6000 ha), changing from 10 large blocks separated by rivers to 1084 pieces—nearly half of these sites are less than 1 ha, and only 14 sites are larger than 100 ha.

(Adapted from Webb and Haskins 1980.)

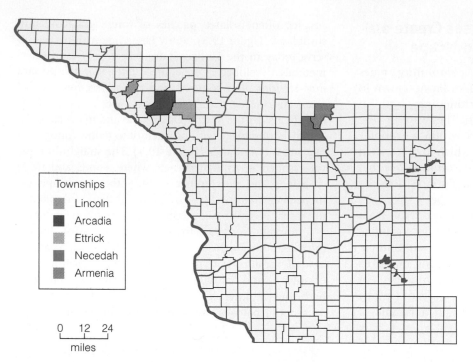

Figure 19.4 | Township boundaries in southwestern Wisconsin. Although the boundaries of some townships are defined by natural features on the landscape (see labeled townships), most boundaries were established by the U.S. Rectangular Survey System used in the 18th and 19th centuries.

(Adapted from Johnson 1976.)

Townships
- Lincoln
- Arcadia
- Ettrick
- Necedah
- Armenia

0   12   24
miles

They may be square, elongate, round, or convoluted, covering tens of square kilometers or only a few meters. The area, shape, and orientation of landscape patches all determine their suitability as habitats for plants and animals and influence many ecological processes, such as wind flow, dispersal of seeds, and movement of animals. In the following section we look at special zones formed where patches meet. These areas are often inhabited by species common to each adjacent community as well as by species unique to transition areas between patches.

19.2 | Transition Zones Offer Diverse Conditions and Habitats

Among the landscape's most conspicuous features are the edges or borders, which mark the perimeter of each patch. Some edges indicate an abrupt change in the physical conditions—topography, substrate, soil type, or microclimate—between communities. An example is a large rock outcrop in a forest. Where long-term natural features underlie adjoining vegetation, edges are usually stable and permanent, and we call them **inherent edges.** Other edges, however, result from natural disturbances such as fires, storms, and flood or from human-related activities such as timber harvesting, livestock grazing, and housing developments. Such edges, subject to successional changes over time, are called **induced edges.** The place where the edge of one patch meets the edge of another is called a **border** (Figure 19.5), an area of contact, separation, or transition between patches. Some borders between landscape patches are narrow and abrupt with a sharp contrast between the adjoining patches, such as between a forest and an adjacent agricultural field. Others are wide with width best defined as the distance between the border

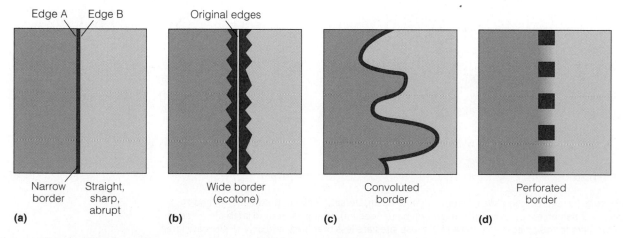

Figure 19.5 | Types of borders. **(a)** Narrow, sharp, abrupt border created where the edges of two patches meet; **(b)** wide border creating an ecotone between two adjacent patches; **(c)** convoluted border; and **(d)** perforated border.

and the point where the physical conditions and vegetation do not differ significantly from those of the interior of the patches. Wide borders form a transition zone, often called an **ecotone,** between the adjoining patches (Figure 19.5b). Borders may be tight or perforated (Figure 19.5d); they may be straight (Figure 19.5a) or convoluted (Figure 19.5c) and vary in length. In addition to width, borders have a vertical structure, the height of which influences the steepness of the physical gradient between patches.

Functionally, borders connect patches through fluxes or flows of material, energy, and organisms. The height, width, and porosity of borders influence the gradients of wind flow, moisture, temperature, and solar radiation between the adjoining patches (Figure 19.6). Borders can differentially restrict or facilitate the dispersal of seed and the movements of animals across the landscape.

Environmental conditions imposed by the physical environment in transition zones between patches enable certain plant and animal species to colonize border environments. Plant species found in such areas tend to be more shade-intolerant (sun adapted) and can tolerate the dry conditions caused by higher air temperatures and rates of evapotranspiration (see Section 6.9). Animal species inhabiting border environments are usually those that require two or more habitat types (plant communities) within their home range or territory. For example, the ruffed grouse (*Bonasa umbellatus*) requires new forest openings with an abundance of herbaceous plants and low shrubs for feeding its young, a dense stand of sapling trees for nesting cover, and mature forests for winter food and cover. Because the home range size of a ruffed grouse is from 4 to 8 ha, the whole range of habitats must be contained within this single area. Other species, such as the indigo bunting (*Passerina cyanea*), are restricted exclusively to the edge environment (Figure 19.7) and are referred to as **edge species.**

Because borders blend elements from all adjacent patches, their structure and composition are often very different from those in any single patch. Thus, borders offer unique habitats with relatively easy access to the adjacent communities. These diverse conditions enable boundaries to support plant and animal species from adjacent patches, as well as those species adapted to the edge environment. As a result, borders are often populated by a rich diversity of life. This phenomenon, called the **edge effect,** is influenced by the area of border available (length and width) and by the degree of contrast between adjoining plant communities. In general, the greater the contrast between adjoining patches, the greater the diversity of species. Therefore, a border between forest and grassland should support more species than would a border between a young and a mature forest.

Although edge effect may increase species diversity, it can also create problems. Narrow, abrupt borders appear to be attractive to predators. Raccoons, opossums, and foxes may use edges as travel lanes. Avian predators such as crows and jays rob nests of smaller birds inhabiting more

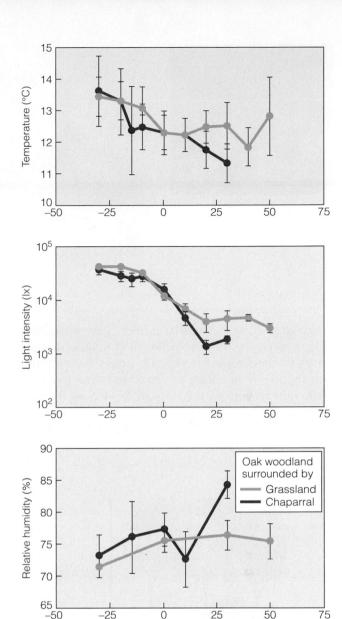

Figure 19.6 | Microhabitat variation (temperature, light intensity, and relative humidity) across edges between oak woodland patches and two different matrix habitats, chaparral and grassland, in the Santa Cruz Mountains of central California. For the *x*-axis, positive values indicate the distance from an edge into the woodland habitat; negative values indicate the distance into the matrix habitat. Bars represent standard errors of the means. Light intensity: ($lx \cdot 0.0185$ = PAR(μmol/m²/s)).

Interpreting Ecological Data

Q1. How does air temperature (top graph) change as you move out of an oak woodland patch into a surrounding chaparral habitat?

Q2. How does light intensity (center graph) change as you move from either chaparral or grassland habitats into woodland patches?

Q3. In general, how does relative humidity (bottom graph) change as you move from the more open grassland and chaparral habitats into oak woodland patches?

(Adapted from Sisk 1997.)

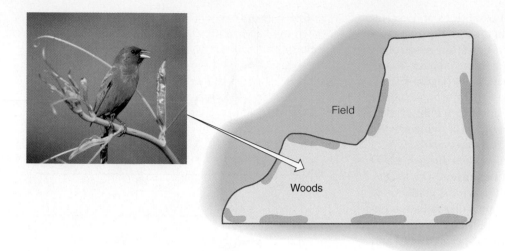

Figure 19.7 | Map of territories of a true edge species, the indigo bunting (*Passerina cyanea*), which inhabits woodland edges, hedgerows, roadside thickets, and large gaps in forests that create edge conditions. The male requires tall, open song perches and the female a dense thicket in which to build a nest.

(Adapted from Whitcomb et al. 1976.)

open edge habitat. Borders further alter interactions among species by either restricting or facilitating dispersal across the landscape. For example, dense, thorny shrubs growing in the boundary of forest and field can block some animals from passing through. Borders are dynamic, changing in space and time. In the example illustrated in Figure 19.8, the border between forest and field is characterized by an abrupt difference in the vertical structure of the vegetation. As the vegetation within the border grows taller, the vertical gap between the forest crown and border diminishes, forming a continuous vertical profile of vegetation. Barring disturbance, the border expands horizontally as vegetation characteristic of the border encroaches into the patch. Plant species change as size and environmental conditions within the border change and competitive interactions among border species increase.

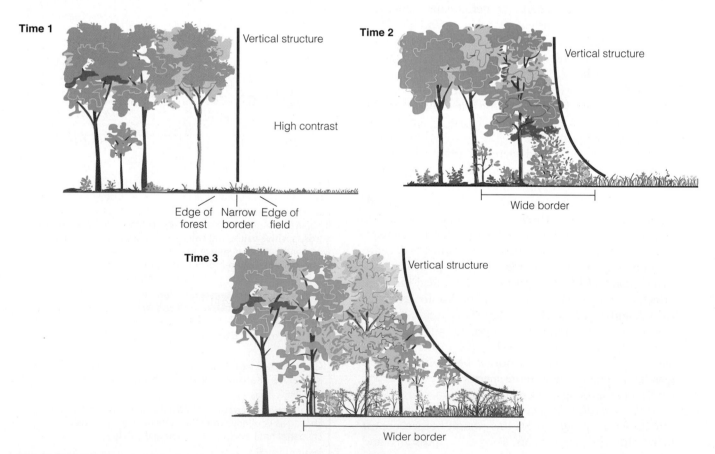

Figure 19.8 | Changes in vertical and horizontal structure of a border through time.

19.3 | Patch Size and Shape Are Crucial to Species Diversity

Patch size has a crucial influence on community structure, species diversity, and the presence and absence of species. As a general rule, large patches of habitat contain a greater number of individuals (population size) and species (species richness) than do small patches. The increase in population size for a given species with increasing area is simply a function of increasing the carrying capacity for the species (see Chapter 11). The greater the area, the greater will be the number of potential home ranges (or territories) that can be supported within the patch. Within the animal community, there is a general relationship between body size (weight) and the size of an animal's home range (see Figure 11.17). In addition, for a given body size, the home range of carnivores is greater than that of herbivores. Thus, large predators such as grizzly bears, wolves, and mountain lions will be limited to much larger, contiguous patches of habitat.

The relationship between patch size and species richness is more complex. Larger patches are more likely to contain variations in topography and soils that give rise to a greater diversity of plant life (both taxonomic and structural), which in turn will create a wide array of habitats for animal species (see Section 17.6).

Another feature of patch size relates to the difference between the habitats provided by edge and interior environments. The size and shape of patches affect the relative abundance of edge (or perimeter) and interior environments. Only when a patch becomes large enough to be deeper than its border can it develop interior conditions (Figure 19.9a). For example, at one extreme a very small patch is all border or edge habitat; but as patch size increases, the ratio of border to interior decreases (Figure 19.9b). Altering the shape of a patch can change this border-to-interior relationship. For example, the long, thin habitat in Figure 19.9c is all border or edge. Such long, narrow patches of woodland, whose depth does not exceed the width of the border, are border (or edge) communities regardless of the total patch area.

In contrast to edge species (see Section 19.2 and Figure 19.7), other species, termed **interior species,** require environmental conditions characteristic of interior habitats and stay away from the abrupt changes associated with border environments. Figure 19.10 shows the relationship between forest area (patch size) and the probability of occurrence of six different bird species of eastern North America. Edge species, such as the gray catbird (*Dumetella carolinensis*) and American robin (*Turdus migratorius*), have a high probability of occurring in small forest patches dominated by edge environments. As

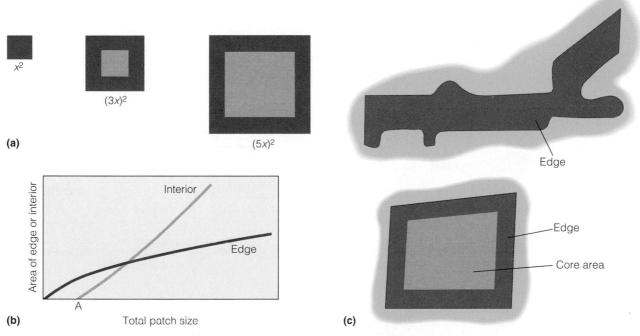

Figure 19.9 | Relationship of habitat patch size to edge and interior conditions. All habitat patches are surrounded by edge. **(a)** Assuming that the depth of the edge remains constant, the ratio of edge to interior decreases as the habitat size increases. When the patch is large enough to maintain shaded, moist conditions, an interior begins to develop. **(b)** The general relationship between patch size and area of edge and interior. Below point A, the habitat is all edge. As size increases, interior area increases, and the ratio of edge to interior decreases. **(c)** This relationship of size to edge holds for a square or circular habitat patch. Long, narrow woodland islands whose widths do not exceed the depth of the edge would be edge communities, even though the area may be the same as that of square or circular ones.

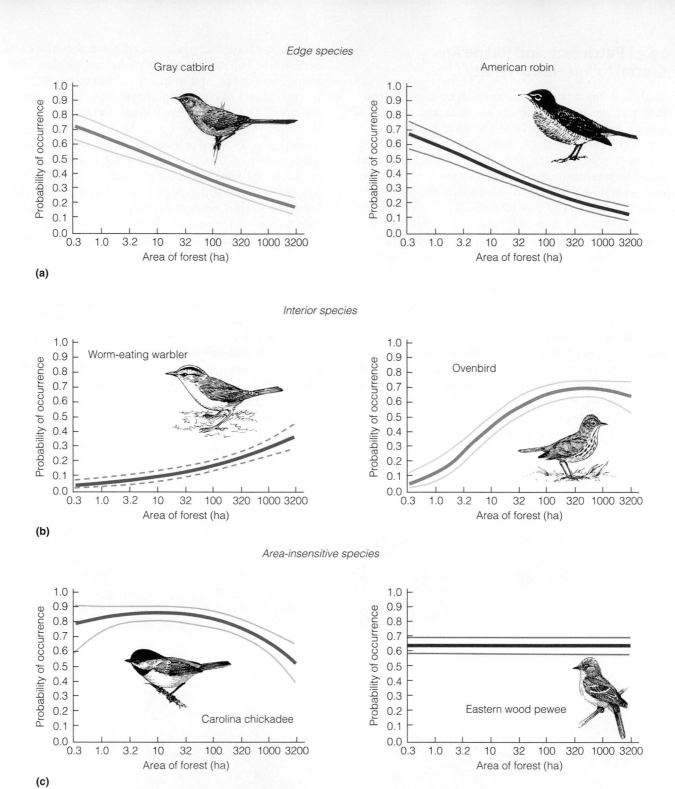

Figure 19.10 | Difference in habitat responses between edge species and area-sensitive or interior species. The graphs indicate the probability of detecting these species from a random point in patches of various sizes. Dashed lines indicate the 95 percent confidence intervals for the predicted probabilities. **(a)** The catbird and the robin are familiar edge species. As patch size increases, the probability of finding them decreases. **(b)** The worm-eating warbler and the ovenbird are ground-nesting birds of the forest interior. The probability of finding them in small patches is low. **(c)** In contrast to edge and interior species, other species—such as the Carolina chickadee and Eastern wood pewee—are insensitive to patch area (probability of encountering species is independent of patch size).

(Adapted from Robbins et al. 1989.)

patch size increases, the probability of finding these birds decreases. In contrast, the ovenbird (*Seiurus aurocapillus*) and the worm-eating warbler (*Helmitheros vermivorus*) are species adapted to the interior of older forest stands, which are characterized by large trees and sparse shrub cover in the understory layer. Accordingly, the probability is low that they will occur in small patches. Intermediate to these two groups are area-insensitive species, such as the Carolina chickadee (*Parus carolinensis*) and the Eastern wood pewee (*Contopus virens*).

The minimum size of habitat needed to maintain interior species differs between plants and animals. For plants, patch size per se is less important to the persistence of a species than are environmental conditions. For many shade-tolerant plant species found in the forest interior, the minimum size is the area needed to sustain moisture and light conditions typical of the interior. That area depends in part on the ratio of edge to interior and on the nature of the surrounding border habitat. If the stand is too small or too open, sunlight and wind will penetrate and dry the interior environment, eliminating herbaceous and woody species that require moister soil conditions. For example, in the northeastern United States, forest fragmentation can result in the decline of moisture-requiring (mesic) species such as sugar maple and beech while encouraging the growth of more xeric species such as oaks.

Several studies that have examined bird species diversity in both forest and grassland patches reveal a pattern of increasing species richness with patch size (Figure 19.11), but only up to a point. R. F. Whitcomb and colleagues studied patterns of species diversity in forest patches in western New Jersey. Their findings suggest that maximum bird diversity is achieved with woodlands 24 ha in size. Similar patterns were observed in studies investigating the species composition of bird communities occupying forest patches in agricultural regions in Illinois and Ontario, Canada. With patches of intermediate size, a general pattern of maximum species diversity results from the negative correlation between edge species and the size of habitat patches, combined with the positive correlation between interior species and increased area (see Figure 19.10).

These studies suggest that two or more small forest patches will support more species than will an equivalent area of contiguous forest. However, smaller woodlands did not support true forest interior species, such as the ovenbird (*Seiurus aurocapillus*), which requires extensive wooded areas (see Figure 19.10). Therefore, estimates of species diversity do not present the complete picture of how forest fragmentation is affecting the biological diversity in the landscape. Large forest tracts with a high degree of heterogeneity are required to support the range of bird species characteristic of both edge and interior habitats.

Although the relation of patch size to species diversity focuses strongly on forests, the same concept applies to

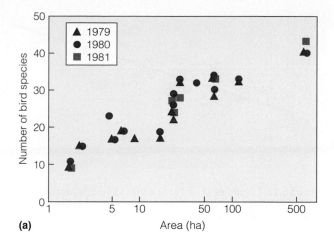

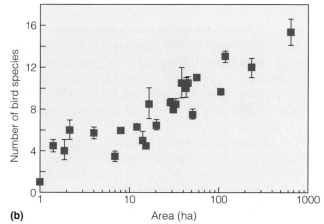

Figure 19.11 | The number of bird species (species richness) plotted as a function of the area of **(a)** woodland or **(b)** grassland habitat. Area (*x*-axis) in both graphs is presented on a log scale. Different symbols in **(a)** refer to results from surveys conducted during three different time periods. Squares in **(b)** represent mean values, while associated bars represent ; 1 standard error.

other landscapes—such as grasslands, shrublands, and marshes—that are all highly fragmented by cropland, grazing, sagebrush eradication, and housing developments. Many grassland species such as grasshopper sparrows (*Ammodramus savannarum*), western meadowlarks (*Sturnella neglecta*), and prairie grouse (*Tympanuchus* spp.) and shrubland species such as sage grouse (*Centrocercus urophasianus*), all of them interior species, are experiencing serious decline as these landscapes become fragmented and patch size decreases.

19.4 | The Theory of Island Biogeography Applies to Landscape Patches

The influence of area (patch size) on species richness did not escape the notice of early naturalist–explorers and biogeographers, who noted that large islands hold more species than small islands do (Figure 19.12). Johann

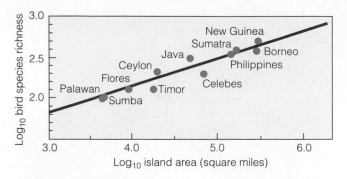

Figure 19.12 | Number of bird species on various islands of the East Indies in relation to area (island size). The x- and y-axes are plotted on a log₁₀ scale.

(Adapted from Preston 1962.)

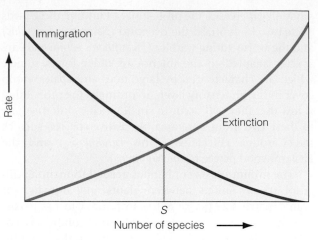

Figure 19.13 | According to the theory of island biogeography, immigration rate declines with increasing species richness (x-axis) while extinction rate increases. The balance between rates of extinction and immigration (immigration rate = extinction rate) defines the equilibrium number of species (S) on the island.

Reinhold Forster, a naturalist on Captain Cook's second voyage to the Southern Hemisphere (1772–75), noted that the number of different species found on islands depended on the island's size. The zoogeographer P. Darlington offered a rule of thumb: On islands, a tenfold increase in land area leads to a doubling of the number of species.

The various patches, large and small, that form the vegetation patterns across the landscape suggest islands of different sizes. Some are near each other; others are remote and isolated. A patch of forest, for example, may sit within a sea of cropland or housing developments, isolated from other forest patches on the landscape. The sizes of these patches and their distances from each other on the landscape have a pronounced influence on the nature and diversity of life they hold.

The similarity between islands and isolated patches of habitat has led ecologists to apply the **theory of island biogeography** to the study of terrestrial landscapes. First developed by Robert MacArthur (formerly of Princeton University) and Edward O. Wilson (Harvard University) in 1963, the theory is quite simple: the number of species established on an island represents a dynamic equilibrium between the immigration of new colonizing species and the extinction of previously established ones (Figure 19.13). Go to **GRAPHit!** at **www.ecologyplace.com** to explore island biogeography theory.

Consider an uninhabited island off the mainland. The species on the mainland make up the species pool of possible colonists. The species with the greatest ability to disperse from the mainland will be the first to colonize the island. As the number of species on the island increases, the immigration rate of new species to the island will decline. The decline results because the more mainland species that successfully colonize the island, the fewer potentially new species for colonization remain on the mainland (the source of immigrating species). When all mainland species exist on the island, the rate of immigration will be zero.

If we assume that extinctions occur at random, the rate of species extinction on the island will increase with the number of species occupying the island based purely on chance. Other factors, however, will amplify this effect. Later immigrants may be unable to establish populations because earlier arrivals will already have used available habitats and resources. As the number of species increases, competition among the species will most likely increase, causing a progressive increase in the extinction rate. An equilibrium species richness (S) is achieved when the immigration rate equals the extinction rate (see Figure 19.13). If the number of species inhabiting the island exceeds this value, the extinction rate is greater than the immigration rate, resulting in a decline in species richness. If the number of species is below this value, the immigration rate is greater than extinction and the number of species increases. At equilibrium, the number of species residing on the island remains stable, although the composition of species may change. The rate at which one species is lost and a replacement species is gained is the turnover rate.

The distance of the island from the mainland and the size of the island both affect equilibrium species richness (Figure 19.14). The greater the distance between the island and the mainland, the less likelihood that many immigrants will successfully complete the journey. The result is a decrease in the equilibrium number of species (Figure 19.14a). On larger islands, extinction rates, which vary with area, are lower because a greater area generally contains a wider array of resources and habitats. For this reason, large islands can support more individuals of each species as well as meet the needs of a wider variety of species. The lower rate of extinction on larger islands results in a higher equilibrium number of species as compared to smaller islands (Figure 19.14b).

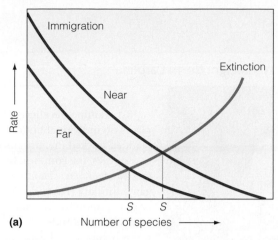

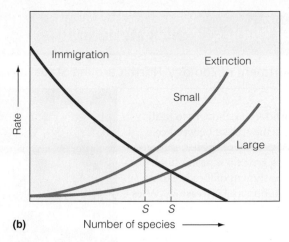

Figure 19.14 | **(a)** Immigration rates are distance related. Islands near a mainland have a higher immigration rate and associated equilibrium species richness (*S*) than do islands distant from a mainland. **(b)** Extinction rates relate to area and are higher on small islands than on large ones. The equilibrium number of species varies according to island size, and larger islands have greater equilibrium species richness than do smaller islands.

Although the theory of island biogeography was applied initially to oceanic islands, there are many other types of "islands." Mountaintops, bogs, ponds, dunes, areas fragmented by human land use, and individual hosts of parasites are all essentially island habitats. As Daniel Simberloff of the University of Tennessee, one of the first ecologists to experimentally test the predictions of this theory, put it: "Any patch of habitat isolated from similar habitat by different, relatively inhospitable terrain traversed only with difficulty by organisms of the habitat patch may be considered an island."

Exactly how does island theory apply to landscape patches? Landscape patches or islands differ considerably from the oceanic islands for which the theory was developed. Oceanic islands are terrestrial environments surrounded by an aquatic barrier to dispersal. They are inhabited by organisms of various species that arrived there by chance dispersal over a long period of time or represent remnant populations that existed in the area long before isolation. By contrast, the organisms that inhabit landscape patches are samples of populations that extend over a much wider area. These patch communities contain fewer species than are found in the larger area, and only a few individuals may represent each species. However, unlike oceanic islands, terrestrial landscape patches are associated with other terrestrial environments, which often present fewer barriers to movement and dispersal among patches.

19.5 | Corridors Permit Movement between Patches

The fragmentation of previously large contiguous areas of habitats, be they forest, grassland, or water, results in a mosaic of patches—islands between which the movement of species is limited or even made dangerous or impossible by the inhospitable nature of the intervening environment. In some situations, **corridors** connect patches of similar habitat, enhancing the ability of organisms to move among patches of their habitat. Typically, corridors are strips of vegetation similar to the patches they connect but different from the surrounding matrix in which they are set. Many corridors are of human origin. Some may be narrow-line corridors, such as hedgerows and lines of trees planted as windbreaks, bridges over fast-flowing streams, highway median strips, and drainage ditches (Figure 19.15a on p. 404). Wider bands of vegetation, called strip corridors, can consist of both interior and edge environments. Such corridors may be broad strips of woodlands left between housing developments, power-line rights-of-way, and belts of vegetation along streams and rivers (Figure 19.15b). An important aspect of any corridor is its **connectivity,** the extent to which a species or a population can move among patches within the matrix (see Field Studies: Nick M. Haddad).

Corridors probably function best as travel lanes for individuals moving within the bounds of their home range; but when corridors interconnect to form networks, they offer dispersal routes for species traveling between habitat patches. They enhance the movement of organisms beyond what is possible through the adjacent matrix. By facilitating the movement among different patches, corridors can encourage gene flow between subpopulations and help reestablish species in habitats that have experienced local extinction (see Chapter 12). Corridors also act as filters, providing dispersal routes for some species but not others. Different-sized gaps in corridors allow certain organisms to cross while restricting others; this is the **filter effect.**

Corridors can negatively affect some populations as well. For example, corridors offer scouting positions for

Department of Zoology, North Carolina State University, Raleigh, North Carolina

Corridors are thought to facilitate movement between connected patches of habitat, thus increasing gene flow, promoting reestablishment of locally extinct populations, and increasing species diversity within otherwise isolated areas. The potential utility of corridors in the conservation of biological diversity has attracted much attention, but the proposed value of corridors is based more on intuition than on empirical evidence. However, recent studies by ecologists, like Nick Haddad of North Carolina State University, are providing valuable data on the role of corridors in facilitating dispersal among otherwise isolated habitat patches on the larger landscape.

Nick Haddad studies the influence of corridors on the dispersal and population dynamics of butterfly species. He has chosen butterflies as a focal organism because their life histories are well defined, they are generally associated with a narrow range of resources, and they are relatively easy to study at large spatial scales.

To examine the effects of corridors on the dispersal and population dynamics of butterfly species, in cooperation with the U.S. Forest Service, Haddad and colleagues established eight 50-ha landscapes on the 1240 km² Savannah River Site, a National Environmental Research Park in South Carolina. All landscapes were composed of mature (40- to 50-year-old) forest dominated by loblolly pine (*Pinus taeda*) and longleaf pine (*Pinus palustris*). Within each landscape, five early successional habitat patches were created by cutting and removing all trees and then burning the cleared areas.

Haddad compared movement rates from a 1-ha central patch, created at the center of each landscape, to four surrounding peripheral patches, created at the same time, each 150 m from the central patch (Figure 1). A corridor 25 m wide connected the central patch to one of the peripheral patches ("connected patches"). All other peripheral patches ("unconnected patches") were equal in size to the area of the connected patch plus the area of the corridor (1.375 ha), thus controlling for effects of increased patch area in the connected patches. In unconnected patches, the corridor's area was added either as 75-m "wings" projecting from the sides of patches ("winged patches") or as additional habitat added to the backs of patches ("rectangular" patches). By comparing the rates of movement from the central patch into the connected patches versus movement into unconnected patches, the biologists were able to test the hypothesis that corridors function as conduits of movement.

To examine the effects of corridors on individual butterflies, Haddad tracked the movement of two species, the common buckeye (*Junonia coenia*) and the variegated fritillary (*Euptoieta claudia*), both common in early successional habitats and rare in mature forest habitats on the Savannah River Site. He marked naturally occurring butterflies in the central patches and recaptured marked individuals in the peripheral patches. All captured butterflies were marked, and the locations of initial and subsequent recaptures were recorded.

The common buckeye was 3 to 4 times more likely to move from center patches to connected patches than to unconnected patches, and the variegated fritillary was

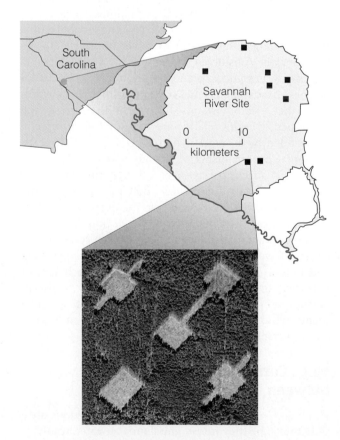

Figure 1 | Map of experimental landscape locations and aerial photograph of one landscape, showing patch configuration.

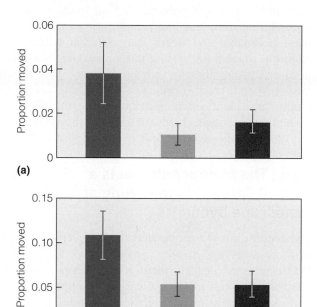

(a)

(b)

Connected   Winged   Rectangular

Figure 2 | Movement rates of butterfly species between connected and isolated patches. *J. coenia* **(a)** and *E. claudia* **(b)** both moved between connected patches more often than between isolated patches. Data in both panels are means ±1 standard error for proportion of individuals marked in the central patch and recaptured in connected, winged, and rectangular peripheral patches. Go to **QUANTIFYit!** at **www.ecologyplace.com** to perform confidence intervals and t-tests.

(Haddad 1999.)

twice as likely to move down corridors than through forests when moving from the center patch (Figure 2). Neither butterfly was more likely to move to winged patches than to rectangular patches.

The results clearly showed that the presence of a corridor facilitated the movement of both butterfly species between connected patches, even after controlling for patch size and shape. But in this experiment, the length of the corridor was fixed at 150 m. How might the length of the corridor influence patterns of dispersal? To examine the influence of corridor length on patterns of butterfly dispersal, Haddad once again created experimental patches and corridors at the Savannah River Site. The experiment consisted of square patches of equal size (1.64 ha). The size of the square patches was fixed at 128 m on a side, and all patches were oriented in the same direction. Two characteristics of the patches were varied: interpatch distance and the presence or absence of a connecting corridor. Distances between patches were 64, 128, 256, or 384 m (0.5, 1, 2, or 3 times the width of the patch).

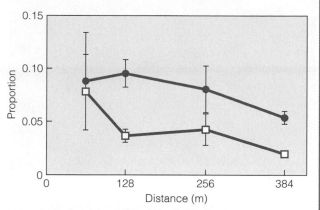

Figure 3 | Mean proportion (±1 standard error) of individuals (males) *Junonia* marked in a patch who moved 1 of 4 distances to an adjacent patch. Circles indicate mean proportions moving between patches connected by a corridor, and squares indicate mean proportions moving between unconnected patches.

(Haddad 1999.)

In this study, Haddad examined the movement patterns of the same two butterfly species, the common buckeye (*J. coenia*) and the variegated fritillary (*E. claudia*). Results of the mark-recapture studies showed that individuals of both species moved more frequently between patches connected by corridors than between unconnected patches, just as was observed in the previous study. In addition, however, interpatch movement was negatively related to interpatch distance (Figure 3), and the density of both butterfly species was significantly higher in connected patches.

The studies by Haddad and colleagues have expanded our understanding of the utility of corridors in aiding patterns of dispersal, a key factor in the persistence of populations in a fragmented landscape. The unique design of these experiments has allowed us to test specific hypotheses regarding the influence of the size and shape of corridors on their utility in conservation efforts.

Bibliography

Haddad, N. 1999a. Corridor and distance effects on interpatch movements: A landscape experiment with butterflies. *Ecological Applications* 9:612–622.

Haddad, N., and K. Baum. 1999b. An experimental test of corridor effects on butterfly densities. *Ecological Applications* 9:623–633.

Tewksbury, J., D. Levey, N. Haddad, S. Sargent, J. Orrock, A. Weldon, B. Danielson, J. Brinkerhoff, E. Damschen, and P. Townsend. 2002. Corridors affect plants, animals, and their interactions in fragmented landscapes. *Proceedings of the National Academy of Sciences of the United States of America* 99:12923–12926.

1. What life history characteristics might you expect for a species that benefits from the presence of corridors?

(a)

(b)

Figure 19.15 | Examples of corridors: **(a)** hedgerow and **(b)** riverine vegetation.

predators that need to remain concealed while hunting in adjacent patches. They can also create avenues for the spread of disease between patches and provide a pathway for the invasion or spread of exotic species from the matrix to other patches. If they are too narrow, corridors can inhibit the movement of social groups.

Corridors may also provide habitats in their own right. Corridors along streams and rivers provide important riparian habitat for animal life. In Europe, the long history of hedgerows in the rural landscape has encouraged the development of typical hedgerow animal and plant communities (see Figure 19.15a). In suburban and urban settings, corridors provide habitat for edge species and act as stopover habitat for migrating birds.

Roads—corridors designed as dispersal routes for humans—dissect the landscape and have effects on adjacent patches of land. Two- to four-lane high-speed roads are a major source of mortality for wildlife ranging in size from large mammals to tiny insects, and such roads effec-

tively divide populations of many species. All types of roads in some way affect roadside vegetation. Salt spread on highways during snow removal, particulate matter from tires and diesel exhaust, and chemical pollutants from trucks and automobiles affect all roadside vegetation. During storms and snowmelt, water runoff carries these pollutants and debris into adjacent patches. In addition, noise from passing traffic discourages wildlife from occupying otherwise suitable habitat. Most important, perhaps, road corridors allow people access to remote areas, with often-disastrous ecological effects, as exemplified by logging roads cut through tropical rain forest. Where roads invade, people and development follow.

19.6 | The Metapopulation Is a Central Concept in the Study of Landscape Dynamics

The distribution of a species over the landscape is limited to those areas that provide a suitable habitat. In many landscapes, habitats are often so fragmented that the species exist as distinct, partially isolated subpopulations, each possessing its own population dynamics. Each subpopulation has its own birthrate, death rate, and probability of extinction. Linking these subpopulations are individuals moving among patches. Such separated populations interconnected by the movement of individuals are called metapopulations (see Chapter 12).

The concept of metapopulations has become central to the study of landscape dynamics because it provides a framework for examining the dynamics of species that are distributed as discrete populations on the larger landscape. You may have noted the similarity in the structure of the basic model of metapopulation dynamics developed by Richard Levins (see Section 12.2) and the model of island biogeography presented in this chapter (see Section 19.4). The theory of island biogeography views the equilibrium number of species (species richness) occupying a given patch (island) as the balance between the processes of colonization and extinction. The model of metapopulation dynamics views the equilibrium number of patches occupied by a given species as the balance between colonization and extinction of local populations. In addition, both models explicitly address how rates of colonization and extinction are influenced by the size and isolation of habitat patches (islands) on the larger landscape (see Section 12.3).

In a way, the model of island biogeography can be viewed as a consequence of the metapopulation dynamics of the set of species that occupy the larger landscape. Some researchers have extended metapopulation models to interactive and noninteractive multispecies communities. Models of interacting metapopulations have been used to explain patterns in species succession, species richness and composition, and the food web structure of communities.

Community models of noninteracting species are constructed simply by summing the predicted patterns of individual species using a metapopulation framework. Metapopulation models can be used to predict the incidence of a species on a given habitat patch (or island), and by summing across all species, the expected number of species on a habitat patch can be predicted. Using this approach, Ilkka Hanski and colleagues have shown that the species–area relationship presented in Figures 19.11 and 19.12 can be derived from a metapopulation model with area-dependent extinction rate and density-dependent colonization rate.

In turn, landscape studies are expanding our understanding of metapopulation dynamics. Studies exploring how landscape pattern functions to constrain patterns of species dispersal (see Field Studies: Nick M. Haddad) yield a richer understanding of this key element of metapopulation dynamics. Likewise, studies examining how patch geometry (patch size and shape) controls the local environment of patches (see Section 19.3) provide a basis for predicting the probability of successful colonization and extinction based on species life history characteristics and habitat needs.

19.7 | Frequency, Intensity, and Scale Determine the Impact of Disturbances

The distinct patterns of communities that we see in the landscape, as well as the plant and animal species inhabiting them, are heavily influenced by disturbances both past and present. A **disturbance** is any relatively discrete event—such as fire, windstorm, flood, extremely cold temperatures, drought, or epidemic—that disrupts community structure and function. Disturbances both create and are influenced by patterns on the landscape. For example, communities on ridgetops are more susceptible to damage from wind and ice storms, and bottomland communities along streams and rivers are more susceptible to flooding. In turn, these disturbances result in a new pattern of patches on the landscape.

Ecologists often distinguish between a particular disturbance event—a single storm or fire—and the disturbance regime, or pattern of disturbance, that characterizes a landscape over a longer period of time. The disturbance regime has spatial and temporal characteristics including intensity, frequency, and spatial extent or scale. **Intensity** is measured by the proportion of total biomass, or population of a species, that the disturbance kills or eliminates. It is influenced by the magnitude of the physical force involved, such as the strength of the wind or the amount of energy released during a fire. **Scale** refers to the spatial extent of the impact of the disturbance relative to the size of the affected landscape. **Frequency** is the mean number of disturbances that occur within a particular time interval

(Figure 19.16). The mean time between disturbances for a given area is the return interval.

Often, a disturbance's frequency is linked to its intensity and scale. For example, natural disturbances taking place on a small scale, such as the death of an individual canopy tree in a temperate or tropical forest, occur quite frequently. By contrast, large-scale disturbances, such as fire, are rarer and occur on average once every 50 to 200 years. In the intertidal zone of the ocean, the abrasive action of waves tears away mussels and algae from rocks along the shore. In grasslands, badgers and groundhogs dig burrows and expose small patches of mineral soil (see Field Studies: Katherine N. Suding, p. 264 in Chapter 13). The outcome of such small-scale disturbances is the creation of a **gap,** an opening that becomes a site of localized regeneration and growth within the community. (Plant ecologist A. S. Watt first applied the term *gap* in 1947.) Gap formation within the community does more than provide access to physical space for colonization by new individuals. Within the gap, the physical environment often differs substantially from conditions in the surrounding area. For example, the microclimate in a gap created by the death of a canopy tree differs from the climate of the rest of the forest. Light and soil temperature increase, while soil moisture and relative humidity decrease. These new conditions encourage the germination and growth of sun-adapted species as well as stimulate the growth of smaller trees already established in the understory (see Figure 6.14 and Field Studies: Kaoru Kitajima, p. 112). The race is on as individuals grow in height; their fate will be determined by how rapidly crowns of trees surrounding the gap can expand to fill the opening.

Community responses to larger-scale disturbances, such as fire, logging, or other forms of land clearing, result in responses that go beyond reorganizing the populations already occupying the site. Large-scale disturbances result in substantial reduction or even elimination of local populations and significantly modify the site's physical environment. In these situations, a period of colonization follows. Some species may already be present but dormant, as seeds of woody and herbaceous plants, roots and rhizomes, stump sprouts, and surviving seedlings and saplings. Other colonizing species are carried to the site by the wind or animals. Long-term recovery involves the process of secondary succession (discussed in Chapter 18) in which species characteristic of the original community eventually replace the early colonizing species.

19.8 | Various Natural Processes Function as Disturbances

Many disturbances arise from natural causes, such as wind and ice storms, lightning fires, hurricanes, floods, grazing animals, and insect outbreaks. Although disturbance usually results in the death of organisms and the loss of

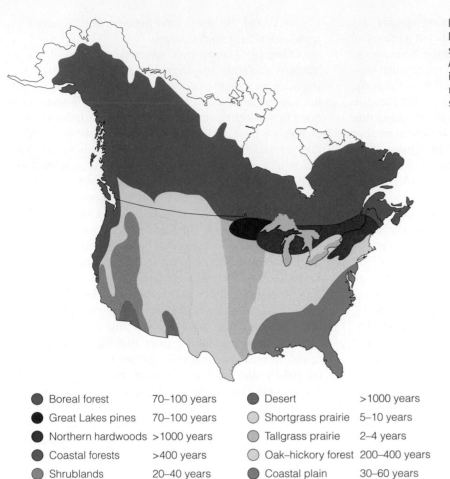

● Boreal forest	70–100 years	● Desert	>1000 years
● Great Lakes pines	70–100 years	○ Shortgrass prairie	5–10 years
● Northern hardwoods	>1000 years	◐ Tallgrass prairie	2–4 years
● Coastal forests	>400 years	○ Oak–hickory forest	200–400 years
◐ Shrublands	20–40 years	● Coastal plain	30–60 years
○ Dry conifer forests			
Surface	5–40 years		
Crown	200–400+ years		

biomass, it can also be a powerful force for change in the physical environment. Storm floods cut away banks, change the course of streams and rivers, scour stream bottoms, redeposit sediments, and carry away aquatic organisms. High winds and tides break down barrier dunes and allow seawater to invade the shoreline, changing the geomorphology of barrier islands (Figure 19.17). By modifying an ecosystem in ways that favor survival of some species and eliminate others, disturbance can either reduce or foster species diversity.

Wind and moving water are two powerful agents of disturbance. Large, inflexible forest canopy trees, as well as trees growing in open fields and along forest edges, are vulnerable to windthrow. So, too, are trees growing on poorly drained soils; their shallow roots are not well anchored. On rocky intertidal and subtidal shores, powerful waves overturn boulders and dislodge sessile organisms. This action clears patches of hard substrate, making them available for recolonization. Hurricanes represent large-scale wind- and water-related disturbances that have a devastating impact on ecosystems. In the Caribbean region, the frequency (average time interval between occurrences) of hurricanes with the intensity of Hurricane Hugo, which

struck the southeastern United States in 1989 (wind speeds more than 166 km/hr), is once every 50 to 60 years.

Fire is a major agent of disturbance, altering both biological and physical environments. In many regions, fire is

Figure 19.17 | Erosion of coastal dunes by storm surges creates breaks in the front dunes, resulting in areas of inundation.

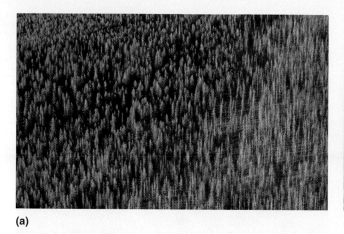

(a)   (b)

Figure 19.18 | **(a)** Crown fire resulting in a landscape mosaic of patches of burned and unburned forest. **(b)** Fires of great intensity can profoundly influence ecosystems. After a spruce forest located on the Alleghany Plateau area of West Virginia was cut in the mid-1800s, intense ground fires burned, fueled by piles of debris from logging. Fire consumed the ground layer, exposing bedrock and mineral soil. The forest never recovered.

a major determinant of landscape patterns. Losses of plants and animals can be significant, reducing biomass as well as species diversity. Fire also consumes standing dead material, accumulated plant litter, and soil organic matter; the nutrients contained in these substances are released through the process of pyromineralization (see Section 27.5 and Field Studies: Deborah Lawrence, p. 585). Fire prepares the seedbed for some species of trees by exposing mineral soil. By removing some or all of the previous vegetation, fire can lead to an increase in the availability of light, water, and nutrients to both remaining (surviving) and newly colonizing plants. Its return rate is influenced by the occurrence of droughts, accumulation and flammability of the fuel (biomass), the resulting intensity of the burn, and human interference. Before the European settlement of North America, fires occurred about every 3 years in the grasslands of the Midwest (see Figure 19.16). In forest ecosystems, the frequency of fires varies greatly, depending on the type of forest. Frequent light fires that burn only the surface layer may have a return interval of 1 to 25 years, whereas fires that destroy canopy trees may have a return interval of 50, 100, or even 300 years (Figure 19.18).

Grazing by domestic animals is a common cause of disturbance. For example, in the southwestern United States, domestic cattle disperse seeds of mesquite (*Prosopis* spp.) and other shrubs through their droppings, encouraging those species to invade already overgrazed grassland. The disturbance of herbivores is not limited to domestic species. In many parts of eastern North America, overpopulations of white-tailed deer (*Odocoileus virginianus*) have decimated herbaceous plants and shrubs in the forest understory (Figure 19.19), eliminated forest regeneration, and destroyed habitat for forest understory wildlife. The African elephant (*Loxodonta africana*) has long been considered a major influence on the development of savanna communities. The combination of high density and re-

stricted movement, limited to the confines of national parks and conservation areas, can result in the large-scale destruction of woodlands (see Figure 16.2).

Beavers (*Castor canadensis*) modify many forested areas in North America and Europe. By damming streams,

Figure 19.19 | High population density of white-tailed deer in the eastern United States can result in elimination of forest understory plants and creation of a "browse line" in the forest trees. The browse line is the highest point that deer can reach for food.

Figure 19.20 | A small beaver dam about 2 m high along a stream in the Rocky Mountains. The reservoir of water behind the dam alters the stream flow.

Figure 19.21 | Block clear-cutting in a coniferous forest in the western United States. Such cutting fragments the forest.

beavers alter the structure and dynamics of flowing water ecosystems. When dammed streams flood lowland areas, beavers convert forested stands into wetlands. Pools behind dams become catchments for sediments. By feeding on aspen, willow, and birch, beavers maintain stands of these trees, which otherwise would be replaced by later successional species. Thus, the action of beavers creates a diversity of patches—pools, open meadows, and thickets of willow and aspen—within the larger landscape (Figure 19.20).

Birds may seem an unlikely cause of major vegetation changes. But in the lowlands along the west coast of Hudson Bay, large numbers of the lesser snow goose (*Chen caerulescens caerulescens*) have affected the brackish and freshwater marshes. Snow geese grub for roots and rhizomes of graminoid plants in early spring and graze intensively on leaves of grasses and sedges in summer. The dramatic increase in the number of geese has stripped large areas of their vegetation, resulting in the erosion of peat and the exposure of underlying glacial gravels. Plant species colonizing these patches differ from the surrounding marsh vegetation, giving rise to a mosaic of patches on the landscape.

Outbreaks of insects such as gypsy moths and spruce budworms defoliate large areas of forest and cause the death or reduced growth of affected trees. Gypsy moth infestations of a hardwood forest may result in a mortality rate of 10–50 percent (see Ecological Issues: Human-Assisted Dispersal, p. 195). Infestation of spruce and fir stands by spruce budworms may result in 100 percent tree mortality. The impact of forest insects is most intense in large expanses of homogeneous forest occupied by older trees.

19.9 | Human Disturbance Creates Some of the Most Long-Lasting Effects

Some of the most lasting disturbances to the landscape are human induced. Because human activity is ongoing and involves continuous management of an ecosystem, it affects

ecosystems more profoundly than natural disturbances do. One of the more permanent and radical changes in vegetation communities occurs when we remove natural communities and replace them with cultivated cropland and pastures. Many of us have the idea that agriculture is a relatively recent disturbance. In fact, prehistoric human populations were cultivating crops and converting land to pasture as early as 5000 years B.P. Their activities changed the landscape pattern by extending or reducing the ranges of woody and herbaceous plants, allowing the invasion and spread of opportunistic weedy species, and altering the dominance structure in woodlands.

Another large-scale disturbance is timber harvesting (see Chapter 27). Disturbance to the forest depends on the logging methods employed, which range from removing only selected trees to clearing entire blocks of timber. In selection cutting, foresters remove only mature single trees or groups of trees scattered through the forest. Selection cutting produces only gaps in the forest canopy (see Section 27.9). Therefore, while this form of timber removal can minimize the scale of disturbance, it can also change the species composition and diversity because it favors shade-tolerant over shade-intolerant species of trees.

Clear-cutting involves removal of wide blocks of trees (Figure 19.21), favoring reproduction of shade-intolerant species of trees. However, unless foresters manage these cleared areas carefully, erosion can badly disturb the ecosystem, affecting recovery of the site as well as adjacent aquatic communities.

19.10 | The Landscape Represents a Shifting Mosaic of Changing Communities

Unlike the artist's mosaic shown in Figure 19.2 that has survived unchanged for centuries, the mosaic of communities defining the landscape is ever-changing through

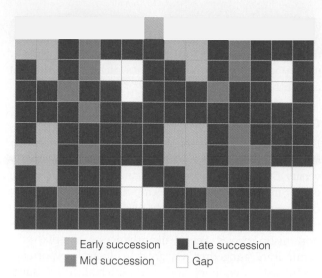

Early succession   Late succession
Mid succession   Gap

Figure 19.22 | Representation of a forested landscape as a mosaic of patches in various stages of successional development. Although each patch is continuously changing, the average characteristic of the forest may remain relatively constant—in a steady state.

time. Disturbances—large and small, frequent and infrequent—alter the biological and physical structures of communities making up the landscape, giving way to the process of succession. This view of the landscape suggests a **shifting mosaic** composed of patches, each in a phase of successional development (Figure 19.22). The ecologists F. Herbert Borman and Gene Likens applied this concept to describe the process of succession in forested landscapes, using the term *shifting-mosaic steady-state*. The term *steady state* is a statistical description of the collection of patches and thus refers to the average state of the forest. In other words, the mosaic of patches shown in Figure 19.22 is not static. Each patch is continuously changing. Disturbance

causes the patches in the mosaic that are currently classified as late successional to revert to early successional. Patches currently classified as early successional undergo shifts in species composition, and later successional species will come to dominate. Although the overall mosaic is continuously changing, the average composition of the forest (average over all patches) may remain fairly constant—in a steady state. This example of a continuously changing population of patches that remains fairly constant when viewed collectively rather than individually is very similar to the concept of a stable age distribution presented in Chapter 10 (Section 10.7). In a population with a stable age distribution, the proportion of individuals in each age class remains constant even though individuals are continuously entering and leaving the population through births and deaths.

Returning to the Virginia countryside in Figure 19.01, we can now see the mosaic of patches—forest, fields, golf course, hedgerows, pine plantation, pond, and human habitations—not as a static image but as a dynamic landscape. Most of the forested lands were once fields and pastureland. During the late 19th and early 20th centuries, when agriculture in the region declined, farmers abandoned their fields and the land reverted to forest (see Ecological Issues: American Forests, p. 370). The current mosaic of land cover is maintained by active processes, many of which are forms of human-induced disturbance. Within these patches, the communities function as islands, some bridged by corridors. Their populations are part of larger metapopulations linked by dispersal of individuals. As time passes, the landscape will continue to change. Patterns of land use will shift boundaries; succession will alter the structure of communities; and natural disturbances such as fire and storms will form new dynamic patches within the mosaic.

Summary

Landscapes | 19.1

Landscapes consist of mosaics of patches related to and interacting with each other. The study of the causes and consequences of these spatial patterns is landscape ecology. Various patches in the landscape have their own unique origin, from remnant patches of original vegetation to introduced patches requiring human maintenance. Several environmental factors interact to create landscape pattern, including geology, topography, soils, climate, biotic processes, and disturbance.

Borders and Edges | 19.2

The place where the edges of two different patches meet is a border. A border may be inherent—produced by a sharp environmental change, such as a topographical feature or a

shift in soil type—or induced, meaning it is created by some form of disturbance that is limited in extent and changes through time. Some borders are narrow and abrupt; others are wide and form a transition zone, or ecotone, between adjoining patches. A border also has a vertical structure that influences physical gradient between patches. Functionally, a border connects patches through fluxes or flows of material, energy, and organisms. Typically, transition zones between patches have high species richness because they support selected species of adjoining communities as well as a group of opportunistic species adapted to edges—a phenomenon called edge effect.

Patch Size, Shape, and Diversity | 19.3

A positive relationship exists between area and species diversity. Generally, large areas support more species than small areas do. The increase in species diversity with increasing patch size is related to several factors. Many species are area sensitive—they require large, unbroken blocks of habitat. Larger areas typically encompass a greater number of microhabitats and thus will support a greater array of animal species. Another feature of patch size relates to differences between the habitats provided by border and interior environments. In contrast to edge species, interior species require environmental conditions found in the interior of large habitat patches, away from the abrupt changes in environmental conditions associated with edge environments.

Island Biogeography | 19.4

The theory of island biogeography proposes that the number of species an island holds represents a balance between immigration and extinction. The island's distance from a mainland or source of potential immigrants influences immigration rates. Thus, islands farthest from a mainland would receive fewer immigrants than would islands closer to the mainland. The area of an island influences extinction rates. Because they hold fewer individuals of a species and their habitat varies less, small islands have higher extinction rates than large islands do. In habitat patches, as in islands, large areas support more species than do small areas.

Corridors | 19.5

Linking one patch to another are corridors, the strips of habitat similar to a patch but unlike the surrounding matrix. Corridors act as conduits or travel lanes, function as filters or barriers, and provide dispersal routes among patches.

Metapopulations | 19.6

Habitat fragmentation and human exploitation have reduced many species to isolated or semi-isolated populations. These subpopulations inhabiting fragmented habitats form metapopulations. Maintaining them depends on movement of individuals among habitat patches. The theory of metapopulation dynamics is a mechanism for understanding patterns of species richness in habitat patches.

Characteristics of Disturbance | 19.7

Disturbance, a discrete event that disrupts communities and populations, also initiates succession and creates diversity. Disturbances have spatial and temporal characteristics. Small-scale disturbances make gaps in the substrate or vegetation, creating patches of different composition or successional stage. Large-scale disturbances favor opportunistic species. Severe disturbances can replace the community altogether. Of great ecological importance are intensity, frequency, and return interval. Too frequent disturbances can eliminate certain species.

Natural Disturbances | 19.8

Fire is a natural large-scale disturbance. It has both beneficial and adverse effects. Other major natural disturbance regimes include wind, floods, storms, and animals.

Human Disturbances | 19.9

Major human-induced disturbances include logging, mining, agriculture, and development. These produce profound, often permanent, changes in the landscape.

Landscape as a Shifting Mosaic | 19.10

The landscape is dynamic, and patches are in various stages of development and disturbance. This process suggests a landscape pattern that represents a shifting mosaic. The term *steady state* is a statistical description of the collection of patches; it describes the average state of the landscape.

Study Questions

1. In what way is the study of spatial pattern the defining characteristic of landscape ecology?
2. How do variations in the physical environment (geology, topography, soils, and climate) give rise to the landscape patterns in your region? Contrast these patterns to the influence of human activities in defining the mosaic of patches that form the surrounding landscape.
3. Distinguish between edge, border, and ecotone.
4. Why do edges and ecotones often support a greater diversity of species than do the adjoining communities?
5. How does the ratio of the proportion of edge to interior habitat change with increasing patch size?

How does species diversity change with increasing patch size?
6. The theory of island biogeography envisions the species richness of island communities as a balance between the processes of colonization and local extinction. In what way can this theory be applied to isolated habitat patches in terrestrial environments?
7. How do island size and isolation (distance to neighboring islands) influence patterns of species diversity?
8. How does the theory of island biogeography relate to the model of metapopulation dynamics developed by Richard Levins and presented in Chapter 12 (Section 12.2)?

9. How do corridors aid in maintaining the diversity of habitat patches?

10. Distinguish among these terms as they apply to landscape disturbances: scale, intensity, frequency.

11. How has the landscape around your home or neighborhood changed since you were a young child? What processes have been responsible for the changes?

Further Readings

Cadenasso, M. L., S. T. A. Pickett, K. E. Weathers, and C. D. Jones. 2003. A framework for a theory of ecological boundaries. *BioScience* 53:750–758.

This paper provides a framework for the study of ecological boundaries. The framework focuses on flows of organisms, materials, and energy in heterogeneous landscape mosaics.

Forman, R. T. T. 1995. *Land mosaics: The ecology of landscapes and regions.* New York: Cambridge University Press.

A state-of-the-art synthesis exploring the ecology of heterogeneous land areas, where natural processes and human activities spatially interact to produce a continually changing mosaic.

Harris, L. D. 1988. Edge effects and the conservation of biotic diversity. *Conservation Biology* 2:330–332.

This short paper is a good overview of the development of the concept of edge effect in ecology. It introduces a series of articles in this journal issue that examine the role of edge communities in conservation ecology.

Hilty, J. A., W. Z. Lidicker, Jr., and A. M. Merenlender. 2006. *Corridor ecology: The science and practice of linking landscapes for biodiversity conservation.* Washington, DC: Island Press.

An excellent review of corridor ecology and its application in creating, enhancing, and maintaining connectivity between natural areas.

Johnson, H. B. 1976. *Order upon the land.* Oxford: Oxford University Press.

This book provides an historical perspective and analysis of the impacts of early survey and settlement practices on the American landscape.

Lindenmayer, D. B., and J. Fischer. 2006. *Habitat fragmentation and landscape change: An ecological and conservation synthesis.* Washington, DC: Island Press.

Reviews ecological problems caused by landscape change and discusses the relationships among landscape change, habitat fragmentation, and biodiversity.

McDonnell, M. J., and S. T. A. Pickett. 1990. Ecosystem structure and function along urban-rural gradients: An unexploited opportunity for ecology. *Ecology* 71:1232–1237.

This excellent paper develops a framework for studying the transition from rural to urban environments to better understand the influence of human activity on ecological patterns and processes on the larger landscape.

Romme, W. H., and D. H. Knight. 1982. Landscape diversity: The concept applied to Yellowstone Park. *BioScience* 32:664–670.

An early application of the principles of landscape ecology to Yellowstone National Park. The authors provide a practical application of landscape theory to understanding pattern and process across this expansive landscape.

Rosenberg, D. K., B. R. Noon, and E. C. Meslow. 1997. Biological corridors: Forms, function and efficiency. *BioScience* 47:677–687.

An excellent introduction to the topic of landscape corridors. This review article presents numerous examples of field research addressing the role of corridors in conservation ecology.

Strayer, D. L., M. E. Power, W. F. Fagen, S. T. Pickett, and J. Belnap. 2003. A classification of ecological boundaries. *BioScience* 53:723–729.

This paper presents a classification of the attributes of ecological boundaries as related to their origin and maintenance, their spatial structure, their function, and their temporal dynamics.

Turner, M. 1989. Landscape ecology: The effects of pattern and process. *Annual Review of Ecology and Systematics* 20:171–197.

One of the earliest reviews of the emerging field of landscape ecology. This paper provides an overview of terms and concepts that are central to the theory of ecological landscapes.

Turner, M. 1998. Landscape ecology. In *Ecology* (S. I. Dodson et al., eds.). Oxford: Oxford University Press.

This well-written chapter by one of the field's leading researchers is an excellent introduction to the ecology of landscapes.

Urban, D., R. V. O'Neill, and H. H. Shugart. 1987. Landscape ecology. *BioScience* 37:119–127.

An early and important paper in the development of landscape ecology. This well-written and illustrated article is a good introduction to many of the central concepts in landscape ecology.

Wuerthner, G. (ed.). 2006. *Wildfire: A century of failed forest policy.* Washington, DC: Island Press.

Covers the topic of wildfire from ecological, economic, and sociopolitical perspectives.

Ecosystem Ecology

Hawaii Volcanoes National Park, on the island of Hawaii, is one of the most dynamic places on Earth—volcanic activity at the park is continuously forming new land as lava flows toward the ocean. Exposed to the sun and wind, the smooth, hard surface of the cooled lava flows is a harsh environment for the establishment and growth of plants. Where there are cracks, however, plants may find a more hospitable environment to colonize. Certain plant species, such as lichens, ferns, and the native tree species (*ohia-lehua*), can tolerate these harsh conditions and appear first on the flows. Temperatures are more moderate in the shaded crevices, moisture may accumulate by running off the adjacent surfaces, and wind-blown dust and organic matter can accumulate in these sheltered areas where the material is less likely to be blown away. These early colonists establish themselves in the most sheltered cracks and crevices, where they have access to the scarce moisture and nutrients. The lichen *Stereocaulon volcani* and the evergreen tree *Myrica faya* are able to fix atmospheric nitrogen (see Section 15.11). As these plants become established, they further contribute to the process of environmental change by shading the substrate and contributing organic matter, which is important for the retention of soil moisture and nutrient availability. As the plant roots grow outward, they add to the process of substrate breakdown by mechanically cracking the lava. Soon a sparse organic layer appears, and over time other plant species become established beneath the protective cover of the original colonizers. This process continues through time as the colonizing plants modify the environmental conditions, facilitating the successful arrival of other species until eventually a forest is established.

As we have discussed in previous chapters, the distribution and abundance of species and the biological structure of the community vary in response to environmental conditions. However, as we can see from the example of plant succession on the lava flows of Hawaii, it is equally true that the organisms themselves, in part, define the abiotic environment. It is this inseparable link between the biotic environment (the community) and the abiotic environment that led A. G. Tansley to coin the term *ecosystem* in an article in the journal *Ecology* in 1935. Tansley wrote:

> The more fundamental conception is . . . the whole system (in the sense of physics) including not only the organism-complex, but also the whole complex of physical factors forming what we call the environment. . . . We cannot separate them [the organisms] from their special environment with which they form one physical system. . . . It is the systems so formed which . . . [are] the basic units of nature on the face of the earth. . . . These ecosystems, as we may call them, are of the most various kinds and sizes.

In the concept of the ecosystem, the biological and physical components of the environment are a single interactive system.

The ecosystem concept brought with it a new way of approaching the study of nature. Given the diversity of life within any given community, the taxonomic perspective had to give way to a more functional approach of viewing nature. In discussing the ecology of a forest with a population or community ecologist, one will hear a story of species—the dynamics of populations, their interactions, food webs, and patterns of diversity. Discuss the same forest with an ecosystem ecologist, however, and a more abstract picture emerges: a story of energy and matter, where the boundary between the biotic and abiotic components of the forest is often blurred. To the ecosystem ecologist, the forest is a system composed of autotrophs, heterotrophs, and the abiotic environment, each component processing and exchanging energy and matter.

The autotrophs, or primary producers, are predominantly green plants and algae. These organisms use the energy of the Sun in photosynthesis (see Section 6.2) to transform inorganic compounds into simple organic compounds. The heterotrophs, or consumers, use the organic compounds produced by the autotrophs as a source of food. Through decomposition, heterotrophs eventually transform these complex organic compounds into simple inorganic compounds that are once again used by the primary producers. The heterotrophic component of the ecosystem is often subdivided into two subsystems: consumers and decomposers. The consumers feed mostly on living tissue, and the decomposers break down dead matter into inorganic substances.

The abiotic component consists of the air, water, soil, sediments, particulate matter, dissolved organic matter in aquatic ecosystems, and dead organic matter. All of the dead organic matter is derived from plant and consumer remains and is acted upon by the decomposers. Such dead organic matter is crucial to the internal cycling of nutrients in the ecosystem (see Chapter 21).

The driving force of the ecosystem is the energy of the Sun. This energy, harnessed by the primary producers, flows from producers to consumers to decomposers and eventually dissipates as heat.

Like the community, the ecosystem is a spatial concept; it has boundaries. Like the community, these boundaries are often difficult to define. At first examination, a pond ecosystem is clearly separate and distinct from the surrounding terrestrial environment. A closer inspection, however, reveals a less distinct boundary between aquatic and adjacent terrestrial ecosystems. Some plants along the shoreline, such as cattails, may be either partially submerged or rooted in the surrounding land and able to tap the shallow water table with their roots. Amphibians move between the shoreline and the water. Surrounding trees drop leaves into the pond, adding to the dead organic matter that feeds the decomposer community on the pond bottom.

Regardless of these difficulties, ecosystems theoretically have spatial boundaries; and having defined the boundaries, we can view our ecosystem in the context of its surrounding environment. Exchanges from the surrounding environment into the ecosystem are inputs. Exchanges from inside the ecosystem to the surrounding environment are outputs. An ecosystem with no inputs is called a **closed ecosystem;** one with inputs is an **open ecosystem.** Inputs and outputs, together with exchanges of energy and matter among components within the ecosystem, will form the basis of our discussion in the following three chapters on ecosystem energetics (Chapter 20), nutrient cycling (Chapter 21), and biogeochemical cycles (Chapter 22).

As you will see in the following chapters, the study of ecosystems draws upon the concepts and understanding that we have developed thus far. In a way, it provides a framework to integrate our accumulated understanding of adaptation, populations, communities, and the abiotic environment.

Plant species characteristic of the early stages of plant succession on the lava flows of Hawaii. (a) *Metrosideros polymorpha* (Hawaiian: 'ohi'a) are well adapted to the environment of the lava flows and are a common plant in the early stages of colonization. (b) Litter breakdown is associated with the common lichen, *Stereocaulon volcani*.

(a)

(b)

CHAPTER 20

Ecosystem Energetics

A Rocky Mountain alpine tundra ecosystem carpeted with cotton grass (*Eriophorum angustifolium*) in full bloom.

The sunlight that floods Earth is the ultimate source of energy that keeps the planet functioning. Solar energy arrives as light in particles of energy known as photons. When these photons reach the atmosphere, land, and water, some are transformed into another form of energy—heat—that warms Earth, warms the atmosphere (see Chapter 2), drives the water cycle (see Chapter 3), and causes the currents of air (winds) and water (see Chapters 2 and 3). Some of those photons that reach plants are transformed into photochemical energy used in photosynthesis (see Chapter 6). That energy, stored in the chemical bonds of carbohydrates and other carbon-based compounds, becomes the source of energy for other living organisms. In this way, the story of energy within an ecosystem is in large part a story of carbon in the form of organic matter—the living and dead tissues of plants and animals.

All ecological processes involve the transfer of energy, and ecosystems are no different from physical systems like the atmosphere because they are subjected to the same physical laws. In this chapter we will explore the pathways, efficiencies, and constraints that characterize energy flow through the ecosystem. But first, we must examine the physical laws governing the flow of energy.

20.1 | The Laws of Thermodynamics Govern Energy Flow

Energy exists in two forms: potential and kinetic. **Potential energy** is stored energy—it is capable of and available for performing work. **Kinetic energy** is energy in motion. It performs work at the expense of potential energy. Work is of at least two kinds: the storage of energy and the arranging or ordering of matter.

Two laws of thermodynamics govern the expenditure and storage of energy. The **first law of thermodynamics** states that energy is neither created nor destroyed. It may change form, pass from one place to another, or act upon matter in various ways. Regardless of what transfers and transformations take place, however, no gain or loss in total energy occurs. Energy is simply transferred from one form or place to another. When wood burns, the potential energy lost from the molecular bonds of the wood equals the kinetic energy released as heat. When a chemical reaction results in the loss of energy from the system, the reaction is **exothermic.**

On the other hand, some chemical reactions must absorb energy in order to proceed. These are **endothermic** reactions. Here, too, the first law of thermodynamics holds true. In photosynthesis, for example, the molecules of the products (simple sugars) store more energy than do the reactants that combined to form the products. The extra energy stored in the products is acquired from the sunlight harnessed by the chlorophyll within the leaf (see Chapter 6). Again, there is no gain or loss in total energy. Although the total amount of energy in any reaction, such as burning wood, does not increase or decrease,

much of the potential energy degrades into a form incapable of doing further work. It is transferred to the surrounding environment as heat. This reduction in potential energy is commonly referred to as **entropy.** The transfer of energy involves the **second law of thermodynamics.** This law states that when energy is transferred or transformed, part of the energy assumes a form that cannot pass on any further. Entropy increases. When coal is burned in a boiler to produce steam, some of the energy creates steam, and some of the energy is dispersed as heat to the surrounding air. The same thing happens to energy in the ecosystem. As energy is transferred from one organism to another in the form of food, a portion is stored as energy in living tissue, whereas a large part of that energy is dissipated as heat—entropy increases.

At first, biological systems do not seem to conform to the second law of thermodynamics. The tendency of life is to produce order out of disorder, to decrease rather than increase entropy. The second law theoretically applies to closed systems in which no energy or matter is exchanged with the surrounding environment. With the passage of time, closed systems tend toward maximum entropy; eventually, no energy is available to do work. Living systems, however, are open systems with a constant input of energy in the form of solar radiation, providing the means to counteract entropy.

20.2 | Energy Fixed in the Process of Photosynthesis Is Primary Production

The flow of energy through a terrestrial ecosystem starts with the harnessing of sunlight by autotrophs. The rate at which radiant energy is converted by photosynthesis to organic compounds is referred to as primary productivity because it is the first and basic form of energy storage.

Gross primary productivity is the total rate of photosynthesis, or energy assimilated by the autotrophs. Like all other organisms, autotrophs must expend energy in the process of respiration (see Chapter 6). The rate of energy storage as organic matter after respiration is **net primary productivity (NPP).** NPP can be described by the following equation:

Net primary productivity (NPP)	=	Gross primary productivity (GPP)	−	Respiration by autotrophs (R)

Productivity is usually expressed in units of energy per unit area per unit time: kilocalories per square meter per year ($kcal/m^2/yr$). However, productivity may also be expressed in units of dry organic matter: ($g/m^2/yr$). As pointed out by the ecologist Eugene Odum, in all these definitions, the term *productivity* and the phrase *rate of production* may be used interchangeably. Even when the word *production* is used, a time element is always assumed or understood, so one should always state the time interval.

The amount of accumulated organic matter found in an area at a given time is the **standing crop biomass.** Biomass is usually expressed as grams of organic matter per square meter (g/m^2) or some other appropriate unit of area. Biomass differs from productivity. Productivity is the rate at which organic matter is created by photosynthesis. Biomass is the amount of organic matter present at any given time.

The simplest and most common method of measuring net primary production in terrestrial ecosystems is to estimate the change in standing crop biomass (SCB) over a given time interval $(t_2 - t_1)$: $\Delta SCB = SCB(t_2) - SCB(t_1)$. Two possible losses of biomass over the time period must also be recognized: loss of biomass due to the death of plants (D), and loss of biomass due to consumption by consumer organisms (C). The estimate of net primary productivity is then $NPP = (\Delta SCB) + D + C$.

In aquatic ecosystems, the most common method of estimating NPP is the light/dark bottle method (Figure 20.1). Because oxygen is one of the most easily measured products of both photosynthesis and respiration (see Section 6.1), a good way to gauge primary productivity in

Light bottle

Water sample containing phytoplankton

O_2 produced by photosynthesis

O_2 consumed in respiration

(Net primary production)

Dark bottle

Water sample containing phytoplankton

O_2 consumed in respiration

(Respiration)

O_2 produced by photosynthesis

(Gross primary production)

Figure 20.1 | Paired light and dark bottles are used to measure photosynthesis (gross production), respiration, and net primary production by phytoplankton in aquatic ecosystems. A sample of water containing phytoplankton (primary producers) is placed in both bottles and allowed to incubate for a period of time. In the light (clear) bottle, O_2 is produced in photosynthesis and consumed in respiration. The resulting change (increase) in O_2 concentration represents the difference in the rates at which these two processes occur: net primary productivity. Lacking light to drive the process of photosynthesis, only respiration occurs in the dark bottle. As a result, O_2 concentration declines. The difference between the O_2 concentrations of the water from the light and dark bottles at the end of the incubation period represents the rate of O_2 produced in photosynthesis: gross primary productivity.

an aquatic ecosystem is to measure the concentration of dissolved oxygen. In one set of clear glass "light bottles," a water sample from the aquatic ecosystem (and associated autotrophic organisms—phytoplankton) is allowed to incubate for a defined time period. If photosynthesis is greater than respiration, oxygen will accumulate in the water, providing an estimate of NPP. Water is also incubated over the same time period in another set of "dark bottles" (painted dark to prevent light from reaching the water). Because the lack of light will prevent photosynthesis, the oxygen content of the water will decline as a function of respiration. The difference between the values of oxygen in the light (photosynthesis + respiration) and dark (respiration) bottles at the end of the time period therefore provides an estimate of total photosynthesis, or gross primary productivity.

20.3 | Temperature, Water, and Nutrients Control Primary Production in Terrestrial Ecosystems

An array of environmental factors, including climate, influence the productivity of terrestrial ecosystems. Measured estimates of net primary productivity (NPP) for various terrestrial ecosystems are plotted in Figure 20.2 against the mean annual precipitation (Figure 20.2a) and mean annual temperature (Figure 20.2b) for each site. NPP increases with increasing mean annual temperature and rainfall. Increasing mean annual temperature is directly related to the annual intercepted solar radiation at the site, reflecting both an increase in mean daily temperature and the length of the growing season (see Chapter 2, Section 2.2). The length of the growing season is defined as the period (number of days) during which temperatures are warm enough to support photosynthesis. As a result, sites with a higher mean annual temperature typically support higher rates of photosynthesis and are associated with a longer time period over which photosynthesis can occur (Figure 20.3).

As we discussed in Chapter 6, for photosynthesis and productivity to occur, the plant must open the stomata to take in CO_2. When the stomata are open, water is lost from the leaf to the surrounding air. For the plant to keep the stomata open, roots must replace the lost water. The higher the rainfall, the more water is available for transpiration. The amount of water available to the plant will therefore limit both the rate of photosynthesis and the amount of leaves (surface area that is transpiring) that can be supported (see Section 6.10). The combination of these factors determines the rate of primary productivity.

Although the two graphs in Figure 20.2 show independent effects of temperature and precipitation on primary productivity, in reality the influence of these two factors is closely related. Warm air temperatures increase the potential for evaporation and therefore increase rates

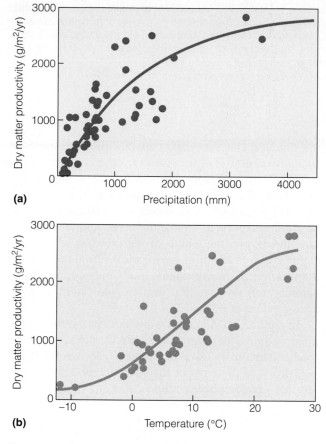

(a)

(b)

Figure 20.2 | Net primary productivity for a variety of terrestrial ecosystems **(a)** as a function of mean annual precipitation and **(b)** as a function of mean annual temperature.

(Adapted from Reichle 1970.)

teraction between temperature and water on the process of NPP explains the high degree of variation in NPP observed in both Figures 20.2a and 20.2b with increasing values of mean annual temperature and precipitation. For example, in Figure 20.2a, values of NPP for sites having a mean annual temperature of approximately 12°C range from a low of 900 to a high of more than 2500 g/m²/yr. This range of values reflects differences in the corresponding mean annual precipitation at these sites, with the low values of productivity associated with low rainfall sites and high values with high precipitation sites. Similarly, in Figure 20.2b, variation in values of NPP for sites receiving approximately the same annual precipitation reflects differences in the mean annual temperature.

It is the combination of warm temperatures and an adequate water supply for transpiration that gives the highest primary productivity. This pattern is reflected in Figure 20.4, which relates the NPP of various ecosystems to estimates of actual evapotranspiration (AET). Actual evapotranspiration is the combined value of surface evaporation and transpiration (see Section 3.1). It reflects both the demand and the supply of water to the ecosystem. The demand is a function of incoming radiation and temperature, whereas the supply is a function of precipitation.

The influence of climate on primary productivity in terrestrial ecosystems is reflected in the global patterns presented in Figure 20.5. These patterns of productivity reflect the global patterns of temperature and precipitation presented in Chapter 2 (see Figures 2.8 and 2.17). In addition, measured estimates of NPP for a variety of ecosystems are summarized in Table 20.1. The regions of highest NPP are located in the equatorial zone where the combination of year-round warm temperatures and precipitation supports high rates of photosynthesis and leaf area (tropical rain forest). Moving north and south from

of transpiration and plant water demand (see Section 2.6). If temperatures are warm but water availability is low, productivity will also be low. Conversely, if temperatures are low, rates of photosynthesis and productivity will be low regardless of the availability of water. This in-

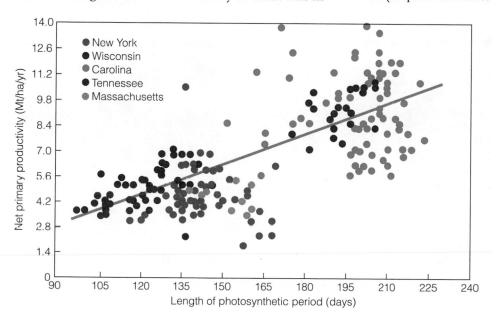

Figure 20.3 | Relationship between net primary productivity and the length of the growing season for deciduous forest stands in eastern North America. Each point represents a single forest site. The (regression) line represents the general trend of increasing productivity with increasing length of the growing season (largely a function of latitude).

(Adapted from Leith 1975.)

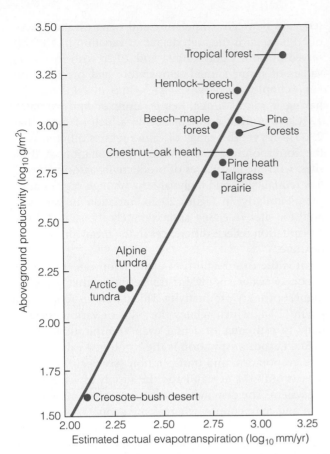

Figure 20.4 | Relationship between aboveground net primary productivity and (actual) evapotranspiration for a range of terrestrial ecosystems. Evapotranspiration—the combination of evaporation and transpiration at a site—depends on both precipitation and temperature (see Chapter 3).

(Adapted from MacArthur and Connell 1966.)

the Equator, the seasonality of precipitation increases (see discussion of intertropical convergence zone, Section 2.7), decreasing the growing season and, subsequently, values of NPP. Continuing into the temperate regions (midlatitudes), an increasing seasonality of temperature functions to reduce the mean annual temperature and restrict the length of the growing season (see Figure 20.3). In addition, as one moves from the coast to the interior of the continents, both mean annual temperature and precipitation decline (see Chapter 2), reducing values of NPP.

In addition to climate, the availability of essential nutrients required for plant growth directly affects ecosystem productivity. As discussed in Chapter 6 (Section 6.12), the availability of nutrients in the soil influences the rate of nutrient uptake, photosynthesis, and plant growth; the net result is a general pattern of increasing NPP with increasing soil nutrient availability.

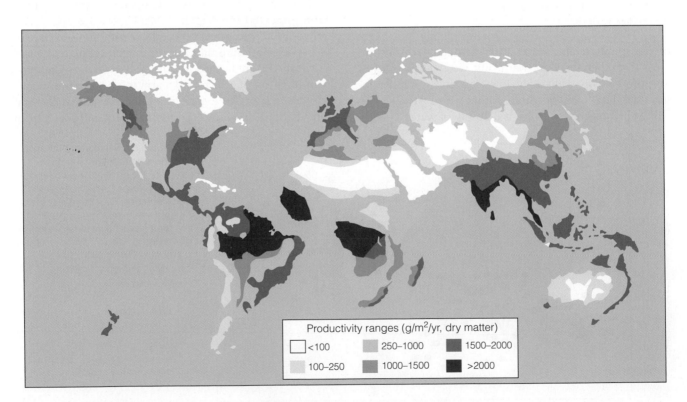

Figure 20.5 | A global map of primary productivity for the terrestrial surface (ecosystems).

(Adapted from Golley and Leigh 1972.)

Table 20.1 | Net Primary Production and Plant Biomass of World Ecosystems

Ecosystems (in Order of Productivity)	Area (10^6 km^2)	Mean Net Primary Production per Unit Area (g/m^2/yr)	World Net Primary Production (10^9 Mt/yr)	Mean Biomass per Unit Area (kg/m^2)
Continental				
Tropical rain forest	17.0	2000.0	34.00	44.00
Tropical seasonal forest	7.5	1500.0	11.30	36.00
Temperate evergreen forest	5.0	1300.0	6.40	36.00
Temperate deciduous forest	7.0	1200.0	8.40	30.00
Boreal forest	12.0	800.0	9.50	20.00
Savanna	15.0	700.0	10.40	4.00
Cultivated land	14.0	644.0	9.10	1.10
Woodland and shrubland	8.0	600.0	4.90	6.80
Temperate grassland	9.0	500.0	4.40	1.60
Tundra and alpine meadow	8.0	144.0	1.10	0.67
Desert shrub	18.0	71.0	1.30	0.67
Rock, ice, sand	24.0	3.3	0.09	0.02
Swamp and marsh	2.0	2500.0	4.90	15.00
Lake and stream	2.5	500.0	1.30	0.02
Total continental	149.0	720.0	107.09	12.30
Marine				
Algal beds and reefs	0.6	2000.0	1.10	2.00
Estuaries	1.4	1800.0	2.40	1.00
Upwelling zones	0.4	500.0	0.22	0.02
Continental shelf	26.6	360.0	9.60	0.01
Open ocean	332.0	127.0	42.00	0.003
Total marine	361.0	153.0	55.32	0.01
World total	510.0	320.0	162.41	3.62

Source: Adapted from Whittaker 1975.

Relative net primary productivity (RNPP) is calculated by dividing the value of net primary production (column 3) by the corresponding value of mean biomass (column 5). Values of mean biomass must first be converted to units of g/m^2. The resulting units for RNPP are g/g/yr.

John Pastor of the University of Minnesota, together with colleagues, examined the role of nitrogen availability on patterns of primary productivity in different forest types on Blackhawk Island, Wisconsin. Their results clearly show the relationship between soil nitrogen availability and aboveground primary productivity (Figure 20.6). A similar response of primary productivity to nutrient availability has been reported for oak savannas that form the transition from the forest ecosystems of eastern North America to the western grasslands of the Great Plains. Peter Reich and colleagues at the University of Minnesota examined the relationship between soil nitrogen availability and aboveground NPP in 20 mature oak savanna stands in Minnesota. Their results show a pattern of increasing primary productivity with available nitrogen in these mixed tree–grass ecosystems (Figure 20.7).

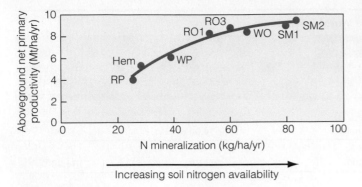

Figure 20.6 | Relationship between net primary production and nutrient availability. Aboveground productivity increases with increasing nitrogen availability (N mineralization rate) for a variety of forest ecosystems on Blackhawk Island, Wisconsin. Abbreviations refer to the dominant trees in each stand: Hem, hemlock; RP, red pine; RO, red oak; WO, white oak; SM, sugar maple; WP, white pine.

(Adapted from Pastor et al. 1984.)

20.4 | Temperature, Light, and Nutrients Control Primary Production in Aquatic Ecosystems

Light is a primary factor limiting productivity in aquatic ecosystems, and the depth to which light penetrates in a lake or ocean is crucial in determining the zone of primary productivity. Recall from Chapter 3 (see Figure 3.7) that photosynthetically active radiation (PAR) declines exponentially with water depth (Figure 20.8). The photosynthetic rate and subsequently the gross productivity of phytoplankton are highest at intermediate levels of PAR (see Figure 6.2). On the other hand, the respiration rate does not change much with depth. This means that as the phytoplankton go deeper in the water column, the photosynthetic rate declines as the light intensity decreases until at some point the rate of photosynthesis (gross production) is equal to the rate of respiration, and net primary productivity (NPP) is zero (Figure 20.8). This zone is referred to as the **compensation depth** and corre-

sponds to the depth at which the availability of light is equal to the light compensation point discussed in Chapter 6 (see Figure 6.2).

In the oceans, nutrients in the deeper waters must be transported to the surface waters, where light (PAR) is sufficient to support photosynthesis. As a result, nutrients—particularly nitrogen, phosphorus, and iron—are a major limitation on primary productivity in the oceans (see Chapters 21 and 24). John Downing, an ecologist at Iowa State University, together with his colleagues Craig Osenburg and Orlando Sarnelle, examined results from more than 300 nutrient-enrichment experiments conducted in marine habitats around the world (Figure 20.9a). The authors found that nitrogen (N) addition stimulated phytoplankton growth the greatest, followed closely by the addition of iron (Fe). In contrast, the addition of phosphorus (P) did not, on average, stimulate phytoplankton growth. These results confirm the prevailing view among marine ecologists that N and Fe are the two most limiting nutrients in marine environments. These

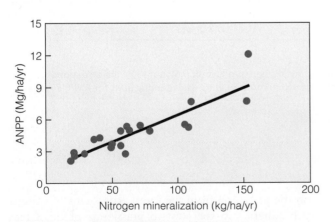

Figure 20.7 | Relationship between aboveground net primary productivity (ANPP) and nitrogen availability (nitrogen mineralization rate) for 20 oak savanna sites in Minnesota.
Go to QUANTIFYit! at **www.ecologyplace.com** to perform regression analysis.

(Adapted from Reich et al. 2001.)

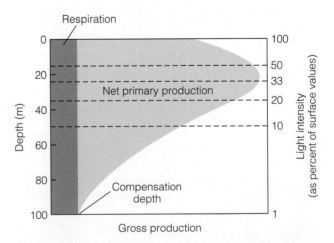

Figure 20.8 | Changes in available light, gross productivity, respiration, and net primary productivity (gross productivity 2 respiration) with water depth. Respiration rate is relatively constant with depth, whereas gross productivity (photosynthesis) declines with depth as a function of declining available light. The depth at which gross productivity is equal to respiration (net photosynthesis equal to zero) is called the compensation depth.

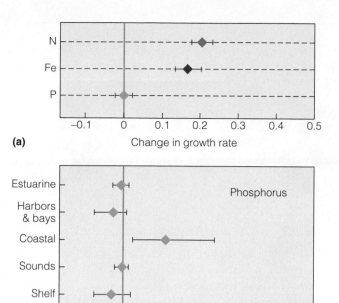

(a)

(b)

Figure 20.9 | **(a)** Effects of nutrient addition (nitrogen, iron, and phosphorus) on marine phytoplankton growth rates in 303 experiments. Diamonds represent mean values from the experiments, and the bars represent 95 percent confidence intervals. Solid red line represents zero effect. Mean response values are based on 148 (N), 114 (P), and 35 (Fe) experiments. **(b)** The change in phytoplankton growth rate due to phosphorus addition varied among different marine environments.
Go to **QUANTIFYit!** at **www.ecologyplace.com** to perform confidence intervals and t-tests.

(Adapted from Downing et al. 1999.)

results, however, are average responses and do not account for differences among habitats. The magnitude of response to nutrient enrichment varied significantly among marine environments, particularly for phosphorus (Figure 20.9b). Growth response to phosphorus addition in the more polluted waters of the nearshore environments (bays, estuaries, and harbors) was largely negative. In contrast, in the more pristine coastal and oceanic environments, the positive response was nearly as great as that observed for nitrogen.

The role of environmental constraints on the primary productivity of the world's oceans can be seen in the map presented in Figure 20.10 (see also Table 20.1). For two reasons, the most productive waters of the oceans are the shallow waters of the coastal environments. First, shallow waters allow for a greater transport of nutrients from the bottom sediments to the surface waters, aided by wave action and the changing tides. Second, coastal waters receive a large input of nutrients carried from terrestrial ecosystems by rivers and streams (see Chapters 21 and 24).

The constraints on NPP in freshwater ecosystems are not always as easy to interpret as those operating in marine ecosystems. Solar radiation limits primary productivity in lake ecosystems, but the close link between light intensity and temperature (see Section 3.4) makes it difficult to evaluate these two factors independently. The role of nutrient availability on primary productivity in lake ecosystems, however, has been well established. Ecologists P. J. Dillon and F. H. Rigler of the University of Toronto examined the relationship between summer chlorophyll and spring total phosphorus concentration for 19 lakes in southern Ontario. Water chlorophyll concentration provides a simple and accurate estimate of phytoplankton standing biomass and productivity. The researchers combined their results with data reported in the literature for other North American lakes. The results of their analysis, presented in Figure 20.11, show a clear pattern of increasing primary productivity with phosphorus concentration. Similar results have been reported for studies in which the nutrient concentrations of lake water have been manipulated experimentally through fertilization.

Although experimental manipulation of water nutrient concentrations in stream and river ecosystems have shown an increase in NPP with nutrient concentration, primary productivity in these ecosystems is typically low in comparison to standing water (lentic ecosystems). The primary source of organic matter in most flowing-water ecosystems (lotic ecosystems) is the input of dead organic matter from adjacent terrestrial ecosystems (see Section 21.10).

20.5 | Energy Allocation and Plant Life-Form Influence Primary Production

In Chapter 6, we examined the implications of how plants allocate the carbon fixed in photosynthesis (carbon allocation; see Section 6.7) to the process of plant growth. You will recall that the process of plant growth functions as a positive feedback system. For a given rate of photosynthesis, the greater the allocation of carbon (energy) to photosynthetic tissues (leaves) relative to non-photosynthetic tissues (stems and roots), the greater the net carbon gain and plant growth.

As discussed in Section 20.3, the pattern of decreasing net primary productivity (NPP) with declining precipitation is partially due to the changing pattern of carbon–energy allocation of plants within the ecosystem. Reduced moisture conditions result in an increased allocation to roots at the expense of leaves, thus reducing leaf area and rates of net carbon gain. Although plant species within an ecosystem exhibit a wide range of characteristics and adaptation to microclimatic conditions (such as the case of shade-tolerant and shade-intolerant species; see Section 6.9), the average patterns of carbon allocation for different ecosystems reflect the general pattern of carbon

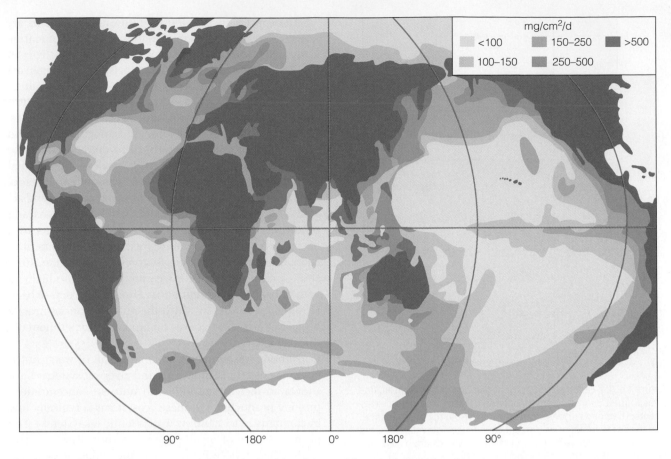

Figure 20.10 | Geographic variation in primary productivity of the world's oceans. Note that the highest productivity is along the coastal regions, and areas of lowest productivity are in the open ocean (see Chapter 27 for detailed discussion).

allocation exhibited by individual plants in response to such environmental gradients as moisture availability (see Section 6.10). Estimates of the ratio of belowground to aboveground biomass (root-to-shoot ratio; R:S) range

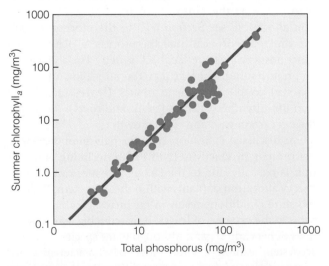

Figure 20.11 | Relationship between summer average chlorophyll (an estimate of phytoplankton net primary productivity) content (*y*-axis) and spring total phosphorus concentration (*x*-axis) for northern temperate lake ecosystems (each point represents a single lake).

(Adapted from Dillon and Rigler 1974.)

from a low of 0.20 in tropical rain forest ecosystems to 1.2 for arid shrublands and to a high of 4.5 in desert ecosystems. These differences reflect the greater allocation to roots relative to aboveground tissues (leaves and stem) with decreasing annual precipitation.

The decline in NPP from mesic to more xeric environments shown in Figure 20.2 is paralleled with a reduction in standing crop biomass, the accumulation of NPP over time (Figure 20.12). Those ecosystems with a greater NPP are those with the greater standing biomass. This relationship of increasing standing biomass with increasing NPP is seen in both terrestrial and marine environments.

Recall from our discussion of the growth of individual plants in Chapter 6 (see Quantifying Ecology 6.1: Relative Growth Rate) that larger plants generally have a greater net carbon gain or absolute growth rate (grams per unit time) than do smaller plants. This situation often changes, however, when we examine the relative growth rate, or biomass gain per unit of plant mass (grams per gram plant mass per unit time). The same holds true for the collective growth of plants within the ecosystem. The ratio of NPP to standing biomass from Table 20.1, relative net primary productivity (RNPP: units of g/g/yr), represents a similar index to that of relative growth rate for individual plants: the

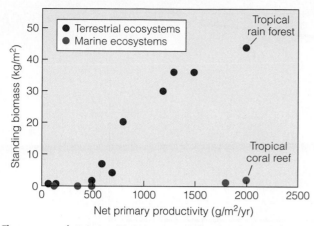

Figure 20.12 | Relationship between standing biomass and net primary productivity for ecosystems in Table 20.1. Note the difference in the standing biomass per unit of net primary productivity between terrestrial and aquatic ecosystems (see text for discussion).

rate of biomass accumulation per unit of plant biomass present. A comparison of RNPP with the average standing biomass in each of the terrestrial ecosystems shows a pattern inverse to that discussed for Figure 20.12. For example, although the NPP of a temperate forest (0.04 g/g/yr) is more than twofold greater than that of a temperate grassland, if we calculate the productivity per unit of standing biomass, the grassland ecosystem is almost an order of magnitude higher than the forest (0.31 g/g/yr). This reflects the general pattern of higher relative growth rate for grasses when compared to trees.

The same inverse relationship between standing biomass and RNPP observed for terrestrial ecosystems also occurs in marine ecosystems. The interpretation, however, is quite different. In the terrestrial ecosystems represented in Table 20.1, the longevity of dominant plant species is typically much greater than the period over which net primary productivity is measured (annual NPP). This is not, however, the case for most marine ecosystems. Phytoplankton (microscopic algae) are the dominant net primary producers in open-water ecosystems. These species are short-lived (weeks) with high rates of reproduction. As a result, there is a constant turnover of the populations, with many generations occurring during the period over which annual NPP is measured. As a result of the fast turnover of the populations, the standing biomass at any time interval is low as compared to the accumulated NPP over the course of the year. This accounts for the extremely high value of RNPP for the open ocean (42.3 g/g/yr) as compared to all terrestrial ecosystems (see Table 20.1).

20.6 | Primary Production Varies with Time

Primary production also varies within an ecosystem with time and age. Both photosynthesis and plant growth are directly influenced by seasonal variations in environmental conditions. Regions with cold winters or distinct wet

and dry seasons have a period of plant dormancy when primary productivity ceases. In the wet regions of the tropics, where conditions are favorable for plant growth year-round, there is little seasonal variation in primary productivity.

Year-to-year variations in primary productivity within an area can occur as a result of climatic variation. In a long-term experiment, known as Park Grass, at the Rothamsted Experimental Station in Hertfordshire (England), grass yields have been recorded since 1856. Yields have been recorded, with constant treatments, using standard measurement methods, since 1965 (Figure 20.13). Yields in late summer are depressed in hot, dry summers. The three lowest-yielding years were 1976, 1990, and 1995—all years with hot, dry summers.

Disturbances such as herbivory and fire can also lead to year-to-year variations in net primary productivity (NPP) at a site (see Section 19.8). Overgrazing of grasslands by cattle and sheep or defoliation of forests by such insects as the gypsy moth can significantly reduce NPP. Fire in grasslands may increase productivity in wet years but reduce it in dry years.

Net primary productivity also varies with stand age, particularly in ecosystems that are dominated by woody vegetation. Trees and woody shrubs can survive for a long

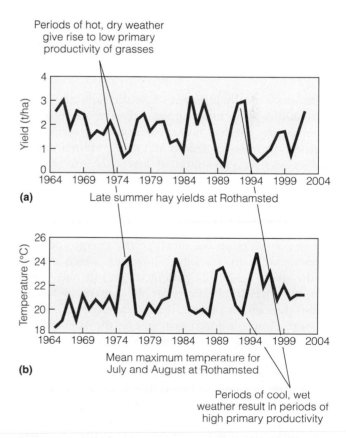

Figure 20.13 | Seasonal variations in **(a)** grass productivity and **(b)** mean maximum temperature for the period of July and August at the Rothamsted Experimental Station in Hertfordshire (England) for the period 1965–2004.

(Adapted from Sparks and Potts 2004.)

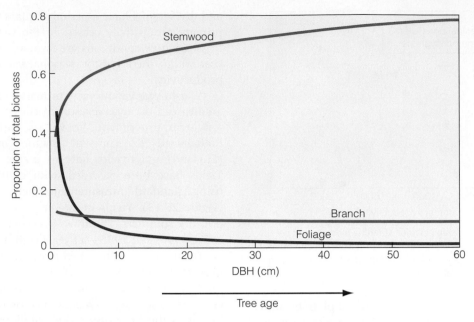

Figure 20.14 | Changes in the proportion of biomass in foliage (leaves), branch, and stemwood (bole) for white oak trees (*Quercus alba*) as a function of tree size (diameter of the trunk at 1.5 m above the ground, or DBH).

Interpreting Ecological Data

Q1. Go to Quantifying Ecology 6.1: Relative Growth Rate on page 109. Assuming that leaf area (cm²) increases linearly with leaf mass (g), how does the leaf area ratio of the oak tree change with age?

Q2. Based on the discussion of relative growth rate presented in Quantifying Ecology 6.1, hypothesize how relative growth rate might change as the tree increases in size (DBH) and age.

time, which greatly influences how they allocate energy. Early in life, leaves make up more than one-half of their biomass (dry weight); but as trees age, they accumulate more woody growth. Trunks and stems become thicker and heavier, and the ratio of leaves to woody tissue changes (Figure 20.14). Eventually, leaves account for only 1 to 5 percent of the total mass of the tree. The production system (leaf mass) that supplies the energy is considerably less than the biomass it supports. Thus, as the woody plant grows, much of the energy goes into support and maintenance (respiration), which increase as the plant ages. This pattern of growth and energy allocation has implications for the pattern of NPP of forests through the process of stand development.

Stith Gower, a forest ecologist at the University of Wisconsin, examined the potential causes of declining productivity with stand age with colleagues Ross McMurtrie and Danuse Murty of Australian National University. The authors found that as the age of a forest stand increases, more and more of the living biomass is in woody tissue while the leaf area remains relatively constant or declines (Figure 20.15a). As the stand ages, rates of both photosynthesis and respiration decline. In addition, more of the gross production (photosynthates) goes for maintenance (respiration of woody tis-

sues) and less remains for growth. The result is a pattern of increasing primary productivity during the early stages of stand development followed by a decline as the forest ages and the standing biomass increases (Figure 20.15b).

20.7 | Primary Productivity Limits Secondary Production

Net primary production (NPP) is the energy available to the heterotrophic component of the ecosystem. Either herbivores or decomposers eventually consume all plant productivity, but often it is not all used within the same ecosystem. Humans or other agents, such as wind or water currents, may disperse the NPP of any given ecosystem to another food chain outside the ecosystem (see Ecological Issues: Human Appropriation of Net Primary Productivity, p. 428). For example, about 45 percent of the net production of a salt marsh is lost to adjacent estuarine waters (see Chapter 25).

Some energy in the form of plant material, once consumed, passes from the body as waste products (such as feces and urine). Of the energy assimilated, part is used as heat for metabolism (see Section 7.7). The remainder is available for maintenance—capturing or harvesting

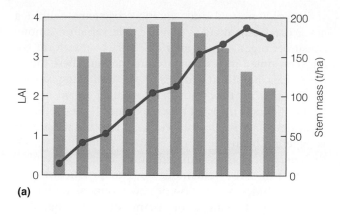

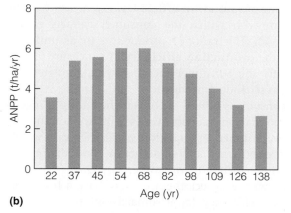

Figure 20.15 | (a) Aboveground stem mass (filled circles), leaf area index (LAI; green bars), and (b) aboveground net primary productivity (ANPP) for stands of the boreal needle-leaf evergreen conifer *Picea abies* of differing age.

(Adapted from Gower et al. 1996.)

food, performing muscular work, and keeping up with wear and tear on the animal's body. The energy used for maintenance is eventually lost to the surrounding environment as heat. The energy left over from maintenance and respiration goes into production, including the growth of new tissues and the production of young (allocation to reproduction; see Chapter 8). This net energy of production is called **secondary production.** As with primary production, secondary production per unit time (grams per unit area per unit time) is referred to as **secondary productivity.** Secondary productivity is greatest when the birthrate of the population and the growth rate of individuals are highest.

Secondary production depends on primary production for energy, and therefore primary productivity should function as a constraint on secondary productivity within the ecosystem. Ecologist Sam McNaughton of Syracuse University compiled data from 69 studies that reported both net primary and secondary productivity for terrestrial ecosystems ranging from Arctic tundra to tropical forests. A number of general patterns emerge. As expected, both herbivore biomass (Figure 20.16a) and consumption of primary productivity by herbivores (Figure 20.16b) increase with primary productivity. Likewise, secondary pro-

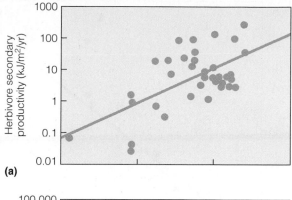

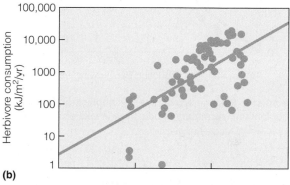

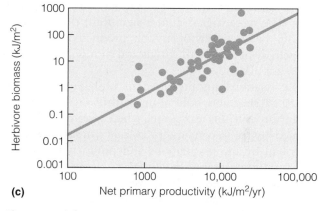

Figure 20.16 | Relationship between aboveground net primary productivity and (a) net secondary productivity of herbivores, (b) consumption, and (c) herbivore biomass. Units are kJ/m²/yr except for biomass, which is kJ/m². Each point represents a different terrestrial ecosystem.

(Adapted from McNaughton et al. 1989; Nature Publishing Group.)

ductivity of herbivores increased with primary productivity (Figure 20.16c). Inspection of the relationship between consumption and NPP (Figure 20.16b) revealed that forests have less consumption per unit of primary productivity than do grasslands. Restricting the estimates of NPP to foliage only (leaves), however, reduced these differences.

A similar relationship to that observed by McNaughton for terrestrial ecosystems has been observed between phytoplankton production (primary productivity) and zooplankton production (secondary productivity) in lake ecosystems. M. Brylinsky and K. H. Mann of Dalhousie University (Canada) examined the relationship between

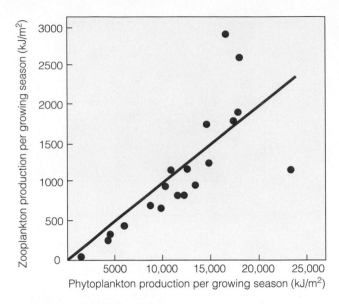

Figure 20.17 | Relationship between phytoplankton (primary) and zooplankton (secondary) productivity in lake ecosystems. (Adapted from Brylinsky and Mann 1973.)

phytoplankton and zooplankton productivity in 43 lakes and 12 reservoirs distributed from the tropics to the Arctic. The researchers found a significant positive relationship between phytoplankton productivity and the productivity of both herbivorous and carnivorous (Figure 20.17) zooplankton.

The relationships presented in both Figures 20.16 and 20.17 suggest a bottom-up control on the flow of energy through the ecosystem, where populations and productivity of secondary producers (herbivores) are controlled by the populations and productivity of primary producers (plants and algae). However, in our discussion of food webs in Part Five, we saw a more complex picture of species interactions between plant, herbivore, and carnivore populations. For example, recall "the world is green" proposition of Hairston and colleagues presented in Chapter 17 (Section 17.4) that suggests a top-down control on primary productivity and standing biomass. Hairston and colleagues proposed that the world is green (plant biomass accumulates) because predators keep herbivore populations in check. A growing body of experimental data suggests that top-down controls are important in many ecosystems and that patterns of NPP are influenced not only by abiotic conditions but also by controls on herbivore populations (and rates of consumption of primary productivity) by predators (see Section 17.4 and Field Studies: Brian Silliman).

20.8 | Consumers Vary in Efficiency of Production

Although there is a general relationship between the availability of primary productivity and the productivity of consumer organisms (secondary productivity) across a variety of terrestrial and aquatic ecosystems, within a given ecosystem there is considerable variation among consumer organisms in their efficiency to transform energy consumed into secondary production. These differences can be illustrated using the following simple model of energy flow through a consumer organism.

Of the food ingested by a consumer (I), a portion is assimilated across the gut wall (A), and the remainder is expelled from the body as waste products (W). Of the energy that is assimilated, some is used in respiration (R) and the remainder goes to production (P), which includes production of new tissues as well as reproduction. The ratio of assimilation to ingestion (A/I), the **assimilation efficiency,** is a measure of the efficiency with which the consumer extracts energy from food. The ratio of production to assimilation (P/A), the **production efficiency,** is a measure of how efficiently the consumer incorporates assimilated energy into secondary production.

A consumer's ability to convert the energy it ingests into secondary production varies with species and the type of consumer. Assimilation efficiencies vary widely among ectotherms and endotherms. Endotherms are much more efficient than ectotherms. However, carnivorous animals, even ectothermic ones, have a higher assimilation efficiency (approximately 80 percent) than herbivores (20 percent to 50 percent). Predatory spiders feeding on invertebrates have assimilation efficiencies of more than 90 percent.

Production efficiency varies mainly according to taxonomic class (Table 20.2). Invertebrates in general have high efficiencies (30 percent to 40 percent), losing

Table 20.2 | Production Efficiency ($P/A \times 100$) of Various Animal Groups

Group	P/A(%)
Mice	4.10
Voles	2.63
Other mammals	2.92
Birds	1.26
Fish	9.74
Social insects	8.31
Orthoptera	41.67
Hemiptera	41.90
All other insects	41.23
Mollusca	21.59
Crustacea	24.96
All other noninsect invertebrates	27.68
Noninsect invertebrates	
Herbivores	18.81
Carnivores	25.05

Source: Data from Humphreys 1979.

relatively little energy in respiratory heat and converting more assimilated energy into production. Among the vertebrates, ectotherms have intermediate values of production efficiency (approximately 10 percent). In contrast, endotherms, with their high energy expenditure associated with maintaining a constant body temperature (see Section 7.11), convert only 1 to 2 percent of their assimilated energy into production.

For endotherms, body size also influences production efficiency. You may recall from Chapter 7 (Section 7.10) that the mass-specific metabolic rate (kcal/g body weight/hr) increases exponentially with decreasing body mass. An increase in mass-specific metabolic rate lowers production efficiency.

20.9 | Ecosystems Have Two Major Food Chains

Energy fixed by plants is the base that the rest of life on Earth depends on. This energy stored by plants is passed along through the ecosystem in a series of steps of eating and being eaten—known as a food chain (see Section 16.4). Feeding relationships within a food chain are defined in terms of trophic or consumer levels. From a functional rather than a species viewpoint, all organisms that obtain their energy in the same number of steps from the autotrophs or primary producers belong to the same trophic level in the ecosystem. The first trophic level belongs to the primary producers, the second level to the herbivores (first-level consumers), and the higher levels to the carnivores (second-level consumers). Some consumers occupy a single trophic level; but many others, such as omnivores, occupy more than one trophic level (see Section 7.1).

Food chains are descriptive. They represent a more abstract expression of the food webs presented in Chapters 16 and 17. Major feeding groups are defined based on a common source of energy, such as autotrophs, herbivores, and carnivores. Each feeding group is then linked to others in a manner that represents the flow of energy. A simple food chain was presented in Figure 17.6. Boxes represent the three feeding groups: autotrophs, herbivores, and carnivores. The arrows linking the boxes represent the direction of energy flow.

Within any ecosystem there are two major food chains: the grazing food chain and the detrital food chain (Figure 20.18). The distinction between these two food chains is the source of energy for the first-level consumers, the herbivores. In the grazing food chain, the source of energy is living plant biomass or net primary production. In the detrital food chain, the source of energy is dead organic matter or detritus. In turn, the herbivores in each food chain are the source of energy for the carnivores, and so on. Cattle grazing on pastureland, deer browsing in the forest, insects feeding on leaves in the forest canopy, or zooplankton feeding on phytoplankton in the water column all represent first-level consumers of the grazing food chain. In contrast, a variety of invertebrates—such as

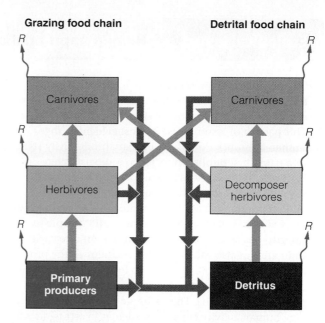

Figure 20.18 | Two parts of any ecosystem: a grazing food chain and a detrital food chain. Orange arrows linking trophic levels represent the flow of energy associated with ingestion. The blue arrows from each trophic level represent the loss of energy through respiration. The brown arrows represent a combination of dead organic matter (unconsumed biomass) and waste products (feces and urine).

snails, beetles, millipedes, and earthworms as well as fungi and bacteria—represent first-level consumers of the detrital food chain (see Chapter 21).

Figure 20.18 combines the two food chains to produce a generalized model of trophic structure and energy flow through an ecosystem. The two food chains are linked. The initial source of energy for the detrital food chain is the input of dead organic matter and waste materials from the grazing food chain. This linkage appears as a series of brown arrows from each of the trophic levels in the grazing food chain, leading to the box designated as detritus or dead organic matter. There is one notable difference in the flow of energy between trophic levels in the grazing and decomposer food chains. In the grazing food chain, the flow is unidirectional, with net primary production providing the energy source for herbivores, herbivores providing the energy for carnivores, and so on. In the decomposer food chain, the flow of energy is not unidirectional. The waste materials and dead organic matter (organisms) in each of the consumer trophic levels are "recycled," returning as an input to the dead organic matter box at the base of the detrital food chain. In addition, the distinction between the grazer and consumer food chains is often blurred at the higher trophic levels (carnivores), as predators rarely distinguish whether prey draw their resources from primary producers or detritus. For example, the diet of an insectivorous bird might include beetle species that feed on detritus, as well as species that feed on green leaf tissues.

Although we represent only 1 of more than 1.5 million known species inhabiting our planet, humans use a vastly disproportionate share of Earth's resources. This pattern of resource use is most evident in the human appropriation of net primary productivity (NPP). In a paper first published in 1986, ecologist Peter Vitousek and colleagues at Stanford University used three different approaches to estimate the fraction of NPP that humans have appropriated. A low estimate was derived by calculating the amount of NPP people use directly—as food, fuel, fiber, or timber. An intermediate estimate was also calculated that includes all the productivity of lands devoted entirely to human activities (such as the NPP of croplands, as opposed to the portion of crops actually eaten). This estimate also included the energy human activity consumes, such as in setting fires to clear land. A third approach provided a high estimate, which further included productive capacity lost as a result of converting open land to cities and forests to pastures or because of desertification or overuse (overgrazing, excessive erosion). Results of the three approaches (low, intermediate, and high) yielded a wide range of estimates: 3 percent, 19 percent, and 40 percent, respectively. These estimates are a remarkable level of use for a species that represents approximately 0.5 percent of the total heterotrophic biomass on Earth.

Since the initial analyses conducted by Vitousek and colleagues in the mid-1980s, advances in satellite technology have greatly enhanced scientists' ability to monitor patterns of land use and primary productivity at a continental to global scale. Marc Imhoff and Lahouari Bounoua, researchers at NASA's Goddard Space Flight Center (GSFC; Greenbelt, Maryland), and colleagues have recently undertaken an analysis of patterns of human use of NPP at a global scale using satellite-derived data. To map terrestrial NPP at a global scale, researchers employed the technique presented in Quantifying Ecology 20.1: Estimating Net Primary Productivity Using Satellite Data, using a global database of NDVI (normalized vegetation index) and climate for the period 1982–98.

The researchers defined human appropriation of terrestrial NPP (HANPP) as the amount of terrestrial NPP required to derive food and fiber products consumed by humans, including the organic matter that is lost during harvesting and processing of whole plants into end products. Using data compiled by the Food and Agricultural Organization (FAO) of the United Nations on products consumed in 1995 for 230 countries in 7 categories (plant foods, meat, milk, eggs, wood, paper, and fiber), a per capita value of HANPP was calculated for each country. The per capita estimate of HANPP was then applied to a gridded database of the human population at a spatial resolution of 0.25° (latitude and longitude). This spatial scale was chosen to match the spatial resolution of NPP derived from the satellite-derived vegetation index (NDVI). The resulting map depicts the spatial pattern of HANPP, showing where the products of terrestrial photosynthesis are consumed (Figure 1a). By combining the maps of global NPP and HANPP, the researchers were able to map HANPP as a percentage of NPP, providing a spatially explicit balance of NPP "supply" and "demand" (Figure 1b). The resulting map reveals a great deal of regional variation in the appropriation of NPP.

Summing for the globe, the researchers estimate annual HANPP to be 24.2 Pg organic matter (1Pg = 1015 g), or approximately 20 percent of terrestrial annual NPP (with high and low estimates of 14–26 percent). Some regions, however, such as Western Europe and south central Asia, consume more than 70 percent of their NPP. In contrast, HANPP in other regions (typically in the wet tropics) is less than 15 percent. The lowest value (approximately 6 percent) is found in South America.

Both population and per capita consumption interact to determine the human ecological impact at a regional scale. The role of population is clear from Figure 1b despite vast differences in consumption among nations. For example, east and south central Asia, with almost half of the world's population, appropriates 72 percent of its regional NPP despite having the lowest per capita consumption of any region. Affluence also plays an important role. The average annual per capita HANPP for industrialized countries is almost double that of developing nations, which are home to 83 percent of the global population. If the per capita HANPP

20.10 | Energy Flows through Trophic Levels Can Be Quantified

To quantify the flux of energy through the ecosystem using the conceptual model of the food chains just discussed, we need to return to the processes involved in secondary production discussed in Section 20.8: consumption, ingestion, assimilation, respiration, and production. We will examine a single trophic compartment (Figure 20.19a). The energy available to a given trophic level (designated as n) is the production of the next-lower level ($n - 1$); for example, net primary production (P_1) is the available energy for grazing herbivores (trophic level 2). Following the simple model of energy of the developing nations increased to match that of the industrialized countries, global HANPP would increase by 75 percent to a value of 35 percent of the current global NPP.

1. Use Figure 20.5 together with Figure 1 to answer the following question. Why is the HANPP expressed as a percentage (Figure 1b) low for the wet tropical regions of the world (regions of tropical rain forest)?

2. Since the mid-20th century, the efficiency of agricultural production (the quantity of crops produced per hectare of land under cultivation) has increased dramatically. How would this increase have changed the values of HANPP shown in Figures 1a and 1b?

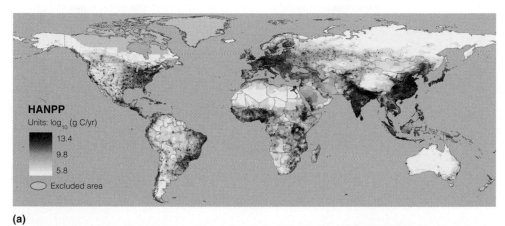

(a)

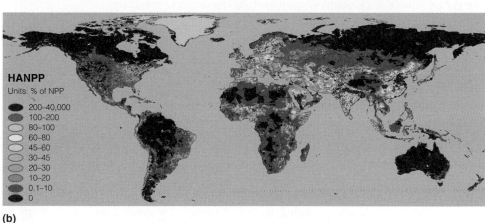

(b)

Figure 1 | Spatial distribution of the annual NPP resources required by the human population, as measured by **(a)** HANPP, and **(b)** HANPP as a percentage of local NPP.

(Nature Publishing Group.)

flow through a consumer organism presented in Section 20.8, some proportion of that productivity is consumed or ingested (I); the remainder makes its way to the dead organic matter of the detrital food chain. Some portion of the energy consumed is assimilated by the organisms (A), and the remainder is lost as waste materials (W) to the detrital food chain. Of the energy assimilated, some is lost to respiration, shown as the arrow labeled R that is leaving the upper left corner of the box, and the remainder goes to production (P). We quantify this flow with the energetic efficiencies defined in Section 20.8: the assimilation efficiency (A/I) and the production efficiency (P/A). One additional index of energetic efficiency, however, must be introduced:

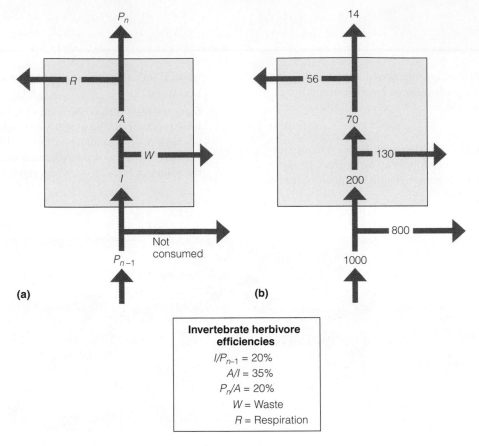

(a)

(b)

Invertebrate herbivore efficiencies

$I/P_{n-1} = 20\%$
$A/I = 35\%$
$P_n/A = 20\%$
W = Waste
R = Respiration

Figure 20.19 | **(a)** Energy flow within a single trophic compartment. **(b)** An example of quantifying energy flow for an invertebrate herbivore using estimates of efficiencies provided in table. Values are kilocalories (kcal).

Interpreting Ecological Data

Q1. Assume that the assimilation efficiency is changed from 35% to 25%; what is the new value for respiration?

Q2. What is the trophic efficiency (TE) for the example presented in the figure above? (See Section 20.12.)

consumption efficiency. The **consumption efficiency,** the ratio of ingestion to production (I_n/P_{n-1}), defines the amount of available energy being consumed. Sample values of these efficiencies for an invertebrate herbivore in the grazing food chain are provided in Figure 20.19b. Using these efficiency values, we can track the fate of a given amount of energy (1000 kcal) available to herbivores in the form of NPP through the herbivore trophic level.

If we apply efficiency values for each trophic level in the grazing and detrital food chains, we can calculate the flow of energy through the whole ecosystem. The production from each trophic level provides the input to the next-higher level, and unconsumed production (dead individuals) and waste products from each trophic level provide input into the dead organic matter compartment. The entire flow of energy through the ecosystem is a function of the initial transformation of solar energy into NPP. All energy entering the ecosystem as NPP eventually is lost through respiration.

20.11 | Consumption Efficiency Determines the Pathway of Energy Flow through the Ecosystem

Although the general model of energy flow presented in Figure 20.18 pertains to all ecosystems, the relative importance of the two major food chains and the rate of energy flow through the various trophic levels can vary widely among different types of ecosystems. The consumption efficiency (I_n/P_{n-1}) defines the amount of available energy produced by any given trophic level (P_{n-1}) that is consumed by the next-higher level (I_n). Values of consumption efficiency for the various consumer trophic levels therefore determine the pathway of energy flow through the food chain, providing a basis for comparison of energy flow through different ecosystems.

Despite its conspicuousness, the grazing food chain is not the major one in most terrestrial and many aquatic ecosystems. Only in some open-water aquatic ecosystems do the grazing herbivores play the dominant role in energy flow. Ecologists Helene Cyr (University of Toronto, Canada) and Michael Pace (University of Virginia) compiled published measurements of herbivore consumption rates (herbivore consumption efficiency), herbivore biomass, and primary productivity for a wide range of aquatic and terrestrial ecosystems (Figure 20.20). Although there is

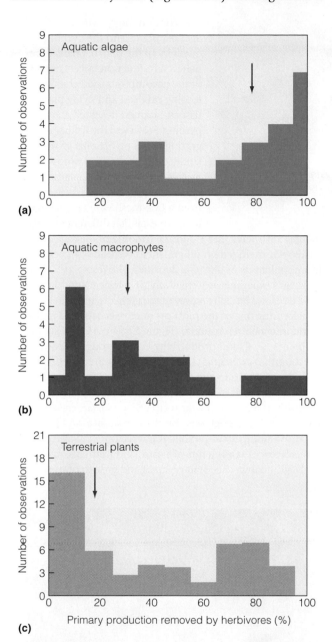

(a)

(b)

(c)

Primary production removed by herbivores (%)

Figure 20.20 | Results from a review of studies that examined rates of herbivory in different ecosystems. Histograms represent the percentage of net primary productivity consumed by herbivores in ecosystems dominated by **(a)** algae (phytoplankton), **(b)** rooted aquatic plants, and **(c)** terrestrial plants. Number of observations refers to the number of experiments having a given level of consumption. Red arrows indicate the median value. Note that herbivores consume a significantly greater proportion of phytoplankton productivity than do either aquatic or terrestrial plants. Go to QUANTIFYit! at **www.ecologyplace.com** to perform a chi-square test. (Nature Publishing Group.)

(Adapted from Cyr and Pace 1993.)

considerable variation in both environments, some generalizations do emerge from their analysis. Aquatic ecosystems dominated by phytoplankton have higher rates of herbivory (median value of 79 percent) than do those in which vascular plants (submerged and emergent) dominate (median value of 30 percent). In contrast, only 17 percent of primary productivity (median value) is removed by herbivores in terrestrial ecosystems. Therefore, in most terrestrial and shallow-water ecosystems, with their high standing biomass and relatively low harvest of primary production by herbivores, the detrital food chain is dominant. In deep-water aquatic ecosystems, with their low standing biomass, rapid turnover of organisms, and high rate of harvest, the grazing food chain may be dominant.

In terrestrial ecosystems, distinct differences in consumption efficiency and energy flow exist between forest and grassland ecosystems. Nelson Hairston of Cornell University reviewed a wide range of studies that examined patterns of energy flow through terrestrial ecosystems, providing a comparison of consumption efficiencies for herbivores (primary producer → herbivore) and their predators (herbivore → carnivore). The author found an average consumption efficiency of 3.7 percent for herbivores inhabiting deciduous forest ecosystems, whereas herbivores inhabiting grassland ecosystems had a value of 9.3 percent (both values lower than the average for terrestrial ecosystems reported by Cyr and Pace). Much smaller differences were observed for the consumption efficiency of predators inhabiting the two ecosystem types. Predators inhabiting forests had a value of 89.9 percent, whereas predators inhabiting grassland ecosystems had an average value of 77 percent.

Patterns of energy flow through flowing-water ecosystems (streams and rivers) differ markedly from both terrestrial and standing-water ecosystems (lakes and oceans). By comparison, stream and river ecosystems have extremely low NPP, and the grazing food chain is minor (see Chapter 24). The detrital food chain dominates and depends on inputs of dead organic matter from adjacent terrestrial ecosystems.

Figure 20.21 (on p. 434) graphically represents the different patterns of energy transfer in the four different ecosystems just discussed: forest, grassland, standing water, and running water.

20.12 | Energy Decreases in Each Successive Trophic Level

Based on the preceding discussion and the analysis presented in Figure 20.19, we can conclude that the quantity of energy flowing into a trophic level decreases with each successive trophic level in the food chain. This pattern occurs because not all energy is used for production. An ecological rule of thumb is that only 10 percent of the energy stored as biomass in a given trophic level is converted to biomass at the next-higher trophic level. If, for

Department of Zoology, University of Florida, Gainesville, Florida

The salt marshes that fringe the coastline of eastern North America are among the most productive ecosystems in the world. For the past half century, the prevailing theory found in the ecological literature and textbooks has been that productivity in these coastal ecosystems is controlled by physical conditions (water depth, frequency of inundation, salinity, etc.) and nutrient availability, referred to as "bottom-up" control (see Sections 17.4 and 20.4). But as a result of research conducted by ecologist Brian Silliman of the University of Florida, this long-held view is being brought into question. The focus of Silliman's research is the role of consumer organisms in the salt marshes of the southeastern United States. What is emerging from his work is a rich, complex picture of salt marshes that involves the interactions of marsh plants, "fungus-farming" snails, and an array of predators, including the blue crab.

The salt-marsh tidal zones of eastern North America are dominated by salt-marsh cordgrass, *Spartina alterniflora*. The most abundant and widespread grazer in these communities is the marsh periwinkle (*Littoraria irrorata*; Figure 1). The marsh periwinkle is a small snail, reaching 2.5 cm in length, with population densities upward of 500 individuals per square meter within the tidal zone dominated by *Spartina*.

While he was a graduate student working in the salt marshes along the eastern shore of Virginia, Silliman explored the role of the herbivory on patterns of net primary productivity (NPP) and standing biomass. Before Silliman's work, it was assumed that the grazers had little influence on the growth of *Spartina* plants and the overall productivity of the marsh. Snails were believed to function largely as part of the decomposer food chain, feeding on dead and dying plant tissues. Silliman designed an experiment to assess the influence of grazing snails on *Spartina* growth (individual plants) and productivity (collective growth of plants). He used cages, 1 m² in size, made from a fine-mesh wire fencing material, to establish different experimental treatments (see Figure 2). In some of the cages, snails were excluded; in others, snails were added to establish populations of differing densities. The results of Silliman's experiments were dramatic. Moderate to high snail densities led to runaway grazing effects, ultimately transforming one of the most productive ecosystems in the world into a barren mudflat (Figure 2). The effect of snails on plant growth and productivity is not through the direct consumption of green plant tissues, but by preparing the leaf tissue for colonization by their preferred food: fungus. As the snail crawls along the leaf surface, it scrapes the surface with its band of sawlike teeth called *radulae*, creating wounds that run lengthwise on the leaf surface and kill the surrounding tissues. While it travels, the snail also deposits feces containing fungal spores and nutrients, effectively stimulating the establishment and growth of fungus.

Figure 1 | The marsh periwinkle on *Spartina* leaf.

Figure 2 | Effects of snail grazing on *Spartina* standing crop and canopy structure after 8 months. Photographs show experimental exclosure in area of high snail density.

In a series of follow-up experiments with colleague Steven Newell of the University of Georgia Marine Institute, Silliman has demonstrated a mutualistic relationship between snails and fungi at the expense of the *Spartina* plants, the resource upon which both depend. The snails employ a low-level food production strategy whereby they prepare a favorable environment for fungal growth, provide substrate to promote growth, add supplemental nutrients and propagules, and consume fungus. Although this type of facilitation, known as "fungus farming," has been reported for some beetle, termite, and ant species, the work by Silliman and his colleague is the first reported case of this type of cultivation behavior outside of insects.

Given the potential of the snails to so dramatically reduce plant growth and marsh primary productivity at even moderate population densities, how do the salt marshes remain so productive? To answer this question, Silliman has turned his attention to the structure of the marsh's food chain. The periwinkle has several potential predators, including the terrapin, mud crab, and blue crab. Silliman hypothesized that predation maintains snail populations below the densities at which they have a devastating impact on *Spartina* plants. This type of control on plant productivity, in which predators limit populations of herbivores, is called "top-down" control.

To establish the role of predators in determining the distribution of snails within the marsh, Silliman once again used wire mesh cages—this time not to exclude the snails, but to exclude their predators. Other plots (areas of marsh) without cages were monitored for comparison (control plots). The experiment ran for a year, after which the snail populations were estimated in both the exclosure and control plots. The experiment results supported the hypothesis that predators control the distribution and abundance of snail populations (density of adults as well as recruitment of juveniles into the population). Once again, the results were dramatic. The density of juvenile snails was 100 times greater in the plots where predators were excluded as compared to the control plots. Predators such as the blue crab do indeed control snail populations in the marsh.

Thanks to Silliman's research, a new, more complex picture of the salt marshes of the eastern U.S. shorelines has emerged. Rather than the classical view of a bottom-up control on NPP, where plant growth and productivity are controlled primarily by the physical conditions and nutrient availability, a new top-down model of control on plant growth and productivity has emerged in which consumers exert a strong control on plant structure and productivity within the ecosystem. This phenomenon is referred to as a *trophic cascade*, because the effects of predators cascade down the food web to influence primary productivity (Figure 3).

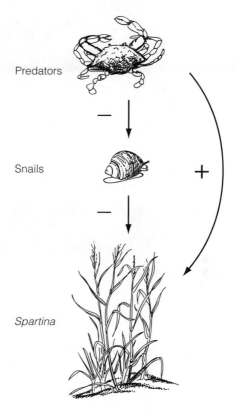

Figure 3 | The proposed mechanism of the marsh *trophic cascade*. Grazing by snails has a negative impact on *Spartina* populations. Likewise, predators have a negative impact on snail populations. The net effect is that predators have a positive indirect effect on *Spartina* populations by reducing rates of herbivory by snails.

Bibliography

Silliman, B. R., and J. C. Zieman. 2001. Top-down control on *Spartina alterniflora* production by periwinkle grazing in a Virginia salt marsh. *Ecology* 82:2830–2845.

Silliman, B. R., and M. D. Bertness. 2002. A trophic cascade regulates salt marsh primary production. *Proceedings of the National Academy of Sciences USA* 99:10500–10505.

Silliman, B. R., and S. Y. Newell. 2003. Fungal farming in a snail. *Proceedings of the National Academy of Sciences USA* 100:15643–15648.

1. How does the model of trophic cascade presented in Figure 3 relate to the concept of indirect mutualism developed in Chapter 17 (Section 17.3)?

2. Suppose we were to introduce a top predator to this ecosystem that feeds on blue crabs. How might this action change the nature of the trophic cascade outlined in Figure 3?

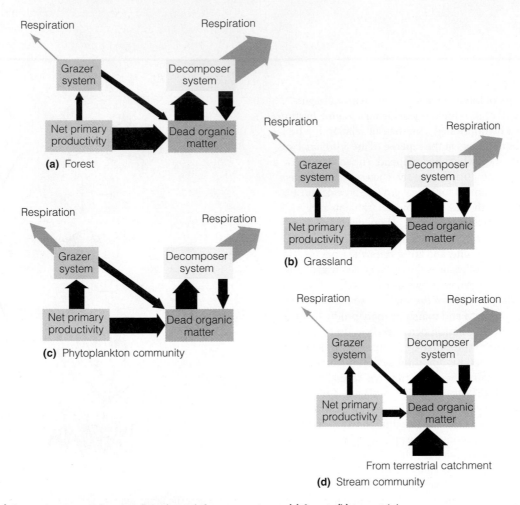

Figure 20.21 | General patterns of energy flow through four ecosystems: **(a)** forest, **(b)** terrestrial grassland, **(c)** ocean (phytoplankton community), and **(d)** stream. Relative sizes of boxes and arrows are proportional to the relative magnitude of compartments and flow.

(Adapted from Begon et al. 1986.)

example, herbivores eat 1000 kcal of plant energy, only about 100 kcal is converted into herbivore tissue, 10 kcal into first-level carnivore production, and 1 kcal into second-level carnivore production. However, ecosystems are not governed by some simple principle that regulates a constant proportion of energy reaching successive trophic levels.

As we have seen thus far in our discussion, differences in the consumption efficiency as well as the efficiency of energy conversion (assimilation and production efficiencies) exist among different feeding groups (see Table 20.2). These differences will directly influence the rate of energy transfer from one trophic level to the next-higher level. A measure of efficiency used to describe the transfer of energy between trophic levels is called the trophic efficiency. The **trophic efficiency** (*TE*) is the ratio of productivity in a given trophic level (P_n) to the trophic level it feeds on (P_{n-1}): $TE = P_n/P_{n-1}$.

Daniel Pauly and Villy Christensen of the University of British Columbia examined the energy transfer efficiency reported in 48 different studies of aquatic ecosystems. There is considerable variation among studies and trophic levels, but the mean value of 10.13 percent is close to the general rule of 10 percent transfer between trophic levels.

An important consequence of decreasing energy transfers through the food web is a corresponding decrease in the standing biomass of organisms within each successive trophic level. If we sum all of the biomass or energy contained in each trophic level, we can construct pyramids for the ecosystem (Figure 20.22). The pyramid of biomass indicates by weight, or other means of measuring living material, the total bulk of organisms or fixed energy present at any one time—the standing crop. Because some energy or material is lost at each successive trophic level, the total mass supported

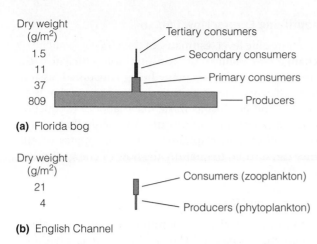

Dry weight (g/m²)
1.5 — Tertiary consumers
11 — Secondary consumers
37 — Primary consumers
809 — Producers

(a) Florida bog

Dry weight (g/m²)
21 — Consumers (zooplankton)
4 — Producers (phytoplankton)

(b) English Channel

Figure 20.22 | Biomass pyramids for the consumer food chain of **(a)** a bog ecosystem in Florida and **(b)** the marine ecosystem of the English Channel. The pyramid for the marine ecosystem is inverted due to the high productivity but fast turnover of phytoplankton populations (short life span and high rate of consumption by zooplankton).

at each level is limited by the rate at which energy is being stored at the next-lower level. In general, the biomass of producers must be greater than that of the herbivores they support, and the biomass of herbivores must be greater than that of carnivores. That circumstance results in a narrowing pyramid for most ecosystems (Figure 20.22a).

This arrangement does not hold for all ecosystems. In such ecosystems as lakes and open seas, primary production is concentrated in the phytoplankton. These microscopic organisms have a short life cycle and rapid reproduction. They are heavily grazed by herbivorous zooplankton that are larger and longer-lived. Thus, despite the high productivity of algae, their biomass is low compared to that of zooplankton herbivores (Figure 20.22b). The result is an inverted pyramid, with a lower standing biomass of primary producers (phytoplankton) and herbivores (zooplankton).

Summary

Laws of Thermodynamics | 20.1

Energy flow in ecosystems supports life. Energy is governed by the laws of thermodynamics. The first law states that although energy can be transferred, it cannot be created or destroyed. The second law states that as energy is transferred, a portion ceases to be usable. As energy moves through an ecosystem, much of it is lost as heat of respiration. Energy is degraded from a more organized to a less organized state, or entropy. However, a continuous flux of energy from the Sun prevents ecosystems from running down.

Primary Production | 20.2

The flow of energy through an ecosystem starts with the harnessing of sunlight by green plants through a process referred to as primary production. The total amount of energy fixed by plants is gross primary production. The amount of energy remaining after plants have met their respiratory need is net primary production in the form of plant biomass. The rate of primary production is net primary productivity, which is measured in units of weight per unit area per unit time.

Terrestrial Ecosystems | 20.3

Productivity of terrestrial ecosystems is influenced by climate, especially temperature and precipitation. Temperature influences the photosynthetic rate, and the amount of available water limits photosynthesis and the amount of leaves that can be supported. Warm, wet conditions make tropical rain forest the most productive terrestrial ecosystem. Nutrient availability also directly influences rates of primary productivity.

Aquatic Ecosystems | 20.4

Light is a primary factor limiting productivity in aquatic ecosystems, and the depth to which light penetrates is crucial to determining the zone of primary productivity. Nutrient availability is the most pervasive influence on the productivity of oceans. The most productive ecosystems are shallow coastal waters, coral reefs, and estuaries, where nutrients are more available. Nutrient availability is also a dominant factor limiting net primary productivity in lake ecosystems. In rivers and streams, net primary productivity is low, with inputs of dead organic matter from adjacent terrestrial ecosystems being the dominant source of energy input.

Energy Allocation | 20.5

Energy fixed by plants is allocated to different parts of the plant and to reproduction. How much is allocated to each component is a function of the plant life-form as well as the environmental conditions. The pattern of allocation will directly influence standing biomass and productivity rate.

Temporal Variation | 20.6

Primary production in an ecosystem varies with time. Seasonal and yearly variations in moisture and temperature directly influence primary production. In ecosystems dominated by woody vegetation, net primary production declines with age. As the ratio of woody biomass to foliage increases, more of gross production goes into maintenance.

Secondary Production | 20.7

Net primary production is available to consumers directly as plant tissue or indirectly through animal tissue. Once consumed and assimilated, energy is diverted to maintenance, growth, and reproduction and to feces, urine, and gas. Change in biomass, including weight change and reproduction, is secondary production. Secondary production depends upon primary production. Any environmental constraint on primary production will constrain secondary production in the ecosystem.

Efficiency of Energy Use | 20.8

Efficiency of production varies. Homeotherms have high assimilation efficiency but low production efficiency because they have to expend so much energy in maintenance. Poikilotherms have low assimilation efficiency but high production efficiency; they put more energy into growth.

Food Chains and Energy Flow | 20.9

A basic function of the ecosystem is to move energy from the Sun through various consumers to its final dissipation in a series of energy transfers known as the food chain. The various members of a food web can be grouped into categories called trophic or feeding levels. Autotrophs occupy the first trophic level. Herbivores that feed on autotrophs make up the next trophic level. Carnivores that feed on herbivores make up the third and higher trophic levels.

Energy flow in ecosystems takes two routes: one through the grazing food chain, the other through the detrital food chain. The bulk of production is used by organisms that feed on dead organic matter. The two food chains are linked by the input of dead organic matter and wastes from the consumer food chain functioning as input into the detrital food chain.

Quantifying Energy Flow | 20.10

At each trophic level, estimates of the efficiency of energy exchange are defined as consumption efficiency, the proportion of available energy being consumed; assimilation efficiency, the portion of energy ingested that is assimilated and not lost as waste material; and production efficiency, the portion of assimilated energy that goes to growth rather than respiration. These estimates of efficiency can be used to quantify the flow of energy through the food chain.

Consumption Efficiency | 20.11

Consumption efficiency determines the flow of energy through the ecosystem. The detrital food chain dominates in terrestrial ecosystems, with only a small proportion of net primary productivity being consumed by herbivores. In open-water ecosystems, such as lakes and oceans, a greater proportion of primary productivity is consumed by herbivores. Consumption efficiency of predators is more similar among these ecosystems.

Energy Pyramids | 20.12

The quantity of energy flowing into a trophic level decreases with each successive trophic level in the food chain. This pattern occurs because not all energy is used for production. An ecological rule of thumb is that only 10 percent of the energy stored as biomass in a given trophic level is converted to biomass at the next-higher trophic level. A plot of the total weight of individuals at each successive level produces a tapering pyramid. In aquatic ecosystems, however, where there is a rapid turnover of small aquatic producers, the pyramid of biomass becomes inverted.

Study Questions

1. How do the concepts of community and ecosystem differ?
2. Relate the following terms: gross primary productivity, autotrophic respiration, and net primary productivity.
3. Contrast net primary productivity and standing biomass for an ecosystem.
4. How do temperature and precipitation interact to influence net primary productivity in terrestrial ecosystems?
5. How does net primary productivity vary with water depth in standing-water ecosystems (lakes and oceans)? What is the basis for the vertical profile of net primary productivity in these ecosystems?
6. What environmental factors might influence the (light) compensation depth of a lake ecosystem?
7. How does primary productivity function as a constraint on secondary productivity in ecosystems?
8. What does the top-down model of food chain structure imply about the role of secondary producers in controlling net primary productivity and standing biomass within ecosystems?
9. How do assimilation efficiency and production efficiency relate to the flow of energy through a trophic level?
10. What is the difference in energy allocation and production efficiency between homeotherms and poikilotherms?
11. What are the two major food chains, and how are they related?
12. How does consumption efficiency differ between terrestrial and marine ecosystems?

Further Readings

Aber, J. D., and J. M. Melillo. 1991. *Terrestrial ecosystems.* Philadelphia: Saunders College Publishing.

This well-written and illustrated text gives readers an excellent introduction to ecosystem ecology.

Chapin, F. S., P. A. Matson, and H. A. Mooney. 2002. *Principles of terrestrial ecosystem ecology.* New York: Springer-Verlag.

This well-written and illustrated text is aimed at a more advanced readership but provides an excellent reference for exploring topics covered in Part Six in more depth.

Gates, D. M. 1985. *Energy and ecology.* Sunderland, MA: Sinauer Associates.

This text shows the function of ecological systems in terms of energy flow and covers a range of topics from photosynthesis to ecosystem productivity.

Golley, F. B. 1994. *A history of the ecosystem concept in ecology.* New Haven, CT: Yale University Press.

This book presents a historical overview of the development of the study of ecosystems in the broader framework of ecology.

Gosz, J. R., R. T. Holmes, G. E. Likens, and F. H. Bormann. 1978. The flow of energy in a forest ecosystem. *Scientific American* 238:92–102.

This paper summarizes one of the most detailed analyses of energy flow through an ecosystem and provides a contrast to the more aggregated view based on food chains presented in this chapter.

Howarth, R. W. 1988. Nutrient limitation of net primary production in marine ecosystems. *Annual Review of Ecology and Systematics* 19:89–110.

This paper reviews research on the role of nutrients in limiting net primary productivity of the world's oceans.

Leith, H., and R. H. Whittaker, eds. 1975. *Primary productivity in the biosphere.* Ecological Studies Vol. 14. New York: Springer-Verlag.

A classic volume exploring patterns of primary productivity at a global scale. An excellent resource for comparing primary productivity patterns in the various terrestrial and aquatic ecosystems on Earth.

Lindeman, R. L. 1942. The trophic-dynamics aspect of ecology. *Ecology* 23:399–418.

The classic paper that developed the concept of viewing ecosystems in terms of productivity and energy flow.

National Academy of Science. 1975. *Productivity of world ecosystems.* Washington, DC: National Academy of Science.

This volume provides an excellent summary of world primary production.

Wiegert, R. C., ed. 1976. *Ecological energetics. Benchmark papers.* Stroudsburg, PA: Dowden, Hutchinson & Ross.

A collection of papers surveying the development of the concept of energy flow through ecological systems. Gives readers an excellent historical overview of the development of ideas in ecosystem ecology.

CHAPTER 21

Decomposition and Nutrient Cycling

Colorful decomposers such as honey mushrooms (*Armillaria mellea*) reside on the forest floor throughout much of continental North America.

As we saw in Chapter 20, the flow of energy through ecosystems is the story of the element carbon. Beginning with the fixing of carbon dioxide into simple organic carbon compounds in the process of primary productivity, carbon moves through the food chain, ultimately returning to the atmosphere through the process of cellular respiration. Primary productivity, however, depends on the uptake of an array of essential mineral (inorganic) nutrients by plants (see the introduction to Part Two) and their incorporation into living tissues. The source of carbon is carbon dioxide in the atmosphere, but what is the source of the array of other essential elements life depends on? Each element has its own story of origin and movement through the ecosystem, and we shall explore these pathways, known as biogeochemical cycles, in Chapter 22. In general, however, the source of these essential nutrients is either the atmosphere, as in the case of carbon, or the weathering of rocks and minerals (a topic discussed in Chapter 4). Once in the soil or water, the nutrients are taken up by plants and then move through the food chain. In fact, nutrients in organic form, stored in living tissues, represent a significant proportion of nutrients within ecosystems. So what is the fate of these nutrients once they make their way into the food chain? As these living tissues senesce, the nutrients are returned to the soil or sediments in the form of dead organic matter, where they make their way through the decomposer food chain. But unlike carbon, most of the nutrients are recycled within the ecosystem. Various microbial decomposers transform the organic nutrients into a mineral form, and the nutrients are once again available to the plants for uptake and incorporation into new tissues. This process, called **internal cycling** (or nutrient cycling), is an essential feature of all ecosystems. It represents a recycling of nutrients within the ecosystem.

In this chapter, we will examine the processes involved in the recycling of nutrients within the ecosystem.

A central focus of our discussion will be the processes of decomposition and nutrient mineralization and the environmental factors that control the rate at which these processes proceed. We also will explore how this general process, common to all ecosystems, varies between terrestrial and aquatic ecosystems—thus setting the stage for discussing specific biogeochemical cycles in Chapter 22.

21.1 | Most Essential Nutrients Are Recycled within the Ecosystem

The cycling of nutrients within a terrestrial ecosystem is represented in Figure 21.1. We use the essential element nitrogen as an example to follow the pathway taken by nutrients from the soil to the vegetation and back again to the soil.

As with all essential nutrients, plants require nitrogen in an inorganic or mineral form. Nitrogen in the soil solution (in the form of ammonium and nitrate) is taken up through plant roots and used to produce proteins, enzymes, and a variety of other nitrogen-based compounds. This step represents a transformation of nitrogen from inorganic to organic form—namely, nitrogen contained in living plant tissues. As we discussed in Section 6.12, the availability of nitrogen and other nutrients in the soil solution will limit the rate of uptake by plants and subsequently the rate of net primary productivity (see Section 20.3). In the case of nitrogen, the rate of plant uptake will directly influence the rate of photosynthesis (see Figure 6.24).

As plant tissues senesce, nutrients are returned to the soil surface in the form of dead organic matter. Before senescence occurs, however, plants absorb some of the nutrients from senescing tissues to be stored and used in producing new tissues. This process of recycling nutrients within the plant is called **retranslocation** or **reabsorption.** For example, as the days of autumn become shorter in the deciduous forests of temperate regions, the synthesis of chlorophyll (the

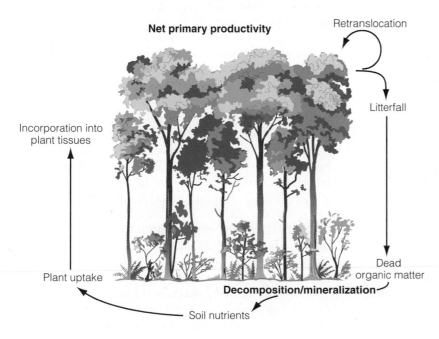

Net primary productivity

Retranslocation

Litterfall

Incorporation into plant tissues

Dead organic matter

Plant uptake

Decomposition/mineralization

Soil nutrients

Figure 21.1 | A generalized model of nutrient cycling in a terrestrial ecosystem.

Table 21.1 | Nitrogen Content of Living and Senescent Leaves for Nine Tree Species Found in Central Virginia*

Species	Green Leaf % N	Senescent Leaf % N	Retranslocation % N
White oak	2.08	0.82	60.6
Scarlet oak	2.14	0.85	60.3
Southern red oak	1.88	0.60	68.1
Red maple	1.96	0.76	61.2
Tulip poplar	2.55	0.90	64.7
Virginia pine	1.62	0.54	66.7
American hornbeam	2.20	1.16	47.3
Sweetgum	1.90	0.59	68.9
Sycamore	2.10	0.90	57.1

*All values are expressed as a percentage of dry weight. Percent nitrogen retranslocation is the difference between the nitrogen content of green and senescent leaves expressed as a percentage of green leaf nitrogen content.

light-harvesting pigment that gives leaves their green color; see Section 6.1) by plants begins to decline. The yellow and orange pigments (carotinoids and xanthophylls), always present within the leaf, begin to show. The senescing cells of the leaves also produce other chemicals, particularly anthocyanins, that are responsible for red and purple colors. The leaves of some species, particularly the oaks, contain high quantities of tannins that produce brown colors. As senescence occurs, water and nutrients are being drawn into the stems and away from the leaves. The plant can recover as much as 70 percent of the nitrogen in the green leaves before they senesce and fall to the forest floor (Table 21.1), thus reducing the amount of nutrients returned to the soil as dead organic matter.

Once on the forest floor, various decomposer organisms break down and consume the dead plant tissues, transforming the organic nutrients into a mineral form through the process of mineralization (discussed in detail later in Section 21.5). The cycle is now complete, and the nutrients are once again available to the plants for uptake and incorporation into plant tissues.

21.2 | Decomposition Is a Complex Process Involving a Variety of Organisms

The key process in the recycling of nutrients within ecosystems is decomposition. Decomposition is the breakdown of chemical bonds formed during the construction of plant and animal tissues. Whereas photosynthesis involves the incorporation of solar energy, carbon dioxide, water, and inorganic nutrients into organic compounds (living matter), decomposition involves the release of energy originally fixed by photosynthesis, carbon dioxide, and water and ultimately the conversion of organic compounds into inorganic nutrients. Decomposition is a complex of many processes, including leaching, fragmentation, changes in physical and chemical structure, ingestion, and excretion of waste products. These processes are accomplished by a variety of decomposer organisms. All heterotrophs function to some degree as decomposers. As they digest food, they break down organic matter, alter it structurally and chemically, and release it partially in the form of waste products. However, what we typically refer to as decomposers are organisms that feed on dead organic matter or detritus. These organisms include microbial decomposers, a group made up primarily of bacteria and fungi, and detritivores, animals that feed on dead material including dung (Figure 21.2).

The innumerable organisms involved in decomposition are categorized into several major groups based on their size and function. Organisms most commonly associated with the process of decomposition are the microflora, composed of the bacteria and fungi. Bacteria may be aerobic, requiring oxygen for metabolism (respiration), or they may be anaerobic, able to carry on their metabolic functions without oxygen by using inorganic compounds. This type of respiration by anaerobic bacteria, commonly found in the mud and sediments of aquatic habitats and in the rumen of ungulate herbivores, is fermentation (see Section 7.2).

Bacteria are the dominant decomposers of dead animal matter, whereas fungi are the major decomposers of plant material. Fungi extend their hyphae into the organic material to withdraw nutrients. Fungi range in type from species that feed on highly soluble, organic compounds such as glucose to the more complex hyphal fungi that invade tissues with their hyphae (Figure 21.2a).

Bacteria and fungi secrete enzymes into plant and animal tissues to break down the complex organic compounds. Some of the resulting products are then absorbed as food. After one group has exploited the material to the extent of its ability, a different group of bacteria and fungi able to use the remaining material moves in. Thus, a succession of microflora are involved in decomposing the organic matter until it is finally reduced to inorganic nutrients.

Decomposition is aided by the fragmentation of leaves, twigs, and other dead organic matter (detritus) by invertebrate detritivores. These organisms fall into four major groups, classified by body width: (1) microfauna and microflora ($<100\ \mu m$) include protozoans and nematodes inhabiting the water in soil pores; (2) mesofauna (between $100\ \mu m$ and 2 mm) include mites (Figure 21.2b), potworms, and springtails that live in soil air spaces; (3) macrofauna (2–20 mm); and (4) megafauna (>20 mm). The last two categories are represented by millipedes, earthworms (Figure 21.2c), and snails in terrestrial habitats and by annelid worms, smaller crustaceans—especially amphipods and isopods—and mollusks and crabs in aquatic habitats (Figure 21.2d). Earthworms and snails dominate the megafauna. The macrofauna and megafauna can burrow into the soil or

Figure 21.2 | **(a)** Fungi and bacteria are major decomposers of plant and animal tissues. **(b)** Mites and springtails are among the most abundant of small detritivores. **(c)** Earthworms and millipedes are large detritivores in terrestrial ecosystems, and **(d)** mollusks and crabs play a similar role in aquatic ecosystems.

substrate to create their own space, and megafauna, such as earthworms, have major influences on soil structure (see Chapter 4). These detritivores feed on plant and animal remains and on fecal material.

Energy and nutrients incorporated into bacterial and fungal biomass do not go unexploited in the decomposer world. Feeding on bacteria and fungi are the microbivores. Making up this group are protozoans such as amoebas, springtails (Collembola), nematodes, larval forms of beetles (Coleoptera), and mites (Acari). Smaller forms feed only on bacteria and fungal hyphae. Because larger forms feed on both microflora and detritus, members of this group are often difficult to separate from detritivores.

21.3 | Studying Decomposition Involves Following the Fate of Dead Organic Matter

Decomposers, like all heterotrophs, derive their energy and most of their nutrients from the consumption of organic compounds. Energy is obtained through the oxidation of carbon compounds, carbohydrates (such as glucose), in the process of respiration (see the introduction to Part Two). Ecologists study the process of decomposition by designing experiments that follow the decay of dead plant and animal tissues through time. The most widely used approach is the use of litterbags to examine the decomposition of dead plant tissues (called plant litter). Litterbags are mesh bags constructed of synthetic material that does not readily decompose (Figure 21.3). The holes in the bag must be large enough to allow decomposer organisms to enter and feed on the litter but small enough to prevent decomposing plant material from falling out of the bag (holes typically 1–2 mm), although this compromise in mesh size typically restricts access to the larger decomposer organisms.

A fixed amount of litter material is placed in each bag. In the experiments presented in Figure 21.4, for each of the 3 tree species, 30 litterbags were filled with 5 g of leaf litter each. These bags were buried in the litter layer of the forest. At six intervals during the course of a year, five bags were collected for each species and their contents dried and weighed in the laboratory. From these data, the decomposition rate can be determined for each of the species (see Quantifying Ecology 21.1: Estimating the Rate of Decomposition).

Note in Figure 21.4 that the mass of litter remaining in the bags decreases continuously as time progresses. As decomposer organisms consume the litter, using the carbon as a source of energy, the carbon is eventually lost to the atmosphere as CO_2 in the process of respiration. It is important to note, however, that in this approach of studying decomposition, as time passes, the mass of organic matter remaining in the litterbag includes original plant material as well as the bacteria and fungi (microbial decomposers) that have colonized and grown on the plant litter. Because of the difficulty in doing so, very few studies have quantified the

Figure 21.3 | Litterbag experiment. In this example, a known quantity of senescent leaves is placed in mesh bags on the forest floor. Bags are retrieved at various intervals, and the mass loss due to consumption by decomposers is tracked through time.

changing contribution of primary (original plant material) and secondary (microbial biomass) organic matter in the decomposing litter (remaining mass).

The microbial ecologist Martin Swift of the University of Zimbabwe was able to estimate the growth of fungi during decomposition of wood sawdust by measuring the change in chitin content, which is restricted to the fungal cell walls. By the end of the experiment, a

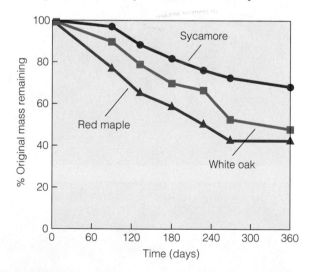

Figure 21.4 | Results of a litterbag experiment in central Virginia designed to examine the decomposition of fallen leaves from red maple, white oak, and sycamore trees. Decomposition is expressed as the percentage of the original mass remaining at different times during the first year of the experiment.
(Adapted from Smith 2002.)

Interpreting Ecological Data

Q1. What proportion of the original leaf mass has been lost (consumed) for each of the three species at day 180 in the experiment?

Q2. Which of the three species has the slowest rate of decomposition?

Estimating the Rate of Decomposition

Litterbag experiments such as the one presented in Figures 21.4 and 21.16 are the primary means by which ecologists study the process of decomposition. By collecting multiple (replicate) litterbags at regular intervals during the process of decay, researchers can plot the proportion of mass loss (or proportion of mass remaining) through time. These data, in turn, can be used to estimate the rate of decomposition. Each point in Figure 1 represents the average of five replicate litterbags collected on the given day during the experiment. The x-axis is time (weeks), and the y-axis is the percentage of the original mass remaining.

The mass loss through time is generally expressed as a negative exponential function:

$$\text{Original mass remaining} = e^{-kt}$$

Here, e is the natural logarithm, t is the time unit used (years, months, weeks, or days), and k is the decomposition coefficient, which defines the slope of the negative exponential curve. The decomposition coefficient can be estimated using regression techniques (fitting an exponential regression model to the data set: $y = e^{-kx}$, where y is the original mass remaining and x is the corresponding time in units of weeks). This technique was used for the data presented in Figure 1, and the resulting equations are presented. The estimates of k for red maple and Virginia pine are 0.0167 and 0.0097, respectively, in units of proportion mass loss per week. In the current example, the values are multiplied by 100 to convert from proportion to percentage. The higher the value of k, the faster the rate of decomposition.

Data from the same litterbag study for a third tree species, sycamore (*Platanus occidentalis*), is presented in Table 1. Each value represents the remaining dry mass of the original 7 grams dry weight of leaf litter that was placed in the bag at the beginning of the experiment (day 0). The percent original mass remaining is therefore calculated by dividing each value by 7

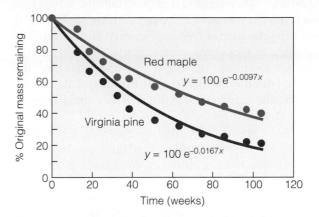

Figure 1 | Data from two litterbag experiments that examined the rate of decomposition for red maple (*Acer rubra*) and Virginia pine (*Pinus virginiana*) leaf litter over a period of two years. Each point represents the average mass remaining in five replicate litterbags sampled during that period. The lines represent the predicted relationship based on the solution for the decomposition coefficient k using the negative exponential model discussed in the text. Values of k were estimated using a nonlinear regression technique.

and multiplying by 100 to convert the proportion to a percentage.

1. Using the data in Table 1, calculate the mean value for the five replicate samples at each time period, and then convert the mean value to percent original mass remaining (%OMR). Now plot the resulting mean values of %OMR (y-axis) as a function of time (x-axis), as shown in Figure 1. For purposes of comparison, you will need to convert days to weeks.

2. How does the general pattern of decomposition (%OMR) through time compare to that of red maple and Virginia pine presented in Figure 1?

3. How do you think the lignin concentration of the sycamore leaves compares to that of red maple or Virginia pine? Why?

Table 1 | Litterbag Study after the Decomposition of Sycamore Leaf Litter*

Litterbag	Day 0	90	131	179	228	269	360	445	525	603	680	730
Replicate1	7	6.92	5.82	5.69	5.29	4.89	4.66	4.33	4.06	3.78	3.84	3.71
Replicate2	7	6.84	6.08	5.59	5.28	4.99	4.87	4.24	3.99	4.09	3.71	3.83
Replicate3	7	6.91	5.98	5.83	5.38	5.18	4.75	4.38	4.26	4.01	3.60	3.78
Replicate4	7	6.75	5.74	5.88	5.44	5.13	4.92	4.50	4.21	3.88	3.92	3.58
Replicate5	7	6.72	5.88	5.78	5.41	5.21	4.72	4.19	4.08	3.87	3.82	3.65

*All values in grams dry weight. Initial weight of leaf litter in each bag was 7.0 g.

39 percent weight loss was recorded for the sawdust, but the biomass estimate showed that 58 percent of the remaining mass was composed of living and dead fungal biomass. The apparent decomposition rate (k) of the sawdust was 0.04/wk, but the rate more than doubled—to 0.09/wk—when recalculated to exclude the fungal biomass (for a discussion of k, see Quantifying Ecology 21.1: Estimating the Rate of Decomposition). This shift in the proportion of remaining organic matter in plant and decomposer biomass is crucial to understanding the dynamics of other nutrients, such as nitrogen, during the decomposition process; that topic is examined later, in Section 21.5.

A similar approach to litterbag experiments is used to evaluate the decomposition of plant litter in stream ecosystems. Leaf litter subsidies to aquatic ecosystems provide large inputs of energy and nutrients into streams that typically exhibit low primary productivity (see Section 20.11). Leaf litter often accumulates in areas of active deposition, forming leaf packs (Figure 21.5a). To

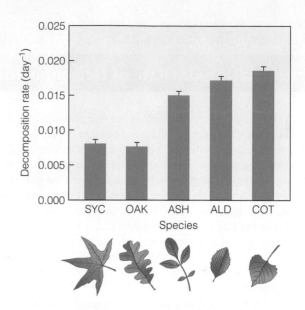

Figure 21.6 | Decomposition rates for leaf litter from five tree species submerged in stream. Experiments used submerged litter bags (leaf packs) that were sampled at five intervals over a period of 83 days. Decomposition rates expressed as mass loss per day (see Quantifying Ecology 21.1). SYC: *Platanus wrightii*; OAK: *Quercus gambelii*; ASH: *Fraxinus velutina*; ALD: *Alnus oblongifolia*; COT: *Populus fremontii*.

(Adapted from Leroy and Marks 2006.)

Figure 21.5 | **(a)** Inputs of plant litter from the surrounding terrestrial environment can form areas of deposition, known as leaf packs, in stream ecosystems. **(b)** Much like the use of litterbag experiments by terrestrial ecologists, stream ecologists use mesh bags to simulate natural leaf packs and examine the processes of decomposition. **(c)** Smaller mesh bags allow access to only microbial decomposers, while larger mesh bags allow access to the diversity of invertebrate decomposers that inhabit streams.

quantify the process of decomposition, plant litter is placed in mesh bags (Figure 21.5b), called **leaf packs,** which are anchored in place in areas where active deposition occurs. By varying the mesh size, investigators can determine the relative importance of microbial and macroinvertebrates on the decomposition of leaves. Fine-mesh screens allow only microbes to colonize and decompose the leaves, while coarse screens also allow the larger organisms to enter and feed on the material (Figure 21.5c). As with terrestrial litterbags, the leaf packs are placed in the stream for several weeks to measure the weight loss and chemical changes of the leaves during the process of decomposition (Figure 21.6).

Unlike terrestrial litterbag experiments, leaf pack studies typically quantify the diversity of micro- and macro-decomposers during decomposition. A detailed discussion of decomposition in stream ecosystems is presented in Chapter 24 (Section 24.6).

Litterbags and leaf packs are also used to examine the decomposition of plant litter in wetland and marsh ecosystems. The mesh bags are tethered to stakes to prevent them from being displaced by the tidal currents.

21.4 | Several Factors Influence the Rate of Decomposition

Not all organic matter decomposes at the same rate. For example, note the differences in decomposition rate for the leaf litter of the three tree species reported in Figure

21.4. From the wealth of studies completed over the past 50 years, some generalizations concerning factors that influence decomposition rate have emerged. The rate of decay (mass loss) is related to (1) the quality of plant litter as a substrate for microorganisms and soil fauna active in the decomposition process and (2) features of the physical environment that directly influence decomposer populations, namely soil properties (e.g., texture and pH) and climate (temperature and precipitation).

Characteristics that influence the quality of plant litter as an energy source relate directly to the types and quantities of carbon compounds present; that is, the types of chemical bonds present and the size and three-dimensional structure of the molecules in which these bonds are formed. Carbon is plentiful in plant remains, typically making up 45–60 percent of the total dry weight of plant tissues, but not all carbon compounds are of equal quality as an energy source for microbial decomposers. Glucose and other simple sugars, the first products of photosynthesis, are high-quality sources of carbon. These molecules are physically small. The breakage of their chemical bonds yields much more energy than required to synthesize the enzymes needed to break them. Cellulose and hemicellulose are the main constituents of cell walls. These compounds are more complex in structure and therefore more difficult to decompose than simple carbohydrates. They are of moderate quality as a substrate for microbial decay. The much larger lignin molecules are among the most complex and variable carbon compounds in nature. There is no precise chemical description of lignin; rather, it represents a class of compounds. These compounds possess very large molecules, intricately folded into complex three-dimensional structures that effectively shield much of the internal structure from attack by enzyme systems. As such, lignins, major components of wood, are among the slowest components of plant tissue to decompose. Lignin compounds are of such low quality as a source of energy that they yield almost no net gain of energy to microbes during decomposition. Bacteria do not degrade lignins; they are broken down by only a single group of fungi, the basidiomycetes (which includes the mushrooms; see Figure 21.2a).

Variation in the consumption rates of different carbon compounds is revealed in an experiment examining the rate at which carbon was consumed during the decomposition of straw that was placed on a soil surface (Figure 21.7). The total carbon content of the straw, expressed as a percentage of the original mass, declined exponentially during the period of the 80-day study. However, when the total carbon was partitioned into various classes of carbon compounds, the decomposition rates of these compounds varied widely. Proteins, simple sugars, and other soluble compounds made up some 15 percent of the original total carbon content. These compounds decomposed very quickly, disappearing com-

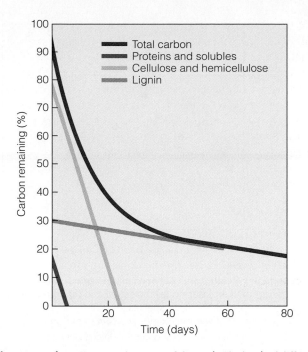

Figure 21.7 | Variation in the rates of decay (mass loss) of different classes of carbon compounds in an experiment examining the decomposition of straw on the soil surface. At any time, the sum of the three classes of carbon compounds is equal to the value for total carbon.
(Adapted from Smith, Heal, and Anderson 1979.)

pletely within the first few days of the experiment. Cellulose and hemicellulose made up some 60 percent of the original carbon content. Although these compounds decomposed more slowly than the proteins and simple sugars, by three weeks into the experiment they had been completely broken down. The third category of carbon compounds examined, the lignins, made up some 20 percent of the total original carbon. These compounds were broken down very slowly during the course of the experiment, and the vast majority of lignins remained intact by day 80. As decomposition proceeded during the experiment, the quality of the carbon resource declined. High- and intermediate-quality carbon compounds declined at a relatively rapid rate. Thus, the proportion of total carbon remaining as lignin compounds continually increased with time. The increasing component of lignin lowered the overall quality of the remaining litter as an energy source for microbial decomposers, therefore slowing the decomposition rate.

Because of its low quality as an energy source, the proportion of carbon contained in lignin-based compounds is used as an index of litter quality for decomposer organisms. The difference in the decomposition rates of the three species shown in Figure 21.4 is a direct result of their initial lignin content. The freshly fallen leaves of red maple (*Acer rubrum*) have a lignin content of 11.7 percent as compared to 17.7 percent for white oak (*Quercus alba*) and 36.4 percent for sycamore (*Platanus occidentalis*) leaves. In general, there is an inverse relationship between

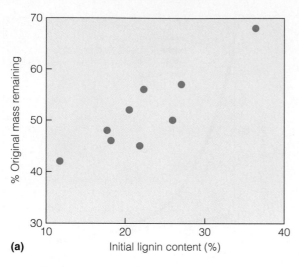

(a)
Initial lignin content (%)

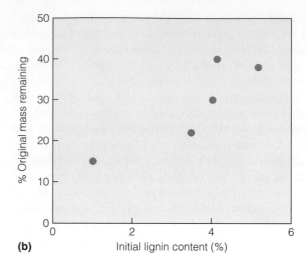

(b)
Initial lignin content (%)

Figure 21.8 | Relationship between initial lignin content of litter material and rate of decomposition for a variety of plant litters in **(a)** terrestrial and **(b)** aquatic environments. Each point on the graphs represents an individual plant species.
Go to **QUANTIFYit!** at **www.ecologyplace.com** to perform regression analysis.

(a) (Adapted from Smith 2002.) (b) (Adapted from Klap et al. 1999.)

the decomposition rate for plant litter and its lignin content at the start of decomposition. This inverse relationship between decomposition rate and the carbon quality of plant litter as a source of energy for decomposers has been reported for a wide range of plant species inhabiting both terrestrial and aquatic environments (Figure 21.8). Carbon quality of plant litters can have a particularly important influence on decomposition in coastal marine environments. Phytoplankton have a low lignin concentration and therefore decompose rather quickly. However, vascular plants, such as sea grasses, marsh grasses, and reeds that inhabit estuarine and marsh ecosystems, can have lignin concentrations that approach those of terrestrial plants. Decomposition of these plant litters is dependent on the oxygen content of the water. In the mud and sediments of aquatic habitats, where oxygen levels can be extremely low, anaerobic bacteria carry out most of the decomposition (see Section 21.2). The absence of fungi, which require oxygen for respiration, hinders the decomposition of lignin compounds, therefore slowing the overall rate of decomposition (Figure 21.9).

Besides the quality of dead organic matter as a food source, the physical environment also directly affects both macro- and micro-decomposers and, therefore, the rate of decomposition. Temperature and moisture greatly influence microbial activity. Low temperatures reduce or inhibit microbial activity, as do dry conditions. The optimum environment for microbes is warm and moist. As a result, decomposition rates are highest in warm, wet climates (see Field Studies: Edward A. G. (Ted) Schuur, p. 447). Alternate wetting and drying and continuous dry spells tend to reduce microflora activity and populations.

This effect of climate on the decomposition of red maple leaves at three sites in eastern North America

(New Hampshire, West Virginia, and Virginia) can be seen in Figure 21.10. Although the lignin content of red maple leaves does not differ significantly at the sites, the decomposition rate increases as you move southward from New Hampshire to West Virginia and Virginia. These observed differences can be attributed directly to climate differences at the sites. Mean daily temperature at the New Hampshire site is 7.2°C, and mean annual potential evaporation (see Chapter 4, Section 4.9) is 621 mm; mean daily temperature at the West Virginia

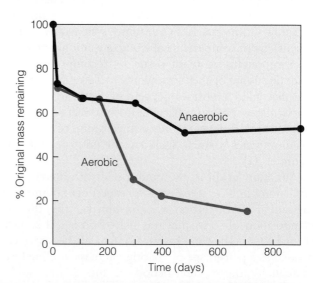

Figure 21.9 | Results of a litterbag experiment designed to examine the decomposition of *Spartina alternifolia* litter exposed to aerobic (litterbags on the marsh surface) and anaerobic (buried 5–10 cm below the marsh surface) conditions. See Figure 21.18 for example of experimental design of litterbag experiment in marsh environment.

(Adapted from Valiela 1984.)

Department of Botany, University of Florida, Gainesville, Florida

The warm, wet environments of the tropical rain forest support the highest rates of net primary productivity and decomposition of any terrestrial ecosystem on Earth. Or so we thought before the work of ecologist Ted Schuur of the University of Florida. The focus of Schuur's research is how species characteristics and features of the physical environment interact to control patterns of net primary productivity (NPP) and nutrient cycling in terrestrial ecosystems. Much of his work on this subject has been done in the montane forests of Maui in the Hawaiian Islands.

The island of Maui provides a unique environment in which to explore the interactions between biotic and abiotic controls on ecosystem processes. Ecologists are drawn to it for two reasons. First, the interaction of topography and prevailing easterly trade winds results in a diversity of microclimatic conditions on the island. Second, the Hawaiian Islands flora is relatively species poor; thus, a few species and genera occupy the broad range of environmental conditions.

The rain shadow (see Section 3.10) created by the 3055-m Haleakala volcano allowed Schuur to establish a series of six experimental plots at a constant altitude (1300 m) while mean annual precipitation ranged systematically from 2200 mm/yr (mesic) to more than 5000 mm/yr (wet) as a function of aspect relative to the prevailing trade winds. Other environmental factors—such as temperature regime, parent material, ecosystem age, vegetation, and topographic relief—were similar among the sites, allowing Schuur to focus on the role of precipitation on ecosystem processes. The forest canopy at all sites on this gradient was consistently dominated by the native evergreen tree *Metrosideros polymorpha* (Myrtaceae; Figure 1).

Initial experiments designed to estimate the aboveground NPP at the sites revealed an unexpected result. The prevailing conceptual (and empirical) model of how net primary productivity relates to precipitation is shown in Chapter 20 (Figure 20.2). Net primary productivity increases with increasing precipitation, leveling off under the conditions of high annual precipitation typical of that found in the wet tropics. What Ted Schuur found in the forests of Maui, however, was a distinct pattern of decreasing NPP with increasing annual precipitation.

Aboveground NPP decreased by a factor of 2.2 over the gradient (1000 g/m^2 to <500 g/m^2).

What mechanism could be at play? How could increasing precipitation result in a decrease in forest productivity? Measurements of the chemical composition of leaves collected from trees at the six sites along the mountainside provided some clues. Two characteristics varied systematically across the gradient of rainfall. Leaf nitrogen concentration decreased, whereas the concentration of lignin increased with increasing annual precipitation at the sites. Both of these characteristics are known to influence the processes of decomposition and mineralization (see Figures 21.9 and 21.16). This evidence pointed to the possibility that systematic changes in plant characteristics along this gradient may reduce rates of decomposition and nutrient cycling and subsequently limit nutrient availability for NPP.

To test this hypothesis, Schuur conducted a set of experiments designed to examine how both plant and site (physical environment) characteristics influenced the process of nutrient cycling across the six study sites. The dynamics of carbon and nitrogen during decomposition

Figure 1 | Typical forest of the Makawao and Koolau Forest Reserves.

continued on page 448

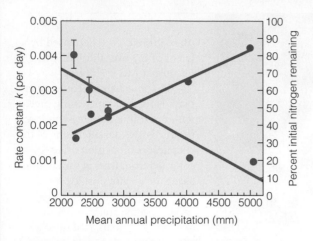

Figure 2 | Rate of decomposition (*k*: blue circles) and nitrogen loss (green circles) as a function of rainfall for the six study sites.

(Adapted from Schuur 2001.)

were examined using litterbag experiments (see Sections 21.3 and 21.5). In each litterbag experiment, replicate samples (bags) were collected at each of five time intervals during a 15-month period (1, 3, 6, 9, and 15 months).

The litterbag experiment results revealed that the decomposition rate of leaves decreased by a factor of 6.4 across the precipitation gradient, while the rate of nitrogen release from decomposing litter declined by a factor of 2.2 across the gradient (Figure 2). The litterbag experiments, therefore, presented a consistent picture of declining decomposition and nitrogen cycling across the gradient of mean annual rainfall.

Differences in the rate of nitrogen release from decomposing litter were reflected in the availability of nitrogen in the soil. Available (inorganic) soil nitrogen declined with increasing annual precipitation at the sites, as did oxygen availability in the soil.

The results clearly showed that reduced nitrogen availability was responsible for the decline in NPP across the rainfall gradient. But what factor (or factors) was controlling the rate of nutrient cycling in these forest ecosystems?

In these forests, the litter decomposition rates and nutrient release slow with increasing rainfall, apparently due to decreased soil oxygen availability and the production of low-quality litter at wetter sites. These two factors are interrelated, however. Because of high precipitation, low soil oxygen reduces rates of decomposition and nitrogen release (mineralization). In turn, low soil nitrogen potentially leads to lower leaf concentrations and litter quality, which in turn further decreases decomposition rates (see Figure 20.22).

Schuur's work in the forests of Maui has important implications for the current debate over how climate change will influence terrestrial ecosystems (see Chapter 29). According to current empirical models, precipitation has little effect on NPP and decomposition in the humid tropical ecosystems (see Figure 20.2). When this relationship is extended to wetter environments within the tropics using Schuur's data, however, an inverse relationship emerges (Figure 3). Because tropical forest is the largest terrestrial biome and accounts for one-third of the potential terrestrial NPP, the nature of this relationship is crucial to understanding how projected climate change will influence the global carbon balance.

Bibliography

Schuur, E. A. G. 2001. The effect of water on decomposition dynamics in mesic to wet Hawaiian montane forests. *Ecosystems* 4:259–273.

Schuur, E. A. G. 2003. Productivity and global climate revisited: The sensitivity of tropical forest growth to precipitation. *Ecology* 84:1165–1170.

Schuur, E. A. G., and P. A. Matson. 2001. Net primary productivity and nutrient cycling across a mesic to wet precipitation gradient in Hawaiian montane forest. *Oecologia* 128:431–442.

1. What type of experiment could Schuur undertake at his research site in Hawaii to determine if nitrogen availability in the soil is directly limiting the nutrient concentration of leaves and subsequently the rates of net primary productivity?
2. Why would reduced oxygen concentrations in the soil function to reduce rates of decomposition?

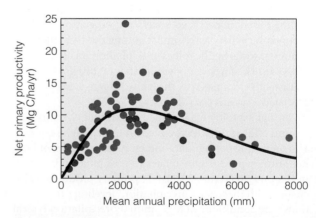

Figure 3 | Relationship between net primary productivity and mean annual precipitation.

(Adapted from Schuur 2003.)

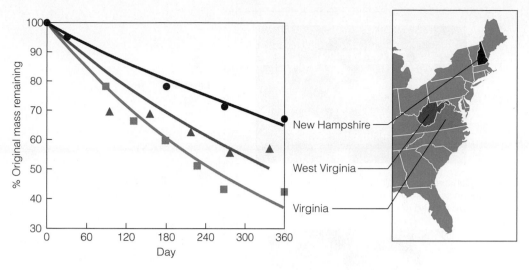

Figure 21.10 | Decomposition of red maple litter at three sites in eastern North America: New Hampshire (circles), West Virginia (triangles), and Virginia (squares). Mass loss through time was estimated using litterbag experiments at each site. The decline in decomposition rate from north to south is a direct result of changes in climate, primarily temperature.

(Adapted from Melillo et al. 1982, New Hampshire; Mudrick et al. 1994, West Virginia; Smith 2002, Virginia.)

site is 12.2°C, and mean annual potential evaporation is 720 mm; mean daily temperature at the Virginia site is 14.4°C, and mean potential evaporation is 806 mm.

The direct influence of temperature on decomposers results in a distinct diurnal pattern of microbial activity as measured by microbial respiration from the soil (Figure 21.11). The daily temperature pattern is closely paralleled by the release of CO_2 from the soil due to the respiration of microbial decomposers.

21.5 | Nutrients in Organic Matter Are Mineralized During Decomposition

Just as dead organic matter varies in the quality of carbon compounds as an energy source for microbial decomposers, likewise the nutrient quality of dead organic material varies greatly. The macronutrient nitrogen can serve as an example. Most dead leaf material has a nitrogen content in the range of 0.5–1.5 percent (see Table 21.1). The higher the nitrogen content of the dead leaf, the higher the nutrient value for the microbes and fungi that feed upon the leaf.

As the dead organic matter is consumed, the microbial decomposers—bacteria and fungi—transform nitrogen and other elements contained in organic compounds into inorganic (or mineral) forms (Figure 21.12 on page 450). This process is called **mineralization.** For example, the inorganic form of nitrogen, ammonia, is a waste product of microbial metabolism. The same decomposers that are responsible for mineralization also require nitrogen for their own growth and reproduction. Therefore, whenever mineralization occurs, **immobilization**—the uptake and assimilation of mineral nitrogen by microbial decomposers—runs counter to it (Figure 21.12). Because both of these processes, mineralization and immobilization, are taking place as decomposer organisms are

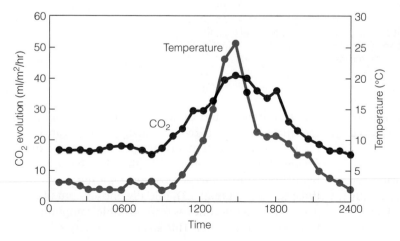

Figure 21.11 | Diurnal changes in air temperature and decomposition in a temperate deciduous forest. Decomposition rate is measured indirectly as the release (evolution) of CO_2 from decomposing litter on the forest floor. The release of CO_2 is a measure of the respiration of decomposer organisms.

(Adapted from Whitkamp and Frank 1969.)

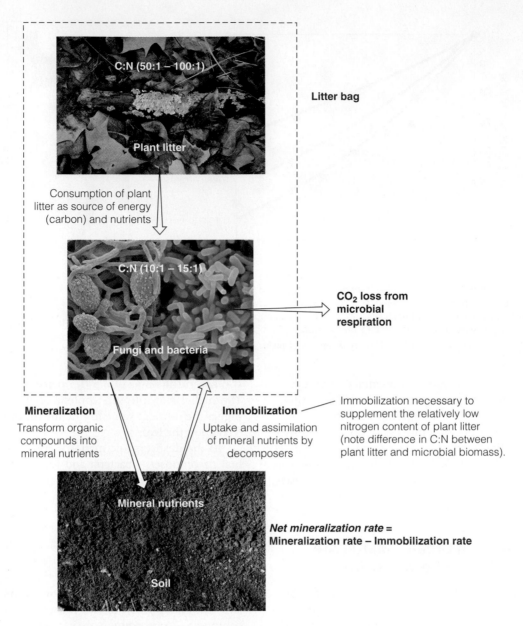

Figure 21.12 | Diagram illustrating the exchanges between litterbag (delineated by the dashed line) and soil in a standard litterbag experiment used to quantify the process of decomposition. Decomposer organisms (bacteria and fungi) colonize the plant litter. As litter is consumed, a significant proportion of carbon is respired, and nutrients bound in organic matter are mineralized and released to the soil. To convert plant carbon to microbial biomass, mineral nutrients are taken up in the process of immobilization. The difference in the rates of mineral nutrient release (mineralization) and immobilization is the rate of net mineralization. Note that as time progresses, the residual organic matter in the litterbag is composed of a growing proportion of microbial biomass as the original plant material is consumed, respired, and converted into microbial biomass.

consuming the litter, the supply rate of mineral nutrients to the soil during the process of decomposition, the **net mineralization rate,** is the difference between the rates of mineralization and immobilization.

Changes in the nutrient content of litter during decomposition are typically examined concurrently with changes in mass and carbon content in the litterbag experiments described in Section 21.3. As with litter mass

(see Figure 21.4) and carbon content (see Figure 21.9), the nitrogen content of the remaining litter can be expressed as a percentage of the nitrogen content of the original litter mass. Changes in the nitrogen content of plant litter during decomposition typically conform to three stages (Figure 21.13). Initially, the amount of nitrogen in the litter declines as water-soluble compounds are leached from the litter. This stage can be very short and,

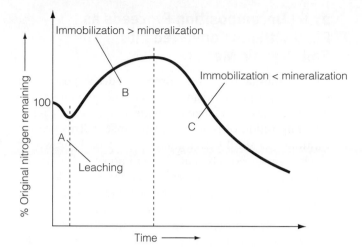

Immobilization > mineralization

Immobilization < mineralization

B

A

Leaching

C

% Original nitrogen remaining

Time

Figure 21.13 | Idealized graph showing the change in nitrogen content of plant litter during decomposition. The initial phase (A) corresponds to the leaching of soluble compounds. Nitrogen content then increases above initial concentrations (phase B) as the rate of immobilization exceeds the rate of mineralization. As decomposition proceeds, the rate of nitrogen mineralization exceeds that of immobilization, and there is a net release of nitrogen from the litter (phase C).

in terrestrial environments, depends on the soil moisture environment.

After the initial period of leaching, nitrogen content typically increases as microbial decomposers immobilize nitrogen from outside the litter. For this reason, nitrogen concentrations can rise above 100 percent, actually exceeding the initial nitrogen content of the litter material. To understand how this can occur, first recall from our discussion that the remaining organic matter in the litterbags includes the original dead leaf material as well as the living and dead microorganisms (see Figures 21.12 and 21.14). Second, the nitrogen content of the fungi and bacteria is considerably higher than that of the plant material they are feeding upon. The ratio of carbon to nitrogen (C:N; grams of carbon per gram of nitrogen) is a widely used index to characterize the nitrogen content of different tissues. The C:N of plant biomass is typically in the range of 50:1 to 100:1, whereas the C:N of bacteria and fungi is in the range of 10:1 to 15:1. As the plant material is consumed, and nitrogen is immobilized to meet the metabolic demands of the decomposers, the C:N will decline (Figure 21.14), reflecting the changing proportion of plant and microbial biomass remaining in the litterbag.

As decomposition proceeds and carbon quality declines (due to a higher proportional fraction of lignin), the mineralization rate exceeds the immobilization rate. The result is a net release of nitrogen to the soil (or water).

The pattern presented in Figure 21.13 is idealized. The actual pattern of nitrogen dynamics during decomposition depends on the initial nutrient content of the

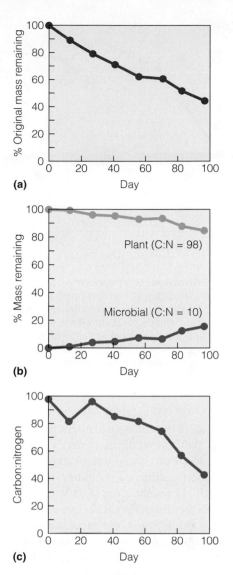

Figure 21.14 | Results from a litterbag experiment designed to examine the changing composition of decomposing winter rye in an agricultural field. **(a)** Mass loss continued throughout the 100 days of the experiment. **(b)** The proportion of the remaining mass in plant and microbial (fungal) biomass (living and dead). **(c)** Because the ratio of carbon to nitrogen (C:N) of the microbial biomass is much lower than that of the remaining plant litter, there is a general pattern of decline in the C:N during decomposition.

(Adapted from Beare et al. 1992.)

Interpreting Ecological Data

Q1. In the above experiment, assume that microbial decomposers assimilate 100 units of carbon in the form of plant litter, of which 40 units are converted into microbial biomass while the remaining 60 are respired as CO_2. Since the C:N ratio of the decomposers is 10:1, the production of 40 units of biomass would require 4 units of nitrogen (40:4). Assuming the C:N ratio of the 100 units of plant litter is 98:1, and that the decomposers retained all the nitrogen in the litter they consumed, how many units of nitrogen must be immobilized to produce the 40 units of microbial biomass?

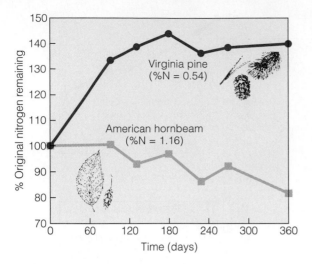

Figure 21.15 | Change in the nitrogen content of leaf litter from two tree species inhabiting the forests of central Virginia: American hornbeam and Virginia pine. Nitrogen content is expressed as the percentage of the original mass of nitrogen remaining at different times during the first year of the experiment. All data are from litterbag experiments. Note the difference between the two species in the initial nitrogen content of the leaf litter and the subsequent rates of immobilization.

(Adapted from Smith 2002.)

litter material. If the nitrogen content of the litter material is high, then mineralization may exceed the rate of immobilization from the onset of decomposition, and nitrogen concentration of the litter will not increase above the initial concentration (Figure 21.15).

Although the discussion and examples just presented have focused on nitrogen, the same pattern of immobilization and mineralization as a function of litter nutrient content applies to all essential nutrients (Figure 21.16). As with nitrogen, the exact pattern of dynamics during decomposition is a function of the nutrient content of the litter and the demand for the nutrient by the microbial population.

21.6 | Decomposition Proceeds as Plant Litter Is Converted into Soil Organic Matter

We have thus far discussed a process of decomposition in which decomposer organisms consume plant litter. In doing so, they mineralize nutrients and alter the chemical composition of residual organic matter. This process continues as the litter degrades into a dark brown or black homogeneous organic matter known as humus. As this organic matter becomes embedded in the soil matrix, it is referred to as soil organic matter (see Section 4.7). Most field studies, including those presented earlier, are relatively short term, examining the initial stages of decomposition over a period of 1 or 2 years at most. The work of Bjorn Berg and colleagues at the Swedish University of Agricultural Sciences, however, provides one of the most complete pictures of long-term decomposition of plant litter.

Berg and colleagues examined the decomposition of leaf (needle) litter in a Scots pine forest in central Sweden over a period of five years. Mass loss continued over the five-year study, and carbon continued to decline as carbon dioxide was lost to the atmosphere through microbial respiration (Figure 21.17). Simultaneously, the need for decomposers to convert plant organic matter with a carbon to nitrogen (C:N) ratio of 134:1 into microbial biomass (C:N of approximately 10:1) results in a prolonged period of immobilization (Figure 21.17a) and increasing nitrogen content of the residual organic matter (Figure 21.17b). The net result is an increasing nitrogen concentration and a declining C:N of the residual organic matter (Figure 21.17c).

As the litter decomposition progresses, the rate of mass loss (decomposition rate) declines. This decline is a result of the preferential consumption of easily digested, high-energy-yielding carbon compounds (termed labile organic matter), leaving increasingly

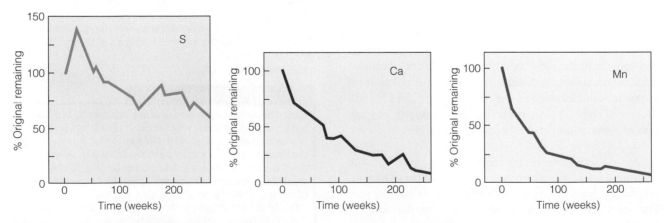

Figure 21.16 | Patterns of immobilization and mineralization for sulfur (S), calcium (Ca), and manganese (Mn) in decomposing needles of Scots pine. Results are from a litterbag experiment during a period of five years.

(Adapted from Staff and Berg 1982.)

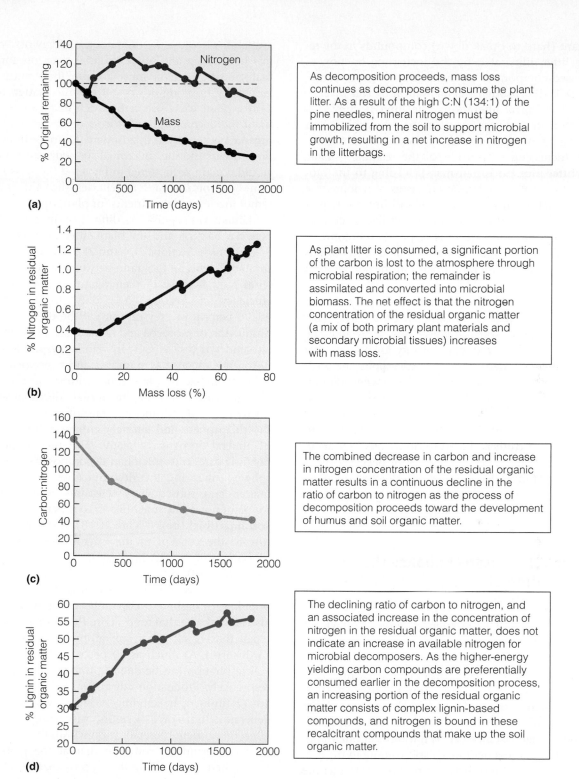

(a)

As decomposition proceeds, mass loss continues as decomposers consume the plant litter. As a result of the high C:N (134:1) of the pine needles, mineral nitrogen must be immobilized from the soil to support microbial growth, resulting in a net increase in nitrogen in the litterbags.

(b)

As plant litter is consumed, a significant portion of the carbon is lost to the atmosphere through microbial respiration; the remainder is assimilated and converted into microbial biomass. The net effect is that the nitrogen concentration of the residual organic matter (a mix of both primary plant materials and secondary microbial tissues) increases with mass loss.

(c)

The combined decrease in carbon and increase in nitrogen concentration of the residual organic matter results in a continuous decline in the ratio of carbon to nitrogen as the process of decomposition proceeds toward the development of humus and soil organic matter.

(d)

The declining ratio of carbon to nitrogen, and an associated increase in the concentration of nitrogen in the residual organic matter, does not indicate an increase in available nitrogen for microbial decomposers. As the higher-energy yielding carbon compounds are preferentially consumed earlier in the decomposition process, an increasing portion of the residual organic matter consists of complex lignin-based compounds, and nitrogen is bound in these recalcitrant compounds that make up the soil organic matter.

Figure 21.17 | Patterns of **(a)** mass loss and nitrogen dynamics, **(b)** changes in nitrogen concentration of residual organic matter, **(c)** ratio of carbon to nitrogen, and **(d)** concentration of lignin in residual organic matter during a five-year experiment examining the decomposition of Scots pine leaf litter in central Sweden.

(Data from Berg et al. 1982.)

recalcitrant (hard to break down) compounds in the remaining litter, illustrated by the increasing fraction of lignin-based compounds in the residual organic matter (Figure 21.17d). Through fragmentation by soil invertebrates and chemical alterations, the litter is converted into soil organic matter. As microbes die, chitin and other recalcitrant components of their cell walls comprise an increasing proportion of the residual organic matter (litter plus microbial mass), leading to the production of humus. All of these processes contribute to a gradual reduction in the quality of soil organic matter as it ages. The C:N ratio continues to decline as decomposition proceeds; the carbon is respired, and some of the mineralized nitrogen is incorporated into humus and combines with mineral particles in the soil to form colloids.

The decline in C:N ratio is not, however, an indicator of increased nitrogen availability, because the nitrogen becomes incorporated into chemical structures that are recalcitrant—carbon compounds of very low quality that yield little energy to decomposers. At that point the original plant litter has been degraded to a form where decomposition proceeds very slowly; the rate is largely a function of carbon quality. Soil organic matter typically has a residence time of 20 to 50 years, although it can range from 1 or 2 years in a cultivated field to thousands of years in environments that support slow rates of decomposition (dry or cold). Humus decomposes very slowly; but due to its abundance, it represents a significant portion of the carbon and nutrients released from soils.

21.7 | Plant Processes Enhance the Decomposition of Soil Organic Matter in the Rhizosphere

Over a century ago, the soil scientist Lorenz Hiltner put forward the term **rhizosphere** to describe a region of the soil where plant roots function. It is an active zone of root growth and death, characterized by intense microbial and fungal activity. Decomposition in the rhizosphere is more rapid than in the bulk soil. The rhizosphere makes up virtually all of the soil in fine-rooted grasslands, where the average distance between roots is about 1 mm. Root density is far less per unit of soil volume in forested ecosystems (often 10 mm between roots). Roots alter the chemistry of the rhizosphere by secreting carbohydrates into the soil. These carbohydrates may account for as much as 40 percent of dry matter production of plants. This large expenditure of carbon must be of fundamental importance for plants to justify the significant trade-off in carbon allocation.

The growth of bacteria in the rhizosphere is supported by the abundant source of high-quality, energy-rich carbon of the root exudates (exuded matter). Bacterial growth is limited most strongly by nutrient availability because the exudates are energy rich, but very low in nitrogen and other essential nutrients for microbial growth. Bacteria must acquire their nutrients for growth by breaking down soil organic matter. In other words, plant roots use carbon-rich exudates to supplement the decomposition process of low carbon quality organic matter in the rhizosphere. Nutrient immobilization occurs during microbial growth, and nutrients would remain sequestered in bacterial biomass if predation by protozoa and nematodes did not constantly remobilize essential nutrients for plant uptake.

Unlike the significant difference in the C:N ratio between bacteria and the plant litter they feed on (see discussions in Sections 21.4 and 21.5), there is a relatively small difference in C:N ratio between predators and bacterial prey. Because of their relatively low assimilation efficiency (see Section 20.8), only 10–40 percent and 50–70 percent of prey carbon will be used for biomass production of protozoa and nematodes, respectively. The remainder is lost as CO_2 in cellular respiration. Excess nitrogen is excreted as ammonia and hence readily available for uptake by plant roots within the rhizosphere. The interplay between microbial decomposers and microbivores determines the rate of nutrient cycling in the rhizosphere and strongly enhances the availability of mineral nutrients to plants. This process of supplementing carbon to microbial decomposers in the rhizosphere, enhancing the decomposition of soil organic matter, and subsequently releasing mineral nutrients for plant uptake by microbial grazers is referred to as the **soil microbial loop** (Figure 21.18). It is similar in structure to the concept of the microbial loop in aquatic ecosystems discussed in Section 21.8 and Chapter 24 (Section 24.10).

Populations of protozoa and nematodes fluctuate. As populations decline, their readily decomposable tissues also enter the detrital food chain. Protozoan biomass can equal that of all other soil animal groups except for earthworms. As much as 70 percent and 15 percent of soil respiration can be due to protozoa and nematodes, respectively. Production rates of microbivores can reach 10–12 times their standing biomass with a minimum generation time of 2–4 hours, suggesting a significant effect on nutrient mineralization within the soil.

Some estimates conclude that at the global scale, rhizosphere processes utilize approximately 50 percent of the energy fixed by photosynthesis in terrestrial ecosystems, contribute roughly 50 percent of the total CO_2 emitted from terrestrial ecosystems, and mediate virtually all aspects of nutrient cycling.

21.8 | Decomposition Occurs in Aquatic Environments

Decomposition in aquatic ecosystems follows a pattern similar to that in terrestrial ecosystems but with some

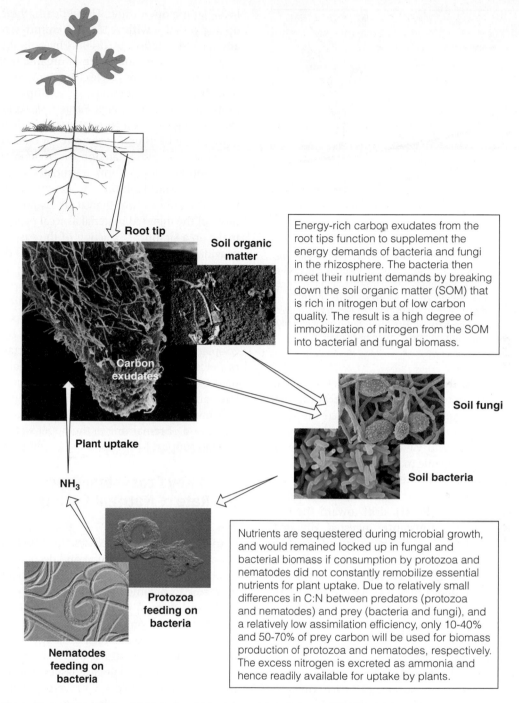

Root tip

Soil organic matter

Carbon exudates

Energy-rich carbon exudates from the root tips function to supplement the energy demands of bacteria and fungi in the rhizosphere. The bacteria then meet their nutrient demands by breaking down the soil organic matter (SOM) that is rich in nitrogen but of low carbon quality. The result is a high degree of immobilization of nitrogen from the SOM into bacterial and fungal biomass.

Plant uptake

Soil fungi

Soil bacteria

NH_3

Protozoa feeding on bacteria

Nematodes feeding on bacteria

Nutrients are sequestered during microbial growth, and would remained locked up in fungal and bacterial biomass if consumption by protozoa and nematodes did not constantly remobilize essential nutrients for plant uptake. Due to relatively small differences in C:N between predators (protozoa and nematodes) and prey (bacteria and fungi), and a relatively low assimilation efficiency, only 10-40% and 50-70% of prey carbon will be used for biomass production of protozoa and nematodes, respectively. The excess nitrogen is excreted as ammonia and hence readily available for uptake by plants.

Figure 21.18 | Illustration of the soil microbial loop in which energy-rich carbon exudates from the plant roots within the rhizosphere enhance the growth of microbial populations and the breakdown of soil organic matter. Nutrients immobilized in microbial biomass are then liberated to the soil through predation by microbivores, providing increased mineral nutrients to support plant growth.

major differences influenced by the watery environment. As in terrestrial environments, decomposition involves leaching, fragmentation, colonization of detrital particles by bacteria and fungi, and consumption by detritivores and microbivores. In coastal environments, permanently submerged plant litters decompose more rapidly than do those on the marsh surface because (1) they are more accessible to detritivores and (2) the stable physical envi-

ronment is more favorable to microbial decomposers than are the alternative periods of wetting and drying that characterize tidal environments (Figure 21.19).

In flowing-water ecosystems, aquatic fungi colonize leaves, twigs, and other particulate matter. One group of aquatic arthropods, called shredders, fragment the organic particles, in the process eating bacteria and fungi on the surface of the litter (see Chapter 24). Downstream, another

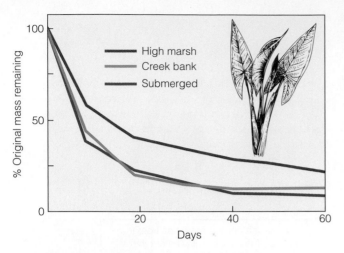

Figure 21.19 | Decomposition of leaves of arrow arum (*Peltandra virginica*) in a tidal freshwater marsh. Decomposition is a measure of the percent of original mass remaining in litterbags under three conditions: irregularly flooded high marsh exposed to alternate periods of wetting and drying, creek bed flooded two times daily (tidal), and permanently submerged. The litter that is consistently wet (high marsh) has the highest rate of decomposition.

(Adapted from Odum and Haywood 1978.)

group of invertebrates, the filtering and gathering collectors, filter from the water fine particles and fecal material left by the shredders. Grazers and scrapers feed on algae, bacteria, fungi, and organic matter collected on rocks and large debris (all discussed in Chapter 24). Algae take up nutrients and dissolved organic matter from the water.

In still, open water of ponds and lakes and in the ocean, dead organisms and other organic material, called particulate organic matter (POM), drift toward the bottom. On its way, POM is constantly ingested, digested, and mineralized until much of the organic matter settles on the bottom in the form of humic compounds (see discussion of humus formation in Chapter 4). How much depends partly on the depth of the water the particulate matter falls through. In shallow water, much of it may arrive in relatively large packages to be further fragmented and digested by bottom-dwelling detritivores such as crabs, snails, and mollusks.

Bacteria work on the bottom, or benthic, organic matter. Bacteria living on the surface can carry on aerobic respiration; but within a few centimeters below the surface of the sediment, the oxygen supply is exhausted. Under this anoxic condition, a variety of bacteria capable of anaerobic respiration take over decomposition, which proceeds at a much slower rate than in the aerobic environment of the surface and shallow sediments (see Figure 21.9).

Aerobic and anaerobic decomposition in the benthic environment form only a part of the decomposition process. Dissolved organic matter (DOM) in the water column also provides a source of fixed carbon for decomposition. Major sources of DOM are the free-floating macroalgae, phytoplankton, and zooplankton in-

habiting the open water. Upon death, their bodies break up and dissolve within 15 to 30 minutes, too rapidly for any bacterial action to occur. Phytoplankton and other algae also excrete quantities of organic matter at certain stages of their life cycles, particularly during rapid growth and reproduction. For example, during photosynthesis the marine alga *Fucus vesiculosus* produces an exudate high in carbon content. This dissolved organic matter then becomes a substrate for bacterial growth.

Ciliates and zooplankton eat bacteria and in turn excrete nutrients in the form of exudates and fecal pellets in the water. Zooplankton, in the presence of abundant food, consumes more than it needs and excretes half or more of the ingested material as fecal pellets. These pellets make up a significant fraction of the suspended material, providing a substrate for further bacterial decomposition. An important component of the aquatic nutrient cycle is the microbial loop. The microbial loop is the trophic pathway through which dissolved organic matter is reintroduced to the food web by being incorporated into bacteria, which in turn are consumed by ciliates and zooplankton (see Section 24.10 for a detailed discussion of the microbial loop in aquatic food chains). The lighter nature of dissolved organic matter (in contrast to particulate) allows it to remain longer in the upper waters, so that nutrients entering the microbial loop have a greater likelihood of remaining in the upper waters of the photic zone to support further primary productivity.

21.9 | Key Ecosystem Processes Influence the Rate of Nutrient Cycling

You can see from Figure 21.1 that the internal cycling of nutrients through the ecosystem depends on the processes of primary production and decomposition. Primary productivity determines the rate of nutrient transfer from inorganic to organic form (nutrient uptake), and decomposition determines the rate of transformation of organic nutrients into inorganic form (net mineralization rate). Therefore, the rates at which these two processes occur directly influence the rates at which nutrients cycle through the ecosystem (see Ecological Issues: Nitrogen Fertilizers). But how do these two key processes interact to limit the rate of the internal cycling of nutrients through the ecosystem? The answer lies in their interdependence.

For example, consider the cycling of nitrogen, an essential nutrient for plant growth. The direct link between soil nitrogen availability, rate of nitrogen uptake by plant roots, and the resulting leaf nitrogen concentrations was discussed in Chapter 6 (see Figure 6.25). The maximum rate of photosynthesis is strongly correlated with nitrogen concentrations in the leaves (see Figure 6.24) because certain compounds directly involved in photosynthesis (e.g., rubisco and chlorophyll) contain a large portion of leaf nitrogen. Thus, availability of nitrogen in the soil (sediments or water in the case of aquatic ecosystems)

In natural ecosystems, there is a tight link between the processes of net primary productivity (NPP) and decomposition. NPP determines the quantity and quality of organic matter available to decomposer populations. In turn, the process of decomposition determines the rate of nutrient release to the soil, constraining plant uptake and NPP. In agriculture, this balance is disrupted. Plants, and the nutrients they contain, are harvested as crops; and the organic matter does not return to the soil surface to undergo the process of decomposition and mineralization. As a result, to maintain an adequate supply of soil nutrients for continued crop production, nutrient supplements (fertilizers) must be added to the soil.

Natural fertilizers such as manures and ground animal bones have been used since ancient times, but it was not until the development of the modern study of plants, soils, and the chemical requirements for growth in the mid-19th century that an understanding of the specific elemental needs of plants began to emerge. And with that understanding came the first production of fertilizers from inorganic chemical sources.

Early scientific study established three elements as necessary in large quantities for plant growth: potassium (K), phosphorus (P), and nitrogen (N). These inorganic fertilizer elements were all originally mineral deposits. From the mid-1800s to the present day, various deposits of phosphate rocks and potash (potassium hydroxide: KOH) have been found to provide adequate sources of the elements phosphorus and potassium. Such was not the case with nitrogen. One source of mineral nitrogen, saltpeter ($NaNO_3$), found in deposits in Chile (South America), accounted for more than 60 percent of the world's supply for most of the 19th century. Although other sources of nitrogen were discovered periodically, they were usually depleted within a few years. Ammonia (NH_4^+) and nitrates (NO_3^-) were also produced from the distillation of coal and as industrial by-products of other chemical processes; however, these quantities were too limited to provide a significant source of nitrogen fertilizer.

As world population increased, the growing demand for food was accompanied by an increasing need for fertilizers. Using fertilizer with a high nitrogen content provided good crop yield, but by the late 1800s there was growing concern about depletion of the sources of nitrogen for chemical fertilizer. What was needed was an effective means of tapping the virtually limitless source of nitrogen contained in the atmosphere: nitrogen gas (N_2).

Although the reaction between nitrogen gas and hydrogen gas (H_2) to produce ammonia gas (NH_3) had been known for many years, the yields were very small and the reaction very slow. In the early 20th century, however, the German physical chemist Fritz Haber developed the synthetic ammonia process, which made the manufacture of ammonia economically feasible. Carl Bosch, an industrial chemist, soon translated the process developed by Haber for synthesizing liquid ammonia into a large-scale process using a catalyst and high-pressure methods. By 1913, a chemical plant was operating in Germany, producing ammonia using what has now become known as the Haber–Bosch process.

In the 1920s, the first ammonia-producing plants based on the Haber–Bosch process were built in the United States and in Europe outside of Germany. There was a general consolidation of smaller chemical companies into larger ones, which seemed to provide cheaper and more efficient methods of producing ammonia. In the 1930s, American agriculture developed methods for adding ammonia directly to the soil as fertilizers. In addition, better processes of carrying on the chemical reaction continued to be developed. Although ammonia from these plants was still more expensive to use in fertilizers than the ammonia from by-products of other reactions, the advent of World War II increased demand and led to still cheaper and more efficient methods.

The Haber–Bosch process has changed the way nitrogen fertilizers are produced and used, and it was an important part of the "Green Revolution" of the 20th century. The use of ammonia in fertilizer has made it the second most important chemical in the United States. It is the most important source of nitrogen in fertilizers today. Because of the explosive growth in world population over the past century, the demand for nitrogen-based fertilizers has grown exponentially—an increase of more than 400 percent since the 1940s. It is estimated that the nitrogen content of more than one-third of the protein consumed by the world's population is a product of the Haber–Bosch process.

The bounty of food produced as a result of the synthesis of nitrogen fertilizers, however, does not come without an environmental cost. Nitrates pollute drinking water; and through farm runoff, nitrogen also makes its way into streams and rivers and eventually to the oceans. There, it disrupts the normal constraints on primary productivity that result from the seasonal cycling of nitrogen in these ecosystems (see Section 21.8) and leads to the explosive growth of algae by a process known as eutrophication (see Section 24.4). The high inputs of organic matter to the waters result in a corresponding increase in decomposition and heterotrophic respiration, dramatically reducing the oxygen content of the water. The result is a decline in other organisms, particularly fish, that cannot survive under the low-oxygen conditions.

Almost a century after the development of the Haber–Bosch process, in June 2003, the Pew Oceans

continued on page 458

Commission released its final report to Congress and the nation. Among the conclusions presented in the report is that nitrogen fertilizer has become the main source of pollution in the ocean. The human ingenuity used in developing the process that enables us to feed the world's population is now faced with an equally difficult task—reducing the environmental consequences created by the abundant production and use of this most essential element for plant growth.

1. How do chemical fertilizers, such as ammonium and nitrate, differ from natural fertilizers such as manures, ground animal bone, or compost?
2. Using the graphic presented in Figure 21.1 as a guide, diagram the steps involved in the cycling of nutrients (such as nitrogen) in an agricultural ecosystem.

directly affects rates of ecosystem primary productivity via the influence of nitrogen on photosynthesis and carbon uptake.

A low availability of soil nitrogen reduces net primary production (the total production of plant tissues) as well as the nitrogen concentration of the plant tissues produced (again, see Figure 6.25). Thus, the reduced availability of soil nitrogen influences the input of dead organic matter to the decomposer food chain (see Section 20.9) by reducing both the total quantity of dead organic matter produced and its nutrient concentration. The net effect is a lower return of nitrogen in the form of dead organic matter.

Both the quantity and quality of organic matter as a food source for decomposers directly influence the rates of decomposition and nitrogen mineralization (nutrient release). Lower nutrient concentrations in the dead organic matter promote immobilization of nutrients from the soil and water to meet the nutrient demands of decomposer populations (see Figure 21.15). This immobilization effectively reduces nutrient availability to the plants (reduces the rate of net mineralization; see Section 21.5), adversely affecting primary productivity.

One can now appreciate the feedback system that exists in the internal cycling of nutrients within an ecosystem (Figure 21.20). Reduced nutrient availability can have the combined effect of reducing both the nutrient concentration of plant tissues (primarily leaf tissues) and net primary productivity (NPP). This reduction lowers the total amount of nutrients returned to the soil in dead organic matter. The reduced quantity and quality (nutrient concentration) of organic matter entering the decomposer food chain increase immobilization and reduce the availability of nutrients for uptake by plants. In effect, low nutrient availability begets low nutrient availability. Conversely, high nutrient availability encourages high plant tissue concentrations and high NPP. In turn, the associated high quantity and quality of dead organic matter encourage high rates of net mineralization and nutrient supply in the soil.

The feedback between litter quality, nutrient cycling, and NPP is illustrated in the work of John Pastor of the University of Minnesota. Pastor and colleagues examined aboveground production and nutrient cycling (nitrogen and phosphorus) in a series of forest stands along a gradient of soil texture on Blackhawk Island, Wisconsin. Tree species producing higher-quality litter (lower C:N ratio) dominated sites with a progressively finer soil texture (silt and clay content; see Chapter 4). The higher-quality litter resulted in a higher rate of nutrient mineralization (release; Figure 21.21a). Higher rates of nutrient availability in turn resulted in a higher rate of primary productivity (see Figure 21.20) and nutrient return in litterfall (Figure 21.21b). The net effect was to increase the rate at which nitrogen and phosphorus cycle through the forest stands. The changes in species composition and litter quality along the soil gradient were directly related to the influence of soil texture on plant available moisture (see Section 4.7).

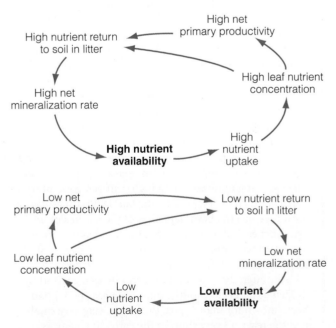

Figure 21.20 | Feedback that occurs between nutrient availability, net primary productivity, and nutrient release in decomposition for initial conditions of low and high nutrient availability. (Adapted from Chapin 1980.)

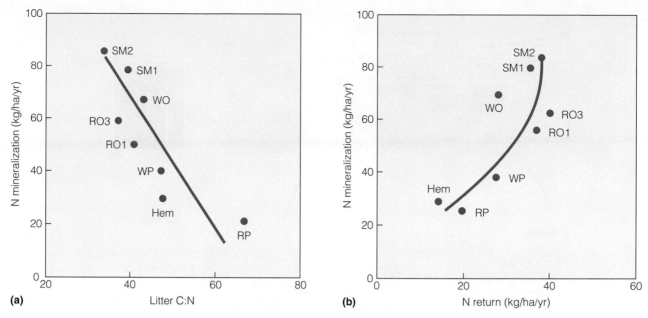

Figure 21.21 | Relationship between **(a)** litter quality (C:N) and nitrogen mineralization rate (N availability) and **(b)** nitrogen mineralization rate and nitrogen returned in annual litterfall for a variety of forest ecosystems on Blackhawk Island, Wisconsin. Abbreviations refer to the dominant trees in each stand: Hem, hemlock; RP, red pine; RO, red oak; WO, white oak; SM, sugar maple; WP, white pine.

(Adapted from Pastor et al. 1984.)

21.10 | Nutrient Cycling Differs between Terrestrial and Open-Water Aquatic Ecosystems

Nutrient cycling is an essential process in all ecosystems and represents a direct (cyclic) link between net primary productivity (NPP) and decomposition. However, the nature of this link varies among ecosystems, particularly between terrestrial and aquatic ecosystems.

In virtually all ecosystems, there is a vertical separation between the zones of production (photosynthesis) and decomposition (Figure 21.22). In terrestrial ecosystems, the plants themselves bridge this physical separation between the decomposition zone at the soil surface and the productivity zone in the plant canopy; the plants physically exist in both zones. The root systems provide access to the nutrients made available in the soil through decomposition, and the vascular system

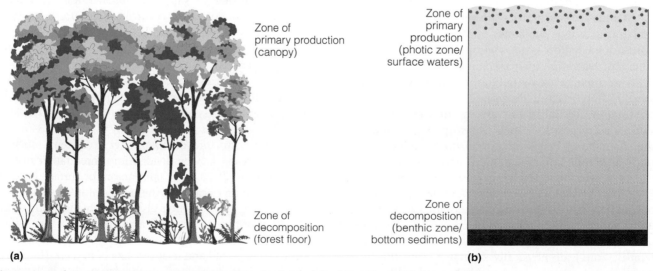

Figure 21.22 | Comparison of the vertical zones of production and decomposition in **(a)** a terrestrial (forest) and **(b)** an open-water (lake) ecosystem. In the terrestrial ecosystem, the two zones are linked by the vegetation (trees). However, this is not the case in the lake ecosystem.

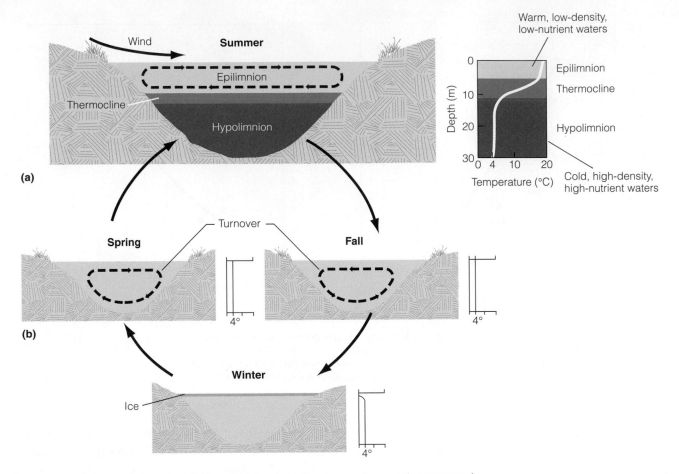

Figure 21.23 | Seasonal dynamics in the vertical structure of an open-water aquatic ecosystem in the temperate zone. Solid arrows track seasonal changes, and dashed arrows show the circulation of waters. Winds mix the waters within the epilimnion during the summer **(a)**, but the thermocline isolates this mixing to the surface waters. With the breakdown of the thermocline during the fall and spring months, turnover occurs, allowing the entire water column to become mixed **(b)**. This mixing allows nutrients in the epilimnion to be brought up to the surface waters.

within the plant transports these nutrients to the sites of production (canopy).

In aquatic ecosystems, however, plants do not always function to bridge the zones of production and decomposition. In shallow-water environments of the shoreline, emergent vegetation such as cattails, cordgrasses (*Spartina* spp.), and sedges are rooted in the sediments. Here, as in terrestrial ecosystems, the zones of decomposition and production are linked directly by the plants. Likewise, submerged vegetation, such as sea grasses, is rooted in the sediments. The plants extend up the water column into the photic zone (see Figure 24.16)—the shallower waters where light levels support higher productivity. However, as water depths increase, primary production is dominated by free-floating phytoplankton within the upper waters (photic zone). In deeper water, the zones of decomposition in the bottom sediments and waters (the benthic zone) are physically separated from the surface waters, where temperatures and light availability support primary productivity. This physical separation between the zones where nutrients become available through

decomposition and the zone of productivity where nutrients are needed to support photosynthesis and plant growth is a major factor controlling the productivity of open-water ecosystems (e.g., lakes and oceans).

To understand how nutrients are transported vertically from deeper waters to the surface, where temperature and light conditions can support primary productivity, we must examine the vertical structure of the physical environment in open-water ecosystems as first presented in Chapter 3 (see Section 3.4). As presented briefly in Chapters 3 and 16, the vertical structure of open-water ecosystems, such as lakes or oceans, can be divided into three rather distinct zones (Figure 21.23). The epilimnion, or surface water, is relatively warm because of the interception of solar radiation. The oxygen content is also relatively high, due to the diffusion of oxygen from the atmosphere into the surface waters (see Section 3.6). In contrast, the hypolimnion, or deep water, is cold and relatively low in oxygen. The transition zone between the surface and deep waters is characterized by a steep temperature gradient called the thermocline. In effect, the vertical structure can be represented as a warm, low-density surface layer of water on top of a denser, colder layer of deep water; these layers are separated by the rather thin zone of the thermocline. This vertical structure and physical separation of the epilimnion and hypolimnion have an important influence on the distribution of nutrients and subsequent patterns of primary productivity in aquatic ecosystems. The colder, deeper waters where decomposition occurs are relatively nutrient rich, but temperature and light conditions cannot support high productivity. In contrast, the surface waters are relatively nutrient poor; however, this is the zone where temperatures and light can support high productivity.

Although winds blowing over the water surface cause turbulence that mixes the waters of the epilimnion, this mixing does not extend into the colder, deeper waters because of the thermocline. As autumn and winter approach in the temperate and polar zones, the amount of solar radiation reaching the water surface decreases, and the temperature of the surface water declines. As the water temperature of the epilimnion approaches that of the hypolimnion, the thermocline breaks down, and mixing throughout the profile can take place (Figure 21.23). If surface waters become cooler than the deeper waters, they will begin to sink, displacing deep waters to the surface. This process is called turnover (see Section 3.4). With the breakdown of the thermocline and mixing of the water column, nutrients are brought up from the bottom to the surface waters. With the onset of spring, increasing temperatures and light in the epilimnion give rise to a peak in productivity due to the increased availability of nutrients in the surface waters. As the spring and summer progress, the nutrients in the surface water are used, reducing the nutrient content of the water, and a subsequent decline in productivity occurs. The annual cycle of productivity in these ecosystems (Figure 21.24) is a direct result of thermocline dynamics and the consequent behavior of the vertical distribution of nutrients.

21.11 | Water Flow Influences Nutrient Cycling in Streams and Rivers

Inputs in the form of dead organic matter from adjacent terrestrial ecosystems (leaves and woody debris), rainwater, and subsurface seepage bring nutrients into streams. Although the internal cycling of nutrients follows the same general pathway as that discussed for terrestrial and open-water ecosystems (see Figure 21.1), the continuous, directional movement of water affects nutrient cycling in stream ecosystems. Jack Webster of the University of Georgia was the first to note that because nutrients are continuously being transported downstream, a spiral rather than a cycle better represents the cycling of nutrients. He coined the term **nutrient spiraling** to describe this process.

Nutrients in terrestrial and open-water ecosystems are recycled more or less in place. An atom of nutrients passes from the soil or water column to plants and consumers

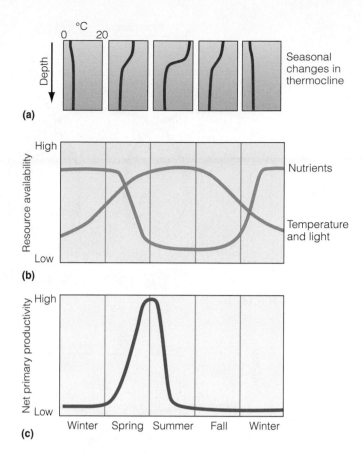

Figure 21.24 | Seasonal dynamics of **(a)** the thermocline and associated changes in **(b)** the availability of light and nutrients, and **(c)** net primary productivity of the surface waters.

and passes back to the soil or water in the form of dead organic matter. Then it is recycled within the same location within the ecosystem, although losses do occur. Cycling essentially involves time. Flowing water has an added element: a spatial cycle. Nutrients in the form of organic matter are constantly being carried downstream. How quickly these materials are carried downstream depends on how fast the water moves and what physical and biological factors hold nutrients in place. Physical retention involves storage in wood detritus such as logs and snags, in debris caught in pools behind logs and boulders, in sediments, and in patches of aquatic vegetation. Biological retention occurs through the uptake and storage of nutrients in animal and plant tissue.

The processes of recycling, retention, and downstream transport may be pictured as a spiral lying horizontally (longitudinally) in a stream (Figure 21.25). One cycle in the spiral is the uptake of an atom of nutrient, its passage through the food chain, and its return to water, where it is available for reuse. Spiraling is measured as the distance needed to complete one cycle. The longer the distance required, the more open the spiral; the shorter the distance, the tighter the spiral. If dead leaves and other debris are physically held in place long enough to allow organisms to process the organic matter, the

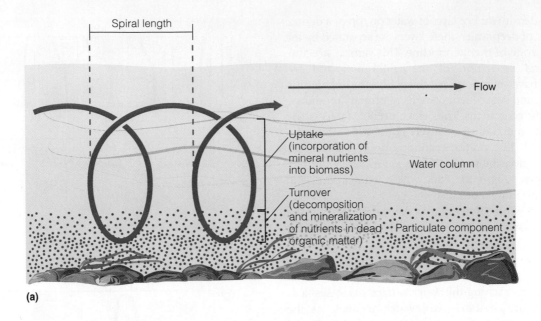

(a)

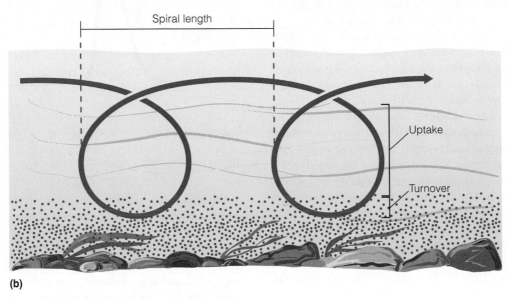

(b)

Figure 21.25 | Nutrient spiraling between organic matter and the water column in a stream ecosystem. Uptake and turnover take place as nutrients flow downstream. The tighter the spiraling, the longer the nutrients remain in place. **(a)** Tight spiraling; **(b)** open spiraling.

(Adapted from Newbold et al. 1982.)

spiral is tight. This type of physical retention is especially important in fast headwater streams, which can rapidly lose organic matter downstream. Organisms can function to both open and tighten the spiral. Organisms that shred and fragment the organic matter can open the spiral by facilitating the transport of organic materials downstream. Other organisms tighten the spiral by physically storing dead organic matter.

J. D. Newbold and colleagues at Oak Ridge National Laboratory in Tennessee experimentally determined how quickly phosphorus moved downstream in a small woodland brook on the Walker Branch watershed. They determined that phosphorus moved downstream at the rate of 10.4 m a day and cycled once every 18.4 days. The average downstream distance of 1 cycle (spiral) was 190 m. In other words, 1 atom of phosphorus on the average completed 1 cycle from the water compartment and back again for every 190 m of downstream travel.

21.12 | Land and Marine Environments Influence Nutrient Cycling in Coastal Ecosystems

Coastal ecosystems are among the most productive environments. Water from most streams and rivers eventually drains into the oceans, and the place where this freshwater

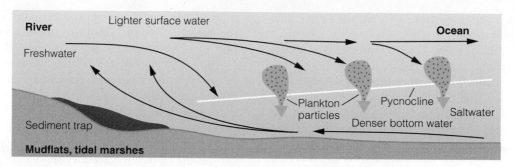

Figure 21.26 | Circulation of freshwater and saltwater in an estuary creates a nutrient trap. A salty wedge of intruding seawater on the bottom produces a surface flow of lighter freshwater and a counterflow of heavier brackish water. These layers are physically separated by variations in water density arising from both salt concentration and temperature differences. The zone of maximum vertical difference in density, the pycnocline, functions much like the thermocline in lake ecosystems. Living and dead particles settle through the pycnocline into the countercurrent and are carried up-estuary along with their nutrient content, conserving the nutrients within the estuary rather than being flushed out to sea.

(Adapted from Correll 1978.)

joins saltwater is called an estuary (see Section 3.10). Estuaries are semi-enclosed parts of the coastal ocean where seawater is diluted and partially mixed with water coming from the land. As the rivers meet the ocean, the current velocity drops, and sediments are deposited within a short distance (referred to as a sediment trap; Figure 21.26). The buildup of sediments creates alluvial plains about the estuary, giving rise to mudflats and salt marshes that are dominated by grasses and small shrubs rooted in the mud and sediments (see Chapter 24 for detailed descriptions of these ecosystems). Nutrient cycling in these ecosystems differs from that of terrestrial, open-water, and stream ecosystems discussed thus far. In a way, it combines some features of each. As in terrestrial ecosystems, the dominant plants are rooted in the sediments and therefore function to link the zones of decomposition and primary production. Submerged plants take up nutrients from the sediments as well as directly from the water column. As in streams and rivers, the directional (horizontal) flow of water transports organic matter (energy) and nutrients both into (inputs) and out of (outputs) the ecosystem.

Nutrients are carried into the coastal marshes by precipitation, surface water (streams and rivers), and groundwater. In addition, the rise and fall of water depth with the tidal cycle serves to flush out salts and other toxins from the marshes and brings in nutrients from the coastal waters by a process referred to as the **tidal subsidy.** The tidal cycle also serves to replace oxygen-depleted waters within the surface sediments with oxygenated water.

The salt marsh is a detrital system with only a small portion of its primary production being consumed by herbivores. Almost three-quarters of the detritus (dead organic matter) produced in the salt-marsh ecosystem is broken down by bacteria and fungi. Nearly 50 percent of the total NPP is lost through respiration by microbial decomposers. The low oxygen content of the sediments favors anaerobic bacteria. They can carry on their metabolic functions without oxygen by using inorganic compounds such as sulfates rather than oxygen in the process of fermentation (see Sections 6.1 and 7.2).

A substantial portion (usually 20 to 45 percent) of the NPP of a salt marsh is exported to adjacent estuaries. The exact nature of this exchange depends on the geomorphology of the basin (its shape and nature of the opening to the sea) and the magnitude of tidal and freshwater flows (fluxes). Each salt marsh apparently differs in the way carbon and other nutrients move through the food web as well as in the route taken and amount of nutrients exported. Some salt marshes are dependent on tidal exchanges and import more than they export, whereas others export more than they import. Some material may be exported to the estuary as mineral nutrients or physically as detritus; as bacteria; or as fish, crabs, and intertidal organisms within the food web. Although inflowing water from rivers and coastal marshes carries mineral nutrients into the estuary, primary production is regulated more by internal nutrient cycling than by external sources. This internal cycling involves the release of nutrients through decomposition and mineralization within the bottom sediments, as well as excretion of mineralized nutrients by herbivorous zooplankton.

As in the tidal marshes, nutrients and oxygen are carried into the estuary by the tides. Typical estuaries maintain a "salt wedge" of intruding seawater on the bottom, producing a surface flow of freshwater and a counterflow of more brackish, heavier water (see Figure 21.26). These layers are physically separated by variations in water density arising from both salt concentration and temperature differences. The zone of maximum vertical difference in density, the **pycnocline,** functions much like the thermocline in lake ecosystems. Living and dead particles settle through the pycnocline into the countercurrent and are carried up the estuary along with their nutrient content, thus conserving the nutrients within the estuary rather than being flushed out to sea.

The regular movement of freshwater and saltwater into the estuary, coupled with the shallowness and turbulence, generally allows for sufficient vertical mixing. In deeper estuaries, a thermocline can form during the summer months. In this case, the seasonal pattern of vertical mixing and nutrient cycling will be similar to the pattern discussed for open-water ecosystems in Section 21.10.

21.13 | Surface Ocean Currents Bring About Vertical Transport of Nutrients

The global pattern of ocean surface currents presented in Chapter 2 (Figure 2.15) influences patterns of surface-water temperature, productivity, and nutrient cycling. The Coriolis effect drives the patterns of surface currents. But how deep does this lateral movement of water extend vertically into the water column? In general, the lateral flow is limited to the upper 100 m, but in certain regions the lateral movement can bring about a vertical circulation or upwelling of water. As presented in Chapter 3 (Section 3.9), along the western margins of the continents, the surface currents flow along the coastline and toward the equator (see Figure 2.15). At the same time, these surface waters are pushed offshore by the Coriolis effect. The movement of surface waters offshore results in deeper, more nutrient-rich waters being transported vertically to the surface (Figure 3.17a).

Surface currents give rise to a similar pattern of upwelling in the equatorial waters. As the two equatorial currents flow west, they are deflected to the right north of the equator and to the left south of the equator. Where this occurs, subsurface water is transported vertically, bringing cold waters, rich in nutrients, to the surface (see Figure 3.17b). These regions of nutrient-rich waters are highly productive (see Figure 20.10) and support some of the world's most important fisheries.

Summary

Nutrient Cycling | 21.1

As plants take up nutrients from the soil or water, they become incorporated into living tissues: organic matter. As the tissues senesce, the dead organic matter returns to the soil or sediment surface. Various decomposers transform the organic nutrients into mineral form, and they are once again available for uptake by plants. This process is called internal cycling.

Decomposition Processes | 21.2

Decomposition is the breakdown of chemical bonds formed during the construction of plant and animal tissues. Decomposition involves an array of processes including leaching, fragmentation, digestion, and excretion. An essential function of decomposers is the release of organically bound nutrients into an inorganic form. The wide variety of organisms involved in decomposition are classified into groups based on both function and size. The microflora, comprising the bacteria and fungi, are the group most commonly associated with decomposition. Bacteria can be grouped as either aerobic or anaerobic based on their requirement of oxygen to carry out respiration. Invertebrate detritivores are classified based on body size. Microbivores feed on bacteria and fungi.

Litterbag Studies | 21.3

Decomposers derive their energy and most of their nutrients by consuming organic compounds. Ecologists study the process of decomposition by designing experiments that follow the decay of dead plant and animal tissues through time. The most widely used approach is litterbags. A fixed amount of dead organic matter is placed in mesh bags, and the rate of loss is followed through time.

Factors Influencing Decomposition | 21.4

Microbial decomposers use carbon compounds contained in dead organic matter as a source of energy. Various carbon compounds differ in their quality as an energy source for decomposers. Glucose and other simple sugars are easily broken down and provide a high-quality source of carbon. Cellulose and hemicellulose, the main constituents of cell walls, are intermediate in quality, whereas the lignins are of very low quality and therefore decompose the slowest. The quality of dead organic matter as a food source for decomposers is influenced by the types of carbon compounds present and by the nutrient content. Because of the low quality of lignins as an energy source for decomposers, there is an inverse relationship between the lignin content of plant litter and its decomposition rate. The physical environment directly influences both macro- and micro-decomposers. Temperature and moisture greatly influence microbial activity. Low temperatures and moisture inhibit microbial activity. As a result, the highest decomposition rates occur during warm, wet conditions. The influence of temperature and moisture on decomposer activity results in geographic variation in rates of decomposition that relate directly to climate.

Nutrient Mineralization | 21.5

The nutrient quality of dead organic matter depends on its nutrient content. As microbial decomposers break down dead organic matter, they transform nutrients tied up in organic compounds into an inorganic form. This process is called nutrient mineralization. The same organisms responsible for mineralization reuse some of the nutrients they have produced, incorporating the inorganic nutrients into an organic form. This process is called nutrient immobilization. The difference between the rates of mineralization and immobilization is the net mineralization rate, which represents the net release of nutrients to the soil or water during decomposition. The rates of mineralization and immobilization during decomposition are related to the nutrient content of the dead organic matter being consumed.

Soil Organic Matter | 21.6

The process of decomposition continues as the litter degrades into a dark brown or black homogeneous organic matter known as humus. As this organic matter becomes embedded in the soil matrix, it is referred to as soil organic matter. Although high in nitrogen, humus decomposes very slowly due to the low quality of carbon compounds. Despite the slow rate of decomposition, due to its sheer abundance, the decomposition of humus represents a significant portion of the carbon and nutrients released from soils.

Rhizosphere | 21.7

The rhizosphere is an active zone of root growth and death, characterized by intense microbial and fungal activity. Decomposition in the rhizosphere is more rapid than in the bulk soil because plant roots use carbon-rich exudates to supplement the decomposition process of low carbon quality organic matter. Nutrients immobilized in bacteria biomass are then released to the soil as microbivores (protozoa and nematodes) that feed on the bacteria.

Decomposition in Aquatic Environments | 21.8

Decomposition in aquatic ecosystems varies as a function of water depth and flow rate. In flowing waters (streams and rivers), various specialized detritivores are involved in the breakdown of plant litter imported from adjacent terrestrial ecosystems. In open-water environments, dead organisms and other organic matter, called particulate organic matter (POM), drift downward to the bottom. On its way, POM is constantly being ingested, digested, and mineralized until much of the organic matter is in the form of humic compounds by the time it reaches the bottom sediments. Bacteria decompose organic matter on the bottom sediments, using aerobic or anaerobic respiration depending on the supply of oxygen. Dissolved organic matter (DOM) in the water column also provides a source of carbon for decomposers.

Rate of Nutrient Cycling | 21.9

The rate at which nutrients cycle through the ecosystem is directly related to the rates of primary productivity (nutrient uptake) and decomposition (nutrient release).

Environmental factors that influence these two processes will affect the rate at which nutrients cycle through the ecosystem.

Comparison between Terrestrial and Aquatic Ecosystems | 21.10

There is typically a vertical separation between the zones of primary production and decomposition. In terrestrial and shallow-water ecosystems, plants function to bridge this gap. In open-water ecosystems, there is a physical separation between these zones that limits nutrient availability in the surface waters. The thermocline functions to limit the movement of nutrients from the bottom (benthic) zone (cold) to the surface (warm) waters. During the winter season, the thermocline breaks down, allowing for a mixing of the water column and the movement of nutrients into the surface waters. This seasonality of the thermocline and mixing of the water column controls seasonal patterns of productivity in these ecosystems.

Stream Ecosystems | 21.11

The continuous, directional movement of water affects nutrient cycling in stream ecosystems. Because nutrients are continuously being transported downstream, a spiral rather than a cycle better represents the cycling of nutrients. One cycle in the spiral is the uptake of an atom of nutrient, its passage through the food chain, and its return to water, where it is available for reuse. The cycle length involved is related to the flow rate of the stream and to the physical and biological mechanisms available for nutrient retention.

Coastal Ecosystems | 21.12

Water from most streams and rivers eventually drains into the oceans, giving rise to estuary and salt-marsh ecosystems along the coastal environment. As in streams and rivers, the directional flow of water functions to transport both organic matter and nutrients both into and out of the ecosystem. The rise and fall of water depth with the tidal cycle serves to flush out salts and other toxins from the marshes and brings in nutrients from the coastal waters. The combined effect of the inward (toward the coast) movement of saltwater and the outward flow of freshwater develops a countercurrent that carries both living and dead particles and the nutrients they contain back toward the coastline. This mechanism conserves nutrients within the estuary and salt-marsh ecosystems.

Vertical Transport in Oceans | 21.13

The global pattern of surface currents brings about the transport of deep, nutrient-rich waters to the surface in the coastal regions. As surface currents move waters away from the western coastal margins, deep water moves to the surface, carrying nutrients with it. A similar pattern of upwelling occurs in the equatorial regions of the oceans, where surface currents move to the north and south.

Study Questions

1. Define decomposition. What are the major microbial decomposers of plant and animal material?
2. (a) How does the type of carbon compounds present in dead organic matter influence its quality as an energy source for decomposers? (b) How does lignin concentration influence the decomposition of plant litter?
3. How does the concentration of lignin-based compounds in the residual (remaining) organic matter change during the decomposition of plant litter?
4. (a) Contrast the processes of mineralization and immobilization. (b) What does an increase in the nitrogen content of decomposing plant tissues imply about the relative rates of mineralization and immobilization?
5. How does the initial ratio of carbon to nitrogen (C:N) of plant litter influence the relative rates of nitrogen mineralization and immobilization during the initial stages of decomposition?
6. What is the difference in the ratio of carbon to nitrogen (C:N) of plant litter and the tissues of microbial decomposers, and how does the difference influence the nutrient dynamics during decomposition?
7. Tree species that inhabit the forests of the far north (boreal forests) are characterized by low concentrations of nitrogen in their leaves. How might t his factor influence nitrogen cycling in these forests?
8. Contrast nutrient cycling in terrestrial and open-water aquatic ecosystems. What is the outstanding difference?
9. How do changes in the thermocline influence seasonal patterns of net primary productivity in lake ecosystems?
10. Go back to the global map of primary productivity of marine ecosystems in Figure 20.10. Identify areas of high productivity in the equatorial region that might be related to the process of upwelling.
11. How does the continuous, directional flow of water influence the cycling of nutrients in stream ecosystems?
12. What mechanism functions to conserve nutrients in estuary ecosystems?

Further Readings

Aber, J. D., and J. M. Melillo. 1991. *Terrestrial ecosystems.* Philadelphia: Saunders College Publishing.

This text provides an excellent introduction to the topic of nutrient cycling in terrestrial ecosystems.

Bertness, M. D. 1992. The ecology of a New England salt marsh. *American Scientist* 80:260–268.

This paper presents a fine overview of nutrient cycling in these important coastal ecosystems.

Hooper, D. U., and P. M. Vitousek. 1998. Effects of plant composition and diversity on nutrient cycling. *Ecological Monographs* 68:121–149.

Although technical, this paper presents a good example of the influences of plant species characteristics on the collective process of internal nutrient cycling within terrestrial ecosystems.

Likens, G. E., and F. H. Bormann. 1995. *Biogeochemistry of a forest ecosystem.* 2nd ed. New York: Springer-Verlag.

This now-classic text provides a 32-year record of the structure, dynamics, and nutrient cycling of the Hubbard Brook ecosystem.

McNaughton, S. J., R. M. Ruess, and S. W. Seagle. 1988. Large mammals and process dynamics in African ecosystems. *BioScience* 38:794–800.

An interesting perspective on the role of consumer organisms in the cycling of nutrients in savanna ecosystems.

Newbold, J. D., J. W. Elwood, R. V. O'Neill, and A. L. Sheldon. 1983. Phosphorus dynamics in a woodland stream ecosystem: A study of nutrient spiraling. *Ecology* 64:1249–1265.

This early study helped change ecologists' view of nutrient cycling in flowing-water ecosystems.

Pastor, J., and S. D. Bridgham. 1999. Nutrient efficiency along nutrient availability gradients. *Oecologia* 118:50–58.

This paper provides an excellent discussion of variation in the processes involved in nutrient cycling along an environmental gradient of nutrient availability.

Swift, M. J., O. W. Heal, and J. M. Anderson. 1979. *Decomposition in terrestrial ecosystems.* Berkeley: University of California Press.

An older, but extensive reference on decomposition processes in terrestrial ecosystems.

Wagener, S. M., M. W. Oswood, and J. P. Schimel. 1998. Rivers and soils: Parallels in carbon and nutrient processing. *BioScience* 48:104–108.

An excellent discussion of the similarities and differences in the processes of decomposition and nutrient cycling in terrestrial and stream ecosystems.

CHAPTER 22

Biogeochemical Cycles

Impala (*Aepyceros melampus*) standing in the shade of acacia trees. Their urine and droppings make the impala important contributors to the internal nitrogen cycle of these trees.

In Chapter 21 we examined the internal cycling of nutrients within the ecosystem, driven by the processes of net primary productivity and decomposition. The internal cycling of nutrients within the ecosystem is a story of biological processes. But not every transformation of elements in the ecosystem is biologically mediated. Many chemical reactions take place in abiotic components of the ecosystem: the atmosphere, water, soil, and parent material. The weathering of rocks and minerals releases certain elements into the soil and water, making them available for uptake by plants. The energy from lightning produces small amounts of ammonia (NH_3) from molecular nitrogen and water in the atmosphere, providing an input of nitrogen to aquatic and terrestrial ecosystems. Other processes, such as the sedimentation of calcium carbonate in marine environments (see Section 3.5), remove elements from the active process of internal cycling.

Each element has its own story, but all nutrients flow from the nonliving to the living and back to the nonliving components of the ecosystem in a more or less cyclic path known as the **biogeochemical cycle** (from *bio*, "living"; *geo* for the rocks and soil; and *chemical* for the processes involved). In this chapter, we will expand our view of nutrient flow through the ecosystem from our previous discussion of internal cycling, which was dominated by the biological processes of uptake and decomposition, to include a wider array of both biotic and abiotic processes. We will also examine in detail some of the major biogeochemical cycles.

22.1 | There Are Two Major Types of Biogeochemical Cycles

There are two basic types of biogeochemical cycles: gaseous and sedimentary. This classification is based on the primary source of nutrient input to the ecosystem. In gaseous cycles, the main pools of nutrients are the atmosphere and the oceans. For this reason, gaseous cycles are distinctly global. The gases most important for life are nitrogen, oxygen, and carbon dioxide. These three gases—in stable quantities of 78 percent, 21 percent, and 0.03 percent, respectively—are the dominant components of Earth's atmosphere.

In sedimentary cycles, the main pool is the soil, rocks, and minerals. The mineral elements required by living organisms come initially from inorganic sources. Available forms occur as salts dissolved in soil water or in lakes, streams, and seas. The mineral cycle varies from one element to another, but essentially it consists of two phases: the rock phase and the salt solution phase. Mineral salts come directly from Earth's crust through weathering (see Section 4.4). The soluble salts then enter the water cycle. With water, the salts move through the soil to streams and lakes and eventually reach the seas, where they remain indefinitely. Other salts return to Earth's crust through sedimentation. They become incor-

porated into salt beds, silts, and limestone. After weathering, they enter the cycle again.

There are many different kinds of sedimentary cycles. Cycles such as the sulfur cycle are a hybrid of the gaseous and the sedimentary because they have major pools in Earth's crust as well as in the atmosphere. Other cycles, such as the phosphorus cycle, have no significant gaseous pool; the element is released from rock and deposited in both the shallow and deep sediments of the sea.

Gaseous and sedimentary cycles involve biological and nonbiological processes. Both cycles are driven by the flow of energy through the ecosystem, and both are tied to the water cycle (see Section 3.1). Water is the medium that moves elements and other materials through the ecosystem. Without the cycling of water, biogeochemical cycles would cease.

Although the biogeochemical cycles of the various essential nutrients required by autotrophs and heterotrophs differ in detail, from the perspective of the ecosystem all biogeochemical cycles have a common structure. They share three basic components: inputs, internal cycling, and outputs (Figure 22.1).

22.2 | Nutrients Enter the Ecosystem via Inputs

The input of nutrients to the ecosystem depends on the type of biogeochemical cycle. Nutrients with a gaseous cycle, such as carbon and nitrogen, enter the ecosystem via the atmosphere. In contrast, nutrients such as calcium and phosphorus have sedimentary cycles, with inputs dependent on the weathering of rocks and minerals (see Section 4.4). The process of soil formation and the resulting soil characteristics have a major influence on processes involved in nutrient release and retention (see Chapter 4).

Supplementing nutrients in the soil are nutrients carried by rain, snow, air currents, and animals. Precipitation brings appreciable quantities of nutrients, called **wetfall.** Some of these nutrients, such as tiny dust particles of calcium and sea salt, form the nuclei of raindrops; others wash out of the atmosphere as the rain falls. Some nutrients are brought in by airborne particles and aerosols, collectively called **dryfall.** Between 70 percent and 90 percent of rainfall striking the forest canopy reaches the forest floor. As it drips through the canopy (throughfall) and runs down the stems (stemflow), rainwater picks up and carries with it nutrients deposited as dust on leaves and stems together with nutrients leached from them. Therefore, rainfall reaching the forest floor is richer in calcium, sodium, potassium, and other nutrients than rain falling in the open at the same time.

The major sources of nutrients for aquatic life are inputs from the surrounding land in the form of drainage water, detritus, and sediment and from the atmosphere in the form of precipitation. Flowing-water aquatic

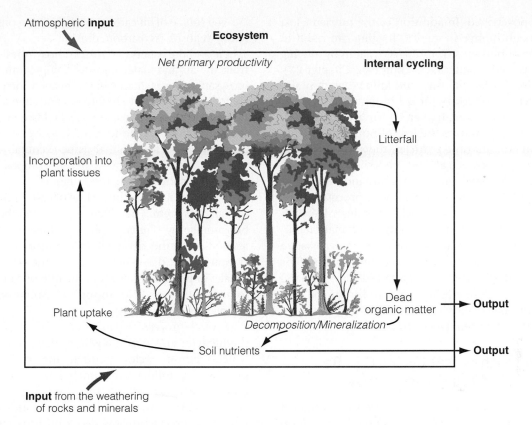

Atmospheric **input**

Ecosystem

Net primary productivity　　　　**Internal cycling**

Litterfall

Incorporation into
plant tissues

Dead
organic matter　**Output**

Plant uptake

Decomposition/Mineralization

Soil nutrients　　**Output**

Input from the weathering
of rocks and minerals

Figure 22.1 | A generalized representation of the biogeochemical cycle of an ecosystem. The three common components of inputs, internal cycling, and outputs are shown in bold. Key processes involved in the internal cycling of nutrients within ecosystems, net primary productivity and decomposition, are shown in italics.

systems (streams and rivers) are highly dependent on a steady input of dead organic matter from the watersheds they flow through (see Chapters 23 and 24).

22.3 | Outputs Represent a Loss of Nutrients from the Ecosystem

The export of nutrients from the ecosystem represents a loss that must be offset by inputs if a net decline is not to occur. Export can occur in a variety of ways depending on the specific biogeochemical cycle. Carbon is exported to the atmosphere in the form of CO_2 via the process of respiration by all living organisms (see the introduction to Part Two). Likewise, various microbial and plant processes result in the transformation of nutrients to a gaseous phase that can subsequently be transported from the ecosystem in the atmosphere. Examples of these processes are provided in the following sections, which examine specific biogeochemical cycles.

Transport of nutrients from the ecosystem can also occur in the form of organic matter. Organic matter from a forested watershed can be carried from the ecosystem through surface flow of water in streams and rivers. The input of organic carbon from terrestrial ecosystems constitutes most of the energy input into stream ecosystems (see

Chapter 24). Organic matter can also be transferred between ecosystems by herbivores. Moose feeding on aquatic plants can transport and deposit nutrients to adjacent terrestrial ecosystems in the form of feces. Conversely, the hippopotamus (*Hippopotamus amphibius*) feeds at night on herbaceous vegetation near the body of water where it lives. Large quantities of nutrients are then transported in the form of feces and other wastes to the water.

Although the transport of organic matter can be a significant source of nutrient loss from an ecosystem, organic matter plays a key role in recycling nutrients because it prevents rapid losses from the system. Large quantities of nutrients are bound tightly in organic matter structure; they are not readily available until released by activities of decomposers.

Some nutrients are leached from the soil and carried out of the ecosystem by underground water flow to streams. These losses may be balanced by inputs to the ecosystem, such as the weathering of rocks and minerals (see Chapter 4).

Considerable quantities of nutrients are withdrawn permanently from ecosystems by harvesting, especially in farming and logging, as biomass is directly removed from the ecosystem. In such ecosystems, these losses must be replaced by applying fertilizer; otherwise, the ecosystem becomes impoverished. In addition to the nutrients lost directly through biomass removal, logging can result in the transport of nutrients from the ecosystem by altering processes involved in internal cycling (see Chapter 27).

Depending on its intensity, fire kills vegetation and converts varying proportions of the biomass and soil organic matter to ash (see Chapter 19 for a discussion of fire as disturbance). Besides the loss of nutrients through volatilization and airborne particles, the addition of ash changes the soil's chemical and biological properties. Many nutrients become readily available, and nitrogen in ash is subject to rapid mineralization—a process known as pyromineralization (see Field Studies: Deborah Lawrence, Chapter 27). If not taken up by vegetation during recovery, nutrients may be lost from the ecosystem through leaching and erosion. Stream-water runoff is often greatest after fire because of reduced water demand for transpiration. High nutrient availability in the soil, coupled with high runoff, can lead to large nutrient losses from the ecosystem (see Section 27.9).

22.4 | Biogeochemical Cycles Can Be Viewed from a Global Perspective

The cycling of nutrients and energy occurs within all ecosystems, and it is most often studied as a local process; that is, the internal cycling of nutrients within the ecosystem and the identification of exchanges both to (inputs) and from (outputs) the ecosystem. Through these processes of exchange, the biogeochemical cycles of differing ecosystems are linked.

Often, the output from one ecosystem represents an input to another, as in the case of exporting nutrients from terrestrial to aquatic ecosystems. The processes of exchanging nutrients among ecosystems requires viewing the biogeochemical cycles from a much broader spatial framework than that of a single ecosystem. This is particularly true of nutrients that go through a gaseous cycle. Because the main pools of these nutrients are the atmosphere and the ocean, they have distinctly global circulation patterns. In the following sections, we explore the cycling of carbon, nitrogen, phosphorus, sulfur, and oxygen and examine the specific processes involved in their movement through the ecosystem. We will then expand our model of biogeochemical cycling to provide a framework for understanding the global cycling of these elements, which are crucial to life.

22.5 | The Carbon Cycle Is Closely Tied to Energy Flow

Carbon, a basic constituent of all organic compounds, is involved in the fixation of energy by photosynthesis (see Chapter 6). Carbon is so closely tied to energy flow that the two are inseparable. In fact, we often express ecosystem productivity in terms of grams of carbon fixed per square meter per year (see Section 20.1).

The source of all carbon, both in living organisms and fossil deposits, is carbon dioxide (CO_2) in the atmosphere and the waters of Earth. Photosynthesis draws CO_2 from the air and water into the living component of the ecosystem. Just as energy flows through the grazing food chain, carbon passes to herbivores and then to carnivores. Primary producers and consumers release carbon back to the atmosphere in the form of CO_2 by respiration. The carbon in plant and animal tissue eventually goes to the reservoir of dead organic matter. Decomposers release it to the atmosphere through respiration.

Figure 22.2 shows the cycling of carbon through a terrestrial ecosystem. The difference between the rate of carbon uptake by plants in photosynthesis and release by respiration is the net primary productivity (in units of carbon). The difference between the rate of carbon uptake in photosynthesis and the rate of carbon loss due to autotrophic and heterotrophic respiration is the **net ecosystem productivity.**

Several processes, particularly the rates of primary productivity and decomposition, determine the rate at which carbon cycles through the ecosystem. Both processes are influenced strongly by environmental conditions such as temperature and precipitation (see Section 20.3). In warm, wet ecosystems such as a tropical rain forest, production and decomposition rates are high, and carbon cycles through the ecosystem quickly. In cool, dry ecosystems, the process is slower. In ecosystems where temperatures are very low, decomposition is slow, and dead organic matter accumulates (see Chapter 23). In swamps and marshes, where dead material falls into the water, organic material does not completely decompose. When stored as raw humus or peat (see Chapter 25), carbon circulates slowly. Over geologic time, this buildup of partially decomposed organic matter in swamps and marshes has formed fossil fuels (oil, coal, and natural gas).

Similar cycling takes place in freshwater and marine environments (see Figure 22.2). Phytoplankton uses the carbon dioxide that diffuses into the upper layers of water or is present as carbonates and converts it into plant tissue. The carbon then passes from the primary producers through the aquatic food chain. Carbon dioxide produced through respiration is either reused or reintroduced to the atmosphere by diffusion from the water surface to the surrounding air (see Section 3.7 for a discussion of the carbon dioxide–carbonate system in aquatic ecosystems).

Significant amounts of carbon can be bound as carbonates in the bodies of mollusks and foraminifers and incorporated into their exoskeletons (shells, etc.). Some of these carbonates dissolve back into solution, and some become buried in the bottom mud at varying depths when the organisms die. Because it is isolated from biotic activity, this carbon is removed from cycling. Upon incorporation into bottom sediments

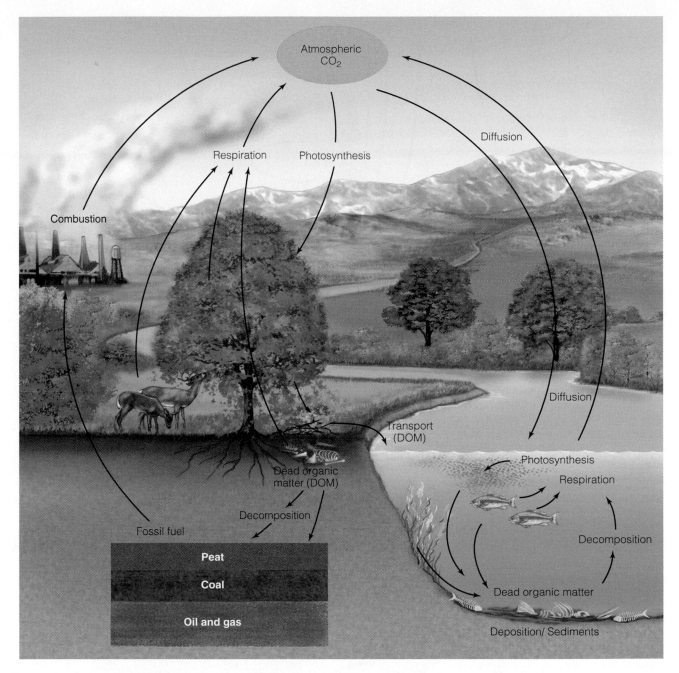

Figure 22.2 | The carbon cycle as it occurs in both terrestrial and aquatic ecosystems.

over geologic time, it may appear in coral reefs and limestone rocks.

22.6 | Carbon Cycling Varies Daily and Seasonally

If you were to measure the concentration of carbon dioxide in the atmosphere above and within a forest on a summer day, you would discover that it fluctuates throughout the day (Figure 22.3). At daylight when photosynthesis begins, plants start to withdraw carbon dioxide from the air, and the concentration declines sharply.

By afternoon when the temperature is increasing and relative humidity is decreasing, the rate of photosynthesis declines, and the concentration of carbon dioxide in the air surrounding the canopy increases. By sunset, photosynthesis ceases (carbon dioxide is no longer being withdrawn from the atmosphere), respiration increases, and the atmospheric concentration of carbon dioxide increases sharply. A similar diurnal fluctuation occurs in aquatic ecosystems.

Likewise, the production and use of carbon dioxide undergoes a seasonal fluctuation relating both to the temperature and the timing of the growing and dormant

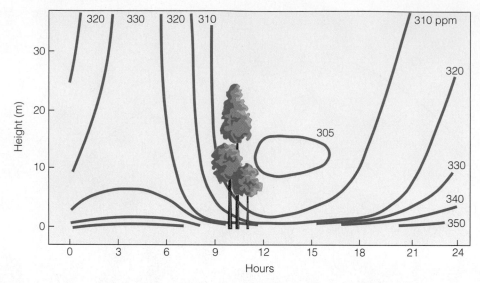

Figure 22.3 | Daily flux of CO_2 in a forest. Isopleths (lines) define concentration gradients. Note the consistently high level of CO_2 on the forest floor—the site of microbial respiration. Atmospheric CO_2 in the forest is lowest from midmorning to late afternoon. CO_2 levels are highest at night, when photosynthesis shuts down and respiration pumps CO_2 into the atmosphere.

(Adapted from Baumgartner 1968.)

seasons (Figure 22.4). With the onset of the growing season when the landscape is greening, the atmospheric concentration begins to drop as plants withdraw carbon dioxide through photosynthesis. As the growing season reaches its end, photosynthesis declines or ceases, respiration is the dominant process, and atmospheric concentrations of carbon dioxide rise. Although these patterns of seasonal rise and decline occur in both aquatic and terrestrial ecosystems, the fluctuations are much greater in terrestrial environments. As a result, these fluctuations in atmospheric concentrations of carbon dioxide are more pronounced in the Northern Hemisphere with its much larger land area.

22.7 | The Global Carbon Cycle Involves Exchanges among the Atmosphere, Oceans, and Land

The carbon budget of Earth is closely linked to the atmosphere, land, and oceans and to the mass movements of air around the planet (see Chapter 2). Earth contains about 10^{23} g of carbon, or 100 million Gt [Gt is a gigaton, equal to 1 billion (10^9) metric tons, or (10^{15}) g]. All but a small fraction of carbon is buried in sedimentary rocks and is not actively involved in the global carbon cycle. The carbon pool involved in the global carbon cycle (Figure 22.5) amounts

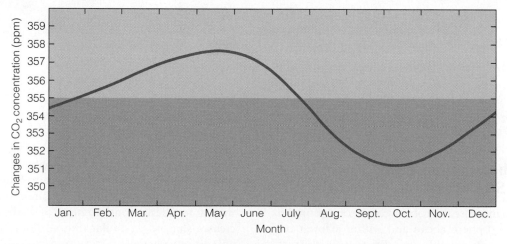

Figure 22.4 | Variation in atmospheric concentration of CO_2 during a typical year at Barrow, Alaska. Concentrations increase during the winter months, declining with the onset of photosynthesis during the growing season (May–June).

(Adapted from Pearman and Hyson 1981.)

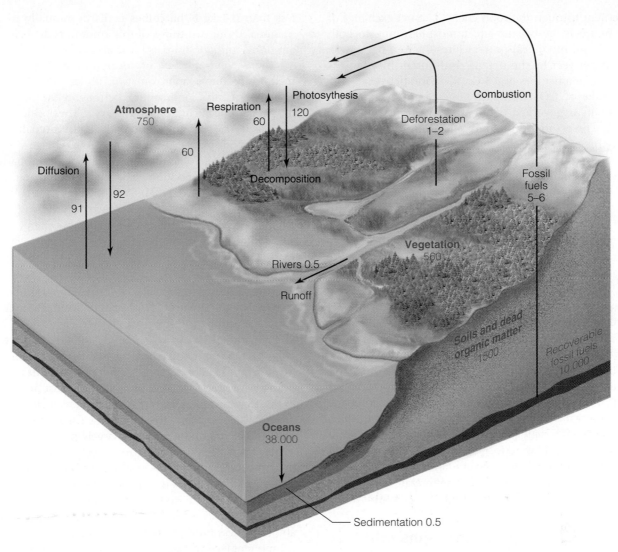

Figure 22.5 | The global carbon cycle. The sizes of the major pools of carbon are labelled in red, and arrows indicate the major exchanges (fluxes) among them. All values are in gigatons (Gt) of carbon, and exchanges are on an annual timescale. The largest pool of carbon, geologic, is not included due to the slow rates (geologic timescale) of transfer with other active pools.

(Adapted from Edmonds 1992.)

Interpreting Ecological Data

Q1. Which two fluxes in Figure 22.5 provide an estimate of global net primary productivity (NPP)? What is the estimate of global NPP in the figure?

Q2. Which of the major pools identified in the figure represent an estimate of global standing crop biomass?

Q3. Which fluxes can be used to provide an estimate of global net ecosystem productivity? What is the estimate of global net ecosystem productivity in the figure?

to an estimated 55,000 Gt. Fossil fuels, created by the burial of partially decomposed organic matter, account for an estimated 10,000 Gt. The oceans contain the vast majority of the active carbon pool, about 38,000 Gt, mostly as bicarbonate and carbonate ions (see Section 3.7). Dead organic matter in the oceans accounts for 1650 Gt of carbon; and living matter, mostly phytoplankton (primary producers), accounts for 3 Gt. The terrestrial biosphere (all terrestrial ecosystems) contains an estimated 1500 Gt of carbon as dead organic matter and 560 Gt as living matter (biomass). The atmosphere holds about 750 Gt of carbon.

In the ocean, the surface water acts as the site of main exchange of carbon between atmosphere and ocean. The ability of the surface waters to take up CO_2 is governed largely by the reaction of CO_2 with the carbonate ion to form bicarbonates (see Section 3.7). In the surface water, carbon circulates physically by means of currents and biologically through photosynthesis by phytoplankton and movement through the food chain. The net exchange of CO_2 between the oceans and atmosphere due to both physical and biological processes results in a net uptake of 1 Gt per year by the oceans, and burial in sediments accounts for a net loss of 0.5 Gt of carbon per year.

Uptake of CO_2 from the atmosphere by terrestrial ecosystems is governed by gross production (photosynthesis). Losses are a function of autotrophic and heterotrophic respiration, the latter being dominated by microbial decomposers. Until recently, exchanges of CO_2 between the landmass and the atmosphere (uptake in photosynthesis and release by respiration and decomposition) were believed to be nearly in equilibrium (Figure 22.5). However, more recent research suggests that the terrestrial surface is acting as a carbon sink, with a net uptake of CO_2 from the atmosphere (see Chapter 29).

Of considerable importance in the terrestrial carbon cycle are the relative proportions of carbon stored in soils and in living vegetation (biomass). Carbon stored in soils includes dead organic matter on the soil surface and in the underlying mineral soil. Estimates place the amount of soil carbon at 1500 Gt compared with 560 Gt in living biomass.

The average amount of carbon per volume of soil increases from the tropical regions poleward to the boreal forest and tundra (see Chapter 23). Low values for the tropical forest reflect high rates of decomposition, which compensate for high productivity and litterfall. Frozen tundra soil and waterlogged soils of swamps and marshes have the greatest accumulation of dead organic matter because factors such as low temperature, saturated soils, and anaerobic conditions function to inhibit decay.

22.8 | The Nitrogen Cycle Begins with Fixing Atmospheric Nitrogen

Nitrogen is an essential constituent of protein, which is a building block of all living tissue. Nitrogen is generally available to plants in only two chemical forms: ammonium (NH_4^+) and nitrate (NO_3^-). Thus, although Earth's atmosphere is almost 80 percent nitrogen gas, it is in a form (N_2) that is not available for uptake (assimilation) by plants. Nitrogen enters the ecosystem via two pathways, and the relative importance of each varies greatly among ecosystems (Figure 22.6). The first pathway is atmospheric deposition. This can be in wetfall—such as rain, snow, or even cloud and fog droplets—and in dryfall, such as aerosols and particulates. Regardless of the form of atmospheric deposition, nitrogen in this pathway is supplied in a form that is already available for uptake by plants.

The second pathway for nitrogen to enter ecosystems is via nitrogen fixation. This fixation comes about in two ways. One is high-energy fixation. Cosmic radiation, meteorite trails, and lightning provide the high energy needed to combine nitrogen with the oxygen and hydrogen of water. The resulting ammonia and nitrates are carried to Earth's surface in rainwater. Estimates suggest that

less than 0.4 kg N/ha comes to Earth annually in this manner. About two-thirds of this amount is deposited as ammonia and one-third as nitric acid (HNO_3).

The second method of fixation is biological. This method produces approximately 10 kg N/yr for each hectare of Earth's land surface, or roughly 90 percent of the fixed nitrogen contributed each year. This fixation is accomplished by symbiotic bacteria living in mutualistic association with plants (see Section 15.11 and Figure 15.7), by free-living aerobic bacteria, and by cyanobacteria (blue-green algae). Fixation splits molecular nitrogen (N_2) into two atoms of free N. The free N atoms then combine with hydrogen to form two molecules of ammonia (NH_3). The process of fixation requires considerable energy. To fix 1 g of nitrogen, nitrogen-fixing bacteria associated with the root system of a plant must expend about 10 g of glucose, a simple sugar produced by the plant in photosynthesis.

In agricultural ecosystems, *Rhizobium* bacteria associated with approximately 200 species of leguminous plants are the preeminent nitrogen fixers (see Figure 15.10). In nonagricultural systems, some 12,000 species, from cyanobacteria to nodule-bearing plants, are responsible for nitrogen fixation. Also contributing to the fixation of nitrogen are free-living soil bacteria. The most prominent of the 15 known genera are the aerobic *Azotobacter* and the anaerobic *Clostridium*. Cyanobacteria (blue-green algae) are another important group of largely nonsymbiotic nitrogen fixers. Of some 40 known species, the most common are in the genera *Nostoc* and *Calothrix*, which are found in soil as well as in aquatic habitats. Certain lichens (*Collema tunaeforme* and *Peltigera rufescens*) are also involved in nitrogen fixation. Lichens with nitrogen-fixing ability possess nitrogen-fixing cyanobacteria as their algal component (see Section 15.9).

Ammonium in the soil can be used directly by plants. In addition to atmospheric deposition, NH_4^+ occurs in the soil as a product of microbial decomposition of organic matter (see Section 21.5), wherein NH_3 is released as a waste product of microbial activity. This process is called **ammonification** (Figure 22.7). Most soils have an excess of H^+ (slightly acidic; see Section 3.7), and the NH_3 is rapidly converted to ammonium (NH_4^+). Interestingly, because NH_3 is a gas, the transfer of nitrogen back to the atmosphere (volatilization) can occur in soils with a pH close to 7 (neutral)—a low concentration of H^+ ions to convert ammonia to ammonium. Volatilization can be especially pronounced in agricultural areas where both nitrogen fertilizers and lime (to decrease soil acidity) are used extensively.

In some ecosystems, plant roots must compete for NH_4^+ with two groups of aerobic bacteria, which use it as part of their metabolism (see Figure 22.6). The first group (*Nitrosomonas*) oxidizes NH_4^+ to NO_2^-, and a second group (*Nitrobacter*) oxidizes NO_2^- to NO_3^-. This process is called **nitrification.** Once nitrate is produced, several things can happen to it. First, plant roots can take it up.

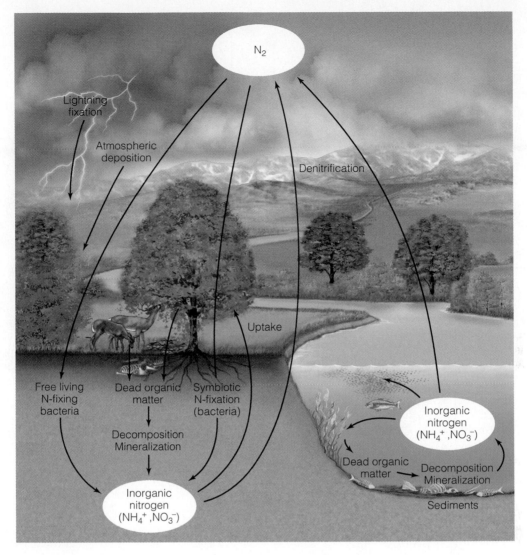

Figure 22.6 | The nitrogen cycle in terrestrial and aquatic ecosystems.

Second, **denitrification** can occur under anaerobic ("lacking oxygen") conditions, when another group of bacteria (*Pseudomonas*) chemically reduces NO_3^- to N_2O and N_2. These gases are then returned to the atmosphere. The anaerobic conditions necessary for denitrification are

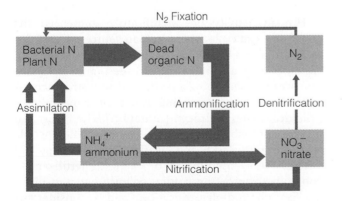

Figure 22.7 | Bacterial processes involved in nitrogen cycling.

generally rare in most terrestrial ecosystems (but can occur seasonally). These conditions, however, are common in wetland ecosystems and in the bottom sediments of open-water aquatic ecosystems (see Chapters 24 and 25).

Finally, nitrate is the most common form of nitrogen exported from terrestrial ecosystems in stream water (see Chapter 27), although in undisturbed ecosystems the amount is usually quite small because of the great demand for nitrogen. Indeed, the amount of nitrogen recycled within the ecosystem is usually much greater than the amount either entering or leaving the ecosystem through inputs and outputs.

Because both nitrogen fixation and nitrification are processes mediated by bacteria, they are influenced by various environmental conditions, such as temperature and moisture. However, one of the more important factors is soil pH. Both processes are usually greatly limited in extremely acidic soils due to the inhibition of bacteria under those conditions.

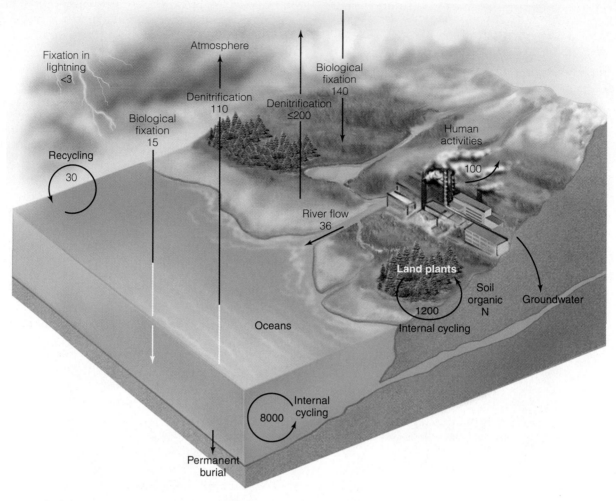

Figure 22.8 | The global nitrogen cycle. Each flux is shown in units of 10^{12} g N/yr. (Adapted from Schlesinger 1997.)

Inputs of nitrogen can vary, but the internal cycling of nitrogen is fairly similar from ecosystem to ecosystem. The process involves the assimilation of ammonium and nitrate by plants and the return of nitrogen to the soil, sediments, and water via the decomposition of dead organic matter.

The global nitrogen cycle follows the pathway of the local nitrogen cycle presented earlier, only on a grander scale (Figure 22.8). The atmosphere is the largest pool, containing 3.9×10^{21} g. Comparatively small amounts of nitrogen are found in the biomass (3.5×10^{15} g) and soils (95×10^{15} to 140×10^{15} g) of terrestrial ecosystems. Global estimates of denitrification in terrestrial ecosystems vary widely but are of the order 200×10^{12} g/yr, and more than half of that total occurs in wetland ecosystems. Major sources of nitrogen to the world's oceans are dissolved forms in the freshwater drainage from rivers (36×10^{12} g/yr) and inputs in precipitation (30×10^{12} g/yr). Biological fixation accounts for another 15×10^{12} g/yr. Denitrification accounts for an estimated

flux of 110×10^{12} g N/yr from the world's oceans to the atmosphere.

There are small but steady losses from the biosphere to the deep sediments of the ocean and to sedimentary rocks. In return, there is a small addition of new nitrogen from the weathering of igneous rocks and from volcanic activity.

Human activity has significantly influenced the global nitrogen cycle. Major human sources of nitrogen input are agriculture (see Chapter 27), industry, and automobiles; and in recent decades, anthropogenic inputs of nitrogen into aquatic and terrestrial ecosystems have been a growing cause of concern (see Ecological Issues: Nitrogen Saturation). The first major intrusion probably came from agriculture, when people began burning forests and clearing land for crops and pasture. Heavy application of chemical fertilizers to croplands disturbs the natural balance between nitrogen fixation and denitrification, and a considerable portion of nitrogen fertilizers is lost as nitrates to groundwater and runoff that find their way into aquatic ecosystems (see Chapter 25).

Automobile exhaust and industrial high-temperature combustion add nitrous oxide (N_2O), nitric oxide (NO), and nitrogen dioxide (NO_2) to the atmosphere. These oxides can reside in the atmosphere for up to 20 years, drifting slowly up to the stratosphere. There, ultraviolet light reduces nitrous oxide to nitric oxide and atomic oxygen (O). Atomic oxygen reacts with oxygen (O_2) to form ozone (O_3) (see Section 22.12).

22.9 | The Phosphorus Cycle Has No Atmospheric Pool

Phosphorus occurs in only minute amounts in the atmosphere. Therefore, the phosphorus cycle can follow the water (hydrological) cycle only part of the way—from land to sea (Figure 22.9). Because phosphorus lost from the ecosystem in this way is not returned via the biogeochemical cycle, phosphorus is in short supply under undisturbed natural conditions. The natural scarcity of phosphorus in aquatic ecosystems is emphasized by the explosive growth of algae in water receiving heavy discharges of phosphorus-rich wastes.

The main reservoirs of phosphorus are rock and natural phosphate deposits. Phosphorus is released from these rocks and minerals by weathering, leaching, erosion, and mining for use as agricultural fertilizers. Nearly all of the phosphorus in terrestrial ecosystems comes from the weathering of calcium phosphate minerals. In most soils, only a small fraction of the

total phosphorus is available to plants. The major process regulating phosphorus availability for net primary production is the internal cycling of phosphorus from organic to inorganic forms (nutrient cycling; see Chapter 21). Some of the available phosphorus in terrestrial ecosystems escapes and is exported to lakes and seas.

In marine and freshwater ecosystems, the phosphorus cycle moves through three states: particulate organic phosphorus, dissolved organic phosphates, and inorganic phosphates. Organic phosphates are taken up quickly by all forms of phytoplankton, which are eaten in turn by zooplankton and detritus-feeding organisms. Zooplankton may excrete as much phosphorus daily as it stores in its biomass, returning it to the cycle. More than half of the phosphorus zooplankton excretes is inorganic phosphate, which is taken up by phytoplankton. The remaining phosphorus in aquatic ecosystems exists in organic compounds that may be used by bacteria, which fail to regenerate much dissolved inorganic phosphate. Bacteria are consumed by the microbial grazers, which then excrete the phosphate they ingest. Part of the phosphate is deposited in shallow sediments and part in deep water. In the process of ocean upwelling (see Sections 3.8 and 21.13), the movement of deep waters to the surface brings some phosphates from the dark depths to shallow waters, where light is available to drive photosynthesis. These phosphates are taken up by phytoplankton. Part of the phosphorus contained in the bodies of plants and animals sinks to the bottom and is deposited in the sediments. As a result, surface waters may become depleted

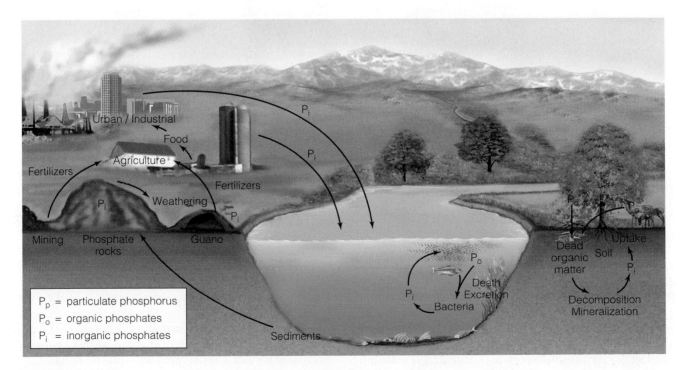

Figure 22.9 | The phosphorus cycle in aquatic and terrestrial ecosystems.

Net primary productivity in most terrestrial forest ecosystems is limited by the availability of soil nitrogen, yet in recent decades, human activities have caused a dramatic increase in the rates of nitrogen deposition. In North America, anthropogenic activities such as fossil fuel combustion (see Chapter 29) and high-intensity agriculture (see Chapter 27) have increased the inputs of nitrogen oxides in the atmosphere far above natural inputs. Nitrogen oxides quickly undergo a variety of chemical reactions in the atmosphere and therefore do not reside in the atmosphere for long periods, but rather are deposited in the region where the emissions originated. The result is that deposition rates vary widely among geographic regions (Figure 1).

As discussed in Chapter 6 (see Section 6.12), the concentration of nitrogen in the soil solution influences the rate of plant uptake and plant tissue concentration. In turn, there is a strong relationship between photosynthetic capacity and leaf nitrogen due to a greater concentration of enzymes and pigments used in photosynthesis (see Figure 6.24). The net result is that initially nitrogen deposition acts as a fertilizer, increasing rates of net primary productivity. However, as water and other nutrients become more limiting relative to nitrogen, these ecosystems may approach "nitrogen saturation." If the nitrogen supply continues to increase, a complex series of changes to soil and plant processes may ultimately lead to forest decline and soil acidification (Figure 2).

Most nitrogen deposited is in the form of either nitrate or ammonium, though ammonium is dominant. In the early stages of nitrogen saturation, plants will take up most of the nitrogen, and forest productivity and growth are stimulated by nitrogen inputs. As the limitations on plant growth shift from nitrogen to other resources, more ammonium will be available in the soil. High concentrations of ammonium may cause the release of other cations to the soil solution by overwhelming cation exchange sites on soil particles (see Section 4.9).

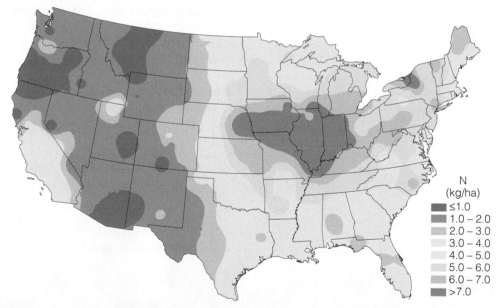

Figure 1 | Estimated inorganic nitrogen deposition from nitrate and ammonium in 1998 (National Atmospheric Deposition Program).

of phosphorus, and the deep waters become saturated. Much of this phosphorus becomes locked up for long periods of time in the hypolimnion and bottom sediments, while some is returned to the surface waters by upwelling.

The global phosphorus cycle (Figure 22.10) is unique among the major biogeochemical cycles in having no significant atmospheric component, although airborne transport of P in soil dust and sea spray is of the order 1×10^{12} g P/yr.

Rivers transport approximately 21×10^{12} g P/yr to the oceans, but only about 10 percent of this amount is available for net primary productivity. The remainder is deposited in sediments. The concentration of phosphorus

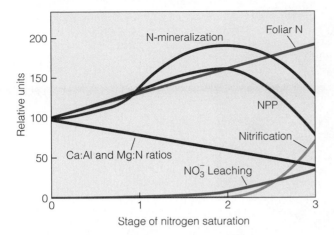

Figure 2 | Hypothesized response of temperate forest ecosystems to long-term nitrogen additions. In stage 1, N-mineralization increases, which results in increased NPP. In stage 2, NPP and N-mineralization decline due to decreasing Ca:Al and Mg:N ratios and to soil acidification. Nitrification also increases as excess ammonium is available. Finally, in stage 3, nitrate leaching increases dramatically.

(Adapted from Aber et al. 1998.)

Microbes may also use some of the excess ammonium to increase their own populations, thus increasing other microbial processes such as decomposition, but much of the ammonium will be nitrified to nitrate.

In contrast to ammonium, nitrate is highly mobile in soils, as it is not strongly adsorbed to soil particles through ion exchange (see Section 4.9). Nitrate may be taken up by plants or used by microbes in the process of denitrification, reducing nitrate to N_2 gas and thus completing the nitrogen cycle (see Section 22.8, Figure 22.7). Nitrate that is not used by plants or microbes may be leached to groundwater and surface water, resulting in eutrophication of streams (see Section 24.4).

When nitrate is leached from the soil, a loss of base cations and soil acidification may result. As the pH of the soil decreases, the buffering ability of cation exchange may be exhausted, resulting in the release of Al ions to the soils solution (see Section 4.9). Aluminum can be leached to aquatic ecosystems, where it is toxic. In addition, high Al concentrations in the soil solution can have detrimental effects on forest ecosystems.

In the later stages of nitrogen saturation, productivity is expected to decrease, and plant mortality may increase. This is partly due to nutrient imbalances resulting from the overwhelming availability of nitrogen. The health of plants is affected primarily by the relative concentrations of nutrients as opposed to their absolute abundances. As Al concentration in the soil solution increases due to soil acidification, the Ca:Al and Mg:Al ratios decrease. This decrease is partly due to the higher affinity of Al during the passive uptake of nutrients in soil solution by roots. These relative nutrient proportions have been correlated with declining spruce (*Picea*) populations in Europe and the United States. Calcium has several important roles in plant functioning. For example, Ca is essential for the production of new sapwood—the outer wood on the bole of the tree that contains the active vascular tissues for water transport. Decreased Ca availability will restrict functioning sapwood and water transport, which would result in decreasing the leaf area that the tree can support. Magnesium is an important element in plant enzymes, particularly chlorophyll. Limitations on chlorophyll production will function to limit photosynthesis. The combined effect is reduced net uptake of carbon by the plant. The yellowing of leaves often identifies trees suffering from the effects of nitrogen saturation. This condition is believed to be caused by the retranslocation of nutrients, such as Ca and Mg, during periods of new growth. Because high nitrogen levels stimulate new growth, the plant mobilizes the necessary Ca and Mg from leaves, regardless of its relative availability.

1. Using the information provided by the map in Figure 1, in which areas in the United States would you expect nitrogen saturation to occur?

2. Increased concentrations of ammonium can result in increased microbial processes, such as decomposition. How might the increase in microbial decomposition influence the availability of other mineral nutrients (in soil) to plants?

in the ocean waters is low, but the large volume of water results in a significant global pool of phosphorus. The turnover of organic phosphorus in the surface waters occurs within days, and the vast majority of phosphorus taken up in primary production is decomposed and mineralized (internally cycled) in the surface waters. However, approximately 2×10^{12} g/yr is deposited in the ocean sediments or transported to the deep waters. In the deep waters, organic phosphorus converted into inorganic, soluble forms remains unavailable to phytoplankton in the surface waters until transported by upwelling (see Section 3.8). On a geological timescale, uplifting and subsequent weathering return this phosphorus to the active cycle.

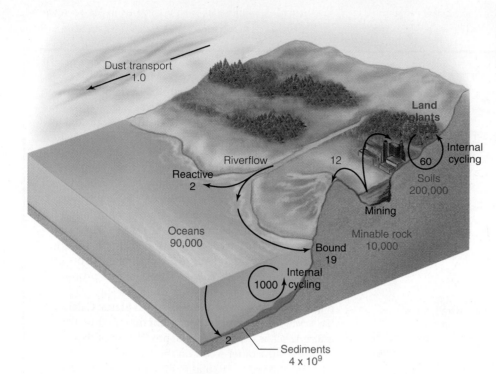

Figure 22.10 | The global phosphorus cycle. Each flux is shown in units of 10^{12} g P/yr.

(Adapted from Schlesinger 1997.)

22.10 | The Sulfur Cycle Is Both Sedimentary and Gaseous

The sulfur cycle has both sedimentary and gaseous phases (Figure 22.11). In the long-term sedimentary phase, sulfur is tied up in organic and inorganic deposits, released by weathering and decomposition, and carried to terrestrial ecosystems in salt solution. The gaseous phase of the cycle permits sulfur circulation on a global scale.

Sulfur enters the atmosphere from several sources: the combustion of fossil fuels, volcanic eruptions, exchange at the surface of the oceans, and gases released by decomposition. It enters the atmosphere initially as hydrogen sulfide (H_2S), which quickly interacts with oxygen to form sulfur dioxide (SO_2). Atmospheric sulfur dioxide, soluble in water, is carried back to the surface in rainwater as weak sulfuric acid (H_2SO_4). Whatever the source, sulfur in a soluble form is taken up by plants and

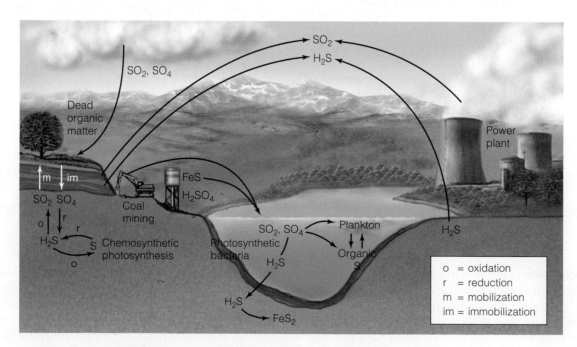

Figure 22.11 | The sulfur cycle. Note the two components: sedimentary and gaseous. Major sources from human activity are the burning of fossil fuels and acidic drainage from coal mines.

incorporated through a series of metabolic processes, starting with photosynthesis, into sulfur-bearing amino acids. From the primary producers, sulfur in amino acid is transferred to consumers.

Excretion and death carry sulfur from living material back to the soil and to the bottom of ponds, lakes, and seas, where bacteria release it as hydrogen sulfite or sulfate. One group, the colorless sulfur bacteria, reduces hydrogen sulfide to elemental sulfur and then oxidizes it to sulfuric acid. Green and purple bacteria, in the presence of light, use hydrogen sulfide during photosynthesis. Best known are the purple bacteria found in salt marshes and in the mudflats of estuaries. These organisms can transform hydrogen sulfide into sulfate, which is then recirculated and taken up by producers or used by bacteria that further transform the sulfates. Green bacteria can transform hydrogen sulfide into elemental sulfur.

Sulfur, in the presence of iron and under anaerobic conditions, will precipitate as ferrous sulfide (FeS_2). This compound is highly insoluble in neutral and low pH (acidic) conditions, and it is firmly held in mud and wet soil. Sedimentary rocks containing ferrous sulfide, called pyritic rocks, may overlie coal deposits. When exposed to air during deep and surface mining for coal, the ferrous sulfide reacts with oxygen. In the presence of water, it produces ferrous sulfate ($FeSO_4$) and sulfuric acid.

In this manner, sulfur in pyritic rocks, suddenly exposed to weathering by human activities, discharges sulfuric acid, ferrous sulfate, and other sulfur compounds into aquatic ecosystems. These compounds destroy aquatic life. They have converted hundreds of kilometers of streams in the eastern United States to lifeless, highly acidic water.

22.11 | The Global Sulfur Cycle Is Poorly Understood

The global sulfur cycle is presented in Figure 22.12. Although a great deal of research now focuses on the sulfur cycle, particularly the role of human inputs, our understanding of the global sulfur cycle is primitive.

The gaseous phase of the sulfur cycle permits circulation on a global scale. The annual flux of sulfur compounds through the atmosphere is of the order 300×10^{12} g. The atmosphere contains not only sulfur dioxide and hydrogen sulfide, but sulfate particles. The sulfate particles become part of dry deposition (dryfall); the gaseous forms combine with moisture and are transported in precipitation (wetfall).

The oceans are a large source of aerosols that contain sulfate (SO_4); however, most are redeposited in the oceans as precipitation and dryfall (see Figure 22.11). Dimethylsulfide [$(CH_3)_2S$] is the major gas emitted from the oceans that is generated by biological processes. Estimates of 16×10^{12} g S/yr make it the largest natural source of sulfur gases released to the atmosphere.

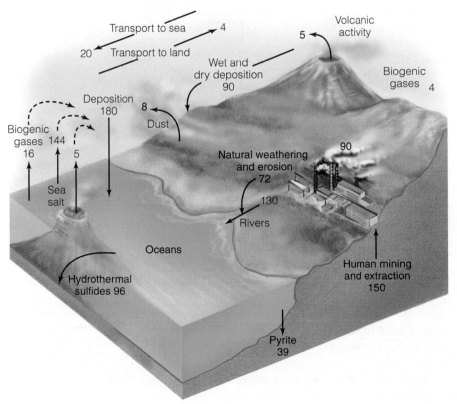

Figure 22.12 | The global sulfur cycle. Each flux is shown in units of 10^{12} g S/yr.

(Adapted from Schlesinger 1997.)

Transport to sea 4
20 Transport to land
Volcanic activity 5
Wet and dry deposition 90
Biogenic gases 4
Deposition 180
8 Dust
Biogenic gases 144
16 5
Sea salt
Natural weathering and erosion 72
90
130
Rivers
Oceans
Hydrothermal sulfides 96
Human mining and extraction 150
Pyrite 39

Various biological sources of sulfur emissions from terrestrial ecosystems exist, but collectively they cause only a minor flux to the atmosphere. The dominant sulfur gas emitted from freshwater wetlands and anoxic (oxygen-depleted) soils is hydrogen sulfide (H_2S). Emissions from plants are poorly understood, but forest fires emit on the order of 3×10^{12} g S annually. It is almost impossible to estimate the biological turnover of sulfur dioxide, due to the complicated cycling within the biosphere. Estimates of the net annual assimilation of sulfur by marine plants are of the order 130×10^{12} g. Adding the anaerobic oxidation of organic matter (see Section 21.4) brings the total to an estimated 200×10^{12} g.

Volcanic activity also contributes to the global biogeochemical cycle of sulfur. Major events, such as the eruption of Mt. Pinatubo in 1991, release on the order of 5×10^{12} to 10×10^{12} g S. When volcanic activity is averaged over a long period, the annual global flux is of the order 10×10^{12} g S.

Human activity plays a dominant role in the biogeochemical cycle of sulfur. Thus, to complete the picture of the global sulfur cycle, we must include the inputs due to industrial activity.

22.12 | The Oxygen Cycle Is Largely under Biological Control

The major source of free oxygen (O_2) that supports life is the atmosphere. There are two significant sources of atmospheric oxygen. One is the breakup of water vapor through a process driven by sunlight. In this reaction, the water molecules (H_2O) are disassociated to produce hydrogen and oxygen. Most of the hydrogen escapes into space. If the hydrogen did not escape, it would recombine with the oxygen to form water vapor again.

The other source of oxygen is photosynthesis, active only since life began on Earth (Figure 22.13). Oxygen is produced by photosynthetic autotrophs (green plants, algae, and photosynthetic bacteria; see Section 6.1) and consumed by both autotrophs and heterotrophs in the process of cellular respiration. Because photosynthesis and aerobic respiration involve the alternate release and use of oxygen, one would seem to balance the other, so no significant quantity of oxygen would accumulate in the atmosphere. Nevertheless, at some time in Earth's history, the amount of oxygen introduced into the atmosphere had to exceed the amount taken up in respiration (including the decay of organic matter) and geological processes, such as the oxidation of sedimentary rocks. Part of the oxygen present in the atmosphere is from the past imbalance between photosynthesis and respiration in plants. Undecomposed organic matter in the form of fossil fuels and carbon in sedimentary rocks represent a net positive flux of oxygen to the atmosphere (see Section 22.7).

The other main reservoirs of oxygen are water and carbon dioxide. All the reservoirs are linked through pho-

tosynthesis. Oxygen is also biologically exchangeable in such compounds from as nitrates and sulfates, which organisms transform from ammonia and hydrogen sulfide (see Sections 22.8 and 22.10).

Because oxygen is so reactive, its cycling in the ecosystem is complex. As a constituent of carbon dioxide, it circulates throughout the ecosystem. Some carbon dioxide combines with calcium to form carbonates. Oxygen combines with nitrogen compounds to form nitrates, with iron to form ferric oxides, and with other minerals to form various oxides. In these states, oxygen is temporarily withdrawn from circulation. In photosynthesis, the freed oxygen is split from the water molecule. The oxygen is then reconstituted into water during cellular respiration in both plants and animals. Part of the atmospheric oxygen is reduced to ozone (O_3) by high-energy ultraviolet radiation.

Ozone (O_3) is an ambivalent atmospheric gas. In the stratosphere, 10 to 40 km above Earth, it shields the planet from biologically harmful ultraviolet radiation. Close to the ground, ozone is a damaging pollutant, cutting visibility, irritating eyes and respiratory systems, and injuring or killing plant life. In the stratosphere, ozone is diminished by its reaction with human-caused pollutants. In the troposphere, ozone is born from the union of nitrogen oxides with oxygen in the presence of sunlight.

A cycling reaction requiring sunlight maintains ozone in the stratosphere. Solar radiation breaks the O–O bond in O_2. Freed oxygen atoms rapidly combine with O_2 to form O_3. At the same time, a reverse reaction consumes

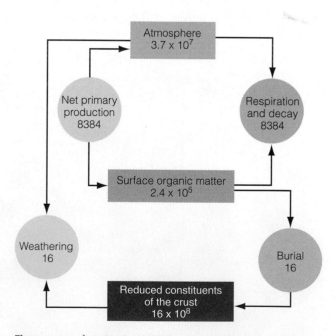

Figure 22.13 | A simple model for the global biogeochemical cycle of O_2. Data are expressed in units of 10^{12} moles of O_2 per year or the equivalent amount of reduced compounds. Note that a small misbalance in the ratio of photosynthesis to respiration can result in a net storage of reduced organic materials in the crust and an accumulation of O_2 in the atmosphere.

(Adapted from Schlesinger 1997.)

ozone to form O and O_2. Under natural conditions in the stratosphere, a balance exists between the rates of ozone formation and destruction. In recent times, however, catalysts—some human-caused and some biologically derived—injected into the stratosphere have been reactive enough to reduce stratospheric ozone. Among them are chlorofluorocarbons (CFCs); methane (CH_4), both natural and human caused; and nitrous oxide from denitrification and synthetic nitrogen fertilizer. Of particular concern is chlorine monoxide (ClO) derived from CFCs and used in aerosol spray propellants (banned in the United States), refrigerants, solvents, and other sources. This form of chlorine can break down ozone.

22.13 | The Various Biogeochemical Cycles Are Linked

Although we have introduced each of the major biogeochemical cycles independently, they are all linked in various ways. In specific cases, they are linked through their common membership in compounds that form an important component of their cycles. Examples are the links between calcium and phosphorus in the mineral apatite, a phosphate of calcium, and the link between nitrogen and oxygen in nitrate. In general, cycled nutrients are all components of living organisms and constituents of organic matter. As a result, they travel together in their odyssey through internal cycling.

Autotrophs and heterotrophs require nutrients in different proportions for different processes. For example (as indicated in the equation for photosynthesis in Chapter 6), photosynthesis uses 6 moles of water (H_2O) and produces 6 moles of oxygen (O_2) for every 6 moles of CO_2 that is transformed into 1 mole of sugar ($CH_2O)_6$. The proportions of hydrogen, oxygen, and carbon involved in photosynthesis are fixed. Likewise, a fixed quantity of nitrogen is required to produce a mole of rubisco, the enzyme that catalyzes the fixation of CO_2 in photosynthesis (see Section 6.1). Therefore, the nitrogen content of a rubisco molecule is the same in every plant, independent of species or environment. The same is true for the variety of amino acids, proteins, and other nitrogenous compounds that are essential for the synthesis of plant cells and tissues. The branch of chemistry dealing with the quantitative relationships of elements in combination is called **stoichiometry.** The stoichiometric relationships among various elements involved in processes related to carbon uptake and plant growth have an important influence on the cycling of nutrients in ecosystems.

Because of similar relationships among the variety of macronutrients and micronutrients required by plants, the limitation of one nutrient can affect the cycling of all the others. As an example, consider the link between carbon and nitrogen presented earlier in Chapter 21. Although the nitrogen content of a rubisco molecule is the same in every plant independent of species or environment, plants can differ in the concentration of rubisco found in their leaves and, therefore, in their concentration of nitrogen (grams N per gram dry weight). Plants growing under low nitrogen availability will have a lower rate of nitrogen uptake and less nitrogen for the production of rubisco and other essential nitrogen-based compounds. In turn, the lower concentrations of rubisco result in lower rates of photosynthesis and carbon gain (see Section 6.12). In turn, the lower concentrations of nitrogen in the leaf litter influence the relative rates of immobilization and mineralization and subsequent nitrogen release to the soil in decomposition (see Section 21.5). In this way, nitrogen availability and uptake by plants influence the rate at which carbon and other essential plant nutrients cycle through the ecosystem.

Conversely, the variety of other essential nutrients and environmental factors that directly influence primary productivity, and thus the demand for nitrogen, influence the rate of nitrogen cycling through the ecosystem. In fact, the cycles of all essential nutrients for plant and animal growth are linked due to the stoichiometric relationships defining the mixture of chemicals that make up all living matter.

Summary

Biogeochemical Cycles | 22.1

Nutrients flow from the living to the nonliving components of the ecosystem and back, in a perpetual cycle. Through these cycles, plants and animals obtain nutrients necessary for their survival and growth. There are two basic types of biogeochemical cycles: the gaseous, represented by oxygen, carbon, and nitrogen, whose major pools are in the atmosphere; and the sedimentary, represented by the sulfur and phosphorus cycles, whose major pools are in Earth's crust. The sedimentary cycles involve two phases: salt solution and rock. Minerals become available through the weathering of Earth's crust, enter the water cycle as a salt solution, take diverse pathways through the ecosystem, and ultimately return to Earth's crust through sedimentation. All nutrient cycles have a common structure and share three basic components: inputs, internal cycling, and outputs.

Inputs | 22.2

The input of nutrients to the ecosystem depends on the type of biogeochemical cycle: gaseous or sedimentary. The availability of essential nutrients in terrestrial ecosystems depends heavily on the nature of the soil. Supplementing the nutrients in soil are the nutrients carried by rain, snow, air currents, and animals. The major sources of nutrients for aquatic life are inputs from the surrounding land (in the form of drainage water, detritus, and sediments).

Outputs | 22.3

The export of nutrients from the ecosystem represents a loss that must be offset by inputs so that a net decline does not occur. Export can occur in a variety of ways, depending on the specific biogeochemical cycle. A major means of transportation is in the form of organic matter carried by surface flow of water in streams and rivers. Leaching of dissolved nutrients from soils into surface and groundwater also represents a significant export in some ecosystems. Harvesting of biomass in forestry and agriculture represents a permanent withdrawal from the ecosystem. Fire is also a major source of nutrient export in some terrestrial ecosystems.

Global Cycles | 22.4

The biogeochemical cycles of various ecosystems are linked. As such, it is important to view the biogeochemical cycles of many elements from a global perspective.

The Carbon Cycle | 22.5

The carbon cycle is inseparable from energy flow. Carbon is assimilated as carbon dioxide by plants, consumed in the form of plant and animal tissue by heterotrophs, released through respiration, mineralized by decomposers, accumulated in standing biomass, and withdrawn into long-term reserves. Carbon cycles through the ecosystem at a rate that depends on the rates of primary productivity and decomposition. Both processes are faster in warm, wet ecosystems. In swamps and marshes, organic material stored as raw humus or peat circulates slowly, forming oil, coal, and natural gas. Similar cycling takes place in freshwater and marine environments.

Temporal Variation in the Carbon Cycle | 22.6

Cycling of carbon exhibits daily and seasonal fluctuations. Carbon dioxide builds up at night, when respiration increases. During the day, plants withdraw carbon dioxide from the air, and CO_2 concentration drops sharply. During the growing season, atmospheric concentration drops.

The Global Carbon Cycle | 22.7

Earth's carbon budget is closely linked to the atmosphere, land, and oceans and to the mass movements of air around the planet. In the ocean, surface water acts as the main site of carbon exchange between the atmosphere and ocean. The ability of the surface waters to take up CO_2 is governed largely by the reaction of CO_2 with the carbonate ion to form bicarbonates. The uptake of CO_2 from the atmosphere by terrestrial ecosystems is governed by gross production (photosynthesis). Losses are a function of

autotrophic and heterotrophic respiration, the latter being dominated by microbial decomposers.

The Nitrogen Cycle | 22.8

The nitrogen cycle is characterized by the fixation of atmospheric nitrogen by mutualistic nitrogen-fixing bacteria associated with roots of many plants, largely legumes, and cyanobacteria. Other processes are ammonification, the breakdown of amino acids by decomposer organisms to produce ammonia; nitrification, the bacterial oxidation of ammonia to nitrate and nitrates; and denitrification, the reduction of nitrates to gaseous nitrogen. The global nitrogen cycle follows the pathway of the local nitrogen cycle just described, only on a grander scale. The atmosphere is the largest pool, with comparatively small amounts of nitrogen found in the biomass and soils of terrestrial ecosystems. Major sources of nitrogen to the world's oceans are dissolved forms in the freshwater drainage from rivers and inputs in precipitation.

The Phosphorus Cycle | 22.9

The phosphorus cycle has no significant atmospheric pool. The main pools of phosphorus are rock and natural phosphate deposits. The terrestrial phosphorus cycle follows the typical biogeochemical pathways. In marine and freshwater ecosystems, however, the phosphorus cycle moves through three states: particulate organic phosphorus, dissolved organic phosphates, and inorganic phosphates. Involved in the cycling are phytoplankton, zooplankton, bacteria, and microbial grazers. The global phosphorus cycle is unique among the major biogeochemical cycles in having no significant atmospheric component, although airborne transport of P occurs in the form of soil dust and sea spray. Nearly all of the phosphorus in terrestrial ecosystems is derived from the weathering of calcium phosphate minerals. The transfer of phosphorus from terrestrial to aquatic ecosystems is low under natural conditions; however, the large-scale application of phosphate fertilizers and the disposal of sewage and wastewater to aquatic ecosystems result in a large input of phosphorus to aquatic ecosystems.

The Sulfur Cycle | 22.10

Sulfur has both gaseous and sedimentary phases. Sedimentary sulfur comes from the weathering of rocks, runoff, and decomposition of organic matter. Sources of gaseous sulfur are decomposition of organic matter, evaporation of oceans, and volcanic eruptions. Much of the sulfur released to the atmosphere is a by-product of the burning of fossil fuels. Sulfur enters the atmosphere mostly as hydrogen sulfide, which quickly oxidizes to sulfur dioxide, SO_2. Sulfur dioxide reacts with moisture in the atmosphere to form sulfuric acid that is carried to Earth in precipitation. Plants incorporate this acid into sulfur-bearing amino acids. Consumption, excretion, and death carry sulfur back to soil and aquatic sediments, where bacteria release it in inorganic form.

The Global Sulfur Cycle | 22.11

The global sulfur cycle is a combination of gaseous and sedimentary cycles, because sulfur has reservoirs in Earth's crust and in the atmosphere. The sulfur cycle involves a long-term sedimentary phase in which sulfur is tied up in organic and inorganic deposits, is released by weathering and decomposition, and is carried to terrestrial and aquatic ecosystems in salt solution. The bulk of sulfur first appears in the gaseous phase as a volatile gas, hydrogen sulfide (H_2S), in the atmosphere, which quickly oxidizes to form sulfur dioxide. Once in soluble form, sulfur is taken up by plants and incorporated into organic compounds. Excretion and death carry sulfur in living material back to the soil and to the bottoms of ponds, lakes, and seas, where sulfate-reducing bacteria release it as hydrogen sulfide or as a sulfate.

The Oxygen Cycle | 22.12

Oxygen, the by-product of photosynthesis, is very active chemically. It combines with a wide range of chemicals in Earth's crust, and it reacts spontaneously with organic compounds and reduced substances. It is involved in oxidizing carbohydrates in the process of respiration to release energy, carbon dioxide, and water. The current atmospheric pool of oxygen is maintained in a dynamic equilibrium between the production of oxygen in photosynthesis and its consumption in respiration. An important constituent of the atmospheric reservoir of oxygen is ozone (O_3).

Biogeochemical Cycles Are Linked | 22.13

All of the major biogeochemical cycles are linked; the nutrients that cycle are all components of living organisms, constituents of organic matter. Stoichiometric relationships among the various elements involved in plant processes related to carbon uptake and plant growth have an important influence on the cycling of nutrients in ecosystems.

Study Questions

1. How are the processes of photosynthesis and decomposition involved in the carbon cycle?
2. In the temperate zone, is the atmospheric concentration of carbon dioxide higher during the day or night? Why?
3. Characterize the following processes in the nitrogen cycle: fixation, ammonification, nitrification, and denitrification.
4. What biological and nonbiological mechanisms are responsible for nitrogen fixation?
5. What is the source of sulfur in the sulfur cycle? Why does the sulfur cycle have characteristics of both sedimentary and gaseous cycles?
6. What is the source of phosphorus in the phosphorus cycle?
7. What is the major source of phosphorus input into aquatic ecosystems?
8. What is the role of photosynthesis and decomposition in the oxygen cycle?
9. How would the development of large deposits of fossil fuels have influenced the oxygen concentration of the atmosphere over geologic time?

Further Readings

Aber, J. D. 1992. Nitrogen cycling and nitrogen saturation in temperate forest ecosystems. *Trends in Ecology and Evolution* 7:220–223.

> This paper is an excellent introduction to the concept and consequences of nitrogen saturation in terrestrial ecosystems.

Post, W. M., T. H. Peng, W. R. Emanuel, A. W. King, V. H. Dale, and D. L. DeAngelis. 1990. The global carbon cycle. *American Scientist* 78:310–326.

> An excellent updated review of the global carbon cycle—what we know and do not know about it.

Sarmiento, J. L., and Gruber, N. 2006. *Ocean biogeochemical dynamics.* Princeton, NJ: Princeton University Press.

> This text is an excellent overview of the circulation and interactions of the major elements in the oceans.

Schlesinger, W. H. 1997. *Biogeochemistry: An analysis of global change,* 2nd ed. London: Academic Press.

> A superb text on the subject of biogeochemical cycles. Provides an excellent discussion of nutrient cycling in both terrestrial and aquatic environments.

Sprent, J. I. 1988. *The ecology of the nitrogen cycle.* New York: Cambridge University Press.

> This text provides an ecological perspective to the processes involved in the nitrogen cycle.

Vitousek, P. M., J. Aber, R. W. Howarth, G. E. Likens, P. A. Matson, D. W. Schindler, W. H. Schlesinger, and G. D. Tilman. 1997. Human alteration of the global nitrogen cycle: Sources and consequences. *Ecological Applications* 7:737–750.

> An excellent discussion of how human activities have changed the nitrogen cycle.

Ecological Biogeography

The 19th century witnessed the golden age of the naturalist-explorer, and the tropics were the frontier of natural science. Aspiring young naturalists such as Alfred Russel Wallace, Henry Walter Bates, Joseph Hooker, Alexander Von Humboldt, and Charles Darwin traveled the tropics, cataloging its amazing array of plants and animals. In 11 years in Brazil, Henry Bates collected some 14,712 species of animals, of which more than 8000 were previously unknown to science. To someone who had been trained in the temperate zone, the biological diversity of the tropics was awe inspiring.

Besides being struck by the diversity of life encountered, these young naturalists were also impressed by the similar appearance and nature of organisms found in geographically different regions. The similarity appeared to link form and function—for example, the flying squirrel of eastern North America (a rodent) and the Australian sugar glider (a marsupial) are unrelated species inhabiting the canopies of forests half a world apart, yet they are remarkably similar in appearance (Figure 1). Both species have a flat, bushy tail and an extension of the skin between the foreleg and the hindleg. This body form enables them to glide through the air from one tree limb to another.

The observed link between form and function made by early naturalists was not limited to animals. One of the most striking patterns of similarity observed by these early explorers was that geographically different regions characterized by similar climates support similar types of plant communities. Consider the photos of two desert regions in southern Africa and Australia shown in Figure 2. Both regions have a similar climate with low, seasonal precipitation. Although the plant species occupying these two shrub–desert ecosystems are quite distinct, the dwarf shrubs and the physical structure of the plant communities are very similar in appearance and function. The early naturalists were therefore faced with two conflicting observations in need of explanation. On the one hand, they needed a way to explain the amazing variation—the diversity—

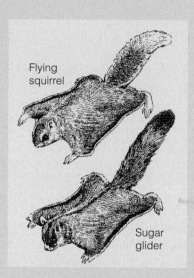

Figure 1 | Convergent evolution of unrelated species showing similar relationships between form and function. A North American rodent, the flying squirrel, and an Australian marsupial, the sugar glider, both have a flat, bushy tail and an extension of the skin between the foreleg and the hindleg that enables them to glide down from one tree limb to another.

observed among plants and animals. On the other hand, they needed an explanation for the observed similarity among different, unrelated species that exhibited similar behaviors (such as the flying squirrel and sugar glider) or inhabited similar types of environments (such as the dwarf shrubs of the shrub deserts).

Although the primary focus of Darwin's theory of natural selection presented in Chapter 5 is to explain the mechanism by which diversity (differences among species) can arise from common ancestry, it offers a second explanation for the similar characteristics observed in plants and animals found in geographically distinct regions—that of convergent evolution. Evolutionary biologists use the term **convergent evolution** to describe the independent evolution of a similar characteristic in two different species not

(a)

(b)

Figure 2 | Shrub desert ecosystems: **(a)** Karoo Desert in southern Africa; **(b)** western New South Wales, Australia. Both sites are characterized by low, seasonal precipitation.

derived from a recent, common ancestor. Recall that natural selection is essentially a two-step process: (1) the production of variation among individuals within the population in some characteristic, (2) which results in differences among individuals in their survival and reproduction. The random processes of genetic mutation and recombination give rise to variation, but the necessity of adaptation to the prevailing (and often changing) environmental conditions guides the invisible hand of design.

The physical processes discussed in Chapters 2–4 give rise to the broad-scale patterns of climate and abiotic environment on Earth. These processes transcend oceans, continents, mountain ranges, and the geographic barriers that serve as mechanisms of isolation in the process of evolution. The similarity between shrub deserts in southern Africa and Australia

shown in Figure 2 reflects the similar form and function of the plant species that have adapted to these distant ecosystems—species that evolved independently but under similar constraints imposed by the physical environment.

The observed similarity in the structure of plant communities found in geographically distant yet physically similar environments led 19th-century plant geographers J. F. Schouw, A. deCandolle, and A. F. W. Schimper to correlate the distribution of vegetation to climate. First, they noted that the world could be divided into zones representing broad categories of vegetation, categories defined on the basis of a similar physical appearance (physiognomy): deserts; grasslands; and coniferous, temperate, and tropical forests. The geographers called these categories **formations.** Second, they noted that regions of the globe occupied by a given type of vegetation formation were characterized by a similar climate. Their observations led to the development of a general understanding of the factors controlling the distribution of vegetation at a global scale.

These early studies relating the global distribution of plant life—and later animal life—with climate developed into the field of biogeography. **Biogeography** is the study of the spatial or geographical distribution of organisms, both past and present. Its goal is to describe and understand the processes responsible for the many patterns in the distribution of species and larger taxonomic groups (genus, family, etc.). Historical biogeography is concerned with reconstructing the origin, dispersal, and extinction of various taxonomic groups. Ecological biogeography is concerned with studying the distribution of contemporary organisms, and it is the focus of Part Seven. We will first explore the large-scale distribution of what the early plant geographers called formations—now generally referred to as *biomes* (see the introduction to Chapter 23). To frame this discussion, we can return to the dichotomy proposed in Part One of dividing Earth's environments into categories of terrestrial (Chapter 23) and aquatic (Chapter 24). To this we will add a category to include the coastal and wetland environments that form the transition zone between land and water (Chapter 25). These chapters reflect ecology of place by examining how the processes and patterns discussed thus far in the text are reflected in different regions of the world.

We will finish our discussion of ecological biogeography by exploring patterns of biological diversity on Earth, both in time and space (Chapter 26).

CHAPTER 23

Terrestrial Ecosystems

Spectacular fall color is a hallmark of the eastern deciduous mixed hardwood forest (Harvey Pond, Madrid, Maine).

n 1939, the ecologists F. E. Clements and V. E. Shelford introduced an approach for combining the broad-scale distribution of plants and associated animals into a single classification system. Clements and Shelford called these biotic units **biomes.** Biomes are classified according to the predominant plant types. Depending on how finely biomes are classified, there are at least eight major terrestrial biome types: tropical forest, temperate forest, conifer forest (taiga or boreal forest), tropical savanna, temperate grasslands, chaparral (shrublands), tundra, and desert (Figure 23.1). These broad categories reflect the relative contribution of three general plant life-forms: trees, shrubs, and grasses. A closed canopy of trees characterizes forest ecosystems. Woodland and savanna ecosystems are characterized by the codominance of grasses and trees (or shrubs). As the names imply, shrubs are the dominant plant form in shrublands, and grasses dominate in grasslands. Desert is a general category used to describe the scarcity of plant cover.

When the plant ecologist Robert Whittaker of Cornell University plotted these biome types on gradients of mean annual temperature and mean annual precipitation, he found they formed a distinctive climatic pattern, as graphed in Figure 23.2. As the graph indicates, boundaries between biomes are broad and often indistinct as they blend into each other. Besides climate, other factors such as topography, soils, and exposure to disturbances such as fire can influence which of several biome types occupies an area.

If we plot the relationship between mean annual temperature and precipitation for locations on the land surface, another general pattern emerges from Whittaker's analysis of the relationship between biomes and climate. The range of observed values for mean annual precipitation declines with decreasing mean annual temperature (note that the range of biomes defined by precipitation along the y-axis decreases with declining temperatures along the x-axis). In geographic terms, this relationship indicates a decrease in the range of environmental conditions defined by moisture availability as one moves from the tropics to the temperate and arctic regions (see label on x-axis of Figure 23.2 and Figure 2.17). This relationship between climate and geography reflects the systematic latitudinal pattern of environmental conditions discussed in Chapter 2 that result directly from seasonal variations in the influx of solar radiation to Earth's surface. Mean annual temperature decreases from the equator to the poles, while seasonal variation in temperatures (and day length;

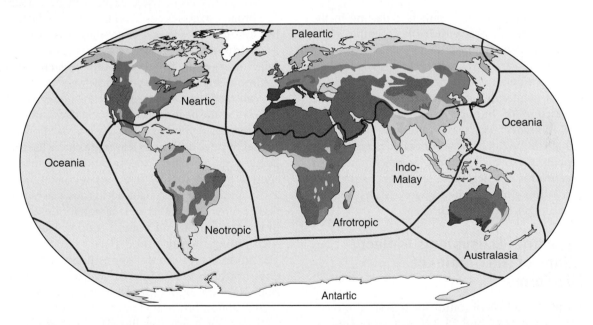

- ◯ Tropical and subtropical moist broadleaf forests
- ◯ Tropical and subtropical dry broadleaf forests
- ◯ Tropical and subtropical coniferous forests
- ● Temperate broadleaf and Mixed forests
- ● Temperate coniferous forests
- ◯ Boreal forests/taiga
- ● Tropical and subtropical grasslands, savannas, and shrublands
- ◯ Temperate grasslands, savannas, and shrublands
- ◯ Flooded grasslands and savannas
- ◐ Montane grasslands and shrublands
- ◯ Tundra
- ● Mediterranean forests, woodlands, and scrub
- ● Deserts and xeric shrublands
- ◐ Mangroves

Figure 23.1 | Major biomes and biogeographical realms of the world.
(Adapted from Olson et al. 2001.)

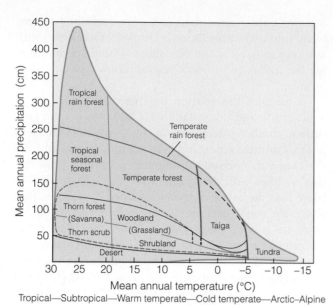

Figure 23.2 | The pattern of terrestrial biomes in relation to temperature and moisture. Where the climate varies, soil can shift the balance between types. The dashed line encloses environments in which either grassland or one of the types dominated by woody plants may prevail.

(Adapted from Whittaker 1970.)

see Figure 2.5) increases. The result is a decline in the growing season (period over which photosynthesis and plant growth can be maintained).

The systematic variation in climate with latitude is not limited to temperature. Average annual precipitation decreases with increasing latitude (see Figure 2.18) due to the interaction of humidity and temperature (see Section 2.6). With declining temperatures, the amount of moisture that can be held in the air declines (see Figure 2.16), reducing the overall amount of precipitation. As we shall see in this chapter, these systematic patterns of climate across the globe dictate the general distribution of terrestrial biomes on Earth's surface.

23.1 | Terrestrial Ecosystems Reflect Adaptations of the Dominant Plant Life-Forms

Given that the broad classification of terrestrial biomes presented in Figures 23.1 and 23.2 (forest, woodland/savanna, shrubland, and grassland) reflects the relative contribution of three general plant life-forms (trees, shrubs, and grasses), the question of what controls the distribution of biomes relative to climate becomes: Why are there consistent patterns in the distribution and abundance of these three dominant plant life-forms that relate to climate and the physical environment? The answer to this question lies in the adaptations that these three very different plant life-forms represent as well as the advantages and constraints arising from these adaptations under different environmental conditions.

Although the broad categories of grasses, shrubs, and trees each represent a diverse range of species and characteristics, they have fundamentally different patterns of carbon allocation and morphology (see Chapter 6). Grasses allocate less carbon to the production of supportive tissues (stems) than do woody plants (shrubs and trees), enabling grasses to maintain a higher proportion of their biomass in photosynthetic tissues (leaves). For woody plants, shrubs allocate a lower percentage of their resources to stems than do trees. The production of woody tissue gives the advantage of height and access to light, but it also has the associated cost of maintenance and respiration. If this cost cannot be offset by carbon gain through photosynthesis, the plant is unable to maintain a positive carbon balance and will die (see Chapter 6, Figure 6.8). As a result, as environmental conditions become adverse for photosynthesis (dry, low nutrient concentrations, or short growing season and cold temperatures), trees will decline in both stature and density until they can no longer persist as part of the plant community.

Within the broad classes of forest and woodland ecosystems in which trees are dominant or codominant, leaf form is another plant characteristic that ecologists use to classify ecosystems. Leaves can be classified into two broad categories based on their longevity. Leaves that live for only a single year or growing season are classified as **deciduous,** whereas those that live beyond a year are called **evergreen.** The deciduous leaf is characteristic of environments with a distinct growing season. Leaves are typically shed at the end of the growing season and then regrown at the beginning of the next. Deciduous leaf type is further divided into two categories based on dormancy period. Winter-deciduous leaves are characteristic of temperate regions, where the period of dormancy corresponds to low (below freezing) temperatures (Figures 23.3a and 23.3b). Drought-deciduous leaves are characteristic of environments with seasonal rainfall, especially in the subtropical and tropical regions, where leaves are shed during the dry period (Figures 23.3c and 23.3d). The advantage of the deciduous habit is that the plant does not have the additional cost of maintenance and respiration during the period of the year when environmental conditions restrict photosynthesis.

Evergreen leaves can likewise be classified into two broad categories. The broadleaf evergreen leaf type (Figure 23.4a) is characteristic of environments with no distinct growing season where photosynthesis and growth continue year-round, such as tropical rain forests. The needle-leaf evergreen form (Figure 23.4b) is characteristic of environments where the growing season is very short (northern latitudes) or nutrient availability severely constrains photosynthesis and plant growth.

A simple economic model has been proposed to explain the adaptation of this leaf form (see Chapter 6). The production of a leaf has a "cost" to the plant that can be

Figure 23.3 | Examples of winter and drought-deciduous trees. Temperate deciduous forest in central Virginia during **(a)** summer and **(b)** winter seasons. Semiarid savanna/woodland in Zimbabwe, Africa, during **(c)** rainy and **(d)** dry seasons.

defined in terms of the carbon and other nutrients required to construct the leaf. The time required to pay back the cost of production (carbon) will be a function of the rate of net photosynthesis (carbon gain). If environmental conditions result in low rates of net photosynthesis, the period of

time required to pay back the cost of production will be longer. If the rate of photosynthesis is low enough, it may not be possible to pay back the cost over the period of a single growing season. A plant adapted to such environmental conditions cannot "afford" a deciduous leaf form,

Figure 23.4 | Examples of evergreen trees. **(a)** Broadleaf evergreen trees dominate the canopy of this tropical rain forest in Queensland, Australia. **(b)** Needle-leaf evergreen trees (foxtail pine) inhabit the high-altitude zone of the Sierra Nevada in western North America.

Climate Diagrams

As illustrated in Figure 23.2, the distribution of terrestrial biomes is closely related to climate. The measures of regional climate used in Whittaker's graph (Figure 23.2) are mean annual temperature and precipitation. Yet, as we will see in the discussion of the various biome types, the distribution of terrestrial ecosystems is influenced by other aspects of climate as well, namely seasonality of both temperature and precipitation. Topographic features such as mountains and valleys also influence the climate of a region.

To help understand the relationship between regional climate and the distribution of terrestrial ecosystems, for each biome discussed in this chapter (tropical forest, savanna, etc.) we present a map showing its global distribution. Accompanying the map is a series of climate diagrams. The diagrams describe the local climate at representative locations around the world where that particular biome type is found. See Figure 1 for a representative climate diagram, which we have labeled to help you interpret the information it presents. As you study the diagram, take particular note of the patterns of seasonality.

1. In Figure 23.12, what is the distinctive feature of the climate diagrams for these tropical savanna ecosystems? How do the patterns differ between sites in the Northern and Southern Hemispheres? What feature of Earth's climate system discussed in Chapter 3 is responsible for these distinctive patterns?

2. In Figure 23.23, what feature of the climate is common to all mediterranean ecosystems?

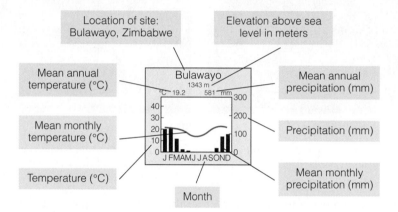

Figure 1 | Climate diagram for Bulawayo, Zimbabwe. This city is in the Southern Hemisphere, where the cooler winter season occurs during the period May–August. Note the distinct dry season during the winter months, with the rainy season beginning in October (spring) and lasting through the summer months.

which requires producing new leaves every year. The leaves of evergreens, however, may survive for several years. So, under this model, we can view the needle-leaf evergreen as a plant adapted for survival in an environment with a distinct growing season, where conditions limit the plant's ability to produce enough carbon through photosynthesis during the growing season to pay for the cost of producing the leaves.

Upon combining the simple classification of plant life-forms and leaf type with the large-scale patterns of climate presented earlier, we can begin to understand the distribution of biome types relative to the axes of temperature and precipitation shown in Figure 23.2. Ecosystems characteristic of warm, wet climates with no distinct seasonality are dominated by broadleaf evergreen trees and are called tropical (and subtropical) rain forest. As conditions become drier, with a distinct dry season, the broadleaf evergreen habit gives way to drought-deciduous trees that characterize the seasonal tropical forests. As precipitation declines further, the stature and density of these trees declines, giving rise to the woodlands and savannas that are characterized by the coexistence of trees (shrubs) and grasses. As precipitation further declines, trees can no longer be supported, giving rise to the arid shrublands (thorn scrub) and desert.

The temperature axis represents the latitudinal gradient from the equator to the poles (see geographical labels on x-axis of Figure 23.2). Moving from the broadleaf evergreen forests of the wet tropics into the cooler, seasonal environments of the temperate regions, the dominant trees are winter-deciduous. These are the regions of temperate deciduous forest. In areas of the temperate region where precipitation is insufficient to support trees, grasses dominate and give rise to the prairies of North America, the steppes of Eurasia, and the pampas of Argentina. Moving poleward, the temperate deciduous forests give way to the needle-leaf–dominated forests of the boreal region (conifer forest or taiga). As temperatures become more extreme and the growing season shorter, trees can no longer be supported, and the short-stature shrubs and grasses characteristic of the tundra dominate the landscape ecosystems of the arctic region.

In the following sections, we will examine the eight major categories of terrestrial biomes outlined in Figure 23.2. We begin each section by relating their geographic distribution to the broad-scale constraints of regional climate, as outlined in Figure 23.2, as well as to associated patterns of seasonality in temperature and precipitation (see Quantifying Ecology 23.1: Climate Diagrams) that function as constraints on the dominant plant lifeforms and patterns of primary and secondary productivity. In our discussion, we emphasize the unique physical and biological characteristics defining these broad categories of terrestrial ecosystems (biomes).

23.2 | Tropical Forests Characterize the Equatorial Zone

The tropical rain forests are restricted primarily to the equatorial zone between latitudes 10°N and 10°S (Figure 23.5), where the temperatures are warm throughout the year and rainfall occurs almost daily. The largest and most continuous region of rain forest in the world is in the Amazon basin of South America (Figure 23.6). The second largest is located in Southeast Asia, and the third largest is in West Africa around the Gulf of Guinea and in the Congo basin. Smaller rain forests occur along the northeastern coast of Australia, the windward side of the Hawaiian Islands, the South Pacific Islands, the east coast of Madagascar, northern South America, and southern Central America.

The climate of tropical rain forest regions varies geographically but is typically characterized by a mean temperature of all months exceeding 18°C and minimum monthly precipitation above 60 mm (see climate diagrams for representative tropical rain forest sites in Figure 23.6). Within the lowland forest zone, mean annual temperatures typically exceed 25°C with an annual range less than 5°C.

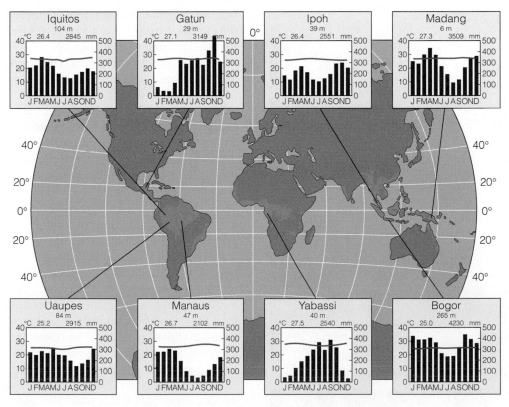

Figure 23.5 | Geographic distribution of Earth's tropical forest ecosystems and associated climate diagrams showing long-term patterns of monthly temperature and precipitation for selected locations. Note the lack of seasonality in mean monthly temperatures, which are above 20°C. Although the rainfall in some regions is seasonal, note that minimum monthly precipitation is typically above 60 mm, and total annual precipitation above 2000 mm.

(Adapted from Archibold 1995.)

(a)

(b)

Figure 23.6 | Tropical rain forests in **(a)** Amazon Basin (South America) and **(b)** Malaysia (Southeast Asia). Despite being taxonomically distinct, these two tropical rain forest regions are dominated by broadleaf evergreen trees and support vigorous plant growth year-round. Tropical rain forests represent the most diverse and productive terrestrial ecosystems on our planet.

Tropical rain forests have a high diversity of plant and animal life. Covering only 6 percent of the land surface, tropical rain forests account for more than 50 percent of all known plant and animal species. Tree species number in the thousands. A 10-km^2 area of tropical rain forest may contain 1500 species of flowering plants and up to 750 species of trees. The richest area is the lowland tropical forest of peninsular Malaysia, which contains some 7900 species.

Nearly 90 percent of all nonhuman primate species live in the tropical rain forests of the world (Figure 23.7). Sixty-four species of New World primates, small mammals with prehensile tails, live in the trees. The Indo-Malaysian forests are inhabited by a number of primates, many with a limited distribution within the region. The orangutan, an arboreal ape, is confined to the island of Borneo. Peninsular Malaysia has seven species of primates, including three gibbons, two langurs, and two macaques. The long-tailed macaque is common in disturbed or secondary forests, and the pig-tailed macaque is a terrestrial species adaptable to human settlements. The tropical rain forest of Africa is home to mountain gorillas and chimpanzees. The diminished rain forest of Madagascar holds 39 species of lemurs.

Tropical rain forests may be divided into five vertical layers (Figure 23.8): emergent trees, upper canopy, lower canopy, shrub understory, and a ground layer of herbs and ferns. Conspicuous in the rain forest are lianas—climbing vines—growing upward into the canopy, epiphytes growing on the trunks and branches, and strangler figs (*Ficus* spp.) that grow downward from the canopy to the ground. Many large trees develop plank-like outgrowths called buttresses (Figure 23.9). They function as prop roots to support trees rooted in shallow soil that offers poor anchorage. The floor of a tropical rain forest is thickly laced with roots, both large and small, forming a dense mat on the ground.

The continually warm, moist conditions in rain forests promote strong chemical weathering and rapid leaching of soluble materials. The characteristic soils are Oxisols, which are deeply weathered with no distinct horizons. Ultisols may develop in areas with more seasonal precipitation regimes and are typically associated with forested regions that exhibit seasonal soil moisture deficits. Areas of volcanic activity in parts of Central and Southeast Asia, where recent ash deposits quickly weather, are characterized by Andosols (see Figure 4.12).

The warmer, wetter conditions of the tropical rain forest result in high rates of net primary productivity and subsequent high annual rates of litter input to the forest floor. Little litter accumulates, however, because decomposers

(a)

(b)

Figure 23.7 | Examples of primate species that inhabit the tropical rain forests of the world: the chimpanzee (*Pan troglodytes*) inhabits the tropical rain forests of Central Africa and (b) the orangutan (*Pongo pygmaeus*) inhabits the tropical rain forests of Borneo (Southeast Asia).

Emergent canopy
(trees widely spaced)

Upper canopy
(medium-spaced
crowns)

Lower canopy

Understory
(shrubs and saplings)

Ground cover
(herbs and ferns)

Height above ground (m)

40
35
30
25
20
15
10
5
0

Figure 23.8 | Vertical stratification of a tropical rain forest.

Figure 23.9 | Plank-like buttresses help to support tall rain forest trees.

(a)

(b)

Figure 23.10 | A tropical dry forest in Costa Rica during the (a) rainy and (b) dry season. Most of the tropical dry forests in Central America have disappeared from land-clearing for agriculture.

consume the dead organic matter almost as rapidly as it falls to the forest floor. Most of the nutrients available for uptake by plants are a result of the rapidly decomposed organic matter that is continuously falling to the surface. Growing plants, however, rapidly absorb these nutrients. The average time for leaf litter to decompose is 24 weeks.

Moving from the equatorial zone to the regions of the tropics that are characterized by greater seasonality in precipitation, the broadleaf evergreen forests are replaced by the dry tropical forests (Figure 23.10). Dry tropical forests undergo a dry season whose length is based on latitude. The more distant the forest is from the equator, the longer is the dry season—in some areas, up to 8 months. During the dry season, the drought-deciduous trees and shrubs drop their leaves. Before the start of the rainy season, which may be much wetter than the wettest time in the rain forest, the trees begin to leaf. During the wet season, the landscape becomes uniformly green.

The largest proportion of tropical dry forest is found in Africa and South America, to the south of the zones dominated by rain forest. These regions are influenced by the seasonal migration of the Intertropical Convergence Zone (see Figure 2.19). In addition, areas of Central America, northern Australia, India, and Southeast Asia are also classified as dry tropical forest. Much of the original forest, especially in Central America and India, has been converted to agricultural and grazing land.

23.3 | Tropical Savannas Are Characteristic of Semiarid Regions with Seasonal Rainfall

The term *savanna* was originally used to describe the treeless areas of South America. Now it is generally applied to a range of vegetation types in the drier tropics and subtropics characterized by a ground cover of grasses with scattered shrubs or trees. Savanna includes an array of veg-

etation types representing a continuum of increasing cover of woody vegetation, from open grassland to widely spaced shrubs or trees and to woodland (Figure 23.11). In South America, the more densely wooded areas are referred to as *cerrado*. The campos and llano are characterized by a more open appearance (lower density of trees), and thorn scrub is the dominant cover of the caatinga. In Africa, the miombo, mopane, and *Acacia* woodlands can be distinguished from the more open and parklike bushveld. Scattered individuals of *Acacia* and *Eucalyptus* dominate the mulga and brigalow of Australia.

The physiognomic diversity of the savanna vegetation reflects the different climate conditions occurring throughout this widely distributed ecosystem (Figure 23.12). Moisture appears to control the density of woody vegetation, a function of both rainfall (amount and distribution) and soil—its texture, structure, and water-holding capacity (Figure 23.13; also see Chapter 4).

(a) **(b)**

Figure 23.11 | Savanna ecosystems, such as the **(a)** cerrano of South America and **(b)** mulga woodlands of central Australia are characterized by a ground cover of grasses with scattered shrubs or trees.

Savannas are associated with a warm continental climate with distinct seasonality in precipitation and a large interannual (year to year) variation in total precipitation (see climate diagrams for representative savanna sites in Figure 23.12). Mean monthly temperatures typically do not fall below 18°C, although during the coldest months in highland areas, temperatures can be considerably lower. There is seasonality in temperatures, and maximum temperatures occur at the end of the wet season. The nature of the vegetation cover, however, is more closely determined by the amount and seasonality of precipitation than by temperature.

Savannas, despite their differences in vegetation, exhibit a certain set of characteristics. Savannas occur on land surfaces of little relief—often on old plateaus, interrupted by escarpments and dissected by rivers. Continuous weathering in these regions has produced nutrient-poor Oxisols, which are particularly deficient in phosphorus. Alfisols are common in the drier savannas, whereas Entisols are associated with the driest savannas (see Figure 4.12). Subject to recurrent fires, the dominant vegetation is fire adapted. Grass cover with or without woody vegetation is always present, and the woody component is short-lived—individuals seldom survive for

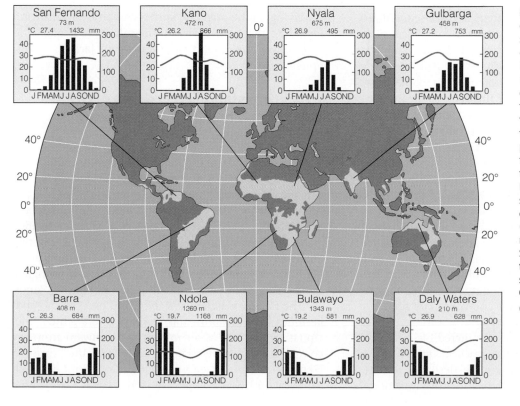

Figure 23.12 | Geographic distribution of Earth's tropical savanna ecosystems and associated climate diagrams showing long-term patterns of monthly temperature and precipitation for selected locations. Seasonal patterns of temperature are similar to those of the tropical forest sites presented in Figure 23.5, reflecting the tropical and subtropical climates at these sites. Note, however, the distinct seasonality of precipitation that characterizes these ecosystems (also note the shift in timing of rainy season for Northern and Southern Hemisphere locations, reflecting the seasonal shift of the Intertropical Convergence Zone).

(Adapted from Archibold 1995.)

Figure 23.13 | Diagram showing the interaction between annual precipitation and soil texture in defining the transition from woodland to savanna and grassland in southern Africa. Plants have more limited access to soil moisture on the heavily textured soils (clays) than on the coarser sands, so annually more precipitation is needed to support the woody plants.

more than several decades. Savannas are characterized by a two-layer vertical structure due to the ground cover of grasses and the presence of shrubs or trees.

The yearly cycle of plant activity and subsequent productivity in tropical savannas is largely controlled by the markedly seasonal precipitation and corresponding changes in available soil moisture. Most leaf litter is decomposed during the wet season, and most woody debris is consumed by termites during the dry season.

The microenvironments associated with tree canopies can influence species distribution, productivity, and soil characteristics. Stem flow and associated litter accumulation result in higher soil nutrients and moisture under tree canopies, often encouraging increased productivity and the establishment of species adapted to the more shaded environments.

Savannas can support a large and varied assemblage of herbivores—invertebrate and vertebrate, grazing and browsing. The African savanna, visually at least, is dominated by a large and diverse ungulate fauna of at least 60 species that share the vegetative resources. Some species, such as the wildebeest and zebra, are migratory during the dry season.

Savanna vegetation supports an incredible number of insects: flies, grasshoppers, locusts, crickets, carabid beetles, ants, and detritus-feeding dung beetles and termites. Mound-building termites excavate and move tons of soil,

mixing mineral soil with organic matter. Some species construct extensive subterranean galleries, and others accumulate organic matter.

Living on the ungulate fauna is an array of carnivores including the lion, leopard, cheetah, hyena, and wild dog. Scavengers, including vultures and jackals, subsist on leftover prey.

23.4 | Grassland Ecosystems of the Temperate Zone Vary with Climate and Geography

Natural grasslands occupy regions where rainfall is between 25 and 80 cm a year, but they are not exclusively climatic. Many exist through the intervention of fire and human activity. Conversions of forests into agricultural lands and the planting of hay and pasturelands extended grasslands into once forested regions. Formerly covering about 42 percent of the land surface of Earth, natural grasslands have shrunk to less than 12 percent of their original size because of conversion to cropland and grazing lands.

The natural grasslands of the world occur in the midlatitudes in midcontinental regions, where annual precipitation declines as air masses move inward from the coastal environments (Figure 23.14). In the Northern Hemisphere, these regions include the prairies of North America and the steppes of central Eurasia. In the

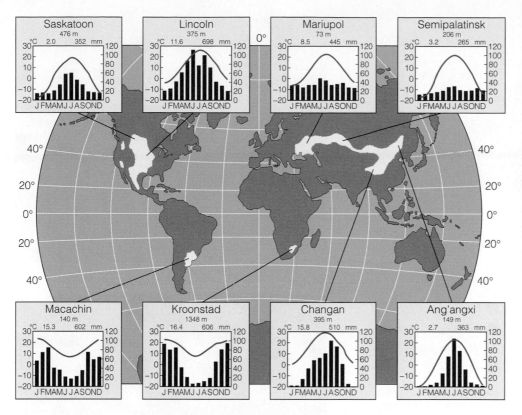

Figure 23.14 | Geographic distribution of Earth's temperate grassland ecosystems and associated climate diagrams showing long-term patterns of monthly temperature and precipitation for selected locations. Most of the major grassland regions are midcontinental with a distinct seasonality in temperature. Mean annual precipitation is typically well below 1000 mm, comparable in range to that observed for tropical and subtropical savannas but less than that of temperate deciduous forests (compare with Figure 23.27).

(Adapted from Archibold 1995.)

Southern Hemisphere, grasslands are represented by the pampas of Argentina and the veld of the high plateaus of southern Africa. Smaller areas occur in southeastern Australia and the drier parts of New Zealand.

The temperate grassland climate is one of recurring drought, and much of the diversity of vegetation cover reflects differences in the amount and reliability of precipitation. Grasslands do the poorest where precipitation is lowest and the temperatures are high. They are tallest in stature and the most productive where mean annual precipitation is greater than 800 mm and mean annual temperature is above 15°C. Thus, native grasslands of North America, influenced by declining precipitation from east to west, consist of three main types distinguished by the height of the dominant species: tallgrass, mixed-grass, and shortgrass prairie (Figure 23.15). **Tallgrass prairie** (Figure 23.16a) is dominated by big bluestem (*Andropogon gerardi*), growing 1 m tall with flowering stalks 1 to 3 1/2 m tall. **Mixed-grass prairie** (Figure 23.16b), typical of the Great Plains, is composed largely of needlegrass–grama grass (*Bouteloua–Stipa*). South and west of the mixed prairie and grading into the desert regions is the **shortgrass prairie** (Figure 23.16c), dominated by sod-forming blue grama (*Bouteloua gracilis*) and buffalo grass (*Buchloe dactyloides*), which has remained somewhat intact, and desert grasslands. From southeastern Texas to southern Arizona and south into Mexico lies the **desert grassland**, similar in many respects to the shortgrass plains, except that three-awn grass (*Aristida* spp.) replaces buffalo grass. Confined largely to the Central Valley of

California is **annual grassland.** It is associated with a mediterranean climate (see Section 23.6) characterized by rainy winters and hot, dry summers. Growth occurs during early spring, and most plants are dormant in summer, turning the hills a dry tan color accented by the deep green foliage of scattered California oaks.

At one time, the great grasslands of the Eurasian continent extended from eastern Europe to western Siberia south to Kazakhstan. These **steppes,** treeless except for ribbons and patches of forest, are divided into four belts of latitude, from the mesic meadow steppes in the north to semiarid grasslands in the south.

In the Southern Hemisphere, the major grasslands exist in southern Africa and southern South America. Known as **pampas,** the South American grasslands extend westward in a large semicircle from Buenos Aires to cover about 15 percent of Argentina. These pampas have been modified by the introduction of European forage grasses and alfalfa (*Medicago sativa*), and the eastern tallgrass pampas have been converted to wheat and corn. In Patagonia, where annual rainfall averages about 25 cm, the pampas change to open steppe.

The **velds** of southern Africa (not to be confused with savanna) occupy the eastern part of a high plateau 1500 to 2000 m above sea level in the Transvaal and the Orange Free State.

Australia has four types of grasslands: arid tussock grassland in the northern part of the continent, where the rainfall averages between 20 and 50 cm, mostly in the summer; arid hummock grasslands in areas with less

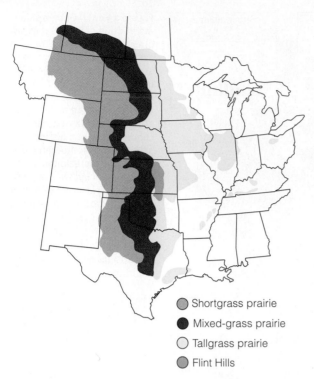

- ● Shortgrass prairie
- ● Mixed-grass prairie
- ○ Tallgrass prairie
- ● Flint Hills

Figure 23.15 | Map showing the original extent of shortgrass, mixed-grass, and tallgrass prairies in North America before the arrival of Europeans.
(After Reichman 1987.)

than 20 cm rainfall; coastal grasslands in the tropical summer rainfall region; and subhumid grasslands along coastal areas where annual rainfall is between 50 and 100 cm. However, the introduction of fertilizers, nonnative grasses, legumes, and sheep grazing have changed most of these grasslands.

Grasslands support a diversity of animal life dominated by herbivorous species, both invertebrate and vertebrate. Large grazing ungulates and burrowing mammals are the most conspicuous vertebrates (Figure 23.17). The North American grasslands once were dominated by huge migratory herds of millions of bison (*Bison bison*) and the forb-consuming pronghorn antelope (*Antilocarpa americana*). The most common burrowing rodent was the prairie dog (*Cynomys* spp.), which along with gophers (*Thomomys* and *Geomys* spp.) and the mound-building harvester ants (*Pogonomyrex* spp.) appeared to be instrumental in developing and maintaining the ecological structure of the shortgrass prairie.

The Eurasian steppes and the Argentine pampas lack herds of large ungulates. On the pampas, the two major large herbivores are the pampas deer (*Ozotoceros bezoarticus*) and, farther south, the guanaco (*Lama guanicoe*), a small relative of the camel. These species, however, are greatly reduced in number compared to historical times.

The African grassveld once supported great migratory herds of wildebeest (*Connochaetes taurinus*) and zebra

(a)

(b)

(c)

Figure 23.16 | North American grasslands. **(a)** A remnant tall-grass prairie in Iowa; **(b)** the mixed-grass prairie has been called "daisyland" for the diversity of its wildflowers; **(c)** shortgrass steppe in western Wyoming.

(a)

(b)

Figure 23.17 | North American grasslands were once dominated by **(a)** large grazing ungulates such as bison and **(b)** burrowing mammals such as the prairie dog.

(*Equus* spp.) along with their associated carnivores, the lion (*Panthera leo*), leopard (*Panthera pardus*), and hyena (*Crocuta crocuta*). The great ungulate herds have been destroyed and replaced with sheep, cattle, and horses.

The Australian marsupial mammals evolved many forms that are the ecological equivalents of placental grassland mammals. The dominant grazing animals are several kangaroo species, especially the red kangaroo (*Macropus rufus*) and the gray kangaroo (*Macropus giganteus*).

Grasslands evolved under the selective pressure of grazing. Thus, up to a point, grazing stimulates primary production (see Section 14.13). Although the most conspicuous grazers are large herbivores, the major consumers in grassland ecosystems are invertebrates. The heaviest consumption takes place belowground, where the dominant herbivores are nematodes.

The most visible feature of grassland is the tall, green, ephemeral herbaceous growth that develops in spring and dies back in autumn. One of the three strata in the grassland, it arises from the crowns, nodes, and rosettes of plants hugging the soil. The ground layer and the belowground root layer are the other two major strata of grasslands. The highly developed root layer can make up more than half the total plant biomass and typically extends fairly deep into the soil.

Depending on their history of fire and degree of grazing and mowing, grasslands accumulate a layer of mulch that retains moisture and, with continuous turnover of fine roots, adds organic matter to the mineral soil. Dominant soils of the grasslands are Mollisols (see Figure 4.12) with a relatively thick, dark-brown to black surface horizon that is rich in organic matter. Soils typically become thinner and paler in the drier regions because less organic material is incorporated into the surface horizon.

The productivity of temperate grassland ecosystems is primarily related to annual precipitation (Figure 23.18), yet temperature can complicate this relationship. Increasing temperatures have a positive effect on photo-

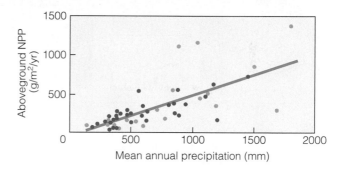

Figure 23.18 | Relationship between aboveground net primary production (NPP) and mean annual precipitation for 52 grassland sites around the world. Each point represents a different grassland site. North American grasslands are indicated by dark-green dots.

(Adapted from Lauenroth 1979.)

synthesis but can actually reduce productivity by increasing the demand for water.

23.5 | Deserts Represent a Diverse Group of Ecosystems

The arid regions of the world occupy from 25 to 35 percent of Earth's landmass (Figure 23.19). The wide range reflects the various approaches used to define desert ecosystems based on climate conditions and vegetation types. Much of this land lies between 15° and 30° latitude, where the air that is carried aloft along the Intertropical Convergence Zone subsides to form the semipermanent high-pressure cells that dominate the climate of tropical deserts (see Figure 2.17). The warming of the air as it descends, and cloudless skies, result in intense radiation heat during the summer months.

Temperate deserts lie in the rain shadow of mountain barriers or are located far inland, where moist maritime air rarely penetrates. Here, temperatures are high during the summer but can drop to below freezing during the winter months. Thus the lack of precipitation, rather

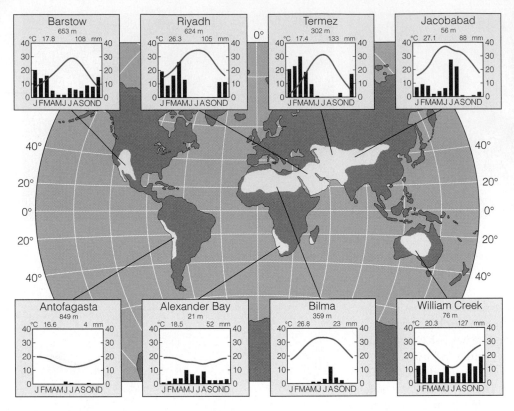

Figure 23.19 | Geographic distribution of Earth's desert (arid) ecosystems and associated climate diagrams showing long-term patterns of monthly temperature and precipitation for selected locations. Seasonal variations in temperature differ among these representative sites; but at all sites, annual precipitation is well below the potential evaporative demand, resulting in low soil moisture to support primary productivity.

(Adapted from Archibold 1995.)

than continually high temperatures, is the distinctive characteristic of all deserts.

Most of the arid environments are found in the Northern Hemisphere. The Sahara, the world's largest desert, covers approximately 9 million km^2 of North Africa. It extends the breadth of the African continent to the deserts of the Arabian Peninsula, continuing eastward to Afghanistan and Pakistan and finally terminating in the Thar Desert of northwest India. The temperate deserts of Central Asia lie to the north. The most westerly of these is the Kara Kum desert region of Turkmenistan. Eastward lie the high-elevation deserts of western China and the high plateau of the Gobi Desert.

A similar transition to temperate desert occurs in western North America. Here, the Sierra Nevada effectively blocks the passage of moist air into the interior of the Southwest. Mountain ranges run parallel to the Sierras throughout the northern part of this region, and desert basins occur on the eastern sides of these ranges.

Apart from the drier parts of southern Argentina, the deserts of the Southern Hemisphere all lie within the subtropical high-pressure belt that mirrors that of the Northern Hemisphere (see preceding discussion). Cold ocean currents also contribute to the development of arid coastal regions (see Section 2.5). Drought conditions are severe along a narrow strip of the coast that includes Chile and Peru. The drier parts of Argentina lie in the rain shadow of the Andes.

The deserts of southern Africa include three regions. The Namib Desert occupies a narrow strip of land that runs along the west coast of Africa from southern Angola to the border of the cape region of South Africa. This region continues south and east across South Africa as the Karoo, which merges with the Kalahari Desert to the north in Botswana. The most extensive region of arid land in the Southern Hemisphere is found in Australia, where more than 40 percent of the land is classified as desert.

Deserts are not the same everywhere. Differences in moisture, temperature, soil drainage, topography, alkalinity, and salinity create variations in vegetation cover, dominant plants, and groups of associated species. There are hot deserts and cold deserts, extreme deserts and semideserts, ones with enough moisture to verge on being grasslands or shrublands, and gradations between those extremes within continental deserts.

Cool deserts—including the Great Basin of North America, the Gobi, Takla Makan, and Turkestan deserts of Asia—and high elevations of hot deserts are dominated by *Artemisia* and chenopod shrubs (Figure 23.20). They may be considered shrub steppes or desert scrub. In the Great Basin of North America, the northern, cool, arid region lying west of the Rocky Mountains is the northern desert scrub. The climate is continental, with warm summers and prolonged cold winters. The vegetation falls into two main associations: one is sagebrush, dominated by *Artemisia tridentata*, which often forms pure stands; the other is shadscale (*Atriplex confertifolia*), a C$_4$ species, and other chenopods (halophytes—tolerant of saline soils).

(a)

(b)

Figure 23.20 | Two examples of desert scrub. **(a)** Northern desert shrubland in Wyoming is dominated by sagebrush (*Artemisia*). Although classified as a cool desert plant, sagebrush forms one of the most important shrub types in North America. **(b)** Saltbrush shrubland in Victoria, Australia, is dominated by *Atriplex* and is an ecological equivalent of the Great Basin shrublands in North America.

A similar type of desert scrub exists in the semiarid inland of southwestern Australia. Many chenopod species, particularly the saltbushes of the genera *Atriplex* and *Maireana*, form extensive low shrublands on low riverine plains.

The hot deserts range from those lacking vegetation to ones with some combination of chenopods, dwarf shrubs, and succulents (Figure 23.21). Creosote bush (*Larrea divaricata*) and bur sage (*Franseria* spp.) dominate the deserts of southwestern North America—the Mojave, the Sonoran, and the Chihuahuan. Areas of favorable moisture support tall growths of *Acacia* spp., saguaro (*Cereus giganteus*), palo verde (*Cercidium* spp.), ocotillo (*Fouquieria* spp.), yucca (*Yucca* spp.), and ephemeral plants.

Both plants and animals adapt to the scarcity of water by either drought evasion or drought resistance. Drought-evading plants flower only when moisture is present. They persist as seeds during drought periods, ready to sprout, flower, and produce seeds when moisture and temperature are favorable. If no rains come, these ephemeral species do not germinate and grow.

Drought-evading animals, like their plant counterparts, adopt an annual cycle of activities or go into estivation or some other dormant stage during the dry season. For example, the spadefoot toad (*Scaphiopus*; Figure 23.22) remains underground in a gel-lined underground cell, making brief reproductive appearances during periods of winter and summer rains. If extreme drought develops during the breeding season, many animals such as lizards and birds do not reproduce.

(a)

(b)

Figure 23.21 | Two examples of hot deserts. **(a)** The Chihuahuan Desert in Nuevo Leon, Mexico. The substrate of this desert is sand-sized particles of gypsum. **(b)** Dunes in the Saudi Arabian desert near Riyadh. Note the extreme sparseness of vegetation.

Figure 23.22 | A spadefoot toad, named for the black, sharp-edged "spades" on its hind feet, emerges from its desert burrow to breed when the rains come.

Desert plants may be deep-rooted woody shrubs, such as mesquite (*Prosopis* spp.) and *Tamarix*, whose tap-roots reach the water table, rendering them independent of water supplied by rainfall. Some plants, such as *Larrea* and *Atriplex*, are deep-rooted perennials with superficial laterals that extend as far as 15 to 30 m from the stems. Other perennials, such as the various species of cactus, have shallow roots that often extend no more than a few centimeters below the surface.

Despite their aridity, desert ecosystems support a surprising diversity of animal life, including a wide assortment of beetles, ants, locusts, lizards, snakes, birds, and mammals. The mammals are mostly herbivorous species. Grazing herbivores of the desert tend to be generalists and opportunists in their mode of feeding. They consume a wide range of species, plant types, and parts. Desert rodents, particularly the family Heteromyidae, and ants feed largely on seeds and are important in the dynamics of desert ecosystems. Seed-eating herbivores can eat up to 90 percent of the available seeds. That consumption can distinctly affect plant composition and plant populations. Desert carnivores, such as foxes and coyotes, have mixed diets that include leaves and fruits; even insectivorous birds and rodents eat some plant material. Omnivory, rather than carnivory and complex food webs, seems to be the rule in desert ecosystems.

The infrequent rainfall coupled with high rates of evaporation limit the availability of water to plants, so primary productivity is low. Most desert soils are poorly developed Aridisols and Entisols (see Figure 4.12), and the sparse cover of arid lands limits the ability of vegetation to heavily modify the soil environment. Underneath established plants, however, "islands of fertility" can develop due to higher litter input and the enrichment by wastes from animals that seek shade, particularly under shrubs.

23.6 | Mediterranean Climates Support Temperate Shrublands

Shrublands—plant communities where the shrub growth form is either dominant or codominant—are difficult types of ecosystems to categorize, largely because of the difficulty in characterizing the term *shrub* itself. In general, a shrub is a plant with multiple woody, persistent stems but no central trunk and a height from 4.5 to 8 m. However, under severe environmental conditions, even many trees do not exceed that size. Some trees—particularly individuals that coppice (resprout from the stump) after destruction of the aboveground tissues by fire, browsing, or cutting—are multistemmed, and some shrubs can have large, single stems. In addition, the shrub growth form can be a dominant component of a variety of tropical and temperate ecosystems, including the tropical savannas and scrub desert communities (see Sections 23.2 and 23.3, respectively). However, in five widely disjunct regions along the western margins of the continents, between 30° and 40° latitude, are found the mediterranean ecosystems dominated by evergreen shrubs and sclerophyllous trees that have adapted to the distinctive climate of summer drought and cool, moist winters.

The five regions of mediterranean ecosystems include the semiarid regions of western North America, the regions bordering the Mediterranean Sea, central Chile, the cape region of South Africa, and southwestern and southern Australia (Figure 23.23). The mediterranean climate has hot, dry summers, with at least 1 month of protracted drought, and cool, moist winters (see representative climate diagrams in Figure 23.23). About 65 percent of the annual precipitation falls during the winter months. Winter temperatures typically average 10–12°C with a risk of frost. The hot, dry summer climates of the mediterranean regions arise from the seasonal change in the semipermanent high-pressure zones that are centered over the tropical deserts at about 20°N and 20°S (see discussion in Section 23.5). The persistent flow of dry air out of these regions during the summer brings several months of hot, dry weather. Fire is a frequent hazard during these periods.

All five regions support similar-looking communities of xeric broadleaf evergreen shrubs and dwarf trees known as sclerophyllous (*scleros*, "hard"; *phyll*, "leaf") vegetation with a herbaceous understory. Sclerophyllous vegetation possesses small leaves, thickened cuticles, glandular hairs, and sunken stomata, all characteristics that function to reduce water loss during the hot, dry summer period (Figure 23.24). Vegetation in each of the mediterranean systems also shares adaptations to fire and to low nutrient levels in the soil.

The largest area of mediterranean ecosystem forms a discontinuous belt around the Mediterranean Sea in

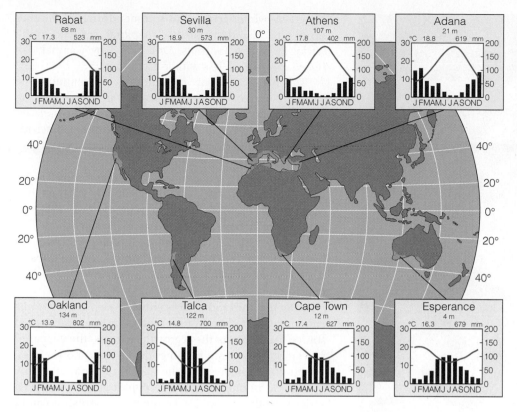

Figure 23.23 | Geographic distribution of Earth's mediterranean ecosystems and associated climate diagrams showing long-term patterns of monthly temperature and precipitation for selected locations. Note that these ecosystems are characterized by a winter rainy season and dry summers.

(Adapted from Archibold 1995.)

southern Europe and North Africa. Much of the area is currently or was once dominated by mixed evergreen woodland supporting species such as holm oak (*Quercus ilex*) and cork oak (*Quercus suber*). Often, these two species grow in mixed stands in association with strawberry tree (*Arbutus unedo*) and various species of shrubs. The easternmost limit of these ecosystems is in the coastal areas of Syria, Lebanon, and Israel, where they grade into the arid lands of the Middle East. Here, deciduous oak species are more abundant. Desert vegetation extends across North Africa as far as Tunisia, with mediterranean shrub and woodland extending through the northern coastal areas of Algeria and Morocco.

The mediterranean zone in southern Africa is restricted to the mountainous region of the Cape Province, where the vegetation is known as *fynbos*. The vegetation is composed primarily of broadleaf proteoid shrubs that grow to a height of 1.5 to 2.5 m (Figure 23.25). In southwest Australia, the mediterranean shrub community known as *mallee* is dominated by low-growing *Eucalyptus*, 5 to 8 m in height, with broad sclerophyllous leaves.

In North America, the sclerophyllous shrub community is known as chaparral, a word of Spanish origin meaning a thicket of shrubby evergreen oaks (Figure 23.26). California chaparral, dominated by scrub oak (*Quercus berberidifolia*) and chamise (*Adenostoma fasciculatum*), is

(a) **(b)** **(c)**

Figure 23.24 | Sclerophyllous leaves of some tree and shrub species inhabiting mediterranean shrublands (chaparral) of California: **(a)** chamise (*Adenostoma fasciculatum*), **(b)** scrub oak (*Quercus dumosa*), and **(c)** chinquapin (*Chrysolepis sempervirens*).

Figure 23.25 | Mediterranean vegetation (fynbos) of the Western Cape region of South Africa.

evergreen, winter-active, and summer-dormant. Another shrub type, also designated as chaparral, is found in the Rocky Mountain foothills. Dominated by Gambel oak (*Quercus gambelii*), it is winter-deciduous.

The matorral shrub communities of central Chile occur in the coastal lowlands and on the west-facing slopes of the Andes. Most of the matorral species are evergreen shrubs 1 to 3 m in height with small sclerophyllous leaves, although drought-deciduous shrubs are also found.

For the most part, mediterranean shrublands lack an understory and ground litter and are highly inflammable. Many species have seeds that require the heat and scarring action of fire to induce germination. Without fire, chaparral grows taller and denser, building up large fuel loads of leaves and twigs on the ground. In the dry season the shrubs, even though alive, nearly explode when ignited.

After fire, the land returns either to lush green sprouts coming up from buried root crowns or to grass if a seed source is nearby. As the regrowth matures, the chaparral vegetation once again becomes dense, the canopy closes, the litter accumulates, and the stage is set for another fire.

Shrub communities have a complex of animal life that varies with the region. In the mediterranean shrublands, similarity in habitat structure has resulted in pronounced parallel and convergent evolution (see Chapter 26) among bird species and some lizard species, especially between the Chilean mattoral and the California chaparral. In North America, chaparral and sagebrush communities support mule deer (*Odocoileus hemionus*), coyotes (*Canis latrans*), a variety of rodents, jackrabbits (*Lepus* spp.), and sage grouse (*Centrocercus urophasianus*). The Australian mallee is rich in birds, including the endemic mallee fowl (*Leipoa ocellata*), which incubates its eggs in a large mound. Among the mammalian life are the gray kangaroo (*Macropus giganteus*) and various species of wallaby.

Figure 23.26 | Chaparral is the dominant mediterranean shrub vegetation of southern California.

The diverse topography and geology of the mediterranean environments give rise to a diversity of soil conditions, but soils are typically classified as Alfisols (see Figure 4.12). The soils of the regions are generally deficient in nutrients, and litter decomposition is limited by low temperatures during the winter and low soil moisture during the summer months. These ecosystems vary in productivity depending on the annual precipitation and the severity of summer drought.

23.7 | Forest Ecosystems Dominate the Wetter Regions of the Temperate Zone

Climatic conditions in the humid midlatitude regions give rise to the development of forests dominated by broadleaf deciduous trees (Figure 23.27). But in the mild, moist climates of the Southern Hemisphere, temperate evergreen forests become predominant. Deciduous forest once covered large areas of Europe and China, parts of North and South America, and the highlands of Central America. The deciduous forests of Europe and Asia, however, have largely disappeared, cleared over the centuries for agriculture. In eastern North America, the deciduous forest consists of several forest types or associations (Figure 23.28) including the mixed mesophytic forest of the unglaciated Appalachian plateau; the beech–maple and northern hardwood forests (with pine and hemlock) in northern regions that eventually grade into the boreal forest (see Section 23.8); the maple–basswood forests of the Great

Lakes states; the oak–chestnut (now oak since the die-off of the American chestnut) or central hardwood forests, which cover most of the Appalachian Mountains; the magnolia–oak forests of the Gulf Coast states; and the oak–hickory forests of the Ozarks. In North America, temperate deciduous forests reach their greatest development in the mesic forests of the central Appalachians, where the number of tree species is unsurpassed by any other temperate area in the world.

The Asiatic broadleaf forest, found in eastern China, Japan, Taiwan, and Korea, is similar to the North American deciduous forest and contains several plant species of the same genera as those found in North America and western Europe. However, broadleaf evergreen species become increasingly present in Japan, South Korea, and southern China and in the wet foothills of the Himalayas. In southern Europe, their presence reflects the transition into the mediterranean region. Evergreen oaks and pines are also widely distributed in the southeastern United States, where they are usually associated with poorly developed sandy or swampy soils.

In the Southern Hemisphere, temperate deciduous forests are found only in the drier parts of the southern Andes. In southern Chile, broadleaf evergreen rain forests have developed in an oceanic climate that is virtually frost free. Evergreen forests are also found in New Zealand, Tasmania, and parts of southeastern Australia where the winter temperatures are moderated by the coastal environment. Climate regions in these areas are

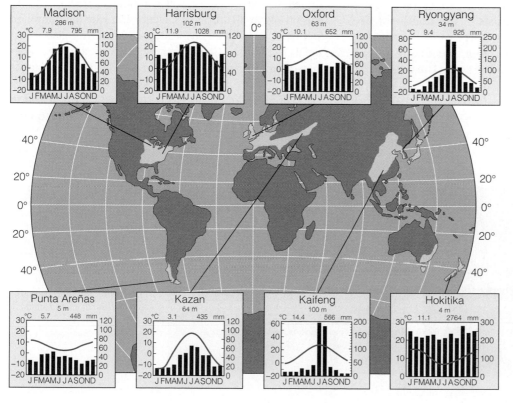

Figure 23.27 | Geographic distribution of Earth's temperate forest ecosystems and associated climate diagrams showing long-term patterns of monthly temperature and precipitation for selected locations. These ecosystems are characterized by distinct seasonality in temperature with adequate precipitation during the growing season to support a closed canopy of trees.

(Adapted from Archibold 1995.)

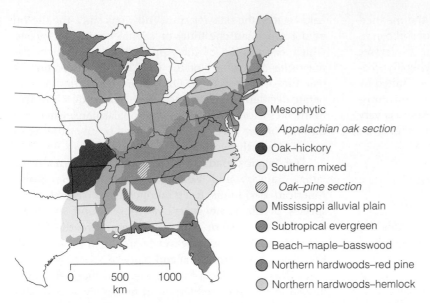

Figure 23.28 | Large-scale distribution of temperate forest communities in the eastern United States, derived from contemporary data. Compare with Figure 16.11, depicting the original forest pattern.

(Adapted from Dyer 2006.)

- Mesophytic
- *Appalachian oak section*
- Oak–hickory
- Southern mixed
- *Oak–pine section*
- Mississippi alluvial plain
- Subtropical evergreen
- Beach–maple–basswood
- Northern hardwoods–red pine
- Northern hardwoods–hemlock

similar to those of the Pacific Northwest of North America, but here the predominant species are conifers.

In the broadleaf deciduous forests of the temperate region, the end of the growing season is marked by the autumn colors of foliage shortly before the trees enter into their leafless winter period (Figure 23.29). The trees resume growth in the spring in response to increasing temperatures and longer day lengths. Many herbaceous species flower at this time before the developing canopy casts a heavy shade on the forest floor.

Highly developed, unevenly aged deciduous forests usually have four vertical layers or strata (see Figure 16.5). The upper canopy consists of the dominant tree species, below which is the lower tree canopy, or understory. Next is the shrub layer, followed finally by the ground layer of herbs, ferns, and mosses. The diversity of animal life is associated with this vertical stratification and the growth forms of plants (see Chapter 16). Some animals, particularly forest arthropods, spend most of their lives in a single stratum; others range over two or more strata. The greatest concentration and variety of life in the forest occurs on and just below the ground layer. Many animals, the soil and litter invertebrates in particular, remain in the subterranean stratum. Others, such as mice, shrews, ground squirrels, and forest salamanders, burrow into the soil or litter for shelter and food. Larger mammals live on the ground layer and feed on herbs, shrubs, and low trees. Birds move rather freely among several strata but typically favor one layer over another (see Figure 16.6).

Differences in climate, bedrock, and drainage are reflected in the variety of soil conditions present. Alfisols, Inceptisols, and Ultisols (see Figure 4.12) are the dominant soil types with Alfisols typically associated with glacial materials in more northern regions. Primary productivity varies geographically and is influenced largely by temperatures and the length of the growing season (see Section 20.3). Leaf fall in deciduous forests occurs over a short period in autumn, and the availability of nutrients is related to rates of decomposition and mineralization (see Chapter 21).

(a)

(b)

Figure 23.29 | A temperate forest of the Appalachian region: **(a)** the canopy during autumn and **(b)** interior of the forest during spring. The forest is dominated by oaks (*Quercus* spp.) and yellow poplar (*Liriodendron tulipifera*), with an understory of redbud (*Cercis canadensis*) in bloom.

23.8 | Conifer Forests Dominate the Cool Temperate and Boreal Zones

Conifer forests, dominated by needle-leaf evergreen trees, are found primarily in a broad circumpolar belt across the Northern Hemisphere and on mountain ranges, where low temperatures limit the growing season to a few months each year (Figure 23.30). The variable composition and structure of these forests reflect the wide range of climatic conditions in which they grow. In central Europe, extensive coniferous forests, dominated by Norway spruce (*Picea abies*), cover the slopes up to the subalpine zone in the Carpathian Mountains and the Alps (Figure 23.31a). In North America, several coniferous forests blanket the Rocky, Wasatch, Sierra Nevada, and Cascade mountains. At high elevations in the Rocky Mountains grows a subalpine forest dominated by Engelmann spruce (*Picea engelmannii*) and subalpine fir (*Abies lasiocarpa*). Middle elevations have stands of Douglas-fir, and lower elevations are dominated by open stands of ponderosa pine (*Pinus ponderosa*) (Figure 23.31b) and dense stands of the early successional conifer, lodgepole pine (*Pinus contorta*). The largest tree of all, the giant sequoia (*Sequoiadendron giganteum*), grows in scattered groves on the western slopes of the California Sierra. In addition, the mild, moist climate of the Pacific Northwest supports a highly productive coastal forest extending along the coastal strip from Alaska to northern California.

The largest expanse of conifer forest, in fact the largest vegetation formation on Earth, is the boreal forest, or taiga (Russian for "land of little sticks"). This belt of coniferous forest, encompassing the high latitudes of the Northern Hemisphere, covers about 11 percent of Earth's terrestrial surface (see Figure 23.30). In North America, the boreal forest covers much of Alaska and Canada and spills into northern New England, with fingers extending down the western mountain ranges and into the Appalachians. In Eurasia, the boreal forest begins in Scotland and Scandinavia and extends across the continent, covering much of Siberia, to northern Japan.

Three major vegetation zones make up the taiga (Figure 23.32): (1) the forest–tundra ecotone with open stands of stunted spruce, lichens, and moss; (2) the open lichen woodland with stands of lichens and black spruce; and (3) the main boreal forest (Figure 23.33) with continuous stands of spruce and pine broken by poplar and birch on disturbed areas. This boreal–mixed forest grades into the temperate forest of southern Canada and the northern United States. Primarily occupying formerly glaciated land, the taiga is also a region of cold lakes, bogs, rivers, and alder thickets.

A cold continental climate with strong seasonal variation dominates the taiga. The summers are short, cool, and moist; the winters are long, harsh, and dry, with a prolonged period of snowfall. The driest winters and the greatest seasonal fluctuations are in interior Alaska and

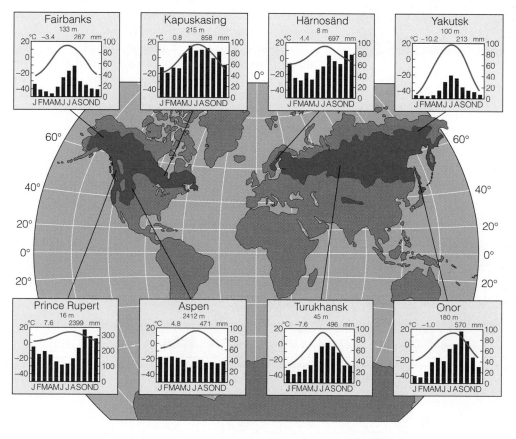

Figure 23.30 | Geographic distribution of Earth's conifer forest ecosystems and associated climate diagrams showing long-term patterns of monthly temperature and precipitation for selected locations. Regions supporting conifer forest ecosystems have a lower mean annual temperature and shorter growing season than areas that support temperate deciduous forest (see Figure 23.27).

(Adapted from Archibold 1995.)

(a)

(b)

Figure 23.31 | Two coniferous forest types. **(a)** A Norway spruce forest in the Carpathian Mountains of central Europe. **(b)** A montane coniferous forest in the Rocky Mountains. The dry, lower slopes support ponderosa pine; the upper slopes are cloaked with Douglas-fir.

central Siberia, which experience seasonal temperature extremes (difference between minimum and maximum annual temperatures) of as much as 100°C.

Much of the taiga is under the controlling influence of permafrost, which impedes infiltration and maintains high soil moisture. Permafrost is the perennially frozen subsurface that may be hundreds of meters deep. It develops where the ground temperatures remain below 0°C for extended periods of time. Its upper layers may thaw in summer and refreeze in winter. Because the permafrost is impervious to water, it forces all water to remain and move above it. Thus, the ground stays soggy even though precipitation is low, enabling plants to exist in the driest parts of the Arctic.

Fires are recurring events in the taiga. During periods of drought, fires can sweep over hundreds of thousands of hectares. All of the boreal species, both broadleaf trees and conifers, are well adapted to fire. Unless too severe, fire provides a seedbed for regeneration of trees. Light surface burns favor early successional hardwood species. More severe fires eliminate hardwood competition and favor spruce and jack pine regeneration.

Because of the global demand for timber and pulp, vast areas of the boreal forest across North America and Siberia are being clear-cut with little concern for their future. This exploitation can alter the nature of and threaten the survival of the boreal forest.

The boreal forest has a unique animal community. Caribou (*Rangifer tarandus*), wide-ranging and feeding on grasses, sedges, and especially lichens, inhabit open spruce–lichen woodlands. Joining the caribou is the moose (*Alces*

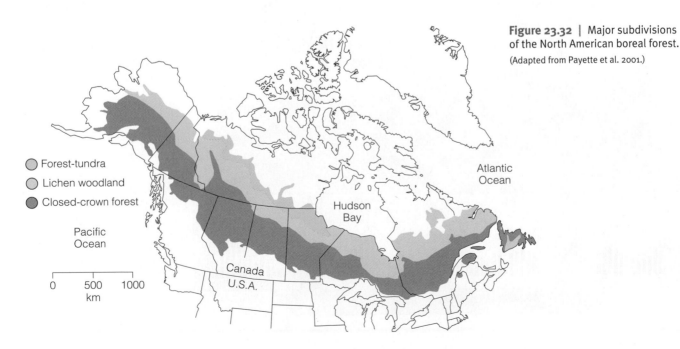

Figure 23.32 | Major subdivisions of the North American boreal forest.

(Adapted from Payette et al. 2001.)

Forest-tundra
Lichen woodland
Closed-crown forest

Pacific
Ocean

0   500   1000
km

Canada
U.S.A.

Hudson
Bay

Atlantic
Ocean

Figure 23.33 | Black spruce is a dominant conifer in the North American taiga.

alces), called elk in Eurasia, the largest of all deer. It is a lowland mammal feeding on aquatic and emergent vegetation as well as alder and willow. Competing with moose for browse is the cyclic snowshoe hare (*Lepus americanus*). The arboreal red squirrel (*Sciurus hudsonicus*) inhabits the conifers and feeds on young pollen-bearing cones and seeds of spruce and fir; and the quill-bearing porcupine (*Erethizon dorsatum*) feeds on leaves, twigs, and the inner bark of trees. Preying on these are an assortment of predators including the wolf, lynx (*Lynx canadensis*), pine martin (*Martes americana*), and owls. The taiga is also the nesting ground of migratory neotropical birds and the habitat of

northern seed-eating birds such as crossbills (*Loxia* spp.), grosbeaks (*Coccothraustes* spp.), and siskins (*Carduelis* spp.).

Of great ecological and economic importance are major herbivorous insects such as the spruce budworm (*Choristoneura fumiferana*). Although they are major food items for insectivorous summer birds, these insects experience periodic outbreaks during which they defoliate and kill large expanses of forest.

Compared to more temperate forests boreal forests have generally low net primary productivity; they are limited by low nutrients, cooler temperatures, and the short growing season. Likewise, inputs of plant litter are low compared to the forests of the warmer temperate zone. However, rates of decomposition are slow under the cold, wet conditions, resulting in the accumulation of organic matter. Soils are primarily Spodosols (see Figure 4.12) characterized by a thick organic layer. The mineral soils beneath mature coniferous forests are comparatively infertile, and growth is often limited by the rate at which mineral nutrients are recycled through the ecosystem.

23.9 | Low Precipitation and Cold Temperatures Define the Arctic Tundra

Encircling the top of the Northern Hemisphere is a frozen plain, clothed in sedges, heaths, and willows; dotted with lakes; and crossed by streams (Figure 23.34). Called tundra, its name comes from the Finnish *tunturi*, meaning "a

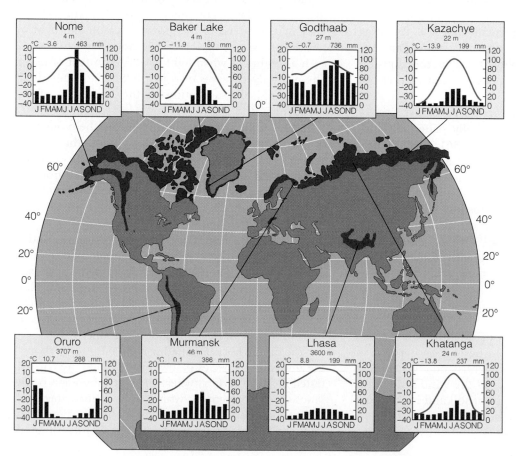

Figure 23.34 | Geographic distribution of Earth's tundra ecosystems and associated climate diagrams showing long-term patterns of monthly temperature and precipitation for selected locations. Tundra ecosystems are characterized by lower mean temperatures, a shorter growing season, and lower annual precipitation than regions supporting conifer forest (see Figure 23.30 for comparison).

(Adapted from Archibold 1995.)

(a) **(b)**

Figure 23.35 | (a) The plant cover that characterizes the wide expanse of the Arctic tundra in the Northwest Territories of Canada is in stark contrast to **(b)** the polar desert which is characterized by dry soils and sparse plant cover.

treeless plain." The arctic tundra falls into two broad types: *tundra* with up to 100 percent plant cover and wet to moist soil (Figure 23.35), and *polar desert* with less than 5 percent plant cover and dry soil.

Conditions unique to the Arctic tundra are a product of at least three interacting forces: (1) the permanently frozen deep layer of permafrost; (2) the overlying active layer of organic matter and mineral soil that thaws each summer and freezes the following winter; and (3) vegetation that reduces warming and retards thawing in summer. Permafrost chills the soil, retarding the general growth of plant parts both above- and belowground, limiting the activity of soil microorganisms, and diminishing the aeration and nutrient content of the soil.

Alternate freezing and thawing of the upper layer of soil creates the unique, symmetrically patterned landforms typical of the tundra (Figure 23.36). This action of frost pushes stones and other material upward and outward from the mass to form a patterned surface of frost hummocks, frost boils, earth stripes, and stone polygons. On sloping ground, creep, frost thrusting, and downward flow of supersaturated soil over the permafrost form *solifluction terraces*, or "flowing soil." This gradual downward creep of soils and rocks eventually rounds off ridges and other irregularities in topography. Such molding of the landscape by frost action, called *cryoplanation*, is far more important than erosion in wearing down the Arctic landscape.

Structurally, the vegetation of the tundra is simple. The number of species tends to be few, and growth is slow. Only those species able to withstand constant disturbance of the soil, buffeting by the wind, and abrasion from wind-carried particles of soil and ice can survive. Low ground is covered with a complex of cotton grasses, sedges, and *Sphagnum*. Well-drained sites sup-

port heath shrubs, dwarf willows and birches, herbs, mosses, and lichens. The driest and most exposed sites support scattered heaths and crustose and foliose lichens growing on the rock. Arctic plants propagate themselves almost entirely by vegetative means, although viable seeds many hundreds of years old exist in the soil.

Plants are photosynthetically active on the Arctic tundra about 3 months out of the year. As snow cover disappears, plants commence photosynthetic activity. They maximize use of the growing season and light by photosynthesizing during the 24-hour daylight period, even at midnight when light is one-tenth that of noon. The nearly erect leaves of some Arctic plants permit almost complete interception of the low angle of the Arctic sun.

Much of the photosynthate goes into the production of new growth; but about 1 month before the growing season ends, plants cease to allocate photosynthate to aboveground biomass. They withdraw nutrients from the leaves and move them to roots and belowground biomass, sequestering 10 times the amount stored by temperate grasslands.

Structurally, most of the tundra vegetation is underground. Root-to-shoot ratios of vascular plants range from 3:1 to 10:1. Roots are concentrated in the upper soil that thaws during the summer, and aboveground parts seldom grow taller than 30 cm. It is not surprising, then, that the belowground net annual production is typically three times that of the aboveground productivity.

The tundra hosts fascinating animal life, even though the diversity of species is low. Invertebrates are concentrated near the surface, where there are abundant populations of segmented whiteworms (Enchytraeidae), collembolas, and flies (Diptera), chiefly crane flies. Summer in the Arctic tundra brings hordes of black

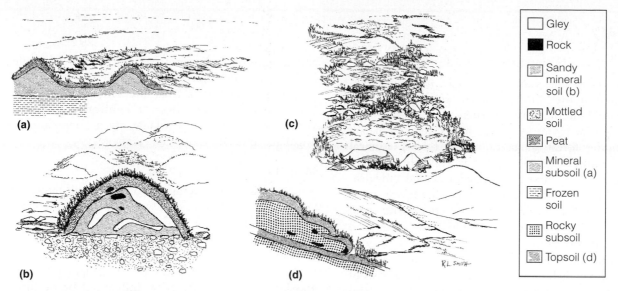

Figure 23.36 | Patterned landforms typical of the tundra region: **(a)** unsorted earth stripes; **(b)** frost hummocks; **(c)** sorted stone nets and polygons; **(d)** a solifluction terrace.

Legend:
- Gley
- Rock
- Sandy mineral soil (b)
- Mottled soil
- Peat
- Mineral subsoil (a)
- Frozen soil
- Rocky subsoil
- Topsoil (d)

flies (*Simulium* spp.), deer flies (*Chrysops* spp.), and mosquitoes.

Dominant vertebrates on the Arctic tundra are herbivores, including lemmings, Arctic hare, caribou, and musk ox (*Ovibos moschatus*). Although caribou have the greatest herbivore biomass, lemmings, which breed throughout the year, may reach densities as great as 125 to 250 per hectare; and they consume 3 to 6 times as much forage as caribou do. Arctic hares that feed on willows disperse over the range in winter and congregate in more restricted areas in summer. Caribou are extensive grazers, spreading out over the tundra in summer to feed on sedges. Musk oxen are more intensive grazers, restricted to more localized areas where they feed on sedges, grasses, and dwarf willow. Herbivorous birds are few, dominated by ptarmigan and migratory geese.

The major Arctic carnivore is the wolf, which preys on musk ox, caribou, and, when they are abundant, lemmings. Medium-sized to small predators include the Arctic fox (*Alopex lagopus*), which preys on Arctic hare, and several species of weasel, which prey on lemmings. Also feeding on lemmings are snowy owls (*Nyctea scandiaca*) and the hawk-like jaegers (*Stercorarius* spp.). Sandpipers (*Tringa* spp.), plovers (*Pluvialis* spp.), longspurs (*Calcarius* spp.), and waterfowl, which nest on the wide expanse of ponds and boggy ground, feed heavily on insects.

At lower latitudes, alpine tundra occurs in the higher mountains of the world. The alpine tundra is a severe environment of rock-strewn slopes, bogs, meadows, and shrubby thickets (Figure 23.37). It is a land of strong winds, snow, cold, and widely fluctuating temperatures. During summer, the temperature on the surface of the soil ranges from 40°C to 0°C. The atmosphere is thin; so light intensity, especially ultraviolet, is high on clear days. Alpine tundras have little permafrost, and it is confined mostly to very high elevations. Lacking permafrost, soils are drier. Only in alpine wet meadows and bogs do soil moisture conditions compare with those of the Arctic. Precipitation, especially snowfall and humidity, is higher in the alpine regions than in the Arctic tundra, but steep topography induces a rapid runoff of water.

Figure 23.37 | Rocky Mountains alpine tundra.

Summary

Ecosystem Distribution and Plant Adaptations | 23.1

Terrestrial ecosystems can be grouped into broad categories called biomes. Biomes are classified according to the predominant plant types. There are at least eight major terrestrial biome types: tropical forest, temperate forest, conifer forest (taiga or boreal forest), tropical savanna, temperate grasslands, chaparral (shrublands), tundra, and desert. These broad categories reflect the relative contribution of three general plant life-forms: trees, shrubs, and grasses. Interaction between moisture and temperature is the primary factor limiting the nature and geographic distribution of terrestrial ecosystems.

Terrestrial ecosystems are classified by their vegetation structure, which is typically defined on the basis of the dominant plant life-forms (grasses, shrubs, and trees). Constraints imposed on the adaptations of these major plant life-forms to features of the physical environment (climate and soils) determine their patterns of dominance along gradients of temperature and moisture. These patterns of plant life-forms determine the corresponding distribution of ecosystems.

Tropical Forests | 23.2

Seasonality of rainfall determines the types of tropical forests. Rain forests, associated with high aseasonal rainfall, are dominated by broadleaf evergreen trees. They are noted for their enormous diversity of plant and animal life. The vertical structure of the forest is divided into five general layers: emergent trees, high upper canopy, low tree stratum, shrub understory, and a ground layer of herbs and ferns. Conspicuous in the rain forest are the lianas or climbing vines, epiphytes growing up in the trees, and stranglers growing downward from the canopy to the ground. Many large trees develop buttresses for support. Nearly 90 percent of nonhuman primate species live in tropical rain forests.

Tropical rain forests support high levels of primary productivity. The high rainfall and consistently warm temperatures also result in high rates of decomposition and nutrient cycling.

Dry tropical forests undergo varying lengths of dry season, during which trees and shrubs drop their leaves (drought-deciduous). New leaves are grown at the onset of the rainy season. Most dry tropical forests have been lost to agriculture and grazing and other disturbances.

Tropical Savannas | 23.3

Savannas are characterized by a codominance of grasses and woody plants. Such vegetation is characteristic of regions with alternating wet and dry seasons. Savannas range from grass with occasional trees to shrubs to communities where trees form an almost continuous canopy as a function of precipitation and soil texture. Productivity and decomposition in savanna ecosystems are closely tied to the seasonality of precipitation.

Savannas support a large and varied assemblage of both invertebrate and vertebrate herbivores. The African savanna is dominated by a large, diverse population of ungulate fauna and associated carnivores.

Temperate Grasslands | 23.4

Natural grasslands occupy regions where rainfall is between 250 and 800 mm a year. Once covering extensive areas of the globe, natural grasslands have shrunk to a fraction of their original size because of conversion to cropland and grazing lands.

Grasslands vary with climate and geography. Native grasslands of North America, influenced by declining precipitation from east to west, consist of tallgrass prairie, mixed-grass prairie, shortgrass prairie, and desert grasslands. Eurasia has steppes, South America the pampas, and southern Africa the veldt. Grassland consists of an ephemeral herbaceous layer that arises from crowns, nodes, and rosettes of plants hugging the ground. It also has a ground layer and a highly developed root layer. Depending on the history of fire and degree of grazing and mowing, grasslands accumulate a layer of mulch.

Grasslands support a diversity of animal life dominated by herbivorous species, both invertebrate and vertebrate. Grasslands once supported herds of large grazing ungulates such as bison in North America, migratory herds of wildebeest in Africa, and marsupial kangaroos in Australia. Grasslands evolved under the selective pressure of grazing. Although the most conspicuous grazers are large herbivores, the major consumers are invertebrates. The heaviest consumption takes place belowground, where the dominant herbivores are nematodes.

Deserts | 23.5

Deserts occupy about one-seventh of Earth's land surface and are largely confined to two worldwide belts between 15° and 30° north and south latitude. Deserts result from dry descending air masses within these regions, the rain shadows of coastal mountain ranges, and remoteness from oceanic moisture. Two broad types of deserts exist: cool deserts, exemplified by the Great Basin of North America, and hot deserts, like the Sahara. Deserts are structurally simple—scattered shrubs, ephemeral plants, and open, stark topography. In this harsh environment, plants and animals have evolved ways of circumventing aridity and high temperatures by either evading or resisting drought. Despite their aridity, deserts support a diversity of animal life, notably opportunistic herbivorous species and carnivores.

Shrubland | 23.6

Shrubs have a densely branched, woody structure and low height. Shrublands are difficult to classify due to the variety of climates in which shrubs can be a dominant or codominant component of the plant community. But in five widely disjunct regions along the western margins of the continents between 30° and 40° latitude are found the mediterranean ecosystems. Dominated by evergreen shrubs and sclerophyll trees, these biomes have adapted to the distinctive climate of summer drought and cool, moist winters. These shrublands are fire adapted and highly flammable.

Temperate Forests | 23.7

Broadleaf deciduous forests are found in the wetter environments of the warm temperate region. They once covered large areas of Europe and China, but their distribution has been reduced by human activity. In North America, deciduous forests are still widespread. They include various types such as beech–maple and oak–hickory forest; the greatest development is in the mixed mesophytic forest of the unglaciated Appalachians. Well-developed deciduous forests have four strata: upper canopy, lower canopy, shrub layer, and ground layer. Vertical structure influences the diversity and distribution of life in the forest. Certain species are associated with each stratum.

Conifer Forests | 23.8

Coniferous forests of temperate regions include the montane pine forests and lower-elevation pine forests of Eurasia and North America and the temperate rain forests of the Pacific Northwest.

North of the temperate coniferous forest is the circumpolar taiga, or boreal forest, the largest biome on Earth. Characterized by a cold continental climate, the taiga consists of four major zones: the forest ecotone, open boreal woodland, main boreal forest, and boreal–mixed forest ecotone.

Permafrost, the maintenance of which is influenced by tree and ground cover, strongly influences the pattern of vegetation, as do recurring fires. Spruces and pines dominate boreal forest with successional communities of birch and poplar. Ground cover below spruce is mostly moss; in open spruce and pine stands, the cover is mostly lichens.

Major herbivores of the boreal region include caribou, moose, and snowshoe hare. Predators include the wolf, lynx, and pine martin.

Tundra | 23.9

The Arctic tundra extends beyond the tree line at the far north of the Northern Hemisphere. It is characterized by low temperature, low precipitation, a short growing season, a perpetually frozen subsurface (the permafrost), and a frost-molded landscape. Plant species are few, growth forms are low, and growth rates are slow. Over much of the Arctic, the dominant vegetation is cotton grass, sedge, and dwarf heaths. These plants exploit the long days of summer by photosynthesizing during the 24-hour daylight period. Most plant growth occurs underground. The animal community is low in diversity but unique. Summer in the Arctic brings hordes of insects, providing a rich food source for shorebirds. Dominant vertebrates are lemming, Arctic hare, caribou, and musk ox. Major carnivores are the wolf, Arctic fox, and snowy owl.

Alpine tundras occur in the mountains of the world. They are characterized by widely fluctuating temperatures, strong winds, snow, and a thin atmosphere.

Study Questions

1. What are the main differences between tropical rain forest and tropical dry forest?
2. What are the major strata in the tropical rain forest?
3. How does the warm, wet environment of tropical rain forest influence rates of net primary productivity and decomposition?
4. What types of trees characterize tropical rain forest (leaf type)?
5. What distinguishes savannas from grassland ecosystems?
6. Under what climate conditions do you find tropical savanna ecosystems?
7. How does seasonality influence rates of net primary productivity and decomposition in savanna ecosystems?
8. What features of regional climate lead to the formation of desert ecosystems?
9. What climate characterizes mediterranean ecosystems?
10. What type of leaves characterize mediterranean plants?
11. What types of leaves characterize the trees of temperate forest ecosystems?
12. How does seasonality of temperature influence the structure and productivity of temperate forest ecosystems?
13. What climate is characteristic of temperate grasslands?
14. How does annual precipitation influence the structure and productivity of grassland ecosystems?
15. What type of trees characterize boreal forest?
16. What is permafrost, and how does it influence the structure and productivity of boreal forest ecosystems?
17. What physical and biological features characterize Arctic tundra?
18. How does alpine tundra differ from Arctic tundra?

Further Readings

Archibold, O. W. 1995. *Ecology of world vegetation.* London: Chapman & Hall.

An outstanding reference for those interested in the geography and ecology of terrestrial ecosystems.

Bliss, L. C., O. H. Heal, and J. J. Moore, eds. 1981. *Tundra ecosystems: A comparative analysis.* New York: Cambridge University Press.

A major reference on the geography, structure, and function of these high-latitude ecosystems.

Bonan, G. B., and H. H. Shugart. 1989. Environmental factors and ecological processes in boreal forests. *Annual Review of Ecology and Systematics* 20:1–18.

An excellent review paper providing a good introduction to boreal forest ecosystems.

Evenardi, M., I. Noy-Meir, and D. Goodall, eds. 1986. *Hot deserts and arid shrublands of the world.* Ecosystems of the World 12A and 12B. Amsterdam: Elsevier Scientific.

A major reference work on the geography and ecology of the world's deserts.

French, N., ed. 1979. *Perspectives on grassland ecology.* New York: Springer-Verlag.

This text provides a good summary of grassland ecology.

Murphy, P. G., and A. E. Lugo. 1986. Ecology of tropical dry forests. *Annual Review of Ecology and Systematics* 17:67–88.

This paper provides an overview of the distribution and ecology of tropical dry forests, one of the most endangered terrestrial ecosystems.

Quinn, R. D., and S. C. Keeley. 2006. *Introduction to California chaparral.* Berkeley: University of California Press.

An overview of the chaparral ecosystem with an excellent discussion of the role and impact of fire.

Reichle, D. E., ed. 1981. *Dynamic properties of forest ecosystems.* Cambridge, UK: Cambridge University Press.

The major reference source on the ecology and function of forest ecosystems throughout the world.

Reichman, O. J. 1987. *Konza prairie: A tallgrass natural history.* Lawrence: University Press of Kansas.

An excellent introduction to the tallgrass prairie ecosystem; emphasizes the interrelationships of plants, animals, and landscape.

Richards, P. W. 1996. *The tropical rain forest: An ecological study,* 2nd ed. New York: Cambridge University Press.

A thoroughly revised edition of a classic book on the ecology of tropical rain forests.

Sinclair, A. R. E., and P. Arcese, eds. 1995. *Serengeti II: Dynamics, management, and conservation of an ecosystem.* Chicago: University of Chicago Press.

A masterful study of an ecosystem. A valuable reference on the ecology of tropical savanna ecosystems.

CHAPTER 24

Aquatic Ecosystems

Waves heavily influence the pattern of life on rocky shores along the California coastline.

Whereas scientists classify terrestrial ecosystems according to their dominant plant life-forms, classification of aquatic ecosystems is largely based on features of the physical environment. A major feature influencing the adaptations of organisms to the aquatic environment is water salinity (see Section 3.5). For this reason, aquatic ecosystems fall into two major categories: freshwater or saltwater (or marine). These categories are further divided into several ecosystem types based on substrate, depth and flow of water, and type of dominant organisms (typically plants).

Ecologists subdivide marine ecosystems into two broad categories of coastal and open-water systems. Freshwater ecosystems are classified on the basis of water depth and flow. Flowing-water, or **lotic**, ecosystems include rivers and streams. Nonflowing-water, or **lentic** ecosystems, include ponds, lakes, and inland wetlands.

All aquatic ecosystems, both freshwater and marine, are linked directly or indirectly as components of the hydrological cycle (Figure 24.1; also see Section 3.1). Water evaporated from oceans and terrestrial environments falls as precipitation. The precipitation that remains on the land surface (i.e., does not infiltrate into the soil or evaporate) follows a path determined by gravity and topography—more specifically, geomorphology. Flowing-water ecosystems begin as streams. These streams, in turn, coalesce into rivers as they follow the topography of the landscape, or they collect in basins and floodplains to form nonflowing-water ecosystems such as ponds, lakes, and inland wetlands. Rivers eventually flow to the coast and form estuaries, which represent the transition from freshwater to marine.

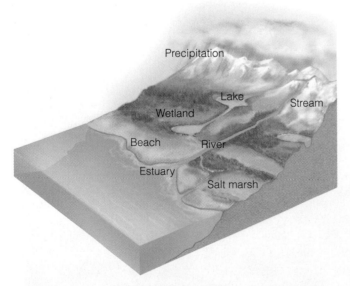

Figure 24.1 | Idealized landscape/seascape showing the linkages among the various types of aquatic ecosystems via the water cycle.
Go to GRAPHiT! at **www.ecologyplace.com** to learn about global freshwater resources.

In this chapter, we will examine the basic characteristics of aquatic ecosystems, both freshwater and marine. Beginning first with freshwater ecosystems, we will examine lentic ecosystems (lakes and ponds). We then turn our attention to lotic ecosystems, following the path and changing characteristics of streams as they coalesce to form rivers, eventually flowing to coastal environments. After examining estuarine environments, we conclude by examining the marine environments that cover over 70 percent of Earth's surface.

24.1 | Lakes Have Many Origins

Lakes and ponds are inland depressions containing standing water (Figure 24.2). They vary in depth from 1 m to more than 2000 m. They range in size from small ponds of less than a hectare to large lakes covering thousands of square kilometers. Ponds are small bodies of water so shallow that rooted plants can grow over much of the bottom. Some lakes are so large that they mimic marine environments. Most ponds and lakes have outlet streams, and both may be more or less temporary features on the landscape, geologically speaking.

Some lakes have formed by glacial erosion and deposition. Abrading slopes in high mountain valleys, glaciers carved basins that filled with water from rain and melting snow to form tarns. Retreating valley glaciers left behind crescent-shaped ridges of rock debris, called moraines, that dammed up water behind them. Many shallow kettle lakes and potholes were left behind by the glaciers that covered much of northern North America and northern Eurasia.

Lakes also form when silt, driftwood, and other debris deposited in beds of slow-moving streams dam up water behind them. Loops of streams that meander over flat valleys and floodplains often become cut off by sediments, forming crescent-shaped oxbow lakes.

Shifts in Earth's crust, uplifting mountains or displacing rock strata, sometimes develop water-filled depressions. Craters of some extinct volcanoes have also become lakes. Landslides may block streams and valleys to form new lakes and ponds.

Many lakes and ponds form through nongeological activity. Beavers dam streams to make shallow but often extensive ponds. Humans create huge lakes by damming rivers and streams for power, irrigation, or water storage and construct smaller ponds and marshes for recreation, fishing, and wildlife (see Ecological Issues: Dams: Regulating the Flow of River Ecosystems). Quarries and surface mines can also form ponds.

24.2 | Lakes Have Well-Defined Physical Characteristics

All lentic ecosystems share certain characteristics. Life in still-water ecosystems depends on light. The amount of light penetrating the water is influenced by natural

Figure 24.2 | Lakes and ponds fill basins or depressions in the land. **(a)** A rock basin glacial lake, or tarn, in the Rocky Mountains. **(b)** Swampy tundra in Siberia is dotted with numerous ponds and lakes. **(c)** An oxbow lake formed when a bend in the river was cut off from the main channel. **(d)** A human-constructed, old New England millpond. Note the floating vegetation.

attenuation, by silt and other material carried into the lake, and by the growth of phytoplankton. Temperatures vary seasonally and with depth (Figure 3.8). Oxygen can be limiting, especially in summer, because only a small proportion of the water is in direct contact with air, and respiration by decomposers on the bottom consumes large quantities (see Figure 3.12). Thus, variation in oxygen, temperature, and light strongly influences the distribution and adaptations of life in lakes and ponds (see Chapter 3 for more detailed discussion).

Ponds and lakes may be divided into both vertical and horizontal strata based on penetration of light and photosynthetic activity (Figure 24.3 on page 522). The horizontal zones are obvious to the eye; the vertical ones, influenced by depth of light penetration, are not. Surrounding most lakes and ponds and engulfing some ponds completely is the **littoral zone,** or shallow-water zone, in which light

reaches the bottom, stimulating the growth of rooted plants. Beyond the littoral is open water, the **limnetic zone,** which extends to the depth of light penetration. Inhabiting this zone are microscopic phytoplankton (autotrophs) and zooplankton (heterotrophs) as well as **nekton,** free-swimming organisms such as fish. Beyond the depth of effective light penetration is the **profundal zone.** Its beginning is marked by the compensation depth of light, the point at which respiration balances photosynthesis (see Figure 20.8). The profundal zone depends on a rain of organic material from the limnetic zone for energy. Common to both the littoral and profundal zones is the third vertical stratum—the **benthic zone,** or bottom region, which is the primary place of decomposition. Although these zones are named and often described separately, all are closely dependent on one another in the dynamics of lake ecosystems.

Dams constructed across rivers and steams interrupt and regulate the natural flow of water (Figure 1), profoundly affecting the river's hydrology and ecology. Dams change the environment in which lotic organisms live, most often to their detriment.

Under normal conditions, free-flowing streams and rivers experience seasonal fluctuation in flow. Snow-melt and early spring rains bring scouring high water; summer brings low water levels that expose some of the streambed and speed the decomposition of organic matter along the edges. Life has adapted to these seasonal changes. Damming a river or stream interrupts both nutrient spiraling and the river continuum.

Figure 1 | A dammed river.

Downstream flow is greatly reduced as a pool of water fills behind the dam, developing characteristics similar to those of a natural lake yet retaining some features of the lotic system, such as a constant inflow of water. Heavily fertilized by decaying material on the newly flooded land, the lake develops a heavy bloom of phytoplankton and, in tropical regions, dense growths of floating plants. Species of fish, often introduced exotics adapted to lake-like conditions, replace fish adapted to flowing water.

The type of pool allowed to develop depends on the purpose of the dam and strongly affects downstream conditions. Single-purpose dams serve only for flood control or water storage; multipurpose dams provide hydroelectric power, irrigation water, and recreation, among other uses. A flood-control dam has a minimum pool; the dam fills only during a flood, at which time inflow exceeds outflow. Engineers release the water slowly to minimize downstream flooding. In time, the water behind the dam recedes to the original pool depth. During and after floods, the river below carries a strong flow for some time, scouring the riverbed. During normal times, flow below the dam is stabilized. If the dam is for water storage, the reservoir holds its maximum pool; but during periods of water shortage and drought, drawdown of the pool can be considerable, exposing large expanses of shoreline for a long time and stressing or killing life in the littoral zone. Only a minimal quantity of water is released downstream—usually an amount required by law. Hydroelectric and multipurpose dams hold a variable amount of water,

24.3 | The Nature of Life Varies in the Different Zones

Aquatic life is richest and most abundant in the shallow water about the edges of lakes and ponds as well as in other places where sediments have accumulated on the bottom and decreased the water depth (Figure 24.4). Dominating these areas is emergent vegetation such as cattails and sedges—plants whose roots are anchored in the bottom mud, lower stems are immersed in water, and upper stems and leaves stand above water. Beyond the emergents and occupying even deeper water is a zone of floating plants such as pondweed (*Potamogeton*) and pond lily (*Nuphar* spp.). In depths too great for floating plants live submerged plants, such as species of pondweed with their finely dissected or ribbonlike leaves.

Associated with the emergents and floating plants is a rich community of organisms, among them hydras, snails, protozoans, dragonflies and diving insects, pickerel (*Esox* spp.), sunfish (*Lepomis* spp.), herons (Ardeidae), and blackbirds (*Agelaius* spp. and *Xanthocephalus xanthocephalus*). Many species of pond fish have compressed bodies, permitting them to move easily through the masses of aquatic plants. The littoral zone contributes heavily to the large input of organic matter into the system.

The main forms of life in the limnetic zone are phytoplankton and zooplankton (Figure 24.5). Phytoplankton, including desmids, diatoms, and filamentous algae, are the primary producers in open-water ecosystems and form the base on which the rest of life in open water depends. Also suspended in the water column are small grazing animals, mostly tiny crustaceans that feed on the phytoplankton. These animals form an important link in energy flow in the limnetic zone.

During the spring and fall turnovers (see Section 21.10 and Figure 21.25), plankton are carried downward, determined by consumer needs. During periods of power production, pulsed releases are strong enough to wipe out or dislodge benthic life downstream, which under the best of conditions has a difficult time becoming established.

Reservoirs with large pools of water become stratified, with well-developed vertical structure—epilimnion, thermocline, and hypolimnion (see Figure 3.9). If water is discharged from the upper layer of the reservoir, the effect of the flow downstream is similar to that of a natural lake. Warm, nutrient-rich, well-oxygenated water creates highly favorable conditions for some species of fish below the spillway and on downstream. If the discharge is from the cold hypolimnion, downstream receives cold, oxygen-poor water carrying an accumulation of iron and other minerals and a concentration of soluble organic materials. Such adverse conditions to stream life may persist for hundreds of kilometers downstream before the river reaches anything near normal conditions. Gated selective withdrawal structures or induced artificial circulation to increase oxygen concentration reduce such problems at some dams.

The impact of dams is compounded when a series of multipurpose dams are built on a river. The amount of water released and moving downstream decreases with each dam, until eventually all available water is consumed and the river simply dries up. This describes the situation on the Colorado River, the most regulated river in the world. The river is nearly dry by the time it reaches Mexico and disappears before reaching the mouth of the Gulf of California.

For many years the impetus has been to build more and more dams on our rivers and streams. Now the focus has shifted to removing dams to restore degraded rivers and streams, eliminate costly repairs, and improve public safety. Over 500 old sediment-filled dams, between 5 and 10 meters high, that do not generate power or control floods have been removed. The positive ecological effects of removing dams include eliminating barriers to upstream movement of fish, restoring spawning areas, and shifting macroinvertebrates from lentic to lotic species upstream. The negative effects are short-term increases of sediment loads with their contaminants downstream, upstream movement of invasive species, and loss of any backwater wetland habitat.

1. How might the development of dams in coastal regions affect migratory fish species, such as the populations of sockeye salmon along the Pacific Northwest coast of North America discussed in the introduction to Part Three?

2. What is the source of freshwater (residential, commercial, and agricultural) for your community? Does water storage involve a reservoir? If so, what watershed (streams and rivers) are affected by the dam(s)?

3. Some people advocate removing dams in their area; others want to repair and retain them. Discuss the pros and cons of dam removal from the social, economic, legal, and ecological viewpoints.

but at the same time nutrients released by decomposition on the bottom are carried upward to the impoverished surface layers. In spring when surface waters warm and stratification develops, phytoplankton have access to both nutrients and light. A spring bloom develops, followed by rapid depletion of nutrients and a reduction in planktonic populations, especially in shallow water.

Fish make up most of the nekton in the limnetic zone. Their distribution is influenced primarily by food supply, oxygen, and temperature. During the summer, large predatory fish such as largemouth bass, pike, and muskellunge inhabit the warmer epilimnion waters, where food is abundant. In winter, these species retreat to deeper water. In contrast, lake trout require colder water temperatures and move to greater depths as summer advances. During the spring and fall turnover, when oxygen and temperature are fairly uniform throughout, both warm-water and cold-water species occupy all levels.

Life in the profundal zone depends not only on the supply of energy and nutrients from the limnetic zone above but also on the temperature and availability of oxygen. In highly productive waters, oxygen may be limiting because the decomposer organisms so deplete it that little aerobic life can survive. Only during spring and fall turnovers, when organisms from the upper layers enter this zone, is life abundant in profundal waters.

Easily decomposed substances drifting down through the profundal zone are partly mineralized while sinking. The remaining organic debris—dead bodies of plants and animals of the open water, and decomposing plant matter from shallow-water areas—settles on the bottom. Together with quantities of material washed in, these substances make up the bottom sediments, the habitat of benthic organisms.

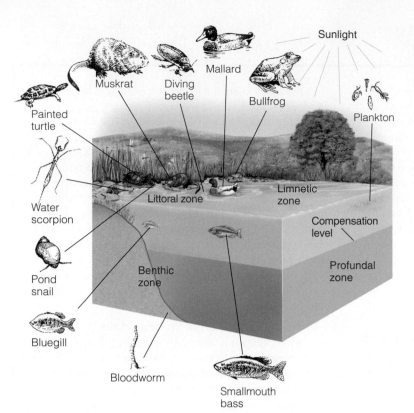

Figure 24.3 | Major zones of a lake in midsummer: littoral, limnetic, profundal, and benthic. The compensation level is the depth where light levels are such that gross production in photosynthesis is equal to respiration, so net production (primary) is zero. The organisms shown are typical of the various zones in a lake community.

The bottom ooze is a region of great biological activity—so great, in fact, that the oxygen curves for lakes and ponds show a sharp drop in the profundal water just above the bottom. Because the organic muck is so low in oxygen, the dominant organisms there are anaerobic bacteria. Under anaerobic conditions, however, decomposition cannot proceed to inorganic end products. When the amounts of organic matter reaching the bottom are greater than can be used by bottom fauna, they form a muck that is rich in hydrogen sulfide and methane.

As the water becomes shallower, the benthos changes. The action of water, plant growth, and recent organic

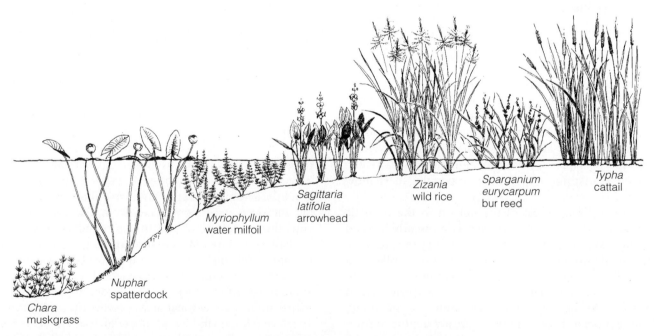

Figure 24.4 | Zonation of emergent, floating, and submerged vegetation at the edge of a lake or pond. Such zonation does not necessarily reflect successional stages but rather a response to water depth (see Chapter 13, Figure 13.12).

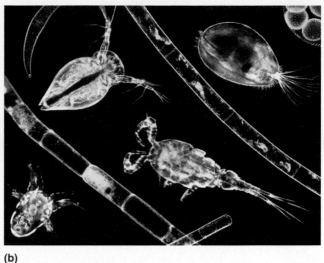

Figure 24.5 | **(a)** Phytoplankton and **(b)** zooplankton.

deposits modifies the bottom material, typically consisting of stones, rubble, gravel, marl, and clay. Increased oxygen, light, and food encourage a richness of life not found on the bottom of the profundal zone.

Closely associated with the benthic community are organisms collectively called **periphyton** or **aufwuchs.** They are attached to or move on a submerged substrate but do not penetrate it. Small aufwuchs communities colonize the leaves of submerged aquatic plants, sticks, rocks, and debris. Periphyton, mostly algae and diatoms living on plants, are fast growing and lightly attached. Aufwuchs on stones, wood, and debris form a more crustlike growth of cyanobacteria, diatoms, water moss, and sponges.

24.4 | The Character of a Lake Reflects Its Surrounding Landscape

Due to the close relationship between land and water ecosystems, lakes reflect the character of the landscape in which they occur. Water that falls on land flows over the surface or moves through the soil to enter springs, streams, and lakes. The water transports with it silt and nutrients in solution. Human activities including road construction, logging, mining, construction, and agriculture add another heavy load of silt and nutrients—especially nitrogen, phosphorus, and organic matter. These inputs enrich aquatic systems by a process called **eutrophication.** The term **eutrophy** (from the Greek *eutrophos,* "well nourished") means a condition of being rich in nutrients.

A typical eutrophic lake (Figure 24.6a) has a high surface-to-volume ratio; that is, the surface area is large relative to depth. Nutrient-rich deciduous forest and farmland often surround it. An abundance of nutrients, especially nitrogen and phosphorus, flowing into the

lake stimulates a heavy growth of algae and other aquatic plants. Increased photosynthetic production leads to increased recycling of nutrients and organic compounds, stimulating even further growth.

Phytoplankton concentrates in the warm upper layer of the water, giving it a murky green cast. Algae, inflowing organic debris and sediment, and remains of rooted plants drift to the bottom, where bacteria feed on this dead organic matter. Their activities deplete the oxygen supply of the bottom sediments and deep water until this region of the lake cannot support aerobic life. The number of bottom species declines, although the biomass and numbers of organisms remain high. In extreme cases, oxygen depletion (anoxic conditions) can result in the die-off of invertebrate and fish populations.

In contrast to eutrophic lakes and ponds are oligotrophic bodies of water (Figure 24.6b). **Oligotrophy** is the condition of being poor in nutrients. Oligotrophic lakes have a low surface-to-volume ratio. The water is clear and appears blue to blue-green in the sunlight. The nutrient content of the water is low; nitrogen may be abundant, but phosphorus is highly limited. A low input of nutrients from surrounding terrestrial ecosystems and other external sources is mostly responsible for this condition. Low availability of nutrients causes low production of organic matter that leaves little for decomposers, so oxygen concentration remains high in the hypolimnion. The bottom sediments are largely inorganic. Although the numbers of organisms in oligotrophic lakes and ponds may be low, species diversity is often high.

Lakes that receive large amounts of organic matter from surrounding land, particularly in the form of humic materials that stain the water brown, are called **dystrophic** (from *dystrophos,* "ill-nourished"). These bodies of water occur generally on peaty substrates, or in contact with

(a)

(b)

Figure 24.6 | **(a)** A eutrophic lake. Note the floating algal mats on the water surface. **(b)** An oligotrophic lake in Montana.

peaty substrates in bogs or heathlands that are usually highly acidic (see Section 25.6). Dystrophic lakes generally have highly productive littoral zones. This littoral vegetation dominates the lake's metabolism, providing a source of both dissolved and particulate organic matter.

24.5 | Flowing-Water Ecosystems Vary in Structure and Types of Habitats

Even the largest rivers begin somewhere back in the hinterlands as springs or seepage areas that become headwater streams, or they arise as outlets of ponds or lakes. A few rivers emerge fully formed from glaciers. As a stream drains away from its source, it flows in a direction and manner dictated by the lay of the land and the underlying rock formations. Joining the new stream are other small streams, spring seeps, and surface water.

Just below its source the stream may be small, straight, and swift, with waterfalls and rapids. Farther downstream, where the gradient is less steep, velocity decreases and the stream begins to meander, depositing its load of sediment as silt, sand, or mud. At flood time, a stream drops its load of sediment on surrounding level land, over which floodwaters spread to form floodplain deposits.

Where a stream flows into a lake or a river into the sea, the velocity of water is suddenly checked. The river then is forced to deposit its load of sediment in a fan-shaped area about its mouth to form a delta (Figure 24.7). Here, its course is carved into several channels, which are blocked or opened with subsequent deposits. As a result, the delta becomes an area of small lakes, swamps, and marshy islands. Material the river fails to deposit in the delta is carried out to open water and deposited on the bottom.

Because streams become larger on their course to rivers and are joined along the way by many others, we

Figure 24.7 | River delta formed by the deposition of sediments.

can classify them according to order (Figure 24.8). A small headwater stream with no tributaries is a first-order stream. Where two streams of the same order join, the stream becomes one of higher order. If two first-order streams unite, the resulting stream becomes a second-order one; and when two second-order streams unite, the stream becomes a third-order one. The order of a stream can increase only when a stream of the same order joins it. It cannot increase with the entry of a lower-order stream. In general, headwater streams are orders 1 to 3; medium-sized streams, 4 to 6; and rivers, greater than 6.

The velocity of a current molds the character and structure of a stream (see Quantifying Ecology 24.1: Streamflow). The shape and steepness of the stream channel; its width, depth, and roughness of the bottom;

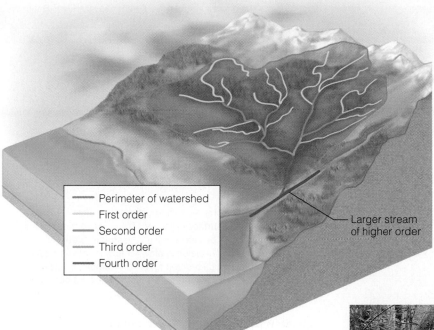

Figure 24.8 | Stream orders within a watershed.

Perimeter of watershed
First order
Second order
Third order
Fourth order

Larger stream
of higher order

and the intensity of rainfall and rapidity of snowmelt all affect velocity. Fast streams are those whose velocity of flow is 50 cm/s or higher. At this velocity, the current removes all particles less than 5 mm in diameter and leaves behind a stony bottom. High water increases the velocity; it moves bottom stones and rubble, scours the streambed, and cuts new banks and channels. As the gradient decreases and the width, depth, and volume of water increase, silt and decaying organic matter accumulate on the bottom. The character of the stream changes from fast water to slow (Figure 24.9), with an associated change in species composition.

Flowing-water ecosystems often alternate between two different but related habitats: the turbulent riffle and the quiet pool (Figure 24.10). Processes occurring in the rapids above influence the waters of the pool; and in turn, the waters of the rapids are influenced by events in the pools upstream.

Riffles are the sites of primary production in the stream. Here the periphyton or aufwuchs, organisms that are attached to or move on submerged rocks and logs, assume dominance. Periphyton, which occupy a position of the same importance as phytoplankton in lakes and ponds, consist chiefly of diatoms, cyanobacteria, and water moss.

Above and below the riffles are the pools. Here, the environment differs in chemistry, intensity of current, and depth. Just as the riffles are the sites of organic production, so the pools are the sites of decomposition. Here, the velocity of the current slows enough for organic matter to settle. Pools, the major sites of carbon dioxide production during the summer and fall, are necessary for maintaining

(a)

(b)

Figure 24.9 | **(a)** A fast mountain stream. The gradient is steep and the bottom is largely bedrock. **(b)** A slow stream is deeper and has a lower slope gradient.

The ecology of a stream ecosystem is determined largely by its streamflow—the water discharge occurring within the natural streambed or channel. The rate at which water flows through the stream channel influences the water temperature, oxygen content, rate of nutrient spiraling, physical structure of the benthic environment, and subsequently the types of organisms inhabiting the stream. As such, streamflow is an important parameter used by ecologists to characterize the stream environment.

Flow is defined simply as the volume of water moving past a given point in the stream per unit time. As such, it can be estimated from the cross-sectional area of the stream channel and the velocity of the flow as follows:

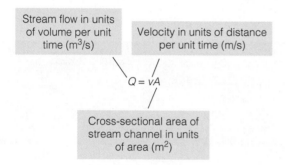

Stream flow in units of volume per unit time (m^3/s)

Velocity in units of distance per unit time (m/s)

$$Q = vA$$

Cross-sectional area of stream channel in units of area (m^2)

As shown in Figure 1, the cross-sectional area (A) can be calculated by measuring the depth (d) and width (w) of the stream and multiplying the two ($A = w \times d$). Estimates of depth and width can be easily made for a point along the stream channel by using a tape measure and meter stick. Velocity (v) can generally be thought of as distance (z) traveled over time (t); see Figure 1. The velocity can be estimated using a "current" or "flow" meter. One flow meter commonly used in streams and rivers is an apparatus with rotating cups (Figure 2a). The flow causes the cups to rotate (Figure 2b), and the number of rotations is monitored electronically. Number of

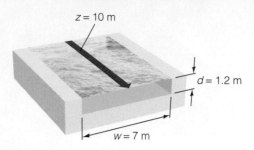

Figure 1 | Idealized representation of a stream channel (water shown in blue, land surface in yellow). The cross-sectional area (A) of the stream can be calculated as the stream width (w) times the depth (d). The velocity of water flow is calculated as the distance traveled (z) per unit time. For example, if a position on the water surface travels downstream 10 m over a period (t) of 20 s, the velocity of the water is z/t or 10 m/20 s = 0.5 m/s.

rotations (distance traveled) per unit time provides a measure of velocity.

For example, in the simple representation of a stream channel depicted in Figure 1, let us assume that the stream depth (d) is 1.2 m and the stream width (w) is 7 m. The cross-sectional area (A) of the stream is then 8.4 m^2(1.2 m × 7 m = 8.4 m^2). If the measured velocity (v) is 0.5 m/s, then the streamflow (Q) is 4.2 m^3/s(8.4 m^2 × 0.5 m/s = 4.2 m^3/s). In reality, the profile of the stream channel is never as simple as the rectangular profile presented in Figure 1, and the water velocity varies as a function of depth and position relative to the stream bank. For this reason, multiple measurements of depth and velocity are taken across the stream profile. For example, the stream profile in Figure 3 is 6 m wide; however, the depth varies across the width of the stream channel. A simple approach to estimating flow for this stream would be to sample water depth and velocity at several locations along the width of the stream, using the average values of water depth and

a constant supply of bicarbonate in solution (see Section 3.7). Without pools, photosynthesis in the riffles would deplete the bicarbonates and result in smaller and smaller quantities of available carbon dioxide downstream.

24.6 | Life Is Highly Adapted to Flowing Water

Living in moving water, inhabitants of streams and rivers have a major problem with remaining in place and not being swept downstream. They have evolved unique adaptations for dealing with life in the current (Figure 24.11a).

Figure 24.10 | Two different but related habitats in a stream: a riffle (background) and a pool (foreground).

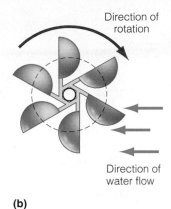

Direction of
rotation

Direction of
water flow

(a)   **(b)**

Figure 2 | **(a)** Stream velocity can be easily measured using a cup-type flow meter. **(b)** As the water flows, the cups rotate about a fixed point on the instrument. The number of rotations per unit time provides an estimate of water velocity.

velocity to calculate the value of streamflow. In most cases, however, stream ecologists will use a much more

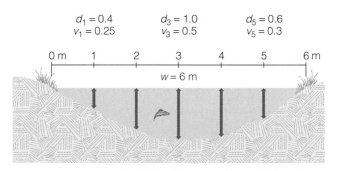

$d_1 = 0.4$ $d_3 = 1.0$ $d_5 = 0.6$
$v_1 = 0.25$ $v_3 = 0.5$ $v_5 = 0.3$

0 m   1   2   3   4   5   6 m

$w = 6$ m

Figure 3 | Cross-sectional view of a stream channel. To estimate the cross-sectional area (A), measurements of depth (d) and velocity (v) are taken at several locations along the width of the stream. By averaging these values of depth and velocity, a more accurate estimate of cross-sectional area and velocity can be obtained for calculating streamflow (Q).

elaborate sampling scheme, estimating water depth at regular intervals along the stream profile and water velocity at several depths at each interval.

The cross-sectional area and velocity of a stream will vary through time based on the amount of water being discharged from the surrounding watershed. In turn, the amount of water discharged will reflect the input of water to the surrounding watershed from precipitation. As a result, an accurate picture of streamflow requires a systematic sampling of the stream morphology (width and depth) and velocity through time.

1. In Figure 3, the water depths at 1, 3, and 5 m from the left stream bank are given as 0.4, 1.0, and 0.6 m, respectively. Estimates of velocity for the three locations are 0.25, 0.5, and 0.3 m/s, respectively. Estimate streamflow by representing depth and velocity as the simple mean of the three samples.

2. Why might the water velocity (v) decrease from the center of the stream channel to the banks?

A streamlined form, which offers less resistance to water current, is typical of many animals found in fast water. Larval forms of many insect species have extremely flattened bodies and broad, flat limbs that enable them to cling to the undersurfaces of stones where the current is weak. The larvae of certain species of caddisflies (Trichoptera) construct protective cases of sand or small pebbles and cement them to the bottoms of stones. Sticky undersurfaces help snails and planarians cling tightly and move about on stones and rubble in the current.

Among the plants, water moss (*Fontinalis*) and heavily branched, filamentous algae cling to rocks by strong holdfasts. Other algae grow in cushionlike colonies or form

sheets—covered with a slippery, gelatinous coating—that lie flat against the surfaces of stones and rocks.

All animal inhabitants of fast-water streams require high, near-saturation concentrations of oxygen and moving water to keep their absorbing and respiratory surfaces in continuous contact with oxygenated water. Otherwise, a closely adhering film of liquid, impoverished of oxygen, forms a cloak about their bodies.

In slow-flowing streams where current is at a minimum, streamlined forms of fish give way to fish species such as smallmouth bass, whose compressed bodies enable them to move through beds of aquatic vegetation. Pulmonate snails (Lymnaeacea) and burrowing mayflies

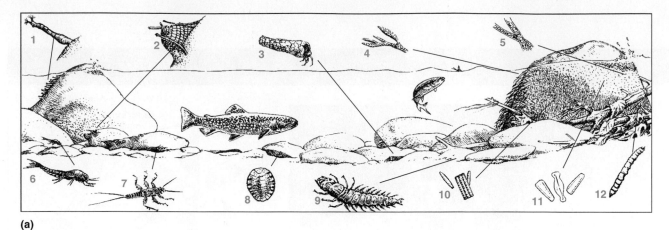

(a)

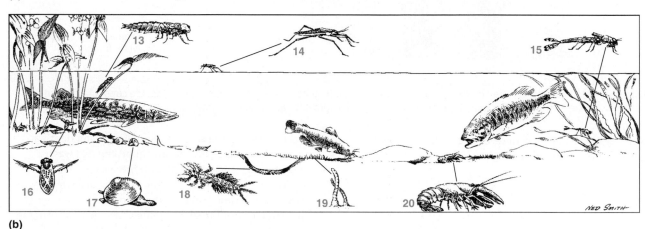

(b)

Figure 24.11 | Life in **(a)** a fast stream and **(b)** a slow stream. Fast stream: (1) black-fly larva (Simuliidae); (2) net-spinning caddisfly (*Hydropsyche* spp.); (3) stone case of caddisfly; (4) water moss (*Fontinalis*); (5) algae (*Ulothrix*); (6) mayfly nymph (*Isonychia*); (7) stonefly nymph (*Perla* spp.); (8) water penny (*Psephenus*); (9) hellgrammite (dobsonfly larva, *Corydalis cornuta*); (10) diatoms (*Diatoma*); (11) diatoms (*Gomphonema*); (12) crane-fly larva (Tipulidae). Slow stream: (13) dragonfly nymph (Odonata, *Anisoptera*); (14) water strider (*Gerris*); (15) damselfly larva (Odonata, *Zygoptera*); (16) water boatman (Corixidae); (17) fingernail clam (*Sphaerium*); (18) burrowing mayfly nymph (*Hexegenia*); (19) bloodworm (Oligochaeta, *Tubifex* spp.); (20) crayfish (*Cambarus* spp.). The fish in the fast stream is a brook trout (*Salvelinus fontinalis*). The fish in the slow stream are, from left to right: northern pike (*Esox lucius*), bullhead (*Ameiurus melas*), and smallmouth bass (*Micropterus dolomieu*).

(Ephemeroptera) replace rubble-dwelling insect larvae. Bottom-feeding fish, such as catfish (Akysidae), feed on life in the silty bottom; and back-swimmers and water striders inhabit sluggish stretches and still backwaters (Figure 24.11b).

Invertebrate inhabitants are classified into four major groups based on their feeding habits (Figure 24.12). **Shredders,** such as caddisflies (Trichoptera) and stoneflies (Plecoptera), make up a large group of insect larvae. They feed on coarse particulate organic matter (CPOM: > 1 mm diameter), mostly leaves that fall into the stream. The shredders break down the CPOM, feeding on the material not so much for the energy it contains as for the bacteria and fungi growing on it. Shredders assimilate about 40 percent of the material they ingest and pass off 60 percent as feces.

When broken up by the shredders and partially decomposed by microbes, the leaves, along with invertebrate feces, become part of the fine particulate organic matter (FPOM: < 1 mm but > 0.45 μm diameter). Drifting downstream and settling on the stream bottom, FPOM is picked up by another feeding group of stream invertebrates, the **filtering collectors** and **gathering collectors.** The filtering collectors include, among others, the larvae of black flies (Simuliidae) with filtering fans and the net-spinning caddisflies. Gathering collectors, such as the larvae of midges, pick up particles from stream-bottom sediments. Collectors obtain much of their nutrition from bacteria associated with the fine detrital particles.

While shredders and collectors feed on detrital material, another group, the **grazers,** feeds on the algal coating of stones and rubble. This group includes the

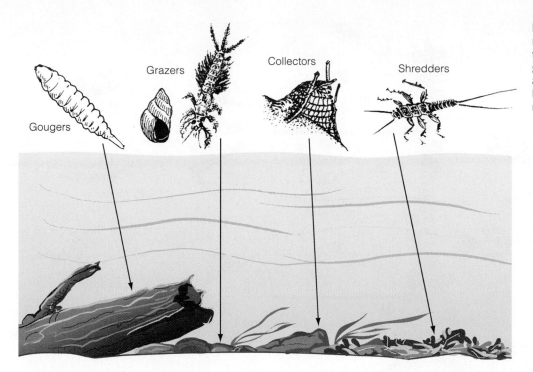

Grazers

Gougers

Collectors

Shredders

beetle larvae, water penny (*Psephenus* spp.), and a number of mobile caddisfly larvae. Much of the material they scrape loose enters the drift as FPOM. Another group, associated with woody debris, are the **gougers**—invertebrates that burrow into waterlogged limbs and trunks of fallen trees.

Feeding on the detrital feeders and grazers are predaceous insect larvae and fish such as the sculpin (*Cottus*) and trout. Even these predators do not depend solely on aquatic insects; they also feed heavily on terrestrial invertebrates that fall or wash into the stream.

Because of the current, quantities of CPOM, FPOM, and invertebrates tend to drift downstream to form a traveling benthos. This drift is a normal process in streams, even in the absence of high water and abnormally fast currents. Drift is so characteristic of streams that a mean rate of drift can serve as an index of a stream's production rate.

24.7 | The Flowing-Water Ecosystem Is a Continuum of Changing Environments

From its headwaters to its mouth, the flowing-water ecosystem is a continuum of changing environmental conditions (Figure 24.13). Headwater streams (orders 1 to 3) are usually swift, cold, and in shaded forested regions. Primary productivity in these streams is typically low, and they depend heavily on the input of detritus from terrestrial streamside vegetation, which contributes more than 90 percent of the organic input. Even when headwater streams are exposed to sunlight and autotrophic production exceeds inputs from adjacent terrestrial ecosystems, organic matter produced invariably enters the detrital food chain. Dominant organisms are shredders, processing large-sized litter and feeding on

CPOM; and collectors, processors of FPOM. Populations of grazers are minimal, reflecting the small amount of autotrophic production. Predators are mostly small fish—sculpins, darters, and trout. Headwater streams, then, are accumulators, processors, and transporters of particulate organic matter of terrestrial origin. As a result, the ratio of gross primary production to respiration is less than 1.

As streams increase in width to medium-sized creeks and rivers (orders 4 to 6), the importance of riparian vegetation and its detrital input decreases. With more surface water exposed to the sun, water temperature increases; and as the elevation gradient declines, the current slows. These changes bring about a shift from dependence on terrestrial input of particulate organic matter to primary production by algae and rooted aquatic plants. Gross primary production now exceeds community respiration. Due to the lack of CPOM, shredders disappear. Collectors, feeding on FPOM transported downstream, and grazers, feeding on autotrophic production, become the dominant consumers. Predators show little increase in biomass but shift from cold-water species to warm-water species, including bottom-feeding fish such as suckers (Catostomidae) and catfish.

As the stream order increases from 6 through 10 and higher, riverine conditions develop. The channel is wider and deeper. The flow volume increases, and the current becomes slower. Sediments accumulate on the bottom. Both riparian and autotrophic production decrease. A basic energy source is FPOM, used by bottom-dwelling collectors that are now the dominant consumers. However, slow, deep water and DOM (dissolved organic matter: $< 0.45 \mu m$ diameter) support a minimal phytoplankton and associated zooplankton population.

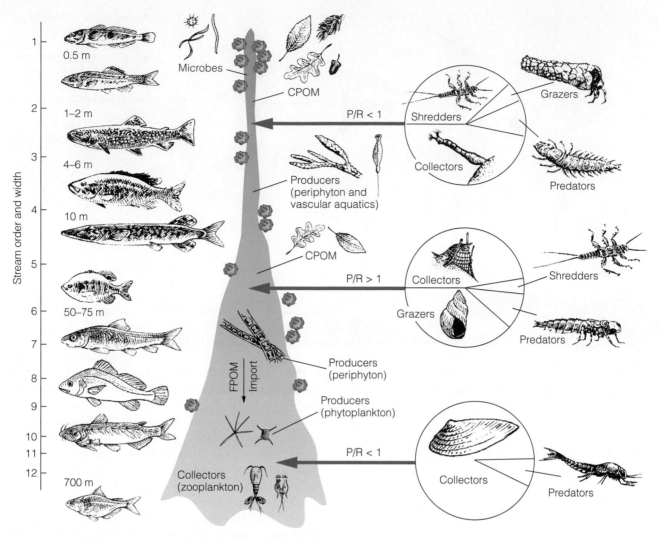

Figure 24.13 | Changes in consumer groups along the river continuum. Stream order and width (m) are shown on the axis to the left of the figure. The headwater stream is strongly heterotrophic, dependent on terrestrial input of detritus. The dominant consumers are shredders and collectors. As stream size increases, the input of organic matter shifts to primary production by algae and rooted vascular plants. The major consumers are now collectors and grazers. As the stream grows into a river, the lotic system shifts back to heterotrophy. A phytoplankton population may develop. The consumers are mostly bottom-dwelling collectors. The fish community likewise changes as one moves downstream (from top to bottom as shown: sculpin, darter, brook trout, smallmouth bass, pickerel, sunfish, sucker, freshwater drum, catfish, and shad).

Throughout the downstream continuum, the community capitalizes on upstream feeding inefficiency. Downstream adjustments in production and the physical environment are reflected in changes in consumer groups.

24.8 | Rivers Flow into the Sea, Forming Estuaries

Waters of most streams and rivers eventually drain into the sea. The place where freshwater joins saltwater is called an estuary. Estuaries are semienclosed parts of the coastal ocean where seawater is diluted and partially mixed with freshwater coming from the land (Figure 24.14). Here, the one-way flow of freshwater streams and rivers into an estuary meets the inflowing and outflowing saltwater tides. This meeting sets up a complex of currents that vary with the structure of the estuary (size, shape, and volume), season, tidal oscillations, and winds.

Mixing waters of different salinities and temperatures creates a counterflow that works as a nutrient trap (see Figure 21.28). Inflowing river waters most often impoverish rather than fertilize the estuary, with the possible exception of phosphorous. Instead, nutrients and oxygen are carried into the estuary by the tides. If vertical mixing occurs, these nutrients are not swept back out to sea but circulate up and down among organisms, water, and bottom sediments (see Figure 21.28).

Organisms inhabiting the estuary face two problems: maintaining their position and adjusting to changing salinity. Most estuarine organisms are benthic. They attach

Figure 24.14 | Estuary on the east coast of Australia.

Figure 24.15 | Oyster reef in estuarine environment.

to the bottom, bury themselves in the mud, or occupy crevices and crannies. Mobile inhabitants are mostly crustaceans and fish, largely young of species that spawn offshore in high-salinity water. Planktonic organisms are wholly at the mercy of the currents. Seaward movements of streamflow and ebb tide transport plankton out to sea, and the rate of water movement determines the size of the plankton population.

Salinity dictates the distribution of life in the estuary. The vast majority of the organisms inhabiting an estuary are marine, able to withstand full seawater. Some estuarine inhabitants cannot withstand lowered salinities, and these species decline along a salinity gradient from the ocean to the river's mouth. Sessile and slightly motile organisms have an optimum salinity range within which they grow best. When salinities vary on either side of this range, populations decline.

Anadromous fish are those that live most of their lives in saltwater and return to freshwater to spawn. These fish are highly specialized to endure the changes in salinity. Some species of fish, such as the striped bass (*Morone saxatilis*), spawn near the interface of fresh and low-salinity water. The larvae and young fish move downstream to more saline waters as they mature. Thus, for the striped bass, an estuary serves as both a nursery and as a feeding ground for the young. Anadromous species such as the shad (*Alosa*) spawn in freshwater; but the young fish spend their first summer in an estuary and then move out to the open sea. Species such as the croaker (Sciaenidae) spawn at the mouth of the estuary, but the larvae are transported upstream to feed in plankton-rich, low-salinity areas.

The oyster bed and oyster reef are the outstanding communities of the estuary (Figure 24.15). The oyster is the dominant organism about which life revolves. Oysters may be attached to every hard object in the intertidal zone; or they may form reefs, areas where clusters of living organisms grow cemented to the almost buried shells of past generations. Oyster reefs usually lie at right angles to tidal currents, which bring planktonic food, carry away wastes, and sweep the oysters clean of sediment and debris. Closely associated with oysters are encrusting organisms such as sponges, barnacles, and bryozoans, which attach themselves to oyster shells and depend on the oyster or algae for food.

In shallow estuarine waters, rooted aquatics such as the sea grasses widgeongrass (*Ruppia maritima*) and eelgrass (*Zostera marina*) assume major importance (Figure 24.16). These aquatic plants are complex systems supporting many epiphytic organisms. Such communities are important to certain vertebrate grazers, such as geese, swans, and sea turtles, and they provide a nursery ground for shrimp and bay scallops.

24.9 | Oceans Exhibit Zonation and Stratification

The marine environment is marked by several differences from the freshwater world. It is large, occupying 70 percent of Earth's surface; and it is deep, in places over 10 km. The surface area lighted by the sun is small compared to the

Figure 24.16 | Sea-grass meadow in the Chesapeake Bay dominated by eelgrass (*Zostera marina*).

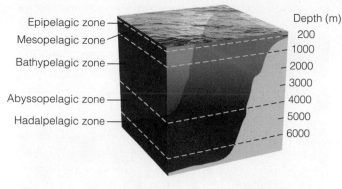

Figure 24.17 | Major regions of the ocean.

total volume of water. This small volume of sunlit water and the dilute solution of nutrients limit primary production. All of the seas are interconnected by currents, influenced by wave actions and tides, and characterized by salinity (see Chapter 3).

Just as lakes exhibit stratification and zonation, so do the seas. The ocean itself has two main divisions: the **pelagic,** or whole body of water, and the **benthic zone,** or bottom region. The pelagic is further divided into two provinces: the **neritic** province, water that overlies the continental shelf, and the **oceanic** province. Because conditions change with depth, the pelagic is divided into several distinct vertical layers or zones (Figure 24.17). From the surface to about 200 m is the **epipelagic zone,** or **photic zone,** in which there are sharp gradients in illumination, temperature, and salinity (see Chapter 4). From 200 to 1000 m is the **mesopelagic zone,** where little light penetrates and the temperature gradient is more even and gradual, without much seasonal variation. This zone contains an oxygen-minimum layer and often the maximum concentration of nutrients (nitrate and phosphate). Below the mesopelagic is the **bathypelagic zone,** where darkness is virtually complete, except for bioluminescent organisms; temperature is low; and water pressure is great. The **abyssopelagic zone** (Greek meaning "no bottom") extends from about 4000 m to the sea floor. The only zone deeper than this is the **hadalpelagic zone,** which includes areas found in deep-sea trenches and canyons.

24.10 | Pelagic Communities Vary among the Vertical Zones

When viewed from the deck of a ship or from an airplane, the open sea appears to be monotonously the same. Nowhere can you detect any strong pattern of life or well-defined communities, as you can over land. The reason is that pelagic ecosystems lack the supporting structures and framework of large, dominant plant life.

The dominant autotrophs are phytoplankton, and their major herbivores are tiny zooplankton.

There is a reason for the smallness of phytoplankton. Surrounded by a chemical medium that contains in varying quantities the nutrients necessary for life, they absorb nutrients directly from the water. The smaller the organism, the greater is the surface-to-volume ratio (see Section 7.1). More surface area is exposed for the absorption of nutrients and solar energy. Seawater is so dense that there is little need for supporting structures (see Section 3.2).

Because they require light, autotrophs are restricted to the upper surface waters where light penetration varies from tens to hundreds of meters (see Section 20.4). In shallow coastal waters, the dominant marine autotrophs are attached algae—restricted by light requirements to a maximum depth of about 120 m. The brown algae (Phaeophyceae) are the most abundant, associated with the rocky shoreline. Included in this group are the large kelps—such as *Macrocystis*, which grows to a length of 50 m and forms dense subtidal forests in the tropical and subtropical regions (see Figure 4.1). The red algae (Rhodophyceae) are the most widely distributed of the larger marine plants. They occur most abundantly in the tropical oceans, where some species grow to depths of 120 m.

The dominant autotrophs of the open water are phytoplankton (see Figure 24.5a). Each ocean or region within an ocean appears to have its own dominant forms. Littoral and neritic waters and regions of upwelling are richer in plankton than the mid-oceans are. In regions of downwelling, the dinoflagellates—a large, diverse group characterized by two whiplike flagella—concentrate near the surface in areas of low turbulence. These organisms attain their greatest abundance in warmer waters. In summer, they may concentrate in the surface waters in such numbers that they color it red or brown. Often toxic to vertebrates, such concentrations of dinoflagellates are responsible for "red tides." In regions of upwelling, the dominant forms of phytoplankton are diatoms. Enclosed in a silica case, diatoms are particularly abundant in Arctic waters.

The **nanoplankton,** smaller than diatoms, make up the largest biomass in temperate and tropical waters. Most abundant are the tiny cyanobacteria. The haptophytes—a group of primarily unicellular, photosynthetic algae that includes more than 500 species—are distributed in all waters except the polar seas. The most important members of this group, the coccolithophores, are a major source of primary production in the oceans. Coccolithophores have an armored appearance due to the calcium carbonate platelets, called coccoliths, covering the exterior of the cell (Figure 24.18).

Converting primary production into animal tissue is the task of herbivorous zooplankton, the most important

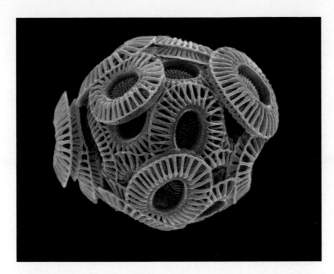

Figure 24.18 | Coccolithophores have an armored appearance due to the calcium carbonate platelets, called coccoliths, covering the exterior of the cell.

of which are the copepods (see Figure 24.5b). To feed on the minute phytoplankton, most of the grazing herbivores must also be small—between 0.5 and 5 mm. Most grazing herbivores in the ocean are copepods, which are probably the most abundant animals in the world. In the Antarctic, the shrimplike euphausiids, or krill (Figure 24.19), fed on by baleen whales and penguins, are the dominant herbivores. Feeding on the herbivorous zooplankton are the carnivorous zooplankton, which include such organisms as the larval forms of comb jellies (Ctenophora) and arrow worms (Chaetognatha).

However, part of the food chain begins not with the phytoplankton, but with organisms even smaller. Bacteria and protists, both heterotrophic and photosynthetic, make up one-half of the biomass of the sea and are responsible for the largest part of energy flow in pelagic systems. Photosynthetic nanoflagellates

(2–20 μm) and cyanobacteria (1–2 μm), responsible for a large part of photosynthesis in the sea, excrete a substantial fraction of their photosynthate in the form of dissolved organic material that heterotrophic bacteria use. In turn, heterotrophic nanoflagellates consume heterotrophic bacteria. This interaction introduces a feeding loop, the **microbial loop** (Figure 24.20), and adds several trophic levels to the plankton food chain.

Like phytoplankton, zooplankton live mainly at the mercy of the currents; but many forms of zooplankton have enough swimming power to exercise some control. Some species migrate vertically each day to arrive at a preferred level of light intensity. As darkness falls, zooplankton rapidly rise to the surface to feed on phytoplankton. At dawn, they move back down to preferred depths.

Feeding on zooplankton and passing energy along to higher trophic levels are the nekton, swimming organisms that can move at will in the water column. They range in size from small fish to large predatory sharks and whales, seals, and marine birds such as penguins. Some predatory fish, such as tuna, are more or less restricted to the photic zone. Others are found in deeper mesopelagic and bathypelagic zones or move between them, as the sperm whale does. Although the ratio in size of predator to prey falls within limits, some of the largest

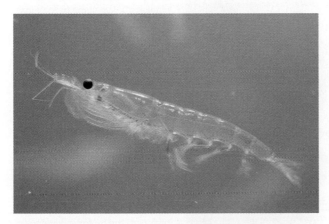

Figure 24.19 | Small euphausiid shrimps, called krill, are eaten by baleen whales and are essential to the marine food chain.

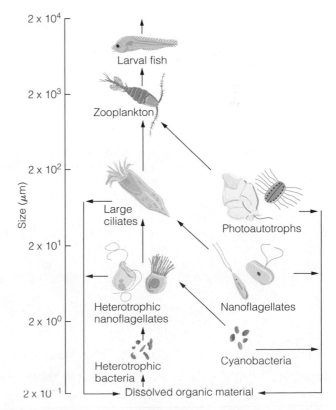

Figure 24.20 | A representation of the *microbial loop* and its relationship to the plankton food web. Autotrophs are on the right side of the diagram, and heterotrophs are on the left.

nekton organisms in the sea—the baleen whales—feed on disproportionately small prey, euphausiids, or krill (see Figure 24.19). By contrast, the sperm whale attacks very large prey such as the giant squid.

Residents of the deep have special adaptations for securing food. Darkly pigmented and weak bodied, many deep-sea fish depend on luminescent lures, mimicry of prey, extensible jaws, and expandable abdomens (enabling them to consume large items of food). In the mesopelagic region, bioluminescence reaches its greatest development—two-thirds of the species produce light. Fish have rows of luminous organs along their sides and lighted lures that enable them to bait prey and recognize other individuals of the same species. Bioluminescence is not restricted to fish. Squid and euphausiid shrimp possess searchlight-like structures complete with lens and iris, and some discharge luminous clouds to escape predators.

24.11 | Benthos Is a World of Its Own

The term *benthic* refers to the floor of the sea, and **benthos** refers to plants and animals that live there. In a world of darkness, no photosynthesis takes place, so the bottom community is strictly heterotrophic (except in vent areas), depending entirely on the rain of organic matter drifting to the bottom. Patches of dead phytoplankton as well as the bodies of dead whales, seals, birds, fish, and invertebrates all provide an array of foods for different feeding groups and species. Despite the darkness and depth, benthic communities support a high diversity of species. In shallow benthic regions, the polychaete worms may exceed 250 species and the pericarid crustaceans well over 100. But the deep-sea benthos supports a surprisingly higher diversity. The number of species collected in more than 500 samples—of which the total surface area sampled was only 50 m² —was 707 species of polychaetes and 426 species of pericarid crustaceans.

Important organisms in the benthic food chain are the bacteria of the sediments. Commonly found where large quantities of organic matter are present, bacteria may reach several tenths of a gram per square meter in the topmost layer of silt. Bacteria synthesize protein from dissolved nutrients and in turn become a source of protein, fat, and oils for other organisms.

In 1977, oceanographers first discovered high-temperature, deep-sea hydrothermal vents along volcanic ridges in the ocean floor of the Pacific near the Galápagos Islands. These vents spew jets of superheated fluids that heat the surrounding water to 8°C to 16°C, considerably higher than the 2°C ambient water. Since then, oceanographers have discovered similar vents on other volcanic ridges along fast-spreading centers of the ocean floor, particularly in the mid-Atlantic and eastern Pacific.

Vents form when cold seawater flows down through fissures and cracks in the basaltic lava floor deep into the underlying crust. The waters react chemically with the hot basalt, giving up some minerals but becoming enriched with others such as copper, iron, sulfur, and zinc. The water, heated to a high temperature, reemerges through mineralized chimneys rising up to 13 m above the sea floor. Among the chimneys are white smokers and black smokers (Figure 24.21). White-smoker chimneys rich in zinc sulfides issue a milky fluid with a temperature of under 300°C. Black smokers, narrower chimneys rich in copper sulfides, issue jets of clear water from 300°C to more than

Figure 24.21 | Black smoker. Deep under the oceans, next to hydrothermal vents issuing from cracked volcanic rocks ("black smokers"), live some of the most unusual animals ever seen. The photograph shows a "black smoker" in full flow, some 2250 m down on the ocean floor west of Vancouver Island, on the Juan de Fuca Ridge. The water temperatures on the chimneys exceed 400°C. Surrounding the vents are tubeworms, a bizarre group of animals found only in conditions where high levels of hydrogen sulfide can fuel their internal symbionts—bacteria that manufacture food for them. Distribution of this fauna is related to the history of the tectonic development of the ridges, thus linking the geology and biology of this peculiar ecosystem.

Figure 24.22 | Pictured is a colony of giant tubeworms with vent fish and crabs, all highly specialized for and found only in the extreme environment of the hydrothermal vent ecosystem.

(Courtesy of Richard Lutz, Rutgers University, Stephen Low Productions, and Woods Hole Oceanographic Institution.)

450°C that are soon blackened by precipitation of fine-grained sulfur–mineral particles.

Associated with these vents is a rich diversity of unique deep-sea life, confined to within a few meters of the vent system. The primary producers are chemosynthetic bacteria that oxidize reduced sulfur compounds such as H_2S to release energy used to form organic matter from carbon dioxide. Primary consumers include giant clams, mussels, and polychaete worms that filter bacteria from water and graze on bacterial film on rocks (Figure 24.22).

24.12 | Coral Reefs Are Complex Ecosystems Built by Colonies of Coral Animals

Lying in the warm, shallow waters about tropical islands and continental landmasses are coral reefs—colorful, rich oases within the nutrient-poor seas (Figure 24.23). They are a unique accumulation of dead skeletal material built up by carbonate-secreting organisms, mostly living coral (Cnidaria, Anthozoa) but also coralline red algae (Rhodophyta, Corallinaceae), green calcerous algae (*Halimeda*), foraminifera, and mollusks. Although various types of corals can be found from the water's surface to depths of 6000 m, reef-building corals are generally found at depths of less than 45 m. Because reef-building corals have a symbiotic relationship with algal cells (zooxanthellae; see Section 15.10 and Figure 15.8), their distribution is limited to depths where sufficient solar radiation (photosynthetically active radiation) is available to support photosynthesis. Precipitation of calcium from the water is necessary to form the coral skeleton. This precipitation occurs when water temperature and salinity are high and carbon dioxide concentrations are low. These requirements limit the distribution of reef-building corals to the shallow, warm tropical waters (20°C to 28°C).

Coral reefs are of three basic types: (1) *Fringing reefs* grow seaward from the rocky shores of islands and conti-nents. (2) *Barrier reefs* parallel shorelines of continents and islands and are separated from land by shallow lagoons. (3) *Atolls* are rings of coral reefs and islands surrounding a lagoon, formed when a volcanic mountain subsides beneath the surface. Such lagoons are about 40 m deep, usually connect to the open sea by breaks in the reef, and may have small islands of patch reefs. Reefs build up to sea level.

Coral reefs are complex ecosystems that begin with the complexity of the corals themselves. Corals are modular animals, anemone-like cylindrical polyps, with prey-capturing tentacles surrounding the opening or mouth. Most corals form sessile colonies supported on the tops of dead ancestors and cease growth when they reach the surface of the water. In the tissues of the gastrodermal layer live zooxanthellae—symbiotic, photosynthetically active, endozoic dinoflagellate algae that coral depend on for most efficient growth (see Chapter 15). On the calcareous skeletons live still other kinds of algae—the encrusting red and green coralline species and the filamentous

Figure 24.23 | A rich diversity of coral species, algae, and colorful fish occupy this reef in Fiji (South Pacific Ocean).

species, including turf algae—and a large bacterial population. Also associated with coral growth are mollusks, such as giant clams (*Tridacna*, *Hippopus*), echinoderms, crustaceans, polychaete worms, sponges, and a diverse array of fishes, both herbivorous and predatory.

Because the coralline community acts as a nutrient trap (see Section 21.10), offshore coral reefs are oases of productivity (1500 to 5000 g C/m^2/yr) within the relatively nutrient-poor, lower-productivity sea (15 to 50 g C/m^2/yr). This productivity and the varied habitats within the reef support a great diversity of life—thousands of species of invertebrates (such as sea urchins, which feed on coral and algae). Many kinds of herbivorous fish graze on algae, including zooxanthellae within the coral tissues; and hundreds of predatory species feed upon both invertebrate and vertebrate prey. Some of these predators, such as puffers (Tetraodontidae) and filefish (Monacanthidae), are corallivores that feed on coral polyps. Other predators lie in ambush for prey in coralline caverns. In addition, there is a wide array of symbionts, such as cleaning fish and crustaceans that pick parasites and detritus from larger fish and invertebrates.

24.13 | Productivity of the Oceans Is Governed by Light and Nutrients

As we discussed in Chapters 20 and 21, primary productivity in marine environments is limited to regions where the availability of light and nutrients can support photosynthesis and plant growth. The vertical attenuation of light in water limits productivity to the shallower waters of the photic zone. The presence of a thermocline (see Section 21.10), however, limits the movement of nutrients from the deeper to the surface waters where light is adequate to support photosynthesis—especially in the tropics, where the thermocline is permanent. The rate at which nutrients are returned to the surface, and therefore productivity, is controlled by two processes: (1) the seasonal breakdown of the thermocline and subsequent turnover (see Section 21.10) and (2) upwelling of deeper nutrient-rich waters to the surface (see Section 21.13). As a result, the highest primary productivity is found in coastal regions (see Figure 20.10). There, the shallower waters of the continental shelf allow for turbulence and seasonal turnover (where it occurs) to increase vertical mixing, and coastal upwelling (see Figure 3.17) brings deeper, colder, nutrient-rich water to the surface (Figure 24.24).

In open waters, productivity is low in most tropical oceans because the permanent nature of the thermocline slows the upward diffusion of nutrients. In these regions, phytoplankton growth is essentially controlled by the cycling of nutrients within the photic zone. Production rates remain more or less constant throughout the year (Figure 24.25a). The highest production in the open waters of the tropical oceans occurs where nutrient-rich water is brought to the surface in the equatorial regions, where upwelling occurs as surface currents diverge (see Figures 3.14 and 24.24).

Productivity is also low in the Arctic, mainly because of light limitations. A considerable amount of light energy is lost through reflection due to the low sun angle, or it is absorbed by the snow-covered sea ice that covers as much as 60 percent of the Arctic Ocean during the summer.

In contrast, the waters of the Antarctic are noted for their high productivity as a result of the continuous upwelling of nutrient-rich water around the continent (Figure 24.25b). The growing season is limited by the short summer period. Primary productivity in temperate oceans (Figure 24.25c) is strongly related to seasonal variation in nutrient supply, driven by the seasonal dynamics of the thermocline (see Section 21.10).

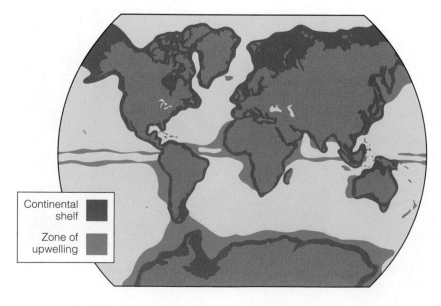

Figure 24.24 | Map of continental shelf and upwellings.

(Adapted from Archibold 1995.)

Continental shelf

Zone of upwelling

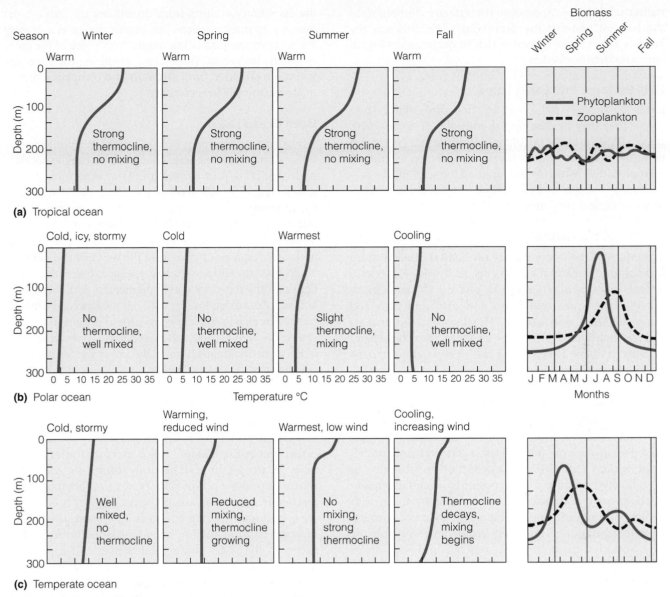

Figure 24.25 | Thermal profiles, extent of vertical mixing, and associated patterns of productivity in **(a)** tropical, **(b)** polar, and **(c)** temperate oceans during the four seasons of the year.

Summary

Lakes Defined | 24.1

Lake and pond ecosystems are bodies of water that fill a depression in the landscape. They are formed by many processes, ranging from glacial and geological to human activities. Geologically speaking, lakes and ponds are successional features. In time, most of them fill, get smaller, and finally may be replaced by a terrestrial ecosystem.

Lake Stratification | 24.2

As a nearly self-contained ecosystem, a lake exhibits both vertical and horizontal gradients. Seasonal stratification

in light, temperature, and dissolved gases influences the distribution of life in the lake.

Zonation of Life in Lakes | 24.3

The area where light penetrates to the bottom of the lake, called the littoral zone, is occupied by rooted plants. Beyond this is the open-water or limnetic zone inhabited by plant and animal plankton and fish. Below the depth of effective light penetration is the profundal region, where the diversity of life varies with temperature and oxygen supply. The bottom or benthic zone is a place of intense biological activity where decomposition of organic matter takes place. Anaerobic bacteria are dominant on the bottom beneath the profundal water, whereas the benthic zone of the littoral is rich in decomposer organisms and detritus feeders.

Nutrient Input into Lakes | 24.4

Lakes are strongly influenced by their surrounding landscape. They may be classified as eutrophic (nutrient rich), oligotrophic (nutrient poor), or dystrophic (acidic and rich in humic material). Most lakes are subjected to cultural eutrophication, which is the rapid addition of nutrients—especially nitrogen and phosphorus—from sewage, industrial wastes, and agricultural runoff.

Flowing-Water Habitat | 24.5

Currents and their dependence on detrital material from surrounding terrestrial ecosystems set flowing-water ecosystems apart from other aquatic systems. Currents shape the life in streams and rivers and carry nutrients and other materials downstream. Flowing-water ecosystems change longitudinally in flow and size from headwater streams to rivers. They may be fast or slow, characterized by a series of riffles and pools.

Adaptations to Flowing Water | 24.6

Organisms well-adapted to living in the current inhabit fast-water streams. They may be streamlined in shape, flattened to conceal themselves in crevices and underneath rocks, or attached to rocks and other substrates. In slow-flowing streams where current is minimal, streamlined forms of fish tend to be replaced by those with compressed bodies that enable them to move through aquatic vegetation. Burrowing invertebrates inhabit the silty bottom. Stream invertebrates fall into four major groups that feed on detrital material: shredders, collectors, grazers, and gougers.

River Continuum | 24.7

Life in flowing water reflects a continuum of changing environmental conditions from headwater streams to the river mouth. Headwater streams depend on inputs of detrital material. As stream size increases, algae and rooted plants become important energy sources as reflected in the changing species composition of fish and invertebrate life. Large rivers depend on fine particulate matter and dissolved organic matter as sources of energy and nutrients. River life is dominated by filter feeders and bottom-feeding fish.

Estuaries | 24.8

Rivers eventually reach the sea. The place where the inflowing freshwater meets the incoming and outgoing tidal water is an estuary. The intermingling of freshwater and tides creates a nutrient trap exploited by estuarine life. Salinity determines the nature and distribution of estuarine life. As salinity declines from the estuary up through the river, so do marine species. An estuary serves as a nursery for many marine organisms, particularly some of the commercially important finfish and shellfish, for here the young are protected from predators and competing species unable to tolerate lower salinity.

Open Ocean | 24.9

The marine environment is characterized by salinity, waves, tides, depth, and vastness. Like lakes, oceans experience both stratification of temperature (and other physical parameters) and stratification of the organisms that inhabit the differing vertical strata. The open sea can be divided into several vertical zones. The hadalpelagic zone includes areas found in the deep-sea trenches and canyons. The abyssopelagic zone extends from the sea floor to a depth of about 4000 m. Above is the bathypelagic zone, void of sunlight and inhabited by darkly pigmented, bioluminescent animals. Above that lies the dimly lit mesopelagic zone, inhabited by characteristic species, such as certain sharks and squid. The bathypelagic and mesopelagic zones depend on a rain of detrital material from the upper lighted zone, the epipelagic zone, for their energy.

Ocean Life | 24.10

Phytoplankton dominate the surface waters. The littoral and neritic zones are richer in plankton than the open ocean. Tiny nanoplankton, which make up the largest biomass in temperate and tropical waters, are the major source of primary production. Feeding on phytoplankton are herbivorous zooplankters, especially copepods. They are preyed upon by carnivorous zooplankton. The greatest diversity of zooplankton, including larval forms of fish, occurs in the water over coastal shelves and upwellings; the least diversity occurs in the open ocean. Making up the larger life-forms are free-swimming nekton, ranging from small fish to sharks and whales. Benthic organisms (those living on the floor of the deep ocean) vary with depth and substrate. They are strictly heterotrophic and depend on organic matter that drifts to the bottom. They include filter feeders, collectors, deposit feeders, and predators.

Hydrothermal Vents | 24.11

Along volcanic ridges are hydrothermal vents inhabited by unique and newly discovered life-forms, including crabs, clams, and worms. Chemosynthetic bacteria that use sulfates as an energy source account for primary production in these hydrothermal vent communities.

Coral Reefs | 24.12

Coral reefs are nutrient-rich oases in nutrient-poor tropical waters. They are complex ecosystems based on anthozoan coral and coralline algae. Their productive and varied habitats support a high diversity of invertebrate and vertebrate life.

Ocean Productivity | 24.13

Primary productivity in marine environments is limited to regions where the availability of light and nutrients can support photosynthesis and plant growth. The areas of highest productivity are coastal regions and areas of upwelling. In open oceans, especially in tropical areas, productivity is low because the permanent nature of the thermocline slows the upward diffusion of nutrients. Primary productivity in temperate oceans is strongly related to seasonal variation in nutrient supply, driven by the seasonal dynamics of the thermocline.

Study Questions

1. What distinguishes the littoral zone from the limnetic zone in a lake? What distinguishes the limnetic zone from the profundal zone?
2. What conditions distinguish the benthic zone from the other strata in lake ecosystems? What is the dominant role of the benthic zone?
3. Distinguish between oligotrophy, eutrophy, and dystrophy.
4. What physical characteristics are unique to flowing-water ecosystems? Contrast these conditions in a fast- and slow-flowing stream.
5. How do environmental conditions change along a river continuum?
6. What is an estuary?
7. Characterize the major life zones of the ocean, both vertical and horizontal.
8. How does temperature stratification in tropical seas differ from that of temperate regions of the ocean? How do these differences influence patterns of primary productivity?
9. What are hydrothermal vents, and what makes life around them unique?
10. What are coral reefs, and how do they form?

Further Readings

Allan, J. D. 1995. *Stream ecology: Structure and functioning of running waters*. Dordrecht: Kluwer Academic Press.
　An extensive reference on the ecology of stream ecosystems.

Grassle, J. F. 1991. Deep-sea benthic diversity. *Bioscience* 41:464–469.
　An excellent and well-written introduction for those interested in the strange and wonderful world of the deep ocean.

Gross, M. G., and E. Gross. 1995. *Oceanography: A view of Earth*, 7th ed. Englewood Cliffs, NJ: Prentice-Hall.
　An excellent reference on the physical aspects of the sea.

Hart, D. D., and N. L. Poff (eds.). 2002. A special section on dam removal and river restoration. *BioScience* 52:653–747.
　An in-depth appraisal of the ecological, social, economic, and legal aspects of dam removal.

Jackson, J. B. C. 1991. Adaptation and diversity of reef corals. *Bioscience* 41:475–482.
　This excellent review paper relates patterns of species distribution to life history and disturbance.

Nybakken, J. W., and M. D. Bertness. 2005. *Marine biology: An ecological approach*, 6th ed. Menlo Park, CA: Benjamin Cummings.
　An excellent reference on marine life and ecosystems. Well written and illustrated.

Wetzel, R. G. 2001. *Limnology: Lake and river ecosystems*, 3rd ed. San Diego: Academic Press.
　An outstanding reference that provides an introduction to the ecology of freshwater ecosystems.

CHAPTER 25

Coastal and Wetland Ecosystems

Pond pine (*Pinus serotina*) and a diversity of floating and emergent aquatic plants dominate this area of cypress swamp in spring (Charleston, South Carolina).

Wherever land and water meet, there is a transitional zone that gives rise to a diverse array of unique ecosystems. In coastal environments, this zone is between the terrestrial and marine environments. These environments are classified based on their underlying geology and substrate—sediment type, size, and shape. The product of marine erosion, the rocky shore, is the most primitive type of coast because it has been altered the least. Sandy beaches, found in wave-dominated, depositional settings, are highly dynamic environments subjected to continuous and often extreme change. Associated with estuarine environments or in the protected regions of coastal dunes are tidal mudflats, salt marshes, and mangrove forests.

The transitional zones between freshwater and land are characterized by terrestrial wetlands dominated by specialized plants that occur where the soil conditions remain saturated for most or all of the year. These are the marshes, swamps, bogs, and zones of emergent vegetation along rivers and lakes.

In this chapter, we will examine these unique environments and the organisms that inhabit them.

25.1 | The Intertidal Zone Is the Transition between Terrestrial and Marine Environments

Rocky, sandy, muddy, and either protected or pounded by incoming swells, all intertidal shores have one feature in common—they are alternately exposed and submerged by the tides. Roughly, the region of the seashore is bounded on one side by the height of extreme high tide and on the other by the height of extreme low tide. Within these confines, conditions change hourly with the ebb and flow of the tides (see Section 3.9). At flood tide, the seashore is a water world; at ebb tide, it belongs to the terrestrial environment, with its extremes in temperature, moisture, and solar radiation. Despite these changes, seashore inhabitants are essentially marine organisms adapted to withstand some degree of exposure to the air for varying periods of time.

At ebb tide, the uppermost layers of intertidal life are exposed to air, wide temperature fluctuations, intense solar radiation, and desiccation for a considerable period, whereas the lowest fringes on the intertidal shore may be exposed only briefly before the rising tide submerges them again. These varying conditions result in one of the most striking features of the coastal shoreline: the zonation of life.

25.2 | Rocky Shorelines Have a Distinct Pattern of Zonation

All rocky shores have three basic zones (Figure 25.1), each characterized by dominant organisms (Figure 25.2). The approach to a rocky shore from the landward side is marked by a gradual transition from lichens and land plants to marine life dependent at least partly on the tidal

(a)

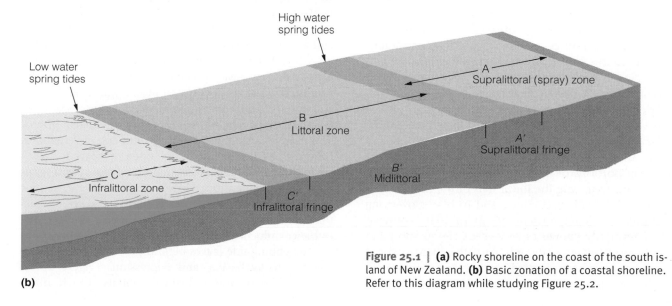

(b)

Figure 25.1 | (a) Rocky shoreline on the coast of the south island of New Zealand. (b) Basic zonation of a coastal shoreline. Refer to this diagram while studying Figure 25.2.

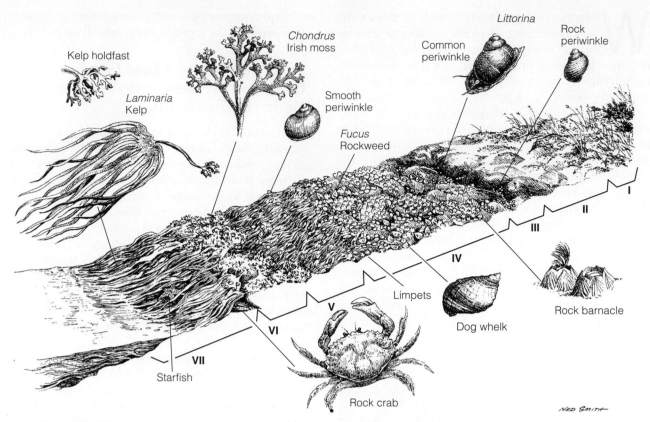

Figure 25.2 | Zonation on a rocky shore along the North Atlantic. Compare with Figure 25.1b. I, land: lichens, herbs, grasses; II, bare rock; III, black algae and rock periwinkle (*Littorina*) zone; IV, barnacle (*Balanus*) zone: barnacles, dog whelks, common periwinkles, mussels, limpets; V, fucoid zone: rockweed (*Fucus*) and smooth periwinkles; VI, Irish moss (*Chondrus*) zone; VII, kelp (*Laminaria*) zone.

waters. Moving from the terrestrial or **supralittoral or supratidal zone,** the first major change from the adjacent terrestrial environment appears at the **supralittoral fringe,** where saltwater comes only once every 2 weeks on the spring tides. It is marked by the black zone, named for the thin black layer of cyanobacteria (*Calothrix*) growing on the rock together with lichens (*Verrucaria*) and green alga (*Entophysalis*) above the high-tide waterline. Living under conditions that few plants could survive, these organisms represent an essentially nonmarine community. Common to this black zone are periwinkles, small snails of the genus *Littorina* (from which the term *littoral zone* is derived) that graze on wet algae covering the rocks.

Below the black zone lies the **littoral or intertidal zone,** which is covered and uncovered daily by the tides. In the upper reaches, barnacles are most abundant. Oysters, blue mussels, and limpets appear in the middle and lower portions of the littoral, as does the common periwinkle. Occupying the lower half of the littoral zone (midlittoral) of colder climates, and in places overlying the barnacles, is an ancient group of plants—brown algae, commonly known as rockweeds (*Fucus* spp.) and wrack (*Ascophyllum nodosum*). On the hard-surfaced shores that have been covered partly by sand and mud, blue mussels may replace the brown algae.

The lowest part of the littoral zone, uncovered only at the spring tides and not even then if wave action is strong, is the **infralittoral fringe.** This zone, exposed for short periods of time, consists of forests of large brown alga—*Laminaria* (one of the kelps)—with a rich undergrowth of smaller plants and animals among the holdfasts. Below the infralittoral fringe is the **infralittoral or subtidal zone.**

Grazing, predation, competition, larval settlement, and action of waves heavily influence the pattern of life on rocky shores. Waves bring in a steady supply of nutrients and carry away organic material. They keep the fronds of seaweeds in constant motion, moving them in and out of shadow and sunlight, allowing more even distribution of incident light and thus more efficient photosynthesis. By dislodging both plants and invertebrates from the rocky substrate, waves open up space for colonization by algae and invertebrates and reduce strong interspecific competition. Heavy wave action can reduce the activity of such predators as starfish and sea urchins that feed on sessile intertidal invertebrates. In effect, disturbance influences community structure.

The ebbing tide leaves behind pools of water in rock crevices, rocky basins, and depressions (Figure 25.3). These pools represent distinct habitats, which differ

Figure 25.3 | Tidal pools fill depressions along this length of rocky coastline in Maine.

considerably from exposed rock and the open sea and even differ among themselves. At low tide, all pools are subjected to wide and sudden fluctuations in temperature and salinity. Changes are most marked in shallow pools. Under the summer Sun, temperatures may rise above the maximum many organisms can tolerate. As water evaporates, especially in the shallower pools, salt crystals may appear around the edges. When rain or drainage from the surrounding land brings freshwater to the pools, salinity may decrease. In deep pools, this freshwater tends to form a layer on top, developing a strong salinity stratification in which the bottom layer and its inhabitants are little affected. If algal growth is considerable, oxygen will be high during the daylight hours but low at night, a situation that rarely occurs at sea. The rise of CO_2 at night lowers the pH (see Section 3.7). Most pools suddenly return to sea conditions on the rising tide and experience drastic and sudden changes in temperature, salinity, and pH. Life in the tidal pools must be able to withstand these extreme fluctuations.

25.3 | Sandy and Muddy Shores Are Harsh Environments

Sandy and muddy shores often appear devoid of marine life at low tide (Figure 25.4), in sharp contrast to the life-filled rocky shore; but sand and black mud are not as barren as they seem, for beneath them life lurks, waiting for the next high tide.

The sandy shore is a harsh environment, a product of the harsh and relentless weathering of rock, both inland and along the shore. Rivers and waves carry the products of rock weathering and deposit them as sand along the edge of the sea. The size of the sand particles deposited influences the nature of the sandy beach, water retention during low tide, and the ability of animals to burrow through it. In sheltered areas of the coast, the slope of the beach may be so gradual that the surface appears to be flat. Because of the flatness, the outgoing tidal currents are slow, leaving behind a residue of organic material settled from the water. In these situations, mudflats develop.

Life on sand is almost impossible. Sand provides no surface for attachment of seaweeds and their associated fauna; and the crabs, worms, and snails characteristic of

Figure 25.4 | A stretch of sandy beach washed by waves on the southern New Zealand coast. Although the beach appears barren, life is abundant beneath the sand.

rocky crevices find no protection there. Life, then, is forced to live beneath the sand.

Life on sandy and muddy beaches (see Figure 16.9 for an illustration of patterns of zonation on a sandy beach) consists of **epifauna** (organisms living on the sediment surface) and **infauna** (organisms living in the sediments). Most infauna occupy either permanent or semipermanent tubes within the sand or mud and are able to burrow rapidly into the substrate. Other infauna live between particles of sand and mud. These tiny organisms, referred to as **meiofauna,** range in size from 0.05 to 0.5 mm and include copepods, ostracods, nematodes, and gastrotrichs.

Sandy beaches also exhibit zonation related to the tides (see Figure 16.9), but you must discover it by digging. Pale, sand-colored ghost crabs and beach hoppers occupy the upper beach, the supralittoral. The intertidal beach, the littoral, is a zone where true marine life appears. Although sandy shores lack the variety found on rocky shores, the populations of individual species of largely burrowing animals often are enormous. An array of animals, among them starfish and the related sand dollar, can be found above the low-tide line in the littoral zone.

Organisms living within the sand and mud do not experience the same extreme fluctuations in temperature as do those on rocky shores. Although the surface temperature of the sand at midday may be 10 °C (or more) higher than the returning seawater, the temperature a few centimeters below the sand remains almost constant throughout the year. Nor is there a great fluctuation in salinity, even when freshwater runs over the surface of the sand. Below 25 cm, salinity is little affected.

Near and below the low-tide line live predatory gastropods, which prey on bivalves beneath the sand. In the same area lurk predatory portunid crabs such as the blue crab and green crab that feed on mole crabs, clams, and other organisms. These species move back and forth with the tides. The incoming tides also bring small predatory fish, such as killifish and silversides. As the tide recedes, gulls and shorebirds scurry across the sand and mudflats to hunt for food.

The energy base for life on the sandy shore is an accumulation of organic matter. Most sandy beaches contain a certain amount of detritus from seaweeds, dead animals, and feces brought in by the tides. This organic matter accumulates within the sand in sheltered areas. It is subject to bacterial decomposition, which is most rapid at low tide. Some detrital-feeding organisms ingest organic matter largely as a means of obtaining bacteria. Prominent among them are many nematodes and copepods (Harpacticoida), polychaete worms (*Nereis*), gastropod mollusks, and lugworms (*Arenicola*), which are responsible for the conspicuous coiled and cone-shaped casts on the beach. Other sandy beach animals are filter feeders that obtain their food by sorting particles of organic matter from tidal water. Two of these, alternately advancing and retreating with the tide, are the mole crab (*Emerita*) and the coquina clam (*Donax*).

25.4 | Tides and Salinity Dictate the Structure of Salt Marshes

Salt or tidal marshes occur in temperate latitudes where coastlines are protected from the action of waves within estuaries, deltas, and by barrier islands and dunes (Figure 25.5). The structure of a salt marsh is dictated by tides and salinity, which create a complex of distinctive and clearly demarked plant communities.

From the edge of the sea to the high land, zones of vegetation distinctive in form and color develop, reflecting a microtopography that lifts the plants to various heights within and above high tide (see Figure 16.8 for an illustration of vegetation zonation in a coastal salt marsh). Commonly found on the seaward edge of marshes and along tidal creeks of the Eastern coastline in North America are tall, deep green growths of salt-marsh cordgrass, *Spartina alterniflora*. Cordgrass forms a marginal strip between the open mud to the front and the high marsh behind. It has a high tolerance for salinity and is able to live in a semi-submerged state. To get air to its roots, which are buried in anaerobic mud, cordgrass has hollow tubes leading from the leaf to the root through which oxygen diffuses.

Above and behind the low marsh is the high marsh, standing at the level of mean high water. At this level, tall salt-marsh cordgrass gives way rather abruptly to a short form. This shorter form of *Spartina* is yellowish, in contrast to the tall, dark green form. This short form is an example of phenotypic plasticity in response to environmental conditions of the high marsh (see Chapter 5). The high marsh has a higher salinity and a decreased input of nutrients that result from a lower tidal exchange rate than in the low marsh. Here also grow the fleshy, translucent glassworts (*Salicornia* spp.; Figure 25.6) that turn bright red in fall, sea lavender (*Limonium carolinianum*), spearscale (*Atriplex patula*), and sea blite (*Suaeda maritima*).

Where the microelevation is about 5 cm above mean high water, short *Spartina alterniflora* and its associates are replaced by salt meadow cordgrass (*Spartina patens*) and an associate spikegrass or saltgrass (*Distichlis spicata*). As

Figure 25.5 | Coastal salt marsh.

Figure 25.6 | Glasswort dominates highly saline areas on the salt marsh. The plant, which turns red in fall, is a major food of overwintering geese.

Figure 25.7 | A tidal creek at high tide on the high marsh. Tall *Spartina* grows along the banks.

the microelevation rises several more centimeters above mean high tide, and if there is some intrusion of freshwater, *Spartina* and *Distichlis* may be replaced by two species of black needlerush or black grass (*Juncus roemerianus* and *Juncus gerardi*), so called because their dark green color becomes almost black in the fall. Beyond the black grass and often replacing it is a shrubby growth of marsh elder (*Iva frutescens*) and groundsel (*Baccharis halimifolia*). On the upland fringe grow bayberry (*Myrica pensylvanica*) and the pink-flowering sea holly (*Hibiscus palustris*).

Two conspicuous features of a salt marsh are the salt pans interspersed among meandering creeks. The creeks form an intricate system of drainage channels that carry tidal waters back out to sea (Figure 25.7). Their exposed banks support a dense population of mud algae, diatoms, and dinoflagellates that are photosynthetically active all year. Salt pans are circular to elliptical depressions flooded at high tide. At low tide, they remain filled with saltwater. If the pans are shallow enough, the water may evaporate completely, leaving an accumulating concentration of salt on the mud. The edges of these salt flats may be invaded by glasswort and spikegrass (Figure 25.8).

Although the salt marsh is not noted for its diversity, it is home to several interesting organisms. Some of the inhabitants are permanent residents in the sand and mud; others are seasonal visitors; and most are transients coming to feed at high and low tide.

Three dominant animals of the low marsh are ribbed mussels (*Modiolus demissus*), buried halfway in the mud; fiddler crabs (*Uca* spp.), running across the marsh at low tide; and marsh periwinkles (*Littorina* spp.) that move up and down the stems of *Spartina* and onto the mud to feed on alga. Three conspicuous vertebrate residents of the low marsh of eastern North America are the diamond-backed terrapin (*Malaclemys terrapin*), clapper rail (*Rallus longirostris*), and seaside sparrow (*Ammospiza maritima*).

In the high marsh, animal life changes as abruptly as the vegetation. The small, coffee-colored pulmonate snail (*Melampus*), found by the thousands under the low grass, replaces the marsh periwinkle. The willet (*Catoptrophorus semipalmatus*) and seaside sharp-tailed sparrow (*Ammospiza caudacuta*) replace the clapper rail and seaside sparrow.

Low tide brings a host of predators into the marsh to feed. Herons, egrets, gulls, terns, willets, ibis, raccoons, and others spread over the exposed marsh floor and muddy banks of tidal creeks. At high tide, the food web changes as

Figure 25.8 | A salt pan or pool in the high marsh.

Figure 25.9 | Mangroves replace tidal marshes in tropical regions.

the tide waters flood the marsh. Small predatory fish such as the silversides (*Menidia menidia*), killifish (*Fundulus heteroclitus*), and four-spined stickleback (*Apeltes quadracus*), restricted to channel waters at low tide, spread over the marsh at high tide, as does the blue crab.

25.5 | Mangroves Replace Salt Marshes in Tropical Regions

Replacing salt marshes on tidal flats in tropical regions are **mangrove forests** or **mangals** (Figure 25.9), which cover 60 to 75 percent of the coastline of the tropical regions. Mangrove forests develop where wave action is absent, sediments accumulate, and the muds are anoxic (without oxygen). They extend landward to the highest vertical tidal range, where they may be only periodically flooded. The dominant plants are mangroves, which include 8 families and 12 genera dominated by *Rhizophora*, *Avicennia*, *Bruguiera*, and *Sonneratia*. Growing with them are other salt-tolerant plants, mostly shrubs.

In growth form, mangroves range from short, prostrate forms to timber-size trees 30 m high. All mangroves have shallow, widely spreading roots, and many have prop roots coming from trunk and limbs (Figure 25.10). Many species have root extensions called pneumatophores that take in oxygen for the roots. The tangle of prop roots and pneumatophores slows the movement of tidal waters, allowing sediments to settle out. Land begins to move seaward, followed by colonizing mangroves.

Mangrove forests support a rich fauna, with a unique mix of terrestrial and marine life. Living and nesting in the upper branches are many species of birds, particularly

Figure 25.10 | Interior of a red mangrove stand. Note the prop roots.

herons and egrets. As in salt marshes, *Littorina* snails live on the prop roots and trunks of mangrove trees. Also attached to the stems and prop roots are oysters and barnacles, and on the mud at the base of the roots are detritus-feeding snails. Fiddler crabs and tropical land crabs burrow into the mud during low tide and live on prop roots and high ground during high tide. In the Indo-Malaysian mangrove forests live mudskippers, fish of the genus *Periophthalmus*, with modified eyes set high on the head. They live in burrows in the mud and crawl about on the top of it. In many ways they act more like amphibians than fish. The sheltered waters about the roots provide a nursery and haven for the larvae and young of crabs, shrimp, and fish.

25.6 | Freshwater Wetlands Are a Diverse Group of Ecosystems

The transitional zones between freshwater and land are characterized by terrestrial wetlands. These unique environments form ecotones between terrestrial and adjacent aquatic ecosystems, sharing characteristics of both. Wetlands cover 6 percent of Earth's surface. They are found in every climatic zone but are local in occurrence. Only a few—such as the Everglades in Florida, the Pantanal in Brazil, the Okavango in southern Africa (Figure 25.11), and the Fens of England—cover extensive areas of the landscape (see Ecological Issues: The Continuing Decline of Wetlands).

Wetlands range along a gradient from permanently flooded to periodically saturated soil (Figure 25.12) and support specialized plants that occur where the soil conditions remain saturated for most or all of the year. These hydrophytic (water-adapted) plants are adapted to grow in water or on soil that is periodically anaerobic (lacking oxygen) because of excess water (see Chapter 6). **Hydrophytic plants** are typically classified into one of three groups: (1) obligate wetland plants that require saturated soils: examples are the submerged pondweeds, floating pond lily, emergent cattails and bulrushes, and trees such as bald cypress (*Taxodium distichum*); (2) facultative wetland plants that can grow in either saturated or upland soil and rarely grow elsewhere, such as certain sedges and alders, and trees such as red maple (*Acer rubrum*) and cottonwoods (*Populus* spp.); and (3) occasional wetland plants that are usually found out of wetland environments but can tolerate wetlands. The third group of plants is critical in determining the upper limit of a wetland along a gradient of soil moisture.

Wetlands most commonly occur in three topographic situations (Figure 25.13). Basin wetlands develop in shallow basins, ranging from upland depressions to filled-in lakes and ponds. Riverine wetlands develop along shallow and periodically flooded banks of rivers and streams. A third type, fringe wetlands, occurs along the coasts of large lakes. The three types are partially separated due to the direction of water flow (see Figure 25.13). Water flow in basin wetlands is vertical due to precipitation and the downward infiltration of water into the soil. In riverine wetlands, water flow is unidirectional. In fringe wetlands, flow is in two directions, because it involves rising lake levels or tidal action. These flows transport nutrients and sediments into and out of the wetland.

Wetlands dominated by emergent herbaceous vegetation are **marshes** (Figure 25.14). With their reeds, sedges, grasses, and cattails, marshes are essentially wet grasslands. Forested wetlands are commonly called **swamps** (Figure 25.15). They may be deep-water swamps dominated by cypress (*Taxodium* spp.), tupelo (*Nyssa* spp.), and swamp oaks (*Quercus* spp.); or they may be shrub swamps dominated by alder (*Alnus* spp.) and willows (*Salix* spp.). Along many large river systems are extensive tracts of **bottomland** or **riparian woodlands** (Figure 25.16 on page 588), which are occasionally or seasonally flooded by river waters but are dry for most of the growing season.

Wetlands that are characterized by an accumulation of partially decayed organic matter with time are called **peatlands** or **mires** (Figure 25.17 on page 550). Organic matter accumulates because it is produced faster than it can be decomposed. The water table is at or near the soil surface, which creates anaerobic conditions that slow microbial activity. Mires that are fed by groundwater moving through mineral soil, from which they obtain most of

(a)

(b)

Figure 25.11 | **(a)** Okavango delta as photographed from the space shuttle. **(b)** The dry northwest corner of Botswana is the starting point for the annual summer floods of the Kavango River, which spills down through a vast network of narrow waterways, lagoons, oxbow lakes, and floodplains that covers some 22,000 km² of Botswana.

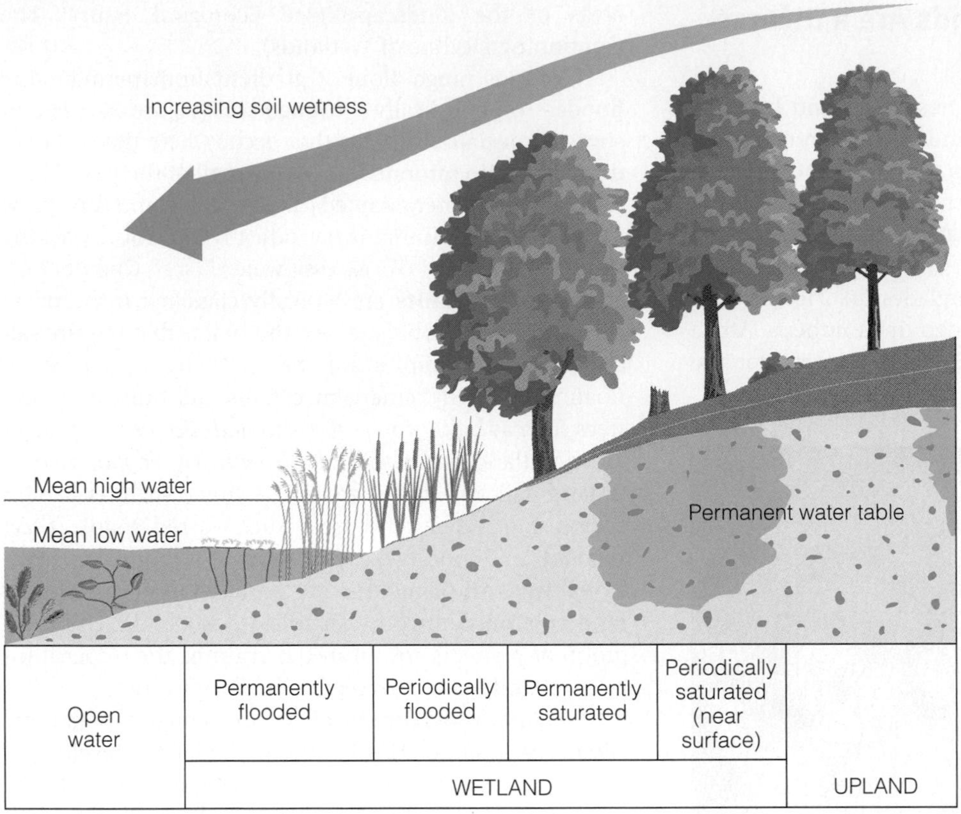

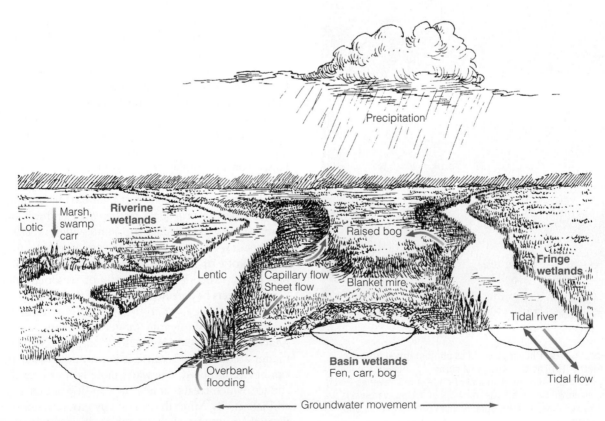

Figure 25.12 | Location of wetlands along a soil moisture gradient.

Increasing soil wetness

Mean high water

Mean low water

Permanent water table

Open water	Permanently flooded	Periodically flooded	Permanently saturated	Periodically saturated (near surface)	
	WETLAND				UPLAND

Precipitation

Lotic

Marsh, swamp carr

Riverine wetlands

Lentic

Overbank flooding

Capillary flow
Sheet flow

Raised bog

Blanket mire

Basin wetlands
Fen, carr, bog

Fringe wetlands

Tidal river

Tidal flow

Groundwater movement

Figure 25.13 | Water flow in various types of freshwater wetlands.

Figure 25.14 | The Horicon marsh in Wisconsin is an outstanding example of a northern marsh with well-developed emergent vegetation and patches of open water—an ideal environment for wildlife.

their nutrients, and dominated by sedges are known as **fens.** Mires dependent largely on precipitation for their water supply and nutrients and dominated by *Sphagnum* are **bogs.** Mires that develop in upland situations—where decomposed, compressed peat forms a barrier to the downward movement of water, resulting in a perched water table (zone of saturation above an impermeable horizon) above mineral soil—are **blanket mires** and **raised bogs** (Figure 25.18). Raised bogs are popularly known as **moors.** Because bogs depend on precipitation for nutrient inputs, they are highly deficient in mineral salts and low in pH. Bogs also develop when a lake basin fills with sediments and organic matter carried by inflowing water. These sediments divert water around the lake basin and raise the surface of the mire above the influence of groundwater. Other bogs form when a lake basin fills in from above rather than from below, creating a floating mat of peat over open water. Such bogs are often termed **quaking bogs** (Figure 25.19).

25.7 | Hydrology Defines the Structure of Freshwater Wetlands

Wetland structure is influenced by the phenomenon that creates it: its hydrology. Hydrology has two components. One involves the physical aspects of water and its movement: precipitation, surface and subsurface flow, direction and kinetic energy of water, and chemistry of the water. The other component is the **hydroperiod,** which involves duration, frequency, depth, and season of flooding. The length of the hydroperiod varies among types of wetlands. Basin wetlands have a longer hydroperiod. They usually flood during periods of high rainfall and draw down during dry periods. Both phenomena appear

to be essential to the long-term existence of wetlands. Riverine wetlands have a short period of flooding associated with peak stream flow. The hydroperiod of fringe wetlands, influenced by wind and lake waves, may be short and regular and does not undergo the seasonal fluctuation characteristic of many basin marshes.

Hydroperiod influences plant composition, for it affects germination, survival, and mortality at various stages of the plants' life cycles. The effect of hydroperiod is most pronounced in basin wetlands, especially those

(a)

(b)

Figure 25.15 | **(a)** A cypress deep-water swamp in the southern United States. **(b)** An alder (*Alnus*) shrub swamp with an herbaceous understory of skunk cabbage (*Symplocarpus foetidus*).

Figure 25.16 | A riparian forest in Alabama.

Figure 25.17 | An upland black spruce–tamarack bog in the Adirondack Mountains of New York.

in the prairie regions of North America. In basins (called potholes in the prairie region) deep enough to have standing water throughout periods of drought, the dominant plants will be submergents (Figure 25.20). If the wetland goes dry annually or during a period of drought, tall or midheight emergent species such as cattails will dominate the marsh. If the pothole is shallow and flooded only briefly in the spring, then grasses, sedges, and forbs will make up a wet-meadow community.

If the basin is deep enough toward its center as well as large enough, then zones of vegetation may develop, ranging from submerged plants to deep-water emergents such as cattails and bulrushes, shallow-water emergents, and wet-ground species such as spike rush. Zonation reflects the response of plants to the hydroperiod. Those areas of wetland subjected to a long hydroperiod will support submerged and deep-water emergents; those with a short hydroperiod and shallow

Figure 25.18 | A raised bog **(a)** develops when an accumulation of peat rises above the surrounding landscape **(b)**.

(a)

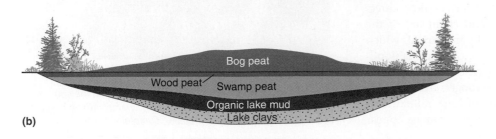

(b)

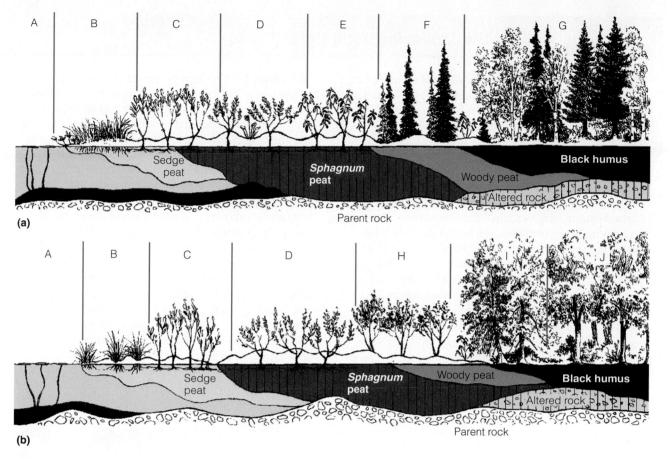

(a)

A B C D E F G

Sedge peat

Sphagnum peat

Woody peat

Black humus

Altered rock

Parent rock

(b)

A B C D H I J

Sedge peat

Sphagnum peat

Woody peat

Black humus

Altered rock

Parent rock

Figure 25.19 | (a) Transect through a quaking bog, showing zones of vegetation, *Sphagnum* mounds, peat deposits, and floating mats. A, pond lily in open water; B, buckbean (*Menyanthes trifoliata*) and sedge; C, sweetgale (*Myrica gale*); D, leatherleaf (*Chamaedaphne calyculata*); E, Labrador tea (*Ledum groenlandicum*); F, black spruce; G, birch–black spruce–balsam fir forest. **(b)** An alternative vegetational sequence. H, alder; I, aspen, red maple; J, mixed deciduous forest.

water are occupied by shallow-water emergents and wet-ground plants.

Periods of drought and wetness can induce vegetation cycles associated with changes in water levels. Periods of above-normal precipitation can raise the water level and drown the emergents to create a lake marsh dominated

Figure 25.20 | Prairie potholes.

by submerged plants. During a drought, the marsh bottom is exposed by receding water, stimulating seed germination in the emergents and annuals characteristic of mudflats. When water levels rise again the mudflat species drown, and the emergents survive and spread vegetatively.

Peatlands differ from other freshwater wetlands in the accumulation of peat that results because organic matter is produced faster than it can be decomposed. In northern regions, acid-forming, water-holding *Sphagnum* add new growth on top of the accumulating remains of past moss generations, and their spongelike ability to hold water increases water retention on the site. As the peat blanket thickens, the water-saturated mat of moss and associated vegetation is raised above and insulated from mineral soil. The peat mat then becomes its own reservoir of water, creating a perched water table.

Peat bogs and mires generally form under oligotrophic and dystrophic conditions (see Section 24.4). Although usually associated with and most abundant in boreal regions of the Northern Hemisphere, peatlands

For centuries, we have looked at wetlands as forbidding, mysterious places: sources of pestilence, home to dangerous and pestiferous insects, and the abode of slimy, sinister creatures that rise out of swamp waters. They have been looked upon as places that should be drained for more productive uses by human standards: agricultural land, solid waste dumps, housing, industrial developments, and roads. The Romans drained the great marshes around the Tiber to make room for the city of Rome. William Byrd described the Great Dismal Swamp on the Virginia–North Carolina border as a "horrible desert, the foul damps ascend without ceasing." Despite the enormous amount of vacant dry land available in 1763, a corporation called the Dismal Swamp Land Company, owned in part by George Washington, failed in an attempt to drain the western end of the swamp for farmland. Although severely affected over the past 200 years, much of the swamp remains as a wildlife refuge.

Rationales for draining wetlands are many. The most persuasive relates to agriculture. Drainage of wetlands opens many hectares of rich organic soil for crop production. In prairie country, agriculturalists viewed the innumerable potholes (see Figure 25.20) as an impediment to efficient farming. Draining them tidies up fields and allows unhindered use of large agricultural machinery. There are other reasons, too. Landowners and local governments view wetlands as an economic liability that produces no economic return and provides little tax revenue. Many regard the wildlife that wetlands support as threats to grain crops. Elsewhere, wetlands are considered valueless lands, at best filled in and used for development. Some major wetlands have been in the way of dam development projects. For example, the large Pymatuning Lake in the states of Pennsylvania and Ohio covers what was once a 4200-hectare *Sphagnum*–tamarack bog. Peat bogs in the northern United States, Canada, Ireland, and northern Europe are excavated for fuel, horticultural peat, and organic soil. In some areas, such exploitation threatens to wipe out peatland ecosystems.

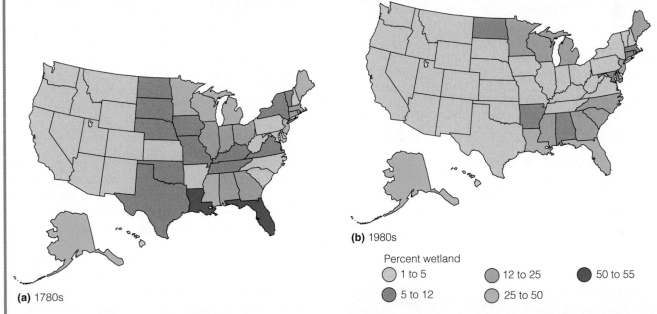

(a) 1780s

(b) 1980s

Percent wetland
- 1 to 5
- 5 to 12
- 12 to 25
- 25 to 50
- 50 to 55

Figure 1 | The loss of wetlands in the United States over the past 200 years.
(Adapted from U.S. Department of Agriculture.)

Many remaining wetlands, especially in the north-central and southwestern United States, are contaminated and degraded by the pesticides and heavy metals carried into them by surface and subsurface drainage and sediments from surrounding croplands. Although inputs of nitrogen and phosphorus increase the productivity of wetlands, a concentration of herbicides, pesticides, and heavy metals poisons the water, destroys invertebrate life, and has debilitating effects on wildlife (including deformities, lowered reproduction, and death). Waterfowl in wetlands scattered throughout agricultural lands are also more exposed to predation, and without access to natural upland vegetation they breed less successfully.

Fifty-one percent of the human population of the contiguous United States, and globally 70 percent of all people, live within 80 km of the coastlines. With so much humanity clustered near the coasts, it is obvious why coastal wetlands are threatened and disappearing rapidly. Despite some efforts at regulation and acquisition at state and federal levels to slow the loss, coastal wetlands in the United States are still disappearing at a rate of 8000 hectares per year. In colonial times, the area now embraced by the 50 United States contained some 160 million hectares of wetlands. Over the past 200 years, that area has decreased to 110 million hectares (Figure 1), and many of these remnants are degraded. Coastal Europe has lost 65 percent of its original tidal marshes, and 75 percent of the remaining are heavily managed. Since the 1980s, 35 percent of tropical mangrove forests have been diked for aquaculture—pond rearing of shrimp and fish—and cut for wood chips and charcoal. Losses are continuing at a rate of 2 percent a year.

Commonly regarded as economic wastelands, salt marshes have been and are still being ditched, drained, and filled for real estate development (everyone likes to live at the water's edge), industrial development, and agriculture. Reclamation of marshes for agriculture is most extensive in Europe, where the high marsh is enclosed within a sea wall and drained. Most of the marshland and tideland in Holland has been reclaimed in this fashion. Many coastal cities such as Boston, Amsterdam, and much of London have been built on filled-in marshes. Salt marshes close to urban and industrial developments occasionally become polluted with spillages of oil, which becomes easily trapped within the vegetation.

Losses of coastal wetlands have a pronounced effect on the salt marsh and associated estuarine ecosystems. They are the nursery ground for commercial and recreational fisheries. There is, for example, a positive correlation between the expanse of coastal marsh and shrimp production in Gulf coastal waters. Oysters and blue crabs are marsh dependent, and the decline of these important species relates to loss of salt marshes. Coastal marshes are major wintering grounds for waterfowl. One-half of the migratory waterfowl of the Mississippi Flyway depend on Gulf Coast wetlands, and the bulk of the snow goose population winters on coastal marshes from the Chesapeake Bay to North Carolina. Through grazing or uprooting, these geese may remove nearly 60 percent of the belowground production of marsh vegetation. Forced concentration of these wintering migratory birds into shrinking salt-marsh habitats could jeopardize marsh vegetation and the future of these salt-marsh ecosystems.

The loss of wetlands has reached a point where both environmental and socioeconomic values—including waterfowl habitat, groundwater supply and quality, floodwater storage, and sediment trapping—are in jeopardy. Although the United States has made some progress toward preserving the remaining wetlands through legislative action and land purchase, the future of freshwater wetlands is not secure. Apathy, hostility toward wetland preservation, political maneuvering, court decisions, and arguments over what constitutes a wetland allow the continued destruction of wetlands at a rate of more than 200,000 hectares per year.

1. The Everglades National Park in South Florida is one of the largest natural wetlands in the world. Over the past century, the draining of lands and the diversion of water to meet growing residential and agricultural needs in this region have threatened this wetland ecosystem. Efforts are currently under way to restore the flow of water that is critical to preserving this unique ecosystem. Information on the history of the Everglades ecosystem and the Comprehensive Everglades Restoration Plan are available online at http://www.evergladesplan.org/.

also exist in tropical and subtropical regions. They develop in mountainous and coastal regions where hydrological conditions encourage an accumulation of partly decayed organic matter. Examples in coastal regions are the Everglades in Florida and the pocosins on the coastal plains of the southeastern United States.

25.8 | Freshwater Wetlands Support a Rich Diversity of Life

Biologically, freshwater wetlands are among the richest and most interesting ecosystems. They support a diverse community of benthic, limnetic, and littoral invertebrates, especially crustaceans and insects. These inverte-brates, along with small fishes, provide a food base for waterfowl, herons, gulls, and other birds and supply the fat-rich nutrients ducks need for egg production and the growth of young. Amphibians and reptiles, notably frogs, toads, and turtles, inhabit the emergent growth, soft mud, and open water of marshes and swamps.

Herbivores make up a conspicuous component of animal life. The dominant herbivore in prairie marshes is the muskrat (*Ondatra zibethicus*). Muskrats are the major prey for mink (*Mustela vison*), the dominant carnivore on the marshes. Other predators, including raccoon, fox, weasel, and skunk, can seriously reduce the reproductive success of waterfowl on small marshes.

Summary

Intertidal Zone | 25.1

Sandy shores and rocky coasts occur where the sea meets the land. The drift line marks the furthest advance of the tide on sandy shores. On rocky shores, a zone of black algal growth marks the tide line.

Rocky Coasts | 25.2

The most striking feature of the rocky shore—zonation of life—results from alternate exposure and submergence by the tides. The black zone marks the supralittoral fringe, the upper part of which is flooded only once every 2 weeks by spring tides. Submerged daily by the tides is the littoral zone, characterized by barnacles, periwinkles, mussels, and fucoid seaweeds. Uncovered only at spring tides is the infralittoral, which is dominated by large brown laminarian seaweeds, Irish moss, and starfish. Distribution and diversity of life across rocky shores are also influenced by wave action, competition herbivory, and predation. Left behind by outgoing tides are tidal pools, distinct habitats subjected over a 24-hour period to wide fluctuations in temperature and salinity and inhabited by varying numbers of organisms, depending on the amount of emergence and exposure.

Sandy Beaches | 25.3

Sandy beaches are a product of weathering of rock. Exposed to wave action, the beaches are subjected to deposition and wearing away of the sandy substrate. Sandy and muddy shores appear barren of life at low tide; but beneath the sand and mud, conditions are more amenable to life than on the rocky shore. Zonation of life is hidden beneath the surface. The energy base for sandy and muddy shores is organic matter carried in by tides and made available by bacterial decomposition. Basic consumers are bacteria, which in turn are a major source of food for both deposit-feeding and filter-feeding organisms.

Salt Marshes | 25.4

The interaction of salinity, tidal flow, and height produces a distinctive zonation of vegetation in salt marshes. Salt-marsh cordgrass dominates marshes flooded by daily tides. Higher microelevations that are shallow, flooded only by spring tides, and subjected to higher salinity support salt meadow cordgrass and spikegrass. Salt-marsh animals are adapted to tidal rhythms. Detrital feeders such as fiddler crabs and their predators are active at low tide; filter-feeding ribbed mussels are active at high tide.

Mangrove Forests | 25.5

In tropical regions, mangrove forests or mangals replace salt marshes and cover up to 70 percent of coastlines. Uniquely adapted to a tidal environment, many mangrove tree species have supporting prop roots that carry oxygen to the roots, and their seeds grow into seedlings on the tree and drop into the water to take root in the mud. Mangroves support a unique mix of terrestrial and marine life. The sheltered water about the prop roots provides a nursery for the larvae and young of crabs, shrimp, and fish.

Freshwater Wetlands | 25.6

Wetlands can be defined as a community of hydrophytic plants occupying a gradient of soil wetness from permanently flooded to periodically saturated during the growing season. Hydrophytic plants are adapted to grow in water or on soil periodically deficient in oxygen. Wetlands dominated by grasses and herbaceous hydrophytes are marshes. Those dominated by wooded vegetation are forested wetlands (riparian forests) or shrub swamps. Wetlands characterized by an accumulation of peat are mires. Mires fed by water moving through the mineral soil and dominated by sedges are fens; those dominated by *Sphagnum* and dependent largely on precipitation for moisture and nutrients are bogs. Bogs are characterized by blocked drainage, an accumulation of peat, and low productivity.

Hydrology Structures Wetlands | 25.7

The structure and function of wetlands are strongly influenced by their hydrology—both the physical movement of water and hydroperiod. Hydroperiod is the depth, frequency, and duration of flooding. Hydroperiod influence on vegetation is most evident in basin wetlands that exhibit zonation from deep-water submerged vegetation to wet-ground emergents.

Diversity of Wetland Life | 25.8

Wetlands support a diversity of wildlife. Freshwater wetlands provide essential habitats for frogs, toads, turtles, and a diversity of invertebrate life. Nesting, migrant, and wintering waterfowl depend on these critical habitats.

Study Questions

1. Describe the three major zones of the rocky shore.
2. What environmental stresses are organisms living in the rocky intertidal zone subjected to?
3. How does life on sandy shores differ from that in the rocky intertidal zone? How is it the same?
4. What influences the major structural features (zonation) of a salt marsh?
5. Compare salt marshes with mangrove forests.
6. What is a freshwater wetland?
7. What are the three major types of wetlands in terms of hydrology (water flow)?
8. What is hydroperiod, and how does it relate to the structure of wetlands?
9. Characterize the types of wetlands by their vegetation.

Further Readings

Bertness, M. D. 1999. *The ecology of Atlantic shorelines.* Sunderland, MA: Sinauer Associates.

A well-written and illustrated introduction to coastal ecology.

Lugo, A. E. 1990. *The forested wetlands.* Amsterdam: Elsevier.
Excellent overview and discussion of the structure and function of these freshwater wetland ecosystems.

Lugo, A. E., and S. C. Snedaker. 1974. The ecology of mangroves. *Annual Review of Ecology and Systematics* 5:39–64.
This review paper provides an excellent introduction to mangrove ecosystems.

Mathieson, A. C., and P. H. Nienhuis, eds. 1991. *Intertidal and littoral ecosystems.* Ecosystems of the world 24. Amsterdam: Elsevier.
A detailed, broad survey of the intertidal and littoral zones of the world.

Mitsch, W. J., and J. C. Gosslink. 2000. *Wetlands,* 3rd ed. New York: John Wiley & Sons.
A pioneering text and major reference of the ecology of wetland ecosystems.

Moore, P. G., and R. Seed, eds. 1986. *The ecology of rocky shores.* New York: Columbia University Press.
A comprehensive, worldwide review of the rocky intertidal zone.

Niering, W. A., and B. Littlehales. 1991. *Wetlands of North America.* Charlottesville, VA: Thomasson-Grant.
An exceptionally illustrated survey of North American wetlands, both freshwater and salt.

Teal, J., and M. Teal. 1969. *Life and death of the salt marsh.* Boston: Little, Brown.
Although out of print, this beautifully written classic is an excellent introduction to salt-marsh ecology.

Valiela, I., J. L. Bowen, and J. K. York. 2001. Mangrove forests: One of the world's threatened major tropical environments. *BioScience* 51:807–815.
An overview of the impact of mariculture on mangroves.

Van der Valk, A., ed. 1989. *Northern prairie wetlands.* Ames, IA: Iowa State University Press.
Detailed studies of this unique wetland ecosystem that is rapidly disappearing.

Williams, M., ed. 1990. *Wetlands: A threatened landscape.* Cambridge, MA: Blackwell Publishers.
This comprehensive appraisal of the world's wetlands covers occurrence and composition, physical and biological dynamics, human impact, and management and preservation.

CHAPTER 26

Large-Scale Patterns of Biological Diversity

Although covering only 6 percent of the land surface, tropical rain forests contain over 50 percent of all known terrestrial species.

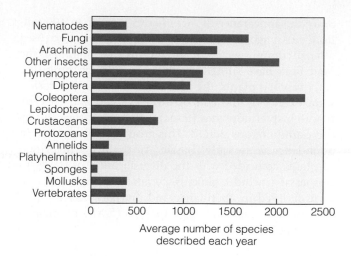

Figure 26.2 | Many new (previously unknown) species are discovered each year. Most of these species belong to taxonomic groups that are relatively small in size.

E arth's ecosystems support an amazing diversity of species. Scientists have identified and named approximately 1.4 million species (Figure 26.1), and the task is not complete. Scientists are continuously discovering new species, and quantifying the actual number of species inhabiting Earth is an ongoing exercise (Figure 26.2). Some scientists, such as Harvard biologist E. O. Wilson, believe that the actual number of species may be close to 10 million.

Regardless of whether the actual number of species is 1.4 million or 10 million, the diversity of our planet is not static. Over evolutionary time, new species evolve while existing species fade away and become extinct (see Chapter 28). The diversity of our planet is a story of constant change.

Neither is biological diversity the same everywhere on Earth's surface. Distinct geographic patterns of diversity relate to environmental conditions that have influenced the evolution of species diversity as well as the ability of local environments to support a diverse community. In this chapter, we will examine these regional and global patterns of Earth's biological diversity, both in time and space.

26.1 | Earth's Biological Diversity Has Changed through Geologic Time

In Chapters 18 and 19, we explored the temporal dynamics of species diversity on a successional timescale. In those examples, however, the local patterns of diversity reflect changes in the local distribution and abundance of species through time. Over geologic time, however, there have been dramatic long-term evolutionary changes in patterns of global diversity.

Over the past 600 million years, the number of different types of organisms has been increasing. Among most groups of organisms for which data exist from the fossil record, the number of species has increased almost continuously since the taxonomic group first appeared in the fossil record. Figure 26.3 represents the estimated species richness of fossilized invertebrates over geologic time. Species richness of this taxonomic group has increased over the past 600 million years with slight decreases during the late Devonian and Permian periods.

The evolution of diversity among vascular land plants presents a particularly interesting pattern (Figure 26.4).

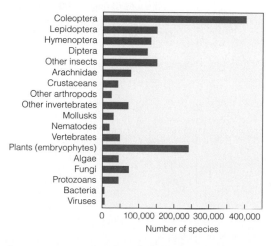

Figure 26.1 | Number of living species of all kinds of organisms currently known. Species are classified by major taxonomic groups. Insects and plants dominate the diversity of living organisms.

(Adapted from Wilson 1999.)

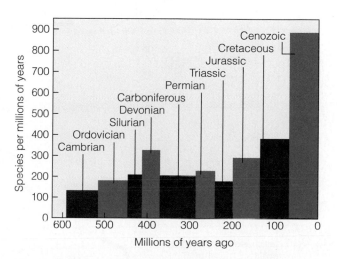

Figure 26.3 | Estimated species richness of fossilized invertebrates over geologic time.

Since the appearance of these plants more than 400 million years ago, the number of species has increased almost continuously, but the groups that dominate the land flora have shifted dramatically through time. The early vascular plants, the rootless and leafless psilopsids, went extinct by the end of the Devonian and were replaced by pteridophytes (ferns), which flourished during the Carboniferous period. This group then decreased in abundance by the early Triassic. The decline of this group of plants coincided with the diversification of the gymnosperms (includes ginkgos, cycads, conifers), which in turn declined in abundance and diversity over the past 100 million years as angiosperms (flowering plants) diversified.

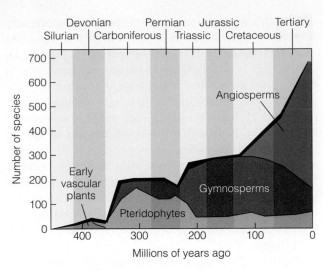

Figure 26.4 | Pattern of expansion and reduction of major terrestrial plant groups during 400 million years of plant evolution. (Adapted from Niklas et al. 1983.)

26.2 | Past Extinctions Have Been Clustered in Time

Although the history of Earth's biological diversity is generally a story of increasing species richness, it has experienced periods of decline. The general pattern of increasing diversity through geologic time has also been accompanied by extinctions, which were not spread evenly across Earth's history (Figure 26.5). Most extinctions are clustered in geologically brief periods of time. One mass extinction occurred at the end of the Permian period, 225 million years ago, when 90 percent of the shallow-water marine invertebrates disappeared. Another

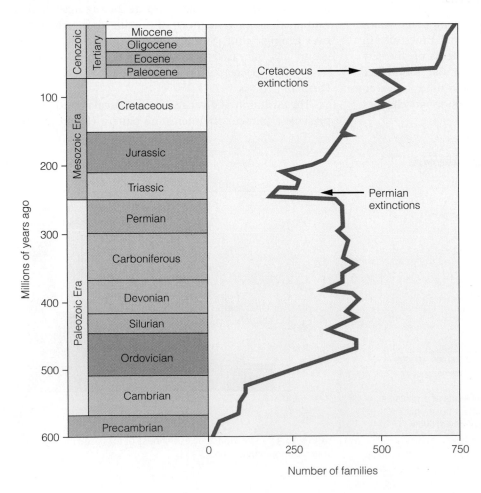

Number of families

Figure 26.5 | The geologic timescale and mass extinctions in the history of life. The fossil record profiles mass extinctions during geological times. The most recent mass extinction occurred during the Cretaceous, which wiped out more than half of all species, including the dinosaurs. The mass extinction event at the end of the Permian resulted in the loss of 96 percent of all marine species and perhaps as many as 50 percent of the total species on Earth.

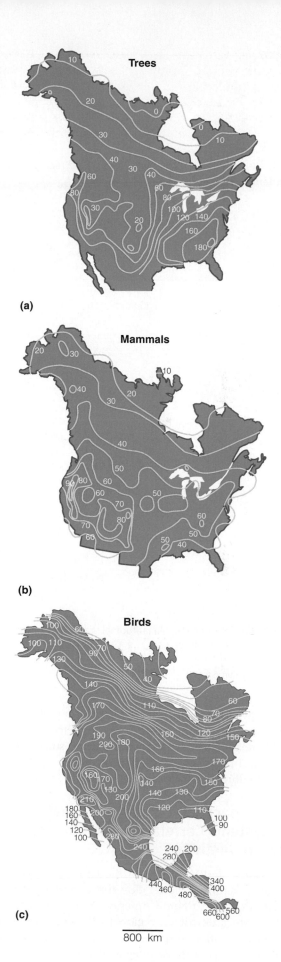

Figure 26.6 | Geographic variation in the distribution of **(a)** trees, **(b)** mammals, and **(c)** birds in North America. Contour lines connect points with about the same number of species.
(Adapted from Currie 1991.)

(a)

(b)

(c)

800 km

occurred at the end of the Cretaceous period, 65 to 125 million years ago, when the dinosaurs vanished. An asteroid striking Earth, interrupting oceanic circulation, altering the climate, and causing volcanic and mountain-building activity is currently believed to have caused that extinction event.

One of the great extinctions of mammalian life took place during the Pleistocene, when such species as the woolly mammoth, giant deer, mastodon, giant sloth, and saber-toothed cat vanished from Earth. Some scientists suggest that climate changes as ice sheets advanced and retreated caused the extinctions. Others argue that Pleistocene hunters overkilled large mammals, especially in North America, as human populations swept through North and South America between 11,550 and 10,000 years ago. Perhaps the large grazing herbivores could not withstand the combined predatory pressure of humans and other large carnivores. The greatest number of present-day extinctions have taken place since A.D. 1600. Humans have caused well over 75 percent of these extinctions, primarily through habitat destruction, but also by introduction of predators and parasites and by exploitative hunting and fishing (see Chapter 28).

26.3 | Regional and Global Patterns of Species Diversity Vary Geographically

The 1.4 million species that have been identified are not distributed equally across Earth's surface. There are distinct geographic patterns of species richness (number of species). In general, the number of terrestrial species decreases as one moves away from the equator toward the poles. The three maps in Figure 26.6 illustrate distinct geographic patterns of species richness for trees, mammals, and birds in North America. Note the overall decrease in species richness as one moves from south to north. This pattern is more obvious if we plot species richness as a function of latitude (Figure 26.7).

The decline in species diversity as you move northward in the North American continent is part of a global pattern of declining diversity (both terrestrial and marine) from the equator northward and southward toward the poles. Although scientists do not know the exact mechanisms underlying the geographic pattern of species diversity, they have postulated various hypotheses related to a range of factors including the age of the community, spatial heterogeneity of the environment, stability of the climate over time, and ecosystem productivity. Of the

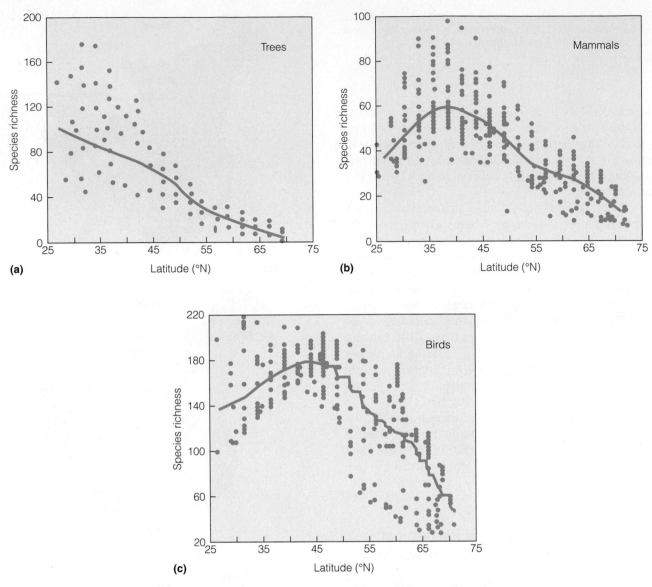

Figure 26.7 | North American latitudinal gradients in species richness for **(a)** trees, **(b)** mammals, and **(c)** birds based on cells of 2.5° × 2.5° latitude/longitude. Species richness per cell is based on range maps for individual species.

(Adapted from Currie 1991.)

various hypotheses proposed to account for global patterns of species diversity, the most easily interpreted are those explicitly relating to environmental features such as climate and availability of essential resources, which are known to directly influence basic plant and animal processes.

26.4 | Species Richness in Terrestrial Ecosystems Correlates with Climate and Productivity

D. J. Currie and V. Paquin of the University of Ottawa, Canada, examined the relationship between patterns of species richness in North American tree species and several variables describing regional differences in climate. Although variation in species richness was correlated with climatic factors such as integrated measures of annual temperature, solar radiation, and precipitation, it correlated most strongly with estimates of actual evapotranspiration (AET) (Figure 26.8).

Recall from earlier discussions in Chapters 3 and 20 that AET is the flux of water from the terrestrial surface to the atmosphere through both evaporation and transpiration. It is a function of both the atmospheric demand for water brought about by the input of solar energy to the surface and the supply of water from precipitation. The pattern of increasing species richness with increasing AET parallels the positive correlation between AET and net primary productivity presented earlier (Figure 20.4), suggesting a relationship between plant diversity and primary productivity. In other words, environmental conditions favorable for photosynthesis and plant growth may well give rise to increased plant diversity over evolutionary time.

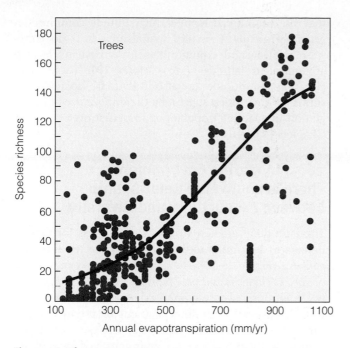

Figure 26.8 | Relationship between annual measure of actual evapotranspiration (AET) and tree species richness for North America.

(From Currie and Paquin 1987.)

Currie also demonstrated a relationship between climate and species richness of vertebrates across North America. He found a positive correlation between potential evapotranspiration (PET), an index of integrated energy availability, and regional patterns of species richness for both mammals and birds across North America (Figure 26.9). Although the correlation between PET and vertebrate species richness offers hope of a mechanistic interpretation other than latitude, PET is correlated with temperature, solar radiation, precipitation, humidity, and a host of other abiotic factors that vary with latitude. Like actual evapotranspiration, PET also correlates with plant productivity. In fact, Currie reported a positive correlation between vertebrate diversity and plant species diversity. Given the correlation between plant diversity and net primary productivity, the correlation between plant and animal diversity may relate to the positive relationship between primary and secondary productivity discussed in Chapter 20 (see Figure 20.16). This suggests a relationship between secondary productivity and animal diversity similar to that observed between primary productivity and plant diversity.

Animal diversity is linked to plant diversity because variety in plant species provides a variety of potential food sources as well as suitable habitat for animals. Increased structural diversity within plant communities, as measured by foliage height diversity (see Figure 17.14), provides a wider range of microhabitats and associated resources and consequently supports a greater variety of animal species. This relationship between habitat heterogeneity and species diversity is not limited to

animals. Environmental heterogeneity also gives rise to increased plant species diversity. For example, the diverse topography of mountainous regions generally supports more species than does the consistent terrain of flatlands. From east to west in North America, the number of species of trees, breeding land birds, and mammals (see Figure 26.6) increases. This increased diversity on an east–west gradient relates to an increased diversity of the environment both horizontally and altitudinally.

Although mountainous regions may support more species than flatlands do, in mountainous regions scientists have observed a pattern of decreasing species richness with increasing elevation. Patterns of species richness with increasing elevation for bird species in New Guinea, and for mammals and vascular plants in the Himalayan Mountains, are shown in Figure 26.10. The mechanisms underlying a decline in species richness with a rise in elevation may be similar to those involved with changing latitude. Variations in temperature, PET, AET, and vegetation structure that occur with increasing elevation parallel those observed with increasing latitude. However, the negative correlation between species richness and elevation

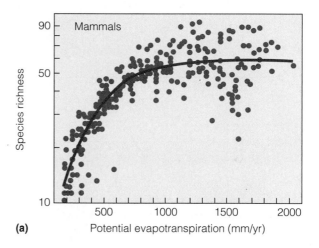

(a)

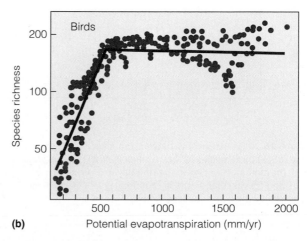

(b)

Figure 26.9 | Relationship between annual estimate of potential evapotranspiration and species richness of **(a)** mammals and **(b)** birds in North America.

(From Currie 1991.)

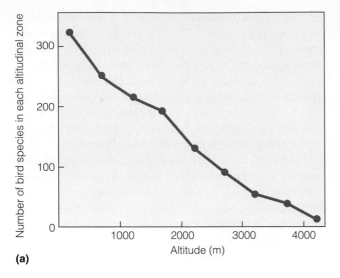

(a)

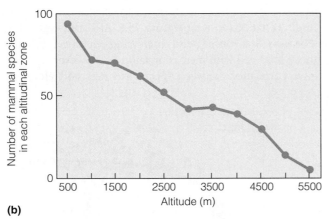

(b)

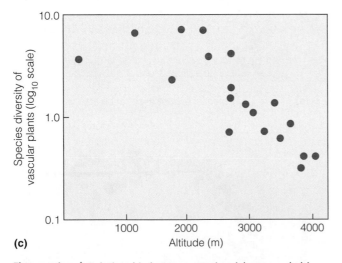

(c)

Figure 26.10 | Relationship between species richness and altitude for **(a)** bird species in New Guinea (from Kikkawa and Williams 1971), **(b)** mammal species in the Himalayas (from Hunter and Yonzon 1992), and **(c)** vascular plants in the Himalayas. (Adapted from Whittaker 1977.)

may be confounded because high-altitude communities generally occupy a smaller spatial area than do corresponding lowland communities in ecosystems located at equivalent latitudes (see Chapter 19). These high-elevation communities are also likely to be isolated from similar communities, suggesting the importance of immigration in enabling populations to persist over time (see Chapters 12 and 19).

26.5 | In Marine Environments, There Is an Inverse Relationship between Productivity and Diversity

The latitudinal gradients of species richness for marine organisms are similar to those observed for terrestrial organisms (Figure 26.11). The correlation between patterns of species richness and productivity are not as straightforward as those observed in terrestrial environments. In fact, the general latitudinal gradient of productivity in the oceans is the reverse of that observed on land (see Figures 20.10 and 24.25). Except in localized areas of upwelling (see Figure 24.24), the primary productivity of oceans increases from the equator to the poles. This relationship suggests an inverse relationship between productivity and diversity—the opposite of that observed for terrestrial environments.

Circumstantial evidence points to the importance of seasonality, rather than total annual productivity, as a factor influencing local patterns of diversity for pelagic and benthic species. Observations show that as the influence of seasonal fluctuations in temperature on primary productivity increases (see Figure 24.25), species richness declines and species dominance increases. Recall from Chapter 21 that primary productivity in the ocean is influenced by seasonal dynamics of the thermocline and vertical transport of nutrients from the deep to surface waters (see Figure 21.26). In northern latitudes, the seasonal formation and breakdown of the thermocline result in the productivity of surface waters ranging from very high (spring and summer) to very low (winter; see Figure 24.25). By contrast, the permanent presence of a thermocline in the tropical ocean waters results in a low but continuous pattern of primary productivity throughout the year. Seasonal variation in the temperature of surface waters functions to increase primary productivity, whereas the lack of seasonal variation functions to support a higher diversity of life.

Scientists have hypothesized that geologic history has been an important factor influencing latitudinal patterns of species richness within oceans. Some suggest that glaciation during the Quaternary period may be responsible for lower diversity observed at higher latitudes (see Section 18.9). During that period, sea ice covered the Norwegian Sea and the northern reaches of the Atlantic. As a result, the poleward decline in regional

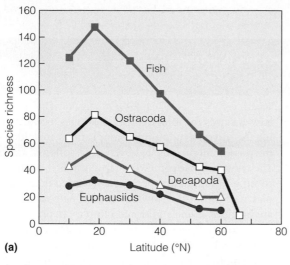
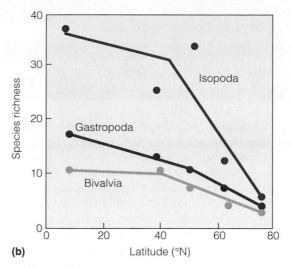

Figure 26.11 | Latitudinal gradient of species richness: **(a)** Four groups of pelagic organisms caught at six stations along 20° W (longitude) in the Northeast Atlantic Ocean. Samples were collected over a period of 14 days (day and night) from the top 2000 m of the vertical water column. **(b)** Three groups of benthic organisms in the North Atlantic. (a) (Adapted from Angel 1991.) (b) (Adapted from Rex et al. 1993.)

diversity may to some extent represent a recovery from the effects of glaciation. The slow northward range expansion of plants and animals suggests that the recovery is still in progress.

26.6 | Species Diversity Is a Function of Processes Operating at Many Scales

The discussion of species diversity, even at the broad geographic scale that has been the focus of this chapter, is complicated by factors that relate directly to topics presented in earlier chapters. For example, in Chapter 16 we discussed species diversity of individual communities. Ecologists define species diversity at this spatial scale as **local (alpha) diversity.** Quantifying local patterns of diversity is hindered by the often difficult task of defining community boundaries. In addition, the relationship between species diversity and area discussed in Chapter 19 makes it difficult to compare patterns of species diversity among communities and ecosystems that differ in size, such as lake ecosystems of varying size (see Quantifying Ecology 26.1: Quantifying Biodiversity: Comparing Species Richness Using Rarefaction Curves). Local patterns of plant and animal diversity also change over time during succession (see Figures 18.12 and 18.16), further compounding the difficulty of comparing communities.

Total species diversity (or species richness) across all communities within a geographic area is called **regional (gamma) diversity.** Diversity at this scale corresponds to the patterns depicted in Figures 26.6 and 26.11. Comparison of even these broad-scale patterns of diversity at a continental or global scale can be confounded by time. Latitudinal patterns of diversity for pelagic species

illustrate this point. Data presented in Figure 26.11 comes from the Institute of Oceanographic Sciences Deacon Laboratory in Surrey, England. At the time of sampling, the major front (boundary) between the South Atlantic Central Water and the North Atlantic Central Water occurred at 18° N. Because these two major regions of the Atlantic Ocean differ in physical characteristics, each supports a different fauna, which come together within this boundary zone. Thus, the peak in species diversity observed for the four groups of organisms at 18° N and graphed in Figure 26.11a represents the rich mix of species characteristic of the edge effect described in Chapter 19. This boundary zone of maximum diversity shifts its geographic position seasonally by several degrees, and thus its location changes with the seasons.

Although not apparent in the map shown in Figure 26.6, regional estimates of bird species diversity in eastern North America are likewise seasonally dependent. More than 50 percent of the bird species that nest and breed in this region during the spring and summer months are migratory; they reside farther south in North, Central, and South America during the fall and winter months. Patterns of species migration alter seasonal patterns of regional diversity for a wide range of taxonomic groups.

Changes in regional diversity also occur over geologic timescales. Over timescales of tens of millions to hundreds of millions of years, evolution drives changes in patterns of diversity through the emergence and extinction of species (see Figures 26.3 through 26.5). On a timescale of thousands to tens of thousands of years, changes in climate have influenced regional patterns of diversity by shifting the geographic ranges of species. In eastern North America, shifts in the distribution of tree

Quantifying Biodiversity: Comparing Species Richness Using Rarefaction Curves

Species richness is the simplest measure of community and regional diversity, but how do ecologists measure species richness within a given area or region? Because they obviously cannot locate and identify each and every individual organism that inhabits a community, no matter how limited in extent, they must use some form of sampling.

Suppose our objective was to quantify patterns of tree species richness within a forested region. A simple approach is first to delineate the area where the estimate of species richness is to be made. A number of smaller sample areas (plots) can then be identified, and a survey can be conducted within the sample areas (plots). The problem is that for a diverse taxonomic group such as trees, as more samples are taken, more species will be recorded (see Section 19.3). Suppose that as we sample the delineated area, we plot the total number of different tree species encountered as a function of the number of samples taken. The result would be an **accumulation curve** similar to the one shown in Figure 1. A species (or higher taxon) accumulation curve records the total number of species revealed, during the survey, as additional sample units are added to the pool of all previously observed or collected samples. The curve will rise somewhat rapidly at first, as the more common (abundant) species are included in the samples, and then more slowly as increasingly rarer species are included in the additional samples. In principle, an asymptote will eventually be reached, and no further species will be added. The point where the curve reaches an asymptote defines the optimal sample size; continued sampling will yield no additional information about the number of species in the area.

Suppose we now wish to compare our findings with other studies of tree species richness that have been reported for other surveys. If the same sampling procedure was used, and the same numbers of samples were collected, a direct statistical comparison would be possible. But generally this is not the case. The areas to be compared may differ in size, or the number of samples collected in the survey might differ. As illustrated by the accumulation curve presented in Figure 1, the estimate of species richness will vary with sample size. For a meaningful comparison of the species richness derived from different surveys, we must obtain values for the same sample size for each survey. Several techniques have been developed to subsample the larger survey to obtain a result that can be compared to the smaller survey. One technique is to construct a rarefaction curve from the samples used in constructing the accumulation curve.

A **rarefaction curve** is produced by repeatedly re-sampling the total pool of samples, defined as N, at random, plotting the average number of species represented by 1, 2, . . . , N samples (see Figure 1). Thus, rarefaction generates the expected number of species in a small collection of n samples drawn at random from the larger total pool of N samples.

In effect, a rarefaction curve is constructed as the opposite of the corresponding accumulation curve. Accumulation curves, in effect, move from left to right, as they are further extended by additional sampling.

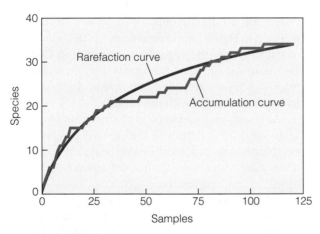

Figure 1 | Examples of accumulation and rarefaction curves constructed from a survey of 125 samples. The sample size is represented by the x-axis, and the y-axis represents the corresponding number of species encountered (species richness).

In contrast, rarefaction curves move from right to left, as the full data set is increasingly "rarefied." Because each sample that forms the corresponding accumulation curve is equally likely to be included in the mean richness value for any level of resampling along the rarefaction curve, the corresponding accumulation and rarefaction curves are closely related to one another. The rarefaction curve can be viewed as the statistical expectation of the corresponding accumulation curve over different reorderings of the samples. As such, by constructing rarefaction curves for both surveys, a comparison can be made for species richness at any given sample size up to the value of N for the smaller of the two surveys.

Table 1 represents a list of tree species identified in each of twenty 20 × 20 m sample plots collected as part of a forest survey undertaken in central Virginia. The samples are presented from left to right in the order they were collected.

1. Using the data provided in Table 1, construct an accumulation curve for the site. The first point would be for the first sample. Plot the value 1 for sample size on the x-axis and 5 as the number of unique tree species surveyed in the first sample. The value of tree species richness for sample size 2 is found by combining the species lists for the first two samples (columns) and counting the number of unique tree species in this combined sample (7 species, as the two samples have 4 species in common). Continue this process until all samples ($N = 20$) are included.

2. How would you use the data in Table 1 to calculate a rarefaction curve for the survey?

3. By selecting five random numbers from 1 to 20 (you can use a random number table, found in any introductory statistics book or online), select 5 of the 20 samples from Table 1 and calculate the species richness of the combined sample. Repeat the procedure three times, and take the resulting average (of the three) as your estimate of species richness for a sample size of 5. How does this value compare with the value from the accumulation curve?

Table 1 | Tree Species Surveyed on Twenty 20 × 20 m Sample Plots

1	2	3	4	5	6	7	8	9	10	11	12	13	14	15	16	17	18	19	20
dw	bg	mh	bg	bc	rc	bg	be	dw	br	ah	bc	be	bh	bg	co	co	ah	bg	bg
rm	dw	rc	dw	mh	rm	mh	dw	hh	rm	hh	bg	br	dw	bh	dw	rm	bw	dw	rm
ro	rc	rm	rc	rc	vp	rc	rb	rm	sg	rm	dw	dw	po	dw	rm	sf	dw	mh	ro
tp	rm	tp	rm	so	wo	rm	rm	sg	sy	sg	wo	tp	rm	ro	sf	sl	sg	rb	tp
wo	vp	wo	wo	vp		sl	ro	tp	wo	sy			wo	so	vp	vp	tp	tp	wo
	wo					so	tp							vp			wo	wo	
						tp													
						vp													

Species codes: dw, dogwood (*Cornus florida*); rm, red maple (*Acer rubrum*); ro, southern red oak (*Quercus falcata*); wo, white oak (*Q. alba*); so, scarlet oak (*Q. coccinea*); co, chestnut oak (*Q. prinus*); po, post oak (*Q. stellata*); tp, tulip poplar (*Liriodendron tulipifera*); bg, black gum (*Nyssa sylvatica*); rc, red cedar (*Juniperus virginiana*); vp, Virginia pine (*Pinus virginiana*); sl, short-leaf pine (*P. echinata*); mh, mockernut hickory (*Carya tomentosa*); bh, bitternut hickory (*C. cordiformis*); be, American beech (*Fagus grandifolia*); rb, red bud (*Cercis canadensis*); sg, sweetgum (*Liquidambar styraciflua*); sy, sycamore (*Platanus occidentalis*); br, river birch (*Betula nigra*); ah, American hornbeam (*Carpinus caroliniana*); hh, hophornbeam (*Ostrya virginiana*); bc, black cherry (*Prunus serotia*); sf, sassafras (*Sassafras albidum*).

species after the last glacial maximum around 20,000 years B.P. represent one such example (see Figures 18.19 and 18.20). Geographic ranges of many species in North America continue to shift, influencing local and regional patterns of diversity. Possible change in the geographic distributions of plant and animal species in response to future changes in Earth's climate is a key area of research on global change (see Chapter 29).

Summary

Temporal Patterns of Species Diversity | 26.1

Earth's biological diversity has changed through evolutionary time. The fossil record suggests a pattern of increasing diversity over the past 600 million years.

Extinctions | 26.2

Despite the overall pattern of increasing diversity through time, Earth's history is marked by periods of large-scale or mass extinctions. Two notable periods are the end of the Permian, when more than 90 percent of marine invertebrates disappeared from the fossil record, and the Cretaceous period, which saw the extinction of dinosaurs.

Geographic Patterns of Species Richness | 26.3

The approximately 1.4 million species identified by scientists are not distributed evenly over Earth's surface. In general, species richness decreases from the equator toward the poles for both aquatic and terrestrial organisms. Various hypotheses, including the role of climate, have been put forth to explain these patterns.

Patterns of Terrestrial Species Richness | 26.4

Regionally, plant species richness is correlated with actual evapotranspiration, suggesting a positive relationship between species richness and net primary productivity. Species richness of terrestrial vertebrates is correlated with potential evapotranspiration, an integrated measure of energy input to the ecosystem.

Patterns of Marine Species Richness | 26.5

Patterns of species diversity for pelagic and benthic organisms appear to be influenced by the seasonality of primary productivity rather than by total productivity per se.

Local and Regional Diversity | 26.6

Ecologists define diversity within a community or ecosystem as local (or alpha) diversity. Quantifying local patterns of diversity is hindered by difficulties in defining community boundaries, the species diversity–area relationship, and changes in diversity during succession. Total diversity (or species richness) across all communities within a geographic area is called regional (or gamma) diversity.

Study Questions

1. How has Earth's biological diversity changed over the past 600 million years?
2. What is a mass extinction?
3. How does species richness vary with latitude? With elevation?
4. How does tree species richness vary with actual evapotranspiration?
5. What does the relationship between tree species richness and actual evapotranspiration presented in Figure 26.8 imply about the relationship between tree species diversity and net primary productivity?
6. How is climate correlated with regional patterns of vertebrate diversity?
7. How does environmental heterogeneity influence patterns of species diversity?
8. How is primary productivity related to species richness in the oceans?
9. Contrast local (alpha) and regional (gamma) diversity.
10. What are some factors that make it difficult to quantify patterns of species diversity/richness?

Further Readings

Brown, J. H. 1995. *Macroecology*. Chicago: University of Chicago Press.

This book provides an excellent discussion of large-scale patterns of biological diversity over geologic time.

Cox, C. B., and P. D. Moore. 2000. *Biogeography: An ecological and evolutionary approach*, 6th ed. Oxford, UK: Blackwell Publishing.

An excellent text; provides an introduction to the geography and ecology of Earth's biological diversity.

Currie, D. J. 1991. Energy and large-scale biogeographical patterns of animal and plant species richness. *American Naturalist* 137:27–49.

Research paper relating regional patterns of diversity to features of climate (as discussed in this chapter).

Huston, M. 1994. *Biological diversity: The coexistence of species on changing landscapes*. New York: Cambridge University Press.

An outstanding review of our current understanding of processes governing patterns of biological diversity, from a local to global scale.

Irigoien, X., J. Huisman, and R. P. Harris. 2004. Global biodiversity patterns of marine phytoplantkton and zooplankton. *Nature* 429:863–867.

This research paper examines global patterns of phytoplankton and zooplankton diversity as they relate to patterns of primary productivity.

Wilson, E. O. 1992. *The diversity of life*. Cambridge, MA: The Belknap Press of Harvard University Press.

A modern-day classic providing an overview of biodiversity over geological and evolutionary time—and human impacts on biodiversity.

Human Ecology

All organisms modify their environment, but perhaps no other organism has affected Earth's environment as much as the human species. As the human population has grown and the power of our technology has expanded, the nature and scope of our modification of Earth's environment have changed dramatically. The story of our species is one of continuously redefining our relationship with the environment—a relationship based on energy.

With the melting of the polar ice cap from 40,000 to 10,000 years B.P. (see Section 18.9), the human species spread over the world's continents, with the Americas and Australia being the last to be inhabited some 25,000 to 10,000 years B.P. (before present). At that time, the global population was approaching 5 million. To this point in human history, dependence on plants and animals for energy was the major constraint on human population growth. Hunter-gatherer societies consisted of small, autonomous bands of a few hundred individuals who depended on the productivity and abundance of plants and animals that make up natural ecosystems and on their own ability to extract and use natural resources. They were nomadic people, tracking their resource in space and time, vulnerable to environmental change. Yet by 10,000 years B.P., a change leading to an era that redefined the relationship between humans and their environment began to take place—the establishment of agriculture.

The Neolithic period (8000–5000 B.C.) saw the development of agriculture—the cultivation of plants and the domestication of animals. Although the shift from hunter-gatherer to agriculture did not change human dependence on primary productivity as the dominant energy source, it shifted the dependency from the productivity of natural ecosystems to that of managed agricultural systems. The result was an increase in the quantity and predictability of food resources, accompanied by the development of permanent villages, division of labor, and rise of a new social structure.

The transition to agricultural society greatly eased the environmental constraints imposed on human carrying capacity. However, the continued dependence on plants and animals as the sole source of energy still set an upper limit on the productivity of human activities. In a burst of effort, the human body can muster 100 watts of power (1 joule per second = 1 watt). The most any society could devote to a given task with humans or animals as the primary source of energy was a few hundred thousand watts. Expanding territory could increase the energy supply, but it could not raise the total that could be applied to a single task. It is impossible to concentrate more than a few thousand bodies on a given project, be it construction or battle. But by the 18th century, the mechanical energy of animals and human labor was replaced by a much more concentrated form of energy: coal.

The Industrial Revolution, beginning in the mid-18th century, saw the development of the steam engine and with it a shift in labor from humans and draft animals to machines. First developed in the late 17th century, the steam engine transformed the heat energy of steam into mechanical energy. At first, steam engines were inefficient; they lost more than 99 percent of their energy. By 1800, however, their efficiency rose to about 5 percent with the capacity of 20 kilowatts of power in a single engine (the equivalent of 200 persons). By 1900, engines were capable of handling high-pressure steam and were 30 times more powerful than those in 1800 (the equivalent of 6000 persons). As important as the power they provided was their portability. Steam engines could be used anywhere (ships and trains), allowing for the large-scale transport of coal and establishing a positive feedback—industrialization.

With mechanization, manufacturing that was once carried out in individual homes and small workshops became centralized in factories. By the latter part of the 19th century, mechanization was changing the nature of agriculture as well. The size of fields increased, and the amount of land dedicated to farm animals decreased greatly. Machines more efficiently accomplished the work once done by the labor of humans and draft animals. A single large farm tractor (150–200 hp) can accomplish the work of more than 1000 workers or 200 draft animals. As a result of mechanization, by 1980, each American farmer fed about 80 people.

The mechanization of farming shifted the labor force, reducing the need for farm labor and providing the workforce needed for industrialization. In 1920, half the U.S. population was involved in the farm labor force; by 1990, the proportion had declined to 2–3 percent. In the 20th century, a transition had occurred from a rural agricultural-based economy to an urban industrial-based economy. In turn, this change triggered a major shift in the distribution of the human population. At the beginning of the 20th century, just over 35 percent of the human population lived in urban areas. As we entered the 21st century, that number had risen to almost 80 percent and continues to grow. Growing urbanization requires an ever-increasing infrastructure of transportation and trade, and with it an increasing demand for energy. By the 1990s, the average global citizen used about 20 "human equivalents" in energy (energy equal to that of a human working 24 hours a day, 365 days a year), primarily in the form of fossil fuels — making economic and population growth possible.

This transformation of energy use has allowed industrial labor efficiency to increase about 200-fold between 1750 and 1990, so that modern workers produce as much in a week as their 18th-century counterparts did in four years. In the 20th century alone, global industrial output grew 40-fold.

During the past 10,000 years, the human race has managed to multiply by a factor of 1000 and at the same time dramatically increase the per capita use of resources. The human population now exceeds 6.5 billion, and our collective "ecological footprint" on the planet continues to grow. Nearly 40 percent of Earth's potential terrestrial net primary productivity is used directly, co-opted, or forgone because of human activity (see Ecological Issues: Human Appropriation of Net Primary Productivity, p. 428). We use more than 50 percent of all freshwater resources, of which 70 percent goes to assist in agricultural production. In all, our activities have transformed 40–50 percent of the terrestrial surface to produce food, fuel, and fiber, and our transformations of the natural environment have led to the extinction of thousands of species.

Although profound, the changes that humans have brought about in our environment are not the product of thoughtless, malicious behavior. They are largely due to meeting the basic needs of a growing human population. Historically, we did not realize the extent of the impact of human activities on our environment. As recently as the 1970s, the mantra was "the solution to pollution is dilution." We had little appreciation of the long-term consequences of our collective actions.

Times have changed, and we are beginning to understand the consequences of our past and current activities. Human ecology is a new field, within the broader discipline of ecology, focusing specifically on interactions between humans and the environment. Like ecology itself, human ecology is an interdisciplinary field that draws upon many of the sciences as well as anthropology, sociology, and history. In Part Eight, we explore three general topics that form the backbone of current environmental issues regarding human impacts on the environment: resource use and environmental sustainability (Chapter 27), the declining biological diversity of our planet (Chapter 28), and the potential for human activity to significantly change Earth's climate (Chapter 29).

CHAPTER 27

Population Growth, Resource Use, and Sustainability

The human population entered the 20th century with 1.6 billion people and left the century with 6.1 billion. Demographers estimate that this number could rise to more than 9 billion in the next 50 years.

n 1890, the superintendent of the U.S. Bureau of Census announced that the frontier was closed. No longer was there a vast expanse of unsettled land waiting to be tamed. Until that time, if you drew a map of where people were and were not living in the United States, the two areas were clearly distinct. But by 1890, there was no longer a visible frontier line on the demographic maps produced by the Census Bureau. From 1790 on, that line had been moving steadily *westward*. Thomas Jefferson had speculated that it would take 100 generations to fill the vast open spaces of America. It took approximately 80 years.

The bland statement that "the frontier was closed" carried immense symbolic meaning, for it suggested to a generation of Americans that the process of exploiting inexhaustible resources had come to an end. In 1890, the year that the Census Bureau declared the frontier closed, the U.S. population was approximately 63 million. Since that time, the U.S. population has increased almost fivefold to over 300 million, and the demand for resources has grown disproportionately with advancements in technology and economic growth. The frontier may have closed, but the exploitation of natural resources continued at an accelerated pace.

Over that same period, the global population has increased sixfold. Earth's current population exceeds 6.5 billion people, who must exploit natural resources to meet the basic human needs of food, water, and shelter. The consumption of resources is driven by two factors: the total number of individuals (population size) and the per capita rate of consumption. Both factors have continued to increase steadily over the past 50 years (Figure 27.1).

The growing human population and the desire for expanded economic growth within the world community have brought the issue of sustainability to the forefront of current economic, political, and environmental

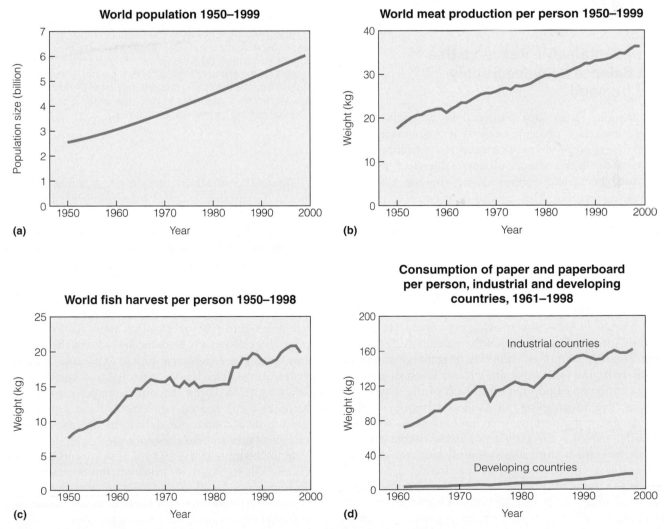

Figure 27.1 | Trends in population **(a)** and per capita resource use **(b–d)** over the past 50 years.
(Source: Census Bureau and FAO.)

discussions. Although the term *environmental sustainability* is broad, covering a range of topics and activities including population growth, energy use, and economic development, most discussions of environmental sustainability focus on the use of *natural capital*. **Natural capital** is the range of natural resources provided by ecosystems—the air, water, soil, nutrients, forests, grasslands, and so forth—that humans draw upon as essential resources. By its very definition, environmental sustainability relates to the exploitation of natural ecosystems by human populations. It is the ability to sustain the exploitation of natural capital to meet the growing human needs.

In this chapter, we will examine the concept of environmental sustainability as it relates to the human activities of agriculture, forestry, and fisheries. All three of these activities involve exploiting populations, plant and animal, to provide the most essential of human resources: food and shelter. We will examine the basic environmental issues related to these activities of resource management and extraction and consider how these issues relate to the ecological processes and patterns we have discussed thus far.

27.1 | Sustainable Resource Use Is a Balance between Supply and Demand

Sustainability is an idea burdened by ambiguity. It is widely used as a concept, rarely is it quantitatively defined. The concept of environmental sustainability seems to be based on the concept of sustainable yield that was first used in German forestry during the late 18th and early 19th centuries. The concept of sustainable yield is one of matching periodic harvests to the rate of biological growth—that is, harvesting without diminishing the forests themselves or undermining their long-term ability to regenerate.

In its simplest form, sustainable resource use is constrained by supply and demand. In Figure 27.2a, the box represents the quantity of a resource—for example, water, trees, or fish—that is being exploited. The arrow leading into the box represents the rate at which the resource is being supplied—the rate of recharge of a lake or reservoir, the rate of tree growth in a forest or plantation, or the rate of population growth of the target fish species. The arrow going out of the box represents the rate at which the resource is being harvested—the rate of water use, the harvest rate of trees, or the rate at which fish are being caught. Simply put, for the exploitation of the resource to be sustainable, the rate at which the resource is being used (consumption or harvest rate) must not exceed the rate at which the resource is being supplied (replacement or regeneration rate). Otherwise, the quantity of resources declines through time (Figure 27.2b).

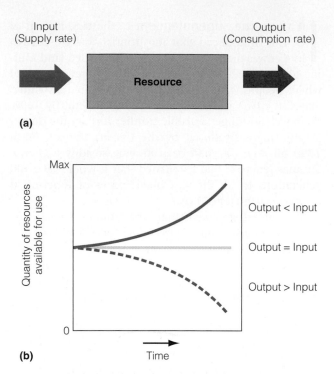

Figure 27.2 | **(a)** A simple model of resource use. The amount of resource available at any time (green box) is a function of the difference between the supply rate (green arrow) and the consumption rate (red arrow). **(b)** If the consumption rate is less than the supply rate, the amount of resource will increase. If the consumption rate exceeds the supply rate, the amount of resources available will decline. Sustainable resource use depends on the consumption rate not exceeding the supply rate.

Figure 27.3 illustrates this simple principle as it applies to the sustainable use of water resources of the Aral Sea in central Asia. In 1963, the surface of the Aral Sea measured 66,100 km². By 1987, 27,000 km² of former sea bottom had become dry land. About 60 percent of the Aral Sea's volume had been lost, and its salt concentration had doubled.

The demise of the Aral Sea occurred primarily because the inflowing Amu Dar'ya and Syr Dar'ya rivers were diverted to provide irrigation water for local croplands. These diversions dramatically reduced the river inflows, causing the Aral Sea to shrink. At the current rate of decline, the Aral Sea has the potential to disappear completely by 2020. Using the simple model presented in Figure 27.2, the harvest rate exceeded the supply rate, resulting in a continuous decline in the resource—an example of unsustainable resource use.

In the example of the Aral Sea, water resources can be continuously harvested for irrigation. But other resources can be harvested only periodically because they require an extended period between harvests to regenerate to a level where they can be harvested again. Trees in a forest plantation are an excellent example. After seedlings are planted, a period of time is required for trees to grow

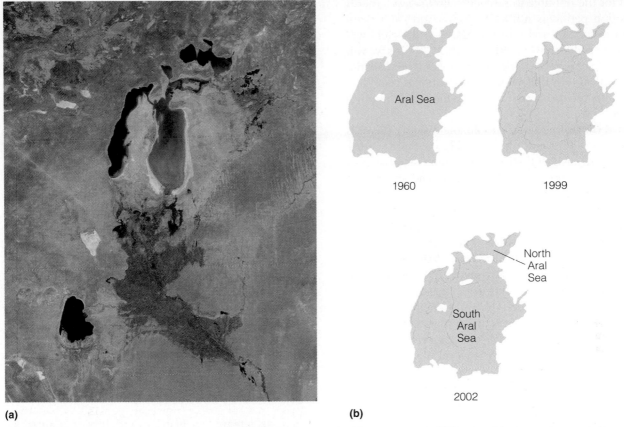

Figure 27.3 | **(a)** The Aral Sea in central Asia is an example of unsustainable resource use. **(b)** The graphic shows changes in the extent of the Aral Sea from 1960 to 2002. Because its waters are diverted for irrigation, the Aral Sea decreased in volume by 60 percent between 1963 and 1987. At the current rate of decline, it could disappear completely by 2020.

(Figure 27.4a). When the biomass of trees reaches a certain level, the forest or plantation is harvested. The amount of resource (tree biomass) harvested per unit time is called the **yield.** After the harvest, a period of time is required for new trees to grow and the amount of re-

source to return to the level of the previous harvest. This period of time is called the **rotation period** (or **harvest interval**). If the objective is to ensure a similar yield at each harvest, termed **sustained yield,** then sufficient time must be allowed between harvests (the rotation

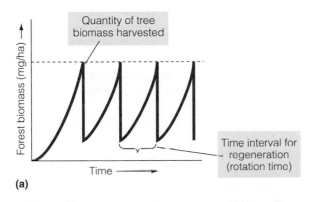

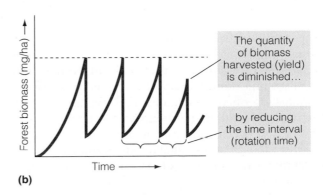

Figure 27.4 | **(a)** To achieve sustainable yield, sufficient time must be allowed between harvests (rotation time) for biomass to return to preharvest levels. Rotation time will depend on the growth rate of the species and site conditions that influence forest productivity. **(b)** If the rotation time is reduced, sufficient time is not allowed for the forest stand to recover (grow) to preharvest levels, and the subsequent yield will decline. The result is that the harvest rate exceeds the resource regeneration rate (as shown in Figure 27.2), and the quantity of resource declines through time.

period) for the resource to recover to preharvest levels. If the rotation period is not sufficient to allow the forest stand to recover to preharvest levels, then the yield will diminish in successive harvests (Figure 27.4b). As we will see in our discussion of agriculture, forestry, and fisheries in the following sections, many of the conflicts over long-term sustainable use of resources center on maintaining a sustainable yield.

In this simple model of sustainable resource use, it is assumed that the resource is renewable—it can be resupplied or regenerated. If the resource is nonrenewable, then by definition its use is not sustainable, and the rate of resource decline is a function of the rates at which the resource is being harvested and used. Mineral resources (such as aluminum, zinc, copper, etc.) are an example of nonrenewable resources. Often, however, resources are classified as nonrenewable even though they are being resupplied, because their resupply rate is virtually nonexistent compared to their consumption rate. Fossil fuels are a good example. Coal, oil, and natural gas are referred to as nonrenewable energy sources because their formation occurs on a timescale of millions of years (see Chapter 29), making their rate of resupply effectively zero on the timescale of human consumption.

Unlike fossil fuel energy, many nonrenewable resources can be recycled, reducing the harvest rate for the initial resource. The effect of recycling is to extend the effective lifetime of the resource.

27.2 | Sustainability Can Be Indirectly Limited by Adverse Consequences of Resource Use

Ecosystem services are the processes by which the environment produces resources (natural capital) such as clean air, water, timber, or fish. Although the rate at which ecosystems supply these essential resources functions as a fundamental constraint on sustainable use, sustainable use can also be constrained indirectly as a consequence of negative impacts on ecosystem services that arise from resource management, extraction, or use. A prime example is waste (unused or unwanted materials). Dealing with domestic, industrial, and agricultural waste is a growing environmental issue with implications for ecosystems and human health. Wastes and by-products of production often contaminate the environment (air, water, and soil) with harmful substances—pollution (see Chapter 22), which can limit or disrupt the ability of ecosystems to provide essential resources and services.

Return to the example of the Aral Sea (see Figure 27.3). Today, about 200,000 metric tons of salt and sand are carried by the wind from the Aral Sea region every day and dumped within a 300-km radius. The salt pollution is decreasing the area available for agriculture, destroying pastures, and creating a shortage of forage for domestic animals. Fishing in the Aral Sea has ceased completely,

shipping and other water-related activities have declined, and the associated economic changes have taken a heavy toll on agricultural production. The quality of drinking water has continued to decline due to increasing salinity, bacteriological contamination, and the presence of pesticides and heavy metals. Diseases like anemia, cancer, and tuberculosis as well as the presence of allergies are on the rise. The incidence of typhoid fever, viral hepatitis, tuberculosis, and throat cancer is three times the national average in some areas.

In discussing sustainable resource use in the following sections, we will focus on the ability to sustain current and future levels of resource yield (the topic of sustainable yield), as well as on the negative impacts arising from the management, extraction, and use of resources. Such impacts may affect an ecosystem's ability to continue providing those natural resources (ecosystem services). In evaluating questions of sustainability, we must focus on the ability to sustain current and future rates of resource consumption as well as on the consequences of managing and consuming those resources for both environmental and human well-being.

27.3 | Sustainability Is a Concept Learned from Natural Ecosystems

When we attempt to manage and harvest natural resources in a sustainable fashion, in many ways we are attempting to mimic the function of natural ecosystems. Natural ecosystems function as sustainable units. This should be clear from our discussion of ecosystems in Chapters 20 and 21. Take as an example the link between primary productivity and decomposition. The uptake of nutrients such as nitrogen by plants is constrained by the rate at which they become available in the soil (see Figure 6.25). In effect, the rate of nutrient harvest by plants, and the subsequent rate of primary productivity within an ecosystem, are constrained by and cannot exceed the rate at which nutrients are being supplied to the soil (see Section 20.3). In turn, nutrients tied up in organic matter are recycled through the processes of microbial decomposition and mineralization (see Section 21.5). As nutrients are mineralized and returned to the soil, they are quickly taken up by plants, thus minimizing their loss through ground and surface waters (see Section 21.9).

To put the functioning of natural ecosystems into the context of the simple graphical model presented in Figure 27.2, the rate of resource use is limited by the resource supply rate. In the case of mineral nutrients such as nitrogen, the rate at which the resource is used is equivalent to the rate at which it is supplied (regenerated), so the size of the box representing the pool of available resources is extremely small.

When the supply rate of a resource varies through time, the rate of resource use must likewise change.

During years of drought, net primary productivity of an ecosystem will decline. If sufficient water is not available to replace water lost through transpiration, stomata will close, leaves will wilt, and the plants could die. Although water stored in the soil may postpone the effects of drought, plant growth and productivity are generally tied closely to annual rates of precipitation (water availability). As we will see in the following discussions of agriculture and forestry, variations in the supply rate of resources necessary to maintain primary productivity are a major constraint on maintaining sustainable yield. Overcoming these constraints is the major source of problems related to sustainable agriculture and forestry.

27.4 | Agricultural Practices Vary in the Level of Energy Input

The exploitation and management of natural populations provides more than 80 percent of fish and shellfish harvested globally on an annual basis, but the vast majority of human food resources are derived from agriculture—the production of crops and the grazing of livestock. Even though botanists estimate that, worldwide, there are as many as 30,000 native plant species with parts (seeds, roots, leaves, fruits, etc.) that can be consumed by humans, only 15 plant and 8 animal species produce 90 percent of our food supply. The seeds of only three annual grasses—wheat, rice, and corn (maize)—constitute more than 80 percent of the cereal crops consumed by the world population. Although initially derived from native plant species, the varieties of the cereal crops cultivated today are the products of intensive selective breeding and genetic modification by plant and agricultural scientists around the world. The same is true for domestic animals used for food production.

Approximately 11 percent of Earth's ice-free land area is under cultivation. Another 25 percent is used as pastureland for grazing livestock (primarily cattle and sheep). No matter what crop is planted or cultivation method is used, agriculture involves replacing diverse natural ecosystems—grassland, forest, shrublands—with a community consisting of a single crop species (**monoculture**) or a mixture of crops (**polyculture**). Although a wide range of agricultural practices are carried out throughout the world, agricultural production can be classified into one of two broad categories: industrialized and traditional.

Industrialized (also termed *mechanized* or *high-input*) agriculture depends on large inputs of energy in the form of fossil fuels (mechanization), chemical fertilizers, irrigation systems, and pesticides. Although energy demanding, this form of agriculture produces large quantities of crops and livestock per unit of land area. Industrialized agriculture is practiced on about 25 percent of all cropland, primarily in developed countries;

however, these practices have spread to developing regions in recent decades.

Traditional agriculture is dominated by subsistence agriculture in which primarily human labor and draft animals are used to produce only enough crops or livestock for a family to survive. Examples of this low-input approach to agriculture are shifting cultivation in tropical forests and nomadic livestock herding.

In reality, these two forms of agricultural production, industrialized and traditional, define two points on a continuum of agricultural methods that are practiced throughout the world. However, we will use these two end points for the purpose of comparison in our discussion of sustainability of agricultural practices.

27.5 | Swidden Agriculture Represents a Dominant Form of Agriculture in the Tropics

A method of subsistence farming that is practiced primarily in the tropical regions is shifting cultivation, or **swidden agriculture.** This method of traditional agriculture involves a rotating cultivation technique in which trees are first cut down and burned in order to clear land for planting (Figure 27.5). The burning of felled trees and brush serves two purposes. First, it removes debris, thus clearing the land for planting and ensuring that the plot is relatively free of weeds. Second, the resulting ash is high in mineral nutrients (see Field Studies: Deborah Lawrence), promoting plant growth (Figure 27.6). The plot is then cultivated and crops are harvested. A characteristic of this type of agriculture is a decline in productivity with each successive harvest (Figure 27.7). The reason for this decline is that each time crops are harvested, nutrients in the form of plant tissues are being removed from the plot. Because even organic fertilizers are rarely used in this form of agriculture, soil nutrients decline. Eventually the site is abandoned as cropland and allowed

Figure 27.5 | In swidden agriculture, a plot of forest is cleared and burned to allow crops to be planted. The ashes are an important source of nutrients. Fertilizers are typically not used.

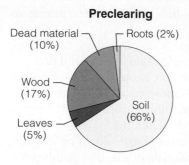
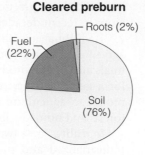
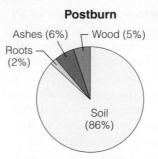

Preclearing

Dead material (10%)
Roots (2%)
Wood (17%)
Leaves (5%)
Soil (66%)

Cleared preburn

Roots (2%)
Fuel (22%)
Soil (76%)

Postburn

Ashes (6%)
Wood (5%)
Roots (2%)
Soil (86%)

Figure 27.6 | Dynamics of total nitrogen during the burning and clearing of a forest site in Turrialba, Costa Rica (values in kg N/ha).

(Adapted from Ewel et al. 1981.)

Losses: Harvesting of wood (100)
Burning (100)
Leaching and wind (340)

to revert to natural vegetation through secondary succession (see Chapter 18).

If left undisturbed or fallow for a sufficient time, the nutrient status of the site recovers to precultivation levels (Figure 27.8). At that point, the site can once again be cleared and planted. In the meantime, other areas have been cleared, burned, and planted. So in effect, this type of agriculture represents a shifting, highly heterogeneous patchwork of plots in various stages of cultivation and secondary vegetation (Figure 27.9).

The swidden cultivation system represents a sustainable form of agriculture when time is permitted for regrowth of natural vegetation and the recovery of soil nutrients. But it also requires sufficient land area to allow for the appropriate rotation period. The problem currently being faced in many parts of the tropics is that growing populations are placing ever-increasing demands on the land, and sufficient recovery periods are not always possible. In these cases the land is quickly degraded, and yield continuously declines.

27.6 | Industrialized Agriculture Dominates the Temperate Zone

Industrialized agriculture is widely practiced in North America, much of Europe, Russia, sections of South America, Australia, and other areas of the world where money and land are available to support this form of agriculture. Machines and fossil fuel energy replace the energy supplied by humans and draft animals in traditional agricultural systems (Figure 27.10). Mechanization requires large expanses of land for machines to operate effectively and economically. Because different crops (wheat, corn, cotton, etc.) require specialized equipment for planting and harvest, farmers typically plant one (monoculture) or few crop varieties season after season (continuous rotation).

With crops such as wheat and corn, most (or all) parts of the plants are removed, so little or no organic matter remains after harvest. Consequently, at each harvest large quantities of nutrients are removed from the

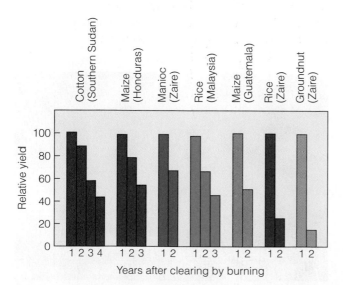

Figure 27.7 | Patterns of declining productivity in successive years in swidden agricultural systems for various crops in several regions of the tropics.

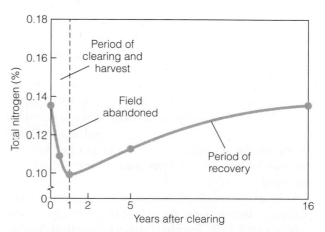

Figure 27.8 | Changes in total soil nitrogen during a cycle of clearing, harvest, abandonment, and recovery for a swidden agricultural system in Costa Rica (Central America). Note that nitrogen levels decline until the plot is abandoned after the first year. Soil nitrogen recovers to original levels by year 16.

(Adapted from Ewel et al. 1981.)

Regenerating forest

One year fallow

Newly cleared for planting

(a)

(b)

Figure 27.10 | Examples of traditional **(a)** and mechanized (industrial) **(b)** agriculture. In industrial agriculture, machines have replaced humans and draft animals as the source of labor (energy).

site in the plant materials (Table 27.1). In addition, tilling the soil to prepare for planting exposes the soil to wind and water erosion, and conventional tilling practices result in annual soil losses totaling upward of 44 tons per hectare (Figure 27.11).

Due to the removal of organic matter, these agricultural ecosystems effectively have no nutrient cycle. Instead, large quantities of chemical fertilizers must be added to maintain productivity. Nutrients are added to the soil in an inorganic (mineral) form that is readily available for

Table 27.1 | Approximate Amounts of Nutrient Elements Contained in Various Crops (values in kg/ha)

Crop	Yield	N	P	K	Ca	Mg
Corn						
Grain	9416	151	26	37	18	22
Stover	10,080	112	18	135	31	19
Rice						
Grain	5380	56	10	9	3	4
Straw	5610	34	6	65	10	6
Wheat						
Grain	2690	56	12	15	1	7
Straw	3360	22	3	33	7	3
Loblolly pine (22 yr)	84,000	135	11	64	85	23
Loblolly pine (60 yr)	234,000	344	31	231	513	80

Figure 27.11 | A field that has recently been tilled. Note the lack of ground cover. Tilled fields have a high rate of erosion from water and wind.

Table 27.3 | Comparison of Energy Inputs in Production and Energy Yields for Corn Harvested Using a Traditional Agricultural System in Mexico and an Industrial Agricultural Production System in the United States

Item	Mexico(kcal/ha)	USA(kcal/ha)
Inputs		
Labor	589,160	5250
Axe and hoe	16,570	
Machinery		1,018,000
Gasoline		400,000
Diesel		855,000
Irrigation		2,250,000
Electricity		100,000
Nitrogen		3,192,000
Phosphorus		730,000
Potassium		240,000
Lime		134,000
Seeds	36,608	520,000
Insecticides		300,000
Herbicides		800,000
Drying		660,000
Transportation		89,000
Total	642,338	11,036,650
Outputs		
Total corn yield	1944 kg	7000 kg
	8,748,000 kcal	31,500,000 kcal
kcal output/kcal input	13.62	2.85

uptake by plants; however, the nutrients are also readily leached to surface and groundwater (Table 27.2).

Because the same crop varieties are being planted over large, contiguous regions, pests and plant diseases spread readily. Pests are typically controlled by chemical means to avoid reduced productivity. Different chemical pesticides target different pests: insecticides (insects), herbicides (weedy herbaceous plants), fungicides (fungi), and rodenticides (rodents). These chemicals bring with them a range of environmental problems.

27.7 | Different Agricultural Methods Represent a Trade-off between Sustainability and Productivity

The two very different agricultural systems just discussed—traditional (swidden) and industrialized—represent a trade-off between energy input in production and energy harvest in food resources. Table 27.3 compares en-

ergy inputs and yields for corn production in Mexico using traditional swidden agriculture and in the United States using large-scale industrial farming techniques. Energy inputs in the traditional agricultural system are dominated by labor (approximately 92 percent of total energy input), with small inputs in the form of tools and seed. Total crop yield is just over 1900 kg/ha. In contrast, labor is only a minor proportion of total energy input in the industrialized agricultural system (approximately 0.05 percent). Major inputs are in the form of farm machinery (3.2%), fossil fuels (4%), irrigation (7.1%), chemical fertilizers (13.6%), and pesticides (3.5%). Total crop yield in this agricultural system is 7000 kg/ha, more than 3.5 times that produced by the traditional agricultural methods used in Mexico. However, the real story emerges when we look at the ratio of energy input in production to energy produced in harvested food. The ratio of kcal output (food) to kcal energy input (energy involved in production) is 13.6 for the traditional agriculture and only 2.8 for the industrialized system. Although industrialized agriculture produces 3.5 times the yield of corn per unit

Table 27.2 | Nutrient Inputs in Precipitation and Fertilizers for a Corn Field in the Central United States (values in kg/ha/yr)

	In Precipitation	Fertilizer	Harvest	Runoff to Streams
Nitrogen	11.0	160.0	60.0	35.0
Calcium	3.2	190.0	1.0	47.0
Phosphorus	0.03	30.0	12.0	3.0
Potassium	0.2	75.0	13.0	15.0

Soil loss due to erosion: 44 t/ha/yr.

of land area under cultivation, it does so at the cost of more than 17 times the energy input for production. In addition, this large input of energy is in the form of materials and services that carry a large environmental cost.

The loss of chemical fertilizers such as nitrates and phosphates from agricultural fields to adjacent streams, lakes, and coastal waters (estuaries and wetlands) has led to nutrient enrichment. This accelerated enrichment causes chemical and environmental changes that result in major shifts in plant and animal life, by a process termed **cultural eutrophication** (see Section 24.4).

Besides the impacts on adjacent natural ecosystems, the widespread use of chemical fertilizers has impacts on human health. Groundwater provides drinking water for more than half of the U.S. population and is the sole source of drinking water for many rural communities. Nitrate (from chemical fertilizers) is one of the most common groundwater contaminants in rural areas. Recent U.S. Environmental Protection Agency (EPA) surveys indicate that 1.2 percent of community water systems and 2.4 percent of rural domestic wells nationwide contain concentrations of nitrate that exceed public health standards. Although this value may seem low, contamination of groundwater is concentrated in agricultural areas, such as the Midwest, where the percentage of wells contaminated is much higher (Figure 27.12). High concentrations of nitrate in drinking water may cause birth defects, cancer, nervous system impairments, and "blue baby syndrome," a condition in which the oxygen content of the infant's blood falls to dangerously

low levels. In addition, high concentration of nitrate can indicate the possible presence of other more serious residential or agricultural contaminants, such as bacteria or pesticides.

Along with the environmental problems caused by the widespread use of chemical fertilizers and pesticides, the large inputs of fossil fuels used for mechanization, irrigation, and chemical fertilizer production add to the growing input of carbon dioxide and other greenhouse gases to the atmosphere (see Chapter 29).

At first it may seem that the solution to sustainable agriculture is to adopt more traditional agricultural practices, such as those used in swidden agriculture. However, agricultural production must be viewed in the context of the growing human population. Increasing agricultural production to meet the growing human demand can be accomplished in two ways: increasing the land under cultivation or increasing the production of food per unit land area. Historically, the total land under cultivation worldwide has risen exponentially, keeping pace with the growing human population. However, in the last half of the 20th century, this trend began to slow (Figure 27.13a), and the per capita land area under production has declined (Figure 27.13b). This decline is due partly to the development of more productive varieties of crops (see Ecological Issues: Genetically Modified Crops) and an expansion of the use of irrigation. But by far the largest factor responsible for the increased productivity per unit land is the growing use of chemical fertilizers, particularly

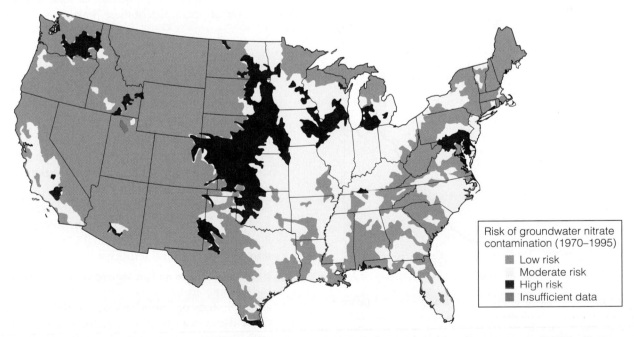

Risk of groundwater nitrate contamination (1970–1995)

■ Low risk
□ Moderate risk
■ High risk
■ Insufficient data

Figure 27.12 | A national risk map designed to present patterns of risk for nitrate contamination at regional or national scales (produced by overlay analysis in a geographic information system). Knowing the site and type of risks to groundwater can alert water-resource managers and private users of the need to protect water supplies. By targeting regions with the highest risk of nitrate contamination, managers can direct resources to areas most likely to benefit from pollution prevention programs and long-term monitoring.

nitrogen (Figure 27.13c). Synthetic nitrogen fertilizer now supplies about half of the nitrogen used annually by the world's crops. At least one-third of the protein in the current global food supply is derived by synthesizing ammonia (NH_3) from hydrogen and nitrogen using the Haber–Bosch process—named after the two engineers, Fritz Haber and Carl Bosch, who would later win the Nobel Prize for developing this process for use in chemical fertilizers (see Ecological Issues: Nitrogen Fertilizers, Chapter 21, p. 457).

So the reality of our situation is that we are dependent on industrialized agriculture to feed the world's growing population. In addition, by slowing the amount of land being cleared for agricultural production, we are reducing the single greatest cause of Earth's declining biological diversity—habitat loss primarily due to the conversion of lands for agriculture (see Chapter 28). The task has now become one of developing methods of large-scale mechanized agriculture that minimize environmental impacts.

27.8 | Sustainable Agriculture Depends on a Variety of Methods

As with the broader concept of environmental sustainability, the term **sustainable agriculture** refers more to a notion of maintaining agriculture production while minimizing environmental impacts than to a quantitative set of criteria. As it pertains to agriculture, *sustainable* describes farming systems that are capable of maintaining their productivity and usefulness to society while simultaneously conserving resources and minimizing the negative impacts that we have discussed.

Today, sustainable farming techniques commonly include a variety of practices designed to reduce soil erosion, reduce the use of chemical fertilizers and pesticides, and conserve and protect the quality of water resources. The following are examples of these practices:

> *Soil conservation methods*: Methods such as contour and strip cropping (Figure 27.14a) and reduced tillage or "no-till" farming (Figure 27.14b) help prevent soil loss due to wind and water erosion. Likewise, planting shrubs and trees as fencerows provides wind barriers, thus reducing soil erosion.
>
> *Reduced use of pesticides*: Crop rotations (such as planting wheat one season and hay the next) and strip cropping with multiple crops or varieties help reduce the spread of pests and disease.
>
> *Alternative sources of soil nutrients*: Increased use of on-farm nutrient sources, such as manure and leguminous cover crops (see Section 15.11), provide alternatives to chemical fertilizers.
>
> *Water conservation and protection*: Water conservation has become an important part of agricultural stewardship. Many practices have been developed to improve the quality of drinking and surface water as well as to protect wetlands. Wetlands play a key role in filtering nutrients and pesticides, in addition to providing wildlife habitat (see Chapter 25).

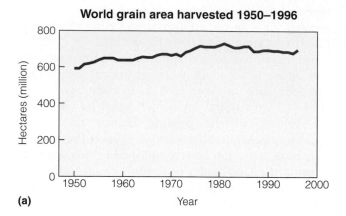

World grain area harvested 1950–1996

(a)

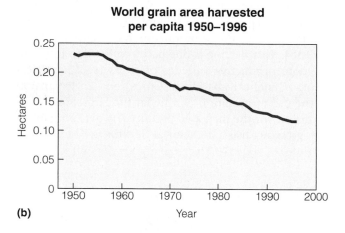

World grain area harvested per capita 1950–1996

(b)

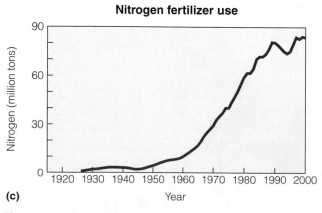

Nitrogen fertilizer use

(c)

Figure 27.13 | The global total of land in crop production **(a)** has begun to stabilize in recent decades, while the per capita area harvested **(b)** has actually declined. This decline is due to an increase in productivity per hectare, resulting largely from increasing use of chemical fertilizers (primarily nitrogen) over the same period **(c)**.
(Source: FAO.)

Over the past 50 years the amount of crops produced per unit of land area under agricultural production has steadily increased. In part, this steady increase is due to increasing use of chemical fertilizers, irrigation, and pesticides (see Figure 27.13). However, increases in production per hectare are also due to improved plant varieties developed in the process of crop breeding.

Crop breeding is a form of selective breeding. Individual plants with desired characteristics are interbred to produce new crop varieties with these traits (see Chapter 5 for discussion of heritability and natural selection). These plants can then be crossbred to introduce the desired traits (genes) of one variety into another. For example, a mildew-resistant variety of corn may be crossed with a high-yielding but susceptible variety, resulting in a plant that has mildew resistance as well as high yield. The seed and planting material of these varieties are then supplied to farmers.

Humans have practiced the approach of selective breeding for literally thousands of years. About 8000 years ago, for example, farmers in Central America crossed two mutant strains of the annual grass *Balsas teosinte* and produced the first ancestor of modern maize (corn; see Ecological Issues: The Life History of Maize, pg. 175). The diversity of modern crops attests to the success of selective breeding; but because this process relies on the random recombination of all of a plant's tens of thousands of genes, the odds of producing a variety with a desired trait are extremely low. Recent innovations in biotechnology, however, allow scientists to select specific genes from one organism and introduce them directly into the chromosome of another to confer a desired trait. Compared to conventional breeding methods, this technology not only produces new varieties of plants or animals more quickly but also introduces traits not possible through traditional techniques. Amazingly, genes from organisms as dissimilar as bacteria and plants can be successfully inserted into each other. Combining genes from different organisms is known as recombinant DNA technology, and the resulting organism is said to be genetically modified, genetically engineered, or transgenic.

The principal application of this technology to date has been the development of genetically modified (GM) crops engineered to tolerate herbicides and resist pests. Crops carrying herbicide-tolerant genes were developed so that farmers could spray their fields to eliminate weeds without damaging the crop. Likewise, pest-resistant crops have been engineered to contain a gene for a toxic protein from the soil bacterium *Bacillus thuringiensis*. This protein, referred to as Bt, is produced by the entire plant, thereby making it resistant to insect

pests like the European corn borer or cotton boll weevil. Other pest-resistant GM crops on the market today have been engineered to contain genes that confer resistance to specific plant viruses.

In 1996, a total of 4.2 million acres in 6 countries were planted with GM crops. By 2003, the numbers had grown to 18 countries and 167.2 million acres, or 25 percent of the total land under agricultural production— a 40-fold increase in eight years (Figure 1). The adoption of GM crops has been the most rapid in the United States, where there has been a 27-fold increase in the area of GM crops planted during the same eight-year period (from 3.7 million acres in 1996 to 105.7 million acres in 2003).

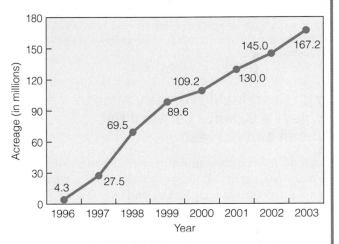

Figure 1 | Increase in global area of genetically modified (GM) crops for the period 1996–2003.
(ISAAA Global Review of Transgenic Crops 2003.)

In the United States the three main GM crops under cultivation are varieties of soybeans, corn, and cotton. By 2004, genetically engineered varieties accounted for 85 percent of all soybeans, 45 percent of all corn, and 76 percent of all cotton produced in the United States.

Other GM crops currently grown in the United States include canola, squash, and papaya. An estimated 54 percent of all canola grown in this country is genetically modified, according to industry estimates. More than 50 percent of the papayas grown in the U.S. (all in Hawaii) are GM.

Other major producers of GM crops are Argentina, which plants primarily GM soybeans; Canada, whose principal GM crop is canola; Brazil, which has recently legalized the planting of GM soybeans; China, where the acreage of GM cotton continues to increase; and South Africa, where cotton is also the principal GM crop.

(a)

(b)

Figure 27.14 | Some methods being adopted in the effort to promote sustainable agricultural practices are **(a)** contour and **(b)** no-till farming. Go to GRAPHit! at **www.ecologyplace.com** to graph food production efficiency.

27.9 | Sustainable Forestry Aims to Achieve a Balance between Net Growth and Harvest

Forest ecosystems cover approximately 35 percent of Earth's surface (see Chapter 23) and provide a wealth of resources, including fuel, building materials, and food (Figure 27.15). Although plantations provide a growing percentage of forest resources, more than 90 percent of global forest resources are still harvested from native forests.

Globally, about half of the forest that was present under modern (post-Pleistocene) climatic conditions (see Section 18.9, Figure 18.20)—and before the spread of human influence—has disappeared (see Figure 28.2), largely through the impact of human activities. The spread of agriculture and animal husbandry, the harvesting of forests for timber and fuel, and the expansion of populated areas have all taken their toll on forests. The causes and timing of forest loss differ between regions and forest types, as do the current trends in change in forest cover. In

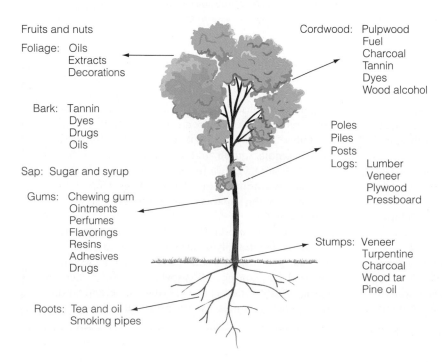

Fruits and nuts

Foliage: Oils
Extracts
Decorations

Bark: Tannin
Dyes
Drugs
Oils

Sap: Sugar and syrup

Gums: Chewing gum
Ointments
Perfumes
Flavorings
Resins
Adhesives
Drugs

Roots: Tea and oil
Smoking pipes

Cordwood: Pulpwood
Fuel
Charcoal
Tannin
Dyes
Wood alcohol

Poles
Piles
Posts
Logs: Lumber
Veneer
Plywood
Pressboard

Stumps: Veneer
Turpentine
Charcoal
Wood tar
Pine oil

Figure 27.15 | A variety of products derived from forests.

the face of increasing demand and declining forest cover, the goal of sustained yield in forestry is to achieve a balance between net growth and harvest. To achieve this end, foresters have an array of silvicultural and harvesting techniques from clear-cutting to selection cutting.

Clear-cutting involves removing the forest and reverting to an early stage of succession (Figure 27.16a). The area harvested can range from thousands of hectares to small patch cuts of a few hectares designed to create habitat for wildlife species that require an opening within the forest (see Sections 19.2 and 19.3). Postharvest management varies widely for clear-cut areas. When natural forest stands are clear-cut, there is generally no follow-up management. Stands are left to regenerate naturally from existing seed and sprouts on the site and the input of seeds from adjacent forest stands. With no follow-up management, clear-cut areas can be badly disturbed by erosion that affects subsequent recovery of the site as well as adjacent aquatic communities.

(a)

(b)

Figure 27.16 | Examples of **(a)** clear-cut and **(b)** shelterwood (seed-tree) forest harvest.

Harvest by clear-cutting is the typical practice on forest plantations; but here, intensive site management follows clearing. Plant materials that are not harvested (branches, leaves, and needles) are typically burned to clear the site for planting. After clearing, seedlings are planted and fertilizer applied to encourage plant growth. Herbicides are often used to discourage the growth of weedy plants that would compete with the seedlings for resources.

The **seed-tree,** or **shelterwood,** system is a method of regenerating a new stand by removing all trees from an area except for a small number of seed-bearing trees (Figure 27.16b). The uncut trees are intended to be the main source of seed for establishing natural regeneration after harvest. Seed trees can be uniformly scattered or left in small clumps, and they may or may not be harvested later.

In many ways, the shelterwood system is similar to a clear-cut because generally not enough trees are left standing to affect the microclimate of the harvested area. The advantage of this harvesting approach is that the seed source for natural regeneration is not limited to adjacent stands. This can result in improved distribution (or stocking) of seedlings as well as a more desirable mix of species.

Like any silviculture system, shelterwood harvesting requires careful planning to be effective. Trees left on the site must be strong enough to withstand winds and capable of producing adequate seed; seedbed conditions must be conducive to seedling establishment (this may require a preparatory treatment during or after harvest); and follow-up management may be required to fully establish the regeneration.

In **selection cutting,** mature single trees or groups of trees scattered through the forest are removed. Selection cutting produces only small openings or gaps in the forest canopy. Although this form of timber harvest can minimize the scale of disturbance within the forest caused by direct removal of trees, the network of trails and roads necessary to provide access can be a major source of disturbance (to both plants and soils). Selective cutting also can cause changes in species composition and diversity because only certain species are selectively removed.

Regardless of the differences in approach, some general principles can be stated regarding sustainable forestry. Forest trees function in the manner discussed for competition in plant populations in Chapter 11 (Section 11.3). Whether a forest is planted as seedlings or grown by natural regeneration, its establishment begins with a population of small individuals (seedlings) that grow and compete for the essential resources of light, water, and nutrients. As biomass in the forest increases, the density of trees decreases; and average tree size increases as a result of self-thinning (Figure 27.17; also see Section 11.4 and Figure 11.10). For a stand to be considered

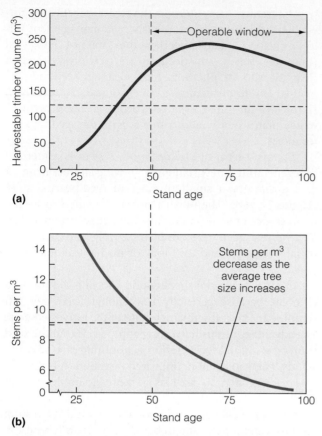

(a)

(b)

Figure 27.17 | Two criteria are used to determine when a stand of trees are suitable for harvest (referred to as the *operable window*): (1) salable volume of wood per hectare (m³/ha) and (2) average tree size as measured by the stems per cubic meter (the number of trees required to make a cubic meter of wood volume). In the example shown, dashed horizontal lines represent constraints of **(a)** minimum salable wood volume of 100 m³/ha and **(b)** minimum average tree size of 9 stems/m³. Dashed vertical lines indicate the earliest stage at which both constraints are satisfied.

Interpreting Ecological Data

Q1. What does graph (a) imply about the change in average tree size (diameter and/or height) with stand age?

Q2. Assume that a decision is made to harvest the stand when it reaches the minimum salable wood volume (100 m³/ha). How many stems (trees) per m³ would be harvested?

(Adapted from Oriens et al. 1986.)

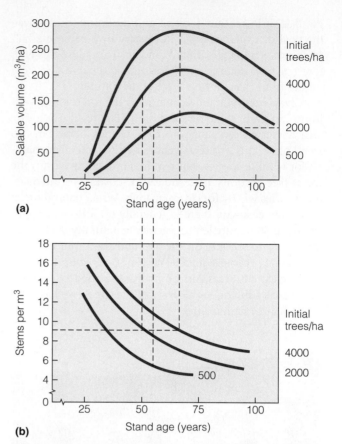

(a)

(b)

Figure 27.18 | Effects of initial stand density on timing of stand availability for harvest (operable window). As in Figure 27.17, dashed horizontal lines represent constraints of **(a)** minimum marketable wood volume of 100 m³/ha and **(b)** a minimum average tree size of 9 stems/m³. Dashed vertical lines indicate the earliest stage at which both constraints are satisfied. Note that the intermediate planting density of 2000 trees per hectare has the earliest operable window.

Interpreting Ecological Data

Q1. The above analysis includes three initial planting densities: 500, 2000, and 4000 trees/ha. At which stand age does each of the three initial planting densities achieve the minimum constraint for average tree size?

Q2. Given the constraints on minimum wood volume and average tree size as defined, which of the initial planting densities reaches these constraints at the earliest stand age (earliest operable window)?

(Adapted from Oriens et al. 1983.)

economically available for harvest (referred to as being in an *operative state*), minimum thresholds must be satisfied for the harvestable volume of timber per hectare and average tree size (see Figure 27.17); and these thresholds will vary depending on the species. In plantation forestry, for a given set of thresholds (timber volume and average tree size), the initial stand density (planting density) can be controlled to influence the timing of the stand's availability for harvest (Figure 27.18).

After trees are harvested, a sufficient time must pass for the forest to regenerate. For sustained yield, the time between harvests must be sufficient for the forest to regain

the level of biomass it had reached at the time of the previous harvest (see Figure 27.4). Rotation time depends on a variety of factors related to the tree species, site conditions, type of management, and intended use of the trees being harvested. Wood for paper products (pulpwood), fence posts, and poles are harvested from fast-growing species, allowing a short rotation period (15–40 years). These species are often grown in highly managed plantations where trees can be spaced to reduce competition and fertilized to maximize growth rates. Trees harvested for timber (saw logs) require a much longer rotation period. Hardwood species used for furniture and cabinetry

Department of Environmental Sciences, University of Virginia, Charlottesville, Virginia

For hundreds of years, the villagers of Kembera in West Kalimantan, Indonesia, have practiced the same form of shifting cultivation. An area of rain forest approximately 1 ha in size is cleared of all vegetation (Figure 1). After drying for several weeks, the debris is burned. With the onset of heavy rains, the field is planted with rice and a few vegetable crops. This type of upland agriculture involves no tillage or mounding, no irrigation, and no chemical inputs. The rice crop is then weeded by hand from one to three times during the 5- or 6-month growing period. One rice crop is grown per year, and although plots are sometimes used for a second year, fields are typically abandoned after a single harvest. During abandonment, the plot naturally regenerates to secondary forest. After a period of recovery (fallow) lasting on average 20 years, the area is once again ready to be cleared and planted.

Are these land-use practices sustainable, and how do they affect the region's biological diversity? These questions are being addressed by ecologist Deborah Lawrence of the Department of Environmental Sciences, University of Virginia.

The area surrounding the Dayak village of Kembera in West Kalimantan has been the location of Lawrence's work, which focuses on the impacts of traditional agriculture on soil resources. Phosphorus (P) availability is often a major limitation on agricultural production in the tropics. Its availability is controlled by the weathering of phosphorus-containing rocks and minerals. Moreover, rapid sorption (absorption and adsorption) of P on abundant iron and aluminum minerals in tropical soils (see Section 4.10) precludes its free movement through the soil solution to plant roots. High amounts of rainfall also cause heavy leaching of nutrients such as P from the surface soil. To understand the limitations of P on the long-term sustainability of agriculture in this region, Lawrence needed to quantify the influence of the process of shifting cultivation on the dynamics of soil P.

Fire transfers nutrients from the vegetation and surface organic matter to the soil. Where temperatures rise to high levels, **pyromineralization** occurs, resulting in the transformation of organic forms of P into inorganic (mineral) forms available for uptake by plants. Depending on the intensity of the fire, Lawrence found that up to 55 percent of the total P stock is lost to the atmosphere as smoke—although much of this may be redistributed across the landscape rather than being lost entirely from the forest. In addition, upward of 50 percent of mineral P that is deposited during the fire may be lost from the site through erosion by water and wind or through leaching. Despite these losses of P from the soil when the site is being prepared for planting, burning results in a large immediate increase in available inorganic P, thus promoting growth of the rice plants. The harvesting of rice, however, represents a further loss of P and depletes soil stocks. These losses of soil P during field preparation and crop harvesting are large relative to the input of P to the soil through the process of weathering. Given these facts, one would hypothesize a decline in total P with each subsequent cultivation cycle that influences the long-term potential for agricultural productivity as well as the process of secondary forest growth (forest regeneration).

To examine this hypothesis, Lawrence designed a study to examine patches of land around the village of Kembera that had been used exclusively for shifting cultivation for up to 200 years. To determine long-term changes in soil P, she sampled the top 30 cm of soil at 24 sites representing a gradient of cultivation history, from zero (primary forest) to 10 or more previous swidden–fallow cultivation cycles. At each site, 30 soil cores were collected, and each core was divided by depth (0–2.5 cm, 2.5–15 cm, and 15–30 cm).

Figure 1 | Aerial view of plot cleared for planting.

continued on page 586

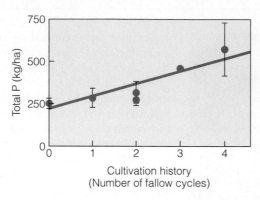

Figure 2 | Changes in phosphorus stocks in the top 30 cm of soil during the first four cycles of cultivation. Values shown are mean ±1 standard deviation for 6 sites on sandy soil.

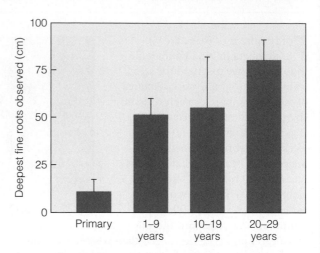

Figure 3 | Maximum depth of fine roots observed as a function of forest age. Deepest roots are defined as the midpoint of soil horizons in which the roots were found. Data are from 2 sites newly cleared from primary forest and 17 sites in secondary forest fallows 1–28 years after clearing.

Results of the analyses were surprising. Rather than declining, total soil P actually increased with each successive cultivation cycle over the first four cycles (approximately 80 years; Figure 2), remaining relatively constant thereafter. How could this be, given the large losses associated with clearing and harvest? To understand how P was increasing in the top 30 cm of soil through time, Lawrence turned to examining the fallow process between periods of cultivation, during which secondary forest growth occurs.

When an area has been abandoned, nutrients accumulate in the secondary forest biomass before soil stocks begin to increase. Soil P taken up by the trees is then returned to the soil surface through litterfall and subsequent decomposition and mineralization. Upon examining the vertical distribution of roots on the study plots, Lawrence began to develop a possible explanation for the observed increase in P in the surface soil through time. Tree species that colonize the secondary forest have more fine roots, deeper in the soil profile, than do species that dominate the primary (uncleared) forest (Figure 3). These fine roots are responsible for most of the nutrient uptake from the soil. The maximum observed depth of fine roots in the secondary forest is 50 cm, as compared to 20 cm in the primary forest. This finding suggests that the increase in total P after cultivation and fallow is a result of "nutrient pumping." Fine roots located deeper in the soil profile take up P, which is incorporated into the plant tissues as the secondary forest grows during the fallow period. Organic P in the form of senescent plant tissues is then deposited on the soil surface, where it is eventually released as inorganic P. Therefore, Lawrence found, profuse deep rooting and periodic death (with each fire) are likely to encourage the transfer of inorganic P to the surface and organic P (in the form of fine roots) to the deeper horizons.

Despite the increase in total P in the top 30 cm of soil observed during the first four cycles of cultivation, potential P losses through the erosion and harvest dictate that future rice production will depend on continued acquisition of P from deeper in the soil profile, hence maintaining the critical fallow periods necessary for regenerating nutrient stocks in the surface soils. In many parts of Indonesia, however, shifting cultivators are having to reduce the length of the fallow period due to the pressures of population growth—potentially jeopardizing this sustainable form of agricultural production that has supported populations of this region for so long.

Bibliography

Lawrence, D., and W. H. Schlesinger. 2001. Changes in soil phosphorus during 200 years of shifting cultivation in Indonesia. *Ecology* 82:2769–2780.

Lawrence, D., D. R. Peart, and M. Leighton. 1998. The impact of shifting cultivation on a rainforest landscape in West Kalimantan: Spatial and temporal dynamics. *Landscape Ecology* 13:135–148.

1. What possible impacts other than altering soil nutrient availability might these agricultural practices have on the tropical forest ecosystems of this region?
2. How does the practice of burning the site to prepare for planting influence the availability of P in the soil in both organic and mineral forms?

are typically slower growing and may have a rotation time of 80 to 120 years. Sustained forestry of these species works best in extensive areas where blocks of land can be maintained in different age classes.

As with agricultural crops, a significant amount of nutrients are lost from the forest when trees are harvested and removed (see Table 27.1). The loss of nutrients in plant biomass is often compounded by further losses from soil erosion and various postharvest management practices—particularly the use of fire (see Field Studies: Deborah Lawrence). The reduction of nutrients will reduce plant growth, requiring a longer rotation period for subsequent harvests or causing reduced forest yield if the rotation period is maintained. Forest managers often counter the loss of nutrients by using chemical fertilizers, which create other environmental problems for adjacent aquatic ecosystems (see Sections 24.4 and 27.7).

In addition to the nutrients removed directly through biomass removal, logging can also result in the transport of nutrients from the ecosystem by altering processes involved in internal cycling. The removal of trees in clear-cutting and other forest management practices increases the amount of radiation (including direct sunlight) reaching the soil surface. The resulting increase in soil temperatures promotes decomposition of remaining soil organic matter (see Section 21.4) and causes an increase in net mineralization rates (see Section 21.5) (Figure 27.19). This increase in nutrient availability in the soil occurs at the same time that demand for nutrients is low because plants have been removed and net primary productivity is low. As a result, there is a dramatic increase in the leaching of nutrients from the soil into ground and surface waters (Figure 27.20). This export of nutrients

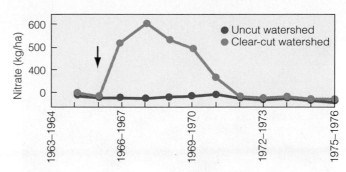

Figure 27.20 | Temporal changes in the nitrate concentration of stream water for two forested watersheds in Hubbard Brook, New Hampshire. The forest on one watershed was clear-cut, and the other forest was undisturbed. Note the large increase in concentrations of nitrate in the stream on the clear-cut watershed. This increase is due to increased decomposition and nitrogen mineralization after clear-cutting (see Figure 27.19). The nitrogen then leached into the surface water and groundwater.

(Adapted from Likens and Borman 1995.)

from the ecosystem results from decoupling the two processes of nutrient release in decomposition and nutrient uptake in net primary productivity.

Sustained yield is a key concept in forestry and is practiced to some degree by large timber companies and federal and state forestry agencies. But all too often, industrial forestry's approach to sustained yield is to grow trees as a crop rather than maintaining a forest ecosystem. Their management approach is a form of agriculture in which trees are grown as crops: they clear-cut, spray herbicides, plant or seed the site to one species, clear-cut, and plant again. Clear-cutting practices in some national forests, especially in the Pacific Northwest and the Tongass National Forest in Alaska, hardly qualify as sustained-yield management when below-cost timber sales are mandated by the government to meet politically determined harvesting quotas. Even more extensive clear-cutting of forests is taking place in the northern forests of Canada, especially British Columbia (Figure 27.21), and in large areas of Siberia. As the timber supply dwindles in the Pacific Northwest, the timber industry that moved west after the depletion of eastern hardwoods and pine forests of the lake states is moving east again to exploit the regrowth of eastern hardwood forests, especially the rich and diverse central hardwood forest. From Virginia and eastern Tennessee to Arkansas and Alabama, timber companies have built more than 140 highly automated chip mills that cut up trees of all sizes into chips for paper pulp and particle board. Feeding the mills requires clear-cutting 500,000 ha annually. The growing demand for timber has boosted timber prices, stimulating more clear-cutting. The rate of harvest is wholly unsustainable. In face of growing timber demands, sustained-yield management has hardly filtered down to smaller parcels of private land.

The problem of sustained-yield forestry is its economic focus on the resource with little concern for the

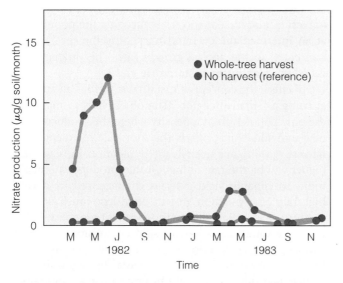

Figure 27.19 | Comparison of nitrate (NO_3^-) production after logging for a loblolly pine (*Pinus taeda*) plantation in the southeastern United States. Data for the reference stand (no harvest) are compared with those of a whole-tree harvest clear-cut.

(Adapted from Vitousek 1992.)

Figure 27.21 | Localized view of forest clearing using aerial photography. Go to **GRAPHit!** at **www.ecologyplace.com** to graph rates of deforestation.

forest as a biological community. A carefully managed stand of trees, often reduced to one or two species, is not a forest in an ecological sense. Rarely will a naturally regenerated forest, and certainly not a planted one, support the diversity of life found in old-growth forests (see Section 17.6 and Figure 17.14). By the time the trees reach economic or financial maturity—based on the type of rotation—they are cut again.

27.10 | Exploitation of Fisheries Has Led to the Need for Management

Although the advent of agriculture some 10,000 years ago has reduced human dependence on natural populations as a food source, more than 80 percent of the world's commercial catches of fish and shellfish is from the harvest of naturally occurring populations in the oceans (71 percent) and inland freshwaters (10 percent).

There are many historical accounts of overexploitation and population declines, yet not until the late 1800s was there any effort to manage fisheries resources to en-

sure their continuance. At that time, wide fluctuations in fish catches in the North Sea began to have an economic impact on the commercial fishing industry. Ensuing debates raged over the cause of the decline and whether commercial harvest was affecting fish populations. Some people argued that removing the fish had no effect on reproduction; others argued that it did. When the Danish fishery biologist C. D. J. Petersen developed a method for estimating population size based on a technique of tagging, releasing, and recapture, biologists could begin assessing fish populations. Combining data from egg surveys and the aging of fish from commercial catches, these studies suggested that overharvesting indeed was the culprit. But the debate continued, and the controversy was not laid to rest until after World War I.

During the war, fishing in the North Sea had stopped. After the war, fishermen experienced sizable increases in their catches. Fishery biologists suggested that the renewed fishing would once again reduce population sizes, and catches would stabilize and eventually decline with overexploitation. Their predictions were correct, and with time, attention turned to the question of sustainable harvest.

The goal of long-term sustainable harvest has been a mainstay of fisheries science for the past 50 years. A central concept of sustainable harvest is the logistic model of population growth presented in Chapter 11 (Section 11.1). Under conditions of the logistic model, growth rate (overall numbers of new organisms produced per year) is low when the population is small (Figure 27.22). It is also low when a population nears its carrying capacity (K) because of density-dependent processes such as competition for limited resources. Intermediate-sized populations have the greatest growth capacity and ability to produce the most harvestable fish per year. The key realization of this model is that fisheries could optimize harvest of a particular species by keeping the population at an intermediate level and harvesting the species at a rate equal to the annual growth rate. This strategy was called the **maximum sustainable yield.**

In effect, the concept of sustainable yield is an attempt at being a "smart predator." The objective is to maintain the prey population at a density where the production of new individuals just offsets the mortality represented by harvest. The higher the rate of population increase, the higher will be the rate of harvest that produces the maximum sustainable yield. Species characterized by a very high rate of population growth often lose much of their production to a high density-independent mortality, influenced by variation in the physical environment such as temperature (see Section 11.12). The management objective for these species is to reduce "waste" by taking all individuals that otherwise would be lost to natural mortality. Such a population is difficult to manage. The stock can be depleted if annual patterns of reproduction are interrupted due to environmental conditions. An example is the Pacific sardine (*Sardinops sagax*). Exploitation of the Pacific

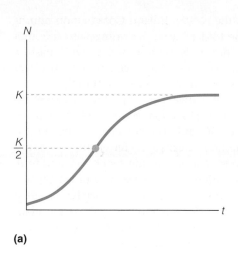

(a)

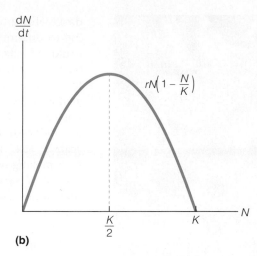

$$rN\left(1 - \frac{N}{K}\right)$$

(b)

Figure 27.22 | Assuming the growth rate of the fish population follows the logistic model presented in Chapter 10 [dN/dt = rN[1 − N/K)], **(a)** in the absence of fishing, the population will grow to carrying capacity, K. **(b)** The relationship between the rate of population growth, dN/dt, and population size, N, takes the form of a parabola, reaching a maximum value at a population size of $N = K/2$.

sardine population in the 1940s and 1950s shifted the age structure of the population to younger age classes. Before exploitation, reproduction was distributed among the first five age classes (years). In the exploited population, this pattern of reproduction shifted, and close to 80 percent of reproduction was associated with the first two age classes. Two consecutive years of environmentally induced reproductive failure (a result of natural climate variations associated with El Niño–Southern Oscillation—ENSO; see Section 2.9) caused a population collapse the species never recovered from (Figure 27.23).

Sustainable yield requires a detailed understanding of the population dynamics of the fish species. Recall from Chapter 10 that the intrinsic rate of population growth, r, is a function of the age-specific birth and mortality rates. Unfortunately, the usual approach to maximum sustained yield fails to consider adequately the size and age classes, differential rates of growth among them, sex ratio, survival, reproduction, and environmental uncertainties—all data difficult to obtain. Adding to the problem is the common-property nature of the resource: because it belongs to no one, it belongs to everyone to use as each of us sees fit.

Perhaps the greatest problem with sustainable harvest models is that they fail to incorporate the most important component of population exploitation: economics (see Section 27.12). Once commercial exploitation begins, the pressure is on to increase it to maintain the underlying economic infrastructure. Attempts to reduce the rate of exploitation meet strong opposition. People argue that reduction will mean unemployment and industrial bankruptcy—that, in fact, the harvest effort should increase. This argument is shortsighted. An overused resource will fail, and the livelihoods it supports will collapse, because in the long run the resource will be depleted. Abandoned fish processing plants and rusting fishing fleets support this view. With conservative exploitation, the resource could be maintained.

27.11 | Fisheries Management Requires an Ecosystem Approach

Another problem with the concept of sustainable harvest is that traditional population management, especially by fisheries, considers stocks of individual species as single biological units rather than components of a larger ecological system. Each stock is managed to bring in a maximum economic return, overlooking the need to leave behind a certain portion to continue its ecological role within the larger community, be it predator or prey. This attitude encourages a tremendous discard problem, euphemistically called *bycatch* (Figure 27.24). Employing large purse seines and midwater gill nets (Figure 27.25) that encompass square kilometers of ocean, fishermen haul in not only commercial species they seek but also a range of other marine life as well, including sea turtles, dolphins, and scores of other species of fish. To harvest bottom-dwelling groundfish, such as halibut, haddock, and flounder, fishermen use bottom trawls that sweep over the ocean floor scraping up all forms of benthic life and destroying the sea bottom habitat. Fishermen discard the unwanted species—at least one-third of the catch—by putting them back into the sea.

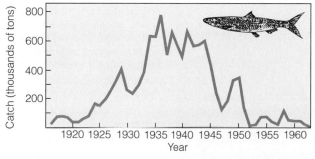

Figure 27.23 | Annual catch of Pacific sardines along the Pacific coast of North America. Overfishing, environmental changes, and an increased population of a competing fish species, the anchovy, resulted in a collapse of the sardine population.

(Adapted from Murphy 1966.)

Figure 27.24 | Bycatch of fish and invertebrates can outweigh the target species (in this case, shrimp) by 5- to 20-fold. This bypass in the Gulf of California includes bottom-dwelling brittlestars, crabs, small fish, and skates.

(Photo by Elliot Norse.)

This practice seriously affects the future of those fisheries, damaging critical habitats and altering the structure of the community at all trophic levels. The cod (*Gadus morhua*) of the North Atlantic provides a case study of these unsustainable practices.

For 500 hundred years the waters of the Atlantic Coast from Newfoundland to Massachusetts supported one of the greatest fisheries in the world. The English explorer John Cabot in 1497 discovered and marveled at the abundance of cod off the Newfoundland Coast. Upon returning to Britain, he told of seas "swarming with fish that could be taken not only with nets but with baskets weighted down with stone." Some cod were 5 to 6 feet long and weighed up to 200 pounds. Cabot's news created a frenzy of exploitative fishery. Portuguese, Spanish, English, and French fishermen sailed to Newfoundland to take between 100,000 and 200,000 metric tons (MT) a year. In 1542 the French sailed no fewer than 60 ships, each making two trips a year. In the 1600s England took control of Newfoundland and its waters and established numerous coastal posts where English merchants salted and dried cod before shipping it to England. So abundant were the fish that the English thought nothing could seriously affect this inexhaustible resource.

Catches remained rather stable until after World War II, when the demand for fish increased dramatically and led to intensified fishing efforts. Large factory trawlers that could harvest and process the catch at sea replaced smaller fishing vessels. Equipped with sonar and satellite navigation, fishing fleets could locate spawning schools. They could engulf schools with huge purse nets and sweep the ocean floor clean of fish and all associated marine life. In the 1950s annual average catch was 500,000 MT of cod, but by the 1960s the catch had almost tripled to 1,475,000 MT. In 15 years from the mid-1950s through the 1960s, two hundred factory ships off Newfoundland took as many northern cod as were caught over the prior 250-year span since Cabot's arrival.

The cod fishery could not endure such intense exploitation. By 1978 the catch had declined to 404,000 MT. To protect their commercial interests in the fishery, the Canadian and U.S. governments excluded all foreign fisheries in a zone extending 200 miles. But instead of capitalizing on this opportunity to allow the stocks to recover, the Canadian government provided the industry with subsidies to build huge factory trawlers. After a brief surge in catches during the 1980s, the catch in the Canadian waters fell to 13,000 MT by the mid 1990s, and in 1997 the North Atlantic Canadian cod fishery collapsed (Figure 27.26).

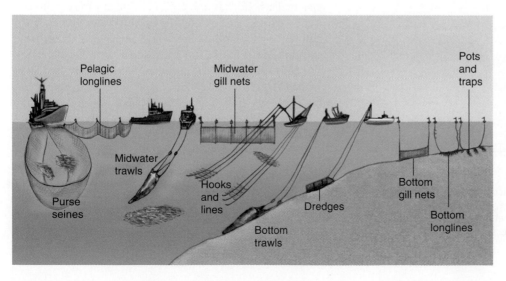

Figure 27.25 | Types of gear employed in taking fish. Purse seines, midwater gill nets, and pelagic longlines cover miles of ocean. Dredges and bottom trawls scrape the seabed.

(Adapted from Chuenpadgee et al. 2003.)

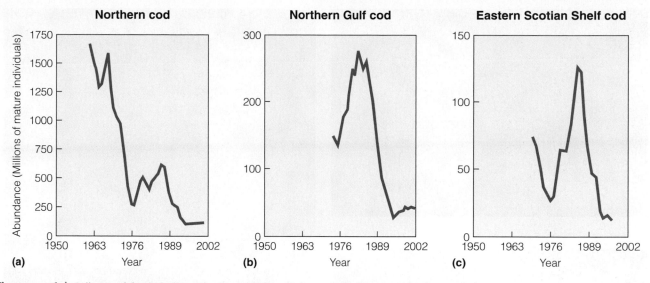

Northern cod   Northern Gulf cod   Eastern Scotian Shelf cod

(a) Year   (b) Year   (c) Year

Figure 27.26 | Collapse of the Canadian Atlantic cod fishery is dramatically illustrated by these graphs showing the harvest of three stocks of cod. Note the spike of overfishing in the 1980s, followed by collapse of the fishery.

(After Hutchings and Reynolds 2004.)

The Gulf of Maine and Grand Banks fisheries controlled by the United States followed a similar trend. The Grand Banks spawning stock biomass reached a record low in the 1990s, and since 2001 has continued to decline (Figure 27.27). Combined total commercial cod landings for Canada and the United States for the Grand Banks and Gulf of Maine in 2005 was 6957 MT, a 17 percent decrease from 8381 MT in 2004. Cod populations are still being overfished, even under a management program designed to reduce fishing mortality to a level allowing for populations to recover. The plan includes time and area closures, gear restrictions, minimum size limits, limits on days at sea, and a moratorium on permits. Despite these restrictions, cheating on catch limits, a substantial trade in illegal landings, and unreported fishing contribute to overfishing.

The collapse of the Atlantic cod fishery created an ecological, economic, and social disaster. In Newfoundland it left 40,000 fishermen and fish processors out of work, and it devastated nearly half of the region's 1300 fishing communities that depended entirely on fishing. It affected the New England fishing industry in the form of fleet reductions, processing plant closures, and loss of jobs.

Fishery managers and the industry blamed the collapse on seal predation of the cod's favorite foods, Atlantic herring (*Clupea harengus*) and capelin (*Mallotus villosus*); on colder water temperatures that drove cod away; and on various other environmental factors. They refused to admit that the real cause was overfishing. After a 10-year moratorium on fishing, the cod population has failed to recover. Many of the fishery scientists who once refused to admit that overfishing was the root cause have now concluded that centuries of overfishing the cod populations and devastating other species through discard are indeed the root cause.

Emphasis on managing commercial fish species using a single-species model, viewing the dynamics of

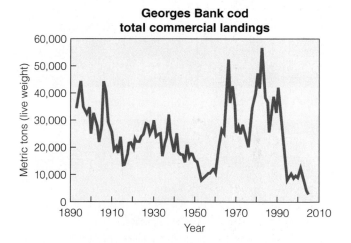

Georges Bank cod total commercial landings

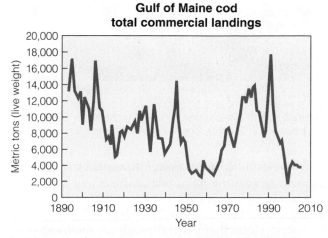

Gulf of Maine cod total commercial landings

Figure 27.27 | Trend of commercial landings of cod in the famous Grand Banks fishery, 1890–2005. Again note the sharp increase in cod take in 1970–1980, followed by a collapse of the stock. (New England Fishery Science Center, National Oceanic and Atmospheric Administration, 2006.)

Figure 27.28 | Major species involved in the North Atlantic harvest: **(a)** Atlantic cod, **(b)** spiny dogfish, **(c)** green sea urchin, **(d)** common periwinkle, and **(e)** rockweed, a dominant brown seaweed.

that species in isolation from other species in the community, ignores the multitude of direct and indirect interactions that influence their populations (see Part Four). As we have discussed in Chapter 17, impact on one species can filter down through the food web to affect other species. Such was the case with the cod fishery. The cod is a top predator (Figure 27.28a). With the collapse of cod populations, its prey—herring, crabs, sea

urchins, shrimp, and lobster—increased dramatically. In addition, the spiny dogfish (*Squalus acanthias*) replaced the cod as a top predator.

With the increase in spiny dogfish populations, the fishery shifted emphasis, fueled by an increasing demand for this species in the European market. A slow-growing, slow-reproducing species of shark, the spiny dogfish (Figure 27.28b) became an important commercial species with the same pattern of overfishing and decline in populations.

In response to the collapse of the cod and dogfish populations, the fishing industry moved to the next-lower trophic level to concentrate on a species that was once the dominant prey of the cod—herring. Intense fishing pressure on herring brought about a sharp decline in that species as well. So to augment this decline in the fishery, the Atlantic Coast fishing industry went further down the food chain to the dominant herbivores, the kelp-grazing spiny green sea urchin (*Strongylocentrotus droebachiensis*; Figure 27.28c), a prey for lobsters, crabs, and seabirds; and the green-algae-feeding periwinkles (*Littorina* spp.; Figure 27.28d), an important food for many invertebrates, fish, and waterfowl. A growing demand for sea urchins for sushi is resulting in unsustainable harvest, accomplished either by towing a dredge on sedimentary bottoms or by scuba divers. Due to these practices, the urchin population has been in decline since 1993 and is now strongly regulated by both federal and state governments. Periwinkles, considered a delicacy in European and Japanese communities, have traditionally been harvested from rocky coasts at low tide. Due to the growing market, however, periwinkles are now collected by diver-operated suction harvesters. These harvesters can collect up to several thousand pounds of periwinkles per day.

Once again facing declining stocks, the industry has now turned to the primary producer level of the food chain, to harvest rockweed (*Fucus* and *Ascophyllum*; Figure 27.28e) along the Atlantic Coast from Newfoundland to Maine. Rockweed is in demand for the production of fertilizer, mulch, processed kelp meal, and gelatinous compounds used as a thickener and stabilizer in thousands of commercial products. Once gathered by hand, rockweed is now harvested by large motorized barges with a vacuum device and cutter blades. As a dominant primary producer of these coastal ecosystems, rockweed functions as the base of the food chain, supporting large populations of invertebrates including periwinkles. It also functions as a critical nursery habitat for many species of fish.

The history of Atlantic fishery emphasizes that managing for a single species alone has direct or indirect impacts on other species in the food web. Fishery removes not just the target species, but scores of other species at all trophic levels as discard, while destroying the benthic habitat. The problem grows more acute as the industry moves down the trophic levels to the primary producer level, as in the harvesting of rockweed. There is a slowly growing realization of the long-term effects of current management and harvesting activities—and a recognition that the long-prevailing model of maximum sustained yield for individual species must be replaced with a model of ecologically sustainable yield that considers the role of the harvested species in the broader context of the community and ecosystem. Such

an approach will require stricter catch limits, and a change in the timing and methods of fishing. Perhaps most important, the implementation of ecologically sustainable yield will require the political, economic, and social changes that involve forgoing short-term profits for long-term sustainability.

27.12 | Economics Are a Key Factor Governing Resource Management

Although a large portion of the human population produces its own food and forest products (such as fuel wood) through subsistence farming, fishing, and wood gathering, these essential human resources are now part of the global marketplace—produced for sale and profit. As such, economic considerations are central to the production and management of natural resources.

Benefit–cost analysis is a critical tool of economics used in making decisions regarding the production and management of natural resources. A **benefit–cost analysis** involves measuring, adding up, and comparing all the benefits and all the costs of a particular project or activity. For example, if it costs a farmer $100 per hectare to produce corn this year, and the expected value of that corn is $200 per hectare, the benefits ($200) outweigh the costs ($100), and the decision will most likely be to plant the corn. If the corn produced per hectare was worth only $80, however, the costs outweigh the benefits, and the farmer is unlikely to plant the fields with corn and incur a loss. In this example, the dollar value of costs and benefits can be compared directly because both costs (expenses) and benefits (revenue) occur during the same time period—the year the corn crop is planted and harvested. When costs and benefits are extended over a much longer period of time, it is necessary to use a procedure called discounting.

Discounting is a technique employed to add and compare costs and benefits that occur at different points in time. It is a major driving force in the economics of natural resource management; unfortunately, discounting often runs counter to the objectives of sustainable resource management. The problem arises because in farming, forestry, and fisheries, there are substantial initial costs in acquiring land, equipment, permits, and so forth. These costs must then be weighed against expected future earnings from the production and harvesting of natural resources (crops, trees, or fish). As we all know, a dollar earned in the future does not have the same value as a dollar in hand today. Inflation depreciates the value of future earnings. In addition, a dollar invested today is worth more in the future because of compounding interest. Therefore, in comparing present-day costs with expected benefits (revenue) in the future, those benefits must be discounted to reflect the reduced value of future dollars. When this comparison is done for activities such as harvesting forests or fisheries, it often leads to the conclusion that it is economically more advantageous to "overexploit" the resource now and invest the resulting profits rather than to harvest the resource in a sustainable way over a much longer period.

In 1973, the economist Colin W. Clark made this point persuasively in the case of the blue whale, *Balaenoptera musculus*. More than 100 feet long and weighing upward of 150 tons at full maturity, the blue whale is the largest animal on either land or sea. It is also among the easiest to hunt and kill. More than 300,000 blue whales were harvested during the 20th century, with a peak harvest of more than 29,000 in the 1930–31 season. By the early 1970s, the population had plummeted to several hundred individuals. Although international negotiations were held to address the problem of overharvest and discuss potential regulatory policies, some countries were especially eager to continue the hunt even at the risk of total extinction. So Clark asked what practice would yield the whalers the most money: cease hunting and let the blue whales recover in numbers and then harvest them sustainably into the future, or kill the rest off as quickly as possible and invest the profits in the stock market? The troubling answer was that if the whalers could achieve an annual rate of return on their investments of 20 percent or more, it would be economically advantageous to harvest all of the blue whales at that time and invest their profits.

The problem with this purely economic approach to evaluation of natural resources is that the value of a blue whale was based only on the measures relevant to the existing market; that is, on the going price per unit weight of whale oil and meat. It does not consider other services provided by the blue whale, such as ecotourism (whale watching), nor does it consider the value of blue whales to future generations. But perhaps most important, it views blue whales, trees, and even whole ecosystems as having no inherent value beyond those calculated in economic terms.

A second important economic concept essential to understanding the sustainable use of natural resources is that of externalities. **Externalities** occur when the actions of one individual (or group of individuals) affect another individual's well-being, but the relevant costs (or benefits) are not reflected in market prices. For example, if a timber company clear-cuts an area of forest, it may well have adverse environmental consequences on adjacent areas. Erosion may transport silt to adjacent streams and rivers. Fertilizers and pesticides used as part of site preparation may likewise find their way to surface water and groundwater. The result may be reduced water quality, affecting drinking water, recreational value of the waterways, and aquatic plants and animals. Although the economic costs of these impacts can often be quantified, these costs are typically not borne by the timber company. These are externalities, or costs that are not reflected in the market prices. The timber company reaps the monetary benefit of the trees that have been harvested, but the environmental costs resulting from the tree harvest are passed on to the public. If these external costs were included in the analysis of costs associated with the timber harvest, the benefits (profits) may no longer outweigh the costs; and the clear-cut would not take place. Another option is to pass on these costs to the consumer to reflect the actual costs of the goods and services.

Recent decades have seen the emergence of a new discipline in the field of economics: **environmental economics.** Environmental economics is the study of environmental problems with the perspective and analytic tools of economics, such as those presented earlier. Incorporating economic principles into the environmental decision-making process is critical. Until the true value of natural resources and the costs of their extraction and use are understood and incorporated into economic decisions, the sustainable management and use of natural resources seems unlikely.

Summary

Sustainable Resource Use | 27.1

In its simplest form, the constraint on sustainable resource use is one of supply and demand. For exploitation of the resource to be sustainable, the rate at which the resource is being used (consumption rate) must not exceed the rate at which the resource is being supplied (regeneration rate). Otherwise, the quantity of resources declines through time. If the resource is nonrenewable, then by definition its use is not sustainable, and the rate of resource decline is a function of the rate at which the resource is being harvested and used.

Negative Impacts of Resource Extraction and Use | 27.2

Sustainable resource use can also be constrained indirectly as a result of negative impacts arising from resource management, extraction, or use. Domestic, industrial, and agricultural waste disposal is a growing environmental issue with implications for ecosystems and human health. Wastes and by-products of production often contaminate the environment (air, water, and soil) with harmful substances—pollution.

Natural Ecosystems | 27.3

When we attempt to manage and harvest natural resources in a sustainable fashion, in many ways we are trying to mimic the function of natural ecosystems. Natural ecosystems function as sustainable units.

Agriculture and Energy | 27.4

No matter what crop is planted or which method of cultivation is used, agriculture involves replacing diverse natural ecosystems with a community consisting of a single crop species or a mixture of crops. A wide range of agricultural practices are carried out worldwide, but agricultural production falls into one of two broad categories: industrialized and traditional.

Industrialized agriculture depends on large inputs of energy in the form of fossil fuels, chemical fertilizers, irrigation systems, and pesticides. Although energy demanding, this form of agriculture produces large quantities of crops and livestock per unit of land area. Traditional agriculture is dominated by subsistence agriculture in which primarily human labor and draft animals are used to produce only enough crops or livestock for a family to survive.

Swidden Agriculture | 27.5

A traditional method of production in the wet tropics is swidden agriculture, a rotating cultivation technique in which trees are first cut down and burned to clear land for planting. Crops are planted, but production declines with each succeeding harvest. The plot is then abandoned, and the forest is allowed to reestablish itself. Eventually, the nutrient status of the site recovers, and it can again be used for crops.

Industrialized Agriculture | 27.6

In industrial agriculture, machines and fossil fuel energy replace the energy supplied by humans and draft animals in traditional agricultural systems. Mechanization requires large expanses of land for machines to operate effectively, and because different crops require specialized equipment for planting and harvest, farmers typically plant one or a few crop varieties season after season. Tilling and removal of organic matter during harvest has the effects of reducing the nutrient status of the soil and encouraging soil erosion. Large quantities of chemical fertilizers and pesticides must be used to maintain productivity.

Trade-offs in Agricultural Production | 27.7

The two very different agricultural systems—traditional (swidden) and industrialized—represent a trade-off between energy input in production and energy harvest in food resources. Industrial agriculture produces high yields per hectare at the expense of large inputs of energy in the form of fossil fuels, fertilizers, and pesticides. Each of these inputs produces serious environmental impacts.

In contrast, yields are lower in traditional agricultural systems; but they are more energy efficient, yielding a greater amount of energy in crops per unit of energy input in crop production.

Sustainable Agriculture | 27.8

The term *sustainable agriculture* refers more to a notion of maintaining agriculture production while minimizing environmental impacts than to a quantitative set of criteria. It involves the use of farming methods that conserve soil and water resources, reduce the use of pesticides, and use alternative (on-site) sources of fertilizers.

Forestry | 27.9

The goal of sustained yield in forestry is to achieve a balance between net growth and harvest. To achieve this end, foresters use an array of silvicultural and harvesting techniques, from clear-cutting to selection cutting. For a stand to be considered economically available for harvest, it must satisfy minimum thresholds for harvestable volume of timber per hectare and average tree size. When trees are harvested, enough time must pass for the forest stand to reach harvestable size again. For sustained yield, the time between harvests must be sufficient for the forest to regain the level of biomass it had reached for the previous harvest.

Tree harvesting, and the alterations to the site after harvesting, results in loss of nutrients from the site. To maintain productivity, it is often necessary to fertilize the site, which can lead to further nutrient loss from the site to adjacent aquatic ecosystems.

Fisheries | 27.10

The goal of long-term sustainable harvest has been a mainstay of fisheries science for the past 50 years. A central concept of sustainable harvest is the logistic model of population growth, in which the population growth rate is highest at intermediate population densities. The key realization of this model is that fisheries could optimize harvest of a particular species by keeping the population at an intermediate level and harvesting the species at a rate equal to the annual growth rate. This strategy was called the maximum sustainable yield. In practice, the concept of maximum sustainable yield is difficult to achieve because it requires a detailed understanding of the population structure and dynamics of the species being harvested.

Ecosystem Approach | 27.11

A problem with the concept of sustainable harvest is that traditional population management, especially by fisheries, considers stocks of individual species as single biological units rather than components of a larger ecological system. Each stock is managed to result in maximum economic return. The catch is dictated by economics and market demands that override considerations of maximum sustainable yield. As target species are depleted, the fisheries exploit lower trophic levels, in turn depleting those stocks. There is a growing realization that the long-term effects of current fishery management are such that the maximum sustained yield concept must be replaced with a model of ecological sustained yield that considers the harvested species within the broader context of community and ecosystem.

Environmental Economics | 27.12

Benefit–cost analysis is a critical tool of economics, used in making decisions regarding the production and management of natural resources. A benefit–cost analysis involves measuring, adding up, and comparing all the benefits and all the costs of a particular project or activity.

When costs and benefits are extended over a much longer time, it is necessary to use a procedure called discounting. In this procedure, the benefits derived in the future must be discounted to reflect the reduced value of future dollars. Discounting often leads to economic choices that run counter to sustainable yield.

Externalities occur when the actions of one individual (or group of individuals) affect another individual's well-being, but the relevant costs (or benefits) are not reflected in market prices. Externalities are important in evaluating methods of sustainable resource extraction because actual costs of production (pollution, habitat degradation, and negative impacts on human health) are typically not built into the price structure, so the true cost of the activity or resource is not being considered.

Study Questions

1. What relationship between supply rate and consumption rate must exist for the consumption of a resource to be sustainable?
2. Contrast renewable and nonrenewable resources in the context of sustainable resource use.
3. How might the sustainable use of resources be limited indirectly by adverse consequences from the management, extraction, and consumption of resources? Provide examples.
4. Contrast traditional and industrialized agricultural methods. What are the major inputs of energy (for production) in each?
5. Which agricultural method (industrial or traditional) produces the greatest yield of crops per unit area? Which produces the most crops per unit of energy input for production?

6. Name some methods or practices that can increase the sustainability of current industrialized agricultural production.
7. What is sustained yield? What is maximum sustainable yield? How do these two concepts differ?
8. Identify and discuss two sources of nutrient loss during forest harvest and management.
9. Why is it important to take an ecosystem approach to fisheries management rather than approaching the management and harvest of each species as a separate issue?
10. Why do economists discount future benefits? What is the effect of discounting on sustainable management of resources?
11. Why is air pollution caused by coal-fired power plants an externality?

Further Readings

Barrett, G. W., and A. Farina (eds.). 2000. Special roundtable section on ecology and economics. *BioScience* 50:311–368.
> An excellent series of eight articles on integrating ecology and economics.

Clover, C. 2006. *End of the line: How overfishing is changing the world and what we eat.* New York: New Press.
> All fisheries are experiencing the cod scenario, as this book points out. An excellent read on the global impacts of overfishing.

Daily, G., ed. 1997. *Nature's services: Societal dependence on natural ecosystems.* Washington, D.C.: Island Press.
> In this excellent book, a group of eminent scientists explain in simple terms the critical role of natural ecosystems in meeting basic human needs.

Farley, J., J. Erickson, and H. E. Daly. 2003. *Ecological economics.* Washington, D.C.: Island Press.
> Places traditional neoclassical economic concepts in the context of economic growth, environmental degradation, and social inequity.

Gliessman, S. R., ed. 1990. *Agroecology: Researching the ecological basis for sustainable agriculture.* Ecological Studies Series no. 78. New York: Springer-Verlag.
> Provides examples of agricultural systems employed in different parts of the world, including tropical and temperate zones, and discusses the ecological issues associated with agricultural production.

Jaeger, W. K. 2005. *Environmental economics for tree huggers and other skeptics*. Washington, D.C.: Island Press.

Explains how to put economics to work for environmental causes.

Jenkins, M. B., ed. 1998. *The business of sustainable forestry: Case studies*. Chicago: J. D. and K. T. MacArthur Foundation.

This excellent volume integrates and analyzes a series of 21 case studies to provide a composite snapshot of the business of sustainable forestry; also presents management practices, techniques, and technologies.

Kurlansky, M. 1997. *Cod: The biography of the fish that changed the world*. New York: Walker and Company.

An outstanding book on the history of the cod, from abundance through scarcity through determined shortsightedness.

McNeeley, J. A., and S. J. Sheerr. 2002. *Ecoagriculture*. Washington, D.C.: Island Press.

Management of landscapes for production of food and conservation of wild diversity. Focuses mainly on tropical regions.

Orians, G., ed. 1986. *Ecological knowledge and environmental problem solving*. Washington, D.C.: National Academy Press.

An excellent introduction to the application of ecological science to addressing environmental problems; excellent case studies dealing with sustainable forestry and fisheries.

Pesek, J., ed. 1989. *Alternative agriculture*. Washington, D.C.: National Academy Press.

This volume includes 11 case studies describing in detail the practices and performance of alternative farming systems in the United States.

CHAPTER 28

Habitat Loss, Biodiversity, and Conservation

Roadways, such as this one through the Tijuca Forest in Brazil, can function as both a barrier to dispersal and a source of mortality for many species.

Scientists believe that approximately 65 million years ago—at the end of the Cretaceous period, in the region of the Yucatán Peninsula—a large meteorite struck Earth's surface, leaving a crater 180 km in diameter under the waters of the Caribbean. Evidence from deep-sea sediment cores reveals a remarkable record of the meteorite's impact and the resulting debris. The debris, blasted high into the atmosphere, may have triggered a major decline in Earth's temperature. Scientists now believe the assault by this massive asteroid or comet was largely responsible for the extinction of 70 percent of all species inhabiting Earth at that time, including the dinosaurs. In the period that followed, the various species that eventually dominated the oceans and land surface changed dramatically from the previous inhabitants.

Paleontologists refer to the loss of species at the end of the Cretaceous as a *mass extinction event*. It was not the only such event; Earth has undergone several mass extinctions, such as the one during the Permian (250 million years ago) when more than 50 percent of Earth's species disappeared from the fossil record, including 96 percent of all marine species (see Section 26.2 and Figure 26.5). The fossil record teaches us that these mass extinction events change the course of evolution, inducing a dramatic shift in the types of organisms inhabiting the planet.

Earth is currently undergoing a mass extinction on par with previous events, with an estimated annual loss of species in the thousands. The current mass extinction event, however, is different from those of the past. The cause is not extraterrestrial, such as from a meteor or comet; nor is it terrestrial, such as a change in sea level or climate. The destruction is due to human activities.

In North America's recent past, for example, hunting for food and other goods has led to the extinction of many mammal and bird species. Hunting has led to the extermination of marine mammals such as Stellar's sea cow (extinct about 1767), the New England sea mink (about 1880), and the Caribbean monk seal (about 1952). At a global scale, overkill has been found to be the main cause in virtually all 46 modern extinctions of large terrestrial mammals. Among birds, about 15 percent of the 88 modern species extinctions and 83 subspecies extinctions have been attributed to overkill. Affected species include the great auk (1844) and the passenger pigeon (1914). In some instances, overkill has resulted from the often mistaken belief that a wild species was a threat to gardens or to domestic animals. Victims of this belief include the Carolina parakeet (Figure 28.1), the only native U.S. parakeet (1914).

When compared to current rates of species extinction, however, the number of species extinguished by hunting and overexploitation is relatively small. By far, the largest and greatest threat to Earth's biological diversity is the alteration and destruction of habitat.

Figure 28.1 | The Carolina parakeet, the only native U.S. parakeet, went extinct in 1914 because of hunting and extermination.

28.1 | Habitat Destruction Is the Leading Cause of Current Species Extinctions

The primary cause of species extinctions is habitat destruction resulting from the expansion of human populations and activities. Historically, the largest cause of land transformation has been the expansion of agricultural lands to meet the needs of a growing human population (see Figure 27.1).

According to the United Nations Food and Agricultural Organization (Global Forest Resources Assessment 2000), the net loss in forest area at the global level during the 1990s was an estimated 94 million ha—an area equivalent to 2.4 percent of the world's total forests (Figure 28.2). However, global statistics tend to obscure significant changes in forest cover among regions and countries. Net deforestation rates were highest in West Africa and South America. This was followed by Asia, particularly in Southeast Asia. The 10 countries that lost the greatest amount of primary forest cover during the decade from 1990 to 2000 are given in Table 28.1.

Global distribution of original and remaining forests

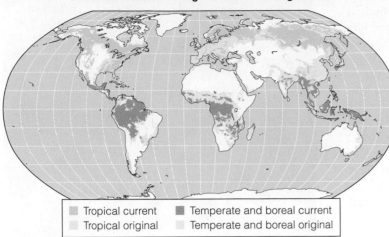

Tropical current   Temperate and boreal current
Tropical original   Temperate and boreal original

Figure 28.2 | Globally, about half of the forest that was present under modern (post-Pleistocene) climatic conditions, and before the spread of human influence, has disappeared—largely because of human activities.

(Adapted from United Nations, FAO.)

Because of their disproportionately high species diversity (see Section 26.3) and the pressures of growing populations and economic development, the world's tropical regions have been the primary focus of attention. The destruction of tropical rain forest has become almost synonymous with declining biodiversity. On a global scale, as much as 140,000 km^2 of tropical forest are being lost each year. In the Amazon region of Brazil, the rate of forest clearing during the 1980s exceeded 1 percent per year. Within this region alone, the extent of forest has decreased from 1 million km^2 to current estimates of just over 50,000 km^2. Forest clearing in Madagascar has resulted in the removal of more than 90 percent of the original forest cover (Figure 28.3). Likewise, 95 percent of the rain forest cover in western Ecuador (west of the Andes Mountains) has been destroyed since 1960.

From a conservation perspective, quantity of forests alone inadequately indicates the status of a forest ecosystem, because much of the world's forests are highly fragmented and face continued pressure from human activities. Although deforestation is widely recognized as a major conservation challenge, the related issue of habitat fragmentation has received less attention. As human pressures increase in both temperate and tropical forests, areas that were once continuously forested have become more fragmented (Figure 28.4; also see Chapter 19).

The changes in land use that bring about this destruction and large-scale extinction are not limited to the wet tropical regions. The clearing of tropical dry forest (see Section 23.3) for crop production and cattle grazing has all but eliminated this ecosystem from large areas of Central and South America, India, and Africa. The current distribution of dry forest cover in the Pacific coastal

Table 28.1 | **Changes in Forest Cover, 1990–2000**

Countries listed represent the top 10 in total area of forest cleared during that period.

Country	Total Forest 1990 ha × 10^6	Total Forest 2000 ha × 10^6	Forest Cover Change (1990–2000) Annual Change ha × 10^6	Annual Rate of Change (%)
Brazil	566,998	543,905	−2309	−0.4
Indonesia	118,110	104,986	−1312	−1.2
Sudan	71,216	61,627	−959	−1.4
Zambia	39,755	31,246	−851	−2.4
Mexico	61,511	55,205	−631	−1.1
Dem. Rep. of the Congo	140,531	135,207	−532	−0.4
Myanmar	39,588	34,419	−517	−1.4
Nigeria	17,501	13,517	−398	−2.6
Zimbabwe	22,239	19,040	−320	−1.5
Argentina	37,499	34,648	−285	−0.8

Source: FAO.

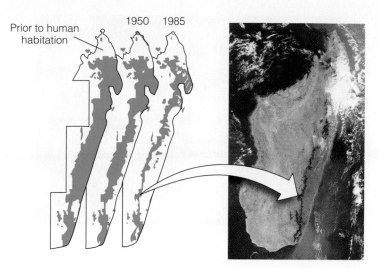

Prior to human habitation   1950   1985

Figure 28.3 | The series of maps represent the decline of rain forest in eastern Madagascar from the arrival of humans to the present. The photograph of Madagascar was taken from the space shuttle. The dark areas shown on the inland regions of the east coast are existing rain forest, which once extended to the coast.

region of Central America is less than 2 percent of its original extent.

Temperate grasslands are another type of ecosystem that has been heavily affected by agricultural activity. Once covering about 42 percent of Earth's land surface, natural grasslands have shrunk to less than 12 percent of their original size because of conversion to cropland and grazing lands. In North America, remaining areas of prairie have been heavily fragmented and widely scattered across the landscape, effectively becoming islands of native grassland within a sea of agricultural land.

Aquatic ecosystems have fared no better. Pollution of our inland waterways, dredging and filling of coastal wetlands, and destruction of coral reefs by pollution and siltation are having an effect similar to forest clearing on Earth's freshwater and coastal environments. Agricultural fertilizers, detergents, sewage, and industrial wastes add large amounts of nitrogen and phosphorus to aquatic ecosystems, resulting in cultural eutrophication (see Section 27.7). Although typically associated with lakes and ponds (enclosed bodies of freshwater), eutrophication is affecting marine coastal ecosystems as well. The

Figure 28.4 | Forest clearing in the Rondonia region of Brazil in the Amazon basin as viewed at several spatial scales. **(a)** The broad-scale view shows a pattern of linear clearings (light-colored areas) within the background of forest cover (dark green). **(b)** The linear pattern of clearing is associated with the development of roads for access to the area. **(c)** The forest is then cleared for the development of agricultural lands (primarily pasture).

(a)

(c)

(b)

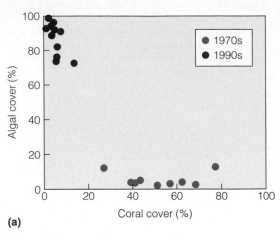

(a)

(b)

Figure 28.5 | **(a)** In the 1970s, coral dominated Jamaica's reef ecosystems; 20 years later, algae has taken over these same areas. The shift is due to water pollution and the overharvesting of algae-eating fish. **(b)** Algae growing on soft corals.

(Adapted from Hughes 1994; as in Primack 1998.)

Caribbean and Mediterranean seas are two examples of marine environments facing severe problems resulting from anthropogenic inputs of nutrients from areas of coastal development (Figure 28.5).

28.2 | Human-Introduced Exotic Species Threaten Many Native Species

Intentionally or unintentionally, humans have functioned as agents of dispersal for countless species of plants and animals, transporting them outside their natural geographic ranges (see Ecological Issues: Human-Assisted Dispersal, p. 195). Although many species fail to survive in their new homes, others flourish in their new environments. Freed from the constraints of their native competitors, predators, and parasites, they successfully establish themselves and spread. These nonnative plants and animals are referred to as exotic, alien, invasive, or nonindigenous species. There are approximately 50,000 nonindigenous species in the United States, and they cause major environmental damage and losses totaling about $137 billion annually.

Animal invaders often cause the extinction of vulnerable native species through predation, grazing, competition, and habitat alteration. Island species suffer the most. In Hawaii, for example, during the past 200 years, 263 native species disappeared; 300 are listed as endangered or threatened; and 1400 life-forms are in trouble or extinct. Among the island's 111 birds, 51 are extinct and 40 are endangered. On the Pacific island of Guam, the brown tree snake (*Boiga irregularis*), a native of New Guinea, accidentally reached the island around 1950, probably aboard military equipment transported there for dismantling. The snake has eliminated 9 of 12 native bird species, 6 of 12 native lizards, and 2 of the 3 native fruit bats.

Plant invaders, many introduced as horticultural plants, outcompete native species and alter fire regimes, nutrient cycling, energy budgets, and hydrology. In the United States some 5000 nonindigenous plant species exist in natural ecosystems that support over 17,000 native species. Introduction of exotics is the cause behind 95 percent of plant species loss and endangerment in Hawaii. Among the 1126 native flowering plants of Hawaii, 93 are extinct and 40 are endangered. On the North American continent, the ornamental perennial herb purple loose-strife (*Lythrum salicaria*; Figure 28.6a), originally introduced from Europe in the mid-1800s, has eliminated native wetland plants to the detriment of wetland wildlife. The Australian paperbark tree (*Melaleuca quinquenervia*; Figure 28.6b), introduced as an ornamental plant in Florida, is displacing cypress, sawgrass, and other native species in the Florida Everglades, drawing down water and fostering more frequent or intense fires. The most notorious plant invader in the United States is cheatgrass (*Bromus tectorum*; Figure 28.7), a winter annual accidentally introduced from Europe into Colorado in the 1800s. It arrived in the form of packing material and possibly crop seeds. It spread explosively across overgrazed rangeland and winter wheat fields in the Pacific Northwest and the Intermountain Region. By 1930 it became the dominant grass, replacing native vegetation. Susceptible to fire, its growth burns rapidly. Cheatgrass is a primary source of fire throughout the western United States.

The problem of invasive species is not restricted to terrestrial environments. Over 139 nonindigenous aquatic species that affect native plant and animal species have invaded the Great Lakes by the way of global shipping. Most

(a)

(b)

Figure 28.6 | Two invasive plant species that have significant negative impacts on native communities in North America: **(a)** the perennial herb purple loosestrife and **(b)** Australian paperbark tree.

notorious is the zebra mussel (*Dreissena polymorpha*; Figure 28.8), introduced from the ballast of ships traversing the St. Lawrence Seaway. Since it first appeared in 1998, the zebra mussel has spread to most eastern river systems. The San Francisco Bay Area is occupied by 96 nonnative invertebrates, from sponges to crustaceans. Exotic fish, introduced purposefully or accidentally, have been responsible for 68 percent of fish extinctions in North America during the past 100 years and for the decline of 70 percent of the fish species listed as endangered.

A classic example of how an invasive animal species can alter community structure was the accidental introduction of peacock bass (*Cichla ocellatus*), a native of the Amazon, into Gatun Lake in the Panama Canal Zone (see the introduction to Part Five). A popular sport and food fish and a voracious predator, the peacock bass escaped from impoundments. Its presence has had a devastating effect on the fish population and profoundly affected community structure (Figure 28.9). In the lake, the peacock bass feeds mostly on adult fish of the genus *Melaniris*, decreasing its populations. Other predatory species that feed on *Melaniris*, such as the Atlantic tarpon, black terns, and herons, have sharply declined. A complex community structure has been highly simplified. Six or eight common fish species within the community have been eliminated or seriously reduced, all by the introduction of a single top-level predator in the lake community.

Figure 28.7 | Cheatgrass (*Bromus tectorum*) grassland, converted from a sagebrush-dominated community by fire, in Grant County, Oregon. Cheatgrass, a nonindigenous invasive species, now covers thousands of acres of rangeland in the western United States.

Figure 28.8 | Zebra mussels, so called because of the striping on the shell, cover solid substrate, water intake pipes, and the shells of native mussels.

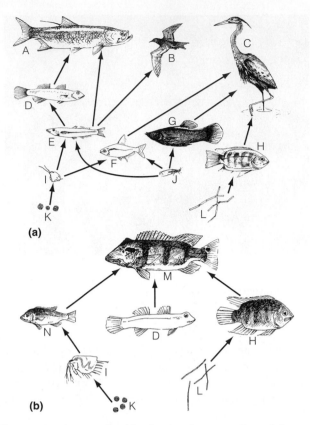

(a)

(b)

Figure 28.9 | Generalized food webs of common Gatun Lake populations contrasting **(a)** regions without or before introduction of *Cichla* (peacock bass) and **(b)** regions with *Cichla*. Key to species: A, *Tarpon atlanticus* (tarpon); B, *Chlidonias niger* (black tern); C, several species of herons and kingfishers; D, *Gobiomorus dormitor*; E, *Melaniris chagresi*; F, Characidae, including four common species; G, Poecillidae, including two common species—the exclusively herbivorous *Poecilia mexicana* and the exclusively insectivorous *Gambusia nicaraguensis*; H, *Cichlasoma maculicauda*; I, zooplankton; J, terrestrial insects; K, nanophytoplankton; L, filamentous green algae; M, adult *Cichla ocellaris*; N, young *Cichla*.

(After Zaret and Paine 1973).

28.3 | Species Differ in Their Susceptibility to Extinction

Not all species are equally susceptible to extinction from human activities. The susceptibility of species to extinction is related to various life history characteristics that influence the species' vulnerability to human activities as well as natural catastrophes.

A species with a geographically widespread distribution is referred to as **ubiquitous,** whereas a species found to occur naturally in a single geographic area and no place else is said to be **endemic** to that location. Endemic species are particularly susceptible to extinction because a loss of habitat in the one geographic region will result in a complete loss of habitat for the species.

Similarly, species with one or few local populations (small metapopulations) may become extinct because of chance factors, such as fire, flood, disease outbreak, or human activity (habitat destruction). Species with many local populations are less vulnerable to these events. This concept is central to metapopulation dynamics, discussed in Chapter 12.

Species that migrate seasonally (see Section 9.7) depend on two or more distinct habitat types in different geographic regions. If either of these habitats is destroyed, the species will not be able to persist. The more than 120 species of neotropical migrant birds that migrate each year between the temperate zone of eastern North America and the tropics of Central and South America (and the islands of the Caribbean) depend on suitable habitat in both locations (Figure 28.10) as well as stopover habitat in between. In addition to destroying habitats, barriers to migration can restrict a species from carrying out its life cycle. Dams along rivers in the Pacific Northwest of North America function as barriers to salmon populations attempting to swim upriver to spawn.

Figure 28.10 | Range map of the Nashville warbler (*Vermivora ruficapilla*). The spring and summer breeding range of the species is colored purple, and the winter habitat is blue.

Figure 28.11 | The shale-barren evening primrose, an endemic species found only on shale outcrops on south-facing slopes of the Allegheny Mountains in the Eastern United States.

Some species have very specialized habitat requirements, making them extremely susceptible to habitat alterations. Often, these habitats are scattered and rare across the landscape or region. For example, the shale-barren evening primrose (Figure 28.11) is a member of the evening primrose family (*Onagraceae*). This species is adapted to hot, shale-barren environments that form when certain types of shale form outcrops on south- to southwest-facing slopes of the Allegheny Mountains. Most members of this group are listed as endangered or threatened because they are found in these specific habitats only from southern Pennsylvania through West Virginia to southern Virginia, where shale barrens are formed.

Species requiring a large home range are often endangered due to habitat fragmentation. Although the habitat they require may be abundant, fragmentation may restrict the availability of contiguous patches that are large enough to support breeding populations.

Species that are hunted or come into conflict with human needs and activities are also vulnerable to extinction. Although hunting and collecting of species for human consumption has led to the endangerment and extinction of a variety of species (see the introduction and Figure 28.1), others have been eradicated because they either threaten human activities or are a direct threat to human lives. A prime example is species of large carnivores. In North America the wolf, grizzly bear, and mountain lion were hunted to near extinction because they posed a threat to livestock and human life. Current plans for reintroducing these species into areas of their former range are meeting resistance from the public (see Ecological Issues: The Wolves of Yellowstone National Park).

28.4 | Identifying Threatened Species Is Critical to Conservation Efforts

To define the status of rare and endangered species for the purpose of conservation, the International Union for the Conservation of Nature (IUCN) developed a quantitative classification based on the probability of extinction. The system has three levels:

1. *Critically endangered species* have a 50 percent or greater probability of extinction within 10 years or 3 generations, whichever is longer.

2. *Endangered species* have a 20 percent probability of extinction within 20 years or 5 generations.

3. *Vulnerable species* have a 10 percent or greater probability of extinction within 100 years.

Assigning a species to one of these categories requires having at least one of the following types of information:

- Observable decline in numbers of individuals
- The geographic area occupied by a species and the number of populations
- The total number of individuals alive and the number of breeding individuals
- The expected decline in the numbers of individuals if current and projected trends in population decline or habitat destruction continue
- The probability of the species going extinct in a certain number of years or generations

Despite the limitation of any such classification system, the advantage of the IUCN system is that it provides a standard, quantitative method of classification that the scientific community can use in reviewing and evaluating decisions. In addition, assigning categories based on habitat loss is particularly useful for species whose life history is largely unknown.

The World Conservation Union of the IUCN, through its Species Survival Commission (SSC), has for four decades been assessing the conservation status of species on a global scale. The SCC highlights those species and taxonomic groups that are threatened with extinction and promotes their conservation. The IUCN Red List of Threatened Species provides conservation status and distribution information on species that have been evaluated using the IUCN classification as just described. Table 28.2 (p. 608) presents the number of threatened species worldwide by major taxonomic group.

When the Yellowstone National Park was established in 1872, the gray wolf (*Canis lupus*; Figure 1) was a key component of the Yellowstone ecosystem. With the mountain lion and grizzly bear, it functioned as one of three top predators in the ecosystem. A social animal living and hunting in packs, the gray wolf fed on the populations of bison, elk, mule deer, and moose. By the early 1900s, however, predator control was being practiced in the park. By the 1940s reports of wolf sightings within the park were rare; and by the 1970s, park scientists found no evidence of wolves in the park.

Outside the park, gray wolves persisted in the lower 48 states only in areas of northern Minnesota and on Isle Royale in Michigan. In contrast, healthy populations continued to flourish in Canada and Alaska. By the 1980s, however, as wolves began to reestablish themselves in northern Montana, Idaho, and Washington states, discussions began about the possibility of reestablishing wolves in the Yellowstone ecosystem.

As early as 1987, as part of the larger Northern Rocky Mountain Wolf Recovery Plan, the U.S. Fish & Wildlife Service had proposed reintroducing an "experimental population" of wolves into Yellowstone and surrounding government lands. National Park Service policy calls for restoring native species when (1) sufficient habitat exists to support a self-perpetuating population, (2) management can prevent serious threats to outside interests, and (3) the original decline of the species resulted from human activities.

In October 1991, when Congress funded the U.S Fish & Wildlife Service in preparing a required environmental impact assessment on restoring wolves to Yellowstone, opposition was immediately voiced. Ranchers argued that the wolves would pose a threat to livestock on adja-

Figure 1 | Gray wolves were reintroduced to Yellowstone National Park in 1995.

cent private and public land (the latter being lands leased for grazing). Hunters objected to the idea because wolves would reduce the region's populations of deer, elk, and moose. Logging and mining companies feared that the presence of this protected species might limit their access to federal lands.

Finally, after several years and a near-record number of public comments, the Secretary of Interior signed the order allowing for the reintroduction of gray wolves to Yellowstone. In 1995, fourteen wolves were released into Yellowstone National Park. In 1996, seventeen more wolves were brought from Canada and released. After the 1996 release, plans to transfer additional wolves were terminated due to reduced funding and to the wolves' unexpected early reproductive success.

As anticipated, the presence of the wolves in Yellowstone is altering many ecological aspects of this ecosystem. Already, wolves have dramatically lowered the coyote population, which had become the top predator in the absence of wolves. The reduction in the coyote population will no doubt influence other species. Other findings indicate that wolves have influenced the behavior and movement patterns of elk, which now congregate in larger herds.

For many years to come, scientists will be studying the effects of reintroducing these important predators to their former habitat. It is clear, however, that wolf restoration in Yellowstone has been successful beyond all expectations. In 2001, the species reached the minimum population size of 30 breeding pairs in the three recovery areas (Yellowstone, central Idaho, northwest Montana). This population size is required for being declassified as "endangered." If wolves maintain these population levels and management plans are approved, they can finally be reclassified under the Endangered Species Act. By 2007 the population had reached 1294 wolves and 86 breeding pairs, but wolves have not been reclassified. If the species were removed from the endangered species list, Montana, Idaho, and Wyoming would have full management authority over wolves. The federal government did cede most wolf management to Idaho and Montana, but not to Wyoming. Nevertheless, attitudes toward the wolf in Idaho threaten to undo wolf recovery there. Strong ill will among ranchers, hunters, and politicians toward wolves in Wyoming prevent any reclassification in that state. The Endangered Species Act not withstanding, societal values will determine the fate of the wolves.

1. Efforts are currently under way to reintroduce the grizzly bear to federal lands in the western United States (see U.S. Fish & Wildlife Service http://mountain-prairie.fws.gov/bitterroot/). What arguments are being posed both in favor of, and in opposition to, the proposed reintroduction?

The number of individuals in a population is generally greater than the number that is actually contributing genes to the next generation. In polygamous populations (see Section 8.3), for example, a few dominant males are responsible for all mating, so the alleles of these males contribute disproportionately to the following generations. From a genetic viewpoint, in such populations the nonbreeding males might as well not exist. For this reason, the actual size of a small population or a subpopulation means little. Of greatest importance is the genetically effective population size that is passing genes to successive generations.

When only a subset of the individuals within a population successfully breed, the effective population size is given by:

Effective population size — Number of breeding females

$$N_e = (4\ N_m N_f)\ /\ (N_m + N_f)$$

Number of breeding males

For example, if we have a population of 200 elk, with 20 breeding males and 180 breeding females, the effective population size is 72 $[(4 \times 20 \times 180)/(20 + 180) = 72]$, rather than 200.

1. What is the effective population size of a population of 80 breeding females and only 20 breeding males?

2. How would the effective population size differ for the same population size of 100, but with 50 breeding females and 50 breeding males (monogamous)?

28.5 | Regions of High Species Diversity Are Crucial to Conservation Efforts

There are more than 1.4 million known and described species on Earth. Many scientists, such as the Harvard biologist E. O. Wilson, believe that the actual number of species may be closer to 10 times that number. As we discussed in Chapter 26 (Section 26.3), Earth's biological diversity is not distributed evenly over the land surface and oceans of the planet. There is a distinct gradient of increasing species richness from the poles to the equator for both terrestrial and marine species. For example, although tropical rain forests cover only 7 percent of the land surface, more than one-half of all known plant and animal species inhabit these ecosystems. In addition to the distinct latitudinal gradient in diversity, regions characterized by topographic variation (such as ridges and valleys) support high numbers of species relative to flatter areas within the same geographic region (see Section 26.3). This pattern is most likely related to the diversity of habitats available in these areas.

Complicating the interpretation of broad-scale patterns of diversity is the fact that most of Earth's species are endemic—they have small, restricted geographic ranges. Out of the world's 10,000 or so bird species, more than 2500 are endemic, being restricted to a range smaller than 50,000 km². Species of flora endemic to a single country represent 46–62 percent of the world flora. Of the thousands of new species being identified each year, virtually all of them are endemic to very restricted regions within the tropics. It is the restricted distribution of these species that makes them extremely vulnerable to human activities that degrade or destroy their habitats. Of the species classified as threatened by the IUCN (see Table 28.2), 91 percent are endemic.

As with general patterns of species richness, endemic species are not distributed evenly over Earth's surface or even within a geographic region (Figure 28.12). Certain regions of the world exhibit both high species richness and endemism. British ecologist Norman Myers defined these regions of unusually high diversity as **hotspots.** Myers developed the concept of biodiversity hotspots in 1988 to address the dilemma that conservationists face: What areas are the most important for preserving species?

The designation of a region as a hotspot of biological diversity is based on two factors: overall diversity of the region and significance of impact from human activities.

Status of Threatened is based on IUCN Classification of Threatened and Endangered Species.

	Number of described species	Number of species evaluated in 2006	Number of threatened species in 2003	Number of threatened species in 2004	Number of threatened species in 2006	Number threatened in 2006, as % of species described	Number threatened in 2006, as % of species evaluated**
Vertebrates							
Mammals	5,416	4,856	1,130	1,101	1,093	20%	23%
Birds	9,934	9,934	1,194	1,213	1,206	12%	12%
Reptiles	8,240	664	293	304	341	4%	51%
Amphibians	5,918	5,918	157	1,770	1,811	31%	31%
Fishes	29,300	2,914	750	800	1,173	4%	40%
Subtotal	**58,808**	**24,284**	**3,524**	**5,188**	**5,624**	**10%**	**23%**
Invertebrates							
Insects	950,000	1,192	553	559	623	0.07%	52%
Mollusks	70,000	2,163	967	974	975	1.39%	45%
Crustaceans	40,000	537	409	429	459	1.15%	85%
Others	130,200	86	30	30	44	0.03%	51%
Subtotal	**1,190,200**	**3,978**	**1,959**	**1,992**	**2,101**	**0.18%**	**53%**
Plants							
Mosses	15,000	93	80	80	80	0.53%	86%
Ferns and allies	13,025	212	111	140	139	1%	66%
Gymnosperms	980	908	304	305	306	31%	34%
Dicotyledons	199,350	9,538	5,768	7,025	7,086	4%	74%
Monocotyledons	59,300	1,150	511	771	779	1%	68%
Subtotal	**287,655**	**11,901**	**6,774**	**8,321**	**8,390**	**3%**	**70%**
Others							
Lichens	10,000	2	2	2	2	0.02%	100%
Mushrooms	16,000	1	—	—	1	0.01%	100%
Subtotal	**26,000**	**3**	**2**	**2**	**3**	**0.01%**	**100%**
TOTAL	**1,562,663**	**40,168**	**12,259**	**15,503**	**16,118**	**1%**	**40%**

Source: IUCN.

Plant diversity is the biological basis for hotspot designation; to qualify as a hotspot, a region must support 1500 or more endemic plant species (0.5 percent of the global total), and the region must have lost more than 70 percent of its original habitat. Plants have been used as qualifiers because they are easy to identify and census as well as the basis of diversity in other taxonomic groups.

The 25 biodiversity regions of Earth that have been designated as hotspots by the IUCN (Figure 28.13) contain 44 percent of all plant species and 35 percent of all terrestrial vertebrate species in only 1.4 percent of the planet's land area. Several hotspots are tropical island archipelagos, like the Caribbean and the Philippines, or relatively large islands, like New Caledonia. However, other hotspots are continental islands. They are effectively isolated, being surrounded by deserts, mountain ranges, and seas. The Cape Floristic Province in South Africa is isolated by the aridity of the Kalahari, Karoo, and Namib deserts. Other landlocked islands are high mountains and mountain ridges. For communities inhabiting mountain ridges in the Tropical Andes (South America) and the Caucasus (central Asia), the lowlands are insurmountable barriers to dispersal.

By far the largest proportion of the 121,000 potentially threatened species in the tropical regions are endemic to countries within the 25 designated biodiversity hotspots. In these areas, high diversity and massive habitat loss coincide.

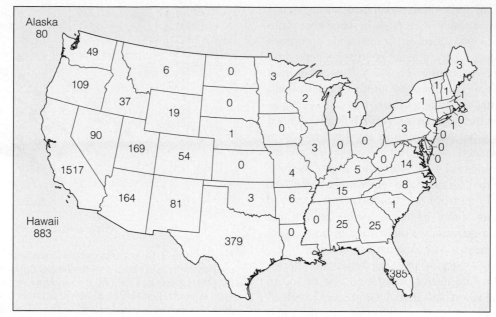

Figure 28.12 | The number of plant species endemic to the different states of the United States. (Adapted from Gentry 1986.)

28.6 | Protecting Populations Is the Key to Conservation Efforts

Because endangered species consist of a few—or even a single—local populations (see Chapter 12), protecting populations is the key to preserving these species. Often, these populations are restricted to protected areas (nature reserves, etc.), and an adequate conservation plan requires the preservation of as many individuals as possible within the greatest possible area. But with limited land and resources, conservation ecologists must address the question: What sizes of populations are needed to save the species?

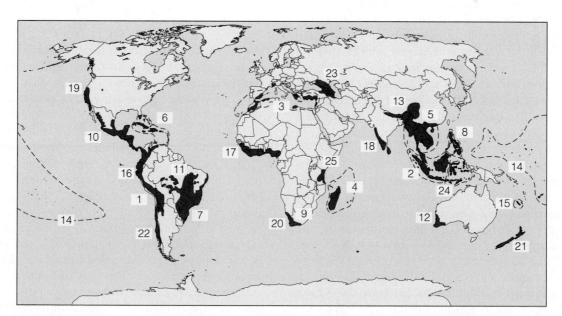

Figure 28.13 | Twenty-five areas of the world that the IUCN has designated as biodiversity hotspots. 1, Tropical Andes; 2, Sundaland; 3, Mediterranean Basin; 4, Madagascar and Indian Ocean Islands; 5, Indo-Burma; 6, Caribbean; 7, Atlantic Forest Region; 8, Philippines; 9, Cape Floristic Province; 10, Mesoamerica; 11, Brazilian Cerrado; 12, Southwest Australia; 13, Mountains of south-central China; 14, Polynesia/Micronesia; 15, New Caledonia; 16, Chocó-Darién of Western Ecuador; 17, Guinean forests of West Africa; 18, Western Ghats and Sri Lanka; 19, California Floristic Province; 20, Succulent Karoo; 21, New Zealand; 22, Central Chile; 23, Caucasus; 24, Wallacea; 25, Eastern Arc Mountains and coastal forests of Tanzania and Kenya.

(Adapted from Myers et al. 1991.)

The number of individuals needed to ensure that a species persists in a viable state must be large enough to cope with chance variations in demographic processes (births and deaths), environmental changes, genetic drift, and catastrophes (see Chapters 5 and 10). The conservation ecologist M. L. Shaffer defined the number of individuals necessary to ensure the long-term survival of a species as the **minimum viable population (MVP).** Shaffer defined the MVP for any given population in any given habitat as the "smallest isolated population having a 99 percent chance of remaining extant for 1000 years despite the fore-seeable effects of demographic and environmental stochasticity, and natural catastrophes." Although Shaffer realized the tentative nature of this definition, the key point is that MVP size allows a quantitative assessment of how large a population must be to assure long-term survival.

Genetic models suggest that for vertebrate species, populations with an effective population size of 100 or less and an actual size of less than 1000 are highly vulnerable to extinction. For species exhibiting extreme variation in population size, such as invertebrates and annual plants, it has been suggested that minimum viable populations of 10,000 individuals would be required.

In fact, the actual MVP for a species depends on the life history of the species (longevity, mating system, etc.) and the ability of individuals to disperse among habitat patches. Despite the difficulty in quantifying MVP for a given species, the concept is of paramount importance in conserving species and maintaining biological diversity.

Once a minimum viable population size has been established for a species, the area required to support that population must be considered. The area of suitable habitat necessary for maintaining the minimum viable population is called the **minimum dynamic area (MDA).** An estimation of MDA for a species begins with an understanding of the home-range size of the individuals, family groups, or colonies. Recall from Chapter 11 (Section 11.9 and Figure 11.17) that the area (home range) requirement of an individual of a species increases with body size. In addition, for a given body size, the home-range requirement of a carnivore is larger than that of an herbivore. With knowledge of the area requirement per individual of a species, and an estimate of MVP, the area necessary to sustain a viable population can be established (Figure 28.14). For large carnivores, the area required to sustain an MVP can be enormous. Wildlife biologist Reed Noss estimated that to preserve a population of 1000 grizzly bears would require an area of 2 million km². This is why most large carnivore populations (such as the African lion, Asian tiger, and gray wolf of North America) are endangered, and why they are restricted to only the largest public lands and nature reserves.

One of the best-documented cases of minimum viable population size comes from J. Berger's study of the persistence of bighorn sheep (*Ovis canadensis*) populations in the deserts of the southwestern United States. The

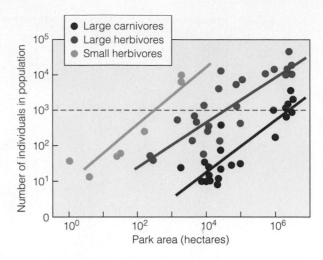

Figure 28.14 | Large parks contain larger populations of each species than small parks; only the largest parks may contain long-term viable populations of many large vertebrate species. Each symbol represents an animal population in a park. If the MVP of a species is 1000 (dashed line), parks of at least 100 ha will be needed to protect small herbivores. Parks of more than 10,000 ha will be needed to support populations of large herbivores, and parks of at least 1,000,000 ha will be needed to protect populations of large carnivores.

(Adapted from Schonewald-Cox 1983.)

Interpreting Ecological Data

Q1. What is the approximate size of a park needed to support populations of large herbivores? To protect populations of large carnivores?

Q2. Why is a larger area required to support a viable population of carnivores than is needed to support a similar-sized population of herbivores?

study, which examined 120 populations, found that all populations with 50 individuals or less went extinct within 50 years. In contrast, virtually all of the populations with 100 or more individuals persisted over that same period (Figure 28.15). No single cause was identified for the local extinctions; rather, a wide variety of factors appear to be responsible for the population declines.

Species rarely occur in a single contiguous population. Given the fragmented nature of most landscapes (see Chapter 19) and the often specific habitat needs of a given species, species frequently consist of a set of semi-isolated subpopulations that are connected by dispersal—metapopulations (see Chapter 12). The persistence of a metapopulation is the result of a complex dynamic among subpopulations. The rates of birth, death, immigration, and emigration of each subpopulation interact with the size and spatial arrangement of habitat patches to determine dynamics of the metapopulation as a whole. In many metapopulations, some areas function as source populations (or patches), and others function as sinks (see Section 12.5). In source patches, the local reproductive rate exceeds the local mortality rate, leading to a surplus of individuals that are able to colonize other

(a)

(b)

Figure 28.15 | Relationship between the size of a population of bighorn sheep (shown in (a)) and the percentage of populations that persist over time. The number on the graph (b) (*N*) represents population size. Almost all populations of 100 or more sheep persisted beyond 50 years, whereas populations with fewer than 50 individuals did not.

(Adapted from Berger 1990.)

patches of habitat. In sink patches, mortality rate exceeds reproduction rate so that the local population will go extinct if it is not regularly recolonized by new immigrants. Dependence on key source patches is a critical feature of such metapopulations. Identifying key source patches and the corridors (see Section 19.5) that link with them is critical to species conservation. Destroying one central core population might result in the extinction of many smaller subpopulations that depend on the core popula-

tion for periodic colonization. The case study of the Iberian lynx (*Lynx pardinus*) in Chapter 12 provides an example (see Ecological Issues: The Metapopulation Concept in Conservation Ecology, p. 243) of the importance of understanding metapopulation structure in relation to species conservation.

28.7 | Reintroduction Is Necessary to Reestablish Populations of Some Species

In some cases, saving species from what appears to be an inevitable decline to extinction requires direct action by conservation ecologists: establishing new populations through transplants and reintroductions. In these cases, conservation biologists move individuals from one location to another (Figure 28.16). The two species of African rhinoceros, the white rhino (*Ceratotherium simum*) and the black rhino (*Diceros bicornis*), provide an example of two efforts at reintroduction that have met with contrasting results.

The southern white rhino was at the brink of extinction—perhaps as few as 50 survived at the beginning of the 20th century. Like other rhinos, farmers and hunters slaughtered the southern white rhino, and poaching continues to be a problem in protected areas. Thanks to conservation efforts, white rhino numbers have recovered to an estimated 7500, partly due to an intensive program of relocating individuals from the few viable populations remaining within the region.

Following the success of its early initiatives, the KwaZulu–Natal Board of South Africa launched an ambitious project dubbed "Operation Rhino" in 1961. This effort involved transporting surplus numbers of white rhinos from the Hluhluwe–Umfolozi Park in KwaZulu–Natal to other protected areas. By the end of 1999, a total

Figure 28.16 | Black rhino captured at Umfolozi Reserve in Natal, South Africa, being released into a newly established reserve at Pilansburg, South Africa.

Figure 28.17 | Map showing the former and current distribution of the African black rhino.

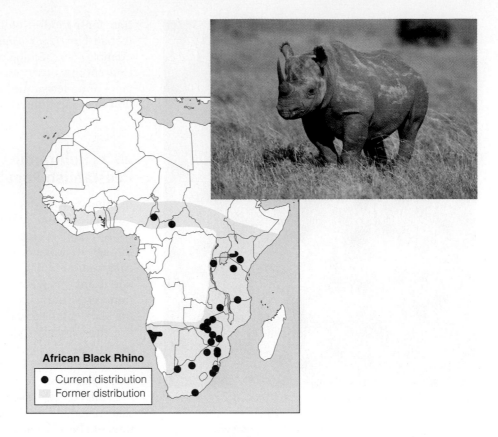

African Black Rhino

● Current distribution
▨ Former distribution

of 2367 white rhinos had been redistributed worldwide, of which 1262 had been rehabilitated in southern African protected areas.

The story of the black rhino is, unfortunately, different. From a total population estimated at around 65,000 in 1970, less than 2500 remained in the mid-1990s (Figure 28.17). Black rhinos are scattered across central and southern Africa in populations too small to be self-sustaining or in large populations that can be vulnerable to outbreaks of poaching, disease, or overpopulation for the size of the reserve.

In a massive program of capture, relocation, and release across the African continent, black rhinos are being moved from parks like Etosha in Namibia and Hluhluwe–Umfolozi in South Africa, which have been their successful safe havens over the last three decades, and sent to new sites on private, public, and communal lands. It is hoped that old and new populations alike will grow quickly and continue to interbreed just as they did before poaching decimated the species.

Unfortunately, black rhinos are reluctant to use unfamiliar habitat. When groups of individuals are introduced into new areas, aggressive interactions often result in deadly fights among individuals. Over half of the deaths after release can be attributed to these fights among newly released individuals, presenting an ongoing difficulty in reestablishment efforts.

Restoration of some species such as whooping crane (*Grus americana*), masked bobwhite (*Colinus virginianus*),

Hawaiian goose or nene (*Branta sandwicensis*), peregrine falcon (*Falco peregrinus*), and California condor (*Gymnogyps californianus*; Figure 28.18)—and, among the mammals, the wolf and European wisent (*Bison bonasus*)—have relied on the introduction of individuals from captive-bred populations. Introducing captive-bred individuals to the wild requires prerelease and post-release conditioning, including the acquiring food, finding shelter, interacting with other individuals of their own species, and learning to fear and avoid humans. Despite these inherent difficulties, reintroduction programs have succeeded in halting, and in some cases reversing, the downward spiral toward extinction.

28.8 | Habitat Conservation Functions to Protect Whole Communities

Despite the necessity of focusing conservation efforts on individual species that are threatened and endangered, the most effective way to preserve overall biological diversity is by protecting habitats or whole ecological communities. In fact, it is most likely the only way we can successfully conserve Earth's biological diversity, given our limited understanding of the natural history of most species and the complex nature of interactions among species within the context of the community (see Chapter 17).

Unlike the population approach to conservation, which focuses on the specific habitat needs and protection of single species, a community-based approach re-

(a)   (b)

Figure 28.18 | **(a)** Staff at the San Diego Wild Animal Park Condor Breeding Facility release adult California condors into a holding area. **(b)** A 12-hour-old, captive-hatched California condor chick that will eventually be released into the wild.

quires an understanding of the relationship between overall patterns of biological diversity and features of the landscape. One key element to designing a program to protect a region's overall species diversity is an understanding of the relationship between area and species richness, as discussed in Chapter 19 (Section 19.3).

As noted first in Section 19.3 (also see Quantifying Ecology 26.1: Quantifying Biodiversity: Comparing Species Richness Using Rarefaction Curves, p. 564), large areas generally contain more species than do smaller areas. There are several reasons for this observed relationship between species richness and area. First, larger areas are often more heterogeneous, encompassing a greater variety of habitats than do smaller areas. Larger areas can thus provide for the needs of a greater variety of species. In addition, as vegetation changes due to natural processes such as succession (Chapter 18), or in response to periodic disturbances (fire, drought, etc.), a more heterogeneous landscape provides a greater probability that a given species can find an area of suitable habitat (see Section 12.4 for an example in the context of metapopulation dynamics).

Second, some species require larger areas to meet their basic resource needs. For example, larger organisms have larger home ranges than do smaller species (see Section 11.9, Figure 11.17, and Figure 28.14), therefore requiring a larger area of habitat to maintain minimum viable populations (see Section 28.6).

As discussed in detail in Section 19.3, smaller areas have a greater amount of edge (greater edge-to-area ratio), and edge environments have unique environmental constraints involving microclimate and contact with predators, pests, and disease. In addition, some species—interior species—require the environmental conditions found only in larger, contiguous tracts of habitat and are particularly sensitive to edge environments (see Figure 19.10).

Finally, many species are locally rare (see Section 10.13) and require a larger area even to be present in small numbers.

Given the general relationship between species diversity and area, it is apparently better for purposes of conserving overall biological diversity to protect as large an area as possible. However, an early debate among conservation ecologists questioned whether species richness would be maximized in one large area of land (preservation area) or in several smaller patches of an equal total area. Proponents of large protected reserves argued that such areas minimize edge effects, encompass the greatest diversity of habitats, and are the only reserves that can contain sufficient numbers of large, low-density species (such as large carnivores) to maintain long-term populations. On the other hand, once an area is larger than a certain size, the number of new species added with each successive increase in area declines (see Quantifying Ecology 26.1: Quantifying Biodiversity: Comparing Species Richness Using Rarefaction Curves, p. 564). In that case, establishing a second area some distance away may be a better strategy for preserving additional species than merely adding area to the existing reserve. In addition, a network of smaller areas positioned over a larger region may be able to include a greater variety of habitat types and more rare species. It also may be less susceptible to single catastrophic events—such as fire, flood, disease, or the introduction of alien species—than would a single, contiguous block of land.

The consensus among conservation ecologists now seems to favor a mixed strategy. Larger areas are required for preservation of larger species, but a network of reserves may well be a better solution to long-term preservation of species. A major force behind this shift in thinking has been the development of metapopulation biology (see Chapter 12).

28.9 | Habitat Conservation Involves Establishing Protected Areas

Given the ever-growing pressures placed on lands by the human population, preservation of biological diversity depends more and more on establishing legally designated protected areas. Protected areas can be established in various ways, but the two most common are through governmental action (at the national, regional, and local levels) and the purchase of lands by private individuals or conservation organizations (such as the Nature Conservancy and the Audubon Society).

Land classified as protected areas covers a wide range of categories. The International Union for the Conservation of Nature (IUCN) has developed a classification system for protected areas that covers a range from minimal to intensive use of the habitat by humans (Table 28.3).

Of these categories, the first five can be considered as truly protected areas, managed primarily for biological diversity. Category VI, managed resource areas, is particularly significant: these areas are often much larger than protected areas and often still contain much of their original biological diversity, despite being managed for multiple purposes.

For example, in the United States (Figure 28.19a), the U.S. Park Service administers approximately 350 areas, including the national parks and covering a total land area of 125,000 square miles. These lands are designated as protected—dedicated to providing for a combination of low-impact recreation and preservation of biological diversity. In addition, there are some 530 national wildlife refuges covering 95 million acres. In contrast, the U.S. Department of Agriculture oversees some 155 national forests covering an area of 300,000 square miles,

---

Table 28.3 | IUCN Protected Area Management Categories

CATEGORY Ia

Strict Nature Reserve: protected area managed mainly for science

Area of land and/or sea possessing some outstanding or representative ecosystems, geological or physiological features, and/or species, available primarily for scientific research and/or environmental monitoring.

CATEGORY Ib

Wilderness Area: protected area managed mainly for wilderness protection

Large area of unmodified or slightly modified land and/or sea, retaining its natural character and influence without permanent or significant habitation, which is protected and managed so as to preserve its natural condition.

CATEGORY II

National Park: protected area managed mainly for ecosystem protection and recreation

Natural area of land and/or sea, designated to (a) protect the ecological integrity of one or more ecosystems for present and future generations, (b) exclude exploitation or occupation inimical to the purposes of designation of the area and (c) provide a foundation for spiritual, scientific, educational, recreational, and visitor opportunities, all of which must be environmentally and culturally compatible.

CATEGORY III

Natural Monument: protected area managed mainly for conservation of specific natural features

Area containing one or more specific natural or natural/cultural features of outstanding or unique value because of their inherent rarity, representative or aesthetic qualities, or cultural significance.

CATEGORY IV

Habitat/Species Management Area: protected area managed mainly for conservation through management intervention

Area of land and/or sea subject to active intervention for management purposes so as to ensure the maintenance of habitats and/or to meet the requirements of specific species.

CATEGORY V

Protected Landscape/Seascape: protected area managed mainly for landscape/seascape conservation and recreation

Area of land, with coast and sea as appropriate, where the interaction of people and nature over time has produced an area of distinct character with significant aesthetic, ecological, and/or cultural value, often with high biological diversity. Safeguarding the integrity of this traditional interaction is vital to the protection, maintenance, and evolution of such an area.

CATEGORY VI

Managed Resource Protected Area: protected area managed mainly for the sustainable use of natural ecosystems

Area containing predominantly unmodified natural systems, managed to ensure long-term protection and maintenance of biological diversity while at the same time providing a sustainable flow of natural products and services to meet community needs.

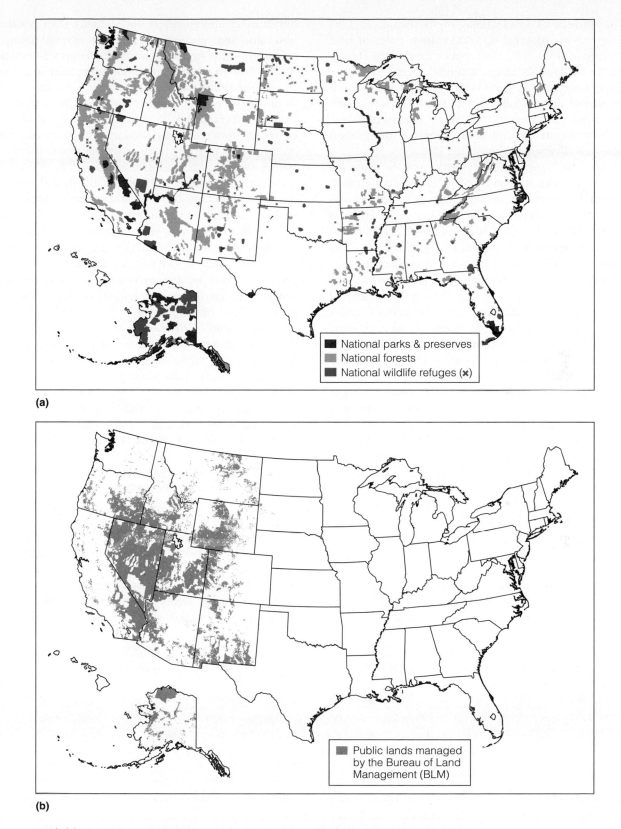

(a)

National parks & preserves
National forests
National wildlife refuges (✖)

(b)

Public lands managed
by the Bureau of Land
Management (BLM)

Figure 28.19 | (a) Map showing the distribution of national parks, wildlife refuges, and forests within the United States. **(b)** Map showing the distribution of U.S. Bureau of Land Management public lands.

and the Bureau of Land Management (Figure 28.17b) oversees the use of some 425,000 square miles of land (11.7 percent of the total land area of the United States), mostly in the Western states. National forests as well as Bureau of Land Management lands are managed for multipurpose use that includes timber harvest, grazing, and the extraction of mineral and water resources. These lands, however, are still critical to the overall preservation of biological diversity in the United States.

As of 2003, approximately 9900 strictly protected areas (categories Ia, Ib, and II of the IUCN classification) have been designated worldwide, covering some 6.46 million km². An additional 54,000 partially protected areas (categories III–V of the IUCN classification) combine to cover an additional 4.35 million km² (Table 28.4). That may seem like a large area of land, but most of the protected areas are relatively small; half of them cover an area of 100 km² or less (Figure 28.20). The combined total of protected lands accounts for only 11.5 percent of Earth's total land surface. Of the total protected land area, some 67 percent has been classified using the IUCN system.

Marine protection efforts have lagged behind conservation of terrestrial environments. Currently, only 1 percent of the marine environment is included in protected areas.

With few exceptions, most large tracts of land that will function as future protected areas already fall within one of the six classes of protected lands as defined by the IUCN. However, new and smaller reserves are being established throughout the world, and lands under limited protection (such as national forests in the United States) are continuously being reclassified into categories of increased protection. For example, in 2002 more than 10,000 acres of the George Washington National Forest (Virginia) were designated as wilderness areas (Priest and Three Ridges), the highest designation of protection for federal lands within the United States.

However, many current conservation efforts focus on working with the existing protected lands, providing buffer zones and corridors that enhance their conservation value. Strategies exist for aggregating small nature reserves and other protected areas into larger conservation blocks. Nature reserves are often embedded in a larger matrix of habitat managed for resource extraction, such as timber harvest, grazing, or farmland (refer to Part 5, Community Ecology). If protecting biological diversity can be incorporated as a secondary priority into the management plan of these lands, a greater representation of species and habitats can be protected.

Whenever possible, a protected area should include a uniform, contiguous block of land or water, such as a watershed, lake, or mountain range. This will allow managers to more effectively control the spread of fire, pests, and destructive outside influences due to human activity.

A new approach to managing a system of nature reserves is to link isolated protected areas into one large system through the use of habitat corridors—areas of protected land running between the reserves (see Section 19.5 for a detailed discussion of corridors). Such corridors can facilitate the dispersal of plants and animals from one reserve to another. Corridors may assist species that migrate seasonally to different habitats to obtain food or breed. This principle was put into practice in Costa Rica to link two wildlife reserves, the Braulio Carillo National Park and La Selva Biological Station. A 7700-ha corridor of forest several kilometers wide, known as La Zona Protectora, was set aside to provide a link that allows at least 35 species of birds to migrate between the two conservation areas.

The idea of corridors is intuitively appealing, but there are some possible drawbacks. Corridors also may facilitate the movement of fire, pest species, or disease.

In some rare cases, actions are being taken to link established protected areas that go beyond the limited use of corridors. Efforts are under way in southern Africa to establish the Great Limpopo Transfrontier Park (Figure 28.21), an international conservation effort that involves linking three

Table 28.4 | **Global Number and Extent of IUCN Classified Protected Areas by Category as of 2003**
(See Table 28.3 for category descriptions.)

Category	No. of sites	Proportion of total no. protected areas (%)	Area covered (km2)	Proportion of total area protected (%)
Ia	4,731	4.6	1,033,888	5.5
Ib	1,302	1.3	1,015,512	5.4
II	3,881	3.8	4,413,142	23.6
III	19,833	19.4	275,432	1.5
IV	27,641	27.1	3,022,515	16.1
V	6,555	6.4	1,056,008	5.6
VI	4,123	4.0	4,377,091	23.3
No category	34,036	33.4	3,569,820	19.0
Total	102,102	100	18,763,407	100

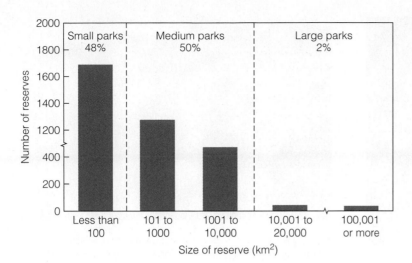

existing national parks in South Africa (Kruger National Park), Zimbabwe (Gonarezhou National Park), and Mozambique (Coutada Wildlife Area, Banhine and Zinave National Parks). Establishing the transfrontier park will create a conservation area measuring 100,000 square kilometers, making it one of the world's largest contiguous nature reserves.

28.10 | Habitat Restoration Is Often Necessary in Conservation Efforts

In recent years, considerable efforts have been under way to restore natural communities affected by human activi-

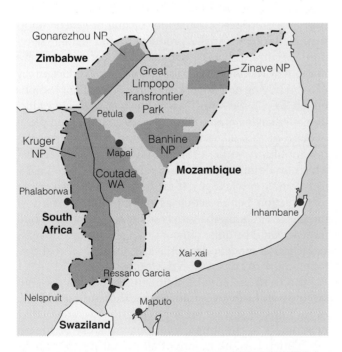

Figure 28.21 | Map of the Great Limpopo Transfrontier Park, an international conservation effort that involves linking three existing national parks in South Africa (Kruger National Park), Zimbabwe (Gonarezhou National Park), and Mozambique (Coutada Wildlife Area, Banhine and Zinave National Parks).

ties. This work has stimulated a new approach to human intervention that is termed **restoration ecology.** The goal of restoration ecology is to return an ecosystem to a close approximation of its condition before disturbance, by applying ecological principles. The approach involves a continuum of approaches ranging from reintroducing species and restoring habitats to attempting to reestablish whole communities as functioning ecosystems.

The least intensive restoration effort involves the rejuvenation of existing communities by eliminating invasive species, replanting native species, and reintroducing natural disturbances such as short-term periodic fires in grasslands and low-intensity ground fires in pine forests. Lake restoration involves reducing inputs of nutrients, especially phosphorus, from the surrounding land that stimulate growth of algae; restoring aquatic plants; and reintroducing fish species native to the lake. Wetland restoration may involve reestablishing the hydrological conditions, so that the wetland is flooded at the appropriate time of year, and the replanting of aquatic plants (Figure 28.22).

More intensive restoration involves re-creating the community from scratch and is best accomplished on relatively small areas. This kind of restoration involves preparing the site, introducing an array of appropriate native species over time, and employing appropriate management to maintain the community, especially against the invasion of nonnative species from adjacent surrounding areas. A classic example is the re-creation of a prairie ecosystem on a 60-acre field near Madison, Wisconsin (Figure 28.23; also see the introduction to Part Five). The previous prairie had been plowed, grazed, and overgrown. The restoration process involved destroying occupying weeds and brush, reseeding and replanting native prairie species, and burning the site once every 2 to 3 years to approximate a natural fire regime. After nearly 60 years, the plant community now resembles the original native prairie.

Much attention is being focused on restoring wetlands that have been lost because water was diverted or

Figure 28.22 | Volunteers help National Oceanic and Atmospheric Administration (NOAA) scientists prepare sea-grass shoots for planting in the Florida Keys. The plantings help enhance recovery of areas where sea-grass communities have been damaged or large-scale die-off has occurred.

Figure 28.23 | Curtis Prairie at the University of Wisconsin Arboretum. Native prairie vegetation has been restored on this 60-acre tract of land that was once used for agriculture.

drained for land development. New techniques are also being developed for constructing wetlands where they did not exist before, often for the purpose of wastewater and stormwater treatment.

28.11 | Environmental Ethics Is at the Core of Conservation

At the United Nations Earth Summit in Rio de Janeiro in 1992, more than 150 nations signed the Convention on Biodiversity (signed by President Clinton in 1993, although the U.S. Senate has yet to give its advice and consent to its ratification). This treaty makes preservation of the world's biodiversity an international priority. Although many reasons are voiced, the arguments regarding the importance of maintaining biodiversity can be grouped into three categories: economic, evolutionary, and ethical.

The economic argument is based largely on self-interest. Many of the products we use come from organisms that we share this planet with. Obviously, the foods we eat are all derived from other organisms. Every time we buy a drug or other pharmaceutical, there is almost a 50–50 chance that we can attribute some of the essential constituents to a wild species. The value of medicinal products derived from such sources now totals more than $40 billion every year. We derive rubber, solvents, and paper from trees. Cotton, flax, leather, and a host of other natural materials are used to clothe us. Modern industrial society owes a lot to Earth's genetic resources that in one way or another contribute to an array of products that better our standard of living.

Although today's benefits from Nature's cornucopia are astonishing enough, they are only the tip of the

iceberg. Scientists have thus far taken only a preliminary look at some 10 percent of the 250,000 plant species, many of which are already proven to have enormous economic importance. Additionally, we have barely scratched the surface regarding the potential of products derived from the animal kingdom. As these species are lost through extinction, so is their potential for human exploitation.

The second argument for preserving biodiversity is based on genetics (see Chapter 5). Current patterns of biodiversity are a product of ecological and evolutionary processes that have acted on species that existed in the past. The processes of mutation, mixing of genetic information through sexual reproduction, and natural selection, together with the essential ingredient of time, give rise to new species. All species eventually go extinct, many of them leaving no traces other than fossilized impressions buried deep in the earth. Others, however, fade into extinction after having given rise to new species. For example, it is believed that all modern birds can trace their evolutionary history to the earliest known bird, *Archaeopteryx*, which lived during the Jurassic period (fossil record dates to 145 million years ago). If *Archaeopteryx* had been driven to extinction before acting as the evolutionary seed of more modern birds, the variety of life at our backyard bird feeders would be quite different today. Likewise, the mass extinction of modern-day species limits the potential evolution of species diversity in the future.

The third category of arguments in support of conserving biodiversity is ethics based. Humans are but one of millions of species inhabiting Earth—we are relative newcomers to the long evolutionary history of life on our planet. It is the nature of all organisms to both respond to and alter their surrounding environment. However, it is unlikely that any other species in the history of Earth has so dramatically affected its environment in such a short time. The fundamental question facing humanity is a moral one. To what degree will we allow human activities to continue resulting in the extinction of tens of thousands of species that we share this planet with? Debate on the value of biodiversity will center on this question. Arguments based on economics will fall to the wayside because technology allows us to synthesize medicines and other products currently made from plant and animal products, and concerns for the evolutionary future of our planet appear all too abstract when balanced against the needs of our growing human population. Science can work to identify and quantify the problem, but its solution lies outside the realm of science. It involves social, economic, and ethical issues that influence all our lives. Unlike so many of the problems facing society that science is called upon to solve, this is a problem that science can identify. But it is up to the members of society—including you—to arrive at the solution.

Summary

Habitat Destruction | 28.1

The primary cause of species extinctions is habitat destruction resulting from the expansion of human populations and activities. Historically, the largest cause of land transformation has been the expansion of agricultural lands to meet the needs of a growing human population. Because of their disproportionately high species diversity (see Section 26.3) and the pressures of growing populations and economic development, the world's tropical regions have been the primary focus of attention.

Quantity of forests alone is an inadequate indicator of the status of a forest ecosystem from a conservation perspective, because many of the world's forests are highly fragmented and face continued pressure from human activities.

Invasive Species | 28.2

Nonnative (nonindigenous) plants and animals, introduced either intentionally or unintentionally, often cause the extinction of vulnerable native species through predation, grazing, competition, and habitat alteration.

Susceptibility to Extinction | 28.3

Not all species are equally susceptible to extinction from human activities. The susceptibility of species to extinction is related to various life history characteristics that influence their vulnerability to human activities as well as natural catastrophes. Endemic species are particularly susceptible to extinction because a loss of habitat in the one geographic region will result in a complete loss of habitat for the species. Likewise, species that migrate seasonally depend on two or more distinct habitat types in different geographic regions. Other characteristics include small populations (or metapopulations), specialized habitats, and species that directly conflict with human activities.

Threatened Species | 28.4

Development of a quantitative method of classification that allows for the definition of threatened and endangered species is critical to conservation efforts. Such a system provides a standard by which decisions can be reviewed and evaluated by the scientific community.

Biodiversity Hotspots | 28.5

The concept of biodiversity hotspots was developed to address the dilemma that conservationists face: What areas are the most important for preserving species? The 25 biodiversity regions of Earth that have been designated as hotspots by the IUCN contain 44 percent of all plant species and 35 percent of all terrestrial vertebrate species in only 1.4 percent of the planet's land area.

Protecting Populations | 28.6

Because endangered species consist of a few populations, or even a single population, protecting populations is the key to preserving these species. These populations are restricted to protected areas, and an adequate conservation plan must preserve as many individuals as possible within the greatest possible area. The minimum viable population for a species is the number of individuals necessary to ensure the long-term survival of a species. The area of suitable habitat necessary for maintaining the minimum viable population is called the minimum dynamic area.

Species rarely occur in a single, contiguous population; instead, species often consist of a set of semi-isolated subpopulations, connected by dispersal—metapopulations. The persistence of a metapopulation is the result of a complex dynamic among subpopulations.

Reintroduction | 28.7

In some cases, saving species from what appears to be an inevitable decline to extinction requires direct action by conservation ecologists—that is, establishing new populations through transplants and reintroductions.

Habitat Conservation | 28.8

Despite the necessity of focusing conservation efforts on individual species that are threatened and endangered, the most effective way to preserve overall biological diversity is by protecting habitats or whole ecological communities. Unlike the population approach to conservation, a community-based approach requires an understanding of the relationship between overall patterns of biological diversity and features of the landscape. Large areas contain a greater number of species than do smaller areas; and other things being equal, it is better for purposes of conserving overall biological diversity to protect as large an area as possible.

Protected Areas | 28.9

Given the ever-growing pressures placed on lands by the human population, the preservation of biological diversity is becoming more and more dependent on establishing legally designated protected areas. By far, the greatest area of protected lands is the category of public lands. Protected lands differ in their degree of protection; many categories serve multiple purposes, including resource extraction. Many current efforts in conservation focus on working with the existing protected lands and providing buffer zones and corridors that enhance their conservation value.

Habitat Restoration | 28.10

Considerable efforts are under way to restore natural communities affected by human activities, using a new approach to human intervention known as *restoration ecology*. The goal of restoration ecology is to return an ecosystem to a close approximation of its conditions before disturbance by applying ecological principles. This process involves a continuum of approaches ranging from reintroducing species and restoring habitats to attempting to reestablish whole communities as functioning ecosystems.

Environmental Ethics | 28.11

Arguments for the importance of maintaining biodiversity can be grouped into three categories: economic, evolutionary, and ethical. The economic argument is based largely on self-interest. It focuses on products derived from natural resources. The evolutionary argument suggests that the extinction of modern species limits the potential evolution of species diversity in the future. Ethical arguments focus on the extent to which human activities cause the extinction of species.

Study Questions

1. What is the leading cause of current species extinctions?
2. Discuss several ways in which introduced, nonnative species of plants or animals may disrupt a community, leading to the decline of native species. Consider the role of species interaction in community structure and dynamics as presented in Parts Five and Six.
3. What is an endemic species?
4. Name three characteristics that might make a species more susceptible to extinction.
5. What is a biodiversity hotspot? What role can hotspots potentially play in conservation efforts?
6. What is the general relationship between population size and the probability of local extinction?
7. What is a minimum viable population? What is a minimum dynamic area? How do these two terms relate to each other?
8. What is restoration ecology?
9. Give three possible explanations for the observed relationship between species diversity and area.
10. Give two reasons why we should be concerned about the current rate of species extinctions.

Further Readings

Baskins, Y. 2000. *A plague of rats and rubbervines*. Washington, D.C.: Island Press.
    A well-written account of the invasion phenomenon.

Cade, T. J., and W. Burnham. 2003. *Return of the peregrine: A North American saga of tenacity and teamwork*. Boise, ID: The Peregrine Fund.
    Story of the restoration of the peregrine falcon to 80 percent of its range despite bureaucratic and other obstacles. An excellent, well-illustrated case history study.

Clark, T. W., A. P. Curlee, S. C. Minta, and P. M. Kareiva (eds.). 1999. *Carnivore ecosystems: The Yellowstone experience*. New Haven: Yale University Press.
    How Yellowstone carnivores and their prey interact with the return of the missing wolf in the ecosystem.

Contreras, G. P., and K. E. Evans (compilers). 1986. *Proceedings—Grizzly bear habitat symposium: Missoula, Montana, April 30–May 2, 1985* (General Technical Report INT-207). Ogden, UT: U. S. Department of Agriculture, Forest Service, Intermountain Research Station.
    State-of-the art information on grizzly bear habitat, management, and cumulative effects of activities on habitat.

Cox, G. 2004. *Alien species and evolution*. Washington, D.C.: Island Press.
    Examines how alien species adapt and evolve in their invaded environment.

Elton, C. 2000 (1958; repr.). *The ecology of invasions by plants and animals*. Chicago: University of Chicago Press.

Reprint of the 1958 classic. Sketches the basic components of invasion biology.

Falk, D. A., M. Palmer, and J. Zedler. 2006. *Foundations of restoration ecology*. Washington, D.C.: Island Press.

Overview of the ecological principles and approaches to restoring natural ecosystems.

Fearn, E. ed. 2008. *State of the Wild 2008–2009. A Global Portrait of Wildlife, Wildlands, and Oceans*. Washington, D.C.: Island Press.

An excellent up-to-date discussion of the critical issues facing the conservation of species and ecosystems; includes a section on emerging diseases, conservation, and human health.

Leopold, A. 1949 (many editions). *A Sand County almanac*. New York: Oxford.

A conservation classic. The essay "Land Ethic" strongly influenced the development of environmental ethics.

Mittermeier, R. A., N. Myers, P. R. Gil, and C. G. Mittermeier. 1999. *Hotspots: Earth's biologically richest and most endangered terrestrial ecoregions*. Mexico City: CEMEX Conservation International.

This fabulous volume provides an atlas of the biodiversity hotspots presented in Figure 28.11. Outstanding photography.

Mittermeier, R. A., P. R. Gil, and M. Hoffman et al. 2004. *Hotspots revisited*. Mexico City: CEMEX Conservation International.

Another magnificent volume with updated information and analysis as well as redefined boundaries. Considers a number of potential hotspots.

Musiani, M., and P. C. Paquet. 2004. *The practice of wolf persecution, protection, and restoration in Canada and United States*. BioScience 54:50–60.

An overview of the wolf problem beyond the Yellowstone.

Noon, B. R., and K. S. McKelvey. 1996. Management of the spotted owl: A case study in conservation biology. *Annual Review of Ecology and Systematics* 27:135–162.

An excellent case study in the conservation of this threatened bird species of the Pacific Northwest.

Pimentel, D., L. Lach, R. Zuniga, and D. Morrison. 2000. *Environmental and economic costs of nonindigenous species in the United States*. BioScience 50:53–64.

An excellent survey of the problems and costs of invasive species.

Primack, R. B. 1998. *Essentials of conservation biology*, 2nd ed. Sunderland, MA: Sinauer Associates.

This is an excellent introductory text for students interested in the broader topic of conservation ecology.

Ray, C., and J. McCormick-Ray. 2004. *Coastal marine conservation: Science and policy*. Madden, MA: Blackwell Publishing.

This text explores coastal-realm issues and conservation tools and linkages between conservation science and policy. Three case studies representing three regions are included: the temperate Chesapeake Bay, the subarctic Bering Sea, and the tropical Bahamas.

Soulé, M. E. 1985. What is conservation biology? *BioScience* 35:727–734.

This paper is an excellent introduction to the philosophy and science of conservation ecology.

Todd, K. 2000. *Tinkering with Eden: A natural history of exotic species in America*. New York: Norton.

Fairly comprehensive treatment of deliberate introductions and the problems they created.

Williams, T. 2007. Back off! *Audubon* (May–June):50–51, 84–87.

Excellent review of the history and current societal problems and controversies surrounding the Yellowstone wolf restoration project.

CHAPTER 29

Global Climate Change

Satellite image of Hurricane Katrina in the Gulf of Mexico (August 2005).

The term *global climate change* is redundant. Change is inherent in Earth's climate system. For example, although the tilt of Earth's axis relative to the Sun is 23.5°, giving rise to the seasons (see Chapter 2), Earth is actually wobbly. In fact, the tilt of Earth's axis varies from 22.5° to 24°. The amount of tilt in Earth's rotation affects the amount of sunlight striking the different parts of the globe, influencing patterns of global climate. This variation in the tilt of Earth takes place during a cycle of 41,000 years and is largely responsible for the ice ages—periods of glacial expansion and retreat (see Section 18.9).

In turn, variations in climate profoundly affect life on Earth. Paleoecology has recorded the response of populations, communities, and ecosystems to climate changes during periods of glacial expansion and retreat over the past 100,000 years (see Figures 18.18 and 18.20). On an even longer timescale, the fossil record recounts a story of evolutionary change resulting from the dynamics of Earth's climate over geologic time. Throughout this text, we have seen countless examples of how climate influences the function of natural ecosystems—from the uptake of carbon dioxide by leaves in the process of photosynthesis to the distribution and productivity of Earth's ecosystems. But now we have entered a new era in the history of life on our planet; one in which a single species—humans—may have the ability to alter Earth's climate.

In this chapter, we examine how human activities are changing the chemistry of the atmosphere and how those changes may alter Earth's climate. We then explore how these predicted changes in Earth's climate have the potential to affect ecological systems by shifting the distribution of species, altering their interactions, and ultimately influencing the distribution and productivity of ecosystems. We also explore whether these changes in Earth's climate and ecosystems will directly affect the health and well-being of the human population.

29.1 | Greenhouse Gases Influence Earth's Energy Balance and Climate

Many chemical compounds naturally present in Earth's atmosphere—principally water vapor (H_2O), carbon dioxide (CO_2), and ozone (O_3)—absorb thermal (longwave) radiation emitted by Earth's surface and atmosphere. The atmosphere is warmed by this mechanism and, in turn, emits thermal radiation; a significant portion of this energy then acts to warm the surface and the lower atmosphere (see Section 2.1 and Figure 2.3). Consequently, the average surface air temperature of the Earth is about 30°C higher than it would be without atmospheric absorption and reradiation of thermal energy. This phenomenon is popularly known as the **greenhouse effect,** and the gases responsible for the effect are likewise referred to as **greenhouse gases.**

As we discussed in Chapter 2 (Section 2.1), over time, the amount of incoming energy from the Sun to Earth's surface should be about the same as the amount of energy radiated back into space, leaving the average temperature of Earth's surface roughly constant. However, since the industrial period began (see the introduction to Part Eight), the concentrations of greenhouse gases in Earth's atmosphere have increased dramatically. Given the role of greenhouse gases in maintaining Earth's energy balance, concerns have risen over the potential impact of rising greenhouse gas concentrations on the global climate system.

29.2 | Atmospheric Concentration of Carbon Dioxide Is Rising

Although human activities have increased the atmospheric concentration of a variety of greenhouse gases, the major concern is focused on CO_2. The atmospheric concentration of CO_2 has increased by more than 25 percent over the past 100 years. The evidence for this rise comes primarily from continuous observations of atmospheric CO_2 started in 1958 at Mauna Loa, Hawaii, by Charles Keeling (Figure 29.1) and from parallel records around the world.

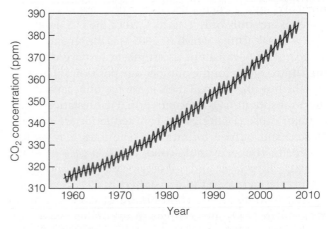

Figure 29.1 | Concentration of atmospheric CO_2 as measured at Mauna Loa Observatory, Hawaii.
(IPCC 2007.)

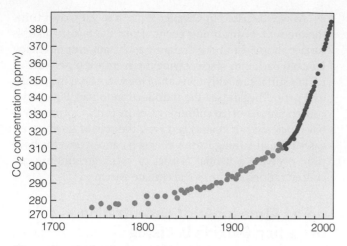

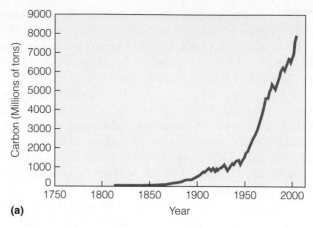

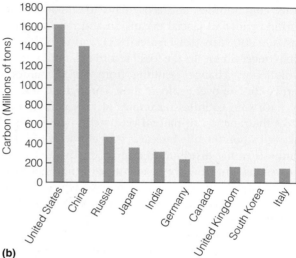

Figure 29.2 | Historical record of atmospheric CO_2 over the past 300 years. Data collected prior to direct observation (1958 to present) are estimated from various techniques including analysis of air trapped in Antarctic ice sheets. (ppmv – parts per million volume) (IPCC 2007.)

Figure 29.3 | (a) Historical record of annual input of CO_2 to the atmosphere from the burning of fossil fuels since 1750. (b) Carbon emissions from the burning of fossil fuels by the top 10 countries in 2005.
(a) (CDIAC; EPI 2007.) (b) (DOE—CDIAC 2007.)

Evidence before the direct observations of 1958 comes from various sources, including the analysis of air bubbles trapped in the ice of glaciers in Greenland and Antarctica.

In reconstructing atmospheric CO_2 concentrations over the past 300 years, we see values that fluctuate between 280 and 290 ppm until the mid-1800s (Figure 29.2). After the onset of the Industrial Revolution, the value increased steadily, rising exponentially by the mid-19th century onward. The change reflects the combustion of fossil fuels (coal, oil, and gas) as an energy source for industrialized nations (Figure 29.3).

In 2005, over 70 percent of the total CO_2 emissions from the burning of fossil fuels came from the developed countries. The United States was the largest single source, accounting for over 22 percent of the total—with carbon emissions per person exceeding 18 tons per year. Over the next few decades, 90 percent of the world's population growth will take place in the developing countries, some of which are also undergoing rapid economic development. Per capita energy use in the developing countries, which is currently only 1/10 to 1/20 of the U.S. level, will also increase. China, which in 2005 was the second-largest source of CO_2 emissions, is estimated to have displaced the United States for the number one position since 2006.

The burning of fossil fuels is not the only cause of rising atmospheric CO_2 concentration. Deforestation is also a major cause (Figure 29.4). Forested lands are typically cleared and burned for farming. The trees may be harvested for timber or pulpwood; but a large part of the biomass, litter layer, and soil organic matter is burned, releasing the carbon to the atmosphere as CO_2.

Calculations of the contribution of land clearing to atmospheric CO_2 are complex. After timber harvest on lands managed for forest production or on lands that have been cultivated and then abandoned, vegetation and soil organic matter become reestablished (see discus-

sion of forestry and swidden agriculture in Chapter 27). We calculate the net contribution to the atmosphere as the difference between CO_2 released during clearing and burning and CO_2 taken up by photosynthesis and the accumulation of biomass during reestablishment. At one time, scientists used regional estimates of population growth and land use (forestry and agriculture) together with simple models of vegetation and soil succession to estimate the contribution due to changing land use. More recent estimates use satellite images to quantify these changes (see Figures 28.3 and 28.4 for examples).

29.3 | Tracking the Fate of CO_2 Emissions

Scientists estimate that the average annual amount of carbon released to the atmosphere during the 1990s was 8.5 gigatons (a gigaton, Gt, is 10^9 metric tons); approximately 6.3 Gt was from combustion of fossil fuels, and 2.2 Gt was from changes in land use (forest clearing). To put this

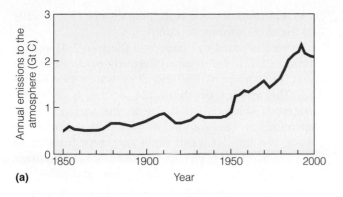

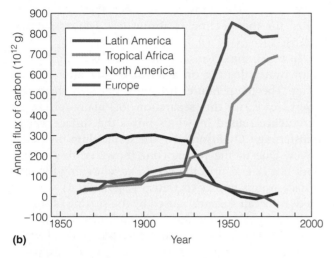

Figure 29.4 | Historical record of annual input of CO_2 to the atmosphere from the clearing and burning of forest **(a)** globally and **(b)** in selected geographic regions: Latin America, Tropical Africa, North America, and Europe.

(Adapted from Houghton 1997.)

number into perspective, if the average weight of a human is 70 kg (approximately 150 pounds), a gigaton would be the weight of more than 14 billion people, or more than two times the world's population.

Direct measurements of atmospheric CO_2 over this same period show an annual accumulation of carbon in the atmosphere of only 3.2 gigatons. The difference, 5.3 gigatons, must have flowed from the atmosphere into the other main pools in the global carbon cycle (see Figure 22.5), the oceans, and terrestrial ecosystems. Determining the fate of CO_2 put into the atmosphere from the burning of fossil fuel requires input from a variety of scientific disciplines, as well as a large dose of detective work.

The process of diffusion controls uptake of carbon dioxide from the atmosphere into the oceans (see Section 3.7). Because physical processes largely control this transfer, scientists are able to make reasonably accurate estimates. The estimate of annual uptake of carbon dioxide by the oceans during the 1990s is 2.4 gigatons. In contrast, although the processes controlling the exchange of carbon between terrestrial ecosystems and the atmosphere are generally well understood, quantifying these processes at a regional to global scale is extremely difficult. As a result, a

simple process of elimination is used to estimate global uptake of carbon by terrestrial ecosystems. Carbon that has been emitted over a specified period of time but cannot be accounted for by measurements of atmospheric carbon concentration or estimates of oceanic absorption is relegated to the terrestrial ecosystems:

**Net uptake by
terrestrial ecosystems =
0.7 Gt**

Emissions from fossil fuels	−	Atmospheric increase	−	Ocean uptake
6.3 Gt		3.2 Gt		2.4 Gt

Using this approach, Earth's terrestrial ecosystems are a net sink of carbon, with an annual net uptake of carbon from the atmosphere of 0.7 Gt. However, as noted earlier, scientists estimate that land clearing (deforestation) resulted in an annual input of 2.2 Gt of carbon to the atmosphere from terrestrial ecosystems during this period, not a net uptake of 0.7 Gt. Therefore, there is a discrepancy of 2.9 Gt per year (2.2 + 0.7 Gt) in this analysis. This discrepancy has been called the problem of the "missing carbon." The terms of the global carbon equation just presented are shown graphically over the period 1850–2000 in Figure 29.5.

Some studies suggest that any possible net uptake of carbon by terrestrial ecosystems may result from reforestation in temperate regions of the Northern Hemisphere. The regrowth of forests followed the large-scale abandonment of lands cleared for agriculture during the latter part of the 19th and early part of the

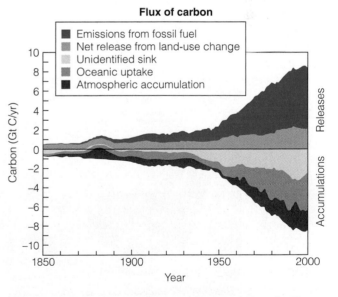

Figure 29.5 | The various releases and accumulation of carbon associated with the global carbon cycle, shown graphically over the period 1850–2000. The unidentified sink is referred to as the missing carbon.

(Adapted from www.whrc.org/carbon/missingc.htm.)

Go to **GRAPHit!** at www.ecologyplace.com to see trends in emission pollution.

20th centuries (see Ecological Issues: American Forests, p. 370). Although reforestation may be a key component in balancing the global carbon cycle, more recent studies suggest that the tropical forests represent a much larger sink of carbon than previously believed, possibly accounting for a significant proportion of the missing carbon. Determining the fate of carbon input to the atmosphere through the burning of fossil fuels, however, requires an understanding of the processes controlling carbon exchange among the major components of the global carbon cycle as well as how these exchanges might be influenced by rising atmospheric concentrations of CO_2.

29.4 | Atmospheric CO_2 Concentrations Affect CO_2 Uptake by Oceans

Carbon dioxide diffuses from the atmosphere into the surface waters of the ocean, where it dissolves and undergoes various chemical reactions, including the transformation to carbonates and bicarbonates (see Section 3.7). The rate of diffusion from the atmosphere to the surface waters of the ocean is a function of the diffusion gradient (difference in concentrations). Therefore, as the concentration of CO_2 in the atmosphere rises, the diffusion of CO_2 into surface waters increases.

Given their volume, the oceans have the potential to absorb most of the carbon that is being transferred to the atmosphere by fossil fuel combustion and land clearing. This potential is not realized, because the oceans do not act as a homogeneous sponge that absorbs CO_2 equally into the entire volume of water.

As we discussed in Chapters 3 (Section 3.4) and 21 (Section 21.10), the oceans effectively function as two layers: the surface waters and deep waters (see Figure 3.8). The average depth of the oceans is 2000 m. Intercepted solar radiation warms the surface waters. Depending on the amount of radiation reaching the surface, the zone of warm water can range from 75 to 200 m in depth. The average temperature of this surface layer is 18°C. The remainder of the vertical profile (200 to 2000 m depth) is deep waters, whose average temperature is 3°C. The transition between these two zones, called the thermocline, is abrupt (see Figures 3.8 and 21.25).

In effect, the oceans can be viewed as a thin layer of warm water floating on a much deeper layer of cold water. The temperature difference between these two layers leads to the separation of many processes. Turbulence caused by winds mixes the surface waters, transferring CO_2 absorbed at the surface into the waters below. Due to the thermocline, however, this mixing does not extend into the deep waters. Mixing between surface waters and deep waters (Figure 29.6) depends on deep ocean currents caused by the sinking of surface waters as they move toward the poles (see Sections 2.5 and 3.8). This process occurs over hundreds of years, limiting the short-term uptake of CO_2 by the deep waters. Consequently, the amount of CO_2 that can be absorbed by the oceans over the short term is limited, despite the large volume of water.

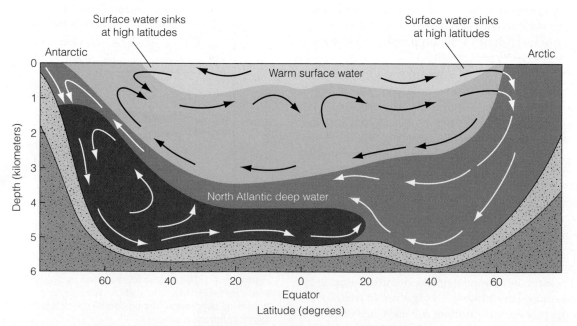

Figure 29.6 | The major pattern of circulation in the Atlantic Ocean. Atlantic surface waters, flowing northward from the tropics, cool and sink when they reach subarctic latitudes. After sinking, these waters become part of the huge, deep, southward countercurrent reaching all the way to the Antarctic.

29.5 | Plants Respond to Increased Atmospheric CO_2

Carbon dioxide flows from the atmosphere into terrestrial ecosystems via photosynthesis (see Section 6.1). To understand how rising atmospheric CO_2 concentrations influence the productivity of terrestrial ecosystems, we must understand how photosynthesis responds in an enriched CO_2 environment.

Elevated atmospheric CO_2 concentrations have two direct, short-term effects on plants. First, they increase the rate of photosynthesis. Recall that CO_2 diffuses from the air into the leaf through the stomatal openings (see Section 6.3). The higher the CO_2 concentration in the outside air, the greater the rate of diffusion into the leaf. A higher rate increases the availability of CO_2 for photosynthesis in the mesophyll cells of the leaf, so it generally results in a higher rate of photosynthesis. The higher rates of diffusion and photosynthesis under elevated atmospheric concentrations of CO_2 have been termed the **CO_2 fertilization effect.**

Second, elevated atmospheric CO_2 concentrations cause stomata to partially close, reducing water loss due to transpiration. Thus, at elevated CO_2 levels, plants increase their water-use efficiency (carbon uptake and water loss; see Section 6.4).

The effects of long-term exposure to elevated CO_2 on plant growth and development, however, may be more complicated. Plant ecologists Hendrik Poorter and Marta Pérez-Soba of Utrecht University in The Netherlands reviewed the results from more than 600 experimental studies examining the growth of plants at elevated carbon dioxide levels. These studies examined a wide variety of plant species representing all three photosynthetic pathways: C_3, C_4, and CAM (see Chapter 6). Their results revealed that C_3 species respond most strongly to elevated CO_2, with an average increase in biomass of 47 percent (Figure 29.7). Data on the response of CAM species were limited, but the mean response for the six species reported was 21 percent. The C_4 species examined also responded positively to elevated CO_2, with an average increase of 11 percent.

Within C_3 species, on average, crop species show the highest biomass enhancement (59 percent) and wild herbaceous plants the lowest (41 percent). Most of the experiments with woody species were conducted with seedlings, therefore covering only a small part of their life cycle. The growth stimulation of woody plants was on average 49 percent.

In some studies, the enhanced effects of elevated CO_2 levels on plant growth have been short-lived (Figure 29.8). Some plants produce less of the photosynthetic enzyme rubisco at elevated CO_2, reducing photosynthesis to rates comparable to those measured at lower CO_2 concentrations; this phenomenon is known as downregulation. Other studies reveal that plants grown at increased CO_2 levels allocate less carbon to producing leaves and more to producing roots. In addition, plants grown at elevated CO_2 levels appear to produce fewer stomata on the leaf surface. The smaller leaf area and lower stomatal density reduce water loss; but they also reduce total carbon uptake and growth rates.

It is uncertain how the results observed for leaves or single plants translate into changes in the net primary productivity (NPP) of terrestrial ecosystems. Availability of water or nutrients in many ecosystems may limit potential increases in plant productivity at elevated CO_2

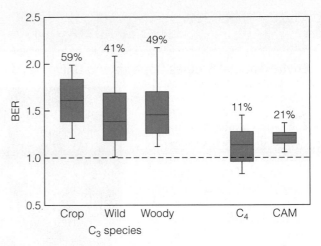

Figure 29.7 | Distribution of biomass enhancement ratio (BER) for several functional types of species. BER is the ratio of biomass growth at elevated and ambient levels of CO_2. Distributions are based on 280 C_3, 30 C_4, and 6 CAM species. C_3 species were separated into three groups: crop, wild herbaceous, and woody species. Boxes indicate the distribution of the range of observation. Line represents median value, lower box 25th percentile, and upper box 75th percentile. Error bars give 10th and 90th percentile.

(Adapted from Poorter and Pérez-Soba 2002.)

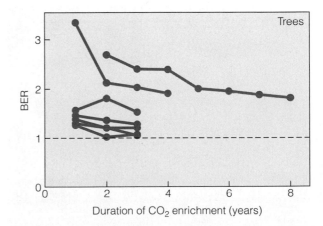

Figure 29.8 | Time course of biomass enhancement ratio (BER) due to elevated CO_2. BER is the ratio of biomass growth at elevated and ambient levels of CO_2. Each line represents the results of an experiment with a different tree species.

(Adapted from Poorter and Pérez-Soba 2002.)

Environmental Studies Department, University of California, Santa Cruz, California

Human activities are altering Earth's atmosphere and climate in a variety of ways. Increases in atmospheric concentrations of CO_2 are contributing to rising global temperatures as well as changes in annual patterns of precipitation (see Section 29.6). Global anthropogenic nitrogen (N) fixation (see Ecological Issues: Nitrogen Fertilizers, p. 457) now exceeds all natural sources of N fixation, and its products include greenhouse gases such as N_2O.

How these global changes in climate and atmospheric chemistry may alter the diversity of plant communities by changing resource availability and affecting individual species performances is a question central to the research of ecologist Erika Zavaleta of the University of California, Santa Cruz. Since the mid-1990s, Zavaleta and her colleagues at Stanford University have been studying the response of California's grassland ecosystems to changes in climate, atmospheric CO_2, and N-deposition based on future scenarios developed for the region.

Zavaleta's studies were conducted in the California grassland at the Jasper Ridge Biological Preserve in the San Francisco Bay Area. The grassland community is composed of annual grasses, annual and biennial forbs, and occasional perennial bunchgrasses, forbs, and shrubs. Annual grasses are the community dominants, contributing the majority of plant biomass during the period of peak primary productivity in the growing season.

In this region's mediterranean-type climate, annual plants (both grasses and forbs) germinate with the onset of the fall–winter rains (see Section 23.6). Plants then set seed and senesce as water limitation becomes severe with the cessation of rain in March–May. The small stature and short life span of the annual plants that dominate these communities make this site an excellent experimental system to examine community response to altered environmental conditions over a period of multiple generations.

In 1997, Zavaleta and colleagues established 32 circular plots 2 m in diameter and surrounded each with a solid belowground partition to 50 cm depth. Each plot was then divided into four 0.78 m^2 quadrats, using solid partitions belowground and mesh partitions aboveground. Four global change treatments were applied to the experimental plots: (1) elevated CO_2 (ambient plus 300 ppm), (2) warming (80 W/m^2 of thermal radiation resulting in soil-surface warming of 0.8–1.0°C), (3) elevated precipitation (increased by 50 percent, including a growing-season extension of 20 days), and (4) N-deposition (increased by 7 $g/m^2/day$).

The experimental design of the plots and application of treatments can be seen in the photograph in Figure 1. CO_2 was elevated by using a free-air system with emitters surrounding each plot and delivering pure CO_2 at the canopy level. Warming was applied with infrared lamps suspended over the center of each plot. Extra precipitation was delivered with overhead sprinklers and drip lines. The growing season extension was delivered in two applications, at 10 and 20 days after the last natural rainfall event. N-deposition was administered with liquid (in autumn) and slow-release (in winter) $Ca(NO_3)_2$ applications each year.

The four treatments were applied in one- to four-way treatment combinations (see Figure 2 caption for description of treatment combinations), each replicated eight times. Treatments were begun in November 1998 and continued for a period of three years.

To evaluate the influence of the global change treatments on the grassland community within the experimental plots, a census was conducted in May of each

Figure 1 | An experimental study plot at the Jasper Ridge Biological Reserve.

year to determine plant species diversity on each plot. Plant diversity was quantified using species richness (total number of species).

After three years, 3 of the 4 global change treatments had altered total plant diversity (Figure 2). N-deposition reduced total plant species diversity by 5 percent, and it elevated CO_2 reduced overall plant diversity by 8 percent. In contrast, elevated precipitation increased plant diversity by 5 percent. The fourth treatment, elevated temperature, had no significant effect on plant species diversity in the experimental plots. The effects of elevated CO_2, N-deposition, and precipitation on total diversity were driven mainly by significant gains and losses of forb species (see Figure 2), which make up most of the native plant diversity in California grasslands. Reduced diversity in the N-deposition treatment was partly a function of the loss of all three N-fixing forb species at the site. In contrast to forb species, annual grass diversity was relatively unresponsive to all individual global change treatments.

All four treatment-combination scenarios produced mean declines in forb diversity of greater than 10 percent (see Figure 2). Diversity of this functional group, which includes many of the remaining native and rare species in California grasslands, seems susceptible to decline regardless of whether N-deposition and precipitation increase. The effects of these four treatment combinations on total plant diversity were not significant, however, because increases in perennial grass diversity partially offset losses of forb species.

Perhaps the most interesting and unexpected result from Zavaleta's experiments emerged from a comparison using only a subset of the treatments: elevated CO_2 (C), warming (T), and CO_2 plus warming (C + T). It is

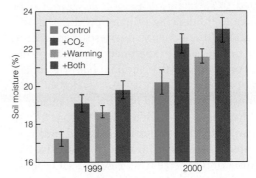

Figure 3 | Warming and elevated CO_2 effects on spring soil moisture for 1999–2000. Values are mean soil moisture from January to July for each year.

generally believed that global warming may increase aridity in water-limited ecosystems, such as the California grasslands, by accelerating evapotranspiration. However, the experiments conducted by Zavaleta and colleagues produced the reverse effect. Simulated warming increased spring soil moisture by 5–10 percent under both ambient and elevated CO_2 (Figure 3). This effect was not caused by decreasing leaf area or plant production under elevated temperatures, but rather by earlier plant senescence (in later May and early June) in the elevated temperature treatments. Lower transpirational water losses resulting from earlier senescence provide a mechanism for the unexpected rise in soil moisture, and this biotic link between warming and water balance may well prove to be an important influence on the response of grassland and savanna communities to climate change.

Bibliography

Zavaleta, E. S., B. D. Thomas, N. R. Chiariello, G. P. Asner, and C. B. Field. 2003. Plants reverse warming effect on ecosystem water balance. *Proceedings of the National Academy of Sciences* USA 100:9892–9893.

Zavaleta, E. S., M. R. Shaw, N. R. Chiariello, H. A. Mooney, and C. B. Field. 2003. Additive effects of simulated climate changes, elevated CO_2, and nitrogen deposition on grassland diversity. *Proceedings of the National Academy of Sciences* USA 100:7650–7654.

1. In the results of the combined treatments presented in Figure 2, how does increased precipitation influence the effects of elevated CO_2 and temperature on the diversity of forb species?

2. Based on the discussion of plant response to elevated CO_2 presented in Section 29.5, how might changes in stomatal conductance (and transpiration) influence soil moisture over the growing season?

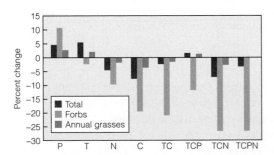

Figure 2 | Changes in the total, forb, and annual grass diversity under single and combined global change treatments. Values are percent difference between controls and elevated levels for each treatment, based on values of mean species richness for each treatment. Treatments: C, CO_2; T, warming; P, precipitation; N, nitrogen; TC, warming and CO_2; TCP, warming, CO_2 and precipitation; TCN, warming, CO_2, and nitrogen; TCPN, warming, CO_2, precipitation, and nitrogen.

concentrations. Several large-scale experiments are currently examining the effects of elevated CO_2 on whole ecosystems. By exposing whole areas of forest and grassland to elevated CO_2, scientists are able to examine the variety of processes influencing primary productivity, decomposition, and nutrient cycling in terrestrial ecosystems (Figure 29.9). One such experiment, at the Duke Experimental Forest in North Carolina (shown in Figure 29.9), has been ongoing since 1996. Elevated CO_2 has resulted in a consistent increase in NPP over that measured in the control (ambient CO_2) forest plots (Figure 29.10).

A comparison of field studies in grassland and agricultural ecosystems reveals an average increase in biomass production of 14 percent at elevated CO_2 levels (double ambient concentrations). However, estimates at individual sites ranged from an increase of 85 percent to a decline of almost 20 percent. These results stress the importance of the interactions of elevated CO_2 with other environmental factors, particularly temperature, moisture, and nutrient availability.

Ecosystems characteristic of low-temperature environments tend to show an initial enhancement of productivity after elevated CO_2, followed by downregulation. A study conducted by Walter Oechel and colleagues examined the response of arctic tundra in Alaska to elevated CO_2. They observed an initial increase in productivity, but primary productivity returned to original levels after three years of continuous exposure to a doubled CO_2 environment.

The largest and most persistent responses to elevated CO_2 have been observed in seasonally dry environments, where primary productivity is enhanced during years of below-average rainfall. In a study of a tallgrass prairie ecosystem in Kansas, C. E. Owensby and colleagues

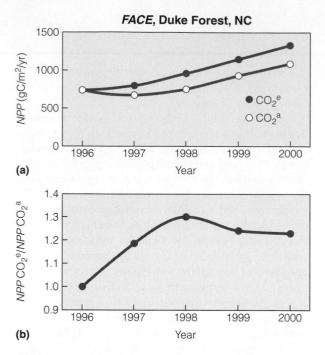

(a)

(b)

Figure 29.10 | **(a)** Net primary productivity (NPP) for the first 5 years of CO_2 enrichment in the Duke Forest FACE experiment since it began in 1996 (experimental forest shown in Figure 29.9). CO_2^e denotes elevated CO_2 conditions; CO_2^a denotes ambient CO_2 conditions. **(b)** Enhancement ratio of NPP under CO_2^e relative to that under CO_2^a since the enrichment began in 1996. The enhancement ratio is the ratio of NPP under elevated CO_2 divided by the NPP under ambient conditions.

(a) (1996 and 1997 data from DeLucia et al. 1999; 1998 data from Hamilton et al. 2002.) (b) (Adapted from Schafer et al. 2003.)

Interpreting Ecological Data

Q1. How did NPP change over the observation period at Duke Forest? Did the general pattern of change differ for the ambient and double CO_2 treatments?

Q2. How does the enhancement ratio change over the observation period? How does the response of NPP to elevated CO_2 for Duke Forest relate to the results of the biomass enhancement experiments presented in Figure 29.8? Are the results of experiments for individual trees in Figure 29.8 consistent with the observed patterns for forest NPP in Figure 29.9?

found no significant increase in aboveground NPP during wet years (greater-than-average rainfall) for experimental plots exposed to a doubled CO_2 concentration when compared with control plots receiving ambient CO_2. In contrast, they observed a 40 percent increase in aboveground NPP during years with average rainfall and an 80 percent increase during years with below-average precipitation. Even though these relative increases in NPP are large, they occur in years of low NPP, so that absolute changes may be quite low.

Enhancement of primary productivity by elevated CO_2 in dry environments arises largely from the small re-

Figure 29.9 | The Free Air CO_2 Experiment (FACE) at Duke Forest in North Carolina. The circle of towers releases carbon dioxide into the surrounding air, allowing scientists to examine the response of the forest ecosystem to elevated concentrations of atmospheric carbon dioxide.

duction in transpiration resulting from partial closure of the stomata. These small reductions have resulted in measurable changes (increases) in soil moisture in grassland ecosystems, particularly during prolonged dry periods (see Field Studies: Erika Zavaleta). Increased soil moisture extends the growing season and increases soil microbial activity, decomposition, and nitrogen mineralization (see Chapter 21).

29.6 | Greenhouse Gases Are Changing the Global Climate

Over the past 100 years (1906–2005), Earth's average surface temperature has increased by approximately 0.74°C (estimates range from 0.56 to 0.92°C). In addition, 11 of the 12 years from 1995–2006 rank among the warmest years in the instrumental record of global surface temperatures since 1850 (see Ecological Issues: Who Turned Up the Heat?). The cause for the observed warming has been the focus of much scientific research and debate over the past two decades, but now a general consensus has begun to emerge. According to the most recent assessment by the Intergovernmental Panel on Climate Change (IPCC 2007), a scientific intergovernmental body set up by the World Meteorological Organization (WMO) and by the United Nations Environment Program (UNEP), most of the observed increase in global average temperature since the mid-20th century is "very likely" due to observed changes in the atmospheric concentrations of greenhouse gases, primarily CO_2.

As human activities continue to increase the atmospheric concentration of CO_2, how will it continue to influence the global climate? Scientists estimate that at current rates of emission, the preindustrial level of 280 ppm of CO_2 in the atmosphere will double by the year 2020. Moreover, CO_2 is not the only greenhouse gas increasing because of human activities (Figure 29.11). Others include methane (CH_4), chlorofluorocarbons (CFCs), hydrogenated chlorofluorocarbons (HCFCs), nitrous oxide (N_2O), ozone (O_3), and sulfur dioxide (SO_2). Although much lower in concentration, some of these gases are much more effective at trapping heat than is CO_2. They are significant components of the total greenhouse effect.

The role of greenhouse gases in warming Earth's surface is well established, but the specific influence that doubling the CO_2 concentration of the atmosphere will exert on the global climate system is much more uncertain. Atmospheric scientists have developed complex computer models of Earth's climate system—called **general circulation models,** or GCMs for short—to help determine how increasing concentrations of greenhouse gases may influence large-scale patterns of global climate. Although all use the same basic physical descriptions of climate processes, GCMs at different research institutions differ in their spatial resolution and in how they describe certain features of Earth's surface and atmosphere. As a result, the models differ in their predictions (Figure 29.12).

Despite these differences, certain patterns consistently emerge. All of the models predict an increase in the average global temperature as well as a corresponding increase in global precipitation. Findings published in 2007 by the IPCC suggest an increase in globally averaged surface temperature in the range of 1.1°C to 6.4°C by the year 2100 (actual range depends on the specific scenario of greenhouse gas emission developed by the IPCC). These changes would not be evenly distributed over Earth's surface. Warming is expected to be greatest during

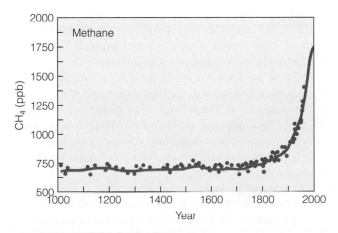
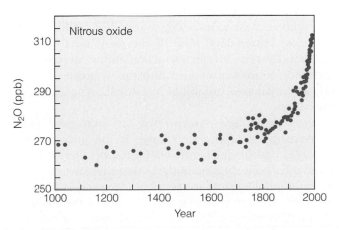

Figure 29.11 | Historic trends in greenhouse gas emissions as illustrated by changes in the atmospheric concentrations of methane and nitrous oxide.

(Adapted from IPCC 2007.)

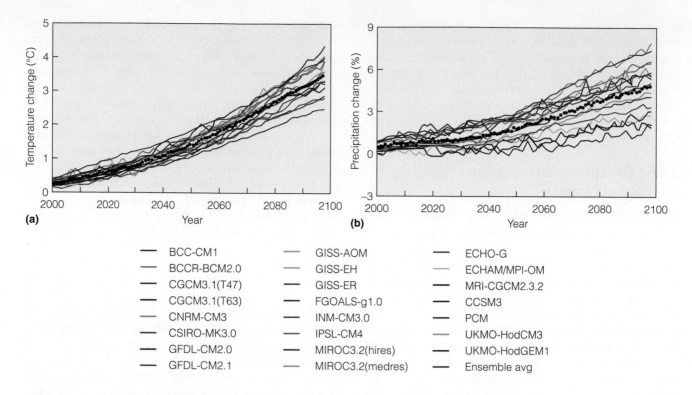

— BCC-CM1	— GISS-AOM	— ECHO-G
— BCCR-BCM2.0	— GISS-EH	— ECHAM/MPI-OM
— CGCM3.1(T47)	— GISS-ER	— MRI-CGCM2.3.2
— CGCM3.1(T63)	— FGOALS-g1.0	— CCSM3
— CNRM-CM3	— INM-CM3.0	— PCM
— CSIRO-MK3.0	— IPSL-CM4	— UKMO-HodCM3
— GFDL-CM2.0	— MIROC3.2(hires)	— UKMO-HodGEM1
— GFDL-CM2.1	— MIROC3.2(medres)	— Ensemble avg

Figure 29.12 | Time series of globally averaged **(a)** surface air temperature change and **(b)** precipitation change from various global circulation models under a scenario (scenario A2) of rising atmospheric concentrations of greenhouse gases developed by the Intergovernmental Panel on Climate Change (IPCC). Changes are relative to the average value for period from 1980 to 1999. Abbreviations refer to the research programs that have developed the various global circulation models used in the analyses. (Adapted from IPCC 2007.)

the winter months and in the northern latitudes. Figure 29.13 shows the spatial variation in changes in mean temperature and precipitation at a global scale for the Northern Hemisphere winter and summer periods based on the most recent (2007) IPCC analyses.

Although in popular speech *greenhouse effect* is synonymous with *global warming*, the models predict more than just hotter days. One of the most notable predictions foretells an increased variability of climate, including more storms and hurricanes; greater snowfall; and increased variability in rainfall, depending on the region.

One recent development that has influenced predicted patterns of climate change is the inclusion of aerosols in calculating Earth's energy balance. Aerosols, or small particles suspended in the atmosphere, absorb radiation from the Sun and scatter it back to space. By scattering solar radiation back to space, they reduce the amount of radiation reaching Earth's surface. Aerosols come from a variety of sources. In desert regions, they originate from winds blowing dust airborne. Over the oceans, aerosols come from sea spray. They also result from the burning of

forests and grasslands (referred to as biomass burning). Occasionally, large quantities of particulates are injected into the upper atmosphere through the eruption of volcanoes, as was the case when Mt. Pinatubo erupted in 1991.

A major source of aerosols resulting from human activities is sulfates and soot from the burning of fossil fuels. Sulfate particles are especially important. They are formed from sulfur dioxide, a gas produced in large quantities by power stations that burn coal (see Chapter 22). These particles remain in the atmosphere for a very short period (on average, five days), so their distribution is concentrated in the regions near their source (Figure 29.14a). In regions of the Northern Hemisphere, their concentration is significant and functions to offset the effects of greenhouse gases, reducing estimates of global warming (Figure 29.14b).

As models of global climate improve, there will no doubt be further changes in the patterns and the severity of changes they predict. However, the physics of greenhouse gases and the consistent qualitative predictions of the GCMs lead scientists to believe that rising concentrations of atmospheric CO_2 will significantly affect global climate.

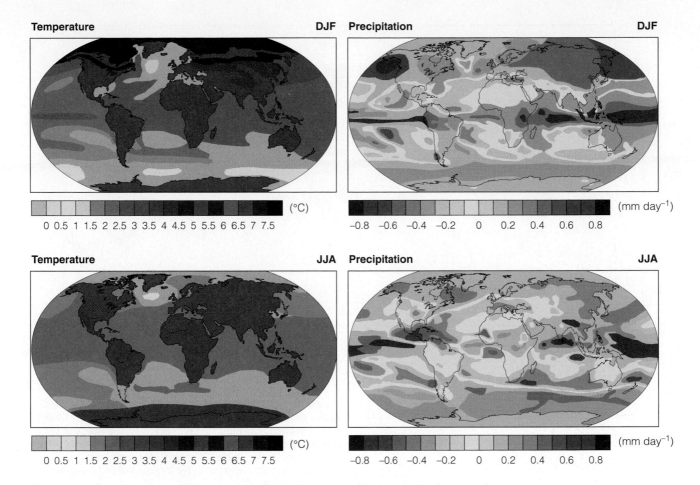

Figure 29.13 | Mean changes in **(a)** surface air temperature (°C) and **(b)** precipitation (mm day^{-1}) for Northern hemisphere winter (DJF—December, January, and February: top) and summer (JJA—June, July, and August: bottom) under a scenario of rising atmospheric concentrations of greenhouse gases developed by the IPCC. Results are an average of the patterns predicted by the various global circulation models in Figure 29.12. Changes are for the period 2080 to 2099 relative to 1980 to 1999. Note that most of the warming is predicted to occur in the more northern latitudes and during the winter months. (IPCC 2007.)

29.7 | Changes in Climate Will Affect Ecosystems at Many Levels

Climate influences almost every aspect of the ecosystem: the physiological and behavioral response of organisms (Chapters 5–8); the birth, death, and growth rates of populations (Chapters 9–12); the relative competitive abilities of species (Chapter 13); community structure (Chapters 16–19); productivity (Chapter 20); and cycling of nutrients (Chapter 21). Current research on greenhouse warming focuses on the response of organisms at all levels of organization: individuals, populations, communities, and ecosystems. Changes in temperature and water availability will directly affect the distribution and abundance of individual species. For example, the relative abundance of three widely distributed European tree

species is plotted as a function of mean annual temperature and rainfall in Figure 29.15. Differing environmental responses determine the distribution and abundance of these three important tree species over the European landscape. Their distribution and abundance will change as regional patterns of temperature and precipitation change.

The potential impact of regional climate change on plant species distribution can be more clearly seen from the work of Anantha Prasad and Louis Iverson of the Northeast Research Station, U.S. Forest Service. Using data from the Inventory and Analysis Program of the U.S. Forest Service, Prasad and Iverson developed statistical models to predict the distribution of 80 different tree species inhabiting the eastern United States. Individual tree species distributions are predicted as a

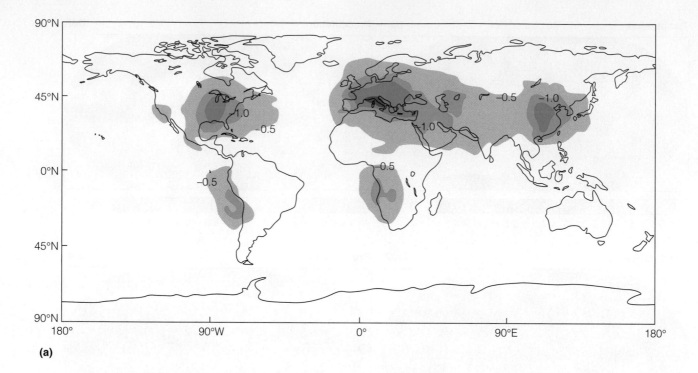

(a)

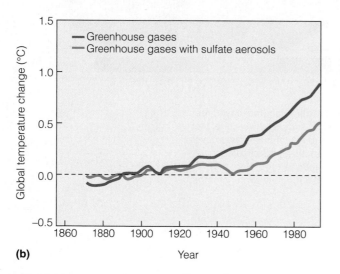

(b)

Figure 29.14 | (a) Estimates of reduction in radiation (watts per square meter) resulting from anthropogenic sulfate aerosols in the atmosphere. The reduction is largest over regions close to the source of the emissions. **(b)** Predicted changes in mean global temperature for the United Kingdom Meteorological Office (UKMO) GCM, with and without the inclusion of sulfate aerosols in the simulation.

(Adapted from Mitchell et al. 1995.)

function of variables describing climate, soils, and topography for any location. This framework allows the investigators to predict shifts in the distribution of these tree species based on changes in temperature and precipitation for the region from a variety of GCM predictions at doubled levels of CO_2. Predicted distributions for three major tree species in the eastern United States under current and doubled CO_2 climate conditions using the National Oceanic and Atmospheric Administration (NOAA) Geophysical Fluid Dynamics Laboratory (GFDL) general circulation model (see Figure 29.12) are presented in Figure 29.16. These predicted changes in temperature and precipitation will dramatically influ-

ence the distribution and abundance of tree species that dominate the forest ecosystems of the eastern United States.

The distribution and abundance of animals are also directly related to features of the climate. For example, the northern limit of the winter range of the Eastern phoebe (*Sayornis phoebe*) is associated with average minimum January temperatures of $-4°C$. The phoebe is not found in areas where temperatures drop below this value. Two lines, or isotherms, defining the region of eastern North America where average minimum January temperatures of $-4°C$ occur are plotted in Figure 29.17. Minimum temperatures drop below $-4°C$ in areas to the

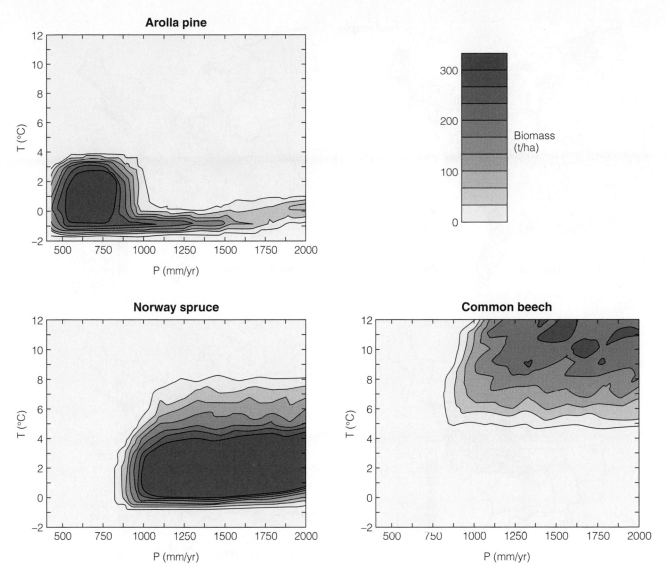

Figure 29.15 | Abundance (biomass t/ha) of three common European tree species as it relates to mean annual temperature (T) and precipitation (P).

(Adapted from Miko et al. 1996.)

north and west of the lines, whereas temperatures are above −4°C to the south and east. The two isotherms show the current −4°C average minimum January temperature isotherm and the −4°C isotherm predicted by the GFDL general circulation model for a doubled atmospheric concentration of CO_2. A change in the isotherm would be expected to result in a northern expansion of the Eastern phoebe's winter range.

Collectively, the shifts in individual species' distributions will have the effect of changing regional patterns of species diversity. Prasad and Iverson used their analysis of tree species distribution in the eastern United States under conditions of climate change (see Figure 29.16) to explore the implications on regional patterns of diversity. By combining the predicted shifts in the distributions of all 80 tree species, they were able to examine the resulting changes in local and regional patterns of tree species richness (Figure 29.18). Under the climate conditions predicted by the GFDL model, there is a marked decline in tree species richness in the southeastern United States.

For most taxonomic groups, however, we do not have sufficient information on the environmental factors controlling the distribution of individual species to allow for an analysis such as that performed by Prasad and Iverson for the trees of eastern North America. For other groups

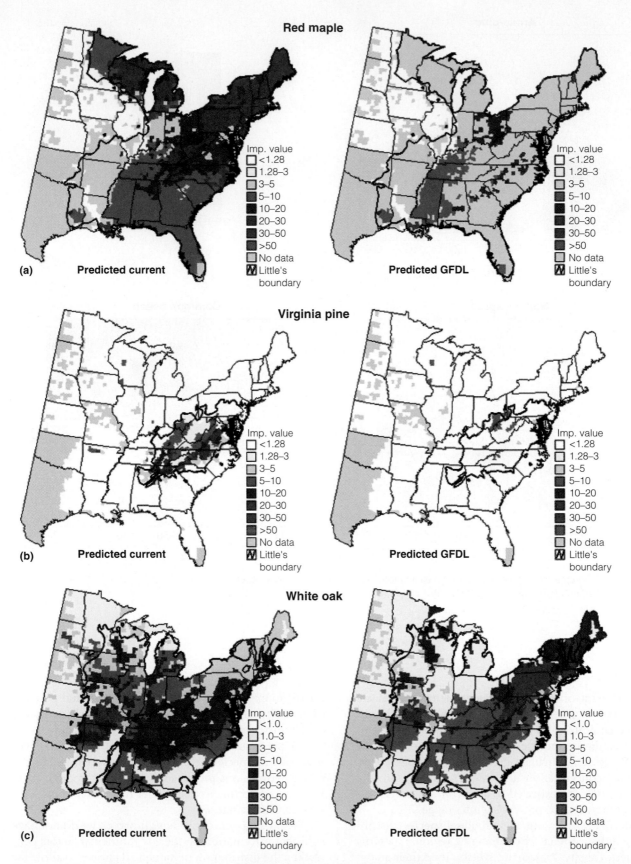

Figure 29.16 | Distributions of **(a)** red maple, **(b)** Virginia pine, and **(c)** white oak under both current climate and a doubled CO_2 climate as predicted by the GFDL general circulation model. Species abundances expressed in terms of importance value (sum of relative density, basal area, and frequency). Little's boundary refers to the observed distribution of the species as reported by Little (1977). See text for description of model used for predicting species distributions based on climate and site factors.

(Adapted from Iverson et al. 1999.)

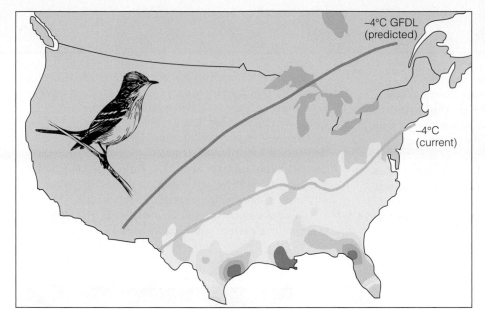

Figure 29.17 | Map showing the existing distribution of the Eastern phoebe along the current average minimum January temperature isotherm, as well as the predicted isotherm under a changed climate. The predicted isotherm is based on temperature changes due to a doubling of atmospheric CO_2 concentration as predicted by the GFDL general circulation model (shown in Figure 29.12a).

(Adapted from Root 1988.)

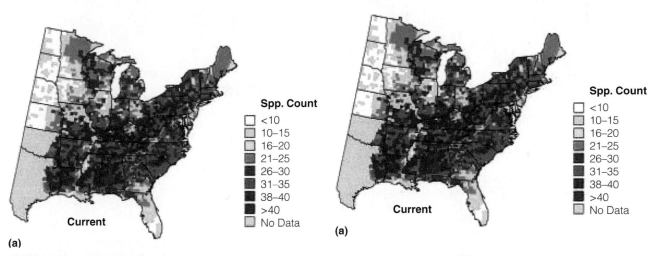

Figure 29.18 | **(a)** Current tree species richness as determined from forest inventory data and **(b)** potential future richness under the climate patterns predicted by the GFDL climate model under conditions of a doubled atmospheric concentration of CO_2.

(Adapted from Iverson and Prasud 2001.)

Interpreting Ecological Data

Q1. What is the general trend in tree species richness for the eastern region of the United States under the GFDL climate change scenario presented in this figure?

Q2. Is there any area of the region included in the maps that is predicted to have a significant increase in tree species richness under the GFDL scenario? If so, what is the dominant ecosystem in this area (or areas) under the current climate? (See Chapter 23 for maps of current ecosystem distribution in the region.)

of organisms, we must depend on more general relationships between features of the environment and overall patterns of diversity. For example, in Chapter 26 we examined the work of ecologist David Currie (University of Ottawa) in correlating the broad-scale patterns of species diversity at the continental scale to features of the physical environment (see Section 26.4). Currie found that the richness of most terrestrial animal groups, including vertebrates, covaries with features of the physical environment related to the energy and water balance of organisms: temperature, evapotranspiration, and incident solar radiation. In a more recent study, Currie has used the relationship between climate (specifically, mean January and July temperature and precipitation) and s Earth's climate changing? According to the Intergovernmental Panel on Climate Change (IPCC), the answer is unequivocally yes. This conclusion is drawn from a suite of observations that allow scientists to track changes in the global climate over the past century (Figure 1). Widespread direct measurements of surface temperature began around the middle of the 19th century. These direct measures from instruments such as thermometers are referred to as the instrumental record. Observations of other surface 'weather' variables, such as precipitation and winds, have been made for about 100 years.

Besides measurements made at the land surface, observations of sea surface temperatures have been made from ships since the mid-19th century. Since the late 1970s, a network of instrumented buoys has supplemented these observations. Measurements of the upper atmosphere have been made systematically only since the late 1940s, but since the late 1970s, Earth-observing satellites have been providing a continuous record of global observations for a wide variety of climate variables.

So what do these climate records reveal? The global average surface temperature has increased by 0.74 °C

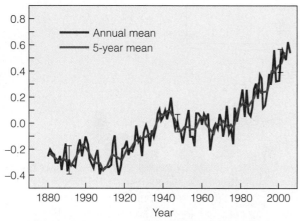

(a) Global-Mean Surface Temperature Anomaly (°C)

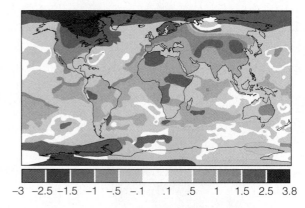

(b) 2006 Surface Temperature Anomaly (°C)

Figure 1 | **(a)** Combined annual land-surface air and sea-surface temperature anomalies (°C) from 1880 to 2006. The anomalies are the difference between annual temperature for any given year and the average annual temperature for the period 1951–1980. **(b)** Global pattern of surface temperature anomalies, as defined in (a), for the year 2006. (NASA GISS 2006.)

species richness at a regional scale to predict changes in bird and mammal diversity for the conterminous United States under conditions of a climate change (Figure 29.19). His analyses predict a northward shift in the regions of highest diversity, with species richness declining in the southern areas of the United States while increasing in New England, the Pacific Northwest, and in the Rocky Mountains and the Sierra Nevada.

These regional analyses of changes in species diversity in response to climate reflect correlations between species distributions and features of the physical environment. We know, however, that the distribution and abundance of species is also a function of species interactions within the community (competition, mutualism, predator–prey). Changes in the growth and reproductive rates of species in response to climate change may well influ-

ence the nature of these species interactions, altering patterns of zonation and succession (see Chapters 16–18). Given the difficulty of experimen-tally changing climate conditions in the field, few studies have examined these effects (see Field Studies: Erika Zavaleta). However, one such experiment was conducted in a meadow community in the Rocky Mountains of Colorado. Using electric heaters suspended 2.6 m above five experimental plots, scientists were able to raise soil temperature and influence soil moisture and the timing of snowmelt. In heated plots, the density of shrubs increased at the expense of grass and forb species. Results suggest that the increased warming expected under an atmosphere with a doubled concentration of CO_2 would shift the dominant vegetation of the widespread mountain meadow habitat. Shrubs would compete better in the altered environment.

$(\pm 0.2\,°C)$ since the early 20th century. The five warmest years in the instrumental record since 1850 are, in descending order, 2005, 1998, 2002, 2003, and 2006. New analyses of daily maximum and minimum land-surface temperatures for 1950 to 1993 show that the diurnal temperature range is decreasing. On average, minimum temperatures are increasing at about twice the rate of maximum temperatures ($0.2\,°C$ versus $0.1\,°C$/decade). In other words, nighttime temperatures (minimum) have increased more than daytime temperatures (maximum) over this period.

New analyses also indicate that global ocean heat content has increased significantly since the late 1950s. More than half of the increase in heat content has occurred in the upper 300 m of the ocean; in this layer the temperature has increased at a rate of about $0.04\,°C$/decade.

Scientists generally agree that the climate has changed significantly over the past century, but debate continues about the answer to the more difficult question: "Why is it changing?" The debate centers on two points. The first relates to the instrumental data measuring the trend in land-surface temperatures. Most weather stations are located in urban areas, which are typically warmer than the surrounding rural areas (see Ecological Issues: Urban Microclimates, p. 34). Recent studies, however, have worked to remove this potential bias from the data. Current findings by the IPCC have established that the warming trend over the past century is independent of the effects of urbanization.

The second point of debate relates to the difficulty in determining a meaningful long-term trend from an instrumental record that covers less than two centuries.

Climate varies on an array of timescales, and Earth has gone through periods of warming and cooling in the past. For example, the Northern Hemisphere is still recovering from the last glacial maximum, some 18,000–20,000 years before the present (see Section 18.9), when surface temperatures were much colder. Climate reconstructions of the more recent past (1000 years ago to the present), however, suggest that the warming trend observed over the past century is consistent with that expected from the rising atmospheric concentrations of greenhouse gases. A consensus has begun to emerge, with the most recent IPCC report (2007) stating that most of the observed increase in global average temperature since the mid-20th century is "very likely" due to observed changes in the atmospheric concentrations of greenhouse gases. The debate will likely continue for years to come; but the real question is, "Will the warming continue?"

1. Go to the website for the U.S. Global Change Research Program (National Climate Change Assessment) at www.usgcrp.gov/usgcrp/nacc/background/regions.htm. The report discusses predicted climate change for various regions of the United States. What changes in climate are predicted to occur in your region because of global climate change?

2. How might these changes affect the natural ecosystems found in your area (forest, grassland, coastal salt marshes, etc.)?

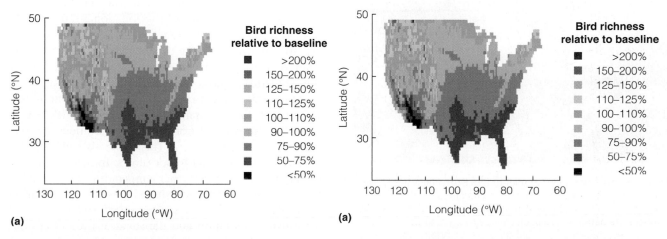

Figure 29.19 | Changes in **(a)** bird and **(b)** mammal species richness, relative to current species richness resulting from the climatic changes associated with doubling of atmospheric CO_2. Richness was projected using five GCMs.
(Adapted from Currie 2001.)

Such shifts have a major impact on plant communities as well as associated animal species.

In a similar approach, the International Tundra Experiment (ITEX) was established in late 1990 as a coordinated group of field experiments aimed at understanding the potential impact of warming at high latitudes on tundra ecosystems. Investigators from 13 countries are applying a range of standard field techniques including passive warming of tundra vegetation using open-top chambers and manipulating snow depth to alter growing season length (Figure 29.20). Studies are examining species-, community-, and ecosystem-level responses to warming in the Arctic region.

Changes in climate also affect vegetation indirectly, through decomposition and nutrient cycling. In terrestrial ecosystems, these processes depend on temperature and available moisture (see Section 21.4). Decomposition proceeds faster under warmer, wetter conditions. An ongoing experiment at Harvard Forest in Massachusetts is examining the effect of elevated soil temperatures on rates of decomposition and nutrient cycling in a forested ecosystem. Buried heating cables raise the soil temperature by 5°C. Initial results show a 60 percent increase in rates of soil respiration (CO_2 emissions), a direct result of increased microbial and root respiration. The results are consistent with patterns of soil respiration observed in other forests in warmer regions around the world. They indicate that greenhouse warming will increase rates of decomposition and microbial respiration, leading to a significant rise in emissions of CO_2 from soils to the atmosphere.

Figure 29.20 | The International Tundra Experiment (ITEX) uses small, passive, clear plastic, open-top chambers to warm the tundra and extend the growing season. The chambers raise the daily temperature of the tundra plant canopy by 1.5°C to 1.7°C, which is in the range predicted by global climate simulations. This experiment, which has been in place in the Barrow Environmental Observatory for 3 years and for 1 year at Atqasuk, Alaska, has already provided new insights on flower and growth responses to warming.

29.8 | Changing Climate Will Shift the Global Distribution of Ecosystems

By studying past climate fluctuations, ecologists have learned a great deal about the responses of terrestrial ecosystems to changing climate conditions. Pollen samples from sediment cores taken in lake beds have allowed paleobotanists to reconstruct the vegetation of many regions existing during the past 20,000 years. The work of Margaret Davis in reconstructing the distribution of tree species in eastern North America since the last glacial maximum (see Section 18.9 and Figure 18.19) is a good example. Tree genera migrated northward at different rates after the retreat of the glaciers. The migration rates depended on how well a species' physiology, dispersal ability, and competitive interactions with other tree species let it respond to changes in climate. Such studies show that the existing forest communities in eastern North America are a recent result of different responses of tree species to changing climate. As Earth's climate has changed in the past, the distribution and abundance of organisms and the communities and ecosystems they compose have changed (see Figure 18.20).

It is virtually impossible to develop experiments in the field to examine the long-term response of terrestrial ecosystems to a future climate change. This limitation means that scientists must base predictions on computer models, such as the one by Prasad and Iverson presented in Figure 29.16. Perhaps the simplest but most telling of these are the biogeographical models that relate the distribution of ecosystems to climate. Since the days of early naturalists, plant ecologists have recognized the link between climate and plant distribution (see the introduction to Part Seven). For example, tropical rain forests are found in the wet tropical regions of Central and South America, Africa, Asia, and Australia. According to the biogeographical model developed by L. R. Holdridge, within these regions the distribution of tropical rain forests is limited to areas where mean annual temperatures are at or above 24°C and annual precipitation is above 2000 mm. Tropical regions that meet these climate restrictions are shown on the map in Figure 29.21a. Under the changed temperature and rainfall patterns predicted by the United Kingdom Meteorological Office (UKMO) GCM for a doubled atmospheric CO_2 concentration, this distribution changes dramatically (Figure 29.21b). The region that can support tropical wet forest under this scenario shrinks by 25 percent. This decline is a direct result of drying due to higher temperatures. In some areas, the drying is caused by increased temperatures accompanied by decreased precipitation. In other areas, precipitation increases—but not enough to meet the increased demand for water (evaporation and transpiration) resulting from the increased temperatures.

Together with the demands of agriculture and forestry (see Chapter 27), this scenario would devastate the

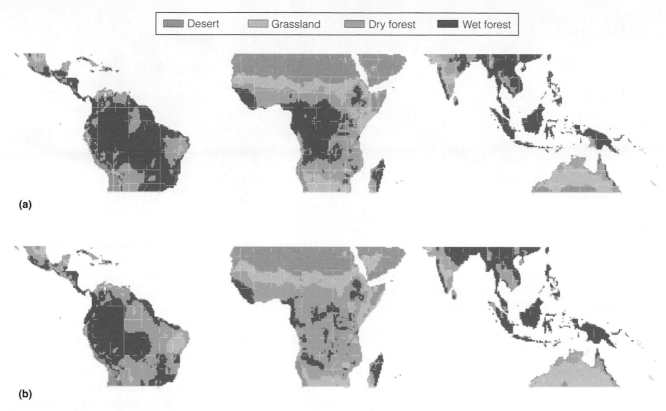

| Desert | Grassland | Dry forest | Wet forest |

(a)

(b)

Figure 29.21 | Maps of the areas in the tropical zone that could possibly support rain forest ecosystems as predicted by the Holdridge biogeographical model of ecosystem distribution. Map **(a)** is the area of tropical rain forest under current climate conditions, and **(b)** is the predicted area under changed climate conditions predicted by the United Kingdom Meteorological Office general circulation model for a doubled atmospheric CO_2 concentration.

(Adapted from Smith et al. 1992.)

tropical rain forest ecosystems as well as the diversity of life they support. Although tropical rain forests cover only 6 percent of the total land area, they are home to more than 50 percent of all terrestrial plant and animal species. Currently, deforestation in the tropics is the single major cause of species extinction, with annual rates of extinction ranging in the thousands of species (see Chapter 28). The loss of tropical rain forest predicted by the UKMO climate model would result in far more extinction.

Changes in global patterns of temperature would also affect the distribution of aquatic ecosystems (see Chapter 24). For instance, the global distribution of coral reefs is limited to the tropical waters where mean surface temperatures are at or above 20°C. Reef development is not possible where the mean minimum temperature is below 18°C. Optimal reef development occurs in waters where the mean annual temperatures are 23°C to 25°C, and some corals can tolerate temperatures up to 36°C to 40°C. A warming of the world's oceans would alter the potential range of waters in which reef development is possible, allowing for reef formation farther up the eastern coast of North America.

Ecologists are far from providing a complete analysis of the potential impacts of a global climate change. There

is little question, however, that changes in patterns of temperature and precipitation of the magnitude predicted by climate models will significantly influence the distribution and functioning of both terrestrial and aquatic ecosystems.

29.9 | Global Warming Would Raise Sea Level and Affect Coastal Environments

During the last glacial maximum some 18,000 years ago, sea level was 100 m lower than today. The highly productive shallow coastal waters, such as the continental shelf of eastern North America, were above sea level and were covered by terrestrial ecosystems (see Figure 18.20). As the climate warmed and the glaciers melted, sea levels rose. Over the past century, sea level has risen at a rate of 1.8 mm per year (Figure 29.22). This is due to the general pattern of global warming over this period and the associated thermal expansion of ocean waters and melting of glaciers. The 2007 IPCC report estimates that global mean sea level will rise by 0.18 to 0.59 m between the years 1990 and 2100 but with considerable regional variation. A rise of this magnitude will have serious effects on coastal environments from the perspectives of both natural ecosystems and human populations.

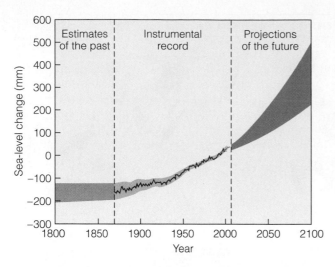

Figure 29.22 | Time series of global mean sea level (deviation from the 1980–1999 mean) in the past and as projected for the future. For the period before 1870, global measurements of sea level are not available. The grey shading shows the uncertainty in the estimated long-term rate of sea-level change. The red line is a reconstruction of global mean sea level from tide gauges, and the red shading denotes the range of variation. The green line shows global mean sea level observed from satellite data. The blue shading represents the range of model projections for a scenario of rising atmospheric concentrations of greenhouse gases developed by the IPCC (scenario A1B). Projected values for the 21st century are relative to the 1980 to 1999 mean.

(IPCC 2007.)

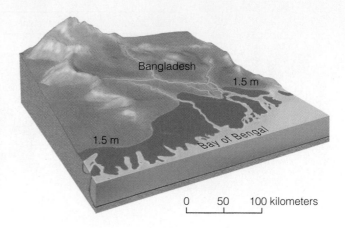

Figure 29.23 | Land area in Bangladesh that would be submerged (dark green area of map) if sea level rose by 1 m.

(Adapted from Nicholls and Leatherman 1995.)

A large portion of the human population lives in coastal areas; in fact, 13 of the world's 20 largest cities are located on coasts. Particularly vulnerable are delta regions, low-lying countries such as The Netherlands, Surinam, and Nigeria, and the smaller low-lying islands of the Pacific and other oceans. Bangladesh, an Asian country of about 120 million inhabitants, is located in the delta region of the Ganges, Brahmaputra, and Meghna Rivers (Figure 29.23). Approximately 25 percent of the country's population lives in areas below 3 m above sea level; about 7 percent of the country's habitable land and 6 million people reside in areas less than 1 m above sea level. Estimates of sea-level rise in this region due to a combination of land subsidence (a result of land collapsing in response to removal of groundwater) and global warming are 1 m by the year 2050 and 2 m by the year 2100. Although there is great uncertainty in these estimates, the effect on Bangladesh would be devastating.

Other coastal regions of Southeast Asia and Africa would be equally affected by the predicted rise in sea level. In Egypt, about 12 percent of the arable land, with a population over 7 million, would be affected by a rise in sea level of 1 m. In the coastal areas of eastern China, a sea-level rise of just half a meter would inundate an area of approximately 40,000 km^2 where more than 30 million people currently live.

Small islands are also particularly vulnerable to a sea-level rise. More than half a million people live in the archipelagos of small islands and coral atolls. Two examples are the Maldives in the Indian Ocean and the Marshall Islands of the Pacific. These island chains lie almost entirely below 3 m above sea level. A half-meter or more rise in sea level would dramatically reduce their land area and have a devastating impact on groundwater (freshwater) supply.

A sea-level rise will also have major effects on coastal ecosystems. Among these effects are direct inundation of low-lying wetlands and dryland areas, erosion of shorelines through loss of sediments, increased salinity of estuaries and aquifers, rising coastal water tables, and increased flooding and storm surges. Estuarine and mangrove ecosystems (see Chapter 25) would be highly susceptible to a sea-level rise of the magnitude predicted. Coastal salt marshes are dependent on the twice-daily tidal inundation of saltwater mixing with the freshwater provided by streams and rivers. Patterns of water depth, temperature, salinity, and turbidity are critical to maintaining these ecosystems. The invasion of saltwater farther into the estuary due to a rise in sea level would be disastrous and could cause salination of land adjacent to the estuary margins. Estuarine and mangrove environments are critical to coastal fisheries. More than two-thirds of the fish caught for human consumption, as well as many birds and animals, depend on the coastal marshes and mangroves for part of their life cycles.

29.10 | Climate Change Will Affect Agricultural Production

Despite technological advances in improved crop varieties and methods of irrigation, climate and weather remain the key factors determining agricultural production.

Changes in the global climate patterns will exacerbate an already increasing problem of feeding the world population, which is predicted to double in size over the next half century.

The major cereal crops that feed people—wheat, corn (maize), and rice—are domesticated species. Like native species, these crops exhibit environmental tolerances to temperature and moisture that control survival, growth, and reproduction. Changes in regional climate conditions will directly influence their suitability and produc-

tivity and therefore current patterns of agricultural production. However, these changes will be complex, with economic and social factors interacting to influence patterns of global food production and distribution.

In examining potential effects of greenhouse warming on agricultural production (Figure 29.24), increasing concentrations of CO_2 as well as climate changes must be considered. The results of many studies suggest that most crop species (and varieties) will benefit from a rise in CO_2 concentration (see Figure 29.7). For

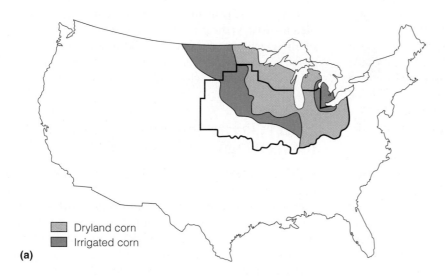

Dryland corn
Irrigated corn

(a)

Figure 29.24 | Regional shifts in areas suitable for crop production under a changed climate as predicted by the Goddard Institute for Space Studies GCM: **(a)** Shift in the region suitable for corn production in the United States. **(b)** Shift in areas suitable for irrigated rice production in northern Japan. Areas in dark green are suitable for irrigated rice production.

(a) (Adapted from Blasing and Solomon 1983.);
(b) (Adapted from Yoshino et al. 1988.)

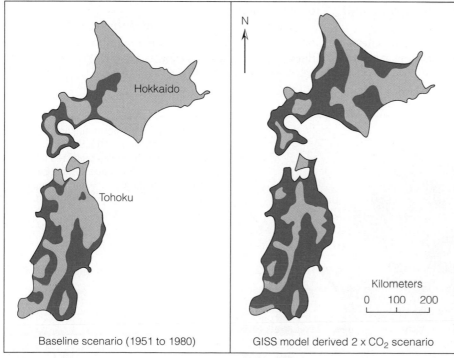

N

Hokkaido

Tohoku

Kilometers
0   100   200

Baseline scenario (1951 to 1980)   GISS model derived 2 x CO_2 scenario

(b)

example, in an experiment in Arizona, cotton and spring wheat were grown in field conditions under elevated CO_2 and irrigation. Cotton yield increased by 60 percent and wheat by more than 10 percent compared with crops grown under identical field conditions and ambient concentrations of CO_2.

One of the simplest ways to evaluate the potential implications of a climate change on agriculture is to examine shifts in the geographic range of certain crop species as they relate directly to climate. For example, an average daily temperature increase of 1°C during the growing season would shift the "corn belt" (region of highest corn production) of the United States significantly to the north (Figure 29.24a). A similar analysis for the shift in suitable regions for irrigated rice production in northern Japan is shown in Figure 29.24b. In both examples, the shifts in agricultural zones imply significant changes in regional land-use patterns, with associated economic and social costs. Although analyses of this type can provide insight into changing patterns of regional agricultural production, evaluating the actual effect on global food production and markets requires a more detailed interdisciplinary approach.

The Environmental Change Unit at Oxford University has carried out a collaborative study with agricultural scientists from 18 countries to examine the regional and global impacts of climate change on world agricultural production. Various assumptions were made about farmers' ability to adapt to changing environmental conditions through shifts in the species or varieties of crops grown or changes in agricultural practices such as irrigation. The analysis also assumes a continuation of current economic growth rates, certain changes in current trade restrictions, and projected estimates of population growth.

A major finding of the study is that the negative effects of climate change are to some extent compensated for by increased productivity resulting from elevated atmospheric concentrations of CO_2. The net effect of a climate change as predicted by the general circulation models, including a doubling of atmospheric CO_2 concentrations, is to reduce the global production of cereal crops by up to 5 percent. An important point is that this reduction is not evenly distributed across the globe, or even within a given region or country (Figure 29.25).

The predicted changes would increase the current disparity in cereal crop production between developed and developing countries. Results of the study tend to show an increase in production in developed countries, particularly in the middle latitudes (temperate regions). In contrast, production in developing nations, as a group, would decline by as much as 10 percent, with an associated increase in the population at risk of hunger. In many of these regions, climatic variability and marginal climatic conditions for agriculture are worsened under the predicted patterns of global climate change.

29.11 | Climate Change Will Directly and Indirectly Affect Human Health

Climatic change will have various direct and indirect effects on human health. Direct effects would include increased heat stress, asthma, and other cardiovascular and respiratory ailments. Indirect health effects are likely to include increased incidence of communicable diseases, increased mortality and injury due to increased natural disasters (floods, hurricanes, etc.), and changes in diet and nutrition due to changed agricultural production.

Several studies have examined the direct relationship between maximum summer temperatures and mortality rates. Climate change is expected to change the frequency

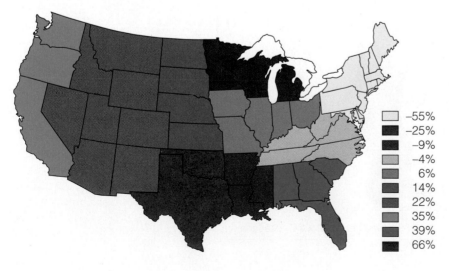

Figure 29.25 | Changes in regional crop production by year 2060 for the United States under a climate change as predicted by the Goddard Institute for Space Studies GCM (assuming an average 3°C increase in temperature, 7 percent increase in precipitation, and 530 ppm).

(Adapted from Adams et al. 1995.)

	−55%
	−25%
	−9%
	−4%
	6%
	14%
	22%
	35%
	39%
	66%

of very hot days. For example, if average July temperature in Chicago, Illinois, were to rise by 3°C, the probability that the heat index will exceed 35°C (95°F) during the month increases from 1 in 20 to 1 in 4. In the United States, warm and humid conditions during summer nights lead to the highest mortality. The greatest death toll in the United States occurred during the summer of 1936, when 4700 excess deaths were recorded due to heat-related causes. In recent decades, 1200 excess deaths occurred in Dallas during the summer of 1980 and 566 in Chicago during July 1995 (Figure 29.26). Analyses of climate change scenarios show a significant rise in heat-related mortality in all regions of the United States over the next several decades (Figure 29.27). During heat waves, cardiovascular and respiratory illnesses are the major causes of mortality. The elderly and children are typically in the greatest danger during these periods.

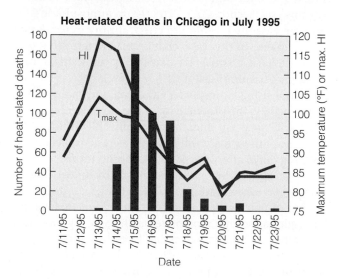

Figure 29.26 | This graph tracks maximum temperature (T$_{max}$), heat index (HI), and heat-related deaths in Chicago each day from July 11 to 23, 1995. The maroon line shows maximum daily temperature, the red line shows the heat index, and the bars indicate number of deaths for the day.

(Adapted from NOAA/NCDC.)

In addition to direct heat-related mortality, the distribution and rates of transmission for various infectious diseases will be influenced by changes in regional climate patterns. Disease consists of agents—such as viruses, bacteria, or protozoa—and host organisms (humans). Some diseases are transmitted to humans by intermediate organisms, or vectors (see Section 15.4). Insects are a primary vector of human disease. Although acting as carriers, the insects themselves are not affected by the disease agent. Insect-borne viruses (referred to as arbovirus, short for arthropod-borne virus) cover a wide variety of diseases. The most common insects involved in transmitting arboviruses are mosquitoes, ticks, and blood flukes (schistosomes). Approximately 102 arboviruses can produce disease in humans. Of that number, approximately 50 percent have been isolated from mosquitoes. Insects that carry these disease agents are adapted to specific ecosystems for survival and reproduction and exhibit specific tolerances to climatic features such as temperature and humidity. Changes in the climate will affect their distribution and abundance, as is true of the Eastern phoebe (see Section 29.7, Figure 29.17).

One insect-borne disease is malaria, a recurring infection produced in humans by protozoan parasites transmitted by the bite of an infected female mosquito of the genus *Anopheles*. The optimal breeding temperature for *Anopheles* is 20°C to 30°C, with relative humidity over 60 percent. Mosquitoes die at temperatures above 35°C and relative humidity less than 25 percent. Forty percent of the world's population is currently at risk for malaria, and more than 2 million people are killed each year by this disease. The current distribution of malaria will be extensively modified under a climate change. Expansion of the *Anopheles* mosquito's geographic range into currently more temperate climates is expected to increase the proportion of the world's population at risk of this infectious disease to more than 60 percent by the latter part of the 21st century.

Dengue and yellow fever, two related viral diseases, are also transmitted by mosquitoes. Their vector is the mosquito *Aedes aegypti*, which is adapted to the urban environment. Colonization by this mosquito is limited to areas with an average daily temperature of 10°C or greater. The virus that causes yellow fever lives in mosquitoes only when temperatures exceed 24°C under high relative humidity. Epidemics occur when mean annual temperatures exceed 20°C, making this a disease of the tropical forested regions. Yellow fever is currently prevalent across Africa and Latin America but has been detected as far north as the midlatitude ports of Bristol, Philadelphia, and Halifax, where the mosquitoes have survived in the water tanks of ships that have traveled from tropical regions. A climate change would directly influence the distribution of both the virus and its vector, the mosquito.

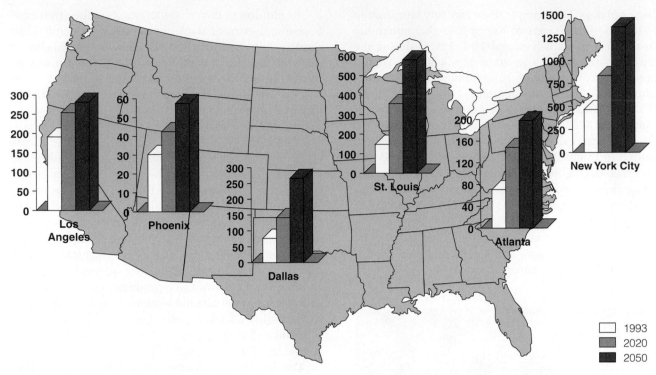

Figure 29.27 | Average annual excess weather-related mortality for the years 1993, 2020, and 2050 in various cities of the United States. Future projections of weather-related mortality are based on changes in climate predicted by the Geophysical Fluid Dynamics Laboratory GCM.

(Adapted from Kalkstan and Green 1997.)

29.12 | Understanding Global Change Requires the Study of Ecology at a Global Scale

Increasing atmospheric concentrations of CO_2 and other greenhouse gases, and the potential changes in global climate patterns that may result, present a new class of ecological problems. To understand the effect of rising CO_2 emissions from fossil fuel burning and land clearing, we have to examine the carbon cycle on a global scale (see Figure 22.4), linking the atmosphere, hydrosphere, biosphere, and lithosphere. The discussion in previous sections focuses on rising CO_2 concentrations and the effects of climate changes on populations, communities, and ecosystems; but the possible effects are not unidirectional. As presented in Section 29.8, ecosystems also influence atmospheric CO_2 and regional climate patterns. For example, if climate does change as shown in Figure 29.20, the global distribution and abundance of tropical rain forests will decline dramatically. Tropical rain forests are the most productive terrestrial ecosystems on the planet. A significant decline in these ecosystems will reduce global primary productivity, the uptake of CO_2 from the atmosphere, and CO_2 storage as organic carbon in biomass. In fact, as tropical rain forests shrink, atmospheric CO_2 will increase. The drying of these regions

will kill trees, increase fires, and transfer carbon stored in the living biomass to the atmosphere as CO_2 in much the same way that forest clearing does in these regions (see Section 29.2). The rise in atmospheric CO_2 will increase the greenhouse effect, further exacerbating the problem. In this case, changes in the terrestrial surface act as a positive feedback loop to rising atmospheric concentrations of CO_2.

But if rising atmospheric CO_2 level and changing climate increase the productivity of the world's ecosystems, they will take up more CO_2 from the atmosphere. Increased productivity will function as negative feedback, drawing down atmospheric CO_2 concentrations.

Besides changing climate indirectly by influencing atmospheric concentrations of CO_2, changes in the distribution of rain forests can affect climate directly by altering regional precipitation patterns. In some regions, such as extensive areas of tropical rain forest, a significant portion of the precipitation is water that has been transpired from the vegetation in the area. In effect, water is being recycled locally through the hydrologic cycle (see Figure 3.2). Removing the forest (either through deforestation or by shifting the distribution of ecosystems as in Figure 29.21) reduces transpiration and increases runoff to rivers that transport water from the area. Experiments using GCMs and regional climate models have examined the potential effects of large-scale deforestation in the

Amazon Basin. Findings suggest that the loss of forest cover would result in a significant reduction in annual precipitation by reducing the internal cycling of water within the forest. This would effectively change the region's climate, making it unlikely that the rain forest could be reestablished.

Gordon Bonan and colleagues at the National Center for Atmospheric Research (NCAR) in Boulder, Colorado, have presented another example of a direct influence of terrestrial ecosystems on regional climate. The largest degree of warming under elevated CO_2 is in the northern latitudes (see Section 29.6). The predicted warming would significantly reduce snow cover in this region and shift the distribution of boreal forests to the north. A major factor influencing the relative absorption and reflection of shortwave radiation (solar radiation) by Earth's surface is its albedo. Albedo is an index of the ability of a surface to reflect solar radiation back to space. Snow has a high albedo, or reflectance, whereas vegetation (with its darker color) has a low albedo. Both reduced snow cover and the northern movement of the boreal forest would reduce the regional albedo, therefore increasing the amount of solar radiation absorbed by Earth's surface. This increase in the absorption of radiation would further increase regional temperatures, functioning as a positive feedback loop.

These are not simple connections. To understand the interactions among the atmosphere, oceans, and terrestrial ecosystems requires that ecologists study Earth as a single, integrated system. It is only by developing a global ecology that ecologists, working with oceanographers and atmospheric scientists, will come to understand the potential consequences of the increasing concentration of CO_2 in the atmosphere over the next century.

Summary

Greenhouse Gases and Climate | 29.1

A variety of chemical compounds in the atmosphere absorb thermal (longwave) radiation emitted by the Earth's surface and atmosphere. These compounds are referred to as greenhouse gases. The atmosphere is warmed by this mechanism, referred to as the greenhouse effect. Concentrations of these greenhouse gases have been rising, causing concern over possible impacts on Earth's climate.

Rising Atmospheric Concentrations of CO_2 | 29.2

Direct observations beginning in 1958 reveal an exponential increase in the atmospheric concentration of CO_2. The rise is a direct result of fossil fuel combustion and the clearing of land for agriculture.

Fate of CO_2 Emissions | 29.3

Of the CO_2 released from fossil fuel combustion and land clearing, only about 60 percent remains in the atmosphere. The remainder is taken up by the oceans and terrestrial ecosystems. Calculations of diffusion of CO_2 into the surface waters provide an estimate of uptake by the oceans. Carbon uptake by terrestrial ecosystems is calculated as the difference between inputs to the atmosphere, atmospheric concentrations, and uptake by oceans.

Ocean Uptake | 29.4

More than 85 percent of the ocean volume is deep water (>200 m depth). Carbon dioxide diffuses from the atmosphere into surface waters. The rise in atmospheric concentrations results in increased uptake of CO_2 into surface waters. The thermocline limits vertical mixing and therefore the rate of CO_2 transfer from the surface to deep waters.

Plant Response | 29.5

In general, plants respond to increased atmospheric CO_2 with higher rates of photosynthesis and partial closure of stomata. These responses increase water-use efficiency. Responses to long-term exposure vary, including increased allocation of carbon to the production of roots, reduced allocation to the production of leaves, and reduction in stomatal density. Scientists are studying long-term effects on net primary productivity.

Global Climate Change | 29.6

Carbon dioxide is a greenhouse gas. It traps longwave radiation emitted from Earth's surface, warming the atmosphere. Rising atmospheric concentrations of CO_2 and other greenhouse gases could raise the global mean temperature by $1.1\,°C$ to $6.4\,°C$ by the year 2100. Warming will not be uniform over Earth. The greatest warming is predicted to come during the winter months and at northern latitudes. Increased variability in climate is predicted, including changes in precipitation and the frequency of storms. The input of sulfates and other aerosols from human sources acts to reduce the input of solar radiation to Earth's surface, thereby reducing warming.

Climate Change and Ecosystems | 29.7

The distribution and abundance of species will shift as temperature and precipitation change. Changes in climate will influence the competitive ability of species and thus change patterns of community zonation and succession. Ecosystem processes such as decomposition and nutrient cycling are sensitive to temperature and moisture, and changing climate will affect them.

Distribution of Ecosystems | 29.8

Changes in climate also will shift the distribution and abundance of both terrestrial and aquatic ecosystems. These changes in ecosystem distribution influence global patterns of plant and animal diversity.

Climate Change and Sea Level | 29.9

Sea level is currently rising globally at an average rate of 1.8 mm per year. It is estimated that global warming will cause sea level to rise 0.18 to 0.59 m by 2100, as the polar ice caps melt and warmer ocean waters expand. A sea-level rise of this magnitude will have major effects on people living in coastal areas. In addition, rising sea level will affect coastal ecosystems such as beaches, estuaries, and mangroves.

Climate Change and Agriculture | 29.10

Climate change will affect global agricultural production. Decreases in crop production from drier conditions will be partially offset by increased rates of photosynthesis under elevated atmospheric CO_2 levels; however, current models project a 5 percent decline in global production of cereal crops. This decline is not distributed evenly. Developed countries in the middle latitudes will realize a slight increase, whereas production in developing countries in the tropics will decline. The result will be increased hunger.

Climate Change and Human Health | 29.11

Climate change will have both direct and indirect effects on human health. Mortality rates are expected to rise as a result of heat-related deaths associated with respiratory and cardiovascular ailments. Indirect health effects include increased mortality and injury from climate-related natural disasters, as well as changes in diet and nutrition resulting from changes in agricultural production. Distribution and transmission rates of various insect-borne infectious diseases that are directly related to climate, such as malaria, will also be affected.

Global Ecology | 29.12

To understand the effect of rising atmospheric concentrations of greenhouse gases and global climate change, we have to study the whole Earth as a single, complex system.

Study Questions

1. Why is CO_2 called a greenhouse gas?
2. What are the major sources of greenhouse gases, especially CO_2?
3. Not all of the CO_2 released to the atmosphere remains there. What happens to the rest?
4. How does elevated CO_2 influence rates of photosynthesis and transpiration?
5. What limits the transfer of CO_2 from the surface waters of the ocean to the deep waters?
6. How might changes in climate (temperature and precipitation) influence the distribution of plant and animal species?
7. How might changes in climate influence the distribution and abundance of terrestrial ecosystems?
8. How is sea level currently changing?
9. How will global warming influence sea levels?
10. How might rising sea levels influence human populations? Coastal environments?
11. How might changes in climate influence agricultural production? How will rising CO_2 levels influence crop production?
12. How might climate change influence human health, both directly and indirectly?

Further Readings

Bazzaz, F. A. 1996. *Plants in changing environments: Linking physiological, population, and community ecology.* New York: Cambridge University Press.
> This book provides several excellent examples of experimental approaches that examine the response of plant species to elevated levels of carbon dioxide.

Graedel, T. E., and P. J. Crutzen. 1995. *Atmosphere, climate, and change.* Scientific American Library. New York: W. H. Freeman.
> An outstanding introduction to the issue of global climate change.

Houghton, J. 1997. *Global warming: The complete briefing*. Cambridge, UK: Cambridge University Press.

A condensed version of the findings of the IPCC report referenced below, written by the chairman of the IPCC.

Intergovernmental Panel on Climate Change (IPCC) Fourth Assessment Report—Climate Change. 2007. Available at http://www.ipcc.ch/.

The summary of the IPCC assessment report provides a comprehensive overview of the climate change issue. Detailed scientific reports are also available at the website.

National Assessment Synthesis Team, U.S. Global Change Research Program. 2000. Climate change impacts on the United States: The potential consequences of climate variability and change. U.S. national assessment. Available at http://www.usgcrp.gov/usgcrp/nacc/default.htm.

This report provides a summary of the U.S. national assessment mandated by Congress. Detailed summaries of regional impacts can be found at the website.

Peters, R. L., and T. E. Lovejoy, eds. 1992. *Global warming and biological diversity*. New Haven: Yale University Press.

A sobering, readable examination of the potential effect of global warming on vegetation, soils, animals, parasites and diseases, and ecosystems.

Schneider, S. H. 1989. *Global warming*. San Francisco: Sierra Club Books.

Highly readable examination of global warming and the debate engendered in politics and the media.

Watson, R. T., M. C. Zinyowera, R. H. Moss, and D. J. Dokken, eds. 1998. The regional impacts of climate change (a special report of IPCC Working Group II). Cambridge, UK: Cambridge University Press.

This short volume provides an excellent overview of the potential impacts of climate change on specific regions of the globe.

Woodward, F. I., ed. 1992. *Global climate change: The ecological consequences*. London: Academic Press.

The emphasis of this series of papers is to provide an overview of climate change research.

References

Chapter 1 | The Nature of Ecology

Bronowski, J. 1956. *Science and human values.* Harper & Row.

Brower, J. E., and J. H. Zar. 1984. *Field and laboratory methods for general ecology.* 2nd ed. Dubuque, IA: Wm. Brown.

Cox, G. W. 1985. *Laboratory manual for general ecology.* 5th ed. Dubuque, IA: Wm. Brown.

Egerton, F. N. (ed.). 1977. *History of American ecology.* New York: Arno Press.

Golley, F. B. 1993. *A history of the ecosystem concept in ecology: More than the sum of its parts.* New Haven, CT: Yale University Press.

Hagen, J. B. 1992. *An entangled bank: The origins of ecosystem ecology.* New Brunswick, NJ: Rutgers University Press.

Kingsland, S. 1995. *Modeling nature.* Chicago: University of Chicago Press.

McIntosh, R. P. 1985. *The background of ecology: Concept and theory.* New York: Cambridge University Press.

Sheall, J. (ed.). 1988. *Seventy-five years of ecology: The British Ecological Society.* Oxford: Blackwell Scientific.

Worster, D. 1977. *Nature's economy.* San Francisco: Sierra Club Books.

Chapter 2 | Climate

Ahrens, C. D. 2003. *Meteorology today.* 7th ed. Pacific Grove, CA: Brooks/Cole.

Berry, R. G., and R. J. Corley. 1998. *Atmosphere, weather, and climate.* 7th ed. New York: Routledge.

Gates, D. M. 1962. *Energy exchange in the biosphere.* New York: Harper & Row.

Geiger, R. 1965. *Climate near the ground.* Cambridge, MA: Harvard University Press.

Graedel, T. E., and P. J. Crutzen. 1997. *Atmosphere, climate and change.* New York: Scientific American Library.

Landsberg, H. E. 1970. Man-made climatic changes. *Science* 170:1265–1274.

Lee, R. 1978. *Forest microclimatology.* New York: Columbia University Press.

Chapter 3 | The Aquatic Environment

Bainbridge, R., G. C. Evans, and O. Rackham (eds.). 1966. *Light as an ecological factor.* Oxford: Blackwell Scientific.

Clark, G. A. 1939. Utilization of solar energy by aquatic organisms. *Problems in Lake Biology* 10:27–38.

Garrison, T. 2005. *Essentials of oceanography.* 4th ed. St. Paul, MN: Brooks/Cole.

Hynes, H. B. N. 1970. *The ecology of running water.* Toronto: University of Toronto Press.

Likens, G. E. (ed.). 1985. *An ecosystem approach to aquatic ecology: Mirror Lake and environment.* New York: Springer-Verlag.

Nybakken, J. W., and M. Bertness. 2004. *Marine biology: An ecological approach.* 6th ed. San Francisco: Benjamin Cummings.

Wetzel, R. G. 1983. *Limnology.* 2nd ed. Philadelphia: Saunders College Publishing.

Chapter 4 | The Terrestrial Environment

Boul, S. W., F. D. Hole, R. J. McCracken, and R. J. Southward. 1997. *Soil genesis and classification.* Ames: Iowa State University Press.

Brady, N. C., and R. W. Weil. 2001. *The nature and properties of soils.* 13th ed. Upper Saddle River, NJ: Prentice Hall.

Brown, L. R., and E. C. Wolf. 1984. *Soil erosion: Quiet crisis in the world economy.* Washington, DC: Worldwatch Institute.

Cernusca, A. 1976. Energy exchange within individual layers of a meadow. *Oecologica* 23:148.

Farb, P. 1959. *The living earth.* New York: Harper & Row.

Hutchinson, B. A., and D. R. Matt. 1977. The distribution of solar radiation within a deciduous forest. *Ecological Monographs* 47:185–207.

Jenny, H. 1980. *The soil resource.* New York: Springer-Verlag.

Killham, K. 1994. *Soil ecology.* New York: Cambridge University Press.

Lutz, H. J., and R. F. Chandler. 1946. *Forest soils.* New York: Wiley.

Patton, T. R. 1996. *Soils: A new global view.* New Haven, CT: Yale University Press.

Pfitsch, W. A., and R. W. Pearcy. 1989. Daily carbon gain by *Adenocaulon bicolor* (Asteraceae), a redwood forest understory herb, in relation to its light environment. *Oecologia* 80:465–470.

Reifsnyder, W. E., and H. W. Lull. 1965. *Radiant energy in relation to forests.* Technical Bulletin No. 1344. Washington, DC: U.S. Department of Agriculture.

Soil Survey Division Staff. 1993. *Soil survey manual.* Handbook 18. Washington, DC: U.S. Government Printing Office.

Chapter 5 | Ecological Genetics: Adaptation and Evolution

Boag, P. T., and P. R. Grant. 1981. Intense natural selection in a population of Darwin's finches (*Geospiza*) in the Galapagos. *Science* 214:82–85.

———. 1984. The classical case of character release: Darwin's finches (*Geospiza*) on Isla Daphne Major, Galapagos. *Biological Journal of the Linnean Society* 22:243–287.

Darwin, C. 1859. *The origin of species.* Philadelphia: McKay. [Reprinted from 6th London edition.]

Desmond, A., and J. Moore. 1991. *Darwin.* New York: Warner Books.

Futuyma, D. J. 1984. *Evolutionary biology.* Sunderland, MA: Sinauer Associates.

Grant, P. 1999. *Ecology and evolution of Darwin's finches.* Princeton, NJ: Princeton University Press.

Grant, V. 1971. *Plant speciation.* New York: Columbia University Press.

———. 1985. *The evolutionary process: A critical review of evolutionary theory.* New York: Columbia University Press.

Hartl, D. 1988. *A primer of population genetics.* Sunderland, MA: Sinauer Associates.

Highton, R. 1995. Speciation in eastern North American salamanders of the genus *Plethodon*. *Annual Review of Ecology and Systematics* 26:579–600.

Lack, D. 1974. *Darwin's finches: An essay on the general biological theory of evolution.* London: Cambridge University Press.

Mayr, E. 1963. *Animal species and evolution.* Cambridge, MA: Harvard University Press.

_____. 1991. *One long argument.* Cambridge, MA: Harvard University Press.

Reznick, D., F. H. Shaw, F. H. Rodd, and R. G. Shaw. 1997. Evaluation of the rate of evolution in natural populations of guppies (*Poecilia reticulata*). *Science* 275:1934–1937.

Sultan, S. E., and F. A. Bazzaz. 1993. Phenotypic plasticity in *Polygonum persicaria*. II. Norms of reaction to soil moisture and the maintenance of genetic diversity. *Evolution* 47:1032–1049.

Chapter 6 | Plant Adaptations to the Environment

Augspurger, C. K. 1982. Light requirements of neotropical tree seedlings: A comparative study of growth and survival. *Journal of Ecology* 72:777–795.

Bjorkman, O. 1973. Comparative studies on photosynthesis in higher plants. Page 53 in A. C. Geise, ed., *Photophysiology.* New York: Academic Press.

Bradshaw, A. D., M. J. Chadwick, D. Jowett, and R. W. Snaydon. 1964. Experimental investigations into the mineral nutrition of several grass species. IV. Nitrogen level. *Journal of Ecology* 52:665–676.

Chapin, F. S., III. 1980. The mineral nutrition of wild plants. *Annual Review of Ecology and Systematics* 11:233–260.

Chazdon, R. L., and R. W. Pearcy. 1991. The importance of sunflecks for forest understory plants. *BioScience* 41:760–765.

Dale, J. E. 1992. How do leaves grow? *BioScience* 42:423–432.

Davies, S. J. 1998. Photosynthesis of nine pioneer *Macaranga* species from Borneo in relation to life-history traits. *Ecology* 79:2292–2308.

Davies, S. J., P. A. Palmiotto, P. S. Ashton, H. S. Lee, and J. V. LaFrankie. 1998. Comparative ecology of 11 sympatric species of *Macaranga* in Borneo: Tree distribution in relation to horizontal and vertical resource heterogeneity. *Journal of Ecology* 86:662–673.

Feldman, L. J. 1988. The habits of roots. *BioScience* 38:612–618.

Field, C. B., and H. Mooney. 1986. The photosynthetic-nitrogen relationship in wild plants. Pages 25–55 in T. J. Givnish, ed., *On the economy of plant form and function.* Cambridge: Cambridge University Press.

Grime, J. 1971. *Plant strategies and vegetative processes.* New York: Wiley.

Kirk, J. T. O. 1983. *Light and photosynthesis in aquatic systems.* New York: Cambridge University Press.

Kozlowski, T. T. 1982. Plant responses to flooding of soil. *BioScience* 34:162–167.

Lambers, H., F. S. Chapin III, and T. L. Pons. 1998. *Plant physiological ecology.* New York: Springer.

Larcher, W. 1996. *Physiological plant ecology.* 3rd ed. New York: Springer-Verlag.

Mooney, H. A. 1986. Photosynthesis. Chapter 11 in M. J. Crawley, ed., *Plant ecology.* New York: Blackwell Scientific.

Mooney, H. A., O. Bjorkman, J. Ehleringer, and J. Berry. 1976. Photosynthetic capacity of in situ Death Valley plants. *Carnegie Institute Yearbook* 75:310–413.

Parsons, T., J. R. Parsons, M. Takahashi, and B. Hargrave. 1984. *Biological oceanographic processes,* 3rd ed. New York: Pergamon Press.

Pearcy, R. W. 1977. Acclimation of photosynthetic and respiratory CO_2 to growth temperature in *Atriplex lentiformis*. *Plant Physiology* 61:484–486.

Reich, P. B., D. S. Ellsworth, and M. B. Walters. 1998. Leaf structure (specific leaf area) regulates photosynthesis-nitrogen relations: Evidence from within and across species and functional groups. *Functional Ecology* 12:948–958.

Reich, P. B., M. G. Tjoelker, M. B. Walters, D. Vanderklein, and C. Buschena. 1998. Close association of RGR, leaf and root morphology, seed mass and shade tolerance in seedlings of nine boreal tree species grown in high and low light. *Functional Ecology* 12:327–338.

Schulze, E. D., R. H. Robichaux, J. Grace, P. W. Rundel, and J. R. Ehleringer. 1987. Plant water balance. *BioScience* 37:30–37.

Turner, N. C., and P. J. Kramer (eds.). 1980. *Adaptation of plants to water and high temperature stress.* New York: Wiley.

Woodward, I., and T. Smith. 1993. Predictions and measurements of the maximum photosynthetic rate, A_{max} at a global scale. Pages 491–508 in E. D. Schultz and M. M. Caldwell, eds., *Ecophysiology of photosynthesis.* Vol. 100. Berlin: Springer.

Chapter 7 | Animal Adaptations to the Environment

Adkisson, P. L. 1966. Internal clocks and insect diapause. *Science* 154:234–241.

Aschoff, J., S. Daan, and G. A. Gross (eds.). 1982. *Vertebrate circadian rhythms.* New York: Springer-Verlag.

Belovsky, G. E., and P. F. Jordan. 1981. Sodium dynamics and adaptations of a moose population. *Journal of Mammalogy* 63:613–621.

Cheatum, E. L., and C. W. Severinghaus. 1950. Variations in fertility of white-tailed deer related to range conditions. *Transactions of the North American Wildlife Conference* 15:170–189.

DeCoursey, P. J. 1960a. Daily light sensitivity rhythm in a rodent. *Science* 131:33–35.

_____. 1960b. Phase control of activity in a rodent. *Cold Spring Harbor Symposium on Quantitative Biology* 25:49–54.

_____. (ed.). 1976. *Biological rhythms in the marine environment.* Columbia: University of South Carolina Press.

Dodd, R. H. 1973. Insect nutrition: Current developments and metabolic implications. *Annual Review of Entomology* 18:381–420.

Edmunds, L. N. 1988. *Cellular and molecular bases of biological clocks.* New York: Springer-Verlag.

Farner, D. S. 1985. Annual rhythms. *Annual Review of Physiology* 47:65–82.

French, A. R. 1992. Mammalian dormancy. Pages 105–121 in T. E. Tomasi and T. H. Horton, eds., *Mammalian energetics.* Ithaca, NY: Comstock.

Gilles, R. (ed.). 1979. *The mechanisms of osmoregulation in animals.* New York: Wiley.

Heinrich, B. (ed.). 1981. *Insect thermoregulation.* New York: Wiley.

Hill, E. P., III. 1972. Litter size in Alabama cottontails as influenced by soil fertility. *Journal of Wildlife Management* 36:1199–1209.

Hill, R. W. 1992. The altricial/precocial contrast in the thermal relations and energetics of small mammals. Pages 122–159 in T. E. Tomasi and T. H. Horton, eds., *Mammalian energetics.* Ithaca, NY: Comstock.

Hill, R. W., and G. A. Wyse. 1989. *Animal physiology.* New York: Harper & Row.

Johnson, C. H., and J. W. Hastings. 1986. The elusive mechanisms of the circadian clock. *American Scientist* 74:29–36.

Jones, R. L., and H. P. Weeks. 1985. Ca, Mg, and P in the annual diet of deer in south-central Indiana. *Journal of Wildlife Management* 49:129–133.

Lee, R. E., Jr. 1989. Insect cold-hardiness: To freeze or not to freeze. *BioScience* 39:308–313.

Leith, H. (ed.). 1974. *Phenology and seasonality modeling.* New York: Springer-Verlag.

Lyman, C. P., A. Malan, J. S. Willis, and L. C. H. Wang. 1982. *Hibernation and torpor in mammals and birds.* New York: Academic Press.

Maloly, C. M. O. (ed.). 1979. *Comparative physiology of osmoregulation in animals.* New York: Academic Press.

Naylor, E. 1985. Tidal rhythmic behaviour of marine animals. *Symposium of Society of Experimental Biology* 39:63–93.

Palmer, J. D. 1990. The rhythmic lives of crabs. *BioScience* 40:352–358.

———. 1996. Time, tide, and living clocks of marine organisms. *American Scientist* 84:570–578.

Schmidt-Neilsen, K. 1997. *Animal physiology: Adaptation and environment.* 5th ed. New York: Cambridge University Press.

Storey, K. B., and J. M. Storey. 1996. Natural freezing survival in animals. *Annual Review of Ecology and Systematics* 27:365–386.

Takahashi, J. S., and M. Hoffman. 1995. Molecular biological clocks. *American Scientist* 83:158–165.

Taylor, C. R. 1972. The desert gazelle: A parody resolved. In G. M. O. Malory, ed., *Comparative physiology of desert animals. Symposium of the Zoological Society of London No. 31.* New York: Academic Press.

Weeks, H. P., Jr., and C. M. Kirkpatrick. 1978. Salt preferences and sodium drive phenology in fox squirrels and woodchuck. *Journal of Mammalogy* 59:531–542.

Weir, J. S. 1972. Spatial distribution of elephants in an African national park in relation to environmental sodium. *Oikos* 23:113.

Chapter 8 | Life History Patterns

Alcock, J. 1995. *Animal behavior: An evolutionary approach.* 4th ed. Sunderland, MA: Sinauer Associates.

Andersson, M., and Y. Iwasa. 1996. Sexual selection. *Trends in Ecology and Evolution* 11:53–58.

Ashmole, N. P. 1963. The regulation of numbers of tropical oceanic birds. *Ibis* 103b:458–473.

Bajema, C. J. (ed.). 1984. *Evolution by sexual selection theory. Benchmark Papers in Systemic and Evolutionary Biology.* New York: Scientific and Academic Editions.

Bierzychudek, P. 1982. The demography of jack-in-the-pulpit, a forest perennial that changes sex. *Ecological Monographs* 52:335–351.

Boyce, M. S. 1984. Restitution of *r* and *K* selection as modes of density-dependent natural selection. *Annual Review of Ecology and Systematics* 15:427–447.

Catchpole, C. K. 1987. Bird song, sexual selection, and female choice. *Trends in Ecology and Evolution* 2:94–97.

Clutton-Brock, T. H., F. E. Guinness, and S. D. Albion. 1982. *Red deer: Behavior and ecology of two sexes.* Chicago: University of Chicago Press.

Cody. M. L. 1966. A general theory of clutch size. *Evolution* 20:174–184.

Darwin, C. 1871. *The descent of man and selection in relation to sex.* London: Murray.

Eis, S., E. H. Garman, and L. F. Ebel. 1965. Relation between cone production and diameter increment of Douglas fir (*Pseudotsuga*), grand fir (*Aibes grandis*), and western white pine (*Pinus monticola*). *Canadian Journal of Botany* 43:1553–1559.

Ellstrand, N. C. 1984. Multiple paternity within the fruits of wild radish *Raphanus sativus. American Naturalist* 123:819–828.

Freeman, C. L., K. T. Harper, and E. L. Charnov. 1980. Sex change in plants: Old and new observations and new hypotheses. *Oecologica* 47:222–232.

Grime, J. P. 1977. Evidence for the existence of three primary strategies in plants and its relevance to ecological and evolutionary theory. *American Naturalist* 111:1169–1194.

———. 1979. *Plant strategies and vegetative processes.* New York: Wiley.

Gubernich, D. J., and P. H. Klopher (eds.). 1981. *Parental care in mammals.* New York: Plenum.

James, F. 1971. Ordination of habitat relationships among birds. *Wilson Bulletin* 83:215–236.

Johnsgard, P. A. 1994. *Arena birds: Sexual selection and behavior.* Washington, DC: Smithsonian Institution Press.

Jones, M. B. 1978. Aspects of the biology of the big handed crab, *Heterozius rohendifrons,* from Kaikoura, New Zealand. *New Zealand Journal of Zoology* 5:783–794.

Krebs, J. R., and N. D. Davies. 1981. *An introduction to behavioral ecology.* Oxford: Blackwell Scientific.

Lack, D. 1954. *The natural regulation of animal numbers.* Oxford: Oxford University Press (Clarendon Press).

MacArthur, R. H., and E. O. Wilson. 1967. *The theory of island biogeography.* Princeton, NJ: Princeton University Press.

McGregor, P. K., J. R. Krebs, and C. M. Perrins. 1981. Song repertoires and lifetime reproductive success in the great tit (*Parus major*). *American Naturalist* 118:49–59.

Mennill, D., L. M. Ratcliffe, and P. T. Boag. 2002. Female eavesdropping on male song contests in songbirds. *Science* 296:873.

Petrie, M. 1994. Improved growth and survival of offspring of peacocks with more elaborate trains. *Nature* 371:598–599.

Pianka, E. 1972. *r* and *K* selections or *b* and *d* selection? *American Naturalist* 100:65–75.

Policansky, D. 1982. Sex change in plants and animals. *Annual Review of Ecology and Systematics* 13:471–495.

Primack, R. B. 1979. Reproductive effort in annual and perennial species of Plantago (*Plantaginaceae*). *American Naturalist* 114:51–62.

Ricklefs, R. 1980. Geographical variation in clutch size in passerine birds: Ashmole's hypothesis. *Auk* 97:38–49.

Shapiro, D. Y. 1980. Serial sex changes after simultaneous removal of males from social groups of coral reef fish. *Science* 209:1136–1137.

Small, M. F. 1992. Female choice in mating. *American Scientist* 80:142–151.

Thornhill, R., and J. Alcock. 1983. *The evolution of insect mating systems.* Cambridge, MA: Harvard University Press.

Warner, R. R. 1988. Sex change and size-advantage model. *Trends in Ecology and Evolution* 3:133–136.

Wasser, S. K. (ed.). 1983. *Social behavior of female vertebrates.* New York: Academic Press.

Wauters, L., and A. A. Dohondt. 1989. Body weight, longevity, and reproductive success in red squirrels (*Sciurus vulgaris*). *Journal of Animal Ecology* 58:637–651.

Werner, P. A., and W. J. Platt. 1976. Ecological relationships of co-occurring goldenrods (*Solidago:Compositae*). *American Naturalist* 110:959–971.

Whitham, T. G. 1980. The theory of habitat selection: Examined and extended using *Pemphigus aphids. American Naturalist* 115:449–466.

Willson, M. F. 1983. *Plant reproductive ecology.* New York: Wiley.

Wilson, E. O. 1980. *Sociobiology: The abridged edition.* Cambridge, MA: Harvard University Press.

Chapter 9 | Properties of Populations

Alexander, M. M. 1958. The place of aging in wildlife management. *American Scientist* 6:123–131.

Begon, M., J. L. Harper, and C. R. Townsend. 1996. *Ecology: Individuals, populations, and communities.* New York: Blackwell Scientific.

Blower, J. G., L. M. Cook, and J. A. Bishop. 1981. *Estimating the size of animal populations.* London: George Allen & Unwin.

Engle, L. G. 1960. Yellow-poplar seedfall pattern. *Central States Forest Experiment Station Note 143.*

Forman, R. T. T. 1964. Growth under controlled conditions to explain the hierarchical distribution of a moss, *Tetraphis pellucida. Ecological Monographs* 34:1–25.

Krebs, C. 2001. *Ecology: The experimental analysis of distribution and abundance.* 5th ed. San Francisco: Benjamin Cummings.

Krebs, C. J. 1989. *Ecological methodology.* 2nd ed. San Francisco: Benjamin Cummings.

Liebhold, A. M., J. A. Halverson, and G. A. Elmes. 1992. Gypsy moth invasion in North America: A quantitative analysis. *Journal of Biogeography* 19:513–520.

MacArthur, R. H., and J. H. Connell. 1966. *The biology of populations.* New York: John Wiley.

Mueller-Dombois, D., and H. Ellenberg. 1974. *Aims and methods of vegetation ecology.* New York: Wiley.

Root, T. 1988. Energy constraints on avian distributions and abundances. *Ecology* 69:330–339.

Southwood, T. R. E. 1978. *Ecological methods.* London: Chapman and Hall.

Wiens, J. 1973. Pattern and process in grassland bird communities. *Ecological Monographs* 43:247–270.

Chapter 10 | Population Growth

Allee, W. C. 1945. *Animal aggregations. A study in general sociology,* Chicago: University of Chicago Press.

Begon, M., and M. Mortimer. 1995. *Population ecology: A unified study of plants and animals.* 2nd ed. Cambridge, MA: Blackwell Scientific.

Binkley, C. S., and R. S. Miller. 1983. Population characteristics of the whooping crane, *Grus americana. Canadian Journal of Zoology* 61:2768–2776.

Campbell, R. W. 1969. Studies on gypsy moth population dynamics. Pages 29–34 in *Forest insect population dynamics.* USDA Research Paper, NE-125.

Cannon, J. R. 1996. Whooping crane recovery: A case study in public and private cooperation in the conservation of an endangered species. *Conservation Biology* 10:813–821.

Caughley, G. 1977. *Analysis of vertebrate populations.* New York: Wiley.

Gotelli, N. J. 1995. *A primer of ecology.* Sunderland, MA: Sinauer Associates.

Hackney, E., and J. B. McGraw. 2001. Experimental demonstration of an allee effect in American ginseng. *Conservation Biology* 15:129–136.

Harper, J. L. 1977. *The population biology of plants.* London: Academic Press.

Johnston, R. F. 1956. Predation by short-eared owls in a *Salicornia* salt marsh. *Wilson Bulletin* 68:91–102.

Krebs, C. 2001. *Ecology: The experimental analysis of distribution and abundance.* 5th ed. San Francisco: Benjamin Cummings.

Lowe, V. P. W. 1969. Population dynamics of red deer (*Cervus elaphus* L.) on the Isle of Rhum. *Journal of Animal Ecology* 38:425–457.

MacLulich, D. A. 1937. Fluctuations in the numbers of varying hare (*Lepus americanus*). *University of Toronto Biological Series No 43.*

Pitelka, F. A. 1957. Some characteristics of microtine cycles in the Arctic. *Proceedings of 18th Biology Colloquium.* Salem: University of Oregon Press.

Scharitz, R. R., and J. R. McCormick. 1973. Population dynamics of two competing plant species. *Ecology* 54:723–740.

Scheffer, V. C. 1951. Rise and fall of a reindeer herd. *Scientific Monthly* 73:356–362.

Stephens P. A., W. J. Sutherland, and R. P. Freckleton. 1999. What is the allee effect? *Oikos* 87:185–190.

Thornhill, N. W. (ed.). 1993. *The natural history of inbreeding and outbreeding: Theoretical and empirical perspectives.* Chicago: University of Chicago Press.

Chapter 11 | Intraspecific Population Regulation

Berteaux, D., and S. Boutin. 2000. Breeding dispersal in female North American red squirrels. *Ecology* 81:1311–1326.

Cahill, J. F. 2000. Investigating the relationship between neighbor root biomass and belowground competition: Field evidence for symmetric competition belowground. *Oikos* 90:311–320.

Chatworthy, J. N. 1960. *Studies on the nature of competition between closely related species.* D. Phil. diss., University of Oxford.

Chepko-Sade, B. D., and Z. T. Halpin. 1987. *Mammalian dispersal patterns.* Chicago: University of Chicago Press.

Dash, M. C., and A. K. Hota. 1980. Density effects on the survival, growth rate, and metamorphosis of *Rana tigrina* tadpoles. *Ecology* 61:1025–1028.

Fery, R. L., and J. Janick. 1971. Response of corn (*Zea mays.* L.) to population pressure. *Crop Science* 11:220–224.

Fowler, C. W. 1981. Density dependence as related to life history strategy. *Ecology* 62:602–610.

Gaines, M. S., and L. R. McClenaghan Jr. 1980. Dispersal in small mammals. *Annual Review of Ecology and Systematics* 11:163–169.

Greenwood, P. J., and P. H. Harvey. 1982. The natal and breeding dispersal in birds. *Annual Review of Ecology and Systematics* 13:1–21.

Grime, J. P. 1979. *Plant strategies and vegetative processes.* New York: Wiley.

Harestad, A. S., and F. L. Bunnell. 1979. Home range and body weight—a reevaluation. *Ecology* 60:389–402.

Harper, J. L. 1977. *Population biology of plants.* New York: Academic Press.

Jones, W. T. 1989. Dispersal distance and the range of nightly movements of Merriam's kangaroo rats. *Journal of Mammalogy* 20:29–34.

King, A. A., and W. C. Schaffer. 2001. The geometry of a population cycles: A mechanistic model of snowshoe demography. *Ecology* 82:814–830.

Krebs, J., and N. B. Davies (eds.). 1991. *Behavioral ecology: An evolutionary approach.* 3rd ed. Oxford: Blackwell Scientific.

Krebs, J. R. 1971. Territory and breeding density in the great tit *Parus major. Ecology* 52:2–22.

Larsen, K. W., and S. Boutin. 1994. Movement, survival, and settlement of red squirrel (*Tamiasciurus hudsonicus*) offspring. *Ecology* 75:214–223.

Lett, P. F., R. K. Mohn, and D. F. Gray. 1981. Density-dependent processes and management strategy for the northwest Atlantic harp seal populations. Pages 135–158 in C. W. Fowler and T. D. Smith, eds., *Dynamics of large mammal populations.* New York: Wiley.

Massey, A., and J. D. Vandenbergh. 1980. Puberty delay by a urinary cue from female house mice in feral populations. *Science* 209:821–822.

McCullough, D. R. 1981. Population dynamics of the Yellowstone grizzly. Pages 173–196 in C. W. Fowler and T. D. Smith, eds., *Dynamics of large mammal populations.* New York: Wiley.

Mech, L. D., R. E. McRoberts, R. O. Peterson, and R. E. Page. 1978. Relationship of deer and moose populations to previous winter's snow. *Journal of Animal Ecology* 56:615–627.

Murdoch, W. W. 1994. Population regulation in theory and practice. *Ecology* 75: 271–287.

Myers, K., C. S. Hale, R. Mykytowycz, and R. L. Hughes. 1971. The effects of varying density and space on sociality and health in animals. Pages 148–187 in A. E. Esser, ed., *Behavior and environment: The use of space by animals and men.* New York: Plenum.

Packard, J. M., and L. D. Mech. 1983. Population regulation in wolves. Pages 151–174 in F. L. Bunnell, D. S. Eastman, and J. M. Peak, eds., *Symposium on natural regulation of wildlife populations.* Moscow: University of Idaho Press.

Searcy, W. A. 1982. The evolutionary effects of mate selection. *Annual Review of Ecology and Systematics* 13:57–85.

Sinclair, A. R. E. 1977. *The African buffalo: A study of resource limitations of populations.* Chicago: University of Chicago Press.

Smith, R. L. 1963. Some ecological notes on the grasshopper sparrow. *Wilson Bulletin* 75:159–165.

Smith, S. M. 1978. The underworld in a territorial adaptive strategy for floaters. *American Naturalist* 112:570–582.

Turchin, P. 1999. Population regulation: A synthetic view. *Oikos* 84:160–163.

Watkinson, A. R., and A. J. Davy. 1985. Population biology of salt marsh and dune annuals. *Vegetatio* 62:487–497.

Wolff, J. O. 1997. Population regulation in mammals: An evolutionary perspective. *Journal of Animal Ecology* 66:1–13.

Yoda, K., T. Kira, H. Ogawa, and K. Hozumi. 1963. Self-thinning in overcrowded pure stands under cultivated and natural conditions. *Journal Biology* 14:107–129.

Zimen, E. 1981. *The wolf: A species in danger.* New York: Delacourt Press.

Chapter 12 | Metapopulations

den Boer, P. 1981. On the survival of populations in a heterogeneous and variable environment. *Oecologia* 50:39–53.

Ehrlich, P., D. Breedlove, and P. Brussard. 1972. Weather and the "regulation" of subalpine populations. *Ecology* 53:243–247.

Ehrlich, P., and D. Murphy. 1987. Conservation lessons from long-term studies of checkerspot butterflies. *Conservation Biology* 1:122–131.

Ehrlich, P., D. Murphy, M. Singer, C. Sherwood, R. White, and I. Brown. 1980. Extinction, reduction, stability and increase: The responses of checkerspot butterfly (*Euphydryas*) populations to the California drought. *Oecologia* 46:101–105.

Hanski, I. 1990. Density dependence, regulation and variability in animal populations. *Philosophical Transactions of the Royal Society London* 330:141–150.

———. 1991. Single-species metapopulation dynamics. *Biological Journal of the Linnean Society* 42:17–38.

———. 1994. Patch occupancy dynamics in fragmented landscapes. *Trends in Ecology and Evolution* 9:131–135.

———. 1999. *Metapopulation ecology.* Oxford: Oxford University Press.

Hanski, I., and M. Gilpin. 1991. Metapopulation dynamics: Brief history and conceptual domain. *Biological Journal of the Linnean Society* 42:3–16.

——— (eds.). 1997. *Metapopulation biology: Ecology, genetics and evolution.* London: Academic Press.

Harrison, S., D. Murphy, and P. Ehrlich. 1988. Distribution of the Bay checkerspot butterfly, *Euphydryas editha bayensis:* Evidence for a metapopulation model. *American Naturalist* 132:360–382.

Harrison, S., and J. Quinn. 1989. Correlated environments and the persistence of metapopulations. *Oikos* 56:293–298.

Hill, J., C. Thomas, and O. Lewis. 1996. Effects of habitat patch size and isolation on dispersal by Hesperia comma butterflies: Implications for metapopulation structure. *Journal of Animal Ecology* 65:725–735.

Kindvall, O. 1996. Habitat heterogeneity and survival in a bush cricket metapopulation. *Ecology* 77:207–214.

Kindvall, O., and I. Ahlen. 1992. Geometrical factors and metapopulation dynamics of the bush cricket, *Metrioptera bicolor. Conservation Biology* 6:520–529.

Murphy, D., and R. White. 1984. Rainfall, resources and dispersal in southern populations of *Euphydryas editha. Pan-Pacific Entomologist* 60:350–354.

Peltonen, and I. Hanski. 1991. Patterns of island occupancy explained by colonization and extinction rates in shrews. *Ecology* 72:1698–1708.

Pettersson, B. 1985. Extinction of an isolated population of the middle spotted woodpecker *Dendrocopos medius* in Sweden and its relation to general theories on extinction. *Biological Conservation* 32:335–353.

Pulliam, R. 1988. Sources, sinks, and population regulation. *American Naturalist* 132:652–661.

Schoener, T., and D. Spiller. 1987. High population persistence in a system with high turnover. *Nature* 330:474–477.

Shaffer, M. 1981. Minimum population sizes for species conservation. *BioScience* 31:131–134.

Stacey, P., and M. Taper. 1992. Environmental variation and the persistence of small populations. *Ecological Applications* 2:18–29.

Sutcliffe, O., C. Thomas, T. Yates, and J. Greatorex-Davies. 1997. Correlated extinctions, colonizations and population fluctuations in a highly connected ringlet butterfly metapopulation. *Oecologia* 109:235–241.

Thomas, C., M. Singer, and D. Boughton. 1996. Catastrophic extinction of population sources in a butterfly metapopulation. *American Naturalist* 148:957–975.

Thomas, C., and T. Jones. 1993. Partial recovery of a skipper butterfly (*Hesperia comma*) from population refuges: Lessons for conservation in a fragmented landscape. *Journal of Animal Ecology* 62:472–481.

Thomas, J. 1983. The ecology and conservation of *Lysandra bellargus* in Britain. *Journal of Applied Ecology* 20:59–83.

Chapter 13 | Interspecific Competition

American Naturalist. 1983. Interspecific competition papers. Vol. 122, no. 5 (November).

Austin, M. P. 1999. A silent clash of paradigms: Some inconsistencies in community ecology. *Oikos* 86:170–178.

Austin, M. P., R. H. Groves, L. M. F. Fresco, and P. E. Laye. 1986. Relative growth of six thistle species along a nutrient gradient with multispecies competition. *Journal of Ecology* 73:667–684.

Bazzaz, F. A. 1996. *Plants in changing environments.* New York: Cambridge University Press.

Brown, J. C., O. J. Reichman, and D. W. Davidson. 1979. Granivory in desert ecosystems. *Annual Review of Ecology and Systematics* 10:210–227.

Connell, J. H. 1975. Some mechanisms producing structure in natural communities: A model and evidence from field experiments. Pages 460–490 in M. L. Cody and J. Diamond, eds., *Ecology and evolution of communities.* Cambridge, MA: Harvard University Press.

———. 1983. On the prevalence and relative importance of interspecific competition: Evidence from field experiments. *American Naturalist* 122:661–696.

Darwin, C. 1859. *The origin of species.* Philadelphia: McKay. [Reprinted from 6th London edition.]

Dayan, T., D. Simberloff, E. Tchernov, and Y. Yom-Tov. 1990. Feline canines: Community-wide character displacement among the small cats of Israel. *American Naturalist* 136:39–60.

Dye, P. J., and P. T. Spear. 1982. The effects of bush clearing and rainfall variability on grass yield and composition in southwest Zimbabwe. *Zimbabwe Journal of Agricultural Research* 20:103–118.

Emery, N. C., P. J. Ewanchuk, and M. D. Bertness. 2001. Competition and salt marsh plant zonation: Stress tolerators may be dominant competitors. *Ecology* 82:2471–2485.

Gause, G. F. 1934. *The struggle for existence.* Baltimore: Williams and Wilkins.

Grace, J. B., and R. G. Wetzel. 1981. Habitat partitioning and competitive displacement in cattails (*Typha*): Experimental field studies. *American Naturalist* 118:463–474.

Grant, P. 1999. *Ecology and evolution of Darwin's finches.* Princeton, NJ: Princeton University Press.

Groves, R. H., and J. D. Williams. 1975. Growth of skeleton weed (*Chondrilla juncea*) as affected by growth of subterranean clover (*Trifolium subterraneum L.*) and infection by *Puccinia chondrilla. Australian Journal of Agricultural Research* 26:975–983.

Gurevitch, J. L. 1986. Competition and the local distribution of the grass *Stipa neomexicana. Ecology* 67:46–57.

Gurevitch, J. L., L. Morrow, A. Wallace, and J. J. Walch. 1992. Meta-analysis of competition in field experiments. *American Naturalist* 140:539–572.

Heller, H. C., and D. Gates. 1971. Altitudinal zonation of chipmunks (*Eutamias*): Energy budgets. *Ecology* 52: 424–443.

Lotka, A. J. 1925. *Elements of physical biology.* Baltimore: Williams and Wilkins.

Park, T. 1954. Experimental studies of interspecies competition: 2. Temperature, humidity and competition in two species of *Trilobium. Physiological Zoology* 27:177–238.

Pianka, E. R. 1978. *Evolutionary ecology.* 3rd ed. New York: HarperCollins.

Pickett, S. T. A., and F. A. Bazzaz. 1978. Organization of an assemblage of early successional species on a soil moisture gradient. *Ecology* 59:1248–1255.

Rice, E. L. 1984. *Allelopathy.* 2nd ed. Orlando, FL: Academic Press.

Schoener, T. W. 1983. Field experiments on interspecific competition. *American Naturalist* 122:240–285.

Tilman, D. M., M. Mattson, and S. Langer. 1981. Competition and nutrient kinetics along a temperature gradient: An experimental test of a mechanistic approach to niche theory. *Limnology and Oceanography* 26:1020–1033.

Volterra, V. 1926. Variation and fluctuations of the numbers of individuals in animal species living together. Reprinted on pages 409–448 in R. M. Chapman (1931), *Animal Ecology.* New York: McGraw-Hill.

Werner, E. E., and J. D. Hall. 1976. Niche shifts in sunfishes: Experimental evidence and significance. *Science* 191:404–406.

———. 1979. Foraging efficiency and habitat switching in competing sunfish. *Ecology* 60:256–264.

Wiens, J. A. 1977. On competition and variable environments. *American Scientist* 65:590–597.

Chapter 14 | Predation

Agrawal, A. A. 2004. Resistance and susceptibility of milkweed to herbivore attack: Consequences of competition, root herbivory, and plant genetic variation. *Ecology* 85:2118–2133.

Charnov, E. L. 1976. Optimal foraging: the marginal value theorem. *Theoretical Population Biology* 9:129–136.

Cooper, S. M., and N. Owen-Smith. 1986. Effects of plant spinescence on large mammalian herbivores. *Oecologica* 68:446–455.

Davies, N. B. 1977. Prey selection and social behavior in wagtails (*Aves monticillidae*). *Journal of Animal Ecology* 46:37–57.

Fox, L. R. 1975. Cannibalism in natural populations. *Annual Review of Ecology and Systematics* 6:87–106.

Godfray, H. C. J. 1994. *Parasitoids: Behavioral and evolutionary ecology.* Princeton, NJ: Princeton University Press.

Hassell, M. P., and R. M. May. 1974. Aggregation in predator sand insect parasites and its effect on stability. *Journal of Animal Ecology* 43:567–594.

Holling, C. S. 1959. The components of predation as revealed by a study of small mammal predation of the European sawfly. *Canadian Entomologist* 91:293–320.

———. 1966. The functional response of invertebrate predators to prey density. *Memoirs Entomological Society of Canada* 48:1–86.

Irons, D. B., R. G. Anthony, and J. A. Estes. 1986. Foraging strategies of glaucous-winged gulls in a rocky intertidal community. *Ecology* 67:1460–1474.

Jedrzejewski, W., B. Jedrzejewski, and L. Szymura. 1995. Weasel population response, home range, and predation on rodents in a deciduous forest in Poland. *Ecology* 76:179–195.

Keith, L. B., J. R. Cary, O. J. Rongstad, and M. C. Brittingham. 1984. Demography and ecology of a declining snowshoe hare population. *Journal of Wildlife Management* 90:1–43.

Korpimaki, E., and K. Norrdahl. 1991. Numerical and functional responses of kestrels, short-eared owls, and long-eared owls to vole densities. *Ecology* 72:814–826.

Krebs, C. J., R. Boonstra, S. Boutin, and A. R. E. Sinclair. 2001. What drives the 10-year cycle of snowshoe hares? *BioScience* 51:25–35.

Krebs, J. R., and N. B. Davies (eds.). 1984. *Behavioral ecology: An evolutionary approach.* 2nd ed. Oxford: Blackwell Scientific.

Leonard, G. H., M. D. Bertness, and P. O. Yund. 1999. Crab predation, water-borne cues, and inducible defenses in the blue mussel *Mytilus edulis. Ecology* 80:1–14.

Lima, S. L. 1998. Nonlethal effects in the ecology of predator-prey interactions. *BioScience* 48:25–34.

Mazncourt, de C., M. Loreau, and U. Dieckmann. 2001. Can the evolution of plant defense lead to plant-herbivore mutualism? *American Naturalist* 158:109–123.

Mook, L. J. 1963. Birds and the spruce budworm. In R. F. Morris, ed., The dynamics of epidemic spruce budworm populations. *Memoirs of the Entomological Society of Canada,* 269–291.

Mooney, H. A., and Gulmon, S. L. 1982. Constraints on leaf structure and function in reference to herbivory. *BioScience* 32:198–206.

Nelson, E. H., C. E. Matthews, and J. A. Roenheim. 2004. Predators reduce prey population growth by inducing change in behavior. *Ecology* 85:1853–1858.

O'Donoghue, M., S. Boutin, C. J. Krebs, G. Zuleta, D. L. Murray, and E. J. Hofer. 1998. Functional response of coyotes and lynx to the snowshoe hare cycle. *Ecology* 79:1193–1208.

Polis, G. 1981. The evolution and dynamics of intraspecific predation. *Annual Review of Ecology and Systematics* 12:225–251.

Polis, G., and R. D. Holt. 1992. Intraguild predation: The dynamics of complex trophic interactions. *Trends in Ecology and Evolution* 7:151–154.

Proulx, M., and A. Mazumder. 1998. Reversal of grazing impact on plant species richness in nutrient-poor vs. nutrient-rich ecosystems. *Ecology* 79:2581–2592.

Relyea, R. A. 2001. The relationship between predation risk and antipredator responses in larval anurans. *Ecology* 82:541–554.

Schnell, D. E. 1976. *Carnivorous plants of the United States and Canada.* Winston-Salem, NC: John F. Blair.

Sinclair, A. R. E., C. J. Krebs, J. M. N. Smith, and S. Boutin. 1988. Population biology of snowshoe hares. III: Nutrition, plant secondary compounds, and food limitation. *Journal of Animal Ecology* 57:787–806.

Stephens, D. W., and J. W. Krebs. 1987. *Foraging theory.* Princeton, NJ: Princeton University Press.

Taylor, R. J. 1984. *Predation.* New York: Chapman and Hall.

Van Valen, L. 1973. A new evolutionary law. *Evolutionary Theory* 1:1–30.

Wajnberg E., X. Fauvergue, and O. Pons. 2000. Patch leaving decision rules and the Marginal Value Theorem: An experimental analysis and a simulation model. *Behavioral Ecology* 11:577–586.

Wagner, J. D., and D. H. Wise. 1996. Cannibalism regulates densities of young wolf spiders: Evidence from field and laboratory experiments. *Ecology* 77:639–652.

Williams, K., K. G. Smith, and F. M. Stevens. 1993. Emergence of 13-year periodical cicada (*Cicadidae: Magicicada*): Phenology, mortality, and predator satiation. *Ecology* 74:1143–1152.

Chapter 15 | Parasitism and Mutualism

Addicott, J. F. 1986. Variations in the costs and benefits of mutualism: The interaction between yucca and yucca moths. *Oeocologica* 70:486–494.

Anderson, R. C. 1963. The incidence, development, and experimental transmission of *Pneumostrongylus tenuis* Dougherty (*Metastrongyloidae: Protostrongyliidae*) of the menges of the white-tailed deer (*Odocoilus virginianus borealis*) in Ontario. *Canadian Journal of Zoology* 41:775–792.

Anderson, R. C., and A. K. Prestwood. 1981. Lungworms. In W. R. Davidson, ed., *Diseases and parasites of white-tailed deer.* Miscellaneous Publication No. 7. Tallahassee, FL: Tall Timbers Research Station.

Bacon, P. J. 1985. *Population dynamics of rabies in wildlife.* New York: Academic Press.

Ball, D. M., J. F. Pedersen, and G. D. Lacefield. 1993. The tall fescue endophyte. *American Scientist* 81:370–379.

Barbour, A. G., and D. Fish. 1993. The biological and social phenomenon of Lyme disease. *Science* 260:1610–1616.

Barrett, J. A. 1983. Plant-fungus symbioses. In D. Futuyma and M. Slakin, eds., *Coevolution.* Sunderland, MA: Sinauer Associates.

Barth, F. G. 1991. *Insects and flowers: The biology of a partnership.* Princeton, NJ: Princeton University Press.

Beattie, A. J., and D. C. Culver. 1981. The guild of myrmeco-hores in the herbaceous flora and West Virginia forests. *Ecology* 62:107–115.

Bentley, B. L. 1977. Extrafloral nectarines and protection by pugnacious bodyguards. *Annual Review of Ecology and Systematics* 8:407–427.

Blount, J. D., N. B. Metcalfe, T. R. Birkhead, and P. F. Surai. 2003. Carotenoid modulation of immune function and sexual attractiveness in zebra finches. *Science* 300:125–127.

Boucher, D. H. (ed.). 1985. *The biology of mutualism.* Oxford: Oxford University Press.

Boucher, D. H., S. James, and H. D. Keller. 1982. The ecology of mutualism. *Annual Review of Ecology and Systematics* 13:315–347.

Burdon, J. J. 1987. *Diseases and plant population biology.* Cambridge: Cambridge University Press.

Buskirk, J. V., and R. S. Ostfeld. 1995. Controlling Lyme disease by modifying the density and species composition of tick hosts. *Ecological Applications* 5:1133–1140.

Clay, K. 1988. Fungal endophytes of grasses: A defensive mutualism between plants and fungi. *Ecology* 60:10–16.

———. 1990. Fungal endophytes of grasses. *Annual Review of Ecology and Systematics* 21:275–297.

Croll, N. A. 1966. *Ecology of parasites.* Cambridge, MA: Harvard University Press.

Dobson, A. P., and E. R. Carper. 1996. Infectious diseases and human population history. *BioScience* 46:115–125.

Doebler, S. A. 2000. The rise and fall of the honeybee. *BioScience* 50:738–732.

Edwards, M. A., and U. McDonald. 1982. *Animal diseases in relation to animal conservation.* New York: Academic Press.

Feinsinger, P. 1983. Coevolution and pollination. Pages 282–310 in D. Futuyma and M. Slatkin, eds., *Coevolution.* Sunderland, MA: Sinauer Associates.

Fenner, F., and F. N. Ratcliffe. 1965. *Myxamatosis.* Cambridge: Cambridge University Press.

Futuyma, D. J., and M. Slatkin (eds.). 1983. *Coevolution.* Sunderland, MA: Sinauer Associates.

Garnett, G. P., and E. C. Holmes. 1996. The ecology of emergent infectious disease. *BioScience* 46:127–135.

Grenfell, B. T. 1988. Gastrointestinal nematode parasites and the stability and productivity of intensive ruminal grazing systems. Philosophical Transactions Royal Society, London. *British Biological Science* 321:541–563.

Hamilton, W. D., and M. Zuk. 1982. Heritable true fitness and bright birds: A role for parasites? *Science* 218:384–387.

Hanzawa, F. M., A. J. Beattie, and D. C. Culvar. 1988. Directed dispersal: Demographic analysis of an ant-seed mutualism. *American Naturalist* 131:1–13.

Harley, J. L., and S. E. Smith. *The biology of mycorrhizae.* London: Academic Press.

Heithaus, E. R. 1981. Seed predation by rodents on three ant-dispersed plants. *Ecology* 63:136–145.

Hocking, B. 1975. Ant-plant mutualism: Evolution and energy. Pages 78–90 in L. E. Gilbert and P. H. Raven, eds., *Coevolution of animals and plants.* Austin: University of Texas Press.

Holmes, J. 1983. Evolutionary relationships between parasitic helminths and their hosts. In D. J. Futuyma and M. Slakin, eds., *Coevolution.* Sunderland, MA: Sinauer Associates.

Hougen-Eitzman, D., and M. D. Rausher. 1994. Interactions between herbivorous insects and plant-insect coevolution. *American Naturalist* 143:677–697.

Howe, H. F. 1986. Seed dispersal by fruit-eating birds and mammals. Pages 123–190 in D. R. Murra, ed., *Seed dispersal.* Sydney: Academic Press.

Howe, H. F., and L. C. Westley. 1988. *Ecological relationships of plants and animals.* New York: Oxford University Press.

Hutchins, H. E. 1994. Role of various animals in the dispersal and establishment of white-barked pine in the Rocky Mountains, USA. Pages 163–171 in W. C. Schmidt and F-K. Holtmeier, comps., Proceedings—International workshop on subalpine stone pines and their environment: The status of our knowledge; 1992 September 5–11; St. Moritz, Switzerland. *General Technical Report INT-GTR-309.* Ogden, UT: USDA Forest Service, Intermountain Research Station.

Iowa, K., and M. D. Rausher. 1997. Evolution of plant resistance to multiple herbivores: Quantifying diffuse evolution. *American Naturalist* 149:316–335.

Janzen, D. H. 1966. Coevolution of mutualism between ants and acacias in Central America. *Evolution* 20:249–275.

Korstian, C. F., and P. W. Stickel. 1927. The natural replacement of blight-killed chestnut. *USDA Miscellaneous Circular 100.* Washington, DC: U.S. Government Printing Office.

Lafferty, K. D., and A. K. Morris. 1996. Altered behavior of parasitized killifish increases susceptibility to predation by bird final hosts. *Ecology* 77:1390–1397.

Lindstrom, E. R., et al. 1994. Disease reveals the predator: Sarcoptic mange, red fox predation, and prey populations. *Ecology* 74:1041–1049.

Margulis, L. 1982. *Early life.* Boston: Science Books International.

———. 1998. *Symbiotic planet.* New York: Basic Books.

Maser, C., J. M. Trappe, and R. A. Nussbaum. 1978. Fungal–small mammal interrelationship with emphasis on Oregon coniferous forests. *Ecology* 59:799–809.

Mattes, H. 1994. Coevolutional aspects of stone pines and nutcrackers. Pages 31–35 in W. C. Schmidt and F-K. Holtmeier, comps., Proceedings–International workshop on subalpine stone pines and their environment: the status of our knowledge; 1992 September 5–11; St. Moritz, Switzerland. *General Technical Report INT-GTR-309.* Ogden, UT: USDA Forest Service, Intermountain Research Station.

May, R. M. 1983. Parasitic infections as regulators of animal populations. *American Scientist* 71:36–45.

May, R. M., and R. M. Anderson. 1979. Population biology of infective diseases: Part II. *Nature* 280:455–461.

McNeill, W. H. 1976. *Plagues and peoples.* New York: Doubleday.

Moore, J. 1995. The behavior of parasitized animals. *BioScience* 45:89–96.

Muscatine, L., and J. W. Porter. 1977. Reef corals: Mutualistic symbioses adapted to nutrient-poor environments. *BioScience* 27:454–460.

Ostfeld, R. S. 1997. The ecology of Lyme disease risk. *American Scientist* 85:338–346.

Randolph, S. E. 1975. Patterns of the distribution of the tick *Ixodes trianguliceps birula* on its host. *Journal of Animal Ecology* 44:451–474.

Real, L. A. (ed.). 1983. *Pollination biology.* Orlando, FL: Academic Press.

———. 1996. Sustainability and the ecology of infectious disease. *BioScience* 46:88–97.

Schall, J. J. 1983. Lizard malaria: Cost to vertebrate host's reproductive success. *Parasitology* 87:1–6.

Sheldon, B. C., and S. Verhulst. 1996. Ecological immunity: Costly parasite defenses and trade-offs in evolutionary ecology. *Trends in Ecology and Evolution* 11:317–321.

Spielman, A., M. I. Wilson, J. F. Levine, and J. Piesman. 1985. Ecology of *Ixodes dammini*–borne human babesiosis and Lyme disease. *Annual Review of Entomology* 30:439–460.

Tomback, D. F. 1982. Dispersal of white-barked pine seeds by Clark's nutcracker: A mutualism hypothesis. *Journal of Animal Ecology* 51:451–457.

———. 1994. Ecological relationship between Clark's nutcracker and four wingless-seed *Strobus* pines of western North America. Pages 221–224 in W. C. Schmidt and F-K. Holtmeier, comps., Proceedings–International workshop on subalpine stone pines and their environment: the status of our knowledge; 1992 September 5–11; St. Moritz, Switzerland. *General Technical Report INT-GTR-309.* Ogden, UT: USDA Forest Service, Intermountain Research Station.

Tomback, D. F., and Y. B. Linhart. 1990. The evolution of bird-dispersed pines. *Evolutionary Ecology* 4:185–219.

Wakelin, D. 1997. Parasites and the immune system. *BioScience* 47:32–40.

Warner, R. E. 1968. The role of introduced diseases in the extinction of endemic Hawaiian avifauna. *Condor* 70:101–120.

Wilson, E. O. 1975. *Sociobiology: The new synthesis.* Cambridge, MA: Harvard University Press.

Woodard, T. N., R. J. Gutierrez, and W. H. Rutherford. 1974. Bighorn lamb production, survival and mortality in south-central Colorado. *Journal of Wildlife Management* 28:381–391.

Zuk, M. 1991. Parasites and bright birds: New data and new predictions. Pages 317–327 in J. E. Loge and M. Zuk, eds., *Bird–parasite interactions.* Oxford: Oxford University Press.

Chapter 16 | Community Structure

Anderson, S., and H. H. Shugart. 1974. Avian community analyses of Walker Branch Watershed. *ORNL Publication No. 623.* Oak Ridge, TN: Environmental Sciences Division, Oak Ridge National Laboratory.

Braun, E. L. 1950. *Deciduous forests of eastern North America.* New York: McGraw-Hill.

Clements, F. E. 1916. Plant succession: An analysis of the development of vegetation. *Carnegie Inst. Wash. Publ. 242.*

Estes, J., M. Tinker, T. Williams, and D. Doak. 1998. Killer whale predation on sea otters linking oceanic and nearshore ecosystems. *Science* 282:473–476.

Gleason, H. A. 1926. The individualistic concept of the plant association. *Bulletin Torrey Botanical Club* 53:1–20.

Morin, P. J. 1999. *Community ecology.* Oxford: Blackwell Scientific.

Pimm, S. L. 1982. *Food webs.* New York: Chapman & Hall.

Power, M. E., D. Tilman, J. Estes, B. Menge, W. Bond, L. Mills, G. Daily, J. Castilla, J. Lubchenco, and R. Paine. 1996. Challenges in the quest for keystones. *BioScience* 46:609–620.

Shannon, C. E., and W. Wiener. 1963. *The mathematical theory of communications.* Urbana: University of Illinois Press.

Simpson, E. H. 1949. Measurement of diversity. *Nature* 163:688.

Whittaker, R. H. 1956. Vegetation of the Great Smoky Mountains. *Ecological Monographs* 26:1–80.

Chapter 17 | Factors Influencing the Structure of Communities

Bazzaz, F. A., and J. L. Harper. 1976. Relationship between plant weight and numbers in mixed populations of *Sinapis arvensis* (L.) Rabenh. and *Lepidium sativum* L. *Journal of Applied Ecology* 13:211–216.

Berger, J., P. B. Stacey, L. Bellis, and M. P. Johnson. A mammalian predator-prey imbalance: Grizzly bear and wolf extinction affect avian neotropical migrants. *Ecological Applications* 11:947–960.

Cahill, J. F. 1999. Fertilization effects on interactions between above- and belowground competition in an old field. *Ecology* 80:466–480.

Cohen, J. E. 1989. Food webs and community structure. Pages 181–202 in J. Roughgarden, R. May, and S. Levin, eds., *Perspectives in ecological theory.* Princeton, NJ: Princeton University Press.

Dodd, M., J. Silvertown, K. McConway, J. Potts, and M. Crawley. (1994) Biomass stability in the plant communities of the Park Grass Experiment: The influence of species richness, soil pH and biomass. *Philosophical Transactions of the Royal Society* B.346:185–193.

Dodson, S. I. 1970. Complementary feeding niches sustained by size-selective predation. *Limnology and Oceanography* 15:131–137.

Ehrlich, P. R., and A. Ehrlich. 1981. *Extinction.* New York: Random House.

Emery, N. C., P. J. Ewanchuk, and M. D. Bertness. 2001. Competition and salt marsh plant zonation: Stress tolerators may be dominant competitors. *Ecology* 82:2471–2485.

Fowler, N. L. 1981. Competition and coexistence in a North Carolina grassland II: The effects of the experimental removal of species. *Journal of Ecology* 69:843–854.

Gotelli, N. J. 2001. Research frontiers in null model analysis. *Global Ecology and Biogeography* 10:337–343.

Gotelli, N. J., and G. R. Graves. 1996. *Null models in ecology.* Herndon, VA: Smithsonian Institution.

Hairston, N. G., F. E. Smith, and L. B. Slobodkin. 1960. Community structure, population control, and competition. *American Naturalist* 94:421–425.

Holt, R. D. 1977. Predation, apparent competition and the structure of prey communities. *Theoretical Population Biology* 12:197–229.

Huston, M. 1979. A general hypothesis of species diversity. *American Naturalist* 113:81–101.

Huston, M. A. 1980. Soil nutrients and tree species richness in Costa Rican forests. *Journal of Biogeography* 7:147–157.

Krebs, C. 2001. *Ecology.* 5th ed. San Francisco: Benjamin Cummings.

Krebs, C. J., S. Boutin, and R. Boonstra (eds.). 2001. *Vertebrate community dynamics in the Kluane Boreal Forest.* Oxford: Oxford University Press.

Krebs, C. J., S. Boutin, R. Boonstra, A. R. E. Sinclair, J. N. M. Smith, M. R. T. Dale, K. Martin, and R. Turkington. 1995. Impact of food and predation on the snowshoe hare cycle. *Science* 269:1112–1115.

MacArthur, R. H., and J. W. MacArthur. 1961. On bird species diversity. *Ecology* 42:594–598.

Paine, R. T. 1966. Food web complexity and species diversity. *American Naturalist* 100:65–75.

———. 1969. The Pisaster-Tegula interaction: Prey patches, predator food preference and intertidal community structure. *Ecology* 50:950–961.

Pimm, S. L. 1982. *Food webs.* New York: Chapman & Hall.

Power, M. E., M. J. Matthews, and A. J. Stewart. 1985. Grazing minnows, piscivorous bass, and stream algae: Dynamics of a strong interaction. *Ecology* 66:1448–1456.

Ricklefs, R. E., and D. Schluter (eds.). 1993. *Ecological communities: Historical and geographical perspectives.* Chicago: University of Chicago Press.

Robertson, G. P., M. A. Huston, F. C. Evans, and J. M. Tiedje. 1988. Spatial variability in a successional plant community: Patterns of nitrogen availability. *Ecology* 69:1517–1524.

Silvertown, J., M. Dodd, K. McConway, J. Potts, and M. Crawley. 1994. Rainfall, biomass variation, and community composition in the Park Grass Experiment. *Ecology* 75:2430–2437.

Smith, T. M., and M. Huston. 1987. A theory of spatial and temporal dynamics of plant communities. *Vegetatio* 3:49–69.

Smith, T. M., F. I. Woodward, and H. H. Shugart (eds.). 1996. *Plant functional types: Their relevance to ecosystem properties and global change.* Cambridge: Cambridge University Press.

Walker, B. H. 1992. Biodiversity and ecological redundancy. *Conservation Biology* 6:18–23.

Whittaker, R. H. 1960. Vegetation of the Siskiyou Mountains, Oregon and California. *Ecological Monographs* 23:41–78.

———. 1975. *Communities and ecosystems.* New York: Macmillan.

Chapter 18 | Community Dynamics

Austin, M. P., and T. M. Smith. 1989. A new model of the continuum concept. *Vegetatio* 83:35–47.

Bazzaz, F. A. 1979. The physiological ecology of plant succession. *Annual Review of Ecology and Systematics* 10:351–371.

Billings, W. D. 1938. The structure and development of oldfield shortleaf pine stands and certain associated physical properties of the soil. *Ecological Monographs* 8:437–499.

Bormann, F., and G. E. Likens. 1979. *Patterns and process in forested ecosystems.* New York: Springer-Verlag.

Clements, F. E. 1916. Plant succession: An analysis of the development of vegetation. *Carnegie Inst. Wash. Publ. 242.*

———. 1936. Nature and structure of the climax. *Journal of Ecology* 24:252–284.

Connell, J. H. 1978. Diversity in tropical rain forests and coral reefs. *Science* 199:1302–1310.

Connell, J. H., and R. O. Slatyer. 1977. Mechanisms of succession in natural communities and their role in community stability and organization. *American Naturalist* 111: 1119–1144.

Cowles, H. C. 1899. The ecological relations of the vegetation on the sand dunes of Lake Michigan. *Botanical Gazette* 27:95–117, 167–202, 281–308, 361–391.

Crocker, R. L., and J. Major. 1955. Soil development in relation to vegetation and surface age at Glacier Bay, Alaska. *Journal of Ecology* 43:427–488.

Davis, M. B. 1983. Holocene vegetational history of the eastern United States. Pages 166–188 in H. E. Wright Jr., ed., *Late quaternary environments of the United States.* Vol. II: *The Holocene.* Minneapolis: University of Minnesota Press.

Delcourt, P. A., and H. R. Delcourt. 1981. Vegetation maps for eastern North America, 40,000 yr bp to present. Pages 123–166 in R. Romans, ed., *Geobotany.* New York: Plenum Press.

Duggins, D. O. 1980. Kelp beds and sea otters: An experimental approach. *Ecology* 61:447–453.

Egler, F. 1954. Vegetation science concepts. I. Initial floristic composition, a factor in old field vegetation development. *Vegetatio* 14:412–417.

Golley, F. (ed.). 1978. *Ecological succession. Benchmark Papers.* Stroudsburg, PA: Dowden, Hutchinson, and Ross.

Hobbie, E. A. 1994. *Nitrogen cycling during succession in Glacier Bay, Alaska.* Masters thesis. University of Virginia.

Huston, M. A. 1979. A general hypothesis of species diversity. *American Naturalist* 113:81–101.

———. 1994. *Biological diversity: The coexistence of species on changing landscapes.* New York: Cambridge University Press.

Huston, M., and T. M. Smith. 1987. Plant succession: Life history and competition. *American Naturalist* 130:168–198.

Maser, C., and J. M. Trappe (eds.). 1984. The seen and unseen world of the fallen tree. *USDA Forest Service General Technical Report PNW-164.*

Olson, J. S. 1958. Rates of succession and soil changes in southern Lake Michigan sand dunes. *Botanical Gazette* 119: 125–170.

Oosting, H. J. 1942. An ecological analysis of the plant communities of Piedmont, North Carolina. *American Midland Naturalist* 28:1–26.

Pastor, J., R. J. Naiman, B. Dewey, and P. McInnes. 1988. Moose, microbes, and boreal forest. *BioScience* 38:770–777.

Smith, R. L. 1959. Conifer plantations as wildlife habitat. *New York Fish and Game Journal* 5:101–132.

Sousa, W. P. 1979. Disturbance in marine intertidal boulder fields: The nonequilibrium maintenance of species diversity. *Ecology* 60:1225–1239.

———. 1984. Intertidal mosaics: Patch size, propagule availability, and spatially variable patterns of succession. *Ecology* 65:1918–1935.

Sprugle, D. G. 1976. Dynamic structure of wave-generated *Abies balsamea* forests in northeastern United States. *Journal of Ecology* 64:889–911.

Warming, J. E. B. 1909. *Oecology of plants.* Oxford: Oxford University Press.

Watts, M. T. 1976. *Reading the American landscape.* New York: Macmillan.

West, D. C., H. H. Shugart, and D. B. Botkin (eds.). 1981. *Forest succession: Concepts and application.* New York: Springer-Verlag.

Whittaker, R. H. 1956. Vegetation of the Great Smoky Mountains. *Ecological Monographs* 26:1–80.

Whittaker, R. H., and G. M. Woodwell. 1968. Dimension and production relations of trees and shrubs in the Brookhaven forest, New York. *Ecology* 56:1–25.

———. 1969. Structure and function, production, and diversity of the oak-pine forest at Brookhaven, New York. *Journal of Ecology* 57:155–174.

Williams, M. 1989. *Americans and their forests: A historical geography.* Oxford: Oxford University Press.

Zieman, J. C. 1982. The ecology of seagrasses in south Florida: A community profile. *U.S. Fish and Wildlife Service FWS/OBS-82/25.*

Chapter 19 | Landscape Ecology

Aber, J. D., and J. M. Melillo. 1991. *Terrestrial ecosystems.* Philadelphia: Saunders College Publishing.

Bormann, F. H., and G. E. Likens. 1979. *Pattern and process in a forested ecosystem.* New York: Springer-Verlag.

Brown, J. H., and A. Kodrich-Brown. 1977. Turnover rates in insular biogeography: Effect of immigration on extinction. *Ecology* 58:445–449.

Chepko-Sade, B. D., and Z. T. Halpin (eds.). 1987. *Mammalian dispersal patterns: The effect of social structure on population genetics.* Chicago: University of Chicago Press.

Christensen, N. L. 1977. Fire and soil-plant nutrient relations in a pine-wiregrass savanna on the coastal plain of North Carolina. *Oecologia* 31:27–44.

Curtis, J. T. 1959. *The vegetation of Wisconsin.* Madison: University of Wisconsin Press.

Delcourt, H. R. 1987. The impact of prehistoric agriculture and land occupation on natural vegetation. *Trends in Ecology and Evolution* 2:39–44.

Fahrig, L., and G. Merriam. 1994. Conservation of fragmented populations. *Conservation Biology* 8:50–59.

Forman, R. T. T. 1995. *Land mosaics: The ecology of landscapes and regions.* New York: Cambridge University Press.

Forman, R. T. T., and M. Gordon. 1981. Patches and structural components for a landscape ecology. *BioScience* 31:733–740.

Frankel, O. H., and M. E. Soule. 1981. *Conservation and evolution.* Cambridge: Cambridge University Press.

Freemark, K. E., and H. G. Merriam. 1986. Importance of area and habitat heterogeneity to bird assemblages in temperate forest fragments. *Biological Conservation* 36:115–141.

Galli, A. E., C. F. Leck, and R. T. T. Forman. 1976. Avian distribution patterns in forest islands of different sizes in central New Jersey. *Auk* 93:356–364.

Gray, J. T., and W. H. Schlesinger. 1981. Nutrient cycling in Mediterranean-type ecosystems. In P. C. Miller, ed., *Resource use by chaparral and matorral.* New York: Springer-Verlag.

Guthery, F. S., and R. L. Bingham. 1992. On Leopold's principle of edge. *Wildlife Society Bulletin* 20:340–344.

Hanski, I. A., and M. E. Gilpin (eds.). 1997. *Metapopulation biology: Ecology, genetics, and evolution.* San Diego: Academic Press.

Harris, L. D. 1984. *The fragmented forest.* Chicago: University of Chicago Press.

———. 1988. Edge effects and the conservation of biological diversity. *Conservation Biology* 2:212–215.

Heinselman, M. L. 1973. Fire in the virgin forest of the Boundary Waters Canoe Area, Minnesota. *Quaternary Research* 3:329–382.

Heinselman, M. L., and T. M. Casey. 1981. Fire and succession in the conifer forests of northern North America. Pages 374–405 in D. E. West, H. H. Shugart, and D. B. Botkin, eds., *Forest succession: Concepts and applications.* New York: Springer-Verlag.

Johnson, H. B. 1976. *Order upon the land.* Oxford: Oxford University Press.

Kerbes, R. H., P. M. Kotanen, and R. L. Jeffries. 1990. Destruction of wetland habitats by lesser snow geese: A keystone species on the west coast of Hudson Bay. *Journal of Applied Ecology* 27:242–258.

Leopold, A. 1933. *Game management.* New York: Scribner.

Levenson, J. B. 1981. Woodlots as biogeographic islands in southeastern Wisconsin. Pages 13–39 in R. L. Burgess and D. M. Sharpe, eds., *Forest island dynamics in man-dominated landscapes.* New York: Springer-Verlag.

MacArthur, R. H., and E. O. Wilson. 1967. *The theory of island biogeography.* Princeton, NJ: Princeton University Press.

Marquis, D. A. 1974. The impact of deer browsing on Allegheny hardwood regeneration. *USDA Forest Service Research Paper NE-308.*

———. 1981. Effect of deer browsing on timber production in Allegheny hardwood forests of northwestern Pennsylvania. *USDA Forest Service Research Paper NE-475.*

Mooney, H. A., et al. (eds.). 1981. Proceedings: Symposia on fire regimes and ecosystem properties. *General Technical Report WO-26.* Washington, DC: USDA Forest Service.

Naiman, R. J., J. M. Melillo, and J. E. Hobbie. 1986. Ecosystem alteration of boreal forest streams by beaver (*Castor canadensis*). *Ecology* 67:1254–1269.

Preston, F. W. 1962. The canonical distribution of commonness and rarity. *Ecology* 43:185–215, 410–432

Ranney, J. W. 1977. Forest island edges—their structure, development, and importance to regional forest ecosystem dynamics. *EDFB/IBP Cont. No. 77/1,* Oak Ridge National Laboratory, Oak Ridge, TN.

Ranney, J. W., M. C. Brunner, and J. B. Levenson. 1981. The importance of edge in the structure and dynamics of forest lands. Pages 67–91 in R. L. Burgess and D. M. Sharpe, eds., *Forest island dynamics in man-dominated landscapes* (Ecological Studies No. 41). New York: Springer-Verlag.

Robbins, C. S., D. K. Dawson, and B. A. Dowell. 1989. Habitat area requirements of breeding forest birds of the Middle Atlantic States. *Wildlife Monographs* 103.

Romme, W. H., and D. H. Knight. 1982. Landscape diversity: The concept applied to Yellowstone Park. *BioScience* 32:664–670.

Rosenberg, D. K., B. R. Noon, and E. C. Meslow. 1997. Biological corridors: Forms, function and efficiency. *BioScience* 47:677–687.

Runkle, J. R. 1985. Disturbance regimes in temperate forests. Pages 17–34 in S. T. A. Pickett and P. S. White, eds., *Ecology of natural disturbance and patch dynamics.* Orlando, FL: Academic Press.

Schonewald-Cox, C. M., S. M. Chambers, B. MacBryde, and W. L. Thomas (eds.). 1983. *Genetics and conservation.* Menlo Park, CA: Benjamin Cummings.

Schwartz, M. W. 1997. *Conservation in highly fragmented landscapes.* New York: Chapman & Hall.

Simberloff, D. S. 1974. Equilibrium theory of island biogeography and ecology. *Annual Review of Ecology and Systematics* 5:161–182.

Simberloff, D. S., and E. O Wilson. 1969. Experimental zoogeography of islands. The colonization of empty islands. *Ecology* 50:278–296.

Temple, S. A. 1986. Predicting impacts of habitat fragmentation on forest birds: A comparison of two models. Pages 301–304 in J. Verner, M. L. Morrison, and C. T. Ralph, eds., *Wildlife 2000: Modeling habitat relations of terrestrial vertebrates.* Madison: University of Wisconsin Press.

Thomas, J. W., R. G. Anderson, C. Master, and E. L. Bull. 1979. Snags. Pages 60–77 in J. W. Thomas, ed., *Wildlife habitats in managed forests (the Blue Mountains of Oregon and Washington), USDA Forest Service Agricultural Handbook No. 553.*

Turner, M. 1989. Landscape ecology: The effects of pattern and process. *Annual Review of Ecology and Systematics* 20: 171–197.

———. 1998. Landscape ecology. In S. I. Dodson, T. F. H. Allen, S. R. Carpenter, A. R. Ives, R. L. Jeanne, J. F. Kitchell, N. E. Langston, and M. G. Turner, *Ecology.* Oxford: Oxford University Press.

Urban, D., R. V. O'Neill, and H. H. Shugart. 1987. Landscape ecology. *BioScience* 37:119–127.

Walker, B. 1989. Diversity and stability in ecosystem conservation. Pages 125–130 in D. Western and M. C. Pearl, eds., *Conservation for the twenty-first century.* New York: Oxford.

Watt, A. S. 1947. Pattern and process in the plant community. *Journal of Ecology* 35:1–22.

Whitcomb, R. E., J. F. Lynch, M. K. Klimkiewicz, C. S. Robbins, B. L. Whitcomb, and D. Bystrak. 1981. Effects of forest fragmentation on avifauna of the eastern deciduous forest. Pages 125–205 in R. L. Burgess and D. M. Sharpe, eds., *Forest island dynamics in man-dominated landscapes.* New York: Springer-Verlag.

Chapter 20 | Ecosystem Energetics

Aber, J. D., and J. M. Melillo. 1991. *Terrestrial ecosystems.* Philadelphia: Saunders College Publishing.

Anderson, J. M., and A. MacFadyen (eds.). 1976. *The role of terrestrial and aquatic organisms in the decomposition process. Seventeenth Symposium, British Ecological Society.* Oxford: Blackwell Scientific.

Andrews, R. D., D. C. Coleman, J. E. Ellis, and J. S. Singh. 1975. Energy flow relationships in a shortgrass prairie ecosystem. Pages 22–28 in *Proceedings 1st International Congress of Ecology.* The Hague: W. Junk.

Begon, M., J. L. Harper, and C. R. Townsend. 1996. *Ecology: Individuals, populations, and communities.* New York: Blackwell Scientific.

Brylinsky, M. 1980. Estimating the productivity of lakes and reservoirs. Pages 411–418 in E. D. Le Cren and R. H. Lowe-McConnell, eds., *The functioning of freshwater ecosystems. International Biological Programme No. 22.* Cambridge: Cambridge University Press.

Brylinsky, M., and K. H. Mann. 1973. An analysis of factors governing productivity in lakes and reservoirs. *Limnology and Oceanography* 18:1–14.

Coe, M. J., D. H. Cummings, and J. Phillipson. 1976. Biomass and production of large African herbivores in relation to rainfall and primary production. *Oecologica* 22:341–354.

Cooper, J. P. 1975. *Photosynthesis and productivity in different environments.* New York: Cambridge University Press.

Cowan, R. L. 1962. Physiology of nutrition of deer. Pages 1–8 in *Proceedings 1st National White-tailed Deer Disease Symposium.*

Cry, H., and M. L. Pace. 1993. Magnitude and patterns of herbivory in aquatic and terrestrial ecosystems. *Nature* 361:148–150.

DeAngelis, D. L., R. H. Gardner, and H. H. Shugart. 1980. Productivity of forest ecosystems studies during the IBP: The woodlands data set. In D. E. Reichle, ed., *Dynamic properties of forest ecosystems. International Biological Programme 23.* Cambridge: Cambridge University Press.

Dillon, P. J., and F. H. Rigler. 1974. The phosphorus–chlorophyll relationship in lakes. *Limnology and Oceanography* 19:767–773.

Downing, J. A., C. W. Osenberg, and O. Sarnelle. 1999. Meta-analysis of marine nutrient-enrichment experiments: Systematic variation in the magnitude of nutrient limitation. *Ecology* 80:1157–1167.

Elton, C. 1927. *Animal ecology.* London: Sidgwick & Jackson.

Gates, D. M. 1985. *Energy and ecology.* Sunderland, MA: Sinauer Associates.

Golley, F. B., and H. Leith. 1972. Basis of organic production in the tropics. Pages 1–26 in P. M. Golley and F. H. Golley, eds., *Tropical ecology with an emphasis on organic production.* Athens: University of Georgia Press.

Gower, S. T., R. E. McMurtie, and D. Murty. 1996. Aboveground net primary productivity declines with stand age: Potential causes. *Trends in Ecology and Evolution* 11:378–383.

Hairston, N. J., Jr., and N. G. Hairston Sr. 1993. Cause-effect relationships in energy flow, trophic structure, and interspecific interactions. *American Naturalist* 143:379–411.

Imhoff, M., L. Bounoua, T. Ricketts, C. Loucks, R. Harriss, and W. T. Lawrence. 2004. Global patterns in human consumption of net primary productivity. *Nature* 439:370–373.

Leith, H. 1973. Primary production: Terrestrial ecosystems. *Human Ecology* 1:303–332.

———. 1975. Primary productivity in ecosystems: Comparative analysis of global patterns. Pages 67–88 in W. H. van Dobben and R. H. Lowe-McConnell, eds., *Unifying concepts in ecology.* The Hague: W. Junk.

Leith, H., and R. H. Whittaker (eds.). 1975. Primary productivity in the biosphere. *Ecological Studies* vol. 14. New York: Springer-Verlag.

MacArthur, R. H., and J. H. Connell. 1966. *The biology of populations.* New York: Wiley.

McNaughton, S. J., M. Oesterheid, D. A. Frank, and K. J. Williams. 1989. Ecosystem-level patterns of primary productivity and herbivory in terrestrial habitats. *Nature* 341:142–144.

National Academy of Science. 1975. *Productivity of world ecosystems.* Washington, DC: National Academy of Science.

Odum, E. P. 1983. *Basic Ecology.* Philadelphia: Saunders College Publishing.

Pastor, J., J. D. Aber, C. A. McClaugherty, and J. M. Melillo. 1984. Aboveground production and N and P cycling along a nitrogen mineralization gradient on Blackhawk Island, Wisconsin. *Ecology* 65:256–268.

Pauly, D., and V. Christensen. 1995. Primary production required to sustain global fisheries. *Nature* 374:255–257.

Petrusewicz, K. (ed.). 1967. *Secondary productivity of terrestrial ecosystems.* Warsaw, Poland: Pantsworve Wydawnictwo Naukowe.

Phillipson, J. J. 1966. *Ecological energetics.* New York: St. Martin's Press.

Reich, P. B., D. A. Peterson, K. Wrage, and D. Wedin. 2001. Fire and vegetation effects on productivity and nitrogen cycling across a forest-grassland continuum. *Ecology* 82:1703–1719.

Reichle, D. E. 1970. *Analysis of temperate forest ecosystems.* New York: Springer-Verlag.

———. (ed.). 1981. *Dynamic properties of forest ecosystems.* Cambridge: Cambridge University Press.

Ryan, M. G., D. Binkley, and J. H. Fownes. 1997. Age-related decline in forest productivity: Pattern and process. *Advances in Ecological Research* 27:213–262.

Schindler, D. W. 1977. Evolution of phosphorus limitation in lakes. *Science* 195:260–262.

———. 1978. Factors regulating phytoplankton production and standing crop in the world's freshwaters. *Limnology and Oceanography* 23:478–486.

Transeau, E. N. 1926. The accumulation of energy by plants. *Ohio Journal of Science* 26:1–10.

Whittaker, R. H., and G. E. Likens. 1973. Carbon in the biota. In G. M. Woodwell and E. V. Pecan, eds., *Carbon and the biosphere conference 72501.* Springfield, VA: National Technical Information Service.

Wiegert, R. C. (ed.). 1976. *Ecological energetics. Benchmark Papers.* Stroudsburg, PA: Dowden, Hutchinson, and Ross. Collection of papers surveying the development of the concept.

Chapter 21 | Decomposition and Nutrient Cycling

Aber, J., and J. Melillo. 1982. Nitrogen immobilization in decaying hardwood leaf litter as a function of initial nitrogen and lignin content. *Canadian Journal of Botany* 60: 2263–2269.

_____. 1991. *Terrestrial ecosystems.* Philadelphia: Saunders College Publishing.

Anderson, J. M., and A. MacFadyen (eds.). 1976. The role of terrestrial and aquatic organisms in the decomposition process. *Seventeenth Symposium, British Ecological Society.* Oxford: Blackwell Scientific.

Austin, A. T., and P. M. Vitousek. 2000. Precipitation, decomposition and litter decomposability of *Metrosideros polymorpha* in native forests in Hawaii. *Journal of Ecology* 88:129–138.

Beare, M. H., R. W. Parmelee, P. F. Hendrix, W. Cheng, D. C. Coleman, and D. A. Crossley. 1992. Microbial and fungal interactions and effects on litter nitrogen and decomposition in agroecosystems. *Ecological Monographs* 62:569–591.

Chapin, F. S. 1990. The mineral nutrition of wild plants. *Annual Review of Ecology and Systematics* 11:233–260.

Correll, D. L. 1978. Estuarine productivity. *BioScience* 28: 646–650.

Fenchel, T. 1988. Marine plankton food chains. *Annual Review of Ecology and Systematics* 19:19–38.

Fenchel, T., and T. H. Blackburn. 1979. *Bacteria and mineral cycling.* London: Academic Press.

Fenchel, T., and P. Harrison. 1976. The significance of bacterial grazing and mineral cycling for the decomposition of particulate detritus. Pages 285–321 in J. M. Anderson and A. MacFayden, eds., *The role of terrestrial and aquatic organisms in the decomposition process.* New York: Blackwell Scientific.

Fletcher, M., G. R. Gray, and J. G. Jones. 1987. *Ecology of microbial communities.* New York: Cambridge University Press.

Klap, V., P. Louchouarn, J. J. Boon, M. A. Hemminga, and J. van Soelen. 1999. Decomposition dynamics of six salt marsh halophytes as determined by cupric oxide oxidation and direct temperature-resolved mass spectrometry. *Limnology and Oceanography* 44:1458–1476.

Likens, G. E., and F. H. Bormann. 1974. Linkages between terrestrial and aquatic ecosystems. *BioScience* 24(8): 447–456.

Lovelock, J. 1979. *GAIA: A new look at life on Earth.* Oxford: Oxford University Press.

Meentemeyer, V. 1978. Macroclimate and lignin control of decomposition. *Ecology* 59:465–472.

Melillo, J., J. Aber, and J. Muratore. 1982. Nitrogen and lignin control of hardwood leaf litter decomposition dynamics. *Ecology* 63:621–626.

Mudrick, D., M. Hoosein, R. Hicks, and E. Townsend. 1994. Decomposition of leaf litter in an Appalachian forest: Effects of leaf species, aspect, slope position and time. *Forest Ecology and Management* 68:231–250.

Newbold, J. D., R. V. O'Neill, J. W. Elwood, and W. Van Winkle. 1982. Nutrient spiraling in streams: Implications for nutrient and invertebrate activity. *American Naturalist* 20:628–652.

Nybakken, J. W. 1988. *Marine biology: An ecological approach.* 2nd ed. New York: Harper & Row.

Odum, W. E., and M. A. Haywood. 1978. Decomposition of intertidal freshwater marsh plants. Pages 89–97 in R. E. Good, D. F. Whigham, and R. L. Simpson, eds., *Freshwater wetlands.* New York: Academic Press.

Pastor, J., J. D. Aber, C. A. McClaugherty, and J. M. Melillo. 1984. Aboveground production and N and P cycling along a nitrogen mineralization gradient on Blackhawk Island, Wisconsin. *Ecology* 65:256–268.

Smith, T. M. 2005. Spatial variation in leaf-litter production and decomposition within a temperate forest: The influence of species composition and diversity. *Journal of Ecology.* In review.

Staff, H., and B. Berg. 1982. Accumulation and release of plant nutrients in decomposing Scots pine needle litter. Long-term decomposition in a Scots pine forest. II. *Canadian Journal of Botany* 60:1516–1568.

Swift, M. J., O. W. Heal, and J. M. Anderson. 1979. *Decomposition in terrestrial ecosystems.* Berkeley: University of California Press.

Valiela, I. 1984. *Marine ecological processes.* New York: Springer-Verlag.

Whitkamp, M., and M. L. Frank. 1969. Evolution of carbon dioxide from litter, humus and subsoil of a pine stand. *Pedobiologia* 9:358–365.

Vitousek, P. M., S. W. Andariese, P. A. Matson, L. Morris, and R. L. Sanford. 1992. Effects of harvest intensity, site preparation, and herbicide use on soil nitrogen transformations in a young loblolly pine plantation. *Forest Ecology and Management* 49:277–292.

Webster, J. R. 1975. *Analysis of potassium and calcium dynamics in stream ecosystems on three Appalachian watersheds of contrasting vegetation.* Ph.D. diss. University of Georgia.

Webster, J. R., D. J. D'angelo, and G. T. Peters. 1991. Nitrate and phosphate uptake in streams at Coweeta Hydrological Laboratory. *Internationale Verein Limnologie* 24:1681–1686.

Witherspoon, J. P. 1964. Cycling of cesium-134 in white oak trees. *Ecological Monographs* 34:403–420.

Chapter 22 | Biogeochemical Cycles

Aber, J. D., and A. H. Magill. 2004. Chronic nitrogen additions at the Harvard Forest: The first fifteen years of a nitrogen saturation experiment. *Forest Ecology and Management* 196:1–6.

Aber, J. D., K. N. Nadelhoffer, P. Steudler, and J. M. Melillo. 1989. Nitrogen saturation in northern forest ecosystems. *BioScience* 39:378–386.

Alexander, M. (ed.). 1980. *Biological nitrogen fixation.* New York: Plenum.

Binkley, D., C. T. Driscoll, H. L. Allen, P. Schoeneberger, and D. McAvoy. 1989. *Acidic deposition and forest soils: Context and case studies in southeastern United States.* New York: Springer-Verlag.

Bolin, B., E. T. Degens, S. Kempe, and P. Ketner (eds.). 1979. *The global carbon cycle. SCOPE 13.* New York: John Wiley.

Bormann, F. H., and G. E. Likens. 1979. *Pattern and process in a forested ecosystem.* New York: Springer-Verlag.

Pomeroy, L. R. 1974. Cycles of essential elements. *Benchmark Papers in Ecology.* Stroudsburg, PA: Dowden, Hutchinson, and Ross.

Post, W. M., T. H. Peng, W. R. Emanuel, A. W. King, V. H. Dale, and D. L. DeAngelis. 1990. The global carbon cycle. *American Scientist* 78:310–326.

Schlesinger, W. H. 1997. *Biogeochemistry: An analysis of global change.* 2nd ed. London: Academic Press.

Schulze, E. D., O. L. Lange, and O. Oren. 1989. *Forest decline and air pollution: A study of spruce* (Picea abies) *on acid soils.* New York: Springer-Verlag.

Smith, W. H. 1990. *Air pollution and forests: Interaction between air contaminants and forest ecosystems.* 2nd ed. New York: Springer-Verlag.

Sprent, J. I. 1988. *The ecology of the nitrogen cycle*. New York: Cambridge University Press.

Wellburn, A. 1988. *Air pollution and acid rain: The biological impact*. Essex, England: Longman Scientific and Technical; New York: Wiley.

Chapter 23 | Terrestrial Ecosystems

Archibold, O. W. 1995. *Ecology of world vegetation*. London: Chapman and Hall.

Bailey, R. G. 1996. *Ecosystem geography*. New York: Springer.

Tropical Forests

Ashton, P. S. 1988. Dipterocarp biology as a window to understanding of tropical forest structure. *Annual Review of Ecology and Systematics* 19:347–370.

Bawa, K. S. 1990. Plant-pollinator interactions in tropical rain forests. *Annual Review of Ecology and Systematics* 21:399–422.

Collins, M. (ed.). 1990. *The last rain forests: A world conservation atlas*. New York: Oxford University Press.

Denslow, J. S. 1987. Tropical rainforest gaps and tree species diversity. *Annual Review of Ecology and Systematics* 18:431–451.

Erwin, T. L. 1988. The tropical forest canopy: The heart of biotic diversity. In E. O. Wilson and F. M. Peters, eds., *Biodiversity*. Washington, DC: National Academy Press.

Fleming, T. H., R. Breitwisch, and G. H. Whitesides. 1987. Patterns of tropical vertebrate frugivore diversity. *Annual Review of Ecology and Systematics* 18:71–90.

Golley, F. B. (ed.). 1983. *Tropical forest ecosystems: Structure and function*. Ecosystems of the World No. 14A. Amsterdam: Elsevier Scientific.

Jordan, C. F., and J. R. Kline. 1972. Mineral cycling: Some basic concepts and their application in a tropical rain forest. *Annual Review of Ecology and Systematics* 3:33–50.

Lathwell, D. J., and T. L. Grove. 1986. Soil-plant relations in the tropics. *Annual Review of Ecology and Systematics* 17:1–16.

Leigh, E. G., Jr. 1975. Structure and climate in tropical rain forest. *Annual Review of Ecology and Systematics* 6:67–86.

Murphy, P. G., and A. E. Lugo. 1986. Ecology of tropical dry forests. *Annual Review of Ecology and Systematics* 17:67–88.

Richards, P. W. 1996. *The tropical rain forest: An ecological study*. 2nd ed. New York: Cambridge University Press.

Simpson, B. B., and J. Haffer. 1978. Spatial patterns in the Amazonian rain forest. *Annual Review of Ecology and Systematics* 9:497–518.

Sutton, S. L., T. C. Whitmore, and A. C. Chadwick (eds.). 1983. *Tropical rain forests: Ecology and management*. Oxford: Blackwell Scientific.

Tomlinson, P. B. 1987. Architecture of tropical plants. *Annual Review of Ecology and Systematics* 18:1–21.

Walter, H. 1971. *Ecology of tropical and subtropical vegetation*. Edinburgh: Oliver & Boyd.

Whitmore, T. C. 1984. *Tropical rainforests of the Far East*. Oxford: Clarendon Press.

———. 1990. *An introduction to tropical rain forests*. New York: Oxford University Press.

Temperate Forests

Bormann, F. H., and G. E. Likens. 1979. *Pattern and process in a forest ecosystem*. New York: Springer-Verlag.

Brinson, M. M., B. L. Swift, R. C. Plantico, and J. S. Barclay. 1981. *Riparian ecosystems: Their ecology and status*. FWS/OBS-81/17. Kearneysville, WV: Eastern Energy and Land Use Team, U.S. Fish and Wildlife Service.

Curtis, J. T. 1959. *The vegetation of Wisconsin*. Madison: University of Wisconsin Press.

Davis, M. B. (ed.). 1996. *Eastern old-growth forests: Prospects for rediscovery and recovery*. Covelo, CA: Island Press.

Duvigneaud, P. (ed.). 1971. *Productivity of forest ecosystems*. Paris: UNESCO.

Lang, G. E., W. A. Reiners, and L. H. Pike. 1980. Structure and biomass dynamics of epiphytic lichen communities of balsam fir forests in New Hampshire. *Ecology* 63: 541–550.

Peterken, G. F. 1996. *Natural woodlands: Ecology and conservation in north temperate regions*. New York: Cambridge University Press.

Polunin, O., and M. Walters. 1985. *A guide to the vegetation of Britain and Europe*. Oxford: Oxford University Press.

Reichle, D. E. (ed.). 1981. *Dynamic properties of forest ecosystems*. Cambridge: Cambridge University Press.

Conifer Forest

Bonan, G. B., and H. H. Shugart. 1989. Environmental factors and ecological processes in boreal forests. *Annual Review of Ecology and Systematics* 20:1–18.

Edmonds, R. L. (ed.). 1981. *Analysis of coniferous forest ecosystems in the western United States*. Stroudsburg, PA: Dowden, Hutchinson & Ross.

Franklin, J. F., and C. T. Dyrness. 1973. Natural vegetation of Oregon and Washington. *USDA Forest Service Gen. Tech. Rept. PNW8*. Corvallis, OR: USDA Forest Service.

Knystautas, A. 1987. *The natural history of the USSR*. New York: McGraw-Hill.

Larsen, J. A. 1980. *The boreal ecosystem*. New York: Academic Press.

———. 1989. *The northern forest border in Canada and Alaska*. Ecological Studies 70. New York: Springer-Verlag.

Polunin, O., and M. Walters. 1985. *A guide to the vegetation of Britain and Europe*. Oxford: Oxford University Press.

Woodland/Savanna

Bourliere, F. (ed.). 1983. *Tropical savannas: Ecosystems of the world 13*. Amsterdam: Elsevier Scientific.

Plumb, T. R. 1979. Proceedings of the symposium on the ecology, management, and utilization of California oaks. *General Technical Report PSW-44*. Berkeley, CA: Pacific Southwest Forest and Range Experiment Station, Forest Service, U.S. Dept. of Agriculture.

Sarimiento, G. 1984. *Ecology of neotropical savannas*. Cambridge, MA: Harvard University Press.

Sinclair, A. R. E., and P. Arcese (eds.). 1995. *Serengeti II: Dynamics, management, and conservation of an ecosystem*. Chicago: University of Chicago Press.

Sinclair, A. R. E., and M. Norton-Griffiths (eds.). 1979. *Serengeti: Dynamics of an ecosystem*.

Tothill, J. C., and J. J. Mott (eds.). 1985. *Ecology and management of the world's ecosystems*. Canberra: Australian Academy of Science.

Grasslands

Breymeyer, A., and G. Van Dyne (eds.). 1980. *Grasslands, systems analysis and man*. Cambridge: Cambridge University Press.

Callenback, E. 1996. *Bring back the buffalo: A sustainable future for America's Great Plains*. Covelo, CA: Island Press.

Coupland, R. T. (ed.) 1979. *Grassland ecosystems of the world: Analysis of grasslands and their uses*. Cambridge: Cambridge University Press.

Duffy, E. 1974. *Grassland ecology and wildlife management.* London: Chapman and Hall.

French, N. (ed.). 1979. *Perspectives on grassland ecology.* New York: Springer-Verlag.

Hodgson, J., and A. W. Illius (eds.). 1966. *Ecology and management of grazing systems.* New York: CAB International.

Levin, S. A. (ed.). 1993. Forum: Grazing theory and rangeland management. *Ecological Applications* 3:1–38.

Manning, D. 1995. *Grassland: History, biology, politics, and promise of the American prairie.* New York: Penguin Books.

Reichman, O. J. 1987. *Konza prairie: Tallgrass natural history.* Lawrence: University Press of Kansas.

Risser, P. G., et al. 1981. *The true prairie ecosystem.* Stroudsburg, PA: Dowden, Hutchinson & Ross.

Weaver, J. E. 1954. *North American prairie.* Lincoln, NE: Johnson.

Shrublands

Castri, F. Di, D. W. Goodall, and R. L. Specht (eds.). 1981. *Mediterranean-type shrublands. Ecosystems of the World No. 11.* Amsterdam: Elsevier Scientific.

Castri, F. Di, and H. A. Mooney (eds.). 1973. *Mediterranean-type ecosystems. Ecological Studies No. 7.* New York: Springer-Verlag.

Groves, R. H. 1981. *Australian vegetation.* Cambridge: Cambridge University Press.

McKell, C. M. (ed.). 1983. *The biology and utilization of shrubs.* Orlando, FL: Academic Press.

Polunin, O., and M. Walters. 1985. *A guide to the vegetation of Britain and Europe.* Oxford: Oxford University Press.

Specht, R. L. (ed.). 1979, 1981. *Heathlands and related shrublands. Ecosystems of the World 9A and 9B.* Amsterdam: Elsevier Scientific.

Desert

Brown, G. W., Jr. (ed.). 1976–1977. *Desert biology,* 2 vols. New York: Academic Press.

Evenardi, M., I. Noy-Meir, and D. Goodall (eds.). 1985–1986. *Hot deserts and arid shrublands of the world. Ecosystems of the World 12A and 12B.* Amsterdam: Elsevier Scientific.

Wagner, F. H. 1980. *Wildlife of the deserts.* New York: Harry W. Abrams.

Tundra

Bliss, L. C., O. H. Heal, and J. J. Moore (eds.). 1981. *Tundra ecosystems: A comparative analysis.* New York: Cambridge University Press.

Brown, J., P. C. Miller, L. L. Tieszen, and F. L. Bunnell. 1980. *An arctic ecosystem: The coastal tundra at Barrow, Alaska.* Stroudsburg, PA: Dowden, Hutchinson & Ross.

Furley, P. A., and W. A. Newey. 1983. *Geography of the biosphere.* London: Butterworths.

Rosswall, T., and O. W. Heal (eds.). 1975. *Structure and function of tundra ecosystems. Ecological Bulletins 20.* Stockholm: Swedish Natural Sciences Research Council.

Smith, A. P., and T. P. Young. 1987. Tropical alpine plant ecology. *Annual Review of Ecology and Systematics* 18:137–158.

Sonesson, M. (ed.). 1980. *Ecology of a subarctic mire. Ecological Bulletins 30.* Stockholm: Swedish Natural Sciences Research Council.

Wielgolaski, F. E. (ed.). 1975. *Fennoscandian tundra ecosystems.* Part I: *Plants and microorganisms.* Part II: *Animals and systems analysis.* New York: Springer-Verlag.

Zwinger, A. H., and B. E. Willard. 1972. *Land above the trees.* New York: Harper & Row.

Chapter 24 | Aquatic Ecosystems

Freshwater and Estuarine Ecosystems

Barnes, R. S. K., and K. H. Mann. 1980. *Fundamentals of aquatic ecosystems.* New York: Blackwell Scientific.

Brock, T. D. 1985. *A eutrophic lake: Lake Mendota, Wisconsin.* New York: Springer-Verlag.

Carpenter, S. A. 1980. Enrichment of Lake Wingra, Wisconsin, by submerged macrophyte decay. *Ecology* 61:1145–1155.

Carpenter, S. A., and J. F. Kitchell. 1984. Plankton community structure and limnetic primary production. *American Naturalist* 124:159–172.

Carpenter, S. A., J. F. Kitchell, and J. Hodgson. 1985. Cascading trophic interactions and lake productivity. *BioScience* 35:634–639.

Cummins, K. W. 1974. Structure and function of stream ecosystems. *BioScience* 24:631–641.

———. 1979. Feeding ecology of stream invertebrates. *Annual Review of Ecology and Systematics* 10:147–172.

Gore, J. A., and G. E. Petts. 1989. *Alternatives in regulated river management.* Boca Raton, FL: CRC Press.

Hutchinson, G. E. 1957–1967. *A treatise on limnology.* Vol. 1: *Geography, physics, and chemistry.* Vol. 2: *Introduction to lake biology and limnoplankton.* New York: Wiley.

Hynes, H. B. N. 1970. *The ecology of running water.* Toronto: University of Toronto Press.

Ketchum, B. H. (ed.). 1983. *Estuaries and enclosed seas. Ecosystems of the World 26.* Amsterdam: Elsevier Scientific.

Likens, G. E. (ed.). 1985. *An ecosystem approach to aquatic ecology: Mirror Lake and environment.* New York: Springer-Verlag.

Macan, T. T. 1970. *Biological studies of English lakes.* New York: Elsevier Scientific.

———. 1973. *Ponds and lakes.* New York: Crane, Russak.

———. 1974. *Freshwater ecology.* 2nd ed. New York: John Wiley & Sons.

Maitland, P. S. 1978. *Biology of fresh waters.* New York: John Wiley & Sons.

McLusky, D. S. 1989. *The estuarine ecosystem.* 2nd ed. New York: Chapman & Hall.

Meyer, J. L. 1990. A blackwater perspective on riverine ecosystems. *BioScience* 40:643–651.

Petts, G. E. 1984. *Impounded rivers: Perspectives for ecological management.* New York: Wiley.

Stanford, J. A., and A. P. Covich (eds.). 1988. Community structure and function in temperate and tropical streams. *Journal of North American Benthological Society* 7:261–529.

Vannote, R. L., G. W. Minshall, K. W. Cummins, J. R. Sedell, and C. E. Cushing. 1980. The river continuum concept. *Canadian Journal of Fisheries and Aquatic Science* 37:130–137.

Wiley, M. 1976. *Estuarine processes.* New York: Academic Press.

Oceans

Grassle, J. F. 1985. Hydrothermal vent animals: Distribution and biology. *Science* 229:713–717.

———. 1989. Species diversity in deep-sea communities. *Trends in Ecology and Evolution* 4:12–15.

———. 1991. Deep-sea benthic diversity. *BioScience* 41:464–469.

Gross, M. G. 1982. *Oceanography: A view of Earth.* 3rd ed. Englewood Cliffs, NJ: Prentice-Hall.

Hardy, A. 1971. *The open sea: Its natural history.* Boston: Houghton Mifflin.

Hayman, R. M., and R. C. McDonald. 1985. The ecology of deep sea hot springs. *American Scientist* 73:441–449.

Kinne, O. (ed.). 1978. *Marine ecology.* 5 vols.

Marshall, N. B. 1980. *Deep-sea biology: Developments and perspectives.* New York: Garland STMP Press.

Nybakken, J. W. 1997. *Marine biology: An ecological approach.* 4th ed. Menlo Park, CA: Benjamin Cummings.

Rex, M. A. 1981. Community structure in the deep-sea benthos. *Annual Review of Ecology and Systematics* 12:331–353.

Steele, J. 1974. *The structure of a marine ecosystem.* Cambridge, MA: Harvard University Press.

Hiatt, R. W., and D. W. Strasburg. 1960. Ecological relationships of the fish fauna on coral reefs of the Marshall Islands. *Ecological Monographs* 30:66–120.

Huston, M. 1985. Patterns of species diversity on coral reefs. *Annual Review of Ecology and Systematics.* 16:149–177.

Jackson, J. B. C. 1991. Adaptation and diversity of reef corals. *BioScience* 41:475–482.

Jones, O. A., and R. Endean (eds.). 1973, 1976. *Biology and geology of coral reefs.* Vols. II, III. New York: Academic Press.

Nybakken, J. W. 1997. *Marine biology: An ecological approach.* 4th ed. Menlo Park, CA: Benjamin Cummings.

Pomeroy, L. R., and E. J. Kuenzler. 1969. Phosphorus turnover by coral reef animals. Pages 478–483 in D. J. Nelson and F. E. Evans, eds., *Symposium on radioecology conference.* 670503. Springfield, VA: National Technical Information Services.

Reaka, M. J. (ed.). 1985. *Ecology of coral reefs.* Symposia Series for Undersea Research, NOAA Underseas Research Program 3. Washington DC: US Department of Commerce.

Sale, P. F. 1980. The ecology of fishes on coral reefs. *Annual Review of Oceanography and Marine Biology* 18:367–421.

Valiela, I. 1984. *Marine biology processes.* New York: Springer-Verlag.

Wellington, G. W. 1982. Depth zonation of corals in the Gulf of Panama: Control and facilitation by resident reef fishes. *Ecological Monographs* 52:223–241.

Wilson, R., and J. Q. Wilson. 1985. *Watching fishes: Life and behavior on coral reefs.* New York: Harper & Row.

Chapter 25 | Land-Water Margins
Coastal Ecosystems

Bertness, M. D. 1999. *The ecology of Atlantic shorelines.* Sunderland, MA: Sinauer Associates.

Carson, R. 1955. *The edge of the sea.* Boston: Houghton Mifflin.

Eltringham, S. K. 1971. *Life in mud and sand.* New York: Crane, Russak.

Leigh, E. G., Jr. 1987. Wave energy and intertidal productivity. *Proceedings of the National Academy of Sciences* 84:1314.

Lubchenco, J. 1978. Algal zonation in the New England rocky intertidal community: An experimental analysis. *Ecology* 61:333–344.

Mathieson, A. C., and P. H. Nienhuis (eds.). 1991. *Intertidal and littoral ecosystems. Ecosystems of the world 24.* Amsterdam: Elsevier Scientific.

Moore, P. G., and R. Seed (eds.). 1986. *The ecology of rocky shores.* New York: Columbia University Press.

Newell, R. C. 1970. *Biology of intertidal animals.* New York: Elsevier Scientific.

Stephenson, T. A., and A. Stephenson. 1973. *Life between the tidemarks on rocky shores.* San Francisco: Freeman.

Underwood, A. J., E. J. Denley, and M. J. Moran. 1983. Experimental analyses of the structure and dynamics of midshore rocky intertidal communities in New South Wales. *Oecologica* 56:202–219.

Yonge, C. M. 1949. *The seashore.* London: Collins.

Wetlands

Bertness, M. D. 1984. Ribbed mussels and *Spartina alterniflora* production on a New England marsh. *Ecology* 65: 1794–1807.

_____. 1999. *The ecology of Atlantic shorelines.* Sunderland, MA: Sinauer Associates.

Bildstein, K. L., G. T. Bancroft, P. J. Dugan et al. 1991. Approaches to the conservation of coastal wetlands in the Western Hemisphere. *Wilson Bulletin* 103:218–254.

Chapman, V. J. 1976. *Mangrove vegetation.* Leutershausen, Germany: J. Cramer.

_____. (ed.). 1977. *Wet coastal ecosystems.* Amsterdam: Elsevier Scientific.

Clark, J. 1974. *Coastal ecosystems: Ecological considerations for the management of the coastal zone.* Washington, DC: Conservation Foundation.

Cowardin, L. M., V. Carter, and E. C. Golet. 1979. *Classification of wetlands and deepwater habitats of the United States.* U.S. Department of Interior, Fish and Wildlife Service FWS/OBS-79/31.

Dahl, T. E. 1990. *Wetland losses in the United States, 1780's to 1980's.* U.S. Department of Interior, Fish and Wildlife Service.

Dugan, P. (ed.). 1993. *Wetlands in danger: A world conservation atlas.* New York: Oxford University Press.

Ewel, K. C. 1990. Multiple demands on wetlands. *BioScience* 40:660–666.

Ewel, K. C., and H. T. Odum (eds.). 1986. *Cypress swamps.* Gainesville: University Presses of Florida.

Gore, A. P. J. (ed.). 1983. *Mire, swamp, bog, fen, and moor. Ecosystems of the world 4A and 4B.* Amsterdam: Elsevier Scientific.

Haines, B. L., and E. L. Dunn. 1985. Coastal marshes. Pages 323–347 in B. F. Chabot and H. A. Mooney, eds., *Physiological ecology of North American plant communities.* New York: Chapman and Hall.

Hopkinson, C. S., and J. P. Schubauer. 1984. Static and dynamic aspects of nitrogen cycling in the salt marsh graminoid *Spartina alterniflora. Ecology* 65:961–969.

Howarth, R. W., and J. Teal. 1979. Sulfate reduction in a New England salt marsh. *Limnology and Oceanography* 24:999–1013.

Jefferies, R. L., and A. J. Davy (eds.). 1979. *Ecological processes in coastal environments.* Oxford: Blackwell Scientific.

Long, S. P., and C. F. Mason. 1983. *Salt marsh ecology.* New York: Chapman & Hall.

Lugo, A. E. 1990. *The forested wetlands.* Amsterdam: Elsevier Scientific.

Lugo, A. E., and S. C. Snedaker. 1974. The ecology of mangroves. *Annual Review of Ecology and Systematics* 5:39–64.

Mitsch, W. J., and J. C. Gosslink. 1993. *Wetlands.* 2nd ed. New York: Van Nostrand Reinhold.

Moore, P. D., and D. J. Bellemany. 1974. *Peatlands.* New York: Springer-Verlag.

Niering, W. A., and B. Hales. 1991. *Wetlands of North America.* Charlottesville, VA: Thomasson-Grant.

Nixon, S. W., and C. A. Oviatt. 1973. Ecology of a New England salt marsh. *Ecological Monographs* 43:463–498.

Nybakken, J. W. 1997. *Marine biology: An ecological approach.* 4th ed. Menlo Park, CA: Benjamin Cummings.

Pomeroy, L. R., and R. G. Wiegert (eds.). 1981. *The ecology of a salt marsh.* New York: Springer-Verlag.

Teal, J. 1962. Energy flow in a salt marsh ecosystem of Georgia. *Ecology* 43:614–624.

Tiner, R. W. 1991. The concept of a hydrophyte for wetland identification. *BioScience* 41:236–247.

Valiela, I., and J. M. Teal. 1979. The nitrogen budget of a salt marsh ecosystem. *Nature* 47:337–371.

Valiela, I., J. M. Teal, and W. G. Denser. 1978. The nature of growth forms in salt marsh grass *Spartina alternifolia. American Naturalist* 112:461–370.

Van der Valk, A. (ed.). 1989. *Northern prairie wetlands.* Ames: Iowa State University Press.

Weller, M. W. 1981. *Freshwater wetlands: Ecology and wildlife management.* Minneapolis: University of Minnesota Press.

Zedler, J., T. Winfield, and D. Mauriello. 1992. *The ecology of Southern California coastal marshes: A community profile.* U.S. Fish and Wildlife Service Office of Biological Services FWS/OBS 81/54.

Chapter 26 | Large-Scale Patterns of Biological Diversity

Angel, M. V. 1991. Variations in time and space: Is biogeography relevant to studies of long-time scale change? *Journal of the Marine Biological Association of the United Kingdom* 71:191–206.

Archibold, O. W. 1995. *Ecology of world vegetation.* London: Chapman and Hall.

Bailey, R. G. 1996. *Ecosystem geography.* New York: Springer.

Bailey, R. W. 1978. *Description of the ecoregions of the United States.* Ogden, UT: USDA Forest Service Intermountain Region.

Bartlein, P. J., T. Webb, and E. C. Fleri. 1984. Climate response surfaces from pollen data for some eastern North American taxa. *Journal of Biogeography* 13:35–57.

Clements, F. E., and V. E. Shelford. 1939. *Bio-ecology.* New York: McGraw-Hill.

Currie, D. J. 1991. Energy and large-scale biogeographical patterns of animal and plant species richness. *American Naturalist* 137:27–49.

Currie, D. J., and V. Paquin. 1987. Large-scale biogeographical patterns of species richness in trees. *Nature* (London) 329:326–327.

DeCandolle, A. P. A. 1874. *Constitution dans le Regne Vegetal de Groupes Physiologiques Applicables a la Geographie Ancienne et Moderne.* Archives des Sciences Physiques et Naturelles. May, Geneva.

Holdridge, L. R. 1947. Determination of wild plant formations from simple climatic data. *Science* 105:367–368.

_____. 1967. Determination of world plant formation from simple climatic data. *Science* 130:572.

Hunter, M. L., and P. Yonzon. 1992. Altitudinal distributions of birds, mammals, people, forests and parks in Nepal. *Conservation Biology* 7:420–423.

Huston, M. 1994. *Biological diversity: The coexistence of species on changing landscapes.* New York: Cambridge University Press.

Kikkawa, J., and W. T. Williams. 1971. Altitudinal distribution of land birds in New Guinea. *Search* 2:64–69.

Koppen, W. 1918. Klassifikation der Klimate nach Temperatur, Niederschlag, und Jahres lauf. *Petermann's Mitteilungen* 64:193–203, 243–248.

Merriam, C. H. 1890. Results of a biological survey of the San Francisco mountain region and the desert of the Little Colorado, Arizona. *North American Fauna* 1–136.

_____. 1899. Zone temperatures. *Science* 9:116.

Niklas, K. J., B. H. Tiffney, and A. H. Knoll. 1983. Patterns in land plant diversification. *Nature* 303:293–299.

Prentice, I. C., P. J. Bartlein, and T. Webb. 1991. Vegetation and climate change in eastern North America since the last glacial maximum: A response to continuous climate forcing. *Ecology* 72:2038–2056.

Rex, M. A., C. T. Stuart, R. R. Hessler, J. A. Allen, H. A. Sanders, and G. D. F. Wilson. 1993. Global-scale latitudinal patterns of species diversity in the deep-sea benthos. *Nature* 365:636–639.

Schimper, A. F. W. 1903. *Plant-geography upon a physiological basis.* Oxford: Clarendon.

Whittaker, R. H. 1975. *Communities and ecosystems.* London: Collier-Macmillan.

_____. 1977. Evolution of species diversity in land communities. *Evolutionary Biology* 10:1–67.

Wilson, E. O. 1999. *The diversity of life.* New York: Norton.

Chapter 27 | Population Growth, Resource Use and Sustainability

Bormann, F. H., D. Balmori, and G. T. Geballe. 1993. *Redesigning the American lawn.* New Haven, CT: Yale University Press.

Bricklemeyer, E. C., Jr., S. Ludicello, and H. J. Hartmann. 1989. Discarded catch in U.S. commercial marine fisheries. Pages 259–295 in *Audubon wildlife report 1989/1990.* San Diego: Academic Press.

Buschbacker, R. J. 1986. Tropical deforestation and pasture development. *BioScience* 36:22–28.

Daily, G. (ed.). 1997. *Nature's Services: Societal Dependence on Natural Ecosystems.* Washington, DC: Island Press.

Ellis, R. 1991. *Men and whales.* New York: Knopf.

Ewel, J., C. Berish, B. Brown, N. Price, and J. Raich. 1981. Slash and burn impacts on a Costa Rican wet forest site. *Ecology* 62:816–829.

Gambell, R. 1976. Population biology and the management of whales. *Applied Biology* 1:237–343.

Gliessman, S. R. (ed.) 1990. *Agroecology: Researching the ecological basis for sustainable agriculture. Ecological Studies Series No. 78.* New York: Springer-Verlag.

Harris, L. D. 1984. *The fragmented forest.* Chicago: University of Chicago Press.

Hoffman, C. A. 1990. Ecological risks of genetic engineering of crop plants. *BioScience* 40:434–436.

Heal, G. 2000. *Nature and the marketplace.* Washington, DC: Island Press.

Horn, D. J. 1988. *Ecological approach to pest management.* New York: Guilford.

Jenkins, M. B. (ed.). 1998. *The business of sustainable forestry: Case studies.* Chicago: J. D. and K. T. MacArthur Foundation.

Lansky, M. 1992. *Beyond the beauty strip: Saving what's left of our forests.* Gardiner, ME: Tilbury House.

Livi-Bacci, M. 2001. *A concise history of world population.* 3rd ed. Malden, MA: Blackwell Scientific.

Mather, G. A. S. 1990. *Global forest resources.* Portland, OR: Timber Press.

Orians, G. H., J. Buckley, W. Clark, M. Gilpin, J. Lehman, R. May, G. Robilliard, and D. Simberloff. 1986. *Ecological knowledge and environmental problem-solving.* Washington, DC: National Academy Press.

Perlan, J. 1991. *A forest journey: The role of wood in the development of civilization.* Cambridge, MA: Harvard University Press.

Pesek, J. (ed.). 1989. *Alternative agriculture.* Washington, DC: National Academy Press.

Pimentel, D. 1984. Energy flow in agroecosystems. Pages 121–132 in R. Lawrence, B. Stinner, and G. J. House, eds., *Agricultural ecosystems: Unifying concepts.* New York: John Wiley & Sons.

Pimentel, D., and C. A. Edwards. 1982. Pesticides and ecosystems. *BioScience* 32:595–600.

Regier, H. A., and G. L. Baskerville. 1986. Sustainable development of regional ecosystems degraded by exploitive development. Pages 74–103 in W. E. Clark and R. E. Munn, eds., *Sustainable development of the biosphere.* Cambridge: Cambridge University Press.

Regier, H. A., and W. L. Hartman. 1973. Lake Erie's fish community: 150 years of cultural stress. *Science* 180: 1248–1255.

Russell, C. S. 2001. *Applying economics to the environment.* Oxford: Oxford University Press.

Russell, E. 2001. *War and nature.* Cambridge: Cambridge University Press.

Small, G. 1976. *The blue whale.* New York: Columbia University Press.

USDA Forest Service. 1980. *Environmental consequences of timber harvesting in Rocky Mountain coniferous forests. USDA Forest Service Gen. Tech. Rept. INT-90.* Ogden, UT: Intermountain Forest and Range Experiment Station.

Walters, C. 1986. *Adaptive management of renewable resources.* New York: Macmillan.

Chapter 28 | Habitat Loss, Biodiversity, and Conservation

Baden, J. A., and D. Leal. 1990. *The Yellowstone primer.* San Francisco: Pacific Research Institute for Public Policy.

Berger, J. 1990. Persistence of different-sized populations: An empirical assessment of rapid extinctions in bighorn sheep. *Conservation Biology* 4:91–98.

Carson, R. 1962. *Silent spring.* Boston: Houghton Mifflin.

Caughley, G. 1976. Wildlife management and the dynamics of ungulate populations. *Applied Biology* 1:183–246.

Cottam, G. 1990. Community dynamics on an artificial prairie. Pages 257–270 in W. R. Jordan III, M. E. Gilpin, and J. D. Aber, eds., *Restoration ecology: A synthetic approach to ecological research.* Cambridge: Cambridge University Press.

Culotta, E. 1995. Many suspects to blame in Madagascar extinctions. *Science* 268:156–159.

DiSilvestro, R. L. 1989. *The endangered kingdom: The struggle to save America's wildlife.* New York: Wiley.

Fitzgerald, S. 1989. *International wildlife trade: Whose business is it?* Washington, DC: World Wildlife Fund.

Green, G. N., and R. W. Sussman. 1990. Deforestation history of the eastern rain forests of Madagascar from satellite images. *Science* 248:212–215.

Hughes, T. P. 1994. Catastrophes, phase shifts and large-scale degradation of a Caribbean coral reef. *Science* 265: 1547–1551.

Jordan, W. R., III, M. E. Gilpin, and J. D. Aber (eds.). 1990. *Restoration ecology: A synthetic approach to ecological research.* Cambridge: Cambridge University Press.

Kleiman, D. G. 1989. Reintroduction of captive mammals for conservation. *BioScience* 39:152–161.

Kot, M. 2001. *Elements of mathematical ecology.* Cambridge: Cambridge University Press.

McCollough, D. R. (ed.). 1996. *Metapopulations and wildlife conservation.* Washington, DC: Island Press.

Meffe, G. K., and C. R. Carroll. 1997. *Principles of conservation biology.* 2nd ed. Sunderland, MA: Sinauer Associates.

Mittermeier, R. A., N. Myers, P. R. Gil, and C. G. Mittermeier. 1999. *Hotspots: Earth's biologically richest and most endangered terrestrial ecoregions.* Mexico City: CEMEX Conservation International.

Myers, N. 1986. Tropical deforestation and a mega-extinction spasm. Pages 349—409 in M. E. Soule, ed., *Conservation biology: The science of scarcity and diversity.* Sunderland, MA: Sinauer Associates.

———. 1991a. Tropical deforestation: The latest situation. *BioScience* 41:282.

———. 1991b. The biodiversity challenge: Expanded hotspots analysis. *Environmentalist* 10:243–256.

———. 1998. Threatened biotas: "Hotspots" in tropical forests. *Environmentalist* 8:1–20.

Pimentel, D., et al. 1992. Environmental and economic costs of pesticide use. *BioScience* 42:750–760.

Prescott-Allen, C., and R. Prescott-Allen. 1986. *The first resource: Wild species in North American economy.* New Haven, CT: Yale University Press.

Primack, R. B. 1998. *Essentials of conservation biology.* 2nd ed. Sunderland, MA: Sinauer Associates.

Schonewald-Cox, C. M. 1983. Conclusions: Guidelines to management: A beginning attempt. Pages 414–445 in C. M. Schonewald-Cox, S. M. Chambers, B. MacBryde, and L. Thomas, eds., *Genetics and conservation: A reference for managing wild animal and plant populations.* Menlo Park, CA: Benjamin Cummings.

Soule, M. E. 1985. What is conservation biology? *BioScience* 35:727–734.

——— (ed.). 1986a. *Conservation biology: The science of scarcity and diversity.* Sunderland, MA: Sinauer Associates.

———. 1986b. *Viable populations for conservation.* Cambridge: Cambridge University Press.

Soule, M. E., and B. A. Wilcox (eds.). 1980. *Conservation biology: An evolutionary ecological perspective.* Sunderland, MA: Sinauer Associates.

Tattersal, I. 1993. Madagascar's lemurs. *Scientific American* 268:110—117.

Thatcher, R. C., J. L. Searcy, J. E. Coster, and G. D. Hertel (eds.). 1986. *The southern pine beetle. USDA Forest Service Science and Education Tech. Bull. 1631.* Washington, DC: U.S. Department of Agriculture.

Trefethen, J. B. 1975. *An American crusade for wildlife.* New York: Winchester Press.

Western, D., and M. C. Pearl. 1988. *Conservation for the twenty-first century.* New York: Oxford University Press.

Whitcomb, R. F., J. F. Lynch, P. A. Opler, and C. S. Robbins. 1976. Island biogeography and conservation: Strategy and limitations. *Science* 193:1030–1032.

Wilson, E. O. 1992. *The diversity of life.* Cambridge, MA: The Belknap Press of Harvard University Press.

Wilson, E. O., and F. M. Peter (eds.). 1988. *Biodiversity.* Washington, DC: National Academy Press.

Zaret, T. M., and R. T. Paine. 1973. Species introduction in a tropical lake. *Science* 182:449–455.

Chapter 29 | Global Climate Change

Adams, R. M., R. Alig, J. M. Callaway, B. A. McCarl, and S. M. Winnet. 1995. *The economic effects of climate change on U.S. agriculture.* Final report. Climate Change Impacts Program. Palo Alto, CA: EPRI.

Adams, R. M., R. A. Fleming, C. C. Chang, and B. A. McCarl. 1995. A reassessment of the economic effects of global climate change on U.S. agriculture. *Climate Change* 30:147–167.

Amthor, J. S. 1995. Terrestrial higher-plant response to increasing atmospheric CO_2 in relation to the global carbon cycle. *Global Change Biology* 1:243–274.

Bazzaz, F. A. 1996. *Plants in changing environments: Linking physiological, population, and community ecology.* New York: Cambridge University Press.

Butcher, S. S., R. J. Charlson, G. H. Orians, and G. V. Wolfe (eds.). 1992. *Global biogeochemical cycles.* New York: Academic Press.

Currie D. J. 2001. Projected effects of climate change on patterns of vertebrate and tree species richness in the conterminous United States. *Ecosystems* 4:216–225.

Drake, B. G., and P. W. Leadley. 1991. Canopy photosynthesis of crops and native plant communities exposed to long-term elevated carbon dioxide. *Plant Cell Environment* 14:853–860.

Drake, B. G., G. Peresta, E. Beugeling, and R. Matamala. 1996. Long-term elevated CO_2 exposure in a Chesapeake Bay wetland: Ecosystem gas exchange, primary productivity, and tissue nitrogen. In G. Koch and H. A. Mooney, eds., *Carbon dioxide and terrestrial ecosystems.* San Diego: Academic Press.

Edmonds, J. 1992. Why understanding the natural sinks and sources of CO_2 is important: A policy analysis perspective. *Water, Air and Soil Pollution* 64:11–21.

Field, C. B., R. B. Jackson, and H. A. Mooney. 1995. Stomatal responses to CO_2: Implications from the plant to global scale. *Plant Cell Environment* 18:1214–1225.

Gates, D. 1993. *Climate change and its biological consequences.* Sunderland, MA: Sinauer Associates.

Harte, J., and R. Shaw. 1995. Shifting dominance with a montane vegetation community: Results of a climate-warming experiment. *Science* 267:876–880.

Heimann, M. (ed.). 1993. *The global carbon cycle.* Berlin: Springer-Verlag.

Holdridge, L. R. 1947. Determination of world formulations from simple climatic data. *Science* 105:367–368.

Houghton, J. T. 1997. *Global warming: The complete briefing.* Cambridge: Cambridge University Press.

Houghton, J. T., L. G. Meira Filho, J. Bruce, H. Lee, B. A. Callander, E. Haites, N. Harris, and K. Maskell (eds). 1995. *Climate change 1994.* Cambridge: Cambridge University Press.

Houghton, J. T., L. G. Meira Filho, B. A. Callander, N. Harris, A. Kattenberg, and K. Maskell (eds.). 1996. *Climate change 1995: The science of climate change.*

Houghton, R. A. 1995. Land-use change and the carbon cycle. *Global Change Biology* 1:275–287.

Intergovernmental Panel on Climate Change. 2007. *Climate Change 2007: Fourth Assessment (AR4).* Cambridge: Cambridge University Press. http://www.ipcc.ch/

Iverson, L. R., and A. M. Prasad. 2001. Potential changes in tree species richness and forest community type following climate change. *Ecosystems* 4:186–199.

Kalkstein, L. S., and J. S. Green. 1997. An evaluation of climate/mortality relationships in large U.S. cities and possible impacts of a climate change. *Environmental Health Perspectives* 105:84–93.

Kalkstein, L. S., and G. Tan. 1995. Human health. In K. Strzepek and J. Smith, eds., *As climate changes: International impacts and implications.* New York: Cambridge University Press.

Kattenberg, A., F. Giorgi, H. Grassl, G. A. Meehl, J. F. B. Mitchell, R. J. Stouffer, T. Tokioka, A. J. Weaver, and T. M. L. Wigley. 1996. Climate models projections of future climate. Pages 285–357 in J. T. Houghton et al., eds., *Climate change 1995. The Science of Climate Change.* Intergovernmental Panel on Climate Change. Cambridge: Cambridge University Press.

Keeling, C. D., T. P. Whorf, M. Wahlen, and J. van der Plicht. 1995. Interannual extremes in the rate of rise of atmospheric carbon dioxide since 1980. *Nature* 375:666–670.

Korner, C., M. Diemer, B. Schappi, P. Niklaus, and J. Arnone. 1997. The response of alpine grassland to four seasons of CO_2 enrichment: A synthesis. *Acta Oecologia* 18:165–175.

Marland, G., and T. Boden. 1993. The magnitude and distribution of fossil-fuel related carbon releases. Pages 117–138 in M. Heimann, ed., *The global carbon cycle.* New York: Springer-Verlag.

Miko, U. F. 1996. Climate change impacts on forests. In R. T. Watson, M. C. Zinyowera, and R. H. Moss, eds., *Climate change 1995: Impacts, adaptations and mitigation of climate change.* New York: Cambridge University Press.

Mitchell, J. F. B., R. A. Davis, W. J. Ingram, and C. A. Senior. 1995. On surface temperature, greenhouse gases and aerosols: Models and observations. *Journal of Climatology* 10:2364–2386.

Mitchell, J. F. B., T. J. Johns, J. M. Gregory, and S. B. F. Tett. 1995. Climate response to increasing levels of greenhouse gases and sulfate aerosols. *Nature* 376:501–504.

Nicholls, R. J., and S. P. Leatherman. 1995. Global sea-level rise. In K. Strzepek and J. B. Smith, eds., *As climate changes: International impacts and implications.* Cambridge: Cambridge University Press.

Oechel, W. C., and G. L. Vourlitis. 1996. Direct effects of elevated CO_2 on arctic plant and ecosystem function. Pages 163–176 in G. Koch and H. A. Mooney, eds., *Carbon dioxide and terrestrial ecosystems.* San Diego: Academic Press.

Owensby, C. E., J. M. Ham, A. Knapp, C. W. Rice, P. I. Coyne, and L. M. Auen. 1996. Ecosystem-level responses of tallgrass prairie to elevated CO_2 Pages 147–162 in G. Koch and H. A. Mooney, eds., *Carbon dioxide and terrestrial ecosystems.* San Diego: Academic Press.

Peterjohn, W. T., J. M. Melillo, F. P. Bowles, and P. A. Steudler. 1993. Soil warming and trace gas fluxes: Experimental design and preliminary flux results. *Oecologia* 93:18–24.

Peters, R. L., and T. E. Lovejoy (eds.). 1992. *Global warming and biological diversity.* New Haven, CT: Yale University Press.

Poorter, H., and M. Perez-Soba. 2002. Plant growth at elevated CO_2 Pages 489–496 in H. A. Mooney and J. G. Canadell, eds., *Encyclopedia of global change.* Vol. 2: *The Earth system: Biological and ecological dimensions of global environmental change.* Chichester, England: John Wiley and Sons, Ltd.

Root, T. 1988. Energy constraints on avian distributions and abundances. *Ecology* 69:330–339.

Schneider, S. H. 1989. *Global warming.* San Francisco: Sierra Club Books.

Smith, T. M., P. N. Halpin, H. H. Shugart, and C. Secrett. 1994. Global forests. In K. Strzpeck and J. Smith, eds., *As climate changes: International impacts and implications.* Cambridge: Cambridge University Press.

Smith, T. M., R. Leemans, and H. H. Shugart. 1992. Sensitivity of terrestrial carbon storage to CO_2-induced climate change: Comparison of four scenarios based on general circulation models. *Climatic Change* 21:367–384.

Strain, B. R., and J. D. Cure (eds.). 1985. *Direct effects of increasing carbon dioxide on vegetation.* Washington, DC: U.S. Department of Energy Publication DOE/ER-0238.

Trabalka, J. R. (ed.). 1985. *Atmospheric carbon dioxide and the global carbon cycle.* Washington, DC: U.S. Department of Energy Report DOE/ER-0239.

Trabalka, J. R., and D. E. Reichle (eds.). 1994. *The changing carbon cycle: A global analysis.* New York: Springer-Verlag.

VEMAP Participants. 1995. Vegetation/ecosystem modeling and analysis project: Comparing biogeography and biogeochemistry models in a continental-scale study of terrestrial ecosystem responses to climate change and CO_2 doubling. *Global Biogeochemical Cycles* 9:407–437.

Watson, R. T., M. C. Zinyowera, R. H. Moss, and D. J. Dokken (eds.). 1998. *The regional impacts of climate change (a special report of IPCC Working Group II).* Cambridge: Cambridge University Press.

Woodward, F. I. (ed.). 1992. *Global climate change: The ecological consequences.* London: Academic Press.

Yoshino, M., T. Horie, H. Seino, H. Tsujii, T. Uchijima, and Z. Uchijima. 1988. The effects of climate variations on agriculture in Japan. In M. Parry, T. R. Carter, and N. T. Konijn, eds., *The impacts of climate variation on agriculture.* Vol. 1: *Assessments in cool temperate and cold regions.* Dordrecht, The Netherlands: Kluwer.

Glossary

A horizon Surface stratum of mineral soil, characterized by maximum accumulation of organic matter, maximum biological activity, and loss of such materials as iron, aluminum oxides, and clays.

abiotic Nonliving; the abiotic component of the environment includes soil, water, air, light, nutrients, and the like.

abundance The number of individuals of a species in a given area.

abyssal Relating to the bottom waters of oceans, usually below 1000 m.

abyssopelagic zone Oceanic depth from 4000 m to ocean floor.

acclimatization Changes in physiological state or tolerance that appear in a species after long exposure to different natural environments.

acid deposition Wet and dry atmospheric fallout with an extremely low pH, brought about when water vapor in the atmosphere combines with hydrogen sulfide and nitrous oxide vapors released by burning fossil fuels; the sulfuric and nitric acid in rain, fog, snow, gases, and particulate matter.

adaptation A genetically determined characteristic (behavioral, morphological, or physiological) that improves an organism's ability to survive and reproduce under prevailing environmental conditions.

adaptive radiation Evolution from a common ancestor of divergent forms adapted to distinct ways of life.

adiabatic cooling A decrease in air temperature when a rising parcel of warm air cools by expanding (which uses energy) rather than losing heat to the surrounding air; the rate of cooling is approximately $1\,°C/100$ m for dry air and $0.6\,°C/100$ m for moist air.

adiabatic lapse rate Rate at which a parcel of air loses temperature with elevation if no heat is gained from or lost to an external source.

adiabatic process A process in which heat is neither lost to nor gained from the outside.

aerobic Living or occurring only in the presence of free uncombined molecular oxygen, either as a gas in the atmosphere or dissolved in water.

age distribution The proportion of individuals in various age classes for any one year. See also age structure.

age-specific mortality rate The proportion of deaths occurring per unit time in each age group within a population.

age-specific schedule of birth Average number of offspring produced per individual per unit time as a function of age class.

age structure The number or proportion of individuals in each age group within a population.

aggregative response Movement of predators into areas of high prey density.

aggressive mimicry Resemblance of a predator or parasite to a harmless species to deceive potential prey.

Alfisol Soil characterized by an accumulation of iron and aluminum in the B horizon.

alien species A species not native or endemic to an area.

Allee effect Reduction in reproduction or survival under low population densities.

allele One of two or more alternative forms of a gene that occupies the same relative position or locus on homologous chromosomes.

allelopathy Effect of metabolic products of plants (excluding microorganisms) on the growth and development of other nearby plants.

allogenic Refers to successional change brought about by a change in the physical environment.

allogenic succession Ecological change or development of species structure and community composition brought about by some external force, such as fire or storms.

allopatric Having different areas of geographical distribution; possessing nonoverlapping ranges.

alpha diversity The variety of organisms occurring in a given place or habitat; compare beta diversity, gamma diversity.

altricial Condition among birds and mammals of being hatched or born usually blind and too weak to support their own weight.

ambient Surrounding, external, or unconfined in condition.

ammonification Breakdown of proteins and amino acids, especially by fungi and bacteria, with ammonia as the excretory by-product.

anaerobic Adapted to environmental conditions devoid of oxygen.

Andisol Soil derived from volcanic ejecta; not highly weathered, with a dark upper layer.

anion Ion carrying a negative charge.

annual grassland Grassland in California dominated by exotic annual grasses that reseed every year, replacing native perennial grasses.

apoematism See warning coloration.

aphotic zone A deepwater area of marine ecosystems below the depth of effective light penetration.

apparent competition Occurs when a single species of predator feeds on two prey species supporting a higher predator density, increasing the rate at which prey are consumed. As a result, the density of both prey species is lowered, suggesting a competitive interaction.

aqueous solution Solution in which water is the solvent.

area-insensitive species Species that are at home in any size habitat patch.

Aridisol Desert soil characterized by little organic matter and high base content.

asexual reproduction Any form of reproduction, such as budding, that does not involve the fusion of gametes.

assimilation Transformation or incorporation of a substance by organisms; absorption and conversion of energy and nutrients into constituents of an organism.

assimilation efficiency Ratio of assimilation to ingestion; a measure of efficiency with which a consumer extracts energy from food.

atmospheric pressure The downward force exerted by the weight of the overlying atmosphere.

atoll A ring-shaped coral reef that encloses or almost encloses a lagoon and is surrounded by open sea.

ATP Adenosine triphosphate; major energy-transferring molecules in all biological systems.

aufwuchs Community of plants and animals attached to or moving about on submerged surfaces; compare periphyton.

autogenic Self-generated.

autogenic succession Succession driven by environmental changes brought about by the organisms themselves.

autotrophic community Community whose energy source is photosynthesis, thus based on primary producers.

autotrophic succession Succession in a predominantly inorganic environment with early and continued dominance of green plants (autotrophs).

autotrophs Organisms that produce organic material from inorganic chemicals and some source of energy.

available water capacity Supply of water available to plants in a well-drained soil.

B horizon Soil stratum beneath the A horizon, characterized by an accumulation of silica, clay, and iron and aluminum oxides and possessing blocky or prismatic structure.

basal metabolic rate The minimal amount of energy expenditure needed by an animal to maintain vital processes.

Batesian mimicry Resemblance of a palatable or harmless species, the mimic, to an unpalatable or dangerous species, the model.

bathyal Pertaining to anything, but especially organisms, in the deep sea, below the photic or lighted zone, and above 4000 m.

bathypelagic zone Lightless zone of the open ocean, lying above the abyssal or bottom water, usually above 4000 m.

behavioral defenses Aggressive and submissive postures or actions that threaten or deter enemies.

behavioral ecology The study of the behavior of an organism in its natural habitat.

benefit–cost analysis An economic analysis in which the benefits of an activity are compared to the associated costs.

benthic zone The area of the sea bottom.

benthos Animals and plants living on the bottom of a lake or sea, from the high-water mark to the greatest depth.

beta diversity Variety of organisms occupying different habitats over a region; regional diversity; compare alpha diversity, gamma diversity.

biennial Plant that requires two years to complete a life cycle, with vegetative growth the first year and reproductive growth (flowers and seeds) the second.

biochemical oxygen demand (BOD) A measure of the oxygen needed in a specified volume of water to decompose organic materials; the greater the amount of organic matter in water, the higher the BOD.

biodiversity A measure of the different kinds of organisms within a given region.

biogeochemical cycle Movement of elements or compounds through living organisms.

biogeography The study of past and present geographical distribution of plants and animals, and their ecological relationships.

biological clock The internal mechanism of an organism that controls circadian rhythms without external time cues.

biological magnification Process by which pesticides and other substances become more concentrated in each link of the food chain.

biological species A group of potentially interbreeding populations reproductively isolated from all other populations.

bioluminescence Production of light by living organisms.

biomass Weight of living material, usually expressed as dry weight per unit area.

biome Major regional ecological community of plants and animals; usually corresponds to plant ecologists' and European ecologists' classification of plant formations and life zones.

biosphere Thin layer about Earth in which all living organisms exist.

biotic Applied to the living component of an ecosystem.

biotic community Any assemblage of populations living in a prescribed area or physical habitat.

blanket mire Large area of upland dominated by sphagnum moss and dependent upon precipitation for a water supply; a moor.

bog Wetland ecosystem characterized by an accumulation of peat, acid conditions, and dominance of sphagnum moss.

border The place where the edge of one vegetation patch meets the edge of another.

bottleneck An evolutionary term for any stressful situation that greatly reduces a population.

bottomland Land bordering a river that floods when the river overflows its banks.

bottom-up control Influence of producers on the trophic levels above them in a food web.

boundary layer A layer of still air close to or at the surface of an object.

buoyancy The power of a fluid to exert an upward force on a body placed in it.

browse The part of current leaf and twig growth of shrubs, woody vines, and trees available for animal consumption.

bryophyte Member of the division in the plant kingdom of nonflowering plants comprising mosses (Musci), liverworts (Hepaticae), and hornworts (Anthocerotae).

buffer A chemical solution that resists or dampens change in pH upon addition of acids or bases.

bundle sheath cells Cells surrounding small vascular bundles in the leaves of vascular plants.

C horizon Soil stratum beneath the solum (A and B horizons), little affected by biological activity or soil-forming processes.

C3 plant Any plant that produces as its first step in photosynthesis the three-carbon compound phosphoglyceric acid.

C4 plant Any plant that produces as its first step in photosynthesis a four-carbon compound, malic or aspartic acid.

calcification Process of soil formation characterized by accumulation of calcium in lower horizons.

calorie Amount of heat needed to raise 1 g of water 1°C, usually from 15°C to 16°C.

CAM plant (crassulacean acid metabolism) Plant (cactus or other succulent) that separates the processes of carbon dioxide uptake and fixation when growing under arid conditions; it takes up gaseous carbon dioxide at night, when stomata are open, and uses it during the day, when stomata are closed.

cannibalism Killing and consuming one's own kind; intraspecific predation.

canopy Uppermost layer of vegetation formed by trees; also the uppermost layer of vegetation in shrub communities or in any terrestrial plant community where the upper layer forms a distinct habitat.

capillary water That portion of water in the soil held by capillary forces between soil particles.

carbon balance Balance between uptake of CO_2 in photosynthesis and its loss through the process of transpiration.

carnivore Organism that feeds on animal tissue; taxonomically, a member of the order Carnivora (Mammalia).

carnivory The killing and eating of animals by another animal.

carrying capacity (K) Number of individual organisms the resources of a given area can support, usually through the most unfavorable period of the year.

cation Part of a dissociated molecule carrying a positive electrical charge.

cation exchange capacity Ability of a soil particle to absorb positively charged ions.

chaparral Vegetation consisting of broadleaved evergreen shrubs, found in regions of mediterranean-type climate.

character displacement The principle that two species are more different where they occur together than where they are separated geographically.

chemical defense The use by organisms of bitter, distasteful, or toxic secretions that deter potential enemies.

chemical ecology Study of the nature and use of chemical substances produced by plants and animals.

chemical weathering The action of a set of chemical processes such as oxidation, hydrolysis, and reduction operating at the atomic and molecular level that break down and reform rocks and minerals.

chilling tolerance Ability of a plant to carry on photosynthesis within a range of $+5°C$ to $+10°C$.

chromosome One of a group of threadlike structures of different lengths and sizes in the nuclei of cells of eukaryote organisms.

chronosequences Groups of sites within the same area that are in different stages of succession.

circadian rhythm Endogenous rhythm of physiological or behavioral activity of approximately 24 hours duration.

clear-cutting Forest harvesting procedure in which all trees on the site are cut and removed.

climate Long-term average pattern of local, regional, or global weather.

climax Stable end community of succession that is capable of self-perpetuation under prevailing environmental conditions.

clone A population of genetically identical individuals resulting from asexual reproduction.

closed system A system that neither receives inputs from nor contributes output to the external environment.

Sorensen's coefficient of community Index of similarity between two stands or communities. The index ranges from 0 to indicate communities with no species in common to 100 to indicate communities with identical species composition.

coevolution Joint evolution of two or more non-interbreeding species that have a close ecological relationship; through reciprocal selective pressures, the evolution of one species in the relationship is partially dependent on the evolution of the other.

coexistence Two or more species living together in the same habitat, usually with some form of competitive interaction.

cohesion The ability of water molecules, because of hydrogen bonding, to stick firmly to each other, restricting external forces to break these bonds.

cohort A group of individuals of the same age.

cold resistance Ability of a plant to resist low-temperature stress without injury.

collectors Feeding group of stream invertebrates that filter fine organic particles from flowing water or pick up particles from the stream bottom.

colloid Negatively charged particles in the soil that provide surfaces with high cation exchange capacity.

commensalism Relationship between species that is beneficial to one, but neutral or of no benefit to the other.

community A group of interacting plants and animals inhabiting a given area.

community ecology Study of the living component of ecosystems; description and analysis of patterns and processes within the community.

compartment Major reservoir or component of an ecosystem.

compensation depth In aquatic ecosystems, the depth of the water column at which light intensity reaching plants is just sufficient for the rate of photosynthesis to balance the rate of respiration.

compensation level Light intensity at which photosynthesis and respiration balance each other, so that net production is 0; in aquatic systems, usually the depth of light penetration at which oxygen utilized in respiration equals oxygen produced by photosynthesis.

competition Any interaction that is mutually detrimental to both participants, occurring between species that share limited resources.

competitive exclusion principle Hypothesis that when two or more species coexist using the same resource, one must displace or exclude the other.

competitive release Niche expansion in response to reduced interspecific competition.

conduction Direct transfer of heat energy from one substance to another.

conductivity Ability to exchange heat with the surrounding environment.

conservation ecology A synthetic field that applies principles of ecology, biogeography, population genetics, economics, sociology, and other fields to the maintenance of biological diversity. Also called conservation biology.

constitutive defense Fixed feature of an organism, such as object resemblance, that deters predators.

consumer Any organism that lives on other organisms, dead or alive.

consumption efficiency Ratio of ingestion to production or energy available; defines the amount of available energy being consumed.

contest competition Competition in which a limited resource is shared only by dominant individuals; a relatively constant number of individuals survive, regardless of initial density.

continental shelf Gently seaward-sloping surface of a continent that extends to a depth of about 200 m.

continuum A gradient of environmental characteristics or changes in community composition.

convection Transfer of heat by the circulation of a liquid or gas.

convergent evolution Development of similar characteristics in different species living in different areas under similar environmental conditions.

coprophagy Feeding on feces.

Coriolis effect Physical consequence of the law of conservation of angular momentum; as a result of Earth's rotation, a moving object veers to the right in the Northern Hemisphere and to the left in the Southern Hemisphere relative to Earth's surface.

corridor A strip of a particular type of vegetation that differs from the land on both sides.

countercurrent heat exchange An anatomical and physiological arrangement by which heat exchange takes place between outgoing warm arterial blood and cool venous blood returning to the body core; important in maintaining temperature homeostasis in many vertebrates.

covalence Sharing of a pair of electrons between two atoms.

critical day length The period of daylight, specific for any given species, that triggers a long-day or a short-day response in organisms.

critical thermal maximum Temperature at which an animal's capacity to move is so reduced that it cannot escape from thermal conditions that will lead to death.

crown fire Fire that sweeps through the canopy of a forest.

crude birthrate The number of young produced per unit of population.

crude density The number of individuals per unit area; compare ecological density.

cryptic coloration Coloration of organisms that makes them resemble or blend into their habitat or background.

cultural eutrophication Accelerated nutrient enrichment of aquatic ecosystems by a heavy influx of pollutants that causes major shifts in plant and animal life.

currents Water movements that result in the horizontal transport of water masses.

day-neutral plant A plant that does not require any particular photoperiod to flower.

death rate Number of individuals in a population dying in a given time interval divided by the number alive at the midpoint of the time interval.

deciduous Of leaves, shed during a certain season (winter in temperate regions; dry season in the tropics); of trees, having deciduous parts.

decomposer Organism that obtains energy from the breakdown of dead organic matter to simpler substances; most precisely refers to bacteria and fungi.

decomposition Breakdown of complex organic substances into simpler ones.

defensive mutualism Relationship in which one of the mutualists seems to protect the other from harm.

definitive host Host in which a parasite becomes an adult and reaches maturity.

demography The statistical study of the size and structure of populations and changes within them.

demographic stochasticity Random variations in birth and death rates that occur in a population from year to year.

denitrification Reduction of nitrates and nitrites to nitrogen by microorganisms.

density Size of a population in relation to a definite unit of space; see crude density, ecological density.

density dependence Regulation of population growth by mechanisms controlled by the size of the population; effect increases as population size increases.

density independence Being unaffected by population density; regulation of growth is not tied to population density.

dependent variable Variable y, the second of two numbers in an ordered pair (x, y); the set of all values taken on by the dependent variable is called the range of the function; compare independent variable.

desert grassland Grassland of hot, dry climates, with rainfall varying between 200 and 500 mm, dominated by bunchgrasses and widely interspersed with other desert vegetation.

desertification Process of desert expansion or formation as a consequence of climatic change, poor land management, or both.

detrital food chain Food chain in which detritivores consume detritus or litter, mostly from plants, with subsequent transfer of energy to various trophic levels; ties into the grazing food chain; compare grazing food chain.

detritivore Organism that feeds on dead organic matter; usually applies to detritus-feeding organisms other than bacteria and fungi.

detritus Fresh to partly decomposed plant and animal matter.

dew point temperature Temperature at which condensation of water in the atmosphere begins.

diameter at breast height (dbh) Diameter of a tree measured at 1.4 m (4 feet, 6 inches) from ground level.

diapause A period of dormancy, usually seasonal, in the life cycle of an insect, in which growth and development cease and metabolism greatly decreases.

diffuse competition Competition in which a species experiences interference from many other species that deplete the same resources.

diffusion Spontaneous movement of particles of gases or liquids from an area of high concentration to an area of low concentration.

dimorphism Existing in two structural forms, two color forms, two sexes, and the like.

dioecious Having male and female reproductive organs on separate plants; compare monoecious.

diploid Having chromosomes in homologous pairs, or twice the haploid number of chromosomes.

directional selection Selection favoring individuals at one extreme of the phenotype in a population.

discounting Adding and comparing costs and benefits that occur at different times; a major driving force in the economics of resource management; often runs counter to the objectives of sustainable resource management.

disease Any deviation from a normal state of health.

dispersal Leaving an area of birth or activity for another area.

dispersion Distribution of organisms within a population over an area.

disruptive selection Selection in which two extreme phenotypes in the population leave more offspring than the intermediate phenotype, which has lower fitness.

distribution Arrangement of organisms within an area.

disturbance A discrete event in time that disrupts an ecosystem, community, or population, changing substrates and resource availability.

diversity Abundance of different species in a given location; species richness.

diversity index The mathematical expression of species richness of a given community or area.

DNA A nucleic acid (deoxyribonucleic acid) mainly found in the chromosomes; the hereditary material of all organisms except certain viruses.

domestic grassland Grasslands that are seeded and maintained by human efforts; includes hayfields, pastures, golf courses, and lawns.

dominance In a community, control over environmental conditions influencing associated species by one or several species, plant or animal, enforced by number, density, or growth form; in a population, behavioral, hierarchical order that gives high-ranking individuals priority of access to essential resources; in genetics, ability of an allele to mask the expression of an alternative form of the same gene in a heterozygous condition.

dominant Population possessing ecological dominance in a given community and thereby governing type and abundance of other species in the community.

dominant gene An allele that is expressed in either the homozygous or heterozygous state.

dormancy State of cessation of growth and suspended biological activity, during which life is maintained.

drought avoidance Ability of a plant to escape dry periods by becoming dormant or surviving as a seed.

drought resistance Sum of drought tolerance and drought avoidance.

drought tolerance Ability of plants to maintain physiological activity despite the lack of water or to survive the drying of tissues.

dry deposition Pollutants—mainly sulfur dioxides and nitrogen oxides—that return to Earth in the form of particulate matter and airborne gases, thus introducing acidic material to the ground and surface waters.

dryfall Nutrients brought into an ecosystem by airborne particles and aerosols.

dynamic composite life table Pooled cohort of individuals born over several time periods instead of just one.

dynamic life table Fate of a group of individuals born at the same time and followed from birth to death.

dystrophic Term applied to a body of water with a high content of humic or organic matter, often with high littoral productivity and low plankton productivity.

E horizon Mineral horizon characterized by the loss of clay, iron, or aluminum and a concentration of quartz and other resistant minerals in sand and silt sizes; light in color.

early successional species Plant species characterized by high dispersal rates, ability to colonize disturbed sites, short life span, and shade intolerance.

easterlies A system of broad, steady, prevailing winds around Earth over the equatorial regions created by the westward deflection of air that follows the barometric pressure gradients from subtropical high to equatorial low; also called trade winds.

ecocline A geographical gradient of communities or ecosystems produced by responses of vegetation to environmental gradients of rainfall, temperature, nutrient concentrations, and other factors.

ecological density Density measured in terms of the number of individuals per area of available living space; compare crude density.

ecological efficiency Percentage of biomass produced by one trophic level that is incorporated into biomass of the next highest trophic level.

ecological pyramid A graphical representation of the trophic structure and function of an ecosystem.

ecological release Expansion of habitat or increase in food availability resulting from release of a species from interspecific competition.

ecology The study of relations between organisms and their natural environment, living and nonliving.

ecosystem The biotic community and its abiotic environment, functioning as a system.

ecosystem ecology The study of natural systems with emphasis on energy flow and nutrient cycling.

ecosystem services Processes by which the environment produces resources such as air, water, timber, or fish.

ecotone Transition zone between two structurally different communities; wide borders that form a transition zone between adjoining patches.

ecotype Subspecies or race adapted to a particular set of environmental conditions.

ectomycorrhizae Mutualistic association between fungi and roots in which the fungi form sheaths around the outside of the roots.

ectoparasite Parasite, such as a flea, that lives in the fur, feathers, or skin of the host.

ectothermy Determination of body temperature primarily by external thermal conditions.

edaphic Relating to soil.

edge Place where two or more vegetation types meet; see border.

edge effect Response of organisms, animals in particular, to environmental conditions created by the edge.

edge species Species that are restricted exclusively to the edge or border environment.

effective population size The size of an ideal population that would undergo the same amount of random genetic drift as the actual population; sometimes used to measure the amount of inbreeding in a finite, randomly mating population.

egestion Elimination of undigested food material.

El Niño-Southern Oscillation (ENSO) Global event arising from large-scale interactions between ocean and atmosphere.

elaiosome Shiny, oil-containing, ant-attracting tissue on the seed coats of many plants.

emigration Movement of part of a population permanently out of an area.

endemic Restricted to a given region.

endoparasite Parasite that lives within the body of the host.

endothermic reaction Chemical reaction that gains energy from the environment.

endothermy Regulation of body temperature by internal heat production; allows maintenance of appreciable difference between body temperature and external temperature.

energy Capacity to do work.

Entisols Embryonic mineral soils whose profile is just beginning to develop; common on recent floodplains and wind deposits, they lack distinct horizons.

entropy Transformation of matter and energy to a more random, more disorganized state.

environment Total surroundings of an organism, including other plants and animals and those of its own kind.

environmental economics Study of environmental problems and the incorporation of economic principles into the environmental decision-making process.

environmental lapse rate Rate at which temperature decreases with altitude; see adiabatic lapse rate.

environmental stochasticity Random variations in the environment that directly affect birth and death rates.

epidemic Rapid spread of a bacterial or viral disease in a human population; compare epizootic.

epifauna Benthic animals that live on or move over the surface of a substrate.

epiflora Benthic plants that live on the surface of a substrate.

epilimnion Warm, oxygen-rich upper layer of water in a lake or other body of water, usually seasonal.

epipelagic zone The lighted area of the ocean; also called photic zone.

epiphyte Plant that lives wholly on the surface of other plants, deriving support but not nutrients from them.

epizootic Rapid spread of a bacterial or viral disease in a dense population of animals.

equilibrium species Species whose population exists in equilibrium with resources and at a stable density.

equilibrium turnover rate Change in species composition per unit time when immigration equals extinction.

equitability Evenness of distribution of species abundance patterns; maximum equitability is the same number of individuals among all species in the community.

estivation Dormancy in animals during a period of drought or a dry season.

estuary A partially enclosed embayment where freshwater and seawater meet and mix.

euphotic zone Surface layer of water to the depth of light penetration where photosynthetic production equals respiration.

eutrophic Term applied to a body of water with high nutrient content and high productivity.

eutrophication Nutrient enrichment of a body of water; called cultural eutrophication when accelerated by introduction of massive amounts of nutrients from human activity.

eutrophy Condition of being nutrient rich.

evaporation Loss of water vapor from soil or open water or another exposed surface.

evapotranspiration Sum of the loss of moisture by evaporation from land and water surfaces and by transpiration from plants.

evenness Degree of equitability in the distribution of individuals among a group of species; see equitability.

evergreen Applied to trees and shrubs for which there is no complete seasonal loss of leaves; two types, broadleaf and needle-leaf.

evolution Change in gene frequency through time resulting from natural selection and producing cumulative changes in characteristics of a population.

evolutionary ecology Integrated study of evolution, genetics, natural selection, and adaptations within an ecological context; evolutionary interpretation of population, community, and ecosystem ecology.

exothermic reaction Chemical reaction that releases heat to the environment.

exotic species A species that is not native to the place where it is found.

exploitative competition Competition by a group or groups of organisms that reduces a resource to a point that adversely affects other organisms.

exponential growth (r) Instantaneous rate of population growth, expressed as proportional increase per unit of time.

externalities When the actions of one individual or a group affect another individual's well-being, but relevant costs are not reflected in market prices.

extinction coefficient Point at which the intensity of light reaching a certain depth is insufficient for photosynthesis; ratio of intensity of light at a given depth to intensity at the surface.

facilitation A situation where one species benefits from the presence or action of another.

facilitation model A model of succession in which a community prepares or "facilitates" the way for a succeeding community.

facultative Able to adjust optimally to different environmental conditions.

fecundity Potential ability of an organism to produce eggs or young; rate of production of young by a female.

fecundity table Shows the number of offspring produced per unit time; constructed by using the survivorship column from the life table and the age-specific birthrates; the mean number of females born in each age group of females.

fen Slightly acidic wetland, dominated by sedges, in which peat accumulates.

fermentation Breakdown of carbohydrates and other organic matter under anaerobic conditions.

field capacity Amount of water held by soil against the force of gravity.

field study A controlled experimental study carried out in a natural environment rather than in the laboratory.

filter effect Corridors that provide dispersal routes for some species but restrict the movement of others.

filtering collectors Stream insect larvae that feed on fine particulate matter by filtering it from water flowing past them.

finite multiplication rate Expressed as lambda, λ, the geometric rate of increase by discrete time intervals; given a stable age distribution, lambda can be used as a multiplier to project population size.

finite population growth rate Geometric rate of population increase by discrete time intervals.

first law of thermodynamics Energy is neither created nor destroyed; in any transfer or transformation, no gain or loss of total energy occurs.

first-level carnivores Organisms that feed on first-level consumers or plant eaters.

first-level consumers Organisms that feed on plants.

first trophic level Producers; organisms that fix energy that becomes the basic source of energy for consumers.

fitness Genetic contribution by an individual's descendants to future generations.

fixed quota Harvest removal of a certain percentage of a population, based on maximum sustained yield estimates.

flashing coloration Hidden markings on animals that, when quickly exposed, startle or divert the attention of a potential predator.

floating reserve Individuals in a population of a territorial species that do not hold territories and remain unmated, but are available to refill territories vacated by death of an owner.

flux Flow of energy from a source to a sink or receiver.

foliage height diversity Measure of the degree of layering or vertical stratification of foliage in a forest.

food chain Movement of energy and nutrients from one feeding group of organisms to another in a series that begins with plants and ends with carnivores, detrital feeders, and decomposers.

food web Interlocking pattern formed by a series of interconnecting food chains.

foraging strategy Manner in which animals seek food and allocate their time and effort in obtaining it.

forb Herbaceous plant other than grass, sedge, or rush.

forest floor Term describing the ground layer of leaves and detritus; site of decomposition.

founder effect Effect of starting a population with a small number of colonists, which contain only a small and often biased sample of genetic variation of the parent population; a markedly different new population may arise.

fragmentation Reduction of a large habitat area into small, scattered remnants; reduction of leaves and other organic matter into smaller particles.

frequency In landscape ecology, the mean number of disturbances that occur within a time interval. In community ecology, the proportion of sample plots in which a species occurs relative to the total number of sample plots; the probability of finding the species in any one sample plot.

frugivore Organism that feeds on fruit.

functional type (group) A collection of species that exploit the same array of resources or perform similar functions within the community.

functional response Change in rate of exploitation of a prey species by a predator in relation to changing prey density.

fundamental niche Total range of environmental conditions under which a species can survive.

gamma diversity Differences among similar habitats in widely separated regions.

gap Opening made in a forest canopy by some small disturbance such as windfall; death of an individual tree or group of trees that influences the development of vegetation beneath.

gaseous cycle A biogeochemical cycle with the main reservoir or pool of nutrients in the atmosphere and ocean.

gathering collectors Stream insect larvae that pick up and feed on particles from stream-bottom sediment.

Gelisol Soil that contains permafrost within 200 cm of the ground surface; it is characterized by perennial coldness rather than diagnostic horizons.

gene Unit material of inheritance; more specifically, a small unit of a DNA molecule, coded for a specific protein to produce one of the many attributes of a species.

gene flow Exchange of genetic material between populations.

gene frequency Relative abundance of different alleles carried by an individual or a population; allele frequency.

gene pool The sum of all the genes of all individuals in a population.

genet A genetic individual that arises from a single fertilized egg.

genetic drift Random fluctuation in allele frequency over time, due to chance alone without any influence by natural selection; important in small populations.

genotype Genetic constitution of an organism.

genotypic frequency The proportion of various genotypes in a population; compare gene frequency.

geographic isolates Groups of populations that are semi-isolated from one another by some extrinsic barrier; compare subspecies.

geometric rate of increase (*l*) Factor by which the size of a population increases over a period of time.

gleization A process in waterlogged soils in which iron, because of an inadequate supply of oxygen, is reduced to a ferrous compound, giving dull gray or bluish mottles and color to the horizons.

global ecology The study of ecological systems on a global scale.

gougers Stream insect larvae that burrow into waterlogged limbs and trunks of fallen trees.

grazers Stream invertebrates that feed on algal coating on rocks and other substrates.

grazing food chain Food chain in which primary producers (green plants) are eaten by grazing herbivores, with subsequent energy transfers to other trophic levels. Compare detrital food chain.

greenhouse effect Selective energy absorption by carbon dioxide in the atmosphere, which allows short wavelength energy to pass through but absorbs longer wavelengths and reflects heat back to Earth.

greenhouse gas A gas that absorbs longwave radiation and thus contributes to the greenhouse effect when present in the atmosphere; includes water vapor, carbon dioxide, methane, nitrous oxides, and ozone.

gross primary production Energy fixed per unit area by photosynthetic activity of plants before respiration; total energy flow at the secondary level is not gross production, but rather assimilation, because consumers use material already produced with respiratory losses.

gross reproductive rate Sum of the mean number of females born to each female age group.

ground fire Fire that consumes organic matter down to the mineral substrate or bare rock.

groundwater Water that occurs below Earth's surface in pore spaces within bedrock and soil, free to move under the influence of gravity.

guild A group of populations that utilize a gradient of resources in a similar way.

gyre Circular motion of water in major ocean basins.

habitat Place where a plant or animal lives.

habitat selection Behavioral responses of individuals of a species involving certain environmental cues used to choose a potentially suitable environment.

hadalpelagic zone That part of the ocean below 6000 m.

halophyte Terrestrial plant adapted morphologically or physiologically to grow in salt-rich soil.

haploid Having a single set of unpaired chromosomes in each cell nucleus.

Hardy–Weinberg principle The proposition that genotypic ratios resulting from random mating remain unchanged from one generation to another, provided natural selection, genetic drift, and mutation are absent.

harvest effort Approach to harvesting populations by manipulating or controlling hunting efforts, by means such as setting seasons and bag limits.

harvest interval Time period between harvests. Compare to rotation period.

heat Energy in the process of being transferred from one object to another because of temperature differences between the two.

hemiparasite Plant parasite that has chlorophyll, carries on photosynthesis, yet derives some nutrients from its host.

herb layer Lichens, moss, ferns, herbaceous plants, and small woody seedlings growing on the forest floor.

herbivore Organism that feeds on plant tissue.

herbivory Feeding on plants.

hermaphrodite Organism possessing the reproductive organs of both sexes.

heterogeneity State of being mixed in composition; can refer to genetic or environmental conditions.

heterotherm An organism that during part of its life history becomes either endothermic or ectothermic; hibernating endotherms become ectothermic, and foraging insects such as bees become endothermic during periods of activity; they are characterized by rapid, drastic, repeated changes in body temperature.

heterotrophic community Community that is dependent upon and supported by energy already fixed by the autotrophic community.

heterotrophic succession Succession that occurs on dead organic matter; detritivores feed in sequence, each group releasing nutrients used by the next group, until resources are exhausted.

heterotrophs Organisms that are unable to manufacture their own food from inorganic materials and thus rely on other organisms, living and dead, as a source of energy and nutrients.

heterozygous Containing two different alleles of a gene, one from each parent, at the corresponding loci of a pair of chromosomes.

hibernation Winter dormancy in animals, characterized by a great decrease in metabolism.

Histosol Soil characterized by high organic matter content.

home range Area over which an animal ranges throughout the year.

homeostasis Maintenance of a nearly constant internal environment in the midst of a varying external environment; more generally, the tendency of a biological system to maintain itself in a state of stable equilibrium.

homeostatic plateau Limited range of maximum and minimum values of physiological tolerances within which an organism operates.

homeotherm Animal with a fairly constant body temperature; also spelled *homoiotherm* and *homotherm*.

homeothermy Regulation of body temperature by physiological means.

homologous chromosomes Corresponding chromosomes from male and female parents that pair during meiosis.

homozygous Containing two identical alleles of a gene at the corresponding loci of a pair of chromosomes.

horizon Major zone or layer of soil, with its own particular structure and characteristics.

horse latitudes Subtropical latitudes coinciding with a major anticyclonic belt, characterized by generally settled weather and a light or moderate wind.

host Organism that provides food or other benefit to another organism of a different species; usually refers to an organism exploited by a parasite.

humus Organic material derived from partial decay of plant and animal matter.

hybrid Plant or animal resulting from a cross between genetically different parents.

hydrogen bonding A type of bond occurring between an atom of oxygen or nitrogen and a hydrogen atom joined to oxygen or nitrogen on another molecule; responsible for the properties of water.

hydrological cycle See water cycle.

hydroperiod In wetlands, the duration, frequency, depth, and season of flooding.

hydrothermal vent Place on ocean floor where water, heated by molten rock, issues from fissures; vent water contains sulfides oxidized by chemosynthetic bacteria, providing support for carnivores and detritivores.

hyperosmotic Having a higher concentration of salts in the body tissue than does the surrounding water.

hyperthermia Rise in body temperature to reduce thermal differences between an animal and a hot environment, thus reducing the rate of heat flow into the body.

hypertrophic Condition of lakes that have received excessive amounts of nutrients, making them highly and unnaturally eutrophic; compare eutrophic.

hypervolume The multidimensional space of a species niche; compare niche.

hypha Filament of a fungus thallus or vegetative body.

hypolimnion Cold, oxygen-poor zone of a lake, below the thermocline.

hypoosmotic Having a lower concentration of salts in the body tissue than does the surrounding water.

hypothesis Proposed explanation for a phenomenon; we should be able to test it, accepting or rejecting it based on experimentation.

immigration Arrival of new individuals into a habitat or population.

immobilization Conversion of an element from inorganic to organic form in microbial or plant tissue, rendering the nutrient unavailable to other organisms.

importance value Sum of relative density, relative dominance, and relative frequency of a species in a community.

inbreeding Mating among close relatives.

inbreeding depression Detrimental effects of inbreeding.

Inceptisol Mineral soil that has one or more horizons in which mineral materials have been weathered or removed and that is only beginning to develop a distinctive soil profile.

incomplete dominance Physical expression of the heterozygous individuals is intermediate between those of heterozygotes.

independent variable Variable x, the first of two numbers of an ordered pair (x, y); the set of all values taken on by the independent variable is called the domain of the function; compare dependent variable.

indeterminate growth Organism continues to grow throughout its adult life; has no characteristic adult size.

index of abundance Estimates of animal populations derived from counts of animal sign, call, and number of animals observed along a prescribed route; useful in indicating trends of populations from year to year or habitat to habitat.

indirect commensalism Indirect interaction that is positive to one of the species involved, while neutral to the other. See indirect interaction.

Indirect interaction When one species does not interact with a second species directly, but instead influences a third species that does have a direct interaction with the second.

indirect mutualism Situation in which one species indirectly benefits another species by reducing the population size of its strong competitor.

individualistic concept The view, first proposed by H. A. Gleason, that vegetation is a continuous variable in a continuously changing environment; therefore, no two vegetational communities are identical, and associations of species arise only from similarities in requirements.

induced edge Edge that results from some disturbance; adjoining vegetation types are successional, changing or disappearing with time, maintained only by periodic disturbances.

induced defense Defense response brought about or induced by the presence or action of a predator; for example, alarm pheromones.

infauna Organisms living within a substrate.

infection Diseased condition arising when pathogenic microorganisms enter a body, become established, and multiply.

infiltration Downward movement of water into the soil.

infralittoral fringe Region below the littoral region of the sea.

inherent edge Stable, permanent edge determined by long-term natural features and conditions.

inhibition model Model of succession proposing that the dominant vegetation occupying a site prevents colonization of that site by other plants of the next successional community.

inputs Flow of energy and nutrients from the surrounding environment into an ecosystem.

integrated pest management Holistic approach to pest control that considers biological, ecological, economic, and social aspects; the object is to control pests before outbreaks can occur.

intensity A measure of the proportion of the total biomass or population of a species that a disturbance kills or eliminates.

interception The capture of rainwater by vegetation, from which the water evaporates and does not reach the ground.

interference competition Competition in which access to a resource is limited by the presence of a competitor.

interior species Organisms that require large areas of habitat, even though their home ranges may be small.

intermediate disturbance hypothesis The concept that species diversity is greatest in those habitats experiencing a moderate amount of disturbance, allowing the coexistence of early and late successional species.

intermediate host Host that harbors a developmental phase of a parasite; the infective stage or stages can develop only when the parasite is independent of its definitive host; compare definitive host.

internal cycling Movement or cycling of nutrients through components of ecosystems.

intersexual selection Choice of a mate, usually by the female.

interspecific Between individuals of different species.

intertidal zone Area lying between the lines of high and low tide.

intertropical convergence zone (ITCZ) The boundary zone separating the northeast trade winds of the Northern Hemisphere from the southeast trade winds of the Southern Hemisphere.

intrasexual selection Competition among members of the same sex for a mate; most common among males and characterized by fighting and display.

intraspecific Between individuals of the same species.

intrinsic rate of increase The per capita rate of growth of a population that has reached a stable age distribution and is free of competition and other growth restraints.

invasive species A nonnative species that successfully colonizes a disturbed area or empty niche, spreads, and outcompetes associated native species.

ion An atom that is electrically charged as a result of a loss of one or more electrons or a gain of electrons.

ion exchange capacity Total number of charged sites on soil particles within a volume of soil

island biogeography Study of distribution of organisms and community structure on islands.

isolating mechanism Any structural, behavioral, or physiological mechanism that blocks or inhibits gene exchange between two populations.

iteroparous Having multiple broods over a lifetime.

***K*-strategist** A competitive species with a stable population with a slow growth rate, long-lived individuals; they produce relatively few young or seeds, and provide parental care.

keystone predation Predation that is central to the organization of a community; the predator enhances one or more inferior competitors by reducing the abundance of the superior competitor.

keystone species A species whose activities have a significant role in determining community structure.

kinetic energy Energy associated with motion; performs work at the expense of potential energy.

Krummholz Stunted form of trees characteristic of transition zone between alpine tundra and subalpine coniferous forest.

landscape ecology Study of structure, function, and change in a heterogeneous landscape composed of interacting ecosystems.

La Niña A global climate phenomenon characterized by strong winds and cool ocean currents flowing westward from the coastal waters of South America to the tropical Pacific Ocean.

lapse rate The rate at which the temperature decreases for each unit of increase of height in the atmosphere.

late successional species Long-lived, shade-tolerant plant species that supplant early successional species.

latent heat Amount of heat given up when a unit mass of a substance converts from a liquid to a solid state, or the amount of heat absorbed when a substance converts from the solid to liquid state.

latent heat of evaporation The amount of heat absorbed per gram of a liquid to convert it to a gas.

laterization Soil-forming process in hot, humid climates, characterized by intense oxidation; results in loss of bases and in a deeply weathered soil composed of silica.

leaching Dissolving and washing of nutrients out of soil, litter, and organic matter.

leaf area index The total leaf area of a plant exposed to incoming light energy relative to the ground surface area beneath the plant.

leaf area ratio (LAR) Total area of leaves per total plant weight.

lek Communal courtship area males use to attract and mate with females.

lentic Pertaining to standing water, such as lakes and ponds; a population is limited by the lowest amount needed of an essential nutrient.

life expectancy The average number of years to be lived in the future by members of a population.

life table Tabulation of mortality and survivorship of a population; static, time-specific, or vertical life tables are based on a cross section of a population at a given time; dynamic, cohort, or horizontal life tables are based on a cohort followed throughout life.

life zone Major area of plant and animal life, equivalent to a biome; transcontinental region or belt characterized by particular plants and animals and distinguished by temperature differences; applies best to mountainous regions where temperature changes accompany changes in altitude.

light compensation point Depth of water or level of light at which photosynthesis and respiration balance each other.

light saturation point Amount of light at which plants achieve the maximum rate of photosynthesis.

limiting resource Resource or environmental condition that limits the abundance and distribution of an organism.

limnetic Pertaining to or living in the open water of a pond or lake.

limnetic zone Shallow-water zone of a lake or sea, in which light penetrates to the bottom.

link In a food web, the arrows leading from a consumer to a species being consumed.

Lithosol Soil showing little or no evidence of soil development and consisting mainly of partly weathered rock fragments or nearly barren rock.

littoral zone Shallow water of a lake, in which light penetrates to the bottom, permitting submerged, floating, and emergent vegetative growth; also shore zone of tidal water between high-water and low-water marks.

local diversity Number of species in a small area of homogeneous habitat. Also called alpha diversity.

local subpopulation A subpopulation associated with a restricted patch of habitat.

locus Site on a chromosome occupied by a specific gene.

logistic curve S-shaped curve of population growth that slows at first, steepens, and then flattens out at asymptote, determined by carrying capacity.

logistic model of population growth (logistic equation) Mathematical expression for the population growth curve in which rate of increase decreases linearly as population size increases.

long-day organism Plant or animal that requires long days—days with more than a certain minimum of daylight—to flower or come into reproductive condition.

longwave radiation Infrared radiation that occurs as wavelengths longer than 3 or 4 microns.

lotic Pertaining to flowing water.

macromutation Mutation at the level of the chromosome.

macronutrients Essential nutrients plants and animals need in large amounts.

macroparasite Any of the parasitic worms, lice, fungi, and the like that have comparatively long generation times, spread by direct or indirect transmission, and may involve intermediate hosts or vectors.

mainland–island metapopulation structure A single habitat patch (the mainland) is the dominant source of individuals emigrating to other habitat patches (island) within a metapopulation network.

mallee Sclerophyllous shrub community in Australia; most of the species are Eucalyptus.

mangal A mangrove swamp.

mangrove forest Tropical inshore communities dominated by several species of mangrove trees and shrubs capable of growth and reproduction in areas inundated daily by seawater.

marsh Wetland dominated by grassy vegetation such as cattails and sedges.

material cycling See nutrient cycle.

mating system Pattern of mating between individuals in a population.

matric potential Tendency of water to adhere to surfaces.

matrix The background land-use type in a mosaic, characterized by extensive cover and high connectivity.

maximum sustainable yield The maximum rate at which individuals can be harvested from a population without reducing its size; recruitment balances harvesting.

mechanical weathering Breakdown of rocks and minerals in place by disintegration processes such as freezing, thawing, and pressure that do not involve chemical reactions.

mediterranean-type climate Semiarid climate characterized by a hot, dry summer and a wet, mild winter.

meiofauna Benthic organisms within the size range from 1 to 0.1 mm; interstitial fauna.

meiosis Two successive divisions by a gametic cell, with only one duplication of chromosomes, so the number of chromosomes in daughter cells is one-half the diploid number.

melatonin Special hormone in animals that serves to measure time; associated with the biological clock.

mesic Moderately moist.

mesopelagic Uppermost lightless pelagic zone.

mesophyll Specialized tissue located between the epidermal layers of a leaf; palisade mesophyll consists of cylindrical cells at right angles to upper epidermis and contains many chloroplasts; spongy mesophyll lies next to the lower epidermis and has interconnecting, irregularly shaped cells with large intercellular spaces.

metabolism The chemical reactions in cells responsible for breaking down molecules to provide energy (catabolism) and building more complex molecules from simpler molecules (anabolism).

metalimnion Transition zone in a lake between hypolimnion and epilimnion; region of rapid temperature decline.

metapopulation A population broken into sets of subpopulations held together by dispersal or movements of individuals among them.

metatrophic Having a moderate amount of nutrients; stage in a nutrient-poor lake becoming eutrophic.

microbial decomposition Decomposition processes performed by bacteria and fungi; involved in immobilization and mineralization of nutrients.

microbial loop Feeding loop in which bacteria take up dissolved organic matter produced by plankton and nanoplankton consume the bacteria; adds several trophic levels to the plankton food chain.

microbivore Organism that feed on microbes, especially in the soil and litter.

microclimate Climate on a very local scale, which differs from the general climate of the area; influences the presence and distribution of organisms.

microflora Bacteria and certain fungi inhabiting the soil.

microhabitat That part of the general habitat utilized by an organism.

micromutation A mutation at the level of the gene; point mutation.

micronutrient Essential nutrient needed in very small quantities by plants and animals.

microparasite Any of the viruses, bacteria, and protozoans, characterized by small size, short generation time, and rapid multiplication.

migration Intentional, directional, usually seasonal movement of animals between two regions or habitats; involves departure and return of the same individual; a round-trip movement.

mimicry Resemblance of one organism to another or to an object in the environment, evolved to deceive predators.

mineralization Microbial breakdown of humus and other organic matter in soil to inorganic substances.

minimum base temperature The temperature at which net photosynthesis is at or approaching zero; used to determine an index of degree days.

minimum dynamic area (MDA) Area of suitable habitat necessary for maintaining a minimum viable population.

minimum viable population (MVP) Size of a population that, with a given probability, will ensure the population's existence for a stated period of time.

mire Wetland characterized by an accumulation of peat.

mitosis Cell division involving chromosome duplication, resulting in two daughter cells with the full complement of chromosomes, genetically the same as the parent cell.

mixed-grass prairie Grassland in mid North America, characterized by great variation in precipitation and a mixture of largely cool-season shortgrass and tallgrass species.

model In theoretical and systems ecology, an abstraction or simplification of a natural phenomenon, developed to predict a new phenomenon or to provide insight into existing ones; in mimetic association, the organism mimicked by a different organism.

modular organism Organism that grows by repeated iteration of parts, such as branches or shoots of a plant; some parts may separate and become physically and physiologically independent.

Mollisol Soil formed by calcification, characterized by accumulation of calcium carbonate in lower horizons and high organic content in upper horizons.

monoculture Planting of a single plant species.

monoecious Having male and female reproductive organs separated in different floral structures on the same plant; compare hermaphrodite, dioecious.

monogamy In animals, mating and maintenance of a pair bond with only one member of the opposite sex at a time.

moor A blanket bog or peatland.

morphological species Species described as monotypic, possessing little variation in color pattern, structure, proportion, and other features; the "field guide" species.

mortality rate The probability of dying; the ratio of number dying in a given time interval to the number alive at the beginning of the time interval.

mosaic A pattern of patches, corridors, and matrices in the landscape.

mutation Transmissible changes in the structure of a gene or chromosome.

mutualism Relationship between two species in which both benefit.

mycelium Mass of hyphae that make up the vegetative portion of a fungus.

mycorrhizae Association of fungus with roots of higher plants, which improves the plants' uptake of nutrients from the soil.

myrmecochory Dispersal by ants.

myrmecochores Plants that possess ant-attracting substances on their seed coats.

nanoplankton Plankton with a size range from 2 to 20 mm.

natality Production of new individuals in a population.

natural capital Range of natural resources provided by ecosystems.

natural selection Differential success (reproduction and survival) of individuals that results in elimination of maladaptive traits from a population.

neap tide Tide of small range that occurs at the first and last quarters of the moon when Earth, Moon, and Sun are at right angles.

negative feedback Homeostatic control in which an increase in some substance or activity ultimately inhibits or reverses the direction of the processes leading to the increase.

nekton Aquatic animals that are able to move at will through the water.

neritic Marine environment embracing the regions where landmasses extend outward as a continental shelf.

net ecosystem productivity Difference between the rates of net primary productivity and carbon lost through consumer and decomposer respiration.

net mineralization rate Difference between the rates of mineralization and immobilization.

net photosynthesis Difference between the rate of carbon uptake in photosynthesis and carbon loss in respiration.

net primary productivity (NPP) The rate of energy storage as organic matter after respiration.

net production Accumulation of total biomass over a given period of time after respiration is deducted from gross production in plants and from assimilated energy in consumer organisms.

net reproductive rate Average number of female offspring produced by an average female during her lifetime.

niche Functional role of a species in the community, including activities and relationships.

niche breadth Range of a single niche dimension occupied by a population.

niche compression Restriction of the use of a resource such as food or space because of intense competition.

niche overlap Sharing of niche space by two or more species.

niche preemption Procurement by a species of a portion of available resources, leaving less for the next.

nitrification Breakdown of nitrogen-containing organic compounds into nitrates and nitrites.

nitrogen fixation Conversion of atmospheric nitrogen to forms usable by organisms.

nucleotide A compound formed by the condensation of a nitrogenous base with a sugar and phosphoric acid; structural unit of DNA.

null hypothesis A statement of no difference between sets of values formulated for statistical testing.

numerical response Change in size of a population of predators in response to change in density of its prey.

nutrient Substance an organism requires for normal growth and activity.

nutrient cycle Pathway of an element or nutrient through the ecosystem, from assimilation by organisms to release by decomposition.

nutrient spiraling In flowing-water ecosystems, the combined processes of nutrient cycling and downstream transport.

object resemblance A prey species assumes the appearance of some feature in the environment, such as a leaf, to avoid detection.

obligate Having no alternative in response to a particular condition or in way of life.

obligatory relationship A symbiotic relationship in which one symbiont cannot survive and reproduce without the other.

oceanic Referring to regions of the sea with depths greater than 200 m that lie beyond the continental shelf.

old-growth forest Forest that has not been disturbed by humans for hundreds of years.

oligotrophic Term applied to a body of water low in nutrients and in productivity.

oligotrophy Nutrient-poor condition.

omnivore An animal (heterotroph) that feeds on both plant and animal matter.

open system System with exchanges of materials and energy to the surrounding environment.

operative temperature range Range of body temperatures at which poikilotherms carry out daily activity.

opportunistic species Organisms able to exploit temporary habitats or conditions.

optimal foraging theory Tendency of animals to harvest food efficiently, selecting food sizes or food patches that supply maximum food intake for energy expended.

optimum yield Amount of material that can be removed from a population to produce maximum biomass based on sustained yield.

organismic concept of community Idea that species, especially plant species, are integrated into an internally interdependent unit; upon maturity and death of the community, another identical plant community replaces it.

osmosis Movement of water molecules across a differentially permeable membrane in response to a concentration or pressure gradient.

osmotic potential The attraction of water across a membrane; the more concentrated a solution, the lower is its osmotic potential.

osmotic pressure Pressure needed to prevent passage of water or another solvent through a semipermeable membrane separating a solvent from a solution.

outbreeding Production of offspring by the fusion of distantly related gametes.

outbreeding depression Hybridization between two populations, each adapted to different environments, that destroys coadapted gene complexes, making the offspring maladapted to local conditions.

outputs Export of nutrients and energy from an ecosystem to the surrounding environment.

overdispersion Situation in which the distribution of organisms is random but clumped, with some areas empty and some heavily overpopulated; contagious distribution.

overturn Vertical mixing of layers in a body of water, brought about by seasonal changes in temperature.

Oxisol Soil developed under humid semitropical and tropical conditions, characterized by silicates and hydrous oxides, clays, residual quartz, deficiency in bases, and low plant nutrients; formed by laterization.

paleoecology Study of ecology of past communities by means of the fossil record.

pampas Temperate South American grassland, dominated by bunchgrasses; much of the moister pampas are under cultivation.

parasitism Relationship between two species in which one benefits while the other is harmed (although not usually killed directly).

parasitoid Insect larva that kills its host by consuming the host's soft tissues before pupation or metamorphosis into an adult.

parthenogenesis Development of an individual from an egg that did not undergo fertilization.

patch An area of habitat that differs from its surroundings and has sufficient resources to allow a population to persist.

peat Unconsolidated material consisting of un-decomposed and only slightly decomposed organic matter under conditions of excessive moisture.

peatland Any ecosystem dominated by peat; compare bog, mire, and fen.

pelagic Referring to the open sea.

PEP carboxylase The enzyme phosphoenolpyruvate carboxylase that catalyzes the fixation of CO_2 into four-carbon acids, malate, and aspartate.

percent base saturation The extent to which the exchange sites of soil particles are occupied by exchangeable base cations or by cations other than hydrogen and aluminum, expressed as percentage of total cation exchange capacity; compare cation exchange capacity.

percent similarity (PS) An index of proportional similarity that considers the number of species in each community, the species common to both communities, and the abundance of species.

percolation The movement of water downward and outward through subsurface soil, often continuing down to groundwater.

periphyton In freshwater ecosystems, organisms that are attached to submerged plant stems and leaves; see aufwuchs.

permafrost Permanently frozen soil in arctic regions.

permanent wilting point Point at which water potential in the soil and conductivity assume such low values that the plant is unable to extract sufficient water to survive and wilts permanently.

pest An animal that humans consider undesirable; what constitutes a pest varies with time, place, circumstance, and individual attitudes.

phenotype Physical expression of a characteristic of an organism, determined by both genetic constitution and environment.

phenotypic plasticity Ability to change form under different environmental conditions.

pheromone Chemical substance released by an animal that influences behavior of others of the same species.

photic zone Lighted water column of a lake or ocean, inhabited by plankton.

photoinhibition The slowing or stopping of a plant process by light.

photoperiodism Response of plants and animals to changes in relative duration of light and dark.

photosynthesis Use of light energy by plants to convert carbon dioxide and water into simple sugars.

photosynthetically active radiation (PAR) That part of the light spectrum between wavelengths of 400 and 700 nm that is used by plants in photosynthesis.

phreatophyte Type of plant that habitually obtains its water supply from groundwater.

physiological ecology Study of the physiological functioning of organisms in relation to their environment.

phytoplankton Small, floating plant life in aquatic ecosystems; planktonic plants.

pioneer species Plants that invade disturbed sites or appear in early stages of succession.

plankton Small, floating or weakly swimming plants and animals in freshwater and marine ecosystems.

Pleistocene Geological epoch extending from about 2 million to 10,000 years ago, characterized by recurring glaciers; the Ice Age.

pneumatophore An erect respiratory root that protrudes above waterlogged soils; typical of bald cypress and mangroves.

podzolization Soil-forming process in which acid leaches the A horizon and iron, aluminum, silica, and clays accumulate in lower horizons.

poikilotherm An organism whose body temperature varies according to the temperature of its surroundings.

poikilothermy Variation of body temperature with external conditions.

polar easterlies Easterly wind located at high latitudes poleward of the subpolar low.

polyandry Mating of one female with several males.

polyculture Planting of several plant species.

polygamy Acquisition by an individual of two or more mates, none of which is mated to other individuals.

polygyny Mating of one male with several females.

polyploid Having three or more times the haploid number of chromosomes.

polyploidy The condition of a cell or an organism that has more than its normal set of chromosomes.

population A group of individuals of the same species living in a given area at a given time.

population cycles Recurrent oscillations between high and low points in a population in a manner more regular than we would expect to occur by chance. The two most common intervals between peaks are 3 to 4 in arctic voles and 9 to 10 in snowshoe hares.

population density The number of individuals in a population per unit area.

population dynamics Study of the factors that influence the number and density of populations in time and space.

population ecology Study of how populations grow, fluctuate, spread, and interact intraspecifically and interspecifically.

population genetics The study of changes in gene frequency and genotypes in populations.

population projection table Chart of growth of a population developed by calculating the births and mortality of each age group over time.

population regulation Mechanisms or factors within a population that cause it to decrease when density is high and increase when density is low.

positive feedback Control in a system that reinforces a process in the same direction.

potential energy Energy available to do work.

potential evapotranspiration Amount of water that would be transpired under constantly optimal conditions of soil moisture and plant cover.

practical salinity unit (psu) The total amount of dissolved material in seawater, expressed as parts per thousand (‰).

precipitation All the forms of water that fall to Earth; includes rain, snow, hail, sleet, fog, mist, drizzle, and the measured amounts of each.

precocial In birds, hatched with down, open eyes, and ability to move about; in mammals, born with open eyes and ability to follow the mother, as fawns and calves can.

predation Relationship in which one living organism serves as a food source for another.

predator defenses Evolved characteristics that help prey avoid detection or capture.

predator satiation A predator defense mechanism involving the physiological timing of reproduction by a prey species, plant or animal, to produce a maximum number of seeds or young within a short period—more than predators can possibly consume—thus allowing a greater percentage of offspring to escape.

preferred temperature Range of temperatures within which poikilotherms function most efficiently.

premating isolating mechanisms Behavioral, habitat, structural or temporal means that prevent mating between individuals of different species.

primary producer Green plant or chemosynthetic bacterium that converts light or chemical energy into organismal tissue.

primary production Production by green plants over time.

primary productivity Rate at which plants produce biomass per unit area per unit time.

primary succession Vegetational development starting on a new site never before colonized by life.

production Amount of energy formed by an individual, population, or community per unit time.

production efficiency The ratio of production to assimilation; a measure of the efficiency of a consumer to incorporate assimilated energy into secondary production.

productivity Rate of energy fixation or storage per unit time; not to be confused with production.

profundal zone Deep zone in aquatic ecosystems, below the limnetic zone.

promiscuity Member of one sex mates with more than one member of the opposite sex, and the relationship terminates after mating.

protective armor Hard outer covering of an animal body, such as shells of turtles and spines of porcupines, that deters or makes the owner somewhat invulnerable to most enemies.

pycnocline Area in the water column where the highest rate of change in density occurs for a given change in depth.

pyramid of biomass Diagrammatic representation of biomass at different trophic levels in an ecosystem.

pyramid of energy Diagrammatic representation of the flow of energy through different trophic levels.

pyramid of numbers Diagrammatic representation of the number of individual organisms present at each trophic level in an ecosystem; the least useful pyramid.

pyromineralization Mineralization of nutrients bound in organic compounds by fire.

quaking bog Bog characterized by a floating mat of peat and vegetation over water.

***r*-strategist** Short-lived species with a high reproductive rate at low population densities; characterized by relatively small body size, rapid growth rate, large number of offspring, and lack of parental care.

radiation Transfer of energy through electromagnetic rays.

rain forest Permanently wet forest of the tropics; also the wet coniferous forest of the Pacific Northwest of the United States.

rain shadow A dry region on the leeward side of a mountain range resulting from a reduction in rainfall.

raised bog A bog in which the accumulation of peat has raised the surface above both the surrounding landscape and the water table; it develops its own perched water table.

ramet An individual member of a plant clone.

random distribution Distribution lacking pattern or order; placement of each individual is independent of all other individuals.

random sample Each object in a population has an equal chance of being selected.

rank-abundance diagram Plots of relative abundance of each species against rank, defined as the order of species from the most to the least abundant.

rate of increase Factor by which a population changes over a given period of time; compare exponential growth, geometric rate of increase, intrinsic rate of increase.

reabsorption See retranslocation.

realized niche Portion of fundamental niche space occupied by a population facing competition from populations of other species; environmental conditions under which a population survives and reproduces in nature.

recessive gene Applies to an allele whose phenotypic effect is expressed in the homozygous state and masked in the presence of an allele in organisms heterozygous for that gene.

recombination Exchange of genetic material by independent assortment of chromosomes and their genes during gamete production, allowing a random mix of different sets of genes at fertilization.

recruitment Addition of new individuals to a population by reproduction.

redundancy model Relates to the effect of loss of species on ecosystem stability. Losing some species may have little effect because other species can expand their role and take up functions vacated by the lost species.

regional diversity Total number of species observed in all habitats within a geographic area. Also called gamma diversity.

regular distribution A pattern in which individuals are more widely separated from each other than would be expected by chance; underdispersion.

relative abundance Proportional representation of a species in a community or sample of a community.

relative growth rate (RGR) Weight gained during a specified period of time.

relative humidity Water vapor content of air at a given temperature, expressed as a percentage of the water vapor needed for saturation at that temperature.

reproductive effort Proportion of its resources an organism expends on reproduction.

reproductive isolation Separation of one population from another by inability to produce viable offspring when the two populations mate.

reproductive value Potential reproductive output of an individual at a particular age relative to that of a newborn individual at the same time.

rescue effect An increase in population size and a decrease in the risk of extinction brought about by an increase of immigration into a population.

resource Environmental component used by a living organism.

resource allocation Apportioning the supply of a resource to specific uses.

respiration Metabolic assimilation of oxygen, accompanied by production of carbon dioxide and water, release of energy, and breakdown of organic compounds.

restoration ecology Applying principles of ecosystem development and function to the restoration and management of disturbed lands.

rete A large network or discrete vascular bundle of intermingling small blood vessels carrying arterial and venous blood that acts as a heat exchanger in mammals and certain fish and sharks.

retranslocation Recycling of nutrients within a plant.

rhizobia Bacteria capable of living mutualistically with higher plants.

rhizome A horizontal underground stem that branches and gives rise to vegetative structures.

richness A component of species diversity; the number of species present in an area.

riffle Stretch of shallow, fast, rough water flowing between pools in a stream.

riparian woodland Woodland along the bank of a river or stream; riverbank forests are often called gallery forests.

rivet model Idea that losing a species in an ecosystem is analogous to losing a rivet from an airplane. When the loss reaches a certain threshold, there are major catastrophic effects.

root-to-shoot ratio Ratio of the weight of roots to the weight of shoots of a plant.

rotation period Interval between the recurrence of a disturbance event; or interval between harvests of a crop, such as trees.

rubisco Enzyme in photosynthesis that catalyzes the initial transformation of into sugar.

ruminant Ungulate with a three-chamber or four-chamber stomach; in the large first chamber, or rumen, bacteria ferment plant matter.

salinity A measure of the total quantity of dissolved substances in water in parts per thousand (0/00) by weight.

salinization The process of accumulation of soluble salts in soil, usually by upward capillary movement from a salty groundwater source. Also called salination.

salt marsh Communities of emergent vegetation rooted in a soil alternately flooded and drained by tidal action.

sample That part of a population actually observed.

saprophyte Plant that draws its nourishment from dead plant and animal matter, mostly the former.

saturated Refers to air that contains the maximum amount of water vapor it can hold at a given temperature and pressure.

saturation vapor pressure Maximum amount of water vapor a volume of air can hold at a given temperature.

savanna Tropical grassland, usually with scattered trees or shrubs.

scale Level of resolution within the dimensions of time and space; spatial proportion as a ratio of length on a map to actual length.

scavenger Animal that feeds on dead animals or on animal products, such as dung.

scramble competition Intraspecific competition in which limited resources are shared to the point that no individual survives.

scrapers Aquatic invertebrates that feed on algal coating on stones and rubble in streams; also called grazers.

search image Mental image formed in predators, enabling them to find prey more quickly and to concentrate on a common type of prey.

seasonality Recurrence of biological events with the seasons.

second law of thermodynamics In any energy transfer or transformation, part of the energy assumes a form that cannot be passed on any further.

second-level carnivores Organisms that feed on first-level carnivores or second-level consumers.

second-level consumers Organisms that feed on first-level consumers or herbivores; carnivores.

secondary producers Organisms that derive energy from consuming plant or animal tissue and breaking down assimilated carbon compounds.

secondary production Production by consumer organisms over time.

secondary succession Development of vegetation after a disturbance.

sedimentary cycle Weathering of rock and leaching of its minerals, transport, deposition, and burial.

seed-tree harvesting Method of forest harvest in which a small number of trees are left on the site (uncut) to provide a source of seeds for natural regeneration of the population.

selection Differential survival or reproduction of individuals in a population because of phenotypic differences among them.

selection cutting Method of forest harvesting in which only selected individual trees of high commercial value are removed from the forest stand.

selective breeding Process in which humans select individuals of animals or plants with a desired trait and then mate them with other individuals exhibiting the same traits, resulting in populations of organisms with specific characteristics; analogous with natural selection.

selective pressure Any force acting on individuals in a population that determines which individuals leave more descendants than others; gives direction to the evolutionary process.

self-thinning Progressive decline in density of plants associated with the increasing size of individuals.

semelparity Having only a single reproductive effort in a lifetime, over one short period of time.

semiarid Fairly dry in climate, with precipitation between 25 and 60 cm a year and with an evapotranspiration rate high enough that the potential loss of water to the environment exceeds inputs.

sequential hermaphrodite An individual organism that changes sex from female to male or male to female at some time in its life cycle.

seral stage Following a series of stages; a point in a continuum of vegetation through time.

sere The series of successional stages on a given site that lead to a terminal community.

serotiny Release by heat of seeds tightly held in the cones of some species of coniferous trees.

serpentine soil Soil derived from ultrabasic rocks that are high in iron, magnesium, nickel, chromium, and cobalt and low in calcium, potassium, sodium, and aluminum; supports distinctive communities.

sessile Not free to move about; permanently attached to a substrate.

sex ratio The relative number of males to females in a population.

sex reversal A change in functioning so that a member of one sex behaves like the other.

sexual dimorphism The occurrence of morphological differences other than primary sexual characteristics that distinguish males from females in a species.

sexual reproduction Two individuals produce haploid gametes (egg and sperm) that combine to form a diploid cell.

sexual selection Selection by one sex for an individual of the other sex based on some specific characteristic or characteristics; usually takes place through courtship behavior.

shade-intolerant Growing and reproducing best under high light conditions; growing poorly and failing to reproduce under low light conditions.

shade-tolerant Able to grow and reproduce under low light conditions.

sheet erosion Transport of soil material from slopes by a thin, mobile sheet of water.

shelterwood cutting See seed-tree cut.

shifting mosaic Constantly changing pattern of patches as each patch passes through successive stages of development.

short-day organisms Plants and animals that come into reproductive condition under conditions of short days—days with less than a certain maximum length.

shortgrass prairie Westernmost grasslands of the Great Plains, characterized by infrequent rainfall, low humidity, and high winds; dominated by shallow-rooted, sod-forming grasses.

shredders Aquatic invertebrates that feed on coarse particulate matter in streams.

siblicide Killing of an offspring by another offspring of the same parents.

sigmoid curve S-shaped curve of logistic growth.

Simpson's index (D) A measure of the probability that two individuals randomly selected from a sample will belong to the same species.

simultaneous hermaphrodite An individual organism that possesses both male and female sex organs at the same time in its life cycle.

sink An unfilled, submarginal, or marginal habitat or patch where the population can persist only by immigration from other habitats because it experiences low reproduction or high mortality.

sink habitat A habitat area that receives immigrants from a source habitat, but in which the subpopulation would continually decrease in size because of mortality and poor reproductive success without continual immigration from excess individuals in a source habitat.

sink patch Area where population of a species can be maintained only by immigration. See sink habitat.

site Combination of biotic, climatic, and soil conditions that determine an area's capacity to produce vegetation.

snag Dead or partially dead tree of at least 10.2 cm diameter at breast height (dbh) and 1.8 m tall; important habitat for cavity-nesting birds and mammals.

social dominance Physical dominance of one individual over another, usually maintained by some manifestation of aggressive behavior.

soil horizon Developmental layer in the soil with characteristic thickness, color, texture, structure, acidity, nutrient concentration, and the like.

soil profile Distinctive layering of horizons in the soil.

soil series Basic unit of soil classification, consisting of soils that are alike in all major profile characteristics except texture of the A horizon; soil series are usually named for the locality where the typical soil was first recorded.

soil structure Arrangement of soil particles and aggregates.

soil texture Relative proportions of the three particle sizes (sand, silt, and clay) in the soil.

solar constant Rate at which solar energy is received on a surface just outside of Earth's atmosphere; current value is 0.140 watt/cm^2.

solifluction The downhill movement of soil that has been saturated with water.

solute A substance that is dissolved in solution.

solution Liquid that is a homogeneous mixture of two or more substances.

solvent Dissolving agent of a solution.

source habitat Area of habitat in which a subpopulation of a species produces more individuals than needed for self-maintenance, thus contributing to emigration.

source patch Area where a population of a species reproductively produces more individuals than needed for replacement; these individuals then emigrate into other areas. See source habitat.

speciation Separation of a population into two or more reproductively isolated populations.

species diversity Measurement that relates density of organisms of each type present in a habitat to the number of species in a habitat.

species evenness A component of species diversity index; a measure of the distribution of individuals among total species occupying a given area.

species richness Number of species in a given area.

specific heat Amount of energy that must be added or removed to raise or lower the temperature of a substance by a specific amount.

Sphagnum A genus of mosses that are most abundant in wet, acidic habitats; the dead cells rapidly fill with water, allowing the plant to hold many times its own weight in water.

spiraling Mechanism of retention of nutrients in flowing-water ecosystems, involving the interdependent processes of nutrient recycling and downstream transport.

Spodosol Soil characterized by the presence of a horizon in which organic matter and amorphous oxides of aluminum and iron have precipitated; includes podzolic soils.

spring tide A tide of greater than mean range that occurs every two weeks, when the Moon is full or new; maximum spring tides occur when Sun and Moon are in the same plane as Earth; compare neap tide.

stabilizing selection Selection favoring the middle in the distribution of phenotypes.

stable age distribution Constant proportion of individuals of various age classes in a population though population changes.

stand Unit of vegetation that is essentially homogeneous in all layers and differs from adjacent types qualitatively and quantitatively.

standard deviation Statistical measure defining the dispersion of values about the mean in a normal distribution.

standing crop biomass Total amount of biomass per unit area at a given time.

static life table See life table.

stationary age distribution Special form of stable age distribution in which the birthrate equals the death rate, and age distribution remains fixed.

statistical population Set of objects about which inferences can be drawn.

steppe Name given to Eurasian grasslands that extend from eastern Europe to western Siberia and China.

stoichiometry Branch of chemistry dealing with the quantitative relationships of elements in combination.

stomata Pores in the leaf or stem of a plant that allow gaseous exchange between the internal tissues and the atmosphere.

stratification Division of an aquatic or terrestrial community into distinguishable layers based on temperature, moisture, light, vegetative structure, and other such factors; stratification creates zones for different plant and animal types.

stratified random sample Random samples taken from each of a defined subpopulation.

sublittoral zone Lower division of the sea, from about 40 m to below 200 m.

subspecies Geographical unit of a species population, distinguishable by morphological, behavioral, or physiological characteristics.

succession Replacement of one community by another; often progresses to a stable terminal community called the climax.

successional sequence Pattern of colonization and extinction of plants on a given area over time; compare sere.

supercooling In ectotherms, lowering body temperature below freezing without freezing body tissue, by means of solutes (particularly glycerol).

supralittoral fringe The highest zone on the intertidal shore, bounded below by the upper limit of barnacles and above by the upper limit of *Littorina* snails.

surface fire Fire that feeds on the litter layer of forests and grasslands.

surface tension Elastic film across the surface of a liquid, caused by the attractive forces between molecules at the surface of the liquid.

survivorship The probability that a representative newborn individual in a cohort will survive to various ages.

survivorship curve A graph describing the survival of a cohort of individuals in a population from birth to the maximum age reached by any one member of the cohort.

sustainable agriculture Farming practices that provide a secure living for farm families while at the same time maintaining the natural environment and resources.

sustained yield Yield per unit time equal to production per unit time in an exploited population.

swamp Wooded wetland in which water is near or above ground level.

swidden agriculture Farming systems that alternate periods of annual cropping with extended fallow periods. Also referred to as shifting cultivation; fire is used to clear fallow areas for cropping.

switching Changing the diet from a less abundant to a more abundant prey species.

symbiosis Situation in which two dissimilar organisms live together in close association.

sympatric Living in the same area; usually refers to overlapping populations.

sympatric speciation Production of a new species within a population or the dispersal range of a population.

system Set or collection of interdependent parts or subsystems enclosed within a defined boundary; the outside environment may provide inputs and receive outputs.

systematic sampling Type of sampling used to determine variations within a population. Application of general systems theory and methods to ecology.

systems ecology Sets of compartments linked by fluxes of energy and nutrients.

taiga The northern circumpolar boreal forest.

tallgrass prairie A narrow belt of tall grasses dominated by big bluestem that once ran north and south adjacent to the deciduous forest of eastern North America; presence maintained by fire; largely destroyed by cultivation.

temperate rain forest Forest in regions characterized by mild climate and heavy rainfall that produces lush vegetative growth; one example is the coniferous forest of the Pacific Northwest of North America.

temperature A measure of the average speed or kinetic energy of atoms and molecules in a substance.

territory Area defended by an animal; varies among animal species according to social behavior, social organization, and resource requirements.

theory of island biogeography Theory stating that the number of species established on an island represents a dynamic equilibrium between the immigration of new colonizing species and the extinction of previously established ones.

thermal conductance Rate at which heat flows through a substance.

thermal neutral zone Among homeotherms, the range of temperatures at which metabolic rate does not vary with temperature.

thermal radiation Heat transfer by longwave or infrared radiation.

thermal tolerance Range of temperatures in which an aquatic poikilotherm is most at home.

thermocline Layer in a thermally stratified body of water in which temperature changes rapidly relative to the remainder of the body.

thinning law Self-thinning plant populations, sown at sufficiently high densities, approach and follow a thinning line with a slope of roughly -3.2; therefore, in a growing population, plant weight increases faster than density decreases to a point where the slope changes to -1.

throughfall That part of precipitation that falls through vegetation to the ground.

tidal overmixing Mixing of freshwater and seawater when a tidal wedge of seawater moves upstream in a tidal river faster than freshwater moves seaward; seawater on the surface tends to sink as lighter freshwater rises to the surface.

tidal subsidy Nutrients carried to coastal ecosystems and wastes carried away by tidal cycles.

time lag Delay in a response to change.

time-specific life table A population sampled in some manner to obtain the distribution of age classes during a single time period.

tolerance model Model proposing that succession leads to a community composed of species most efficient in exploiting resources; colonists neither increase nor decrease the rate of recruitment or growth of later colonists.

top-down control Influence of predators on the structure of lower trophic levels in a food web.

topography Physical structure of the landscape.

torpor Temporary great reduction in an animal's respiration, with loss of motion and feeling; reduces energy expenditure in response to some unfavorable environmental condition, such as heat or cold.

trace element Element occurring and needed in small quantities; see micronutrient.

trade winds Tropical easterly winds that blow in a steady direction from the subtropical high-pressure areas to the equatorial low-pressure areas between the latitudes 30° and 40° north and south; these winds are generally northeasterly in the Northern Hemisphere and southeasterly in the Southern Hemisphere.

translocation Transport of materials within a plant; absorption of minerals from soil into roots and their movement throughout the plant.

transpiration Loss of water vapor from a plant to the outside atmosphere.

trophic Related to feeding.

trophic efficiency Ratio of productivity in a given trophic level with the trophic level on which it feeds.

trophic level Functional classification of organisms in an ecosystem according to feeding relationships, ranging from first-level autotrophs through succeeding levels of herbivores and carnivores.

trophic structure Organization of a community based on the number of feeding or energy transfer levels.

tundra Area in an arctic or alpine (high mountain) region, characterized by bare ground, absence of trees, and growth of mosses, lichens, sedges, forbs, and low shrubs.

turgor pressure The state in a plant cell in which the protoplast is exerting pressure on the cell wall due to intake of water by osmosis.

turnover rate Rate of species lost and others gained.

type I functional response Rate of prey mortality due to predation is constant, a function of the efficiency of predators.

type II functional response Per capita rate of predation increases in a decelerating fashion only up to a maximum rate that is attained at some high prey density.

type III functional response Rate of prey consumed; slow at first and then increasing in an S-shaped (sigmoid) fashion as the rate of predation reaches a maximum.

ubiquitous Having widespread geographic distribution.

Ultisol Low base soil associated with warm, humid climate and old terrain, taking on a reddish color from secondary iron oxides.

understory Growth of medium-height and small trees beneath the canopy of a forest; sometimes includes a shrub layer as well.

ungulate Any hoofed grazing mammal; usually refers to ruminants, such as cattle and deer.

unitary organism An organism, such as an arthropod or vertebrate, whose growth to adult form follows a determinate pathway, unlike modular organisms whose growth involves indeterminate repetition of units of structure.

upwelling In oceans and large lakes, a water current or movement of surface waters produced by wind that brings nutrient-loaded colder water to the surface; in open oceans, regions where surface currents diverge deep waters, which rise to the surface to replace departing waters.

vapor pressure The amount of pressure water vapor exerts independent of dry air.

vapor pressure deficit The difference between saturation vapor pressure and the actual vapor pressure at any given temperature.

vector Organism that transmits a pathogen from one organism to another.

vegetative reproduction Asexual reproduction in plants by means of specialized multicellular organs, such as bulbs, corms, rhizomes, stems, and the like.

veld Extensive grasslands in the east of the interior of South Africa, largely confined to high terrain.

vertical stratification Layering of physical conditions and life in a community.

vertical structure The arrangement of layers of vegetation in terrestrial and aquatic communities.

Vertisol Mineral soil that contains more than 30 percent of swelling clays that expand when wet and contract when dry; associated with seasonal wet and dry environments.

vesicular arbuscular mycorrhizae (VAM) A form of endomycorrhizae in which the fungus enters and grows within the host's cells and extends widely into the surrounding soil.

viscosity Property of a fluid that resists the force that causes it to flow.

visible light Light comprising wavelengths of 3400 to 740 nanometers.

warning coloration Conspicuous color or markings on an animal that serve to discourage potential predators.

water balance Maintenance of the balance of water between organisms and their surrounding environment.

water cycle Movement of water between atmosphere and Earth by way of precipitation and evaporation.

water potential Measure of energy needed to move water molecules across a semipermeable membrane; water tends to move from areas of high or less negative potential to areas of low or more negative potential.

water-use efficiency Ratio of net primary production to transpiration of water by a plant.

watershed Entire region drained by a waterway into a lake or reservoir; total area above a given point on a stream that contributes water to the flow at that point; the topographic dividing line from which surface streams flow in two different directions.

weather The combination of temperature, humidity, precipitation, wind, cloudiness, and other atmospheric conditions at a specific place and time.

weathering Physical and chemical breakdown of rock and its components at and below Earth's surface.

weed Plant possessing a high rate of dispersal, occurring opportunistically on land or water disturbed by human activity, and competing for resources with cultivated plants; a plant growing in the wrong place.

westerlies The dominant east-to-west motion of the winds centered over the middle latitudes of both hemispheres.

wet deposition Sulfur dioxide and nitrogen oxides that return as dilute solutions of nitric acid and sulfuric acid dissolved in rain and snow.

wetfall Component of acid deposition that reaches Earth by some form of precipitation; wet deposition.

wetland A general term applied to open-water habitats and seasonally or permanently waterlogged land areas; defining the extent of a wetland is controversial because of conflicting land-use demands.

wilting point Moisture content of soil at which plants wilt and fail to recover their turgidity when placed in a dark, humid atmosphere; measured by oven drying.

xeric Dry, especially in soil.

yield Individuals or biomass removed or harvested from a population per unit time.

zero-growth isocline An isocline along which the net population growth rate is zero.

zonation Characteristic distribution of vegetation along an environmental gradient; this gradient may form latitudinal, altitudinal, or horizontal belts within an ecosystem.

zooplankton Floating or weakly swimming animals in freshwater and marine ecosystems; planktonic animals.

Credits

Text/Illustration Credits

Chapter 2 2.7: Barry & Chorley, *Atmosphere, Weather and Climate, 6/e.* Copyright © 1992. Reprinted by permission of Taylor & Francis Group, LLC. 2.11: T.E. Graedel and Paul J. Crutzen, *Atmosphere, Climate, And Change.* Copyright © 1995, 1997 by Lucent Technologies. Reprinted by permission of Henry Holt and Company, LLC. 2.15: From *This Great and Wide Sea* by Robert E. Coker. Copyright 1949 by the University of North Carolina Press, renewed © 1977 by Robert M. Coker. Used by permission of the publisher. 2.21: From *USDA Agricultural Handbook 360*, Schroeder & Buck.

Chapter 3 3.7: R.G. Wetzel, *Limnology, 2/e.* 3.10: From *Biological Survey of the Raquette Watershed.*

Chapter 4 4.5: William A. Pfitsch, "Steady-state and dynamic photosynthetic response of *Adenocaulon bicolor* (Asteraceæ) in its redwood forest habitat," *Oecologia* 80(4):471–476, 2004. Copyright © 2004 Springer Berlin/Heidelberg. Used with permission. 4.6: Adapted from B.A. Hutchinson and D.R. Matt, "The distribution of solar radiation within a deciduous forest," *Ecological Monographs* 47:205, 1972. Copyright © 1972 Ecological Society of America. 4.7: Adapted from USGS, Soil Conservation Service.

Chapter 5 5.4: *Ecology and Evolution of Darwin's Finches*, Peter Grant, p. 155, fig. 59. 5.5a: *Ecology and Evolution of Darwin's Finches*, Peter Grant, p. 182, fig. 55. 5.5b: P.T. Boag, "The heritability of external morphololgy in Darwin's Ground Finches (geospiza) on Isla Daphne Major, Galapagos, *Evolution.* Vol. 37, No. 5, Sept. 1983, pp. 877–894. 5.6: P.T. Boag and P.R. Grant, "The classical case of character release: *Biological Journal of the Linnean Societ.* 5.7: P.T. Boag and P.R. Grant, "Intense Natural Selection in a population of Darwin's Finches in the Galapagos" *Science* 214:82-85, 1981. Copyright © 1981. Reprinted by permission of AAAS. 5.8a: Grant, Peter R., *Ecology and Evolution of Darwin's Finches.* © 1986 Princeton University Press. Reprinted by permission. 5.8a: From Boag and Grant, "Intense natural selection in a natural population of Darwin's Finches," *Science* 214, 1981, p. 83. Copyright © 1981 AAAS. Used with permission. 5.9: *Ecology and Evolution of Darwin's Finches*, Peter Grant, p. 205, fig. 59b. 5.11b: Adapted from Smith, *Nature* 363:618, 1993. 5.12: R.H. Baker, *Origin, Classification, and Distribution* in L.K. Halls, *White-tailed deer: Ecology and Management.* 5.15: D. Schluter, "Seed and patch selection by Galapagos ground finches," *Ecology* and Peter Grant, *Ecology and Evolution of Darwin's Finches.* 5.16a,b: From *Evolution of Darwin's Finches*, Peter Grant. 5.16c: *Morphological Differentiation and adaptation in the Galapagos Finches* by R.I. Bowman 1961. 5.17: Patel, "How to build a longer beak," *Nature* 442:515–516, p. 515. 5.19: S.E. Sultan and F.A Bazzaz, "Phenotypic plasticity in polygonumpersicaria," *Ecology* fig 3. Graphs h & I, and legend, p. 27. FS5.1: Courtesy of Laura Nagel. Used with permission. FS5.2: Adapted from Beren W. Robinson, *Behavior* 137, 2000. © Brill Academic Publishers, Leiden, The Netherlands 2000.

Chapter 6 6.4: *Handbook of Plant and Crop Physiology*, M. Pessarakli, editor. 6.9: Adapted from Larcher, *Physiological Plant Ecology*, 3/e, p. 96, 108. Copyright © 1995. Reprinted by permission of Springer-Verlag. 6.10: Adapted from Davies, *Ecology* 79:2292–2308, 1998. Copyright © 1998 Ecological Society of America. Used with permission. 6.13: Reich et al., *Functional Ecology.* 6.14: Adapted from C.K. Ausperger. 6.16: J.A. Teeri, "Climatic patterns and the distribution of C_4 grasses in North America," *Oecologia* 23(1):1-12, 2004. Copyright © 2004 Springer Berlin/Heidelberg. Used with permission. 6.20: Adapted from Mooney et al., "Photosynthetic capacity of Death Valley plants" *Carnegie Institute Yearbook* 75:310–413, 1976. 6.22: Data from Pearcy, *Plant Physiology* 61:484–486, 1977. 6.24: Thomas Givnish, *On the Economy of Plant Formand Function: Proceedings of the Sixth Maria Moors Cabot.* Copyright © 1986. Reprinted by permission of Cambridge University Press. 6.27: Adapted from Reich et al, *Ecological Monographs* 62, 1982. Copyright © 1982 Ecological Society of America. Used with permission. FS6.2: Adapted from Kitajima, "The relative importance of photosynthetic traits and allocation patterns as correlates of seedling shade tolerance of 13 tropical trees," *Oecologia* 98:419–428, 1994. Copyright © 1994. Reprinted by permission of Springer Berlin, Heidelberg.

Chapter 7 7.12: Hilly and Wyse, ANIMAL PHYSIOLOGY 2/e. Copyright © 1989. Reprinted by permission of Pearson Education. 7.15: K. Schmidt-Nielsen, *Ani al Physiology 5/3.* 7.25: Edwin Bunning, "Circulation rhythms and the time measurement in photoperiodism," *Cold Spring Harbor Symposium on Quantitative Biology*, Vol. 25. FS7.2: Adapted from Wikelski, *Ecology* 78(7):2204–2217, 1997. Copyright © 1997 Ecological Society of America. Used with permission. FS7.3: Adapted from Wikelski, *Ecology* 51(3):922–938, 1997. Copyright © 1997 Ecological Society of America. Used with permission. FS7.4: M. Wikelski & C. Thom, "marine iguanas shrink to survive El Nino," *Nature* 403:37-38, 2000. Copyright © 2000 Nature Publishing Group. Used with permission.

Chapter 8 8.11: Begon et al., *Ecology 2/e.* 8.12: P. Werner and W. Platt, "Ecological relationships of co-occurring goldenrods," *The American Naturalist* 110;959–971. Copyright © 1976 The University of Chicago Press. Used with permission. 8.13a: Adapted from Jones, *New Zealand Journal of Zoology* 5:783, 1978. Used with permission of the Royal Society of New Zealand. 8.13b: Adapted from L. Wauters and A.A. Dohondt, "Body weight, longevity, and reproductive success in red squirrels (*Sciurusvulgaris*)," *Journal of Animal Ecology* 58:637–651, 1989. Copyright © 1989 Blackwell Science. FS8.2: Adapted from A.L. Basolo, "Female preference for male sword length in the green swordtail *Xiphophorushelleri*," *Animal Behaviour* 40:332–338, 1990. Copyright © 1990, with permission from Elsevier. FS8.3: Basolo & Alcaraz, "The turn of the sword" *Proceedings of the Royal Society of London* Vol. 270, pp. 1631–1636, 2003. Copyright © 2003 The Royal Society. Used with permission.

Chapter 9 9.14: H. Hett and O. Loucks, "Age Structure Models of Balsam Fir and Eastern Hemlock," *Journal of Ecololgy* 64:1035, 1976. Copyright © 1976. Reprinted by permission of Blackwell Science, Ltd. **9.15:** Adapted from Engle, "Yellow Poplar seed fall pattern," *Central States Forest Experimental Station Note 143*, 1960.

Chapter 10 10.6: Based on R.R. Sharitz and J.R. McCormick, "Population Dynamics of Two Competing Plant Species," *Ecology* 54, 1973. 10.13b: E. Hackney and J. McGraw, *Conservation Biology*.

Chapter 11 11.6: J.N. Chatworthy, *Studies on the Nature of Competition between Closely Related Species*, D. Phil. Thesis, University of Oxford. 11.7: L. Wang et al., "Effects of intraspecific competition on growth and photosynthesis of *Atriplexprostrata*," *Aquatic Botany*. 11.8: M.C. Dash and A.R. Hota, "Density Effects on Survival Growth Rate and metamorphosis on *Ran tigrina* tadpoles," *Ecology*. 11.9: R.A. Relyea, fig. 1, p. 529 from "Competitor-induced plasticity in tadpoles," *Ecological Monographs*. 11.11: R.N. Hughes & C.L. Griffiths, "Self-Thinning in barnacles and mussels," *American Naturalist* 132:484–491, 1988. Copyright © 1988 The University of Chicago Press. Used with permission. 11.12: T.M. Jenkins et al., "Effects of population density on individual growth of Brown Trout in streams," *Ecology* fig. 4, p. 948. 11.13: E.R. Keeley, "Demographic responses to food and space competition by juvenile steelhead trout," *Ecology*. 11.14: From C.W. Fowler and T.D. Smith, *Dynamics of Large Mammal Populations*. Copyright © 1981. Reprinted by permission of the author. 11.15: A. M. Schueller et al., "Density dependence of Walleye maturity and fecundity in Big Crooked Lake, Wisconsin, 1997–2003," *North American Journal of Fisheries Management*. Copyright © 2005. Reprinted by permission from American Fisheries Society. 11.18: Adapted from R.L. Smith, "Some Ecological Notes on the Grasshopper Sparrow," *Wilson Bulletin 75*, 1963. 11.20: Adapted from Cahill, *Oikos* Vol. 90, pp. 311–320. 11.21: Mech et al., "Relationships of Deer and Moose populations to previous winters' snow," *Journal of Animal Ecology*. FS11.2: Sillett et al., "Impacts of a global climate cycle on population dynamics of a migratory songbird," *Science* 288:2040–2042, 2000. Copyright © 2000. Reprinted by permission of AAAS. FS11.3: Sillett et al., "Impacts of a global climate cycle on population dynamics of a migratory songbird," *Science* 288:2040–2042, 2000. Copyright © 2000. Reprinted by permission of AAAS.

Chapter 12 12.2b-d: Ehrlich & D. Murphy, *Conservation Biology*. 12.4b,c: O. Kindvall and I. Ahlen, *Conservation Biology* 6: 520–529, 1992. 12.5: O. Kindvall and I. Ahlen, *Conservation Biology* 6: 520–529, 1992. 12.6b: C. Thomas and T. Jones, *Journal of Animal Ecology* 62:472–481, 1993. 12.8: Adapted from O. Kindvall, *Ecology* 77:207–214. Copyright © 1996 Ecological Society of America. Used with permission. 12.9: Susan Harrison et al., "Distribution of the Bay Checkerspot Butterfly, *Euphydryasedithabayensis*: Evidence for a metapopulation model," *The American Naturalist*, 9/1/1988. Copyright © 1988 The University of Chicago Press. 12.10: C. Thomas and T. Jones, *Journal of Animal Ecology* 62:472–481, 1993. E1.12.1b,c: Adapted from A. Rodriguez and M. Delibes, "Current range and status of the Iberian lynx *Felisparadina* in Spain," *Biological Conservation* 61: 189–196. Used with permission.

Chapter 13 13.3: Tilman et al., "Competition between species of diatoms for silica," *Limnology Oceanopgraphy*. 13.8: M.P. Austin et al., *Journal of Ecology*. 13.9: After Groves & Williams, *Australian Journal of Agricultural Research* 26:977–983, 1975. Copyright © 1975. 13.10: N.C. Emery et a., *Ecology*. 13.11: C. Heller and D. Gates, *Ecology*. 13.15: N.K. Wieland and F.A. Bazzaz, *Ecology*. 13.16: Tamar Dayan et al., "Feline canines: community-wide character displacement among the small cats of Israel," *The American Naturalist* 136, pp. 39-60, 1990. Copyright © 1990 The University of Chicago Press. Used with permission. 13.20: Grant, Peter R., *Ecology and Evolution of Darwin's Finches*. © 1986 Princeton University Press. Reprinted by permission. FS13.1: Katherine Suding, "The effect of spring burning on competitive rankings of prairie species," *Journal of Vegetative Science* 12:849–856, 2001. Copyright © 2001 Opulus Press. Used with permission. FS13.2: Adapted from K. Suding and D. Goldberg, "Do disturbances alter competitive hierarchies?" Mechanisms of change following gap formation," *Ecology* 82:2133–2149. Copyright © 2001 Ecological Society of America. Used with permission.

Chapter 14 14.4a: Adapted from Korpimake and Norrdahl, "Numerical functional responses of kestrels, short-eared owls, and long-eared owls to vole densities," *Ecology* 72:814–826, 1991. 14.4b: Adapted from Jedrzejewski et al., "Weasel population response, home range and predation on rodents in a deciduous forest in Poland," *Ecology* 76:179–195, 1995. 14.4c: Mook & Morris, "Birds and Spruce Budworm," *Entomological Society of Canada Memoirs*. 14.8: Mook & Morris, "Birds and Spruce Budworm," *Entomological Society of Canada Memoirs*. 14.9: Adapted from Jedrzejewski et al., "Weasel population response, home range and predation on rodents in a deciduous forest in Poland," *Ecology* 76:179–195, 1995. 14.10: Adapted from Davies, "Prey selection and social behavior in wagtails," *Journal of Animal Ecology* 46:37-57, 1997. 14.21: Adapted from Williams et al., *Ecology* 74:1143–1152. Copyright © 1993 Ecological Society of America. Used with permission. 14.27: Garry J. Scrimgeour, "Feeding while evading predators by a lotic mayfly: linking short-term foraging behaviours to long-term fitness consequences," *Oecologia* 100(1):128–134, 2004. Copyright © 2004, Springer Berlin/Heidelberg. 14.28: Adapted form Nelson et al., *Ecology* 85:1855. Copyright © 2004 Ecological Society of America. Used with permission. FS14.2: Adapted from R. Relyea, "The many faces of predation: how induction, selections, and thinning combine to alter prey phenotypes," *Ecology* 83:1953–1964. Copyright © 2002 Ecological Society of America. Used with permission. FS14.3: Adapted from R. Relyea, "The many faces of predation: how induction, selections, and thinning combine to alter prey phenotypes," *Ecology* 83:1953–1964. Copyright © 2002 Ecological Society of America. Used with permission. FS14.4: Adapted from R. Relyea, "The many faces of predation: how induction, selections, and thinning combine to alter prey phenotypes," *Ecology* 83:1953–1964. Copyright © 2002 Ecological Society of America. Used with permission.

Chapter 15 15.6: Adapted from K. Lafferty and K. Morris, *Ecology* 77:1390–1397, 1996. Copyright © 1996 Ecological Society of America. Used with permission. FS15.1: J. Stachowicz & M. Hay, "Mutualism and coral persistence," *Ecology* September 1999. FS15.2: J. Stachowicz, *BioScience*, 2001.

Chapter 16   16.10: R.H. Whittaker, "Vegetation of the Great Smoky Mountains" *Ecological Monographs*. 16.11: Henski et al., *Metapopulation Biology*, 1996. EI.16.1: Adapted from Keith Langdon, GRSM 2004.

Chapter 17   17.9: T. Smith and M. Huston, "A theory of spatial and temporal dynamics of plant communities," *Plant Ecology* 83:46-69. © 1989 with kind permission of Kluwer. 17.12: N.C. Emery et al., "Competition and salt marsh plant zonation: stress tolerators may be dominant competitors" *Ecology*. 17.14: Huston. FS17.1: S.D. Hacker and M.D. Bertness, "Trophic Consequences of a Positive Plant Interaction," *American Naturalist* 144:363–372, 1994. Copyright © 1994. Reprinted by permission of the University of Chicago Press. FS17.2: S.D. Hacker and M.D. Bertness, "Trophic Consequences of a Positive Plant Interaction," *American Naturalist* 144:363–372, 1994. Copyright © 1994. Reprinted by permission of the University of Chicago Press. FS17.3: S.D. Hacker and M.D. Bertness, "Trophic Consequences of a Positive Plant Interaction," *American Naturalist* 144:363–372, 1994. Copyright © 1994. Reprinted by permission of the University of Chicago Press.

Chapter 18   18.3: *Pattern and Process in a Forested Ecosystem* by F. Bormann and G.E. Likens. 18.11: H.J. Oosting, 1942. 18.14: From University of Virginia, 1994, Master's Thesis. 18.19: R.H. Whitaker, "Quaternary history and the stability of forest communities". F18.20: "Vegetation Maps for Eastern North America, 40,000 yr. BP to Present." *Geobotany*, P.A. Delacourt. F18.22b: Adapted from Whittaker 1954. F18.22c: Adapted from Whittaker 1956. EI.18.1: Michael Williams, *Americans and their Forests*. EI.18.2: Michael Williams, *Americans and their Forests*.

Chapter 19   19.3: N.R. Webb and L.E. Haskins "An ecological survey of the heathlands in the Poole Basin, Dorset, England in 1978," *Biological Conservation*. 19.4: H.B. Johnson, *Order Upon the Land*. Copyright © Oxford University Press. 19.6: *Elements of Ecology 7/e* by Smith & Smith. 19.12: F.W. Preston. FS19.1: J. Tewksbury et al., "Corridors affect plants, animals, and their interaction in fragmented landscapes," *PNAS*. FS19.2: J. Tewksbury et al., "Corridors affect plants, animals, and their interaction in fragmented landscapes," *PNAS*. FS19.3: Haddad.

Chapter 20   20.2: H. Leith, "Primary production terrestrial ecosystems," *Human Ecology* vol. 1q7, 1973, p. 303. 20.3: H. Leith, "Modeling primary productivity of the world". 20.4: R.H. MacArthur & J.H. Connell, *The biology of Populations* 1966. 20.6: Pastor et al, "Above-ground production and N&P cycling along a nitrogen mineralization gradient on Blackhawk Island, Wisconsin," *Ecology*. 20.7: Reich et al., *Ecology*. 20.9: Downing et la., *Ecology*. 20.11: "The phosphorus-chlorophyll relationship in lakes," P.J. Dilon and F.H. Rigler, *Limnology and Oceanography*. 20.15: S.T. Gower et al., *Trends in Ecology and Evolution*. 20.16: Nature fig. 1, vol. 14, pp. 142–144. 20.20: Cyr & Pace, Nature Vol. 361, pp. 148–150, 1993. 20.21: Begon, *Ecology: Individuals, Populations and Communities*. FS20.3: Brian Silliman Nature Vol. 429, pp. 870–873. EI.20.1: Nature fig. 1, vol. 429, pp. 87-873.

Chapter 21   21.6: C.J. Leroy and J.C. Makrs, "Litter quality, stream characteristics, and litter diversity influence decomposition rats and microinvertebrates. 21.10: Ivan Valiela, *Marine Biology Processes*, P. 301. Copyright © 1984 Ivan Valiela. Used by permission. 21.17: Aber and Melillo. 21.19: B. Berg et al, "Changes in organic chemical components of needle litter during decomposition," *Canadian Journal of Botany*. 21.21: *Freshwater Wetlands: Ecological Processes and Management Potential* R.E. Good et al. 21.23: J. Pastor et al., "Above-ground production and N&P cycling along a nitrogen mineralization gradient on Blackhawk Island, Wisconsin," *Ecology*. FS21.2: Ted Schuur. FS21.3: Ted Schuur.

Chapter 22   22.8: Schlessinger, *Biochemistry: An analysis of global change*. 22.12: Schlessinger, *Biochemistry: An analysis of global change*. 22.13: Schlessinger, *Biochemistry: An analysis of global change*.

Chapter 23   23.1: D.M. Olson et al., "Terrestrial Ecoregions of the Worlds; A New Map of Life on Earth," *BioScience*. 23.15: O.J. Reichman, *Konza Prairie: A Tallgrass Natural History*. 23.28: J.M. Dyer, *BioScience*. 23.32: S. Payette, "The Subarctic Forest-Tundra: the structure of a biome in a changing climate," *BioSciencei*.

Chapter 26   26.4: Huston & Niklas. 26.8: Currie & Paquin, "Large scale biogeographical patterns of species richness in Trees" Nature Vol. 329, fig. 2, p. 327, 1987. 26.9: Currie, "Energy and Large Scale Biographical Patterns of Species Richness in Trees," *American Naturalist*. Copyright © 1991 The University of Chicago Press.

Chapter 27   27.6: Ewel et al., *Ecology*. 27.16: *Ecological Knowledge and Environmental Problem Solving: Concepts and Cast Studies*. 27.19: Vitousek & Matson, *Ecology*. 27.26: J.A. Reynolds & J.D. Reynolds, "Marine fish population collapse: consequences for recovery and extinction risk," *BioScience*. EI27.1: Figure 1, p. 5 from ISAAA 2003 Report. FS27.2: Lawrence & Schelsinger, *Ecology* 28:2769–2780, 2001. Copyright © 2001 Ecological Society of America. Used with permission. FS27.3: Lawrence & Schelsinger, *Ecology* 28:2769–2780, 2001. Copyright © 2001 Ecological Society of America. Used with permission.

Chapter 28   28.12: Gentry, *Essentials of Conservation Biology*, f16.7. 28.15: J. Berger, *Conservation Biology* vol. 4, pp. 91–98, 1990. T28.3: IUCN. T28.4: IUCN.

Chapter 29   29.3: Data from CDIAC. 29.5: www.whrc.org. 29.10: K.V.R. Schafter, "Exposure to an enriched CO2 atmosphere alters carbon assimilation and allocation in a pine forest ecosystem," *Global Change Biology*. 29.18: Iverson & Prasud, 2001. 29.19: Currie. FS29.2: Erika Zavaleta, *PNAS* 100:7650–7654, figure 1. FS29.3: Erika Zavaleta.

Photo Credits

Chapter 1   CO: Franz Lanting/Minden Pictures. 1.1: NASA/Johnson Space Center. 1.2: Tom Smith (author). 1.3a: Bjorn Forsberg/naturepl.com. 1.3b: Elizabeth Whiting & Associates/Alamy. 1.3c: John Lemker/Animals Animals. 1.3d: Dennis Frates/Alamy. 1.3e: Jim Richardson/National Geographic Image Collection. 1.3g: NASA/Goddard Space Flight Center. 1.6: David Tilman, Department of Ecology, Evolution and Behavior, University of Minnesota, 100 Ecology Building, 1987 Upper Buford Circle St., Paul, MN 55108 (612) 625-5740 larso106@umn.edu. 1.7: Tom Smith (author).

Animals/Earth Scenes 21.2d: (L) Gary Braasch/Corbis; (R) Bill Curtsinger. 21.3: R. L. Smith. 21.5a: (left) Susan Gottberg/ iStockphoto, (right) Steve McBeath/iStockphoto. 21.5c: Wayne Wurtsbaugh, Utah State University and the American Society of Limnology and Oceanography. 21.FS1: Ted Schur. 21.FS2: Micheal T. Sedam/Corbis. 21.11a: Cecelia Spitznas/iStockPhoto. 21.11b: Dennis Kunkel/Phototake. 21.11c: Kristin Piljay/Pearson Science. 21.16a: Thomas Bredenfeld/iStockPhoto. 21.16b,c: Kristin Piljay/Pearson Science. 21.16d: Dorling Kindersley Media Library. 21.18b: Photo Researchers, Inc. 21.18c: Kristin Piljay/Pearson Science. 21.18d,e: Dennis Kunkel/Phototake. 21.18f: Wim van Egmond/Visuals Unlimited, Inc.. 21.18g: Sinclair Stammers/Photo Researchers, Inc.

Chapter 22 CO22: R. L. Smith.

Chapter 23 CO23: George Ranalli/Photo Researchers, Inc. 23.3a-d: Thomas M. Smith. 23.4a,b: Thomas M. Smith. 23.6a: David Tipling/Photolibrary. 23.6b: Nick Garbutt/Nature Picture Library. 23.7a: Anup Shah/Nature Picture Library. 23.7b: Gerry Ellis/Minden Pictures. 23.9: Tom Stack & Associates, Inc. 23.10a,b: Erin Bohman. 23.11a: Luiz C. Marigo/Peter Arnold. 23.11b: Robin Smith/Photolibrary. 23.13: (TL) Gerry Ellis/Minden Pictures. 23.13: (TR) Nigel Dennis/age footstock. 23.13: (BL) Tom Smith (author). 23.16a,c: DRK Photo 23.16b: Tom Stack & Associates, Inc. 23.17a: Jim Parkin/Shutterstock. 23.17b: Galina Dreyzina/Shutterstock. 23.20a: NASA Earth Observing System. 23.20b: Australia Department of Lands. 23.21a, b: R. L. Smith. 23.21c: Ray Ellis/Photo Researchers, Inc. 23.22: Chris Mattison/agefotostock. 23.24a: Lynn Watson. 23.24b: Charles E. Jones. 23.24c: Br. Alfred Brousseau, Saint Mary's College. 23.25: Martin Harvey / Peter Arnold, Inc. 23.26a: Bob Gibbons/Alamy. 23.26b: Dan Suzio Photography. 23.29a: Royalty-Free/Corbis. 23.29b: R. L. Smith. 23.31a: Bruce Coleman Inc. 23.29b,c: Tom Stack & Associates, Inc. 23.31a: Bruce Coleman Inc. 23.31b: Tom Stack & Associates, Inc. 23.33: R. L. Smith. 23.35a: Paul Nicklen/National Geographic Image Collection. 23.35b: Neil Nightingale/naturepl.com. 23.37: John Shaw/Tom Stack & Associates, Inc.

Chapter 24 CO24: Gary Randall/FPG International. 24.2a. Bob and Ira Spring. 24.2b: Jack S. Grove. 24.2c: Doug Drake/Tom Stack & Associates, Inc. 24.2d: R. L. Smith. 24.EI1: Gary Braasch. 24.5a: Darlyne A. Murawski/Peter Arnold. 24.5b: Laguna Design/Photo Researchers, Inc. 24.6a: Brian Parker/ Tom Stack & Associates, Inc. 24.6b: R. L. Smith. 24.7: Geoff Higgins/ Photolibrary.Com. 24.9a,b: John Shaw/Tom Stack & Associates, Inc. 24.10: John Shaw/Tom Stack & Associates, Inc. 24.QE1: http://www.brown.edu/Courses/ GEO0158/measuringstreamflow. 24.14: T. M. Smith. 24.15: South Carolina Department of Natural Resources. 24.16: Mareveision/age footstock. 24.18: Steve Gschmeissner/Photo Researchers, Inc. 24.19: National Ocean Service/NOAA. 24.21: Verena Tunnicliffe. 24.22: Richard A. Lutz. 24.23: Tammy Peluso/Tom Stack & Associates, Inc.

Chapter 25 CO5: Sisse Birmberg/National Geographic Image Collection. 25.1: Tom Smith (author). 25.3: Tom Bean/ Allstock, Inc./Tony Stone Images. 25.4: Ron Dahlquist/Pacific Stock. 25.5: McDonald Wildlife Photography/Animals Animals/Earth Scenes. 25.6: Florida Images/Alamy. 25.7: Joe Arrington/Visuals Unlimited. 25.8: R. L. Smith. 25.9: Gary Braasch. 25.10: Larry Lipske/Tom Stack & Associates, Inc. 25.11a: Richard Du Toit/Minden Pictures. 25.11b: NASA. 25.14: Terry Donnelly/Tom Stack & Associates, Inc. 25.15a: USDA Forest Service. 25.15b: Gary Braasch. 25.16: R. L. Smith. 25.17: Mark Rollo/Photo Researchers, Inc. 25.18a: G. Schwabe/Arco Images/Peter Arnold. 25.20: Doug Wechsler/Animals Animals/Earth Scenes.

Chapter 26 CO26: Jean E. Roche/naturepl.com.

Chapter 27 CO27: Mira/Alamy Images. 27.3a: NASA. 27.5: Aflo/naturepl.com. 27.9: Tom Smith (author). 27.10a: Santokh Kochar/Getty Images. 27.10b: Jeremy Bright/ Robert Harding World Imagery. 27.11: Andreas Stirnberg/Getty Images. 27.14a: USDA/NRCS/Natural Resources Conservation Service. 27.11b: James Strawser/Grant Heilman Photography. 27.14: USDA/ NRCS/Natural Resources Conservation Service. 27.16a: Joe Sohm/ Alamy. 27.16b: Getty Images/Photodisc Collection. 27.FS1: Deb Lawrence. 27.FS2: Tom Smith (author). 27.21: Image courtesy of Earth Sciences and Image Analysis Laboratory, NASA Johnson Space Center. 27.24: Eliott Norse, Marine Conservation Biology Institute/Marine Photobank. 27.28a: blickwinkel/Alamy. 27.28b: Doug Perrine/naturepl.com. 27.28c: Peter Scoones/naturepl.com. 27.28d: Gary K. Smith/naturepl.com. 27.28e: Scott W. Smith/ Animals Animals/Earth Scenes.

Chapter 28 28CO: Silvestre Machado/Getty Images. 28.1: Stephen J. Krasemann/DRK Photo. 28.3: NASA/http:// visibleearth.nasa.gov/. 28.4a,b: Image courtesy of Earth Sciences and Image Analysis Laboratory, NASA Johnson Space Center. 28.4c: Paul Franklin/OSF/Earth Sciences. 28.5b: Corbis. 28.6a: Stephen G. Maka/DRK Photo. 28.6b: Wolfgang Kaehler/Corbis. 28.7: Raymond Gehman/Corbis. 28.8: Runk/Schoenberger/Grant Heilman Photography, Inc. 28.10a: Neil Julian/Alamy. 28.10b: Kathleen Brown/Corbis. 28.11: Carl S. Keener. 28.EI1: AP Photo/U.S. Fish and Wildlife Service. 28.15a: Tom Ulrich/photolibrary. 28.16: Thomas M. Smith. 28.17: Juniors Bildarchiv/Alamy. 28.18a: Y. Galindo/Zoological Society of San Diego via Getty Images. 28.18b: AP Photo/San Diego Wild Animal Park. 28.22: NOAA Restoration Center. 28.23: University of Wisconsin-Madison Arboretum.

Chapter 29 CO29: NOAA. 29.1: Simon Fraser/Mauna Loa Observatory/Photo Researchers, Inc. 29.FS1: Robert Houser. 29.FS2: Nona Chiariello. 29.9: Will Owens Photography. 29.20: Greg Henry. 29.26: Comstock Images.

Part Openers PO1: NASA/Johnson Space Center. PO2: (L) Mark Moffett/Minden Pictures, (R) Michael Fogden/Animals Animals/Earth Scenes. PO3: Tom & Pat Leeson. PO4: Ingo Bartussek/naturepl.com. PO5: Reidar Hahn/Fermilab. PO6: Gary Braasch/Corbis; Frans Lanting/Minder Pictures; David Muench/Corbis. PO7: Thomas M. Smith. PO8: Jeremy Walker/Getty Images.

Index

Note: Boldface page numbers refer to material contained in tables, boxes, figures, and legends.

A

A horizon, **64**, 65
Abies spp. (firs)
    A. lasiocarpa (subalpine fir), 509
    postglacial migration of, 384
Abiotic component of ecosystem, 3, 413
Abiotic factors, in competition, 278
Abomasum, 131, **131**, 132
Abundance, population density and distribution and, 186–187
Abutilon theophrasti (Indian mallow), resource partitioning and, 273, **274**
Abyssopelagic zone, of ocean, 532, **532**
Acacia tortillis trees, spatial distribution of, 187, **188**
Acacia trees, nitrogen cycle of, impala in, **467**
Acacia woodlands, 496
Acclimation, 93, **94**
Accumulation curve, **564**
Acer spp. (maples). *See* Maples (*Acer* spp.)
Achillea millefolium (yarrow), ecotypes of, 85, **85**
Acidity
    of aquatic environments, 48–49
    measures of, 49
    of soils, 67
Acorn woodpecker (*Melanerpes formicivorus*), territoriality of, 263
Acyrthosiphon pisum (pea aphids), predators and, in alfalfa, 304–305, **305**
Adaptation(s)
    animal, 126–157
    constraints and, 86–91
    definition of, 76
    natural selection and, 75–76
    plant, 97–125
        environmental constraints and, 106–107
        link between water demand and temperature influencing, 114–117
        to wetland environments, 121–123
    as product of evolution by natural selection, 78–80
    for stress tolerance and competitive ability, 360–363
    in terrestrial ecosystems, 490–493, **491**
    as trade-offs, 86, **87–88**, 89, **90**, 91
Adaptive radiation, 89, **90**
Adenosine diphosphate (ADP), in ATP synthesis, 98
Adenosine triphosphate (ATP), synthesis of, 98
Adenostoma fasciculatum (chamise), 505, **505**
Adiabatic cooling, 25
Adiabetic lapse rate, 25
Aedes aegypti (mosquitoes), dengue and yellow fever transmission by, climate change and, 645

Aepyceros melampus (impala), in acacia tree nitrogen cycle, **467**
Aerenchyma, in aquatic plants, **122**, 122
Aerosols, global warming and, 632, 634
African elephants (*Loxodonta africana*), **127**, **199**
    disturbances from, 407
    as keystone species, 336, **336**
    sodium deficiency in, 136
Age
    fecundity and, 170–171
    sex ratios in populations and, 191–192
Age distribution, calculation of, population projection table in, 209–210, **210**
Age pyramids, 190, **191**
Age structure, of populations, 189–190
Age-specific birthrates, 208
    in population growth projection, **209**, 209–211, **210**
Age-specific mortality rates, 204
    in population growth projection, **209**, 209–211, **210**
Aggregative response, **288**, 288–289
Agrawal, Anurag, 299
Agriculture
    carbon dioxide concentrations and, 624
    climate change and, 642–644, **643**
    decline in, **370**, 371
    development of, in Neolithic Period, 568
    disturbances from, 408
    energy inputs and, 575
    fields for, abandoned, colonization of, **368**, 369
    genetically modified crops in, **581**
    industrialized, 575, **577**
        in temperate zone, 576–578, **577**, **578**
    loss of temperate zone grasslands to, 601
    mechanization of, 568–569
    nitrogen cycle and, 476–477
    old-field communities and, **370–371**
    shifting cultivation in, **585–586**
    sustainable, 580, **582**
    swidden, **575**, 575–576, **576**, **577**
    trade-offs in, **578**, 578–580, **579**
    traditional, 575, **577**
Agrostis spp. (bentgrasses), responses of, to nitrogen in soil, 121
Ahlen, Ingemar, 244
Air
    circulation of, global, **25**, 25–26
    density, 24, **24**
    moisture content of, temperature influences on, 27–28
    quality of, in urban areas, heat island effect and, **34**
    pressure, 24, **24**
    temperature, altitude and, 23–25
Alaria fistulosa (kelp), in secondary succession, 375, **375**
Alarm calls, as predator defense, 297

Alces alces (moose/elk)
    in boreal forest, 510–511
    sodium deficiency in, 136
Alfalfa (*Medicago sativa*), herbivorous-predatory insect interactions and, 304–305, **305**
Alfisols, *507*
    description of, **67**
    world distribution of, **68**
Algae
    blue-green, in nitrogen fixation, 474
    in lichens, 320, **320**
Alkaline solutions, definition of, 49
Allee, W. C., 213
Allee effect, 213, **213**
Allegheny Mountain salamander (*Desmognathus ochrophaeus*), reproductive trade-offs in, 167
Alleles, 76
    codominant, **76**, 76–77
    dominant, 76, **76**
    frequency of, as measure of genetic variation, 77
    recessive, 76, **76**
Alligator snapping turtles, cryptic coloration and mimicry in, **298**
Allogenic environmental change, 377
Alopex lagopus (arctic fox), 513
Alpha diversity, 563
Alpine chipmunk (*Tamias slpinus*), 270, **271**
Alsophila pometaria (fall cankerworms), defoliation of trees by, 299
Altitude
    air temperature and, 23–25
    species richness and, **562**
Altricial organisms, 170
Aluminum, concentrations of, in solutions, 49
Aluminum oxides, in soils formed by laterization, 68
Amazon Basin, South America, 494
Amazon region of Brazil, deforestation in, 600
Amblyrhynchus cristatus (marine iguana), size variations in, survival and, **134–135**
Ambrosia arterisiifolia (ragweed), in secondary succession, 374
Ambush hunting, 298
Ambystoma maculatum (spotted salamander), life history strategy of, 176, **176**
American beech tree (*Fagus grandifolia*), age assessment in, **191**
American chestnut (*Castanea dentata*)
    chestnut blight and, 316
    as dominant species, 336
American elm (*Ulmus americana*), Dutch elm disease and, 316
American ginseng (*Panax quinquefolius*), reproductive limitations of, 213, **213**
American robins (*Turdus migratorius*)
    as edge species, 397, **398**
    parental care of offspring in, 170

Amino acids, in animal nutrition, 132
Ammadramus savannarum (grasshopper sparrow), territories of, **233**
Ammonification, 474, **475**
Amphibians
    as poikilotherms, 142
    thermoregulation in, 143
    water balance maintenance in, 149
Anadromous fish, 531
Anax longipes (dragonfly), larvae of, induction in prey caused by, **300–301**
Anders, William A., 2
Andisols
    description of, **67**
    world distribution of, **68**
Andosols, 494
Andropogon gerardii, **265–266**
Andropogon virginicus (broomsedge), in secondary succession, 374–375
Animals
    adaptations of, 126–157
    aquatic
        buoyancy of, 150–151, **151**
        nonnative, environmental damage from, 602–603, **603, 604**
        water balance maintenance in, 149–150
    energy acquisition by, 130–132
    energy exchange with environment in, 138–141
    evolution of, size as constraint on, 127–129
    hermaphroditic reproduction in, 161
    homeostasis and feedback in, 138, **139**
    light/dark cycles influencing, 151–152
    nutrient acquisition by, 130–132
    nutritional needs of, varied, 132–136
    oxygen needs of, for energy release from food, 136–138
    processes common to, 127
    seasonal responses in, critical day lengths triggering, **152**, 152–153
    size of, structure and, 128–129
    in succession, **381**, 381–382, **382**
    terrestrial
        nonnative, environmental damage from, 602
        water balance maintenance in, 148–149
    thermoregulation in, **139, 140–141**
    torpor in, 147
Anions, definition of, 66
Annual crabgrass *(Digitaria sanguinalis)*, in secondary succession, 374
Annual grassland, 499
Anopheles spp. (mosquitoes)
    distribution of, climate change and, 644–645
    in malaria transmission, 311, **311**
Antibiotic resistance, ecology of, **92**
Antigens, in parasitic invasion, 313
Ants *(Pseudomyrmex* spp.), mutualism in, 322
Aphantopus hyperantus (ringlet butterfly), synchronization of local population extinction for, 249
Aphids, reproduction in, 159
Aphotic zone, of aquatic ecosystem, 340
Apodemus flavicollis (yellow-necked mouse), 256
    density of, numerical response of weasels to, 289, **289**

Apoematism, as predator defense, 295, **296**
Apollo 8, 2, **2**
Apparent competion, 356–357, **357**
Appalachian Mountains, southern, salamander diversity in, 85, **86**
Aquatic ecosystem(s), 37–54, 517–539
    acidity in, influence of, 48–49
    carbon cycle in, 470–471, **471**
    decomposition in, 456, **456**
    distribution of, global climate change and, 641
    eutrophication in, 601–602
    fish harvesting from, 588–593. *See also* Fisheries management
    flowing-water, 524–531. *See also* Streams/rivers
    freshwater. *See also* Lakes/ponds; Streams/rivers
        phosphorus cycle in, 477
        and saltwater, transition zone between, **52**, 52–53
        saltwater environment and, transition zone between, **52**, 52–53
    light in, by depth, **43**, 43–44
    marine coastal, tides and, **51**, 51–52
    nonnative species invading, environmental damage from, 602–603, **603, 604**
    oceans as, 531–536. *See also* Oceans
    open-water, nutrient cycling in, 459–460, **460, 461, 462**
    oxygen stratification in, 47–48, **48**
    pH of, 49
    phosphorus cycle in, **477**, 477–479
    primary production in, control of, 420–421
    stream and river. *See* Streams/rivers
    structure of, 339, **339**, 340
    temperature in, by depth, **44**, 44–45
    water movements shaping, **50**, 50–51, **51**
Aquatic nutrient cycle, 456
Aquatic plants, carbon uptake process for, 103–104
Aqueous solution, definition of, 45
Aquifer, High Plains–Ogallala, 39
Aral Sea, demise of, 572, **573**, 574
Arbuscules, 321, **321**
Arbutus unedo (strawberry tree), 505
Arctic fox *(Alopex lagopus)*, 513
Arctic tundra, **511**, 511–513, **512, 513**
Arenicola cristata (lugworm), 342
Aridisols, 504
    description of, **67**
    world distribution of, **68**
Arisaema triphyllum (jack-in-the-pulpit), sex change in, **161**, 161–162
Armadillidium vulgare, reproductive trade-offs in, 167
Armillaria mellea (honey mushrooms), **438**
Arolla pine, geographic distribution of, climate change and, **635**
Arrow arum *(Peltandra virginica)*, decomposition of leaves of, **456**
Artemia salina (brine shrimp), dispersal of, 192
Artemisia (sagebrush), 502, **503**
Artiplex confertifolia (shadescale), 502, **503**
Artyroxiphium spp. (Hawaiian silverswords), semelparity in, 169
Ascaris (roundworms), direct transmission of, 310–311

Asclepias spp. (milkweed)
    impact of herbivory on, 299
    seed dispersal in, 192
Asexual reproduction, 159–160
Ash *(Fraxinus* spp.), postglacial migration of, 384
Ashmole, N. Philip, 173
Aspens. *See* Cottonwoods *(Populus* spp.)
Aspidosperma cruenta, **113**
Assimilation efficiency, 426
Association, 344
Assortative mating, 82
Aster ericoides (white aster), in secondary succession, 374
Asterionella formosa, competition of, with *Synedra ulna*, 260–261, **261**
Asynchronous hatching, 171
Atmosphere
    in global carbon cycle, **473**
    in oxygen cycle, 482, **482**
    oxygen diffusion to surface waters from, **47**, 47–48, **48**
    regions in, 24, **25**
    water cycles between earth and, 38–40, **39**
Atmospheric pressure, 24
Atolls, 535
Atriplex prostrata (spear-leaved orache), intraspecific competition in, 223, **223**
Atriplex spp. (saltbushes), 116, **119**
Aufwuchs, 523
Augspurger, Caroline, 114
Austin, Mike, 264, 269
Australian magpie *(Gymnorhina tibicen)*, reserve of potential breeders in, 233
Australian paperbark tree *(Melaleuca quinquenervia)*, invading Florida, environmental damage from, 602, **603**
Autogenic environmental change, succession associated with, 377–379
Autotrophs, 73, 98
    in ecosystem, 413
    herbivores preying on, 298–299, 302
Autumnal equinox, 21
    circle of illumination at, **22**
Available water capacity (AWC), of soil, **65**, 65–66
Azotobacter, in nitrogen fixation, 474

B

B horizon, **64**, 65
Bacteria
    in decomposition
        in aquatic environments, 456
        in terrestrial ecosystems, 440, 441
    growth of, in rhizosphere, 453
    in herbivore digestion, 130
    in nitrogen cycling, 474–475, **475**
    Rhizobium, as nitrogen fixers, 474
Baetis tricaudatus (mayflies), prey-predator interactions and, 304, **304**
Balaenoptera musculus (blue whale), **127**
    economic approach to resource management and, 594
Bald cypress *(Taxodium distichum)*
    adaptations of, to aquatic environment, 122, **123**
    in wetlands, 547

Grime, J. Phillip, 249–250
　　life history variation model of, for
　　　　plants, 176, **177**
Grizzly bears *(Ursus arctos)*, local extinction
　　of, effects of, 358
Grosbeaks *(Coccothraustes* spp.), 511
Gross primary productivity, 415
Gross reproductive rate, 208
Ground finches *(Geospiza* spp.)
　　G. fortis (medium ground finch)
　　　　beak size of, climate change and, **78,**
　　　　　　78–80, **79, 80,** 86, 89, **89, 90**
　　　　character displacement in, 276–278,
　　　　　　277
　　G. fuliginosa (small ground finch), 86,
　　　　89, **89, 90**
　　　　character displacement in, 276–277,
　　　　　　277
　　G. magnirostris (large ground finch), 86,
　　　　89, **89, 90**
　　　　character displacement and,
　　　　　　277–278
Groundhogs *(Marmota monax)*, hibernation
　　in, 147
Groundwater, **38,** 38–39, **39, 40**
　　nitrate contamination of, risks of, 579,
　　　　579
Groves, R. H., 263–264, 360–361
Growing season, length of
　　definition of, 416
　　net primary productivity and, 416, **417**
Growth
　　density-dependent, 222
　　effects of intraspecific competition on,
　　　　223, 223–224
　　indeterminate, 170
　　intraspecific competition affecting,
　　　　222–224
Growth rate
　　animal, mineral availability and, 136
　　plant
　　　　relative, **109–110**
　　　　slow, low-nutrient conditions and,
　　　　　　121
Growth rings, in age determination, 190, **191**
Grus americana (whooping crane), expo-
　　nential growth of, 203, **203**
Guilds, in species classification, 338
Gurevitch, Jessica, 272
Gymnogyps californianus (California con-
　　dor), restoration of, 612, **613**
Gypsy moths *(Lymantria dispar)*
　　defoliation of trees by, 299
　　disturbances from, 408
　　spread of, 194, **194, 195**
Gyres, 26–27, **27**

H

Haber, Fritz, 580
Haber-Bosch process, in nitrogen fertilizer
　　production, 457, 580
Habitability, 16
Habitat(s)
　　conservation of
　　　　establishing protected areas in, **614,**
　　　　　　614–617, **615, 616, 617**
　　　　in whole community protection,
　　　　　　612–613
　　corridors as, 404
　　destruction of, current species extinc-
　　　　tion from, 599–602

heterogeneity of, local population per-
　　sistence and, 247, **247**
loss of, population extinction and, 212
restoration of, in conservation efforts,
　　617–618, **618**
Habitat patches
　　as emigrant source, 248, **248**
　　in metapopulation dynamics, 244–247,
　　　　45, 246, 247
Habitat selection, reproductive success and,
　　173–174
Hacker, Sally D., **352–353**
Hackney, Erine, 213
Hadalpelagic zone, of ocean, 532, **532**
Haddad, Nick M., **402–403**
Hadley, George, 26
Hadley cells, 26, **26**
Haeckel, Ernst, 2
Hairston, Nelson, 359, 426, 431
Haldane, J. B. S., 75
*Halimeda,*in secondary succession, 375, **376**
*Halodule wrightii,*in secondary succession,
　　375, **376**
Halophytes, *123*
Hanski, Ilkka, 240, 250, 405
Hard-shelled clam *(Mercenaria merce-
　　naria)*, 342
Hardy, Godfrey, 82
Hardy-Weinberg principle, 82, **83–84**
Harems, 162, **163**
Harp seals *(Phoca groenlandica)*, sexual
　　maturity of, weight and, 226, **230**
Harper, John, 365
Harrison, Susan, 248
Hartig's net, **321**
Harvest interval, 573
Harvesting
　　in boreal forests, 510
　　in withdrawal of nutrients from ecosys-
　　　　tem, 469–470
Hatching, asynchronous, 171
Haustorium, 309
Hawaiian goose *(Branta sandwicensis)*,
　　restoration of, 612
Hawaiian silverswords *(Argyroxiphium*
　　spp.), semelparity in, 169
Hawthorn *(Crataegus* spp.), in succession,
　　369
Heat
　　latent, 27
　　specific, of water, 41
Heat exchange
　　countercurrent, in thermoregulation,
　　　　147–148, **148**
　　temperature regulation and, **140–141**
Heat islands, urban, 34
Heat transfer coefficient, 141
Helmintheros vermivorus (worm-eating war-
　　bler**)**, as interior pecies, **398,** 399
Hemiparasites, 309, **310**
Hemlock trees *(Tsuga* spp.), reproduction
　　in, 160
Herb layer, **339,** 340
Herbivores
　　in Arctin tundra, 513
　　definition of, 130
　　desert, 504
　　digestive systems of, **131**
　　energy and nutrient acquisition by,
　　　　130–132
　　in freshwater wetlands, 554

grassland, 500–501
interaction of, with plants and carni-
　　vores, 303–304
plant characteristics to deter, **302,**
　　302–303
as predators, 282
preying on autotrophs, 298–299, 302
in savannas, 498
sodium deficiency in, 136
Hermaphrodites
　　sequential, 161–162
　　simultaneous, 161
Hermaphroditic organisms, **160,** 160–161
Herring gulls *(Larus argentatus)*, habitat
　　selection in, 174
Hesperia comma (skipper butterfly)
　　decline of, 249, **249**
　　patterns of extinction and colonization
　　　　in, 246, **246**
*Heteropogon contortus,*competition of, with
　　U. mosambicensis, rainfall and,
　　262, **263**
Heterotherms, 142
　　characteristics of, 146
　　temporal, 146
Heterotrophs, 73
　　in ecosystem, 413
　　in succession, **381,** 381–382, **382**
Heterozius rohendifrons (big-handed crabs),
　　fecundity and size in, 171, **171**
Heterozygous, definition of, 76
Hibernation, 147
Hickory *(Carya* spp.), 341
　　in secondary succession, **375**
Hickory leaf gall, **313**
Hiesey, W. M., 85
High Plains-Ogallala aquifer, **39**
Highlands salamander *(Plethodon jordani
　　melaventris)*, **86**
Hiltner, Lorenz, 453
Hippopotamus amphibius (hippopotamus),
　　in transport of nutrients from
　　ecosystem, 469
Hippopotamus *(Hippopotamus amphibius)*,
　　in transport of nutrients from
　　ecosystem, 469
Histogram, **10**
Histosols
　　description of, **67**
　　world distribution of, **68**
Holdridge, L. R., 640
Holling, 288
Holm oak *(Quercus ilex)*, 505
Holoparasites, 309, **310**
　　direct transmission of, 311
Holt, Robert, 356
Home range
　　definition of, 232
　　size of, body size and, 232, **232**
Homeostasis, definition of, 138
Homeostatic plateaus, 138
Homeotherms
　　countercurrent heat exchange in, 148
　　hibernation in, 147
　　thermoregulation in, 143–145, **144**
Homeothermy, 142
Homozygous, definition of, 76
Honey mushrooms *(Armillaria mellea)*, **438**
Hooded warbler *(Wilsonia citrina)*
　　forest strata occupied by, 364
　　habitat of, **173**

Hooker, Joseph, 486
Horizons, soil, **64**, 64–65
Hornbeam *(Carpinus caroliniana)*, 341
   in understory, 339
Horned owl *(Bubo virginianus)*, in three-level interaction, **303**, 303–304
Horseweed *(Erigeron canadensis)*, intraspecific competition in, mortality rates and, 224, **225**
Horseweed *(Lactuca canadensis)*, in secondary succession, 374
Hosts
   definitive, 312
   direct transmission of parasites between, 310–311
   intermediate, 312
   multiple, in parasite transmission, 311–312, **312**
   parasitic invasion of, 310
   populations of, regulation of, parasites in, 314–316
   reproduction in, parasites affecting, 313–314
   response of, to parasites, 312–313, **313**
   survival of, parasites affecting, 313–314
Hota, Ashok, 224
Hotspots, biodiversity, 607–608, **609**
House mouse *(Mus musculus)*, pheromones in, stress and, 231
How Many People Can the Earth Support? (Cohen), **219**
Hulett, H. R., **219**
Human activity
   agricultural. *See* Agriculture
   disturbances from, long-lasting effects of, 408
   exotic species introduced by, native species threatened by, 602–603, **603**, **604**
   landscape patches and, 393
   net primary productivity and, **428–429**
   nitrogen cycle and, 476–477
   resource use and, 569
   sulfur cycle and, **480**, 481, 482
   sustainability and, 571–575. *See also* Sustainability
Human factor, in ecology, 12, **13**
Human health, climate change affecting, 644–645, **645**, **646**
Human population growth, environmental problems related to, 12
Humidity, relative, 27
Hummingbirds, as heterotherms, 142
Humpback whales *(Megaptera novaeangliae)*, migration of, 193
Humus, 451
Hunting tactics, evolved by predators, 298
Hurricane Katrina, **622**
Hurricanes, disturbances from, 406
Huston, Michael, 365, 380–381
Hutchinson, G. E., 275
Hydra, freshwater, asexual reproduction in, 159, **159**
Hydrogen bonding, in water, 41, **41**, 42
Hydrogen (H)
   in animal nutrition, **133**
   as macronutrient, 119, **120**
Hydrogen ions, in solution, acidity and, 49
Hydrologic cycle, 38. *See also* Water cycle(s)
Hydroperiod, of wetlands, 549–551
Hydrophytic plants, 547

Hyla crucifer (spring peepers), supercooling in, 147
Hyla versicolor (gray tree frogs)
   supercooling in, 147
   tadpoles of, induction in, **300–301**
Hylurgopinus rufipes (elm bark beetles), in parasite transmission, 311
Hyperosmotic organisms, 149
Hypervolume, niche as, 275, **276**
Hypolimnion, 44, **44**, **45**
Hypoosmotic organisms, marine fish as, 150
Hypothesis
   as model, 9
   in scientific method, **6**, 6–7
   testing of, in scientific method, **6**, 7–9

I

Iberian lynx *(Lynx pardinus)*, habitat fragmentation in, **243–244**
Ice Age, Little, 31
Ice sheets. *See* Glaciers
Icefields, world distribution of, **68**
Ictalurus nebulosus (bullhead catfish), temperature acclimation in, 93, **94**
Iguana, parietal eye of, 152
Immigration, 192
Immobilization, mineralization and, 449–451, **452**
Immune response, in defense against parasites, 313
Impala *(Aepyceros melampus)*, in acacia tree nitrogen cycle, **467**
Inbreeding, 82
Inbreeding depression, 82
Inceptisols
   description of, **67**
   world distribution of, **68**
Indeterminate growth, 170
Indian mallow *(Abutilon theophrasti)*, resource partitioning and, 273, **274**
Indigo bunting *(Passerina cyanea)*, as edge species, 395, **396**
Indirect commensalism, 358
Indirect mutualism, 358
Individual(s), 4
   as basic unit of ecology, 13–14
   high density as stressful to, 227, 230–231
   patterns and processes in, study of, 5
   in population, spatial distribution patterns for, 186–187, **187**
   response of, to environmental variation, 91–94
Individualistic concept of community, 346, **346**
Induced defenses, predator, 298
Induced edges, 394
Induction, predators and, **300**
Industrial Revolution, 568–569
Infauna, 544
Infection, parasitic, 309
Infiltration, of precipitation, 38, **38**
   soil texture and, 65, **65**
Infralittoral fringe, 542
Infralittoral zone, 542
Inherent edges, 394
Inheritance, units of, genes as, 76
Inhibition model, of succession, 377
Initial floristic composition, in succession, 377
Inputs, of nutrients to ecosystem, 468–469, **469**

Insect(s)
   adult, as heterothermic, 146
   as poikilotherms, 142
   respiratory system of, **137**
   thermoregulation in, 143
Insect vectors, in parasite transmission, 311, **311**
Insects, in savannas, 498
Insulation, in homeotherms, 144–145
Intensity, of disturbance, impact and, 405
Interception, of water, 38, **38**
Interference, competition as, 222
Intergovernmental Panel on Climate Change (IPCC), 631–632
Interior species, 397
Intermediate disturbance hypothesis, 381
Intermediate species, in food web, 337, **338**
Internal cycling, of nutrients, 439
   feedback system in, **458**, 458–459
International Tundra Experiment (ITEX), 640, **640**
International Union for the Conservation of Nature (IUCN)
   biodiversity hotspots designated by, 608, **609**
   classification of threatened species by, 605, **608**
   debate over causes and future of climate change and, **638–639**
   protected area management categories of, 614, **614**
   global number and extent of, in 2003, 616, **616**
Intersexual selection, 163, 166
Interspecific competition, 256–280
   along environmental gradients, 264, **267–268**, **269**, 269–270, **270**, **271**
   biotic and abiotic factors in, 278
   competitive exclusion principle and, studies supporting, 261–262
   environmental influences on, temporal variation in, 262–263
   Lotka–Volterra model of, 258, **259**
      application of, **267–268**
      laboratory experiments supporting, 260–261, **261**
   for multiple resources, 263–264
   multiple species in, 257
   natural selection and, 275–278
   niche of species and, 270–273, **272**, **273**, **274**
   nonresource factors influencing, 262, **265–266**
   outcomes of, 257–258, **259**, 260
Intertidal organisms, activity rhythms of, tidal cycles and, **153**, 153–154
Intertidal zone, 52, 542
   description of, 541
Intertropical Convergence Zone (ITCZ), 28, 501
   precipitation and, 28–30, **30**
Intrasexual selection, 163
Intraspecific competition
   definition of, 222
   growth and development and, **222**, 222–224, **223**, **224**
   mortality rates and, 224–226, **225**, **226**
   reducing reproduction, 226–227, **228–229**, 230, **231**
   as social behaviors, 231–232

Light
attenuation of
Beer's law and, 59–60
by water depth, 43, 43–44
interspecific competition and, 262
productivity of oceans and, 536
vertical distribution of, in terrestrial
environment, plant cover and,
57–60, 58
visible, 20, 21
Light compensation point (LCP), 99, 108,
108
Light environments
alterations in, succession and, 377–378,
378
different, plant adaptations for,
107–114
Light extinction coefficient, for leaf area, 60
Light saturation point, 99–100
Likens, Gene, 409
Lilies (Lilium spp.), reproduction in, 160
Lilium spp. (lilies), reproduction in, 160
Limnetic zone, 519, 520, 522
Links, in food web, 337
Lipid storage, as buoyancy adaptation,
150–151, 151
Liquidambar styraciflua (sweet gum), 341
Liriodendron tulipifera (yellow poplar), 62,
508
seed dispersal in, 192
Litterbag experiments, 442, 442–444, 443,
444
Little Ice Age, 31
Littoral zone, 519, 520, 522, 542
Littoraria irrorata (marsh periwinkle), in salt
marsh trophic cascade, 432–433
Lizards (Lacerta vivipara)
reproductive trade-offs in, 167
thermoregulation in, 143
Llampropeltis pyromelana (mountain
kingsnake), mimicry in, 295
Llano, 496
Local (alpha) diversity, 563
Local populations, 250
Local scale, of metapopulation dynamics,
241
Locus, definition of, 76
Lodgepole chipmunk (Tamias speciosus),
270, 271
Lodgepole pine (Pinus contorta), 509
Loggerhead sea turtles (Caretta caretta),
fecundity and size in, 171
Logistic model, of population growth, 218,
218, 220
Lonchocarpus latifolius, 113
Long-day organisms, 152
Longhorn beetles (Tetraopes spp.), her-
bivory by, 299
Longwave radiation, 19
Lotic ecosystems, 518
Lotka, Alfred, 257–258, 259,
282–285, 284
Lotka-Volterra model
of interspecific competition, 258, 259
application of, 267–268
laboratory experiments supporting,
260–261, 261
of predator-prey interactions, 283–285,
284
mutual population regulation and,
285

Loxia spp. (crossbills), 511
Loxodonta africana (African elephants), 127,
199. See also African elephants
(Loxodonta africana)
Lugworm (Arenicola cristata), 342
Lumbering, in boreal forests, 510
Lymantria dispar (gypsy moths)
defoliation of trees by, 299
disturbances from, 408
spread of, 194, 194, 195
Lyme disease, transmission of, 311
Lynx (Felis lynx), in three-level interaction,
303, 303–304
Lynx spp. (lynxes)
L. canadensis (lynx), 511
L. pardinus (Iberian lynx), habitat frag-
mentation in, 243–244
Lythrum salicaria (purple loosestrife), invad-
ing North America, environmen-
tal damage from, 602, 603

M
Macaques, 494
Macaranga spp., 108, 108, 110
MacArthur, Robert, 174, 364, 400
Mackerel shark (Isurus tigris), countercur-
rent heat exchange in, 148
Macrocystis pyrifera (giant kelp), 56, 57
Macronutrients, 119
Macroparasites, 309
direct transmission of, 310–311
life cycle of, 312, 312
Madagascar, 126
forest clearing in, 600, 601
Magicicada spp. (periodic cicadas), predator
satiation and, 297, 298
Magnesium (Mg)
in animal nutrition, 133
deficiency of, in ruminants, 136
as macronutrient, 119, 120
Mainland-island metapopulation structure,
248
Maize. See Corn (Zea mays)
Malacoctenus (blennies), daily and tidal
cycles of, 153
Malaria, distribution of, climate change
and, 645
Malaria parasites, 311, 311
Malayasia, 494
Maldives, sea-level rise from global warming and, 642
Mallee, 505
Mallee fowl (Leipoa ocellata), 506
Mammals
extinctions of, in Pleistocene period,
559
geographic distribution of, 559
home range of, body size and, 232, 232
as homeotherms, 142
thermoregulation in, 143–145, 144
Mangals, 546
Manganese (Mn)
in animal nutrition, 133
as micronutrient, 120
Manganese oxides, in soil, 63
Mangroves
adaptation of, to aquatic environment,
122–123
in tropical regions, 546, 546
Mann, K. H., 425–426
Maples (Acer spp.)

A. rubrum (red maple)
geographic distribution of, CO_2
increases and, 636
geographic range of, 183, 185, 185
litter from, decomposition rate for,
443, 445
root systems of, 122
in wetlands, 547
postglacial migration of, 385
seed dispersal in, 192
in succession, 369
Marcopus gigantus (gray kangaroo), 506
Marginal value theorem, optimal foraging
and, 291, 292
Margulis, Lynn, 309
Marine ecosystems. See Oceans
Marine iguanas (Amblyrhynchus cristatus),
size variations in, survival and,
134–135
Marine invertebrates, mass extinction of, in
Permian period, 558, 558
Mark-recapture methods of sampling,
188–189
Marmota monax (groundhogs; woodchucks)
hibernation in, 147
stress effects on, 227
Mars, search for water on, rationale for, 16
Mars Exploration Rovers, 16
Marsh elder (Iva frutescens), 342
interactions of, with black grass,
352–353
Marsh periwinkle (Littoraria irrorata), in
salt marsh trophic cascade,
432–433
Marshall Islands, sea-level rise from global
warming and, 642
Marshes, 547, 549
Martes americana (pine martin), 511
Maser, Chris, 325
Masked bobwhite (Colinus virginianus),
restoration of, 612
Masked booby (Sula dactylatra), siblicide
in, 172
Mass extinction event, 599
Massey, Adrianne, 230–231
Mates, acquisition of
by females based on resources, 166,
166–167, 167
sexual selection in, 163–166
Mating, assortative, 82
Mating systems, 162–163
Matorral shrub communities, 506
Matric potential, 101–102
Maximum sustainable yield, for fisheries,
588
Mayflies (Baetis tricaudatus), prey-predator
interactions and, 304, 304
Mayflies (Ephemeroptera), 42
semelparity in, 169
Mayr, Ernst, on adaptations, 75
McGraw, James, 213
McKibben, Bill, 13
McMurtrie, Ross, 424
McNaughton, Sam, 425
Meadow voles (Microtus pennsylvanicus),
dispersal of, 193
Mech, L. D., 235
Mechanical weathering, in soil formation, 61
Medicago sativa (alfalfa), herbivorous-
predatory insect interactions
and, 304–305, 305

Peregrine falcon (*Falco peregrinus*), restoration of, 612

Pérez-Soba, Marta, 627

Periodic cicadas (*Magicicada* spp.), predator satiation and, 297, **98**

Periphyton, 523

Permafrost, 510, 511

Permian extinctions, 558, **558**, 599

Peromyscus maniculatus (deer mice), dispersal of, 193

Pesticides, reduced use of, for sustainable farming, 580

Petersen, C. D. J., 588

Petrie, Marion, 166

Petrinvich, Lewis, 233

pH, definition of, 49

Phagocytes vernalis (flatworm), water balance maintenance in, 149

Phalaropes (*Phalaropus* spp.), polyandry in, 162

Phalaropus spp. (phalaropes), polyandry in, 162

Phenotype, **76**, 76–77, **77**

Phenotypic plasticity, 91–93

Pheromones, stress and, 230–231

Phoca groenlandica (harp seals), sexual maturity of, weight and, 226, **230**

Phocaena spp. (porpoises), countercurrent heat exchange in, 148, **148**

Phoradendron spp. (mistletoe), as hemiparasite, **310**, 311

Phosphorus cycle, **477**, 477–479, **480**

Phosphorus (P)
in animal nutrition, **133**
availability of, in agricultural production in tropics, **586–587**
as macronutrient, 119, **120**

Photic zone
of aquatic ecosystem, 340
of ocean, 532

Photoautotrophs, 98

Photoinhibition, 100

Photosensory organs, circadian rhythm and, 151, **152**

Photosynthesis
affect of light received by plant on, 99–100, **100**
carbon gained in, allocation of, 105–106, **106**
definition of, 98
effects of intraspecific competition on, **223**, 223–224
gas exchange and, 100
increased, with elevated atmospheric CO_2, 627
net, 99
maximum rate of, light levels and, 108, **108**
as oxygen source, 482
process of, 98–99, **99**
rate of
leaf temperature and, 104, **104**
light availability and, 107, 107–108
temperature response curves for, 117–118, **118**

Photosynthetic electron transport, 98

Photosynthetically active radiation (PAR), 20–21
in forest over year, 60, **62**
rate of photosynthesis and, **107**, 107–108

Physeter catodon (sperm whale), 42, **43**

Physical environment, 16–71
aquatic, 37–54. *See also* Aquatic environment
climate and, 18–36. *See also* Climate
terrestrial, 55–71. *See also* Terrestrial environment

Phytoplankton
in net primary production, 423
in ocean, 532
in open-water ecosystems, 520

Pianka, E., 174

Picea spp. (spruces). *See* Spruces (*Picea* spp.)

Pickett, Steward, 264, 269

Picoides pubescens (downy woodpecker), goldenrod ball gall attracting, 313

Pied wagtail (*Motacilla alba*), optimal foraging by, 290, **290**

Pieris proptodice (checkered white butterfly), source-sink metapopulation of, 248

Pigmy shrew, 127, **127**

Pine cone gall, **313**

Pine martin (*Martes americana*), 511

Pines (*Pinus* spp.)
freezing temperatures and, 119
P. albicaulis (whitebark pine), seed dispersal for, 324
*P. banksiana,*shade tolerance of, **111**
P. contorta (lodgepole pine), 509
P. echinata (shortleaf pine), in secondary succession, **375**
P. ponderosa (ponderosa pine), 509
P. serotina (pond pine), **540**
P. strobus (white pine)
root systems of, 122
shade tolerance of, **111**
P. virginiana (Virginia pine), litter from, decomposition rate for, **443**
postglacial migration of, 384, **385**, **386**
in succession, 369

Pinus spp. (pines). *See* Pines (*Pinus* spp.)

Pioneer species, 372

Piranga olivacea (scarlet tanager), forest strata occupied by, 364

Pisaster spp. (starfish), keystone predation and, 356, **356**

*Pituophis melanoleucus,*mimicry in, 296

Plague, **315**

Planktivores, as predators, 282

Plant(s)
adaptations by, 97–125
environmental constraints and, 106–107
biomass of, in world ecosystems, 419
C_3, 99
diversity of, in community, resource availability and, 364–366, **365**
evolution of, over geologic time, 557–558, **558**
growth rate of, relative, **109–110**
heat exchange by, 104
herbivory and, 298–299, 302
interaction of, with herbivores and carnivores, 303–304
in movement of water from soil to atmosphere, 100–103
nutrient recycling in, 439–449
nutrient uptake in, mutualism in, 320–322, **321**

parasitic, 309
direct transmission of, 311
in preempting space and resources, 234, **234**
response of, to increased atmospheric CO_2, **627**, 627–631, **628–629**, **630**
roots of. *See* Roots
seeds of
dispersal of, mutualisms in, **324**, 324–325, **325**
size of, light environment and, **112–113**
sexually reproductive forms of, **160**, 160–161
succession of, 369. *See also* Succession
temperature of, energy balance with environment and, 104–105
terrestrial, nonnative, environmental damage from, 602

Plant litter
conversion of, into soil organic matter, 451–453, **453**
quality of
decomposition rate and, 445, **446**
nutrient cycling and, **458**, 458–459, **459**

Plasmodium spp., causing malaria, 311

Plasticity
developmental, 93
phenotypic, 91–93

Platanus occidentalis (sycamore), 341
litter from, decomposition rate for, **443**

*Platypodium elegans,***113**

Pleistocene epoch, community structure changes in, 383–384, **384**, **385**, **386**

Pleistocene extinctions, 559

Plethodon spp. (salamanders). *See* Salamanders (*Plethodon* spp.)

Pneumatophores, 122, **123**

Podzolization, in soil formation, soils resulting from, 68–69

Poikilotherms
amphibious, behavioral thermoregulation in, 143, **143**
aquatic, thermoregulatory adjustments in, 142–143
countercurrent heat exchange in, 148
dependence of, on environmental temperatures, **142**, 142–143
fecundity in, size and age and, 171
terrestrial, behavioral thermoregulation in, 143

Poikilothermy, 142

Polar cell, 26, **26**

Polar covalent bond, in water, 41, **41**

Polar desert, 512

Polar easterlies, 26, **26**

Polar front, 26

Pollination, mutualism in, 323–324, **324**

Polyandry, **162**, 162–163

Polyculture, 575

Polygamous females, mate acquisition by, 166–167, **167**

Polygamy, **162**, 162–163

Polygonum pensylvanicum (smartweed), resource partitioning and, 273, **274**

Polygonum persicaria (spotted Lady's thumb), 93, **93**

Polygyny, 162

Primates, rain forest, 494
Production efficiency, 426
    of consumers, **426**, 426–427
Productivity
    diversity and, inverse relationships of,
        in marine environments,
        562–563, **563**
    net ecosystem, 470
    net primary, 415. *See also* Net primary
        productivity (NPP)
    of oceans, light and nutrients govern-
        ing, 536, **537**
    primary, 415
    species richness in terrestrial ecosystems
        and, 561, 561–562, **562**
    trade-offs between sustainability and, in
        agriculture, 578, 578–580, **579**,
        580
Profundal zone, 519, 521–522, **522**
Promiscuity, 162
Protected areas, establishing, in habitat
    conservation, 614, 614–617,
    615, 616, 617
Protective armor, as predator defense, 297
Protozoa
    in herbivore digestion, 130
    in soil microbial loop, 454, 455, **455**
Prunus pennsylvanica (fire cherry), in succes-
    sion, 369
*Pseudobombax septenatum,*113
Pseudomyrmex spp. (ants), mutualism in, 322
Pseudotsuga menziesii (Douglas-fir trees),
    reproductive trade-offs in, 167,
    167
Pulliam, Ronald, 248
Purple loosestrife *(Lythrum salicaria)*, invad-
    ing North America, environmen-
    tal damage from, 602, **603**
Pursuit hunting, 298
Pycnocline, 463, **464**
Pygmy sweep *(Parapriacanthus ransonetti)*,
    216
Pyrenestes ostrinus (black-bellied seedcrack-
    er), disruptive selection in, 80, **81**
Pyromineralization, of phosphorus,
    585–586

Q

Quaking aspen *(Populus tremuloides)*, mod-
    ular growth of, 183
Quaking bogs, 549, **551**
Qualitative inhibitors, of herbivory,
    302–303
Qualitative traits, 77
Quantitative traits, 77
Quantitatve inhibitors, of herbivory, 302
Quercus spp. (oaks). *See also* Oaks *(Quercus*
    spp.)
Question, in scientific method, 6, **6**
Quiscalus quiscula (common grackle), asyn-
    chronous hatching in, 171

R

Rabbits *(Oryctolagus cuniculus)*
    sodium deficiency in, 136
    stress effects on, 227
    tularemia in, predation and, 314
Rabun bald salamander *(Plethodon jordani
    rabunensis)*, 86
Radiation
    adaptive, 89, **90**

longwave, 19
photosynthetically active, 20–21
shortwave, 19
solar. *See* Solar radiation
thermal, wavelengths of, **21**
wavelength of, temperature of object
    and, 19, **20**
Radulae, 432
Ragweed *(Ambrosia arterisiifolia)*
    in secondary succession, 374
    semelparity in, 169
Rain shadow, 30, **31**
Rainfall. *See* Precipitation
Rainstorm, over ocean, 37
Raised bogs, 549, **550**
Ramets, definition of, 183
Rana sylvatica (wood frogs)
    supercooling in, 147
    tadpoles of, intraspecific competition
        in, 224, **224**
Rangifer tarandus (caribou, St. Paul reindeer).
    See Caribou *(Rangifer tarandus)*
Rank-abundance diagram, 334, **334**
Ranunculus spp. (buttercups), reproduction
    in, 160
Rarefaction curves, in comparing species
    richness, **564–565**
*Ratibida pinnata,*265–266
Rattus rattus (black rat), in plague transmis-
    sion, **315**
Ravenstein, E. G., **219**
Reabsorption, 439–440
Reaction, norm of, 91, **91, 93**
Realized niche, 270–272, **272, 273**
Recessive allele, 76, **76**
Red fox *(Vulpes vulpes)*, food sources for, 132
Red grouse *(Lagopus lagopus)*, reserve of
    potential breeders in, 233
Red mangrove *(Rhizophora mangle)*,
    aerenchyma in, 122
Red maple *(Acer rubrum). See under* Maples
    (Acer spp.)
Red oak *(Quercus rubra)*, **110**
Red squirrel *(Sciurus hudsonicus)*, 511
Redback salamander *(Plethodon cinereus)*,
    life history strategy of, 176, **176**
Red-billed oxpecker, mutualism and,
    322–323, **323**
Redbud *(Cercis canadensis)*, **508**
Red-cheeked salamander *(Plethodon jordani
    jordani)*, 86
Redds, description of, 180
Red-legged salamander *(Plethodon jordani
    shermani)*, 86
Redshank *(Tringa totanus)*, aggregative
    response in, **288**
Redwood tree *(Sequoia sempervirens)*, **57**
Reef-forming corals, mutualism and, 317
Reefs, coral, **535**, 535–536
Regional (gamma) diversity, 563
Regional scale, of population dynamics, 241
Regolith, 61
Regression model, 9, **9**
Reich, Peter, 419
Reindeer. *See* Caribou *(Rangifer tarandus)*
Relative abundance, of species, 333, **333, 334**
Relative growth rate (RGR), plant, **109–110**
Relative humidity, 27
Relyea, Rick A., 224, **300–301**
Reproduction. *See also* Fecundity
    animal, mineral availability and, 136

asexual, 159–160
budgeting time and energy to, 167–168
of hosts, parasites affecting, 313–314
mode of, dispersal and colonization
    rates and, 250
parental investment in, number and
    size of young and, 170
reduction of, intraspecific competition
    and, 226–227
sexual, 159–162
    forms of, 160–162
timing of, species differences in, 168–170
Reproductive effort, 167–168
    latitude and, **172**, 172–173
Reproductive rate
    gross, 208
    net, birthrate and survivorship and,
        208–209
Reptiles
    as poikilotherms, 142
    thermoregulation in, 143, **143**
Rescue effect, 248
Resistance, antibiotic, ecology of, **92**
Resources
    acquisition of, by parasites,
        309–310, **310**
    availability of, plant diversity in com-
        munity and, 364–366, **365**
    limited, competition and, 222
    management of, economics and,
        593–594
    multiple, interspecific competition for,
        263–264
    partitioning of, species coexistence
        involving, 273–275, **275, 276**
    plants preempting, 234, **234**
    severe shortage of, population extinc-
        tion and, 212, **212**
    use of
        adverse consequences of, limiting
            sustainability, 574
        sustainable, **572**, 572–574, **573**
Respiration
    leaf, rate of, light levels and, 108, **108**
    rate of, leaf temperature and, 104, **104**
Respiratory systems, **137**
Restoration ecology, 617
Reticulum, 131, **131**
Retranslocation, 439–440
Rhizobium bacteria
    in legumes, 320–321, **321**
    as nitrogen fixers, 474
Rhizophora mangle (red mangrove),
    aerenchyma in, 122
Rhizosphere, decomposition in, 454, 455,
    455
Rhopalosiphum padi (grass aphid), in appar-
    ent competition, 357
Rhus spp. (sumac), in succession, 369
Riffles, in streams, 525, **526**
Ringlet butterfly *(Aphantopus hyperantus)*,
    synchronization of local
    population extinction for, 249
Riparian woodlands, 547, **550**
Rivers. *See* Streams/rivers
Roads, effects of, on wildlife, 404
Robber flies *(Laphria* spp.), mimicry in,
    298, **299**
Robertson, Philip, 363
Robinia pseudoacacia (black locust), modu-
    lar growth of, 183

Robinson, Beren, **87–88**
Rock, as phosphorus reservoir, 477
Rocky intertidal zone, succession in, 372, **372**
Rocky soil, world distribution of, 68
Rodents
    density of, weasels and, 286, **286**
    in shrublands, 506
Roff, Derek, 250
Roots
    in aquatic plants, 122
    production of
        light environment and, 110, **111**
        nutrient availability and, 121
        water availability and, 117, **117**
    in soil formation, 62
Rotation period, **573**, 573–574
Rotation time, in timber management, 584, 587
Roundworms *(Ascaris)*, direct transmission of, 310–311
r-selection, 174, 176
r-strategists, 174, 176
Rubisco
    in carboxylation, 98
    levels of, light levels and, 108
Rubus spp. (blackberries), in succession, 369
Ruffed grouse *(Bonasa umbellatus)*, in border environments, 395
Rumen, 131, **131**
Ruminants
    digestive tract of, **131**
        mutualism in, 320
    energy/nutrient acquisition by, 131–132
Runoff, surface, 38, **38**
Ruppia maritima (widgeon grass), 342

S

Sage grouse *(Centrocercus urophasianus)*, 506
    mate selection in, 167, **167**
Sagebrush *(Artemisia)*, 502, **503**
Sahara Desert, 502
St. Paul reindeer *(Rangifer tarandus). See* Caribou *(Rangifer tarandus)*
Salamanders *(Plethodon* spp.), 85, **86**
    diversity of, 85, **86**
    geographic isolates of, 85, **86**
    P. cinereus (redback salamander), life history strategy of, 176, **176**
    P. jordani, geographic isolates of, *85, 86*
    P. jordani clemsonae (Clemson salamander), 86
    P. jordani jordani (red-cheeked salamander), 86
    P. jordani melaventris (highlands salamander), **86**
    P. jordani metcalfi (Metcalf's salamander), **86**
    P. jordani rabunensis (Rabun bald salamander), 86
    P. jordani shermani (red-legged salamander), 86
    P. jordani teyahalee (Tayahalee salamander), 86
    subspecies of, 85
Salicornia spp. (glassworts), **342**, **544**, **545**
    density-dependent control on fecundity in, 227, **231**
    salinity tolerance of, 123

Saline habitats, plants of, 123
Salinity
    of estuarine environment, **52**, 52–53
    salt marsh structure and, 544–546
    of seawater, 46–47, **47**
    tolerance of plants for, 123
Salinization, in soil formation, soils resulting from, 68, **69**
Salmo trutta (brown trout), self-thinning of, 225, **226**
Salt marsh cordgrass *(Spartina alterniflora)*, **342**, 544
    litter from, decomposition of, 446
    salinity tolerance of, 123
    in salt marsh trophic cascade, 432–433
    stress tolerance in, plant zonation and, 361–363, **362**
Salt marsh hay grass *(Spartina patens)*, salinity tolerance of, 123
Salt marshes
    nutrient cycling in, 462
    plants of, 123
    role of consumer organisms in, **432–433**
    structure of, tides and salinity dictating, **544**, 544–546, **545**
    zonation in, 341–342, **342**
        plant-physical environment interactions in, **352–353**
Salt meadow cordgrass *(Spartina patens)*, **342**
    stress tolerance in, plant zonation and, 361–363, **362**
Salt pans, in salt marshes, 545, **545**
Saltbushes *(Atriplex* spp.), 116, **119**
Saltwater environment, freshwater environment and, transition zone between, **52**, 52–53
Sample, definition of, 7
Sampling, in population density determination, **188**, 188–189
Sand, 63, **63**
Sand dunes, colonization of, succession in, 373–374
Sander vitreus (walleyes), fecundity of, population density and, 226, **230**
Sandpipers *(Scolopacidae)*, polyandry in, 162
Sands, shifting, world distribution of, **68**
Sandy beaches, zonation in, 343
Sandy soil, field capacity of, 65, **65**
Sardinops sagax (Pacific sardine), population management for, 588–589, **589**
Sarnelle, Orlando, 420–421
Satellite populations, in metapopulations, 240
Satiation, predator, 297
Saturation, of soil, 65, **65**
Saturation vapor pressure, 27, **28**
Savannas
    fauna in, 498
    geographic distribution of, 497
    precipitation and, 497
    tropical, 496–498, **497**, 498
    vegetation of, 496, 497, 497–498
Sayornis phoebe (Eastern phoebe), winter range of, climate change and, 634–635, **637**
Scale, of disturbance, impact and, 405
Scaling
    definition of, 127
    isometric, 128, **128**

Scaphiopus couchi (spadefoot toad), 503, **504**
    water balance maintenance in, 149
Scaridae (parrotfishes), sex change in, **161**
Scarlet king snake *(Lampropeltis triangulum)*, mimicry in, 295, **296**
Scarlet tanager *(Piranga olivacea)*, forest strata occupied by, 364
Scatter plots, 7, **8**, **10**, 11
Sceloporus occidentalis (western fence lizard), impact of malaria on, 313
Schall, J. J., 313
Schimper, A. F. W., 487
Schoener, Thomas, 257
Schouw, J. F., 487
Schuur, Edward A. G., **447–448**
Science
    environmental, 12
    uncertainty in, 9, 12
Scientific method, **6**, 6–9
Sciurus hudsonicus (red squirrel), 511
Sciurus vulgaris (European red squirrels), fecundity and size in, 171, **171**
Sclerophyllous vegetation, 504–506, **505**
Scolopacidae (sandpipers), polyandry in, 162
Scolytus multistriatus (elm bark beetles), in parasite transmission, 311
Scramble competition, definition of, 222
Scrub, desert, 502, **503**
Scrub oak *(Quercus berberidifolia)*, 505, **505**
Sea otters *(Enhydra lutris)*, as keystone species, 336
Seagrass communities, secondary succession in, 375, **376**
Search image, of predator, 287–288
Search time, in foraging, 291
Seasonal variations
    in carbon cycling, 471–472, **472**
    in primary productivity, 423, **423**
    in solar radiation, 21–23
    in water temperatures, 44–45, **45**
Seawater composition, 46–47, **47**
Second law of thermodynamics, 415
Secondary producers, 73
Secondary production
    definition of, 425
    primary production and, 424–426, **425**
Secondary productivity, definition of, 425
Sedge warblers *(Acrocephalus schoenobaenus)*, mate selection in, 166
Sedimentary cycles, 468
    sulfur cycle as, **480**, 480–481
Sedum smallii (elf orpine)
    life table for, 206, **206**
    mortality curve for, **206**, 207
    survivorship curve for, **207**
Seed predators, 282
Seeds
    dispersal of, mutualisms in, **324**, 324–325, **325**
    size of, light environments and, **112–113**
Seed-tree system, of forest management, 583
Seiurus aurocapillus (ovenbirds)
    forest strata occupied by, 364
    habitat of, **173**
    as interior species, **398**, 399

Water
in air, 27–28
capillary, in soil, 65
cohesion in, 42
density of, 41–43
density-temperature relationship of, 41–42, **42**
depth of
light attenuation by, **43**, 43–44
temperature and, 44, 44–45
flow of, in freshwater wetlands, **548**
flowing, temperature changes in, 45, **46**
hydrogen bonding in, 41, **41**, 42
in living cells, 38
on Mars, search for, rationale for, 16
movement of, through plant, 100–103
movements of, **50**, 50–51, **51**
moving, disturbances from, 406
in oxygen cycle, 482
physical properties of, 40–43
polarity of, 41, **41**
as solvent, 45–47
specific heat of, 41
structure of, 40–41, **41**
surface, oxygen diffusion from atmosphere to, **47**, 47–48, **48**
surface tension of, 42, **42**
viscosity of, 42
volume of, on Earth, 40
Water balance
maintenance of
for aquatic animals, 149–150
for terrestrial animals, 148–149
in organism, 56
Water conservation and protection, 580
Water cycle(s), 38–40, **39**
global, 40, **40**
local, **38**
precipitation in, 38, **38**, 40
Water potential, 101–103
Water spiders (*Dolomedes* spp.), 42
Water striders (*Gerridae* spp.), 42, **42**
Water tupelo (*Nyssa aquatica*), 122
Water-use efficiency, 103
increased, with elevated atmospheric CO$_2$, 627
Watt, A. S., 405
Wavelengths
of solar radiation, 20–21, **21**
of thermal radiation, **21**
Waves
breaking, 50, **50**
wind generating, 50
Wax myrtle (*Myrica cerifera*), **342**
Weasels (*Mustela* spp.)
monogamy in, 162
numerical response of, to rodent density, 289, **289**
rodent density and, 286, **286**
Weather, definition of, 19
Weathering, in soil formation, 61–62
Webster, Jack, 460–461
Weeks, Paul, 322–323
Weinberg, Wilhelm, 82
Westerlies, 26, **26**
Western coral snake (*Microuroides euryxanthus*), 295
Western fence lizard (*Sceloporus occidentalis*), impact of malaria on, 313

Wetfall, 468
Wetlands
freshwater, 547–554
decline of, **552–553**
diversity of life in, 554
location of, along soil moisture gradient, **548**
structure of, hydrology defining, 549–554
water flow in, **548**
plant adaptations in, 121–123
Whitcomb, R. F., 399
White aster (*Aster ericoides*), in secondary succession, 374
White clover (*Trifolium repens*), growth of, population density and, 222, **222**
White mustard (*Sinapis alba*), soil fertility and, 365
White oak (*Quercus alba*). *See under* Oaks (*Quercus* spp.)
White pine (*Pinus strobus*), root systems of, 122
White rhino (*Ceratotherium simum*), reestablishing populations of, 611–612
Whitebark pine (*Pinus albicaulis*), seed dispersal for, 324
White-crowned sparrow (*Zonotrichia leucophrys*), reserve of potential breeders in, 233
White-fronted brown lemur (*Eulemur fulvus albifrons*), **126**
White-smoker chimneys, 534
White-tailed deer (*Odocoileus virginianus*)
browse lines due to, 407, **407**
clinal differences in, 84–85, **85**
polygamy in, 162
population growth in, winter snow accumulation and, 235, **235**
sodium deficiency in, 136
Whitham, Thomas, 174
Whittaker, Robert, 360, 379, 489–490
Whooping cranes (*Grus americana*)
exponential growth of, 203, **203**
restoration of, 612
Widgeon grass (*Ruppia maritima*), **342**
Wikelski, Martin, **134–135**
Williams, J. D., 263–264, 360–361
Williams, Kathy, 297
Willow tits (*Parus montanus*), foraging behavior of, predation risk and, 292, 294
Willows, 122
Wilson, Edward O., 174, 400
Wilsonia citrina (hooded warbler)
forest strata occupied by, 364
habitat of, **173**
Wilting point (WP), of soil, 65, **65**
texture and, **65**, 66
Winds
disturbances from, 406, **406**
ocean currents and, 51, **51**
waves generated by, 50
Winter deciduous plants, 119
Winter solstice, 22
circle of illumination at, **22**
Winter-deciduous trees, 490, **491**
Wisconsin township borders, 394

Witch hobble (*Viburnum alnifolium*), in understory, 339
Within-patch scale, of metapopulation dynamics, 241
Wolves (*Canis lupus*)
local extinction of, effects of, 358
pack behavior of, 232
of Yellowtone National Park, **606**
Wood frogs (*Rana sylvatica*)
supercooling in, 147
tadpoles of, intraspecific competition in, 224, **224**
Wood pewee (*Contopus virens*)
as area-insensitive species, 398, 399
forest strata occupied by, 364
Woodchucks (*Marmota monax*), stress effects on, 227
Woodlands, riparian, 547, **550**
Woodward, Thomas, 316
Worm-eating warbler (*Helmintheros vermivorus*), as interior species, **398**, 399

X

Xiphophorus spp. (swordtails), sexual selection in, **164–165**

Y

Yarrow (*Achillea millefolium*), ecotypes of, 85, **85**
Yellow fever, distribution of, climate change and, 645
Yellow poplar (*Liriodendron tulipifera*), 62, **508**
seed dispersal in, **192**
Yellow-necked mouse (*Apodemus flavicollis*), 256
density of, numerical response of weasels to, 289, **289**
Yellow-pine chipmunk (*Tamias amoenus*), 270, **271**
Yellowthroats (*Geothlypis trichas*), habitat of, **173**
Yersinia pestis, causing plague, **315**
Yield, 573, **573**, 574
Yoda, Kyoji, 224

Z

Zavaleta, Erika, **628–629**
Zea mays (corn). *See* Corn (*Zea mays*)
Zebra mussel (*Dreissena polymorpha*), invading Great Lakes, environmental damage from, 603, **603**
Zero-growth isoclines, 258, **259**
Zieman, Jay, 375
Zinc (Zn)
in animal nutrition, **133**
as micronutrient, **120**
Zonation, in community structure, 340–342, **341**, **342**, **343**
Zonotrichia leucophrys (white-crowned sparrow), reserve of potential breeders in, 233
Zooplankton
carnivorous, 533
in decomposition, 456
herbivorous, 532–533
Zooxanthellae, 317, **320**
Zostera marina (eelgrass), 531, **531**